SERIES ON CONCRETE AND APPLICABLE MATHEMATICS

ISSN: 1793-1142

Series Editor: Professor George A. Anastassiou
Department of Mathematical Sciences
University of Memphis
Memphis, TN 38152, USA

---

*Published**

Vol. 16 Frontiers in Approximation Theory
by George A. Anastassiou

Vol. 17 Frontiers in Time Scales and Inequalities
by George A. Anastassiou

Vol. 18 Differential Sheaves and Connections: A Natural Approach to Physical Geometry
by Anastasios Mallios & Elias Zafiris

Vol. 19 Stochastic Models with Applications to Genetics, Cancers, AIDS and Other Biomedical Systems (Second Edition)
by Wai-Yuan Tan

Vol. 20 Discrete Approximation Theory
by George A. Anastassiou & Merve Kester

Vol. 21 Functional Equations and Inequalities: Solutions and Stability Results
by John Michael Rassias, E. Thandapani, K. Ravi & B. V. Senthil Kumar

Vol. 22 Bloch-Type Periodic Functions: Theory and Applications to Evolution Equations
by Yong-Kui Chang, Gaston Mandata N'Guérékata & Rodrigo Ponce

Vol. 23 Introduction to Matrix Theory: with Applications in Economics and Engineering
Second Edition
by Ferenc Szidarovszky, Sandor Molnar & Mark Molnar

Vol. 24 Trigonometric and Hyperbolic Generated Approximation Theory
by George A. Anastassiou

*To view the complete list of the published volumes in the series, please visit:
http://www.worldscientific/series/scam

Trigonometric and Hyperbolic Generated Approximation Theory

Series on Concrete and Applicable Mathematics – Vol. 24

Trigonometric and Hyperbolic Generated Approximation Theory

George A. Anastassiou

The University of Memphis, USA

World Scientific

NEW JERSEY • LONDON • SINGAPORE • BEIJING • SHANGHAI • HONG KONG • TAIPEI • CHENNAI • TOKYO

Published by

World Scientific Publishing Co. Pte. Ltd.
5 Toh Tuck Link, Singapore 596224
USA office: 27 Warren Street, Suite 401-402, Hackensack, NJ 07601
UK office: 57 Shelton Street, Covent Garden, London WC2H 9HE

Library of Congress Control Number: 2024024467

British Library Cataloguing-in-Publication Data
A catalogue record for this book is available from the British Library.

Series on Concrete and Applicable Mathematics — Vol. 24
TRIGONOMETRIC AND HYPERBOLIC GENERATED APPROXIMATION THEORY

ISBN 978-981-12-9385-6 (hardcover)
ISBN 978-981-12-9386-3 (ebook for institutions)
ISBN 978-981-12-9387-0 (ebook for individuals)

For any available supplementary material, please visit
https://www.worldscientific.com/worldscibooks/10.1142/13857#t=suppl

Desk Editors: Balasubramanian Shanmugam/Kwong Lai Fun

Typeset by Stallion Press
Email: enquiries@stallionpress.com

Dedicated to all those I love.

Preface

This monograph is made possible because of the following two articles, where new types of Taylor's formulae are presented, of the trigonometric and hyperbolic forms:

Ali Hasan Ali, Zsolt Pales, Taylor-type expansions in terms of exponential polynomials, *Mathematical Inequalities and Applications*, 25 (4) (2022), 1123–1141.
G.A. Anastassiou, Opial and Ostrowski type inequalities based on trigonometric and hyperbolic type Taylor formulae, *Malaya J. Mathematik*, 11(S) (2023), 1–26.

Differentiability of functions is a strong component of our work.

We cover perturbed neural networks approximation by themselves and their connection to Brownian motion and stochastic processes, univariate and multivariate analytical inequalities ordinary and fractional, Korovkin theory, and approximation by singular integrals univariate and multivariate.

This monograph is the natural and expected evolution of recent author's research work put in a book form for the first time. The presented approaches are original, and chapters are self-contained and can be read independently. This monograph is suitable to be used in related graduate classes and research projects. We exhibit to the maximum our computational and estimating methods to all possible directions. All presented here are under the umbrella of computational analysis and its applications.

The list of topics covered in this monograph is as follows:

- Opial and Ostrowski-type inequalities based on trigonometric and hyperbolic-type Taylor formulae,
- q-deformed and L-parametrized A-generalized logistic function-based complex-valued trigonometric and hyperbolic neural network high order approximations,

- q-deformed and L-parametrized A-generalized logistic function-based complex-valued multivariate trigonometric and hyperbolic neural network approximations,
- q-deformed and L-parametrized hyperbolic tangent function-relied complex-valued trigonometric and hyperbolic neural network high order approximations,
- q-deformed and L-parametrized hyperbolic tangent function-relied complex-valued multivariate trigonometric and hyperbolic neural network approximations,
- general sigmoid function-based complex-valued trigonometric and hyperbolic neural network high order approximations,
- general multiple sigmoid functions relied complex-valued multivariate trigonometric and hyperbolic neural network approximations,
- approximation of multiple time separating random functions by neural networks,
- Brownian motion approximation by parametrized and deformed neural networks,
- approximation of Brownian motion on simple graphs,
- multivariate fuzzy-random and perturbed neural network approximation,
- trigonometric and hyperbolic Korovkin theory,
- integral inequalities involving new conformable derivatives,
- toward proportional fractional calculus and inequalities,
- updated radial Ostrowski inequalities over a ball,
- updated Ostrowski inequalities over a spherical shell,
- trigonometric and hyperbolic Polya-type inequalities,
- new Opial and Polya-type inequalities over a spherical shell,
- trigonometric and hyperbolic Poincaré, Sobolev and Hilbert–Pachpatte-type inequalities,
- trigonometric generated rate of convergence of smooth Picard singular integral operators,
- trigonometric generated L_p degree of approximation by smooth Picard singular integral operators,
- parametrized and trigonometric-generated quantitative convergence of smooth Picard singular integral operators,
- parametrized and trigonometric L_p quantitative convergence of smooth Picard singular integral operators,
- uniform approximation by smooth Picard multivariate singular integral operators revisited,
- trigonometric-based multivariate smooth Picard singular integrals L_p approximation,

- trigonometric-derived rate of convergence of various smooth singular integral operators,
- trigonometric-deduced L_p degree of approximation by various smooth singular integral operators,
- parametrized and trigonometric-derived uniform approximation by various smooth singular integral operators,
- parametrized trigonometric-derived L_p degree of approximation by various smooth singular integral operators,
- trigonometric-induced multivariate smooth Gauss–Weierstrass singular integrals approximation,
- trigonometric background multivariate smooth Poisson–Cauchy singular integrals approximation,
- trigonometric background multivariate smooth trigonometric singular integrals approximations.

The above are contained in 32 chapters, and at the end of each chapter, a detailed list of references is included.

This book's results are expected to find applications in the many areas of Pure and Applied Mathematics, Computer Science and Engineering, especially in Artificial Intelligence, Machine Learning and Deep Learning, Analytical Inequalities, and Approximation Theory. Other possible applications can be in applied sciences like Statistics, Economics, etc. All in all, what is presented here is a valuable tool for a large range of applications. Therefore, this monograph is suitable for researchers, graduate students, practitioners and seminars of the above disciplines, also to be in all Science and Engineering libraries.

The preparation of this book took place during 2023–2024 at the University of Memphis.

The author likes to thank Professor Alina Lupas of University of Oradea, Romania, for checking and reading the manuscript. Also he thanks his family for tolerating his math habit. Last but not least, the author likes to thank Professor Dimitra Kouloumpou of Hellenic Naval Academy, Greece, for fruitful discussions during the course of the writing of this monograph.

George A. Anastassiou
Department of Mathematical Sciences
University of Memphis
Memphis, TN 38152, USA

About the Author

George A. Anastassiou received his BSc degree in Mathematics from Athens University, Greece, in 1975. He received his diploma in Operations Research from Southampton University, UK, in 1976. He also received his MA in Mathematics from the University of Rochester, USA, in 1981. He was awarded his PhD in Mathematics from the University of Rochester, USA, in 1984. From 1984 to 1986, he served as a Visiting Assistant Professor at the University of Rhode Island, USA. He has been a Faculty Member at the University of Memphis, USA, since 1986. Also, he is currently a Full Professor of Mathematics since 1994. His research area is Computational Analysis, in a very broad sense. He has published over 700 research articles in international mathematics journals and over 52 monographs, proceedings and textbooks in well-known publishing houses. Several awards have been conferred to George Anastassiou. In 2007, he received an Honorary Doctoral Degree from the University of Oradea, Romania. He is Associate Editor in over 80 international journals in mathematics and Editor-in-Chief in 3 journals: most notably in the well-known *Journal of Computational Analysis and Applications.*

Contents

Preface vii

About the Author xi

1. Trigonometric and Hyperbolic Taylor Formulae and Opial and Ostrowski Inequalities **1**

1.1 Taylor Formulae Based on Linear Differential Operators 1
1.2 Main Results 4
Bibliography 35

2. Perturbed A-Generalized Logistic Function-Activated Complex-Valued Trigonometric and Hyperbolic Neural Network High Order Approximation **37**

2.1 Introduction 37
2.2 Preliminaries 39
2.3 Main Results 46
Bibliography 75

3. Perturbed A-Generalized Logistic Function-Activated Complex-Valued Multivariate Trigonometric and Hyperbolic Neural Network Approximation **77**

3.1 Introduction 77
3.2 Preliminaries 78
3.3 Main Results 94
Bibliography 109

4. Perturbed Hyperbolic Tangent Function-Activated Complex-Valued Trigonometric and Hyperbolic Neural Network High Order Approximation **111**

4.1 Introduction 111
4.2 About q-Deformed and λ-Parametrized Hyperbolic Tangent Function $g_{q,\lambda}$ 113
4.3 Main Results 120
Bibliography 148

5. Perturbed Hyperbolic Tangent Function-Activated Complex-Valued Multivariate Trigonometric and Hyperbolic Neural Network Approximation **149**

5.1 Introduction 149
5.2 About q-Deformed and λ-Parametrized Hyperbolic Tangent Function $g_{q,\lambda}$ 150
5.3 Main Results 166
Bibliography 180

6. General Sigmoid Function-Activated Complex-Valued Trigonometric and Hyperbolic Neural Network High Order Approximation **183**

6.1 Introduction 183
6.2 Basics 184
6.3 Main Results 187
Bibliography 216

7. General Multiple Sigmoid Functions Activated Complex Valued Multivariate Trigonometric and Hyperbolic Neural Network Approximation **217**

7.1 Introduction 217
7.2 Basics 218
7.2.1 General neural network background 218
7.2.2 Multivariate new Taylor formulae 227
7.3 Main Results 232
Bibliography 246

8. Approximation of Multiple Time Separating Random Functions by Neural Networks 249

8.1 Introduction 249
8.2 Background 251
8.2.1 About the logistic sigmoid activation function 251
8.2.2 About the hyperbolic tangent sigmoid activation function 259
8.2.3 Combining 2.1 and 2.2 266
8.3 Multiple Time Separating Random Functions 270
8.4 Main Results 271
8.5 Applications 274
8.6 Specific Applications 277
Bibliography 282

9. Brownian Motion Approximation by Perturbed Neural Networks 285

9.1 Introduction 285
9.2 About Parametrized and Deformed Neural Networks 286
9.2.1 About the q-deformed and λ-parametrized logistic activation function 286
9.2.2 About q-deformed and λ-parametrized hyperbolic tangent activation function $g_{q,\lambda}$ 292
9.3 Combine 2.1 and 2.2 297
9.4 About Brownian Motion on Three-dimensional Sphere 298
9.4.1 Explicit form of the heat kernel of S^3 300
9.5 Main Results 302
9.6 Applications 304
Bibliography 314

10. Approximation of Brownian Motion over Simple Graphs 317

10.1 Introduction 317
10.2 Basics 318
10.2.1 About the parametrized (Gauss) error special activation function 318

10.2.2 About q-deformed and β-parametrized half-hyperbolic tangent function φ_q 324
10.3 Combine 2.1 and 2.2 330
10.4 About Random Motion on Simple Graphs 333
10.5 Main Results 337
10.6 Applications 340
Bibliography 355

11. Multivariate Fuzzy-Random with Perturbed Neural Network Approximation 359

11.1 Fuzzy-Random Functions Background 359
11.2 About Perturbed Neural Network Background 363
11.2.1 About q-Deformed and λ-parametrized A-generalized logistic function-induced real space-valued multivariate multilayer neural network approximation 363
11.2.2 About q-deformed and λ-parametrized hyperbolic tangent function $g_{q,\lambda}$ 367
11.2.3 About q-deformed and parametrized half-hyperbolic tangent function ϑ_q 371
11.3 Main Result 376
11.3.1 About p-mean approximation by fuzzy-random perturbed quasi-interpolation neural network operators 376
Bibliography 380

12. Trigonometric and Hyperbolic Korovkin Properties 383

12.1 Introduction 383
12.2 Main Results 384
12.2.1 Trigonometric approximation by positive linear operators 384
12.2.2 Hyperbolic approximation by positive linear operators 390
12.3 Applications 395
Bibliography 397

13. Integral Inequalities Using the New Conformable Derivatives **399**

13.1 Background 399
13.2 Main Results 404
Bibliography 416

14. About Proportional Fractional Calculus and Inequalities **417**

14.1 Background 417
14.1.1 From left fractional calculus and more 417
14.1.2 From right fractional calculus and more 418
14.1.3 Conclusion 419
14.1.4 About new conformable calculus 420
14.1.5 Proportional left Caputo fractional calculus . . . 423
14.1.6 Proportional right Caputo fractional calculus 425
14.2 Main Results 426
Bibliography 442

15. New Radial Ostrowski Inequalities on the Ball **445**

15.1 Introduction 445
15.2 Main Results 446
Bibliography 456

16. New Ostrowski Inequalities on a Spherical Shell **457**

16.1 Introduction 457
16.2 Main Results 458
Bibliography 473

17. Trigonometric and Hyperbolic Polya Inequalities **475**

17.1 Main Results 475
Bibliography 481

18. New Opial and Polya Inequalities Over a Spherical Shell **483**

18.1 Background 483
18.2 Results 486
Bibliography 494

19. Trigonometric and Hyperbolic Poincaré, Sobolev and Hilbert-Pachpatte Inequalities **495**

19.1 Main Results 495
Bibliography 505

20. Trigonometric-Induced Degree of Approximation by Smooth Picard Singular Integral Operators **507**

20.1 Introduction 507
20.2 Results 508
Bibliography 515

21. Trigonometric-Derived L_p Degree of Approximation by Smooth Picard Singular Integral Operators **517**

21.1 Introduction 517
21.2 Results 517
Bibliography 531

22. Parametrized and Trigonometric-Induced Quantitative Convergence of Smooth Picard Singular Integral Operators **533**

22.1 Introduction 533
22.2 Results 533
Bibliography 547

23. Parametrized and Trigonometric L_p Approximation with Rates of Smooth Picard Singular Integral Operators **549**

23.1 Introduction 549
23.2 Results 549
Bibliography 567

24. Update on Uniform Approximation by Smooth Picard Multivariate Singular Integral Operators Revisited **569**

24.1 Introduction 569
24.2 Results 570
Bibliography 584

25. Trigonometric-Derived Multivariate Smooth Picard Singular Integrals L_p Quantitative Approximation **587**

25.1 Introduction 587
25.2 Results 588
Bibliography 612

26. Trigonometric Produced Rate of Approximation of Various Smooth Singular Integral Operators **613**

26.1 Introduction 613
26.2 Results 614
Bibliography 631

27. Trigonometric Derived L_p Quantitative Approximation by Various Smooth Singular Integral Operators **633**

27.1 Introduction 633
27.2 Basics 634
27.3 Main Results 641
Bibliography 667

28. Parametrized and Trigonometric-Produced Uniform Quantitative Approximation by Various Smooth Singular Integral Operators **669**

28.1 Introduction 669
28.2 About Part I: On Gauss–Weierstrass Smooth Singular Integrals 670
28.3 Main Results 684
Bibliography 692

29. Parametrized Trigonometric-Implied L_p Degree of Quantitative Approximation by Various Smooth Singular Integral Operators **693**

29.1 Introduction 693
29.2 Part I: On Gauss–Weierstrass Smooth Singular Integrals 694
29.3 Part II: On the Smooth Poisson–Cauchy Singular Integral Operators 709

29.4 Part III: On the Smooth Trigonometric Singular Integral Operators 719
Bibliography 731

30. Trigonometric-Implied Multivariate Smooth Gauss–Weierstrass Singular Integrals Quantitative Approximation 733

30.1 Introduction 733
30.2 Background on General Theory 734
30.3 Auxiliary Essential Results 742
30.4 Main Results 749
Bibliography 755

31. Trigonometric-Based Multivariate Smooth Poisson–Cauchy Singular Integrals Quantitative Approximation 757

31.1 Introduction 757
31.2 Background of General Theory 758
31.3 Auxiliary Results 766
31.4 Main Results 777
Bibliography 783

32. Trigonometric-Implied Multivariate Smooth Trigonometric Singular Integrals Quantitative Approximation 785

32.1 Introduction 785
32.2 Background of General Theory 786
32.3 Auxiliary Results 794
32.4 Main Results 805
Bibliography 811

Conclusion 813

Index 815

Chapter 1

Trigonometric and Hyperbolic Taylor Formulae and Opial and Ostrowski Inequalities

Based on Ali and Pales (2022), we produce a collection of interesting trigonometric and hyperbolic Taylor formulae with integral remainders. Using these we derive Opial and Ostrowski-type corresponding inequalities of various kinds and norms. It follows Anastassiou (2023).

1.1 Taylor Formulae Based on Linear Differential Operators

This section is based entirely on Ali and Pales (2022). Here K denotes $\mathbb{R}$ or $\mathbb{C}$.

Let I be an interval subset of $\mathbb{R}$, and $c = (c_0, \ldots, c_n) \in \mathbb{R}^{n+1}$, $c_n = 1$, $n \in \mathbb{N}$, and the nth-order linear differential operator D_c from $C_K^n(I)$ (n-times continuously differentiable K-valued functions defined on I) into $C_K(I)$ (continuous functions from I to K), where $D_c(f) := c_n f^{(n)} + \cdots + c_1 f' + c_0 f$, with $f \in C_K^n(I)$. Let $\omega_c \in C_{\mathbb{C}}^n(\mathbb{R})$ be the unique solution of initial value problem:

$$D_c(\omega_c) = 0, \quad \omega_c^{(i)}(0) = \delta_{l,n-1} \quad (l \in \{0, 1, \ldots, n-1\}).$$

By Theorem 3.2 of Ali and Pales (2022), p. 1131, we have that

$$f(x) = (T_{a,c}f)(x) + \int_a^x D_c(f)(t)\, \omega_c(x-t)\, dt, \tag{1.1}$$

where

$$(T_{a,c}f)(x) := \sum_{j=0}^{n-1} \left(f^{(j)}(a) \sum_{i=0}^{n-1-j} c_{i+j+1} \omega_c^{(i)}(x-a) \right), \tag{1.2}$$

for all $x, a \in I$.

Next, let $k \in \mathbb{N}_0$, $k < n$, $\gamma \in \mathbb{R}$, and we consider $\mathcal{J}_{n,k,\gamma}$ from $\mathbb{R}$ into $\mathbb{R}$, given by

$$\mathcal{J}_{n,k,\gamma}(t) := \sum_{i=0}^{\infty} \frac{\gamma^i t^{i(n-k)+n}}{(i(n-k)+n)!}, \tag{1.3}$$

which converges for all t.

A further interpretation of (1.1), see Theorem 3.3 of Ali and Pales (2022, p. 1132), gives us

$$f(x) = \sum_{j=0}^{k-1} f^{(j)}(a) \frac{(x-a)^j}{j!} + \sum_{j=k}^{n-1} f^{(j)}(a) \mathcal{J}_{n,k,\gamma}^{(n-j)}(x-a)$$
$$+ \int_a^x \left(f^{(n)}(t) - \gamma f^{(k)}(t)\right) \mathcal{J}'_{n,k,\gamma}(x-t)\, dt, \tag{1.4}$$

for all $f \in C_K^n(I)$, $n \in \mathbb{N}$ and $x, a \in I$.

Applications of (1.4) provide us the following interesting Taylor formulae of trigonometric and hyperbolic types:

Theorem 1.1 (Ali and Pales 2022). *For all $f \in C_K^2(I)$ and $a, x \in I$, we have*

$$f(x) = f(a)\cos(x-a) + f'(a)\sin(x-a)$$
$$+ \int_a^x (f''(t) + f(t))\sin(x-t)\, dt. \tag{1.5}$$

Theorem 1.2 (Ali and Pales 2022). *For all $f \in C_K^2(I)$ and $a, x \in I$, we have*

$$f(x) = f(a)\cosh(x-a) + f'(a)\sinh(x-a)$$
$$+ \int_a^x (f''(t) - f(t))\sinh(x-t)\, dt. \tag{1.6}$$

Theorem 1.3 (Ali and Pales 2022). *For all $f \in C_K^4(I)$ and $a, x \in I$, we have*

$$f(x) = f(a)\left(\frac{\cosh(x-a) + \cos(x-a)}{2}\right) + f'(a)\left(\frac{\sinh(x-a) + \sin(x-a)}{2}\right)$$
$$+ f''(a)\left(\frac{\cosh(x-a) - \cos(x-a)}{2}\right) + f'''(a)\left(\frac{\sinh(x-a) - \sin(x-a)}{2}\right)$$
$$+ \int_a^x (f''''(t) - f(t))\left(\frac{\sinh(x-t) - \sin(x-t)}{2}\right) dt. \tag{1.7}$$

Theorem 1.4 (Ali and Pales 2022). *Let $\alpha, \beta \in \mathbb{R}$ with $\alpha\beta(\alpha^2 - \beta^2) \neq 0$. Then, for all $f \in C_K^4(I)$ and $a, x \in I$, we have*

$$\begin{aligned} f(x) &= f(a)\left(\frac{\beta^2 \cos(\alpha(x-a)) - \alpha^2 \cos(\beta(x-a))}{\beta^2 - \alpha^2}\right) \\ &\quad + f'(a)\left(\frac{\beta^3 \sin(\alpha(x-a)) - \alpha^3 \sin(\beta(x-a))}{\alpha\beta(\beta^2 - \alpha^2)}\right) \\ &\quad + f''(a)\left(\frac{\cos(\alpha(x-a)) - \cos(\beta(x-a))}{\beta^2 - \alpha^2}\right) \\ &\quad + f'''(a)\left(\frac{\beta \sin(\alpha(x-a)) - \alpha \sin(\beta(x-a))}{\alpha\beta(\beta^2 - \alpha^2)}\right) \\ &\quad + \int_a^x \left(f''''(t) + (\alpha^2 + \beta^2) f''(t) + \alpha^2\beta^2 f(t)\right) \\ &\quad \times \left(\frac{\beta \sin(\alpha(x-t)) - \alpha \sin(\beta(x-t))}{\alpha\beta(\beta^2 - \alpha^2)}\right) dt. \end{aligned} \tag{1.8}$$

Finally, we include the following result.

Theorem 1.5 (Ali and Pales 2022). *Let $\alpha, \beta \in \mathbb{R}$ with $\alpha\beta(\alpha^2 - \beta^2) \neq 0$. Then, for all $f \in C_K^4(I)$ and $a, x \in I$, we have*

$$\begin{aligned} f(x) &= f(a)\left(\frac{\beta^2 \cosh(\alpha(x-a)) - \alpha^2 \cosh(\beta(x-a))}{\beta^2 - \alpha^2}\right) \\ &\quad + f'(a)\left(\frac{\beta^3 \sinh(\alpha(x-a)) - \alpha^3 \sinh(\beta(x-a))}{\alpha\beta(\beta^2 - \alpha^2)}\right) \\ &\quad + f''(a)\left(\frac{\cosh(\beta(x-a)) - \cosh(\alpha(x-a))}{\beta^2 - \alpha^2}\right) \\ &\quad + f'''(a)\left(\frac{\alpha \sinh(\beta(x-a)) - \beta \sinh(\alpha(x-a))}{\alpha\beta(\beta^2 - \alpha^2)}\right) \\ &\quad + \int_a^x \left(f''''(t) - (\alpha^2 + \beta^2) f''(t) + \alpha^2\beta^2 f(t)\right) \\ &\quad \times \left(\frac{\alpha \sinh(\beta(x-t)) - \beta \sinh(\alpha(x-t))}{\alpha\beta(\beta^2 - \alpha^2)}\right) dt. \end{aligned} \tag{1.9}$$

In this chapter, first, we give the needed variants to Theorems 1.1–1.5, then based on these modified Taylor formulae, we derive several Ostrowski-type inequalities. In between, we deal with Opial-type inequalities.

1.2 Main Results

We are inspired by Anastassiou (2010)–Anastassiou (2016).

We give the following Taylor formula results of trigonometric and hyperbolic types.

Theorem 1.6. *For all $f \in C_K^2(I)$ and $a, x \in I$, we have*

$$\begin{aligned} f(x) - f(a) &= f'(a)\sin(x-a) + 2f''(a)\sin^2\left(\frac{x-a}{2}\right) \\ &\quad + \int_a^x \left[(f''(t) + f(t)) - (f''(a) + f(a))\right]\sin(x-t)\,dt \\ &= f'(a)\sin(x-a) + 2f''(a)\sin^2\left(\frac{x-a}{2}\right) \\ &\quad + \int_a^x \left[(f''(t) - f''(a)) + (f(t) - f(a))\right]\sin(x-t)\,dt. \end{aligned} \tag{1.10}$$

Proof. Here we use Theorem 1.1.

We have by (1.5) that

$$\begin{aligned} f(x) - f(a) = f(a)(\cos(x-a) - 1) + f'(a)\sin(x-a) \\ + \int_a^x (f''(t) + f(t))\sin(x-t)\,dt. \end{aligned} \tag{1.11}$$

(By $\cos 2x = 1 - 2\sin^2 x$, $\cos 2x - 1 = -2\sin^2 x$, and $\cos(x-a) - 1 = \cos 2\left(\frac{x-a}{2}\right) - 1 = -2\sin^2\left(\frac{x-a}{2}\right)$.)

Therefore, it holds

$$\begin{aligned} f(x) - f(a) &= -2f(a)\sin^2\left(\frac{x-a}{2}\right) + f'(a)\sin(x-a) \\ &\quad + \int_a^x (f''(t) + f(t))\sin(x-t)\,dt \end{aligned} \tag{1.12}$$

$$\begin{aligned} &= -2f(a)\sin^2\left(\frac{x-a}{2}\right) + f'(a)\sin(x-a) + (f''(a) + f(a))(1 - \cos(x-a)) \\ &\quad + \int_a^x \left[(f''(t) + f(t)) - (f''(a) + f(a))\right]\sin(x-t)\,dt \\ &= -2f(a)\sin^2\left(\frac{x-a}{2}\right) + f'(a)\sin(x-a) + (f''(a) + f(a))\,2\sin^2\left(\frac{x-a}{2}\right) \end{aligned} \tag{1.13}$$

$$+\int_a^x \left[(f''(t)+f(t))-(f''(a)+f(a))\right]\sin(x-t)\,dt$$

$$=f'(a)\sin(x-a)+2f''(a)\sin^2\left(\frac{x-a}{2}\right)$$

$$+\int_a^x \left[(f''(t)+f(t))-(f''(a)+f(a))\right]\sin(x-t)\,dt.$$

Note above that $(\cos(x-t))'=\sin(x-t)$.

The claim is proved. □

Theorem 1.7. *For all $f\in C_K^2(I)$ and $a,x\in I$, we have*

$$f(x)-f(a)=f'(a)\sinh(x-a)+2f''(a)\sinh^2\left(\frac{x-a}{2}\right)$$

$$+\int_a^x \left[(f''(t)-f(t))-(f''(a)-f(a))\right]\sinh(x-t)\,dt$$

$$=f'(a)\sinh(x-a)+2f''(a)\sinh^2\left(\frac{x-a}{2}\right)$$

$$+\int_a^x \left[(f''(t)-f''(a))-(f(t)-f(a))\right]\sinh(x-t)\,dt. \tag{1.14}$$

Proof. Here, we use Theorem 1.2 (1.6).

Note that $(-\cosh(x-t))'=\sinh(x-t)$.

By $\cosh 2x-1=2\sinh^2 x$, we get that $\cosh(x-a)-1=\cosh 2\left(\frac{x-a}{2}\right)-1=2\sinh^2\left(\frac{x-a}{2}\right)$.

By (1.6), we obtain

$$f(x)-f(a)=f(a)(\cosh(x-a)-1)+f'(a)\sinh(x-a)$$

$$+\int_a^x (f''(t)-f(t))\sinh(x-t)\,dt$$

$$=2f(a)\sinh^2\left(\frac{x-a}{2}\right)+f'(a)\sinh(x-a) \tag{1.15}$$

$$+\int_a^x (f''(t)-f(t))\sinh(x-t)\,dt$$

$$=2f(a)\sinh^2\left(\frac{x-a}{2}\right)+f'(a)\sinh(x-a)$$

$$+(f''(a)-f(a))(\cosh(x-a)-1)$$

$$+\int_a^x \left[(f''(t)-f(t))-(f''(a)-f(a))\right]\sinh(x-t)\,dt$$

$$= 2f(a)\sinh^2\left(\frac{x-a}{2}\right) + f'(a)\sinh(x-a)$$

$$+ (f''(a) - f(a))\left(2\sinh^2\left(\frac{x-a}{2}\right)\right) \tag{1.16}$$

$$+ \int_a^x [(f''(t) - f(t)) - (f''(a) - f(a))]\sinh(x-t)\,dt$$

$$= f'(a)\sinh(x-a) + 2f''(a)\sinh^2\left(\frac{x-a}{2}\right)$$

$$+ \int_a^x [(f''(t) - f(t)) - (f''(a) - f(a))]\sinh(x-t)\,dt.$$

The claim is proved. □

Theorem 1.8. *For all $f \in C_K^4(I)$ and $a, x \in I$, we have*

$$f(x) - f(a) = f'(a)\left(\frac{\sinh(x-a) + \sin(x-a)}{2}\right)$$

$$+ f''(a)\left(\frac{\cosh(x-a) - \cos(x-a)}{2}\right)$$

$$+ f'''(a)\left(\frac{\sinh(x-a) - \sin(x-a)}{2}\right)$$

$$+ f''''(a)\left(\sinh^2\left(\frac{x-a}{2}\right) - \sin^2\left(\frac{x-a}{2}\right)\right)$$

$$+ \int_a^x [(f''''(t) - f''''(a)) - (f(t) - f(a))]\left(\frac{\sinh(x-t) - \sin(x-t)}{2}\right)dt. \tag{1.17}$$

Proof. Here, we use Theorem 1.3 (1.7).

We have that

$$f(x) - f(a) = f(a)\left(\frac{(\cosh(x-a) - 1) + (\cos(x-a) - 1)}{2}\right)$$

$$+ f'(a)\left(\frac{\sinh(x-a) + \sin(x-a)}{2}\right)$$

$$+ f''(a)\left(\frac{\cosh(x-a) - \cos(x-a)}{2}\right) + f'''(a)\left(\frac{\sinh(x-a) - \sin(x-a)}{2}\right) \tag{1.18}$$

$$+\int_a^x \left(f''''(t) - f(t)\right)\left(\frac{\sinh(x-t) - \sin(x-t)}{2}\right) dt$$

$$= f(a)\left(\sinh^2\left(\frac{x-a}{2}\right) - \sin^2\left(\frac{x-a}{2}\right)\right)$$

$$+ f'(a)\left(\frac{\sinh(x-a) + \sin(x-a)}{2}\right) \tag{1.19}$$

$$+ f''(a)\left(\frac{\cosh(x-a) - \cos(x-a)}{2}\right) + f'''(a)\left(\frac{\sinh(x-a) - \sin(x-a)}{2}\right)$$

$$+ \left(f''''(a) - f(a)\right)\left(\sinh^2\left(\frac{x-a}{2}\right) - \sin^2\left(\frac{x-a}{2}\right)\right)$$

$$+ \int_a^x \left[\left(f''''(t) - f(t)\right) - \left(f''''(a) - f(a)\right)\right]\left(\frac{\sinh(x-t) - \sin(x-t)}{2}\right) dt$$

$$= f'(a)\left(\frac{\sinh(x-a) + \sin(x-a)}{2}\right)$$

$$+ f''(a)\left(\frac{\cosh(x-a) - \cos(x-a)}{2}\right) + f'''(a)\left(\frac{\sinh(x-a) - \sin(x-a)}{2}\right)$$

$$+ f''''(a)\left(\sinh^2\left(\frac{x-a}{2}\right) - \sin^2\left(\frac{x-a}{2}\right)\right)$$

$$+ \int_a^x \left[\left(f''''(t) - f''''(a)\right) - \left(f(t) - f(a)\right)\right]\left(\frac{\sinh(x-t) - \sin(x-t)}{2}\right) dt. \tag{1.20}$$

The claim is proved. ☐

We continue with

Theorem 1.9. *Let $\alpha, \beta \in \mathbb{R}$ with $\alpha\beta\left(\alpha^2 - \beta^2\right) \neq 0$. Then, for all $f \in C_K^4(I)$ and $a, x \in I$, we have*

$$f(x) - f(a) = f'(a)\left(\frac{\beta^3 \sin(\alpha(x-a)) - \alpha^3 \sin(\beta(x-a))}{\alpha\beta\left(\beta^2 - \alpha^2\right)}\right)$$

$$+ f''(a)\left(\frac{\cos(\alpha(x-a)) - \cos(\beta(x-a))}{\beta^2 - \alpha^2}\right)$$

$$+ f'''(a)\left(\frac{\beta \sin(\alpha(x-a)) - \alpha \sin(\beta(x-a))}{\alpha\beta\left(\beta^2 - \alpha^2\right)}\right)$$

$$+ \frac{2\left(f''''(a) + \left(\alpha^2 + \beta^2\right) f''(a)\right)}{(\alpha\beta)^2\left(\beta^2 - \alpha^2\right)}\left(\beta^2 \sin^2\left(\frac{\alpha(x-a)}{2}\right)\right.$$

$$\left. - \alpha^2 \sin^2\left(\frac{\beta(x-a)}{2}\right)\right) + \frac{1}{\alpha\beta\left(\beta^2 - \alpha^2\right)} \int_a^x \left[\left(f''''(t) + \left(\alpha^2 + \beta^2\right) f''(t)\right.\right.$$

$$+\alpha^2\beta^2 f(t)) - \left(f''''(a) + \left(\alpha^2+\beta^2\right) f''(a) + \alpha^2\beta^2 f(a)\right)]$$
$$\times \left[\beta \sin(\alpha(x-t)) - \alpha \sin(\beta(x-t))\right] dt. \tag{1.21}$$

Proof. We see that

$$\begin{aligned}
I^* &:= f(a)\left(\frac{\beta^2(\cos(\alpha(x-a))-1) - \alpha^2(\cos(\beta(x-a))-1)}{\beta^2-\alpha^2}\right) \\
&= f(a)\left(\frac{\beta^2(\cos(\alpha(x-a))) - \beta^2 - \alpha^2(\cos(\beta(x-a))) + \alpha^2}{\beta^2-\alpha^2}\right) \\
&= f(a)\left[\left(\frac{\beta^2(\cos(\alpha(x-a))) - \alpha^2(\cos(\beta(x-a)))}{\beta^2-\alpha^2}\right) - \left(\frac{\beta^2-\alpha^2}{\beta^2-\alpha^2}\right)\right] \\
&= f(a)\left[\left(\frac{\beta^2(\cos(\alpha(x-a))) - \alpha^2(\cos(\beta(x-a)))}{\beta^2-\alpha^2}\right) - 1\right] \\
&= f(a)\left(\frac{\beta^2(\cos(\alpha(x-a))) - \alpha^2(\cos(\beta(x-a)))}{\beta^2-\alpha^2}\right) - f(a).
\end{aligned} \tag{1.22}$$

That is,

$$\begin{aligned}
I^* &= f(a)\left(\frac{-2\beta^2 \sin^2\left(\frac{\alpha(x-a)}{2}\right) + 2\alpha^2 \sin^2\left(\frac{\beta(x-a)}{2}\right)}{\beta^2-\alpha^2}\right) \\
&= f(a)\left(\frac{\beta^2(\cos(\alpha(x-a))) - \alpha^2(\cos(\beta(x-a)))}{\beta^2-\alpha^2}\right) - f(a).
\end{aligned} \tag{1.23}$$

By Theorem 1.4, we obtain

$$\begin{aligned}
f(x) - f(a) &= f(a)\left(\frac{-2\beta^2 \sin^2\left(\frac{\alpha(x-a)}{2}\right) + 2\alpha^2 \sin^2\left(\frac{\beta(x-a)}{2}\right)}{\beta^2-\alpha^2}\right) \\
&\quad + f'(a)\left(\frac{\beta^3 \sin(\alpha(x-a)) - \alpha^3 \sin(\beta(x-a))}{\alpha\beta\left(\beta^2-\alpha^2\right)}\right) \\
&\quad + f''(a)\left(\frac{\cos(\alpha(x-a)) - \cos(\beta(x-a))}{\beta^2-\alpha^2}\right) \\
&\quad + f'''(a)\left(\frac{\beta \sin(\alpha(x-a)) - \alpha \sin(\beta(x-a))}{\alpha\beta\left(\beta^2-\alpha^2\right)}\right) \\
&\quad + \frac{\left(f''''(a) + \left(\alpha^2+\beta^2\right) f''(a) + \alpha^2\beta^2 f(a)\right)}{\left(\beta^2-\alpha^2\right)}
\end{aligned}$$

$$\left(\frac{2\beta}{\alpha(\alpha\beta)}\sin^2\left(\frac{\alpha(x-a)}{2}\right)-\frac{2\alpha}{\beta(\alpha\beta)}\sin^2\left(\frac{\beta(x-a)}{2}\right)\right)$$
$$+\frac{1}{\alpha\beta(\beta^2-\alpha^2)}\int_a^x\left[\left(f''''(t)+\left(\alpha^2+\beta^2\right)f''(t)+\alpha^2\beta^2f(t)\right)\right.$$
$$-\left.\left(f''''(a)+\left(\alpha^2+\beta^2\right)f''(a)+\alpha^2\beta^2f(a)\right)\right]\left[\beta\sin(\alpha(x-t))\right.$$
$$\left.-\alpha\sin(\beta(x-t))\right]dt. \tag{1.24}$$

Note that

$$(\cos(\alpha(x-t)))'=-\sin(\alpha(x-t))(-\alpha)=\alpha\sin(\alpha(x-t)), \tag{1.25}$$

and

$$\left(\frac{1}{\alpha}\cos(\alpha(x-t))\right)l'=\sin(\alpha(x-t)), \tag{1.26}$$

and

$$\int_a^x\beta(\sin(\alpha(x-t)))\,dt=\frac{\beta}{\alpha}(\cos(\alpha(x-t)))\,|_a^x=\frac{\beta}{\alpha}(1-\cos(\alpha(x-a)))$$
$$=\frac{2\beta}{\alpha}\left(\sin^2\left(\frac{\alpha(x-a)}{2}\right)\right). \tag{1.27}$$

Similarly, we get

$$-\int_a^x\alpha\sin(\beta(x-t))\,dt=-\frac{2\alpha}{\beta}\left(\sin^2\left(\frac{\beta(x-a)}{2}\right)\right). \tag{1.28}$$

Furthermore, we see that

$$\frac{\alpha^2\beta^2f(a)}{(\beta^2-\alpha^2)}\left(\frac{2}{\alpha^2}\sin^2\left(\frac{\alpha(x-a)}{2}\right)-\frac{2}{\beta^2}\sin^2\left(\frac{\beta(x-a)}{2}\right)\right)$$
$$=\frac{f(a)}{(\beta^2-\alpha^2)}\left(2\beta^2\sin^2\left(\frac{\alpha(x-a)}{2}\right)-2\alpha^2\sin^2\left(\frac{\beta(x-a)}{2}\right)\right). \tag{1.29}$$

The claim is proved. □

We give

Theorem 1.10. *Let $\alpha,\beta\in\mathbb{R}$ and $\alpha\beta\left(\alpha^2-\beta^2\right)\neq 0$. Then, for all $f\in C_K^4(I)$ and $a,x\in I$, we have*

$$f(x)-f(a)=f'(a)\left(\frac{\beta^3\sinh(\alpha(x-a))-\alpha^3\sinh(\beta(x-a))}{\alpha\beta(\beta^2-\alpha^2)}\right)$$

$$+ f''(a)\left(\frac{\cosh(\beta(x-a)) - \cosh(\alpha(x-a))}{\beta^2-\alpha^2}\right)$$
$$+ f'''(a)\left(\frac{\alpha\sinh(\beta(x-a)) - \beta\sinh(\alpha(x-a))}{\alpha\beta(\beta^2-\alpha^2)}\right)$$
$$+ \frac{2\left(f''''(a) - \left(\alpha^2+\beta^2\right)f''(a)\right)}{(\alpha\beta)^2(\beta^2-\alpha^2)}\left(\alpha^2\sinh^2\left(\frac{\beta(x-a)}{2}\right)\right.$$
$$\left.-\ \beta^2\sinh^2\left(\frac{\alpha(x-a)}{2}\right)\right) + \frac{1}{\alpha\beta(\beta^2-\alpha^2)}\int_a^x [(f''''(t)$$
$$-\ \left(\alpha^2+\beta^2\right)f''(t) + \alpha^2\beta^2 f(t)) - (f''''(a) - \left(\alpha^2+\beta^2\right)f''(a)$$
$$+\ \alpha^2\beta^2 f(a))]\left[\alpha\sinh(\beta(x-t)) - \beta\sinh(\alpha(x-t))\right]dt. \tag{1.30}$$

Proof. We see that

$$J := f(a)\left(\frac{\beta^2(\cosh(\alpha(x-a))-1) - \alpha^2(\cosh(\beta(x-a))-1)}{\beta^2-\alpha^2}\right)$$
$$= f(a)\left(\frac{\beta^2(\cosh(\alpha(x-a))) - \alpha^2(\cosh(\beta(x-a)))}{\beta^2-\alpha^2}\right) - f(a). \tag{1.31}$$

Hence,

$$J = f(a)\left(\frac{2\beta^2\sinh^2\left(\frac{\alpha(x-a)}{2}\right) - 2\alpha^2\sinh^2\left(\frac{\beta(x-a)}{2}\right)}{\beta^2-\alpha^2}\right)$$
$$= f(a)\left(\frac{\beta^2\cosh(\alpha(x-a)) - \alpha^2\cosh(\beta(x-a))}{\beta^2-\alpha^2}\right) - f(a). \tag{1.32}$$

By Theorem 1.5, we obtain

$$f(x) - f(a) = f(a)\left(\frac{2\beta^2\sinh^2\left(\frac{\alpha(x-a)}{2}\right) - 2\alpha^2\sinh^2\left(\frac{\beta(x-a)}{2}\right)}{\beta^2-\alpha^2}\right)$$
$$+ f'(a)\left(\frac{\beta^3\sinh(\alpha(x-a)) - \alpha^3\sinh(\beta(x-a))}{\alpha\beta(\beta^2-\alpha^2)}\right)$$
$$+ f''(a)\left(\frac{\cosh(\beta(x-a)) - \cosh(\alpha(x-a))}{\beta^2-\alpha^2}\right)$$
$$+ f'''(a)\left(\frac{\alpha\sinh(\beta(x-a)) - \beta\sinh(\alpha(x-a))}{\alpha\beta(\beta^2-\alpha^2)}\right)$$

$$+\frac{\left(f''''(a)-\left(\alpha^2+\beta^2\right)f''(a)+\alpha^2\beta^2 f(a)\right)}{\left(\beta^2-\alpha^2\right)}$$
$$\left(\frac{2\alpha}{\beta(\alpha\beta)}\sinh^2\left(\frac{\beta(x-a)}{2}\right)-\frac{2\beta}{\alpha(\alpha\beta)}\sinh^2\left(\frac{\alpha(x-a)}{2}\right)\right)$$
$$+\frac{1}{\alpha\beta\left(\beta^2-\alpha^2\right)}\int_a^x\left[\left(f''''(t)-\left(\alpha^2+\beta^2\right)f''(t)+\alpha^2\beta^2 f(t)\right)\right.$$
$$\left.-\left(f''''(a)-\left(\alpha^2+\beta^2\right)f''(a)+\alpha^2\beta^2 f(a)\right)\right]\left(\alpha\sinh\left(\beta(x-t)\right)\right.$$
$$\left.-\beta\sinh\left(\alpha(x-t)\right)\right)dt. \tag{1.33}$$

Note that

$$\begin{aligned}(\cosh(\beta(x-t)))' &= \sinh(\beta(x-t))(\beta x-\beta t)'\\ &= \sinh(\beta(x-t))(-\beta) = -\beta\sinh(\beta(x-t)),\end{aligned} \tag{1.34}$$

that is,

$$\left(-\frac{1}{\beta}\cosh(\beta(x-t))\right)' = \sinh(\beta(x-t)) \tag{1.35}$$

and

$$\left(-\frac{\alpha}{\beta}\cosh(\beta(x-t))\right)' = \alpha\sinh(\beta(x-t)). \tag{1.36}$$

Thus, it holds

$$\begin{aligned}\int_a^x \alpha(\sinh(\beta(x-t)))\,dt &= -\frac{\alpha}{\beta}(\cosh(\beta(x-t)))\,|_a^x\\ &= -\frac{\alpha}{\beta}(1-\cosh(\beta(x-a))) = -\frac{\alpha}{\beta}(-2)\sinh^2\left(\frac{\beta(x-a)}{2}\right)\\ &= \frac{2\alpha}{\beta}\sinh^2\left(\frac{\beta(x-a)}{2}\right).\end{aligned} \tag{1.37}$$

That is,

$$\int_a^x \alpha\sinh(\beta(x-t))\,dt = \frac{2\alpha}{\beta}\sinh^2\left(\frac{\beta(x-a)}{2}\right) \tag{1.38}$$

and

$$\int_a^x (-\beta)\sinh(\alpha(x-t))\,dt = -\frac{2\beta}{\alpha}\sinh^2\left(\frac{\alpha(x-a)}{2}\right). \tag{1.39}$$

Furthermore,

$$\frac{\alpha^2\beta^2 f(a)}{(\beta^2-\alpha^2)}\left(\frac{2}{\beta^2}\sinh^2\left(\frac{\beta(x-a)}{2}\right)-\frac{2}{\alpha^2}\sinh^2\left(\frac{\alpha(x-a)}{2}\right)\right)$$
$$=\frac{f(a)}{(\beta^2-\alpha^2)}\left(2\alpha^2\sinh^2\left(\frac{\beta(x-a)}{2}\right)-2\beta^2\sinh^2\left(\frac{\alpha(x-a)}{2}\right)\right). \quad (1.40)$$

The claim is proved. □

Next come Opial-type inequalities, for basics see Anastassiou (2011).

Theorem 1.11. *Let $f \in C_K^2(I)$, with interval $I \subset \mathbb{R}$, $a, x \in I$, $a < x$, and $f(a) = f'(a) = 0$, with $p, q > 1 : \frac{1}{p} + \frac{1}{q} = 1$. Then*

$$\int_a^x |f(w)|\,|f''(w)+f(w)|\,dw$$
$$\leq 2^{-\frac{1}{q}}\left(\int_a^x\left(\int_a^w |\sin(w-t)|^p\,dt\right)dw\right)^{\frac{1}{p}}\left(\int_a^x |f''(w)+f(w)|^q\,dw\right)^{\frac{2}{q}}. \quad (1.41)$$

Proof. By Theorem 1.1, we have

$$f(x)=\int_a^x (f''(t)+f(t))\sin(x-t)\,dt$$
and
$$f(w)=\int_a^w (f''(t)+f(t))\sin(w-t)\,dt, \quad (1.42)$$

for $a \leq w \leq x$.

By Hölder's inequality, we have

$$|f(w)| \leq \int_a^w |f''(t)+f(t)|\,|\sin(w-t)|\,dt$$
$$\leq \left(\int_a^w |\sin(w-t)|^p\,dt\right)^{\frac{1}{p}}\left(\int_a^w |f''(t)+f(t)|^q\,dt\right)^{\frac{1}{q}}. \quad (1.43)$$

Call

$$z(w) := \int_a^w |f''(t)+f(t)|^q\,dt, \quad z(a)=0, \quad a \leq w \leq x. \quad (1.44)$$

Then

$$z'(w) = |f''(w)+f(w)|^q, \quad (1.45)$$

and

$$|f''(w) + f(w)| = (z'(w))^{\frac{1}{q}}, \quad \text{all } a \le w \le x.$$

Therefore, we have (all $a \le w \le x$)

$$|f(w)|\,|f''(w) + f(w)| \le \left(\int_a^w |\sin(w-t)|^p\, dt\right)^{\frac{1}{p}} (z(w)\, z'(w))^{\frac{1}{q}}. \tag{1.46}$$

Hence it holds

$$\begin{aligned}
&\int_a^x |f(w)|\,|f''(w) + f(w)|\, dw \\
&\le \int_a^x \left(\int_a^w |\sin(w-t)|^p\, dt\right)^{\frac{1}{p}} (z(w)\, z'(w))^{\frac{1}{q}}\, dw \\
&\le \left(\int_a^x \left(\int_a^w |\sin(w-t)|^p\, dt\right) dw\right)^{\frac{1}{p}} \left(\int_a^x z(w)\, z'(w)\, dw\right)^{\frac{1}{q}} \\
&= \left(\int_a^x \left(\int_a^w |\sin(w-t)|^p\, dt\right) dw\right)^{\frac{1}{p}} \left(\int_a^x z(w)\, dz(w)\right)^{\frac{1}{q}} \qquad (1.47) \\
&= \left(\int_a^x \left(\int_a^w |\sin(w-t)|^p\, dt\right) dw\right)^{\frac{1}{p}} \frac{z(x)^{\frac{2}{q}}}{2^{\frac{1}{q}}} \\
&= \left(\int_a^x \left(\int_a^w |\sin(w-t)|^p\, dt\right) dw\right)^{\frac{1}{p}} \frac{\left(\int_a^x |f''(t) + f(t)|^q\, dt\right)^{\frac{2}{q}}}{2^{\frac{1}{q}}}.
\end{aligned} \tag{1.48}$$

The claim is proved. □

Next come several similar results.

Theorem 1.12. *Let $f \in C_K^2(I)$, $a, x \in I$, $a < x$, and $f(a) = f'(a) = 0$, with $p, q > 1 : \frac{1}{p} + \frac{1}{q} = 1$. Then*

$$\begin{aligned}
&\int_a^x |f(w)|\,|f''(w) - f(w)|\, dw \\
&\le 2^{-\frac{1}{q}} \left(\int_a^x \left(\int_a^w |\sinh(w-t)|^p\, dt\right) dw\right)^{\frac{1}{p}} \left(\int_a^x |f''(w) - f(w)|^q\, dw\right)^{\frac{2}{q}}.
\end{aligned} \tag{1.49}$$

Proof. By Theorem 1.11, use of (1.6). □

Theorem 1.13. *Let $f \in C_K^4(I)$, interval $I \subset \mathbb{R}$, let $a, x \in I$, $a < x$, $f(a) = f'(a) = f''(a) = f'''(a) = 0$, with $p, q > 1 : \frac{1}{p} + \frac{1}{q} = 1$. Then*

$$\int_a^x |f(w)| \left| f^{(iv)}(w) - f(w) \right| dw$$
$$\leq 2^{-\left(1+\frac{1}{q}\right)} \left(\int_a^x \left(\int_a^w |\sinh(w-t) - \sin(w-t)|^p \, dt \right) dw \right)^{\frac{1}{p}}$$
$$\left(\int_a^x \left| f^{(iv)}(w) - f(w) \right|^q dw \right)^{\frac{2}{q}}. \tag{1.50}$$

Proof. As in Theorem 1.11, use of (1.7). □

Theorem 1.14. *All as in Theorem 1.13. Let $\alpha, \beta \in \mathbb{R} : \alpha\beta\left(\alpha^2 - \beta^2\right) \neq 0$. Then*

$$\int_a^x |f(w)| \left| f^{(4)}(w) + \left(\alpha^2 + \beta^2\right) f''(w) + \alpha^2\beta^2 f(w) \right| dw$$
$$\leq \frac{1}{2^{\frac{1}{q}} \left|\alpha\beta\left(\beta^2 - \alpha^2\right)\right|} \left(\int_a^x \left(\int_a^w |\beta \sin(\alpha(w-t)) \right.\right.$$
$$\left.\left. -\alpha \sin(\beta(w-t))|^p \, dt \right) dw \right)^{\frac{1}{p}}$$
$$\left(\int_a^x \left| f^{(4)}(w) + \left(\alpha^2 + \beta^2\right) f''(w) + \alpha^2\beta^2 f(w) \right|^q dw \right)^{\frac{2}{q}}. \tag{1.51}$$

Proof. As in Theorem 1.11, use of (1.8). □

Theorem 1.15. *All as in Theorem 1.13. Let $\alpha \in \mathbb{R}$, $\alpha \neq 0$. Then*

$$\int_a^x |f(w)| \left| f^{(4)}(w) + 2\alpha^2 f''(w) + \alpha^4 f(w) \right| dw$$
$$\leq \frac{1}{2^{\frac{1}{q}+1} |\alpha|^3} \left(\int_a^x \left(\int_a^w |\sin(\alpha(w-t)) - \alpha(w-t)\cos(\alpha(w-t))|^p \, dt \right) dw \right)^{\frac{1}{p}} \tag{1.52}$$
$$\left(\int_a^x \left| f^{(4)}(w) + 2\alpha^2 f''(w) + \alpha^4 f(w) \right|^q dw \right)^{\frac{2}{q}}.$$

Proof. As in Theorem 1.11, use of Corollary 3.8 of Ali and Pales (2022), p. 1135. □

Theorem 1.16. *All as in Theorem 1.14. Then*

$$\int_a^x |f(w)| \left| f^{(4)}(w) - \left(\alpha^2+\beta^2\right) f''(w) + \alpha^2\beta^2 f(w) \right| dw$$

$$\leq \frac{1}{2^{\frac{1}{q}} |\alpha\beta(\beta^2-\alpha^2)|} \left(\int_a^x \left(\int_a^w |\alpha \sinh(\beta(w-t))\right.\right.$$

$$\left.\left. - \beta \sinh(\alpha(w-t))|^p dt \right) dw \right)^{\frac{1}{p}}$$

$$\left(\int_a^x \left| f^{(4)}(w) - \left(\alpha^2+\beta^2\right) f''(w) + \alpha^2\beta^2 f(w) \right|^q dw \right)^{\frac{2}{q}}. \tag{1.53}$$

Proof. As in Theorem 1.11, use of (1.9). □

Theorem 1.17. *All as in Theorem 1.15. Then*

$$\int_a^x |f(w)| \left| f^{(4)}(w) - 2\alpha^2 f''(w) + \alpha^4 f(w) \right| dw$$

$$\leq \frac{1}{2^{\frac{1}{q}+1} |\alpha|^3} \left(\int_a^x \left(\int_a^w |\alpha(w-t)\cosh(\alpha(w-t))\right.\right.$$

$$\left.\left. - \sinh(\alpha(w-t))|^p dt \right) dw \right)^{\frac{1}{p}} \tag{1.54}$$

$$\left(\int_a^x \left| f^{(4)}(w) - 2\alpha^2 f''(w) + \alpha^4 f(w) \right|^q dw \right)^{\frac{2}{q}}.$$

Proof. As in Theorem 1.11, use of Corollary 3.10 of Ali and Pales (2022), p. 1135. □

We finish Opial-type inequalities with the following general result.

Theorem 1.18. *Let $n \in \mathbb{N}$, $c = (c_0, \ldots, c_n) \in K^{n+1}$ with $c_n = 1$, $f \in C_K^n(I)$ and $x, a \in I$, $a < x$, with interval $I \subset \mathbb{R}$. Let also $p, q > 1$ such that $\frac{1}{p} + \frac{1}{q} = 1$. Assume further that $f^{(j)}(a) = 0$, $j = 0, 1, \ldots, n-1$. Then*

$$\int_a^x |f(w)| \, |D_c(f)(w)| \, dw$$

$$\leq 2^{-\frac{1}{q}} \left(\int_a^x \left(\int_a^w |\omega_c(w-t)|^p \, dt \right) dw \right)^{\frac{1}{p}} \left(\int_a^x |D_c(f)(w)|^q \, dw \right)^{\frac{2}{q}}. \tag{1.55}$$

Proof. By Theorem 3.2 of Ali and Pales (2022, p. 1131), for $x, a \in I$, we have

$$f(x) = (T_{a,c}f)(x) + \int_a^x D_c(f)(t)\,\omega_c(x-t)\,dt, \tag{1.56}$$

where

$$(T_{a,c}f)(x) := \sum_{j=0}^{n-1}\left(f^{(j)}(a)\sum_{i=0}^{n-1-j} c_{i+j+1}\omega_c^{(i)}(x-a)\right). \tag{1.57}$$

Because $f^{(j)}(a) = 0$, $j = 0, 1, \ldots, n-1$, we get

$$f(x) = \int_a^x D_c(f)(t)\,\omega_c(x-t)\,dt. \tag{1.58}$$

The rest of the proof similar to Theorem 1.11 is omitted. □

Next we present Ostrowski-type inequalities involving $\|\cdot\|_\infty$. For basics, see Anastassiou (2011).

Theorem 1.19. *Let $f \in C_K^3([c,d])$, $a \in [c,d]$, such that $f'(a) = f''(a) = 0$. Then*

(1)

$$\left|\frac{1}{d-c}\int_c^d f(x)\,dx - f(a)\right| \leq \|f''' + f'\|_\infty \frac{\left[(d-a)^3 + (a-c)^3\right]}{6(d-c)}, \tag{1.59}$$

(2) *when $f'\left(\frac{c+d}{2}\right) = f''\left(\frac{c+d}{2}\right) = 0$, and $a = \frac{c+d}{2}$, we get*

$$\left|\frac{1}{d-c}\int_c^d f(x)\,dx - f\left(\frac{c+d}{2}\right)\right| \leq \|f' + f'''\|_\infty \frac{(d-c)^2}{24}. \tag{1.60}$$

Proof. By Theorem 1.6 (1.10), we have

$$f(x) - f(a) = \int_a^x \left[(f''(t) + f(t)) - (f''(a) + f(a))\right]\sin(x-t)\,dt. \tag{1.61}$$

Let $x \geq a$, then

$$\begin{aligned}|f(x) - f(a)| &\leq \int_a^x |(f''(t) + f(t)) - (f''(a) + f(a))|\,|\sin(x-t)|\,dt \\ &\leq \int_a^x |(f''(t) + f(t)) - (f''(a) + f(a))|\,dt \\ &\leq \|f''' + f'\|_\infty \int_a^x (t-a)\,dt = \|f''' + f'\|_\infty \frac{(x-a)^2}{2}.\end{aligned} \tag{1.62}$$

Let $x < a$, then

$$-\left(f\left(x\right)-f\left(a\right)\right)=\int_{x}^{a}\left[\left(f''\left(a\right)+f\left(a\right)\right)-\left(f''\left(t\right)+f\left(t\right)\right)\right]\sin\left(x-t\right)dt.$$

Hence,

$$\begin{aligned}\left|f\left(x\right)-f\left(a\right)\right| &\leq \int_{x}^{a}\left|\left(f''\left(a\right)+f\left(a\right)\right)-\left(f''\left(t\right)+f\left(t\right)\right)\right|\left|\sin\left(x-t\right)\right|dt\\ &\leq \left\|f'''+f'\right\|_{\infty}\int_{x}^{a}\left(a-t\right)dt=\left\|f'''+f'\right\|_{\infty}\frac{\left(a-t\right)^{2}}{2}\Big|_{a}^{x}\\ &= \left\|f'''+f'\right\|_{\infty}\frac{\left(a-x\right)^{2}}{2}. \qquad (1.63)\end{aligned}$$

Therefore,

$$\left|f\left(x\right)-f\left(a\right)\right| \leq \frac{\left\|f'''+f'\right\|_{\infty}}{2}\left(x-a\right)^{2}, \qquad (1.64)$$

$\forall\, x \in [c,d]$.

We observe that

$$\begin{aligned}\left|\frac{1}{d-c}\int_{c}^{d}f\left(x\right)dx-f\left(a\right)\right| &= \left|\frac{1}{d-c}\int_{c}^{d}\left(f\left(x\right)-f\left(a\right)\right)dx\right|\\ &\leq \frac{1}{d-c}\int_{c}^{d}\left|f\left(x\right)-f\left(a\right)\right|dx \leq \frac{1}{d-c}\left(\frac{\left\|f'''+f'\right\|_{\infty}}{2}\right)\int_{c}^{d}\left(x-a\right)^{2}dx\\ &= \frac{\left\|f'''+f'\right\|_{\infty}}{2\left(d-c\right)}\left[\int_{c}^{a}\left(a-x\right)^{2}dx+\int_{a}^{d}\left(x-a\right)^{2}dx\right]\\ &= \frac{\left\|f'''+f'\right\|_{\infty}}{6\left(d-c\right)}\left[\left(a-x\right)^{3}\Big|_{a}^{c}+\left(x-a\right)^{3}\Big|_{a}^{d}\right]\\ &= \frac{\left\|f'''+f'\right\|_{\infty}}{6\left(d-c\right)}\left[\left(a-c\right)^{3}+\left(d-a\right)^{3}\right]\\ &= \left(\frac{\left\|f'''+f'\right\|_{\infty}}{6\left(d-c\right)}\right)\left[\left(d-a\right)^{3}+\left(a-c\right)^{3}\right]. \qquad (1.65)\end{aligned}$$

The claim is proved. □

Using $|\sin x| \leq |x|$, we obtain an alternative Ostrowski-type inequality.

Theorem 1.20. *All as in Theorem 1.19. Then*

(1)

$$\left| \frac{1}{d-c} \int_c^d f(x)\,dx - f(a) \right| \leq \|f''' + f'\|_\infty \frac{\left[(d-a)^4 + (a-c)^4 \right]}{24(d-c)}, \quad (1.66)$$

(2) *when* $f'\left(\frac{c+d}{2}\right) = f''\left(\frac{c+d}{2}\right) = 0$, $a = \frac{c+d}{2}$, *we get that*

$$\left| \frac{1}{d-c} \int_c^d f(x)\,dx - f\left(\frac{c+d}{2}\right) \right| \leq \|f' + f'''\|_\infty \frac{(d-c)^3}{192}. \quad (1.67)$$

Proof. Again by Theorem 1.6 (1.10) we have

$$f(x) - f(a) = \int_a^x \left[(f''(t) + f(t)) - (f''(a) + f(a)) \right] \sin(x-t)\,dt. \quad (1.68)$$

Let $x \geq a$, then

$$\begin{aligned} |f(x) - f(a)| &\leq \int_a^x \left| (f''(t) + f(t)) - (f''(a) + f(a)) \right| |\sin(x-t)|\,dt \\ &\leq \|f''' + f'\|_\infty \int_a^x (t-a) |\sin(x-t)|\,dt \\ &\leq \|f''' + f'\|_\infty \int_a^x (x-t)^{2-1} (t-a)^{2-1}\,dt \quad (1.69) \end{aligned}$$

$$\|f''' + f'\|_\infty \frac{\Gamma(2)\Gamma(2)}{\Gamma(4)} (x-a)^3 = \|f''' + f'\|_\infty \frac{(x-a)^3}{3!}.$$

That is,

$$|f(x) - f(a)| \leq \|f''' + f'\|_\infty \frac{(x-a)^3}{3!}. \quad (1.70)$$

Let $x < a$, then

$$-(f(x) - f(a)) = \int_x^a \left[(f''(a) + f(a)) - (f''(t) + f(t)) \right] \sin(x-t)\,dt. \quad (1.71)$$

Hence,

$$\begin{aligned} |f(x) - f(a)| &\leq \int_x^a \left| (f''(a) + f(a)) - (f''(t) + f(t)) \right| |\sin(x-t)|\,dt \\ &\leq \|f''' + f'\|_\infty \int_x^a (a-t) |x-t|\,dt \end{aligned}$$

$$= \|f''' + f'\|_\infty \int_x^a (a-t)^{2-1} (t-x)^{2-1} \, dt$$

$$= \|f''' + f'\|_\infty \frac{\Gamma(2)\Gamma(2)}{\Gamma(4)} (a-x)^3 = \|f''' + f'\|_\infty \frac{(a-x)^3}{3!}. \tag{1.72}$$

That is,

$$|f(x) - f(a)| \leq \|f''' + f'\|_\infty \frac{(a-x)^3}{3!}. \tag{1.73}$$

Next we observe that

$$\left| \frac{1}{d-c} \int_c^d f(x) \, dx - f(a) \right| \leq \frac{1}{d-c} \int_c^d |f(x) - f(a)| \, dx$$

$$\leq \frac{\|f''' + f'\|_\infty}{3!\,(d-c)} \left[\int_c^a (a-x)^3 \, dx + \int_a^d (x-a)^3 \, dx \right]$$

$$= \frac{\|f''' + f'\|_\infty}{4!\,(d-c)} \left[(a-c)^4 + (d-a)^4 \right]. \tag{1.74}$$

The claim is proved. □

We continue with more involved Ostrowski-type inequalities.

Theorem 1.21. *Let $f \in C_K^5([c,d])$, $a, x \in [c,d]$, and assume that $f^{(i)}(a) = 0$, $i = 1,2,3,4$. Here $\alpha, \beta \in \mathbb{R} : \alpha\beta\left(\alpha^2 - \beta^2\right) \neq 0$. Then*

(1)

$$\left| \frac{1}{d-c} \int_c^d f(x) \, dx - f(a) \right|$$

$$\leq \frac{(|\alpha| + |\beta|) \left\| f^{(5)} + \left(\alpha^2 + \beta^2\right) f^{(3)} + \alpha^2 \beta^2 f' \right\|_\infty}{6 \left| \alpha\beta \left(\beta^2 - \alpha^2\right) \right| (d-c)} \left[(d-a)^3 + (a-c)^3 \right], \tag{1.75}$$

(2) *when $f^{(i)}\left(\frac{c+d}{2}\right) = 0$, $i = 1,2,3,4$, we have*

$$\left| \frac{1}{d-c} \int_c^d f(x) \, dx - f\left(\frac{c+d}{2}\right) \right|$$

$$\leq \frac{(|\alpha| + |\beta|) \left\| f^{(5)} + \left(\alpha^2 + \beta^2\right) f^{(3)} + \alpha^2 \beta^2 f' \right\|_\infty}{24 \left| \alpha\beta \left(\beta^2 - \alpha^2\right) \right|} (d-c)^2. \tag{1.76}$$

Example 1.1. (to (1.75)) Let $\alpha = 1$, $\beta = 10$, then

$$\left| \frac{1}{d-c} \int_c^d f(x)\,dx - f(a) \right| \le \frac{\left\| f^{(5)} + 101 f^{(3)} + 100 f' \right\|_\infty}{540(d-c)} \left[(d-a)^3 + (a-c)^3 \right]. \tag{1.77}$$

Example 1.2. (to (1.76)) When $\alpha = 2$, $\beta = 1$, we get

$$\left| \frac{1}{d-c} \int_c^d f(x)\,dx - f\left(\frac{c+d}{2} \right) \right| \le \frac{\left\| f^{(5)} + 5 f^{(3)} + 4 f' \right\|_\infty}{48} (d-c)^2. \tag{1.78}$$

Proof (of Theorem 1.21). By Theorem 1.9 (1.21), we have

$$f(x) - f(a) = \frac{1}{\alpha\beta(\beta^2 - \alpha^2)} \int_a^x \left[\left(f''''(t) + (\alpha^2 + \beta^2) f''(t) + \alpha^2\beta^2 f(t) \right) - \left(f''''(a) + (\alpha^2+\beta^2) f''(a) + \alpha^2\beta^2 f(a) \right) \right] \left[\beta \sin(\alpha(x-t)) - \alpha \sin(\beta(x-t)) \right] dt. \tag{1.79}$$

Let $x \ge a$, then

$$\begin{aligned} |f(x) - f(a)| &\le \frac{(|\alpha| + |\beta|)}{|\alpha\beta(\beta^2 - \alpha^2)|} \left\| f^{(5)} + (\alpha^2+\beta^2) f^{(3)} + \alpha^2\beta^2 f' \right\|_\infty \int_a^x (t-a)\,dt \\ &= \frac{(|\alpha| + |\beta|)}{2|\alpha\beta(\beta^2 - \alpha^2)|} \left\| f^{(5)} + (\alpha^2+\beta^2) f^{(3)} + \alpha^2\beta^2 f' \right\|_\infty (x-a)^2. \end{aligned} \tag{1.80}$$

Let $x < a$, then

$$-(f(x) - f(a)) = \frac{1}{\alpha\beta(\beta^2 - \alpha^2)} \int_x^a \left[\left(f''''(t) + (\alpha^2 + \beta^2) f''(t) + \alpha^2\beta^2 f(t) \right) - \left(f''''(a) + (\alpha^2+\beta^2) f''(a) + \alpha^2\beta^2 f(a) \right) \right] \left[\beta \sin(\alpha(x-t)) - \alpha \sin(\beta(x-t)) \right] dt. \tag{1.81}$$

Hence we have

$$
\begin{aligned}
&|f(x)-f(a)| \\
&\leq \frac{(|\alpha|+|\beta|)}{|\alpha\beta(\beta^2-\alpha^2)|}\left\|f^{(5)}+\left(\alpha^2+\beta^2\right)f^{(3)}+\alpha^2\beta^2 f'\right\|_\infty \int_x^a (a-t)\,dt \\
&= \frac{(|\alpha|+|\beta|)}{|\alpha\beta(\beta^2-\alpha^2)|}\left\|f^{(5)}+\left(\alpha^2+\beta^2\right)f^{(3)}+\alpha^2\beta^2 f'\right\|_\infty \frac{(a-x)^2}{2}. \quad (1.82)
\end{aligned}
$$

Therefore, it holds

$$
|f(x)-f(a)| \leq \frac{(|\alpha|+|\beta|)\left\|f^{(5)}+\left(\alpha^2+\beta^2\right)f^{(3)}+\alpha^2\beta^2 f'\right\|_\infty}{2|\alpha\beta(\beta^2-\alpha^2)|}(x-a)^2, \quad (1.83)
$$

$\forall\, x \in [c,d]$.

We observe that

$$
\begin{aligned}
&\left|\frac{1}{d-c}\int_c^d f(x)\,dx - f(a)\right| \leq \frac{1}{d-c}\int_c^d |f(x)-f(a)|\,dx \\
&\leq \frac{(|\alpha|+|\beta|)\left\|f^{(5)}+\left(\alpha^2+\beta^2\right)f^{(3)}+\alpha^2\beta^2 f'\right\|_\infty}{2|\alpha\beta(\beta^2-\alpha^2)|(d-c)}\int_c^d (x-a)^2\,dx \\
&= \frac{(|\alpha|+|\beta|)\left\|f^{(5)}+\left(\alpha^2+\beta^2\right)f^{(3)}+\alpha^2\beta^2 f'\right\|_\infty}{6|\alpha\beta(\beta^2-\alpha^2)|(d-c)}\left[(a-c)^3+(d-a)^3\right]. \quad (1.84)
\end{aligned}
$$

The claim is proved. □

A long alternative Ostrowski-type inequality follows.

Theorem 1.22. *Let $f \in C_K^5([c,d])$, $a, x \in [c,d]$, and assume that $f^{(i)}(a) = 0$, $i = 1,2,3,4$. Here $\alpha, \beta \in \mathbb{R} : \alpha\beta\left(\alpha^2-\beta^2\right) \neq 0$. Then*

(1)

$$
\begin{aligned}
&\left|\frac{1}{d-c}\int_c^d f(x)\,dx - f(a)\right| \\
&\leq \frac{\left\|f^{(5)}+\left(\alpha^2+\beta^2\right)f^{(3)}+\alpha^2\beta^2 f'\right\|_\infty}{12|\beta^2-\alpha^2|(d-c)}\left[(d-a)^4+(a-c)^4\right], \quad (1.85)
\end{aligned}
$$

(2) *above let* $a = \frac{c+d}{2}$, *then*

$$\left|\frac{1}{d-c}\int_c^d f(x)\,dx - f\left(\frac{c+d}{2}\right)\right| \le \frac{\left\|f^{(5)} + \left(\alpha^2+\beta^2\right) f^{(3)} + \alpha^2\beta^2 f'\right\|_\infty}{96\left|\beta^2-\alpha^2\right|}(d-c)^3. \tag{1.86}$$

Example 1.3 (to (1.85)). Let $\alpha = 1$, $\beta = 10$, then

$$\left|\frac{1}{d-c}\int_c^d f(x)\,dx - f(a)\right| \le \frac{\left\|f^{(5)} + 101 f^{(3)} + 100 f'\right\|_\infty}{1188\,(d-c)}\left[(d-a)^4 + (a-c)^4\right]. \tag{1.87}$$

Example 1.4 (to (1.86)). When $\alpha = 2$, $\beta = 1$, we get

$$\left|\frac{1}{d-c}\int_c^d f(x)\,dx - f\left(\frac{c+d}{2}\right)\right| \le \frac{\left\|f^{(5)} + 5 f^{(3)} + 4 f'\right\|_\infty}{288}(d-c)^3. \tag{1.88}$$

Proof (of Theorem 1.22). By Theorem 1.9 (1.21), we have

$$f(x) - f(a) = \frac{1}{\alpha\beta\left(\beta^2-\alpha^2\right)} \int_a^x \left[\left(f''''(t) + \left(\alpha^2+\beta^2\right) f''(t) + \alpha^2\beta^2 f(t)\right) - \left(f''''(a) + \left(\alpha^2+\beta^2\right) f''(a) + \alpha^2\beta^2 f(a)\right)\right]\left[\beta\sin\left(\alpha(x-t)\right) - \alpha\sin\left(\beta(x-t)\right)\right] dt. \tag{1.89}$$

Let $x \ge a$ (by $|\sin x| \le |x|$), then

$$|f(x) - f(a)| \le \frac{1}{\left|\alpha\beta\left(\beta^2-\alpha^2\right)\right|}\left\|f^{(5)} + \left(\alpha^2+\beta^2\right) f^{(3)} + \alpha^2\beta^2 f'\right\|_\infty \int_a^x (t-a)\left[|\beta|\left|\alpha(x-t)\right| + |\alpha|\left|\beta(x-t)\right|\right] dt \tag{1.90}$$

$$= \frac{2|\alpha\beta|}{\left|\alpha\beta\left(\beta^2-\alpha^2\right)\right|}\left\|f^{(5)} + \left(\alpha^2+\beta^2\right) f^{(3)} + \alpha^2\beta^2 f'\right\|_\infty \int_a^x (x-t)(t-a)\,dt$$

$$= \frac{2}{|\beta^2 - \alpha^2|} \left\| f^{(5)} + (\alpha^2 + \beta^2) f^{(3)} + \alpha^2 \beta^2 f' \right\|_\infty \frac{(x-a)^3}{3!}$$

$$= \frac{1}{3|\beta^2 - \alpha^2|} \left\| f^{(5)} + (\alpha^2 + \beta^2) f^{(3)} + \alpha^2 \beta^2 f' \right\|_\infty (x-a)^3. \quad (1.91)$$

So, when $x \geq a$, we get that

$$|f(x) - f(a)| \leq \frac{1}{3|\beta^2 - \alpha^2|} \left\| f^{(5)} + (\alpha^2 + \beta^2) f^{(3)} + \alpha^2 \beta^2 f' \right\|_\infty (x-a)^3. \quad (1.92)$$

Let $x < a$, then

$$-(f(x) - f(a)) = \frac{1}{\alpha\beta(\beta^2 - \alpha^2)}$$

$$\int_x^a \left[\left(f''''(t) + (\alpha^2 + \beta^2) f''(t) + \alpha^2 \beta^2 f(t) \right) \right.$$

$$- \left. \left(f''''(a) + (\alpha^2 + \beta^2) f''(a) + \alpha^2 \beta^2 f(a) \right) \right] [\beta \sin(\alpha(x-t))$$

$$-\alpha \sin(\beta(x-t))] \, dt. \quad (1.93)$$

Hence (by $|\sin x| \leq |x|$), we get that

$$|f(x) - f(a)| \leq \frac{1}{|\alpha\beta(\beta^2 - \alpha^2)|} \left\| f^{(5)} + (\alpha^2 + \beta^2) f^{(3)} + \alpha^2 \beta^2 f' \right\|_\infty$$

$$\int_x^a (a-t) \left[|\beta| \, |\alpha| \, (t-x) + |\alpha| \, |\beta| \, (t-x) \right] dt$$

$$= \frac{2}{|\beta^2 - \alpha^2|} \left\| f^{(5)} + (\alpha^2 + \beta^2) f^{(3)} + \alpha^2 \beta^2 f' \right\|_\infty \int_x^a (a-t)(t-x) \, dt$$

$$= \frac{2}{|\beta^2 - \alpha^2|} \left\| f^{(5)} + (\alpha^2 + \beta^2) f^{(3)} + \alpha^2 \beta^2 f' \right\|_\infty \frac{(a-x)^3}{3!}$$

$$= \frac{1}{3|\beta^2 - \alpha^2|} \left\| f^{(5)} + (\alpha^2 + \beta^2) f^{(3)} + \alpha^2 \beta^2 f' \right\|_\infty (a-x)^3. \quad (1.94)$$

So, when $x < a$, we got that

$$|f(x) - f(a)| \leq \frac{1}{3|\beta^2 - \alpha^2|} \left\| f^{(5)} + (\alpha^2 + \beta^2) f^{(3)} + \alpha^2 \beta^2 f' \right\|_\infty (a-x)^3. \quad (1.95)$$

We observe that

$$\left|\frac{1}{d-c}\int_c^d f(x)\,dx - f(a)\right| \leq \frac{1}{d-c}\int_c^d |f(x) - f(a)|\,dx$$

$$\leq \frac{1}{3\left|\beta^2-\alpha^2\right|(d-c)}\left\|f^{(5)} + \left(\alpha^2+\beta^2\right)f^{(3)} + \alpha^2\beta^2 f'\right\|_\infty$$

$$\left[\int_c^a (a-x)^3\,dx + \int_a^d (x-a)^3\,dx\right]$$

$$= \frac{\left\|f^{(5)} + \left(\alpha^2+\beta^2\right)f^{(3)} + \alpha^2\beta^2 f'\right\|_\infty}{12\left|\beta^2-\alpha^2\right|(d-c)}\left[(d-a)^4 + (a-c)^4\right]. \quad (1.96)$$

The claim is proved. □

More Ostrowski-type inequalities follow regarding $\|\cdot\|_p$, $p \geq 1$.

Theorem 1.23. *Let $f \in C_K^2([c,d])$, $a \in [c,d]$, and assume that $f'(a) = f''(a) = 0$. Then*

$$\left|\frac{1}{d-c}\int_c^d f(x)\,dx - f(a)\right|$$

$$\leq \frac{\left[(a-c)\left\|f''+f-f(a)\right\|_{L_1([c,a])} + (d-a)\left\|f''+f-f(a)\right\|_{L_1([a,d])}\right]}{d-c}. \quad (1.97)$$

Proof. As in the proof of Theorem 1.19 (1.62), let $x \geq a$, then

$$|f(x) - f(a)| \leq \int_a^x |f''(t) + f(t) - f(a)|\,dt$$

$$\leq \int_a^d |f''(t) + f(t) - f(a)|\,dt = \|f'' + f - f(a)\|_{L_1([a,d])}. \quad (1.98)$$

Next let $x < a$, by (1.63) we get that

$$|f(x) - f(a)| \leq \int_x^a |f(a) - f''(t) - f(t)|\,dt$$

$$\leq \int_c^a |f''(t) + f(t) - f(a)|\,dt = \|f'' + f - f(a)\|_{L_1([c,a])}. \quad (1.99)$$

Hence it holds

$$\left|\frac{1}{d-c}\int_c^d f(x)\,dx - f(a)\right| = \frac{1}{d-c}\left|\int_c^d (f(x)-f(a))\,dx\right|$$
$$\leq \frac{1}{d-c}\int_c^d |f(x)-f(a)|\,dx$$
$$= \frac{1}{d-c}\left[\int_c^a |f(x)-f(a)|\,dx + \int_a^d |f(x)-f(a)|\,dx\right]$$
$$\leq \frac{1}{d-c}\left[\|f''+f-f(a)\|_{L_1([c,a])}(a-c)\right.$$
$$\left. + \|f''+f-f(a)\|_{L_1([a,d])}(d-a)\right], \tag{1.100}$$

proving the claim. □

Theorem 1.24. *All as in Theorem 1.23. Then*

$$\left|\frac{1}{d-c}\int_c^d f(x)\,dx - f(a)\right|$$
$$\leq \frac{\left[(a-c)^2\|f''+f-f(a)\|_{L_1([c,a])} + (d-a)^2\|f''+f-f(a)\|_{L_1([a,d])}\right]}{2(d-c)}. \tag{1.101}$$

Proof. As in (1.69) for $x \geq a$, we get

$$|f(x)-f(a)| \leq \int_a^x |(f''(t)+f(t))-f(a)|\,|\sin(x-t)|dt$$
$$\leq \int_a^d |(f''(t)+f(t))-f(a)|\,(x-t)\,dt$$
$$\leq (x-a)\|f''+f-f(a)\|_{L_1([a,d])}. \tag{1.102}$$

Also, as in (1.72) for $x < a$, we have

$$|f(x)-f(a)| \leq \int_x^a |f(a)-f''(t)-f(t)|\,|\sin(x-t)|dt$$
$$\leq \int_x^a |f(a)-f''(t)-f(t)|\,(t-x)\,dt \leq (a-x)\|f''+f-f(a)\|_{L_1([c,a])}. \tag{1.103}$$

Hence it holds (see (1.100))

$$\begin{aligned}
&\left|\frac{1}{d-c}\int_c^d f(x)\,dx - f(a)\right| \\
&\leq \frac{1}{d-c}\left[\int_c^a |f(x)-f(a)|\,dx + \int_a^d |f(x)-f(a)|\,dx\right] \\
&\leq \frac{1}{d-c}\left[\|f''+f-f(a)\|_{L_1([c,a])}\int_c^a (a-x)\,dx\right. \\
&\quad \left. + \|f''+f-f(a)\|_{L_1([a,d])}\int_a^d (x-a)\,dx\right] \\
&= \frac{1}{d-c}\left[\|f''+f-f(a)\|_{L_1([c,a])}\frac{(a-c)^2}{2}\right. \\
&\quad \left. + \|f''+f-f(a)\|_{L_1([a,d])}\frac{(d-a)^2}{2}\right], \qquad (1.104)
\end{aligned}$$

proving the claim. □

Theorem 1.25. *All as in Theorem 1.23. Let also* $p, q > 1 : \frac{1}{p} + \frac{1}{q} = 1$. *Then*

$$\left|\frac{1}{d-c}\int_c^d f(x)\,dx - f(a)\right| \leq \frac{\left[\|f''+f-f(a)\|_{L_p([c,a])}(a-c)^{1+\frac{1}{q}} + \|f''+f-f(a)\|_{L_p([a,d])}(d-a)^{1+\frac{1}{q}}\right]}{\left(1+\frac{1}{q}\right)(d-c)}. \qquad (1.105)$$

Proof. Let $x \geq a$, as in (1.98), we get

$$\begin{aligned}
|f(x)-f(a)| &\leq \int_a^x |(f''(t)+f(t)) - f(a)|\,dt \\
&\leq \left(\int_a^x |f''(t)+f(t)-f(a)|^p\,dt\right)^{\frac{1}{p}}\left(\int_a^x 1dt\right)^{\frac{1}{q}} \\
&\leq \|f''+f-f(a)\|_{L_p([a,d])}(x-a)^{\frac{1}{q}}. \qquad (1.106)
\end{aligned}$$

Let $x < a$, as in (1.99), we get

$$\begin{aligned}|f(x) - f(a)| &\leq \int_x^a |f''(t) + f(t) - f(a)|\, dt \\ &\leq \left(\int_x^a |f''(t) + f(t) - f(a)|^p\, dt\right)^{\frac{1}{p}} \left(\int_x^a 1dt\right)^{\frac{1}{q}} \\ &\leq \|f'' + f - f(a)\|_{L_p([c,a])} (a-x)^{\frac{1}{q}}. \end{aligned} \tag{1.107}$$

Acting as in (1.100) we obtain

$$\begin{aligned}&\left|\frac{1}{d-c}\int_c^d f(x)\,dx - f(a)\right| \\ &\leq \frac{1}{d-c}\left[\int_c^a |f(x) - f(a)|\,dx + \int_a^d |f(x) - f(a)|\,dx\right] \\ &\leq \frac{1}{d-c}\left[\|f'' + f - f(a)\|_{L_p([c,a])}\int_c^a (a-x)^{\frac{1}{q}}\,dx\right. \\ &\quad \left. + \|f'' + f - f(a)\|_{L_p([a,d])}\int_a^d (x-a)^{\frac{1}{q}}\,dx\right] \\ &= \frac{1}{d-c}\left[\|f'' + f - f(a)\|_{L_p([c,a])}\frac{(a-c)^{1+\frac{1}{q}}}{\left(1+\frac{1}{q}\right)}\right. \\ &\quad \left. + \|f'' + f - f(a)\|_{L_p([a,d])}\frac{(d-a)^{1+\frac{1}{q}}}{\left(1+\frac{1}{q}\right)}\right], \end{aligned} \tag{1.108}$$

proving the claim. □

Theorem 1.26. *All as in Theorem 1.25. Then*

$$\begin{aligned}&\left|\frac{1}{d-c}\int_c^d f(x)\,dx - f(a)\right| \\ &\leq \frac{\left[(a-c)^{2+\frac{1}{q}}\|f'' + f - f(a)\|_{L_p([c,a])} + (d-a)^{2+\frac{1}{q}}\|f'' + f - f(a)\|_{L_p([a,d])}\right]}{(q+1)^{\frac{1}{q}}\left(2+\frac{1}{q}\right)(d-c)}. \end{aligned} \tag{1.109}$$

Proof. Let $x \geq a$, as in (1.102), we get

$$\begin{aligned} |f(x) - f(a)| &\leq \int_a^x |(f''(t) + f(t)) - f(a)| (x-t)\, dt \\ &\leq \left(\int_a^x |f''(t) + f(t) - f(a)|^p\, dt \right)^{\frac{1}{p}} \left(\int_a^x (x-t)^q\, dt \right)^{\frac{1}{q}} \\ &\leq \|f'' + f - f(a)\|_{L_p([a,d])} \frac{(x-a)^{\frac{q+1}{q}}}{(q+1)^{\frac{1}{q}}}. \end{aligned} \tag{1.110}$$

Also, when $x < a$, by (1.103) we obtain

$$\begin{aligned} |f(x) - f(a)| &\leq \int_x^a |f''(t) + f(t) - f(a)| (t-x)\, dt \\ &\leq \left(\int_x^a |f''(t) + f(t) - f(a)|^p\, dt \right)^{\frac{1}{p}} \left(\int_x^a (t-x)^q\, dt \right)^{\frac{1}{q}} \\ &\leq \|f'' + f - f(a)\|_{L_p([c,a])} \frac{(a-x)^{\frac{q+1}{q}}}{(q+1)^{\frac{1}{q}}}. \end{aligned} \tag{1.111}$$

As in (1.104) we derive

$$\begin{aligned} &\left| \frac{1}{d-c} \int_c^d f(x)\, dx - f(a) \right| \\ &\leq \frac{1}{d-c} \left[\int_c^a |f(x) - f(a)|\, dx + \int_a^d |f(x) - f(a)|\, dx \right] \\ &\leq \frac{1}{d-c} \left[\|f'' + f - f(a)\|_{L_p([c,a])} \frac{\int_c^a (a-x)^{\frac{q+1}{q}}\, dx}{(q+1)^{\frac{1}{q}}} \right. \\ &\quad \left. + \|f'' + f - f(a)\|_{L_p([a,d])} \frac{\int_a^d (x-a)^{\frac{q+1}{q}}\, dx}{(q+1)^{\frac{1}{q}}} \right] \qquad (1.112) \\ &= \frac{1}{d-c} \left[\|f'' + f - f(a)\|_{L_p([c,a])} \frac{(a-c)^{\frac{q+1}{q}+1}}{(q+1)^{\frac{1}{q}} \left(\frac{q+1}{q} + 1 \right)} \right. \end{aligned}$$

$$+ \left\| f'' + f - f(a) \right\|_{L_p([a,d])} \frac{(d-a)^{\frac{q+1}{q}+1}}{(q+1)^{\frac{1}{q}} \left(\frac{q+1}{q}+1\right)} \Bigg]$$

$$= \frac{\left[\|f'' + f - f(a)\|_{L_p([c,a])} (a-c)^{2+\frac{1}{q}} + \|f'' + f - f(a)\|_{L_p([a,d])} (d-a)^{2+\frac{1}{q}} \right]}{(q+1)^{\frac{1}{q}} \left(2+\frac{1}{q}\right)(d-c)}, \tag{1.113}$$

proving the claim. □

We continue with more involved L_p, $p \geq 1$, Ostrowski-type inequalities.

Theorem 1.27. *Let $f \in C_K^4([c,d])$; $a \in [c,d] : f^{(i)}(a) = 0$, $i = 1,2,3,4$. Here $\alpha, \beta \in \mathbb{R} : \alpha\beta\left(\alpha^2 - \beta^2\right) \neq 0$. Then*

$$\left| \frac{1}{d-c} \int_c^d f(x)\,dx - f(a) \right| \leq \left(\frac{(|\alpha| + |\beta|)}{|\alpha\beta(\beta^2 - \alpha^2)|(d-c)} \right) \tag{1.114}$$

$$\left[(a-c) \left\| f'''' + \left(\alpha^2 + \beta^2\right) f'' + \alpha^2\beta^2 f - \alpha^2\beta^2 f(a) \right\|_{L_1([c,a])} \right.$$

$$\left. + (d-a) \left\| f'''' + \left(\alpha^2 + \beta^2\right) f'' + \alpha^2\beta^2 f - \alpha^2\beta^2 f(a) \right\|_{L_1([a,d])} \right].$$

Proof. By (1.79), we obtain ($x \geq a$)

$$|f(x) - f(a)| \leq \left(\frac{(|\alpha| + |\beta|)}{|\alpha\beta(\beta^2 - \alpha^2)|} \right)$$

$$\int_a^x \left| \left[\left(f''''(t) + \left(\alpha^2 + \beta^2\right) f''(t) + \alpha^2\beta^2 f(t) \right) \right. \right.$$

$$\left. \left. - \left(f''''(a) + \left(\alpha^2 + \beta^2\right) f''(a) + \alpha^2\beta^2 f(a) \right) \right] \right| dt$$

$$\leq \left(\frac{(|\alpha| + |\beta|)}{|\alpha\beta(\beta^2 - \alpha^2)|} \right) \int_a^d \left| f''''(t) + \left(\alpha^2 + \beta^2\right) f''(t) + \alpha^2\beta^2 f(t) \right.$$

$$\left. -\alpha^2\beta^2 f(a) \right| dt$$

$$= \left(\frac{(|\alpha| + |\beta|)}{|\alpha\beta(\beta^2 - \alpha^2)|} \right) \left\| f'''' + \left(\alpha^2 + \beta^2\right) f'' + \alpha^2\beta^2 f \right.$$

$$\left. - \alpha^2\beta^2 f(a) \right\|_{L_1([a,d])}. \tag{1.115}$$

From (1.81), we get $(x < a)$

$$\begin{aligned} &|f(x) - f(a)| \\ &\leq \left(\frac{(|\alpha| + |\beta|)}{|\alpha\beta(\beta^2 - \alpha^2)|}\right) \int_x^a \left|f''''(t) + \left(\alpha^2 + \beta^2\right) f''(t) + \alpha^2\beta^2 f(t) - \alpha^2\beta^2 f(a)\right| dt \\ &\leq \left(\frac{(|\alpha| + |\beta|)}{|\alpha\beta(\beta^2 - \alpha^2)|}\right) \int_c^a \left|f''''(t) + \left(\alpha^2 + \beta^2\right) f''(t) + \alpha^2\beta^2 f(t) - \alpha^2\beta^2 f(a)\right| dt \\ &= \left(\frac{(|\alpha| + |\beta|)}{|\alpha\beta(\beta^2 - \alpha^2)|}\right) \left\|f'''' + \left(\alpha^2 + \beta^2\right) f'' + \alpha^2\beta^2 f - \alpha^2\beta^2 f(a)\right\|_{L_1([c,a])}. \end{aligned} \tag{1.116}$$

By (1.100), we get that

$$\begin{aligned} &\left|\frac{1}{d-c}\int_c^d f(x)\,dx - f(a)\right| \\ &\leq \frac{1}{d-c}\left[\int_c^a |f(x) - f(a)|\,dx + \int_a^d |f(x) - f(a)|\,dx\right] \\ &\leq \left(\frac{(|\alpha| + |\beta|)}{|\alpha\beta(\beta^2 - \alpha^2)|(d-c)}\right) \\ &\quad \left[\left\|f'''' + \left(\alpha^2 + \beta^2\right) f'' + \alpha^2\beta^2 f - \alpha^2\beta^2 f(a)\right\|_{L_1([c,a])} (a-c)\right. \\ &\quad \left. + \left\|f'''' + \left(\alpha^2 + \beta^2\right) f'' + \alpha^2\beta^2 f - \alpha^2\beta^2 f(a)\right\|_{L_1([a,d])} (d-a)\right]. \end{aligned} \tag{1.117}$$

The claim is proved. □

The counterpart of (1.114) follows.

Theorem 1.28. *All as in Theorem 1.27. Then*

$$\begin{aligned} &\left|\frac{1}{d-c}\int_a^d f(x)\,dx - f(a)\right| \leq \frac{1}{(d-c)\left|\beta^2 - \alpha^2\right|} \\ &\quad \left[(a-c)^2 \left\|f'''' + \left(\alpha^2 + \beta^2\right) f'' + \alpha^2\beta^2 f - \alpha^2\beta^2 f(a)\right\|_{L_1([c,a])}\right. \\ &\quad \left. + (d-a)^2 \left\|f'''' + \left(\alpha^2 + \beta^2\right) f'' + \alpha^2\beta^2 f - \alpha^2\beta^2 f(a)\right\|_{L_1([a,d])}\right]. \end{aligned} \tag{1.118}$$

Proof. By (1.89), we have ($x \geq a$)

$$|f(x) - f(a)| \leq \frac{1}{|\alpha\beta(\beta^2 - \alpha^2)|}$$

$$\int_a^x \left| f''''(t) + (\alpha^2 + \beta^2) f''(t) + \alpha^2\beta^2 f(t) - \alpha^2\beta^2 f(a) \right| 2|\alpha\beta|(x - t)\, dt$$

$$\leq \frac{2}{|\beta^2 - \alpha^2|} \int_a^x \left| f''''(t) + (\alpha^2 + \beta^2) f''(t) + \alpha^2\beta^2 f(t) - \alpha^2\beta^2 f(a) \right| (x - t)\, dt$$

$$\leq \frac{2(x - a)}{|\beta^2 - \alpha^2|} \left\| f'''' + (\alpha^2 + \beta^2) f'' + \alpha^2\beta^2 f - \alpha^2\beta^2 f(a) \right\|_{L_1([a,d])}. \tag{1.119}$$

When $x < a$, by (1.93), we obtain

$$|f(x) - f(a)| \leq \frac{1}{|\alpha\beta(\beta^2 - \alpha^2)|}$$

$$\int_x^a \left| f''''(t) + (\alpha^2 + \beta^2) f''(t) + \alpha^2\beta^2 f(t) - \alpha^2\beta^2 f(a) \right| 2|\alpha\beta|(t - x)\, dt$$

$$\leq \frac{2}{|\beta^2 - \alpha^2|} \int_x^a \left| f''''(t) + (\alpha^2 + \beta^2) f''(t) + \alpha^2\beta^2 f(t) - \alpha^2\beta^2 f(a) \right| (t - x)\, dt$$

$$\leq \frac{2(a - x)}{|\beta^2 - \alpha^2|} \left\| f'''' + (\alpha^2 + \beta^2) f'' + \alpha^2\beta^2 f - \alpha^2\beta^2 f(a) \right\|_{L_1([c,a])}. \tag{1.120}$$

By (1.100), we get that

$$\left| \frac{1}{d - c} \int_c^d f(x)\, dx - f(a) \right|$$

$$\leq \frac{1}{d - c} \left[\int_c^a |f(x) - f(a)|\, dx + \int_a^d |f(x) - f(a)|\, dx \right]$$

$$\leq \frac{2}{(d - c)|\beta^2 - \alpha^2|} \left[\left\| f'''' + (\alpha^2 + \beta^2) f'' + \alpha^2\beta^2 f \right.\right.$$

$$\left. - \alpha^2\beta^2 f(a) \right\|_{L_1([c,a])} \int_c^a (a - x)\, dx + \left\| f'''' + (\alpha^2 + \beta^2) f'' \right.$$

$$\left.\left. + \alpha^2\beta^2 f - \alpha^2\beta^2 f(a) \right\|_{L_1([a,d])} \int_a^d (x - a)\, dx \right]$$

$$= \frac{1}{(d-c)\left|\beta^2-\alpha^2\right|}\Big[(a-c)^2\left\|f'''' + \left(\alpha^2+\beta^2\right)f'' + \alpha^2\beta^2 f\right.$$
$$\left.- \alpha^2\beta^2 f(a)\right\|_{L_1([c,a])} + (d-a)^2\left\|f'''' + \left(\alpha^2+\beta^2\right)f'' + \alpha^2\beta^2 f\right.$$
$$\left.- \alpha^2\beta^2 f(a)\right\|_{L_1([a,d])}\Big]. \tag{1.121}$$

The claim is proved. □

The counterparts of the last two theorems, Theorems 1.27 and 1.28, for $p > 1$, follow.

Theorem 1.29. *Let* $f \in C_K^4([c,d])$, $a \in [c,d] : f^{(i)}(a) = 0$, $i = 1,2,3,4$. *Here* $\alpha, \beta \in \mathbb{R} : \alpha\beta\left(\alpha^2-\beta^2\right) \neq 0$. *Let also* $p, q > 1 : \frac{1}{p} + \frac{1}{q} = 1$. *Then*

$$\left|\frac{1}{d-c}\int_c^d f(x)\,dx - f(a)\right| \leq \left(\frac{(|\alpha|+|\beta|)}{\left|\alpha\beta\left(\beta^2-\alpha^2\right)\right|(d-c)\left(1+\frac{1}{q}\right)}\right)$$
$$\Big[(a-c)^{1+\frac{1}{q}}\left\|f'''' + \left(\alpha^2+\beta^2\right)f'' + \alpha^2\beta^2 f - \alpha^2\beta^2 f(a)\right\|_{L_p([c,a])}$$
$$+ (d-a)^{1+\frac{1}{q}}\left\|f'''' + \left(\alpha^2+\beta^2\right)f'' + \alpha^2\beta^2 f - \alpha^2\beta^2 f(a)\right\|_{L_p([a,d])}\Big]. \tag{1.122}$$

Proof. Let $x \geq a$, by (1.115), we obtain

$$|f(x) - f(a)| \leq \left(\frac{(|\alpha|+|\beta|)}{\left|\alpha\beta\left(\beta^2-\alpha^2\right)\right|}\right)$$
$$\int_a^x \left|\left[\left(f''''(t) + \left(\alpha^2+\beta^2\right)f''(t) + \alpha^2\beta^2 f(t)\right) - \alpha^2\beta^2 f(a)\right]\right| dt$$
$$\leq \left(\frac{(|\alpha|+|\beta|)}{\left|\alpha\beta\left(\beta^2-\alpha^2\right)\right|}\right)$$
$$\left(\int_a^x \left|f''''(t) + \left(\alpha^2+\beta^2\right)f''(t) + \alpha^2\beta^2 f(t) - \alpha^2\beta^2 f(a)\right|^p dt\right)^{\frac{1}{p}} (x-a)^{\frac{1}{q}}$$
$$\leq \left(\frac{(|\alpha|+|\beta|)}{\left|\alpha\beta\left(\beta^2-\alpha^2\right)\right|}\right)\left\|f'''' + \left(\alpha^2+\beta^2\right)f'' + \alpha^2\beta^2 f\right.$$
$$\left.- \alpha^2\beta^2 f(a)\right\|_{L_p([a,d])}(x-a)^{\frac{1}{q}}. \tag{1.123}$$

Let $x < a$, by (1.116), we get

$$\left|f(x) - f(a)\right| \leq \left(\frac{(|\alpha| + |\beta|)}{|\alpha\beta(\beta^2 - \alpha^2)|}\right)$$
$$\int_x^a \left|\left(f''''(t) + \left(\alpha^2 + \beta^2\right) f''(t) + \alpha^2\beta^2 f(t)\right) - \alpha^2\beta^2 f(a)\right| dt$$
$$\leq \left(\frac{(|\alpha| + |\beta|)}{|\alpha\beta(\beta^2 - \alpha^2)|}\right)$$
$$\left(\int_x^a \left|f''''(t) + \left(\alpha^2 + \beta^2\right) f''(t) + \alpha^2\beta^2 f(t) - \alpha^2\beta^2 f(a)\right|^p dt\right)^{\frac{1}{p}} (a - x)^{\frac{1}{q}}$$
$$\leq \left(\frac{(|\alpha| + |\beta|)}{|\alpha\beta(\beta^2 - \alpha^2)|}\right) \Big\|f'''' + \left(\alpha^2 + \beta^2\right) f'' + \alpha^2\beta^2 f$$
$$- \alpha^2\beta^2 f(a)\Big\|_{L_p([c,a])} (a - x)^{\frac{1}{q}}. \tag{1.124}$$

By (1.100), we get that

$$\left|\frac{1}{d-c}\int_c^d f(x)\,dx - f(a)\right|$$
$$\leq \frac{1}{d-c}\left[\int_c^a |f(x) - f(a)|\,dx + \int_a^d |f(x) - f(a)|\,dx\right]$$
$$\leq \left(\frac{(|\alpha| + |\beta|)}{|\alpha\beta(\beta^2 - \alpha^2)|(d-c)}\right)$$
$$\Bigg[\left\|f'''' + \left(\alpha^2 + \beta^2\right) f'' + \alpha^2\beta^2 f - \alpha^2\beta^2 f(a)\right\|_{L_p([c,a])} \left(\int_c^a (a-x)^{\frac{1}{q}}\,dx\right)$$
$$+ \left\|f'''' + \left(\alpha^2 + \beta^2\right) f'' + \alpha^2\beta^2 f - \alpha^2\beta^2 f(a)\right\|_{L_p([a,d])} \left(\int_a^d (x-a)^{\frac{1}{q}}\,dx\right)\Bigg]$$
$$= \left(\frac{(|\alpha| + |\beta|)}{|\alpha\beta(\beta^2 - \alpha^2)|(d-c)\left(1 + \frac{1}{q}\right)}\right)$$
$$\Big[(a-c)^{1+\frac{1}{q}} \left\|f'''' + \left(\alpha^2 + \beta^2\right) f'' + \alpha^2\beta^2 f - \alpha^2\beta^2 f(a)\right\|_{L_p([c,a])}$$
$$+ (d-a)^{1+\frac{1}{q}} \left\|f'''' + \left(\alpha^2 + \beta^2\right) f'' + \alpha^2\beta^2 f - \alpha^2\beta^2 f(a)\right\|_{L_p([a,d])}\Big]. \tag{1.125}$$

The claim is proved. □

At last we give

Theorem 1.30. *All as in Theorem 1.29. Then*

$$\left|\frac{1}{d-c}\int_c^d f(x)\,dx - f(a)\right| \leq \frac{2}{\left|\beta^2-\alpha^2\right|(q+1)^{\frac{1}{q}}\left(2+\frac{1}{q}\right)(d-c)}$$

$$\left[(a-c)^{2+\frac{1}{q}}\left\|f'''' + \left(\alpha^2+\beta^2\right)f'' + \alpha^2\beta^2 f - \alpha^2\beta^2 f(a)\right\|_{L_p([c,a])}\right.$$

$$\left. + (d-a)^{2+\frac{1}{q}}\left\|f'''' + \left(\alpha^2+\beta^2\right)f'' + \alpha^2\beta^2 f - \alpha^2\beta^2 f(a)\right\|_{L_p([a,d])}\right]. \tag{1.126}$$

Proof. Let $x \geq a$, by (1.119), we get that

$$|f(x) - f(a)| \leq \frac{2}{\left|\beta^2-\alpha^2\right|}$$

$$\int_a^x \left|f''''(t) + \left(\alpha^2+\beta^2\right)f''(t) + \alpha^2\beta^2 f(t) - \alpha^2\beta^2 f(a)\right|(x-t)\,dt$$

$$\leq \frac{2}{\left|\beta^2-\alpha^2\right|}\left(\int_a^x \left|f''''(t) + \left(\alpha^2+\beta^2\right)f''(t) + \alpha^2\beta^2 f(t) - \alpha^2\beta^2 f(a)\right|^p dt\right)^{\frac{1}{p}}$$

$$\left(\int_a^x (x-t)^q\,dt\right)^{\frac{1}{q}}$$

$$\leq \frac{2}{\left|\beta^2-\alpha^2\right|}\left\|f'''' + \left(\alpha^2+\beta^2\right)f'' + \alpha^2\beta^2 f - \alpha^2\beta^2 f(a)\right\|_{L_p([a,d])}\frac{(x-a)^{\frac{q+1}{q}}}{(q+1)^{\frac{1}{q}}}. \tag{1.127}$$

Let $x < a$, by (1.120), we derive

$$|f(x) - f(a)| \leq \frac{2}{\left|\beta^2-\alpha^2\right|}$$

$$\int_x^a \left|f''''(t) + \left(\alpha^2+\beta^2\right)f''(t) + \alpha^2\beta^2 f(t) - \alpha^2\beta^2 f(a)\right|(t-x)\,dt$$

$$\leq \frac{2}{\left|\beta^2-\alpha^2\right|}\left(\int_x^a \left|f''''(t) + \left(\alpha^2+\beta^2\right)f''(t) + \alpha^2\beta^2 f(t) - \alpha^2\beta^2 f(a)\right|^p dt\right)^{\frac{1}{p}}$$

$$\left(\int_x^a (t-x)^q\,dt\right)^{\frac{1}{q}}$$

$$\leq \frac{2}{\left|\beta^2-\alpha^2\right|}\left\|f'''' + \left(\alpha^2+\beta^2\right)f'' + \alpha^2\beta^2 f - \alpha^2\beta^2 f(a)\right\|_{L_p([c,a])}\frac{(a-x)^{\frac{q+1}{q}}}{(q+1)^{\frac{1}{q}}}. \tag{1.128}$$

Finally, by (1.100), we get that

$$\left|\frac{1}{d-c}\int_c^d f(x)\,dx - f(a)\right|$$

$$\leq \frac{1}{d-c}\left[\int_c^a |f(x)-f(a)|\,dx + \int_a^d |f(x)-f(a)|\,dx\right]$$

$$\leq \frac{2}{|\beta^2-\alpha^2|(q+1)^{\frac{1}{q}}(d-c)}$$

$$\left[\left(\int_c^a (a-x)^{\frac{q+1}{q}}\,dx\right)\left\|f'''' + \left(\alpha^2+\beta^2\right)f'' + \alpha^2\beta^2 f - \alpha^2\beta^2 f(a)\right\|_{L_9([c,a])}\right.$$

$$\left. + \left(\int_a^d (x-a)^{\frac{q+1}{q}}\,dx\right)\left\|f'''' + \left(\alpha^2+\beta^2\right)f'' + \alpha^2\beta^2 f - \alpha^2\beta^2 f(a)\right\|_{L_p([a,d])}\right]$$

$$= \frac{2}{|\beta^2-\alpha^2|(q+1)^{\frac{1}{q}}\left(2+\frac{1}{q}\right)(d-c)}$$

$$\left[(a-c)^{2+\frac{1}{q}}\left\|f'''' + \left(\alpha^2+\beta^2\right)f'' + \alpha^2\beta^2 f - \alpha^2\beta^2 f(a)\right\|_{L_p([c,a])}\right.$$

$$\left. + (d-a)^{2+\frac{1}{q}}\left\|f'''' + \left(\alpha^2+\beta^2\right)f'' + \alpha^2\beta^2 f - \alpha^2\beta^2 f(a)\right\|_{L_p([a,d])}\right]. \tag{1.129}$$

The claim is proved. □

Bibliography

Ali H.A. and Zsolt P., Taylor-type expansions in terms of exponential polynomials, *Mathematical Inequalities and Applications*, 25(4) (2022), 1123–1141.

G.A. Anastassiou, *Probabilistic Inequalities*, World Scientific, New Jersey, Singapore, 2010.

G.A. Anastassiou, *Advanced Inequalities*, World Scientific, New Jersey, Singapore, 2011.

G.A. Anastassiou, *Intelligent Comparisons: Analytic Inequalities*, Springer, Heidelberg, New York, 2016.

G.A. Anastassiou, Opial and Ostrowski type inequalities based on trigonometric and hyperbolic type Taylor formulae, *Malaya Journal of Matematik*, 11(S) 2023, 1–26.

Chapter 2

Perturbed A-Generalized Logistic Function-Activated Complex-Valued Trigonometric and Hyperbolic Neural Network High Order Approximation

Here, we research the univariate quantitative approximation of complex-valued continuous functions on a compact interval by complex-valued neural network operators. These approximations are derived by establishing Jackson-type inequalities involving the modulus of continuity of the engaged function's high order derivatives. The nature of our approximations are trigonometric and hyperbolic. Our operators are defined by using a density function generated by a q-deformed and λ-parametrized A-generalized logistic function, which is a sigmoid function. The approximations are pointwise and of the uniform norm. The related complex-valued feed-forward neural networks are with one hidden layer. It follows Anastassiou (2023d).

2.1 Introduction

The author in Anastassiou (1997, 2001), see Chapters 2–5, was the first to establish neural network approximation to continuous functions with rates by very specifically defined neural network operators of Cardaliaguet–Euvrard and "Squashing" types, by employing the modulus of continuity of the engaged function or its high order derivative, and producing very tight Jackson-type inequalities. He treats there both the univariate and multivariate cases. The definitions of these operators' "bell-shaped" and "squashing" functions are assumed to be of compact support.

Again, the author, inspired by Chen and Cao (2009), continued his studies on neural network approximation by introducing and using the proper quasi-interpolation operators of sigmoidal and hyperbolic tangent type which resulted into Anastassiou (2011a)–Anastassiou (2012), by treating both the univariate and multivariate cases.

Let h be a general sigmoid activation function with $h(0) = 0$, and $y = \pm 1$ the horizontal asymptotes. Of course, h is strictly increasing over $\mathbb{R}$. Let the parameter $0 < r < 1$ and $x > 0$. Then clearly, $-x < x$ and $-x < -rx < rx < x$, furthermore, it holds $h(-x) < h(-rx) < h(rx) < h(x)$. Consequently, the sigmoid $y = h(rx)$ has a graph inside the graph of $y = h(x)$, of course with the same asymptotes $y = \pm 1$. Therefore, $h(rx)$ has derivatives (gradients) at more points x than $h(x)$ has different than zero or not as close to zero, thus killing less number of neurons! And of course, $h(rx)$ is more distant from $y = \pm 1$ than it is from $h(x)$. A highly desired fact in neural networks theory.

Different activation functions allow for different nonlinearities which might work better for solving a specific function. So the need to use neural networks with various activation functions is vivid. Thus, performing neural network approximations using different activation functions is not only necessary but also fully justified.

Also brain asymmetry has been observed in animals and humans in terms of structure, function and behaviour. This lateralization is thought to reflect evolutionary, hereditary, developmental, experiential and pathological factors. Therefore, it is natural to consider for our study deformed neural network activation functions and operators. So this chapter is a specific study under this philosophy of approaching reality as close as possible.

Consequently, the author here performs q-deformed and λ-parametrized A-generalized logistic function-activated high order neural network approximations to continuous functions over compact intervals of the real line with complex values. All convergences are with rates expressed via the modulus of continuity of the involved functions' high order derivatives, deriving by very tight Jackson-type inequalities.

The basis of our higher-order approximations, here are some trigonometric and hyperbolic-type Taylor's formulae newly discovered by the author.

Our compact intervals are not necessarily symmetric to the origin. In preparation to prove our results, we describe important properties of the basic density function defining our operators which is induced by a q-deformed and λ-parametrized A-generalized logistic function, which is a sigmoid activation function.

Feed-forward X-valued neural networks (FNNs) with one hidden layer, the only type of network we deal with in this chapter, are mathematically expressed as

$$N_n(x) = \sum_{j=0}^{n} c_j \sigma(\langle a_j \cdot x \rangle + b_j), \quad x \in \mathbb{R}^s, \quad s \in \mathbb{N},$$

where for $0 \le j \le n$, $b_j \in \mathbb{R}$ are the thresholds, $a_j \in \mathbb{R}^s$ are the connection weights, $c_j \in \mathbb{C}$ are the coefficients, $\langle a_j \cdot x \rangle$ is the inner product of a_j and x, and σ is the activation function of the network. About neural networks in general, read Haykin (1998), McCulloch and Pitts (1943) and Mitchell (1997).

2.2 Preliminaries

The following background comes from Anastassiou (2022).

We consider here the q-deformed and λ-parametrized function

$$\varphi_{q,\lambda}(x) = \frac{1}{1 + qA^{-\lambda x}}, \quad x \in \mathbb{R}, \quad \text{where } q, \lambda > 0, \quad A > 1. \tag{2.1}$$

This is an A-generalized logistic-type function.

We easily observe that

$$\varphi_{q,\lambda}(+\infty) = 1, \quad \varphi_{q,\lambda}(-\infty) = 0. \tag{2.2}$$

Furthermore, we have

$$\begin{aligned} 1 - \varphi_{\frac{1}{q},\lambda}(-x) &= 1 - \frac{1}{1 + \frac{1}{q}A^{\lambda x}} = \frac{1 + \frac{1}{q}A^{\lambda x} - 1}{1 + \frac{1}{q}A^{\lambda x}} \\ &= \frac{\frac{1}{q}A^{\lambda x}}{1 + \frac{1}{q}A^{\lambda x}} = \frac{1}{\frac{1}{\frac{1}{q}A^{\lambda x}} + 1} = \frac{1}{1 + qA^{-\lambda x}} = \varphi_{q,\lambda}(x), \end{aligned}$$

proving

$$\varphi_{q,\lambda}(x) = 1 - \varphi_{\frac{1}{q},\lambda}(-x). \tag{2.3}$$

We also have that

$$\varphi_{q,\lambda}(0) = \frac{1}{1+q}. \tag{2.4}$$

Consider the activation function

$$G_{q,\lambda}(x) := \frac{1}{2}(\varphi_{q,\lambda}(x+1) - \varphi_{q,\lambda}(x-1)), \quad x \in \mathbb{R}. \tag{2.5}$$

Then

$$\begin{aligned} G_{q,\lambda}(-x) &= \frac{1}{2}\left(\varphi_{q,\lambda}(-x+1) - \varphi_{q,\lambda}(-x-1)\right) \\ &= \frac{1}{2}\left(1 - \varphi_{\frac{1}{q},\lambda}(x-1) - 1 + \varphi_{\frac{1}{q},\lambda}(x+1)\right) \\ &= \frac{1}{2}\left(\varphi_{\frac{1}{q},\lambda}(x+1) - \varphi_{\frac{1}{q},\lambda}(x-1)\right) = G_{\frac{1}{q},\lambda}(x). \end{aligned} \tag{2.6}$$

That is,

$$G_{q,\lambda}(-x) = G_{\frac{1}{q},\lambda}(x), \quad \forall\, x \in \mathbb{R}. \tag{2.7}$$

We have

$$\begin{aligned} \varphi'_{q,\lambda}(x) &= \left(\left(1 + qA^{-\lambda x}\right)^{-1}\right)' = -1\left(1 + qA^{-\lambda x}\right)^{-2} q(\ln A)A^{-\lambda x}(-\lambda) \\ &= q\lambda(\ln A)\left(1 + qA^{-\lambda x}\right)^{-2} A^{-\lambda x} > 0. \end{aligned} \tag{2.8}$$

So that $\varphi_{q,\lambda}$ is a strictly increasing function over $\mathbb{R}$.

Hence it holds

$$\begin{aligned} \varphi'_{q,\lambda}(x) &= \frac{q\lambda(\ln A)}{\left(1 + qA^{-\lambda x}\right)^2 A^{\lambda x}} = \frac{q\lambda(\ln A)}{\left(1 + q^2 A^{-2\lambda x} + 2qA^{-\lambda x}\right) A^{\lambda x}} \\ &= \frac{q\lambda(\ln A)}{\left(A^{\lambda x} + q^2 A^{-\lambda x} + 2q\right)}. \end{aligned} \tag{2.9}$$

That is,

$$\varphi'_{q,\lambda}(x) = q\lambda(\ln A)\left(A^{\lambda x} + q^2 A^{-\lambda x} + 2q\right)^{-1}. \tag{2.10}$$

Therefore, it holds

$$\begin{aligned} \varphi''_{q,\lambda}(x) &= q\lambda(\ln A)(-1)\left(A^{\lambda x} + q^2 A^{-\lambda x} + 2q\right)^{-2} \\ &\quad \left((\ln A)A^{\lambda x}\lambda + q^2(\ln A)A^{-\lambda x}(-\lambda)\right) \\ &= q\lambda^2(\ln A)^2\left(A^{\lambda x} + q^2 A^{-\lambda x} + 2q\right)^{-2}\left(q^2 A^{-\lambda x} - A^{\lambda x}\right). \end{aligned} \tag{2.11}$$

That is,

$$\varphi''_{q,\lambda}(x) = q\lambda^2(\ln A)^2\left(A^{\lambda x} + q^2 A^{-\lambda x} + 2q\right)^{-2}\left(q^2 A^{-\lambda x} - A^{\lambda x}\right) \in C(\mathbb{R}). \tag{2.12}$$

We have

$$q^2 A^{-\lambda x} - A^{\lambda x} > 0, \text{ iff } q^2 A^{-\lambda x} > A^{\lambda x}, \text{ iff } q^2 > A^{2\lambda x}, \text{ iff } q > A^{\lambda x},$$

$$\text{iff } \log_A q > \lambda x, \text{ iff } x < \frac{\log_A q}{\lambda}.$$

So, $\varphi''_{q,\lambda}(x) > 0$, for $x < \frac{\log_A q}{\lambda}$ and there $\varphi_{q,\lambda}$ is concave up.

When $x > \frac{\log_A q}{\lambda}$, we have $\varphi''_{q,\lambda}(x) < 0$ and $\varphi_{q,\lambda}$ is concave down.

Of course,

$$\varphi''_{q,\lambda}\left(\frac{\log_A q}{\lambda}\right) = 0.$$

So, $\varphi_{q,\lambda}$ is a sigmoid function, see Anastassiou (2023a).

We have that

$$G'_{q,\lambda}(x) = \frac{1}{2}\left(\varphi'_{q,\lambda}(x+1) - \varphi'_{q,\lambda}(x-1)\right).$$

We got that $\varphi'_{q,\lambda}$ is strictly increasing for $x < \frac{\log_A q}{\lambda}$. Let $x < \frac{\log_A q}{\lambda} - 1$, then $x - 1 < x + 1 < \frac{\log_A q}{\lambda}$. Hence $\varphi'_{q,\lambda}(x+1) > \varphi'_{q,\lambda}(x-1)$. Thus $G'_{q,\lambda} > 0$, i.e. $G_{q,\lambda}$ is strictly increasing over $\left(-\infty, \frac{\log_A q}{\lambda} - 1\right)$.

Let now $x > \frac{\log_A q}{\lambda} + 1$, then $x + 1 > x - 1 > \frac{\log_A q}{\lambda}$, and $\varphi'_{q,\lambda}(x+1) < \varphi'_{q,\lambda}(x-1)$, by $\varphi'_{q,\lambda}$ being strictly decreasing over $\left(\frac{\log_A q}{\lambda}, +\infty\right)$. Hence $G'_{q,\lambda} < 0$, and $G_{q,\lambda}$ is strictly decreasing over $\left(\frac{\log_A q}{\lambda}, +\infty\right)$.

Let now $\frac{\log_A q}{\lambda} - 1 \le x \le \frac{\log_A q}{\lambda} + 1$. We have that

$$G''_{q,\lambda}(x) = \frac{1}{2}\left(\varphi''_{q,\lambda}(x+1) - \varphi''_{q,\lambda}(x-1)\right) = \frac{q\lambda^2(\ln A)^2}{2}$$

$$\left[\frac{\left(q^2A^{-\lambda(x+1)} - A^{\lambda(x+1)}\right)}{\left(A^{\lambda(x+1)} + q^2A^{-\lambda(x+1)} + 2q\right)^2} - \frac{\left(q^2A^{-\lambda(x-1)} - A^{\lambda(x-1)}\right)}{\left(A^{\lambda(x-1)} + q^2A^{-\lambda(x-1)} + 2q\right)^2}\right]$$

$$= \frac{q\lambda^2(\ln A)^2}{2}\left[\frac{\left(q^2 - A^{2\lambda(x+1)}\right)}{\left(A^{\lambda(x+1)} + q^2A^{-\lambda(x+1)} + 2q\right)^2 A^{\lambda(x+1)}}\right.$$

$$\left.- \frac{\left(q^2 - A^{2\lambda(x-1)}\right)}{\left(A^{\lambda(x-1)} + q^2A^{-\lambda(x-1)} + 2q\right)^2 A^{\lambda(x-1)}}\right]$$

$$= \frac{q\lambda^2(\ln A)^2}{2}\left[\frac{\left(q - A^{\lambda(x+1)}\right)\left(q + A^{\lambda(x+1)}\right)}{\left(A^{\lambda(x+1)} + q^2A^{-\lambda(x+1)} + 2q\right)^2 A^{\lambda(x+1)}}\right.$$

$$\left.- \frac{\left(q - A^{\lambda(x-1)}\right)\left(q + A^{\lambda(x-1)}\right)}{\left(A^{\lambda(x-1)} + q^2A^{-\lambda(x-1)} + 2q\right)^2 A^{\lambda(x-1)}}\right]. \tag{2.13}$$

By $\frac{\log_A q}{\lambda} \leq x+1 \Leftrightarrow \log_A q \leq \lambda(x+1) \Leftrightarrow q \leq A^{\lambda(x+1)} \Leftrightarrow q - A^{\lambda(x+1)} \leq 0$.

By $x \leq \frac{\log_A q}{\lambda} + 1 \Leftrightarrow x - 1 \leq \frac{\log_A q}{\lambda} \Leftrightarrow \lambda(x-1) \leq \log_A q \Leftrightarrow A^{\lambda(x-1)} \leq q \Leftrightarrow q - A^{\lambda(x-1)} \geq 0$.

Clearly, when $\frac{\log_A q}{\lambda} - 1 \leq x \leq \frac{\log_A q}{\lambda} + 1$, by the above, we get that $G''_{q,\lambda}(x) \leq 0$, that is, $G''_{q,\lambda}$ is concave down there.

Clearly, $G_{q,\lambda}$ is strictly concave down over $\left(\frac{\log_A q}{\lambda} - 1, \frac{\log_A q}{\lambda} + 1\right)$.

Overall, $G_{q,\lambda}$ is a bell-shaped function over $\mathbb{R}$.

Of course, it holds $G''_{q,\lambda}\left(\frac{\log_A q}{\lambda}\right) < 0$.

We have that

$$G'_{q,\lambda}\left(\frac{\log_A q}{\lambda}\right) = \frac{1}{2}\left(\varphi'_{q,\lambda}\left(\frac{\log_A q}{\lambda} + 1\right) - \varphi'_{q,\lambda}\left(\frac{\log_A q}{\lambda} - 1\right)\right)$$

$$= \frac{q\lambda(\ln A)}{2}\left[\frac{1}{A^{\lambda\left(\frac{\log_A q}{\lambda}+1\right)} + q^2 A^{-\lambda\left(\frac{\log_A q}{\lambda}+1\right)} + 2q} - \frac{1}{A^{\lambda\left(\frac{\log_A q}{\lambda}-1\right)} + q^2 A^{-\lambda\left(\frac{\log_A q}{\lambda}-1\right)} + 2q}\right] \tag{2.14}$$

$$= \frac{q\lambda(\ln A)}{2}\left[\frac{A^{\lambda\left(\frac{\log_A q}{\lambda}-1\right)} + q^2 A^{-\lambda\left(\frac{\log_A q}{\lambda}-1\right)} - A^{\lambda\left(\frac{\log_A q}{\lambda}+1\right)} - q^2 A^{-\lambda\left(\frac{\log_A q}{\lambda}+1\right)}}{\left(A^{\lambda\left(\frac{\log_A q}{\lambda}+1\right)} + q^2 A^{-\lambda\left(\frac{\log_A q}{\lambda}+1\right)} + 2q\right) \times \left(A^{\lambda\left(\frac{\log_A q}{\lambda}-1\right)} + q^2 A^{-\lambda\left(\frac{\log_A q}{\lambda}-1\right)} + 2q\right)}\right] \tag{2.15}$$

$$= \frac{q\lambda(\ln A)}{2}\left[\frac{qA^{-\lambda} + q^2 q^{-1} A^{\lambda} - qA^{\lambda} - q^2 q^{-1} A^{-\lambda}}{(qA^{\lambda} + q^2 q^{-1} A^{-\lambda} + 2q)(qA^{\lambda} + q^2 q^{-1} A^{-\lambda} + 2q)}\right]$$

$$= \frac{q\lambda(\ln A)}{2}\left[\frac{qA^{-\lambda} + qA^{\lambda} - qA^{\lambda} - qA^{-\lambda}}{(qA^{\lambda} + qA^{-\lambda} + 2q)(qA^{\lambda} + qA^{-\lambda} + 2q)}\right] = 0. \tag{2.16}$$

So $\frac{\log_A q}{\lambda}$ is the only critical number of $G_{q,\lambda}$ over $\mathbb{R}$. Therefore, $G_{q,\lambda}\left(\frac{\log_A q}{\lambda}\right)$ is the maximum of $G_{q,\lambda}$.

We calculate it:

We have that

$$G_{q,\lambda}\left(\frac{\log_A q}{\lambda}\right) = \frac{1}{2}\left(\varphi_{q,\lambda}\left(\frac{\log_A q}{\lambda}+1\right) - \varphi_{q,\lambda}\left(\frac{\log_A q}{\lambda}-1\right)\right)$$

$$= \frac{1}{2}\left(\frac{1}{1+qA^{-\lambda\left(\frac{\log_A q}{\lambda}+1\right)}} - \frac{1}{1+qA^{-\lambda\left(\frac{\log_A q}{\lambda}-1\right)}}\right)$$

$$= \frac{1}{2}\left(\frac{1}{1+qq^{-1}A^{-\lambda}} - \frac{1}{1+qq^{-1}A^{\lambda}}\right) = \frac{1}{2}\left(\frac{1}{1+A^{-\lambda}} - \frac{1}{1+A^{\lambda}}\right)$$

$$= \frac{1}{2}\left(\frac{A^{\lambda}-A^{-\lambda}}{\left(1+A^{-\lambda}\right)\left(1+A^{\lambda}\right)}\right) = \frac{A^{\lambda}-1}{2\left(A^{\lambda}+1\right)}. \tag{2.17}$$

The global maximum of $G_{q,\lambda}$ is

$$G_{q,\lambda}\left(\frac{\log_A q}{\lambda}\right) = \frac{A^{\lambda}-1}{2\left(A^{\lambda}+1\right)}. \tag{2.18}$$

Finally, we have that

$$\lim_{x\to+\infty} G_{q,\lambda}(x) = \frac{1}{2}\left(\varphi_{q,\lambda}(+\infty) - \varphi_{q,\lambda}(+\infty)\right) = 0 \tag{2.19}$$

and

$$\lim_{x\to-\infty} G_{q,\lambda}(x) = \frac{1}{2}\left(\varphi_{q,\lambda}(-\infty) - \varphi_{q,\lambda}(-\infty)\right) = 0. \tag{2.20}$$

Consequently, the x-axis is the horizontal asymptote of $G_{q,\lambda}$. Of course, $G_{q,\lambda}(x) > 0$, $\forall\, x \in \mathbb{R}$.

We need

Theorem 2.1 (Anastassiou 2022). *It holds*

$$\sum_{i=-\infty}^{\infty} G_{q,\lambda}(x-i) = 1, \quad \forall\, x \in \mathbb{R}, \quad \forall\, q, \lambda > 0, \quad A > 1. \tag{2.21}$$

Thus,

$$\sum_{i=-\infty}^{\infty} G_{q,\lambda}(nx-i) = 1, \quad \forall\, n \in \mathbb{N}, \quad \forall\, x \in \mathbb{R}. \tag{2.22}$$

Similarly, it holds

$$\sum_{i=-\infty}^{\infty} G_{\frac{1}{q},\lambda}(x-i) = 1, \quad \forall\, x \in \mathbb{R}. \tag{2.23}$$

But $G_{\frac{1}{q},\lambda}(x-i) \overset{(2.7)}{=} G_{q,\lambda}(i-x)$, $\forall\, x \in \mathbb{R}$.

Hence,

$$\sum_{i=-\infty}^{\infty} G_{q,\lambda}(i-x) = 1, \quad \forall\, x \in \mathbb{R} \tag{2.24}$$

and

$$\sum_{i=-\infty}^{\infty} G_{q,\lambda}(i+x) = 1, \quad \forall\, x \in \mathbb{R}. \tag{2.25}$$

It follows

Theorem 2.2 (Anastassiou 2022). *It holds*

$$\int_{-\infty}^{\infty} G_{q,\lambda}(x)\,dx = 1, \quad \lambda, q > 0, \quad A > 1. \tag{2.26}$$

So $G_{q,\lambda}$ is a density function on $\mathbb{R}$; $\lambda, q > 0$, $A > 1$.

We need the following result.

Theorem 2.3 (Anastassiou 2022). *Let* $0 < \alpha < 1$, *and* $n \in \mathbb{N}$ *with* $n^{1-\alpha} > 2$. *Then*

$$\sum_{\substack{k=-\infty \\ \{:|nx-k| \geq n^{1-\alpha}}}^{\infty} G_{q,\lambda}(nx-k) < 2\max\left\{q, \frac{1}{q}\right\} \frac{1}{A^{\lambda(n^{1-\alpha}-2)}} = \gamma A^{-\lambda(n^{1-\alpha}-2)}, \tag{2.27}$$

where $q, \lambda > 0$, $A > 1$; $\gamma := 2\max\left\{q, \frac{1}{q}\right\}$.

Let $\lceil \cdot \rceil$ be the ceiling of the number, and $\lfloor \cdot \rfloor$ be the integral part of the number.

We also need

Theorem 2.4 (Anastassiou 2022). *Let* $x \in [a,b] \subset \mathbb{R}$ *and* $n \in \mathbb{N}$ *so that* $\lceil na \rceil \leq \lfloor nb \rfloor$. *For* $q > 0$, $\lambda > 0$, $A > 1$, *we consider the number* $\lambda_q > z_0 > 0$ *with* $G_{q,\lambda}(z_0) = G_{q,\lambda}(0)$ *and* $\lambda_q > 1$. *Then*

$$\frac{1}{\sum_{k=\lceil na \rceil}^{\lfloor nb \rfloor} G_{q,\lambda}(nx-k)} < max\left\{\frac{1}{G_{q,\lambda}(\lambda_q)}, \frac{1}{G_{\frac{1}{q},\lambda}\left(\lambda_{\frac{1}{q}}\right)}\right\} =: K(q). \tag{2.28}$$

We finally mention

Remark 2.1 (Anastassiou 2022). (i) We have that

$$\lim_{n\to+\infty}\sum_{k=\lceil na\rceil}^{\lfloor nb\rfloor} G_{q,\lambda}(nx-k)\neq 1,\quad \text{for at least some } x\in[a,b], \tag{2.29}$$

where $\lambda, q > 0$.

(ii) Let $[a,b]\subset\mathbb{R}$. For large n, we always have $\lceil na\rceil\leq\lfloor nb\rfloor$. Also, $a\leq\frac{k}{n}\leq b$, iff $\lceil na\rceil\leq k\leq\lfloor nb\rfloor$. In general, it holds

$$\sum_{k=\lceil na\rceil}^{\lfloor nb\rfloor} G_{q,\lambda}(nx-k)\leq 1. \tag{2.30}$$

Let $(\mathbb{C},|\cdot|)$ be the Banach space of the complex numbers over the reals.

Definition 2.1. Let $f\in C([a,b],\mathbb{C})$ and $n\in\mathbb{N}:\lceil na\rceil\leq\lfloor nb\rfloor$. We introduce and define the $\mathbb{C}$-valued linear neural network operators

$$L_n(f,x):=\frac{\sum_{k=\lceil na\rceil}^{\lfloor nb\rfloor} f\left(\frac{k}{n}\right)G_{q,\lambda}(nx-k)}{\sum_{k=\lceil na\rceil}^{\lfloor nb\rfloor} G_{q,\lambda}(nx-k)},\quad x\in[a,b]. \tag{2.31}$$

Clearly, here $L_n(f,x)\in C([a,b],\mathbb{C})$.

For convenience, we use the same L_n for real-valued functions when needed. We study here the high order pointwise and uniform convergence of $L_n(f,x)$ to $f(x)$ with rates.

For convenience, also we call

$$L_n^*(f,x):=\sum_{k=\lceil na\rceil}^{\lfloor nb\rfloor} f\left(\frac{k}{n}\right)G_{q,\lambda}(nx-k) \tag{2.32}$$

(similarly, L_n^* can be defined for real-valued functions), that is,

$$L_n(f,x):=\frac{L_n^*(f,x)}{\sum_{k=\lceil na\rceil}^{\lfloor nb\rfloor} G_{q,\lambda}(nx-k)}, \tag{2.33}$$

so that

$$\begin{aligned}L_n(f,x)-f(x)&=\frac{L_n^*(f,x)}{\sum_{k=\lceil na\rceil}^{\lfloor nb\rfloor} G_{q,\lambda}(nx-k)}-f(x)\\&=\frac{L_n^*(f,x)-f(x)\left(\sum_{k=\lceil na\rceil}^{\lfloor nb\rfloor} G_{q,\lambda}(nx-k)\right)}{\sum_{k=\lceil na\rceil}^{\lfloor nb\rfloor} G_{q,\lambda}(nx-k)}.\end{aligned} \tag{2.34}$$

Consequently, we derive that

$$\begin{aligned}|L_n(f,x)-f(x)| &\leq K(q)\left|L_n^*(f,x)-f(x)\left(\sum_{k=\lceil na\rceil}^{\lfloor nb\rfloor}G_{q,\lambda}(nx-k)\right)\right| \\ &= K(q)\left|\sum_{k=\lceil na\rceil}^{\lfloor nb\rfloor}\left(f\left(\frac{k}{n}\right)-f(x)\right)G_{q,\lambda}(nx-k)\right|.\end{aligned} \tag{2.35}$$

We use in our estimations the right-hand side of the last quantity. For that we need, for $f \in C([a,b],\mathbb{C})$, the first modulus of continuity

$$\omega_1(f,\delta) := \sup_{\substack{x,y\in[a,b]\\ |x-y|\leq\delta}} |f(x)-f(y)|, \quad \delta>0. \tag{2.36}$$

The fact $f \in C([a,b],\mathbb{C})$ is equivalent to $\lim_{\delta\to 0}\omega_1(f,\delta) = 0$, see Anastassiou (2017).

2.3 Main Results

We present $\mathbb{C}$-valued neural network high order approximations to a function given with rates. We start with a trigonometric approximation.

Theorem 2.5. *Let $f \in C^3([a,b],\mathbb{C})$, $0<\alpha<1$, $n\in\mathbb{N} : n^{1-\alpha}>2$, $x\in[a,b]$; denote $\delta := \frac{(b-a)^2}{2}\min\left\{1,\frac{b-a}{3}\right\}$. Then*

(1)

$$\begin{aligned}|L_n(f,x)-f(x)| \leq K(q)\Bigg[&|f'(x)|\left(\frac{1}{n^\alpha}+(b-a)\gamma A^{-\lambda\left(n^{1-\alpha}-2\right)}\right) \\ &+\frac{|f''(x)|}{2}\left(\frac{1}{n^{2\alpha}}+(b-a)^2\gamma A^{-\lambda\left(n^{1-\alpha}-2\right)}\right) \\ &+\left(\frac{\omega_1\left(f''+f,\frac{1}{n^\alpha}\right)}{2n^{2\alpha}}+\|f'''+f'\|_\infty\delta\gamma A^{-\lambda\left(n^{1-\alpha}-2\right)}\right)\Bigg],\end{aligned} \tag{2.37}$$

(2) *if $f'(x)=f''(x)=0$, we obtain*

$$|L_n(f,x)-f(x)| \leq K(q)\left[\frac{\omega_1\left(f''+f,\frac{1}{n^\alpha}\right)}{2n^{2\alpha}}+\|f'''+f'\|_\infty\delta\gamma A^{-\lambda\left(n^{1-\alpha}-2\right)}\right], \tag{2.38}$$

note here the high rate of convergence at $n^{-3\alpha}$,

(3) *furthermore we get*

$$\|L_n f - f\|_\infty \leq K(q) \Bigg[\|f'\|_\infty \left(\frac{1}{n^\alpha} + (b-a)\gamma A^{-\lambda(n^{1-\alpha}-2)} \right) + \frac{\|f''\|_\infty}{2} \left(\frac{1}{n^{2\alpha}} + (b-a)^2 \gamma A^{-\lambda(n^{1-\alpha}-2)} \right) + \left(\frac{\omega_1\left(f''+f, \frac{1}{n^\alpha}\right)}{2n^{2\alpha}} + \|f'''+f'\|_\infty \delta\gamma A^{-\lambda(n^{1-\alpha}-2)} \right) \Bigg], \tag{2.39}$$

i.e. $\lim_{n\to+\infty} L_n(f) = f$, *pointwise and uniformly,*

(4) *and finally, it holds*

$$\left| L_n(f,x) - f'(x) L_n(\sin(\cdot - x), x) - 2f''(x) L_n\left(\sin^2\left(\frac{\cdot - x}{2} \right), x \right) - f(x) \right| \leq K(q) \left[\frac{\omega_1\left(f''+f, \frac{1}{n^\alpha}\right)}{2n^{2\alpha}} + \|f'''+f'\|_\infty \delta\gamma A^{-\lambda(n^{1-\alpha}-2)} \right], \tag{2.40}$$

again here we achieve high speed of convergence at $n^{-3\alpha}$.

Proof. Here $f \in C^3([a,b], \mathbb{C})$, and we apply the trigonometric Taylor's formula for $f \in C^2([a,b], \mathbb{C})$, see Theorem 6 of Anastassiou (2023b).

Let $\frac{k}{n}, x \in [a,b]$, then

$$f\left(\frac{k}{n}\right) = f(x) + f'(x)\sin\left(\frac{k}{n} - x\right) + 2f''(x)\sin^2\left(\frac{\frac{k}{n} - x}{2}\right) + \int_x^{\frac{k}{n}} \left[(f''(t) + f(t)) - (f''(x) + f(x)) \right] \sin\left(\frac{k}{n} - t\right) dt. \tag{2.41}$$

Hence it holds

$$f\left(\frac{k}{n}\right) G_{q,\lambda}(nx-k) = f(x) G_{q,\lambda}(nx-k) + f'(x)\sin\left(\frac{k}{n} - x\right) G_{q,\lambda}(nx-k) + 2f''(x)\sin^2\left(\frac{\frac{k}{n} - x}{2}\right) G_{q,\lambda}(nx-k) + G_{q,\lambda}(nx-k)\left(\int_x^{\frac{k}{n}} \left[(f''(t) + f(t)) - (f''(x) + f(x)) \right] \sin\left(\frac{k}{n} - t\right) dt \right) \tag{2.42}$$

so that we have

$$\sum_{k=\lceil na\rceil}^{\lfloor nb\rfloor} f\left(\frac{k}{n}\right) G_{q,\lambda}(nx-k) - f(x) \sum_{k=\lceil na\rceil}^{\lfloor nb\rfloor} G_{q,\lambda}(nx-k)$$

$$= f'(x) \sum_{k=\lceil na\rceil}^{\lfloor nb\rfloor} G_{q,\lambda}(nx-k) \sin\left(\frac{k}{n}-x\right)$$

$$+ 2f''(x) \sum_{k=\lceil na\rceil}^{\lfloor nb\rfloor} G_{q,\lambda}(nx-k) \sin^2\left(\frac{\frac{k}{n}-x}{2}\right) + \sum_{k=\lceil na\rceil}^{\lfloor nb\rfloor} G_{q,\lambda}(nx-k)$$

$$\times \left(\int_x^{\frac{k}{n}} \left[(f''(t)+f(t)) - (f''(x)+f(x))\right] \sin\left(\frac{k}{n}-t\right) dt\right). \quad (2.43)$$

Thus, we obtain

$$L_n^*(f,x) - f(x) \sum_{k=\lceil na\rceil}^{\lfloor nb\rfloor} G_{q,\lambda}(nx-k)$$

$$= f'(x) L_n^*(\sin(\cdot - x), x) + 2f''(x) L_n^*\left(\sin^2\left(\frac{\cdot - x}{2}\right), x\right) + \Lambda_n(x), \quad (2.44)$$

where

$$\Lambda_n(x) := \sum_{k=\lceil na\rceil}^{\lfloor nb\rfloor} G_{q,\lambda}(nx-k)$$

$$\times \left(\int_x^{\frac{k}{n}} \left[(f''(t)+f(t)) - (f''(x)+f(x))\right] \sin\left(\frac{k}{n}-t\right) dt\right). \quad (2.45)$$

We call

$$R_2(n) := \int_x^{\frac{k}{n}} \left[(f''(t)+f(t)) - (f''(x)+f(x))\right] \sin\left(\frac{k}{n}-t\right) dt. \quad (2.46)$$

We assume that $b-a > \frac{1}{n^\alpha}$, which is always the case for large enough $n \in \mathbb{N}$, that is, when $n > \left\lceil (b-a)^{-\frac{1}{\alpha}} \right\rceil$.

Thus, $\left|\frac{k}{n}-x\right| \le \frac{1}{n^\alpha}$ or $\left|\frac{k}{n}-x\right| > \frac{1}{n^\alpha}$.

In case of $\left|\frac{k}{n}-x\right| \leq \frac{1}{n^{\alpha}}$, we have the following cases:

(i) If $\frac{k}{n} \geq x$, then

$$|R_2(n)| = \left|\int_x^{\frac{k}{n}} \left[(f''(t)+f(t)) - (f''(x)+f(x))\right] \sin\left(\frac{k}{n}-t\right) dt\right|$$

$$\leq \int_x^{\frac{k}{n}} \omega_1(f''+f, t-x) \left|\sin\left(\frac{k}{n}-t\right)\right| dt \leq \tag{2.47}$$

(by $|\sin x| \leq |x|$, $\forall\, x \in \mathbb{R}$)

$$\int_x^{\frac{k}{n}} \omega_1(f''+f, t-x)\left(\frac{k}{n}-t\right) dt \leq \omega_1\left(f''+f, \frac{k}{n}-x\right) \frac{\left(\frac{k}{n}-x\right)^2}{2}$$
$$\leq \frac{\omega_1\left(f''+f, \frac{1}{n^{\alpha}}\right)}{2n^{2\alpha}},$$

that is,

$$|R_2(n)| \leq \frac{\omega_1\left(f''+f, \frac{1}{n^{\alpha}}\right)}{2n^{2\alpha}}. \tag{2.48}$$

(ii) If $\frac{k}{n} < x$, then

$$|R_2(n)| = \left|\int_x^{\frac{k}{n}} \left[(f''(t)+f(t)) - (f''(x)+f(x))\right] \sin\left(\frac{k}{n}-t\right) dt\right|$$
$$= \left|\int_{\frac{k}{n}}^{x} \left[(f''(t)+f(t)) - (f''(x)+f(x))\right] \sin\left(\frac{k}{n}-t\right) dt\right|$$
$$\leq \int_{\frac{k}{n}}^{x} \left|(f''(t)+f(t)) - (f''(x)+f(x))\right| \left|\sin\left(\frac{k}{n}-t\right)\right| dt$$
$$\leq \int_{\frac{k}{n}}^{x} \omega_1\left(f''+f, x-\frac{k}{n}\right)\left(t-\frac{k}{n}\right) dt$$
$$\leq \omega_1\left(f''+f, x-\frac{k}{n}\right) \frac{\left(x-\frac{k}{n}\right)^2}{2} \leq \frac{\omega_1\left(f''+f, \frac{1}{n^{\alpha}}\right)}{2n^{2\alpha}}. \tag{2.49}$$

That is,

$$|R_2(n)| \leq \frac{\omega_1\left(f''+f, \frac{1}{n^{\alpha}}\right)}{2n^{2\alpha}}. \tag{2.50}$$

So, we have proved when $\left|\frac{k}{n} - x\right| \leq \frac{1}{n^{\alpha}}$, always it holds

$$|R_2(n)| \leq \frac{\omega_1\left(f'' + f, \frac{1}{n^{\alpha}}\right)}{2n^{2\alpha}}. \tag{2.51}$$

Next assume again $\frac{k}{n} \geq x$, then

$$|R_2(n)| = \left|\int_x^{\frac{k}{n}} \left[\left(f''(t) + f(t)\right) - \left(f''(x) + f(x)\right)\right] \sin\left(\frac{k}{n} - t\right) dt\right|$$

$$\leq \int_x^{\frac{k}{n}} \left|\left(f''(t) + f(t)\right) - \left(f''(x) + f(x)\right)\right| \left|\sin\left(\frac{k}{n} - t\right)\right| dt \leq$$

(by Lakshmikantham and Ladas (1972), p. 3, abstract mean value theorem, and $|\sin x| \leq 1$ along with $|\sin x| \leq |x|$, $\forall\, x \in \mathbb{R}$)

$$\|f''' + f'\|_{\infty} \min\left\{\int_x^{\frac{k}{n}} (t - x)\, dt,\ \int_x^{\frac{k}{n}} \left(\frac{k}{n} - t\right)(t - x)\, dt\right\}$$

$$= \|f''' + f'\|_{\infty} \min\left\{\frac{\left(\frac{k}{n} - x\right)^2}{2},\ \frac{\left(\frac{k}{n} - x\right)^3}{6}\right\}$$

$$\leq \|f''' + f'\|_{\infty} \min\left\{\frac{(b - a)^2}{2},\ \frac{(b - a)^3}{6}\right\}$$

$$= \|f''' + f'\|_{\infty} \frac{(b - a)^2}{2} \min\left\{1,\ \frac{(b - a)}{3}\right\}. \tag{2.52}$$

Hence,

$$|R_2(n)| \leq \|f''' + f'\|_{\infty} \frac{(b - a)^2}{2} \min\left\{1,\ \frac{(b - a)}{3}\right\}. \tag{2.53}$$

When $\frac{k}{n} < x$, we have

$$|R_2(n)| = \left|\int_x^{\frac{k}{n}} \left[\left(f''(t) + f(t)\right) - \left(f''(x) + f(x)\right)\right] \sin\left(\frac{k}{n} - t\right) dt\right|$$

$$= \left|\int_{\frac{k}{n}}^{x} \left[\left(f''(x) + f(x)\right) - \left(f''(t) + f(t)\right)\right] \sin\left(\frac{k}{n} - t\right) dt\right|$$

$$\leq \int_{\frac{k}{n}}^{x} \left|\left(f''(x) + f(x)\right) - \left(f''(t) + f(t)\right)\right| \left|\sin\left(\frac{k}{n} - t\right)\right| dt \leq$$

(as in (2.52))

$$\|f''' + f'\|_\infty \min\left\{\int_{\frac{k}{n}}^{x} (x-t)\,dt,\ \int_{\frac{k}{n}}^{x} (x-t)\left(t-\frac{k}{n}\right)dt\right\}$$

$$= \|f''' + f'\|_\infty \min\left\{\frac{\left(x-\frac{k}{n}\right)^2}{2},\ \frac{\left(x-\frac{k}{n}\right)^3}{6}\right\}$$

$$\leq \|f''' + f'\|_\infty \min\left\{\frac{(b-a)^2}{2},\ \frac{(b-a)^3}{6}\right\}$$

$$= \|f''' + f'\|_\infty \frac{(b-a)^2}{2} \min\left\{1,\ \frac{(b-a)}{3}\right\}. \tag{2.54}$$

Therefore, always it holds

$$|R_2(n)| \leq \|f''' + f'\|_\infty \frac{(b-a)^2}{2} \min\left\{1,\ \frac{(b-a)}{3}\right\}. \tag{2.55}$$

And we have

$$\Lambda_n(x) = \sum_{\left\{\substack{k=\lceil na\rceil \\ :\left|\frac{k}{n}-x\right|\leq\frac{1}{n^\alpha}}\right.}^{\lfloor nb\rfloor} G_{q,\lambda}(nx-k)\, R_2(n)$$

$$+ \sum_{\left\{\substack{k=\lceil na\rceil \\ :\left|\frac{k}{n}-x\right|>\frac{1}{n^\alpha}}\right.}^{\lfloor nb\rfloor} G_{q,\lambda}(nx-k)\, R_2(n). \tag{2.56}$$

Hence it holds

$$-|\Lambda_n(x)| \leq \sum_{\left\{\substack{k=\lceil na\rceil \\ :\left|\frac{k}{n}-x\right|\leq\frac{1}{n^\alpha}}\right.}^{\lfloor nb\rfloor} G_{q,\lambda}(nx-k)\, |R_2(n)|$$

$$+ \sum_{\left\{\substack{k=\lceil na\rceil \\ :\left|\frac{k}{n}-x\right|>\frac{1}{n^\alpha}}\right.}^{\lfloor nb\rfloor} G_{q,\lambda}(nx-k)\, |R_2(n)| \tag{2.57}$$

$$\leq \left(\sum_{\substack{k=\lceil na \rceil \\ \left\{ : \left| \frac{k}{n} - x \right| \leq \frac{1}{n^{\alpha}} \right.}}^{\lfloor nb \rfloor} G_{q,\lambda}(nx-k) \right) \frac{\omega_1\left(f''+f, \frac{1}{n^{\alpha}}\right)}{2n^{2\alpha}}$$

$$+ \sum_{\substack{k=\lceil na \rceil \\ \left\{ : \left| \frac{k}{n} - x \right| > \frac{1}{n^{\alpha}} \right.}}^{\lfloor nb \rfloor} G_{q,\lambda}(nx-k) \left\| f''' + f' \right\|_{\infty} \frac{(b-a)^2}{2} \min\left\{1, \frac{(b-a)}{3}\right\}$$

$$\stackrel{\text{(by (2.30))}}{\leq} \frac{\omega_1\left(f''+f, \frac{1}{n^{\alpha}}\right)}{2n^{2\alpha}} + \left\| f''' + f' \right\|_{\infty} \frac{(b-a)^2}{2} \min\left\{1, \frac{(b-a)}{3}\right\}$$

$$\left(\sum_{\substack{k=\lceil na \rceil \\ \left\{ : \left| \frac{k}{n} - x \right| > \frac{1}{n^{\alpha}} \right.}}^{\lfloor nb \rfloor} G_{q,\lambda}(nx-k) \right) \stackrel{\text{(by Theorem 2.3)}}{\leq} \frac{\omega_1\left(f''+f, \frac{1}{n^{\alpha}}\right)}{2n^{2\alpha}}$$

$$+ \left\| f''' + f' \right\|_{\infty} \frac{(b-a)^2}{2} \min\left\{1, \frac{(b-a)}{3}\right\} \gamma A^{-\lambda\left(n^{1-\alpha}-2\right)}. \tag{2.58}$$

From now on for convenience we call

$$\delta := \frac{(b-a)^2}{2} \min\left\{1, \frac{(b-a)}{3}\right\}. \tag{2.59}$$

Consequently, we have derived that

$$\left|\Lambda_n(x)\right| \leq \frac{\omega_1\left(f''+f, \frac{1}{n^{\alpha}}\right)}{2n^{2\alpha}} + \left\| f''' + f' \right\|_{\infty} \delta \gamma A^{-\lambda\left(n^{1-\alpha}-2\right)}. \tag{2.60}$$

Next we use again $|\sin x| \leq |x|$, $\forall\, x \in \mathbb{R}$.
We have that

$$L_n^*(\sin(\cdot - x), x) = \sum_{k=\lceil na \rceil}^{\lfloor nb \rfloor} G_{q,\lambda}(nx-k) \sin\left(\frac{k}{n} - x\right) \tag{2.61}$$

and

$$\left|L_n^*\left(\sin\left(\cdot-x\right),x\right)\right| \leq \sum_{k=\lceil na\rceil}^{\lfloor nb\rfloor} G_{q,\lambda}\left(nx-k\right)\left|\sin\left(\frac{k}{n}-x\right)\right|$$

$$= \sum_{\begin{cases} k=\lceil na\rceil \\ :\left|\frac{k}{n}-x\right|\leq\frac{1}{n^{\alpha}}\end{cases}}^{\lfloor nb\rfloor} G_{q,\lambda}\left(nx-k\right)\left|\sin\left(\frac{k}{n}-x\right)\right|$$

$$+ \sum_{\begin{cases} k=\lceil na\rceil \\ :\left|\frac{k}{n}-x\right|>\frac{1}{n^{\alpha}}\end{cases}}^{\lfloor nb\rfloor} G_{q,\lambda}\left(nx-k\right)\left|\sin\left(\frac{k}{n}-x\right)\right| \quad (2.62)$$

$$\leq \sum_{\begin{cases} k=\lceil na\rceil \\ :\left|\frac{k}{n}-x\right|\leq\frac{1}{n^{\alpha}}\end{cases}}^{\lfloor nb\rfloor} G_{q,\lambda}\left(nx-k\right)\left|\frac{k}{n}-x\right|$$

$$+ \sum_{\begin{cases} k=\lceil na\rceil \\ :\left|\frac{k}{n}-x\right|>\frac{1}{n^{\alpha}}\end{cases}}^{\lfloor nb\rfloor} G_{q,\lambda}\left(nx-k\right)\left|\frac{k}{n}-x\right|$$

$$\leq \frac{1}{n^{\alpha}} + (b-a)\left(\sum_{\begin{cases} k=\lceil na\rceil \\ :\left|\frac{k}{n}-x\right|>\frac{1}{n^{\alpha}}\end{cases}}^{\lfloor nb\rfloor} G_{q,\lambda}\left(nx-k\right)\right)$$

$$\overset{\text{(by (2.27))}}{\leq} \frac{1}{n^{\alpha}} + (b-a)\,\gamma A^{-\lambda\left(n^{1-\alpha}-2\right)}. \quad (2.63)$$

We found that

$$\left|L_n^*\left(\sin\left(\cdot-x\right),x\right)\right| \leq \frac{1}{n^{\alpha}} + (b-a)\,\gamma A^{-\lambda\left(n^{1-\alpha}-2\right)}. \quad (2.64)$$

Next we estimate

$$L_n^*\left(\sin^2\left(\frac{\cdot-x}{2}\right),x\right) = \sum_{k=\lceil na\rceil}^{\lfloor nb\rfloor} G_{q,\lambda}\left(nx-k\right)\sin^2\left(\frac{\frac{k}{n}-x}{2}\right). \quad (2.65)$$

We have that (by $|\sin x| \leq |x|$, $\forall\, x \in \mathbb{R}$)

$$L_n^*\left(\sin^2\left(\frac{\cdot - x}{2}\right), x\right) = \sum_{k=\lceil na\rceil}^{\lfloor nb\rfloor} G_{q,\lambda}(nx-k)\left|\sin\left(\frac{\frac{k}{n}-x}{2}\right)\right|^2$$

$$\leq \frac{1}{4}\sum_{k=\lceil na\rceil}^{\lfloor nb\rfloor} G_{q,\lambda}(nx-k)\left|\frac{k}{n}-x\right|^2$$

$$= \frac{1}{4}\left[\sum_{\substack{k=\lceil na\rceil \\ \left\{:\left|\frac{k}{n}-x\right|\leq\frac{1}{n^\alpha}\right.}}^{\lfloor nb\rfloor} G_{q,\lambda}(nx-k)\left|\frac{k}{n}-x\right|^2\right.$$

$$\left. + \sum_{\substack{k=\lceil na\rceil \\ \left\{:\left|\frac{k}{n}-x\right|>\frac{1}{n^\alpha}\right.}}^{\lfloor nb\rfloor} G_{q,\lambda}(nx-k)\left|\frac{k}{n}-x\right|^2\right]$$

$$\leq \frac{1}{4}\left[\frac{1}{n^{2\alpha}} + (b-a)^2\gamma A^{-\lambda\left(n^{1-\alpha}-2\right)}\right]. \tag{2.66}$$

That is,

$$L_n^*\left(\sin^2\left(\frac{\cdot - x}{2}\right), x\right) \leq \frac{1}{4}\left[\frac{1}{n^{2\alpha}} + (b-a)^2\gamma A^{-\lambda\left(n^{1-\alpha}-2\right)}\right]. \tag{2.67}$$

Consequently, we have derived the following:

(1)

$$|L_n(f,x) - f(x)| \leq K(q)\left[|f'(x)|\left(\frac{1}{n^\alpha} + (b-a)\gamma A^{-\lambda\left(n^{1-\alpha}-2\right)}\right)\right.$$

$$+ \frac{|f''(x)|}{2}\left(\frac{1}{n^{2\alpha}} + (b-a)^2\gamma A^{-\lambda\left(n^{1-\alpha}-2\right)}\right)$$

$$\left. + \left(\frac{\omega_1\left(f''+f, \frac{1}{n^\alpha}\right)}{2n^{2\alpha}} + \|f'''+f'\|_\infty\,\delta\gamma A^{-\lambda\left(n^{1-\alpha}-2\right)}\right)\right]. \tag{2.68}$$

(2) If $f'(x) = f''(x) = 0$, by (2.68), we obtain

$$|L_n(f,x) - f(x)| \leq K(q)\left[\frac{\omega_1\left(f''+f,\frac{1}{n^{\alpha}}\right)}{2n^{2\alpha}} + \|f'''+f'\|_{\infty}\,\delta\gamma A^{-\lambda\left(n^{1-\alpha}-2\right)}\right], \tag{2.69}$$

note here the high rate of convergence at $n^{-3\alpha}$.

(3) Furthermore, by (2.68), we get

$$\begin{aligned} \|L_n f - f\|_{\infty} \leq K(q) \Bigg[& \|f'\|_{\infty}\left(\frac{1}{n^{\alpha}} + (b-a)\,\gamma A^{-\lambda\left(n^{1-\alpha}-2\right)}\right) \\ & + \frac{\|f''\|_{\infty}}{2}\left(\frac{1}{n^{2\alpha}} + (b-a)^2\,\gamma A^{-\lambda\left(n^{1-\alpha}-2\right)}\right) \\ & + \left(\frac{\omega_1\left(f''+f,\frac{1}{n^{\alpha}}\right)}{2n^{2\alpha}} + \|f'''+f'\|_{\infty}\,\delta\gamma A^{-\lambda\left(n^{1-\alpha}-2\right)}\right)\Bigg]. \end{aligned} \tag{2.70}$$

We derive that $\lim_{n\to+\infty} L_n(f) = f$, pointwise and uniformly.

We observe that

$$\begin{aligned} & L_n(f,x) - f'(x)\,L_n(\sin(\cdot - x),x) - 2f''(x)\,L_n\left(\sin^2\left(\frac{\cdot - x}{2}\right),x\right) - f(x) \\ & = \frac{L_n^*(f,x)}{\sum_{k=\lceil na\rceil}^{\lfloor nb\rfloor} G_{q,\lambda}(nx-k)} - f'(x)\,\frac{L_n^*(\sin(\cdot - x),x)}{\sum_{k=\lceil na\rceil}^{\lfloor nb\rfloor} G_{q,\lambda}(nx-k)} \\ & \quad - 2f''(x)\frac{L_n^*\left(\sin^2\left(\frac{\cdot - x}{2}\right),x\right)}{\sum_{k=\lceil na\rceil}^{\lfloor nb\rfloor} G_{q,\lambda}(nx-k)} - f(x)\left(\frac{\sum_{k=\lceil na\rceil}^{\lfloor nb\rfloor} G_{q,\lambda}(nx-k)}{\sum_{k=\lceil na\rceil}^{\lfloor nb\rfloor} G_{q,\lambda}(nx-k)}\right) \end{aligned} \tag{2.71}$$

$$\begin{aligned} & = \frac{1}{\sum_{k=\lceil na\rceil}^{\lfloor nb\rfloor} G_{q,\lambda}(nx-k)}\Bigg[L_n^*(f,x) - f'(x)L_n^*(\sin(\cdot - x),x) \\ & \quad - 2f''(x)\,L_n^*\left(\sin^2\left(\frac{\cdot - x}{2}\right),x\right) - f(x)\sum_{k=\lceil na\rceil}^{\lfloor nb\rfloor} G_{q,\lambda}(nx-k)\Bigg] \\ & = \frac{1}{\sum_{k=\lceil na\rceil}^{\lfloor nb\rfloor} G_{q,\lambda}(nx-k)}\,(\Lambda_n(x)). \end{aligned} \tag{2.72}$$

Finally, we obtain ($\forall\, x \in [a,b]$, $n \in \mathbb{N}$):
(4)

$$\left| L_n(f,x) - f'(x) L_n(\sin(\cdot - x), x) - 2f''(x) L_n\left(\sin^2\left(\frac{\cdot - x}{2}\right), x\right) - f(x) \right|$$

$$\overset{(2.28)}{\leq} K(q) \left|\Lambda_n(x)\right|$$

$$\overset{(2.60)}{\leq} K(q) \left[\frac{\omega_1\left(f'' + f, \frac{1}{n^\alpha}\right)}{2n^{2\alpha}} + \left\| f''' + f' \right\|_\infty \delta \gamma A^{-\lambda\left(n^{1-\alpha}-2\right)} \right]. \tag{2.73}$$

The theorem is proved. □

We continue with a hyperbolic high order neural network approximation.

Theorem 2.6. *Let $f \in C^3([a,b],\mathbb{C})$, $0 < \alpha < 1$, $n \in \mathbb{N} : n^{1-\alpha} > 2$, $x \in [a,b]$. Then*

(1)

$$|L_n(f,x) - f(x)| \leq K(q) \cosh(b-a) \left[|f'(x)| \left(\frac{1}{n^\alpha} + (b-a)\gamma A^{-\lambda\left(n^{1-\alpha}-2\right)} \right) \right.$$

$$\frac{|f''(x)|}{2} \left(\frac{1}{n^{2\alpha}} + (b-a)^2 \gamma A^{-\lambda\left(n^{1-\alpha}-2\right)} \right)$$

$$\left. + \frac{1}{2} \left(\frac{\omega_1\left(f'' - f, \frac{1}{n^\alpha}\right)}{n^{2\alpha}} + \frac{\left\| f''' - f' \right\|_\infty (b-a)^3}{3} \gamma A^{-\lambda\left(n^{1-\alpha}-2\right)} \right) \right], \tag{2.74}$$

(2) *if $f'(x) = f''(x) = 0$, we obtain*

$$|L_n(f,x) - f(x)| \leq \frac{K(q) \cosh(b-a)}{2}$$

$$\left[\frac{\omega_1\left(f'' - f, \frac{1}{n^\alpha}\right)}{n^{2\alpha}} + \frac{\left\| f''' - f' \right\|_\infty (b-a)^3}{3} \gamma A^{-\lambda\left(n^{1-\alpha}-2\right)} \right], \tag{2.75}$$

note here the high rate of convergence at $n^{-3\alpha}$,

(3) *furthermore, we get*

$$\|L_n f - f\|_\infty \leq K(q)\cosh(b-a)\left[\|f'\|_\infty\left(\frac{1}{n^\alpha} + (b-a)\gamma A^{-\lambda\left(n^{1-\alpha}-2\right)}\right)\right.$$

$$\frac{\|f''\|_\infty}{2}\left(\frac{1}{n^{2\alpha}} + (b-a)^2\gamma A^{-\lambda\left(n^{1-\alpha}-2\right)}\right)$$

$$\left.+\frac{1}{2}\left(\frac{\omega_1\left(f''-f,\frac{1}{n^\alpha}\right)}{n^{2\alpha}} + \frac{\|f'''-f'\|_\infty (b-a)^3}{3}\gamma A^{-\lambda\left(n^{1-\alpha}-2\right)}\right)\right], \tag{2.76}$$

it follows that $\lim_{n\to+\infty} L_n(f) = f$, *pointwise and uniformly,*

and

(4)

$$\left|L_n(f,x) - f'(x)L_n(\sinh(\cdot - x), x) - 2f''(x)L_n\left(\sinh^2\left(\frac{\cdot - x}{2}\right), x\right) - f(x)\right|$$

$$\leq \frac{K(q)\cosh(b-a)}{2}\left[\frac{\omega_1\left(f''-f,\frac{1}{n^\alpha}\right)}{n^{2\alpha}} + \frac{\|f'''-f'\|_\infty (b-a)^3}{3}\gamma A^{-\lambda\left(n^{1-\alpha}-2\right)}\right], \tag{2.77}$$

again here we achieve high speed of convergence at $n^{-3\alpha}$.

Proof. By the mean value theorem, we have that

$$\sinh x = \sinh x - \sinh 0 = (\cosh\xi)(x-0),$$

for some ξ between $\{0, x\}$, for any $x \in \mathbb{R}$.

Hence,

$$|\sinh x| \leq \|\cosh\|_{\infty,[-(b-a),b-a]}|x|, \quad \forall x \in [-(b-a), b-a]. \tag{2.78}$$

That is, there exists $M \geq 1$ such that

$$|\sinh x| \leq M|x|, \quad \forall x \in [-(b-a), b-a], \tag{2.79}$$

where $M := \|\cosh\|_{\infty,[-(b-a),b-a]} = \cosh(b-a)$.

Here $f \in C^3([a,b],\mathbb{C})$, and we apply the hyperbolic Taylor's formula for $f \in C^2([a,b],\mathbb{C})$, see Theorem 7 of Anastassiou (2023b).

Let $\frac{k}{n}, x \in [a, b]$, then

$$f\left(\frac{k}{n}\right) = f(x) + f'(x)\sinh\left(\frac{k}{n} - x\right) + 2f''(x)\sinh^2\left(\frac{\frac{k}{n} - x}{2}\right)$$
$$+ \int_x^{\frac{k}{n}} \left[(f''(t) - f(t)) - (f''(x) - f(x))\right]\sinh\left(\frac{k}{n} - t\right) dt. \tag{2.80}$$

Hence it holds

$$f\left(\frac{k}{n}\right) G_{q,\lambda}(nx - k) = f(x) G_{q,\lambda}(nx - k)$$
$$+ f'(x)\sinh\left(\frac{k}{n} - x\right) G_{q,\lambda}(nx - k) + 2f''(x)\sinh^2\left(\frac{\frac{k}{n} - x}{2}\right) G_{q,\lambda}(nx - k)$$
$$+ G_{q,\lambda}(nx - k)\left(\int_x^{\frac{k}{n}} \left[(f''(t) - f(t)) - (f''(x) - f(x))\right]\sinh\left(\frac{k}{n} - t\right) dt\right) \tag{2.81}$$

so that we have

$$\sum_{k=\lceil na \rceil}^{\lfloor nb \rfloor} f\left(\frac{k}{n}\right) G_{q,\lambda}(nx - k) - f(x) \sum_{k=\lceil na \rceil}^{\lfloor nb \rfloor} G_{q,\lambda}(nx - k)$$
$$= f'(x) \sum_{k=\lceil na \rceil}^{\lfloor nb \rfloor} G_{q,\lambda}(nx - k)\sinh\left(\frac{k}{n} - x\right)$$
$$+ 2f''(x) \sum_{k=\lceil na \rceil}^{\lfloor nb \rfloor} G_{q,\lambda}(nx - k)\sinh^2\left(\frac{\frac{k}{n} - x}{2}\right) + \sum_{k=\lceil na \rceil}^{\lfloor nb \rfloor} G_{q,\lambda}(nx - k)$$
$$\times \left(\int_x^{\frac{k}{n}} \left[(f''(t) - f(t)) - (f''(x) - f(x))\right]\sinh\left(\frac{k}{n} - t\right) dt\right). \tag{2.82}$$

Thus, we obtain

$$L_n^*(f,x) - f(x) \sum_{k=\lceil na \rceil}^{\lfloor nb \rfloor} G_{q,\lambda}(nx-k)$$
$$= f'(x) L_n^*(\sinh(\cdot - x), x) + 2f''(x) L_n^*\left(\sinh^2\left(\frac{\cdot - x}{2}\right), x\right) + \Lambda_n(x), \tag{2.83}$$

where

$$\Lambda_n(x) := \sum_{k=\lceil na \rceil}^{\lfloor nb \rfloor} G_{q,\lambda}(nx-k)$$
$$\times \left(\int_x^{\frac{k}{n}} \left[(f''(t) - f(t)) - (f''(x) - f(x))\right] \sinh\left(\frac{k}{n} - t\right) dt\right). \tag{2.84}$$

We call

$$R_2(n) := \int_x^{\frac{k}{n}} \left[(f''(t) - f(t)) - (f''(x) - f(x))\right] \sinh\left(\frac{k}{n} - t\right) dt. \tag{2.85}$$

We assume that $b - a > \frac{1}{n^\alpha}$, which is always the case for large enough $n \in \mathbb{N}$, that is, when $n > \lceil (b-a)^{-\frac{1}{\alpha}} \rceil$.

Thus, $\left|\frac{k}{n} - x\right| \leq \frac{1}{n^\alpha}$ or $\left|\frac{k}{n} - x\right| > \frac{1}{n^\alpha}$.

In case of $\left|\frac{k}{n} - x\right| \leq \frac{1}{n^\alpha}$, we have the following cases:

(i) If $\frac{k}{n} \geq x$, then

$$|R_2(n)| = \left|\int_x^{\frac{k}{n}} \left[(f''(t) - f(t)) - (f''(x) - f(x))\right] \sinh\left(\frac{k}{n} - t\right) dt\right|$$
$$\leq \int_x^{\frac{k}{n}} \omega_1(f'' - f, t - x) \left|\sinh\left(\frac{k}{n} - t\right)\right| dt$$
$$\overset{(2.79)}{\leq} \int_x^{\frac{k}{n}} \omega_1(f'' - f, t - x) M \left(\frac{k}{n} - t\right) dt$$
$$\leq M\omega_1\left(f'' - f, \frac{k}{n} - x\right) \int_x^{\frac{k}{n}} \left(\frac{k}{n} - t\right) dt$$
$$= M\omega_1\left(f'' - f, \frac{k}{n} - x\right) \frac{\left(\frac{k}{n} - x\right)^2}{2} \leq \frac{M\omega_1\left(f'' - f, \frac{1}{n^\alpha}\right)}{2n^{2\alpha}}, \tag{2.86}$$

that is,

$$|R_2(n)| \leq \frac{M\omega_1\left(f''-f,\frac{1}{n^\alpha}\right)}{2n^{2\alpha}}. \tag{2.87}$$

(ii) If $\frac{k}{n} < x$, then

$$\begin{aligned}
|R_2(n)| &= \left|\int_x^{\frac{k}{n}} \left[\left(f''(t)-f(t)\right)-\left(f''(x)-f(x)\right)\right]\sinh\left(\frac{k}{n}-t\right)dt\right| \\
&= \left|\int_{\frac{k}{n}}^{x} \left[\left(f''(t)-f(t)\right)-\left(f''(x)-f(x)\right)\right]\sinh\left(\frac{k}{n}-t\right)dt\right| \\
&\leq \int_{\frac{k}{n}}^{x} \left|\left(f''(t)-f(t)\right)-\left(f''(x)-f(x)\right)\right|\left|\sinh\left(\frac{k}{n}-t\right)\right|dt \\
&\leq M\omega_1\left(f''-f,x-\frac{k}{n}\right)\int_{\frac{k}{n}}^{x}\left(t-\frac{k}{n}\right)dt \\
&= M\omega_1\left(f''-f,x-\frac{k}{n}\right)\frac{\left(x-\frac{k}{n}\right)^2}{2} \leq \frac{M\omega_1\left(f''-f,\frac{1}{n^\alpha}\right)}{2n^{2\alpha}},
\end{aligned} \tag{2.88}$$

that is,

$$|R_2(n)| \leq \frac{M\omega_1\left(f''-f,\frac{1}{n^\alpha}\right)}{2n^{2\alpha}}. \tag{2.89}$$

So, we have proved when $\left|\frac{k}{n}-x\right| \leq \frac{1}{n^\alpha}$, always it holds

$$|R_2(n)| \leq \frac{M\omega_1\left(f''-f,\frac{1}{n^\alpha}\right)}{2n^{2\alpha}}. \tag{2.90}$$

Next assume again $\frac{k}{n} \geq x$, then

$$\begin{aligned}
|R_2(n)| &= \left|\int_x^{\frac{k}{n}} \left[\left(f''(t)-f(t)\right)-\left(f''(x)-f(x)\right)\right]\sinh\left(\frac{k}{n}-t\right)dt\right| \\
&\leq \int_x^{\frac{k}{n}} \left|\left(f''(t)-f(t)\right)-\left(f''(x)-f(x)\right)\right|\left|\sinh\left(\frac{k}{n}-t\right)\right|dt \leq
\end{aligned}$$

(by Lakshmikantham and Ladas (1972), p. 3, abstract mean value theorem)

$$M\left\|f'''-f'\right\|_{\infty}\int_{x}^{\frac{k}{n}}\left(\frac{k}{n}-t\right)(t-x)\,dt$$

$$=M\left\|f'''-f'\right\|_{\infty}\frac{\left(\frac{k}{n}-x\right)^{3}}{6}\leq M\left\|f'''-f'\right\|_{\infty}\frac{(b-a)^{3}}{6}. \tag{2.91}$$

Hence,

$$\left|R_{2}(n)\right|\leq\frac{M\left\|f'''-f'\right\|_{\infty}(b-a)^{3}}{6}. \tag{2.92}$$

When $\frac{k}{n}<x$, we have

$$\left|R_{2}(n)\right|=\left|\int_{\frac{k}{n}}^{x}\left[\left(f''(t)-f(t)\right)-\left(f''(x)-f(x)\right)\right]\sinh\left(\frac{k}{n}-t\right)dt\right|$$

$$\leq\int_{\frac{k}{n}}^{x}\left|\left(f''(t)-f(t)\right)-\left(f''(x)-f(x)\right)\right|\left|\sinh\left(\frac{k}{n}-t\right)\right|dt\leq$$

(as in (2.91))

$$M\left\|f'''-f'\right\|_{\infty}\int_{\frac{k}{n}}^{x}(x-t)\left(t-\frac{k}{n}\right)dt$$

$$=M\left\|f'''-f'\right\|_{\infty}\frac{\left(x-\frac{k}{n}\right)^{3}}{6}\leq M\left\|f'''-f'\right\|_{\infty}\frac{(b-a)^{3}}{6}. \tag{2.93}$$

Therefore, always it holds

$$\left|R_{2}(n)\right|\leq\frac{M\left\|f'''-f'\right\|_{\infty}(b-a)^{3}}{6}. \tag{2.94}$$

And we have

$$\Lambda_{n}(x)=\sum_{\substack{k=\lceil na\rceil\\ \left\{:\left|\frac{k}{n}-x\right|\leq\frac{1}{n^{\alpha}}\right.}}^{\lfloor nb\rfloor}G_{q,\lambda}(nx-k)\,R_{2}(n)$$

$$+\sum_{\substack{k=\lceil na\rceil\\ \left\{:\left|\frac{k}{n}-x\right|>\frac{1}{n^{\alpha}}\right.}}^{\lfloor nb\rfloor}G_{q,\lambda}(nx-k)\,R_{2}(n). \tag{2.95}$$

Hence it holds

$$|\Lambda_n(x)| \leq \sum_{\left\{\substack{k=\lceil na\rceil \\ :\left|\frac{k}{n}-x\right|\leq\frac{1}{n^\alpha}}\right.}^{\lfloor nb\rfloor} G_{q,\lambda}(nx-k)\,|R_2(n)| \tag{2.96}$$

$$+ \sum_{\left\{\substack{k=\lceil na\rceil \\ :\left|\frac{k}{n}-x\right|>\frac{1}{n^\alpha}}\right.}^{\lfloor nb\rfloor} G_{q,\lambda}(nx-k)\,|R_2(n)|$$

$$\leq \left(\sum_{\left\{\substack{k=\lceil na\rceil \\ :\left|\frac{k}{n}-x\right|\leq\frac{1}{n^\alpha}}\right.}^{\lfloor nb\rfloor} G_{q,\lambda}(nx-k)\right)\frac{\omega_1\left(f''-f,\frac{1}{n^\alpha}\right)M}{2n^{2\alpha}}$$

$$+\left(\sum_{\left\{\substack{k=\lceil na\rceil \\ :\left|\frac{k}{n}-x\right|>\frac{1}{n^\alpha}}\right.}^{\lfloor nb\rfloor} G_{q,\lambda}(nx-k)\right)\frac{M\left\|f'''-f'\right\|_\infty(b-a)^3}{6}$$

$$\overset{((2.30))}{\leq} \frac{M\omega_1\left(f''-f,\frac{1}{n^\alpha}\right)}{2n^{2\alpha}}+\frac{M\left\|f'''-f'\right\|_\infty(b-a)^3}{6}$$

$$\left(\sum_{\left\{\substack{k=\lceil na\rceil \\ :\left|\frac{k}{n}-x\right|>\frac{1}{n^\alpha}}\right.}^{\lfloor nb\rfloor} G_{q,\lambda}(nx-k)\right)$$

$$\overset{\text{(by Theorem 2.3)}}{\leq} M\frac{\omega_1\left(f''-f,\frac{1}{n^\alpha}\right)}{2n^{2\alpha}}+\frac{M\left\|f'''-f'\right\|_\infty(b-a)^3}{6}\gamma A^{-\lambda\left(n^{1-\alpha}-2\right)}. \tag{2.97}$$

Consequently, we have derived that

$$|\Lambda_n(x)| \leq \frac{M}{2}\left[\frac{\omega_1\left(f''-f,\frac{1}{n^\alpha}\right)}{n^{2\alpha}}+\frac{\left\|f'''-f'\right\|_\infty(b-a)^3}{3}\gamma A^{-\lambda\left(n^{1-\alpha}-2\right)}\right]. \tag{2.98}$$

We have that

$$L_n^*(\sinh(\cdot-x),x)=\sum_{k=\lceil na\rceil}^{\lfloor nb\rfloor} G_{q,\lambda}(nx-k)\sinh\left(\frac{k}{n}-x\right), \tag{2.99}$$

and

$$
\begin{aligned}
\left|L_n^*\left(\sinh\left(\cdot-x\right),x\right)\right| &\leq \sum_{k=\lceil na\rceil}^{\lfloor nb\rfloor} G_{q,\lambda}\left(nx-k\right)\left|\sinh\left(\frac{k}{n}-x\right)\right| \\
&= \sum_{\begin{cases} k=\lceil na\rceil \\ :\left|\frac{k}{n}-x\right|\leq\frac{1}{n^{\alpha}} \end{cases}}^{\lfloor nb\rfloor} G_{q,\lambda}\left(nx-k\right)\left|\sinh\left(\frac{k}{n}-x\right)\right| \\
&\quad + \sum_{\begin{cases} k=\lceil na\rceil \\ :\left|\frac{k}{n}-x\right|>\frac{1}{n^{\alpha}} \end{cases}}^{\lfloor nb\rfloor} G_{q,\lambda}\left(nx-k\right)\left|\sinh\left(\frac{k}{n}-x\right)\right|
\end{aligned} \tag{2.100}
$$

$$
\begin{aligned}
&\leq M\left[\sum_{\begin{cases} k=\lceil na\rceil \\ :\left|\frac{k}{n}-x\right|\leq\frac{1}{n^{\alpha}} \end{cases}}^{\lfloor nb\rfloor} G_{q,\lambda}\left(nx-k\right)\left|\frac{k}{n}-x\right| \right. \\
&\quad \left. + \sum_{\begin{cases} k=\lceil na\rceil \\ :\left|\frac{k}{n}-x\right|>\frac{1}{n^{\alpha}} \end{cases}}^{\lfloor nb\rfloor} G_{q,\lambda}\left(nx-k\right)\left|\frac{k}{n}-x\right|\right] \\
&\leq M\left[\frac{1}{n^{\alpha}}+(b-a)\left(\sum_{\begin{cases} k=\lceil na\rceil \\ :\left|\frac{k}{n}-x\right|>\frac{1}{n^{\alpha}} \end{cases}}^{\lfloor nb\rfloor} G_{q,\lambda}\left(nx-k\right)\right)\right] \\
&\overset{\text{(by (2.27))}}{\leq} M\left[\frac{1}{n^{\alpha}}+(b-a)\,\gamma A^{-\lambda\left(n^{1-\alpha}-2\right)}\right].
\end{aligned} \tag{2.101}
$$

We found that

$$
\left|L_n^*\left(\sinh\left(\cdot-x\right),x\right)\right| \leq M\left[\frac{1}{n^{\alpha}}+(b-a)\,\gamma A^{-\lambda\left(n^{1-\alpha}-2\right)}\right]. \tag{2.102}
$$

Next we estimate

$$
L_n^*\left(\sinh^2\left(\frac{\cdot-x}{2}\right),x\right) = \sum_{k=\lceil na\rceil}^{\lfloor nb\rfloor} G_{q,\lambda}\left(nx-k\right)\sinh^2\left(\frac{\frac{k}{n}-x}{2}\right). \tag{2.103}
$$

We have that

$$L_n^*\left(\sinh^2\left(\frac{\cdot - x}{2}\right), x\right) = \sum_{k=\lceil na\rceil}^{\lfloor nb\rfloor} G_{q,\lambda}(nx-k)\left|\sinh\left(\frac{\frac{k}{n}-x}{2}\right)\right|^2$$

$$\leq \frac{M}{4}\sum_{k=\lceil na\rceil}^{\lfloor nb\rfloor} G_{q,\lambda}(nx-k)\left|\frac{k}{n}-x\right|^2$$

$$= \frac{M}{4}\left[\sum_{\substack{k=\lceil na\rceil \\ \left\{:\left|\frac{k}{n}-x\right|\leq\frac{1}{n^\alpha}\right.}}^{\lfloor nb\rfloor} G_{q,\lambda}(nx-k)\left|\frac{k}{n}-x\right|^2\right.$$

$$\left.+\sum_{\substack{k=\lceil na\rceil \\ \left\{:\left|\frac{k}{n}-x\right|>\frac{1}{n^\alpha}\right.}}^{\lfloor nb\rfloor} G_{q,\lambda}(nx-k)\left|\frac{k}{n}-x\right|^2\right]$$

$$\leq \frac{M}{4}\left[\frac{1}{n^{2\alpha}}+(b-a)^2\gamma A^{-\lambda\left(n^{1-\alpha}-2\right)}\right]. \tag{2.104}$$

That is,

$$L_n^*\left(\sinh^2\left(\frac{\cdot - x}{2}\right), x\right) \leq \frac{M}{4}\left[\frac{1}{n^{2\alpha}}+(b-a)^2\gamma A^{-\lambda\left(n^{1-\alpha}-2\right)}\right]. \tag{2.105}$$

By (2.35) and putting together (2.83), (2.98), (2.102) and (2.105), we derive the following:

(1)

$$|L_n(f,x)-f(x)| \leq K(q)M\left[|f'(x)|\left(\frac{1}{n^\alpha}+(b-a)\gamma A^{-\lambda\left(n^{1-\alpha}-2\right)}\right)\right.$$

$$+\frac{|f''(x)|}{2}\left(\frac{1}{n^{2\alpha}}+(b-a)^2\gamma A^{-\lambda\left(n^{1-\alpha}-2\right)}\right)$$

$$\left.+\frac{1}{2}\left(\frac{\omega_1\left(f''-f,\frac{1}{n^\alpha}\right)}{n^{2\alpha}}+\frac{\|f'''-f'\|_\infty(b-a)^3}{3}\gamma A^{-\lambda\left(n^{1-\alpha}-2\right)}\right)\right]. \tag{2.106}$$

(2) If $f'(x) = f''(x) = 0$, by (2.106), we obtain

$$|L_n(f,x) - f(x)| \leq \frac{MK(q)}{2}\left[\frac{\omega_1\left(f''-f,\frac{1}{n^\alpha}\right)}{n^{2\alpha}} + \frac{\|f'''-f'\|_\infty (b-a)^3}{3}\gamma A^{-\lambda\left(n^{1-\alpha}-2\right)}\right], \tag{2.107}$$

note here the high rate of convergence at $n^{-3\alpha}$.

(3) Furthermore, by (2.106), we get

$$\|L_n f - f\|_\infty \leq K(q) M\left[\|f'\|_\infty\left(\frac{1}{n^\alpha} + (b-a)\gamma A^{-\lambda\left(n^{1-\alpha}-2\right)}\right)\right.$$
$$\frac{\|f''\|_\infty}{2}\left(\frac{1}{n^{2\alpha}} + (b-a)^2\gamma A^{-\lambda\left(n^{1-\alpha}-2\right)}\right)$$
$$\left. + \frac{1}{2}\left(\frac{\omega_1\left(f''-f,\frac{1}{n^\alpha}\right)}{n^{2\alpha}} + \frac{\|f'''-f'\|_\infty (b-a)^3}{3}\gamma A^{-\lambda\left(n^{1-\alpha}-2\right)}\right)\right]. \tag{2.108}$$

It follows that $\lim_{n\to+\infty} L_n(f) = f$, pointwise and uniformly.

We observe that

$$L_n(f,x) - f'(x) L_n(\sinh(\cdot - x), x) - 2f''(x) L_n\left(\sinh^2\left(\frac{\cdot - x}{2}\right), x\right) - f(x)$$
$$= \frac{L_n^*(f,x)}{\sum_{k=\lceil na\rceil}^{\lfloor nb\rfloor} G_{q,\lambda}(nx-k)} - f'(x)\frac{L_n^*(\sinh(\cdot - x), x)}{\sum_{k=\lceil na\rceil}^{\lfloor nb\rfloor} G_{q,\lambda}(nx-k)}$$
$$- 2f''(x)\frac{L_n^*\left(\sinh^2\left(\frac{\cdot - x}{2}\right), x\right)}{\sum_{k=\lceil na\rceil}^{\lfloor nb\rfloor} G_{q,\lambda}(nx-k)} - f(x)\left(\frac{\sum_{k=\lceil na\rceil}^{\lfloor nb\rfloor} G_{q,\lambda}(nx-k)}{\sum_{k=\lceil na\rceil}^{\lfloor nb\rfloor} G_{q,\lambda}(nx-k)}\right) \tag{2.109}$$

$$= \frac{1}{\sum_{k=\lceil na\rceil}^{\lfloor nb\rfloor} G_{q,\lambda}(nx-k)}\left[L_n^*(f,x) - f'(x) L_n^*(\sinh(\cdot - x), x)\right.$$
$$\left. - 2f''(x) L_n^*\left(\sinh^2\left(\frac{\cdot - x}{2}\right), x\right) - f(x)\sum_{k=\lceil na\rceil}^{\lfloor nb\rfloor} G_{q,\lambda}(nx-k)\right]$$
$$= \frac{1}{\sum_{k=\lceil na\rceil}^{\lfloor nb\rfloor} G_{q,\lambda}(nx-k)}(\Lambda_n(x)). \tag{2.110}$$

Finally, we obtain ($\forall\ x \in [a,b]$, $n \in \mathbb{N}$):
(4)

$$\left| L_n(f,x) - f'(x) L_n(\sinh(\cdot - x), x) - 2f''(x) L_n\left(\sinh^2\left(\frac{\cdot - x}{2}\right), x\right) - f(x) \right|$$

$$\overset{(2.28)}{\leq} K(q) |\Lambda_n(x)|$$

$$\overset{(2.98)}{\leq} \frac{K(q) M}{2} \left[\frac{\omega_1\left(f'' - f, \frac{1}{n^\alpha}\right)}{n^{2\alpha}} + \frac{\|f''' - f'\|_\infty (b-a)^3}{3} \gamma A^{-\lambda(n^{1-\alpha}-2)} \right]. \tag{2.111}$$

The theorem is established. □

Next follows a mixed hyperbolic-trigonometric high order neural network approximation.

Theorem 2.7. *Let* $f \in C^5([a,b], \mathbb{C})$, $0 < \alpha < 1$, $n \in \mathbb{N} : n^{1-\alpha} > 2$, $x \in [a,b]$. *Then*

(1)

$$\left| L_n(f,x) - f(x) - \frac{f'(x)}{2} L_n((\sinh(\cdot - x) + \sin(\cdot - x)), x) \right.$$

$$- \frac{f''(x)}{2} L_n((\cosh(\cdot - x) - \cos(\cdot - x)), x) \tag{2.112}$$

$$- \frac{f^{(3)}(x)}{2} L_n((\sinh(\cdot - x) - \sin(\cdot - x)), x)$$

$$\left. - f^{(4)}(x) L_n\left(\left(\sinh^2\left(\frac{\cdot - x}{2}\right) - \sin^2\left(\frac{\cdot - x}{2}\right)\right), x\right) \right|$$

$$\leq \frac{K(q)(\cosh(b-a)+1)}{4}$$

$$\left[\frac{\omega_1\left(f^{(4)} - f, \frac{1}{n^\alpha}\right)}{n^{2\alpha}} + \frac{\left\|f^{(5)} - f'\right\|_\infty (b-a)^3}{3} \gamma A^{-\lambda(n^{1-\alpha}-2)} \right], \tag{2.113}$$

(2) *if* $f^{(i)}(x) = 0$, $i = 1,2,3,4$, *we get*

$$|L_n(f,x) - f(x)| \leq \frac{K(q)(\cosh(b-a)+1)}{4}$$

$$\left[\frac{\omega_1\left(f^{(4)} - f, \frac{1}{n^\alpha}\right)}{n^{2\alpha}} + \frac{\left\|f^{(5)} - f'\right\|_\infty (b-a)^3}{3} \gamma A^{-\lambda(n^{1-\alpha}-2)} \right], \tag{2.114}$$

in the last (2.114) *observe the high speed of convergence at* $n^{-3\alpha}$.

Proof. Here $f \in C^5([a,b],\mathbb{C})$, and we apply the hyperbolic-trigonometric Taylor's formula for $f \in C^4([a,b],\mathbb{C})$, see Theorem 8 of Anastassiou (2023b).

Let $\frac{k}{n}, x \in [a,b]$, then

$$\begin{aligned} & f\left(\frac{k}{n}\right) - f(x) - f'(x)\left(\frac{\sinh\left(\frac{k}{n}-x\right)+\sin\left(\frac{k}{n}-x\right)}{2}\right) \\ & \quad - f''(x)\left(\frac{\cosh\left(\frac{k}{n}-x\right)-\cos\left(\frac{k}{n}-x\right)}{2}\right) \\ & \quad - f^{(3)}(x)\left(\frac{\sinh\left(\frac{k}{n}-x\right)-\sin\left(\frac{k}{n}-x\right)}{2}\right) \\ & \quad - f^{(4)}(x)\left(\sinh^2\left(\frac{\frac{k}{n}-x}{2}\right)-\sin^2\left(\frac{\frac{k}{n}-x}{2}\right)\right) \\ & \quad = \int_x^{\frac{k}{n}} \left[\left(f^{(4)}(t)-f(t)\right)-\left(f^{(4)}(x)-f(x)\right)\right] \\ & \quad \times \left(\frac{\sinh\left(\frac{k}{n}-t\right)-\sin\left(\frac{k}{n}-t\right)}{2}\right)dt \\ & \quad =: R_4\left(\frac{k}{n},x\right). \end{aligned} \tag{2.115}$$

As in Theorems 2.5 and 2.6, we derive

$$\begin{aligned} & L_n^*(f,x) - f(x)\sum_{k=\lceil na\rceil}^{\lfloor nb\rfloor} G_{q,\lambda}(nx-k) \\ & \quad - \frac{f'(x)}{2}L_n^*\left(\left(\sinh(\cdot-x)+\sin(\cdot-x)\right),x\right) \\ & \quad - \frac{f''(x)}{2}L_n^*\left(\left(\cosh(\cdot-x)-\cos(\cdot-x)\right),x\right) \\ & \quad - \frac{f^{(3)}(x)}{2}L_n^*\left(\left(\sinh(\cdot-x)-\sin(\cdot-x)\right),x\right) \\ & \quad - \frac{f^{(4)}(x)}{2}L_n^*\left(\left(\sinh^2\left(\frac{\cdot-x}{2}\right)-\sin^2\left(\frac{\cdot-x}{2}\right)\right),x\right) = \Phi_n(x), \end{aligned} \tag{2.116}$$

where

$$\Phi_n(x) := \sum_{k=\lceil na \rceil}^{\lfloor nb \rfloor} G_{q,\lambda}(nx-k) R_4\left(\frac{k}{n}, x\right). \tag{2.117}$$

Without loss of generality, we can assume that $n > \left\lceil (b-a)^{-\frac{1}{\alpha}} \right\rceil$.

Thus, $\left|\frac{k}{n} - x\right| \leq \frac{1}{n^\alpha}$ or $\left|\frac{k}{n} - x\right| > \frac{1}{n^\alpha}$.

In case of $\left|\frac{k}{n} - x\right| \leq \frac{1}{n^\alpha}$, we have the following cases:

(i) If $\frac{k}{n} \geq x$, then

$$\begin{aligned}
&\left|R_4\left(\frac{k}{n}, x\right)\right| \\
&= \left|\frac{1}{2}\int_x^{\frac{k}{n}} \left[\left(f^{(4)}(t) - f(t)\right) - \left(f^{(4)}(x) - f(x)\right)\right]\left(\sinh\left(\frac{k}{n} - t\right) - \sin\left(\frac{k}{n} - t\right)\right) dt\right| \\
&\leq \frac{1}{2}\int_x^{\frac{k}{n}} \left|\left(f^{(4)}(t) - f(t)\right) - \left(f^{(4)}(x) - f(x)\right)\right| \left|\sinh\left(\frac{k}{n} - t\right) - \sin\left(\frac{k}{n} - t\right)\right| dt
\end{aligned} \tag{2.118}$$

$$\begin{aligned}
&\leq \frac{1}{2}\int_x^{\frac{k}{n}} \omega_1\left(f^{(4)} - f, t - x\right)\left(\left|\sinh\left(\frac{k}{n} - t\right)\right| + \left|\sin\left(\frac{k}{n} - t\right)\right|\right) dt \\
&\leq \frac{\omega_1\left(f^{(4)} - f, \frac{k}{n} - x\right)}{2}\int_x^{\frac{k}{n}} \left(\cosh(b-a)\left(\frac{k}{n} - t\right) + \left(\frac{k}{n} - t\right)\right) dt \\
&= \frac{(\cosh(b-a)+1)\,\omega_1\left(f^{(4)} - f, \frac{k}{n} - x\right)}{2}\int_x^{\frac{k}{n}} \left(\frac{k}{n} - t\right) dt \\
&= \frac{(\cosh(b-a)+1)\,\omega_1\left(f^{(4)} - f, \frac{k}{n} - x\right)}{4}\left(\frac{k}{n} - x\right)^2 \\
&\leq \frac{(\cosh(b-a)+1)\,\omega_1\left(f^{(4)} - f, \frac{1}{n^\alpha}\right)}{4n^{2\alpha}}.
\end{aligned} \tag{2.119}$$

That is, when $\frac{k}{n} \geq x$, then

$$\left|R_4\left(\frac{k}{n}, x\right)\right| \leq \frac{(\cosh(b-a)+1)\,\omega_1\left(f^{(4)} - f, \frac{1}{n^\alpha}\right)}{4n^{2\alpha}}. \tag{2.120}$$

(ii) If $\frac{k}{n} < x$, then

$$\left|R_4\left(\frac{k}{n}, x\right)\right|$$

$$= \left|\frac{1}{2}\int_{\frac{k}{n}}^{x}\left[\left(f^{(4)}(t) - f(t)\right) - \left(f^{(4)}(x) - f(x)\right)\right]\left(\sinh\left(\frac{k}{n} - t\right) - \sin\left(\frac{k}{n} - t\right)\right)dt\right|$$

$$\leq \frac{\omega_1\left(f^{(4)} - f, x - \frac{k}{n}\right)}{2}\int_{\frac{k}{n}}^{x}\left(\cosh(b-a)\left(t - \frac{k}{n}\right) + \left(t - \frac{k}{n}\right)\right)dt$$

$$= \frac{(\cosh(b-a) + 1)\,\omega_1\left(f^{(4)} - f, x - \frac{k}{n}\right)}{2}\int_{\frac{k}{n}}^{x}\left(t - \frac{k}{n}\right)dt$$

$$= \frac{(\cosh(b-a) + 1)\,\omega_1\left(f^{(4)} - f, x - \frac{k}{n}\right)}{4}\left(x - \frac{k}{n}\right)^2$$

$$\leq \frac{(\cosh(b-a) + 1)\,\omega_1\left(f^{(4)} - f, \frac{1}{n^\alpha}\right)}{4n^{2\alpha}}. \tag{2.121}$$

Consequently, when $\left|\frac{k}{n} - x\right| \leq \frac{1}{n^\alpha}$, we always obtain that

$$\left|R_4\left(\frac{k}{n}, x\right)\right| \leq \frac{(\cosh(b-a) + 1)\,\omega_1\left(f^{(4)} - f, \frac{1}{n^\alpha}\right)}{4n^{2\alpha}}. \tag{2.122}$$

Next assume again $\frac{k}{n} \geq x$, then

$$\left|R_4\left(\frac{k}{n}, x\right)\right| \leq \frac{1}{2}\int_{x}^{\frac{k}{n}}\left|\left(f^{(4)}(t) - f(t)\right) - \left(f^{(4)}(x) - f(x)\right)\right|$$

$$\times\left|\sinh\left(\frac{k}{n} - t\right) - \sin\left(\frac{k}{n} - t\right)\right|dt \leq$$

(by Lakshmikantham and Ladas (1972), p. 3, abstract mean value theorem)

$$\frac{\left\|f^{(5)} - f'\right\|_\infty}{2}\int_{x}^{\frac{k}{n}}(t - x)\left[\cosh(b-a)\left(\frac{k}{n} - t\right) + \left(\frac{k}{n} - t\right)\right]dt$$

$$= \frac{\left\|f^{(5)} - f'\right\|_\infty}{2}(\cosh(b-a) + 1)\int_{x}^{\frac{k}{n}}\left(\frac{k}{n} - t\right)(t - x)\,dt$$

$$= \frac{\left\|f^{(5)} - f'\right\|_\infty(\cosh(b-a) + 1)}{12}\left(\frac{k}{n} - x\right)^3$$

$$\leq \frac{\left\|f^{(5)} - f'\right\|_\infty(\cosh(b-a) + 1)(b-a)^3}{12}. \tag{2.123}$$

Hence,

$$\left|R_4\left(\frac{k}{n},x\right)\right| \leq \frac{\left\|f^{(5)}-f'\right\|_\infty\left(\cosh\left(b-a\right)+1\right)\left(b-a\right)^3}{12}. \tag{2.124}$$

When $\frac{k}{n} < x$, we have

$$\begin{aligned}
&\left|R_4\left(\frac{k}{n},x\right)\right| \\
&\leq \frac{1}{2}\int_{\frac{k}{n}}^{x}\left|\left(f^{(4)}\left(t\right)-f\left(t\right)\right)-\left(f^{(4)}\left(x\right)-f\left(x\right)\right)\right| \\
&\quad \times\left|\sinh\left(\frac{k}{n}-t\right)-\sin\left(\frac{k}{n}-t\right)\right|dt \\
&\leq \frac{\left\|f^{(5)}-f'\right\|_\infty}{2}\int_{\frac{k}{n}}^{x}\left(x-t\right)\left[\cosh\left(b-a\right)\left(t-\frac{k}{n}\right)+\left(t-\frac{k}{n}\right)\right]dt \\
&= \frac{\left\|f^{(5)}-f'\right\|_\infty\left(\cosh\left(b-a\right)+1\right)}{2}\int_{\frac{k}{n}}^{x}\left(x-t\right)\left(t-\frac{k}{n}\right)dt \\
&= \frac{\left\|f^{(5)}-f'\right\|_\infty\left(\cosh\left(b-a\right)+1\right)}{12}\left(x-\frac{k}{n}\right)^3 \\
&\leq \frac{\left\|f^{(5)}-f'\right\|_\infty\left(\cosh\left(b-a\right)+1\right)\left(b-a\right)^3}{12}.
\end{aligned} \tag{2.125}$$

So, it is always true that

$$\left|R_4\left(\frac{k}{n},x\right)\right| \leq \frac{\left\|f^{(5)}-f'\right\|_\infty\left(\cosh\left(b-a\right)+1\right)\left(b-a\right)^3}{12}. \tag{2.126}$$

Thus,

$$\begin{aligned}
\left|\Phi_n\left(x\right)\right| &\leq \sum_{k=\lceil na\rceil}^{\lfloor nb\rfloor} G_{q,\lambda}\left(nx-k\right)\left|R_4\left(\frac{k}{n},x\right)\right| \\
&= \sum_{\substack{k=\lceil na\rceil \\ :\left|\frac{k}{n}-x\right|\leq\frac{1}{n^\alpha}}}^{\lfloor nb\rfloor} G_{q,\lambda}\left(nx-k\right)\left|R_4\left(\frac{k}{n},x\right)\right|
\end{aligned} \tag{2.127}$$

$$+ \sum_{\substack{k=\lceil na \rceil \\ \left\{ : \left| \frac{k}{n} - x \right| > \frac{1}{n^{\alpha}} \right.}}^{\lfloor nb \rfloor} G_{q,\lambda}\left(nx-k\right) \left| R_4\left(\frac{k}{n}, x\right) \right|$$

$$\leq \left(\sum_{\substack{k=\lceil na \rceil \\ \left\{ : \left| \frac{k}{n} - x \right| \leq \frac{1}{n^{\alpha}} \right.}}^{\lfloor nb \rfloor} G_{q,\lambda}\left(nx-k\right) \right) \frac{\left(\cosh\left(b-a\right)+1\right) \omega_1\left(f^{(4)}-f, \frac{1}{n^{\alpha}}\right)}{4n^{2\alpha}} \tag{2.128}$$

$$+ \left(\sum_{\substack{k=\lceil na \rceil \\ \left\{ : \left| \frac{k}{n} - x \right| > \frac{1}{n^{\alpha}} \right.}}^{\lfloor nb \rfloor} G_{q,\lambda}\left(nx-k\right) \right)$$

$$\left(\frac{\left\| f^{(5)} - f' \right\|_{\infty} \left(\cosh\left(b-a\right)+1\right) \left(b-a\right)^3}{12} \right)$$

$$\overset{\text{(by Theorem 2.3)}}{\leq} \frac{\left(\cosh\left(b-a\right)+1\right) \omega_1\left(f^{(4)}-f, \frac{1}{n^{\alpha}}\right)}{4n^{2\alpha}}$$

$$+ \frac{\left\| f^{(5)} - f' \right\|_{\infty} \left(\cosh\left(b-a\right)+1\right) \left(b-a\right)^3}{12} \gamma A^{-\lambda\left(n^{1-\alpha}-2\right)}. \tag{2.129}$$

We have proved that

$$\left|\Phi_n\left(x\right)\right| \leq \frac{\left(\cosh\left(b-a\right)+1\right)}{4}$$

$$\left[\frac{\omega_1\left(f^{(4)}-f, \frac{1}{n^{\alpha}}\right)}{n^{2\alpha}} + \frac{\left\| f^{(5)} - f' \right\|_{\infty} \left(b-a\right)^3}{3} \gamma A^{-\lambda\left(n^{1-\alpha}-2\right)} \right]. \tag{2.130}$$

We observe that

$$L_n\left(f,x\right) - f\left(x\right) - \frac{f'\left(x\right)}{2} L_n\left(\left(\sinh\left(\cdot - x\right) + \sin\left(\cdot - x\right)\right), x\right)$$

$$- \frac{f''\left(x\right)}{2} L_n\left(\left(\cosh\left(\cdot - x\right) - \cos\left(\cdot - x\right)\right), x\right) \tag{2.131}$$

$$- \frac{f^{(3)}\left(x\right)}{2} L_n\left(\left(\sinh\left(\cdot - x\right) - \sin\left(\cdot - x\right)\right), x\right)$$

$$- f^{(4)}\left(x\right) L_n\left(\left(\sinh^2\left(\frac{\cdot - x}{2}\right) - \sin^2\left(\frac{\cdot - x}{2}\right)\right), x\right)$$

$$= \Bigg[L_n^* (f, x) - f(x) \sum_{k=\lceil na \rceil}^{\lfloor nb \rfloor} G_{q,\lambda}(nx - k)$$

$$- \frac{f'(x)}{2} L_n^* ((\sinh(\cdot - x) + \sin(\cdot - x)), x)$$

$$- \frac{f''(x)}{2} L_n^* ((\cosh(\cdot - x) - \cos(\cdot - x)), x)$$

$$- \frac{f^{(3)}(x)}{2} L_n^* ((\sinh(\cdot - x) - \sin(\cdot - x)), x)$$

$$- f^{(4)}(x) L_n^* \left(\left(\sinh^2\left(\frac{\cdot - x}{2}\right) - \sin^2\left(\frac{\cdot - x}{2}\right) \right), x \right) \Bigg]$$

$$\times \frac{1}{\sum_{k=\lceil na \rceil}^{\lfloor nb \rfloor} G_{q,\lambda}(nx - k)}$$

$$= \frac{\Phi_n(x)}{\sum_{k=\lceil na \rceil}^{\lfloor nb \rfloor} G_{q,\lambda}(nx - k)}. \tag{2.132}$$

Finally, we obtain ($\forall\, x \in [a, b]$, $n \in \mathbb{N}$):

$$\Bigg| L_n(f, x) - f(x) - \frac{f'(x)}{2} L_n ((\sinh(\cdot - x) + \sin(\cdot - x)), x)$$

$$- \frac{f''(x)}{2} L_n ((\cosh(\cdot - x) - \cos(\cdot - x)), x) \tag{2.133}$$

$$- \frac{f^{(3)}(x)}{2} L_n ((\sinh(\cdot - x) - \sin(\cdot - x)), x)$$

$$- f^{(4)}(x) L_n \left(\left(\sinh^2\left(\frac{\cdot - x}{2}\right) - \sin^2\left(\frac{\cdot - x}{2}\right) \right), x \right) \Bigg|$$

$$= \frac{|\Phi_n(x)|}{\sum_{k=\lceil na \rceil}^{\lfloor nb \rfloor} G_{q,\lambda}(nx - k)} \leq K(q) |\Phi_n(x)| \overset{(by(??))}{\leq} \frac{K(q)(\cosh(b-a)+1)}{4}$$

$$\left[\frac{\omega_1\left(f^{(4)} - f, \frac{1}{n^\alpha}\right)}{n^{2\alpha}} + \frac{\left\| f^{(5)} - f' \right\|_\infty (b-a)^3}{3} \gamma A^{-\lambda\left(n^{1-\alpha} - 2\right)} \right]. \tag{2.134}$$

The theorem is proved. ∎

We continue with a general trigonometric result.

Theorem 2.8. *Let* $f \in C^5([a, b], \mathbb{C})$, $0 < \alpha < 1$, $n \in \mathbb{N} : n^{1-\alpha} > 2$, $x \in [a, b]$. *Let also* $\overline{\alpha}, \overline{\beta} \in \mathbb{R}$ *with* $\overline{\alpha}\overline{\beta}\left(\overline{\alpha}^2 - \overline{\beta}^2\right) \neq 0$. *Then*

(1)

$$\left| L_n(f,x) - f(x) - \frac{f'(x)}{\overline{\alpha}\overline{\beta}\left(\overline{\beta}^2 - \overline{\alpha}^2\right)} \right.$$
$$\times L_n\left(\left(\overline{\beta}^3 \sin\left(\overline{\alpha}(\cdot - x)\right) - \overline{\alpha}^3 \sin\left(\overline{\beta}(\cdot - x)\right)\right), x\right)$$
$$- \frac{f''(x)}{\left(\overline{\beta}^2 - \overline{\alpha}^2\right)} L_n\left(\left(\cos\left(\overline{\alpha}(\cdot - x)\right) - \cos\left(\overline{\beta}(\cdot - x)\right)\right), x\right)$$
$$- \frac{f'''(x)}{\overline{\alpha}\overline{\beta}\left(\overline{\beta}^2 - \overline{\alpha}^2\right)} L_n\left(\left(\overline{\beta}\sin\left(\overline{\alpha}(\cdot - x)\right) - \overline{\alpha}\sin\left(\overline{\beta}(\cdot - x)\right)\right), x\right)$$
$$- \left(\frac{2f^{(4)}(x) + \left(\overline{\alpha}^2 + \overline{\beta}^2\right) f''(x)}{\left(\overline{\alpha}\overline{\beta}\right)^2 \left(\overline{\beta}^2 - \overline{\alpha}^2\right)} \right)$$
$$\left. \times L_n\left(\left(\overline{\beta}^2 \sin^2\left(\frac{\overline{\alpha}(\cdot - x)}{2}\right) - \overline{\alpha}^2 \sin^2\left(\frac{\overline{\beta}(\cdot - x)}{2}\right)\right), x\right) \right|$$
$$\leq \frac{K(q)}{\left|\overline{\beta}^2 - \overline{\alpha}^2\right|} \left[\frac{\omega_1\left(\left(f^{(4)} + \left(\overline{\alpha}^2 + \overline{\beta}^2\right) f'' + \overline{\alpha}^2\overline{\beta}^2 f\right), \frac{1}{n^\alpha}\right)}{n^{2\alpha}} \right.$$
$$\left. + \frac{\left\| f^{(5)} + \left(\overline{\alpha}^2 + \overline{\beta}^2\right) f''' + \overline{\alpha}^2\overline{\beta}^2 f' \right\|_\infty (b-a)^3}{3} \gamma A^{-\lambda\left(n^{1-\alpha} - 2\right)} \right], \tag{2.135}$$

(2) *if* $f^{(i)}(x) = 0$, $i = 1, 2, 3, 4$, *we get*

$$|L_n(f,x) - f(x)| \leq \frac{K(q)}{\left|\overline{\beta}^2 - \overline{\alpha}^2\right|}$$
$$\left[\frac{\omega_1\left(\left(f^{(4)} + \left(\overline{\alpha}^2 + \overline{\beta}^2\right) f'' + \overline{\alpha}^2\overline{\beta}^2 f\right), \frac{1}{n^\alpha}\right)}{n^{2\alpha}} \right.$$
$$\left. + \frac{\left\| f^{(5)} + \left(\overline{\alpha}^2 + \overline{\beta}^2\right) f''' + \overline{\alpha}^2\overline{\beta}^2 f' \right\|_\infty (b-a)^3}{3} \gamma A^{-\lambda\left(n^{1-\alpha} - 2\right)} \right]. \tag{2.136}$$

The high speed of convergence in (1) *and* (2) *is* $n^{-3\alpha}$.

Proof. As similar to Theorem 2.7 is omitted. It is based on Theorem 9 of Anastassiou (2023b). □

We finish with a general hyperbolic result.

Theorem 2.9. *Let $f \in C^5([a,b],\mathbb{C})$, $0<\alpha<1$, $n \in \mathbb{N} : n^{1-\alpha} > 2$, $x \in [a,b]$. Let also $\overline{\alpha}, \overline{\beta} \in \mathbb{R}$ with $\overline{\alpha}\overline{\beta}\left(\overline{\alpha}^2 - \overline{\beta}^2\right) \neq 0$. Then*

(1)

$$\Bigg| L_n(f,x) - f(x) - \frac{f'(x)}{\overline{\alpha}\overline{\beta}\left(\overline{\beta}^2 - \overline{\alpha}^2\right)}$$

$$\times L_n\left(\left(\overline{\beta}^3 \sinh\left(\overline{\alpha}(\cdot - x)\right) - \overline{\alpha}^3 \sinh\left(\overline{\beta}(\cdot - x)\right)\right), x\right)$$

$$- \frac{f''(x)}{\overline{\beta}^2 - \overline{\alpha}^2} L_n\left(\left(\cosh\left(\overline{\beta}(\cdot - x)\right) - \cosh\left(\overline{\alpha}(\cdot - x)\right)\right), x\right)$$

$$- \frac{f'''(x)}{\overline{\alpha}\overline{\beta}\left(\overline{\beta}^2 - \overline{\alpha}^2\right)} L_n\left(\left(\overline{\alpha}\sinh\left(\overline{\beta}(\cdot - x)\right) - \overline{\beta}\sinh\left(\overline{\alpha}(\cdot - x)\right)\right), x\right)$$

$$- \left(\frac{2(f^{(4)}(x) - \left(\overline{\alpha}^2 + \overline{\beta}^2\right) f''(x))}{\left(\overline{\alpha}\overline{\beta}\right)^2 \left(\overline{\beta}^2 - \overline{\alpha}^2\right)} \right)$$

$$\times L_n\left(\left(\overline{\alpha}^2 \sinh^2\left(\frac{\overline{\beta}(\cdot - x)}{2}\right) - \overline{\beta}^2 \sinh^2\left(\frac{\overline{\alpha}(\cdot - x)}{2}\right)\right), x\right)\Bigg|$$

$$\leq \frac{K(q)\cosh(b-a)}{\left|\overline{\beta}^2 - \overline{\alpha}^2\right|} \left[\frac{\omega_1\left(\left(f^{(4)} - \left(\overline{\alpha}^2 + \overline{\beta}^2\right) f'' + \overline{\alpha}^2\overline{\beta}^2 f\right), \frac{1}{n^\alpha}\right)}{n^{2\alpha}} \right.$$

$$\left. + \frac{\left\| f^{(5)} - \left(\overline{\alpha}^2 + \overline{\beta}^2\right) f''' + \overline{\alpha}^2\overline{\beta}^2 f' \right\|_\infty (b-a)^3}{3} \gamma A^{-\lambda\left(n^{1-\alpha} - 2\right)} \right], \tag{2.137}$$

(2) *if* $f^{(i)}(x) = 0$, $i = 1, 2, 3, 4$, *we get*

$$|L_n(f, x) - f(x)| \leq \frac{K(q) \cosh(b-a)}{\left|\overline{\beta}^2 - \overline{\alpha}^2\right|}$$

$$\left[\frac{\omega_1\left(\left(f^{(4)} - \left(\overline{\alpha}^2 + \overline{\beta}^2\right) f'' + \overline{\alpha}^2 \overline{\beta}^2 f\right), \frac{1}{n^{\alpha}}\right)}{n^{2\alpha}} \right.$$

$$\left. + \frac{\left\|f^{(5)} - \left(\overline{\alpha}^2 + \overline{\beta}^2\right) f''' + \overline{\alpha}^2 \overline{\beta}^2 f'\right\|_{\infty} (b-a)^3}{3} \gamma A^{-\lambda\left(n^{1-\alpha}-2\right)} \right]. \tag{2.138}$$

The high speed of convergence in (1) *and* (2) *is* $n^{-3\alpha}$.

Proof. As similar to Theorem 2.7 is omitted. It is based on Theorem 10 of Anastassiou (2023b). □

Bibliography

G.A. Anastassiou, Rate of convergence of some neural network operators to the unit-univariate case, *Journal of Mathematical Analysis and Applications,* 212 (1997), 237–262.

G.A. Anastassiou, *Quantitative Approximations*, Chapman & Hall / CRC, Boca Raton, New York, 2001.

G.A. Anastassiou, Univariate hyperbolic tangent neural network approximation, *Mathematics and Computer Modelling*, 53 (2011a), 1111–1132.

G.A. Anastassiou, Multivariate hyperbolic tangent neural network approximation, *Computers and Mathematics*, 61 (2011b), 809–821.

G.A. Anastassiou, Multivariate sigmoidal neural network approximation, *Neural Networks*, 24 (2011c), 378–386.

G.A. Anastassiou, *Inteligent Systems: Approximation by Artificial Neural Networks*, Intelligent Systems Reference Library, Vol. 19, Springer, Heidelberg, 2011d.

G.A. Anastassiou, Univariate sigmoidal neural network approximation, *J. of Computational Analysis and Applications*, 14(4) (2012), 659–690.

G.A. Anastassiou, Vector fractional Korovkin type Approximations, *Dynamic Systems and Applications*, 26(2017), 81–104.

G.A. Anastassiou, General sigmoid based Banach space valued neural network approximation, *J. of Computational Analysis and Applications*, 31(4) (2023a), 520–534.

G.A. Anastassiou, *q-Deformed and λ-parametrized A-generalized logistic function induced Banach space valued multivariate multi layer neural network approximations*, submitted, 2022.

G.A. Anastassiou, Opial and Ostrowski type inequalities based on trigonometric and hyperbolic type Taylor formulae, *Malaya J. Mathematik,* 11(S) (2023b), 1–26.

G.A. Anastassiou, q-Deformed and λ-*parametrized* A-generalized logistic function based complex valued trigonometric and hyperbolic neural network high order approximations, *J. of Mathematical Analysis*, accepted, 2023d.

Z. Chen and F. Cao, The approximation operators with sigmoidal functions, *Computers and Mathematics with Applications*, 58 (2009), 758–765.

S. Haykin, *Neural Networks: A Comprehensive Foundation* 2nd ed., Prentice Hall, New York, 1998.

V. Lakshmikantham, G.E. Ladas, *Differential Equations in Abstract Spaces*, Academic Press, San Diego, New York, 1972.

W. McCulloch and W. Pitts, A logical calculus of the ideas immanent in nervous activity, *Bulletin of Mathematical Biophysis*, 7 (1943), 115–133.

T.M. Mitchell, *Machine Learning*, WCB-McGraw-Hill, New York, 1997.

Chapter 3

Perturbed A-Generalized Logistic Function-Activated Complex-Valued Multivariate Trigonometric and Hyperbolic Neural Network Approximation

Here we research the multivariate quantitative approximation of complex-valued continuous functions on a box of $\mathbb{R}^N$, $N \in \mathbb{N}$, by the multivariate normalized-type neural network operators. We investigate also the case of approximation by iterated multilayer neural network operators. These approximations are achieved by establishing multidimensional Jackson-type inequalities involving the multivariate moduli of continuity of the engaged function and its partial derivatives. Our multivariate operators are defined by using a multidimensional density function induced by a q-deformed and λ-parametrized A-generalized logistic function, which is a sigmoid function. The approximations are pointwise and uniform. The related feed-forward neural networks are with one or multiple hidden layers. The basis of our theory is the introduced multivariate Taylor formulae of trigonometric and hyperbolic type. It follows Anastassiou (2023d).

3.1 Introduction

The author in Anastassiou (1997, 2001), see Chapters 2–5, was the first to establish neural network approximations to continuous functions with rates by very specifically defined neural network operators of Cardaliaguet–Euvrard and "Squashing" types, by employing the modulus of continuity of the engaged function or its high order derivative, and producing very tight Jackson-type inequalities. He treats there both the univariate and multivariate cases. The definitions of these operators' "bell-shaped" and "squashing" functions are assumed to be of compact support.

Motivations for this work include articles by Chen and Cao (2009), Anastassiou (2011a)–Anastassiou (2023b), and Costarelli and Spigler (2013), Costarelli and Spigler (2013).

Here we perform q-deformed and λ-parametrized, $q, \lambda > 0$, $A > 1$, A-generalized logistic sigmoid function-based trigonometric and hyperbolic neural network approximations to complex-valued continuous functions over boxes in $\mathbb{R}^N$, $N \in \mathbb{N}$, and also iterated, multilayer approximations. All convergences here are with rates expressed via the multivariate moduli of continuity of the involved function and its partial derivatives and given by very tight multidimensional Jackson-type inequalities.

We come up with the "right" precisely defined multivariate normalized, quasi-interpolation neural network operators based on boxes of $\mathbb{R}^N$. Our boxes are not necessarily symmetric to the origin. In preparation to prove our results, we mention important properties of the basic multivariate density function induced by the q-deformed and λ-parametrized A-generalized logistic sigmoid function.

Feed-forward neural networks (FNNs) with one hidden layer here are mathematically expressed as

$$N_n(x) = \sum_{j=0}^{n} c_j \sigma(\langle a_j \cdot x \rangle + b_j), \quad x \in \mathbb{R}^s, \quad s \in \mathbb{N},$$

where for $0 \le j \le n$, $b_j \in \mathbb{R}$ are the thresholds, $a_j \in \mathbb{R}^s$ are the connection weights, $c_j \in \mathbb{C}$ are the coefficients, $\langle a_j \cdot x \rangle$ is the inner product of a_j and x, and σ is the activation function of the network. In many fundamental network models, the activation function is a kind of logistic sigmoid function. For more information about neural networks, read Haykin (1998)–Mitchell (1997).

3.2 Preliminaries

(I) About neural network operators

Here all come from Anastassiou (2023b, Ch. 15).

We consider the q-deformed and λ-parametrized activation function

$$\varphi_{q,\lambda}(x) = \frac{1}{1 + qA^{-\lambda x}}, \quad x \in \mathbb{R}, \text{ where } q, \lambda > 0, \ A > 1. \tag{3.1}$$

This is a generalized logistic-type function.

We easily observe that

$$\varphi_{q,\lambda}(+\infty) = 1, \ \ \varphi_{q,\lambda}(-\infty) = 0. \tag{3.2}$$

Furthermore, we have

$$1-\varphi_{\frac{1}{q},\lambda}(-x)=1-\frac{1}{1+\frac{1}{q}A^{\lambda x}}=\frac{1+\frac{1}{q}A^{\lambda x}-1}{1+\frac{1}{q}A^{\lambda x}}$$
$$=\frac{\frac{1}{q}A^{\lambda x}}{1+\frac{1}{q}A^{\lambda x}}=\frac{1}{\frac{1}{\frac{1}{q}A^{\lambda x}}+1}=\frac{1}{1+qA^{-\lambda x}}=\varphi_{q,\lambda}(x),$$

proving

$$\varphi_{q,\lambda}(x)=1-\varphi_{\frac{1}{q},\lambda}(-x). \tag{3.3}$$

We also have that

$$\varphi_{q,\lambda}(0)=\frac{1}{1+q}. \tag{3.4}$$

Consider the basic function

$$G_{q,\lambda}(x):=\frac{1}{2}\left(\varphi_{q,\lambda}(x+1)-\varphi_{q,\lambda}(x-1)\right),\quad x\in\mathbb{R}. \tag{3.5}$$

Then,

$$G_{q,\lambda}(-x)=\frac{1}{2}\left(\varphi_{q,\lambda}(-x+1)-\varphi_{q,\lambda}(-x-1)\right)$$
$$=\frac{1}{2}\left(1-\varphi_{\frac{1}{q},\lambda}(x-1)-1+\varphi_{\frac{1}{q},\lambda}(x+1)\right)$$
$$=\frac{1}{2}\left(\varphi_{\frac{1}{q},\lambda}(x+1)-\varphi_{\frac{1}{q},\lambda}(x-1)\right)=G_{\frac{1}{q},\lambda}(x). \tag{3.6}$$

That is,

$$G_{q,\lambda}(-x)=G_{\frac{1}{q},\lambda}(x),\quad\forall\, x\in\mathbb{R}. \tag{3.7}$$

We have

$$\varphi'_{q,\lambda}(x)=\left(\left(1+qA^{-\lambda x}\right)^{-1}\right)'$$
$$=-1\left(1+qA^{-\lambda x}\right)^{-2}q(\ln A)A^{-\lambda x}(-\lambda)$$
$$=q\lambda(\ln A)\left(1+qA^{-\lambda x}\right)^{-2}A^{-\lambda x}>0, \tag{3.8}$$

so that $\varphi_{q,\lambda}$ is a strictly increasing function over $\mathbb{R}$.

Hence it holds

$$\varphi_{q,\lambda}'(x) = \frac{q\lambda(\ln A)}{(1+qA^{-\lambda x})^2 A^{\lambda x}}$$

$$= \frac{q\lambda(\ln A)}{(1+q^2A^{-2\lambda x}+2qA^{-\lambda x})A^{\lambda x}} = \frac{q\lambda(\ln A)}{(A^{\lambda x}+q^2A^{-\lambda x}+2q)}. \quad (3.9)$$

That is,

$$\varphi_{q,\lambda}'(x) = q\lambda(\ln A)\left(A^{\lambda x}+q^2A^{-\lambda x}+2q\right)^{-1}. \quad (3.10)$$

Therefore, it holds

$$\varphi_{q,\lambda}''(x) = q\lambda(\ln A)(-1)\left(A^{\lambda x}+q^2A^{-\lambda x}+2q\right)^{-2}$$

$$\left((\ln A)A^{\lambda x}\lambda + q^2(\ln A)A^{-\lambda x}(-\lambda)\right)$$

$$= q\lambda^2(\ln A)^2\left(A^{\lambda x}+q^2A^{-\lambda x}+2q\right)^{-2}\left(q^2A^{-\lambda x}-A^{\lambda x}\right). \quad (3.11)$$

That is,

$$\varphi_{q,\lambda}''(x) = q\lambda^2(\ln A)^2\left(A^{\lambda x}+q^2A^{-\lambda x}+2q\right)^{-2}\left(q^2A^{-\lambda x}-A^{\lambda x}\right) \in C(\mathbb{R}). \quad (3.12)$$

We have

$$q^2A^{-\lambda x}-A^{\lambda x}>0, \text{ iff } q^2A^{-\lambda x}>A^{\lambda x}, \text{ iff } q^2>A^{2\lambda x}, \text{ iff } q>A^{\lambda x},$$

$$\text{iff } \log_A q > \lambda x, \text{ iff } x < \frac{\log_A q}{\lambda}.$$

So, $\varphi_{q,\lambda}''(x)>0$, for $x<\frac{\log_A q}{\lambda}$ and there $\varphi_{q,\lambda}$ is concave up.

When $x>\frac{\log_A q}{\lambda}$, we have $\varphi_{q,\lambda}''(x)<0$ and $\varphi_{q,\lambda}$ is concave down.

Of course,

$$\varphi_{q,\lambda}''\left(\frac{\log_A q}{\lambda}\right)=0.$$

So, $\varphi_{q,\lambda}$ is a sigmoid function, see Anastassiou (2023a).

We have that

$$G_{q,\lambda}'(x) = \frac{1}{2}\left(\varphi_{q,\lambda}'(x+1)-\varphi_{q,\lambda}'(x-1)\right).$$

We got that $\varphi_{q,\lambda}'$ is strictly increasing for $x<\frac{\log_A q}{\lambda}$. Let $x<\frac{\log_A q}{\lambda}-1$, then $x-1<x+1<\frac{\log_A q}{\lambda}$. Hence $\varphi_{q,\lambda}'(x+1)>\varphi_{q,\lambda}'(x-1)$. Thus $G_{q,\lambda}'>0$, i.e. $G_{q,\lambda}$ is strictly increasing over $\left(-\infty,\frac{\log_A q}{\lambda}-1\right)$.

Let now $x>\frac{\log_A q}{\lambda}+1$, then $x+1>x-1>\frac{\log_A q}{\lambda}$, and $\varphi_{q,\lambda}'(x+1)<\varphi_{q,\lambda}'(x-1)$, by $\varphi_{q,\lambda}'$ being strictly decreasing over $\left(\frac{\log_A q}{\lambda},+\infty\right)$. Hence $G_{q,\lambda}'<0$, and $G_{q,\lambda}$ is strictly decreasing over $\left(\frac{\log_A q}{\lambda},+\infty\right)$.

Let now $\frac{\log_A q}{\lambda} - 1 \leq x \leq \frac{\log_A q}{\lambda} + 1$. We have that

$$G''_{q,\lambda}(x) = \frac{1}{2}\left(\varphi''_{q,\lambda}(x+1) - \varphi''_{q,\lambda}(x-1)\right)$$
$$= \frac{q\lambda^2(\ln A)^2}{2}\left[\frac{\left(q^2A^{-\lambda(x+1)} - A^{\lambda(x+1)}\right)}{\left(A^{\lambda(x+1)} + q^2A^{-\lambda(x+1)} + 2q\right)^2}\right.$$
$$\left. - \frac{\left(q^2A^{-\lambda(x-1)} - A^{\lambda(x-1)}\right)}{\left(A^{\lambda(x-1)} + q^2A^{-\lambda(x-1)} + 2q\right)^2}\right]$$
$$= \frac{q\lambda^2(\ln A)^2}{2}\left[\frac{\left(q^2 - A^{2\lambda(x+1)}\right)}{\left(A^{\lambda(x+1)} + q^2A^{-\lambda(x+1)} + 2q\right)^2 A^{\lambda(x+1)}}\right.$$
$$\left. - \frac{\left(q^2 - A^{2\lambda(x-1)}\right)}{\left(A^{\lambda(x-1)} + q^2A^{-\lambda(x-1)} + 2q\right)^2 A^{\lambda(x-1)}}\right]$$
$$= \frac{q\lambda^2(\ln A)^2}{2}\left[\frac{\left(q - A^{\lambda(x+1)}\right)\left(q + A^{\lambda(x+1)}\right)}{\left(A^{\lambda(x+1)} + q^2A^{-\lambda(x+1)} + 2q\right)^2 A^{\lambda(x+1)}}\right.$$
$$\left. - \frac{\left(q - A^{\lambda(x-1)}\right)\left(q + A^{\lambda(x-1)}\right)}{\left(A^{\lambda(x-1)} + q^2A^{-\lambda(x-1)} + 2q\right)^2 A^{\lambda(x-1)}}\right]. \tag{3.13}$$

By $\frac{\log_A q}{\lambda} \leq x+1 \Leftrightarrow \log_A q \leq \lambda(x+1) \Leftrightarrow q \leq A^{\lambda(x+1)} \Leftrightarrow q - A^{\lambda(x+1)} \leq 0$.

By $x \leq \frac{\log_A q}{\lambda} + 1 \Leftrightarrow x - 1 \leq \frac{\log_A q}{\lambda} \Leftrightarrow \lambda(x-1) \leq \log_A q \Leftrightarrow A^{\lambda(x-1)} \leq q \Leftrightarrow q - A^{\lambda(x-1)} \geq 0$.

Clearly, when $\frac{\log_A q}{\lambda} - 1 \leq x \leq \frac{\log_A q}{\lambda} + 1$, by the above we get that $G''_{q,\lambda}(x) \leq 0$, that is, $G''_{q,\lambda}$ is concave down there.

Clearly, $G_{q,\lambda}$ is strictly concave down over $\left(\frac{\log_A q}{\lambda} - 1, \frac{\log_A q}{\lambda} + 1\right)$.

Overall $G_{q,\lambda}$ is a bell-shaped function over $\mathbb{R}$.

Of course it holds $G''_{q,\lambda}\left(\frac{\log_A q}{\lambda}\right) < 0$.

We have that

$$G'_{q,\lambda}\left(\frac{\log_A q}{\lambda}\right) = \frac{1}{2}\left(\varphi'_{q,\lambda}\left(\frac{\log_A q}{\lambda} + 1\right) - \varphi'_{q,\lambda}\left(\frac{\log_A q}{\lambda} - 1\right)\right)$$
$$= \frac{q\lambda(\ln A)}{2}\left[\frac{1}{A^{\lambda\left(\frac{\log_A q}{\lambda}+1\right)} + q^2A^{-\lambda\left(\frac{\log_A q}{\lambda}+1\right)} + 2q}\right.$$

$$- \frac{1}{A^{\lambda\left(\frac{\log_A q}{\lambda}-1\right)} + q^2 A^{-\lambda\left(\frac{\log_A q}{\lambda}-1\right)} + 2q}\Bigg] \tag{3.14}$$

$$= \frac{q\lambda(\ln A)}{2}\left[\frac{A^{\lambda\left(\frac{\log_A q}{\lambda}-1\right)} + q^2 A^{-\lambda\left(\frac{\log_A q}{\lambda}-1\right)} - A^{\lambda\left(\frac{\log_A q}{\lambda}+1\right)} - q^2 A^{-\lambda\left(\frac{\log_A q}{\lambda}+1\right)}}{\left(A^{\lambda\left(\frac{\log_A q}{\lambda}+1\right)} + q^2 A^{-\lambda\left(\frac{\log_A q}{\lambda}+1\right)} + 2q\right) \times \left(A^{\lambda\left(\frac{\log_A q}{\lambda}-1\right)} + q^2 A^{-\lambda\left(\frac{\log_A q}{\lambda}-1\right)} + 2q\right)}\right] \tag{3.15}$$

$$= \frac{q\lambda(\ln A)}{2}\left[\frac{qA^{-\lambda} + q^2q^{-1}A^{\lambda} - qA^{\lambda} - q^2q^{-1}A^{-\lambda}}{(qA^{\lambda} + q^2q^{-1}A^{-\lambda} + 2q)(qA^{-\lambda} + q^2q^{-1}A^{\lambda} + 2q)}\right]$$

$$= \frac{q\lambda(\ln A)}{2}\left[\frac{qA^{-\lambda} + qA^{\lambda} - qA^{\lambda} - qA^{-\lambda}}{(qA^{\lambda} + qA^{-\lambda} + 2q)(qA^{\lambda} + qA^{-\lambda} + 2q)}\right] = 0. \tag{3.16}$$

So $\frac{\log_A q}{\lambda}$ is the only critical number of $G_{q,\lambda}$ over $\mathbb{R}$. Therefore, $G_{q,\lambda}\left(\frac{\log_A q}{\lambda}\right)$ is the maximum of $G_{q,\lambda}$.

We calculate it:

We have that

$$G_{q,\lambda}\left(\frac{\log_A q}{\lambda}\right) = \frac{1}{2}\left(\varphi_{q,\lambda}\left(\frac{\log_A q}{\lambda}+1\right) - \varphi_{q,\lambda}\left(\frac{\log_A q}{\lambda}-1\right)\right)$$

$$= \frac{1}{2}\left(\frac{1}{1+qA^{-\lambda\left(\frac{\log_A q}{\lambda}+1\right)}} - \frac{1}{1+qA^{-\lambda\left(\frac{\log_A q}{\lambda}-1\right)}}\right)$$

$$= \frac{1}{2}\left(\frac{1}{1+qq^{-1}A^{-\lambda}} - \frac{1}{1+qq^{-1}A^{\lambda}}\right) = \frac{1}{2}\left(\frac{1}{1+A^{-\lambda}} - \frac{1}{1+A^{\lambda}}\right)$$

$$= \frac{1}{2}\left(\frac{A^{\lambda} - A^{-\lambda}}{(1+A^{-\lambda})(1+A^{\lambda})}\right) = \frac{A^{\lambda}-1}{2(A^{\lambda}+1)}. \tag{3.17}$$

The global maximum of $G_{q,\lambda}$ is

$$G_{q,\lambda}\left(\frac{\log_A q}{\lambda}\right) = \frac{A^{\lambda}-1}{2(A^{\lambda}+1)}. \tag{3.18}$$

Finally, we have that

$$\lim_{x \to +\infty} G_{q,\lambda}(x) = \frac{1}{2}\left(\varphi_{q,\lambda}(+\infty) - \varphi_{q,\lambda}(+\infty)\right) = 0 \tag{3.19}$$

and

$$\lim_{x \to -\infty} G_{q,\lambda}(x) = \frac{1}{2}\left(\varphi_{q,\lambda}(-\infty) - \varphi_{q,\lambda}(-\infty)\right) = 0. \tag{3.20}$$

Consequently, the x-axis is the horizontal asymptote of $G_{q,\lambda}$. Of course, $G_{q,\lambda}(x) > 0$, $\forall\, x \in \mathbb{R}$.

We need

Theorem 3.1 (Anastassiou 2023b, Ch. 15). *It holds*

$$\sum_{i=-\infty}^{\infty} G_{q,\lambda}(x - i) = 1, \quad \forall\, x \in \mathbb{R}, \forall\, q, \lambda > 0,\ A > 1. \tag{3.21}$$

It follows

Theorem 3.2 (Anastassiou 2023b, Ch. 15). *It holds*

$$\int_{-\infty}^{\infty} G_{q,\lambda}(x)\, dx = 1, \quad \lambda, q > 0,\ A > 1, \tag{3.22}$$

so that $G_{q,\lambda}$ is a density function.

We need the following result.

Theorem 3.3 (Anastassiou 2023b, Ch. 15). *Let $0 < \alpha < 1$, and $n \in \mathbb{N}$ with $n^{1-\alpha} > 2$. Then*

$$\sum_{\substack{k=-\infty \\ :|nx-k| \geq n^{1-\alpha}}}^{\infty} G_{q,\lambda}(nx - k) < 2\max\left\{q, \frac{1}{q}\right\} \frac{1}{A^{\lambda\left(n^{1-\alpha}-2\right)}} = \gamma A^{-\lambda\left(n^{1-\alpha}-2\right)}, \tag{3.23}$$

where $q, \lambda > 0$, $A > 1$; $\gamma := 2\max\left\{q, \frac{1}{q}\right\}$.

Let $\lceil \cdot \rceil$ be the ceiling of the number and $\lfloor \cdot \rfloor$ be the integral part of the number.

Theorem 3.4 (Anastassiou 2023b, Ch. 15). *Let $x \in [a, b] \subset \mathbb{R}$ and $n \in \mathbb{N}$ so that $\lceil na \rceil \leq \lfloor nb \rfloor$. For $q > 0$, $\lambda > 0$, $A > 1$, we consider the number $\lambda_q > z_0 > 0$ with $G_{q,\lambda}(z_0) = G_{q,\lambda}(0)$ and $\lambda_q > 1$. Then*

$$\frac{1}{\sum_{k=\lceil na \rceil}^{\lfloor nb \rfloor} G_{q,\lambda}(nx - k)} < \max\left\{\frac{1}{G_{q,\lambda}(\lambda_q)}, \frac{1}{G_{\frac{1}{q},\lambda}\left(\lambda_{\frac{1}{q}}\right)}\right\} =: K(q). \tag{3.24}$$

We make

Remark 3.1 (Anastassiou 2023b, Ch. 15). (i) We have that

$$\lim_{n\to+\infty}\sum_{k=\lceil na\rceil}^{\lfloor nb\rfloor} G_{q,\lambda}(nx-k)\neq 1, \quad \text{for at least some } x\in[a,b], \tag{3.25}$$

where $\lambda, q>0$.

(ii) Let $[a,b]\subset\mathbb{R}$. For large n, we always have $\lceil na\rceil \leq \lfloor nb\rfloor$. Also $a\leq \frac{k}{n}\leq b$, iff $\lceil na\rceil\leq k\leq\lfloor nb\rfloor$. In general, it holds

$$\sum_{k=\lceil na\rceil}^{\lfloor nb\rfloor} G_{q,\lambda}(nx-k)\leq 1. \tag{3.26}$$

We make

Remark 3.2. We introduce

$$Z_{q,\lambda}(x_1,\ldots,x_N) := Z_{q,\lambda}(x) := \prod_{i=1}^{N} G_{q,\lambda}(x_i), \quad x=(x_1,\ldots,x_N)\in\mathbb{R}^N, \tag{3.27}$$

$\lambda, q>0$, $A>1$, $N\in\mathbb{N}$.

It has the following properties:

(i) $Z_{q,\lambda}(x)>0, \ \forall\, x\in\mathbb{R}^N$,

(ii)

$$\sum_{k=-\infty}^{\infty} Z_{q,\lambda}(x-k) := \sum_{k_1=-\infty}^{\infty}\sum_{k_2=-\infty}^{\infty}\cdots\sum_{k_N=-\infty}^{\infty} Z_{q,\lambda}(x_1-k_1,\ldots,x_N-k_N)=1, \tag{3.28}$$

where $k:=(k_1,\ldots,k_n)\in\mathbb{Z}^N$, $\forall\, x\in\mathbb{R}^N$,

hence

(iii)

$$\sum_{k=-\infty}^{\infty} Z_{q,\lambda}(nx-k)=1, \tag{3.29}$$

$\forall\, x\in\mathbb{R}^N$; $n\in\mathbb{N}$,

and

(iv)

$$\int_{\mathbb{R}^N} Z_{q,\lambda}(x)\,dx=1, \tag{3.30}$$

that is, Z_q is a multivariate density function.

Here denote $\|x\|_{\infty} := \max\{|x_1|, \ldots, |x_N|\}$, $x \in \mathbb{R}^N$, also set $\infty := (\infty, \ldots, \infty)$, $-\infty := (-\infty, \ldots, -\infty)$ upon the multivariate context, and

$$\lceil na \rceil := (\lceil na_1 \rceil, \ldots, \lceil na_N \rceil),$$
$$\lfloor nb \rfloor := (\lfloor nb_1 \rfloor, \ldots, \lfloor nb_N \rfloor), \tag{3.31}$$

where $a := (a_1, \ldots, a_N)$, $b := (b_1, \ldots, b_N)$.

We obviously see that

$$\sum_{k=\lceil na \rceil}^{\lfloor nb \rfloor} Z_{q,\lambda}(nx-k) = \sum_{k=\lceil na \rceil}^{\lfloor nb \rfloor} \left(\prod_{i=1}^{N} G_{q,\lambda}(nx_i - k_i) \right)$$
$$= \sum_{k_1=\lceil na_1 \rceil}^{\lfloor nb_1 \rfloor} \cdots \sum_{k_N=\lceil na_N \rceil}^{\lfloor nb_N \rfloor} \left(\prod_{i=1}^{N} G_{q,\lambda}(nx_i - k_i) \right)$$
$$= \prod_{i=1}^{N} \left(\sum_{k_i=\lceil na_i \rceil}^{\lfloor nb_i \rfloor} G_{q,\lambda}(nx_i - k_i) \right). \tag{3.32}$$

For $0 < \beta < 1$ and $n \in \mathbb{N}$, a fixed $x \in \mathbb{R}^N$, we have that

$$\sum_{k=\lceil na \rceil}^{\lfloor nb \rfloor} Z_{q,\lambda}(nx-k)$$
$$= \sum_{\substack{k=\lceil na \rceil \\ \left\{ \left\| \frac{k}{n} - x \right\|_{\infty} \le \frac{1}{n^{\beta}} \right.}}^{\lfloor nb \rfloor} Z_{q,\lambda}(nx-k) + \sum_{\substack{k=\lceil na \rceil \\ \left\{ \left\| \frac{k}{n} - x \right\|_{\infty} > \frac{1}{n^{\beta}} \right.}}^{\lfloor nb \rfloor} Z_{q,\lambda}(nx-k). \tag{3.33}$$

In the last two sums, the counting is over disjoint vector sets of k's because the condition $\left\| \frac{k}{n} - x \right\|_{\infty} > \frac{1}{n^{\beta}}$ implies that there exists at least one $\left| \frac{k_r}{n} - x_r \right| > \frac{1}{n^{\beta}}$, where $r \in \{1, \ldots, N\}$.

(v) By Theorem 3.3 and as in Anastassiou (2016), pp. 379–380, we derive that

$$\sum_{\substack{k=\lceil na \rceil \\ \left\{ \left\| \frac{k}{n} - x \right\|_{\infty} > \frac{1}{n^{\beta}} \right.}}^{\lfloor nb \rfloor} Z_{q,\lambda}(nx-k) < \gamma A^{-\lambda\left(n^{1-\beta}-2\right)}, \quad 0 < \beta < 1, \tag{3.34}$$

with $n \in \mathbb{N} : n^{1-\beta} > 2$, $x \in \prod_{i=1}^{N} [a_i, b_i]$.

(vi) By Theorem 3.4 we get that

$$0 < \frac{1}{\sum_{k=\lceil na \rceil}^{\lfloor nb \rfloor} Z_{q,\lambda}(nx-k)} < (K(q))^N, \tag{3.35}$$

$\forall\, x \in \left(\prod_{i=1}^{N}[a_i,b_i]\right),\ \ n \in \mathbb{N}.$

It is also clear that

(vii)

$$\sum_{\substack{k=-\infty \\ \left\{\left\|\frac{k}{n}-x\right\|_{\infty}>\frac{1}{n^{\beta}}\right.}}^{\infty} Z_{q,\lambda}(nx-k) < \gamma A^{-\lambda\left(n^{1-\beta}-2\right)}, \tag{3.36}$$

$0<\beta<1,\, n\in\mathbb{N} : n^{1-\beta}>2,\, x\in\mathbb{R}^N.$

Furthermore, it holds

$$\lim_{n\to\infty} \sum_{k=\lceil na\rceil}^{\lfloor nb\rfloor} Z_{q,\lambda}(nx-k) \neq 1, \tag{3.37}$$

for at least some $x \in \left(\prod_{i=1}^{N}[a_i,b_i]\right)$.

Here $(\mathbb{C},|\cdot|)$ is the complex Banach space.

Let $f \in C\left(\prod_{i=1}^{N}[a_i,b_i],\mathbb{C}\right)$, $x=(x_1,\ldots,x_N)\in\prod_{i=1}^{N}[a_i,b_i]$, $n\in\mathbb{N}$ such that $\lceil na_i\rceil \le \lfloor nb_i\rfloor$, $i=1,\ldots,N$.

We introduce and define the following multivariate linear normalized neural network operator ($x:=(x_1,\ldots,x_N)\in\left(\prod_{i=1}^{N}[a_i,b_i]\right)$):

$$A_n(f,x_1,\ldots,x_N) := A_n(f,x) := \frac{\sum_{k=\lceil na\rceil}^{\lfloor nb\rfloor} f\left(\frac{k}{n}\right) Z_{q,\lambda}(nx-k)}{\sum_{k=\lceil na\rceil}^{\lfloor nb\rfloor} Z_{q,\lambda}(nx-k)}$$

$$= \frac{\sum_{k_1=\lceil na_1\rceil}^{\lfloor nb_1\rfloor}\sum_{k_2=\lceil na_2\rceil}^{\lfloor nb_2\rfloor}\cdots\sum_{k_N=\lceil na_N\rceil}^{\lfloor nb_N\rfloor} f\left(\frac{k_1}{n},\ldots,\frac{k_N}{n}\right)\left(\prod_{i=1}^{N} G_{q,\lambda}(nx_i-k_i)\right)}{\prod_{i=1}^{N}\left(\sum_{k_i=\lceil na_i\rceil}^{\lfloor nb_i\rfloor} G_{q,\lambda}(nx_i-k_i)\right)}. \tag{3.38}$$

For large enough $n\in\mathbb{N}$, we always obtain $\lceil na_i\rceil \le \lfloor nb_i\rfloor$, $i=1,\ldots,N$. Also $a_i \le \frac{k_i}{n} \le b_i$, iff $\lceil na_i\rceil \le k_i \le \lfloor nb_i\rfloor$, $i=1,\ldots,N$.

When $g\in C\left(\prod_{i=1}^{N}[a_i,b_i]\right)$, we define the companion operator

$$\widetilde{A}_n(g,x) := \frac{\sum_{k=\lceil na\rceil}^{\lfloor nb\rfloor} g\left(\frac{k}{n}\right) Z_{q,\lambda}(nx-k)}{\sum_{k=\lceil na\rceil}^{\lfloor nb\rfloor} Z_{q,\lambda}(nx-k)}. \tag{3.39}$$

Clearly, $\widetilde{A}_n$ is a positive linear operator. We have that

$$\widetilde{A}_n(1,x)=1, \quad \forall\, x\in\left(\prod_{i=1}^{N}[a_i,b_i]\right).$$

Note that $A_n(f) \in C\left(\prod_{i=1}^N [a_i, b_i], \mathbb{C}\right)$ and $\widetilde{A}_n(g) \in C\left(\prod_{i=1}^N [a_i, b_i]\right)$.

Furthermore, it holds

$$|A_n(f, x)| \leq \frac{\sum_{k=\lceil na \rceil}^{\lfloor nb \rfloor} \left| f\left(\frac{k}{n}\right)\right| Z_{q,\lambda}(nx - k)}{\sum_{k=\lceil na \rceil}^{\lfloor nb \rfloor} Z_{q,\lambda}(nx - k)} = \widetilde{A}_n(|f|, x), \tag{3.40}$$

$\forall\, x \in \prod_{i=1}^N [a_i, b_i]$.

Clearly, $|f| \in C\left(\prod_{i=1}^N [a_i, b_i]\right)$.

So, we have that

$$|A_n(f, x)| \leq \widetilde{A}_n(|f|, x), \tag{3.41}$$

$\forall\, x \in \prod_{i=1}^N [a_i, b_i]$, $\forall\, n \in \mathbb{N}$, $\forall\, f \in C\left(\prod_{i=1}^N [a_i, b_i], \mathbb{C}\right)$.

Let $c \in X$ and $g \in C\left(\prod_{i=1}^N [a_i, b_i]\right)$, then $cg \in C\left(\prod_{i=1}^N [a_i, b_i], \mathbb{C}\right)$.

Furthermore, it holds

$$A_n(cg, x) = c\widetilde{A}_n(g, x), \quad \forall\, x \in \prod_{i=1}^N [a_i, b_i]. \tag{3.42}$$

Since $\widetilde{A}_n(1) = 1$, we get that

$$A_n(c) = c, \quad \forall\, c \in \mathbb{C}. \tag{3.43}$$

We call $\widetilde{A}_n$ the companion operator of A_n.

For convenience we call

$$A_n^*(f, x) := \sum_{k=\lceil na \rceil}^{\lfloor nb \rfloor} f\left(\frac{k}{n}\right) Z_{q,\lambda}(nx - k)$$

$$= \sum_{k_1=\lceil na_1 \rceil}^{\lfloor nb_1 \rfloor} \sum_{k_2=\lceil na_2 \rceil}^{\lfloor nb_2 \rfloor} \cdots \sum_{k_N=\lceil na_N \rceil}^{\lfloor nb_N \rfloor} f\left(\frac{k_1}{n}, \ldots, \frac{k_N}{n}\right)\left(\prod_{i=1}^N G_{q,\lambda}(nx_i - k_i)\right), \tag{3.44}$$

$\forall\, x \in \left(\prod_{i=1}^N [a_i, b_i]\right)$.

That is,

$$A_n(f,x) := \frac{A_n^*(f,x)}{\sum_{k=\lceil na \rceil}^{\lfloor nb \rfloor} Z_{q,\lambda}(nx-k)}, \tag{3.45}$$

$\forall\, x \in \left(\prod_{i=1}^{N}[a_i,b_i]\right)$, $n \in \mathbb{N}$.

Hence,

$$A_n(f,x) - f(x) = \frac{A_n^*(f,x) - f(x)\left(\sum_{k=\lceil na \rceil}^{\lfloor nb \rfloor} Z_{q,\lambda}(nx-k)\right)}{\sum_{k=\lceil na \rceil}^{\lfloor nb \rfloor} Z_{q,\lambda}(nx-k)}. \tag{3.46}$$

Consequently, we derive

$$|A_n(f,x) - f(x)| \overset{(3.35)}{\leq} (K(q))^N \left| A_n^*(f,x) - f(x) \sum_{k=\lceil na \rceil}^{\lfloor nb \rfloor} Z_{q,\lambda}(nx-k) \right|, \tag{3.47}$$

$\forall\, x \in \left(\prod_{i=1}^{N}[a_i,b_i]\right)$.

We estimate the right-hand side of (3.47).

For the last and others, we need

Definition 3.1 (Anastassiou 2023b, Ch. 15). Let M be a convex and compact subset of $\left(\mathbb{R}^N, \|\cdot\|_p\right)$, $p \in [1,\infty]$, and $(\mathbb{C}, |\cdot|)$ be the complex Banach space. Let $f \in C(M,X)$. We define the first modulus of continuity of f as

$$\omega_1(f,\delta) := \sup_{\substack{x,y \in M: \\ \|x-y\|_p \leq \delta}} |f(x) - f(y)|, \quad 0 < \delta \leq \operatorname{diam}(M). \tag{3.48}$$

If $\delta > \operatorname{diam}(M)$, then

$$\omega_1(f,\delta) = \omega_1(f, \operatorname{diam}(M)). \tag{3.49}$$

Note $\omega_1(f,\delta)$ is increasing in $\delta > 0$. For $f \in C_B(M,\mathbb{C})$ (continuous and bounded functions), $\omega_1(f,\delta)$ is defined similarly.

Lemma 3.1 (Anastassiou 2023b, Ch. 15). *We have $\omega_1(f,\delta) \to 0$ as $\delta \downarrow 0$, iff $f \in C(M,\mathbb{C})$, where M is a convex compact subset of $\left(\mathbb{R}^N, \|\cdot\|_p\right)$, $p \in [1,\infty]$.*

In this chapter (see (3.48)), we work with $p = \infty$.

(II) Multivariate Exotic Taylor formulae

We use

Theorem 3.5 (Anastassiou 2023c). *Let* $f \in C^2([c,d], \mathbb{C})$, *where* $a, x \in [c,d]$. *Then*

$$f(x) - f(a) = f'(a)\sin(x-a) + 2f''(a)\sin^2\left(\frac{x-a}{2}\right)$$

$$+ \int_a^x [(f''(t) + f(t)) - (f''(a) + f(a))] \sin(x-t)\, dt. \tag{3.50}$$

We make

Remark 3.3. Let now Q be an open convex subset of $\mathbb{R}^k$, $k \geq 2$; $z = (z_1, \ldots, z_k)$, $x_0 := (x_{01}, \ldots, x_{0k}) \in Q$. We consider $f \in C^2(Q, \mathbb{C})$ each second-order partial derivative is denoted by $f_\alpha := \frac{\partial^\alpha f}{\partial x^\alpha}$, where $\alpha := (\alpha_1, \ldots, \alpha_k)$, $\alpha_i \in \mathbb{Z}^+$, $i = 1, \ldots, k$ and $|\alpha| := \sum_{i=1}^k \alpha_i = 2$. We consider $g_z(t) := f(x_0 + t(z - x_0))$, $0 \leq t \leq 1$. Clearly, $x_0 + t(z - x_0) \in Q$. Then

$$g_z(0) = f(x_0), \quad g_z(1) = f(z),$$

$$g_z'(t) = \sum_{i=1}^k (z_i - x_{0i}) \frac{\partial f}{\partial x_i}(x_{01} + t(z_1 - x_{01}), \ldots, x_{0k} + t(z_k - x_{0k})),$$

$$g_z'(0) = \sum_{i=1}^k (z_i - x_{0i}) \frac{\partial f}{\partial x_i}(x_{01}, \ldots, x_{0k}), \tag{3.51}$$

and

$$g_z''(t) = \left[\left(\sum_{i=1}^k (z_i - x_{0i}) \frac{\partial}{\partial x_i}\right)^2 f\right](x_{01} + t(z_1 - x_{01}), \ldots, x_{0k} + t(z_k - x_{0k})), \tag{3.52}$$

$$g_z''(0) = \left[\left(\sum_{i=1}^k (z_i - x_{0i}) \frac{\partial}{\partial x_i}\right)^2 f\right](x_{01}, \ldots, x_{0k}).$$

Note above the second-order partials commute.

Clearly, $g_z \in C^2([0,1], \mathbb{C})$, and by Theorem 3.5, we obtain

$$f(z_1, \ldots, z_k) - f(x_{01}, \ldots, x_{0k}) = g_z(1) - g_z(0)$$

$$= g_z'(0)\sin(1) + 2g_z''(0)\sin^2\left(\frac{1}{2}\right)$$

$$+ \int_0^1 [(g_z''(t) + g_z(t)) - (g_z''(0) + g_z(0))] \sin(1-t)\, dt. \tag{3.53}$$

We also mention

Theorem 3.6 (Anastassiou 2023c). *Let* $f \in C^2([c,d], \mathbb{C})$, *where* $a, x \in [c,d]$. *Then*

$$f(x) - f(a) = f'(a)\sinh(x-a) + 2f''(a)\sinh^2\left(\frac{x-a}{2}\right)$$

$$+ \int_a^x [(f''(t) - f(t)) - (f''(a) - f(a))]\sinh(x-t)\,dt. \tag{3.54}$$

We make

Remark 3.4. Consequently, we get that

$$f(z_1, \ldots, z_k) - f(x_{01}, \ldots, x_{0k}) = g_z(1) - g_z(0)$$

$$= g_z'(0)\sinh(1) + 2g_z''(0)\sinh^2\left(\frac{1}{2}\right)$$

$$+ \int_0^1 [(g_z''(t) - g_z(t)) - (g_z''(0) - g_z(0))]\sinh(1-t)\,dt. \tag{3.55}$$

We make

Remark 3.5. Let $f \in C^2\left(\prod_{i=1}^N [a_i, b_i], \mathbb{C}\right)$, $N \in \mathbb{N}$. Here f_α denotes a partial derivative of f, $\alpha := (\alpha_1, \ldots, \alpha_N)$, $\alpha_i \in \mathbb{Z}_+$, $i = 1, \ldots, N$, and $|\alpha| := \sum_{i=1}^N \alpha_i = l$, where $l = 0, 1, 2$.

We write also $f_\alpha := \frac{\partial^\alpha f}{\partial x^\alpha}$ and we say it is of order l.

We denote

$$\omega_{1,2}^{\max}(f_\alpha, h) := \max_{\alpha:|\alpha|=2} \omega_1(f_\alpha, h). \tag{3.56}$$

Call also

$$\|f_\alpha\|_{\infty,2}^{\max} := \max_{|\alpha|=2} \{\|f_\alpha\|_\infty\}, \tag{3.57}$$

where $\|\cdot\|_\infty$ is the supremum norm.

Above the mixed partials commute.

Here $\frac{k}{n} := \left(\frac{k_1}{n}, \ldots, \frac{k_N}{n}\right)$, and $x := (x_1, \ldots, x_N)$, with $\frac{k}{n}$, $x \in \left(\prod_{i=1}^N [a_i, b_i]\right)$, then (by (3.53), where $g_{\frac{k}{n}}(t) := f\left(x + t\left(\frac{k}{n} - x\right)\right)$, $0 \le t \le 1$) we have

$$f\left(\frac{k}{n}\right) - f(x) = \left(\sum_{i=1}^N \left(\frac{k_i}{n} - x_i\right) \frac{\partial f}{\partial x_i}(x)\right) \sin(1)$$

$$+ 2\left\{\left[\left(\sum_{i=1}^N \left(\frac{k_i}{n} - x_i\right) \frac{\partial}{\partial x_i}\right)^2 f\right](x)\right\} \sin^2\left(\frac{1}{2}\right)$$

$$+\int_0^1 \left\{\left\{\left[\left(\sum_{i=1}^{N}\left(\frac{k_i}{n}-x_i\right)\frac{\partial}{\partial x_i}\right)^2 f\right]\left(x+t\left(\frac{k}{n}-x\right)\right)\right.\right.$$

$$\left. + f\left(x+t\left(\frac{k}{n}-x\right)\right)\right\}$$

$$\left.-\left\{\left[\left(\sum_{i=1}^{N}\left(\frac{k_i}{n}-x_i\right)\frac{\partial}{\partial x_i}\right)^2 f\right](x)+f(x)\right\}\right\}\sin(1-t)\,dt. \tag{3.58}$$

Denote the remainder

$R :=$

$$\int_0^1 \left\{\left\{\left[\left(\sum_{i=1}^{N}\left(\frac{k_i}{n}-x_i\right)\frac{\partial}{\partial x_i}\right)^2 f\right]\left(x+t\left(\frac{k}{n}-x\right)\right)+f\left(x+t\left(\frac{k}{n}-x\right)\right)\right\}\right.$$

$$\left.-\left\{\left[\left(\sum_{i=1}^{N}\left(\frac{k_i}{n}-x_i\right)\frac{\partial}{\partial x_i}\right)^2 f\right](x)+f(x)\right\}\right\}\sin(1-t)\,dt \tag{3.59}$$

$$=\int_0^1 \left\{ \sum_{\substack{\alpha:=(\alpha_1,\ldots,\alpha_N),\alpha_i\in\mathbb{Z}_+\\ i=1,\ldots,N,|\alpha|:=\sum_{i=1}^N \alpha_i=2}} \left(\frac{2}{\prod_{i=1}^N \alpha_i!}\right)\left(\prod_{i=1}^{N}\left(\frac{k_i}{n}-x_i\right)^{\alpha_i}\right)\right.$$

$$\left[f_\alpha\left(x+t\left(\frac{k}{n}-x\right)\right)-f_\alpha(x)\right]$$

$$\left.+\left(f\left(x+t\left(\frac{k}{n}-x\right)\right)-f(x)\right)\right\}\sin(1-t)\,dt.$$

Therefore it holds

$$|R| \le \int_0^1 \left\{ \sum_{\substack{\alpha:=(\alpha_1,\ldots,\alpha_N),\alpha_i\in\mathbb{Z}_+\\ i=1,\ldots,N,|\alpha|:=\sum_{i=1}^N \alpha_i=2}} \left(\frac{2}{\prod_{i=1}^N \alpha_i!}\right)\right.$$

$$\left(\prod_{i=1}^{N}\left|\frac{k_i}{n}-x_i\right|^{\alpha_i}\right)\left|f_\alpha\left(x+t\left(\frac{k}{n}-x\right)\right)-f_\alpha(x)\right|$$

$$+\left|f\left(x+t\left(\frac{k}{n}-x\right)\right)-f(x)\right|\Bigg\}|\sin(1-t)|\,dt$$

$$\leq\int_0^1\Bigg\{\sum_{\substack{\alpha:=(\alpha_1,\ldots,\alpha_N),\alpha_i\in\mathbb{Z}_+\\ i=1,\ldots,N,|\alpha|:=\sum_{i=1}^N\alpha_i=2}}\left(\frac{2}{\prod_{i=1}^N\alpha_i!}\right)\left(\prod_{i=1}^N\left|\frac{k_i}{n}-x_i\right|^{\alpha_i}\right)$$

$$\times\,\omega_1\left(f_\alpha,t\left\|\frac{k}{n}-x\right\|_\infty\right)+\omega_1\left(f,t\left\|\frac{k}{n}-x\right\|_\infty\right)\Bigg\}|\sin(1-t)|\,dt\leq(*). \tag{3.60}$$

Note here that $(0<\beta<1)$

$$\left\|\frac{k}{n}-x\right\|_\infty\leq\frac{1}{n^\beta}\Leftrightarrow\left|\frac{k_i}{n}-x_i\right|\leq\frac{1}{n^\beta},\quad i=1,\ldots,N. \tag{3.61}$$

We further see that

$$(*)\leq\Bigg\{\omega_{1,2}^{\max}\left(f_\alpha,\frac{1}{n^\beta}\right)\left(\sum_{\substack{\alpha:=(\alpha_1,\ldots,\alpha_N),\alpha_i\in\mathbb{Z}_+\\ i=1,\ldots,N,|\alpha|:=\sum_{i=1}^N\alpha_i=2}}\left(\frac{2}{\prod_{i=1}^N\alpha_i!}\right)\left(\prod_{i=1}^N\frac{1}{n^{\beta\alpha_i}}\right)\right)$$

$$+\,\omega_1\left(f,\frac{1}{n^\beta}\right)\Bigg\}\int_0^1|\sin(1-t)|\,dt$$

$$=\Bigg[\omega_{1,2}^{\max}\left(f_\alpha,\frac{1}{n^\beta}\right)\left(\sum_{\substack{\alpha:=(\alpha_1,\ldots,\alpha_N),\alpha_i\in\mathbb{Z}_+\\ i=1,\ldots,N,|\alpha|:=\sum_{i=1}^N\alpha_i=2}}\left(\frac{2}{\prod_{i=1}^N\alpha_i!}\right)\right)\frac{1}{n^{2\beta}}$$

$$+ \omega_1 \left(f, \frac{1}{n^\beta} \right) \Bigg] (1 - \cos(1))$$

$$= (1 - \cos(1)) \left\{ \frac{\omega_{1,2}^{\max}\left(f_\alpha, \frac{1}{n^\beta}\right) N^2}{n^{2\beta}} + \omega_1 \left(f, \frac{1}{n^\beta} \right) \right\}. \tag{3.62}$$

We have proved that

$$|R| \le (1 - \cos(1)) \left\{ \frac{\omega_{1,2}^{\max}\left(f_\alpha, \frac{1}{n^\beta}\right) N^2}{n^{2\beta}} + \omega_1 \left(f, \frac{1}{n^\beta} \right) \right\}, \tag{3.63}$$

given that $\left\| \frac{k}{n} - x \right\|_\infty \le \frac{1}{n^\beta}$.

We note also that

$$|R| \le \int_0^1 \left\{ \sum_{\substack{\alpha := (\alpha_1, \ldots, \alpha_N), \alpha_i \in \mathbb{Z}_+ \\ i = 1, \ldots, N, |\alpha| := \sum_{i=1}^N \alpha_i = 2}} \left(\frac{2}{\prod_{i=1}^N \alpha_i!} \right) \right.$$

$$\left. \left(\prod_{i=1}^N (b_i - a_i)^{\alpha_i} \right) 2 \|f_\alpha\|_\infty + 2 \|f\|_\infty \right\} |\sin(1 - t)| \, dt \tag{3.64}$$

$$\le \left\{ \left(\sum_{\substack{\alpha := (\alpha_1, \ldots, \alpha_N), \alpha_i \in \mathbb{Z}_+ \\ i = 1, \ldots, N, |\alpha| := \sum_{i=1}^N \alpha_i = 2}} \left(\frac{2}{\prod_{i=1}^N \alpha_i!} \right) \right) \right.$$

$$\left. 2 \|b - a\|_\infty^2 \|f_\alpha\|_{\infty,2}^{\max} + 2 \|f\|_\infty \right\} \left(\int_0^1 |\sin(1 - t)| \, dt \right)$$

$$= \left(2 \|b - a\|_\infty^2 \|f_\alpha\|_{\infty,2}^{\max} N^2 + 2 \|f\|_\infty \right) (1 - \cos(1)),$$

where $a := (a_1, \ldots, a_N)$, $b = (b_1, \ldots, b_N)$.

We have proved that

$$|R| \leq \left(2 \|b-a\|_{\infty}^{2} \|f_{\alpha}\|_{\infty,2}^{\max} N^{2} + 2\|f\|_{\infty}\right)(1-\cos(1)) =: \rho. \tag{3.65}$$

3.3 Main Results

Here we discuss the trigonometric approximation by using the smoothness of f.

Theorem 3.7. *Let $f \in C^{2}\left(\prod_{i=1}^{N}[a_{i},b_{i}],\mathbb{C}\right)$, $0<\beta<1$, $n,N \in \mathbb{N}$, $n^{1-\beta}>2$, $x,x_{0} \in \left(\prod_{i=1}^{N}[a_{i},b_{i}]\right)$, $a:=(a_{1},\ldots,a_{N})$, $b:=(b_{1},\ldots,b_{N})$. Then*

(i)

$$\Bigg| A_{n}(f,x)-f(x)-\left(\sum_{=1}^{N}\frac{\partial f(x)}{\partial x_{i}}A_{n}((\cdot-x_{i}),x)\right)\sin(1)$$

$$-4\left\{\sum_{\substack{\alpha:=(\alpha_{1},\ldots,\alpha_{N}),\alpha_{i}\in\mathbb{Z}_{+}\\ i=1,\ldots,N,|\alpha|:=\sum_{i=1}^{N}\alpha_{i}=2}} f_{\alpha}(x)\left(\frac{1}{\prod_{i=1}^{N}\alpha_{i}!}\right)A_{n}\left(\prod_{i=1}^{N}(\cdot-x_{i})^{\alpha_{i}},x\right)\right\}$$

$$\times\sin^{2}\left(\frac{1}{2}\right)\Bigg| \leq (K(q))^{N}\left\{\left[(1-\cos(1))\left\{\frac{\omega_{1,2}^{\max}\left(f_{\alpha},\frac{1}{n^{\beta}}\right)N^{2}}{n^{2\beta}}+\omega_{1}\left(f,\frac{1}{n^{\beta}}\right)\right\}\right]\right.$$

$$\left.+\left[2\|b-a\|_{\infty}^{2}\|f_{\alpha}\|_{\infty,2}^{\max}N^{2}+2\|f\|_{\infty}\right](1-\cos(1))\gamma A^{-\lambda\left(n^{1-\beta}-2\right)}\right\}, \tag{3.66}$$

(ii) *assume that $\frac{\partial f(x_{0})}{\partial x_{i}}=0$, $i=1,\ldots,N$, and $f_{\alpha}(x_{0})=0$, $\alpha:|\alpha|=2$, we have that*

$$|A_{n}(f,x)-f(x)|$$

$$\leq (K(q))^{N}\left\{\left[(1-\cos(1))\left\{\frac{\omega_{1,2}^{\max}\left(f_{\alpha},\frac{1}{n^{\beta}}\right)N^{2}}{n^{2\beta}}+\omega_{1}\left(f,\frac{1}{n^{\beta}}\right)\right\}\right]\right.$$

$$\left.+\left[2\|b-a\|_{\infty}^{2}\|f_{\alpha}\|_{\infty,2}^{\max}N^{2}+2\|f\|_{\infty}\right](1-\cos(1))\gamma A^{-\lambda\left(n^{1-\beta}-2\right)}\right\}, \tag{3.67}$$

(iii)

$$|A_n(f,x) - f(x)| \leq (K(q))^N$$

$$\left\{\left\{\left\{\sum_{i=1}^{N}\left|\frac{\partial f(x)}{\partial x_i}\right|\left\{\frac{1}{n^\beta} + (b_i - a_i)\gamma A^{-\lambda\left(n^{1-\beta}-2\right)}\right\}\right\}\sin(1)\right.\right.\right.$$

$$+ 4\left\{\sum_{\alpha:|\alpha|=2}|f_\alpha(x)|\left(\frac{1}{\prod_{i=1}^{N}\alpha_i!}\right)\left[\frac{1}{n^{2\beta}} + \left(\prod_{i=1}^{N}(b_i - a_i)^{\alpha_i}\right)\gamma A^{-\lambda\left(n^{1-\beta}-2\right)}\right]\right\}$$

$$\times \sin^2\left(\frac{1}{2}\right) + \left\{\left[(1-\cos(1))\left\{\frac{\omega_{1,2}^{\max}\left(f_\alpha, \frac{1}{n^\beta}\right)N^2}{n^{2\beta}} + \omega_1\left(f, \frac{1}{n^\beta}\right)\right\}\right]\right.$$

$$\left. + \left[2\|b-a\|_\infty^2 \|f_\alpha\|_{\infty,2}^{\max} N^2 + 2\|f\|_\infty\right](1-\cos(1))\gamma A^{-\lambda\left(n^{1-\beta}-2\right)}\right\}, \quad (3.68)$$

and

(iv)

$$\|A_n(f) - f\|_\infty \leq (K(q))^N$$

$$\left\{\left\{\left\{\sum_{i=1}^{N}\left\|\frac{\partial f}{\partial x_i}\right\|_\infty\left\{\frac{1}{n^\beta} + (b_i - a_i)\gamma A^{-\lambda\left(n^{1-\beta}-2\right)}\right\}\right\}\sin(1)\right.\right.\right.$$

$$+ 4\left\{\sum_{\alpha:|\alpha|=2}\|f_\alpha\|_\infty\left(\frac{1}{\prod_{i=1}^{N}\alpha_i!}\right)\left[\frac{1}{n^{2\beta}} + \left(\prod_{i=1}^{N}(b_i - a_i)^{\alpha_i}\right)\gamma A^{-\lambda\left(n^{1-\beta}-2\right)}\right]\right\}$$

$$\left.\times \sin^2\left(\frac{1}{2}\right)\right\} + \left\{\left[(1-\cos(1))\left\{\frac{\omega_{1,2}^{\max}\left(f_\alpha, \frac{1}{n^\beta}\right)N^2}{n^{2\beta}} + \omega_1\left(f, \frac{1}{n^\beta}\right)\right\}\right]\right.$$

$$\left.\left. + \left[2\|b-a\|_\infty^2 \|f_\alpha\|_{\infty,2}^{\max} N^2 + 2\|f\|_\infty\right](1-\cos(1))\gamma A^{-\lambda\left(n^{1-\beta}-2\right)}\right\}\right\} =: \xi_n(f). \quad (3.69)$$

We observe that $A_n \to I$ (unit operator), as $n \to \infty$, pointwise and uniformly.

Proof. Here R is as in (3.59). We see that

$$U_n := \sum_{k=\lceil na\rceil}^{\lfloor nb\rfloor} Z_{q,\lambda}(nx-k)R$$

$$= \sum_{\substack{k=\lceil na\rceil \\ \left\{:\left\|\frac{k}{n}-x\right\|_\infty \leq \frac{1}{n^\beta}\right.}}^{\lfloor nb\rfloor} Z_{q,\lambda}(nx-k)R + \sum_{\substack{k=\lceil na\rceil \\ \left\{:\left\|\frac{k}{n}-x\right\|_\infty > \frac{1}{n^\beta}\right.}}^{\lfloor nb\rfloor} Z_{q,\lambda}(nx-k)R. \quad (3.70)$$

Therefore,

$$|U_n| \leq \left(\sum_{\substack{k=\lceil na \rceil \\ \left\{ : \left\| \frac{k}{n} - x \right\|_\infty \leq \frac{1}{n^\beta} \right.}}^{\lfloor nb \rfloor} Z_{q,\lambda}(nx-k) \right)$$

$$\left[(1-\cos(1)) \left\{ \frac{\omega_{1,2}^{\max}\left(f_\alpha, \frac{1}{n^\beta}\right) N^2}{n^{2\beta}} + \omega_1\left(f, \frac{1}{n^\beta}\right) \right\} \right] + \rho\gamma A^{-\lambda\left(n^{1-\beta}-2\right)}$$

$$\leq \left[(1-\cos(1)) \left\{ \frac{\omega_{1,2}^{\max}\left(f_\alpha, \frac{1}{n^\beta}\right) N^2}{n^{2\beta}} + \omega_1\left(f, \frac{1}{n^\beta}\right) \right\} \right] + \rho\gamma A^{-\lambda\left(n^{1-\beta}-2\right)}. \tag{3.71}$$

We have established that

$$|U_n| \leq \left[(1-\cos(1)) \left\{ \frac{\omega_{1,2}^{\max}\left(f_\alpha, \frac{1}{n^\beta}\right) N^2}{n^{2\beta}} + \omega_1\left(f, \frac{1}{n^\beta}\right) \right\} \right]$$
$$+ \left[2\|b-a\|_\infty^2 \|f_\alpha\|_{\infty,2}^{\max} N^2 + 2\|f\|_\infty \right] (1-\cos(1))\,\gamma A^{-\lambda\left(n^{1-\beta}-2\right)}. \tag{3.72}$$

By (3.58), we observe that

$$\sum_{k=\lceil na \rceil}^{\lfloor nb \rfloor} f\left(\frac{k}{n}\right) Z_{q,\lambda}(nx-k) - f(x) \left(\sum_{k=\lceil na \rceil}^{\lfloor nb \rfloor} Z_{q,\lambda}(nx-k) \right)$$
$$= \left(\sum_{i=1}^{N} \left(\left(\sum_{k=\lceil na \rceil}^{\lfloor nb \rfloor} Z_{q,\lambda}(nx-k) \left(\frac{k_i}{n} - x_i \right) \right) \frac{\partial f}{\partial x_i}(x) \right) \right) \sin(1)$$
$$+ 2 \left\{ \sum_{\substack{\alpha:=(\alpha_1,\ldots,\alpha_N),\alpha_i \in \mathbb{Z}_+ \\ i=1,\ldots,N, |\alpha|:=\sum_{i=1}^N \alpha_i = 2}} f_\alpha(x) \left(\frac{2}{\prod_{i=1}^{N} \alpha_i!} \right) \left(\sum_{k=\lceil na \rceil}^{\lfloor nb \rfloor} Z_{q,\lambda}(nx-k) \right.\right.$$
$$\left.\left. \left(\prod_{i=1}^{N} \left(\frac{k_i}{n} - x_i \right)^{\alpha_i} \right) \right) \right\} \sin^2\left(\frac{1}{2}\right) + U_n. \tag{3.73}$$

The last says

$$A_n^*(f,x) - f(x)\left(\sum_{k=\lceil na\rceil}^{\lfloor nb\rfloor} Z_{q,\lambda}(nx-k)\right)$$
$$-\left(\sum_{=1}^{N}\frac{\partial f(x)}{\partial x_i}A_n^*((\cdot - x_i), x)\right)\sin(1)$$
$$-2\left\{\sum_{\substack{\alpha:=(\alpha_1,\ldots,\alpha_N),\alpha_i\in\mathbb{Z}_+\\ i=1,\ldots,N,|\alpha|:=\sum_{i=1}^N\alpha_i=2}} f_\alpha(x)\left(\frac{2}{\prod_{i=1}^N\alpha_i!}\right)A_n^*\left(\prod_{i=1}^N(\cdot - x_i)^{\alpha_i}, x\right)\right\}$$
$$\times\sin^2\left(\frac{1}{2}\right) = U_n. \tag{3.74}$$

We note that

$$|A_n^*((\cdot - x_i), x)| \leq A_n^*(|\cdot - x_i|, x) = \sum_{k=\lceil na\rceil}^{\lfloor nb\rfloor}\left|\frac{k_i}{n} - x_i\right| Z_{q,\lambda}(nx-k)$$
$$= \sum_{\substack{k=\lceil na\rceil\\ \left\{:\left\|\frac{k}{n}-x\right\|_\infty\leq\frac{1}{n^\beta}\right.}}^{\lfloor nb\rfloor}\left|\frac{k_i}{n} - x_i\right| Z_{q,\lambda}(nx-k)$$
$$+ \sum_{\substack{k=\lceil na\rceil\\ \left\{:\left\|\frac{k}{n}-x\right\|_\infty>\frac{1}{n^\beta}\right.}}^{\lfloor nb\rfloor}\left|\frac{k_i}{n} - x_i\right| Z_{q,\lambda}(nx-k)$$
$$\leq \frac{1}{n^\beta} + (b_i - a_i)\sum_{\substack{k=\lceil na\rceil\\ \left\{:\left\|\frac{k}{n}-x\right\|_\infty>\frac{1}{n^\beta}\right.}}^{\lfloor nb\rfloor} Z_{q,\lambda}(nx-k)$$
$$\leq \frac{1}{n^\beta} + (b_i - a_i)\gamma A^{-\lambda\left(n^{1-\beta}-2\right)}. \tag{3.75}$$

We have proved that

$$|A_n^*((\cdot - x_i), x)| \leq \frac{1}{n^\beta} + (b_i - a_i)\gamma A^{-\lambda\left(n^{1-\beta}-2\right)}, \tag{3.76}$$

$i = 1, \ldots, N$.

Next we see that

$$\left| A_n^* \left(\prod_{i=1}^{N} (\cdot - x_i)^{\alpha_i}, x \right) \right| \leq A_n^* \left(\prod_{i=1}^{N} |\cdot - x_i|^{\alpha_i}, x \right)$$

$$= \sum_{k=\lceil na \rceil}^{\lfloor nb \rfloor} \left(\prod_{i=1}^{N} \left| \frac{k_i}{n} - x_i \right|^{\alpha_i} \right) Z_{q,\lambda} (nx - k)$$

$$= \sum_{\begin{cases} k=\lceil na \rceil \\ : \left\| \frac{k}{n} - x \right\|_\infty \leq \frac{1}{n^\beta} \end{cases}}^{\lfloor nb \rfloor} \left(\prod_{i=1}^{N} \left| \frac{k_i}{n} - x_i \right|^{\alpha_i} \right) Z_{q,\lambda} (nx - k)$$

$$+ \sum_{\begin{cases} k=\lceil na \rceil \\ : \left\| \frac{k}{n} - x \right\|_\infty > \frac{1}{n^\beta} \end{cases}}^{\lfloor nb \rfloor} \left(\prod_{i=1}^{N} \left| \frac{k_i}{n} - x_i \right|^{\alpha_i} \right) Z_{q,\lambda} (nx - k)$$

$$\leq \frac{1}{n^{2\beta}} + \left(\prod_{i=1}^{N} (b_i - a_i)^{\alpha_i} \right) \gamma A^{-\lambda\left(n^{1-\beta}-2\right)}. \tag{3.77}$$

We have proved that

$$\left| A_n^* \left(\prod_{i=1}^{N} (\cdot - x_i)^{\alpha_i}, x \right) \right| \leq \frac{1}{n^{2\beta}} + \left(\prod_{i=1}^{N} (b_i - a_i)^{\alpha_i} \right) \gamma A^{-\lambda\left(n^{1-\beta}-2\right)}. \tag{3.78}$$

At last we observe that

$$\left| A_n (f, x) - f (x) - \left(\sum_{=1}^{N} \frac{\partial f (x)}{\partial x_i} A_n ((\cdot - x_i), x) \right) \sin (1) \right.$$

$$- 4 \left\{ \sum_{\begin{cases} \alpha := (\alpha_1, \ldots, \alpha_N), \alpha_i \in \mathbb{Z}_+ \\ i=1,\ldots,N, |\alpha| := \sum_{i=1}^{N} \alpha_i = 2 \end{cases}} f_\alpha (x) \left(\frac{1}{\prod_{i=1}^{N} \alpha_i!} \right) A_n \right.$$

$$\left. \left. \times \left(\prod_{i=1}^{N} (\cdot - x_i)^{\alpha_i}, x \right) \right\} \sin^2 \left(\frac{1}{2} \right) \right|$$

$$\leq (K(q))^N |U_n|$$

$$= (K(q))^N \left| A_n^*(f,x) - f(x) \left(\sum_{k=\lceil na \rceil}^{\lfloor nb \rfloor} Z_{q,\lambda}(nx-k) \right) \right.$$

$$- \left(\sum_{=1}^{N} \frac{\partial f(x)}{\partial x_i} A_n^*((\cdot - x_i), x) \right) \sin(1)$$

$$- 4 \left\{ \sum_{\substack{\alpha := (\alpha_1, \ldots, \alpha_N), \alpha_i \in \mathbb{Z}_+ \\ i=1,\ldots,N, |\alpha| := \sum_{i=1}^{N} \alpha_i = 2}} f_\alpha(r) \left(\frac{1}{\prod_{i=1}^{N} \alpha_i!} \right) A_n^* \right.$$

$$\left. \left. \times \left(\prod_{i=1}^{N} (\cdot - x_i)^{\alpha_i}, x \right) \right\} \sin^2\left(\frac{1}{2}\right) \right|. \quad (3.79)$$

Putting all of the above together, we prove theorem. □

We make

Remark 3.6. All here are as in Remark 3.5. By the mean value theorem, we have that $\sinh x = \sinh x - \sinh 0 = (\cosh \xi)(x-0)$, for some ξ between $\{0, x\}$, for any $x \in \mathbb{R}$.

Hence,

$$|\sinh x| \leq \|\cosh\|_{\infty,[-1,1]} |x|, \quad \forall\, x \in [-1,1].$$

But

$$\|\cosh\|_{\infty,[-1,1]} = \cosh(1).$$

Thus, we have

$$|\sinh x| \leq \cosh(1)\, |x|, \quad \forall\, x \in [-1,1].$$

Let $\frac{k}{n} := \left(\frac{k_1}{n}, \ldots, \frac{k_N}{n}\right)$, and $x := (x_1, \ldots, x_N)$, with $\frac{k}{n}, x \in \left(\prod_{i=1}^{N} [a_i, b_i]\right)$, then (by (3.55), where $g_{\frac{k}{n}}(t) := f\left(x + t\left(\frac{k}{n} - x\right)\right)$, $0 \leq t \leq 1$), we have

$$
\begin{aligned}
f\left(\frac{k}{n}\right) - f(x) &= \left(\sum_{i=1}^{N}\left(\frac{k_i}{n} - x_i\right)\frac{\partial f}{\partial x_i}(x)\right)\sinh(1) \\
&+ 2\left\{\left[\left(\sum_{i=1}^{N}\left(\frac{k_i}{n} - x_i\right)\frac{\partial}{\partial x_i}\right)^2 f\right](x)\right\}\sinh^2\left(\frac{1}{2}\right) \\
&+ \int_0^1 \left\{\left\{\left[\left(\sum_{i=1}^{N}\left(\frac{k_i}{n} - x_i\right)\frac{\partial}{\partial x_i}\right)^2 f\right]\left(x + t\left(\frac{k}{n} - x\right)\right)\right.\right. \\
&- f\left(x + t\left(\frac{k}{n} - x\right)\right)\Bigg\} \\
&- \left.\left\{\left[\left(\sum_{i=1}^{N}\left(\frac{k_i}{n} - x_i\right)\frac{\partial}{\partial x_i}\right)^2 f\right](x) - f(x)\right\}\right\}\sinh(1-t)\,dt.
\end{aligned}
\tag{3.80}
$$

Denote the remainder

$$
\begin{aligned}
R &:= \int_0^1 \left\{\left\{\left[\left(\sum_{i=1}^{N}\left(\frac{k_i}{n} - x_i\right)\frac{\partial}{\partial x_i}\right)^2 f\right]\left(x + t\left(\frac{k}{n} - x\right)\right)\right.\right. \\
&\quad -f\left(x + t\left(\frac{k}{n} - x\right)\right)\Bigg\} \\
&\quad - \left.\left\{\left[\left(\sum_{i=1}^{N}\left(\frac{k_i}{n} - x_i\right)\frac{\partial}{\partial x_i}\right)^2 f\right](x) - f(x)\right\}\right\}\sinh(1-t)\,dt \\
&= \int_0^1 \left\{ \sum_{\substack{\alpha := (\alpha_1, \ldots, \alpha_N), \alpha_i \in \mathbb{Z}_+ \\ i = 1, \ldots, N, |\alpha| := \sum_{i=1}^{N} \alpha_i = 2}} \left(\frac{2}{\prod_{i=1}^{N} \alpha_i!}\right)\left(\prod_{i=1}^{N}\left(\frac{k_i}{n} - x_i\right)^{\alpha_i}\right) \right.
\end{aligned}
$$

$$\times\left[f_{\alpha}\left(x+t\left(\frac{k}{n}-x\right)\right)-f_{\alpha}(x)\right]$$

$$\left.-\left(f\left(x+t\left(\frac{k}{n}-x\right)\right)-f(x)\right)\right\}\sinh(1-t)\,dt. \quad (3.81)$$

Therefore it holds

$$|R|\leq\int_{0}^{1}\left\{\sum_{\substack{\alpha:=(\alpha_1,\ldots,\alpha_N),\alpha_i\in\mathbb{Z}_+\\ i=1,\ldots,N,|\alpha|:=\sum_{i=1}^{N}\alpha_i=2}}\left(\frac{2}{\prod_{i=1}^{N}\alpha_i!}\right)\right.$$

$$\left(\prod_{i=1}^{N}\left|\frac{k_i}{n}-x_i\right|^{\alpha_i}\right)\left|f_{\alpha}\left(x+t\left(\frac{k}{n}-x\right)\right)-f_{\alpha}(x)\right|$$

$$\left.+\left|f\left(x+t\left(\frac{k}{n}-x\right)\right)-f(x)\right|\right\}|\sinh(1-t)|\,dt$$

$$\leq\int_{0}^{1}\left\{\sum_{\substack{\alpha:=(\alpha_1,\ldots,\alpha_N),\alpha_i\in\mathbb{Z}_+\\ i=1,\ldots,N,|\alpha|:=\sum_{i=1}^{N}\alpha_i=2}}\left(\frac{2}{\prod_{i=1}^{N}\alpha_i!}\right)\left(\prod_{i=1}^{N}\left|\frac{k_i}{n}-x_i\right|^{\alpha_i}\right)\right.$$

$$\times\,\omega_1\left(f_{\alpha},t\left\|\frac{k}{n}-x\right\|_{\infty}\right)$$

$$\left.+\,\omega_1\left(f,t\left\|\frac{k}{n}-x\right\|_{\infty}\right)\right\}\cosh(1)(1-t)\,dt\leq(*). \quad (3.82)$$

Note here that $(0<\beta<1)$

$$\left\|\frac{k}{n}-x\right\|_{\infty}\leq\frac{1}{n^{\beta}}\Leftrightarrow\left|\frac{k_i}{n}-x_i\right|\leq\frac{1}{n^{\beta}},\quad i=1,\ldots,N. \quad (3.83)$$

We further see that

$$(*) \leq \cosh(1) \left\{ \omega_{1,2}^{\max}\left(f_\alpha, \frac{1}{n^\beta}\right) \left(\sum_{\substack{\alpha:=(\alpha_1,\ldots,\alpha_N),\alpha_i \in \mathbb{Z}_+ \\ i=1,\ldots,N,|\alpha|:=\sum_{i=1}^N \alpha_i = 2}} \left(\frac{2}{\prod_{i=1}^N \alpha_i!} \right) \right.\right.$$

$$\left.\left. \times \left(\prod_{i=1}^N \frac{1}{n^{\beta\alpha_i}} \right) \right) + \omega_1\left(f, \frac{1}{n^\beta}\right) \right\} \int_0^1 (1-t)\,dt$$

$$= \cosh(1) \left\{ \omega_{1,2}^{\max}\left(f_\alpha, \frac{1}{n^\beta}\right) \left(\sum_{\substack{\alpha:=(\alpha_1,\ldots,\alpha_N),\alpha_i \in \mathbb{Z}_+ \\ i=1,\ldots,N,|\alpha|:=\sum_{i=1}^N \alpha_i = 2}} \left(\frac{2}{\prod_{i=1}^N \alpha_i!} \right) \right) \frac{1}{n^{2\beta}} \right.$$

$$\left. + \omega_1\left(f, \frac{1}{n^\beta}\right) \right\} \frac{1}{2}$$

$$= \frac{\cosh(1)}{2} \left\{ \frac{\omega_{1,2}^{\max}\left(f_\alpha, \frac{1}{n^\beta}\right) N^2}{n^{2\beta}} + \omega_1\left(f, \frac{1}{n^\beta}\right) \right\}. \tag{3.84}$$

We have proved that

$$|R| \leq \frac{\cosh(1)}{2} \left\{ \frac{\omega_{1,2}^{\max}\left(f_\alpha, \frac{1}{n^\beta}\right) N^2}{n^{2\beta}} + \omega_1\left(f, \frac{1}{n^\beta}\right) \right\}, \tag{3.85}$$

given that $\left\| \frac{k}{n} - x \right\|_\infty \leq \frac{1}{n^\beta}$.

We note also that

$$|R| \leq \cosh(1) \int_0^1 \left\{ \sum_{\substack{\alpha:=(\alpha_1,\ldots,\alpha_N),\alpha_i \in \mathbb{Z}_+ \\ i=1,\ldots,N,|\alpha|:=\sum_{i=1}^N \alpha_i = 2}} \left(\frac{2}{\prod_{i=1}^N \alpha_i!} \right) \right.$$

$$\left.\left(\prod_{i=1}^{N}(b_i-a_i)^{\alpha_i}\right)2\|f_\alpha\|_\infty+2\|f\|_\infty\right\}(1-t)\,dt$$

$$\leq \cosh(1)\left\{\left(\sum_{\substack{\alpha:=(\alpha_1,\ldots,\alpha_N),\alpha_i\in\mathbb{Z}_+\\ i=1,\ldots,N,|\alpha|:=\sum_{i=1}^{N}\alpha_i=2}}\left(\frac{2}{\prod_{i=1}^{N}\alpha_i!}\right)\right)\right.$$

$$\left.2\|b-a\|_\infty^2\|f_\alpha\|_{\infty,2}^{\max}+2\|f\|_\infty\right\}\left(\int_0^1(1-t)\,dt\right)$$

$$=\cosh(1)\left\{2\|b-a\|_\infty^2\|f_\alpha\|_{\infty,2}^{\max}N^2+2\|f\|_\infty\right\}\frac{1}{2}$$

$$=\cosh(1)\left(\|b-a\|_\infty^2\|f_\alpha\|_{\infty,2}^{\max}N^2+\|f\|_\infty\right), \tag{3.86}$$

where $a:=(a_1,\ldots,a_N)$, $b=(b_1,\ldots,b_N)$.

We have proved that

$$|R|\leq\cosh(1)\left(\|b-a\|_\infty^2\|f_\alpha\|_{\infty,2}^{\max}N^2+\|f\|_\infty\right)=:\rho. \tag{3.87}$$

We continue with the hyperbolic approximation.

Theorem 3.8. *Let* $f\in C^2\left(\prod_{i=1}^{N}[a_i,b_i],\mathbb{C}\right)$, $0<\beta<1$, $n,N\in\mathbb{N}$, $n^{1-\beta}>2$, $x,x_0\in\left(\prod_{i=1}^{N}[a_i,b_i]\right)$, $a:=(a_1,\ldots,a_N)$, $b:=(b_1,\ldots,b_N)$. *Then*

(i)

$$\left|A_n(f,x)-f(x)-\left(\sum_{=1}^{N}\frac{\partial f(x)}{\partial x_i}A_n((\cdot-x_i),x)\right)\sinh(1)\right.$$

$$-4\left\{\sum_{\substack{\alpha:=(\alpha_1,\ldots,\alpha_N),\alpha_i\in\mathbb{Z}_+\\ i=1,\ldots,N,|\alpha|:=\sum_{i=1}^{N}\alpha_i=2}}f_\alpha(x)\left(\frac{1}{\prod_{i=1}^{N}\alpha_i!}\right)A_n\left(\prod_{i=1}^{N}(\cdot-x_i)^{\alpha_i},x\right)\right\}$$

$$\times \sinh^2\left(\frac{1}{2}\right)\Bigg| \le (K(q))^N \cosh(1) \left\{ \left[\frac{1}{2} \left\{ \frac{\omega_{1,2}^{\max}\left(f_\alpha, \frac{1}{n^\beta}\right) N^2}{n^{2\beta}} + \omega_1\left(f, \frac{1}{n^\beta}\right) \right\} \right] \right.$$

$$\left. + \left[\|b-a\|_\infty^2 \|f_\alpha\|_{\infty,2}^{\max} N^2 + \|f\|_\infty \right] \gamma A^{-\lambda\left(n^{1-\beta}-2\right)} \right\}, \tag{3.88}$$

(ii) *assume that* $\frac{\partial f(x_0)}{\partial x_i} = 0,\ i = 1, \ldots, N,$ *and* $f_\alpha(x_0) = 0,\ \alpha : |\alpha| = 2,$ *we have that*

$$|A_n(f,x) - f(x)|$$

$$\le (K(q))^N (\cosh(1)) \left\{ \left[\frac{1}{2} \left\{ \frac{\omega_{1,2}^{\max}\left(f_\alpha, \frac{1}{n^\beta}\right) N^2}{n^{2\beta}} + \omega_1\left(f, \frac{1}{n^\beta}\right) \right\} \right] \right.$$

$$\left. + \left[\|b-a\|_\infty^2 \|f_\alpha\|_{\infty,2}^{\max} N^2 + \|f\|_\infty \right] \gamma A^{-\lambda\left(n^{1-\beta}-2\right)} \right\}, \tag{3.89}$$

(iii)

$$|A_n(f,x) - f(x)| \le (K(q))^N$$

$$\left\{ \left\{ \left\{ \sum_{i=1}^N \left| \frac{\partial f(x)}{\partial x_i} \right| \left\{ \frac{1}{n^\beta} + (b_i - a_i) \gamma A^{-\lambda\left(n^{1-\beta}-2\right)} \right\} \right\} \sinh(1) \right. \right. \right.$$

$$+ 4 \left\{ \sum_{\alpha:|\alpha|=2} |f_\alpha(x)| \left(\frac{1}{\prod_{i=1}^N \alpha_i!} \right) \left[\frac{1}{n^{2\beta}} + \left(\prod_{i=1}^N (b_i - a_i)^{\alpha_i} \right) \gamma A^{-\lambda\left(n^{1-\beta}-2\right)} \right] \right\}$$

$$\times \sinh^2\left(\frac{1}{2}\right) + \cosh(1) \left\{ \left[\frac{1}{2} \left\{ \frac{\omega_{1,2}^{\max}\left(f_\alpha, \frac{1}{n^\beta}\right) N^2}{n^{2\beta}} + \omega_1\left(f, \frac{1}{n^\beta}\right) \right\} \right] \right.$$

$$\left. + \left[\|b-a\|_\infty^2 \|f_\alpha\|_{\infty,2}^{\max} N^2 + \|f\|_\infty \right] \gamma A^{-\lambda\left(n^{1-\beta}-2\right)} \right\}, \tag{3.90}$$

and

(iv)

$$\|A_n(f) - f\|_\infty \le (K(q))^N$$

$$\left\{ \left\{ \left\{ \sum_{i=1}^N \left\| \frac{\partial f}{\partial x_i} \right\|_\infty \left\{ \frac{1}{n^\beta} + (b_i - a_i) \gamma A^{-\lambda\left(n^{1-\beta}-2\right)} \right\} \right\} \sinh(1) \right. \right. \right.$$

$$+4\left\{\sum_{\alpha:|\alpha|=2}\|f_\alpha\|_\infty\left(\frac{1}{\prod_{i=1}^N\alpha_i!}\right)\left[\frac{1}{n^{2\beta}}+\left(\prod_{i=1}^N(b_i-a_i)^{\alpha_i}\right)\gamma A^{-\lambda\left(n^{1-\beta}-2\right)}\right]\right\}$$

$$\times\sinh^2\left(\frac{1}{2}\right)+\cosh(1)\left\{\left[\frac{1}{2}\left\{\frac{\omega_{1,2}^{\max}\left(f_\alpha,\frac{1}{n^\beta}\right)N^2}{n^{2\beta}}+\omega_1\left(f,\frac{1}{n^\beta}\right)\right\}\right]\right.$$

$$\left.+\left[\|b-a\|_\infty^2\|f_\alpha\|_{\infty,2}^{\max}N^2+\|f\|_\infty\right]\gamma A^{-\lambda\left(n^{1-\beta}-2\right)}\right\}=:\psi_n(f). \quad (3.91)$$

We observe that $A_n\to I$ (unit operator), as $n\to\infty$, pointwise and uniformly.

Proof. Here R is as in (3.81). We see that

$$U_n:=\sum_{k=\lceil na\rceil}^{\lfloor nb\rfloor}Z_{q,\lambda}(nx-k)R$$

$$=\sum_{\substack{k=\lceil na\rceil\\ \left\{:\left\|\frac{k}{n}-x\right\|_\infty\le\frac{1}{n^\beta}\right.}}^{\lfloor nb\rfloor}Z_{q,\lambda}(nx-k)R+\sum_{\substack{k=\lceil na\rceil\\ \left\{:\left\|\frac{k}{n}-x\right\|_\infty>\frac{1}{n^\beta}\right.}}^{\lfloor nb\rfloor}Z_{q,\lambda}(nx-k)R. \quad (3.92)$$

Therefore,

$$|U_n|\le\left(\sum_{\substack{k=\lceil na\rceil\\ \left\{:\left\|\frac{k}{n}-x\right\|_\infty\le\frac{1}{n^\beta}\right.}}^{\lfloor nb\rfloor}Z_{q,\lambda}(nx-k)\right)$$

$$\cosh(1)\left[\frac{1}{2}\left\{\frac{\omega_{1,2}^{\max}\left(f_\alpha,\frac{1}{n^\beta}\right)N^2}{n^{2\beta}}+\omega_1\left(f,\frac{1}{n^\beta}\right)\right\}\right]+\rho\gamma A^{-\lambda\left(n^{1-\beta}-2\right)}$$

$$\le\cosh(1)\left[\frac{1}{2}\left\{\frac{\omega_{1,2}^{\max}\left(f_\alpha,\frac{1}{n^\beta}\right)N^2}{n^{2\beta}}+\omega_1\left(f,\frac{1}{n^\beta}\right)\right\}\right]+\rho\gamma A^{-\lambda\left(n^{1-\beta}-2\right)}. \quad (3.93)$$

We have established that

$$|U_n|\le\cosh(1)\left[\frac{1}{2}\left\{\frac{\omega_{1,2}^{\max}\left(f_\alpha,\frac{1}{n^\beta}\right)N^2}{n^{2\beta}}+\omega_1\left(f,\frac{1}{n^\beta}\right)\right\}\right]$$

$$+\cosh(1)\left[\|b-a\|_\infty^2\|f_\alpha\|_{\infty,2}^{\max}N^2+\|f\|_\infty\right]\gamma A^{-\lambda\left(n^{1-\beta}-2\right)}. \quad (3.94)$$

By (3.80), we observe that

$$\sum_{k=\lceil na\rceil}^{\lfloor nb\rfloor} f\left(\frac{k}{n}\right) Z_{q,\lambda}(nx-k) - f(x)\left(\sum_{k=\lceil na\rceil}^{\lfloor nb\rfloor} Z_{q,\lambda}(nx-k)\right)$$

$$= \left(\sum_{i=1}^{N}\left(\left(\sum_{k=\lceil na\rceil}^{\lfloor nb\rfloor} Z_{q,\lambda}(nx-k)\left(\frac{k_i}{n}-x_i\right)\right)\frac{\partial f}{\partial x_i}(x)\right)\right)\sinh(1)$$

$$+2\left\{\sum_{\substack{\alpha:=(\alpha_1,\ldots,\alpha_N),\alpha_i\in\mathbb{Z}_+\\ i=1,\ldots,N,|\alpha|:=\sum_{i=1}^N\alpha_i=2}} f_\alpha(x)\left(\frac{2}{\prod_{i=1}^N \alpha_i!}\right)\right.$$

$$\left.\times\left(\sum_{k=\lceil na\rceil}^{\lfloor nb\rfloor} Z_{q,\lambda}(nx-k)\left(\prod_{i=1}^{N}\left(\frac{k_i}{n}-x_i\right)^{\alpha_i}\right)\right)\right\}\sinh^2\left(\frac{1}{2}\right)+U_n.$$

The last says

$$A_n^*(f,x) - f(x)\left(\sum_{k=\lceil na\rceil}^{\lfloor nb\rfloor} Z_{q,\lambda}(nx-k)\right)$$

$$-\left(\sum_{=1}^{N}\frac{\partial f(x)}{\partial x_i}A_n^*((\cdot-x_i),x)\right)\sinh(1)$$

$$-2\left\{\sum_{\substack{\alpha:=(\alpha_1,\ldots,\alpha_N),\alpha_i\in\mathbb{Z}_+\\ i=1,\ldots,N,|\alpha|:=\sum\limits_{i=1}^N\alpha_i=2}} f_\alpha(x)\left(\frac{2}{\prod\limits_{i=1}^N \alpha_i!}\right)A_n^*\left(\prod_{i=1}^{N}(\cdot-x_i)^{\alpha_i},x\right)\right\}$$

$$\times\sinh^2\left(\frac{1}{2}\right)=U_n. \tag{3.95}$$

We note that

$$|A_n^*((\cdot-x_i),x)| \le A_n^*(|\cdot-x_i|,x) = \sum_{k=\lceil na\rceil}^{\lfloor nb\rfloor}\left|\frac{k_i}{n}-x_i\right| Z_{q,\lambda}(nx-k)$$

$$= \sum_{\substack{k=\lceil na\rceil \\ \left\{:\left\|\frac{k}{n}-x\right\|_{\infty}\leq\frac{1}{n^{\beta}}\right.}}^{\lfloor nb\rfloor} \left|\frac{k_i}{n}-x_i\right| Z_{q,\lambda}(nx-k)$$

$$+ \sum_{\substack{k=\lceil na\rceil \\ \left\{:\left\|\frac{k}{n}-x\right\|_{\infty}>\frac{1}{n^{\beta}}\right.}}^{\lfloor nb\rfloor} \left|\frac{k_i}{n}-x_i\right| Z_{q,\lambda}(nx-k)$$

$$\leq \frac{1}{n^{\beta}} + (b_i - a_i) \sum_{\substack{k=\lceil na\rceil \\ \left\{:\left\|\frac{k}{n}-x\right\|_{\infty}>\frac{1}{n^{\beta}}\right.}}^{\lfloor nb\rfloor} Z_{q,\lambda}(nx-k)$$

$$\leq \frac{1}{n^{\beta}} + (b_i - a_i)\gamma A^{-\lambda\left(n^{1-\beta}-2\right)}. \tag{3.96}$$

We have proved that

$$\left|A_n^*\left((\cdot - x_i), x\right)\right| \leq \frac{1}{n^{\beta}} + (b_i - a_i)\gamma A^{-\lambda\left(n^{1-\beta}-2\right)}, \tag{3.97}$$

$i = 1, \ldots, N$.

Next we see that

$$\left|A_n^*\left(\prod_{i=1}^{N}(\cdot - x_i)^{\alpha_i}, x\right)\right| \leq A_n^*\left(\prod_{i=1}^{N}|\cdot - x_i|^{\alpha_i}, x\right)$$

$$= \sum_{k=\lceil na\rceil}^{\lfloor nb\rfloor}\left(\prod_{i=1}^{N}\left|\frac{k_i}{n}-x_i\right|^{\alpha_i}\right) Z_{q,\lambda}(nx-k)$$

$$= \sum_{\substack{k=\lceil na\rceil \\ \left\{:\left\|\frac{k}{n}-x\right\|_{\infty}\leq\frac{1}{n^{\beta}}\right.}}^{\lfloor nb\rfloor}\left(\prod_{i=1}^{N}\left|\frac{k_i}{n}-x_i\right|^{\alpha_i}\right) Z_{q,\lambda}(nx-k)$$

$$+ \sum_{\substack{k=\lceil na\rceil \\ \left\{:\left\|\frac{k}{n}-x\right\|_{\infty}>\frac{1}{n^{\beta}}\right.}}^{\lfloor nb\rfloor}\left(\prod_{i=1}^{N}\left|\frac{k_i}{n}-x_i\right|^{\alpha_i}\right) Z_{q,\lambda}(nx-k)$$

$$\leq \frac{1}{n^{2\beta}} + \left(\prod_{i=1}^{N}(b_i - a_i)^{\alpha_i}\right)\gamma A^{-\lambda\left(n^{1-\beta}-2\right)}. \tag{3.98}$$

We have proved that

$$\left|A_n^*\left(\prod_{i=1}^{N}(\cdot - x_i)^{\alpha_i}, x\right)\right| \leq \frac{1}{n^{2\beta}} + \left(\prod_{i=1}^{N}(b_i - a_i)^{\alpha_i}\right)\gamma A^{-\lambda\left(n^{1-\beta}-2\right)}. \tag{3.99}$$

At last we observe that

$$\Bigg| A_n(f,x) - f(x) - \left(\sum_{=1}^{N} \frac{\partial f(x)}{\partial x_i} A_n((\cdot - x_i), x)\right) \sinh(1)$$

$$- 4 \left\{ \sum_{\substack{\alpha:=(\alpha_1,\dots,\alpha_N),\alpha_i\in\mathbb{Z}_+\\ i=1,\dots,N,|\alpha|:=\sum_{i=1}^{N}\alpha_i=2}} f_\alpha(x) \left(\frac{1}{\prod_{i=1}^{N}\alpha_i!}\right) A_n \right.$$

$$\left. \times \left(\prod_{i=1}^{N} (\cdot - x_i)^{\alpha_i}, x\right) \right\} \sinh^2\left(\frac{1}{2}\right) \Bigg|$$

$$\leq (K(q))^N |U_n|$$

$$= (K(q))^N \Bigg| A_n^*(f,x) - f(x) \left(\sum_{k=\lceil na\rceil}^{\lfloor nb\rfloor} Z_{q,\lambda}(nx-k)\right)$$

$$- \left(\sum_{=1}^{N} \frac{\partial f(x)}{\partial x_i} A_n^*((\cdot - x_i), x)\right) \sinh(1)$$

$$- 4 \left\{ \sum_{\substack{\alpha:=(\alpha_1,\dots,\alpha_N),\alpha_i\in\mathbb{Z}_+\\ i=1,\dots,N,|\alpha|:=\sum_{i=1}^{N}\alpha_i=2}} f_\alpha(x) \left(\frac{1}{\prod_{i=1}^{N}\alpha_i!}\right) A_n^* \right.$$

$$\left. \times \left(\prod_{i=1}^{N} (\cdot - x_i)^{\alpha_i}, x\right) \right\} \sinh^2\left(\frac{1}{2}\right) \Bigg|. \tag{3.100}$$

Putting all of the above together, we prove theorem. □

We make

Remark 3.7. By (3.38), we get that $\|A_n(f)\|_\infty \leq \|f\|_\infty < \infty$, and $A_n(f) \in C\left(\prod_{i=1}^N [a_i, b_i], \mathbb{C}\right)$, given that $f \in C\left(\prod_{i=1}^N [a_i, b_i], \mathbb{C}\right)$.

Clearly, then

$$\left\|A_n^2(f)\right\|_\infty = \|A_n(A_n(f))\|_\infty \leq \|A_n(f)\|_\infty \leq \|f\|_\infty, \tag{3.101}$$

etc.

Therefore, we get

$$\left\|A_n^k(f)\right\|_\infty \leq \|f\|_\infty, \quad \forall\, k \in \mathbb{N}, \tag{3.102}$$

the contraction property.

Also we see that

$$\left\|A_n^k(f)\right\|_\infty \leq \left\|A_n^{k-1}(f)\right\|_\infty \leq \cdots \leq \|A_n(f)\|_\infty \leq \|f\|_\infty. \tag{3.103}$$

Also $A_n(1) = 1$, $A_n^k(1) = 1$, $\forall\, k \in \mathbb{N}$.

Following 18.14 of Anastassiou (2016), pp. 401–402, similarly we obtain that

$$\|A_n^r f - f\|_\infty \leq r\,\|A_n(f) - f\|_\infty, \quad r \in \mathbb{N}. \tag{3.104}$$

We give

Theorem 3.9. *All as in Theorems 3.7, 3.8. Then*

(i)

$$\|A_n^r f - f\|_\infty \leq r\xi_n(f), \tag{3.105}$$

where $\xi_n(f)$ *as in (3.69),*

(ii)

$$\|A_n^r f - f\|_\infty \leq r\psi_n(f), \tag{3.106}$$

where $\psi_n(f)$ *as in (3.91).*

So the speed of convergence to the unit operator of A_n^r *is not worse than of* A_n.

Bibliography

G.A. Anastassiou, Rate of convergence of some neural network operators to the unit-univariate case, *Journal of Mathematical Analysis and Applications*, 212 (1997), 237–262.

G.A. Anastassiou, *Quantitative Approximations*, Chapman & Hall/CRC, Boca Raton, New York, 2001.

G.A. Anastassiou, *Inteligent Systems: Approximation by Artificial Neural Networks*, Intelligent Systems Reference Library, Vol. 19, Springer, Heidelberg, 2011a.

G.A. Anastassiou, Univariate hyperbolic tangent neural network approximation, *Mathematics and Computer Modelling*, 53 (2011b), 1111–1132.

G.A. Anastassiou, Multivariate hyperbolic tangent neural network approximation, *Computers and Mathematics*, 61 (2011c), 809–821.

G.A. Anastassiou, Multivariate sigmoidal neural network approximation, *Neural Networks,* 24(2011d), 378–386.

G.A. Anastassiou, Univariate sigmoidal neural network approximation, *Journal of Computational Analysis and Applications*, 14(4) (2012), 659–690.

G.A. Anastassiou, Approximation by neural networks iterates, G. Anastassiou, O. Duman. (Eds.), *Advances in Applied Mathematics and Approximation Theory,* pp. 1–20, Springer Proceedings in Math. & Stat., Springer, New York, 2013.

G.A. Anastassiou, *Intelligent Systems II: Complete Approximation by Neural Network Operators*, Springer, Heidelberg, New York, 2016.

G.A. Anastassiou, General sigmoid based Banach space valued neural network approximation, *Journal of Computational Analysis and Applications*, 31(4) (2023a), 520–534.

G.A. Anastassiou, *Parametrized, deformed and general neural networs*, accepted for publication, Springer, Heidelberg, New York, 2023b.

G.A. Anastassiou, Opial and Ostrowski type inequalities based on trigonometric and hyperbolic type Taylor formulae, *Malaya Journal of Matematik*, 11(S) 2023c, 1–26.

G.A. Anastassiou, *q-Deformed and λ-parametrized A-generalized logistic function based complex valued multivariate trigonometric and hyperbolic neural network approximations*, Bulletin of The Allahabad Mathematical Society, accepted, 2023d.

Z. Chen and F. Cao, The approximation operators with sigmoidal functions, *Computers and Mathematics with Applications*, 58 (2009), 758–765.

D. Costarelli, R. Spigler, Approximation results for neural network operators activated by sigmoidal functions, *Neural Networks* , 44 (2013), 101–106.

D. Costarelli, R. Spigler, Multivariate neural network operators with sigmoidal activation functions, *Neural Networks*, 48 (2013), 72–77.

S. Haykin, *Neural Networks: A Comprehensive Foundation* 2nd ed., Prentice Hall, New York, 1998.

W. McCulloch and W. Pitts, A logical calculus of the ideas immanent in nervous activity, *Bulletin of Mathematical Biophysics*, 7 (1943), 115–133.

T.M. Mitchell, *Machine Learning*, WCB-McGraw-Hill, New York, 1997.

Chapter 4

Perturbed Hyperbolic Tangent Function-Activated Complex-Valued Trigonometric and Hyperbolic Neural Network High Order Approximation

Here we study the univariate quantitative approximation of complex-valued continuous functions on a compact interval by complex-valued neural network operators. These approximations are derived by establishing Jackson-type inequalities involving the modulus of continuity of the used function's high order derivatives. The kinds of our approximations are trigonometric and hyperbolic. Our operators are defined by using a density function generated by a q-deformed and λ-parametrized hyperbolic tangent function, which is a sigmoid function. The approximations are pointwise and of the uniform norm. The related complex-valued feed-forward neural networks are with one hidden layer. It follows Anastassiou (2023d).

4.1 Introduction

The author in Anastassiou (1997, 2001), see Chapters 2–5, was the first to establish neural network approximation to continuous functions with rates by very specifically defined neural network operators of Cardaliaguet–Euvrard and "Squashing" types, by employing the modulus of continuity of the engaged function or its high order derivative, and producing very tight Jackson-type inequalities. He treats there both the univariate and multivariate cases. The definitions of these operators' "bell-shaped" and "squashing" functions are assumed to be of compact support.

Again the author inspired by Chen and Cao (2009) continued his studies on neural network approximation by introducing and using the proper quasi-interpolation operators of sigmoidal and hyperbolic tangent-type which resulted into Anastassiou (2011a)–Anastassiou (2012), by treating both the univariate and multivariate cases.

Let h be a general sigmoid activation function with $h(0) = 0$, and $y = \pm 1$ the horizontal asymptotes. Of course, h is strictly increasing over $\mathbb{R}$. Let the parameter $0 < r < 1$ and $x > 0$. Then clearly $-x < x$ and $-x < -rx < rx < x$, furthermore it holds $h(-x) < h(-rx) < h(rx) < h(x)$. Consequently, the sigmoid $y = h(rx)$ has a graph inside the graph of $y = h(x)$, of course with the same asymptotes $y = \pm 1$. Therefore, $h(rx)$ has derivatives (gradients) at more points x than $h(x)$ has different than zero or not as close to zero, thus killing less number of neurons! And of course $h(rx)$ is more distant from $y = \pm 1$ than $h(x)$. A highly desired fact in neural network theory.

Different activation functions allow for different nonlinearities which might work better for solving a specific function. So the need to use neural networks with various activation functions is vivid. Thus, performing neural network approximations using different activation functions is not only necessary but also fully justified.

Also brain asymmetry has been observed in animals and humans in terms of structure, function and behaviour. This lateralization is thought to reflect evolutionary, hereditary, developmental, experiential and pathological factors. Therefore, it is natural to consider for our study deformed neural network activation functions and operators. So this chapter is a specific study under this philosophy of approaching reality as close as possible.

Consequently, the author here performs q-deformed and λ-parametrized hyperbolic tangent function-activated high order neural network approximations to continuous functions over compact intervals of the real line with complex values. All convergences are with rates expressed via the modulus of continuity of the involved functions high order derivatives, deriving by very tight Jackson-type inequalities.

The basis of our higher-order approximations here are some trigonometric and hyperbolic-type Taylor's formulae newly discovered by the author.

Our compact intervals are not necessarily symmetric to the origin. In preparation to prove our results, we describe important properties of the basic density function defining our operators which is induced by a q-deformed and λ-parametrized hyperbolic tangent function, which is a sigmoid activation function.

Feed-forward X-valued neural networks (FNNs) with one hidden layer, the only type of network we deal with in this chapter, are mathematically expressed as

$$N_n(x) = \sum_{j=0}^{n} c_j \sigma(\langle a_j \cdot x \rangle + b_j), \quad x \in \mathbb{R}^s, \;\; s \in \mathbb{N},$$

where for $0 \le j \le n$, $b_j \in \mathbb{R}$ are the thresholds, $a_j \in \mathbb{R}^s$ are the connection weights, $c_j \in \mathbb{C}$ are the coefficients, $\langle a_j \cdot x \rangle$ is the inner product of a_j and x, and σ is the activation function of the network. For more information about neural networks in general, read Haykin (1998), McCulloch and Pitts (1943), and Mitchell (1997).

4.2 About q-Deformed and λ-Parametrized Hyperbolic Tangent Function $g_{q,\lambda}$

Here all this background comes from Anastassiou (2023b, Ch. 17).

We use $g_{q,\lambda}$, see (4.1), and exhibit that it is a sigmoid function and we present several of its properties related to the approximation by neural network operators.

So, let us consider the function

$$g_{q,\lambda}(x) := \frac{e^{\lambda x} - q e^{-\lambda x}}{e^{\lambda x} + q e^{-\lambda x}}, \quad \lambda, q > 0, \quad x \in \mathbb{R}. \tag{4.1}$$

We have that

$$g_{q,\lambda}(0) = \frac{1-q}{1+q}.$$

We note also that

$$g_{q,\lambda}(-x) = \frac{e^{-\lambda x} - q e^{\lambda x}}{e^{-\lambda x} + q e^{\lambda x}} = \frac{\frac{1}{q} e^{-\lambda x} - e^{\lambda x}}{\frac{1}{q} e^{-\lambda x} + e^{\lambda x}} = -\frac{\left(e^{\lambda x} - \frac{1}{q} e^{-\lambda x}\right)}{e^{\lambda x} + \frac{1}{q} e^{-\lambda x}} = -g_{\frac{1}{q},\lambda}(x). \tag{4.2}$$

That is,

$$g_{q,\lambda}(-x) = -g_{\frac{1}{q},\lambda}(x), \quad \forall\, x \in \mathbb{R}, \tag{4.3}$$

and

$$g_{\frac{1}{q},\lambda}(x) = -g_{q,\lambda}(-x),$$

hence

$$g'_{\frac{1}{q},\lambda}(x) = g'_{q,\lambda}(-x). \tag{4.4}$$

It is

$$g_{q,\lambda}(x) = \frac{e^{2\lambda x} - q}{e^{2\lambda x} + q} = \frac{1 - \frac{q}{e^{2lx}}}{1 + \frac{q}{e^{2\lambda x}}} \underset{(x \to +\infty)}{\to} 1,$$

i.e.

$$g_{q,\lambda}(+\infty) = 1, \tag{4.5}$$

Furthermore,

$$g_{q,\lambda}(x) = \frac{e^{2\lambda x} - q}{e^{2\lambda x} + q} \underset{(x \to -\infty)}{\to} \frac{-q}{q} = -1,$$

i.e.

$$g_{q,\lambda}(-\infty) = -1. \tag{4.6}$$

We find that

$$g'_{q,\lambda}(x) = \frac{4q\lambda e^{2\lambda x}}{\left(e^{2\lambda x} + q\right)^2} > 0, \tag{4.7}$$

therefore $g_{q,\lambda}$ is strictly increasing.

Next we obtain ($x \in \mathbb{R}$)

$$g''_{q,\lambda}(x) = 8q\lambda^2 e^{2\lambda x} \left(\frac{q - e^{2\lambda x}}{\left(e^{2\lambda x} + q\right)^3} \right) \in C(\mathbb{R}). \tag{4.8}$$

We observe that

$$q - e^{2\lambda x} \gtreqless 0 \Leftrightarrow q \gtreqless e^{2\lambda x} \Leftrightarrow \ln q \gtreqless 2\lambda x \Leftrightarrow x \lesseqgtr \frac{\ln q}{2\lambda}.$$

So, in case of $x < \frac{\ln q}{2\lambda}$, we have that $g_{q,\lambda}$ is strictly concave up, with $g''_{q,\lambda}\left(\frac{\ln q}{2\lambda}\right) = 0$.

And in case of $x > \frac{\ln q}{2\lambda}$, we have that $g_{q,\lambda}$ is strictly concave down.

Clearly, $g_{q,\lambda}$ is a shifted sigmoid function with $g_{q,\lambda}(0) = \frac{1-q}{1+q}$, and $g_{q,\lambda}(-x) = -g_{q^{-1},\lambda}(x)$ (a semi-odd function), see also Anastassiou (2023a).

By $1 > -1$, $x + 1 > x - 1$, we consider the activation function

$$M_{q.\lambda}(x) := \frac{1}{4}\left(g_{q,\lambda}(x+1) - g_{q,\lambda}(x-1)\right) > 0, \tag{4.9}$$

$\forall x \in \mathbb{R}$; $q, \lambda > 0$. Note that $M_{q,\lambda}(\pm\infty) = 0$, so the x-axis is horizontal asymptote.

We have that

$$\begin{aligned}M_{q,\lambda}(-x) &= \frac{1}{4}\left(g_{q,\lambda}(-x+1) - g_{q,\lambda}(-x-1)\right)\\ &= \frac{1}{4}\left(g_{q,\lambda}(-(x-1)) - g_{q,\lambda}(-(x+1))\right)\\ &= \frac{1}{4}\left(-g_{\frac{1}{q},\lambda}(x-1) + g_{\frac{1}{q},\lambda}(x+1)\right)\\ &= \frac{1}{4}\left(g_{\frac{1}{q},\lambda}(x+1) - g_{\frac{1}{q},\lambda}(x-1)\right) = M_{\frac{1}{q},\lambda}(x), \quad \forall\, x \in \mathbb{R}.\end{aligned} \tag{4.10}$$

Thus,

$$M_{q,\lambda}(-x) = M_{\frac{1}{q},\lambda}(x), \ \ \forall\, x \in \mathbb{R};\ q,\lambda > 0, \tag{4.11}$$

a deformed symmetry.

Next, we have that

$$M'_{q,\lambda}(x) = \frac{1}{4}\left(g'_{q,\lambda}(x+1) - g'_{q,\lambda}(x-1)\right), \ \ \forall\, x \in \mathbb{R}. \tag{4.12}$$

Let $x < \frac{\ln q}{2\lambda} - 1$, then $x - 1 < x + 1 < \frac{\ln q}{2\lambda}$ and $g'_{q,\lambda}(x+1) > g'_{q,\lambda}(x-1)$ (by $g_{q,\lambda}$ being strictly concave up for $x < \frac{\ln q}{2\lambda}$), that is, $M'_{q,\lambda}(x) > 0$. Hence, $M_{q,\lambda}$ is strictly increasing over $\left(-\infty, \frac{\ln q}{2\lambda} - 1\right)$.

Let now $x - 1 > \frac{\ln q}{2\lambda}$, then $x + 1 > x - 1 > \frac{\ln q}{2\lambda}$, and $g'_{q,\lambda}(x+1) < g'_{q,\lambda}(x-1)$, that is, $M'_{q,\lambda}(x) < 0$.

Therefore, $M_{q,\lambda}$ is strictly decreasing over $\left(\frac{\ln q}{2\lambda} + 1, +\infty\right)$.

Let us next consider, $\frac{\ln q}{2\lambda} - 1 \leq x \leq \frac{\ln q}{2\lambda} + 1$. We have that

$$\begin{aligned}M''_{q,\lambda}(x) &= \frac{1}{4}\left(g''_{q,\lambda}(x+1) - g''_{q,\lambda}(x-1)\right)\\ &= 2q\lambda^2\left[e^{2\lambda(x+1)}\left(\frac{q - e^{2\lambda(x+1)}}{\left(e^{2\lambda(x+1)} + q\right)^3}\right) - e^{2\lambda(x-1)}\left(\frac{q - e^{2\lambda(x-1)}}{\left(e^{2\lambda(x-1)} + q\right)^3}\right)\right].\end{aligned} \tag{4.13}$$

By $\frac{\ln q}{2\lambda} - 1 \leq x \Leftrightarrow \frac{\ln q}{2\lambda} \leq x + 1 \Leftrightarrow \ln q \leq 2\lambda(x+1) \Leftrightarrow q \leq e^{2\lambda(x+1)} \Leftrightarrow q - e^{2\lambda(x+1)} \leq 0$.

By $x \leq \frac{\ln q}{2\lambda} + 1 \Leftrightarrow x - 1 \leq \frac{\ln q}{2\lambda} \Leftrightarrow 2\lambda(x-1) \leq \ln q \Leftrightarrow e^{2\lambda(x-1)} \leq q \Leftrightarrow q - e^{2\lambda\beta(x-1)} \geq 0$.

Clearly by (4.13) we get that $M''_{q,\lambda}(x) \leq 0$, for $x \in \left[\frac{\ln q}{2\lambda} - 1, \frac{\ln q}{2\lambda} + 1\right]$.

More precisely, $M_{q,\lambda}$ is concave down over $\left[\frac{\ln q}{2\lambda} - 1, \frac{\ln q}{2\lambda} + 1\right]$ and strictly concave down over $\left(\frac{\ln q}{2\lambda} - 1, \frac{\ln q}{2\lambda} + 1\right)$.

Consequently, $M_{q,\lambda}$ has a bell-type shape over $\mathbb{R}$.

Of course it holds $M''_{q,\lambda}\left(\frac{\ln q}{2\lambda}\right) < 0$.

At $x = \frac{\ln q}{2\lambda}$, we have

$$\begin{aligned} M'_{q,\lambda}(x) &= \frac{1}{4}\left(g'_{q,\lambda}(x+1) - g'_{q,\lambda}(x-1)\right) \\ &= q\lambda\left(\frac{e^{2\lambda(x+1)}}{\left(e^{2\lambda(x+1)} + q\right)^2} - \frac{e^{2\lambda(x-1)}}{\left(e^{2\lambda(x-1)} + q\right)^2}\right). \end{aligned} \tag{4.14}$$

Thus,

$$\begin{aligned} M'_{q,\lambda}\left(\frac{\ln q}{2\lambda}\right) &= q\lambda\left(\frac{e^{2\lambda\left(\frac{\ln q}{2\lambda}+1\right)}}{\left(e^{2\lambda\left(\frac{\ln q}{2\lambda}+1\right)} + q\right)^2} - \frac{e^{2\lambda\left(\frac{\ln q}{2\lambda}-1\right)}}{\left(e^{2\lambda\left(\frac{\ln q}{2\lambda}-1\right)} + q\right)^2}\right) \\ &= q\lambda\left(\frac{qe^{2\lambda}}{\left(qe^{2\lambda} + q\right)^2} - \frac{qe^{-2\lambda}}{\left(qe^{-2\lambda} + q\right)^2}\right) \\ &= \lambda\left(\frac{e^{2\lambda}}{\left(e^{2\lambda} + 1\right)^2} - \frac{e^{-2\lambda}}{\left(e^{-2\lambda} + 1\right)^2}\right) \\ &= \lambda\left(\frac{e^{2\lambda}\left(e^{-2\lambda} + 1\right)^2 - e^{-2\lambda}\left(e^{2\lambda} + 1\right)^2}{\left(e^{2\lambda} + 1\right)^2\left(e^{-2\lambda} + 1\right)^2}\right) = 0. \end{aligned} \tag{4.15}$$

That is, $\frac{\ln q}{2\lambda}$ is the only critical number of $M_{q,\lambda}$ over $\mathbb{R}$. Hence at $x = \frac{\ln q}{2\lambda}$, $M_{q,\lambda}$ achieves its global maximum, which is

$$\begin{aligned} M_{q,\lambda}\left(\frac{\ln q}{2\lambda}\right) &= \frac{1}{4}\left[g_{q,\lambda}\left(\frac{\ln q}{2\lambda} + 1\right) - g_{q,\lambda}\left(\frac{\ln q}{2\lambda} - 1\right)\right] \\ &= \frac{1}{4}\left[\left(\frac{e^{\lambda\left(\frac{\ln q}{2\lambda}+1\right)} - qe^{-\lambda\left(\frac{\ln q}{2\lambda}+1\right)}}{e^{\lambda\left(\frac{\ln q}{2\lambda}+1\right)} + qe^{-\lambda\left(\frac{\ln q}{2\lambda}+1\right)}}\right) - \left(\frac{e^{\lambda\left(\frac{\ln q}{2\lambda}-1\right)} - qe^{-\lambda\left(\frac{\ln q}{2\lambda}-1\right)}}{e^{\lambda\left(\frac{\ln q}{2\lambda}-1\right)} + qe^{-\lambda\left(\frac{\ln q}{2\lambda}-1\right)}}\right)\right] \\ &= \frac{1}{4}\left[\left(\frac{\sqrt{q}e^{\lambda} - qq^{-\frac{1}{2}}e^{-\lambda}}{\sqrt{q}e^{\lambda} + qq^{-\frac{1}{2}}e^{-\lambda}}\right) - \left(\frac{\sqrt{q}e^{-\lambda} - qq^{-\frac{1}{2}}e^{\lambda}}{\sqrt{q}e^{-\lambda} + qq^{-\frac{1}{2}}e^{\lambda}}\right)\right] \end{aligned}$$

$$= \frac{1}{4}\left[\left(\frac{e^{\lambda}-e^{-\lambda}}{e^{\lambda}+e^{-\lambda}}\right) - \left(\frac{e^{-\lambda}-e^{\lambda}}{e^{-\lambda}+e^{\lambda}}\right)\right]$$

$$= \frac{1}{4}\left[\frac{2\left(e^{\lambda}-e^{-\lambda}\right)}{e^{\lambda}+e^{-\lambda}}\right] = \frac{1}{2}\left(\frac{e^{\lambda}-e^{-\lambda}}{e^{\lambda}+e^{-\lambda}}\right) = \frac{\tanh(\lambda)}{2}. \tag{4.17}$$

Conclusion: The maximum value of $M_{q,\lambda}$ is

$$M_{q,\lambda}\left(\frac{\ln q}{2\lambda}\right) = \frac{\tanh(\lambda)}{2}, \quad \lambda > 0. \tag{4.18}$$

We give

Theorem 4.1 (Anastassiou 2023b, Ch. 17). *We have that*

$$\sum_{i=-\infty}^{\infty} M_{q,\lambda}(x-i) = 1, \quad \forall\, x \in \mathbb{R}, \quad \forall\, \lambda, q > 0. \tag{4.19}$$

Thus,

$$\sum_{i=-\infty}^{\infty} M_{q,\lambda}(nx-i) = 1, \quad \forall\, n \in \mathbb{N}, \quad \forall\, x \in \mathbb{R}. \tag{4.20}$$

Similarly, it holds

$$\sum_{i=-\infty}^{\infty} M_{\frac{1}{q},\lambda}(x-i) = 1, \quad \forall\, x \in \mathbb{R}. \tag{4.21}$$

But $M_{\frac{1}{q},\lambda}(x-i) \overset{(4.11)}{=} M_{q,\lambda}(i-x)$, $\forall\, x \in \mathbb{R}$.

Hence,

$$\sum_{i=-\infty}^{\infty} M_{q,\lambda}(i-x) = 1, \quad \forall\, x \in \mathbb{R}, \tag{4.22}$$

and

$$\sum_{i=-\infty}^{\infty} M_{q,\lambda}(i+x) = 1, \quad \forall\, x \in \mathbb{R}. \tag{4.23}$$

It follows

Theorem 4.2 (Anastassiou 2023b, Ch. 17). *It holds*

$$\int_{-\infty}^{\infty} M_{q,\lambda}(x)\, dx = 1, \quad \lambda, q > 0. \tag{4.24}$$

So $M_{q,\lambda}$ is a density function on $\mathbb{R}$; $\lambda, q > 0$.

We need the following result

Theorem 4.3 (Anastassiou 2023b, Ch. 17). *Let* $0 < \alpha < 1$, *and* $n \in \mathbb{N}$ *with* $n^{1-\alpha} > 2$; $q, \lambda > 0$. *Then*

$$\sum_{\substack{k=-\infty \\ \{ : |nx-k| \geq n^{1-\alpha}}}^{\infty} M_{q,\lambda}(nx-k) < 2\max\left\{q, \frac{1}{q}\right\} e^{4\lambda} e^{-2\lambda n^{(1-\alpha)}} = Te^{-2\lambda n^{(1-\alpha)}}, \tag{4.25}$$

where $T := 2\max\left\{q, \frac{1}{q}\right\} e^{4\lambda}$.

Let $\lceil\cdot\rceil$ be the ceiling of the number and $\lfloor\cdot\rfloor$ be the integral part of the number.

Theorem 4.4 (Anastassiou 2023b, Ch. 17). *Let* $x \in [a, b] \subset \mathbb{R}$ *and* $n \in \mathbb{N}$ *so that* $\lceil na\rceil \leq \lfloor nb\rfloor$. *For* $q > 0$, $\lambda > 0$, *we consider the number* $\lambda_q > z_0 > 0$ *with* $M_{q,\lambda}(z_0) = M_{q,\lambda}(0)$ *and* $\lambda_q > 1$. *Then*

$$\frac{1}{\sum_{k=\lceil na\rceil}^{\lfloor nb\rfloor} M_{q,\lambda}(nx-k)} < \max\left\{\frac{1}{M_{q,\lambda}(\lambda_q)}, \frac{1}{M_{\frac{1}{q},\lambda}\left(\lambda_{\frac{1}{q}}\right)}\right\} =: \Delta(q). \tag{4.26}$$

We make

Remark 4.1 (Anastassiou 2023b, Ch. 17). (i) We have that

$$\lim_{n\to+\infty} \sum_{k=\lceil na\rceil}^{\lfloor nb\rfloor} M_{q,\lambda}(nx-k) \neq 1, \quad \text{for at least some } x \in [a, b], \tag{4.27}$$

where $\lambda, q > 0$.

(ii) Let $[a, b] \subset \mathbb{R}$. For large n, we always have $\lceil na\rceil \leq \lfloor nb\rfloor$. Also $a \leq \frac{k}{n} \leq b$, iff $\lceil na\rceil \leq k \leq \lfloor nb\rfloor$. In general, it holds

$$\sum_{k=\lceil na\rceil}^{\lfloor nb\rfloor} M_{q,\lambda}(nx-k) \leq 1. \tag{4.28}$$

Let $(\mathbb{C}, |\cdot|)$ be the Banach space of the complex numbers over the reals.

Definition 4.1. Let $f \in C([a, b], \mathbb{C})$ and $n \in \mathbb{N} : \lceil na\rceil \leq \lfloor nb\rfloor$. We introduce and define the $\mathbb{C}$-valued linear neural network operators

$$H_n(f, x) := \frac{\sum_{k=\lceil na\rceil}^{\lfloor nb\rfloor} f\left(\frac{k}{n}\right) M_{q,\lambda}(nx-k)}{\sum_{k=\lceil na\rceil}^{\lfloor nb\rfloor} M_{q,\lambda}(nx-k)}, \quad x \in [a, b]; \quad q > 0, \ q \neq 1. \tag{4.29}$$

For large enough n, we always obtain $\lceil na \rceil \leq \lfloor nb \rfloor$. Also $a \leq \frac{k}{n} \leq b$, iff $\lceil na \rceil \leq k \leq \lfloor nb \rfloor$. The same H_n is used for real-valued functions. We study here the pointwise and uniform convergence of $H_n(f,x)$ to $f(x)$ with rates.

Clearly here $H_n(f) \in C([a,b],\mathbb{C})$.

For convenience, also we call

$$H_n^*(f,x) := \sum_{k=\lceil na \rceil}^{\lfloor nb \rfloor} f\left(\frac{k}{n}\right) M_{q,\lambda}(nx-k) \tag{4.30}$$

(the same H_n^* can be defined for real-valued functions), that is,

$$H_n(f,x) := \frac{H_n^*(f,x)}{\sum_{k=\lceil na \rceil}^{\lfloor nb \rfloor} M_{q,\lambda}(nx-k)}. \tag{4.31}$$

So

$$\begin{aligned} H_n(f,x) - f(x) &= \frac{H_n^*(f,x)}{\sum_{k=\lceil na \rceil}^{\lfloor nb \rfloor} M_{q,\lambda}(nx-k)} - f(x) \\ &= \frac{H_n^*(f,x) - f(x)\left(\sum_{k=\lceil na \rceil}^{\lfloor nb \rfloor} M_{q,\lambda}(nx-k)\right)}{\sum_{k=\lceil na \rceil}^{\lfloor nb \rfloor} M_{q,\lambda}(nx-k)}. \end{aligned} \tag{4.32}$$

Consequently, we derive that

$$\begin{aligned} |H_n(f,x) - f(x)| &\leq \Delta(q)\left| H_n^*(f,x) - f(x)\left(\sum_{k=\lceil na \rceil}^{\lfloor nb \rfloor} M_{q,\lambda}(nx-k)\right)\right| \\ &= \Delta(q)\left|\sum_{k=\lceil na \rceil}^{\lfloor nb \rfloor}\left(f\left(\frac{k}{n}\right) - f(x)\right) M_{q,\lambda}(nx-k)\right|, \end{aligned} \tag{4.33}$$

where $\Delta(q)$ as in (4.26).

We estimate the right-hand side of the last quantity.

For that we need, for $f \in C([a,b],\mathbb{C})$, the first modulus of continuity

$$\omega_1(f,\delta) := \sup_{\substack{x,y \in [a,b] \\ |x-y| \leq \delta}} |f(x) - f(y)|, \quad \delta > 0. \tag{4.34}$$

The fact $f \in C([a,b],\mathbb{C})$ is equivalent to $\lim_{\delta \to 0} \omega_1(f,\delta) = 0$, see Anastassiou (2017).

4.3 Main Results

We present $\mathbb{C}$-valued neural network high order approximations to a function given with rates. We start with a trigonometric approximation.

Theorem 4.5. *Let $f \in C^2([a,b],\mathbb{C})$, $0 < \alpha < 1$, $n \in \mathbb{N} : n^{1-\alpha} > 2$, $x \in [a,b]$. Then*

(1)

$$|H_n(f,x) - f(x)| \leq \Delta(q)\left[|f'(x)|\left(\frac{1}{n^\alpha} + (b-a)\,Te^{-2\lambda n^{(1-\alpha)}}\right)\right.$$

$$+\frac{|f''(x)|}{2}\left(\frac{1}{n^{2\alpha}} + (b-a)^2\,Te^{-2\lambda n^{(1-\alpha)}}\right)$$

$$\left.+\left(\frac{\omega_1\left(f''+f,\frac{1}{n^\alpha}\right)}{2n^{2\alpha}} + \|f''+f\|_\infty (b-a)^2\,Te^{-2\lambda n^{(1-\alpha)}}\right)\right], \quad (4.35)$$

(2) *if $f'(x) = f''(x) = 0$, we obtain*

$$|H_n(f,x) - f(x)|$$

$$\leq \Delta(q)\left[\frac{\omega_1\left(f''+f,\frac{1}{n^\alpha}\right)}{2n^{2\alpha}} + \|f''+f\|_\infty (b-a)^2\,Te^{-2\lambda n^{(1-\alpha)}}\right], \quad (4.36)$$

note here the high rate of convergence at $n^{-3\alpha}$,

(3) *furthermore we get*

$$\|H_nf - f\|_\infty \leq \Delta(q)\left[\|f'\|_\infty\left(\frac{1}{n^\alpha} + (b-a)\,Te^{-2\lambda n^{(1-\alpha)}}\right)\right.$$

$$+\frac{\|f''\|_\infty}{2}\left(\frac{1}{n^{2\alpha}} + (b-a)^2\,Te^{-2\lambda n^{(1-\alpha)}}\right)$$

$$\left.+\left(\frac{\omega_1\left(f''+f,\frac{1}{n^\alpha}\right)}{2n^{2\alpha}} + \|f''+f\|_\infty (b-a)^2\,Te^{-2\lambda n^{(1-\alpha)}}\right)\right], \quad (4.37)$$

i.e. $\lim_{n\to+\infty} H_n(f) = f$, *pointwise and uniformly,*

(4) *and finally, it holds*

$$\left| H_n(f,x) - f'(x) H_n(\sin(\cdot - x), x) - 2f''(x) H_n\left(\sin^2\left(\frac{\cdot - x}{2}\right), x\right) - f(x) \right|$$

$$\leq \Delta(q) \left[\frac{\omega_1\left(f'' + f, \frac{1}{n^{\alpha}}\right)}{2n^{2\alpha}} + \|f'' + f\|_{\infty} (b-a)^2 T e^{-2\lambda n^{(1-\alpha)}} \right], \quad (4.38)$$

again here we achieve high speed of convergence at $n^{-3\alpha}$.

Proof. Here $f \in C^2([a,b], \mathbb{C})$, and we apply the trigonometric Taylor's formula for $f \in C^2([a,b], \mathbb{C})$, see Theorem 6 of Anastassiou (2023c).

Let $\frac{k}{n}, x \in [a,b]$, then

$$f\left(\frac{k}{n}\right) = f(x) + f'(x) \sin\left(\frac{k}{n} - x\right) + 2f''(x) \sin^2\left(\frac{\frac{k}{n} - x}{2}\right)$$

$$+ \int_x^{\frac{k}{n}} \left[(f''(t) + f(t)) - (f''(x) + f(x)) \right] \sin\left(\frac{k}{n} - t\right) dt. \quad (4.39)$$

Hence it holds

$$f\left(\frac{k}{n}\right) M_{q,\lambda}(nx-k) = f(x) M_{q,\lambda}(nx-k)$$

$$+ f'(x) \sin\left(\frac{k}{n} - x\right) M_{q,\lambda}(nx-k) + 2f''(x) \sin^2\left(\frac{\frac{k}{n} - x}{2}\right) M_{q,\lambda}(nx-k)$$

$$+ M_{q,\lambda}(nx-k) \left(\int_x^{\frac{k}{n}} \left[(f''(t) + f(t)) - (f''(x) + f(x)) \right] \sin\left(\frac{k}{n} - t\right) dt \right). \quad (4.40)$$

So we have

$$\sum_{k=\lceil na \rceil}^{\lfloor nb \rfloor} f\left(\frac{k}{n}\right) M_{q,\lambda}(nx-k) - f(x) \sum_{k=\lceil na \rceil}^{\lfloor nb \rfloor} M_{q,\lambda}(nx-k)$$

$$= f'(x) \sum_{k=\lceil na \rceil}^{\lfloor nb \rfloor} M_{q,\lambda}(nx-k) \sin\left(\frac{k}{n} - x\right)$$

$$+2f''(x)\sum_{k=\lceil na\rceil}^{\lfloor nb\rfloor} M_{q,\lambda}(nx-k)\sin^2\left(\frac{\frac{k}{n}-x}{2}\right)+\sum_{k=\lceil na\rceil}^{\lfloor nb\rfloor} M_{q,\lambda}(nx-k)$$

$$\times\left(\int_x^{\frac{k}{n}}\left[\left(f''(t)+f(t)\right)-\left(f''(x)+f(x)\right)\right]\sin\left(\frac{k}{n}-t\right)dt\right). \quad (4.41)$$

Thus, we obtain

$$H_n^*(f,x)-f(x)\sum_{k=\lceil na\rceil}^{\lfloor nb\rfloor} M_{q,\lambda}(nx-k)$$

$$=f'(x)H_n^*(\sin(\cdot-x),x)+2f''(x)H_n^*\left(\sin^2\left(\frac{\cdot-x}{2}\right),x\right)+\Lambda_n(x), \quad (4.42)$$

where

$$\Lambda_n(x):=\sum_{k=\lceil na\rceil}^{\lfloor nb\rfloor} M_{q,\lambda}(nx-k)$$

$$\times\left(\int_x^{\frac{k}{n}}\left[\left(f''(t)+f(t)\right)-\left(f''(x)+f(x)\right)\right]\sin\left(\frac{k}{n}-t\right)dt\right).$$

We call

$$R_2(n):=\int_x^{\frac{k}{n}}\left[\left(f''(t)+f(t)\right)-\left(f''(x)+f(x)\right)\right]\sin\left(\frac{k}{n}-t\right)dt. \quad (4.43)$$

We assume that $b-a>\frac{1}{n^\alpha}$, which is always the case for large enough $n\in\mathbb{N}$, that is, when $n>\left\lceil(b-a)^{-\frac{1}{\alpha}}\right\rceil$.

Thus, $\left|\frac{k}{n}-x\right|\le\frac{1}{n^\alpha}$ or $\left|\frac{k}{n}-x\right|>\frac{1}{n^\alpha}$.

In case of $\left|\frac{k}{n}-x\right|\le\frac{1}{n^\alpha}$, we have the following cases:

(i) If $\frac{k}{n}\ge x$, then

$$|R_2(n)|=\left|\int_x^{\frac{k}{n}}\left[\left(f''(t)+f(t)\right)-\left(f''(x)+f(x)\right)\right]\sin\left(\frac{k}{n}-t\right)dt\right|$$

$$\le\int_x^{\frac{k}{n}}\omega_1\left(f''+f,t-x\right)\left|\sin\left(\frac{k}{n}-t\right)\right|dt\le \quad (4.44)$$

(by $|\sin x| \leq |x|, \forall\, x \in \mathbb{R}$)

$$\int_x^{\frac{k}{n}} \omega_1 \left(f'' + f, t - x\right) \left(\frac{k}{n} - t\right) dt \leq \omega_1 \left(f'' + f, \frac{k}{n} - x\right) \frac{\left(\frac{k}{n} - x\right)^2}{2} \leq \frac{\omega_1 \left(f'' + f, \frac{1}{n^\alpha}\right)}{2n^{2\alpha}},$$

that is,

$$|R_2(n)| \leq \frac{\omega_1 \left(f'' + f, \frac{1}{n^\alpha}\right)}{2n^{2\alpha}}. \tag{4.45}$$

(ii) If $\frac{k}{n} < x$, then

$$|R_2(n)| = \left|\int_x^{\frac{k}{n}} \left[\left(f''(t) + f(t)\right) - \left(f''(x) + f(x)\right)\right] \sin\left(\frac{k}{n} - t\right) dt\right|$$
$$= \left|\int_{\frac{k}{n}}^{x} \left[\left(f''(t) + f(t)\right) - \left(f''(x) + f(x)\right)\right] \sin\left(\frac{k}{n} - t\right) dt\right|$$
$$\leq \int_{\frac{k}{n}}^{x} \left|\left(f''(t) + f(t)\right) - \left(f''(x) + f(x)\right)\right| \left|\sin\left(\frac{k}{n} - t\right)\right| dt$$
$$\leq \int_{\frac{k}{n}}^{x} \omega_1 \left(f'' + f, x - \frac{k}{n}\right) \left(t - \frac{k}{n}\right) dt$$
$$\leq \omega_1 \left(f'' + f, x - \frac{k}{n}\right) \frac{\left(x - \frac{k}{n}\right)^2}{2} \leq \frac{\omega_1 \left(f'' + f, \frac{1}{n^\alpha}\right)}{2n^{2\alpha}}. \tag{4.46}$$

That is,

$$|R_2(n)| \leq \frac{\omega_1 \left(f'' + f, \frac{1}{n^\alpha}\right)}{2n^{2\alpha}}. \tag{4.47}$$

So, we have proved when $\left|\frac{k}{n} - x\right| \leq \frac{1}{n^\alpha}$, always it holds

$$|R_2(n)| \leq \frac{\omega_1 \left(f'' + f, \frac{1}{n^\alpha}\right)}{2n^{2\alpha}}. \tag{4.48}$$

Next assume again $\frac{k}{n} \geq x$, then

$$|R_2(n)| = \left|\int_x^{\frac{k}{n}} \left[\left(f''(t) + f(t)\right) - \left(f''(x) + f(x)\right)\right] \sin\left(\frac{k}{n} - t\right) dt\right|$$
$$\leq \int_x^{\frac{k}{n}} \left|\left(f''(t) + f(t)\right) - \left(f''(x) + f(x)\right)\right| \left|\sin\left(\frac{k}{n} - t\right)\right| dt \leq$$

(by $|\sin x| \leq |x|$, $\forall\ x \in \mathbb{R}$)

$$\begin{aligned}
&2\left\|f''+f\right\|_{\infty}\left(\int_{x}^{\frac{k}{n}}\left|\sin\left(\frac{k}{n}-t\right)\right|dt\right)\\
&\quad\leq 2\left\|f''+f\right\|_{\infty}\left(\int_{x}^{\frac{k}{n}}\left(\frac{k}{n}-t\right)dt\right)\\
&\quad= 2\left\|f''+f\right\|_{\infty}\frac{\left(\frac{k}{n}-x\right)^{2}}{2}\leq\left\|f''+f\right\|_{\infty}(b-a)^{2}. \qquad (4.49)
\end{aligned}$$

Hence it is

$$|R_2(n)| \leq \left\|f''+f\right\|_{\infty}(b-a)^{2}. \qquad (4.50)$$

When $\frac{k}{n} < x$, we have

$$\begin{aligned}
|R_2(n)| &= \left|\int_{x}^{\frac{k}{n}}\left[\left(f''(t)+f(t)\right)-\left(f''(x)+f(x)\right)\right]\sin\left(\frac{k}{n}-t\right)dt\right|\\
&= \left|\int_{\frac{k}{n}}^{x}\left[\left(f''(x)+f(x)\right)-\left(f''(t)+f(t)\right)\right]\sin\left(\frac{k}{n}-t\right)dt\right|\\
&\leq \int_{\frac{k}{n}}^{x}\left|\left(f''(x)+f(x)\right)-\left(f''(t)+f(t)\right)\right|\left|\sin\left(\frac{k}{n}-t\right)\right|dt\\
&\leq 2\left\|f''+f\right\|_{\infty}\int_{\frac{k}{n}}^{x}\left|\sin\left(\frac{k}{n}-t\right)\right|dt\\
&\leq 2\left\|f''+f\right\|_{\infty}\int_{\frac{k}{n}}^{x}\left(t-\frac{k}{n}\right)dt=\left\|f''+f\right\|_{\infty}\left(x-\frac{k}{n}\right)^{2}\\
&\leq \left\|f''+f\right\|_{\infty}(b-a)^{2}. \qquad (4.51)
\end{aligned}$$

Therefore, always it holds

$$|R_2(n)| \leq \left\|f''+f\right\|_{\infty}(b-a)^{2}. \qquad (4.52)$$

And we have

$$\begin{aligned}
\Lambda_n(x) = &\sum_{\substack{k=\lceil na\rceil\\ \left\{:\left|\frac{k}{n}-x\right|\leq\frac{1}{n^{\alpha}}\right.}}^{\lfloor nb\rfloor} M_{q,\lambda}(nx-k)\,R_2(n)\\
&+\sum_{\substack{k=\lceil na\rceil\\ \left\{:\left|\frac{k}{n}-x\right|>\frac{1}{n^{\alpha}}\right.}}^{\lfloor nb\rfloor} M_{q,\lambda}(nx-k)\,R_2(n). \qquad (4.53)
\end{aligned}$$

Hence it holds

$$\begin{aligned}
|\Lambda_n(x)| \leq & \sum_{\substack{k=\lceil na \rceil \\ \left\{ : \left| \frac{k}{n} - x \right| \leq \frac{1}{n^\alpha} \right.}}^{\lfloor nb \rfloor} M_{q,\lambda}(nx-k) |R_2(n)| \\
& + \sum_{\substack{k=\lceil na \rceil \\ \left\{ : \left| \frac{k}{n} - x \right| > \frac{1}{n^\alpha} \right.}}^{\lfloor nb \rfloor} M_{q,\lambda}(nx-k) |R_2(n)| \qquad (4.54) \\
& \leq \left(\sum_{\substack{k=\lceil na \rceil \\ \left\{ : \left| \frac{k}{n} - x \right| \leq \frac{1}{n^\alpha} \right.}}^{\lfloor nb \rfloor} M_{q,\lambda}(nx-k) \right) \frac{\omega_1\left(f''+f, \frac{1}{n^\alpha}\right)}{2n^{2\alpha}} \\
& + \sum_{\substack{k=\lceil na \rceil \\ \left\{ : \left| \frac{k}{n} - x \right| > \frac{1}{n^\alpha} \right.}}^{\lfloor nb \rfloor} M_{q,\lambda}(nx-k) \|f''+f\|_\infty (b-a)^2 \\
& \overset{\text{(by (4.28))}}{\leq} \frac{\omega_1\left(f''+f, \frac{1}{n^\alpha}\right)}{2n^{2\alpha}} + \|f''+f\|_\infty (b-a)^2 \\
& \left(\sum_{\substack{k=\lceil na \rceil \\ \left\{ : \left| \frac{k}{n} - x \right| > \frac{1}{n^\alpha} \right.}}^{\lfloor nb \rfloor} M_{q,\lambda}(nx-k) \right) \\
& \overset{\text{(by Theorem 4.3)}}{\leq} \frac{\omega_1\left(f''+f, \frac{1}{n^\alpha}\right)}{2n^{2\alpha}} + \|f''+f\|_\infty (b-a)^2 T e^{-2\lambda n^{(1-\alpha)}}. \qquad (4.55)
\end{aligned}$$

Consequently, we have derived that

$$|\Lambda_n(x)| \leq \frac{\omega_1\left(f''+f, \frac{1}{n^\alpha}\right)}{2n^{2\alpha}} + \|f''+f\|_\infty (b-a)^2 T e^{-2\lambda n^{(1-\alpha)}}. \qquad (4.56)$$

Next we use again $|\sin x| \leq |x|$, $\forall\, x \in \mathbb{R}$.

We have that

$$H_n^*(\sin(\cdot - x), x) = \sum_{k=\lceil na \rceil}^{\lfloor nb \rfloor} M_{q,\lambda}(nx-k) \sin\left(\frac{k}{n} - x\right), \qquad (4.57)$$

and

$$
\begin{aligned}
\left|H_n^*\left(\sin\left(\cdot - x\right), x\right)\right| &\leq \sum_{k=\lceil na\rceil}^{\lfloor nb\rfloor} M_{q,\lambda}\left(nx-k\right)\left|\sin\left(\frac{k}{n}-x\right)\right| \\
&= \sum_{\substack{k=\lceil na\rceil \\ \left\{:\left|\frac{k}{n}-x\right|\leq\frac{1}{n^{\alpha}}\right.}}^{\lfloor nb\rfloor} M_{q,\lambda}\left(nx-k\right)\left|\sin\left(\frac{k}{n}-x\right)\right| \\
&\quad + \sum_{\substack{k=\lceil na\rceil \\ \left\{:\left|\frac{k}{n}-x\right|>\frac{1}{n^{\alpha}}\right.}}^{\lfloor nb\rfloor} M_{q,\lambda}\left(nx-k\right)\left|\sin\left(\frac{k}{n}-x\right)\right| \qquad (4.58) \\
&\leq \sum_{\substack{k=\lceil na\rceil \\ \left\{:\left|\frac{k}{n}-x\right|\leq\frac{1}{n^{\alpha}}\right.}}^{\lfloor nb\rfloor} M_{q,\lambda}\left(nx-k\right)\left|\frac{k}{n}-x\right| \\
&\quad + \sum_{\substack{k=\lceil na\rceil \\ \left\{:\left|\frac{k}{n}-x\right|>\frac{1}{n^{\alpha}}\right.}}^{\lfloor nb\rfloor} M_{q,\lambda}\left(nx-k\right)\left|\frac{k}{n}-x\right| \\
&\leq \frac{1}{n^{\alpha}} + (b-a)\left(\sum_{\substack{k=\lceil na\rceil \\ \left\{:\left|\frac{k}{n}-x\right|>\frac{1}{n^{\alpha}}\right.}}^{\lfloor nb\rfloor} M_{q,\lambda}\left(nx-k\right)\right) \\
&\overset{\text{(by (4.25))}}{\leq} \frac{1}{n^{\alpha}} + (b-a)\,Te^{-2\lambda n^{(1-\alpha)}}. \qquad (4.59)
\end{aligned}
$$

We found that

$$
\left|H_n^*\left(\sin\left(\cdot - x\right), x\right)\right| \leq \frac{1}{n^{\alpha}} + (b-a)\,Te^{-2\lambda n^{(1-\alpha)}}. \tag{4.60}
$$

Next we estimate

$$
H_n^*\left(\sin^2\left(\frac{\cdot - x}{2}\right), x\right) = \sum_{k=\lceil na\rceil}^{\lfloor nb\rfloor} M_{q,\lambda}\left(nx-k\right)\sin^2\left(\frac{\frac{k}{n}-x}{2}\right). \tag{4.61}
$$

We have that (by $|\sin x| \leq |x|$, $\forall\, x \in \mathbb{R}$)

$$H_n^*\left(\sin^2\left(\frac{\cdot - x}{2}\right), x\right) = \sum_{k=\lceil na \rceil}^{\lfloor nb \rfloor} M_{q,\lambda}(nx-k)\left|\sin\left(\frac{\frac{k}{n}-x}{2}\right)\right|^2$$

$$\leq \frac{1}{4}\sum_{k=\lceil na \rceil}^{\lfloor nb \rfloor} M_{q,\lambda}(nx-k)\left|\frac{k}{n}-x\right|^2$$

$$= \frac{1}{4}\left[\sum_{\substack{k=\lceil na \rceil \\ \left\{: \left|\frac{k}{n}-x\right| \leq \frac{1}{n^\alpha}\right.}}^{\lfloor nb \rfloor} M_{q,\lambda}(nx-k)\left|\frac{k}{n}-x\right|^2\right.$$

$$\left. + \sum_{\substack{k=\lceil na \rceil \\ \left\{: \left|\frac{k}{n}-x\right| > \frac{1}{n^\alpha}\right.}}^{\lfloor nb \rfloor} M_{q,\lambda}(nx-k)\left|\frac{k}{n}-x\right|^2\right]$$

$$\leq \frac{1}{4}\left[\frac{1}{n^{2\alpha}} + (b-a)^2 T e^{-2\lambda n^{(1-\alpha)}}\right]. \tag{4.62}$$

That is,

$$H_n^*\left(\sin^2\left(\frac{\cdot - x}{2}\right), x\right) \leq \frac{1}{4}\left[\frac{1}{n^{2\alpha}} + (b-a)^2 T e^{-2\lambda n^{(1-\alpha)}}\right]. \tag{4.63}$$

Consequently, we have derived the following:

(1)

$$|H_n(f,x) - f(x)| \leq \Delta(q)\left[|f'(x)|\left(\frac{1}{n^\alpha} + (b-a) T e^{-2\lambda n^{(1-\alpha)}}\right)\right.$$

$$+ \frac{|f''(x)|}{2}\left(\frac{1}{n^{2\alpha}} + (b-a)^2 T e^{-2\lambda n^{(1-\alpha)}}\right)$$

$$\left. + \left(\frac{\omega_1\left(f''+f, \frac{1}{n^\alpha}\right)}{2n^{2\alpha}} + \|f''+f\|_\infty (b-a)^2 T e^{-2\lambda n^{(1-\alpha)}}\right)\right]. \tag{4.64}$$

(2) If $f'(x) = f''(x) = 0$, by (4.64), we obtain

$$|H_n(f,x) - f(x)|$$
$$\leq \Delta(q)\left[\frac{\omega_1\left(f''+f, \frac{1}{n^\alpha}\right)}{2n^{2\alpha}} + \|f''+f\|_\infty (b-a)^2 T e^{-2\lambda n^{(1-\alpha)}}\right], \quad (4.65)$$

note here the high rate of convergence at $n^{-3\alpha}$.

(3) Furthermore, by (4.64), we get

$$\|H_n f - f\|_\infty \leq \Delta(q)\left[\|f'\|_\infty\left(\frac{1}{n^\alpha} + (b-a) T e^{-2\lambda n^{(1-\alpha)}}\right)\right.$$
$$+ \frac{\|f''\|_\infty}{2}\left(\frac{1}{n^{2\alpha}} + (b-a)^2 T e^{-2\lambda n^{(1-\alpha)}}\right)$$
$$\left. + \left(\frac{\omega_1\left(f''+f, \frac{1}{n^\alpha}\right)}{2n^{2\alpha}} + \|f''+f\|_\infty (b-a)^2 T e^{-2\lambda n^{(1-\alpha)}}\right)\right]. \quad (4.66)$$

We derive that $\lim_{n\to+\infty} H_n(f) = f$, pointwise and uniformly.

We observe that

$$H_n(f,x) - f'(x) H_n(\sin(\cdot - x), x) - 2f''(x) H_n\left(\sin^2\left(\frac{\cdot - x}{2}\right), x\right) - f(x)$$
$$= \frac{H_n^*(f,x)}{\sum_{k=\lceil na\rceil}^{\lfloor nb\rfloor} M_{q,\lambda}(nx-k)} - f'(x)\frac{H_n^*(\sin(\cdot - x), x)}{\sum_{k=\lceil na\rceil}^{\lfloor nb\rfloor} M_{q,\lambda}(nx-k)}$$
$$-2f''(x)\frac{H_n^*\left(\sin^2\left(\frac{\cdot - x}{2}\right), x\right)}{\sum_{k=\lceil na\rceil}^{\lfloor nb\rfloor} M_{q,\lambda}(nx-k)} - f(x)\left(\frac{\sum_{k=\lceil na\rceil}^{\lfloor nb\rfloor} M_{q,\lambda}(nx-k)}{\sum_{k=\lceil na\rceil}^{\lfloor nb\rfloor} M_{q,\lambda}(nx-k)}\right) \quad (4.67)$$

$$= \frac{1}{\sum_{k=\lceil na\rceil}^{\lfloor nb\rfloor} M_{q,\lambda}(nx-k)}\left[H_n^*(f,x) - f'(x) H_n^*(\sin(\cdot - x), x)\right.$$
$$\left. -2f''(x) H_n^*\left(\sin^2\left(\frac{\cdot - x}{2}\right), x\right) - f(x)\sum_{k=\lceil na\rceil}^{\lfloor nb\rfloor} M_{q,\lambda}(nx-k)\right]$$
$$= \frac{1}{\sum_{k=\lceil na\rceil}^{\lfloor nb\rfloor} M_{q,\lambda}(nx-k)}(\Lambda_n(x)). \quad (4.68)$$

Finally, we obtain ($\forall\ x \in [a,b]$, $n \in \mathbb{N}$):
(4)

$$\left|H_n(f,x) - f'(x)\, H_n(\sin(\cdot - x), x) - 2f''(x)\, H_n\left(\sin^2\left(\frac{\cdot - x}{2}\right), x\right) - f(x)\right|$$

$$\overset{(4.26)}{\leq} \Delta(q)\, |\Lambda_n(x)|$$

$$\overset{(4.56)}{\leq} \Delta(q) \left[\frac{\omega_1\left(f'' + f, \frac{1}{n^\alpha}\right)}{2n^{2\alpha}} + \|f'' + f\|_\infty (b-a)^2\, T e^{-2\lambda n^{(1-\alpha)}}\right]. \tag{4.69}$$

The theorem is proved. □

We continue with a hyperbolic high order neural network approximation.

Theorem 4.6. *Let $f \in C^2([a,b],\mathbb{C})$, $0 < \alpha < 1$, $n \in \mathbb{N} : n^{1-\alpha} > 2$, $x \in [a,b]$. Then*

(1)

$$|H_n(f,x) - f(x)| \leq \Delta(q) \cosh(b-a) \left[|f'(x)| \left(\frac{1}{n^\alpha} + (b-a)\, T e^{-2\lambda n^{(1-\alpha)}}\right)\right.$$

$$\frac{|f''(x)|}{2}\left(\frac{1}{n^{2\alpha}} + (b-a)^2\, T e^{-2\lambda n^{(1-\alpha)}}\right)$$

$$\left. + \left(\frac{\omega_1\left(f'' - f, \frac{1}{n^\alpha}\right)}{2n^{2\alpha}} + \|f'' - f\|_\infty (b-a)^2\, T e^{-2\lambda n^{(1-\alpha)}}\right)\right], \tag{4.70}$$

(2) *if $f'(x) = f''(x) = 0$, we obtain*

$$|H_n(f,x) - f(x)| \leq \Delta(q) \cosh(b-a)$$

$$\left[\frac{\omega_1\left(f'' - f, \frac{1}{n^\alpha}\right)}{2n^{2\alpha}} + \|f'' - f\|_\infty (b-a)^2\, T e^{-2\lambda n^{(1-\alpha)}}\right], \tag{4.71}$$

note here the high rate of convergence at $n^{-3\alpha}$,

(3) *furthermore, we get*

$$\|H_n f - f\|_\infty \leq \Delta(q)\cosh(b-a)\left[\|f'\|_\infty\left(\frac{1}{n^\alpha} + (b-a)\,Te^{-2\lambda n^{(1-\alpha)}}\right)\right.$$

$$\frac{\|f''\|_\infty}{2}\left(\frac{1}{n^{2\alpha}} + (b-a)^2\,Te^{-2\lambda n^{(1-\alpha)}}\right)$$

$$\left.+\left(\frac{\omega_1\left(f''-f,\frac{1}{n^\alpha}\right)}{2n^{2\alpha}} + \|f''-f\|_\infty(b-a)^2\,Te^{-2\lambda n^{(1-\alpha)}}\right)\right], \tag{4.72}$$

it follows that $\lim_{n\to+\infty} H_n(f) = f$, *pointwise and uniformly,*

and

(4)

$$\left|H_n(f,x) - f'(x)\,H_n(\sinh(\cdot - x), x) - 2f''(x)H_n\left(\sinh^2\left(\frac{\cdot - x}{2}\right), x\right) - f(x)\right|$$

$$\leq \Delta(q)\cosh(b-a)\left[\frac{\omega_1\left(f''-f,\frac{1}{n^\alpha}\right)}{2n^{2\alpha}} + \|f''-f\|_\infty(b-a)^2\,Te^{-2\lambda n^{(1-\alpha)}}\right], \tag{4.73}$$

again here we achieve high speed of convergence at $n^{-3\alpha}$.

Proof. By the mean value theorem, we have that

$$\sinh x = \sinh x - \sinh 0 = (\cosh\xi)(x-0),$$

for some ξ between $\{0, x\}$, for any $x \in \mathbb{R}$.

Hence,

$$|\sinh x| \leq \|\cosh\|_{\infty,[-(b-a),b-a]}\,|x|, \ \ \forall\, x \in [-(b-a), b-a]. \tag{4.74}$$

That is, there exists $M \geq 1$ such that

$$|\sinh x| \leq M\,|x|, \ \ \forall\, x \in [-(b-a), b-a], \tag{4.75}$$

where $M := \|\cosh\|_{\infty,[-(b-a),b-a]} = \cosh(b-a)$.

Here $f \in C^2([a,b],\mathbb{C})$, and we apply the hyperbolic Taylor's formula for $f \in C^2([a,b],\mathbb{C})$, see Theorem 7 of Anastassiou (2023c).

Let $\frac{k}{n}, x \in [a,b]$, then

$$f\left(\frac{k}{n}\right) = f(x) + f'(x)\sinh\left(\frac{k}{n} - x\right) + 2f''(x)\sinh^2\left(\frac{\frac{k}{n} - x}{2}\right)$$
$$+ \int_x^{\frac{k}{n}} \left[(f''(t) - f(t)) - (f''(x) - f(x))\right]\sinh\left(\frac{k}{n} - t\right) dt. \tag{4.76}$$

Hence it holds

$$f\left(\frac{k}{n}\right) M_{q,\lambda}(nx-k) = f(x) M_{q,\lambda}(nx-k)$$
$$+ f'(x)\sinh\left(\frac{k}{n} - x\right) M_{q,\lambda}(nx-k) + 2f''(x)\sinh^2\left(\frac{\frac{k}{n} - x}{2}\right) M_{q,\lambda}(nx-k)$$
$$+ M_{q,\lambda}(nx-k)\left(\int_x^{\frac{k}{n}} \left[(f''(t) - f(t)) - (f''(x) - f(x))\right]\sinh\left(\frac{k}{n} - t\right) dt\right). \tag{4.77}$$

So we have

$$\sum_{k=\lceil na\rceil}^{\lfloor nb\rfloor} f\left(\frac{k}{n}\right) M_{q,\lambda}(nx-k) - f(x)\sum_{k=\lceil na\rceil}^{\lfloor nb\rfloor} M_{q,\lambda}(nx-k)$$
$$= f'(x)\sum_{k=\lceil na\rceil}^{\lfloor nb\rfloor} M_{q,\lambda}(nx-k)\sinh\left(\frac{k}{n} - x\right)$$
$$+ 2f''(x)\sum_{k=\lceil na\rceil}^{\lfloor nb\rfloor} M_{q,\lambda}(nx-k)\sinh^2\left(\frac{\frac{k}{n} - x}{2}\right) + \sum_{k=\lceil na\rceil}^{\lfloor nb\rfloor} M_{q,\lambda}(nx-k)$$
$$\times\left(\int_x^{\frac{k}{n}} \left[(f''(t) - f(t)) - (f''(x) - f(x))\right]\sinh\left(\frac{k}{n} - t\right) dt\right). \tag{4.78}$$

Thus, we obtain

$$H_n^*(f,x) - f(x)\sum_{k=\lceil na\rceil}^{\lfloor nb\rfloor} M_{q,\lambda}(nx-k)$$
$$= f'(x) H_n^*(\sinh(\cdot - x), x) + 2f''(x) H_n^*\left(\sinh^2\left(\frac{\cdot - x}{2}\right), x\right) + \Lambda_n(x), \tag{4.79}$$

where

$$\Lambda_n(x) := \sum_{k=\lceil na \rceil}^{\lfloor nb \rfloor} M_{q,\lambda}(nx-k)$$
$$\times \left(\int_x^{\frac{k}{n}} \left[(f''(t) - f(t)) - (f''(x) - f(x)) \right] \sinh\left(\frac{k}{n} - t\right) dt \right). \tag{4.80}$$

We call

$$R_2(n) := \int_x^{\frac{k}{n}} \left[(f''(t) - f(t)) - (f''(x) - f(x)) \right] \sinh\left(\frac{k}{n} - t\right) dt. \tag{4.81}$$

We assume that $b - a > \frac{1}{n^\alpha}$, which is always the case for large enough $n \in \mathbb{N}$, that is, when $n > \left\lceil (b-a)^{-\frac{1}{\alpha}} \right\rceil$.

Thus, $\left|\frac{k}{n} - x\right| \leq \frac{1}{n^\alpha}$ or $\left|\frac{k}{n} - x\right| > \frac{1}{n^\alpha}$.

In case of $\left|\frac{k}{n} - x\right| \leq \frac{1}{n^\alpha}$, we have the following cases:

(i) If $\frac{k}{n} \geq x$, then

$$\begin{aligned}
|R_2(n)| &= \left| \int_x^{\frac{k}{n}} \left[(f''(t) - f(t)) - (f''(x) - f(x)) \right] \sinh\left(\frac{k}{n} - t\right) dt \right| \\
&\leq \int_x^{\frac{k}{n}} \omega_1(f'' - f, t - x) \left| \sinh\left(\frac{k}{n} - t\right) \right| dt \\
&\overset{(4.75)}{\leq} \int_x^{\frac{k}{n}} \omega_1(f'' - f, t - x) M \left(\frac{k}{n} - t\right) dt \\
&\leq M\omega_1\left(f'' - f, \frac{k}{n} - x\right) \int_x^{\frac{k}{n}} \left(\frac{k}{n} - t\right) dt \\
&= M\omega_1\left(f'' - f, \frac{k}{n} - x\right) \frac{\left(\frac{k}{n} - x\right)^2}{2} \leq \frac{M\omega_1\left(f'' - f, \frac{1}{n^\alpha}\right)}{2n^{2\alpha}},
\end{aligned} \tag{4.82}$$

that is,

$$|R_2(n)| \leq \frac{M\omega_1\left(f'' - f, \frac{1}{n^\alpha}\right)}{2n^{2\alpha}}. \tag{4.83}$$

(ii) If $\frac{k}{n} < x$, then

$$
\begin{aligned}
|R_2(n)| &= \left| \int_x^{\frac{k}{n}} [(f''(t) - f(t)) - (f''(x) - f(x))] \sinh\left(\frac{k}{n} - t\right) dt \right| \\
&= \left| \int_{\frac{k}{n}}^{x} [(f''(t) - f(t)) - (f''(x) - f(x))] \sinh\left(\frac{k}{n} - t\right) dt \right| \\
&\leq \int_{\frac{k}{n}}^{x} |(f''(t) - f(t)) - (f''(x) - f(x))| \left|\sinh\left(\frac{k}{n} - t\right)\right| dt \\
&\leq M\omega_1\left(f'' - f, x - \frac{k}{n}\right) \int_{\frac{k}{n}}^{x} \left(t - \frac{k}{n}\right) dt \\
&= M\omega_1\left(f'' - f, x - \frac{k}{n}\right) \frac{\left(x - \frac{k}{n}\right)^2}{2} \\
&\leq \frac{M\omega_1\left(f'' - f, \frac{1}{n^\alpha}\right)}{2n^{2\alpha}},
\end{aligned} \tag{4.84}
$$

that is,

$$
|R_2(n)| \leq \frac{M\omega_1\left(f'' - f, \frac{1}{n^\alpha}\right)}{2n^{2\alpha}}. \tag{4.85}
$$

So, we have proved when $\left|\frac{k}{n} - x\right| \leq \frac{1}{n^\alpha}$, always it holds

$$
|R_2(n)| \leq \frac{M\omega_1\left(f'' - f, \frac{1}{n^\alpha}\right)}{2n^{2\alpha}}. \tag{4.86}
$$

Next assume again $\frac{k}{n} \geq x$, then

$$
\begin{aligned}
|R_2(n)| &= \left| \int_x^{\frac{k}{n}} [(f''(t) - f(t)) - (f''(x) - f(x))] \sinh\left(\frac{k}{n} - t\right) dt \right| \\
&\leq \int_x^{\frac{k}{n}} |(f''(t) - f(t)) - (f''(x) - f(x))| \left|\sinh\left(\frac{k}{n} - t\right)\right| dt \\
&\leq 2M \|f'' - f\|_\infty \int_x^{\frac{k}{n}} \left(\frac{k}{n} - t\right) dt \\
&= 2M \|f'' - f\|_\infty \frac{\left(\frac{k}{n} - x\right)^2}{2} \leq M \|f'' - f\|_\infty (b - a)^2.
\end{aligned} \tag{4.87}
$$

Hence,

$$|R_2(n)| \leq M \|f'' - f\|_\infty (b-a)^2. \tag{4.88}$$

When $\frac{k}{n} < x$, we have

$$\begin{aligned} |R_2(n)| &= \left| \int_{\frac{k}{n}}^{x} [(f''(t) - f(t)) - (f''(x) - f(x))] \sinh\left(\frac{k}{n} - t\right) dt \right| \\ &\leq \int_{\frac{k}{n}}^{x} |(f''(t) - f(t)) - (f''(x) - f(x))| \left|\sinh\left(\frac{k}{n} - t\right)\right| dt \\ &\leq 2M \|f'' - f\|_\infty \int_{\frac{k}{n}}^{x} \left(t - \frac{k}{n}\right) dt \\ &= 2M \|f'' - f\|_\infty \frac{\left(x - \frac{k}{n}\right)^2}{2} \leq M \|f'' - f\|_\infty (b-a)^2. \end{aligned} \tag{4.89}$$

Therefore, always it holds

$$|R_2(n)| \leq M \|f'' - f\|_\infty (b-a)^2. \tag{4.90}$$

And we have

$$\Lambda_n(x) = \sum_{\left\{\substack{k=\lceil na \rceil \\ :\left|\frac{k}{n} - x\right| \leq \frac{1}{n^\alpha}}\right.}^{\lfloor nb \rfloor} M_{q,\lambda}(nx-k) R_2(n) + \sum_{\left\{\substack{k=\lceil na \rceil \\ :\left|\frac{k}{n} - x\right| > \frac{1}{n^\alpha}}\right.}^{\lfloor nb \rfloor} M_{q,\lambda}(nx-k) R_2(n). \tag{4.91}$$

Hence it holds

$$|\Lambda_n(x)| \leq \sum_{\left\{\substack{k=\lceil na \rceil \\ :\left|\frac{k}{n} - x\right| \leq \frac{1}{n^\alpha}}\right.}^{\lfloor nb \rfloor} M_{q,\lambda}(nx-k) |R_2(n)| \tag{4.92}$$

$$+ \sum_{\left\{\substack{k=\lceil na \rceil \\ :\left|\frac{k}{n} - x\right| > \frac{1}{n^\alpha}}\right.}^{\lfloor nb \rfloor} M_{q,\lambda}(nx-k) |R_2(n)|$$

$$\leq \left(\sum_{\left\{\substack{k=\lceil na \rceil \\ :\left|\frac{k}{n} - x\right| \leq \frac{1}{n^\alpha}}\right.}^{\lfloor nb \rfloor} M_{q,\lambda}(nx-k) \right) \frac{\omega_1\left(f'' - f, \frac{1}{n^\alpha}\right) M}{2n^{2\alpha}}$$

$$+\left(\sum_{\substack{k=\lceil na\rceil \\ \left\{:\left|\frac{k}{n}-x\right|>\frac{1}{n^{\alpha}}\right.}}^{\lfloor nb\rfloor} M_{q,\lambda}(nx-k)\right) M\|f''-f\|_{\infty}(b-a)^2$$

$$\overset{((4.28))}{\leq} \frac{M\omega_1\left(f''-f,\frac{1}{n^{\alpha}}\right)}{2n^{2\alpha}} + M\|f''-f\|_{\infty}(b-a)^2$$

$$\left(\sum_{\substack{k=\lceil na\rceil \\ \left\{:\left|\frac{k}{n}-x\right|>\frac{1}{n^{\alpha}}\right.}}^{\lfloor nb\rfloor} M_{q,\lambda}(nx-k)\right)$$

$$\overset{\text{(by Theorem 4.3)}}{\leq} M\frac{\omega_1\left(f''-f,\frac{1}{n^{\alpha}}\right)}{2n^{2\alpha}} + M\|f''-f\|_{\infty}(b-a)^2 Te^{-2\lambda n^{(1-\alpha)}}. \tag{4.93}$$

Consequently, we have derived that

$$|\Lambda_n(x)| \leq M\left[\frac{\omega_1\left(f''-f,\frac{1}{n^{\alpha}}\right)}{2n^{2\alpha}} + \|f''-f\|_{\infty}(b-a)^2 Te^{-2\lambda n^{(1-\alpha)}}\right]. \tag{4.94}$$

We have that

$$H_n^*(\sinh(\cdot - x), x) = \sum_{k=\lceil na\rceil}^{\lfloor nb\rfloor} M_{q,\lambda}(nx-k)\sinh\left(\frac{k}{n}-x\right), \tag{4.95}$$

and

$$|H_n^*(\sinh(\cdot - x), x)| \leq \sum_{k=\lceil na\rceil}^{\lfloor nb\rfloor} M_{q,\lambda}(nx-k)\left|\sinh\left(\frac{k}{n}-x\right)\right|$$

$$= \sum_{\substack{k=\lceil na\rceil \\ \left\{:\left|\frac{k}{n}-x\right|\leq\frac{1}{n^{\alpha}}\right.}}^{\lfloor nb\rfloor} M_{q,\lambda}(nx-k)\left|\sinh\left(\frac{k}{n}-x\right)\right|$$

$$+ \sum_{\substack{k=\lceil na\rceil \\ \left\{:\left|\frac{k}{n}-x\right|>\frac{1}{n^{\alpha}}\right.}}^{\lfloor nb\rfloor} M_{q,\lambda}(nx-k)\left|\sinh\left(\frac{k}{n}-x\right)\right| \tag{4.96}$$

$$\leq M \left[\sum_{\substack{k=\lceil na \rceil \\ \left\{: \left| \frac{k}{n} - x \right| \leq \frac{1}{n^{\alpha}} \right.}}^{\lfloor nb \rfloor} M_{q,\lambda} (nx - k) \left| \frac{k}{n} - x \right| \right.$$

$$\left. + \sum_{\substack{k=\lceil na \rceil \\ \left\{: \left| \frac{k}{n} - x \right| > \frac{1}{n^{\alpha}} \right.}}^{\lfloor nb \rfloor} M_{q,\lambda} (nx - k) \left| \frac{k}{n} - x \right| \right]$$

$$\leq M \left[\frac{1}{n^{\alpha}} + (b-a) \left(\sum_{\substack{k=\lceil na \rceil \\ \left\{: \left| \frac{k}{n} - x \right| > \frac{1}{n^{\alpha}} \right.}}^{\lfloor nb \rfloor} M_{q,\lambda} (nx - k) \right) \right]$$

$$\overset{\text{(by (4.25))}}{\leq} M \left[\frac{1}{n^{\alpha}} + (b-a) T e^{-2\lambda n^{(1-\alpha)}} \right]. \tag{4.97}$$

We found that

$$\left| H_n^* (\sinh (\cdot - x), x) \right| \leq M \left[\frac{1}{n^{\alpha}} + (b-a) T e^{-2\lambda n^{(1-\alpha)}} \right]. \tag{4.98}$$

Next we estimate

$$H_n^* \left(\sinh^2 \left(\frac{\cdot - x}{2} \right), x \right) = \sum_{k=\lceil na \rceil}^{\lfloor nb \rfloor} M_{q,\lambda} (nx - k) \sinh^2 \left(\frac{\frac{k}{n} - x}{2} \right). \tag{4.99}$$

We have that

$$H_n^* \left(\sinh^2 \left(\frac{\cdot - x}{2} \right), x \right) = \sum_{k=\lceil na \rceil}^{\lfloor nb \rfloor} M_{q,\lambda} (nx - k) \left(\sinh \left(\frac{\frac{k}{n} - x}{2} \right) \right)^2$$

$$\leq \frac{M}{4} \sum_{k=\lceil na \rceil}^{\lfloor nb \rfloor} M_{q,\lambda} (nx - k) \left(\frac{k}{n} - x \right)^2$$

$$= \frac{M}{4}\left[\sum_{\substack{k=\lceil na\rceil \\ \left\{:\left|\frac{k}{n}-x\right|\le\frac{1}{n^\alpha}\right.}}^{\lfloor nb\rfloor} M_{q,\lambda}(nx-k)\left(\frac{k}{n}-x\right)^2\right.$$

$$\left.+\sum_{\substack{k=\lceil na\rceil \\ \left\{:\left|\frac{k}{n}-x\right|>\frac{1}{n^\alpha}\right.}}^{\lfloor nb\rfloor} M_{q,\lambda}(nx-k)\left(\frac{k}{n}-x\right)^2\right]$$

$$\le \frac{M}{4}\left[\frac{1}{n^{2\alpha}}+(b-a)^2\,Te^{-2\lambda n^{(1-\alpha)}}\right]. \quad (4.100)$$

That is,

$$H_n^*\left(\sinh^2\left(\frac{\cdot - x}{2}\right), x\right) \le \frac{M}{4}\left[\frac{1}{n^{2\alpha}}+(b-a)^2\,Te^{-2\lambda n^{(1-\alpha)}}\right]. \quad (4.101)$$

By (4.33) and putting together (4.79), (4.94), (4.98) and (4.101), we derive the following:

(1)

$$|H_n(f,x)-f(x)| \le \Delta(q)\,M\left[|f'(x)|\left(\frac{1}{n^\alpha}+(b-a)\,Te^{-2\lambda n^{(1-\alpha)}}\right)\right.$$

$$+\frac{|f''(x)|}{2}\left(\frac{1}{n^{2\alpha}}+(b-a)^2\,Te^{-2\lambda n^{(1-\alpha)}}\right)$$

$$\left.+\left(\frac{\omega_1\left(f''-f,\frac{1}{n^\alpha}\right)}{2n^{2\alpha}}+\|f''-f\|_\infty (b-a)^2\,Te^{-2\lambda n^{(1-\alpha)}}\right)\right]. \quad (4.102)$$

(2) If $f'(x) = f''(x) = 0$, by (4.102), we obtain

$$|H_n(f,x)-f(x)|$$

$$\le M\Delta(q)\left[\frac{\omega_1\left(f''-f,\frac{1}{n^\alpha}\right)}{2n^{2\alpha}}+\|f''-f\|_\infty (b-a)^2\,Te^{-2\lambda n^{(1-\alpha)}}\right], \quad (4.103)$$

note here the high rate of convergence at $n^{-3\alpha}$.

(3) Furthermore, by (4.102), we get

$$\|H_n f - f\|_\infty \leq \Delta(q) M \left[\|f'\|_\infty \left(\frac{1}{n^\alpha} + (b-a) T e^{-2\lambda n^{(1-\alpha)}} \right) \right.$$
$$+ \frac{\|f''\|_\infty}{2} \left(\frac{1}{n^{2\alpha}} + (b-a)^2 T e^{-2\lambda n^{(1-\alpha)}} \right)$$
$$\left. + \left(\frac{\omega_1 \left(f'' - f, \frac{1}{n^\alpha} \right)}{2n^{2\alpha}} + \|f'' - f\|_\infty (b-a)^2 T e^{-2\lambda n^{(1-\alpha)}} \right) \right]. \quad (4.104)$$

It follows that $\lim_{n \to +\infty} H_n(f) = f$, pointwise and uniformly.

We observe that

$$H_n(f,x) - f'(x) H_n(\sinh(\cdot - x), x) - 2f''(x) H_n\left(\sinh^2\left(\frac{\cdot - x}{2} \right), x \right) - f(x)$$
$$= \frac{H_n^*(f,x)}{\sum_{k=\lceil na \rceil}^{\lfloor nb \rfloor} M_{q,\lambda}(nx-k)} - f'(x) \frac{H_n^*(\sinh(\cdot - x), x)}{\sum_{k=\lceil na \rceil}^{\lfloor nb \rfloor} M_{q,\lambda}(nx-k)}$$
$$- 2f''(x) \frac{H_n^*\left(\sinh^2\left(\frac{\cdot - x}{2} \right), x \right)}{\sum_{k=\lceil na \rceil}^{\lfloor nb \rfloor} M_{q,\lambda}(nx-k)} - f(x) \left(\frac{\sum_{k=\lceil na \rceil}^{\lfloor nb \rfloor} M_{q,\lambda}(nx-k)}{\sum_{k=\lceil na \rceil}^{\lfloor nb \rfloor} M_{q,\lambda}(nx-k)} \right) \quad (4.105)$$

$$= \frac{1}{\sum_{k=\lceil na \rceil}^{\lfloor nb \rfloor} M_{q,\lambda}(nx-k)} \left[H_n^*(f,x) - f'(x) H_n^*(\sinh(\cdot - x), x) \right.$$
$$\left. - 2f''(x) H_n^*\left(\sinh^2\left(\frac{\cdot - x}{2} \right), x \right) - f(x) \sum_{k=\lceil na \rceil}^{\lfloor nb \rfloor} M_{q,\lambda}(nx-k) \right]$$
$$= \frac{1}{\sum_{k=\lceil na \rceil}^{\lfloor nb \rfloor} M_{q,\lambda}(nx-k)} (\Lambda_n(x)). \quad (4.106)$$

Finally, we obtain ($\forall\, x \in [a,b]$, $n \in \mathbb{N}$):

(4)

$$\left| H_n(f,x) - f'(x) H_n(\sinh(\cdot - x), x) - 2f''(x) H_n\left(\sinh^2\left(\frac{\cdot - x}{2} \right), x \right) - f(x) \right|$$
$$\overset{(4.26)}{\leq} \Delta(q) |\Lambda_n(x)|$$
$$\overset{(4.94)}{\leq} \Delta(q) M \left[\frac{\omega_1\left(f'' - f, \frac{1}{n^\alpha} \right)}{2n^{2\alpha}} + \|f'' - f\|_\infty (b-a)^2 T e^{-2\lambda n^{(1-\alpha)}} \right]. \quad (4.107)$$

The theorem is established. □

Next follows a mixed hyperbolic-trigonometric high order neural network approximation.

Theorem 4.7. *Let* $f \in C^4([a,b],\mathbb{C})$, $0 < \alpha < 1$, $n \in \mathbb{N} : n^{1-\alpha} > 2$, $x \in [a,b]$. *Then*

(1)

$$\left| H_n(f,x) - f(x) - \frac{f'(x)}{2} H_n((\sinh(\cdot - x) + \sin(\cdot - x)), x) \right.$$

$$- \frac{f''(x)}{2} H_n((\cosh(\cdot - x) - \cos(\cdot - x)), x) \tag{4.108}$$

$$- \frac{f^{(3)}(x)}{2} H_n((\sinh(\cdot - x) - \sin(\cdot - x)), x)$$

$$\left. - f^{(4)}(x) H_n\left(\left(\sinh^2\left(\frac{\cdot - x}{2}\right) - \sin^2\left(\frac{\cdot - x}{2}\right)\right), x\right) \right|$$

$$\leq \frac{\Delta(q)(\cosh(b-a)+1)}{2}$$

$$\left[\frac{\omega_1\left(f^{(4)} - f, \frac{1}{n^\alpha}\right)}{2n^{2\alpha}} + \left\| f^{(4)} - f \right\|_\infty (b-a)^2 T e^{-2\lambda n^{(1-\alpha)}} \right], \tag{4.109}$$

(2) *if* $f^{(i)}(x) = 0$, $i = 1,2,3,4$, *we get*

$$|H_n(f,x) - f(x)| \leq \frac{\Delta(q)(\cosh(b-a)+1)}{2}$$

$$\left[\frac{\omega_1\left(f^{(4)} - f, \frac{1}{n^\alpha}\right)}{2n^{2\alpha}} + \left\| f^{(4)} - f \right\|_\infty (b-a)^2 T e^{-2\lambda n^{(1-\alpha)}} \right], \tag{4.110}$$

in the last (4.110) observe the high speed of convergence at $n^{-3\alpha}$.

Proof. Here $f \in C^4([a,b],\mathbb{C})$, and we apply the hyperbolic-trigonometric Taylor's formula for $f \in C^4([a,b],\mathbb{C})$, see Theorem 8 of Anastassiou (2023c).

Let $\frac{k}{n}, x \in [a,b]$, then

$$f\left(\frac{k}{n}\right) - f(x) - f'(x)\left(\frac{\sinh\left(\frac{k}{n}-x\right)+\sin\left(\frac{k}{n}-x\right)}{2}\right)$$

$$- f''(x)\left(\frac{\cosh\left(\frac{k}{n}-x\right)-\cos\left(\frac{k}{n}-x\right)}{2}\right)$$

$$- f^{(3)}(x)\left(\frac{\sinh\left(\frac{k}{n}-x\right)-\sin\left(\frac{k}{n}-x\right)}{2}\right)$$

$$- f^{(4)}(x)\left(\sinh^2\left(\frac{\frac{k}{n}-x}{2}\right) - \sin^2\left(\frac{\frac{k}{n}-x}{2}\right)\right)$$

$$= \int_x^{\frac{k}{n}} \left[\left(f^{(4)}(t) - f(t)\right) - \left(f^{(4)}(x) - f(x)\right)\right]$$

$$\times \left(\frac{\sinh\left(\frac{k}{n}-t\right)-\sin\left(\frac{k}{n}-t\right)}{2}\right) dt$$

$$=: R_4\left(\frac{k}{n}, x\right). \tag{4.111}$$

As in Theorems 4.5 and 4.6, we derive

$$H_n^*(f,x) - f(x) \sum_{k=\lceil na \rceil}^{\lfloor nb \rfloor} M_{q,\lambda}(nx-k)$$

$$- \frac{f'(x)}{2} H_n^*\left(\left(\sinh(\cdot - x) + \sin(\cdot - x)\right), x\right)$$

$$- \frac{f''(x)}{2} H_n^*\left(\left(\cosh(\cdot - x) - \cos(\cdot - x)\right), x\right)$$

$$- \frac{f^{(3)}(x)}{2} H_n^*\left(\left(\sinh(\cdot - x) - \sin(\cdot - x)\right), x\right)$$

$$- \frac{f^{(4)}(x)}{2} H_n^*\left(\left(\sinh^2\left(\frac{\cdot - x}{2}\right) - \sin^2\left(\frac{\cdot - x}{2}\right)\right), x\right) = \Phi_n(x), \tag{4.112}$$

where

$$\Phi_n(x) := \sum_{k=\lceil na \rceil}^{\lfloor nb \rfloor} M_{q,\lambda}(nx-k)\, R_4\left(\frac{k}{n}, x\right). \tag{4.113}$$

Without loss of generality, we can assume that $n > \left\lceil (b-a)^{-\frac{1}{\alpha}} \right\rceil$.

Thus, $\left|\frac{k}{n} - x\right| \leq \frac{1}{n^\alpha}$ or $\left|\frac{k}{n} - x\right| > \frac{1}{n^\alpha}$.

In case of $\left|\frac{k}{n} - x\right| \leq \frac{1}{n^\alpha}$, we have the following cases:

(i) If $\frac{k}{n} \geq x$, then

$$\begin{aligned}
&\left|R_4\left(\frac{k}{n}, x\right)\right| \\
&\quad = \left|\frac{1}{2}\int_x^{\frac{k}{n}} \left[\left(f^{(4)}(t) - f(t)\right) - \left(f^{(4)}(x) - f(x)\right)\right]\right. \\
&\qquad \left. \times \left(\sinh\left(\frac{k}{n} - t\right) - \sin\left(\frac{k}{n} - t\right)\right) dt\right| \\
&\quad \leq \frac{1}{2}\int_x^{\frac{k}{n}} \left|\left(f^{(4)}(t) - f(t)\right) - \left(f^{(4)}(x) - f(x)\right)\right| \\
&\qquad \times \left|\sinh\left(\frac{k}{n} - t\right) - \sin\left(\frac{k}{n} - t\right)\right| dt \qquad (4.114) \\
&\quad \leq \frac{1}{2}\int_x^{\frac{k}{n}} \omega_1\left(f^{(4)} - f, t - x\right)\left(\left|\sinh\left(\frac{k}{n} - t\right)\right| + \left|\sin\left(\frac{k}{n} - t\right)\right|\right) dt \\
&\quad \leq \frac{\omega_1\left(f^{(4)} - f, \frac{k}{n} - x\right)}{2}\int_x^{\frac{k}{n}} \left(\cosh(b-a)\left(\frac{k}{n} - t\right) + \left(\frac{k}{n} - t\right)\right) dt \\
&\quad = \frac{(\cosh(b-a)+1)\,\omega_1\left(f^{(4)} - f, \frac{k}{n} - x\right)}{2}\int_x^{\frac{k}{n}} \left(\frac{k}{n} - t\right) dt \\
&\quad = \frac{(\cosh(b-a)+1)\,\omega_1\left(f^{(4)} - f, \frac{k}{n} - x\right)}{4}\left(\frac{k}{n} - x\right)^2 \\
&\quad \leq \frac{(\cosh(b-a)+1)\,\omega_1\left(f^{(4)} - f, \frac{1}{n^\alpha}\right)}{4n^{2\alpha}}. \qquad (4.115)
\end{aligned}$$

That is, when $\frac{k}{n} \geq x$, then

$$\left|R_4\left(\frac{k}{n}, x\right)\right| \leq \frac{(\cosh(b-a)+1)\,\omega_1\left(f^{(4)} - f, \frac{1}{n^\alpha}\right)}{4n^{2\alpha}}. \tag{4.116}$$

(ii) If $\frac{k}{n} < x$, then

$$\left| R_4\left(\frac{k}{n}, x\right) \right|$$

$$= \left| \frac{1}{2} \int_{\frac{k}{n}}^{x} \left[\left(f^{(4)}(t) - f(t) \right) - \left(f^{(4)}(x) - f(x) \right) \right] \right.$$

$$\left. \times \left(\sinh\left(\frac{k}{n} - t\right) - \sin\left(\frac{k}{n} - t\right) \right) dt \right|$$

$$\leq \frac{\omega_1\left(f^{(4)} - f, x - \frac{k}{n}\right)}{2} \int_{\frac{k}{n}}^{x} \left(\cosh(b-a)\left(t - \frac{k}{n}\right) + \left(t - \frac{k}{n}\right) \right) dt$$

$$= \frac{(\cosh(b-a)+1)\,\omega_1\left(f^{(4)} - f, x - \frac{k}{n}\right)}{2} \int_{\frac{k}{n}}^{x} \left(t - \frac{k}{n}\right) dt$$

$$= \frac{(\cosh(b-a)+1)\,\omega_1\left(f^{(4)} - f, x - \frac{k}{n}\right)}{4} \left(x - \frac{k}{n}\right)^2$$

$$\leq \frac{(\cosh(b-a)+1)\,\omega_1\left(f^{(4)} - f, \frac{1}{n^\alpha}\right)}{4n^{2\alpha}}. \tag{4.117}$$

Consequently, when $\left|\frac{k}{n} - x\right| \leq \frac{1}{n^\alpha}$, we always obtain that

$$\left| R_4\left(\frac{k}{n}, x\right) \right| \leq \frac{(\cosh(b-a)+1)\,\omega_1\left(f^{(4)} - f, \frac{1}{n^\alpha}\right)}{4n^{2\alpha}}. \tag{4.118}$$

Next assume again $\frac{k}{n} \geq x$, then

$$\left| R_4\left(\frac{k}{n}, x\right) \right| \leq \frac{1}{2} \int_{x}^{\frac{k}{n}} \left| \left(f^{(4)}(t) - f(t) \right) - \left(f^{(4)}(x) - f(x) \right) \right|$$

$$\times \left| \sinh\left(\frac{k}{n} - t\right) - \sin\left(\frac{k}{n} - t\right) \right| dt$$

$$\leq \left\| f^{(4)} - f \right\|_\infty \int_{x}^{\frac{k}{n}} \left[\cosh(b-a)\left(\frac{k}{n} - t\right) + \left(\frac{k}{n} - t\right) \right] dt$$

$$= \left\| f^{(4)} - f \right\|_\infty (\cosh(b-a)+1) \int_{x}^{\frac{k}{n}} \left(\frac{k}{n} - t\right) dt$$

$$= \frac{\left\|f^{(4)} - f\right\|_\infty (\cosh(b-a)+1)}{2} \left(\frac{k}{n} - x\right)^2$$

$$\leq \frac{\left\|f^{(4)} - f\right\|_\infty (\cosh(b-a)+1)(b-a)^2}{2}. \tag{4.119}$$

Hence,

$$\left| R_4\left(\frac{k}{n}, x\right)\right| \leq \frac{\left\|f^{(4)} - f\right\|_\infty (\cosh(b-a)+1)(b-a)^2}{2}. \tag{4.120}$$

When $\frac{k}{n} < x$, we have

$$\left| R_4\left(\frac{k}{n}, x\right)\right| \leq \frac{1}{2} \int_{\frac{k}{n}}^{x} \left|\left(f^{(4)}(t) - f(t)\right) - \left(f^{(4)}(x) - f(x)\right)\right|$$

$$\times \left|\sinh\left(\frac{k}{n} - t\right) - \sin\left(\frac{k}{n} - t\right)\right| dt$$

$$\leq \left\|f^{(4)} - f\right\|_\infty \int_{\frac{k}{n}}^{x} \left[\cosh(b-a)\left(t - \frac{k}{n}\right) + \left(t - \frac{k}{n}\right)\right] dt$$

$$= \left\|f^{(4)} - f\right\|_\infty (\cosh(b-a)+1) \int_{\frac{k}{n}}^{x} \left(t - \frac{k}{n}\right) dt$$

$$= \frac{\left\|f^{(4)} - f\right\|_\infty (\cosh(b-a)+1)}{2} \left(x - \frac{k}{n}\right)^2$$

$$\leq \frac{\left\|f^{(4)} - f\right\|_\infty (\cosh(b-a)+1)(b-a)^2}{2}. \tag{4.121}$$

So, it is always true that

$$\left| R_4\left(\frac{k}{n}, x\right)\right| \leq \frac{\left\|f^{(4)} - f\right\|_\infty (\cosh(b-a)+1)(b-a)^2}{2}. \tag{4.122}$$

Thus,

$$|\Phi_n(x)| \leq \sum_{k=\lceil na \rceil}^{\lfloor nb \rfloor} M_{q,\lambda}(nx-k) \left| R_4\left(\frac{k}{n}, x\right)\right|$$

$$= \sum_{\left\{\substack{k=\lceil na \rceil \\ :\left|\frac{k}{n} - x\right| \leq \frac{1}{n^\alpha}}\right.}^{\lfloor nb \rfloor} M_{q,\lambda}(nx-k) \left| R_4\left(\frac{k}{n}, x\right)\right| \tag{4.123}$$

$$+\sum_{\left\{\substack{k=\lceil na\rceil \\ :\left|\frac{k}{n}-x\right|>\frac{1}{n^{\alpha}}}\right.}^{\lfloor nb\rfloor} M_{q,\lambda}(nx-k)\left|R_4\left(\frac{k}{n},x\right)\right|$$

$$\leq\left(\sum_{\left\{\substack{k=\lceil na\rceil \\ :\left|\frac{k}{n}-x\right|\leq\frac{1}{n^{\alpha}}}\right.}^{\lfloor nb\rfloor} M_{q,\lambda}(nx-k)\right)\frac{(\cosh(b-a)+1)\,\omega_1\left(f^{(4)}-f,\frac{1}{n^{\alpha}}\right)}{4n^{2\alpha}} \tag{4.124}$$

$$+\left(\sum_{\left\{\substack{k=\lceil na\rceil \\ :\left|\frac{k}{n}-x\right|>\frac{1}{n^{\alpha}}}\right.}^{\lfloor nb\rfloor} M_{q,\lambda}(nx-k)\right)$$

$$\left(\frac{\left\|f^{(4)}-f\right\|_{\infty}(\cosh(b-a)+1)(b-a)^2}{2}\right)$$

$$\overset{\text{(by Theorem 4.3)}}{\leq}\frac{(\cosh(b-a)+1)\,\omega_1\left(f^{(4)}-f,\frac{1}{n^{\alpha}}\right)}{4n^{2\alpha}}$$

$$+\frac{\left\|f^{(4)}-f\right\|_{\infty}(\cosh(b-a)+1)(b-a)^2}{2}Te^{-2\lambda n^{(1-\alpha)}}. \tag{4.125}$$

We have proved that

$$|\Phi_n(x)|\leq\frac{(\cosh(b-a)+1)}{2}$$

$$\left[\frac{\omega_1\left(f^{(4)}-f,\frac{1}{n^{\alpha}}\right)}{2n^{2\alpha}}+\left\|f^{(4)}-f\right\|_{\infty}(b-a)^2\,Te^{-2\lambda n^{(1-\alpha)}}\right]. \tag{4.126}$$

We observe that

$$H_n(f,x)-f(x)-\frac{f'(x)}{2}H_n((\sinh(\cdot-x)+\sin(\cdot-x)),x)$$

$$-\frac{f''(x)}{2}H_n((\cosh(\cdot-x)-\cos(\cdot-x)),x) \tag{4.127}$$

$$-\frac{f^{(3)}(x)}{2}H_n((\sinh(\cdot-x)-\sin(\cdot-x)),x)$$

$$-f^{(4)}(x)\,H_n\left(\left(\sinh^2\left(\frac{\cdot-x}{2}\right)-\sin^2\left(\frac{\cdot-x}{2}\right)\right),x\right)$$

$$= \Bigg[H_n^* (f, x) - f(x) \sum_{k=\lceil na \rceil}^{\lfloor nb \rfloor} M_{q,\lambda} (nx - k)$$

$$- \frac{f'(x)}{2} H_n^* ((\sinh(\cdot - x) + \sin(\cdot - x)), x)$$

$$- \frac{f''(x)}{2} H_n^* ((\cosh(\cdot - x) - \cos(\cdot - x)), x)$$

$$- \frac{f^{(3)}(x)}{2} H_n^* ((\sinh(\cdot - x) - \sin(\cdot - x)), x)$$

$$- f^{(4)}(x) H_n^* \left(\left(\sinh^2 \left(\frac{\cdot - x}{2} \right) - \sin^2 \left(\frac{\cdot - x}{2} \right) \right), x \right) \Bigg]$$

$$\times \frac{1}{\sum_{k=\lceil na \rceil}^{\lfloor nb \rfloor} M_{q,\lambda} (nx - k)}$$

$$= \frac{\Phi_n (x)}{\sum_{k=\lceil na \rceil}^{\lfloor nb \rfloor} M_{q,\lambda} (nx - k)}. \tag{4.128}$$

Finally, we obtain ($\forall\ x \in [a, b]$, $n \in \mathbb{N}$):

$$\Bigg| H_n (f, x) - f(x) - \frac{f'(x)}{2} H_n ((\sinh(\cdot - x) + \sin(\cdot - x)), x)$$

$$- \frac{f''(x)}{2} H_n ((\cosh(\cdot - x) - \cos(\cdot - x)), x) \tag{4.129}$$

$$- \frac{f^{(3)}(x)}{2} H_n ((\sinh(\cdot - x) - \sin(\cdot - x)), x)$$

$$- f^{(4)}(x) H_n \left(\left(\sinh^2 \left(\frac{\cdot - x}{2} \right) - \sin^2 \left(\frac{\cdot - x}{2} \right) \right), x \right) \Bigg|$$

$$= \frac{|\Phi_n (x)|}{\sum_{k=\lceil na \rceil}^{\lfloor nb \rfloor} M_{q,\lambda} (nx - k)} \leq \Delta(q) |\Phi_n (x)|$$

$$\overset{(\text{by } (4.126))}{\leq} \frac{\Delta(q) (\cosh(b - a) + 1)}{2}$$

$$\left[\frac{\omega_1 \left(f^{(4)} - f, \frac{1}{n^\alpha} \right)}{2n^{2\alpha}} + \left\| f^{(4)} - f \right\|_\infty (b - a)^2 T e^{-2\lambda n^{(1-\alpha)}} \right]. \tag{4.130}$$

The theorem is proved. □

We continue with a general trigonometric result.

Theorem 4.8. *Let* $f \in C^4([a,b],\mathbb{C})$, $0 < \alpha < 1$, $n \in \mathbb{N} : n^{1-\alpha} > 2$, $x \in [a,b]$. *Let also* $\overline{\alpha}, \overline{\beta} \in \mathbb{R}$ *with* $\overline{\alpha}\overline{\beta}\left(\overline{\alpha}^2 - \overline{\beta}^2\right) \neq 0$. *Then*

(1)

$$\Bigg| H_n(f,x) - f(x) - \frac{f'(x)}{\overline{\alpha}\overline{\beta}\left(\overline{\beta}^2 - \overline{\alpha}^2\right)}$$

$$\times H_n\left(\left(\overline{\beta}^3 \sin\left(\overline{\alpha}(\cdot - x)\right) - \overline{\alpha}^3 \sin\left(\overline{\beta}(\cdot - x)\right)\right), x\right)$$

$$- \frac{f''(x)}{\left(\overline{\beta}^2 - \overline{\alpha}^2\right)} H_n\left(\left(\cos\left(\overline{\alpha}(\cdot - x)\right) - \cos\left(\overline{\beta}(\cdot - x)\right)\right), x\right)$$

$$- \frac{f'''(x)}{\overline{\alpha}\overline{\beta}\left(\overline{\beta}^2 - \overline{\alpha}^2\right)} H_n\left(\left(\overline{\beta}\sin\left(\overline{\alpha}(\cdot - x)\right) - \overline{\alpha}\sin\left(\overline{\beta}(\cdot - x)\right)\right), x\right)$$

$$- \left(\frac{2f^{(4)}(x) + \left(\overline{\alpha}^2 + \overline{\beta}^2\right) f''(x)}{\left(\overline{\alpha}\overline{\beta}\right)^2 \left(\overline{\beta}^2 - \overline{\alpha}^2\right)}\right)$$

$$\times H_n\left(\left(\overline{\beta}^2 \sin^2\left(\frac{\overline{\alpha}(\cdot - x)}{2}\right) - \overline{\alpha}^2 \sin^2\left(\frac{\overline{\beta}(\cdot - x)}{2}\right)\right), x\right) \Bigg|$$

$$\leq \frac{\Delta(q)}{\left|\overline{\beta}^2 - \overline{\alpha}^2\right|} \left[\frac{\omega_1\left(\left(f^{(4)} + \left(\overline{\alpha}^2 + \overline{\beta}^2\right) f'' + \overline{\alpha}^2\overline{\beta}^2 f\right), \frac{1}{n^\alpha}\right)}{n^{2\alpha}}\right.$$

$$\left. + 2\left\| f^{(4)} + \left(\overline{\alpha}^2 + \overline{\beta}^2\right) f'' + \overline{\alpha}^2\overline{\beta}^2 f \right\|_\infty (b-a)^2 T e^{-2\lambda n^{(1-\alpha)}} \right], \tag{4.131}$$

(2) *if* $f^{(i)}(x) = 0$, $i = 1,2,3,4$, *we get*

$$|H_n(f,x) - f(x)| \leq \frac{\Delta(q)}{\left|\overline{\beta}^2 - \overline{\alpha}^2\right|}$$

$$\left[\frac{\omega_1\left(\left(f^{(4)} + \left(\overline{\alpha}^2 + \overline{\beta}^2\right) f'' + \overline{\alpha}^2\overline{\beta}^2 f\right), \frac{1}{n^\alpha}\right)}{n^{2\alpha}}\right.$$

$$\left. + 2\left\| f^{(4)} + \left(\overline{\alpha}^2 + \overline{\beta}^2\right) f'' + \overline{\alpha}^2\overline{\beta}^2 f \right\|_\infty (b-a)^2 T e^{-2\lambda n^{(1-\alpha)}} \right]. \tag{4.132}$$

The high speed of convergence in (1) and (2) is $n^{-3\alpha}$.

Proof. As similar to Theorem 4.7 is omitted. It is based on Theorem 9 of Anastassiou (2023c). □

We finish with a general hyperbolic result.

Theorem 4.9. *Let* $f \in C^4([a,b],\mathbb{C})$, $0<\alpha<1$, $n \in \mathbb{N} : n^{1-\alpha} > 2$, $x \in [a,b]$. *Let also* $\overline{\alpha}, \overline{\beta} \in \mathbb{R}$ *with* $\overline{\alpha}\overline{\beta}\left(\overline{\alpha}^2 - \overline{\beta}^2\right) \neq 0$. *Then*

(1)

$$\Bigg| H_n(f,x) - f(x) - \frac{f'(x)}{\overline{\alpha}\overline{\beta}\left(\overline{\beta}^2 - \overline{\alpha}^2\right)} \times H_n\left(\left(\overline{\beta}^3 \sinh(\overline{\alpha}(\cdot - x)) - \overline{\alpha}^3 \sinh(\overline{\beta}(\cdot - x))\right), x\right)$$
$$- \frac{f''(x)}{\overline{\beta}^2 - \overline{\alpha}^2} H_n\left(\left(\cosh(\overline{\beta}(\cdot - x)) - \cosh(\overline{\alpha}(\cdot - x))\right), x\right)$$
$$- \frac{f'''(x)}{\overline{\alpha}\overline{\beta}\left(\overline{\beta}^2 - \overline{\alpha}^2\right)} H_n\left(\left(\overline{\alpha}\sinh(\overline{\beta}(\cdot - x)) - \overline{\beta}\sinh(\overline{\alpha}(\cdot - x))\right), x\right)$$
$$- \left(\frac{2(f^{(4)}(x) - \left(\overline{\alpha}^2 + \overline{\beta}^2\right) f''(x))}{\left(\overline{\alpha}\overline{\beta}\right)^2 \left(\overline{\beta}^2 - \overline{\alpha}^2\right)}\right) \times H_n\left(\left(\overline{\alpha}^2 \sinh^2\left(\frac{\overline{\beta}(\cdot - x)}{2}\right) - \overline{\beta}^2 \sinh^2\left(\frac{\overline{\alpha}(\cdot - x)}{2}\right)\right), x\right)\Bigg|$$
$$\leq \frac{\Delta(q)\cosh(b-a)}{\left|\overline{\beta}^2 - \overline{\alpha}^2\right|}\left[\frac{\omega_1\left(\left(f^{(4)} - \left(\overline{\alpha}^2 + \overline{\beta}^2\right) f'' + \overline{\alpha}^2\overline{\beta}^2 f\right), \frac{1}{n^\alpha}\right)}{n^{2\alpha}} + 2\left\|f^{(4)} - \left(\overline{\alpha}^2 + \overline{\beta}^2\right) f'' + \overline{\alpha}^2\overline{\beta}^2 f\right\|_\infty (b-a)^2 T e^{-2\lambda n^{(1-\alpha)}}\right], \tag{4.133}$$

(2) *if* $f^{(i)}(x) = 0$, $i = 1,2,3,4$, *we get*

$$|H_n(f,x) - f(x)| \leq \frac{\Delta(q)\cosh(b-a)}{\left|\overline{\beta}^2 - \overline{\alpha}^2\right|} \left[\frac{\omega_1\left(\left(f^{(4)} - \left(\overline{\alpha}^2 + \overline{\beta}^2\right) f'' + \overline{\alpha}^2\overline{\beta}^2 f\right), \frac{1}{n^\alpha}\right)}{n^{2\alpha}} + 2\left\|f^{(4)} - \left(\overline{\alpha}^2 + \overline{\beta}^2\right) f'' + \overline{\alpha}^2\overline{\beta}^2 f\right\|_\infty (b-a)^2 T e^{-2\lambda n^{(1-\alpha)}}\right]. \tag{4.134}$$

The high speed of convergence in (1) and (2) is $n^{-3\alpha}$.

Proof. As similar to Theorem 4.7 is omitted. It is based on Theorem 10 of Anastassiou (2023c). □

Bibliography

G.A. Anastassiou, Rate of convergence of some neural network operators to the unit-univariate case, *Journal of Mathematical Analysis and Applications,* 212 (1997), 237–262.

G.A. Anastassiou, *Quantitative Approximations*, Chapman & Hall/CRC, Boca Raton, New York, 2001.

G.A. Anastassiou, Univariate hyperbolic tangent neural network approximation, *Mathematics and Computer Modelling*, 53 (2011a), 1111–1132.

G.A. Anastassiou, Multivariate hyperbolic tangent neural network approximation, *Computers and Mathematics*, 61 (2011b), 809–821.

G.A. Anastassiou, Multivariate sigmoidal neural network approximation, *Neural Networks*, 24 (2011c), 378–386.

G.A. Anastassiou, *Inteligent Systems: Approximation by Artificial Neural Networks*, Intelligent Systems Reference Library, Vol. 19, Springer, Heidelberg, 2011d.

G.A. Anastassiou, Univariate sigmoidal neural network approximation, *Journal of Computational Analysis and Applications*, 14(4) 2012, 659–690.

G.A. Anastassiou, Vector fractional Korovkin type Approximations, D*ynamic Systems and Applications*, 26 (2017), 81–104.

G.A. Anastassiou, General sigmoid based Banach space valued neural network approximation, *Journal of Computational Analysis and Applications*, 31(4) (2023a), 520–534.

G.A. Anastassiou, *Parametrized, deformed and general neural networks*, accepted for publication, Springer, Heidelberg, New York, 2023b.

G.A. Anastassiou, Opial and Ostrowski type inequalities based on trigonometric and hyperbolic type Taylor formulae, *Malaya Journal of Matematik*, 11(S) (2023c), 1–26.

G.A. Anastassiou, *q-Deformed and λ-parametrized hyperbolic tangent function relied complex valued trigonometric and hyperbolic neural network high order approximations*, submitted, 2023d.

Z. Chen and F. Cao, The approximation operators with sigmoidal functions, *Computers and Mathematics with Applications*, 58 (2009), 758–765.

S. Haykin, *Neural Networks: A Comprehensive Foundation* (2 ed.), Prentice Hall, New York, 1998.

W. McCulloch and W. Pitts, A logical calculus of the ideas immanent in nervous activity, *Bulletin of Mathematical Biophysis*, 7 (1943), 115–133.

T.M. Mitchell, *Machine Learning*, WCB-McGraw-Hill, New York, 1997.

Chapter 5

Perturbed Hyperbolic Tangent Function-Activated Complex-Valued Multivariate Trigonometric and Hyperbolic Neural Network Approximation

Here we study the multivariate quantitative approximation of complex-valued continuous functions on a box of $\mathbb{R}^N$, $N \in \mathbb{N}$, by the multivariate normalized-type neural network operators. We investigate also the case of approximation by iterated multilayer neural network operators. These approximations are achieved by establishing multidimensional Jackson-type inequalities involving the multivariate moduli of continuity of the engaged function and its partial derivatives. Our multivariate operators are defined by using a multidimensional density function induced by a q-deformed and λ-parametrized hyperbolic tangent function, which is a sigmoid function. The approximations are pointwise and uniform. The related feed-forward neural networks are with one or multiple hidden layers. The basis of our theory is the introduced multivariate Taylor formulae of trigonometric and hyperbolic type. It follows Anastassiou (2023d).

5.1 Introduction

The author in Anastassiou (1997, 2001), see Chapters 2–5, was the first to establish neural network approximations to continuous functions with rates by very specifically defined neural network operators of Cardaliaguet–Euvrard and "Squashing" types, by employing the modulus of continuity of the engaged function or its high order derivative, and producing very tight Jackson-type inequalities. He treats there both the univariate and multivariate cases. The definitions of these operators' "bell-shaped" and "squashing" functions are assumed to be of compact support.

Motivations for this work include articles by Chen and Cao (2009), Anastassiou (2011a)–Anastassiou (2023b), and Costarelli and Spigler (2013a), Costarelli and Spigler (2013b).

Here we perform q-deformed and λ-parametrized, $q, \lambda > 0$, hyperbolic tangent sigmoid function-based trigonometric and hyperbolic neural network approximations to complex-valued continuous functions over boxes in $\mathbb{R}^N$, $N \in \mathbb{N}$, and also iterated, multilayer approximations. All convergences here are with rates expressed via the multivariate moduli of continuity of the involved function and its partial derivatives and given by very tight multidimensional Jackson-type inequalities.

We come up with the "right" precisely defined multivariate normalized, quasi-interpolation neural network operators based on boxes of $\mathbb{R}^N$. Our boxes are not necessarily symmetric to the origin. In preparation to prove our results, we mention important properties of the basic multivariate density function induced by the q-deformed and λ-parametrized hyperbolic tangent sigmoid function.

Feed-forward neural networks (FNNs) with one hidden layer here are mathematically expressed as

$$N_n(x) = \sum_{j=0}^{n} c_j \sigma\left(\langle a_j \cdot x \rangle + b_j\right), \quad x \in \mathbb{R}^s, \quad s \in \mathbb{N},$$

where for $0 \leq j \leq n$, $b_j \in \mathbb{R}$ are the thresholds, $a_j \in \mathbb{R}^s$ are the connection weights, $c_j \in \mathbb{C}$ are the coefficients, $\langle a_j \cdot x \rangle$ is the inner product of a_j and x, and σ is the activation function of the network. In many fundamental network models, the activation function is a kind of hyperbolic tangent sigmoid function. For more information about neural networks, read Haykin (1998)–Mitchell (1997).

5.2 About q-Deformed and λ-Parametrized Hyperbolic Tangent Function $g_{q,\lambda}$

We talk in detail about $g_{q,\lambda}$, see Anastassiou (2023b), ch. 17 and (5.1), and justify that it is an activation sigmoid function and we give several of its properties related to the approximation by neural network operators.

So, let us consider the function

$$g_{q,\lambda}(x) := \frac{e^{\lambda x} - q e^{-\lambda x}}{e^{\lambda x} + q e^{-\lambda x}}, \quad \lambda, q > 0, \quad x \in \mathbb{R}. \tag{5.1}$$

We have that

$$g_{q,\lambda}(0) = \frac{1-q}{1+q}.$$

We note also that

$$g_{q,\lambda}(-x) = \frac{e^{-\lambda x} - qe^{\lambda x}}{e^{-\lambda x} + qe^{\lambda x}} = \frac{\frac{1}{q}e^{-\lambda x} - e^{\lambda x}}{\frac{1}{q}e^{-\lambda x} + e^{\lambda x}} = -\frac{\left(e^{\lambda x} - \frac{1}{q}e^{-\lambda x}\right)}{e^{\lambda x} + \frac{1}{q}e^{-\lambda x}} = -g_{\frac{1}{q},\lambda}(x). \tag{5.2}$$

That is,

$$g_{q,\lambda}(-x) = -g_{\frac{1}{q},\lambda}(x), \ \ \forall\, x \in \mathbb{R}, \tag{5.3}$$

and

$$g_{\frac{1}{q},\lambda}(x) = -g_{q,\lambda}(-x),$$

hence

$$g'_{\frac{1}{q},\lambda}(x) = g'_{q,\lambda}(-x). \tag{5.4}$$

It is

$$g_{q,\lambda}(x) = \frac{e^{2\lambda x} - q}{e^{2\lambda x} + q} = \frac{1 - \frac{q}{e^{2lx}}}{1 + \frac{q}{e^{2\lambda x}}} \underset{(x \to +\infty)}{\to} 1,$$

i.e.

$$g_{q,\lambda}(+\infty) = 1. \tag{5.5}$$

Furthermore,

$$g_{q,\lambda}(x) = \frac{e^{2\lambda x} - q}{e^{2\lambda x} + q} \underset{(x \to -\infty)}{\to} \frac{-q}{q} = -1,$$

i.e.

$$g_{q,\lambda}(-\infty) = -1. \tag{5.6}$$

We find that

$$g'_{q,\lambda}(x) = \frac{4q\lambda e^{2\lambda x}}{\left(e^{2\lambda x} + q\right)^2} > 0, \tag{5.7}$$

therefore $g_{q,\lambda}$ is strictly increasing.

Next we obtain ($x \in \mathbb{R}$)

$$g''_{q,\lambda}(x) = 8q\lambda^2 e^{2\lambda x}\left(\frac{q - e^{2\lambda x}}{\left(e^{2\lambda x} + q\right)^3}\right) \in C(\mathbb{R}). \tag{5.8}$$

We observe that

$$q - e^{2\lambda x} \gtreqless 0 \Leftrightarrow q \gtreqless e^{2\lambda x} \Leftrightarrow \ln q \gtreqless 2\lambda x \Leftrightarrow x \lesseqgtr \frac{\ln q}{2\lambda}.$$

So, in case of $x < \frac{\ln q}{2\lambda}$, we have that $g_{q,\lambda}$ is strictly concave up, with $g''_{q,\lambda}\left(\frac{\ln q}{2\lambda}\right) = 0$.

And in case of $x > \frac{\ln q}{2\lambda}$, we have that $g_{q,\lambda}$ is strictly concave down.

Clearly, $g_{q,\lambda}$ is a shifted sigmoid function with $g_{q,\lambda}(0) = \frac{1-q}{1+q}$, and $g_{q,\lambda}(-x) = -g_{q^{-1},\lambda}(x)$, (a semi-odd function), see also Anastassiou (2023a).

By $1 > -1$, $x + 1 > x - 1$, we consider the function

$$M_{q.\lambda}(x) := \frac{1}{4}\left(g_{q,\lambda}(x+1) - g_{q,\lambda}(x-1)\right) > 0, \tag{5.9}$$

$\forall x \in \mathbb{R}$; $q, \lambda > 0$. Note that $M_{q,\lambda}(\pm\infty) = 0$, so the x-axis is horizontal asymptote.

We have that

$$\begin{aligned} M_{q,\lambda}(-x) &= \frac{1}{4}\left(g_{q,\lambda}(-x+1) - g_{q,\lambda}(-x-1)\right) \\ &= \frac{1}{4}\left(g_{q,\lambda}(-(x-1)) - g_{q,\lambda}(-(x+1))\right) \\ &= \frac{1}{4}\left(-g_{\frac{1}{q},\lambda}(x-1) + g_{\frac{1}{q},\lambda}(x+1)\right) \\ &= \frac{1}{4}\left(g_{\frac{1}{q},\lambda}(x+1) - g_{\frac{1}{q},\lambda}(x-1)\right) = M_{\frac{1}{q},\lambda}(x), \quad \forall\, x \in \mathbb{R}. \end{aligned} \tag{5.10}$$

Thus,

$$M_{q,\lambda}(-x) = M_{\frac{1}{q},\lambda}(x), \quad \forall\, x \in \mathbb{R}; \quad q, \lambda > 0, \tag{5.11}$$

a deformed symmetry.

Next, we have that

$$M'_{q,\lambda}(x) = \frac{1}{4}\left(g'_{q,\lambda}(x+1) - g'_{q,\lambda}(x-1)\right), \quad \forall\, x \in \mathbb{R}. \tag{5.12}$$

Let $x < \frac{\ln q}{2\lambda} - 1$, then $x - 1 < x + 1 < \frac{\ln q}{2\lambda}$ and $g'_{q,\lambda}(x+1) > g'_{q,\lambda}(x-1)$ (by $g_{q,\lambda}$ being strictly concave up for $x < \frac{\ln q}{2\lambda}$), that is, $M'_{q,\lambda}(x) > 0$. Hence, $M_{q,\lambda}$ is strictly increasing over $\left(-\infty, \frac{\ln q}{2\lambda} - 1\right)$.

Let now $x - 1 > \frac{\ln q}{2\lambda}$, then $x + 1 > x - 1 > \frac{\ln q}{2\lambda}$, and $g'_{q,\lambda}(x+1) < g'_{q,\lambda}(x-1)$, that is, $M'_{q,\lambda}(x) < 0$.

Therefore, $M_{q,\lambda}$ is strictly decreasing over $\left(\frac{\ln q}{2\lambda}+1,+\infty\right)$.

Let us next consider $\frac{\ln q}{2\lambda}-1\leq x\leq\frac{\ln q}{2\lambda}+1$. We have that

$$\begin{aligned}M''_{q,\lambda}(x) &= \frac{1}{4}\left(g''_{q,\lambda}(x+1)-g''_{q,\lambda}(x-1)\right)\\ &= 2q\lambda^2\left[e^{2\lambda(x+1)}\left(\frac{q-e^{2\lambda(x+1)}}{\left(e^{2\lambda(x+1)}+q\right)^3}\right)-e^{2\lambda(x-1)}\left(\frac{q-e^{2\lambda(x-1)}}{\left(e^{2\lambda(x-1)}+q\right)^3}\right)\right].\end{aligned} \tag{5.13}$$

By $\frac{\ln q}{2\lambda}-1\leq x\Leftrightarrow\frac{\ln q}{2\lambda}\leq x+1\Leftrightarrow\ln q\leq 2\lambda(x+1)\Leftrightarrow q\leq e^{2\lambda(x+1)}\Leftrightarrow$ $q-e^{2\lambda(x+1)}\leq 0$.

By $x\leq\frac{\ln q}{2\lambda}+1\Leftrightarrow x-1\leq\frac{\ln q}{2\lambda}\Leftrightarrow 2\lambda(x-1)\leq\ln q\Leftrightarrow e^{2\lambda(x-1)}\leq q\Leftrightarrow$ $q-e^{2\lambda\beta(x-1)}\geq 0$.

Clearly by (5.13) we get that $M''_{q,\lambda}(x)\leq 0$, for $x\in\left[\frac{\ln q}{2\lambda}-1,\frac{\ln q}{2\lambda}+1\right]$.

More precisely, $M_{q,\lambda}$ is concave down over $\left[\frac{\ln q}{2\lambda}-1,\frac{\ln q}{2\lambda}+1\right]$ and strictly concave down over $\left(\frac{\ln q}{2\lambda}-1,\frac{\ln q}{2\lambda}+1\right)$.

Consequently, $M_{q,\lambda}$ has a bell-type shape over $\mathbb{R}$.

Of course it holds $M''_{q,\lambda}\left(\frac{\ln q}{2\lambda}\right)<0$.

At $x=\frac{\ln q}{2\lambda}$, we have

$$\begin{aligned}M'_{q,\lambda}(x) &= \frac{1}{4}\left(g'_{q,\lambda}(x+1)-g'_{q,\lambda}(x-1)\right)\\ &= q\lambda\left(\frac{e^{2\lambda(x+1)}}{\left(e^{2\lambda(x+1)}+q\right)^2}-\frac{e^{2\lambda(x-1)}}{\left(e^{2\lambda(x-1)}+q\right)^2}\right).\end{aligned} \tag{5.14}$$

Thus,

$$\begin{aligned}M'_{q,\lambda}\left(\frac{\ln q}{2\lambda}\right) &= q\lambda\left(\frac{e^{2\lambda\left(\frac{\ln q}{2\lambda}+1\right)}}{\left(e^{2\lambda\left(\frac{\ln q}{2\lambda}+1\right)}+q\right)^2}-\frac{e^{2\lambda\left(\frac{\ln q}{2\lambda}-1\right)}}{\left(e^{2\lambda\left(\frac{\ln q}{2\lambda}-1\right)}+q\right)^2}\right)\\ &= q\lambda\left(\frac{qe^{2\lambda}}{\left(qe^{2\lambda}+q\right)^2}-\frac{qe^{-2\lambda}}{\left(qe^{-2\lambda}+q\right)^2}\right)\\ &= \lambda\left(\frac{e^{2\lambda}}{\left(e^{2\lambda}+1\right)^2}-\frac{e^{-2\lambda}}{\left(e^{-2\lambda}+1\right)^2}\right)\\ &= \lambda\left(\frac{e^{2\lambda}\left(e^{-2\lambda}+1\right)^2-e^{-2\lambda}\left(e^{2\lambda}+1\right)^2}{\left(e^{2\lambda}+1\right)^2\left(e^{-2\lambda}+1\right)^2}\right)=0.\end{aligned} \tag{5.15}$$

That is, $\frac{\ln q}{2\lambda}$ is the only critical number of $M_{q,\lambda}$ over $\mathbb{R}$. Hence at $x = \frac{\ln q}{2\lambda}$, $M_{q,\lambda}$ achieves its global maximum, which is

$$\begin{aligned}
M_{q,\lambda}\left(\frac{\ln q}{2\lambda}\right) &= \frac{1}{4}\left[g_{q,\lambda}\left(\frac{\ln q}{2\lambda}+1\right) - g_{q,\lambda}\left(\frac{\ln q}{2\lambda}-1\right)\right] \\
&= \frac{1}{4}\left[\left(\frac{e^{\lambda\left(\frac{\ln q}{2\lambda}+1\right)} - qe^{-\lambda\left(\frac{\ln q}{2\lambda}+1\right)}}{e^{\lambda\left(\frac{\ln q}{2\lambda}+1\right)} + qe^{-\lambda\left(\frac{\ln q}{2\lambda}+1\right)}}\right) - \left(\frac{e^{\lambda\left(\frac{\ln q}{2\lambda}-1\right)} - qe^{-\lambda\left(\frac{\ln q}{2\lambda}-1\right)}}{e^{\lambda\left(\frac{\ln q}{2\lambda}-1\right)} + qe^{-\lambda\left(\frac{\ln q}{2\lambda}-1\right)}}\right)\right] \\
&= \frac{1}{4}\left[\left(\frac{\sqrt{q}e^{\lambda} - qq^{-\frac{1}{2}}e^{-\lambda}}{\sqrt{q}e^{\lambda} + qq^{-\frac{1}{2}}e^{-\lambda}}\right) - \left(\frac{\sqrt{q}e^{-\lambda} - qq^{-\frac{1}{2}}e^{\lambda}}{\sqrt{q}e^{-\lambda} + qq^{-\frac{1}{2}}e^{\lambda}}\right)\right] \qquad (5.16) \\
&= \frac{1}{4}\left[\left(\frac{e^{\lambda} - e^{-\lambda}}{e^{\lambda} + e^{-\lambda}}\right) - \left(\frac{e^{-\lambda} - e^{\lambda}}{e^{-\lambda} + e^{\lambda}}\right)\right] \\
&= \frac{1}{4}\left[\frac{2\left(e^{\lambda} - e^{-\lambda}\right)}{e^{\lambda} + e^{-\lambda}}\right] = \frac{1}{2}\left(\frac{e^{\lambda} - e^{-\lambda}}{e^{\lambda} + e^{-\lambda}}\right) = \frac{\tanh(\lambda)}{2}. \qquad (5.17)
\end{aligned}$$

Conclusion: The maximum value of $M_{q,\lambda}$ is

$$M_{q,\lambda}\left(\frac{\ln q}{2\lambda}\right) = \frac{\tanh(\lambda)}{2}, \quad \lambda > 0. \tag{5.18}$$

We give

Theorem 5.1 (Anastassiou 2023b, Ch. 17). *We have that*

$$\sum_{i=-\infty}^{\infty} M_{q,\lambda}(x-i) = 1, \quad \forall\, x \in \mathbb{R}, \quad \forall\, \lambda, q > 0. \tag{5.19}$$

It follows

Theorem 5.2 (Anastassiou 2023b, Ch. 17). *It holds*

$$\int_{-\infty}^{\infty} M_{q,\lambda}(x)\, dx = 1, \quad \lambda, q > 0. \tag{5.20}$$

So $M_{q,\lambda}$ is a density function on $\mathbb{R}$; $\lambda, q > 0$.

We need the following result

Theorem 5.3 (Anastassiou 2023b, Ch. 17). *Let* $0 < \alpha < 1$*, and* $n \in \mathbb{N}$ *with* $n^{1-\alpha} > 2$*;* $q, \lambda > 0$*. Then*

$$\sum_{\substack{k=-\infty \\ \left\{ :|nx-k| \geq n^{1-\alpha} \right.}}^{\infty} M_{q,\lambda}(nx-k) < 2\max\left\{q, \frac{1}{q}\right\} e^{4\lambda} e^{-2\lambda n^{(1-\alpha)}}$$

$$= Te^{-2\lambda n^{(1-\alpha)}}, \tag{5.21}$$

where $T := 2\max\left\{q, \frac{1}{q}\right\} e^{4\lambda}$.

Let $\lceil \cdot \rceil$ be the ceiling of the number and $\lfloor \cdot \rfloor$ be the integral part of the number.

Theorem 5.4 (Anastassiou 2023b, Ch. 17). *Let* $x \in [a,b] \subset \mathbb{R}$ *and* $n \in \mathbb{N}$ *so that* $\lceil na \rceil \leq \lfloor nb \rfloor$*. For* $q > 0$*,* $\lambda > 0$*, we consider the number* $\lambda_q > z_0 > 0$ *with* $M_{q,\lambda}(z_0) = M_{q,\lambda}(0)$ *and* $\lambda_q > 1$*. Then*

$$\frac{1}{\sum_{k=\lceil na \rceil}^{\lfloor nb \rfloor} M_{q,\lambda}(nx-k)} < max\left\{\frac{1}{M_{q,\lambda}(\lambda_q)}, \frac{1}{M_{\frac{1}{q},\lambda}\left(\lambda_{\frac{1}{q}}\right)}\right\} =: \Delta(q). \tag{5.22}$$

We make

Remark 5.1 (Anastassiou 2023b, Ch. 17). (i) We have that

$$\lim_{n \to +\infty} \sum_{k=\lceil na \rceil}^{\lfloor nb \rfloor} M_{q,\lambda}(nx-k) \neq 1, \quad \text{for at least some } x \in [a,b], \tag{5.23}$$

where $\lambda, q > 0$.

(ii) Let $[a,b] \subset \mathbb{R}$. For large n, we always have $\lceil na \rceil \leq \lfloor nb \rfloor$. Also $a \leq \frac{k}{n} \leq b$, iff $\lceil na \rceil \leq k \leq \lfloor nb \rfloor$. In general, it holds

$$\sum_{k=\lceil na \rceil}^{\lfloor nb \rfloor} M_{q,\lambda}(nx-k) \leq 1. \tag{5.24}$$

We make

Remark 5.2. We introduce

$$Z_{q,\lambda}(x_1,\ldots,x_N) := Z_{q,\lambda}(x) := \prod_{i=1}^{N} M_{q,\lambda}(x_i), \tag{5.25}$$

$x=(x_1,\ldots,x_N)\in\mathbb{R}^N$, $\lambda,q>0$, $N\in\mathbb{N}$.

It has the following properties:

(i) $Z_{q,\lambda}(x)>0,\ \forall\, x\in\mathbb{R}^N$,

(ii)

$$\sum_{k=-\infty}^{\infty} Z_{q,\lambda}(x-k)$$

$$:= \sum_{k_1=-\infty}^{\infty}\sum_{k_2=-\infty}^{\infty}\cdots\sum_{k_N=-\infty}^{\infty} Z_{q,\lambda}(x_1-k_1,\ldots,x_N-k_N)=1, \tag{5.26}$$

where $k:=(k_1,\ldots,k_n)\in\mathbb{Z}^N$, $\forall\, x\in\mathbb{R}^N$,

hence

(iii)

$$\sum_{k=-\infty}^{\infty} Z_{q,\lambda}(nx-k)=1, \tag{5.27}$$

$\forall\, x\in\mathbb{R}^N;\ n\in\mathbb{N}$,

and

(iv)

$$\int_{\mathbb{R}^N} Z_{q,\lambda}(x)\,dx=1, \tag{5.28}$$

that is, Z_q is a multivariate density function.

Here denote $\|x\|_\infty := \max\{|x_1|,\ldots,|x_N|\}$, $x\in\mathbb{R}^N$, also set $\infty := (\infty,\ldots,\infty)$, $-\infty := (-\infty,\ldots,-\infty)$ upon the multivariate context, and

$$\begin{aligned}\lceil na\rceil &:= (\lceil na_1\rceil,\ldots,\lceil na_N\rceil),\\ \lfloor nb\rfloor &:= (\lfloor nb_1\rfloor,\ldots,\lfloor nb_N\rfloor),\end{aligned} \tag{5.29}$$

where $a:=(a_1,\ldots,a_N)$, $b:=(b_1,\ldots,b_N)$.

We obviously see that

$$\sum_{k=\lceil na\rceil}^{\lfloor nb\rfloor} Z_{q,\lambda}(nx-k) = \sum_{k=\lceil na\rceil}^{\lfloor nb\rfloor}\left(\prod_{i=1}^{N} M_{q,\lambda}(nx_i-k_i)\right)$$

$$= \sum_{k_1=\lceil na_1\rceil}^{\lfloor nb_1\rfloor}\cdots\sum_{k_N=\lceil na_N\rceil}^{\lfloor nb_N\rfloor}\left(\prod_{i=1}^{N} M_{q,\lambda}(nx_i-k_i)\right)$$

$$= \prod_{i=1}^{N}\left(\sum_{k_i=\lceil na_i\rceil}^{\lfloor nb_i\rfloor} M_{q,\lambda}(nx_i-k_i)\right). \tag{5.30}$$

For $0<\beta<1$ and $n\in\mathbb{N}$, a fixed $x\in\mathbb{R}^N$, we have that

$$\sum_{k=\lceil na\rceil}^{\lfloor nb\rfloor} Z_{q,\lambda}(nx-k)$$

$$= \sum_{\begin{cases} k=\lceil na\rceil \\ \left\|\frac{k}{n}-x\right\|_\infty \le \frac{1}{n^\beta}\end{cases}}^{\lfloor nb\rfloor} Z_{q,\lambda}(nx-k) + \sum_{\begin{cases} k=\lceil na\rceil \\ \left\|\frac{k}{n}-x\right\|_\infty > \frac{1}{n^\beta}\end{cases}}^{\lfloor nb\rfloor} Z_{q,\lambda}(nx-k). \tag{5.31}$$

In the last two sums, the counting is over disjoint vector sets of k's because the condition $\left\|\frac{k}{n}-x\right\|_\infty > \frac{1}{n^\beta}$ implies that there exists at least one $\left|\frac{k_r}{n}-x_r\right| > \frac{1}{n^\beta}$, where $r\in\{1,\ldots,N\}$.

(v) By Theorem 5.3 and as in Anastassiou (2016), pp. 379–380, we derive that

$$\sum_{\begin{cases} k=\lceil na\rceil \\ \left\|\frac{k}{n}-x\right\|_\infty > \frac{1}{n^\beta}\end{cases}}^{\lfloor nb\rfloor} Z_{q,\lambda}(nx-k) < Te^{-2\lambda n^{(1-\beta)}}, \quad 0<\beta<1, \tag{5.32}$$

with $n\in\mathbb{N}: n^{1-\beta}>2$, $x\in\prod_{i=1}^{N}[a_i,b_i]$.

(vi) By Theorem 5.4, we get that

$$0 < \frac{1}{\sum_{k=\lceil na\rceil}^{\lfloor nb\rfloor} Z_{q,\lambda}(nx-k)} < (\Delta(q))^N, \tag{5.33}$$

$\forall\, x\in\left(\prod_{i=1}^{N}[a_i,b_i]\right)$, $n\in\mathbb{N}$.

It is also clear that

(vii)

$$\sum_{\substack{k=-\infty \\ \left\{\left\|\frac{k}{n}-x\right\|_{\infty}>\frac{1}{n^{\beta}}\right.}}^{\infty} Z_{q,\lambda}(nx-k) < Te^{-2\lambda n^{(1-\beta)}}, \tag{5.34}$$

$0<\beta<1$, $n\in\mathbb{N}: n^{1-\beta}>2$, $x\in\mathbb{R}^N$.

Furthermore it holds

$$\lim_{n\to\infty}\sum_{k=\lceil na\rceil}^{\lfloor nb\rfloor} Z_{q,\lambda}(nx-k) \neq 1, \tag{5.35}$$

for at least some $x\in\left(\prod_{i=1}^{N}[a_i,b_i]\right)$.

Here $(\mathbb{C},|\cdot|)$ is the complex Banach space.

Let $f\in C\left(\prod_{i=1}^{N}[a_i,b_i],\mathbb{C}\right)$, $x=(x_1,\ldots,x_N)\in\prod_{i=1}^{N}[a_i,b_i]$, $n\in\mathbb{N}$ such that $\lceil na_i\rceil\leq\lfloor nb_i\rfloor$, $i=1,\ldots,N$.

We introduce and define the following complex-valued multivariate linear normalized neural network operator ($x:=(x_1,\ldots,x_N)\in(\prod_{i=1}^{N}[a_i,b_i])$):

$$A_n(f,x_1,\ldots,x_N) := A_n(f,x) := \frac{\sum_{k=\lceil na\rceil}^{\lfloor nb\rfloor} f\left(\frac{k}{n}\right) Z_{q,\lambda}(nx-k)}{\sum_{k=\lceil na\rceil}^{\lfloor nb\rfloor} Z_{q,\lambda}(nx-k)}$$

$$= \frac{\sum_{k_1=\lceil na_1\rceil}^{\lfloor nb_1\rfloor}\sum_{k_2=\lceil na_2\rceil}^{\lfloor nb_2\rfloor}\cdots\sum_{k_N=\lceil na_N\rceil}^{\lfloor nb_N\rfloor} f\left(\frac{k_1}{n},\ldots,\frac{k_N}{n}\right)\left(\prod_{i=1}^{N} M_{q,\lambda}(nx_i-k_i)\right)}{\prod_{i=1}^{N}\left(\sum_{k_i=\lceil na_i\rceil}^{\lfloor nb_i\rfloor} M_{q,\lambda}(nx_i-k_i)\right)}. \tag{5.36}$$

For large enough $n\in\mathbb{N}$, we always obtain $\lceil na_i\rceil\leq\lfloor nb_i\rfloor$, $i=1,\ldots,N$. Also $a_i\leq\frac{k_i}{n}\leq b_i$, iff $\lceil na_i\rceil\leq k_i\leq\lfloor nb_i\rfloor$, $i=1,\ldots,N$.

When $g\in C\left(\prod_{i=1}^{N}[a_i,b_i]\right)$, we define the companion operator

$$\widetilde{A}_n(g,x) := \frac{\sum_{k=\lceil na\rceil}^{\lfloor nb\rfloor} g\left(\frac{k}{n}\right) Z_{q,\lambda}(nx-k)}{\sum_{k=\lceil na\rceil}^{\lfloor nb\rfloor} Z_{q,\lambda}(nx-k)}. \tag{5.37}$$

Clearly, $\widetilde{A}_n$ is a positive linear operator. We have that

$$\widetilde{A}_n(1,x)=1, \ \forall\, x\in\left(\prod_{i=1}^{N}[a_i,b_i]\right).$$

Note that $A_n(f)\in C\left(\prod_{i=1}^{N}[a_i,b_i],\mathbb{C}\right)$ and $\widetilde{A}_n(g)\in C\left(\prod_{i=1}^{N}[a_i,b_i]\right)$.

Furthermore it holds

$$|A_n(f,x)| \leq \frac{\sum_{k=\lceil na \rceil}^{\lfloor nb \rfloor} \left| f\left(\frac{k}{n}\right) \right| Z_{q,\lambda}(nx-k)}{\sum_{k=\lceil na \rceil}^{\lfloor nb \rfloor} Z_{q,\lambda}(nx-k)} = \widetilde{A}_n(|f|,x), \tag{5.38}$$

$\forall\, x \in \prod_{i=1}^{N} [a_i, b_i]$.

Clearly, $|f| \in C\left(\prod_{i=1}^{N} [a_i, b_i]\right)$.

So, we have that

$$|A_n(f,x)| \leq \widetilde{A}_n(|f|,x), \tag{5.39}$$

$\forall\, x \in \prod_{i=1}^{N} [a_i, b_i]$, $\forall\, n \in \mathbb{N}$, $\forall\, f \in C\left(\prod_{i=1}^{N} [a_i, b_i], \mathbb{C}\right)$.

Let $c \in \mathbb{C}$ and $g \in C\left(\prod_{i=1}^{N} [a_i, b_i]\right)$, then $cg \in C\left(\prod_{i=1}^{N} [a_i, b_i], \mathbb{C}\right)$.

Furthermore it holds

$$A_n(cg,x) = c\widetilde{A}_n(g,x), \quad \forall\, x \in \prod_{i=1}^{N} [a_i, b_i]. \tag{5.40}$$

Since $\widetilde{A}_n(1) = 1$, we get that

$$A_n(c) = c, \quad \forall\, c \in \mathbb{C}. \tag{5.41}$$

We call $\widetilde{A}_n$ the companion operator of A_n.

For convenience, we call

$$\begin{aligned} A_n^*(f,x) &:= \sum_{k=\lceil na \rceil}^{\lfloor nb \rfloor} f\left(\frac{k}{n}\right) Z_{q,\lambda}(nx-k) \\ &= \sum_{k_1=\lceil na_1 \rceil}^{\lfloor nb_1 \rfloor} \sum_{k_2=\lceil na_2 \rceil}^{\lfloor nb_2 \rfloor} \cdots \sum_{k_N=\lceil na_N \rceil}^{\lfloor nb_N \rfloor} f\left(\frac{k_1}{n}, \ldots, \frac{k_N}{n}\right) \left(\prod_{i=1}^{N} M_{q,\lambda}(nx_i - k_i)\right), \end{aligned} \tag{5.42}$$

$\forall\, x \in \left(\prod_{i=1}^{N} [a_i, b_i]\right)$.

That is,

$$A_n(f,x) := \frac{A_n^*(f,x)}{\sum_{k=\lceil na \rceil}^{\lfloor nb \rfloor} Z_{q,\lambda}(nx-k)}, \tag{5.43}$$

$\forall\, x \in \left(\prod_{i=1}^{N} [a_i, b_i]\right)$, $n \in \mathbb{N}$.

Hence,

$$A_n(f,x) - f(x) = \frac{A_n^*(f,x) - f(x)\left(\sum_{k=\lceil na \rceil}^{\lfloor nb \rfloor} Z_{q,\lambda}(nx-k)\right)}{\sum_{k=\lceil na \rceil}^{\lfloor nb \rfloor} Z_{q,\lambda}(nx-k)}. \quad (5.44)$$

Consequently we derive

$$|A_n(f,x) - f(x)| \overset{(5.33)}{\leq} (\Delta(q))^N \left| A_n^*(f,x) - f(x) \sum_{k=\lceil na \rceil}^{\lfloor nb \rfloor} Z_{q,\lambda}(nx-k) \right|, \quad (5.45)$$

$\forall\, x \in \left(\prod_{i=1}^N [a_i, b_i]\right)$.

We estimate the right-hand side of (5.45).

For the last we need

Definition 5.1 (Anastassiou 2018, p. 274). Let M be a convex and compact subset of $\left(\mathbb{R}^N, \|\cdot\|_p\right)$, $p \in [1,\infty]$, and $(\mathbb{C}, |\cdot|)$ be a Banach space. Let $f \in C(M, \mathbb{C})$. We define the first modulus of continuity of f as

$$\omega_1(f,\delta) := \sup_{\substack{x,y \in M: \\ \|x-y\|_p \leq \delta}} |f(x) - f(y)|, \quad 0 < \delta \leq \operatorname{diam}(M). \quad (5.46)$$

If $\delta > diam(M)$, then

$$\omega_1(f,\delta) = \omega_1(f, \operatorname{diam}(M)). \quad (5.47)$$

Note $\omega_1(f,\delta)$ is increasing in $\delta > 0$.

Lemma 5.1 (Anastassiou 2018, p. 274). *We have $\omega_1(f,\delta) \to 0$ as $\delta \downarrow 0$, iff $f \in C(M,\mathbb{C})$, where M is a convex compact subset of $\left(\mathbb{R}^N, \|\cdot\|_p\right)$, $p \in [1,\infty]$.*

Let now $f \in C^2\left(\prod_{i=1}^N [a_i,b_i], \mathbb{C}\right)$, $N \in \mathbb{N}$. Here f_α denotes a partial derivative of f, $\alpha := (\alpha_1, \ldots, \alpha_N)$, $\alpha_i \in \mathbb{Z}_+$, $i = 1, \ldots, N$, and $|\alpha| := \sum_{i=1}^N \alpha_i = l$, where $l = 0, 1, 2$. We write also $f_\alpha := \frac{\partial^n f}{\partial x^n}$ and we say it is of order l.

We denote

$$\omega_1^{\max}(f_\alpha, h) := \max_{\alpha : |\alpha| = 2} \omega_1(f_\alpha, h). \quad (5.48)$$

Call also

$$\|f_\alpha\|_\infty^{\max} := \max_{|\alpha|=2} \{\|f_\alpha\|_\infty\}, \tag{5.49}$$

where $\|\cdot\|_\infty$ is the supremum norm.

In this chapter (see (5.46)), we work with $p = \infty$.

(II) Multivariate New Taylor formulae

We use

Theorem 5.5 (Anastassiou 2023c). *Let $f \in C^2([c,d],\mathbb{C})$, where $a, x \in [c,d]$. Then*

$$f(x) - f(a) = f'(a)\sin(x-a) + 2f''(a)\sin^2\left(\frac{x-a}{2}\right)$$
$$+ \int_a^x [(f''(t) + f(t)) - (f''(a) + f(a))]\sin(x-t)\,dt. \tag{5.50}$$

We make

Remark 5.3. Let now Q be an open convex subset of $\mathbb{R}^k$, $k \geq 2$; $z = (z_1, \ldots, z_k)$, $x_0 := (x_{01}, \ldots, x_{0k}) \in Q$. We consider $f \in C^2(Q,\mathbb{C})$ each second-order partial derivative is denoted by $f_\alpha := \frac{\partial^\alpha f}{\partial x^\alpha}$, where $\alpha := (\alpha_1, \ldots, \alpha_k)$, $\alpha_i \in \mathbb{Z}^+$, $i = 1, \ldots, k$ and $|\alpha| := \sum_{i=1}^k \alpha_i = 2$. We consider $g_z(t) := f(x_0 + t(z - x_0))$, $0 \leq t \leq 1$. Clearly $x_0 + t(z - x_0) \in Q$. Then

$$g_z(0) = f(x_0), \quad g_z(1) = f(z),$$

$$g_z'(t) = \sum_{i=1}^k (z_i - x_{0i}) \frac{\partial f}{\partial x_i}(x_{01} + t(z_1 - x_{01}), \ldots, x_{0k} + t(z_k - x_{0k})), \tag{5.51}$$

$$g_z'(0) = \sum_{i=1}^k (z_i - x_{0i}) \frac{\partial f}{\partial x_i}(x_{01}, \ldots, x_{0k}),$$

and

$$g_z''(t) = \left[\left(\sum_{i=1}^k (z_i - x_{0i}) \frac{\partial}{\partial x_i}\right)^2 f\right](x_{01} + t(z_1 - x_{01}), \ldots, x_{0k} + t(z_k - x_{0k})), \tag{5.52}$$

$$g_z''(0) = \left[\left(\sum_{i=1}^k (z_i - x_{0i}) \frac{\partial}{\partial x_i}\right)^2 f\right](x_{01}, \ldots, x_{0k}).$$

Note above the second order partials commute.

Clearly, $g_z \in C^2([0,1],\mathbb{C})$, and by Theorem 5.5, we obtain

$$f(z_1,\ldots,z_k) - f(x_{01},\ldots,x_{0k}) = g_z(1) - g_z(0)$$

$$= g_z'(0)\sin(1) + 2g_z''(0)\sin^2\left(\frac{1}{2}\right)$$

$$+ \int_0^1 \left[\left(g_z''(t) + g_z(t)\right) - \left(g_z''(0) + g_z(0)\right)\right]\sin(1-t)\,dt. \tag{5.53}$$

We also mention

Theorem 5.6 (Anastassiou 2023c). *Let $f \in C^2([c,d],\mathbb{C})$, where $a, x \in [c,d]$. Then*

$$f(x) - f(a) = f'(a)\sinh(x-a) + 2f''(a)\sinh^2\left(\frac{x-a}{2}\right)$$

$$+ \int_a^x \left[\left(f''(t) - f(t)\right) - \left(f''(a) - f(a)\right)\right]\sinh(x-t)\,dt. \tag{5.54}$$

We make

Remark 5.4. Consequently, we get that

$$f(z_1,\ldots,z_k) - f(x_{01},\ldots,x_{0k}) = g_z(1) - g_z(0)$$

$$= g_z'(0)\sinh(1) + 2g_z''(0)\sinh^2\left(\frac{1}{2}\right)$$

$$+ \int_0^1 \left[\left(g_z''(t) - g_z(t)\right) - \left(g_z''(0) - g_z(0)\right)\right]\sinh(1-t)\,dt. \tag{5.55}$$

We make

Remark 5.5. Let $f \in C^2\left(\prod_{i=1}^N [a_i,b_i],\mathbb{C}\right)$, $N \in \mathbb{N}$.

Clearly, the mixed partials commute.

Here $\frac{k}{n} := \left(\frac{k_1}{n},\ldots,\frac{k_N}{n}\right)$, and $x := (x_1,\ldots,x_N)$, with $\frac{k}{n}, x \in \left(\prod_{i=1}^N [a_i,b_i]\right)$, then (by (5.53), where $g_{\frac{k}{n}}(t) := f\left(x + t\left(\frac{k}{n} - x\right)\right)$, $0 \le t \le 1$)

we have

$$f\left(\frac{k}{n}\right)-f(x)=\left(\sum_{i=1}^{N}\left(\frac{k_i}{n}-x_i\right)\frac{\partial f}{\partial x_i}(x)\right)\sin(1)$$
$$+2\left\{\left[\left(\sum_{i=1}^{N}\left(\frac{k_i}{n}-x_i\right)\frac{\partial}{\partial x_i}\right)^2 f\right](x)\right\}\sin^2\left(\frac{1}{2}\right)$$
$$+\int_0^1\left\{\left\{\left[\left(\sum_{i=1}^{N}\left(\frac{k_i}{n}-x_i\right)\frac{\partial}{\partial x_i}\right)^2 f\right]\right.\right.$$
$$\left.\times\left(x+t\left(\frac{k}{n}-x\right)\right)+f\left(x+t\left(\frac{k}{n}-x\right)\right)\right\}$$
$$\left.-\left\{\left[\left(\sum_{i=1}^{N}\left(\frac{k_i}{n}-x_i\right)\frac{\partial}{\partial x_i}\right)^2 f\right](x)+f(x)\right\}\right\}\sin(1-t)\,dt. \quad (5.56)$$

Denote the remainder

$$R:=\int_0^1\left\{\left\{\left[\left(\sum_{i=1}^{N}\left(\frac{k_i}{n}-x_i\right)\frac{\partial}{\partial x_i}\right)^2 f\right]\right.\right.$$
$$\left.\times\left(x+t\left(\frac{k}{n}-x\right)\right)+f\left(x+t\left(\frac{k}{n}-x\right)\right)\right\}$$
$$\left.-\left\{\left[\left(\sum_{i=1}^{N}\left(\frac{k_i}{n}-x_i\right)\frac{\partial}{\partial x_i}\right)^2 f\right](x)+f(x)\right\}\right\}\sin(1-t)\,dt$$
$$=\int_0^1\left\{\sum_{\substack{\alpha:=(\alpha_1,\ldots,\alpha_N),\alpha_i\in\mathbb{Z}_+\\ i=1,\ldots,N,|\alpha|:=\sum_{i=1}^{N}\alpha_i=2}}\left(\frac{2}{\prod_{i=1}^{N}\alpha_i!}\right)\left(\prod_{i=1}^{N}\left(\frac{k_i}{n}-x_i\right)^{\alpha_i}\right)\right.$$
$$\left[f_\alpha\left(x+t\left(\frac{k}{n}-x\right)\right)-f_\alpha(x)\right]$$
$$\left.+\left(f\left(x+t\left(\frac{k}{n}-x\right)\right)-f(x)\right)\right\}\sin(1-t)\,dt. \quad (5.57)$$

Therefore it holds

$$|R| \leq \int_0^1 \left\{ \sum_{\substack{\alpha:=(\alpha_1,\ldots,\alpha_N),\alpha_i\in\mathbb{Z}_+\\ i=1,\ldots,N,|\alpha|:=\sum_{i=1}^N \alpha_i=2}} \left(\frac{2}{\prod_{i=1}^N \alpha_i!}\right) \left(\prod_{i=1}^N \left|\frac{k_i}{n}-x_i\right|^{\alpha_i}\right) \left|f_\alpha\left(x+t\left(\frac{k}{n}-x\right)\right)-f_\alpha(x)\right| + \left|f\left(x+t\left(\frac{k}{n}-x\right)\right)-f(x)\right| \right\} |\sin(1-t)|\,dt$$

$$\leq \int_0^1 \left\{ \sum_{\substack{\alpha:=(\alpha_1,\ldots,\alpha_N),\alpha_i\in\mathbb{Z}_+\\ i=1,\ldots,N,|\alpha|:=\sum_{i=1}^N \alpha_i=2}} \left(\frac{2}{\prod_{i=1}^N \alpha_i!}\right) \times \left(\prod_{i=1}^N \left|\frac{k_i}{n}-x_i\right|^{\alpha_i}\right) \omega_1\left(f_\alpha, t\left\|\frac{k}{n}-x\right\|_\infty\right) + \omega_1\left(f, t\left\|\frac{k}{n}-x\right\|_\infty\right) \right\} |\sin(1-t)|\,dt \leq (*). \tag{5.58}$$

Note here that $(0<\beta<1)$

$$\left\|\frac{k}{n}-x\right\|_\infty \leq \frac{1}{n^\beta} \Leftrightarrow \left|\frac{k_i}{n}-x_i\right| \leq \frac{1}{n^\beta}, \quad i=1,\ldots,N. \tag{5.59}$$

We further see that

$$(*) \leq \left\{ \omega_{1,2}^{\max}\left(f_\alpha, \frac{1}{n^\beta}\right) \left(\sum_{\substack{\alpha:=(\alpha_1,\ldots,\alpha_N),\alpha_i\in\mathbb{Z}_+\\ i=1,\ldots,N,|\alpha|:=\sum_{i=1}^N \alpha_i=2}} \left(\frac{2}{\prod_{i=1}^N \alpha_i!}\right) \left(\prod_{i=1}^N \frac{1}{n^{\beta\alpha_i}}\right) \right) + \omega_1\left(f, \frac{1}{n^\beta}\right) \right\} \int_0^1 |\sin(1-t)|\,dt$$

$$= \left[\omega_{1,2}^{\max}\left(f_\alpha, \frac{1}{n^\beta}\right)\left(\sum_{\substack{\alpha:=(\alpha_1,\ldots,\alpha_N),\alpha_i\in\mathbb{Z}_+\\ i=1,\ldots,N,|\alpha|:=\sum_{i=1}^N \alpha_i=2}}\left(\frac{2}{\prod_{i=1}^N \alpha_i!}\right)\right)\frac{1}{n^{2\beta}}\right.$$

$$\left. + \omega_1\left(f, \frac{1}{n^\beta}\right)\right](1-\cos(1))$$

$$= (1-\cos(1))\left\{\frac{\omega_{1,2}^{\max}\left(f_\alpha, \frac{1}{n^\beta}\right)N^2}{n^{2\beta}} + \omega_1\left(f, \frac{1}{n^\beta}\right)\right\}. \tag{5.60}$$

We have proved that

$$|R| \leq (1-\cos(1))\left\{\frac{\omega_{1,2}^{\max}\left(f_\alpha, \frac{1}{n^\beta}\right)N^2}{n^{2\beta}} + \omega_1\left(f, \frac{1}{n^\beta}\right)\right\}, \tag{5.61}$$

given that $\left\|\frac{k}{n} - x\right\|_\infty \leq \frac{1}{n^\beta}$.

We note also that

$$|R| \leq \int_0^1 \left\{\sum_{\substack{\alpha:=(\alpha_1,\ldots,\alpha_N),\alpha_i\in\mathbb{Z}_+\\ i=1,\ldots,N,|\alpha|:=\sum_{i=1}^N \alpha_i=2}}\left(\frac{2}{\prod_{i=1}^N \alpha_i!}\right)\right.$$

$$\left.\left(\prod_{i=1}^N (b_i - a_i)^{\alpha_i}\right) 2\|f_\alpha\|_\infty + 2\|f\|_\infty\right\}|\sin(1-t)|\,dt$$

$$\leq \left\{\left(\sum_{\substack{\alpha:=(\alpha_1,\ldots,\alpha_N),\alpha_i\in\mathbb{Z}_+\\ i=1,\ldots,N,|\alpha|:=\sum_{i=1}^N \alpha_i=2}}\left(\frac{2}{\prod_{i=1}^N \alpha_i!}\right)\right)\right.$$

$$\left. 2\|b-a\|_\infty^2 \|f_\alpha\|_{\infty,2}^{\max} + 2\|f\|_\infty\right\}\left(\int_0^1 |\sin(1-t)|\,dt\right)$$

$$= \left(2\|b-a\|_\infty^2 \|f_\alpha\|_{\infty,2}^{\max} N^2 + 2\|f\|_\infty\right)(1-\cos(1)), \tag{5.62}$$

where $a := (a_1, \ldots, a_N)$, $b = (b_1, \ldots, b_N)$.

We have proved that

$$|R| \leq \left(2\|b-a\|_{\infty}^{2}\|f_{\alpha}\|_{\infty,2}^{\max}N^{2}+2\|f\|_{\infty}\right)(1-\cos(1)) =: \rho. \tag{5.63}$$

5.3 Main Results

Here we discuss the trigonometric approximation by using the smoothness of f.

Theorem 5.7. *Let $f \in C^{2}\left(\prod_{i=1}^{N}[a_i,b_i],\mathbb{C}\right)$, $0<\beta<1$, $n,N\in\mathbb{N}$, $n^{1-\beta}>2$, $x,x_0\in\left(\prod_{i=1}^{N}[a_i,b_i]\right)$, $a:=(a_1,\ldots,a_N)$, $b:=(b_1,\ldots,b_N)$. Then*

(i)

$$\Bigg|A_n(f,x)-f(x)-\left(\sum_{=1}^{N}\frac{\partial f(x)}{\partial x_i}A_n((\cdot-x_i),x)\right)\sin(1)$$

$$-4\left\{\sum_{\substack{\alpha:=(\alpha_1,\ldots,\alpha_N),\alpha_i\in\mathbb{Z}_+\\ i=1,\ldots,N,|\alpha|:=\sum_{i=1}^{N}\alpha_i=2}} f_{\alpha}(x)\left(\frac{1}{\prod_{i=1}^{N}\alpha_i!}\right)A_n\left(\prod_{i=1}^{N}(\cdot-x_i)^{\alpha_i},x\right)\right\}$$

$$\times\sin^{2}\left(\frac{1}{2}\right)\Bigg|$$

$$\leq(\Delta(q))^{N}\left\{\left[(1-\cos(1))\left\{\frac{\omega_{1,2}^{\max}\left(f_{\alpha},\frac{1}{n^{\beta}}\right)N^{2}}{n^{2\beta}}+\omega_1\left(f,\frac{1}{n^{\beta}}\right)\right\}\right]\right.$$

$$\left.+\left[2\|b-a\|_{\infty}^{2}\|f_{\alpha}\|_{\infty,2}^{\max}N^{2}+2\|f\|_{\infty}\right](1-\cos(1))Te^{-2\lambda n^{(1-\beta)}}\right\}, \tag{5.64}$$

(ii) *assume that $\frac{\partial f(x_0)}{\partial x_i}=0$, $i=1,\ldots,N$, and $f_{\alpha}(x_0)=0$, $\alpha:|\alpha|=2$, we have that*

$$|A_n(f,x)-f(x)|$$

$$\leq(\Delta(q))^{N}\left\{\left[(1-\cos(1))\left\{\frac{\omega_{1,2}^{\max}\left(f_{\alpha},\frac{1}{n^{\beta}}\right)N^{2}}{n^{2\beta}}+\omega_1\left(f,\frac{1}{n^{\beta}}\right)\right\}\right]\right.$$

$$\left.+\left[2\|b-a\|_{\infty}^{2}\|f_{\alpha}\|_{\infty,2}^{\max}N^{2}+2\|f\|_{\infty}\right](1-\cos(1))Te^{-2\lambda n^{(1-\beta)}}\right\}, \tag{5.65}$$

(iii)

$$|A_n(f,x) - f(x)| \leq (\Delta(q))^N$$

$$\left\{\left\{\left\{\sum_{i=1}^{N} \left|\frac{\partial f(x)}{\partial x_i}\right| \left\{\frac{1}{n^{\beta}} + (b_i - a_i) T e^{-2\lambda n^{(1-\beta)}}\right\}\right\} \sin(1) \right.\right.\right.$$

$$+ 4 \left\{ \sum_{\alpha:|\alpha|=2} |f_\alpha(x)| \left(\frac{1}{\prod_{i=1}^{N} \alpha_i!}\right) \right.$$

$$\left.\left. \times \left[\frac{1}{n^{2\beta}} + \left(\prod_{i=1}^{N} (b_i - a_i)^{\alpha_i}\right) T e^{-2\lambda n^{(1-\beta)}}\right]\right\} \sin^2\left(\frac{1}{2}\right)\right\}$$

$$+ \left\{\left[(1 - \cos(1)) \left\{\frac{\omega_{1,2}^{\max}\left(f_\alpha, \frac{1}{n^{\beta}}\right) N^2}{n^{2\beta}} + \omega_1\left(f, \frac{1}{n^{\beta}}\right)\right\}\right]\right.$$

$$\left.\left. + \left[2\|b - a\|_\infty^2 \|f_\alpha\|_{\infty,2}^{\max} N^2 + 2\|f\|_\infty\right](1 - \cos(1)) T e^{-2\lambda n^{(1-\beta)}}\right\}\right\}, \tag{5.66}$$

and

(iv)

$$\|A_n(f) - f\|_\infty \leq (\Delta(q))^N$$

$$\left\{\left\{\left\{\sum_{i=1}^{N} \left\|\frac{\partial f}{\partial x_i}\right\|_\infty \left\{\frac{1}{n^{\beta}} + (b_i - a_i) T e^{-2\lambda n^{(1-\beta)}}\right\}\right\} \sin(1) \right.\right.\right.$$

$$+ 4 \left\{ \sum_{\alpha:|\alpha|=2} \|f_\alpha\|_\infty \left(\frac{1}{\prod_{i=1}^{N} \alpha_i!}\right) \right.$$

$$\left.\left. \times \left[\frac{1}{n^{2\beta}} + \left(\prod_{i=1}^{N} (b_i - a_i)^{\alpha_i}\right) T e^{-2\lambda n^{(1-\beta)}}\right]\right\} \sin^2\left(\frac{1}{2}\right)\right\}$$

$$+ \left\{\left[(1 - \cos(1)) \left\{\frac{\omega_{1,2}^{\max}\left(f_\alpha, \frac{1}{n^{\beta}}\right) N^2}{n^{2\beta}} + \omega_1\left(f, \frac{1}{n^{\beta}}\right)\right\}\right]\right.$$

$$\left.\left. + \left[2\|b - a\|_\infty^2 \|f_\alpha\|_{\infty,2}^{\max} N^2 + 2\|f\|_\infty\right](1 - \cos(1)) T e^{-2\lambda n^{(1-\beta)}}\right\}\right\} =: \xi_n(f). \tag{5.67}$$

We observe that $A_n \to I$ (unit operator), as $n \to \infty$, pointwise and uniformly.

Proof. Here R is as in (5.57). We see that

$$U_n := \sum_{k=\lceil na \rceil}^{\lfloor nb \rfloor} Z_{q,\lambda}(nx-k) R$$

$$= \sum_{\substack{k=\lceil na \rceil \\ \left\{ : \left\| \frac{k}{n} - x \right\|_{\infty} \leq \frac{1}{n^{\beta}} \right.}}^{\lfloor nb \rfloor} Z_{q,\lambda}(nx-k) R + \sum_{\substack{k=\lceil na \rceil \\ \left\{ : \left\| \frac{k}{n} - x \right\|_{\infty} > \frac{1}{n^{\beta}} \right.}}^{\lfloor nb \rfloor} Z_{q,\lambda}(nx-k) R. \tag{5.68}$$

Therefore,

$$|U_n| \leq \left(\sum_{\substack{k=\lceil na \rceil \\ \left\{ : \left\| \frac{k}{n} - x \right\|_{\infty} \leq \frac{1}{n^{\beta}} \right.}}^{\lfloor nb \rfloor} Z_{q,\lambda}(nx-k) \right)$$

$$\left[(1-\cos(1)) \left\{ \frac{\omega_{1,2}^{\max}\left(f_{\alpha}, \frac{1}{n^{\beta}}\right) N^2}{n^{2\beta}} + \omega_1\left(f, \frac{1}{n^{\beta}}\right) \right\} \right] + \rho T e^{-2\lambda n^{(1-\beta)}}$$

$$\leq \left[(1-\cos(1)) \left\{ \frac{\omega_{1,2}^{\max}\left(f_{\alpha}, \frac{1}{n^{\beta}}\right) N^2}{n^{2\beta}} + \omega_1\left(f, \frac{1}{n^{\beta}}\right) \right\} \right] + \rho T e^{-2\lambda n^{(1-\beta)}}. \tag{5.69}$$

We have established that

$$|U_n| \leq \left[(1-\cos(1)) \left\{ \frac{\omega_{1,2}^{\max}\left(f_{\alpha}, \frac{1}{n^{\beta}}\right) N^2}{n^{2\beta}} + \omega_1\left(f, \frac{1}{n^{\beta}}\right) \right\} \right]$$

$$+ \left[2 \|b-a\|_{\infty}^2 \|f_{\alpha}\|_{\infty,2}^{\max} N^2 + 2\|f\|_{\infty} \right] (1-\cos(1)) T e^{-2\lambda n^{(1-\beta)}}. \tag{5.70}$$

By (5.56), we observe that

$$\sum_{k=\lceil na \rceil}^{\lfloor nb \rfloor} f\left(\frac{k}{n}\right) Z_{q,\lambda}(nx-k) - f(x) \left(\sum_{k=\lceil na \rceil}^{\lfloor nb \rfloor} Z_{q,\lambda}(nx-k) \right)$$

$$= \left(\sum_{i=1}^{N} \left(\left(\sum_{k=\lceil na \rceil}^{\lfloor nb \rfloor} Z_{q,\lambda}(nx-k) \left(\frac{k_i}{n} - x_i \right) \right) \frac{\partial f}{\partial x_i}(x) \right) \right) \sin(1)$$

$$+2\left\{\sum_{\substack{\alpha:=(\alpha_1,\ldots,\alpha_N),\alpha_i\in\mathbb{Z}_+\\ i=1,\ldots,N,|\alpha|:=\sum_{i=1}^N\alpha_i=2}} f_\alpha(x)\left(\frac{2}{\prod_{i=1}^N\alpha_i!}\right)\left(\sum_{k=\lceil na\rceil}^{\lfloor nb\rfloor} Z_{q,\lambda}(nx-k)\right.\right.$$

$$\left.\left.\left(\prod_{i=1}^N\left(\frac{k_i}{n}-x_i\right)^{\alpha_i}\right)\right)\right\}\sin^2\left(\frac{1}{2}\right)+U_n. \tag{5.71}$$

The last says

$$A_n^*(f,x)-f(x)\left(\sum_{k=\lceil na\rceil}^{\lfloor nb\rfloor} Z_{q,\lambda}(nx-k)\right)$$

$$-\left(\sum_{=1}^{N}\frac{\partial f(x)}{\partial x_i}A_n^*((\cdot-x_i),x)\right)\sin(1)$$

$$-2\left\{\sum_{\substack{\alpha:=(\alpha_1,\ldots,\alpha_N),\alpha_i\in\mathbb{Z}_+\\ i=1,\ldots,N,|\alpha|:=\sum_{i=1}^N\alpha_i=2}} f_\alpha(x)\left(\frac{2}{\prod_{i=1}^N\alpha_i!}\right)A_n^*\left(\prod_{i=1}^N(\cdot-x_i)^{\alpha_i},x\right)\right\}$$

$$\times\sin^2\left(\frac{1}{2}\right)=U_n. \tag{5.72}$$

We note that

$$\begin{aligned}|A_n^*((\cdot-x_i),x)| &\leq A_n^*(|\cdot-x_i|,x)=\sum_{k=\lceil na\rceil}^{\lfloor nb\rfloor}\left|\frac{k_i}{n}-x_i\right|Z_{q,\lambda}(nx-k)\\ &= \sum_{\substack{k=\lceil na\rceil\\ \left\{:\left\|\frac{k}{n}-x\right\|_\infty\leq\frac{1}{n^\beta}\right.}}^{\lfloor nb\rfloor}\left|\frac{k_i}{n}-x_i\right|Z_{q,\lambda}(nx-k)\\ &\quad+\sum_{\substack{k=\lceil na\rceil\\ \left\{:\left\|\frac{k}{n}-x\right\|_\infty>\frac{1}{n^\beta}\right.}}^{\lfloor nb\rfloor}\left|\frac{k_i}{n}-x_i\right|Z_{q,\lambda}(nx-k)\end{aligned}$$

$$\leq \frac{1}{n^{\beta}} + (b_i - a_i) \sum_{\substack{k=\lceil na \rceil \\ \left\{ : \left\| \frac{k}{n} - x \right\|_{\infty} > \frac{1}{n^{\beta}} \right.}}^{\lfloor nb \rfloor} Z_{q,\lambda}(nx - k)$$

$$\leq \frac{1}{n^{\beta}} + (b_i - a_i) T e^{-2\lambda n^{(1-\beta)}}. \tag{5.73}$$

We have proved that

$$|A_n^* ((\cdot - x_i), x)| \leq \frac{1}{n^{\beta}} + (b_i - a_i) T e^{-2\lambda n^{(1-\beta)}}, \tag{5.74}$$

$i = 1, \ldots, N$.

Next we see that

$$\left| A_n^* \left(\prod_{i=1}^{N} (\cdot - x_i)^{\alpha_i}, x \right) \right| \leq A_n^* \left(\prod_{i=1}^{N} |\cdot - x_i|^{\alpha_i}, x \right)$$

$$= \sum_{k=\lceil na \rceil}^{\lfloor nb \rfloor} \left(\prod_{i=1}^{N} \left| \frac{k_i}{n} - x_i \right|^{\alpha_i} \right) Z_{q,\lambda}(nx - k)$$

$$= \sum_{\substack{k=\lceil na \rceil \\ \left\{ : \left\| \frac{k}{n} - x \right\|_{\infty} \leq \frac{1}{n^{\beta}} \right.}}^{\lfloor nb \rfloor} \left(\prod_{i=1}^{N} \left| \frac{k_i}{n} - x_i \right|^{\alpha_i} \right) Z_{q,\lambda}(nx - k)$$

$$+ \sum_{\substack{k=\lceil na \rceil \\ \left\{ : \left\| \frac{k}{n} - x \right\|_{\infty} > \frac{1}{n^{\beta}} \right.}}^{\lfloor nb \rfloor} \left(\prod_{i=1}^{N} \left| \frac{k_i}{n} - x_i \right|^{\alpha_i} \right) Z_{q,\lambda}(nx - k)$$

$$\leq \frac{1}{n^{2\beta}} + \left(\prod_{i=1}^{N} (b_i - a_i)^{\alpha_i} \right) T e^{-2\lambda n^{(1-\beta)}}. \tag{5.75}$$

We have proved that

$$\left| A_n^* \left(\prod_{i=1}^{N} (\cdot - x_i)^{\alpha_i}, x \right) \right| \leq \frac{1}{n^{2\beta}} + \left(\prod_{i=1}^{N} (b_i - a_i)^{\alpha_i} \right) T e^{-2\lambda n^{(1-\beta)}}. \tag{5.76}$$

At last we observe that

$$\left| A_n(f,x) - f(x) - \left(\sum_{=1}^{N} \frac{\partial f(x)}{\partial x_i} A_n((\cdot - x_i), x) \right) \sin(1) \right.$$

$$\left. -4 \left\{ \sum_{\substack{\alpha:=(\alpha_1,\ldots,\alpha_N), \alpha_i \in \mathbb{Z}_+ \\ i=1,\ldots,N, |\alpha|:=\sum_{i=1}^N \alpha_i = 2}} f_\alpha(x) \left(\frac{1}{\prod_{i=1}^N \alpha_i!} \right) A_n \left(\prod_{i=1}^N (\cdot - x_i)^{\alpha_i}, x \right) \right\} \sin^2\left(\frac{1}{2}\right) \right|$$

$$\leq (\Delta(q))^N |U_n|$$

$$= (\Delta(q))^N \left| A_n^*(f,x) - f(x) \left(\sum_{k=\lceil na \rceil}^{\lfloor nb \rfloor} Z_{q,\lambda}(nx-k) \right) \right.$$

$$- \left(\sum_{=1}^{N} \frac{\partial f(x)}{\partial x_i} A_n^*((\cdot - x_i), x) \right) \sin(1)$$

$$\left. -4 \left\{ \sum_{\substack{\alpha:=(\alpha_1,\ldots,\alpha_N), \alpha_i \in \mathbb{Z}_+ \\ i=1,\ldots,N, |\alpha|:=\sum_{i=1}^N \alpha_i = 2}} f_\alpha(x) \left(\frac{1}{\prod_{i=1}^N \alpha_i!} \right) A_n^* \left(\prod_{i=1}^N (\cdot - x_i)^{\alpha_i}, x \right) \right\} \sin^2\left(\frac{1}{2}\right) \right|. \tag{5.77}$$

Putting all of the above together, we prove the theorem. □

We make

Remark 5.6. Let $f \in C^2\left(\prod_{i=1}^N [a_i, b_i], \mathbb{C}\right)$, $N \in \mathbb{N}$. By the mean value theorem, we have that $\sinh x = \sinh x - \sinh 0 = (\cosh \xi)(x - 0)$, for some ξ between $\{0, x\}$, for any $x \in \mathbb{R}$.

Hence

$$|\sinh x| \leq \|\cosh\|_{\infty,[-1,1]} |x|, \quad \forall\, x \in [-1,1].$$

But

$$\|\cosh\|_{\infty,[-1,1]} = \cosh(1).$$

Thus, we have

$$|\sinh x| \leq \cosh(1) |x|, \quad \forall\, x \in [-1,1].$$

Let $\frac{k}{n} := \left(\frac{k_1}{n}, \ldots, \frac{k_N}{n}\right)$, and $x := (x_1, \ldots, x_N)$, with $\frac{k}{n}, x \in \left(\prod_{i=1}^{N} [a_i, b_i]\right)$, then (by (5.55), where $g_{\frac{k}{n}}(t) := f\left(x + t\left(\frac{k}{n} - x\right)\right)$, $0 \leq t \leq 1$) we have

$$f\left(\frac{k}{n}\right) - f(x) = \left(\sum_{i=1}^{N}\left(\frac{k_i}{n} - x_i\right)\frac{\partial f}{\partial x_i}(x)\right)\sinh(1)$$
$$+ 2\left\{\left[\left(\sum_{i=1}^{N}\left(\frac{k_i}{n} - x_i\right)\frac{\partial}{\partial x_i}\right)^2 f\right](x)\right\}\sinh^2\left(\frac{1}{2}\right)$$
$$+ \int_0^1 \left\{\left\{\left[\left(\sum_{i=1}^{N}\left(\frac{k_i}{n} - x_i\right)\frac{\partial}{\partial x_i}\right)^2 f\right]\left(x + t\left(\frac{k}{n} - x\right)\right) - f\left(x + t\left(\frac{k}{n} - x\right)\right)\right\}\right.$$
$$\left. - \left\{\left[\left(\sum_{i=1}^{N}\left(\frac{k_i}{n} - x_i\right)\frac{\partial}{\partial x_i}\right)^2 f\right](x) - f(x)\right\}\right\}\sinh(1-t)\,dt. \tag{5.78}$$

Denote the remainder

$$R := \int_0^1 \left\{\left\{\left[\left(\sum_{i=1}^{N}\left(\frac{k_i}{n} - x_i\right)\frac{\partial}{\partial x_i}\right)^2 f\right]\right.\right.$$
$$\left. \times \left(x + t\left(\frac{k}{n} - x\right)\right) - f\left(x + t\left(\frac{k}{n} - x\right)\right)\right\}$$
$$\left. - \left\{\left[\left(\sum_{i=1}^{N}\left(\frac{k_i}{n} - x_i\right)\frac{\partial}{\partial x_i}\right)^2 f\right](x) - f(x)\right\}\right\}\sinh(1-t)\,dt$$
$$= \int_0^1 \left\{\sum_{\substack{\alpha:=(\alpha_1,\ldots,\alpha_N),\alpha_i\in\mathbb{Z}_+ \\ i=1,\ldots,N,|\alpha|:=\sum_{i=1}^{N}\alpha_i=2}} \left(\frac{2}{\prod_{i=1}^{N}\alpha_i!}\right)\left(\prod_{i=1}^{N}\left(\frac{k_i}{n} - x_i\right)^{\alpha_i}\right)\right.$$
$$\left[f_\alpha\left(x + t\left(\frac{k}{n} - x\right)\right) - f_\alpha(x)\right]$$
$$\left. - \left(f\left(x + t\left(\frac{k}{n} - x\right)\right) - f(x)\right)\right\}\sinh(1-t)\,dt. \tag{5.79}$$

Therefore it holds

$$|R| \leq \int_0^1 \left\{ \sum_{\substack{\alpha:=(\alpha_1,\ldots,\alpha_N),\alpha_i\in\mathbb{Z}_+ \\ i=1,\ldots,N,|\alpha|:=\sum_{i=1}^N \alpha_i=2}} \left(\frac{2}{\prod_{i=1}^N \alpha_i!}\right) \right.$$

$$\left(\prod_{i=1}^N \left|\frac{k_i}{n} - x_i\right|^{\alpha_i}\right) \left|f_\alpha\left(x + t\left(\frac{k}{n} - x\right)\right) - f_\alpha(x)\right|$$

$$\left. + \left|f\left(x + t\left(\frac{k}{n} - x\right)\right) - f(x)\right| \right\} |\sinh(1-t)|\, dt$$

$$\leq \int_0^1 \left\{ \sum_{\substack{\alpha:=(\alpha_1,\ldots,\alpha_N),\alpha_i\in\mathbb{Z}_+ \\ i=1,\ldots,N,|\alpha|:=\sum_{i=1}^N \alpha_i=2}} \left(\frac{2}{\prod_{i=1}^N \alpha_i!}\right) \right.$$

$$\times \left(\prod_{i=1}^N \left|\frac{k_i}{n} - x_i\right|^{\alpha_i}\right) \omega_1\left(f_\alpha, t\left\|\frac{k}{n} - x\right\|_\infty\right)$$

$$\left. + \omega_1\left(f, t\left\|\frac{k}{n} - x\right\|_\infty\right) \right\} \cosh(1)(1-t)\, dt \leq (*). \tag{5.80}$$

Note here that $(0 < \beta < 1)$

$$\left\|\frac{k}{n} - x\right\|_\infty \leq \frac{1}{n^\beta} \Leftrightarrow \left|\frac{k_i}{n} - x_i\right| \leq \frac{1}{n^\beta}, \quad i = 1, \ldots, N. \tag{5.81}$$

We further see that

$$(*) \leq \cosh(1) \left\{ \omega_{1,2}^{\max}\left(f_\alpha, \frac{1}{n^\beta}\right) \left(\sum_{\substack{\alpha:=(\alpha_1,\ldots,\alpha_N),\alpha_i\in\mathbb{Z}_+ \\ i=1,\ldots,N,|\alpha|:=\sum_{i=1}^N \alpha_i=2}} \left(\frac{2}{\prod_{i=1}^N \alpha_i!}\right) \left(\prod_{i=1}^N \frac{1}{n^{\beta\alpha_i}}\right) \right) \right.$$

$$\left. + \omega_1\left(f, \frac{1}{n^\beta}\right) \right\} \int_0^1 (1-t)\, dt$$

$$= \cosh(1) \left\{ \omega_{1,2}^{\max}\left(f_\alpha, \frac{1}{n^\beta}\right) \left(\sum_{\substack{\alpha:=(\alpha_1,\ldots,\alpha_N),\alpha_i\in\mathbb{Z}_+ \\ i=1,\ldots,N,|\alpha|:=\sum_{i=1}^N \alpha_i=2}} \left(\frac{2}{\prod_{i=1}^N \alpha_i!}\right)\right) \frac{1}{n^{2\beta}} \right.$$

$$\left. + \omega_1\left(f, \frac{1}{n^\beta}\right) \right\} \frac{1}{2}$$

$$= \frac{\cosh(1)}{2} \left\{ \frac{\omega_{1,2}^{\max}\left(f_\alpha, \frac{1}{n^\beta}\right) N^2}{n^{2\beta}} + \omega_1\left(f, \frac{1}{n^\beta}\right) \right\}. \tag{5.82}$$

We have proved that

$$|R| \leq \frac{\cosh(1)}{2} \left\{ \frac{\omega_{1,2}^{\max}\left(f_\alpha, \frac{1}{n^\beta}\right) N^2}{n^{2\beta}} + \omega_1\left(f, \frac{1}{n^\beta}\right) \right\}, \tag{5.83}$$

given that $\left\|\frac{k}{n} - x\right\|_\infty \leq \frac{1}{n^\beta}$.

We note also that

$$|R| \leq \cosh(1) \int_0^1 \left\{ \sum_{\substack{\alpha:=(\alpha_1,\ldots,\alpha_N),\alpha_i\in\mathbb{Z}_+ \\ i=1,\ldots,N,|\alpha|:=\sum_{i=1}^N \alpha_i=2}} \left(\frac{2}{\prod_{i=1}^N \alpha_i!}\right) \right.$$

$$\left. \left(\prod_{i=1}^N (b_i - a_i)^{\alpha_i}\right) 2\|f_\alpha\|_\infty + 2\|f\|_\infty \right\} (1-t)\,dt$$

$$\leq \cosh(1) \left\{ \left(\sum_{\substack{\alpha:=(\alpha_1,\ldots,\alpha_N),\alpha_i\in\mathbb{Z}_+ \\ i=1,\ldots,N,|\alpha|:=\sum_{i=1}^N \alpha_i=2}} \left(\frac{2}{\prod_{i=1}^N \alpha_i!}\right)\right) \right.$$

$$\left. 2\|b-a\|_\infty^2 \|f_\alpha\|_{\infty,2}^{\max} + 2\|f\|_\infty \right\} \left(\int_0^1 (1-t)\,dt\right)$$

$$= \cosh(1)\left\{2\|b-a\|_\infty^2 \|f_\alpha\|_{\infty,2}^{\max} N^2 + 2\|f\|_\infty\right\}\frac{1}{2}$$

$$= \cosh(1)\left(\|b-a\|_\infty^2 \|f_\alpha\|_{\infty,2}^{\max} N^2 + \|f\|_\infty\right), \tag{5.84}$$

where $a := (a_1, \ldots, a_N)$, $b = (b_1, \ldots, b_N)$.

We have proved that

$$|R| \leq \cosh(1)\left(\|b-a\|_\infty^2 \|f_\alpha\|_{\infty,2}^{\max} N^2 + \|f\|_\infty\right) =: \rho. \tag{5.85}$$

We continue with the hyperbolic approximation.

Theorem 5.8. *Let* $f \in C^2\left(\prod_{i=1}^N [a_i, b_i], \mathbb{C}\right)$, $0 < \beta < 1$, $n, N \in \mathbb{N}$, $n^{1-\beta} > 2$, $x, x_0 \in \left(\prod_{i=1}^N [a_i, b_i]\right)$, $a := (a_1, \ldots, a_N)$, $b := (b_1, \ldots, b_N)$. *Then*

(i)

$$\Bigg| A_n(f,x) - f(x) - \left(\sum_{=1}^{N} \frac{\partial f(x)}{\partial x_i} A_n((\cdot - x_i), x)\right)\sinh(1)$$

$$-4\left\{\sum_{\substack{\alpha:=(\alpha_1,\ldots,\alpha_N),\alpha_i\in\mathbb{Z}_+\\ i=1,\ldots,N,|\alpha|:=\sum_{i=1}^N \alpha_i=2}} f_\alpha(x)\left(\frac{1}{\prod_{i=1}^N \alpha_i!}\right)A_n\left(\prod_{i=1}^N (\cdot - x_i)^{\alpha_i}, x\right)\right\}\sinh^2\left(\frac{1}{2}\right)\Bigg|$$

$$\leq (\Delta(q))^N \cosh(1)\left\{\left[\frac{1}{2}\left\{\frac{\omega_{1,2}^{\max}\left(f_\alpha, \frac{1}{n^\beta}\right)N^2}{n^{2\beta}} + \omega_1\left(f, \frac{1}{n^\beta}\right)\right\}\right]\right.$$

$$\left. + \left[\|b-a\|_\infty^2 \|f_\alpha\|_{\infty,2}^{\max} N^2 + \|f\|_\infty\right] T e^{-2\lambda n^{(1-\beta)}}\right\}, \tag{5.86}$$

(ii) *assume that* $\frac{\partial f(x_0)}{\partial x_i} = 0$, $i = 1, \ldots, N$, *and* $f_\alpha(x_0) = 0$, $\alpha : |\alpha| = 2$, *we have that*

$$|A_n(f,x) - f(x)|$$

$$\leq (\Delta(q))^N (\cosh(1))\left\{\left[\frac{1}{2}\left\{\frac{\omega_{1,2}^{\max}\left(f_\alpha, \frac{1}{n^\beta}\right)N^2}{n^{2\beta}} + \omega_1\left(f, \frac{1}{n^\beta}\right)\right\}\right]\right.$$

$$\left. + \left[\|b-a\|_\infty^2 \|f_\alpha\|_{\infty,2}^{\max} N^2 + \|f\|_\infty\right] T e^{-2\lambda n^{(1-\beta)}}\right\}, \tag{5.87}$$

(iii)

$$|A_n(f,x)-f(x)| \leq (\Delta(q))^N$$
$$\left\{\left\{\left\{\sum_{i=1}^{N}\left|\frac{\partial f(x)}{\partial x_i}\right|\left\{\frac{1}{n^\beta}+(b_i-a_i)\,Te^{-2\lambda n^{(1-\beta)}}\right\}\right\}\sinh(1)\right.\right.\right.$$
$$+4\left\{\sum_{\alpha:|\alpha|=2}|f_\alpha(x)|\left(\frac{1}{\prod_{i=1}^{N}\alpha_i!}\right)\left[\frac{1}{n^{2\beta}}+\left(\prod_{i=1}^{N}(b_i-a_i)^{\alpha_i}\right)Te^{-2\lambda n^{(1-\beta)}}\right]\right\}\sinh^2\left(\frac{1}{2}\right)\Bigg\}$$
$$+\cosh(1)\left\{\left[\frac{1}{2}\left\{\frac{\omega_{1,2}^{\max}\left(f_\alpha,\frac{1}{n^\beta}\right)N^2}{n^{2\beta}}+\omega_1\left(f,\frac{1}{n^\beta}\right)\right\}\right]\right.$$
$$\left.\left.+\left[\|b-a\|_\infty^2\,\|f_\alpha\|_{\infty,2}^{\max}N^2+\|f\|_\infty\right]Te^{-2\lambda n^{(1-\beta)}}\right\}\right\}, \tag{5.88}$$

and

(iv)

$$\|A_n(f)-f\|_\infty \leq (\Delta(q))^N$$
$$\left\{\left\{\left\{\sum_{i=1}^{N}\left\|\frac{\partial f}{\partial x_i}\right\|_\infty\left\{\frac{1}{n^\beta}+(b_i-a_i)\,Te^{-2\lambda n^{(1-\beta)}}\right\}\right\}\sinh(1)\right.\right.\right.$$
$$+4\left\{\sum_{\alpha:|\alpha|=2}\|f_\alpha\|_\infty\left(\frac{1}{\prod_{i=1}^{N}\alpha_i!}\right)\left[\frac{1}{n^{2\beta}}+\left(\prod_{i=1}^{N}(b_i-a_i)^{\alpha_i}\right)Te^{-2\lambda n^{(1-\beta)}}\right]\right\}\sinh^2\left(\frac{1}{2}\right)\Bigg\}$$
$$+\cosh(1)\left\{\left[\frac{1}{2}\left\{\frac{\omega_{1,2}^{\max}\left(f_\alpha,\frac{1}{n^\beta}\right)N^2}{n^{2\beta}}+\omega_1\left(f,\frac{1}{n^\beta}\right)\right\}\right]\right.$$
$$\left.\left.+\left[\|b-a\|_\infty^2\,\|f_\alpha\|_{\infty,2}^{\max}N^2+\|f\|_\infty\right]Te^{-2\lambda n^{(1-\beta)}}\right\}\right\} =: \psi_n(f). \tag{5.89}$$

We observe that $A_n \to I$ (unit operator), as $n\to\infty$, pointwise and uniformly.

Proof. Here R is as in (5.79). We see that

$$U_n := \sum_{k=\lceil na\rceil}^{\lfloor nb\rfloor} Z_{q,\lambda}(nx-k)\,R$$
$$= \sum_{\substack{k=\lceil na\rceil \\ \left\{:\left\|\frac{k}{n}-x\right\|_\infty\leq\frac{1}{n^\beta}\right.}}^{\lfloor nb\rfloor} Z_{q,\lambda}(nx-k)\,R + \sum_{\substack{k=\lceil na\rceil \\ \left\{:\left\|\frac{k}{n}-x\right\|_\infty>\frac{1}{n^\beta}\right.}}^{\lfloor nb\rfloor} Z_{q,\lambda}(nx-k)\,R. \tag{5.90}$$

Therefore,

$$|U_n| \leq \left(\sum_{\substack{k=\lceil na \rceil \\ \left\{ : \left\| \frac{k}{n} - x \right\|_\infty \leq \frac{1}{n^\beta} \right.}}^{\lfloor nb \rfloor} Z_{q,\lambda}(nx-k) \right)$$

$$\cosh(1) \left[\frac{1}{2} \left\{ \frac{\omega_{1,2}^{\max}\left(f_\alpha, \frac{1}{n^\beta}\right) N^2}{n^{2\beta}} + \omega_1\left(f, \frac{1}{n^\beta}\right) \right\} \right] + \rho T e^{-2\lambda n^{(1-\beta)}}$$

$$\leq \cosh(1) \left[\frac{1}{2} \left\{ \frac{\omega_{1,2}^{\max}\left(f_\alpha, \frac{1}{n^\beta}\right) N^2}{n^{2\beta}} + \omega_1\left(f, \frac{1}{n^\beta}\right) \right\} \right] + \rho T e^{-2\lambda n^{(1-\beta)}}. \tag{5.91}$$

We have established that

$$|U_n| \leq \cosh(1) \left[\frac{1}{2} \left\{ \frac{\omega_{1,2}^{\max}\left(f_\alpha, \frac{1}{n^\beta}\right) N^2}{n^{2\beta}} + \omega_1\left(f, \frac{1}{n^\beta}\right) \right\} \right]$$

$$+ \cosh(1) \left[\|b-a\|_\infty^2 \|f_\alpha\|_{\infty,2}^{\max} N^2 + \|f\|_\infty \right] T e^{-2\lambda n^{(1-\beta)}}. \tag{5.92}$$

By (5.78), we observe that

$$\sum_{k=\lceil na \rceil}^{\lfloor nb \rfloor} f\left(\frac{k}{n}\right) Z_{q,\lambda}(nx-k) - f(x) \left(\sum_{k=\lceil na \rceil}^{\lfloor nb \rfloor} Z_{q,\lambda}(nx-k) \right)$$

$$= \left(\sum_{i=1}^{N} \left(\left(\sum_{k=\lceil na \rceil}^{\lfloor nb \rfloor} Z_{q,\lambda}(nx-k) \left(\frac{k_i}{n} - x_i \right) \right) \frac{\partial f}{\partial x_i}(x) \right) \right) \sinh(1)$$

$$+ 2 \left\{ \sum_{\substack{\alpha := (\alpha_1, \ldots, \alpha_N), \alpha_i \in \mathbb{Z}_+ \\ i=1,\ldots,N, |\alpha| := \sum_{i=1}^N \alpha_i = 2}} f_\alpha(x) \left(\frac{2}{\prod_{i=1}^N \alpha_i!} \right) \left(\sum_{k=\lceil na \rceil}^{\lfloor nb \rfloor} Z_{q,\lambda}(nx-k) \right. \right.$$

$$\left. \left. \left(\prod_{i=1}^N \left(\frac{k_i}{n} - x_i \right)^{\alpha_i} \right) \right) \right\} \sinh^2\left(\frac{1}{2}\right) + U_n.$$

The last says

$$A_n^*(f,x) - f(x)\left(\sum_{k=\lceil na\rceil}^{\lfloor nb\rfloor} Z_{q,\lambda}(nx-k)\right)$$

$$-\left(\sum_{=1}^{N} \frac{\partial f(x)}{\partial x_i} A_n^*((\cdot - x_i), x)\right)\sinh(1)$$

$$-2\left\{\sum_{\substack{\alpha:=(\alpha_1,\ldots,\alpha_N),\alpha_i\in\mathbb{Z}_+\\ i=1,\ldots,N,|\alpha|:=\sum_{i=1}^N \alpha_i=2}} f_\alpha(x)\left(\frac{2}{\prod_{i=1}^N \alpha_i!}\right) A_n^*\left(\prod_{i=1}^N (\cdot - x_i)^{\alpha_i}, x\right)\right\}$$

$$\times \sinh^2\left(\frac{1}{2}\right) = U_n. \tag{5.93}$$

As earlier it holds

$$|A_n^*((\cdot - x_i), x)| \leq \frac{1}{n^\beta} + (b_i - a_i)\, T e^{-2\lambda n^{(1-\beta)}}, \tag{5.94}$$

$i = 1, \ldots, N$.

Also, as earlier we have

$$\left|A_n^*\left(\prod_{i=1}^N (\cdot - x_i)^{\alpha_i}, x\right)\right| \leq \frac{1}{n^{2\beta}} + \left(\prod_{i=1}^N (b_i - a_i)^{\alpha_i}\right) T e^{-2\lambda n^{(1-\beta)}}. \tag{5.95}$$

At last we observe that

$$\Bigg| A_n(f,x) - f(x) - \left(\sum_{=1}^{N} \frac{\partial f(x)}{\partial x_i} A_n((\cdot - x_i), x)\right)\sinh(1)$$

$$-4\left\{\sum_{\substack{\alpha:=(\alpha_1,\ldots,\alpha_N),\alpha_i\in\mathbb{Z}_+\\ i=1,\ldots,N,|\alpha|:=\sum_{i=1}^N \alpha_i=2}} f_\alpha(x)\left(\frac{1}{\prod_{i=1}^N \alpha_i!}\right) A_n\left(\prod_{i=1}^N (\cdot - x_i)^{\alpha_i}, x\right)\right\}\sinh^2\left(\frac{1}{2}\right)\Bigg|$$

$$\leq (\Delta(q))^N |U_n|$$

$$
= (\Delta (q))^N \left| A_n^* (f, x) - f (x) \left(\sum_{k=\lceil na \rceil}^{\lfloor nb \rfloor} Z_{q,\lambda} (nx - k) \right) \right.
$$

$$
- \left(\sum_{=1}^{N} \frac{\partial f (x)}{\partial x_i} A_n^* ((\cdot - x_i), x) \right) \sinh (1)
$$

$$
\left. - 4 \left\{ \sum_{\substack{\alpha:=(\alpha_1,\ldots,\alpha_N),\alpha_i \in \mathbb{Z}_+ \\ i=1,\ldots,N,|\alpha|:=\sum_{i=1}^N \alpha_i = 2}} f_\alpha (x) \left(\frac{1}{\prod_{i=1}^N \alpha_i!} \right) A_n^* \left(\prod_{i=1}^N (\cdot - x_i)^{\alpha_i}, x \right) \right\} \sinh^2 \left(\frac{1}{2} \right) \right|. \tag{5.96}
$$

Putting all of the above together, we prove theorem. □

We make

Remark 5.7. By (5.36) we get that $\|A_n (f)\|_\infty \leq \|f\|_\infty < \infty$, and $A_n (f) \in C \left(\prod_{i=1}^N [a_i, b_i], \mathbb{C} \right)$, given that $f \in C \left(\prod_{i=1}^N [a_i, b_i], \mathbb{C} \right)$.

Clearly then

$$
\left\| A_n^2 (f) \right\|_\infty = \| A_n (A_n (f)) \|_\infty \leq \| A_n (f) \|_\infty \leq \| f \|_\infty, \tag{5.97}
$$

etc.

Therefore we get

$$
\left\| A_n^k (f) \right\|_\infty \leq \| f \|_\infty, \quad \forall \, k \in \mathbb{N}, \tag{5.98}
$$

the contraction property.

Also we see that

$$
\left\| A_n^k (f) \right\|_\infty \leq \left\| A_n^{k-1} (f) \right\|_\infty \leq \cdots \leq \| A_n (f) \|_\infty \leq \| f \|_\infty. \tag{5.99}
$$

Also $A_n (1) = 1$, $A_n^k (1) = 1$, $\forall \, k \in \mathbb{N}$.

Following 18.14 in Anastassiou (2016), pp. 401–402, similarly we obtain that

$$
\| A_n^r f - f \|_\infty \leq r \, \| A_n (f) - f \|_\infty, \quad r \in \mathbb{N}. \tag{5.100}
$$

We give

Theorem 5.9. *All as in Theorems 5.7 and 5.8. Then*

(i)

$$\|A_n^r f - f\|_\infty \leq r\xi_n(f), \tag{5.101}$$

where $\xi_n(f)$ *as in (5.67),*

(ii)

$$\|A_n^r f - f\|_\infty \leq r\psi_n(f), \tag{5.102}$$

where $\psi_n(f)$ *as in (5.89).*

So the speed of convergence to the unit operator of A_n^r *is not worse than of* A_n*, see also Anastassiou (2013).*

Bibliography

G.A. Anastassiou, Rate of convergence of some neural network operators to the unit-univariate case, *Journal of Mathematical Analysis and Applications*, 212 (1997), 237–262.

G.A. Anastassiou, *Quantitative Approximations*, Chapman & Hall/CRC, Boca Raton, New York, 2001.

G.A. Anastassiou, *Inteligent Systems: Approximation by Artificial Neural Networks*, Intelligent Systems Reference Library, Vol. 19, Springer, Heidelberg, 2011a.

G.A. Anastassiou, Univariate hyperbolic tangent neural network approximation, *Mathematics and Computer Modelling*, 53 (2011b), 1111–1132.

G.A. Anastassiou, Multivariate hyperbolic tangent neural network approximation, *Computers and Mathematics*, 61 (2011c), 809–821.

G.A. Anastassiou, Multivariate sigmoidal neural network approximation, *Neural Networks*, 24 (2011d), 378–386.

G.A. Anastassiou, Univariate sigmoidal neural network approximation, *Journal of Computational Analysis and Applications*, 14(4) (2012), 659–690.

G.A. Anastassiou, Approximation by neural networks iterates, G. Anastassiou, O. Duman. (Eds.), *Advances in Applied Mathematics and Approximation Theory*, pp. 1–20, Springer Proceedings in Math. & Stat., Springer, New York, 2013.

G.A. Anastassiou, *Intelligent Systems II: Complete Approximation by Neural Network Operators*, Springer, Heidelberg, New York, 2016.

G.A. Anastassiou, *Intelligent Computations: Abstract Fractional Calculus, Inequalities, Approximations,* Springer, Heidelberg, New York, 2018.

G.A. Anastassiou, General sigmoid based Banach space valued neural network approximation, *Journal of Computational Analysis and Applications*, 31(4) (2023a), 520–534.

G.A. Anastassiou, *Parametrized, Deformed and General Neural Networks*, accepted for publication, Springer, Heidelberg, New York, 2023b.

G.A. Anastassiou, Opial and Ostrowski type inequalities based on trigonometric and hyperbolic type Taylor formulae, *Malaya Journal of Mathematik*, 11(S) (2023c), 1–26.

G.A. Anastassiou, q-Deformed and λ-parametrized hyperbolic tangent function relied complex valued multivariate trigonometric and hyperbolic neural network approximations, *Annals of Communications in Mathematics*, 6(3) (2023d), 141–164.

Z. Chen and F. Cao, The approximation operators with sigmoidal functions, *Computers and Mathematics with Applications*, 58 (2009), 758–765.

D. Costarelli, R. Spigler, Approximation results for neural network operators activated by sigmoidal functions, *Neural Networks*, 44 (2013a), 101–106.

D. Costarelli, R. Spigler, Multivariate neural network operators with sigmoidal activation functions, *Neural Networks* 48 (2013b), 72–77.

S. Haykin, *Neural Networks: A Comprehensive Foundation* 2nd ed., Prentice Hall, New York, 1998.

W. McCulloch and W. Pitts, A logical calculus of the ideas immanent in nervous activity, *Bulletin of Mathematical Biophysics*, 7 (1943), 115–133.

T.M. Mitchell, *Machine Learning*, WCB-McGraw-Hill, New York, 1997.

Chapter 6

General Sigmoid Function-Activated Complex-Valued Trigonometric and Hyperbolic Neural Network High Order Approximation

Here we research the univariate quantitative approximation of complex-valued continuous functions on a compact interval by complex-valued neural network operators. These approximations are derived by establishing Jackson-type inequalities involving the modulus of continuity of the used function's high order derivatives. The kinds of our approximations are trigonometric and hyperbolic. Our operators are defined by using a density function generated by a general sigmoid function. The approximations are pointwise and of the uniform norm. The related complex-valued feed-forward neural networks are with one hidden layer. It follows Anastassiou (2023d).

6.1 Introduction

The author in Anastassiou (1997, 2001), see Chapters 2–5, was the first to establish neural network approximation to continuous functions with rates by very specifically defined neural network operators of Cardaliaguet–Euvrard and "Squashing" types, by employing the modulus of continuity of the engaged function or its high order derivative, and producing very tight Jackson-type inequalities. He treats there both the univariate and multivariate cases. The definitions of these operators' "bell-shaped" and "squashing" functions are assumed to be of compact support.

Again the author inspired by Chen and Cao (2009) continued his studies on neural network approximation by introducing and using the proper quasi-interpolation operators of sigmoidal and hyperbolic tangent types which resulted into Anastassiou (2011a)–Anastassiou (2012), by treating both the univariate and multivariate cases.

Here the author performs general sigmoid function-activated high order neural network approximations to continuous functions over compact intervals of the real line with complex values. All convergences are with rates expressed via the modulus of continuity of the involved functions high order derivatives, deriving by very tight Jackson-type inequalities.

The basis of our higher-order approximations here are some newly discovered by the author trigonometric and hyperbolic type Taylor's formulae.

Our compact intervals are not necessarily symmetric to the origin. In preparation to prove our results, we describe important properties of the basic density function defining our operators which is induced by a general sigmoid function, which is the activation function.

Feed-forward $\mathbb{C}$-valued neural networks (FNNs) with one hidden layer, the only type of networks we deal with in this chapter, are mathematically expressed as

$$N_n(x) = \sum_{j=0}^{n} c_j \sigma\left(\langle a_j \cdot x \rangle + b_j\right), \quad x \in \mathbb{R}^s, \ \ s \in \mathbb{N},$$

where for $0 \leq j \leq n$, $b_j \in \mathbb{R}$ are the thresholds, $a_j \in \mathbb{R}^s$ are the connection weights, $c_j \in \mathbb{C}$ are the coefficients, $\langle a_j \cdot x \rangle$ is the inner product of a_j and x, and σ is the activation function of the network. For more information about neural networks in general, read Haykin (1998), McCulloch and Pitts (1943), and Mitchell (1997).

6.2 Basics

Here we follow Anastassiou (2023a).

Let $h : \mathbb{R} \to [-1, 1]$ be a general sigmoid function, such that it is strictly increasing, $h(0) = 0$, $h(-x) = -h(x)$, $h(+\infty) = 1$, $h(-\infty) = -1$. Also h is strictly convex over $(-\infty, 0]$ and strictly concave over $[0, +\infty)$, with $h^{(2)} \in C(\mathbb{R})$.

We consider the activation function

$$\psi(x) := \frac{1}{4}\left(h(x+1) - h(x-1)\right), \quad x \in \mathbb{R}. \tag{6.1}$$

As in Anastassiou (2011d), p. 34, we get that $\psi(-x) = \psi(x)$, thus ψ is an even function. Since $x+1 > x-1$, then $h(x+1) > h(x-1)$, and $\psi(x) > 0$, all $x \in \mathbb{R}$.

We see that

$$\psi(0) = \frac{h(1)}{2}. \tag{6.2}$$

Let $x > 1$, we have that

$$\psi'(x) = \frac{1}{4}\left(h'(x+1) - h'(x-1)\right) < 0,$$

by h' being strictly decreasing over $[0, +\infty)$.

Let now $0 < x < 1$, then $1 - x > 0$ and $0 < 1 - x < 1 + x$. It holds $h'(x-1) = h'(1-x) > h'(x+1)$ so that again $\psi'(x) < 0$. Consequently, ψ is straitly decreasing on $(0, +\infty)$.

Clearly, ψ is strictly increasing on $(-\infty, 0)$, and $\psi'(0) = 0$.

See that

$$\lim_{x \to +\infty} \psi(x) = \frac{1}{4}\left(h(+\infty) - h(+\infty)\right) = 0 \tag{6.3}$$

and

$$\lim_{x \to -\infty} \psi(x) = \frac{1}{4}\left(h(-\infty) - h(-\infty)\right) = 0. \tag{6.4}$$

That is, the x-axis is the horizontal asymptote on ψ.

Conclusion, ψ is a bell symmetric function with maximum

$$\psi(0) = \frac{h(1)}{2}.$$

We need

Theorem 6.1 (Anastassiou 2023b, Ch. 21). *We have that*

$$\sum_{i=-\infty}^{\infty} \psi(x - i) = 1, \quad \forall\, x \in \mathbb{R}. \tag{6.5}$$

Theorem 6.2 (Anastassiou 2023b, Ch. 21). *It holds*

$$\int_{-\infty}^{\infty} \psi(x)\, dx = 1. \tag{6.6}$$

Thus, $\psi(x)$ is a density function on $\mathbb{R}$.

We give

Theorem 6.3 (Anastassiou 2023b, Ch. 21). *Let $0 < \alpha < 1$, and $n \in \mathbb{N}$ with $n^{1-\alpha} > 2$. It holds*

$$\sum_{\begin{cases} k=-\infty \\ :|nx-k| \geq n^{1-\alpha} \end{cases}}^{\infty} \psi(nx - k) < \left(1 - h\left(n^{1-\alpha} - 2\right)\right). \tag{6.7}$$

Note that

$$\lim_{n\to+\infty}\left(1-h\left(n^{1-\alpha}-2\right)\right)=0.$$

Denote by $\lfloor\cdot\rfloor$ the integral part of the number and by $\lceil\cdot\rceil$ the ceiling of the number.

We further give

Theorem 6.4 (Anastassiou 2023b, Ch. 21). *Let $x\in[a,b]\subset\mathbb{R}$ and $n\in\mathbb{N}$ so that $\lceil na\rceil\leq\lfloor nb\rfloor$. It holds*

$$\frac{1}{\sum_{k=\lceil na\rceil}^{\lfloor nb\rfloor}\psi\left(nx-k\right)}<\frac{1}{\psi\left(1\right)},\quad\forall\,x\in[a,b]. \tag{6.8}$$

Remark 6.1 (Anastassiou 2023b, Ch. 21). We have that

$$\lim_{n\to\infty}\sum_{k=\lceil na\rceil}^{\lfloor nb\rfloor}\psi\left(nx-k\right)\neq1, \tag{6.9}$$

for at least some $x\in[a,b]$.

See Anastassiou (2011d), p. 39, same reasoning.

Note 6.1 (Anastassiou 2023b, Ch. 21). For large enough n, we always obtain $\lceil na\rceil\leq\lfloor nb\rfloor$. Also $a\leq\frac{k}{n}\leq b$, iff $\lceil na\rceil\leq k\leq\lfloor nb\rfloor$. In general, it holds (by (6.5))

$$\sum_{k=\lceil na\rceil}^{\lfloor nb\rfloor}\psi\left(nx-k\right)\leq1. \tag{6.10}$$

Let $(\mathbb{C},\|\cdot\|)$ be the complex number Banach space.

Definition 6.1. Let $f\in C\left([a,b],\mathbb{C}\right)$ and $n\in\mathbb{N}:\lceil na\rceil\leq\lfloor nb\rfloor$. We introduce and define the $\mathbb{C}$-valued linear neural network operators

$$A_n\left(f,x\right):=\frac{\sum_{k=\lceil na\rceil}^{\lfloor nb\rfloor}f\left(\frac{k}{n}\right)\psi\left(nx-k\right)}{\sum_{k=\lceil na\rceil}^{\lfloor nb\rfloor}\psi\left(nx-k\right)},\quad x\in[a,b]. \tag{6.11}$$

Clearly here $A_n\left(f,x\right)\in C\left([a,b],\mathbb{C}\right)$. For conveniencc we use the same A_n for real-valued function when needed. We study here the pointwise and uniform convergence of $A_n\left(f,x\right)$ to $f\left(x\right)$ with rates.

For convenience also we call

$$A_n^*\left(f,x\right):=\sum_{k=\lceil na\rceil}^{\lfloor nb\rfloor}f\left(\frac{k}{n}\right)\psi\left(nx-k\right) \tag{6.12}$$

(similarly A_n^* can be defined for real-valued function), that is,

$$A_n(f,x) = \frac{A_n^*(f,x)}{\sum_{k=\lceil na \rceil}^{\lfloor nb \rfloor} \psi(nx-k)}. \tag{6.13}$$

So

$$\begin{aligned} A_n(f,x) - f(x) &= \frac{A_n^*(f,x)}{\sum_{k=\lceil na \rceil}^{\lfloor nb \rfloor} \psi(nx-k)} - f(x) \\ &= \frac{A_n^*(f,x) - f(x)\left(\sum_{k=\lceil na \rceil}^{\lfloor nb \rfloor} \psi(nx-k)\right)}{\sum_{k=\lceil na \rceil}^{\lfloor nb \rfloor} \psi(nx-k)}. \end{aligned} \tag{6.14}$$

Consequently we derive

$$|A_n(f,x) - f(x)| \leq \frac{1}{\psi(1)} \left| A_n^*(f,x) - f(x) \left(\sum_{k=\lceil na \rceil}^{\lfloor nb \rfloor} \psi(nx-k) \right) \right|. \tag{6.15}$$

That is,

$$|A_n(f,x) - f(x)| \leq \frac{1}{\psi(1)} \left| \sum_{k=\lceil na \rceil}^{\lfloor nb \rfloor} \left(f\left(\frac{k}{n}\right) - f(x) \right) \psi(nx-k) \right|. \tag{6.16}$$

We estimate the right-hand side of (6.19).

For that we need for $f \in C([a,b],\mathbb{C})$ the first modulus of continuity

$$\omega_1(f,\delta)_{[a,b]} := \omega_1(f,\delta) := \sup_{\substack{x,y \in [a,b] \\ |x-y| \leq \delta}} |f(x) - f(y)|, \quad \delta > 0. \tag{6.17}$$

The fact $f \in C([a,b],\mathbb{C})$ is equivalent to $\lim_{\delta \to 0} \omega_1(f,\delta) = 0$, see Anastassiou (2017).

We denote by $\|f\|_\infty := \sup_{x \in [a,b]} \|f(x)\|$, for $f \in C([a,b],\mathbb{C})$.

6.3 Main Results

We present $\mathbb{C}$-valued neural network high order approximations to a function given with rates. We start with a trigonometric approximation.

Theorem 6.5. *Let* $f \in C^2([a,b],\mathbb{C})$, $0 < \alpha < 1$, $n \in \mathbb{N} : n^{1-\alpha} > 2$, $x \in [a,b]$. *Then*

(1)

$$|A_n(f,x)-f(x)| \leq \frac{1}{\psi(1)}\left[|f'(x)|\left(\frac{1}{n^\alpha}+(b-a)\left(1-h\left(n^{1-\alpha}-2\right)\right)\right)\right.$$
$$+\frac{|f''(x)|}{2}\left(\frac{1}{n^{2\alpha}}+(b-a)^2\left(1-h\left(n^{1-\alpha}-2\right)\right)\right)$$
$$\left.+\left(\frac{\omega_1\left(f''+f,\frac{1}{n^\alpha}\right)}{2n^{2\alpha}}+\|f''+f\|_\infty(b-a)^2\left(1-h\left(n^{1-\alpha}-2\right)\right)\right)\right], \tag{6.18}$$

(2) *if* $f'(x)=f''(x)=0$, *we obtain*

$$|A_n(f,x)-f(x)|$$
$$\leq \frac{1}{\psi(1)}\left[\frac{\omega_1\left(f''+f,\frac{1}{n^\alpha}\right)}{2n^{2\alpha}}+\|f''+f\|_\infty(b-a)^2\left(1-h\left(n^{1-\alpha}-2\right)\right)\right], \tag{6.19}$$

note here the high rate of convergence at $n^{-3\alpha}$,
(3) *furthermore we get*

$$\|A_nf-f\|_\infty \leq \frac{1}{\psi(1)}\left[\|f'\|_\infty\left(\frac{1}{n^\alpha}+(b-a)\left(1-h\left(n^{1-\alpha}-2\right)\right)\right)\right.$$
$$+\frac{\|f''\|_\infty}{2}\left(\frac{1}{n^{2\alpha}}+(b-a)^2\left(1-h\left(n^{1-\alpha}-2\right)\right)\right)$$
$$\left.+\left(\frac{\omega_1\left(f''+f,\frac{1}{n^\alpha}\right)}{2n^{2\alpha}}+\|f''+f\|_\infty(b-a)^2\left(1-h\left(n^{1-\alpha}-2\right)\right)\right)\right], \tag{6.20}$$

i.e. $\lim_{n\to+\infty} A_n(f)=f$, *pointwise and uniformly, and*
(4) *finally, it holds*

$$\left|A_n(f,x)-f'(x)A_n(\sin(\cdot-x),x)-2f''(x)A_n\left(\sin^2\left(\frac{\cdot-x}{2}\right),x\right)-f(x)\right|$$
$$\leq \frac{1}{\psi(1)}\left[\frac{\omega_1\left(f''+f,\frac{1}{n^\alpha}\right)}{2n^{2\alpha}}+\|f''+f\|_\infty(b-a)^2\left(1-h\left(n^{1-\alpha}-2\right)\right)\right], \tag{6.21}$$

again here we achieve high speed of convergence at $n^{-3\alpha}$.

Proof. Here $f \in C^2([a,b],\mathbb{C})$, and we apply the trigonometric Taylor's formula for $f \in C^2([a,b],\mathbb{C})$, see Theorem 6 of Anastassiou (2023c).

Let $\frac{k}{n}, x \in [a,b]$, then

$$f\left(\frac{k}{n}\right) = f(x) + f'(x)\sin\left(\frac{k}{n} - x\right) + 2f''(x)\sin^2\left(\frac{\frac{k}{n} - x}{2}\right) + \int_x^{\frac{k}{n}} \left[\left(f''(t) + f(t)\right) - \left(f''(x) + f(x)\right)\right]\sin\left(\frac{k}{n} - t\right)dt. \tag{6.22}$$

Hence it holds

$$f\left(\frac{k}{n}\right)\psi(nx-k) = f(x)\psi(nx-k) + f'(x)\sin\left(\frac{k}{n} - x\right)\psi(nx-k) + 2f''(x)\sin^2\left(\frac{\frac{k}{n} - x}{2}\right)\psi(nx-k) + \psi(nx-k)\left(\int_x^{\frac{k}{n}} \left[\left(f''(t) + f(t)\right) - \left(f''(x) + f(x)\right)\right]\sin\left(\frac{k}{n} - t\right)dt\right). \tag{6.23}$$

So we have

$$\sum_{k=\lceil na\rceil}^{\lfloor nb\rfloor} f\left(\frac{k}{n}\right)\psi(nx-k) - f(x)\sum_{k=\lceil na\rceil}^{\lfloor nb\rfloor}\psi(nx-k) = f'(x)\sum_{k=\lceil na\rceil}^{\lfloor nb\rfloor}\psi(nx-k)\sin\left(\frac{k}{n} - x\right) + 2f''(x)\sum_{k=\lceil na\rceil}^{\lfloor nb\rfloor}\psi(nx-k)\sin^2\left(\frac{\frac{k}{n} - x}{2}\right) + \sum_{k=\lceil na\rceil}^{\lfloor nb\rfloor}\psi(nx-k)\left(\int_x^{\frac{k}{n}} \left[\left(f''(t) + f(t)\right) - \left(f''(x) + f(x)\right)\right]\sin\left(\frac{k}{n} - t\right)dt\right). \tag{6.24}$$

Thus, we obtain

$$A_n^*(f,x) - f(x)\sum_{k=\lceil na\rceil}^{\lfloor nb\rfloor}\psi(nx-k) = f'(x)A_n^*(\sin(\cdot - x), x) + 2f''(x)A_n^*\left(\sin^2\left(\frac{\cdot - x}{2}\right), x\right) + \Lambda_n(x), \tag{6.25}$$

where

$$\Lambda_n(x)$$
$$:= \sum_{k=\lceil na\rceil}^{\lfloor nb\rfloor} \psi(nx-k)\left(\int_x^{\frac{k}{n}} \left[\left(f''(t)+f(t)\right)-\left(f''(x)+f(x)\right)\right]\sin\left(\frac{k}{n}-t\right)dt\right).$$

We call

$$R_2(n) := \int_x^{\frac{k}{n}} \left[\left(f''(t)+f(t)\right)-\left(f''(x)+f(x)\right)\right]\sin\left(\frac{k}{n}-t\right)dt. \tag{6.26}$$

We assume that $b-a>\frac{1}{n^{\alpha}}$, which is always the case for large enough $n\in\mathbb{N}$, that is, when $n>\left\lceil (b-a)^{-\frac{1}{\alpha}}\right\rceil$.

Thus, $\left|\frac{k}{n}-x\right|\leq\frac{1}{n^{\alpha}}$ or $\left|\frac{k}{n}-x\right|>\frac{1}{n^{\alpha}}$.

In case of $\left|\frac{k}{n}-x\right|\leq\frac{1}{n^{\alpha}}$, we have the following cases:

(i) If $\frac{k}{n}\geq x$, then

$$|R_2(n)| = \left|\int_x^{\frac{k}{n}} \left[\left(f''(t)+f(t)\right)-\left(f''(x)+f(x)\right)\right]\sin\left(\frac{k}{n}-t\right)dt\right|$$
$$\leq \int_x^{\frac{k}{n}} \omega_1\left(f''+f,t-x\right)\left|\sin\left(\frac{k}{n}-t\right)\right|dt \leq \tag{6.27}$$

(by $|\sin x|\leq|x|$, $\forall\, x\in\mathbb{R}$)

$$\int_x^{\frac{k}{n}} \omega_1\left(f''+f,t-x\right)\left(\frac{k}{n}-t\right)dt \leq \omega_1\left(f''+f,\frac{k}{n}-x\right)\frac{\left(\frac{k}{n}-x\right)^2}{2}$$
$$\leq \frac{\omega_1\left(f''+f,\frac{1}{n^{\alpha}}\right)}{2n^{2\alpha}},$$

that is,

$$|R_2(n)| \leq \frac{\omega_1\left(f''+f,\frac{1}{n^{\alpha}}\right)}{2n^{2\alpha}}. \tag{6.28}$$

(ii) If $\frac{k}{n} < x$, then

$$\begin{aligned}
|R_2(n)| &= \left| \int_x^{\frac{k}{n}} \left[\left(f''(t) + f(t) \right) - \left(f''(x) + f(x) \right) \right] \sin\left(\frac{k}{n} - t \right) dt \right| \\
&= \left| \int_{\frac{k}{n}}^{x} \left[\left(f''(t) + f(t) \right) - \left(f''(x) + f(x) \right) \right] \sin\left(\frac{k}{n} - t \right) dt \right| \\
&\leq \int_{\frac{k}{n}}^{x} \left| \left(f''(t) + f(t) \right) - \left(f''(x) + f(x) \right) \right| \left| \sin\left(\frac{k}{n} - t \right) \right| dt \\
&\leq \int_{\frac{k}{n}}^{x} \omega_1\left(f'' + f, x - \frac{k}{n} \right) \left(t - \frac{k}{n} \right) dt \\
&\leq \omega_1\left(f'' + f, x - \frac{k}{n} \right) \frac{\left(x - \frac{k}{n} \right)^2}{2} \leq \frac{\omega_1\left(f'' + f, \frac{1}{n^\alpha} \right)}{2n^{2\alpha}}. \qquad (6.29)
\end{aligned}$$

That is,

$$|R_2(n)| \leq \frac{\omega_1\left(f'' + f, \frac{1}{n^\alpha} \right)}{2n^{2\alpha}}. \qquad (6.30)$$

So, we have proved when $\left| \frac{k}{n} - x \right| \leq \frac{1}{n^\alpha}$, always it holds

$$|R_2(n)| \leq \frac{\omega_1\left(f'' + f, \frac{1}{n^\alpha} \right)}{2n^{2\alpha}}. \qquad (6.31)$$

Next assume again $\frac{k}{n} \geq x$, then

$$\begin{aligned}
|R_2(n)| &= \left| \int_x^{\frac{k}{n}} \left[\left(f''(t) + f(t) \right) - \left(f''(x) + f(x) \right) \right] \sin\left(\frac{k}{n} - t \right) dt \right| \\
&\leq \int_x^{\frac{k}{n}} \left| \left(f''(t) + f(t) \right) - \left(f''(x) + f(x) \right) \right| \left| \sin\left(\frac{k}{n} - t \right) \right| dt \leq
\end{aligned}$$

(by $|\sin x| \leq |x|$, $\forall\, x \in \mathbb{R}$)

$$\begin{aligned}
& 2\left\| f'' + f \right\|_\infty \left(\int_x^{\frac{k}{n}} \left| \sin\left(\frac{k}{n} - t \right) \right| dt \right) \\
&\leq 2\left\| f'' + f \right\|_\infty \left(\int_x^{\frac{k}{n}} \left(\frac{k}{n} - t \right) dt \right) \\
&= 2\left\| f'' + f \right\|_\infty \frac{\left(\frac{k}{n} - x \right)^2}{2} \leq \left\| f'' + f \right\|_\infty (b - a)^2. \qquad (6.32)
\end{aligned}$$

Hence it is

$$\left|R_2(n)\right| \leq \left\|f'' + f\right\|_{\infty} (b-a)^2. \tag{6.33}$$

When $\frac{k}{n} < x$, we have

$$\begin{aligned}
\left|R_2(n)\right| &= \left|\int_x^{\frac{k}{n}} \left[\left(f''(t) + f(t)\right) - \left(f''(x) + f(x)\right)\right] \sin\left(\frac{k}{n} - t\right) dt\right| \\
&= \left|\int_{\frac{k}{n}}^{x} \left[\left(f''(x) + f(x)\right) - \left(f''(t) + f(t)\right)\right] \sin\left(\frac{k}{n} - t\right) dt\right| \\
&\leq \int_{\frac{k}{n}}^{x} \left|\left(f''(x) + f(x)\right) - \left(f''(t) + f(t)\right)\right| \left|\sin\left(\frac{k}{n} - t\right)\right| dt \\
&\leq 2\left\|f'' + f\right\|_{\infty} \int_{\frac{k}{n}}^{x} \left|\sin\left(\frac{k}{n} - t\right)\right| dt \\
&\leq 2\left\|f'' + f\right\|_{\infty} \int_{\frac{k}{n}}^{x} \left(t - \frac{k}{n}\right) dt = \left\|f'' + f\right\|_{\infty} \left(x - \frac{k}{n}\right)^2 \\
&\leq \left\|f'' + f\right\|_{\infty} (b-a)^2.
\end{aligned} \tag{6.34}$$

Therefore, always it holds

$$\left|R_2(n)\right| \leq \left\|f'' + f\right\|_{\infty} (b-a)^2. \tag{6.35}$$

And we have

$$\begin{aligned}
\Lambda_n(x) = &\sum_{\substack{k=\lceil na\rceil \\ \left\{:\left|\frac{k}{n} - x\right| \leq \frac{1}{n^{\alpha}}\right.}}^{\lfloor nb\rfloor} \psi(nx - k)\, R_2(n) \\
&+ \sum_{\substack{k=\lceil na\rceil \\ \left\{:\left|\frac{k}{n} - x\right| > \frac{1}{n^{\alpha}}\right.}}^{\lfloor nb\rfloor} \psi(nx - k)\, R_2(n).
\end{aligned} \tag{6.36}$$

Hence it holds

$$
\begin{aligned}
|\Lambda_n(x)| \leq & \sum_{\substack{k=\lceil na\rceil \\ \left\{: \left|\frac{k}{n}-x\right| \leq \frac{1}{n^\alpha}\right.}}^{\lfloor nb\rfloor} \psi(nx-k)\,|R_2(n)| \\
& + \sum_{\substack{k=\lceil na\rceil \\ \left\{: \left|\frac{k}{n}-x\right| > \frac{1}{n^\alpha}\right.}}^{\lfloor nb\rfloor} \psi(nx-k)\,|R_2(n)| \qquad (6.37) \\
\leq & \left(\sum_{\substack{k=\lceil na\rceil \\ \left\{: \left|\frac{k}{n}-x\right| \leq \frac{1}{n^\alpha}\right.}}^{\lfloor nb\rfloor} \psi(nx-k) \right) \frac{\omega_1\left(f''+f, \frac{1}{n^\alpha}\right)}{2n^{2\alpha}} \\
& + \sum_{\substack{k=\lceil na\rceil \\ \left\{: \left|\frac{k}{n}-x\right| > \frac{1}{n^\alpha}\right.}}^{\lfloor nb\rfloor} \psi(nx-k)\,\|f''+f\|_\infty (b-a)^2 \\
\overset{\text{(by (6.10))}}{\leq} & \frac{\omega_1\left(f''+f, \frac{1}{n^\alpha}\right)}{2n^{2\alpha}} + \|f''+f\|_\infty (b-a)^2 \\
& \left(\sum_{\substack{k=\lceil na\rceil \\ \left\{: \left|\frac{k}{n}-x\right| > \frac{1}{n^\alpha}\right.}}^{\lfloor nb\rfloor} \psi(nx-k) \right) \\
\overset{\text{(by Theorem 6.3)}}{\leq} & \frac{\omega_1\left(f''+f, \frac{1}{n^\alpha}\right)}{2n^{2\alpha}} + \|f''+f\|_\infty (b-a)^2 \left(1-h\left(n^{1-\alpha}-2\right)\right).
\end{aligned} \tag{6.38}
$$

Consequently, we have derived that

$$
|\Lambda_n(x)| \leq \frac{\omega_1\left(f''+f, \frac{1}{n^\alpha}\right)}{2n^{2\alpha}} + \|f''+f\|_\infty (b-a)^2 \left(1-h\left(n^{1-\alpha}-2\right)\right). \tag{6.39}
$$

Next we use again $|\sin x| \leq |x|$, $\forall\, x \in \mathbb{R}$.

We have that

$$
A_n^*(\sin(\cdot - x), x) = \sum_{k=\lceil na\rceil}^{\lfloor nb\rfloor} \psi(nx-k) \sin\left(\frac{k}{n}-x\right), \tag{6.40}
$$

and

$$
\begin{aligned}
\left|A_n^*(\sin(\cdot - x), x)\right| &\leq \sum_{k=\lceil na \rceil}^{\lfloor nb \rfloor} \psi(nx-k) \left|\sin\left(\frac{k}{n} - x\right)\right| \\
&= \sum_{\substack{k=\lceil na \rceil \\ \left\{ : \left|\frac{k}{n} - x\right| \leq \frac{1}{n^\alpha} \right.}}^{\lfloor nb \rfloor} \psi(nx-k) \left|\sin\left(\frac{k}{n} - x\right)\right| \\
&\quad + \sum_{\substack{k=\lceil na \rceil \\ \left\{ : \left|\frac{k}{n} - x\right| > \frac{1}{n^\alpha} \right.}}^{\lfloor nb \rfloor} \psi(nx-k) \left|\sin\left(\frac{k}{n} - x\right)\right| \qquad (6.41) \\
&\leq \sum_{\substack{k=\lceil na \rceil \\ \left\{ : \left|\frac{k}{n} - x\right| \leq \frac{1}{n^\alpha} \right.}}^{\lfloor nb \rfloor} \psi(nx-k) \left|\frac{k}{n} - x\right| \\
&\quad + \sum_{\substack{k=\lceil na \rceil \\ \left\{ : \left|\frac{k}{n} - x\right| > \frac{1}{n^\alpha} \right.}}^{\lfloor nb \rfloor} \psi(nx-k) \left|\frac{k}{n} - x\right| \\
&\leq \frac{1}{n^\alpha} + (b-a) \left(\sum_{\substack{k=\lceil na \rceil \\ \left\{ : \left|\frac{k}{n} - x\right| > \frac{1}{n^\alpha} \right.}}^{\lfloor nb \rfloor} \psi(nx-k) \right) \\
&\overset{\text{(by (6.7))}}{\leq} \frac{1}{n^\alpha} + (b-a)\left(1 - h\left(n^{1-\alpha} - 2\right)\right). \qquad (6.42)
\end{aligned}
$$

We found that

$$
\left|A_n^*(\sin(\cdot - x), x)\right| \leq \frac{1}{n^\alpha} + (b-a)\left(1 - h\left(n^{1-\alpha} - 2\right)\right). \tag{6.43}
$$

Next we estimate

$$
A_n^*\left(\sin^2\left(\frac{\cdot - x}{2}\right), x\right) = \sum_{k=\lceil na \rceil}^{\lfloor nb \rfloor} \psi(nx-k) \sin^2\left(\frac{\frac{k}{n} - x}{2}\right). \tag{6.44}
$$

We have that (by $|\sin x| \leq |x|$, $\forall\, x \in \mathbb{R}$)

$$A_n^*\left(\sin^2\left(\frac{\cdot - x}{2}\right), x\right) = \sum_{k=\lceil na \rceil}^{\lfloor nb \rfloor} \psi(nx-k)\left|\sin\left(\frac{\frac{k}{n}-x}{2}\right)\right|^2$$

$$\leq \frac{1}{4}\sum_{k=\lceil na \rceil}^{\lfloor nb \rfloor} \psi(nx-k)\left|\frac{k}{n}-x\right|^2$$

$$= \frac{1}{4}\left[\sum_{\substack{k=\lceil na \rceil \\ \left\{:\left|\frac{k}{n}-x\right| \leq \frac{1}{n^{\alpha}}\right.}}^{\lfloor nb \rfloor} \psi(nx-k)\left|\frac{k}{n}-x\right|^2\right.$$

$$\left. + \sum_{\substack{k=\lceil na \rceil \\ \left\{:\left|\frac{k}{n}-x\right| > \frac{1}{n^{\alpha}}\right.}}^{\lfloor nb \rfloor} \psi(nx-k)\left|\frac{k}{n}-x\right|^2\right]$$

$$\leq \frac{1}{4}\left[\frac{1}{n^{2\alpha}} + (b-a)^2\left(1-h\left(n^{1-\alpha}-2\right)\right)\right]. \quad (6.45)$$

That is,

$$A_n^*\left(\sin^2\left(\frac{\cdot - x}{2}\right), x\right) \leq \frac{1}{4}\left[\frac{1}{n^{2\alpha}} + (b-a)^2\left(1-h\left(n^{1-\alpha}-2\right)\right)\right]. \quad (6.46)$$

Consequently we have derived the following:

(1)

$$|A_n(f,x) - f(x)| \leq \frac{1}{\psi(1)}\left[|f'(x)|\left(\frac{1}{n^{\alpha}} + (b-a)\left(1-h\left(n^{1-\alpha}-2\right)\right)\right)\right.$$

$$+ \frac{|f''(x)|}{2}\left(\frac{1}{n^{2\alpha}} + (b-a)^2\left(1-h\left(n^{1-\alpha}-2\right)\right)\right)$$

$$\left. + \left(\frac{\omega_1\left(f''+f, \frac{1}{n^{\alpha}}\right)}{2n^{2\alpha}} + \|f''+f\|_{\infty}(b-a)^2\left(1-h\left(n^{1-\alpha}-2\right)\right)\right)\right]. \quad (6.47)$$

(2) If $f'(x) = f''(x) = 0$, by (6.47), we obtain

$$|A_n(f,x) - f(x)| \leq \frac{1}{\psi(1)}\left[\frac{\omega_1\left(f''+f,\frac{1}{n^\alpha}\right)}{2n^{2\alpha}} + \|f''+f\|_\infty (b-a)^2\left(1-h\left(n^{1-\alpha}-2\right)\right)\right], \tag{6.48}$$

note here the high rate of convergence at $n^{-3\alpha}$.

(3) Furthermore, by (6.47), we get

$$\|A_nf - f\|_\infty \leq \frac{1}{\psi(1)}\left[\|f'\|_\infty\left(\frac{1}{n^\alpha}+(b-a)\left(1-h\left(n^{1-\alpha}-2\right)\right)\right)\right.$$
$$+\frac{\|f''\|_\infty}{2}\left(\frac{1}{n^{2\alpha}}+(b-a)^2\left(1-h\left(n^{1-\alpha}-2\right)\right)\right)$$
$$\left.+\left(\frac{\omega_1\left(f''+f,\frac{1}{n^\alpha}\right)}{2n^{2\alpha}}+\|f''+f\|_\infty(b-a)^2\left(1-h\left(n^{1-\alpha}-2\right)\right)\right)\right]. \tag{6.49}$$

We derive that $\lim_{n\to+\infty} A_n(f) = f$, pointwise and uniformly.

We observe that

$$A_n(f,x) - f'(x)A_n(\sin(\cdot-x),x) - 2f''(x)A_n\left(\sin^2\left(\frac{\cdot-x}{2}\right),x\right) - f(x)$$
$$= \frac{A_n^*(f,x)}{\sum_{k=\lceil na\rceil}^{\lfloor nb\rfloor}\psi(nx-k)} - f'(x)\frac{A_n^*(\sin(\cdot-x),x)}{\sum_{k=\lceil na\rceil}^{\lfloor nb\rfloor}\psi(nx-k)}$$
$$-2f''(x)\frac{A_n^*\left(\sin^2\left(\frac{\cdot-x}{2}\right),x\right)}{\sum_{k=\lceil na\rceil}^{\lfloor nb\rfloor}\psi(nx-k)} - f(x)\left(\frac{\sum_{k=\lceil na\rceil}^{\lfloor nb\rfloor}\psi(nx-k)}{\sum_{k=\lceil na\rceil}^{\lfloor nb\rfloor}\psi(nx-k)}\right) \tag{6.50}$$

$$= \frac{1}{\sum_{k=\lceil na\rceil}^{\lfloor nb\rfloor}\psi(nx-k)}\left[A_n^*(f,x) - f'(x)A_n^*(\sin(\cdot-x),x)\right.$$
$$\left.-2f''(x)A_n^*\left(\sin^2\left(\frac{\cdot-x}{2}\right),x\right) - f(x)\sum_{k=\lceil na\rceil}^{\lfloor nb\rfloor}\psi(nx-k)\right]$$
$$= \frac{1}{\sum_{k=\lceil na\rceil}^{\lfloor nb\rfloor}\psi(nx-k)}(\Lambda_n(x)). \tag{6.51}$$

Finally, we obtain ($\forall\ x \in [a,b]$, $n \in \mathbb{N}$)
(4)

$$\left| A_n(f,x) - f'(x) A_n(\sin(\cdot - x), x) - 2f''(x) A_n\left(\sin^2\left(\frac{\cdot - x}{2}\right), x\right) - f(x) \right|$$

$$\overset{(6.8)}{\leq} \frac{1}{\psi(1)} |\Lambda_n(x)|$$

$$\overset{(6.39)}{\leq} \frac{1}{\psi(1)} \left[\frac{\omega_1\left(f'' + f, \frac{1}{n^\alpha}\right)}{2n^{2\alpha}} + \|f'' + f\|_\infty (b-a)^2 \left(1 - h\left(n^{1-\alpha} - 2\right)\right) \right]. \tag{6.52}$$

The theorem is proved. □

We continue with a hyperbolic high order neural network approximation.

Theorem 6.6. *Let* $f \in C^2([a,b], \mathbb{C})$, $0 < \alpha < 1$, $n \in \mathbb{N} : n^{1-\alpha} > 2$, $x \in [a,b]$. *Then*

(1)

$$|A_n(f,x) - f(x)|$$

$$\leq \frac{1}{\psi(1)} \cosh(b-a) \left[|f'(x)| \left(\frac{1}{n^\alpha} + (b-a)\left(1 - h\left(n^{1-\alpha} - 2\right)\right) \right) \right.$$

$$+ \frac{|f''(x)|}{2} \left(\frac{1}{n^{2\alpha}} + (b-a)^2 \left(1 - h\left(n^{1-\alpha} - 2\right)\right) \right)$$

$$\left. + \left(\frac{\omega_1\left(f'' - f, \frac{1}{n^\alpha}\right)}{2n^{2\alpha}} + \|f'' - f\|_\infty (b-a)^2 \left(1 - h\left(n^{1-\alpha} - 2\right)\right) \right) \right], \tag{6.53}$$

(2) *if* $f'(x) = f''(x) = 0$, *we obtain*

$$|A_n(f,x) - f(x)| \leq \frac{1}{\psi(1)} \cosh(b-a)$$

$$\left[\frac{\omega_1\left(f'' - f, \frac{1}{n^\alpha}\right)}{2n^{2\alpha}} + \|f'' - f\|_\infty (b-a)^2 \left(1 - h\left(n^{1-\alpha} - 2\right)\right) \right], \tag{6.54}$$

note here the high rate of convergence at $n^{-3\alpha}$,

(3) *furthermore, we get*

$$\|A_n f - f\|_\infty \leq \frac{1}{\psi(1)} \cosh(b-a) \left[\|f'\|_\infty \left(\frac{1}{n^\alpha} + (b-a)\left(1 - h\left(n^{1-\alpha} - 2\right)\right) \right) \right.$$

$$+ \frac{\|f''\|_\infty}{2} \left(\frac{1}{n^{2\alpha}} + (b-a)^2 \left(1 - h\left(n^{1-\alpha} - 2\right)\right) \right)$$

$$\left. + \left(\frac{\omega_1\left(f'' - f, \frac{1}{n^\alpha}\right)}{2n^{2\alpha}} + \|f'' - f\|_\infty (b-a)^2 \left(1 - h\left(n^{1-\alpha} - 2\right)\right) \right) \right], \tag{6.55}$$

it follows that $\lim_{n \to +\infty} A_n(f) = f$*, pointwise and uniformly,*

and

(4)

$$\left| A_n(f,x) - f'(x) A_n(\sinh(\cdot - x), x) - 2f''(x) A_n\left(\sinh^2\left(\frac{\cdot - x}{2}\right), x\right) - f(x) \right|$$

$$\leq \frac{\cosh(b-a)}{\psi(1)} \left[\frac{\omega_1\left(f'' - f, \frac{1}{n^\alpha}\right)}{2n^{2\alpha}} + \|f'' - f\|_\infty (b-a)^2 \left(1 - h\left(n^{1-\alpha} - 2\right)\right) \right], \tag{6.56}$$

again here we achieve high speed of convergence at $n^{-3\alpha}$.

Proof. By the mean value theorem, we have that

$$\sinh x = \sinh x - \sinh 0 = (\cosh \xi)(x - 0),$$

for some ξ between $\{0, x\}$, for any $x \in \mathbb{R}$.

Hence,

$$|\sinh x| \leq \|\cosh\|_{\infty,[-(b-a),b-a]} |x|, \quad \forall\, x \in [-(b-a), b-a]. \tag{6.57}$$

That is, there exists $M \geq 1$ such that

$$|\sinh x| \leq M |x|, \quad \forall\, x \in [-(b-a), b-a], \tag{6.58}$$

where $M := \|\cosh\|_{\infty,[-(b-a),b-a]} = \cosh(b-a)$.

Here $f \in C^2([a,b], \mathbb{C})$, and we apply the hyperbolic Taylor's formula for $f \in C^2([a,b], \mathbb{C})$, see Theorem 7 of Anastassiou (2023c).

Let $\frac{k}{n}, x \in [a, b]$, then

$$f\left(\frac{k}{n}\right) = f(x) + f'(x)\sinh\left(\frac{k}{n} - x\right) + 2f''(x)\sinh^2\left(\frac{\frac{k}{n} - x}{2}\right)$$
$$+ \int_x^{\frac{k}{n}} \left[(f''(t) - f(t)) - (f''(x) - f(x))\right]\sinh\left(\frac{k}{n} - t\right) dt. \tag{6.59}$$

Hence it holds

$$f\left(\frac{k}{n}\right)\psi(nx - k) = f(x)\psi(nx - k)$$
$$+ f'(x)\sinh\left(\frac{k}{n} - x\right)\psi(nx - k) + 2f''(x)\sinh^2\left(\frac{\frac{k}{n} - x}{2}\right)\psi(nx - k)$$
$$+ \psi(nx - k)\left(\int_x^{\frac{k}{n}} \left[(f''(t) - f(t)) - (f''(x) - f(x))\right]\sinh\left(\frac{k}{n} - t\right) dt\right). \tag{6.60}$$

So we have

$$\sum_{k=\lceil na\rceil}^{\lfloor nb\rfloor} f\left(\frac{k}{n}\right)\psi(nx - k) - f(x)\sum_{k=\lceil na\rceil}^{\lfloor nb\rfloor}\psi(nx - k)$$
$$= f'(x)\sum_{k=\lceil na\rceil}^{\lfloor nb\rfloor}\psi(nx - k)\sinh\left(\frac{k}{n} - x\right)$$
$$+ 2f''(x)\sum_{k=\lceil na\rceil}^{\lfloor nb\rfloor}\psi(nx - k)\sinh^2\left(\frac{\frac{k}{n} - x}{2}\right) + \sum_{k=\lceil na\rceil}^{\lfloor nb\rfloor}\psi(nx - k)$$
$$\times\left(\int_x^{\frac{k}{n}} \left[(f''(t) - f(t)) - (f''(x) - f(x))\right]\sinh\left(\frac{k}{n} - t\right) dt\right). \tag{6.61}$$

Thus, we obtain

$$A_n^*(f,x) - f(x) \sum_{k=\lceil na \rceil}^{\lfloor nb \rfloor} \psi(nx-k)$$

$$= f'(x) A_n^*(\sinh(\cdot - x), x) + 2f''(x) A_n^*\left(\sinh^2\left(\frac{\cdot - x}{2}\right), x\right) + \Lambda_n(x), \tag{6.62}$$

where

$$\Lambda_n(x) := \sum_{k=\lceil na \rceil}^{\lfloor nb \rfloor} \psi(nx-k) \times \left(\int_x^{\frac{k}{n}} \left[(f''(t) - f(t)) - (f''(x) - f(x))\right] \sinh\left(\frac{k}{n} - t\right) dt\right). \tag{6.63}$$

We call

$$R_2(n) := \int_x^{\frac{k}{n}} \left[(f''(t) - f(t)) - (f''(x) - f(x))\right] \sinh\left(\frac{k}{n} - t\right) dt. \tag{6.64}$$

We assume that $b - a > \frac{1}{n^\alpha}$, which is always the case for large enough $n \in \mathbb{N}$, that is, when $n > \left\lceil (b-a)^{-\frac{1}{\alpha}} \right\rceil$.

Thus, $\left|\frac{k}{n} - x\right| \leq \frac{1}{n^\alpha}$ or $\left|\frac{k}{n} - x\right| > \frac{1}{n^\alpha}$.

In case of $\left|\frac{k}{n} - x\right| \leq \frac{1}{n^\alpha}$, we have the following cases:

(i) If $\frac{k}{n} \geq x$, then

$$\begin{aligned} |R_2(n)| &= \left|\int_x^{\frac{k}{n}} \left[(f''(t) - f(t)) - (f''(x) - f(x))\right] \sinh\left(\frac{k}{n} - t\right) dt\right| \\ &\leq \int_x^{\frac{k}{n}} \omega_1(f'' - f, t - x) \left|\sinh\left(\frac{k}{n} - t\right)\right| dt \\ &\overset{(6.58)}{\leq} \int_x^{\frac{k}{n}} \omega_1(f'' - f, t - x) M\left(\frac{k}{n} - t\right) dt \\ &\leq M\omega_1\left(f'' - f, \frac{k}{n} - x\right) \int_x^{\frac{k}{n}} \left(\frac{k}{n} - t\right) dt \\ &= M\omega_1\left(f'' - f, \frac{k}{n} - x\right) \frac{\left(\frac{k}{n} - x\right)^2}{2} \leq \frac{M\omega_1\left(f'' - f, \frac{1}{n^\alpha}\right)}{2n^{2\alpha}}, \end{aligned} \tag{6.65}$$

that is,

$$|R_2(n)| \leq \frac{M\omega_1\left(f''-f,\frac{1}{n^\alpha}\right)}{2n^{2\alpha}}. \tag{6.66}$$

(ii) If $\frac{k}{n} < x$, then

$$\begin{aligned}
|R_2(n)| &= \left|\int_x^{\frac{k}{n}} \left[\left(f''(t)-f(t)\right)-\left(f''(x)-f(x)\right)\right]\sinh\left(\frac{k}{n}-t\right)dt\right| \\
&= \left|\int_{\frac{k}{n}}^{x} \left[\left(f''(t)-f(t)\right)-\left(f''(x)-f(x)\right)\right]\sinh\left(\frac{k}{n}-t\right)dt\right| \\
&\leq \int_{\frac{k}{n}}^{x} \left|\left(f''(t)-f(t)\right)-\left(f''(x)-f(x)\right)\right|\left|\sinh\left(\frac{k}{n}-t\right)\right|dt \\
&\leq M\omega_1\left(f''-f,x-\frac{k}{n}\right)\int_{\frac{k}{n}}^{x}\left(t-\frac{k}{n}\right)dt \\
&= M\omega_1\left(f''-f,x-\frac{k}{n}\right)\frac{\left(x-\frac{k}{n}\right)^2}{2} \leq \frac{M\omega_1\left(f''-f,\frac{1}{n^\alpha}\right)}{2n^{2\alpha}},
\end{aligned} \tag{6.67}$$

that is,

$$|R_2(n)| \leq \frac{M\omega_1\left(f''-f,\frac{1}{n^\alpha}\right)}{2n^{2\alpha}}. \tag{6.68}$$

So, we have proved when $\left|\frac{k}{n}-x\right| \leq \frac{1}{n^\alpha}$, always it holds

$$|R_2(n)| \leq \frac{M\omega_1\left(f''-f,\frac{1}{n^\alpha}\right)}{2n^{2\alpha}}. \tag{6.69}$$

Next assume again $\frac{k}{n} \geq x$, then

$$\begin{aligned}
|R_2(n)| &= \left|\int_x^{\frac{k}{n}} \left[\left(f''(t)-f(t)\right)-\left(f''(x)-f(x)\right)\right]\sinh\left(\frac{k}{n}-t\right)dt\right| \\
&\leq \int_x^{\frac{k}{n}} \left|\left(f''(t)-f(t)\right)-\left(f''(x)-f(x)\right)\right|\left|\sinh\left(\frac{k}{n}-t\right)\right|dt \\
&\leq 2M\left\|f''-f\right\|_\infty \int_x^{\frac{k}{n}}\left(\frac{k}{n}-t\right)dt \\
&= 2M\left\|f''-f\right\|_\infty \frac{\left(\frac{k}{n}-x\right)^2}{2} \leq M\left\|f''-f\right\|_\infty (b-a)^2.
\end{aligned} \tag{6.70}$$

Hence,

$$|R_2(n)| \leq M\left\|f''-f\right\|_\infty (b-a)^2. \tag{6.71}$$

When $\frac{k}{n} < x$, we have

$$\left|R_2(n)\right| = \left|\int_{\frac{k}{n}}^{x} \left[\left(f''(t) - f(t)\right) - \left(f''(x) - f(x)\right)\right] \sinh\left(\frac{k}{n} - t\right) dt\right|$$

$$\leq \int_{\frac{k}{n}}^{x} \left|\left(f''(t) - f(t)\right) - \left(f''(x) - f(x)\right)\right| \left|\sinh\left(\frac{k}{n} - t\right)\right| dt$$

$$\leq 2M \left\|f'' - f\right\|_{\infty} \int_{\frac{k}{n}}^{x} \left(t - \frac{k}{n}\right) dt$$

$$= 2M \left\|f'' - f\right\|_{\infty} \frac{\left(x - \frac{k}{n}\right)^2}{2} \leq M \left\|f'' - f\right\|_{\infty} (b-a)^2. \quad (6.72)$$

Therefore, always it holds

$$\left|R_2(n)\right| \leq M \left\|f'' - f\right\|_{\infty} (b-a)^2. \quad (6.73)$$

And we have

$$\Lambda_n(x) = \sum_{\left\{\substack{k=\lceil na\rceil \\ :\left|\frac{k}{n}-x\right| \leq \frac{1}{n^{\alpha}}}\right.}^{\lfloor nb\rfloor} \psi(nx-k) R_2(n)$$

$$+ \sum_{\left\{\substack{k=\lceil na\rceil \\ :\left|\frac{k}{n}-x\right| > \frac{1}{n^{\alpha}}}\right.}^{\lfloor nb\rfloor} \psi(nx-k) R_2(n). \quad (6.74)$$

Hence it holds

$$\left|\Lambda_n(x)\right| \leq \sum_{\left\{\substack{k=\lceil na\rceil \\ :\left|\frac{k}{n}-x\right| \leq \frac{1}{n^{\alpha}}}\right.}^{\lfloor nb\rfloor} \psi(nx-k) \left|R_2(n)\right| \quad (6.75)$$

$$+ \sum_{\left\{\substack{k=\lceil na\rceil \\ :\left|\frac{k}{n}-x\right| > \frac{1}{n^{\alpha}}}\right.}^{\lfloor nb\rfloor} \psi(nx-k) \left|R_2(n)\right|$$

$$\leq \left(\sum_{\left\{\substack{k=\lceil na\rceil \\ :\left|\frac{k}{n}-x\right| \leq \frac{1}{n^{\alpha}}}\right.}^{\lfloor nb\rfloor} \psi(nx-k) \right) \frac{\omega_1\left(f'' - f, \frac{1}{n^{\alpha}}\right) M}{2n^{2\alpha}}$$

$$+\left(\sum_{\substack{k=\lceil na\rceil \\ \left\{:\left|\frac{k}{n}-x\right|>\frac{1}{n^{\alpha}}\right.}}^{\lfloor nb\rfloor} \psi(nx-k)\right) M\left\|f''-f\right\|_{\infty}(b-a)^{2}$$

$$\overset{((6.10))}{\leq} \frac{M\omega_{1}\left(f''-f,\frac{1}{n^{\alpha}}\right)}{2n^{2\alpha}}+M\left\|f''-f\right\|_{\infty}(b-a)^{2}$$

$$\left(\sum_{\substack{k=\lceil na\rceil \\ \left\{:\left|\frac{k}{n}-x\right|>\frac{1}{n^{\alpha}}\right.}}^{\lfloor nb\rfloor} \psi(nx-k)\right) \overset{\text{(by Theorem 6.3)}}{\leq} M\frac{\omega_{1}\left(f''-f,\frac{1}{n^{\alpha}}\right)}{2n^{2\alpha}}$$

$$+M\left\|f''-f\right\|_{\infty}(b-a)^{2}\frac{\left(1-h\left(n^{1-\alpha}-2\right)\right)}{2}. \tag{6.76}$$

Consequently, we have derived that

$$\left|\Lambda_{n}(x)\right| \leq M\left[\frac{\omega_{1}\left(f''-f,\frac{1}{n^{\alpha}}\right)}{2n^{2\alpha}}+\left\|f''-f\right\|_{\infty}(b-a)^{2}\frac{\left(1-h\left(n^{1-\alpha}-2\right)\right)}{2}\right]. \tag{6.77}$$

We have that

$$A_{n}^{*}(\sinh(\cdot-x),x)=\sum_{k=\lceil na\rceil}^{\lfloor nb\rfloor}\psi(nx-k)\sinh\left(\frac{k}{n}-x\right), \tag{6.78}$$

and

$$\left|A_{n}^{*}(\sinh(\cdot-x),x)\right| \leq \sum_{k=\lceil na\rceil}^{\lfloor nb\rfloor}\psi(nx-k)\left|\sinh\left(\frac{k}{n}-x\right)\right|$$

$$=\sum_{\substack{k=\lceil na\rceil \\ \left\{:\left|\frac{k}{n}-x\right|\leq\frac{1}{n^{\alpha}}\right.}}^{\lfloor nb\rfloor}\psi(nx-k)\left|\sinh\left(\frac{k}{n}-x\right)\right|$$

$$+\sum_{\substack{k=\lceil na\rceil \\ \left\{:\left|\frac{k}{n}-x\right|>\frac{1}{n^{\alpha}}\right.}}^{\lfloor nb\rfloor}\psi(nx-k)\left|\sinh\left(\frac{k}{n}-x\right)\right| \tag{6.79}$$

$$\leq M\left[\sum_{\substack{k=\lceil na\rceil \\ \left\{:\left|\frac{k}{n}-x\right|\leq\frac{1}{n^{\alpha}}\right.}}^{\lfloor nb\rfloor}\psi(nx-k)\left|\frac{k}{n}-x\right|\right.$$

$$+ \sum_{\substack{k=\lceil na \rceil \\ \left\{ : \left| \frac{k}{n} - x \right| > \frac{1}{n^{\alpha}} \right.}}^{\lfloor nb \rfloor} \psi(nx-k) \left| \frac{k}{n} - x \right| \Bigg]$$

$$\leq M \left[\frac{1}{n^{\alpha}} + (b-a) \left(\sum_{\substack{k=\lceil na \rceil \\ \left\{ : \left| \frac{k}{n} - x \right| > \frac{1}{n^{\alpha}} \right.}}^{\lfloor nb \rfloor} \psi(nx-k) \right) \right]$$

$$\leq M \left[\frac{1}{n^{\alpha}} + (b-a) \left(1 - h \left(n^{1-\alpha} - 2 \right) \right) \right]. \tag{6.80}$$

We found that

$$\left| A_n^* \left(\sinh(\cdot - x), x \right) \right| \leq M \left[\frac{1}{n^{\alpha}} + (b-a) \left(1 - h \left(n^{1-\alpha} - 2 \right) \right) \right]. \tag{6.81}$$

Next we estimate

$$A_n^* \left(\sinh^2 \left(\frac{\cdot - x}{2} \right), x \right) = \sum_{k=\lceil na \rceil}^{\lfloor nb \rfloor} \psi(nx-k) \sinh^2 \left(\frac{\frac{k}{n} - x}{2} \right). \tag{6.82}$$

We have that

$$A_n^* \left(\sinh^2 \left(\frac{\cdot - x}{2} \right), x \right) = \sum_{k=\lceil na \rceil}^{\lfloor nb \rfloor} \psi(nx-k) \left(\sinh \left(\frac{\frac{k}{n} - x}{2} \right) \right)^2$$

$$\leq \frac{M}{4} \sum_{k=\lceil na \rceil}^{\lfloor nb \rfloor} \psi(nx-k) \left(\frac{k}{n} - x \right)^2$$

$$= \frac{M}{4} \left[\sum_{\substack{k=\lceil na \rceil \\ \left\{ : \left| \frac{k}{n} - x \right| \le \frac{1}{n^{\alpha}} \right.}}^{\lfloor nb \rfloor} \psi (nx - k) \left(\frac{k}{n} - x \right)^2 \right.$$

$$\left. + \sum_{\substack{k=\lceil na \rceil \\ \left\{ : \left| \frac{k}{n} - x \right| > \frac{1}{n^{\alpha}} \right.}}^{\lfloor nb \rfloor} \psi (nx - k) \left(\frac{k}{n} - x \right)^2 \right]$$

$$\le \frac{M}{4} \left[\frac{1}{n^{2\alpha}} + (b-a)^2 \left(1 - h \left(n^{1-\alpha} - 2 \right) \right) \right]. \tag{6.83}$$

That is,

$$A_n^* \left(\sinh^2 \left(\frac{\cdot - x}{2} \right), x \right) \le \frac{M}{4} \left[\frac{1}{n^{2\alpha}} + (b-a)^2 \left(1 - h \left(n^{1-\alpha} - 2 \right) \right) \right]. \tag{6.84}$$

By (6.16) and putting together (6.62), (6.77), (6.81) and (6.84), we derive the following:

(1)

$$\left| A_n (f, x) - f (x) \right| \le \frac{1}{\psi (1)} M \left[\left| f' (x) \right| \left(\frac{1}{n^{\alpha}} + (b-a) \left(1 - h \left(n^{1-\alpha} - 2 \right) \right) \right) \right.$$

$$+ \frac{\left| f'' (x) \right|}{2} \left(\frac{1}{n^{2\alpha}} + (b-a)^2 \left(1 - h \left(n^{1-\alpha} - 2 \right) \right) \right)$$

$$\left. + \left(\frac{\omega_1 \left(f'' - f, \frac{1}{n^{\alpha}} \right)}{2n^{2\alpha}} + \left\| f'' - f \right\|_{\infty} (b-a)^2 \left(1 - h \left(n^{1-\alpha} - 2 \right) \right) \right) \right]. \tag{6.85}$$

(2) If $f' (x) = f'' (x) = 0$, by (6.85), we obtain

$$\left| A_n (f, x) - f (x) \right|$$

$$\le \frac{M}{\psi (1)} \left[\frac{\omega_1 \left(f'' - f, \frac{1}{n^{\alpha}} \right)}{2n^{2\alpha}} + \left\| f'' - f \right\|_{\infty} (b-a)^2 \left(1 - h \left(n^{1-\alpha} - 2 \right) \right) \right], \tag{6.86}$$

note here the high rate of convergence at $n^{-3\alpha}$.

(3) Furthermore, by (6.85), we get

$$\|A_n f - f\|_\infty \leq \frac{M}{\psi(1)} \left[\|f'\|_\infty \left(\frac{1}{n^\alpha} + (b-a)\left(1 - h\left(n^{1-\alpha} - 2\right)\right) \right) \right.$$

$$+ \frac{\|f''\|_\infty}{2} \left(\frac{1}{n^{2\alpha}} + (b-a)^2 \left(1 - h\left(n^{1-\alpha} - 2\right)\right) \right)$$

$$\left. + \left(\frac{\omega_1\left(f'' - f, \frac{1}{n^\alpha}\right)}{2n^{2\alpha}} + \|f'' - f\|_\infty (b-a)^2 \left(1 - h\left(n^{1-\alpha} - 2\right)\right) \right) \right]. \tag{6.87}$$

It follows that $\lim_{n \to +\infty} A_n(f) = f$, pointwise and uniformly.

We observe that

$$A_n(f,x) - f'(x) A_n(\sinh(\cdot - x), x) - 2f''(x) A_n\left(\sinh^2\left(\frac{\cdot - x}{2}\right), x\right) - f(x)$$

$$= \frac{A_n^*(f,x)}{\sum_{k=\lceil na \rceil}^{\lfloor nb \rfloor} \psi(nx-k)} - f'(x) \frac{A_n^*(\sinh(\cdot - x), x)}{\sum_{k=\lceil na \rceil}^{\lfloor nb \rfloor} \psi(nx-k)}$$

$$- 2f''(x) \frac{A_n^*\left(\sinh^2\left(\frac{\cdot - x}{2}\right), x\right)}{\sum_{k=\lceil na \rceil}^{\lfloor nb \rfloor} \psi(nx-k)} - f(x) \left(\frac{\sum_{k=\lceil na \rceil}^{\lfloor nb \rfloor} \psi(nx-k)}{\sum_{k=\lceil na \rceil}^{\lfloor nb \rfloor} \psi(nx-k)} \right) \tag{6.88}$$

$$= \frac{1}{\sum_{k=\lceil na \rceil}^{\lfloor nb \rfloor} \psi(nx-k)} \left[A_n^*(f,x) - f'(x) A_n^*(\sinh(\cdot - x), x) \right.$$

$$\left. - 2f''(x) A_n^*\left(\sinh^2\left(\frac{\cdot - x}{2}\right), x\right) - f(x) \sum_{k=\lceil na \rceil}^{\lfloor nb \rfloor} \psi(nx-k) \right]$$

$$= \frac{1}{\sum_{k=\lceil na \rceil}^{\lfloor nb \rfloor} \psi(nx-k)} \left(\Lambda_n(x)\right). \tag{6.89}$$

Finally, we obtain ($\forall\, x \in [a,b]$, $n \in \mathbb{N}$)

(4)

$$\left| A_n(f,x) - f'(x)\, A_n(\sinh(\cdot - x), x) - 2f''(x)\, A_n\left(\sinh^2\left(\frac{\cdot - x}{2}\right), x\right) - f(x) \right|$$

$$\overset{(6.8)}{\leq} \left|\Lambda_n(x)\right| / \psi(1)$$

$$\overset{(6.77)}{\leq} \frac{M}{\psi(1)} \left[\frac{\omega_1\left(f'' - f, \frac{1}{n^\alpha}\right)}{2n^{2\alpha}} + \|f'' - f\|_\infty (b-a)^2 \frac{\left(1 - h\left(n^{1-\alpha} - 2\right)\right)}{2} \right]. \tag{6.90}$$

The theorem is established. □

Next follows a mixed hyperbolic-trigonometric high order neural network approximation.

Theorem 6.7. *Let* $f \in C^4([a,b], \mathbb{C})$, $0 < \alpha < 1$, $n \in \mathbb{N} : n^{1-\alpha} > 2$, $x \in [a,b]$. *Then*

(1)

$$\begin{aligned} &\left| A_n(f,x) - f(x) - \frac{f'(x)}{2} A_n\left((\sinh(\cdot - x) + \sin(\cdot - x)), x\right) \right. \\ &\quad - \frac{f''(x)}{2} A_n\left((\cosh(\cdot - x) - \cos(\cdot - x)), x\right) \\ &\quad - \frac{f^{(3)}(x)}{2} A_n\left((\sinh(\cdot - x) - \sin(\cdot - x)), x\right) \\ &\quad \left. - f^{(4)}(x)\, A_n\left(\left(\sinh^2\left(\frac{\cdot - x}{2}\right) - \sin^2\left(\frac{\cdot - x}{2}\right)\right), x\right) \right| \end{aligned} \tag{6.91}$$

$$\begin{aligned} &\leq \frac{(\cosh(b-a) + 1)}{2\psi(1)} \\ &\quad \left[\frac{\omega_1\left(f^{(4)} - f, \frac{1}{n^\alpha}\right)}{2n^{2\alpha}} + \left\| f^{(4)} - f \right\|_\infty (b-a)^2 \left(1 - h\left(n^{1-\alpha} - 2\right)\right) \right], \end{aligned} \tag{6.92}$$

(2) *if* $f^{(i)}(x) = 0$, $i = 1, 2, 3, 4$, *we get*

$$|A_n(f, x) - f(x)| \leq \frac{(\cosh(b-a)+1)}{2\psi(1)}$$

$$\left[\frac{\omega_1\left(f^{(4)} - f, \frac{1}{n^{\alpha}}\right)}{2n^{2\alpha}} + \left\|f^{(4)} - f\right\|_{\infty} (b-a)^2 \left(1 - h\left(n^{1-\alpha} - 2\right)\right)\right], \quad (6.93)$$

in the last (6.93) observe the high speed of convergence at $n^{-3\alpha}$.

Proof. Here $f \in C^4([a, b], \mathbb{C})$, and we apply the hyperbolic-trigonometric Taylor's formula for $f \in C^4([a, b], \mathbb{C})$, see Theorem 8 of Anastassiou (2023c).

Let $\frac{k}{n}, x \in [a, b]$, then

$$f\left(\frac{k}{n}\right) - f(x) - f'(x)\left(\frac{\sinh\left(\frac{k}{n} - x\right) + \sin\left(\frac{k}{n} - x\right)}{2}\right)$$

$$- f''(x)\left(\frac{\cosh\left(\frac{k}{n} - x\right) - \cos\left(\frac{k}{n} - x\right)}{2}\right)$$

$$- f^{(3)}(x)\left(\frac{\sinh\left(\frac{k}{n} - x\right) - \sin\left(\frac{k}{n} - x\right)}{2}\right)$$

$$- f^{(4)}(x)\left(\sinh^2\left(\frac{\frac{k}{n} - x}{2}\right) - \sin^2\left(\frac{\frac{k}{n} - x}{2}\right)\right)$$

$$= \int_x^{\frac{k}{n}} \left[\left(f^{(4)}(t) - f(t)\right) - \left(f^{(4)}(x) - f(x)\right)\right]$$

$$\times \left(\frac{\sinh\left(\frac{k}{n} - t\right) - \sin\left(\frac{k}{n} - t\right)}{2}\right) dt$$

$$=: R_4\left(\frac{k}{n}, x\right). \quad (6.94)$$

As in Theorems 6.5 and 6.6, we derive

$$A_n^*(f, x) - f(x) \sum_{k=\lceil na \rceil}^{\lfloor nb \rfloor} \psi(nx - k)$$

$$- \frac{f'(x)}{2} A_n^*((\sinh(\cdot - x) + \sin(\cdot - x)), x)$$

$$- \frac{f''(x)}{2} A_n^* ((\cosh(\cdot - x) - \cos(\cdot - x)), x)$$

$$- \frac{f^{(3)}(x)}{2} A_n^* ((\sinh(\cdot - x) - \sin(\cdot - x)), x)$$

$$- \frac{f^{(4)}(x)}{2} A_n^* \left(\left(\sinh^2 \left(\frac{\cdot - x}{2} \right) - \sin^2 \left(\frac{\cdot - x}{2} \right) \right), x \right) = \Phi_n(x), \quad (6.95)$$

where

$$\Phi_n(x) := \sum_{k=\lceil na \rceil}^{\lfloor nb \rfloor} \psi(nx - k) R_4 \left(\frac{k}{n}, x \right). \quad (6.96)$$

Without loss of generality, we can assume that $n > \left\lceil (b-a)^{-\frac{1}{\alpha}} \right\rceil$.

Thus, $\left| \frac{k}{n} - x \right| \leq \frac{1}{n^\alpha}$ or $\left| \frac{k}{n} - x \right| > \frac{1}{n^\alpha}$.

In case of $\left| \frac{k}{n} - x \right| \leq \frac{1}{n^\alpha}$, we have the following cases:

(i) If $\frac{k}{n} \geq x$, then

$$\left| R_4 \left(\frac{k}{n}, x \right) \right|$$

$$= \left| \frac{1}{2} \int_x^{\frac{k}{n}} \left[\left(f^{(4)}(t) - f(t) \right) - \left(f^{(4)}(x) - f(x) \right) \right] \left(\sinh \left(\frac{k}{n} - t \right) - \sin \left(\frac{k}{n} - t \right) \right) dt \right|$$

$$\leq \frac{1}{2} \int_x^{\frac{k}{n}} \left| \left(f^{(4)}(t) - f(t) \right) - \left(f^{(4)}(x) - f(x) \right) \right| \left| \sinh \left(\frac{k}{n} - t \right) - \sin \left(\frac{k}{n} - t \right) \right| dt \quad (6.97)$$

$$\leq \frac{1}{2} \int_x^{\frac{k}{n}} \omega_1 \left(f^{(4)} - f, t - x \right) \left(\left| \sinh \left(\frac{k}{n} - t \right) \right| + \left| \sin \left(\frac{k}{n} - t \right) \right| \right) dt$$

$$\leq \frac{\omega_1 \left(f^{(4)} - f, \frac{k}{n} - x \right)}{2} \int_x^{\frac{k}{n}} \left(\cosh(b-a) \left(\frac{k}{n} - t \right) + \left(\frac{k}{n} - t \right) \right) dt$$

$$= \frac{(\cosh(b-a) + 1) \omega_1 \left(f^{(4)} - f, \frac{k}{n} - x \right)}{2} \int_x^{\frac{k}{n}} \left(\frac{k}{n} - t \right) dt$$

$$= \frac{(\cosh(b-a) + 1) \omega_1 \left(f^{(4)} - f, \frac{k}{n} - x \right)}{4} \left(\frac{k}{n} - x \right)^2$$

$$\leq \frac{(\cosh(b-a) + 1) \omega_1 \left(f^{(4)} - f, \frac{1}{n^\alpha} \right)}{4n^{2\alpha}}. \quad (6.98)$$

That is , when $\frac{k}{n} \geq x$, then

$$\left| R_4 \left(\frac{k}{n}, x \right) \right| \leq \frac{(\cosh(b-a)+1)\,\omega_1\left(f^{(4)}-f, \frac{1}{n^{\alpha}}\right)}{4n^{2\alpha}}. \tag{6.99}$$

(ii) If $\frac{k}{n} < x$, then

$$\left| R_4 \left(\frac{k}{n}, x \right) \right|$$

$$= \left| \frac{1}{2} \int_{\frac{k}{n}}^{x} \left[\left(f^{(4)}(t) - f(t) \right) - \left(f^{(4)}(x) - f(x) \right) \right] \left(\sinh\left(\frac{k}{n} - t \right) - \sin\left(\frac{k}{n} - t \right) \right) dt \right|$$

$$\leq \frac{\omega_1\left(f^{(4)}-f, x-\frac{k}{n}\right)}{2} \int_{\frac{k}{n}}^{x} \left(\cosh(b-a)\left(t-\frac{k}{n}\right) + \left(t-\frac{k}{n}\right) \right) dt$$

$$= \frac{(\cosh(b-a)+1)\,\omega_1\left(f^{(4)}-f, x-\frac{k}{n}\right)}{2} \int_{\frac{k}{n}}^{x} \left(t-\frac{k}{n}\right) dt$$

$$= \frac{(\cosh(b-a)+1)\,\omega_1\left(f^{(4)}-f, x-\frac{k}{n}\right)}{4} \left(x-\frac{k}{n}\right)^2$$

$$\leq \frac{(\cosh(b-a)+1)\,\omega_1\left(f^{(4)}-f, \frac{1}{n^{\alpha}}\right)}{4n^{2\alpha}}. \tag{6.100}$$

Consequently, when $\left|\frac{k}{n} - x\right| \leq \frac{1}{n^{\alpha}}$, we always obtain that

$$\left| R_4 \left(\frac{k}{n}, x \right) \right| \leq \frac{(\cosh(b-a)+1)\,\omega_1\left(f^{(4)}-f, \frac{1}{n^{\alpha}}\right)}{4n^{2\alpha}}. \tag{6.101}$$

Next assume again $\frac{k}{n} \geq x$, then

$$\begin{aligned} \left| R_4 \left(\frac{k}{n}, x \right) \right| &\leq \frac{1}{2} \int_{x}^{\frac{k}{n}} \left| \left(f^{(4)}(t) - f(t) \right) - \left(f^{(4)}(x) - f(x) \right) \right| \\ &\quad \times \left| \sinh\left(\frac{k}{n} - t \right) - \sin\left(\frac{k}{n} - t \right) \right| dt \\ &\leq \left\| f^{(4)} - f \right\|_{\infty} \int_{x}^{\frac{k}{n}} \left[\cosh(b-a)\left(\frac{k}{n} - t \right) + \left(\frac{k}{n} - t \right) \right] dt \\ &= \left\| f^{(4)} - f \right\|_{\infty} (\cosh(b-a)+1) \int_{x}^{\frac{k}{n}} \left(\frac{k}{n} - t \right) dt \end{aligned}$$

$$= \frac{\left\| f^{(4)} - f \right\|_{\infty} (\cosh(b-a) + 1)}{2} \left(\frac{k}{n} - x \right)^2$$
$$\leq \frac{\left\| f^{(4)} - f \right\|_{\infty} (\cosh(b-a) + 1)(b-a)^2}{2}. \tag{6.102}$$

Hence,

$$\left| R_4 \left(\frac{k}{n}, x \right) \right| \leq \frac{\left\| f^{(4)} - f \right\|_{\infty} (\cosh(b-a) + 1)(b-a)^2}{2}. \tag{6.103}$$

When $\frac{k}{n} < x$, we have

$$\left| R_4 \left(\frac{k}{n}, x \right) \right| \leq \frac{1}{2} \int_{\frac{k}{n}}^{x} \left| \left(f^{(4)}(t) - f(t) \right) - \left(f^{(4)}(x) - f(x) \right) \right|$$
$$\times \left| \sinh \left(\frac{k}{n} - t \right) - \sin \left(\frac{k}{n} - t \right) \right| dt$$
$$\leq \left\| f^{(4)} - f \right\|_{\infty} \int_{\frac{k}{n}}^{x} \left[\cosh(b-a) \left(t - \frac{k}{n} \right) + \left(t - \frac{k}{n} \right) \right] dt$$
$$= \left\| f^{(4)} - f \right\|_{\infty} (\cosh(b-a) + 1) \int_{\frac{k}{n}}^{x} \left(t - \frac{k}{n} \right) dt$$
$$= \frac{\left\| f^{(4)} - f \right\|_{\infty} (\cosh(b-a) + 1)}{2} \left(x - \frac{k}{n} \right)^2$$
$$\leq \frac{\left\| f^{(4)} - f \right\|_{\infty} (\cosh(b-a) + 1)(b-a)^2}{2}. \tag{6.104}$$

So, it is always true that

$$\left| R_4 \left(\frac{k}{n}, x \right) \right| \leq \frac{\left\| f^{(4)} - f \right\|_{\infty} (\cosh(b-a) + 1)(b-a)^2}{2}. \tag{6.105}$$

Thus,

$$|\Phi_n(x)| \leq \sum_{k=\lceil na \rceil}^{\lfloor nb \rfloor} \psi(nx - k) \left| R_4 \left(\frac{k}{n}, x \right) \right|$$
$$= \sum_{\left\{ \substack{k=\lceil na \rceil \\ : \left| \frac{k}{n} - x \right| \leq \frac{1}{n^{\alpha}}} \right.}^{\lfloor nb \rfloor} \psi(nx - k) \left| R_4 \left(\frac{k}{n}, x \right) \right| \tag{6.106}$$
$$+ \sum_{\left\{ \substack{k=\lceil na \rceil \\ : \left| \frac{k}{n} - x \right| > \frac{1}{n^{\alpha}}} \right.}^{\lfloor nb \rfloor} \psi(nx - k) \left| R_4 \left(\frac{k}{n}, x \right) \right|$$

$$\leq \left(\sum_{\substack{k=\lceil na \rceil \\ \left\{ : \left| \frac{k}{n} - x \right| \leq \frac{1}{n^{\alpha}} \right.}}^{\lfloor nb \rfloor} \psi(nx-k) \right) \frac{(\cosh(b-a)+1)\,\omega_1\left(f^{(4)}-f, \frac{1}{n^{\alpha}}\right)}{4n^{2\alpha}} \tag{6.107}$$

$$+ \left(\sum_{\substack{k=\lceil na \rceil \\ \left\{ : \left| \frac{k}{n} - x \right| > \frac{1}{n^{\alpha}} \right.}}^{\lfloor nb \rfloor} \psi(nx-k) \right) \left(\frac{\left\| f^{(4)} - f \right\|_{\infty} (\cosh(b-a)+1)(b-a)^2}{2} \right)$$

$$\overset{\text{(by Theorem 6.3)}}{\leq} \frac{(\cosh(b-a)+1)\,\omega_1\left(f^{(4)}-f, \frac{1}{n^{\alpha}}\right)}{4n^{2\alpha}} + \frac{\left\| f^{(4)} - f \right\|_{\infty} (\cosh(b-a)+1)(b-a)^2}{2} \left(1 - h\left(n^{1-\alpha}-2\right)\right). \tag{6.108}$$

We have proved that

$$|\Phi_n(x)| \leq \frac{(\cosh(b-a)+1)}{2} \left[\frac{\omega_1\left(f^{(4)}-f, \frac{1}{n^{\alpha}}\right)}{2n^{2\alpha}} + \left\| f^{(4)} - f \right\|_{\infty} (b-a)^2 \left(1 - h\left(n^{1-\alpha}-2\right)\right) \right]. \tag{6.109}$$

We observe that

$$A_n(f,x) - f(x) - \frac{f'(x)}{2} A_n((\sinh(\cdot - x) + \sin(\cdot - x)), x)$$
$$- \frac{f''(x)}{2} A_n((\cosh(\cdot - x) - \cos(\cdot - x)), x) \tag{6.110}$$
$$- \frac{f^{(3)}(x)}{2} A_n((\sinh(\cdot - x) - \sin(\cdot - x)), x)$$
$$- f^{(4)}(x) A_n\left(\left(\sinh^2\left(\frac{\cdot - x}{2}\right) - \sin^2\left(\frac{\cdot - x}{2}\right)\right), x\right)$$
$$= \left[A_n^*(f,x) - f(x) \sum_{k=\lceil na \rceil}^{\lfloor nb \rfloor} \psi(nx-k) - \frac{f'(x)}{2} A_n^*((\sinh(\cdot - x) + \sin(\cdot - x)), x) \right.$$
$$- \frac{f''(x)}{2} A_n^*((\cosh(\cdot - x) - \cos(\cdot - x)), x)$$

$$- \frac{f^{(3)}(x)}{2} A_n^* ((\sinh(\cdot - x) - \sin(\cdot - x)), x)$$

$$\left. - f^{(4)}(x) A_n^* \left(\left(\sinh^2 \left(\frac{\cdot - x}{2} \right) - \sin^2 \left(\frac{\cdot - x}{2} \right) \right), x \right) \right] \frac{1}{\sum_{k=\lceil na \rceil}^{\lfloor nb \rfloor} \psi(nx - k)}$$

$$= \frac{\Phi_n(x)}{\sum_{k=\lceil na \rceil}^{\lfloor nb \rfloor} \psi(nx - k)}. \tag{6.111}$$

Finally, we obtain ($\forall\, x \in [a, b]$, $n \in \mathbb{N}$)

$$\left| A_n(f, x) - f(x) - \frac{f'(x)}{2} A_n ((\sinh(\cdot - x) + \sin(\cdot - x)), x) \right.$$

$$- \frac{f''(x)}{2} A_n ((\cosh(\cdot - x) - \cos(\cdot - x)), x) \tag{6.112}$$

$$- \frac{f^{(3)}(x)}{2} A_n ((\sinh(\cdot - x) - \sin(\cdot - x)), x)$$

$$\left. - f^{(4)}(x) A_n \left(\left(\sinh^2 \left(\frac{\cdot - x}{2} \right) - \sin^2 \left(\frac{\cdot - x}{2} \right) \right), x \right) \right|$$

$$= \frac{|\Phi_n(x)|}{\sum_{k=\lceil na \rceil}^{\lfloor nb \rfloor} \psi(nx - k)} \leq \frac{|\Phi_n(x)|}{\psi(1)} \overset{\text{(by (6.109))}}{\leq} \frac{(\cosh(b - a) + 1)}{2\psi(1)}$$

$$\left[\frac{\omega_1 \left(f^{(4)} - f, \frac{1}{n^\alpha} \right)}{2n^{2\alpha}} + \left\| f^{(4)} - f \right\|_\infty (b - a)^2 \left(1 - h \left(n^{1-\alpha} - 2 \right) \right) \right]. \tag{6.113}$$

The theorem is proved. □

We continue with a general trigonometric result.

Theorem 6.8. *Let $f \in C^4([a, b], \mathbb{C})$, $0 < \alpha < 1$, $n \in \mathbb{N} : n^{1-\alpha} > 2$, $x \in [a, b]$. Let also $\overline{\alpha}, \overline{\beta} \in \mathbb{R}$ with $\overline{\alpha}\overline{\beta}\left(\overline{\alpha}^2 - \overline{\beta}^2\right) \neq 0$. Then*

(1)

$$\left| A_n(f, x) - f(x) - \frac{f'(x)}{\overline{\alpha}\overline{\beta}\left(\overline{\beta}^2 - \overline{\alpha}^2\right)} A_n \left(\left(\overline{\beta}^3 \sin(\overline{\alpha}(\cdot - x)) - \overline{\alpha}^3 \sin\left(\overline{\beta}(\cdot - x)\right) \right), x \right) \right.$$

$$- \frac{f''(x)}{\left(\overline{\beta}^2 - \overline{\alpha}^2\right)} A_n \left(\left(\cos(\overline{\alpha}(\cdot - x)) - \cos\left(\overline{\beta}(\cdot - x)\right) \right), x \right)$$

$$- \frac{f'''(x)}{\overline{\alpha}\overline{\beta}\left(\overline{\beta}^2 - \overline{\alpha}^2\right)} A_n\left(\left(\overline{\beta}\sin\left(\overline{\alpha}(\cdot - x)\right) - \overline{\alpha}\sin\left(\overline{\beta}(\cdot - x)\right)\right), x\right)$$

$$- \left(\frac{2f^{(4)}(x) + \left(\overline{\alpha}^2 + \overline{\beta}^2\right) f''(x)}{\left(\overline{\alpha}\overline{\beta}\right)^2 \left(\overline{\beta}^2 - \overline{\alpha}^2\right)}\right)$$

$$A_n\left(\left(\overline{\beta}^2 \sin^2\left(\frac{\overline{\alpha}(\cdot - x)}{2}\right) - \overline{\alpha}^2 \sin^2\left(\frac{\overline{\beta}(\cdot - x)}{2}\right)\right), x\right)\Bigg|$$

$$\leq \frac{1}{\psi(1)\left|\overline{\beta}^2 - \overline{\alpha}^2\right|}\left[\frac{\omega_1\left(\left(f^{(4)} + \left(\overline{\alpha}^2 + \overline{\beta}^2\right) f'' + \overline{\alpha}^2\overline{\beta}^2 f\right), \frac{1}{n^\alpha}\right)}{n^{2\alpha}}\right.$$

$$\left. + 2\left\|f^{(4)} + \left(\overline{\alpha}^2 + \overline{\beta}^2\right) f'' + \overline{\alpha}^2\overline{\beta}^2 f\right\|_\infty (b-a)^2 \left(1 - h\left(n^{1-\alpha} - 2\right)\right)\right], \tag{6.114}$$

(2) *if* $f^{(i)}(x) = 0$, $i = 1, 2, 3, 4$, *we get*

$$|A_n(f, x) - f(x)| \leq \frac{1}{\left|\overline{\beta}^2 - \overline{\alpha}^2\right| \psi(1)}$$

$$\left[\frac{\omega_1\left(\left(f^{(4)} + \left(\overline{\alpha}^2 + \overline{\beta}^2\right) f'' + \overline{\alpha}^2\overline{\beta}^2 f\right), \frac{1}{n^\alpha}\right)}{n^{2\alpha}}\right.$$

$$\left. + 2\left\|f^{(4)} + \left(\overline{\alpha}^2 + \overline{\beta}^2\right) f'' + \overline{\alpha}^2\overline{\beta}^2 f\right\|_\infty (b-a)^2 \left(1 - h\left(n^{1-\alpha} - 2\right)\right)\right]. \tag{6.115}$$

The high speed of convergence in (1) and (2) is $n^{-3\alpha}$.

Proof. As similar to Theorem 8 is omitted. It is based on Theorem 9 of Anastassiou (2023c). □

We finish with a general hyperbolic result.

Theorem 6.9. *Let* $f \in C^4([a, b], \mathbb{C})$, $0 < \alpha < 1$, $n \in \mathbb{N} : n^{1-\alpha} > 2$, $x \in [a, b]$. *Let also* $\overline{\alpha}, \overline{\beta} \in \mathbb{R}$ *with* $\overline{\alpha}\overline{\beta}\left(\overline{\alpha}^2 - \overline{\beta}^2\right) \neq 0$. *Then*

(1)

$$\Bigg| A_n(f,x) - f(x) - \frac{f'(x)}{\overline{\alpha}\overline{\beta}\left(\overline{\beta}^2 - \overline{\alpha}^2\right)} A_n\left(\left(\overline{\beta}^3 \sinh\left(\overline{\alpha}(\cdot - x)\right) - \overline{\alpha}^3 \sinh\left(\overline{\beta}(\cdot - x)\right)\right), x\right)$$

$$- \frac{f''(x)}{\overline{\beta}^2 - \overline{\alpha}^2} A_n\left(\left(\cosh\left(\overline{\beta}(\cdot - x)\right) - \cosh\left(\overline{\alpha}(\cdot - x)\right)\right), x\right)$$

$$- \frac{f'''(x)}{\overline{\alpha}\overline{\beta}\left(\overline{\beta}^2 - \overline{\alpha}^2\right)} A_n\left(\left(\overline{\alpha}\sinh\left(\overline{\beta}(\cdot - x)\right) - \overline{\beta}\sinh\left(\overline{\alpha}(\cdot - x)\right)\right), x\right)$$

$$- \left(\frac{2(f^{(4)}(x) - \left(\overline{\alpha}^2 + \overline{\beta}^2\right) f''(x))}{\left(\overline{\alpha}\overline{\beta}\right)^2 \left(\overline{\beta}^2 - \overline{\alpha}^2\right)} \right)$$

$$A_n\left(\left(\overline{\alpha}^2 \sinh^2\left(\frac{\overline{\beta}(\cdot - x)}{2}\right) - \overline{\beta}^2 \sinh^2\left(\frac{\overline{\alpha}(\cdot - x)}{2}\right)\right), x\right) \Bigg|$$

$$\leq \frac{\cosh(b-a)}{\psi(1)\left|\overline{\beta}^2 - \overline{\alpha}^2\right|} \left[\frac{\omega_1\left(\left(f^{(4)} - \left(\overline{\alpha}^2 + \overline{\beta}^2\right) f'' + \overline{\alpha}^2\overline{\beta}^2 f\right), \frac{1}{n^\alpha}\right)}{n^{2\alpha}} \right.$$

$$\left. + 2\left\| f^{(4)} - \left(\overline{\alpha}^2 + \overline{\beta}^2\right) f'' + \overline{\alpha}^2\overline{\beta}^2 f \right\|_\infty (b-a)^2 \left(1 - h\left(n^{1-\alpha} - 2\right)\right) \right], \tag{6.116}$$

(2) *if* $f^{(i)}(x) = 0$, $i = 1, 2, 3, 4$, *we get*

$$|A_n(f,x) - f(x)| \leq \frac{\cosh(b-a)}{\left|\overline{\beta}^2 - \overline{\alpha}^2\right| \psi(1)}$$

$$\left[\frac{\omega_1\left(\left(f^{(4)} - \left(\overline{\alpha}^2 + \overline{\beta}^2\right) f'' + \overline{\alpha}^2\overline{\beta}^2 f\right), \frac{1}{n^\alpha}\right)}{n^{2\alpha}} \right.$$

$$\left. + 2\left\| f^{(4)} - \left(\overline{\alpha}^2 + \overline{\beta}^2\right) f'' + \overline{\alpha}^2\overline{\beta}^2 f \right\|_\infty (b-a)^2 \left(1 - h\left(n^{1-\alpha} - 2\right)\right) \right]. \tag{6.117}$$

The high speed of convergence in (1) and (2) is $n^{-3\alpha}$.

Proof. As similar to Theorem 8 is omitted. It is based on Theorem 10 of Anastassiou (2023c). ☐

Bibliography

G.A. Anastassiou, Rate of convergence of some neural network operators to the unit-univariate case, *Journal of Mathematical Analysis and Applications*, 212 (1997), 237–262.

G.A. Anastassiou, *Quantitative Approximations*, Chapman & Hall / CRC, Boca Raton, New York, 2001.

G.A. Anastassiou, Univariate hyperbolic tangent neural network approximation, *Mathematics and Computer Modelling*, 53 (2011a), 1111–1132.

G.A. Anastassiou, Multivariate hyperbolic tangent neural network approximation, *Computers and Mathematics*, 61 (2011b), 809–821.

G.A. Anastassiou, Multivariate sigmoidal neural network approximation, *Neural Networks*, 24 (2011c), 378–386.

G.A. Anastassiou, *Inteligent Systems: Approximation by Artificial Neural Networks*, Intelligent Systems Reference Library, Vol. 19, Springer, Heidelberg, 2011d.

G.A. Anastassiou, Univariate sigmoidal neural network approximation, *J. of Computational Analysis and Applications*, 14(4) (2012), 659–690.

G.A. Anastassiou, Vector fractional Korovkin type Approximations, *Dynamic Systems and Applications*, 26 (2017), 81–104.

G.A. Anastassiou, General sigmoid based Banach space valued neural network approximation, *Journal of Computational Analysis and Applications*, 31(4) (2023a), 520–534.

G.A. Anastassiou, *Parametrized, deformed and general neural networks*, accepted for publication, Springer, Heidelberg, New York, 2023b.

G.A. Anastassiou, Opial and Ostrowski type inequalities based on trigonometric and hyperbolic type Taylor formulae, *Malaya Journal of Mathematik*, 11(S) (2023c), 1–26.

G.A. Anastassiou, *General sigmoid function based complex valued trigonometric and hyperbolic neural network high order approximations*, Analele Universitatii Oradea, Fasc. Matematica, accepted, 2023d.

Z. Chen and F. Cao, The approximation operators with sigmoidal functions, *Computers and Mathematics with Applications*, 58 (2009), 758–765.

S. Haykin, *Neural Networks: A Comprehensive Foundation* 2nd ed., Prentice Hall, New York, 1998.

W. McCulloch and W. Pitts, A logical calculus of the ideas immanent in nervous activity, *Bulletin of Mathematical Biophysis*, 7 (1943), 115–133.

T.M. Mitchell, *Machine Learning*, WCB-McGraw-Hill, New York, 1997.

Chapter 7

General Multiple Sigmoid Functions Activated Complex Valued Multivariate Trigonometric and Hyperbolic Neural Network Approximation

Here we research the multivariate quantitative approximation of complex-valued continuous functions on a box of $\mathbb{R}^N$, $N \in \mathbb{N}$, by the multivariate normalized-type neural network operators. We investigate also the case of approximation by iterated multilayer neural network operators. These approximations are achieved by establishing multidimensional Jackson-type inequalities involving the multivariate moduli of continuity of the engaged function and its partial derivatives. Our multivariate operators are defined by using a multidimensional density function induced by general multiple sigmoid functions. The approximations are pointwise and uniform. The related feed-forward neural networks are with one or multiple hidden layers. The basis of our theory is the introduced multivariate Taylor formulae of trigonometric and hyperbolic type. It follows Anastassiou (2023).

7.1 Introduction

The author in Anastassiou (1997, 2001), see Chapters 2–5, was the first to establish neural network approximations to continuous functions with rates by very specifically defined neural network operators of Cardaliaguet–Euvrard and "Squashing" types, by employing the modulus of continuity of the engaged function or its high order derivative, and producing very tight Jackson-type inequalities. He treats there both the univariate and multivariate cases. The definitions of these operators' "bell-shaped" and "squashing" functions are assumed to be of compact support.

Motivations for this work include articles by Chen and Cao (2009), Anastassiou (2011a)–Anastassiou (2023b), and Costarelli and Spigler (2013a), Costarelli and Spigler (2013b).

Here we perform general multiple sigmoid function-based trigonometric and hyperbolic neural network approximations to complex-valued continuous functions over boxes in $\mathbb{R}^N$, $N \in \mathbb{N}$, and also iterated, multilayer approximations. All convergences here are with rates expressed via the multivariate moduli of continuity of the involved function and its partial derivatives and given by very tight multidimensional Jackson-type inequalities.

We come up with the "right" precisely defined multivariate normalized, quasi-interpolation neural network operators based on boxes of $\mathbb{R}^N$. Our boxes are not necessarily symmetric to the origin. In preparation to prove our results, we mention important properties of the basic multivariate density function induced by a set of general multiple sigmoid functions.

Feed-forward neural networks (FNNs) with one hidden layer here are mathematically expressed as

$$N_n(x) = \sum_{j=0}^{n} c_j \sigma\left(\langle a_j \cdot x \rangle + b_j\right), \quad x \in \mathbb{R}^s, \quad s \in \mathbb{N},$$

where for $0 \leq j \leq n$, $b_j \in \mathbb{R}$ are the thresholds, $a_j \in \mathbb{R}^s$ are the connection weights, $c_j \in \mathbb{C}$ are the coefficients, $\langle a_j \cdot x \rangle$ is the inner product of a_j and x, and σ is the activation function of the network. In many fundamental network models, the activation function is a kind of general sigmoid function. For more information about neural networks, read Haykin (1998)–Mitchell (1997).

7.2 Basics

7.2.1 *General neural network background*

The following comes from Anastassiou (2023b, Chapter 27).

Let $i = 1, \ldots, N \in \mathbb{N}$ and $h_i : \mathbb{R} \to [-1, 1]$ be a general sigmoid activation function, such that it is strictly increasing, $h_i(0) = 0$, $h_i(-x) = -h_i(x)$, $h_i(+\infty) = 1$, $h_i(-\infty) = -1$. Also h_i is strictly convex over $(-\infty, 0]$ and strictly concave over $[0, +\infty)$, with $h_i^{(2)} \in C(\mathbb{R})$.

We consider the scaled function

$$\psi_i(x) := \frac{1}{4}(h_i(x+1) - h_i(x-1)), \quad x \in \mathbb{R}, \; i = 1, \ldots, N. \tag{7.1}$$

As in Anastassiou (2018, p. 285), we get that $\psi_i(-x) = \psi_i(x)$, thus ψ_i is an even function. Since $x + 1 > x - 1$, then $h_i(x+1) > h_i(x-1)$, and $\psi_i(x) > 0$, all $x \in \mathbb{R}$.

We see that

$$\psi_i(0) = \frac{h_i(1)}{2}, \quad i = 1, \ldots, N. \tag{7.2}$$

Let $x > 1$, we have that

$$\psi_i'(x) = \frac{1}{4}(h_i'(x+1) - h_i'(x-1)) < 0,$$

by h_i' being strictly decreasing over $[0, +\infty)$.

Let now $0 < x < 1$, then $1 - x > 0$ and $0 < 1 - x < 1 + x$. It holds $h_i'(x-1) = h_i'(1-x) > h_i'(x+1)$ so that again $\psi_i'(x) < 0$. Consequently, ψ_i is strictly decreasing on $(0, +\infty)$.

Clearly, ψ_i is strictly increasing on $(-\infty, 0)$, and $\psi_i'(0) = 0$.

See that

$$\lim_{x \to +\infty} \psi_i(x) = \frac{1}{4}(h_i(+\infty) - h_i(+\infty)) = 0 \tag{7.3}$$

and

$$\lim_{x \to -\infty} \psi_i(x) = \frac{1}{4}(h_i(-\infty) - h_i(-\infty)) = 0. \tag{7.4}$$

That is, the x-axis is the horizontal asymptote on ψ_i.

Conclusion, ψ is a bell symmetric function with maximum

$$\psi_i(0) = \frac{h_i(1)}{2}.$$

We need

Theorem 7.1 (Anastassiou 2023b, Ch. 27). *We have that*

$$\sum_{i=-\infty}^{\infty} \psi_i(x - i) = 1, \quad \forall\, x \in \mathbb{R}, \; i = 1, \ldots, N. \tag{7.5}$$

Theorem 7.2 (Anastassiou 2023b, Ch. 27). *It holds*

$$\int_{-\infty}^{\infty} \psi_i(x)dx = 1, \quad i = 1, \ldots, N. \tag{7.6}$$

Thus, $\psi_i(x)$ is a density function on $\mathbb{R}$, $i = 1, \ldots, N$.

We give

Theorem 7.3 (Anastassiou 2023b, Ch. 27). *Let* $0 < \alpha < 1$, *and* $n \in \mathbb{N}$ *with* $n^{1-\alpha} > 2$. *It holds*

$$\sum_{\left\{\substack{k=-\infty \\ :|nx-k| \geq n^{1-\alpha}}\right.}^{\infty} \psi_i(nx-k) < (1 - h_i(n^{1-\alpha} - 2)),\ i = 1, \ldots, N. \tag{7.7}$$

Note that

$$\lim_{n \to +\infty} (1 - h_i(n^{1-\alpha} - 2)) = 0,\ \ i = 1, \ldots, N.$$

Denote by $\lfloor \cdot \rfloor$ the integral part of the number and by $\lceil \cdot \rceil$ the ceiling of the number.

We further give

Theorem 7.4 (Anastassiou 2023b, Ch. 27). *Let* $x \in [a, b] \subset \mathbb{R}$ *and* $n \in \mathbb{N}$ *so that* $\lceil na \rceil \leq \lfloor nb \rfloor$. *It holds*

$$\frac{1}{\sum_{k=\lceil na \rceil}^{\lfloor nb \rfloor} \psi_i(nx-k)} < \frac{1}{\psi_i(1)},\ \ \forall\, x \in [a, b]\,,\ \ i = 1, \ldots, N. \tag{7.8}$$

Remark 7.1 (Anastassiou 2023b, Ch. 27). We have that

$$\lim_{n \to \infty} \sum_{k=\lceil na \rceil}^{\lfloor nb \rfloor} \psi_i(nx-k) \neq 1,\ \ \ i = 1, \ldots, N, \tag{7.9}$$

for at least some $x \in [a, b]$.

Note 7.1 (Anastassiou 2023b, Ch. 27). For large enough n, we always obtain $\lceil na \rceil \leq \lfloor nb \rfloor$. Also $a \leq \frac{k}{n} \leq b$, iff $\lceil na \rceil \leq k \leq \lfloor nb \rfloor$. In general, it holds (by (7.5))

$$\sum_{k=\lceil na \rceil}^{\lfloor nb \rfloor} \psi_i(nx-k) \leq 1,\ \ \ i = 1, \ldots, N. \tag{7.10}$$

We make

Remark 7.2 (Anastassiou 2023b, Ch. 27). We define

$$Z(x_1, \ldots, x_N) := Z(x) := \prod_{i=1}^{N} \psi_i(x_i),\ \ \ x = (x_1, \ldots, x_N) \in \mathbb{R}^N,\ \ N \in \mathbb{N}.$$

It has the following properties:

(i)

$$Z(x) > 0, \ \forall x \in \mathbb{R}^N, \tag{7.11}$$

(ii)

$$\sum_{k=-\infty}^{\infty} Z(x-k) := \sum_{k_1=-\infty}^{\infty} \sum_{k_2=-\infty}^{\infty} \cdots \sum_{k_N=-\infty}^{\infty} Z(x_1-k_1,\ldots,x_N-k_N) =$$

$$\sum_{k_1=-\infty}^{\infty} \sum_{k_2=-\infty}^{\infty} \cdots \sum_{k_N=-\infty}^{\infty} \prod_{i=1}^{N} \psi_i(x_i-k_i) = \prod_{i=1}^{N} \left(\sum_{k_i=-\infty}^{\infty} \psi_i(x_i-k_i) \right) \overset{(7.5)}{=} 1.$$

Hence,

$$\sum_{k=-\infty}^{\infty} Z(x-k) = 1. \tag{7.12}$$

That is,

(iii)

$$\sum_{k=-\infty}^{\infty} Z(nx-k) = 1, \ \ \forall x \in \mathbb{R}^N; \ n \in \mathbb{N}. \tag{7.13}$$

And

(iv)

$$\int_{\mathbb{R}^N} Z(x)dx = \int_{\mathbb{R}^N} \left(\prod_{i=1}^{N} \psi_i(x_i) \right) dx_1 \ldots dx_N = \prod_{i=1}^{N} \left(\int_{-\infty}^{\infty} \psi_i(x_i)dx_i \right) \overset{(7.6)}{=} 1, \tag{7.14}$$

thus

$$\int_{\mathbb{R}^N} Z(x)dx = 1, \tag{7.15}$$

that is, Z is a multivariate density function.

Here denote $\|x\|_\infty := \max\{|x_1|,\ldots,|x_N|\}$, $x \in \mathbb{R}^N$, also set $\infty := (\infty,\ldots,\infty)$, $-\infty := (-\infty,\ldots,-\infty)$ upon the multivariate context, and

$$\lceil na \rceil := (\lceil na_1 \rceil, \ldots, \lceil na_N \rceil),$$
$$\lfloor nb \rfloor := (\lfloor nb_1 \rfloor, \ldots, \lfloor nb_N \rfloor),$$

where $a := (a_1,\ldots,a_N)$, $b := (b_1,\ldots,b_N)$.

We obviously see that

$$\sum_{k=\lceil na \rceil}^{\lfloor nb \rfloor} Z(nx-k) = \sum_{k=\lceil na \rceil}^{\lfloor nb \rfloor} \left(\prod_{i=1}^{N} \psi_i (nx_i - k_i) \right)$$

$$= \sum_{k_1=\lceil na_1 \rceil}^{\lfloor nb_1 \rfloor} \cdots \sum_{k_N=\lceil na_N \rceil}^{\lfloor nb_N \rfloor} \left(\prod_{i=1}^{N} \psi_i (nx_i - k_i) \right)$$

$$= \prod_{i=1}^{N} \left(\sum_{k_i=\lceil na_i \rceil}^{\lfloor nb_i \rfloor} \psi_i (nx_i - k_i) \right). \tag{7.16}$$

For $0 < \beta < 1$ and $n \in \mathbb{N}$, a fixed $x \in \mathbb{R}^N$, we have that

$$\sum_{k=\lceil na \rceil}^{\lfloor nb \rfloor} Z(nx-k)$$

$$= \sum_{\begin{cases} k=\lceil na \rceil \\ \left\| \frac{k}{n} - x \right\|_\infty \le \frac{1}{n^\beta} \end{cases}}^{\lfloor nb \rfloor} Z(nx-k) + \sum_{\begin{cases} k=\lceil na \rceil \\ \left\| \frac{k}{n} - x \right\|_\infty > \frac{1}{n^\beta} \end{cases}}^{\lfloor nb \rfloor} Z(nx-k). \tag{7.17}$$

In the last two sums, the counting is over disjoint vector sets of k's, because the condition $\| \frac{k}{n} - x \|_\infty > \frac{1}{n^\beta}$ implies that there exists at least one $| \frac{k_r}{n} - x_r | > \frac{1}{n^\beta}$, where $r \in \{1, \ldots, N\}$.

(v) We note that

$$\sum_{\begin{cases} k=\lceil na \rceil \\ \left\| \frac{k}{n} - x \right\|_\infty > \frac{1}{n^\beta} \end{cases}}^{\lfloor nb \rfloor} Z(nx-k) = \sum_{\substack{k_1=\lceil na_1 \rceil \\ \left\| \frac{k}{n} - x \right\|_\infty > \frac{1}{n^\beta}}}^{\lfloor nb_1 \rfloor} \cdots \sum_{k_N=\lceil na_N \rceil}^{\lfloor nb_N \rfloor} \left(\prod_{i=1}^{N} \psi_i (nx_i - k_i) \right)$$

$$= \prod_{i=1}^{N} \left(\sum_{\begin{cases} k_i=\lceil na_i \rceil \\ \left\| \frac{k}{n} - x \right\|_\infty > \frac{1}{n^\beta} \end{cases}}^{\lfloor nb_i \rfloor} \psi_i (nx_i - k_i) \right)$$

$$\le \left(\prod_{\substack{i=1 \\ i \ne r}}^{N} \left(\sum_{k_i=-\infty}^{\infty} \psi_i (nx_i - k_i) \right) \right) \left(\sum_{\begin{cases} k_r=\lceil na_r \rceil \\ \left| \frac{k_r}{n} - x_r \right| > \frac{1}{n^\beta} \end{cases}}^{\lfloor nb_r \rfloor} \psi_r (nx_r - k_r) \right)$$

$$= \left(\sum_{\left\{ \substack{k_r = \lceil na_r \rceil \\ \left| \frac{k_r}{n} - x_r \right| > \frac{1}{n^\beta}} \right.}^{\lfloor nb_r \rfloor} \psi_r (nx_r - k_r) \right)$$

$$\leq \sum_{\left\{ \substack{k_r = -\infty \\ \left| \frac{k_r}{n} - x_r \right| > \frac{1}{n^\beta}} \right.}^{\infty} \psi_r (nx_r - k_r) = \sum_{\left\{ \substack{k_r = -\infty \\ |nx_r - k_r| > n^{1-\beta}} \right.}^{\infty} \psi_r (nx_r - k_r) \overset{(7.7)}{<}$$

$$1 - h_r \left(n^{1-\beta} - 2 \right) \leq \max_{i \in \{1,\ldots,N\}} \left(1 - h_i \left(n^{1-\beta} - 2 \right) \right), \tag{7.18}$$

where $0 < \beta < 1$.

That is, we get

$$\sum_{\left\{ \substack{k = \lceil na \rceil \\ \left\| \frac{k}{n} - x \right\|_\infty > \frac{1}{n^\beta}} \right.}^{\lfloor nb \rfloor} Z(nx - k) < \max_{i \in \{1,\ldots,N\}} \left(1 - h_i \left(n^{1-\beta} - 2 \right) \right), \tag{7.19}$$

$0 < \beta < 1$, with $n \in \mathbb{N} : n^{1-\beta} > 2$, $\forall\, x \in \prod_{i=1}^N [a_i, b_i]$.

(vi) It is clear that

$$\sum_{\left\{ \substack{k = -\infty \\ \left\| \frac{k}{n} - x \right\|_\infty > \frac{1}{n^\beta}} \right.}^{\infty} Z(nx - k) < \max_{i \in \{1,\ldots,N\}} \left(1 - h_i \left(n^{1-\beta} - 2 \right) \right), \tag{7.20}$$

$0 < \beta < 1$, $n \in \mathbb{N} : n^{1-\beta} > 2$, $\forall\, x \in \prod_{i=1}^N [a_i, b_i]$.

(viii) By Theorem 7.4, we get that

$$0 < \frac{1}{\sum_{k=\lceil na \rceil}^{\lfloor nb \rfloor} Z(nx - k)} = \frac{1}{\prod_{i=1}^N \left(\sum_{k_i = \lceil na_i \rceil}^{\lfloor nb_i \rfloor} \psi_i (nx_i - k_i) \right)} < \frac{1}{\prod_{i=1}^N \psi_i(1)},$$

thus

$$0 < \frac{1}{\sum_{k=\lceil na \rceil}^{\lfloor nb \rfloor} Z(nx - k)} < \frac{1}{\prod_{i=1}^N \psi_i(1)}, \tag{7.21}$$

$\forall\, x \in \left(\prod_{i=1}^N [a_i, b_i] \right)$, $n \in \mathbb{N}$.

Furthermore, it holds

$$\lim_{n\to\infty}\sum_{k=\lceil na\rceil}^{\lfloor nb\rfloor} Z(nx-k)=\lim_{n\to\infty}\prod_{i=1}^{N}\left(\sum_{k_i=\lceil na_i\rceil}^{\lfloor nb_i\rfloor}\psi_i(nx_i-k_i)\right)$$

$$=\prod_{i=1}^{N}\left(\lim_{n\to\infty}\sum_{k_i=\lceil na_i\rceil}^{\lfloor nb_i\rfloor}\psi_i(nx_i-k_i)\right)\neq 1, \tag{7.22}$$

for at least some $x\in\left(\prod_{i=1}^{N}[a_i,b_i]\right)$.

We state

Definition 7.1 (Anastassiou 2023b, Ch. 27). We denote by

$$\delta_N(\beta,n):=\max_{i\in\{1,\ldots,N\}}\left(1-h_i\left(n^{1-\beta}-2\right)\right), \tag{7.23}$$

where $0<\beta<1$.

We make

Remark 7.3. Let $f\in C(\prod_{i=1}^{N}[a_i,b_i],\mathbb{C})$, $x=(x_1,\ldots,x_N)\in\prod_{i=1}^{N}[a_i,b_i]$, $n\in\mathbb{N}$ such that $\lceil na_i\rceil\leq\lfloor nb_i\rfloor$, $i=1,\ldots,N$.

We introduce and define the following multivariate linear normalized neural network operator ($x:=(x_1,\ldots,x_N)\in(\prod_{i=1}^{N}[a_i,b_i])$):

$$A_n(f,x_1,\ldots,x_N):=A_n(f,x):=\frac{\sum_{k=\lceil na\rceil}^{\lfloor nb\rfloor}f\left(\frac{k}{n}\right)Z(nx-k)}{\sum_{k=\lceil na\rceil}^{\lfloor nb\rfloor}Z(nx-k)}$$

$$=\frac{\sum_{k_1=\lceil na_1\rceil}^{\lfloor nb_1\rfloor}\sum_{k_2=\lceil na_2\rceil}^{\lfloor nb_2\rfloor}\cdots\sum_{k_N=\lceil na_N\rceil}^{\lfloor nb_N\rfloor}f\left(\frac{k_1}{n},\ldots,\frac{k_N}{n}\right)\left(\prod_{i=1}^{N}\psi_i(nx_i-k_i)\right)}{\prod_{i=1}^{N}\left(\sum_{k_i=\lceil na_i\rceil}^{\lfloor nb_i\rfloor}\psi_i(nx_i-k_i)\right)}. \tag{7.24}$$

For large enough $n\in\mathbb{N}$, we always obtain $\lceil na_i\rceil\leq\lfloor nb_i\rfloor$, $i=1,\ldots,N$. Also $a_i\leq\frac{k_i}{n}\leq b_i$, iff $\lceil na_i\rceil\leq k_i\leq\lfloor nb_i\rfloor$, $i=1,\ldots,N$.

When $g\in C(\prod_{i=1}^{N}[a_i,b_i])$, we define the companion operator

$$\widetilde{A}_n(g,x):=\frac{\sum_{k=\lceil na\rceil}^{\lfloor nb\rfloor}g\left(\frac{k}{n}\right)Z(nx-k)}{\sum_{k=\lceil na\rceil}^{\lfloor nb\rfloor}Z(nx-k)}. \tag{7.25}$$

Clearly, $\widetilde{A}_n$ is a positive linear operator. We have that

$$\widetilde{A}_n(1,x)=1,\ \ \forall\, x\in\left(\prod_{i=1}^{N}[a_i,b_i]\right).$$

Note that $A_n(f) \in C(\prod_{i=1}^N [a_i, b_i], \mathbb{C})$ and $\widetilde{A}_n(g) \in C(\prod_{i=1}^N [a_i, b_i])$.

Furthermore, it holds

$$|A_n(f, x)| \leq \frac{\sum_{k=\lceil na \rceil}^{\lfloor nb \rfloor} \left| f\left(\frac{k}{n}\right) \right| Z(nx - k)}{\sum_{k=\lceil na \rceil}^{\lfloor nb \rfloor} Z(nx - k)} = \widetilde{A}_n(|f|, x), \tag{7.26}$$

$\forall\, x \in \prod_{i=1}^N [a_i, b_i]$.

Clearly, $|f| \in C(\prod_{i=1}^N [a_i, b_i])$.

So, we have that

$$|A_n(f, x)| \leq \widetilde{A}_n(|f|, x), \tag{7.27}$$

$\forall\, x \in \prod_{i=1}^N [a_i, b_i]$, $\forall\, n \in \mathbb{N}$, $\forall\, f \in C\left(\prod_{i=1}^N [a_i, b_i], \mathbb{C}\right)$.

Let $c \in \mathbb{C}$ and $g \in C(\prod_{i=1}^N [a_i, b_i])$, then $cg \in C(\prod_{i=1}^N [a_i, b_i], \mathbb{C})$.

Furthermore, it holds

$$A_n(cg, x) = c\widetilde{A}_n(g, x), \quad \forall\, x \in \prod_{i=1}^N [a_i, b_i]. \tag{7.28}$$

Since $\widetilde{A}_n(1) = 1$, we get that

$$A_n(c) = c, \quad \forall\, c \in \mathbb{C}. \tag{7.29}$$

We call $\widetilde{A}_n$ the companion operator of A_n.

For convenience, we call

$$\begin{aligned} A_n^*(f, x) &:= \sum_{k=\lceil na \rceil}^{\lfloor nb \rfloor} f\left(\frac{k}{n}\right) Z(nx - k) \\ &= \sum_{k_1=\lceil na_1 \rceil}^{\lfloor nb_1 \rfloor} \sum_{k_2=\lceil na_2 \rceil}^{\lfloor nb_2 \rfloor} \cdots \sum_{k_N=\lceil na_N \rceil}^{\lfloor nb_N \rfloor} f\left(\frac{k_1}{n}, \ldots, \frac{k_N}{n}\right) \\ &\quad \times \left(\prod_{i=1}^N \psi_i(nx_i - k_i)\right), \end{aligned} \tag{7.30}$$

$\forall\, x \in (\prod_{i=1}^N [a_i, b_i])$.

That is,

$$A_n(f, x) := \frac{A_n^*(f, x)}{\sum_{k=\lceil na \rceil}^{\lfloor nb \rfloor} Z(nx - k)}, \tag{7.31}$$

$\forall\, x \in (\prod_{i=1}^N [a_i, b_i])$, $n \in \mathbb{N}$.

Hence,

$$A_n(f,x) - f(x) = \frac{A_n^*(f,x) - f(x)\left(\sum_{k=\lceil na \rceil}^{\lfloor nb \rfloor} Z(nx-k)\right)}{\sum_{k=\lceil na \rceil}^{\lfloor nb \rfloor} Z(nx-k)}. \tag{7.32}$$

Consequently, we derive

$$|A_n(f,x) - f(x)| \overset{(7.21)}{\leq} \left(\prod_{i=1}^{N} \psi_i(1)\right)^{-1} \left| A_n^*(f,x) - f(x) \sum_{k=\lceil na \rceil}^{\lfloor nb \rfloor} Z(nx-k) \right|, \tag{7.33}$$

$\forall\, x \in \left(\prod_{i=1}^{N} [a_i, b_i]\right)$.

We estimate the right-hand side of (7.33).

For the last, we need

Definition 7.2 (Anastassiou 2018, p. 274). Let M be a convex and compact subset of $(\mathbb{R}^N, \|\cdot\|_p)$, $p \in [1,\infty]$. Let $f \in C(M,\mathbb{C})$, we define the first modulus of continuity of f as

$$\omega_1(f,\delta) := \sup_{\substack{x,y \in M: \\ \|x-y\|_p \leq \delta}} |f(x) - f(y)|, \quad 0 < \delta \leq diam(M). \tag{7.34}$$

If $\delta > diam(M)$, then

$$\omega_1(f,\delta) = \omega_1(f, diam(M)). \tag{7.35}$$

Note that $\omega_1(f,\delta)$ is increasing in $\delta > 0$.

Lemma 7.1 (Anastassiou 2018, p. 274). *We have* $\omega_1(f,\delta) \to 0$ *as* $\delta \downarrow 0$, *iff* $f \in C(M,\mathbb{C})$, *where* M *is a convex compact subset of* $\left(\mathbb{R}^N, \|\cdot\|_p\right)$, $p \in [1,\infty]$.

In our results, we use $p = \infty$.

Let now $f \in C^2(\prod_{i=1}^{N} [a_i,b_i], \mathbb{C})$, $N \in \mathbb{N}$. Here f_α denotes a partial derivative of f, $\alpha := (\alpha_1, \ldots, \alpha_N)$, $\alpha_i \in \mathbb{Z}_+$, $i = 1, \ldots, N$, and $|\alpha| := \sum_{i=1}^{N} \alpha_i = l$, where $l = 0,1,2$. We write also $f_\alpha := \frac{\partial^n f}{\partial x^n}$ and we say it is of order l.

We denote

$$\omega_1^{\max}(f_\alpha, h) := \max_{\alpha:|\alpha|=2} \omega_1(f_\alpha, h). \tag{7.36}$$

Call also

$$\|f_\alpha\|_\infty^{\max} := \max_{|\alpha|=2} \{\|f_\alpha\|_\infty t\}, \tag{7.37}$$

where $\|\cdot\|_\infty$ is the supremum norm.

7.2.2 Multivariate new Taylor formulae

We use

Theorem 7.5 (Anastassiou 2023c). *Let $f \in C^2([c,d], \mathbb{C})$, where $a, x \in [c,d]$. Then,*

$$f(x) - f(a) = f'(a)\sin(x-a) + 2f''(a)\sin^2\left(\frac{x-a}{2}\right) + \int_a^x [(f''(t) + f(t)) - (f''(a) + f(a))] \sin(x-t)dt. \tag{7.38}$$

We make

Remark 7.4. Let now Q be an open convex subset of $\mathbb{R}^k$, $k \geq 2$; $z = (z_1, \ldots, z_k)$, $x_0 := (x_{01}, \ldots, x_{0k}) \in Q$. We consider $f \in C^2(Q, \mathbb{C})$ each second-order partial derivative is denoted by $f_\alpha := \frac{\partial^\alpha f}{\partial x^\alpha}$, where $\alpha := (\alpha_1, \ldots, \alpha_k)$, $\alpha_i \in \mathbb{Z}^+$, $i = 1, \ldots, k$ and $|\alpha| := \sum_{i=1}^k \alpha_i = 2$. We consider $g_z(t) := f(x_0 + t(z - x_0))$, $0 \leq t \leq 1$. Clearly, $x_0 + t(z - x_0) \in Q$. Then,

$$g_z(0) = f(x_0), \quad g_z(1) = f(z),$$

$$g_z'(t) = \sum_{i=1}^k (z_i - x_{0i}) \frac{\partial f}{\partial x_i} (x_{01} + t(z_1 - x_{01}), \ldots, x_{0k} + t(z_k - x_{0k})),$$

$$g_z'(0) = \sum_{i=1}^k (z_i - x_{0i}) \frac{\partial f}{\partial x_i} (x_{01}, \ldots, x_{0k}), \tag{7.39}$$

and

$$g_z''(t) = \left[\left(\sum_{i=1}^k (z_i - x_{0i}) \frac{\partial}{\partial x_i}\right)^2 f\right] \times (x_{01} + t(z_1 - x_{01}), \ldots, x_{0k} + t(z_k - x_{0k})),$$

$$g_z''(0) = \left[\left(\sum_{i=1}^k (z_i - x_{0i}) \frac{\partial}{\partial x_i}\right)^2 f\right] (x_{01}, \ldots, x_{0k}). \tag{7.40}$$

Note above that the second-order partials commute.

Clearly, $g_z \in C^2([0,1], \mathbb{C})$, and by Theorem 7.5, we obtain

$$f(z_1, \ldots, z_k) - f(x_{01}, \ldots, x_{0k}) = g_z(1) - g_z(0) =$$
$$g_z'(0)\sin(1) + 2g_z''(0)\sin^2\left(\frac{1}{2}\right) +$$
$$\int_0^1 [(g_z''(t) + g_z(t)) - (g_z''(0) + g_z(0))] \sin(1-t)dt. \qquad (7.41)$$

We also mention

Theorem 7.6 (Anastassiou 2023c). *Let $f \in C^2([c,d], \mathbb{C})$, where $a, x \in [c,d]$. Then,*

$$f(x) - f(a) = f'(a)\sinh(x-a) + 2f''(a)\sinh^2\left(\frac{x-a}{2}\right)$$
$$+ \int_a^x [(f''(t) - f(t)) - (f''(a) - f(a))] \sinh(x-t)dt. \qquad (7.42)$$

We make

Remark 7.5. Consequently, we get that

$$f(z_1, \ldots, z_k) - f(x_{01}, \ldots, x_{0k}) = g_z(1) - g_z(0)$$
$$= g_z'(0)\sinh(1) + 2g_z''(0)\sinh^2\left(\frac{1}{2}\right)$$
$$+ \int_0^1 [(g_z''(t) - g_z(t)) - (g_z''(0) - g_z(0))] \sinh(1-t)dt. \qquad (7.43)$$

We make

Remark 7.6. Let $f \in C^2(\prod_{i=1}^N [a_i, b_i], \mathbb{C})$, $N \in \mathbb{N}$.

Clearly, the mixed partials commute.

Here $\frac{k}{n} := (\frac{k_1}{n}, \ldots, \frac{k_N}{n})$, and $x := (x_1, \ldots, x_N)$, with $\frac{k}{n}, x \in (\prod_{i=1}^N [a_i, b_i])$, then (by (7.41), where $g_{\frac{k}{n}}(t) := f(x + t(\frac{k}{n} - x))$, $0 \leq t \leq 1$)

we have

$$
\begin{aligned}
f\left(\frac{k}{n}\right) - f(x) &= \left(\sum_{i=1}^{N}\left(\frac{k_i}{n} - x_i\right)\frac{\partial f}{\partial x_i}(x)\right)\sin(1) \\
&+ 2\left\{\left[\left(\sum_{i=1}^{N}\left(\frac{k_i}{n} - x_i\right)\frac{\partial}{\partial x_i}\right)^2 f\right](x)\right\}\sin^2\left(\frac{1}{2}\right) \\
&+ \int_0^1 \left\{\left\{\left[\left(\sum_{i=1}^{N}\left(\frac{k_i}{n} - x_i\right)\frac{\partial}{\partial x_i}\right)^2 f\right]\right.\right. \\
&\times \left.\left(x + t\left(\frac{k}{n} - x\right)\right) + f\left(x + t\left(\frac{k}{n} - x\right)\right)\right\} \\
&- \left.\left\{\left[\left(\sum_{i=1}^{N}\left(\frac{k_i}{n} - x_i\right)\frac{\partial}{\partial x_i}\right)^2 f\right](x) + f(x)\right\}\right\}\sin(1-t)\,dt. \quad (7.44)
\end{aligned}
$$

Denote the remainder

$$
\begin{aligned}
R &:= \int_0^1 \left\{\left\{\left[\left(\sum_{i=1}^{N}\left(\frac{k_i}{n} - x_i\right)\frac{\partial}{\partial x_i}\right)^2 f\right]\right.\right. \\
&\times \left.\left(x + t\left(\frac{k}{n} - x\right)\right) + f\left(x + t\left(\frac{k}{n} - x\right)\right)\right\} \\
&- \left.\left\{\left[\left(\sum_{i=1}^{N}\left(\frac{k_i}{n} - x_i\right)\frac{\partial}{\partial x_i}\right)^2 f\right](x) + f(x)\right\}\right\}\sin(1-t)dt \\
&= \int_0^1 \left\{ \sum_{\substack{\alpha := (\alpha_1, \ldots, \alpha_N),\, \alpha_i \in \mathbb{Z}_+ \\ i = 1, \ldots, N,\, |\alpha| := \sum_{i=1}^{N} \alpha_i = 2}} \left(\frac{2}{\prod_{i=1}^{N}\alpha_i!}\right)\left(\prod_{i=1}^{N}\left(\frac{k_i}{n} - x_i\right)^{\alpha_i}\right)\right. \\
&\quad \left[f_\alpha\left(x + t\left(\frac{k}{n} - x\right)\right) - f_\alpha(x)\right] \\
&\quad + \left.\left(f\left(x + t\left(\frac{k}{n} - x\right)\right) - f(x)\right)\right\}\sin(1-t)\,dt. \quad (7.45)
\end{aligned}
$$

Therefore, it holds

$$|R| \leq \int_0^1 \left\{ \sum_{\substack{\alpha:=(\alpha_1,\dots,\alpha_N),\alpha_i \in \mathbb{Z}_+ \\ i=1,\dots,N,|\alpha|:=\sum_{i=1}^{N} \alpha_i = 2}} \left(\frac{2}{\prod_{i=1}^N \alpha_i!} \right) \right.$$

$$\left(\prod_{i=1}^N \left| \frac{k_i}{n} - x_i \right|^{\alpha_i} \right) \left| f_\alpha \left(x + t \left(\frac{k}{n} - x \right) \right) - f_\alpha(x) \right|$$

$$\left. + \left| f \left(x + t \left(\frac{k}{n} - x \right) \right) - f(x) \right| \right\} |\sin(1-t)|\, dt$$

$$\leq \int_0^1 \left\{ \sum_{\substack{\alpha:=(\alpha_1,\dots,\alpha_N),\alpha_i \in \mathbb{Z}_+ \\ i=1,\dots,N,|\alpha|:=\sum_{i=1}^{N} \alpha_i = 2}} \left(\frac{2}{\prod_{i=1}^N \alpha_i!} \right) \left(\prod_{i=1}^N \left| \frac{k_i}{n} - x_i \right|^{\alpha_i} \right) \omega_1 \right.$$

$$\left. \left(f_\alpha, t \left\| \frac{k}{n} - x \right\|_\infty \right) + \omega_1 \left(f, t \left\| \frac{k}{n} - x \right\|_\infty \right) \right\} |\sin(1-t)|\, dt \leq (*). \tag{7.46}$$

Note here that $(0 < \beta < 1)$

$$\left\| \frac{k}{n} - x \right\|_\infty \leq \frac{1}{n^\beta} \Leftrightarrow \left| \frac{k_i}{n} - x_i \right| \leq \frac{1}{n^\beta}, \quad i = 1, \dots, N. \tag{7.47}$$

We further see that

$$(*) \leq \left\{ \omega_{1,2}^{\max} \left(f_\alpha, \frac{1}{n^\beta} \right) \left(\sum_{\substack{\alpha:=(\alpha_1,\dots,\alpha_N),\alpha_i \in \mathbb{Z}_+ \\ i=1,\dots,N,|\alpha|:=\sum_{i=1}^{N} \alpha_i = 2}} \left(\frac{2}{\prod_{i=1}^N \alpha_i!} \right) \left(\prod_{i=1}^N \frac{1}{n^{\beta \alpha_i}} \right) \right) \right.$$

$$\left. + \omega_1 \left(f, \frac{1}{n^\beta} \right) \right\} \int_0^1 |\sin(1-t)|\, dt$$

$$= \left[\omega_{1,2}^{\max}\left(f_\alpha, \frac{1}{n^\beta}\right) \left(\sum_{\substack{\alpha:=(\alpha_1,\ldots,\alpha_N),\alpha_i\in\mathbb{Z}_+ \\ i=1,\ldots,N,|\alpha|:=\sum_{i=1}^N \alpha_i=2}} \left(\frac{2}{\prod_{i=1}^N \alpha_i!}\right) \right) \frac{1}{n^{2\beta}} \right.$$

$$\left. + \, \omega_1\left(f, \frac{1}{n^\beta}\right)\right] (1-\cos(1))$$

$$= (1-\cos(1)) \left\{ \frac{\omega_{1,2}^{\max}\left(f_\alpha, \frac{1}{n^\beta}\right) N^2}{n^{2\beta}} + \omega_1\left(f, \frac{1}{n^\beta}\right) \right\}. \tag{7.48}$$

We have proved that

$$|R| \le (1-\cos(1)) \left\{ \frac{\omega_{1,2}^{\max}\left(f_\alpha, \frac{1}{n^\beta}\right) N^2}{n^{2\beta}} + \omega_1\left(f, \frac{1}{n^\beta}\right) \right\}, \tag{7.49}$$

given that $\left\| \frac{k}{n} - x \right\|_\infty \le \frac{1}{n^\beta}$.

We note also that

$$|R| \le \int_0^1 \left\{ \sum_{\substack{\alpha:=(\alpha_1,\ldots,\alpha_N),\alpha_i\in\mathbb{Z}_+ \\ i=1,\ldots,N,|\alpha|:=\sum_{i=1}^N \alpha_i=2}} \left(\frac{2}{\prod_{i=1}^N \alpha_i!}\right) \right.$$

$$\left. \left(\prod_{i=1}^N (b_i-a_i)^{\alpha_i}\right) 2\|f_\alpha\|_\infty + 2\|f\|_\infty \right\} |\sin(1-t)|\, dt$$

$$\le \left\{ \left(\sum_{\substack{\alpha:=(\alpha_1,\ldots,\alpha_N),\alpha_i\in\mathbb{Z}_+ \\ i=1,\ldots,N,|\alpha|:=\sum_{i=1}^N \alpha_i=2}} \left(\frac{2}{\prod_{i=1}^N \alpha_i!}\right) \right) \right. \tag{7.50}$$

$$\left. 2\|b-a\|_\infty^2 \|f_\alpha\|_{\infty,2}^{\max} + 2\|f\|_\infty \right\} \left(\int_0^1 |\sin(1-t)|\, dt \right)$$

$$= \left(2\|b-a\|_\infty^2 \|f_\alpha\|_{\infty,2}^{\max} N^2 + 2\|f\|_\infty \right) (1-\cos(1)),$$

where $a := (a_1, \ldots, a_N)$, $b = (b_1, \ldots, b_N)$.

We have proved that

$$|R| \leq \left(2\|b-a\|_{\infty}^{2}\|f_{\alpha}\|_{\infty,2}^{\max} N^{2}+2\|f\|_{\infty}\right)(1-\cos(1)) =: \rho. \tag{7.51}$$

7.3 Main Results

Here we discuss the trigonometric approximation by using the smoothness of f.

Theorem 7.7. *Let* $f \in C^{2}(\prod_{i=1}^{N}[a_i,b_i],\mathbb{C})$, $0<\beta<1$, $n, N \in \mathbb{N}$, $n^{1-\beta} > 2$, $x, x_0 \in (\prod_{i=1}^{N}[a_i,b_i])$, $a := (a_1,\ldots,a_N)$, $b := (b_1,\ldots,b_N)$. *Then*

(i)

$$\Bigg| A_n(f,x) - f(x) - \left(\sum_{=1}^{N} \frac{\partial f(x)}{\partial x_i} A_n\left(\left(\cdot - x_i\right), x\right)\right)\sin(1)$$

$$-4\left\{\sum_{\substack{\alpha:=(\alpha_1,\ldots,\alpha_N),\alpha_i\in\mathbb{Z}_+\\ i=1,\ldots,N,|\alpha|:=\sum_{i=1}^{N}\alpha_i=2}} f_{\alpha}(x)\left(\frac{1}{\prod_{i=1}^{N}\alpha_i!}\right)\right.$$

$$\left. A_n\left(\prod_{i=1}^{N}\left(\cdot - x_i\right)^{\alpha_i}, x\right)\right\}\sin^2\left(\frac{1}{2}\right)\Bigg|$$

$$\leq \left(\prod_{i=1}^{N}\psi_i(1)\right)^{-1}\left\{\left[(1-\cos(1))\left\{\frac{\omega_{1,2}^{\max}\left(f_{\alpha},\frac{1}{n^{\beta}}\right)N^2}{n^{2\beta}} + \omega_1\left(f,\frac{1}{n^{\beta}}\right)\right\}\right]\right.$$

$$\left. + \left[2\|b-a\|_{\infty}^{2}\|f_{\alpha}\|_{\infty,2}^{\max}N^2 + 2\|f\|_{\infty}\right](1-\cos(1))\,\delta_N(\beta,n)\right\}, \tag{7.52}$$

(ii) assume that $\frac{\partial f(x_0)}{\partial x_i} = 0$, $i = 1,\ldots,N$, *and* $f_{\alpha}(x_0) = 0$, $\alpha : |\alpha| = 2$, *we have that*

$$|A_n(f,x) - f(x)|$$

$$\leq \left(\prod_{i=1}^{N}\psi_i(1)\right)^{-1}\left\{\left[(1-\cos(1))\left\{\frac{\omega_{1,2}^{\max}\left(f_{\alpha},\frac{1}{n^{\beta}}\right)N^2}{n^{2\beta}} + \omega_1\left(f,\frac{1}{n^{\beta}}\right)\right\}\right]\right.$$

$$\left. + \left[2\|b-a\|_{\infty}^{2}\|f_{\alpha}\|_{\infty,2}^{\max}N^2 + 2\|f\|_{\infty}\right](1-\cos(1))\,\delta_N(\beta,n)\right\}, \tag{7.53}$$

(iii)

$$|A_n(f,x) - f(x)| \leq \left(\prod_{i=1}^{N} \psi_i(1)\right)^{-1}$$

$$\left\{\left\{\left\{\sum_{i=1}^{N} \left|\frac{\partial f(x)}{\partial x_i}\right| \left\{\frac{1}{n^\beta} + (b_i - a_i)\,\delta_N(\beta,n)\right\}\right\}\sin(1) + \right.\right.$$

$$\left. 4\left\{\sum_{\alpha:|\alpha|=2} |f_\alpha(x)| \left(\frac{1}{\prod_{i=1}^{N}\alpha_i!}\right)\left[\frac{1}{n^{2\beta}} + \left(\prod_{i=1}^{N}(b_i-a_i)^{\alpha_i}\right)\delta_N(\beta,n)\right]\right\}\sin^2\left(\frac{1}{2}\right)\right\} +$$

$$\left\{\left[(1-\cos(1))\left\{\frac{\omega_{1,2}^{\max}\left(f_\alpha, \frac{1}{n^\beta}\right)N^2}{n^{2\beta}} + \omega_1\left(f, \frac{1}{n^\beta}\right)\right\}\right]\right.$$

$$\left.\left. + \left[2\|b-a\|_\infty^2 \|f_\alpha\|_{\infty,2}^{\max} N^2 + 2\|f\|_\infty\right](1-\cos(1))\,\delta_N(\beta,n)\right\}\right\}, \quad (7.54)$$

and

(iv)

$$\|A_n(f) - f\|_\infty \leq \left(\prod_{i=1}^{N} \psi_i(1)\right)^{-1}$$

$$\left\{\left\{\left\{\sum_{i=1}^{N} \left\|\frac{\partial f}{\partial x_i}\right\|_\infty \left\{\frac{1}{n^\beta} + (b_i - a_i)\,\delta_N(\beta,n)\right\}\right\}\sin(1) + \right.\right.$$

$$\left. 4\left\{\sum_{\alpha:|\alpha|=2} \|f_\alpha\|_\infty \left(\frac{1}{\prod_{i=1}^{N}\alpha_i!}\right)\left[\frac{1}{n^{2\beta}} + \left(\prod_{i=1}^{N}(b_i-a_i)^{\alpha_i}\right)\delta_N(\beta,n)\right]\right\}\sin^2\left(\frac{1}{2}\right)\right\}$$

$$+\left\{\left[(1-\cos(1))\left\{\frac{\omega_{1,2}^{\max}\left(f_\alpha, \frac{1}{n^\beta}\right)N^2}{n^{2\beta}} + \omega_1\left(f, \frac{1}{n^\beta}\right)\right\}\right]\right.$$

$$\left.\left. + \left[2\|b-a\|_\infty^2 \|f_\alpha\|_{\infty,2}^{\max} N^2 + 2\|f\|_\infty\right](1-\cos(1))\,\delta_N(\beta,n)\right\}\right\} =: \xi_n(f). \quad (7.55)$$

We observe that $A_n \to I$ (unit operator), as $n \to \infty$, pointwise and uniformly.

Proof. Here R is as in (7.45). We see that

$$U_n := \sum_{k=\lceil na \rceil}^{\lfloor nb \rfloor} Z(nx-k)R$$

$$= \sum_{\begin{cases} k=\lceil na \rceil \\ :\left\| \frac{k}{n}-x \right\|_\infty \leq \frac{1}{n^\beta} \end{cases}}^{\lfloor nb \rfloor} Z(nx-k)R + \sum_{\begin{cases} k=\lceil na \rceil \\ :\left\| \frac{k}{n}-x \right\|_\infty > \frac{1}{n^\beta} \end{cases}}^{\lfloor nb \rfloor} Z(nx-k)R. \quad (7.56)$$

Therefore,

$$|U_n| \leq \left(\sum_{\begin{cases} k=\lceil na \rceil \\ :\left\| \frac{k}{n}-x \right\|_\infty \leq \frac{1}{n^\beta} \end{cases}}^{\lfloor nb \rfloor} Z(nx-k) \right)$$

$$\left[(1-\cos(1)) \left\{ \frac{\omega_{1,2}^{\max}\left(f_\alpha, \frac{1}{n^\beta}\right) N^2}{n^{2\beta}} + \omega_1\left(f, \frac{1}{n^\beta}\right) \right\} \right] + \rho\delta_N(\beta, n)$$

$$\leq \left[(1-\cos(1)) \left\{ \frac{\omega_{1,2}^{\max}\left(f_\alpha, \frac{1}{n^\beta}\right) N^2}{n^{2\beta}} + \omega_1\left(f, \frac{1}{n^\beta}\right) \right\} \right] + \rho\delta_N(\beta, n). \quad (7.57)$$

We have established that

$$|U_n| \leq \left[(1-\cos(1)) \left\{ \frac{\omega_{1,2}^{\max}\left(f_\alpha, \frac{1}{n^\beta}\right) N^2}{n^{2\beta}} + \omega_1\left(f, \frac{1}{n^\beta}\right) \right\} \right]$$

$$+ \left[2\|b-a\|_\infty^2 \|f_\alpha\|_{\infty,2}^{\max} N^2 + 2\|f\|_\infty \right] (1-\cos(1))\,\delta_N(\beta, n). \quad (7.58)$$

By (7.44), we observe that

$$\sum_{k=\lceil na\rceil}^{\lfloor nb\rfloor} f\left(\frac{k}{n}\right) Z(nx-k) - f(x)\left(\sum_{k=\lceil na\rceil}^{\lfloor nb\rfloor} Z(nx-k)\right)$$

$$= \left(\sum_{i=1}^{N}\left(\left(\sum_{k=\lceil na\rceil}^{\lfloor nb\rfloor} Z(nx-k)\left(\frac{k_i}{n}-x_i\right)\right)\frac{\partial f}{\partial x_i}(x)\right)\right)\sin(1)$$

$$+2\left\{\sum_{\substack{\alpha:=(\alpha_1,\ldots,\alpha_N),\alpha_i\in\mathbb{Z}_+\\ i=1,\ldots,N,|\alpha|:=\sum_{i=1}^{N}\alpha_i=2}} f_\alpha(x)\left(\frac{2}{\prod_{i=1}^{N}\alpha_i!}\right)\left(\sum_{k=\lceil na\rceil}^{\lfloor nb\rfloor} Z(nx-k)\right.\right.$$

$$\left.\left.\left(\prod_{i=1}^{N}\left(\frac{k_i}{n}-x_i\right)^{\alpha_i}\right)\right)\right\}\sin^2\left(\frac{1}{2}\right)+U_n. \tag{7.59}$$

The last says

$$A_n^*(f,x) - f(x)\left(\sum_{k=\lceil na\rceil}^{\lfloor nb\rfloor} Z(nx-k)\right)$$

$$-\left(\sum_{=1}^{N}\frac{\partial f(x)}{\partial x_i}A_n^*\left(\left(\cdot-x_i\right),x\right)\right)\sin(1)$$

$$-2\left\{\sum_{\substack{\alpha:=(\alpha_1,\ldots,\alpha_N),\alpha_i\in\mathbb{Z}_+\\ i=1,\ldots,N,|\alpha|:=\sum_{i=1}^{N}\alpha_i=2}} f_\alpha(x)\left(\frac{2}{\prod_{i=1}^{N}\alpha_i!}\right)A_n^*\left(\prod_{i=1}^{N}\left(\cdot-x_i\right)^{\alpha_i},x\right)\right\}$$

$$\times\sin^2\left(\frac{1}{2}\right)=U_n. \tag{7.60}$$

We note that

$$|A_n^*((\cdot-x_i),x)| \le A_n^*\left(\left|\cdot-x_i\right|,x\right) = \sum_{k=\lceil na\rceil}^{\lfloor nb\rfloor}\left|\frac{k_i}{n}-x_i\right| Z(nx-k)$$

$$= \sum_{\substack{k=\lceil na\rceil\\ :\left\|\frac{k}{n}-x\right\|_\infty\le\frac{1}{n^\beta}}}^{\lfloor nb\rfloor}\left|\frac{k_i}{n}-x_i\right| Z(nx-k)$$

$$+ \sum_{\substack{k=\lceil na \rceil \\ \left\{ : \left\| \frac{k}{n} - x \right\|_\infty > \frac{1}{n^\beta} \right.}}^{\lfloor nb \rfloor} \left| \frac{k_i}{n} - x_i \right| Z(nx - k)$$

$$\leq \frac{1}{n^\beta} + (b_i - a_i) \sum_{\substack{k=\lceil na \rceil \\ \left\{ : \left\| \frac{k}{n} - x \right\|_\infty > \frac{1}{n^\beta} \right.}}^{\lfloor nb \rfloor} Z(nx - k)$$

$$\leq \frac{1}{n^\beta} + (b_i - a_i)\, \delta_N(\beta, n). \tag{7.61}$$

We have proved that

$$|A_n^*((\cdot - x_i), x)| \leq \frac{1}{n^\beta} + (b_i - a_i)\, \delta_N(\beta, n), \tag{7.62}$$

$i = 1, \ldots, N$.

Next, we see that

$$\left| A_n^* \left(\prod_{i=1}^{N} (\cdot - x_i)^{\alpha_i}, x \right) \right| \leq A_n^* \left(\prod_{i=1}^{N} |\cdot - x_i|^{\alpha_i}, x \right)$$

$$= \sum_{k=\lceil na \rceil}^{\lfloor nb \rfloor} \left(\prod_{i=1}^{N} \left| \frac{k_i}{n} - x_i \right|^{\alpha_i} \right) Z(nx - k)$$

$$= \sum_{\substack{k=\lceil na \rceil \\ \left\{ : \left\| \frac{k}{n} - x \right\|_\infty \leq \frac{1}{n^\beta} \right.}}^{\lfloor nb \rfloor} \left(\prod_{i=1}^{N} \left| \frac{k_i}{n} - x_i \right|^{\alpha_i} \right) Z(nx - k)$$

$$+ \sum_{\substack{k=\lceil na \rceil \\ \left\{ : \left\| \frac{k}{n} - x \right\|_\infty > \frac{1}{n^\beta} \right.}}^{\lfloor nb \rfloor} \left(\prod_{i=1}^{N} \left| \frac{k_i}{n} - x_i \right|^{\alpha_i} \right) Z(nx - k)$$

$$\leq \frac{1}{n^{2\beta}} + \left(\prod_{i=1}^{N} (b_i - a_i)^{\alpha_i} \right) \delta_N(\beta, n). \tag{7.63}$$

We have proved that

$$\left| A_n^* \left(\prod_{i=1}^{N} (\cdot - x_i)^{\alpha_i}, x \right) \right| \leq \frac{1}{n^{2\beta}} + \left(\prod_{i=1}^{N} (b_i - a_i)^{\alpha_i} \right) \delta_N(\beta, n). \tag{7.64}$$

At last we observe that

$$\left| A_n(f,x) - f(x) - \left(\sum_{=1}^{N} \frac{\partial f(x)}{\partial x_i} A_n((\cdot - x_i), x) \right) \sin(1) \right.$$

$$-4 \left\{ \sum_{\substack{\alpha:=(\alpha_1,\ldots,\alpha_N),\alpha_i \in \mathbb{Z}_+ \\ i=1,\ldots,N,|\alpha|:=\sum_{i=1}^{N} \alpha_i = 2}} f_\alpha(x) \left(\frac{1}{\prod_{i=1}^{N} \alpha_i!} \right) A_n \left(\prod_{i=1}^{N} (\cdot - x_i)^{\alpha_i}, x \right) \right\}$$

$$\left. \times \sin^2 \left(\frac{1}{2} \right) \right|$$

$$\leq \left(\prod_{i=1}^{N} \psi_i(1) \right)^{-1} |U_n|$$

$$= \left(\prod_{i=1}^{N} \psi_i(1) \right)^{-1} \left| A_n^* (f,x) - f(x) \left(\sum_{k=\lceil na \rceil}^{\lfloor nb \rfloor} Z(nx-k) \right) \right.$$

$$- \left(\sum_{=1}^{N} \frac{\partial f(x)}{\partial x_i} A_n^* ((\cdot - x_i), x) \right) \sin(1)$$

$$-4 \left\{ \sum_{\substack{\alpha:=(\alpha_1,\ldots,\alpha_N),\alpha_i \in \mathbb{Z}_+ \\ i=1,\ldots,N,|\alpha|:=\sum_{i=1}^{N} \alpha_i = 2}} f_\alpha(x) \left(\frac{1}{\prod_{i=1}^{N} \alpha_i!} \right) A_n^* \left(\prod_{i=1}^{N} (\cdot - x_i)^{\alpha_i}, x \right) \right\}$$

$$\left. \times \sin^2 \left(\frac{1}{2} \right) \right|. \tag{7.65}$$

Putting all of the above together, we prove the theorem. □

We make

Remark 7.7. Let $f \in C^2(\prod_{i=1}^{N} [a_i, b_i], \mathbb{C})$, $N \in \mathbb{N}$. By the mean value theorem, we have that $\sinh x = \sinh x - \sinh 0 = (\cosh \xi)(x - 0)$, for some ξ between $\{0, x\}$, for any $x \in \mathbb{R}$.

Hence,

$$|\sinh x| \leq \|\cosh\|_{\infty,[-1,1]} |x|, \ \forall x \in [-1,1].$$

But

$$\|\cosh\|_{\infty,[-1,1]} = \cosh(1).$$

Thus, we have

$$|\sinh x| \leq \cosh(1)\,|x|\,,\ \ \forall\, x \in [-1,1].$$

Let $\frac{k}{n} := \left(\frac{k_1}{n}, \ldots, \frac{k_N}{n}\right)$, and $x := (x_1, \ldots, x_N)$, with $\frac{k}{n}, x \in \left(\prod_{i=1}^{N} [a_i, b_i]\right)$, then (by (7.43), where $g_{\frac{k}{n}}(t) := f\left(x + t\left(\frac{k}{n} - x\right)\right)$, $0 \leq t \leq 1$) we have

$$f\left(\frac{k}{n}\right) - f(x) = \left(\sum_{i=1}^{N}\left(\frac{k_i}{n} - x_i\right)\frac{\partial f}{\partial x_i}(x)\right)\sinh(1) +$$

$$2\left\{\left[\left(\sum_{i=1}^{N}\left(\frac{k_i}{n} - x_i\right)\frac{\partial}{\partial x_i}\right)^2 f\right](x)\right\}\sinh^2\left(\frac{1}{2}\right) +$$

$$\int_0^1 \left\{\left\{\left[\left(\sum_{i=1}^{N}\left(\frac{k_i}{n} - x_i\right)\frac{\partial}{\partial x_i}\right)^2 f\right]\left(x + t\left(\frac{k}{n} - x\right)\right) - f\left(x + t\left(\frac{k}{n} - x\right)\right)\right\}\right.$$

$$\left. - \left\{\left[\left(\sum_{i=1}^{N}\left(\frac{k_i}{n} - x_i\right)\frac{\partial}{\partial x_i}\right)^2 f\right](x) - f(x)\right\}\right\}\sinh(1-t)\,dt. \quad (7.66)$$

Denote the remainder

$$R := \int_0^1 \left\{\left\{\left[\left(\sum_{i=1}^{N}\left(\frac{k_i}{n} - x_i\right)\frac{\partial}{\partial x_i}\right)^2 f\right]\left(x + t\left(\frac{k}{n} - x\right)\right) - f\left(x + t\left(\frac{k}{n} - x\right)\right)\right\}\right.$$

$$\left. - \left\{\left[\left(\sum_{i=1}^{N}\left(\frac{k_i}{n} - x_i\right)\frac{\partial}{\partial x_i}\right)^2 f\right](x) - f(x)\right\}\right\}\sinh(1-t)dt$$

$$= \int_0^1 \left\{ \sum_{\substack{\alpha:=(\alpha_1,\ldots,\alpha_N),\alpha_i\in\mathbb{Z}_+ \\ i=1,\ldots,N,|\alpha|:=\sum_{i=1}^{N}\alpha_i=2}} \left(\frac{2}{\prod_{i=1}^N \alpha_i!} \right) \left(\prod_{i=1}^N \left(\frac{k_i}{n} - x_i \right)^{\alpha_i} \right) \right.$$

$$\left[f_\alpha \left(x + t\left(\frac{k}{n} - x \right) \right) - f_\alpha(x) \right]$$

$$\left. - \left(f\left(x + t\left(\frac{k}{n} - x \right) \right) - f(x) \right) \right\} \sinh(1-t)\, dt. \tag{7.67}$$

Therefore, it holds

$$|R| \le \int_0^1 \left\{ \sum_{\substack{\alpha:=(\alpha_1,\ldots,\alpha_N),\alpha_i\in\mathbb{Z}_+ \\ i=1,\ldots,N,|\alpha|:=\sum_{i=1}^{N}\alpha_i=2}} \left(\frac{2}{\prod_{i=1}^N \alpha_i!} \right) \right.$$

$$\left(\prod_{i=1}^N \left| \frac{k_i}{n} - x_i \right|^{\alpha_i} \right) \left| f_\alpha \left(x + t\left(\frac{k}{n} - x \right) \right) - f_\alpha(x) \right|$$

$$\left. + \left| f\left(x + t\left(\frac{k}{n} - x \right) \right) - f(x) \right| \right\} |\sinh(1-t)|\, dt \le$$

$$\int_0^1 \left\{ \sum_{\substack{\alpha:=(\alpha_1,\ldots,\alpha_N),\alpha_i\in\mathbb{Z}_+ \\ i=1,\ldots,N,|\alpha|:=\sum_{i=1}^{N}\alpha_i=2}} \left(\frac{2}{\prod_{i=1}^N \alpha_i!} \right) \left(\prod_{i=1}^N \left| \frac{k_i}{n} - x_i \right|^{\alpha_i} \right) \omega_1 \left(f_\alpha, t \left\| \frac{k}{n} - x \right\|_\infty \right) \right.$$

$$\left. + \omega_1 \left(f, t \left\| \frac{k}{n} - x \right\|_\infty \right) \right\} \cosh(1)(1-t)\, dt \le (*). \tag{7.68}$$

Note here that $(0 < \beta < 1)$

$$\left\| \frac{k}{n} - x \right\|_\infty \le \frac{1}{n^\beta} \Leftrightarrow \left| \frac{k_i}{n} - x_i \right| \le \frac{1}{n^\beta}, \quad i = 1, \ldots, N. \tag{7.69}$$

We further see that

$$(*) \leq \cosh(1) \left\{ \omega_{1,2}^{\max}\left(f_{\alpha}, \frac{1}{n^{\beta}}\right) \left(\sum_{\substack{\alpha:=(\alpha_1,\ldots,\alpha_N),\alpha_i\in\mathbb{Z}_+ \\ i=1,\ldots,N,|\alpha|:=\sum_{i=1}^N \alpha_i=2}} \left(\frac{2}{\prod_{i=1}^N \alpha_i!}\right) \right.\right.$$

$$\left.\left. \times \left(\prod_{i=1}^N \frac{1}{n^{\beta\alpha_i}}\right)\right) + \omega_1\left(f, \frac{1}{n^{\beta}}\right) \right\} \int_0^1 (1-t)\, dt$$

$$= \cosh(1) \left\{ \omega_{1,2}^{\max}\left(f_{\alpha}, \frac{1}{n^{\beta}}\right) \left(\sum_{\substack{\alpha:=(\alpha_1,\ldots,\alpha_N),\alpha_i\in\mathbb{Z}_+ \\ i=1,\ldots,N,|\alpha|:=\sum_{i=1}^N \alpha_i=2}} \left(\frac{2}{\prod_{i=1}^N \alpha_i!}\right) \right) \frac{1}{n^{2\beta}} \right.$$

$$\left. + \omega_1\left(f, \frac{1}{n^{\beta}}\right) \right\} \frac{1}{2} =$$

$$\frac{\cosh(1)}{2} \left\{ \frac{\omega_{1,2}^{\max}\left(f_{\alpha}, \frac{1}{n^{\beta}}\right) N^2}{n^{2\beta}} + \omega_1\left(f, \frac{1}{n^{\beta}}\right) \right\}. \tag{7.70}$$

We have proved that

$$|R| \leq \frac{\cosh(1)}{2} \left\{ \frac{\omega_{1,2}^{\max}\left(f_{\alpha}, \frac{1}{n^{\beta}}\right) N^2}{n^{2\beta}} + \omega_1\left(f, \frac{1}{n^{\beta}}\right) \right\}, \tag{7.71}$$

given that $\left\|\frac{k}{n} - x\right\|_{\infty} \leq \frac{1}{n^{\beta}}$.

We note also that

$$|R| \leq \cosh(1) \int_0^1 \left\{ \sum_{\substack{\alpha:=(\alpha_1,\ldots,\alpha_N),\alpha_i\in\mathbb{Z}_+ \\ i=1,\ldots,N,|\alpha|:=\sum_{i=1}^N \alpha_i=2}} \left(\frac{2}{\prod_{i=1}^N \alpha_i!}\right) \right.$$

$$\left. \left(\prod_{i=1}^N (b_i - a_i)^{\alpha_i}\right) 2\left\|f_{\alpha}\right\|_{\infty} + 2\left\|f\right\|_{\infty} \right\} (1-t)\, dt$$

$$\leq \cosh(1) \left\{ \left(\sum_{\substack{\alpha:=(\alpha_1,\ldots,\alpha_N),\alpha_i\in\mathbb{Z}_+ \\ i=1,\ldots,N,|\alpha|:=\sum_{i=1}^{N}\alpha_i=2}} \left(\frac{2}{\prod_{i=1}^{N}\alpha_i!} \right) \right) \right.$$

$$\left. 2\left\|b-a\right\|_\infty^2 \left\|f_\alpha\right\|_{\infty,2}^{\max} + 2\left\|f\right\|_\infty \right\} \left(\int_0^1 (1-t)\,dt \right)$$

$$= \cosh(1) \left\{ 2\left\|b-a\right\|_\infty^2 \left\|f_\alpha\right\|_{\infty,2}^{\max} N^2 + 2\left\|f\right\|_\infty \right\} \frac{1}{2}$$

$$= \cosh(1) \left(\left\|b-a\right\|_\infty^2 \left\|f_\alpha\right\|_{\infty,2}^{\max} N^2 + \left\|f\right\|_\infty \right), \tag{7.72}$$

where $a := (a_1, \ldots, a_N)$, $b = (b_1, \ldots, b_N)$.

We have proved that

$$|R| \leq \cosh(1) \left(\left\|b-a\right\|_\infty^2 \left\|f_\alpha\right\|_{\infty,2}^{\max} N^2 + \left\|f\right\|_\infty \right) =: \rho. \tag{7.73}$$

We continue with the hyperbolic approximation.

Theorem 7.8. *Let* $f \in C^2\left(\prod_{i=1}^{N}[a_i,b_i], \mathbb{C}\right)$, $0 < \beta < 1$, $n, N \in \mathbb{N}$, $n^{1-\beta} > 2$, $x, x_0 \in \left(\prod_{i=1}^{N}[a_i,b_i]\right)$, $a := (a_1, \ldots, a_N)$, $b := (b_1, \ldots, b_N)$. *Then*

(i)

$$\left| A_n(f,x) - f(x) - \left(\sum_{=1}^{N} \frac{\partial f(x)}{\partial x_i} A_n((\cdot - x_i), x) \right) \sinh(1) - \right.$$

$$\left. 4 \left\{ \sum_{\substack{\alpha:=(\alpha_1,\ldots,\alpha_N),\alpha_i\in\mathbb{Z}_+ \\ i=1,\ldots,N,|\alpha|:=\sum_{i=1}^{N}\alpha_i=2}} f_\alpha(x) \left(\frac{1}{\prod_{i=1}^{N}\alpha_i!} \right) A_n\left(\prod_{i=1}^{N} (\cdot - x_i)^{\alpha_i}, x \right) \right\} \sinh^2\left(\frac{1}{2}\right) \right| \leq$$

$$\left(\prod_{i=1}^{N} \psi_i(1) \right)^{-1} \cosh(1) \left\{ \left[\frac{1}{2} \left\{ \frac{\omega_{1,2}^{\max}\left(f_\alpha, \frac{1}{n^\beta}\right) N^2}{n^{2\beta}} + \omega_1\left(f, \frac{1}{n^\beta}\right) \right\} \right] + \right.$$

$$\left. \left[\left\|b-a\right\|_\infty^2 \left\|f_\alpha\right\|_{\infty,2}^{\max} N^2 + \left\|f\right\|_\infty \right] \delta_N(\beta, n) \right\}, \tag{7.74}$$

(ii) assume that $\frac{\partial f(x_0)}{\partial x_i}=0$, $i=1,\ldots,N$, *and* $f_\alpha(x_0)=0$, $\alpha:|\alpha|=2$, *we have that*

$$|A_n(f,x)-f(x)| \leq \left(\prod_{i=1}^{N}\psi_i(1)\right)^{-1}(\cosh(1))\left\{\left[\frac{1}{2}\left\{\frac{\omega_{1,2}^{\max}\left(f_\alpha,\frac{1}{n^\beta}\right)N^2}{n^{2\beta}}+\omega_1\left(f,\frac{1}{n^\beta}\right)\right\}\right]\right.$$
$$\left.+\left[\|b-a\|_\infty^2\|f_\alpha\|_{\infty,2}^{\max}N^2+\|f\|_\infty\right]\delta_N(\beta,n)\right\}, \tag{7.75}$$

(iii)

$$|A_n(f,x)-f(x)| \leq \left(\prod_{i=1}^{N}\psi_i(1)\right)^{-1}$$
$$\left\{\left\{\left\{\sum_{i=1}^{N}\left|\frac{\partial f(x)}{\partial x_i}\right|\left\{\frac{1}{n^\beta}+(b_i-a_i)\,\delta_N(\beta,n)\right\}\right\}\sinh(1)\right.\right.\right.$$
$$+4\left\{\sum_{\alpha:|\alpha|=2}|f_\alpha(x)|\left(\frac{1}{\prod_{i=1}^{N}\alpha_i!}\right)\left[\frac{1}{n^{2\beta}}+\left(\prod_{i=1}^{N}(b_i-a_i)^{\alpha_i}\right)\delta_N(\beta,n)\right]\right\}$$
$$\left.\sinh^2\left(\frac{1}{2}\right)\right\}+\cosh(1)\left\{\left[\frac{1}{2}\left\{\frac{\omega_{1,2}^{\max}\left(f_\alpha,\frac{1}{n^\beta}\right)N^2}{n^{2\beta}}+\omega_1\left(f,\frac{1}{n^\beta}\right)\right\}\right]\right.$$
$$\left.\left.+\left[\|b-a\|_\infty^2\|f_\alpha\|_{\infty,2}^{\max}N^2+\|f\|_\infty\right]\delta_N(\beta,n)\right\}\right\}, \tag{7.76}$$

and

(iv)

$$\|A_n(f)-f\|_\infty \leq \left(\prod_{i=1}^{N}\psi_i(1)\right)^{-1}$$
$$\left\{\left\{\left\{\sum_{i=1}^{N}\left\|\frac{\partial f}{\partial x_i}\right\|_\infty\left\{\frac{1}{n^\beta}+(b_i-a_i)\,\delta_N(\beta,n)\right\}\right\}\sinh(1)+\right.\right.\right.$$

$$4\left\{\sum_{\alpha:|\alpha|=2}\|f_\alpha\|_\infty\left(\frac{1}{\prod_{i=1}^{N}\alpha_i!}\right)\left[\frac{1}{n^{2\beta}}+\left(\prod_{i=1}^{N}(b_i-a_i)^{\alpha_i}\right)\delta_N(\beta,n)\right]\right\}\sinh^2\left(\frac{1}{2}\right)\right\}$$

$$+\cosh(1)\left\{\left[\frac{1}{2}\left\{\frac{\omega_{1,2}^{\max}\left(f_\alpha,\frac{1}{n^\beta}\right)N^2}{n^{2\beta}}+\omega_1\left(f,\frac{1}{n^\beta}\right)\right\}\right]+\right.$$

$$\left.\left[\|b-a\|_\infty^2\|f_\alpha\|_{\infty,2}^{\max}N^2+\|f\|_\infty\right]\delta_N(\beta,n)\right\}\right\}=:\psi_n(f). \tag{7.77}$$

We observe that $A_n \to I$ (unit operator), as $n \to \infty$, pointwise and uniformly.

Proof. Here R is as in (7.67). We see that

$$U_n := \sum_{k=\lceil na\rceil}^{\lfloor nb\rfloor} Z(nx-k)\,R$$

$$= \sum_{\substack{k=\lceil na\rceil \\ \left\{:\left\|\frac{k}{n}-x\right\|_\infty\le\frac{1}{n^\beta}\right.}}^{\lfloor nb\rfloor} Z(nx-k)\,R + \sum_{\substack{k=\lceil na\rceil \\ \left\{:\left\|\frac{k}{n}-x\right\|_\infty>\frac{1}{n^\beta}\right.}}^{\lfloor nb\rfloor} Z(nx-k)R. \tag{7.78}$$

Therefore,

$$|U_n| \le \left(\sum_{\substack{k=\lceil na\rceil \\ \left\{:\left\|\frac{k}{n}-x\right\|_\infty\le\frac{1}{n^\beta}\right.}}^{\lfloor nb\rfloor} Z(nx-k)\right)$$

$$\cosh(1)\left[\frac{1}{2}\left\{\frac{\omega_{1,2}^{\max}\left(f_\alpha,\frac{1}{n^\beta}\right)N^2}{n^{2\beta}}+\omega_1\left(f,\frac{1}{n^\beta}\right)\right\}\right]+\rho\delta_N(\beta,n)$$

$$\le \cosh(1)\left[\frac{1}{2}\left\{\frac{\omega_{1,2}^{\max}\left(f_\alpha,\frac{1}{n^\beta}\right)N^2}{n^{2\beta}}+\omega_1\left(f,\frac{1}{n^\beta}\right)\right\}\right]+\rho\delta_N(\beta,n). \tag{7.79}$$

We have established that

$$|U_n| \le \cosh(1)\left[\frac{1}{2}\left\{\frac{\omega_{1,2}^{\max}\left(f_\alpha,\frac{1}{n^\beta}\right)N^2}{n^{2\beta}}+\omega_1\left(f,\frac{1}{n^\beta}\right)\right\}\right]$$

$$+\cosh(1)\left[\|b-a\|_\infty^2\|f_\alpha\|_{\infty,2}^{\max}N^2+\|f\|_\infty\right]\delta_N(\beta,n). \tag{7.80}$$

By ((7.66)), we observe that

$$\sum_{k=\lceil na\rceil}^{\lfloor nb\rfloor} f\left(\frac{k}{n}\right) Z\left(nx-k\right) - f(x)\left(\sum_{k=\lceil na\rceil}^{\lfloor nb\rfloor} Z(nx-k)\right)$$

$$= \left(\sum_{i=1}^{N}\left(\left(\sum_{k=\lceil na\rceil}^{\lfloor nb\rfloor} Z\left(nx-k\right)\left(\frac{k_i}{n}-x_i\right)\right)\frac{\partial f}{\partial x_i}(x)\right)\right)\sinh(1)$$

$$+2\left\{\sum_{\substack{\alpha:=(\alpha_1,\ldots,\alpha_N),\alpha_i\in\mathbb{Z}_+\\ i=1,\ldots,N,|\alpha|:=\sum_{i=1}^{N}\alpha_i=2}} f_\alpha(x)\left(\frac{2}{\prod_{i=1}^{N}\alpha_i!}\right)\left(\sum_{k=\lceil na\rceil}^{\lfloor nb\rfloor} Z(nx-k)\right.\right.$$

$$\left.\left.\left(\prod_{i=1}^{N}\left(\frac{k_i}{n}-x_i\right)^{\alpha_i}\right)\right)\right\}\sinh^2\left(\frac{1}{2}\right)+U_n.$$

The last says

$$A_n^*(f,x) - f(x)\left(\sum_{k=\lceil na\rceil}^{\lfloor nb\rfloor} Z\left(nx-k\right)\right) -$$

$$\left(\sum_{=1}^{N}\frac{\partial f(x)}{\partial x_i}A_n^*\left(\left(\cdot-x_i\right),x\right)\right)\sinh(1) -$$

$$2\left\{\sum_{\substack{\alpha:=(\alpha_1,\ldots,\alpha_N),\alpha_i\in\mathbb{Z}_+\\ i=1,\ldots,N,|\alpha|:=\sum_{i=1}^{N}\alpha_i=2}} f_\alpha(x)\left(\frac{2}{\prod_{i=1}^{N}\alpha_i!}\right)A_n^*\left(\prod_{i=1}^{N}\left(\cdot-x_i\right)^{\alpha_i},x\right)\right\}\sinh^2\left(\frac{1}{2}\right)=U_n. \tag{7.81}$$

As earlier, it holds

$$\left|A_n^*\left(\left(\cdot-x_i\right),x\right)\right| \leq \frac{1}{n^\beta}+\left(b_i-a_i\right)\delta_N(\beta,n), \tag{7.82}$$

$i=1,\ldots,N$.

Also, as earlier, we have

$$\left|A_n^*\left(\prod_{i=1}^{N}\left(\cdot-x_i\right)^{\alpha_i},x\right)\right| \leq \frac{1}{n^{2\beta}}+\left(\prod_{i=1}^{N}\left(b_i-a_i\right)^{\alpha_i}\right)\delta_n(\beta,n). \tag{7.83}$$

At last we observe that

$$\Bigg| A_n(f,x) - f(x) - \left(\sum_{=1}^{N} \frac{\partial f(x)}{\partial x_i} A_n((\cdot - x_i), x) \right) \sinh(1)$$

$$- 4 \left\{ \sum_{\substack{\alpha := (\alpha_1, \ldots, \alpha_N), \alpha_i \in \mathbb{Z}_+ \\ i=1,\ldots,N, |\alpha| := \sum_{i=1}^{N} \alpha_i = 2}} f_\alpha(x) \left(\frac{1}{\prod_{i=1}^{N} \alpha_i!} \right) A_n \left(\prod_{i=1}^{N} (\cdot - x_i)^{\alpha_i}, x \right) \right\}$$

$$\sinh^2\left(\frac{1}{2}\right) \Bigg|$$

$$\leq \left(\prod_{i=1}^{N} \psi_i(1) \right)^{-1} |U_n|$$

$$= \left(\prod_{i=1}^{N} \psi_i(1) \right)^{-1} \Bigg| A_n^*(f,x) - f(x) \left(\sum_{k=\lceil na \rceil}^{\lfloor nb \rfloor} Z(nx-k) \right)$$

$$- \left(\sum_{=1}^{N} \frac{\partial f(x)}{\partial x_i} A_n^*((\cdot - x_i), x) \right) \sinh(1)$$

$$- 4 \left\{ \sum_{\substack{\alpha := (\alpha_1, \ldots, \alpha_N), \alpha_i \in \mathbb{Z}_+ \\ i=1,\ldots,N, |\alpha| := \sum_{i=1}^{N} \alpha_i = 2}} f_\alpha(x) \left(\frac{1}{\prod_{i=1}^{N} \alpha_i!} \right) A_n^* \left(\prod_{i=1}^{N} (\cdot - x_i)^{\alpha_i}, x \right) \right\}$$

$$\times \sinh^2\left(\frac{1}{2}\right) \Bigg|. \tag{7.84}$$

Putting all of the above together, we prove theorem. □

We make

Remark 7.8. By (7.24), we get that $\|A_n(f)\|_\infty \leq \|f\|_\infty < \infty$, and $A_n(f) \in C\left(\prod_{i=1}^{N}[a_i,b_i], \mathbb{C}\right)$, given that $f \in C\left(\prod_{i=1}^{N}[a_i,b_i], \mathbb{C}\right)$.

Clearly, then

$$\left\|A_n^2(f)\right\|_\infty = \|A_n(A_n(f))\|_\infty \leq \|A_n(f)\|_\infty \leq \|f\|_\infty, \tag{7.85}$$

etc.

Therefore, we get

$$\left\|A_n^k(f)\right\|_\infty \leq \|f\|_\infty, \quad \forall\, k \in \mathbb{N}, \tag{7.86}$$

the contraction property.

Also we see that

$$\left\|A_n^k(f)\right\|_\infty \leq \left\|A_n^{k-1}(f)\right\|_\infty \leq \cdots \leq \|A_n(f)\|_\infty \leq \|f\|_\infty. \tag{7.87}$$

Also $A_n(1) = 1$, $A_n^k(1) = 1$, $\forall\, k \in \mathbb{N}$.

Following 18.14, pp. 401–402, of Anastassiou (2016), similarly we obtain that

$$\|A_n^r f - f\|_\infty \leq r\,\|A_n(f) - f\|_\infty, \quad r \in \mathbb{N}. \tag{7.88}$$

We give

Theorem 7.9. *All as in Theorems 7.7 and 7.8. Then*

(i)

$$\|A_n^r f - f\|_\infty \leq r\xi_n(f), \tag{7.89}$$

where $\xi_n(f)$ *as in ((7.55)),*

(ii)

$$\|A_n^r f - f\|_\infty \leq r\psi_n(f), \tag{7.90}$$

where $\psi_n(f)$ *as in (7.77).*

So the speed of convergence to the unit operator of A_n^r *is not worse than that of* A_n*, see also Anastassiou (2013).*

Bibliography

G.A. Anastassiou, Rate of convergence of some neural network operators to the unit-univariate case, *Journal of Mathematical Analysis and Applications,* 212 (1997), 237–262.

G.A. Anastassiou, *Quantitative Approximations*, Chapman & Hall/CRC, Boca Raton, New York, 2001.

G.A. Anastassiou, *Inteligent Systems: Approximation by Artificial Neural Networks,* Intelligent Systems Reference Library, Vol. 19, Springer, Heidelberg, 2011a.

G.A. Anastassiou, Univariate hyperbolic tangent neural network approximation, *Mathematics and Computer Modelling,* 53 (2011b), 1111–1132.

G.A. Anastassiou, Multivariate hyperbolic tangent neural network approximation, *Computers and Mathematics,* 61 (2011c), 809–821.

G.A. Anastassiou, Multivariate sigmoidal neural network approximation, *Neural Networks,* 24 (2011d), 378–386.

G.A. Anastassiou, Univariate sigmoidal neural network approximation, *Journal of Computational Analysis and Applications*, 14(4) 2012, 659–690.

G.A. Anastassiou, Approximation by neural networks iterates. In G. Anastassiou and O. Duman (Eds.), *Advances in Applied Mathematics and Approximation Theory,* pp. 1–20, Springer Proceedings in Math. & Stat., Springer, New York, 2013.

G.A. Anastassiou, *Intelligent Systems II: Complete Approximation by Neural Network Operators*, Springer, Heidelberg, New York, 2016.

G.A. Anastassiou, *Intelligent Computations: Abstract Fractional Calculus, Inequalities, Approximations,* Springer, Heidelberg, New York, 2018.

G.A. Anastassiou, General sigmoid based Banach space valued neural network approximation, *Journal of Computational Analysis and Applications*, 31(4) (2023a), 520–534.

G.A. Anastassiou, *Parametrized, Deformed and General Neural Networks,* accepted for publication, Springer, Heidelberg, New York, 2023b.

G.A. Anastassiou, Opial and Ostrowski type inequalities based on trigonometric and hyperbolic type Taylor formulae, *Malaya Journal of Mathematik,* 11(S) (2023c), 1–26.

G.A. Anastassiou, General multiple sigmoid functions relied complex valued multivariate trigonometric and hyperbolic neural network approximations 2023.

Z. Chen and F. Cao, The approximation operators with sigmoidal functions, *Computers and Mathematics with Applications*, 58 (2009), 758–765.

D. Costarelli and R. Spigler, Approximation results for neural network operators activated by sigmoidal functions, *Neural Networks*, 44 (2013a), 101–106.

D. Costarelli and R. Spigler, Multivariate neural network operators with sigmoidal activation functions, *Neural Networks*, 48 (2013b), 72–77.

S. Haykin, *Neural Networks: A Comprehensive Foundation* 2nd edn., Prentice Hall, New York, 1998.

W. McCulloch and W. Pitts, A logical calculus of the ideas immanent in nervous activity, *Bulletin of Mathematical Biophysics*, 7 (1943), 115–133.

T.M. Mitchell, *Machine Learning,* WCB-McGraw-Hill, New York, 1997.

Chapter 8

Approximation of Multiple Time Separating Random Functions by Neural Networks

Here we study the multivariate approximation of time separating random functions over a box or all of $\mathbb{R}^N$, $N \in \mathbb{N}$, by quasi-interpolation neural network operators. These approximations are derived by establishing Jackson-type inequalities involving the multivariate modulus of continuity of the engaged random function or its high order partial derivatives. Our operators are defined by using density functions induced by the logistic and hyperbolic tangent activation sigmoid functions. The approximations are pointwise and with respect to the uniform norm. The feed-forward neural networks are with one hidden layer. We finish with a lot of interesting applications. It follows Anastassiou and Kouloumpou (2023).

8.1 Introduction

The first author in Anastassiou (1997, 2001), see Chapters 2–5, was the first to establish neural network approximation to continuous functions with rates by very specifically defined neural network operators of Cardaliaguet–Euvrard and "Squashing" types, by employing the modulus of continuity of the engaged function or its high order derivative, and producing very tight Jackson-type inequalities. He treats there both the univariate and multivariate cases. The definitions of these operators' "bell-shaped" and "squashing" functions are assumed to be of compact support. Also in Anastassiou (2001) he gives the Nth-order asymptotic expansion for the error of weak approximation of these two operators to a special natural class of smooth functions, see Chapters 4–5 there. The first author inspired by Chen and Cao (2009) continued his studies on neural network approximation by introducing and using the proper quasi interpolation operators of sigmoidal and hyperbolic tangent types which resulted into Anastassiou (2010)–Anastassiou (2012), by treating both the univariate and multivariate cases. In this chapter, we are also inspired by the related works Costarelli and Spigler (2013a), Costarelli and Spigler (2013b). The authors here use the logistic and hyperbolic tangent sigmoid functions-based neural network quantitative approximations to continuous functions over a box all of $\mathbb{R}^N$, $N \in \mathbb{N}$, with values in $\mathbb{R}$. All convergences here are with rates expressed via the multivariate modulus of continuity of the involved function or its high order partial derivative and given by very tight Jackson-type inequalities. More precisely, here we perform quantitative approximations of time separating random functions by neural networks. We give plenty of varied and interesting applications. Specific motivations came by

(1) stationary Gaussian processes with an explicit representation such as

$$X_t = \cos(\alpha t)\,\xi_1 + \sin(\alpha t)\,\xi_2, \alpha \in \mathbb{R},$$

where ξ_1, ξ_2 are independent random variables with the standard normal distribution, see Kac and Siegert (1947),

(2) the "Fourier model" of a stationary process, see Kozachenko (2016).

Feed-forward neural networks (FNNs) with one hidden layer, the only type of networks we deal with in this chapter, are mathematically expressed as

$$N_n(x) = \sum_{j=0}^{n} c_j \sigma(\langle a_j \cdot x \rangle + b_j), \quad x \in \mathbb{R}^s, \; s \in \mathbb{N},$$

where for $0 \leq j \leq n$, $b_j \in \mathbb{R}$ are the thresholds, $a_j \in \mathbb{R}^s$ are the connection weights, $c_j \in X$ are the coefficients, $\langle a_j \cdot x \rangle$ is the inner product of a_j and x, and σ is the activation function of the network. In many fundamental neural network models, the activation function is given by various specific sigmoid functions. Here we use the logistic and hyperbolic tangent activation functions. For more information about neural networks in general, read Haykin (1998), McCulloch and Pitts (1943), and Mitchell (1997). See also Anastassiou (2016) for a complete study of real-valued approximation by neural network operators.

8.2 Background

8.2.1 *About the logistic sigmoid activation function*

Here we follow Anastassiou (2011c). We consider here the sigmoidal function of logarithmic type

$$s_i\left(x_i\right)=\frac{1}{1+e^{-x_i}},\quad x_i\in\mathbb{R},\ i=1,\ldots,N;\ x:=\left(x_1,\ldots,x_N\right)\in\mathbb{R}^N.$$

Each has the properties $\lim_{x_i\to+\infty}s_i\left(x_i\right)=1$ and $\lim_{x_i\to-\infty}s_i\left(x_i\right)=0$, $i=1,\ldots,N$.

These functions play the role of activation functions in the hidden layer of neural networks and also have applications in biology, demography, etc (Brauer and Castillo-Chavez 2001, Hritonenko and Yatsenko 2000).

As in Chen and Cao (2009), we consider

$$\Phi_i\left(x_i\right):=\frac{1}{2}\left(s_i\left(x_i+1\right)-s_i\left(x_i-1\right)\right),\quad x_i\in\mathbb{R},\ i=1,\ldots,N.$$

We note the following properties:

(i) $\Phi_i\left(x_i\right)>0,\ \forall\, x_i\in\mathbb{R}$.

(ii) $\sum_{k_i=-\infty}^{\infty}\Phi_i\left(x_i-k_i\right)=1,\ \forall\, x_i\in\mathbb{R}$.

(iii) $\sum_{k_i=-\infty}^{\infty}\Phi_i\left(nx_i-k_i\right)=1,\ \forall\, x_i\in\mathbb{R};\ n\in\mathbb{N}$.

(iv) $\int_{-\infty}^{\infty}\Phi_i\left(x_i\right)dx_i=1$.

(v) Φ_i is a density function.

(vi) Φ_i is even: $\Phi_i\left(-x_i\right)=\Phi_i\left(x_i\right),\ x_i\geq 0$, for $i=1,\ldots,N$.

We see that

$$\Phi_i\left(x_i\right)=\left(\frac{e^2-1}{2e^2}\right)\frac{1}{\left(1+e^{x_i-1}\right)\left(1+e^{-x_i-1}\right)},\quad i=1,\ldots,N. \tag{8.1}$$

(vii) Φ_i is decreasing on $\mathbb{R}_+$, and increasing on $\mathbb{R}_-$, $i=1,\ldots,N$.

Let $0<\beta<1$, $n\in\mathbb{N}$. Then as in Anastassiou (2012), we get

(viii)

$$\sum_{\substack{k_i=-\infty\\ \left\{:\left|nx_i-k_i\right|>n^{1-\beta}\right.}}^{\infty}\Phi_i\left(nx_i-k_i\right)=\sum_{\substack{k_i=-\infty\\ \left\{:\left|nx_i-k_i\right|>n^{1-\beta}\right.}}^{\infty}\Phi_i\left(\left|nx_i-k_i\right|\right)$$
$$\leq 6.3984e^{-n^{(1-\beta)}},\quad i=1,\ldots,N. \tag{8.2}$$

Denote by $\lceil\cdot\rceil$ the ceiling of a number and by $\lfloor\cdot\rfloor$ the integral part of a number. Consider here $x \in \left(\prod_{i=1}^{N}[a_i, b_i]\right) \subset \mathbb{R}^N$, $N \in \mathbb{N}$ such that $\lceil na_i\rceil \leq \lfloor nb_i\rfloor$, $i = 1, \ldots, N$; $a := (a_1, \ldots, a_N)$, $b := (b_1, \ldots, b_N)$.

We obtain

(ix)
$$0 < \frac{1}{\sum_{k_i=\lceil na_i\rceil}^{\lfloor nb_i\rfloor} \Phi_i(nx_i - k_i)} < \frac{1}{\Phi_i(1)} = 5.250312578, \tag{8.3}$$
$\forall\, x_i \in [a_i, b_i]$, $i = 1, \ldots, N$.

(x) As in Anastassiou (2012), we see that
$$\lim_{n\to\infty} \sum_{k_i=\lceil na_i\rceil}^{\lfloor nb_i\rfloor} \Phi_i(nx_i - k_i) \neq 1, \tag{8.4}$$
for at least some $x_i \in [a_i, b_i]$, $i = 1, \ldots, N$.

We use here
$$\Phi(x_1, \ldots, x_N) := \Phi(x) := \prod_{i=1}^{N} \Phi_i(x_i) =: T_1(x), \quad x \in \mathbb{R}^N. \tag{8.5}$$

It has the following properties:

(i)′ $\Phi(x) > 0$, $\forall\, x \in \mathbb{R}^N$,

We see that
$$\sum_{k_1=-\infty}^{\infty}\sum_{k_2=-\infty}^{\infty}\cdots\sum_{k_N=-\infty}^{\infty} \Phi(x_1 - k_1, x_2 - k_2, \ldots, x_N - k_N)$$
$$= \sum_{k_1=-\infty}^{\infty}\sum_{k_2=-\infty}^{\infty}\cdots\sum_{k_N=-\infty}^{\infty} \prod_{i=1}^{N} \Phi_i(x_i - k_i)$$
$$= \prod_{i=1}^{N}\left(\sum_{k_i=-\infty}^{\infty} \Phi_i(x_i - k_i)\right) = 1. \tag{8.6}$$

That is,

(ii)′
$$\sum_{k=-\infty}^{\infty} \Phi(x - k)$$
$$:= \sum_{k_1=-\infty}^{\infty}\sum_{k_2=-\infty}^{\infty}\cdots\sum_{k_N=-\infty}^{\infty} \Phi(x_1 - k_1, \ldots, x_N - k_N) = 1, \tag{8.7}$$
$k := (k_1, \ldots, k_n)$, $\forall\, x \in \mathbb{R}^N$.

(iii)′

$$\sum_{k=-\infty}^{\infty} \Phi(nx-k) := \sum_{k_1=-\infty}^{\infty} \sum_{k_2=-\infty}^{\infty} \cdots \sum_{k_N=-\infty}^{\infty} \Phi(nx_1-k_1,\ldots,nx_N-k_N) = 1, \quad (8.8)$$

$\forall\, x \in \mathbb{R}^N$; $n \in \mathbb{N}$.

(iv)′

$$\int_{\mathbb{R}^N} \Phi(x)\,dx = 1, \quad (8.9)$$

that is, Φ is a multivariate density function.

Here $\|x\|_\infty := \max\{|x_1|,\ldots,|x_N|\}$, $x \in \mathbb{R}^N$, also set $\infty := (\infty,\ldots,\infty)$, $-\infty := (-\infty,\ldots,-\infty)$ upon the multivariate context, and

$$\lceil na \rceil := (\lceil na_1 \rceil,\ldots,\lceil na_N \rceil),$$

$$\lfloor nb \rfloor := (\lfloor nb_1 \rfloor,\ldots,\lfloor nb_N \rfloor).$$

For $0 < \beta < 1$ and $n \in \mathbb{N}$, fixed $x \in \mathbb{R}^N$ have that

$$\sum_{k=\lceil na \rceil}^{\lfloor nb \rfloor} \Phi(nx-k) = \sum_{\substack{k=\lceil na \rceil \\ \left\{\left\|\frac{k}{n}-x\right\|_\infty \le \frac{1}{n^\beta}\right.}}^{\lfloor nb \rfloor} \Phi(nx-k) + \sum_{\substack{k=\lceil na \rceil \\ \left\{\left\|\frac{k}{n}-x\right\|_\infty > \frac{1}{n^\beta}\right.}}^{\lfloor nb \rfloor} \Phi(nx-k). \quad (8.10)$$

In the last two sums, the counting is over disjoint vector of k's because the condition $\left\|\frac{k}{n}-x\right\|_\infty > \frac{1}{n^\beta}$ implies that there exists at least one $\left|\frac{k_r}{n}-x_r\right| > \frac{1}{n^\beta}$, $r \in \{1,\ldots,N\}$.

We treat

$$\sum_{\substack{k=\lceil na \rceil \\ \left\{\left\|\frac{k}{n}-x\right\|_\infty > \frac{1}{n^\beta}\right.}}^{\lfloor nb \rfloor} \Phi(nx-k) = \prod_{i=1}^{N} \left(\sum_{\substack{k_i=\lceil na_i \rceil \\ \left\{\left\|\frac{k}{n}-x\right\|_\infty > \frac{1}{n^\beta}\right.}}^{\lfloor nb_i \rfloor} \Phi_i(nx_i-k_i) \right)$$

$$\le \left(\prod_{\substack{i=1 \\ i \ne r}}^{N} \left(\sum_{k_i=-\infty}^{\infty} \Phi_i(nx_i-k_i) \right) \right) \cdot \left(\sum_{\substack{k_r=\lceil na_r \rceil \\ \left\{\left|\frac{k_r}{n}-x_r\right| > \frac{1}{n^\beta}\right.}}^{\lfloor nb_r \rfloor} \Phi_r(nx_r-k_r) \right).$$

$$= \left(\sum_{\substack{k_r = \lceil na_r \rceil \\ \left\{ \left| \frac{k_r}{n} - x_r \right| > \frac{1}{n^\beta} \right.}}^{\lfloor nb_r \rfloor} \Phi_r \left(nx_r - k_r \right) \right)$$

$$\leq \sum_{\substack{k_r = -\infty \\ \left\{ \left| \frac{k_r}{n} - x_r \right| > \frac{1}{n^\beta} \right.}}^{\infty} \Phi_r \left(nx_r - k_r \right) \overset{\text{(by (viii))}}{\leq} 6.3984 e^{-n^{(1-\beta)}}. \tag{8.11}$$

We have proved that

(v)′

$$\sum_{\substack{k = \lceil na \rceil \\ \left\{ \left\| \frac{k}{n} - x \right\|_\infty > \frac{1}{n^\beta} \right.}}^{\lfloor nb \rfloor} \Phi \left(nx - k \right) \leq 6.3984 e^{-n^{(1-\beta)}}, \tag{8.12}$$

$0 < \beta < 1$, $n \in \mathbb{N}$, $x \in \left(\prod_{i=1}^N [a_i, b_i] \right)$.

By (ix) clearly we obtain

$$0 < \frac{1}{\sum_{k=\lceil na \rceil}^{\lfloor nb \rfloor} \Phi \left(nx - k \right)} = \frac{1}{\prod_{i=1}^N \left(\sum_{k_i = \lceil na_i \rceil}^{\lfloor nb_i \rfloor} \Phi_i \left(nx_i - k_i \right) \right)}$$

$$< \frac{1}{\prod_{i=1}^N \Phi_i (1)} = (5.250312578)^N. \tag{8.13}$$

That is,

(vi)′ it holds

$$0 < \frac{1}{\sum_{k=\lceil na \rceil}^{\lfloor nb \rfloor} \Phi \left(nx - k \right)} < (5.250312578)^N, \tag{8.14}$$

$\forall\, x \in \left(\prod_{i=1}^N [a_i, b_i] \right)$, $n \in \mathbb{N}$.

It is clear also that

(vii)′

$$\sum_{\substack{k = -\infty \\ \left\{ \left\| \frac{k}{n} - x \right\|_\infty > \frac{1}{n^\beta} \right.}}^{\infty} \Phi \left(nx - k \right) \leq 6.3984 e^{-n^{(1-\beta)}}, \tag{8.15}$$

$0 < \beta < 1$, $n \in \mathbb{N}$, $x \in \mathbb{R}^N$.

By (x) we obviously see that

(viii)′

$$\lim_{n\to\infty}\sum_{k=\lceil na\rceil}^{\lfloor nb\rfloor}\Phi\left(nx-k\right)\neq 1 \tag{8.16}$$

for at least some $x\in\left(\prod_{i=1}^{N}\left[a_i,b_i\right]\right)$.

Let $f\in C\left(\prod_{i=1}^{N}\left[a_i,b_i\right]\right)$ and $n\in\mathbb{N}$ such that $\lceil na_i\rceil\leq\lfloor nb_i\rfloor$, $i=1,\ldots,N$.

We introduce and define the multivariate positive linear neural network operator ($x:=\left(x_1,\ldots,x_N\right)\in\left(\prod_{i=1}^{N}\left[a_i,b_i\right]\right)$)

$$G_n\left(f,x_1,\ldots,x_N\right):=G_n\left(f,x\right):=\frac{\sum_{k=\lceil na\rceil}^{\lfloor nb\rfloor}f\left(\frac{k}{n}\right)\Phi\left(nx-k\right)}{\sum_{k=\lceil na\rceil}^{\lfloor nb\rfloor}\Phi\left(nx-k\right)}$$

$$:=\frac{\sum_{k_1=\lceil na_1\rceil}^{\lfloor nb_1\rfloor}\sum_{k_2=\lceil na_2\rceil}^{\lfloor nb_2\rfloor}\cdots\sum_{k_N=\lceil na_N\rceil}^{\lfloor nb_N\rfloor}f\left(\frac{k_1}{n},\ldots,\frac{k_N}{n}\right)\left(\prod_{i=1}^{N}\Phi_i\left(nx_i-k_i\right)\right)}{\prod_{i=1}^{N}\left(\sum_{k_i=\lceil na_i\rceil}^{\lfloor nb_i\rfloor}\Phi_i\left(nx_i-k_i\right)\right)}. \tag{8.17}$$

For large enough n, we always obtain $\lceil na_i\rceil\leq\lfloor nb_i\rfloor$, $i=1,\ldots,N$. Also $a_i\leq\frac{k_i}{n}\leq b_i$, iff $\lceil na_i\rceil\leq k_i\leq\lfloor nb_i\rfloor$, $i=1,\ldots,N$.

We study here the pointwise and uniform convergence of $G_n\left(f\right)$ to f with rates.

For convenience we call

$$G_n^*\left(f,x\right):=\sum_{k=\lceil na\rceil}^{\lfloor nb\rfloor}f\left(\frac{k}{n}\right)\Phi\left(nx-k\right)$$

$$:=\sum_{k_1=\lceil na_1\rceil}^{\lfloor nb_1\rfloor}\sum_{k_2=\lceil na_2\rceil}^{\lfloor nb_2\rfloor}\cdots\sum_{k_N=\lceil na_N\rceil}^{\lfloor nb_N\rfloor}f\left(\frac{k_1}{n},\ldots,\frac{k_N}{n}\right)\left(\prod_{i=1}^{N}\Phi_i\left(nx_i-k_i\right)\right), \tag{8.18}$$

$\forall\, x\in\left(\prod_{i=1}^{N}\left[a_i,b_i\right]\right)$.

That is,

$$G_n\left(f,x\right):=\frac{G_n^*\left(f,x\right)}{\sum_{k=\lceil na\rceil}^{\lfloor nb\rfloor}\Phi\left(nx-k\right)},\quad \forall x\in\left(\prod_{i=1}^{N}\left[a_i,b_i\right]\right),\ n\in\mathbb{N}. \tag{8.19}$$

Hence,

$$G_n(f,x) - f(x) = \frac{G_n^*(f,x) - f(x)\sum_{k=\lceil na \rceil}^{\lfloor nb \rfloor} \Phi(nx-k)}{\sum_{k=\lceil na \rceil}^{\lfloor nb \rfloor} \Phi(nx-k)}. \tag{8.20}$$

Consequently we derive

$$|G_n(f,x) - f(x)| \leq (5.250312578)^N \left| G_n^*(f,x) - f(x) \sum_{k=\lceil na \rceil}^{\lfloor nb \rfloor} \Phi(nx-k) \right|, \tag{8.21}$$

$\forall\, x \in \left(\prod_{i=1}^N [a_i, b_i]\right)$.

We estimate the right-hand side of (8.21).

For that we need, for $f \in C\left(\prod_{i=1}^N [a_i, b_i]\right)$, the first multivariate modulus of continuity

$$\omega_1(f,h) := \sup_{\substack{x,y \in \left(\prod_{i=1}^N [a_i,b_i]\right) \\ \|x-y\|_\infty \leq h}} |f(x) - f(y)|, \quad h > 0. \tag{8.22}$$

Similarly, it is defined for $f \in C_B(\mathbb{R}^N)$ (continuous and bounded functions on $\mathbb{R}^N$). We have that $\lim_{h \to 0} \omega_1(f,h) = 0$.

When $f \in C_B(\mathbb{R}^N)$, we define

$$\overline{G}_n(f,x) := \overline{G}_n(f,x_1,\ldots,x_N) := \sum_{k=-\infty}^{\infty} f\left(\frac{k}{n}\right) \Phi(nx-k)$$

$$:= \sum_{k_1=-\infty}^{\infty} \sum_{k_2=-\infty}^{\infty} \cdots \sum_{k_N=-\infty}^{\infty} f\left(\frac{k_1}{n}, \frac{k_2}{n}, \ldots, \frac{k_N}{n}\right) \left(\prod_{i=1}^N \Phi_i(nx_i - k_i)\right), \tag{8.23}$$

$n \in \mathbb{N}$, $\forall\, x \in \mathbb{R}^N$, $N \geq 1$, the multivariate quasi-interpolation neural network operator.

Note here that for large enough $n \in \mathbb{N}$ we get that

$$e^{-n^{(1-\beta)}} < n^{-\beta j}, \quad j = 1, \ldots, m \in \mathbb{N}, \quad 0 < \beta < 1. \tag{8.24}$$

Thus, given fixed $A, B > 0$, for the linear combination $(An^{-\beta j} + Be^{-n^{(1-\beta)}})$ the (dominant) rate of convergence to zero is $n^{-\beta j}$. The closer β is to 1, the faster and better rate of convergence to zero we get.

Let $f \in C^m\left(\prod_{i=1}^N [a_i, b_i]\right)$, $m, N \in \mathbb{N}$. Here f_α denotes a partial derivative of f, $\alpha := (\alpha_1, \ldots, \alpha_N)$, $\alpha_i \in \mathbb{Z}^+$, $i = 1, \ldots, N$, and $|\alpha| := \sum_{i=1}^N \alpha_i = l$, where $l = 0, 1, \ldots, m$. We write also $f_\alpha := \frac{\partial^\alpha f}{\partial x^\alpha}$ and we say it is of order l.

We denote

$$\omega_{1,m}^{\max}(f_\alpha, h) := \max_{\alpha : |\alpha| = m} \omega_1(f_\alpha, h). \tag{8.25}$$

Call also

$$\|f_\alpha\|_{\infty,m}^{\max} := \max_{|\alpha| = m} \{\|f_\alpha\|_\infty\}, \tag{8.26}$$

$\|\cdot\|_\infty$ is the supremum norm.

Next we present a series of multivariate neural network approximations to a function given with rates.

We first give

Theorem 8.1. *Let $f \in C\left(\prod_{i=1}^N [a_i, b_i]\right)$, $0 < \beta < 1$, $x \in \left(\prod_{i=1}^N [a_i, b_i]\right)$, $n, N \in \mathbb{N}$. Then*

(i)

$$|G_n(f, x) - f(x)| \leq (5.250312578)^N \cdot \left\{\omega_1\left(f, \frac{1}{n^\beta}\right) + (12.7968)\|f\|_\infty e^{-n^{(1-\beta)}}\right\} =: \lambda_1, \tag{8.27}$$

(ii)

$$\|G_n(f) - f\|_\infty \leq \lambda_1. \tag{8.28}$$

Next we present

Theorem 8.2. *(Anastassiou 2011c) Let $f \in C_B(\mathbb{R}^N)$, $0 < \beta < 1$, $x \in \mathbb{R}^N$, $n, N \in \mathbb{N}$. Then*

(i)

$$\left|\overline{G}_n(f, x) - f(x)\right| \leq \omega_1\left(f, \frac{1}{n^\beta}\right) + (12.7968)\|f\|_\infty e^{-n^{(1-\beta)}} =: \lambda_2, \tag{8.29}$$

(ii)

$$\left\|\overline{G}_n(f) - f\right\|_\infty \leq \lambda_2. \tag{8.30}$$

In the following, we discuss high order of approximation by using the smoothness of f.

We give

Theorem 8.3 (Anastassiou 2011c). *Let* $f \in C^m\left(\prod_{i=1}^N [a_i,b_i]\right)$, $0 < \beta < 1$, $n,m,N \in \mathbb{N}$, $x \in \left(\prod_{i=1}^N [a_i,b_i]\right)$. *Then*

(i)

$$\left| G_n(f,x) - f(x) - \sum_{j=1}^{m} \left(\sum_{|\alpha|=j} \left(\frac{f_\alpha(x)}{\prod_{i=1}^N \alpha_i!} \right) G_n\left(\prod_{i=1}^N (\cdot - x_i)^{\alpha_i}, x \right) \right) \right|$$
$$\leq (5.250312578)^N \cdot \left\{ \frac{N^m}{m! n^{m\beta}} \omega_{1,m}^{\max}\left(f_\alpha, \frac{1}{n^\beta}\right) + \left(\frac{(12.7968) \|b-a\|_\infty^m \|f_\alpha\|_{\infty,m}^{\max} N^m}{m!} \right) e^{-n^{(1-\beta)}} \right\}, \tag{8.31}$$

(ii)

$$|G_n(f,x) - f(x)| \leq (5.250312578)^N \cdot$$
$$\left\{ \sum_{j=1}^{m} \left(\sum_{|\alpha|=j} \left(\frac{|f_\alpha(x)|}{\prod_{i=1}^N \alpha_i!} \right) \left[\frac{1}{n^{\beta j}} + \left(\prod_{i=1}^N (b_i - a_i)^{\alpha_i} \right) \cdot (6.3984) e^{-n^{(1-\beta)}} \right] \right) + \frac{N^m}{m! n^{m\beta}} \omega_{1,m}^{\max}\left(f_\alpha, \frac{1}{n^\beta}\right) + \left(\frac{(12.7968) \|b-a\|_\infty^m \|f_\alpha\|_{\infty,m}^{\max} N^m}{m!} \right) e^{-n^{(1-\beta)}} \right\}, \tag{8.32}$$

(iii)

$$\|G_n(f) - f\|_\infty \leq (5.250312578)^N \cdot$$
$$\left\{ \sum_{j=1}^{N} \left(\sum_{|\alpha|=j} \left(\frac{\|f_\alpha\|_\infty}{\prod_{i=1}^N \alpha_i!} \right) \left[\frac{1}{n^{\beta j}} + \left(\prod_{i=1}^N (b_i - a_i)^{\alpha_i} \right) (3.1992) e^{-n^{(1-\beta)}} \right] \right) + \frac{N^m}{m! n^{m\beta}} \omega_{1,m}^{\max}\left(f_\alpha, \frac{1}{n^\beta}\right) + \left(\frac{(12.7968) \|b-a\|_\infty^m \|f_\alpha\|_{\infty,m}^{\max} N^m}{m!} \right) e^{-n^{(1-\beta)}} \right\}, \tag{8.33}$$

(iv) *assume* $f_\alpha(x_0) = 0$, *for all* $\alpha : |\alpha| = 1, \ldots, m$; $x_0 \in \left(\prod_{i=1}^N [a_i,b_i]\right)$. *Then*

$$|G_n(f,x_0) - f(x_0)| \leq (5.250312578)^N \cdot$$
$$\left\{ \frac{N^m}{m! n^{m\beta}} \omega_1^{\max}\left(f_\alpha, \frac{1}{n^\beta}\right) + \left(\frac{(12.7968) \|b-a\|_\infty^m \|f_\alpha\|_{\infty,m}^{\max} N^m}{m!} \right) e^{-n^{(1-\beta)}} \right\}, \tag{8.34}$$

note in the last the extremely high rate of convergence at $n^{-\beta(m+1)}$.

8.2.2 *About the hyperbolic tangent sigmoid activation function*

We consider now the hyperbolic tangent function $\tanh x$, $x \in \mathbb{R}$:

$$\tanh x := \frac{e^x - e^{-x}}{e^x + e^{-x}}.$$

It has the properties $\tanh 0 = 0$, $-1 < \tanh x < 1, \forall\, x \in \mathbb{R}$, and $\tanh(-x) = -\tanh x$. Furthermore, $\tanh x \to 1$ as $x \to \infty$, and $\tanh x \to -1$, as $x \to -\infty$, and it is strictly increasing on $\mathbb{R}$.

This function plays the role of an activation function in the hidden layer of neural networks.

We further consider

$$\Psi(x) := \frac{1}{4}\left(\tanh(x+1) - \tanh(x-1)\right) > 0, \quad \forall\, x \in \mathbb{R}.$$

We easily see that $\Psi(-x) = \Psi(x)$, that is Ψ is even on $\mathbb{R}$. Obviously, Ψ is differentiable, thus continuous.

Proposition 8.1 (Anastassiou 2010). *$\Psi(x)$ for $x \geq 0$ is strictly decreasing.*

Obviously, $\Psi(x)$ is strictly increasing for $x \leq 0$. Also it holds $\lim_{x \to -\infty} \Psi(x) = 0 = \lim_{x \to \infty} \Psi(x)$.

In fact Ψ has the bell shape with horizontal asymptote the x-axis. So the maximum of Ψ is zero, $\Psi(0) = 0.3809297$.

Theorem 8.4 (Anastassiou 2010). *We have that $\sum_{i=-\infty}^{\infty} \Psi(x-i) = 1$, $\forall\, x \in \mathbb{R}$.*

Thus,

$$\sum_{i=-\infty}^{\infty} \Psi(nx - i) = 1, \quad \forall\, n \in \mathbb{N},\ \forall\, x \in \mathbb{R}. \tag{8.35}$$

Also it holds

$$\sum_{i=-\infty}^{\infty} \Psi(x + i) = 1, \quad \forall x \in \mathbb{R}. \tag{8.36}$$

Theorem 8.5 (Anastassiou 2010). *It holds*

$$\int_{-\infty}^{\infty} \Psi(x)\, dx = 1. \tag{8.37}$$

So $\Psi(x)$ is a density function on $\mathbb{R}$.

Theorem 8.6 (Anastassiou 2010). *Let $0 < \alpha < 1$ and $n \in \mathbb{N}$. It holds*

$$\sum_{\begin{cases} k=-\infty \\ :|nx-k| \geq n^{1-\alpha} \end{cases}}^{\infty} \Psi(nx-k) \leq 2e^4 \cdot e^{-2n^{(1-\alpha)}}. \tag{8.38}$$

Denote by $\lfloor \cdot \rfloor$ the integral part of the number and by $\lceil \cdot \rceil$ the ceiling of the number.

Theorem 8.7 (Anastassiou 2010). *Let $x \in [a, b] \subset \mathbb{R}$ and $n \in \mathbb{N}$ so that $\lceil na \rceil \leq \lfloor nb \rfloor$. It holds*

$$\frac{1}{\sum_{k=\lceil na \rceil}^{\lfloor nb \rfloor} \Psi(nx-k)} < \frac{1}{\Psi(1)} = 4.1488766. \tag{8.39}$$

Also by Anastassiou (2010) we get that

$$\lim_{n \to \infty} \sum_{k=\lceil na \rceil}^{\lfloor nb \rfloor} \Psi(nx-k) \neq 1, \tag{8.40}$$

for at least some $x \in [a, b]$.

In this chapter, we use

$$\Theta(x_1, \ldots, x_N)$$

$$:= \Theta(x) := \prod_{i=1}^{N} \Psi(x_i), \quad x = (x_1, \ldots, x_N) = T_2(x) \in \mathbb{R}^N, \ N \in \mathbb{N}. \tag{8.41}$$

It has the following properties:

(i) $\Theta(x) > 0, \ \forall\, x \in \mathbb{R}^N$.

(ii)

$$\sum_{k=-\infty}^{\infty} \Theta(x-k)$$

$$:= \sum_{k_1=-\infty}^{\infty} \sum_{k_2=-\infty}^{\infty} \cdots \sum_{k_N=-\infty}^{\infty} \Theta(x_1-k_1, \ldots, x_N-k_N) = 1, \tag{8.42}$$

where $k := (k_1, \ldots, k_n)$, $\forall\, x \in \mathbb{R}^N$.

(iii)

$$\sum_{k=-\infty}^{\infty} \Theta(nx-k)$$

$$:= \sum_{k_1=-\infty}^{\infty} \sum_{k_2=-\infty}^{\infty} \cdots \sum_{k_N=-\infty}^{\infty} \Theta(nx_1-k_1,\ldots,nx_N-k_N) = 1, \tag{8.43}$$

$\forall\, x \in \mathbb{R}^N;\ n \in \mathbb{N}$.

(iv)

$$\int_{\mathbb{R}^N} \Theta(x)\,dx = 1, \tag{8.44}$$

that is, Θ is a multivariate density function.

We obviously see that

$$\sum_{k=\lceil na \rceil}^{\lfloor nb \rfloor} \Theta(nx-k) = \sum_{k=\lceil na \rceil}^{\lfloor nb \rfloor} \prod_{i=1}^{N} \Psi(nx_i - x_i)$$

$$= \sum_{k_1=\lceil na_1 \rceil}^{\lfloor nb_1 \rfloor} \cdots \sum_{k_N=\lceil na_N \rceil}^{\lfloor nb_N \rfloor} \prod_{i=1}^{N} \Psi(nx_i-k_i) = \prod_{i=1}^{N} \left(\sum_{k_i=\lceil na_i \rceil}^{\lfloor nb_i \rfloor} \Psi(nx_i-k_i) \right). \tag{8.45}$$

For $0 < \beta < 1$ and $n \in \mathbb{N}$, fixed $x \in \mathbb{R}^N$, we have that

$$\sum_{k=\lceil na \rceil}^{\lfloor nb \rfloor} \Theta(nx-k)$$

$$= \sum_{\begin{cases} k=\lceil na \rceil \\ \left\| \frac{k}{n}-x \right\|_\infty \le \frac{1}{n^\beta} \end{cases}}^{\lfloor nb \rfloor} \Theta(nx-k) + \sum_{\begin{cases} k=\lceil na \rceil \\ \left\| \frac{k}{n}-x \right\|_\infty > \frac{1}{n^\beta} \end{cases}}^{\lfloor nb \rfloor} \Theta(nx-k). \tag{8.46}$$

In the last two sums, the counting is over disjoint vector sets of k's because the condition $\left\| \frac{k}{n} - x \right\|_\infty > \frac{1}{n^\beta}$ implies that there exists at least one $\left| \frac{k_r}{n} - x_r \right| > \frac{1}{n^\beta}$, $r \in \{1, \ldots, N\}$.

We treat

$$\sum_{\substack{k=\lceil na\rceil \\ \left\{\left\|\frac{k}{n}-x\right\|_{\infty}>\frac{1}{n^{\beta}}\right.}}^{\lfloor nb\rfloor} \Theta(nx-k) = \prod_{i=1}^{N}\left(\sum_{\substack{k_i=\lceil na_i\rceil \\ \left\{\left\|\frac{k}{n}-x\right\|_{\infty}>\frac{1}{n^{\beta}}\right.}}^{\lfloor nb_i\rfloor} \Psi(nx_i-k_i)\right)$$

$$\leq \left(\prod_{\substack{i=1 \\ i\neq r}}^{N}\left(\sum_{k_i=-\infty}^{\infty} \Psi(nx_i-k_i)\right)\right)\cdot\left(\sum_{\substack{k_r=\lceil na_r\rceil \\ \left\{\left|\frac{k_r}{n}-x_r\right|>\frac{1}{n^{\beta}}\right.}}^{\lfloor nb_r\rfloor} \Psi(nx_r-k_r)\right)$$

$$= \left(\sum_{\substack{k_r=\lceil na_r\rceil \\ \left\{\left|\frac{k_r}{n}-x_r\right|>\frac{1}{n^{\beta}}\right.}}^{\lfloor nb_r\rfloor} \Psi(nx_r-k_r)\right)$$

$$\leq \sum_{\substack{k_r=-\infty \\ \left\{\left|\frac{k_r}{n}-x_r\right|>\frac{1}{n^{\beta}}\right.}}^{\infty} \Psi(nx_r-k_r) \overset{\text{(by Theorem 8.6)}}{\leq 2} e^4\cdot e^{-2n^{(1-\beta)}}. \tag{8.47}$$

We have proved that

(v)

$$\sum_{\substack{k=\lceil na\rceil \\ \left\{\left\|\frac{k}{n}-x\right\|_{\infty}>\frac{1}{n^{\beta}}\right.}}^{\lfloor nb\rfloor} \Theta(nx-k) \leq 2e^4\cdot e^{-2n^{(1-\beta)}}, \tag{8.48}$$

$0<\beta<1$, $n\in\mathbb{N}$, $x\in\left(\prod_{i=1}^{N}[a_i,b_i]\right)$.

By Theorem 8.7 clearly we obtain

$$0<\frac{1}{\sum_{k=\lceil na\rceil}^{\lfloor nb\rfloor}\Theta(nx-k)} = \frac{1}{\prod_{i=1}^{N}\left(\sum_{k_i=\lceil na_i\rceil}^{\lfloor nb_i\rfloor}\Psi(nx_i-k_i)\right)}$$

$$<\frac{1}{(\Psi(1))^N} = (4.1488766)^N. \tag{8.49}$$

That is,

(vi) it holds

$$0 < \frac{1}{\sum_{k=\lceil na \rceil}^{\lfloor nb \rfloor} \Theta(nx-k)} < \frac{1}{(\Psi(1))^N} = (4.1488766)^N, \tag{8.50}$$

$\forall\, x \in \left(\prod_{i=1}^N [a_i, b_i]\right),\ n \in \mathbb{N}.$

It is also clear that

(vii)

$$\sum_{\substack{k=-\infty \\ \left\{\left\| \frac{k}{n} - x \right\|_\infty > \frac{1}{n^\beta}\right.}}^{\infty} \Theta(nx-k) \leq 2e^4 \cdot e^{-2n^{(1-\beta)}}, \tag{8.51}$$

$0 < \beta < 1$, $n \in \mathbb{N}$, $x \in \mathbb{R}^N$.

Also we get that

$$\lim_{n\to\infty} \sum_{k=\lceil na \rceil}^{\lfloor nb \rfloor} \Theta(nx-k) \neq 1, \tag{8.52}$$

for at least some $x \in \left(\prod_{i=1}^N [a_i, b_i]\right)$.

Let $f \in C\left(\prod_{i=1}^N [a_i, b_i]\right)$ and $n \in \mathbb{N}$ such that $\lceil na_i \rceil \leq \lfloor nb_i \rfloor$, $i = 1, \ldots, N$.

We introduce and define the multivariate positive linear neural network operator $\left(x := (x_1, \ldots, x_N) \in \left(\prod_{i=1}^N [a_i, b_i]\right)\right)$

$$F_n(f, x_1, \ldots, x_N) := F_n(f, x) := \frac{\sum_{k=\lceil na \rceil}^{\lfloor nb \rfloor} f\left(\frac{k}{n}\right) \Theta(nx-k)}{\sum_{k=\lceil na \rceil}^{\lfloor nb \rfloor} \Theta(nx-k)}$$

$$:= \frac{\sum_{k_1=\lceil na_1 \rceil}^{\lfloor nb_1 \rfloor} \sum_{k_2=\lceil na_2 \rceil}^{\lfloor nb_2 \rfloor} \cdots \sum_{k_N=\lceil na_N \rceil}^{\lfloor nb_N \rfloor} f\left(\frac{k_1}{n}, \ldots, \frac{k_N}{n}\right) \left(\prod_{i=1}^N \Psi(nx_i - k_i)\right)}{\prod_{i=1}^N \left(\sum_{k_i=\lceil na_i \rceil}^{\lfloor nb_i \rfloor} \Psi(nx_i - k_i)\right)}. \tag{8.53}$$

For large enough n, we always obtain $\lceil na_i \rceil \leq \lfloor nb_i \rfloor$, $i = 1, \ldots, N$. Also $a_i \leq \frac{k_i}{n} \leq b_i$, iff $\lceil na_i \rceil \leq k_i \leq \lfloor nb_i \rfloor$, $i = 1, \ldots, N$.

We study here the pointwise and uniform convergence of $F_n(f)$ to f with rates.

For convenience we call

$$F_n^*(f,x) := \sum_{k=\lceil na \rceil}^{\lfloor nb \rfloor} f\left(\frac{k}{n}\right) \Theta(nx-k)$$

$$:= \sum_{k_1=\lceil na_1 \rceil}^{\lfloor nb_1 \rfloor} \sum_{k_2=\lceil na_2 \rceil}^{\lfloor nb_2 \rfloor} \cdots \sum_{k_N=\lceil na_N \rceil}^{\lfloor nb_N \rfloor} f\left(\frac{k_1}{n},\ldots,\frac{k_N}{n}\right)\left(\prod_{i=1}^{N} \Psi(nx_i-k_i)\right), \tag{8.54}$$

$\forall\, x \in \left(\prod_{i=1}^{N} [a_i, b_i]\right)$.

That is,

$$F_n(f,x) := \frac{F_n^*(f,x)}{\sum_{k=\lceil na \rceil}^{\lfloor nb \rfloor} \Theta(nx-k)}, \tag{8.55}$$

$\forall\, x \in \left(\prod_{i=1}^{N} [a_i, b_i]\right)$, $n \in \mathbb{N}$.

Hence,

$$F_n(f,x) - f(x) = \frac{F_n^*(f,x) - f(x)\sum_{k=\lceil na \rceil}^{\lfloor nb \rfloor} \Theta(nx-k)}{\sum_{k=\lceil na \rceil}^{\lfloor nb \rfloor} \Theta(nx-k)}. \tag{8.56}$$

Consequently we derive

$$|F_n(f,x) - f(x)| \leq (4.1488766)^N \left| F_n^*(f,x) - f(x) \sum_{k=\lceil na \rceil}^{\lfloor nb \rfloor} \Theta(nx-k) \right|, \tag{8.57}$$

$\forall\, x \in \left(\prod_{i=1}^{N} [a_i, b_i]\right)$. When $f \in C_B(\mathbb{R}^N)$ we define,

$$\overline{F}_n(f,x) := \overline{F}_n(f, x_1, \ldots, x_N) := \sum_{k=-\infty}^{\infty} f\left(\frac{k}{n}\right) \Theta(nx-k)$$

$$:= \sum_{k_1=-\infty}^{\infty} \sum_{k_2=-\infty}^{\infty} \cdots \sum_{k_N=-\infty}^{\infty} f\left(\frac{k_1}{n},\frac{k_2}{n},\ldots,\frac{k_N}{n}\right)\left(\prod_{i=1}^{N} \Psi(nx_i-k_i)\right), \tag{8.58}$$

$n \in \mathbb{N}$, $\forall\, x \in \mathbb{R}^N$, $N \geq 1$, the multivariate quasi-interpolation neural network operator. Here we present a series of multivariate neural network approximations to a function given with rates. We first give

Theorem 8.8. *Let* $f \in C\left(\prod_{i=1}^{N}[a_i,b_i]\right)$, $0<\beta<1$, $x \in \left(\prod_{i=1}^{N}[a_i,b_i]\right)$, $n, N \in \mathbb{N}$. *Then*

(i)

$$\left|F_n(f,x) - f(x)\right| \leq (4.1488766)^N \cdot \left\{\omega_1\left(f, \frac{1}{n^\beta}\right) + 4e^4 \|f\|_\infty e^{-2n^{(1-\beta)}}\right\} =: \lambda_1, \tag{8.59}$$

(ii)

$$\|F_n(f) - f\|_\infty \leq \lambda_1. \tag{8.60}$$

Next we present

Theorem 8.9. *Let* $f \in C_B\left(\mathbb{R}^N\right)$, $0<\beta<1$, $x \in \mathbb{R}^N$, $n, N \in \mathbb{N}$. *Then*

(i)

$$\left|\overline{F}_n(f,x) - f(x)\right| \leq \omega_1\left(f, \frac{1}{n^\beta}\right) + 2e^4 \|f\|_\infty e^{-2n^{(1-\beta)}} =: \lambda_2, \tag{8.61}$$

(ii)

$$\left\|\overline{F}_n(f) - f\right\|_\infty \leq \lambda_2. \tag{8.62}$$

In the following, we discuss high order of approximation by using the smoothness of f. We give

Theorem 8.10. *Let* $f \in C^m\left(\prod_{i=1}^{N}[a_i,b_i]\right)$, $0<\beta<1$, $n, m, N \in \mathbb{N}$, $x \in \left(\prod_{i=1}^{N}[a_i,b_i]\right)$. *Then*

(i)

$$\left|F_n(f,x) - f(x) - \sum_{j=1}^{m}\left(\sum_{|\alpha|=j}\left(\frac{f_\alpha(x)}{\prod_{i=1}^{N}\alpha_i!}\right) F_n\left(\prod_{i=1}^{N}(\cdot - x_i)^{\alpha_i}, x\right)\right)\right|$$

$$\leq (4.1488766)^N \cdot \left\{\frac{N^m}{m! n^{m\beta}} \omega_{1,m}^{\max}\left(f_\alpha, \frac{1}{n^\beta}\right)\right.$$

$$\left. + \left(\frac{2e^4 \|b-a\|_\infty^m \|f_\alpha\|_{\infty,m}^{\max} N^m}{m!}\right) e^{-2n^{(1-\beta)}}\right\}, \tag{8.63}$$

(ii)

$$|F_n(f,x)-f(x)| \leq (4.1488766)^N \cdot$$
$$\left\{\sum_{j=1}^{m}\left(\sum_{|\alpha|=j}\left(\frac{|f_\alpha(x)|}{\prod_{i=1}^{N}\alpha_i!}\right)\left[\frac{1}{n^{\beta j}}+\left(\prod_{i=1}^{N}(b_i-a_i)^{\alpha_i}\right)\cdot e^4 e^{-2n^{(1-\beta)}}\right]\right)\right.$$
$$\left.+\frac{N^m}{m!n^{m\beta}}\omega_{1,m}^{\max}\left(f_\alpha,\frac{1}{n^\beta}\right)+\left(\frac{2e^4\|b-a\|_\infty^m\|f_\alpha\|_{\infty,m}^{\max}N^m}{m!}\right)e^{-2n^{(1-\beta)}}\right\}, \tag{8.64}$$

(iii)

$$\|F_n(f)-f\|_\infty \leq (4.1488766)^N \cdot$$
$$\left\{\sum_{j=1}^{m}\left(\sum_{|\alpha|=j}\left(\frac{\|f_\alpha\|_\infty}{\prod_{i=1}^{N}\alpha_i!}\right)\left[\frac{1}{n^{\beta j}}+\left(\prod_{i=1}^{N}(b_i-a_i)^{\alpha_i}\right)e^4 e^{-2n^{(1-\beta)}}\right]\right)\right.$$
$$\left.+\frac{N^m}{m!n^{m\beta}}\omega_{1,m}^{\max}\left(f_\alpha,\frac{1}{n^\beta}\right)+\left(\frac{2e^4\|b-a\|_\infty^m\|f_\alpha\|_{\infty,m}^{\max}N^m}{m!}\right)e^{-2n^{(1-\beta)}}\right\}, \tag{8.65}$$

(iv) *assume* $f_\alpha(x_0)=0$, *for all* $\alpha : |\alpha|=1,\ldots,m$; $x_0 \in \left(\prod_{i=1}^{N}[a_i,b_i]\right)$. *Then*

$$|F_n(f,x_0)-f(x_0)| \leq (4.1488766)^N \cdot$$
$$\left\{\frac{N^m}{m!n^{m\beta}}\omega_1^{\max}\left(f_\alpha,\frac{1}{n^\beta}\right)+\left(\frac{2e^4\|b-a\|_\infty^m\|f_\alpha\|_{\infty,m}^{\max}N^m}{m!}\right)e^{-2n^{(1-\beta)}}\right\}, \tag{8.66}$$

note in the last the extremely high rate of convergence at $n^{-\beta(m+1)}$.

8.2.3 *Combining 2.1 and 2.2*

For the next theorems, we call

$$T_1(x) := \Phi(x)$$
$$T_2(x) := \Theta(x), x \in \mathbb{R}^{\mathbb{N}}, N \in \mathbb{N}.$$

We also set

$$\gamma_{1,N} := (5.250312578)^N,$$
$$\gamma_{2,N} := (4.1488766)^N.$$

Furthermore we set

$$\alpha_{1,n}(\beta) := 6.3984e^{-n^{(1-\beta)}},$$

$$\alpha_{2,n}(\beta) := 2e^{4}e^{-2n^{(1-\beta)}},$$

where $0 < \beta < 1, n \in \mathbb{N}$.
We define

$${}_{1}L_{n}\left(f,x\right) := G_{n}\left(f,x\right),$$

$${}_{2}L_{n}\left(f,x\right) := F_{n}\left(f,x\right), x \in \mathbb{R}^{\mathbb{N}}, n \in \mathbb{N}$$

and

$${}_{1}\overline{L}_{n}\left(f,x\right) := \overline{G}_{n}\left(f,x\right),$$

$${}_{2}\overline{L}_{n}\left(f,x\right) := \overline{F}_{n}\left(f,x\right), x \in \mathbb{R}^{\mathbb{N}}, n \in \mathbb{N}.$$

Note that

$$6.3984e^{-n^{(1-\beta)}} = 2a_{1,n}(\beta).$$

Theorem 8.11. *Let* $f \in C\left(\prod_{i=1}^{N}\left[a_i,b_i\right]\right)$, $0 < \beta < 1$, $x \in \left(\prod_{i=1}^{N}\left[a_i,b_i\right]\right)$, $n, N \in \mathbb{N}$ *and* $k = 1, 2$. *Then*

(i)

$$\left|{}_{k}L_{n}\left(f,x\right) - f\left(x\right)\right| \leq \gamma_{k,N}$$
$$\left\{\omega_1\left(f, \frac{1}{n^{\beta}}\right) + 2\alpha_{k,n}(\beta)\left\|f\right\|_{\infty}\right\} =: \lambda_1, \quad (8.67)$$

(ii)

$$\left\|{}_{k}L_{n}\left(f\right) - f\right\|_{\infty} \leq \lambda_1. \quad (8.68)$$

Proof. From Theorems 8.1 and 8.8. □

Next we present

Theorem 8.12. *Let* $f \in C_B\left(\mathbb{R}^N\right)$, $0 < \beta < 1$, $x \in \mathbb{R}^N$, $n, N \in \mathbb{N}$ *and* $k = 1, 2$. *Then*

(i)

$$\left|{}_{k}\overline{L}_{n}\left(f,x\right) - f\left(x\right)\right| \leq \omega_1\left(f, \frac{1}{n^{\beta}}\right) + 2\alpha_{k,n}\left(\beta\right)\left\|f\right\|_{\infty} =: \lambda_2, \quad (8.69)$$

(ii)

$$\left\|{}_{k}\overline{L}_{n}\left(f\right) - f\right\|_{\infty} \leq \lambda_2. \quad (8.70)$$

Proof. From Theorems 8.2 and 8.9. □

In the following, we discuss high order of approximation by using the smoothness of f.

We give

Theorem 8.13. *Let* $f \in C^m\left(\prod_{i=1}^N [a_i, b_i]\right)$, $0 < \beta < 1$, $n, m, N \in \mathbb{N}$, $x \in \left(\prod_{i=1}^N [a_i, b_i]\right)$ *and* $k = 1, 2$. *Then*

(i)

$$\left| {}_kL_n(f,x) - f(x) - \sum_{j=1}^{m}\left(\sum_{|\alpha|=j}\left(\frac{f_\alpha(x)}{\prod_{i=1}^N \alpha_i!}\right) {}_kL_n\left(\prod_{i=1}^N (\cdot - x_i)^{\alpha_i}, x\right)\right)\right|$$

$$\leq \gamma_{k,N} \cdot \left\{ \frac{N^m}{m! n^{m\beta}} \omega_{1,m}^{\max}\left(f_\alpha, \frac{1}{n^\beta}\right) + \left(\frac{2\alpha_{k,n}\|b-a\|_\infty^m \|f_\alpha\|_{\infty,m}^{\max} N^m}{m!}\right)\right\}, \tag{8.71}$$

(ii)

$$|{}_kL_n(f,x) - f(x)| \leq \gamma_{k,N} \cdot$$

$$\left\{\sum_{j=1}^m\left(\sum_{|\alpha|=j}\left(\frac{|f_\alpha(x)|}{\prod_{i=1}^N \alpha_i!}\right)\left[\frac{1}{n^{\beta j}} + \left(\prod_{i=1}^N (b_i - a_i)^{\alpha_i}\right) \cdot \alpha_{k,n}(\beta)\right]\right) + \frac{N^m}{m! n^{m\beta}} \omega_{1,m}^{\max}\left(f_\alpha, \frac{1}{n^\beta}\right) + \left(\frac{2\alpha_{k,n}(\beta)\|b-a\|_\infty^m \|f_\alpha\|_{\infty,m}^{\max} N^m}{m!}\right)\right\}, \tag{8.72}$$

(iii)

$$\|{}_kL_n(f) - f\|_\infty \leq \gamma_{k,N} \cdot$$

$$\left\{\sum_{j=1}^N\left(\sum_{|\alpha|=j}\left(\frac{\|f_\alpha\|_\infty}{\prod_{i=1}^N \alpha_i!}\right)\left[\frac{1}{n^{\beta j}} + \left(\prod_{i=1}^N (b_i - a_i)^{\alpha_i}\right) \alpha_{k,n}(\beta)\right]\right) + \frac{N^m}{m! n^{m\beta}} \omega_{1,m}^{\max}\left(f_\alpha, \frac{1}{n^\beta}\right) + \left(\frac{2\alpha_{k,n}(\beta)\|b-a\|_\infty^m \|f_\alpha\|_{\infty,m}^{\max} N^m}{m!}\right)\right\}, \tag{8.73}$$

(iv) *assume* $f_\alpha(x_0) = 0$, *for all* $\alpha : |\alpha| = 1, \ldots, m$; $x_0 \in \left(\prod_{i=1}^{N}[a_i, b_i]\right)$. *Then*

$$|{}_kL_n(f, x_0) - f(x_0)| \leq \gamma_{k,N} \cdot \left\{ \frac{N^m}{m! n^{m\beta}} \omega_1^{\max}\left(f_\alpha, \frac{1}{n^\beta}\right) + \left(\frac{2\alpha_{k,n}(\beta) \|b-a\|_\infty^m \|f_\alpha\|_{\infty,m}^{\max} N^m}{m!} \right) \right\}, \tag{8.74}$$

note in the last the extremely high rate of convergence at $n^{-\beta(m+1)}$.

Proof. From Theorems 8.3 and 8.10. □

Next, we apply Theorem 8.13 for m = 1.

Corollary 8.1. *Let* $f \in C^1\left(\prod_{i=1}^{N}[a_i, b_i]\right)$, $0 < \beta < 1$, $n, m, N \in \mathbb{N}$, $x \in \left(\prod_{i=1}^{N}[a_i, b_i]\right)$ *and* $k = 1, 2$. *Then*

(i)

$$\left| {}_kL_n(f, x) - f(x) - \sum_{i=1}^{N} \frac{\partial f}{\partial x_i}\, {}_kL_n((\cdot - x_i), x) \right| \leq \gamma_{k,N} \cdot \left\{ \frac{N}{n^\beta} \max_{i \in \{1,\ldots,N\}} \omega_1\left(\frac{\partial f}{\partial x_i}, \frac{1}{n^\beta}\right) + \left(2\alpha_{k,n} N \|b-a\|_\infty \max_{i \in \{1,\ldots,N\}} \left\| \frac{\partial f}{\partial x_i} \right\|_\infty \right) \right\}, \tag{8.75}$$

(ii)

$$|{}_kL_n(f, x) - f(x)| \leq \gamma_{k,N} \cdot \left\{ \sum_{i=1}^{N} \left| \frac{\partial f(x)}{\partial x_i} \right| \left[\frac{1}{n^\beta} + (b_i - a_i) \cdot \alpha_{k,n}(\beta) \right] + \frac{N}{n^\beta} \max_{i \in \{1,\ldots,N\}} \omega_1\left(\frac{\partial f}{\partial x_i}, \frac{1}{n^\beta}\right) + \left(2\alpha_{k,n}(\beta) N \|b-a\|_\infty \max_{i \in \{1,\ldots,N\}} \left\| \frac{\partial f}{\partial x_i} \right\|_\infty \right) \right\}, \tag{8.76}$$

(iii)

$$\|{}_kL_n(f)-f\|_\infty \le \gamma_{k,N}\cdot$$

$$\left\{\sum_{i=1}^{N}\left\|\frac{\partial f}{\partial x_i}\right\|_\infty\left[\frac{1}{n^\beta}+(b_i-a_i)\,\alpha_{k,n}(\beta)\right]\right.$$

$$\left.+\frac{N}{n^\beta}\max_{i\in\{1,\ldots,N\}}\omega_1\left(\frac{\partial f}{\partial x_i},\frac{1}{n^\beta}\right)+\left(2\alpha_{k,n}(\beta)\,N\,\|b-a\|_\infty\max_{i\in\{1,\ldots,N\}}\left\|\frac{\partial f}{\partial x_i}\right\|_\infty\right)\right\}, \tag{8.77}$$

(iv) *assume* $\frac{\partial f}{\partial x_i}(x_0)=0, i=1,\ldots,N$, *where* $x_0\in\left(\prod_{i=1}^{N}[a_i,b_i]\right)$. *Then*

$$|{}_kL_n(f,x_0)-f(x_0)|\le\gamma_{k,N}\cdot$$

$$\left\{\frac{N}{n^\beta}\max_{i\in\{1,\ldots,N\}}\omega_1\left(\frac{\partial f}{\partial x_i},\frac{1}{n^\beta}\right)+\left(2\alpha_{k,n}(\beta)\,N\,\|b-a\|_\infty\max_{i\in\{1,\ldots,N\}}\left\|\frac{\partial f}{\partial x_i}\right\|_\infty\right)\right\}, \tag{8.78}$$

note in the last the extremely high rate of convergence at $n^{-2\beta}$.

8.3 Multiple Time Separating Random Functions

Let $(\Omega,\mathcal{F},P)$ be a probability space, $\omega\in\Omega; Y_1,Y_2,\ldots,Y_m, m\in\mathbb{N}$, be real-valued random variables on Ω with finite expectations and $h_1(t),h_2(t),\ldots h_m(t):\prod_{j=1}^{N}I_j\to\mathbb{R}$, where I_j are infinite subsets of $\mathbb{R}$ for every $j=1,\ldots,N$. Typically, here I_j is an infinite length interval of $\mathbb{R}$, usually $I_j=\mathbb{R}$ or $I_j=\mathbb{R}_+$, for $j=1,\ldots,N$.
Clearly, then

$$Y(t,\omega):=\sum_{i=1}^{m}h_i(t)Y_i(\omega),\ t\in\prod_{j=1}^{N}I_j, \tag{8.79}$$

is a quite common time separating random function.
We can assume that $h_i\in C^r(\prod_{j=1}^{N}I_j), i=1,2,\ldots,m; r\in\mathbb{N}$. Consequently, we have that the expectation

$$(EY)(t)=\sum_{i=1}^{m}h_i(t)EY_i\in C\left(\prod_{j=1}^{N}I_j\right)\quad\text{or}\quad C^r\left(\prod_{j=1}^{N}I_j\right). \tag{8.80}$$

A classical example of multiple time separating process is

$$\left(\sin\left(\prod_{j=1}^{N}t_j\right)\right)Y_1(\omega)+\left(\cos\left(\prod_{j=1}^{N}t_j\right)\right)Y_2(\omega),\ t_j\in I_j,$$

for $j=1,\ldots,N$.

Note that $\left|\sin\left(\prod_{j=1}^{N} t_j\right)\right| \le 1$ and $\left|\cos\left(\prod_{j=1}^{N} t_j\right)\right| \le 1$.
Another typical example is

$$\left(\sinh\left(\prod_{j=1}^{N} t_j\right)\right) Y_1(\omega) + \left(\cosh\left(\prod_{j=1}^{N} t_j\right)\right) Y_2(\omega),\ t_j \in I_j, \quad \text{for } j = 1,\dots,N. \tag{8.81}$$

In this chapter, we apply the main results of Section 2.3, to $f(t) = (EY)(t)$. We finish with several applications.

8.4 Main Results

We present the following stochastic approximation result.

Theorem 8.14. *Let $(EY)(t)$ as in (8.80), $t \in \prod_{j=1}^{N} I_j$. Here $I_j = [t_{1,j}, t_{2,j}]$, where $t_{1,j}, t_{2,j} \in \mathbb{R}$ with $t_{1,j} < t_{2,j}$, for every $j = 1,\dots,N$. Let also $0 < \beta < 1$, $n, N \in \mathbb{N}$ and $k = 1, 2$. Then*

(i)

$$\left|{}_kL_n\left((EY), t\right) - (EY)(t)\right| \le \gamma_{k,N}$$
$$\left\{\omega_1\left((EY), \frac{1}{n^\beta}\right) + 2\alpha_{k,n}(\beta)\left\|(EY)\right\|_\infty\right\} =: \lambda_1, \tag{8.82}$$

(ii)

$$\left\|{}_kL_n(EY) - (EY)\right\|_\infty \le \lambda_1. \tag{8.83}$$

Proof. From Theorem 8.11. □

It follows the counter part of the previous theorem.

Theorem 8.15. *Let $(EY)(t)$ as in (8.80), $h_i \in C_B\left(\mathbb{R}^{\mathbb{N}}\right)$ for every $i = 1,\dots,m$. Let also $0 < \beta < 1$, $n, N \in \mathbb{N}$ and $k = 1, 2$. Then*

(i)

$$\left|{}_k\overline{L}_n\left((EY), t\right) - (EY)(t)\right| \le \omega_1\left((EY), \frac{1}{n^\beta}\right) + 2\alpha_{k,n}(\beta)\left\|(EY)\right\|_\infty =: \lambda_2, \tag{8.84}$$

(ii)

$$\left\|{}_k\overline{L}_n(EY) - (EY)\right\|_\infty \le \lambda_2. \tag{8.85}$$

Proof. From Theorem 8.12. □

We give

Theorem 8.16. *Let* $(EY)(t)$ *as in (8.80),* $(EY)(t) \in C^m\left(\prod_{j=1}^N I_j\right)$, $t \in \prod_{j=1}^N I_j$. *Here* $I_j = [t_{1,j}, t_{2,j}]$, *where* $t_{1,j}, t_{2,j} \in \mathbb{R}$ *with* $t_{1,j} < t_{2,j}$, *for* $j = 1, \ldots, N, t_1 = (t_{1,1}, \ldots, t_{1,N}), t_2 = (t_{2,1}, \ldots, t_{2,N})$. *Let also* $0 < \beta < 1$, $n, N \in \mathbb{N}$ *and* $k = 1, 2$. *Then*

(i)

$$\left| {}_kL_n((EY),t) - (EY)(t) - \sum_{j^*=1}^{m}\left(\sum_{|\alpha|=j^*}\left(\frac{(EY)_\alpha(t)}{\prod_{i=1}^N \alpha_i!}\right) {}_kL_n\left(\prod_{i=1}^N (\cdot - x_i)^{\alpha_i}, x\right)\right)\right|$$
$$\leq \gamma_{k,N} \cdot \left\{ \frac{N^m}{m! n^{m\beta}} \omega_{1,m}^{\max}\left((EY)_\alpha, \frac{1}{n^\beta}\right) + \left(\frac{2\alpha_{k,n} \|t_2 - t_1\|_\infty^m \|(EY)_\alpha\|_{\infty,m}^{\max} N^m}{m!}\right)\right\}, \tag{8.86}$$

(ii)

$$|{}_kL_n((EY),t) - (EY)(t)| \leq \gamma_{k,N} \cdot$$
$$\left\{\sum_{j^*=1}^{m}\left(\sum_{|\alpha|=j^*}\left(\frac{|(EY)_\alpha(t)|}{\prod_{i=1}^N \alpha_i!}\right)\left[\frac{1}{n^{\beta j^*}} + \left(\prod_{i=1}^N (t_{2,i} - t_{1,i})^{\alpha_i}\right) \cdot \alpha_{k,n}(\beta)\right]\right)\right.$$
$$\left. + \frac{N^m}{m! n^{m\beta}} \omega_{1,m}^{\max}\left((EY)_\alpha, \frac{1}{n^\beta}\right) + \left(\frac{2\alpha_{k,n}(\beta) \|t_2 - t_1\|_\infty^m \|(EY)_\alpha\|_{\infty,m}^{\max} N^m}{m!}\right)\right\}, \tag{8.87}$$

(iii)

$$\|{}_kL_n((EY)) - (EY)\|_\infty \leq \gamma_{k,N} \cdot$$
$$\left\{\sum_{j^*=1}^{N}\left(\sum_{|\alpha|=j^*}\left(\frac{\|(EY)_\alpha\|_\infty}{\prod_{i=1}^N \alpha_i!}\right)\left[\frac{1}{n^{\beta j^*}} + \left(\prod_{i=1}^N (t_{2,i} - t_{1,i})^{\alpha_i}\right) \alpha_{k,n}(\beta)\right]\right)\right.$$
$$\left. + \frac{N^m}{m! n^{m\beta}} \omega_{1,m}^{\max}\left(f_\alpha, \frac{1}{n^\beta}\right) + \left(\frac{2\alpha_{k,n}(\beta) \|t_2 - t_1\|_\infty^m \|f_\alpha\|_{\infty,m}^{\max} N^m}{m!}\right)\right\}, \tag{8.88}$$

(iv) *assume* $(EY)_{\alpha}(t_0) = 0$, *for all* $\alpha : |\alpha| = 1, \ldots, m$; $x_0 \in \left(\prod_{i=1}^{N} [t_{1,i}, t_{2,i}]\right)$. *Then*

$$|{}_kL_n((EY), t_0) - (EY)(t_0)| \leq \gamma_{k,N} \cdot$$
$$\left\{\frac{N^m}{m!n^{m\beta}} \omega_1^{\max}\left((EY)_{\alpha}, \frac{1}{n^{\beta}}\right) + \left(\frac{2\alpha_{k,n}(\beta) \|t_2 - t_1\|_{\infty}^m \|f_{\alpha}\|_{\infty,m}^{\max} N^m}{m!}\right)\right\}, \tag{8.89}$$

notice in the last the extremely high rate of convergence at $n^{-\beta(m+1)}$.

Proof. From Theorem 8.13. □

Next, we apply Theorem 8.16 for m = 1.

Corollary 8.2. *Let* $(EY)(t)$ *as in (8.80),* $(EY)(t) \in C^1\left(\prod_{j=1}^{N} I_j\right)$, $t \in \prod_{j=1}^{N} I_j$. *Here* $I_j = [t_{1,j}, t_{2,j}]$, *where* $t_{1,j}, t_{2,j} \in \mathbb{R}$ *with* $t_{1,j} < t_{2,j}$ *for* $j = 1, \ldots, N, t_1 = (t_{1,1}, \ldots, t_{1,N}), t_2 = (t_{2,1}, \ldots, t_{2,N})$. *Let also* $0 < \beta < 1$, $n, m, N \in \mathbb{N}$, *and* $k = 1, 2$. *Then*

(i)

$$\left|{}_kL_n((EY), t) - (EY)(t) - \sum_{i=1}^{N} \frac{\partial (EY)}{\partial t_i} \, {}_kL_n((\cdot - t_i), t)\right|$$
$$\leq \gamma_{k,N} \cdot \left\{\frac{N}{n^{\beta}} \max_{i \in \{1,\ldots,N\}} \omega_1\left(\frac{\partial (EY)}{\partial t_i}, \frac{1}{n^{\beta}}\right)\right.$$
$$\left. + \left(2\alpha_{k,n} N \|t_2 - t_1\|_{\infty} \max_{i \in \{1,\ldots,N\}} \left\|\frac{\partial (EY)}{\partial t_i}\right\|_{\infty}\right)\right\}, \tag{8.90}$$

(ii)

$$|{}_kL_n((EY), t) - (EY)(t)| \leq \gamma_{k,N} \cdot$$
$$\left\{\sum_{i=1}^{N} \left|\frac{\partial (EY)(t)}{\partial t_i}\right| \left[\frac{1}{n^{\beta}} + (t_{2,i} - t_{1,i}) \cdot \alpha_{k,n}(\beta)\right]\right.$$
$$+ \frac{N}{n^{\beta}} \max_{i \in \{1,\ldots,N\}} \omega_1\left(\frac{\partial (EY)}{\partial t_i}, \frac{1}{n^{\beta}}\right)$$
$$\left. + \left(2\alpha_{k,n}(\beta) N \|t_2 - t_1\|_{\infty} \max_{i \in \{1,\ldots,N\}} \left\|\frac{\partial (EY)}{\partial t_i}\right\|_{\infty}\right)\right\}, \tag{8.91}$$

(iii)

$$\left\|{}_kL_n\left((EY)\right)-(EY)\right\|_\infty \leq \gamma_{k,N}\cdot$$

$$\left\{\sum_{i=1}^{N}\left|\frac{\partial(EY)}{\partial t_i}\right|_\infty\left[\frac{1}{n^\beta}+(t_{2,i}-t_{1,i})\,\alpha_{k,n}(\beta)\right]\right.$$

$$+\frac{N}{n^\beta}\max_{i\in\{1,\ldots,N\}}\omega_1\left(\frac{\partial(EY)}{\partial t_i},\frac{1}{n^\beta}\right)$$

$$\left.+\left(2\alpha_{k,n}(\beta)\,N\left\|t_2-t_1\right\|_\infty\max_{i\in\{1,\ldots,N\}}\left\|\frac{\partial(EY)}{\partial t_i}\right\|_\infty\right)\right\}, \tag{8.92}$$

(iv) *assume* $\frac{\partial(EY)}{\partial t_i}(t_0)=0, i=1,\ldots,N,$ *where* $t_0\in\left(\prod_{i=1}^{N}[t_{1,i},t_{2,i}]\right)$. *Then*

$$\left|{}_kL_n\left((EY),t_0\right)-(EY)(t_0)\right|\leq \gamma_{k,N}\cdot$$

$$\left\{\frac{N}{n^\beta}\max_{i\in\{1,\ldots,N\}}\omega_1\left(\frac{\partial(EY)}{\partial t_i},\frac{1}{n^\beta}\right)\right.$$

$$\left.+\left(2\alpha_{k,n}(\beta)\,N\left\|t_2-t_1\right\|_\infty\max_{i\in\{1,\ldots,N\}}\left\|\frac{\partial(EY)}{\partial t_i}\right\|_\infty\right)\right\}, \tag{8.93}$$

note in the last the extremely high rate of convergence at $n^{-2\beta}$.

8.5 Applications

For the next applications, we consider (Ω, F, P) to be a probability space and Y_0, Y_1, Y_2 to be real-valued random variables on Ω with finite expectations. We consider the stochastic processes $Z_i(t,\omega)$ for $i=1,2,\ldots,7$, where $t=(t_1,\ldots,t_N)\in\mathbb{R}^{\mathbb{N}}$ and $\omega\in\Omega$ as follows:

$$Z_1(t,\omega)=\left[\left(\sum_{j=1}^{N}(t_j-t_{j,0})\right)^{\mu+1}+1\right]Y_0(\omega), \tag{8.94}$$

where $t_0=(t_{1,0},\ldots,t_{N,0})\in\mathbb{R}^{\mathbb{N}}$ and $\mu\in\mathbb{N}$ are fixed;

$$Z_2(t,\omega)=\sin\left(\xi\sum_{j=1}^{N}t_j\right)Y_1(\omega)+\cos\left(\xi\sum_{j=1}^{N}t_j\right)Y_2(\omega), \tag{8.95}$$

where $\xi>0$ is fixed;

$$Z_3(t,\omega)=\sinh\left(\mu\sum_{j=1}^{N}t_j\right)Y_1(\omega)+\cosh\left(\mu\sum_{j=1}^{N}t_j\right)Y_2(\omega), \tag{8.96}$$

where $\mu > 0$ is fixed;

$$Z_4(t,\omega) = sech\left(\mu\sum_{j=1}^{N} t_j\right)Y_1(\omega) + \tanh\left(\mu\sum_{j=1}^{N} t_j\right)Y_2(\omega), \tag{8.97}$$

where $\mu > 0$ is fixed.

Here $sechx := \frac{1}{\cosh\left(\sum_{j=1}^{N} x_j\right)} = \frac{2}{\exp\left(\sum_{j=1}^{N} x_j\right)+\exp\left(-\sum_{j=1}^{N} x_j\right)}$, $x = (x_1,\ldots,x_n) \in \mathbb{R}^{\mathbb{N}}$.

$$Z_5(t,\omega) = \exp\left(-\ell_1\prod_{j=1}^{N} t_j\right)Y_1(\omega) + \exp\left(-\ell_2\prod_{j=1}^{N} t_j\right)Y_2(\omega), \tag{8.98}$$

where $\ell_1, \ell_2 > 0$ are fixed;

$$Z_6(t,\omega) = \frac{1}{1+\exp\left(-\ell_1\prod_{j=1}^{N} t_j\right)}Y_1(\omega) + \frac{1}{1+\exp\left(-\ell_2\prod_{j=1}^{N} t_j\right)}Y_2(\omega), \tag{8.99}$$

where $\ell_1, \ell_2 > 0$ are fixed;

$$Z_7(t,\omega) = e^{-e^{-\mu_1\prod_{j=1}^{N} t_j}}Y_1(\omega) + e^{-e^{-\mu_2\prod_{j=1}^{N} t_j}}Y_2(\omega), \tag{8.100}$$

where $\mu_1, \mu_2 > 0$ are fixed.

The expectations of $Z_i, i = 1, 2, \ldots, 7$ are

$$(EZ_1)(t) = \left[\left(\sum_{j=1}^{N}(t_j - t_{j,0})\right)^{\mu+1} + 1\right]E(Y_0), \tag{8.101}$$

$$(EZ_2)(t) = \sin\left(\xi\sum_{j=1}^{N} t_j\right)E(Y_1) + \cos\left(\xi\sum_{j=1}^{N} t_j\right)E(Y_2), \tag{8.102}$$

$$(EZ_3)(t) = \sinh\left(\mu\sum_{j=1}^{N} t_j\right)E(Y_1) + \cosh\left(\mu\sum_{j=1}^{N} t_j\right)E(Y_2), \tag{8.103}$$

$$(EZ_4)(t) = sech\left(\mu\sum_{j=1}^{N} t_j\right)E(Y_1) + \tanh\left(\mu\sum_{j=1}^{N} t_j\right)E(Y_2), \tag{8.104}$$

$$(EZ_5)(t) = \exp\left(-\ell_1\prod_{j=1}^{N} t_j\right)E(Y_1) + \exp\left(-\ell_2\prod_{j=1}^{N} t_j\right)E(Y_2), \tag{8.105}$$

$$(EZ_6)(t) = \frac{1}{1+\exp\left(-\ell_1 \prod_{j=1}^{N} t_j\right)} E(Y_1) + \frac{1}{1+\exp\left(-\ell_2 \prod_{j=1}^{N} t_j\right)} E(Y_2), \tag{8.106}$$

$$(EZ_7)(t) = e^{-e^{-\mu_1 \prod_{j=1}^{N} t_j}} E(Y_1) + e^{-e^{-\mu_2 \prod_{j=1}^{N} t_j}} E(Y_2). \tag{8.107}$$

For the following, $(EZ_i)(t), i = 1,2,\ldots,7$, are as defined in relations between (8.101) and (8.107), respectively.
We present the following result.

Proposition 8.2. *Let $t \in \prod_{j=1}^{N} I_j$. Here $I_j = [t_{1,j}, t_{2,j}]$, where $t_{1,j}, t_{2,j} \in \mathbb{R}$ with $t_{1,j} < t_{2,j}$, for every $j = 1,\ldots,N$. Let also $0 < \beta < 1$, $n, N \in \mathbb{N}$ and $k = 1,2$. Then, for $i = 1,2,\ldots,7$,*

(i)

$$\left|{}_kL_n\left((EZ_i),t\right) - (EZ_i)(t)\right| \le \gamma_{k,N} \left\{\omega_1\left((EZ_i), \frac{1}{n^\beta}\right) + 2\alpha_{k,n}(\beta)\left\|(EZ_i)\right\|_\infty\right\} =: \lambda_1, \tag{8.108}$$

(ii)

$$\left\|{}_kL_n(EZ_i) - (EZ_i)\right\|_\infty \le \lambda_1. \tag{8.109}$$

Proof. From Theorem 8.14. □

Next we present

Proposition 8.3. *Let $0 < \beta < 1$, $n, N \in \mathbb{N}$ and $k = 1,2$. Then, for $i \in \{2,4,6,7\}$,*

(i)

$$\left|{}_k\overline{L}_n\left((EZ_i),t\right) - (EZ_i)(t)\right| \le \omega_1\left((EZ_i), \frac{1}{n^\beta}\right) + 2\alpha_{k,n}(\beta)\left\|(EZ_i)\right\|_\infty =: \lambda_2, \tag{8.110}$$

(ii)

$$\left\|{}_k\overline{L}_n(EZ_i) - (EZ_i)\right\|_\infty \le \lambda_2. \tag{8.111}$$

Proof. Note that for every $t \in \mathbb{R}$ we have that

- for $Z_2(t,\omega)$, $\left|\sin\left(\xi \sum_{i=1}^{N} t_i\right)\right| \le 1$ and $\left|\cos\left(\xi \sum_{i=1}^{N} t_i\right)\right| \le 1$,
- for $Z_4(t,\omega)$, $\left|sech\left(\mu \sum_{i=1}^{N} t_i\right)\right| \le 1$ and $\left|tanh\left(\mu \sum_{i=1}^{N} t_i\right)\right| \le 1$,

- for $Z_6(t,\omega)$, $0 < \frac{1}{1+\exp\left(-\ell_1 \prod_{i=1}^N t_i\right)} < 1$ and $0 < \frac{1}{1+\exp\left(-\ell_2 \prod_{i=1}^N t_i\right)} < 1$,
- for $Z_7(t,\omega)$, $0 < e^{-e^{-\mu_1 \prod_{i=1}^N t_i}} < 1$ and $0 < e^{-e^{-\mu_2 \prod_{i=1}^N t_i}} < 1$.

Thus, the results come from Theorem 8.16. □

Proposition 8.4. *Let $t \in \prod_{i=j}^N I_j$. Here $I_j = [t_{1,j}, t_{2,j}]$, where $t_{1,j}, t_{2,j} \in \mathbb{R}$ with $t_{1,j} < t_{2,j}$ for $j = 1, \ldots, N, t_1 = (t_{1,1}, \ldots, t_{1,N}), t_2 = (t_{2,1}, \ldots, t_{2,N})$. Let also $0 < \beta < 1$, $n, m, N \in \mathbb{N}$, and $k = 1, 2$. Then, for $i = 1, \ldots, 7$,*

(i)

$$\left| {}_kL_n\left(\left(EZ_i\right), t\right) - \left(EZ_i\right)(t) - \sum_{j=1}^N \frac{\partial\left(EZ_i\right)}{\partial t_j}\, {}_kL_n\left(\left(\cdot - t_j\right), t\right) \right|$$
$$\leq \gamma_{k,N} \cdot \left\{ \frac{N}{n^\beta} \max_{j \in \{1,\ldots,N\}} \omega_1\left(\frac{\partial\left(EZ_i\right)}{\partial t_j}, \frac{1}{n^\beta}\right) + \left(2\alpha_{k,n} N \left\|t_2 - t_1\right\|_\infty \max_{j \in \{1,\ldots,N\}} \left\|\frac{\partial\left(EZ_i\right)}{\partial t_j}\right\|_\infty\right)\right\}, \tag{8.112}$$

(ii) *assume $\frac{\partial(EZ_i)}{\partial t_j}(t_0) = 0, j = 1, \ldots, N$, where $t_0 \in \left(\prod_{j=1}^N [t_{1,j}, t_{2,j}]\right)$. Then*

$$\left| {}_kL_n\left(\left(EZ_i\right), t_0\right) - \left(EZ_i\right)(t_0)\right| \leq \gamma_{k,N} \cdot$$
$$\left\{ \frac{N}{n^\beta} \max_{j \in \{1,\ldots,N\}} \omega_1\left(\frac{\partial\left(EZ_i\right)}{\partial t_j}, \frac{1}{n^\beta}\right) + \left(2\alpha_{k,n}(\beta) N \left\|t_2 - t_1\right\|_\infty \max_{j \in \{1,\ldots,N\}} \left\|\frac{\partial\left(EZ_i\right)}{\partial t_j}\right\|_\infty\right)\right\}, \tag{8.113}$$

note in the last the extremely high rate of convergence at $n^{-2\beta}$.

Proof. From Corollary 8.2 □

8.6 Specific Applications

Let $(\Omega, \mathcal{F}, P)$, where Ω is the set of non-negative integers, be a probability space and $Y_{1,1}, Y_{2,1}$ be real-valued random variables on Ω following Poisson distributions with parameters $\lambda_1, \lambda_2 \in (0, \infty)$, respectively.

We consider the stochastic process $W_1(t,\omega)$, where $t=(t_1,\ldots,t_N)\in\mathbb{R}^{\mathrm{N}}$ and $\omega\in\Omega$ as follows:

$$W_1(t,\omega)=\sin\left(\xi\sum_{j=1}^{N}t_j\right)Y_{1,1}(\omega)+\cos\left(\xi\sum_{j=1}^{N}t_j\right)Y_{2,1}(\omega), \tag{8.114}$$

where $\xi>0$ is fixed.

Since $E\left(Y_{1,1}\right)=\lambda_1$ and $E\left(Y_{2,1}\right)=\lambda_2$, the expectations of $Z_{2,1}$ are

$$\left(EW_1\right)(t)=\sin\left(\xi\sum_{j=1}^{N}t_j\right)\lambda_1+\cos\left(\xi\sum_{j=1}^{N}t_j\right)\lambda_2, \tag{8.115}$$

furthermore, for $j^*=1,\ldots,N$ we have that

$$\frac{\partial\left(EW_1\right)(t)}{\partial t_{j^*}}=\xi\left[\cos\left(\xi\sum_{j=1}^{N}t_j\right)\lambda_1-\sin\left(\xi\sum_{j=1}^{N}t_j\right)\lambda_2\right]=:D_1(t). \tag{8.116}$$

Note that

$$\frac{\partial\left(EW_1\right)(t)}{\partial t_{j^*}}=\frac{\partial\left(EW_1\right)(t)}{\partial t_{i^*}}=D_1(t)$$

for every $j^*,i^*=1,\ldots,N$.

For the following, we consider $(\Omega,\mathcal{F},P)$, where $\Omega=\mathbb{R}$, be a probability space, $Y_{1,2},Y_{2,2}$ on Ω following Gaussian distributions with expectations $\hat{\mu}_1,\hat{\mu}_2\in\mathbb{R}$, respectively. We consider the stochastic process $W_2(t,\omega)$, where $t=(t_1,\ldots,t_N)\in\mathbb{R}^{\mathrm{N}}$ and $\omega\in\Omega$ as follows:

$$W_2(t,\omega)=sech\left(\mu\sum_{j=1}^{N}t_j\right)Y_{1,2}(\omega)+\tanh\left(\mu\sum_{j=1}^{N}t_j\right)Y_{2,2}(\omega), \tag{8.117}$$

where $\mu>0$ is fixed.

Since $E\left(Y_{1,1}\right)=\hat{\mu}_1$ and $E\left(Y_{2,1}\right)=\hat{\mu}_2$, the expectations of $Z_{2,1}$ are

$$\left(EW_2\right)(t)=\hat{\mu}_1 sech\left(\mu\sum_{j=1}^{N}t_j\right)+\hat{\mu}_2\tanh\left(\mu\sum_{j=1}^{N}t_j\right), \tag{8.118}$$

furthermore, for $j^* = 1, \ldots, N$ we have that

$$\begin{aligned}
&\frac{\partial (EW_2)(t)}{\partial t_{j^*}} \\
&= -\mu \left[\hat{\mu}_1 sech \left(\mu \sum_{j=1}^{N} t_j \right) \tanh \left(\mu \sum_{j=1}^{N} t_j \right) \right. \\
&\quad \left. + \hat{\mu}_2 \left(\tanh^2 \left(\mu \sum_{j=1}^{N} t_j \right) - 1 \right) \right]. \qquad (8.119) \\
&=: D_2(t)
\end{aligned}$$

Note that

$$\frac{\partial (EW_2)(t)}{\partial t_{j^*}} = \frac{\partial (EW_2)(t)}{\partial t_{i^*}} = D_2(t)$$

for every $j^*, i^* = 1, \ldots, N$.

Next, we consider $(\Omega, \mathcal{F}, P)$, where $\Omega = [0, \infty)$, be a probability space, $Y_{1,3}, Y_{2,3}$ be real-valued random variables on Ω following Weibull distributions with scale parameters 1 and shape parameters $\gamma_1, \gamma_2 \in (0, \infty)$, respectively.

We consider the stochastic process $W_3(t, \omega)$, where $t = (t_1, \ldots, t_N) \in \mathbb{R}^N$ and $\omega \in \Omega$ as follows:

$$W_3(t, \omega) = \frac{1}{1 + \exp\left(-\ell_1 \prod_{j=1}^{N} t_j\right)} Y_{1,3}(\omega) + \frac{1}{1 + \exp\left(-\ell_2 \prod_{j=1}^{N} t_j\right)} Y_{2,3}(\omega), \qquad (8.120)$$

where $\ell_1, \ell_2 > 0$ are fixed.

Since $E(Y_{1,3}) = \Gamma\left(1 + \frac{1}{\gamma_1}\right)$ and $E(Y_{2,3}) = \Gamma\left(1 + \frac{1}{\gamma_2}\right)$, where $\Gamma(\cdot)$ is the Gamma function, the expectations of $Z_{i,3}, i = 1, 2, 3, 5$, are

$$\begin{aligned}
(EW_3)(t) &= \Gamma\left(1 + \frac{1}{\gamma_1}\right) \frac{1}{1 + \exp\left(-\ell_1 \prod_{j=1}^{N} t_j\right)} \\
&\quad + \Gamma\left(1 + \frac{1}{\gamma_2}\right) \frac{1}{1 + \exp\left(-\ell_2 \prod_{j=1}^{N} t_j\right)}, \qquad (8.121)
\end{aligned}$$

furthermore, for $j = 1, \ldots, N$ we have that

$$r\frac{\partial (EW_3)(t)}{\partial t_j} = \left(\prod_{r=1, r \neq j}^{N} t_r \right) \left[\frac{\ell_1 \Gamma\left(1 + \frac{1}{\gamma_1}\right) \exp\left(-\ell_1 \prod_{r=1}^{N} t_r\right)}{\left(1 + \exp\left(-\ell_1 \prod_{r=1}^{N} t_r\right)\right)^2} + \frac{\ell_2 \Gamma\left(1 + \frac{1}{\gamma_2}\right) \exp\left(-\ell_2 \prod_{r=1}^{N} t_r\right)}{\left(1 + \exp\left(-\ell_2 \prod_{r=1}^{N} t_r\right)\right)^2} \right]. \tag{8.122}$$

Note that

$$\frac{\partial (EW_3)(t)}{\partial t_j} = \left(\prod_{i=1, i \neq j}^{N} t_i \right) D_3(t)$$

for every $j = 1, \ldots, N$, where

$$D_3(t) =: \frac{\ell_1 \Gamma\left(1 + \frac{1}{\gamma_1}\right) \exp\left(-\ell_1 \prod_{j=1}^{N} t_j\right)}{\left(1 + \exp\left(-\ell_1 \prod_{j=1}^{N} t_j\right)\right)^2} + \frac{\ell_2 \Gamma\left(1 + \frac{1}{\gamma_2}\right) \exp\left(-\ell_2 \prod_{j=1}^{N} t_j\right)}{\left(1 + \exp\left(-\ell_2 \prod_{j=1}^{N} t_j\right)\right)^2}. \tag{8.123}$$

We present the following result.

Proposition 8.5. *Let $t \in \prod_{j=1}^{N} I_j$. Here $I_j = [t_{1,j}, t_{2,j}]$, where $t_{1,j}, t_{2,j} \in \mathbb{R}$ with $t_{1,j} < t_{2,j}$, for every $j = 1, \ldots, N$. Let also $0 < \beta < 1$, $n, N \in \mathbb{N}$ and $k = 1, 2$. Then, for $i = 1, 2, 3$,*

(i)

$$|_k L_n ((EW_i), t) - (EW_i)(t)| \leq \gamma_{k,N} \left\{ \omega_1 \left((EW_i), \frac{1}{n^\beta} \right) + 2\alpha_{k,n}(\beta) \|(EW_i)\|_\infty \right\} =: \lambda_1, \tag{8.124}$$

(ii)

$$\|_k L_n (EW_i) - (EW_i)\|_\infty \leq \lambda_1. \tag{8.125}$$

Proof. From Proposition 8.2. □

Next we present

Proposition 8.6. *Let* $0 < \beta < 1$, $n, N \in \mathbb{N}$ *and* $k = 1, 2$. *Then, for* $i = 1, 2, 3,$

(i)

$$\left|{}_k\overline{L}_n\left((EW_i), t\right) - (EW_i)(t)\right| \\ \leq \omega_1\left((EW_i), \frac{1}{n^\beta}\right) + 2\alpha_{k,n}(\beta)\left\|(EW_i)\right\|_\infty =: \lambda_2, \quad (8.126)$$

(ii)

$$\left\|{}_k\overline{L}_n(EW_i) - (EW_i)\right\|_\infty \leq \lambda_2. \quad (8.127)$$

Proof. From Proposition 8.3. □

Proposition 8.7. *Let* $t \in \prod_{i=j}^N I_j$. *Here* $I_j = [t_{1,j}, t_{2,j}]$, *where* $t_{1,j}, t_{2,j} \in \mathbb{R}$ *with* $t_{1,j} < t_{2,j}$ *for* $j = 1, \ldots, N, t_1 = (t_{1,1}, \ldots, t_{1,N}), t_2 = (t_{2,1}, \ldots, t_{2,N})$. *Let also* $0 < \beta < 1$, $n, m, N \in \mathbb{N}$, *and* $k = 1, 2$. *Then, for* $i = 1, 2$,

(i)

$$\left|{}_kL_n\left((EW_i), t\right) - (EW_i)(t) - D_i(t)\sum_{j=1}^N {}_kL_n\left((\cdot - t_j), t\right)\right| \\ \leq \gamma_{k,N} \cdot \left\{\frac{N}{n^\beta}\omega_1\left(D_i(t), \frac{1}{n^\beta}\right) + \left(2\alpha_{k,n}N\left\|t_2 - t_1\right\|_\infty\left\|D_i(t)\right\|_\infty\right)\right\}, \quad (8.128)$$

(ii) *assume* $D_i(t_0) = 0$, *where* $t_0 \in \left(\prod_{j=1}^N [t_{1,j}, t_{2,j}]\right)$. *Then*

$$\left|{}_kL_n\left((EW_i), t_0\right) - (EW_i)(t_0)\right| \leq \gamma_{k,N} \cdot \\ \left\{\frac{N}{n^\beta}\omega_1\left(D_i(t), \frac{1}{n^\beta}\right) + \left(2\alpha_{k,n}(\beta)N\left\|t_2 - t_1\right\|_\infty\left\|D_i(t)\right\|_\infty\right)\right\}, \quad (8.129)$$

note in the last the extremely high rate of convergence at $n^{-2\beta}$.

Proof. From Proposition 8.4. □

Proposition 8.8. *Let* $t \in \prod_{i=j}^N I_j$. *Here* $I_j = [t_{1,j}, t_{2,j}]$, *where* $t_{1,j}, t_{2,j} \in \mathbb{R}$ *with* $t_{1,j} < t_{2,j}$ *for* $j = 1, \ldots, N, t_1 = (t_{1,1}, \ldots, t_{1,N}), t_2 = (t_{2,1}, \ldots, t_{2,N})$. *Let also* $0 < \beta < 1$, $n, m, N \in \mathbb{N}$, *and* $k = 1, 2$. *Then*

(i)

$$\left| {}_kL_n\left((EW_3),t\right) - (EW_3)(t) - D_3(t)\sum_{j=1}^{N}\left(\prod_{i=1,i\neq j}^{N} t_i\right) {}_kL_n\left((\cdot - t_j),t\right)\right|$$

$$\leq \gamma_{k,N}\cdot\left\{\frac{N}{n^{\beta}}\max_{j\in\{1,\ldots,N\}}\omega_1\left(D_3(t)\prod_{i=1,i\neq j}^{N} t_i,\frac{1}{n^{\beta}}\right)\right.$$

$$\left.+\left(2\alpha_{k,n}N\left\|t_2-t_1\right\|_{\infty}\max_{j\in\{1,\ldots,N\}}\left\|D_3(t)\prod_{i=1,i\neq j}^{N} t_i\right\|_{\infty}\right)\right\}, \tag{8.130}$$

(ii) *assume that there exist* $t_0 = (t_{0,1}, t_{0,2}, \cdots, t_{0,N})$, *where* $t_0 \in \left(\prod_{j=1}^{N}\left[t_{1,j}, t_{2,j}\right]\right)$, *such that* $t_{0,k} = t_{0,l} = 0$, *for* $k,l \in \{1,\ldots,N\}$, *with* $k \neq l$. *Then*

$$\left|{}_kL_n\left((EW_3),t_0\right) - (EW_3)(t_0)\right| \leq \gamma_{k,N}\cdot$$

$$\left\{\frac{N}{n^{\beta}}\max_{j\in\{1,\ldots,N\}}\omega_1\left(D_3(t)\prod_{i=1,i\neq j}^{N} t_j,\frac{1}{n^{\beta}}\right)\right.$$

$$\left.+\left(2\alpha_{k,n}(\beta)N\left\|t_2-t_1\right\|_{\infty}\max_{j\in\{1,\ldots,N\}}\left\|\frac{\partial(EW_3)}{\partial t_j}\right\|_{\infty}\right)\right\}, \tag{8.131}$$

note in the last the extremely high rate of convergence at $n^{-2\beta}$.

Proof. From Proposition 8.4. □

Bibliography

G.A. Anastassiou, Rate of convergence of some neural network operators to the unit-univariate case, *Journal of Mathematical Analysis and Applications*, 212 (1997), 237–262.

G.A. Anastassiou, *Quantitative Approximations*, Chapman & Hall/CRC, Boca Raton, New York, 2001.

G.A. Anastassiou, *Univariate hyperbolic tangent neural network approximation*, submitted for publication, 2010. pp. 92–93, 2006.

G.A. Anastassiou, Multivariate hyperbolic tangent neural network approximation, *Computers and Mathematics*, 61 (2011a), 809–821.

G.A. Anastassiou, Multivariate sigmoidal neural network approximation, *Neural Networks*, 24 (2011b), 378–386.

G.A. Anastassiou, *Inteligent Systems: Approximation by Artificial Neural Networks*, Intelligent Systems Reference Library, Vol. 19, Springer, Heidelberg, 2011c.

G.A. Anastassiou, Univariate sigmoidal neural network approximation, *Journal of Computational Analysis and Applications*, 14(4) (2012) 659–690.

G.A. Anastassiou, *Intelligent Systems II: Complete Approximation by Neural Network Operators*, Springer, Heidelberg, New York, 2016.

G. Anastassiou and D. Kouloumpou, Neural networks approximation for time splitting random functions, *Mathematics*, 11 (2023), 2183, https://doi.org/10.3390/math11092183.

F. Brauer and C. Castillo-Chavez, *Mathematical Models in Population Biology and Epidemiology*, Springer-Verlag, New York, pp. 8–9, 2001.

Z. Chen and F. Cao, The approximation operators with sigmoidal functions, *Computers and Mathematics with Applications*, 58 (2009), 758–765.

D. Costarelli, R. Spigler, Approximation results for neural network operators activated by sigmoidal functions, *Neural Networks,* 44 (2013a), 101–106.

D. Costarelli, R. Spigler, Multivariate neural network operators with sigmoidal activation functions, *Neural Networks*, 48 (2013b), 72–77.

S. Haykin, *Neural Networks: A Comprehensive Foundation* (2 ed.), Prentice Hall, New York, 1998.

N. Hritonenko and Y. Yatsenko, *Mathematical Modeling in Economics, Ecology and the Environment*, Reprint, Science Press, Beijing, 2000.

M. Kac and A.J.F. Siegert, An explicit representation of a stationary Gaussian process, *The Annals of Mathematical Statistics*, 18(3) (1947), 438–442.

Y. Kozachenko *et al.*, *Simulation of Stochastic Processes with Given Accuracy and Reliability*, Elsevier (2016), pp. 71–104.

W. McCulloch and W. Pitts, A logical calculus of the ideas immanent in nervous activity, *Bulletin of Mathematical Biophysis*, 7 (1943), 115–133.

T.M. Mitchell, *Machine Learning*, WCB-McGraw-Hill, New York, 1997.

Chapter 9

Brownian Motion Approximation by Perturbed Neural Networks

The first author recently derived several approximation results by neural network operators, see the new monograph Anastassiou (2023d). There, the approximation methods are derived from the parametrized and deformed neural networks induced by the q-deformed and λ-parametrized logistic and hyperbolic tangent activation functions. The results we apply here are univariate on a compact interval, regular and fractional. The outcome is the quantitative approximation of Brownian motion over the three-dimensional sphere. We derive several Jackson-type inequalities estimating the degree of convergence of our neural network operators to a general expectation function of Brownian motion. We give a detailed list of approximation applications regarding the expectation of well-known functions of Brownian motion. Smoothness of our functions is taken into account producing higher speeds of approximation. It follows Anastassiou and Kouloumpou (2024).

9.1 Introduction

The first author in Anastassiou (1997, 2001), see Chapters 2–5 , was the first to establish neural network approximation to continuous functions with rates by very specifically defined neural network operators of Cardaliaguet–Euvrard and "Squashing" types, by employing the modulus of continuity of the engaged function or its high order derivative, and producing very tight Jackson-type inequalities. He treats there both the univariate and multivariate cases. The defining these operators "bell-shaped" and "squashing" functions are assumed to be compact support.

Also the first author inspired by Chen and Cao (2009) continued his studies on neural network approximation by introducing and using the proper quasi-interpolation operators of sigmoidal and hyperbolic tangent types which resulted into Anastassiou (2011a), Anastassiou (2011b), Anastassiou (2011c), Anastassiou (2011d) and Anastassiou (2012b), by treating both the univariate and multivariate cases. He did also the corresponding fractional cases Anastassiou (2012a), Anastassiou (2016) and Anastassiou (2018a).

In Anastassiou (2023b), Anastassiou (2023d), the first author continued similar studies for Banach space-valued functions for activation functions deriving from the q-deformed and λ-parametrized logistic and hyperbolic tangent sigmoid functions. The authors based and inspired by Kouloumpou and Papanicolaou (2013) perform here neural network quantitative approximations to Brownian motion over the three-dimensional sphere.

They present a series of Jackson-type inequalities estimating the error of approximation to a general expectation function of the Brownian motion and its derivative. They produce regular and fractional calculus results. They finish with a lot of important applications.

9.2 About Parametrized and Deformed Neural Networks

9.2.1 *About the q-deformed and λ-parametrized logistic activation function*

Here we follow Anastassiou (2023f).

We consider here the q-deformed and λ-parametrized function acting as an activation function

$$\varphi_{q,\lambda}(x) = \frac{1}{1+qA^{-\lambda x}}, \quad x \in \mathbb{R}, \quad \text{where } q,\lambda > 0,\ A > 1. \tag{9.1}$$

This is an A-generalized logistic-type function.

We easily observe that

$$\lim_{x \longrightarrow +\infty} \varphi_{q,\lambda}(x) = 1, \quad \lim_{x \longrightarrow -\infty} \varphi_{q,\lambda}(x) = 0. \tag{9.2}$$

Furthermore we have

$$\varphi_{q,\lambda}(x) = 1 - \varphi_{\frac{1}{q},\lambda}(-x). \tag{9.3}$$

We also have that

$$\varphi_{q,\lambda}(0) = \frac{1}{1+q} \quad \text{and} \quad \varphi_{\frac{1}{q},\lambda}(0) = \frac{q}{1+q}. \tag{9.4}$$

We have

$$\varphi'_{q,\lambda}(x) = q\lambda(\ln A)\left(1+qA^{-\lambda x}\right)^{-2} A^{-\lambda x} > 0. \tag{9.5}$$

So $\varphi_{q,\lambda}$ is a strictly increasing function over $\mathbb{R}$. Furthermore it holds

$$\varphi''_{q,\lambda}(x) = q\lambda^2(\ln A)^2\left(A^{\lambda x}+q^2A^{-\lambda x}+2q\right)^{-2}\left(q^2A^{-\lambda x}-A^{\lambda x}\right) \in C(\mathbb{R}), \tag{9.6}$$

$\varphi''_{q,\lambda}(x) > 0$, for $x < \frac{\log_A q}{\lambda}$ and there $\varphi_{q,\lambda}$ is concave up.

When $x > \frac{\log_A q}{\lambda}$, we have $\varphi''_{q,\lambda}(x) < 0$ and $\varphi_{q,\lambda}$ is concave down.

Of course,

$$\varphi''_{q,\lambda}\left(\frac{\log_A q}{\lambda}\right) = 0.$$

So, $\varphi_{q,\lambda}$ is a sigmoid function, see Anastassiou (2023c).

Consider the function

$$G_{q,\lambda}(x) := \frac{1}{2}\left(\varphi_{q,\lambda}(x+1) - \varphi_{q,\lambda}(x-1)\right), \quad x \in \mathbb{R}. \tag{9.7}$$

Then,

$$G_{q,\lambda}(-x) = G_{\frac{1}{q},\lambda}(x), \quad \forall\, x \in \mathbb{R}. \tag{9.8}$$

We have that

$$G'_{q,\lambda}(x) = \frac{1}{2}\left(\varphi'_{q,\lambda}(x+1) - \varphi'_{q,\lambda}(x-1)\right), \tag{9.9}$$

$G'_{q,\lambda} > 0$, i.e. $G_{q,\lambda}$ is strictly increasing over $\left(-\infty, \frac{\log_A q}{\lambda} - 1\right)$. $G_{q,\lambda}$ is strictly decreasing over $\left(\frac{\log_A q}{\lambda}, +\infty\right)$. $G_{q,\lambda}$ is strictly concave down over $\left(\frac{\log_A q}{\lambda} - 1, \frac{\log_A q}{\lambda} + 1\right)$.

Overall, $G_{q,\lambda}$ is a bell-shaped function over $\mathbb{R}$. $\frac{\log_A q}{\lambda}$ is the only critical number of $G_{q,\lambda}$ over $\mathbb{R}$. Therefore, $G_{q,\lambda}\left(\frac{\log_A q}{\lambda}\right)$ is the maximum of $G_{q,\lambda}$.

The global maximum of $G_{q,\lambda}$ is

$$G_{q,\lambda}\left(\frac{\log_A q}{\lambda}\right) = \frac{A^\lambda - 1}{2\left(A^\lambda + 1\right)}. \tag{9.10}$$

Finally, we have that

$$\lim_{x\to+\infty} G_{q,\lambda}(x) = \frac{1}{2}\left(\varphi_{q,\lambda}(+\infty) - \varphi_{q,\lambda}(+\infty)\right) = 0 \tag{9.11}$$

and

$$\lim_{x\to-\infty} G_{q,\lambda}(x) = \frac{1}{2}\left(\varphi_{q,\lambda}(-\infty) - \varphi_{q,\lambda}(-\infty)\right) = 0. \tag{9.12}$$

Consequently, the x-axis is the horizontal asymptote of $G_{q,\lambda}$. Of course, $G_{q,\lambda}(x) > 0$, $\forall\, x \in \mathbb{R}$. We need

Theorem 9.1 (Anastassiou 2023g). *It holds*

$$\sum_{i=-\infty}^{\infty} G_{q,\lambda}(x-i) = 1, \quad \forall\, x \in \mathbb{R}, \quad \forall\, q, \lambda > 0,\ A > 1. \tag{9.13}$$

Furthermore,

Theorem 9.2 (Anastassiou 2023g). *It holds*

$$\int_{-\infty}^{\infty} G_{q,\lambda}(x)\, dx = 1, \quad \lambda, q > 0,\ A > 1. \tag{9.14}$$

So $G_{q,\lambda}$ is a density function on $\mathbb{R}$; $\lambda, q > 0$, $A > 1$.

We need the following result.

Theorem 9.3 (Anastassiou 2023g). *Let* $0 < \alpha < 1$, *and* $n \in \mathbb{N}$ *with* $n^{1-\alpha} > 2$. *Then*

$$\sum_{\substack{k=-\infty \\ \{: |nx-k| \geq n^{1-\alpha}}}^{\infty} G_{q,\lambda}(nx-k) < 2\max\left\{q, \frac{1}{q}\right\} \frac{1}{A^{\lambda\left(n^{1-\alpha}-2\right)}} = \gamma A^{-\lambda\left(n^{1-\alpha}-2\right)}, \tag{9.15}$$

where $q, \lambda > 0$, $A > 1$; $\gamma := 2\max\left\{q, \frac{1}{q}\right\}$.

Let $\lceil \cdot \rceil$ be the ceiling of the number and $\lfloor \cdot \rfloor$ be the integral part of the number.

We also need

Theorem 9.4 (Anastassiou 2023g). *Let* $x \in [a,b] \subset \mathbb{R}$ *and* $n \in \mathbb{N}$ *so that* $\lceil na \rceil \leq \lfloor nb \rfloor$. *For* $q > 0$, $\lambda > 0$, $A > 1$, *we consider the number* $\lambda_q > z_0 > 0$ *with* $G_{q,\lambda}(z_0) = G_{q,\lambda}(0)$ *and* $\lambda_q > 1$. *Then*

$$\frac{1}{\sum_{k=\lceil na \rceil}^{\lfloor nb \rfloor} G_{q,\lambda}(nx-k)} < max\left\{\frac{1}{G_{q,\lambda}(\lambda_q)}, \frac{1}{G_{\frac{1}{q},\lambda}\left(\lambda_{\frac{1}{q}}\right)}\right\} =: K(q). \tag{9.16}$$

We finally mention

Remark 9.1. (i) We have that

$$\lim_{n\to+\infty}\sum_{k=\lceil na\rceil}^{\lfloor nb\rfloor} G_{q,\lambda}\left(nx-k\right)\neq 1,\quad \text{for at least some } x\in[a,b], \tag{9.17}$$

where $\lambda, q > 0$.

(ii) Let $[a,b]\subset\mathbb{R}$. For large n, we always have $\lceil na\rceil \leq \lfloor nb\rfloor$. Also $a\leq\frac{k}{n}\leq b$, iff $\lceil na\rceil\leq k\leq\lfloor nb\rfloor$. In general, it holds

$$\sum_{k=\lceil na\rceil}^{\lfloor nb\rfloor} G_{q,\lambda}\left(nx-k\right)\leq 1. \tag{9.18}$$

Definition 9.1. Let $f\in C([a,b])$ and $n\in\mathbb{N}:\lceil na\rceil\leq\lfloor nb\rfloor$. We introduce and define the X-valued linear neural network operators

$$L_n\left(f,x\right):=\frac{\sum_{k=\lceil na\rceil}^{\lfloor nb\rfloor} f\left(\frac{k}{n}\right)G_{q,\lambda}\left(nx-k\right)}{\sum_{k=\lceil na\rceil}^{\lfloor nb\rfloor} G_{q,\lambda}\left(nx-k\right)},\quad x\in[a,b]. \tag{9.19}$$

Clearly here $L_n(f,x)\in C([a,b])$. We study here the pointwise and uniform convergence of $L_n(f,x)$ to $f(x)$ with rates.
For convenience, also we call

$$L_n^*\left(f,x\right):=\sum_{k=\lceil na\rceil}^{\lfloor nb\rfloor} f\left(\frac{k}{n}\right)G_{q,\lambda}\left(nx-k\right), \tag{9.20}$$

$$L_n\left(f,x\right):=\frac{L_n^*\left(f,x\right)}{\sum_{k=\lceil na\rceil}^{\lfloor nb\rfloor} G_{q,\lambda}\left(nx-k\right)}. \tag{9.21}$$

So

$$L_n\left(f,x\right)-f\left(x\right)=\frac{L_n^*\left(f,x\right)}{\sum_{k=\lceil na\rceil}^{\lfloor nb\rfloor} G_{q,\lambda}\left(nx-k\right)}-f\left(x\right)$$

$$=\frac{L_n^*\left(f,x\right)-f\left(x\right)\left(\sum_{k=\lceil na\rceil}^{\lfloor nb\rfloor} G_{q,\lambda}\left(nx-k\right)\right)}{\sum_{k=\lceil na\rceil}^{\lfloor nb\rfloor} G_{q,\lambda}\left(nx-k\right)}. \tag{9.22}$$

Consequently, we derive that

$$|L_n(f,x) - f(x)| \le K(q) \left| L_n^*(f,x) - f(x) \left(\sum_{k=\lceil na \rceil}^{\lfloor nb \rfloor} G_{q,\lambda}(nx-k) \right) \right|$$

$$= K(q) \left| \sum_{k=\lceil na \rceil}^{\lfloor nb \rfloor} \left(f\left(\frac{k}{n}\right) - f(x) \right) G_{q,\lambda}(nx-k) \right|. \tag{9.23}$$

We estimate the right-hand side of the last quantity.

For that we need, for $f \in C([a,b])$, the first modulus of continuity

$$\omega_1(f,\delta) := \sup_{\substack{x,y \in [a,b] \\ |x-y| \le \delta}} |f(x) - f(y)|, \quad \delta > 0. \tag{9.24}$$

Similarly, it is defined ω_1 for $f \in C_{uB}(\mathbb{R})$ (uniformly continuous and bounded functions from $\mathbb{R}$ into $\mathbb{R}$), for $f \in C_B(\mathbb{R})$ (continuous and bounded real valued) and for $f \in C_u(\mathbb{R})$ (uniformly continuous).

The fact $f \in C([a,b])$ or $f \in C_u(\mathbb{R})$ is equivalent to $\lim_{\delta \to 0} \omega_1(f,\delta) = 0$, see Anastassiou (2017c).

We present a set of real-valued neural network approximations to a function given with rates.

Theorem 9.5 (Anastassiou 2023f). *Let $f \in C([a,b])$, $0 < \alpha < 1$, $n \in \mathbb{N} : n^{1-\alpha} > 2$, $x \in [a,b]$. Then*

(i)

$$|L_n(f,x) - f(x)| \le K(q) \left[\omega_1\left(f, \frac{1}{n^\alpha}\right) + 2\|f\|_\infty \gamma A^{-\lambda\left(n^{1-\alpha}-2\right)} \right] =: \rho \tag{9.25}$$

and

(ii)

$$\|L_n(f) - f\|_\infty \le \rho. \tag{9.26}$$

We get that $\lim_{n\to\infty} L_n(f) = f$, pointwise and uniformly.

We need

Definition 9.2 (Anastassiou 2017a, Diethelm 2004). Let $[a,b] \subset \mathbb{R}$, $\alpha > 0$; $m = \lceil \alpha \rceil \in \mathbb{N}$ ($\lceil \cdot \rceil$ is the ceiling of the number), $f : [a,b] \to \mathbb{R}$. We assume that $f^{(m)} \in L_1([a,b])$. We call the Caputo left fractional derivative of order α:

$$\left(D_{*a}^{\alpha}f\right)(x) := \frac{1}{\Gamma(m-\alpha)}\int_a^x (x-t)^{m-\alpha-1} f^{(m)}(t)\,dt, \quad \forall\, x \in [a,b]. \tag{9.27}$$

If $\alpha \in \mathbb{N}$, we set $D_{*a}^{\alpha}f := f^{(m)}$ the ordinary real-valued derivative and also set $D_{*a}^{0}f := f$.

By Anastassiou (2017a), $\left(D_{*a}^{\alpha}f\right)(x)$ exists almost everywhere in $x \in [a,b]$ and $D_{*a}^{\alpha}f \in L_1([a,b])$.

If $\left\|f^{(m)}\right\|_{L_\infty([a,b])} < \infty$, then by Anastassiou (2017a), $D_{*a}^{\alpha}f \in C([a,b])$, hence $\left\|D_{*a}^{\alpha}f\right\| \in C([a,b])$.

We mention

Definition 9.3 (Anastassiou 2017b). Let $[a,b] \subset \mathbb{R}$, $\alpha > 0$, $m := \lceil \alpha \rceil$. We assume that $f^{(m)} \in L_1([a,b])$, where $f : [a,b] \to \mathbb{R}$. We call the Caputo right fractional derivative of order α:

$$\left(D_{b-}^{\alpha}f\right)(x) := \frac{(-1)^m}{\Gamma(m-\alpha)}\int_x^b (z-x)^{m-\alpha-1} f^{(m)}(z)\,dz, \quad \forall\, x \in [a,b]. \tag{9.28}$$

We observe that $\left(D_{b-}^{m}f\right)(x) = (-1)^m f^{(m)}(x)$, for $m \in \mathbb{N}$, and $\left(D_{b-}^{0}f\right)(x) = f(x)$.

By Anastassiou (2017b), $\left(D_{b-}^{\alpha}f\right)(x)$ exists almost everywhere on $[a,b]$ and $\left(D_{b-}^{\alpha}f\right) \in L_1([a,b])$.

If $\left\|f^{(m)}\right\|_{L_\infty([a,b],X)} < \infty$, and $\alpha \notin \mathbb{N}$, by Anastassiou (2017b), $D_{b-}^{\alpha}f \in C([a,b])$, hence $\left\|D_{b-}^{\alpha}f\right\| \in C([a,b])$. Next we present

Theorem 9.6 (Anastassiou 2023f). *Let* $0 < \alpha, \beta < 1$, $f \in C^1([a,b])$, $x \in [a,b]$, $n \in \mathbb{N} : n^{1-\beta} > 2$. *Then*

(i)

$$\begin{aligned} &\left|L_n(f,x) - f(x)\right| \\ &\le K(q)\frac{1}{\Gamma(\alpha+1)}\left\{\frac{\left(\omega_1\left(D_{x-}^{\alpha}f, \frac{1}{n^{\beta}}\right)_{[a,x]} + \omega_1\left(D_{*x}^{\alpha}f, \frac{1}{n^{\beta}}\right)_{[x,b]}\right)}{n^{\alpha\beta}}\right. \\ &\quad \left. + \gamma A^{-\lambda\left(n^{1-\beta}-2\right)}\left(\left\|D_{x-}^{\alpha}f\right\|_{\infty,[a,x]}(x-a)^{\alpha} + \left\|D_{*x}^{\alpha}f\right\|_{\infty,[x,b]}(b-x)^{\alpha}\right)\right\} \end{aligned} \tag{9.29}$$

and

(ii)

$$\|L_n f - f\|_\infty \leq K(q) \frac{1}{\Gamma(\alpha+1)} \cdot$$

$$\left\{ \frac{\left(\sup_{x\in[a,b]} \omega_1 \left(D_{x-}^{\alpha} f, \frac{1}{n^\beta} \right)_{[a,x]} + \sup_{x\in[a,b]} \omega_1 \left(D_{*x}^{\alpha} f, \frac{1}{n^\beta} \right)_{[x,b]} \right)}{n^{\alpha\beta}} \right.$$

$$\left. + (b-a)^\alpha \gamma A^{-\lambda\left(n^{1-\beta}-2\right)} \left(\sup_{x\in[a,b]} \|D_{x-}^{\alpha} f\|_{\infty,[a,x]} + \sup_{x\in[a,b]} \|D_{*x}^{\alpha} f\|_{\infty,[x,b]} \right) \right\}. \tag{9.30}$$

9.2.2 *About q-deformed and λ-parametrized hyperbolic tangent activation function $g_{q,\lambda}$*

Here all this background comes from Anastassiou (2023a). We use $g_{q,\lambda}$, see (9.29), and exhibit that it is a sigmoid activation function and we present several of its properties related to the approximation by neural network operators. So, let us consider the function

$$g_{q,\lambda}(x) := \frac{e^{\lambda x} - q e^{-\lambda x}}{e^{\lambda x} + q e^{-\lambda x}}, \quad \lambda, q > 0, \ x \in \mathbb{R}. \tag{9.31}$$

We have that

$$g_{q,\lambda}(0) = \frac{1-q}{1+q}. \tag{9.32}$$

We also have

$$g_{q,\lambda}(-x) = -g_{\frac{1}{q},\lambda}(x), \quad \forall\, x \in \mathbb{R}, \tag{9.33}$$

Furthermore,

$$\lim_{x\to-\infty} g_{q,\lambda}(x) = -1 \quad \text{and} \quad \lim_{x\to+\infty} g_{q,\lambda}(x) = +1. \tag{9.34}$$

We find that

$$g'_{q,\lambda}(x) = \frac{4q\lambda e^{2\lambda x}}{\left(e^{2\lambda x} + q\right)^2} > 0, \tag{9.35}$$

therefore $g_{q,\lambda}$ is strictly increasing.

Next we obtain ($x \in \mathbb{R}$)

$$g''_{q,\lambda}(x) = 8q\lambda^2 e^{2\lambda x}\left(\frac{q - e^{2\lambda x}}{\left(e^{2\lambda x} + q\right)^3}\right) \in C(\mathbb{R}). \tag{9.36}$$

So, in case of $x < \frac{\ln q}{2\lambda}$, we have that $g_{q,\lambda}$ is strictly concave up, with $g''_{q,\lambda}\left(\frac{\ln q}{2\lambda}\right) = 0$. And in case of $x > \frac{\ln q}{2\lambda}$, we have that $g_{q,\lambda}$ is strictly concave down.

Clearly, $g_{q,\lambda}$ is a shifted sigmoid function with $g_{q,\lambda}(0) = \frac{1-q}{1+q}$, and $g_{q,\lambda}(-x) = -g_{q^{-1},\lambda}(x)$ (a semi-odd function), see also Anastassiou (2023c).

By $1 > -1$, $x + 1 > x - 1$, we consider the activation function

$$M_{q.\lambda}(x) := \frac{1}{4}\left(g_{q,\lambda}(x+1) - g_{q,\lambda}(x-1)\right) > 0, \tag{9.37}$$

$\forall x \in \mathbb{R}$; $q, \lambda > 0$. Note that $M_{q,\lambda}(\pm\infty) = 0$, so the x-axis is horizontal asymptote. We have that

$$M_{q,\lambda}(-x) = M_{\frac{1}{q},\lambda}(x), \quad \forall\, x \in \mathbb{R};\ q, \lambda > 0, \tag{9.38}$$

a deformed symmetry.

Furthermore, $M_{q,\lambda}$ is strictly decreasing over $\left(\frac{\ln q}{2\lambda} + 1, +\infty\right)$. $M_{q,\lambda}$ is concave down over $\left[\frac{\ln q}{2\lambda} - 1, \frac{\ln q}{2\lambda} + 1\right]$ and strictly concave down over $\left(\frac{\ln q}{2\lambda} - 1, \frac{\ln q}{2\lambda} + 1\right)$.

Consequently, $M_{q,\lambda}$ has a bell-type shape over $\mathbb{R}$.

At $x = \frac{\ln q}{2\lambda}$, $M_{q,\lambda}$ achieves its global maximum, which is

$$M_{q,\lambda}\left(\frac{\ln q}{2\lambda}\right) = \frac{\tanh(\lambda)}{2}, \quad \lambda > 0. \tag{9.39}$$

We mention

Theorem 9.7 (Anastassiou 2023e). *We have that*

$$\sum_{i=-\infty}^{\infty} M_{q,\lambda}(x - i) = 1, \quad \forall\, x \in \mathbb{R}, \quad \forall\, \lambda, q > 0. \tag{9.40}$$

It follows

Theorem 9.8 (Anastassiou 2023e). *It holds*

$$\int_{-\infty}^{\infty} M_{q,\lambda}(x)\, dx = 1, \quad \lambda, q > 0. \tag{9.41}$$

So $M_{q,\lambda}$ is a density function on $\mathbb{R}$; $\lambda, q > 0$.

We need the following result.

Theorem 9.9 (Anastassiou 2023e). *Let* $0 < \alpha < 1$, *and* $n \in \mathbb{N}$ *with* $n^{1-\alpha} > 2$; $q, \lambda > 0$. *Then*

$$\sum_{\substack{k=-\infty \\ \left\{ :|nx-k| \geq n^{1-\alpha} \right.}}^{\infty} M_{q,\lambda}(nx-k) < 2\max\left\{q, \frac{1}{q}\right\} e^{4\lambda} e^{-2\lambda n^{(1-\alpha)}} = Te^{-2\lambda n^{(1-\alpha)}}, \tag{9.42}$$

where $T := 2\max\left\{q, \frac{1}{q}\right\} e^{4\lambda}$.

Let $\lceil \cdot \rceil$ be the ceiling of the number and $\lfloor \cdot \rfloor$ be the integral part of the number.

Theorem 9.10 (Anastassiou 2023e). *Let* $x \in [a, b] \subset \mathbb{R}$ *and* $n \in \mathbb{N}$ *so that* $\lceil na \rceil \leq \lfloor nb \rfloor$. *For* $q > 0$, $\lambda > 0$, *we consider the number* $\lambda_q > z_0 > 0$ *with* $M_{q,\lambda}(z_0) = M_{q,\lambda}(0)$ *and* $\lambda_q > 1$. *Then*

$$\frac{1}{\sum_{k=\lceil na \rceil}^{\lfloor nb \rfloor} M_{q,\lambda}(nx-k)} < max\left\{ \frac{1}{M_{q,\lambda}(\lambda_q)}, \frac{1}{M_{\frac{1}{q},\lambda}\left(\lambda_{\frac{1}{q}}\right)} \right\} =: \Delta(q). \tag{9.43}$$

We also mention

Remark 9.2. (i) We have that

$$\lim_{n \to +\infty} \sum_{k=\lceil na \rceil}^{\lfloor nb \rfloor} M_{q,\lambda}(nx-k) \neq 1, \quad \text{for at least some } x \in [a, b], \tag{9.44}$$

where $\lambda, q > 0$.

(ii) Let $[a, b] \subset \mathbb{R}$. For large n, we always have $\lceil na \rceil \leq \lfloor nb \rfloor$. Also $a \leq \frac{k}{n} \leq b$, iff $\lceil na \rceil \leq k \leq \lfloor nb \rfloor$. In general, it holds

$$\sum_{k=\lceil na \rceil}^{\lfloor nb \rfloor} M_{q,\lambda}(nx-k) \leq 1. \tag{9.45}$$

Definition 9.4. Let $f \in C([a, b])$ and $n \in \mathbb{N} : \lceil na \rceil \leq \lfloor nb \rfloor$. We introduce and define the real-valued linear neural network operators

$$H_n(f, x) := \frac{\sum_{k=\lceil na \rceil}^{\lfloor nb \rfloor} f\left(\frac{k}{n}\right) M_{q,\lambda}(nx-k)}{\sum_{k=\lceil na \rceil}^{\lfloor nb \rfloor} M_{q,\lambda}(nx-k)}, \quad x \in [a, b]; q > 0, q \neq 1. \tag{9.46}$$

For large enough n, we always obtain $\lceil na \rceil \leq \lfloor nb \rfloor$. Also $a \leq \frac{k}{n} \leq b$, iff $\lceil na \rceil \leq k \leq \lfloor nb \rfloor$.

We study here the pointwise and uniform convergence of $H_n(f,x)$ to $f(x)$ with rates.

For convenience, also we call

$$H_n^*(f,x) := \sum_{k=\lceil na \rceil}^{\lfloor nb \rfloor} f\left(\frac{k}{n}\right) M_{q,\lambda}(nx-k), \tag{9.47}$$

that is,

$$H_n(f,x) := \frac{H_n^*(f,x)}{\sum_{k=\lceil na \rceil}^{\lfloor nb \rfloor} M_{q,\lambda}(nx-k)}. \tag{9.48}$$

So

$$H_n(f,x) - f(x) = \frac{H_n^*(f,x)}{\sum_{k=\lceil na \rceil}^{\lfloor nb \rfloor} M_{q,\lambda}(nx-k)} - f(x)$$

$$= \frac{H_n^*(f,x) - f(x)\left(\sum_{k=\lceil na \rceil}^{\lfloor nb \rfloor} M_{q,\lambda}(nx-k)\right)}{\sum_{k=\lceil na \rceil}^{\lfloor nb \rfloor} M_{q,\lambda}(nx-k)}. \tag{9.49}$$

Consequently, we derive that

$$|H_n(f,x) - f(x)| \leq \Delta(q) \left| H_n^*(f,x) - f(x) \left(\sum_{k=\lceil na \rceil}^{\lfloor nb \rfloor} M_{q,\lambda}(nx-k) \right) \right|$$

$$= \Delta(q) \left| \sum_{k=\lceil na \rceil}^{\lfloor nb \rfloor} \left(f\left(\frac{k}{n}\right) - f(x) \right) M_{q,\lambda}(nx-k) \right|, \tag{9.50}$$

where $\Delta(q)$ as in (9.43). We estimate the right-hand side of the last quantity.

We present a set of real-valued neural network approximations to a function given with rates.

Theorem 9.11 (Anastassiou 2023a). *Let $f \in C([a,b])$, $0 < \alpha < 1$, $n \in \mathbb{N} : n^{1-\alpha} > 2$, $q > 0$, $q \neq 1$, $x \in [a,b]$. Then*

(i)

$$|H_n(f,x) - f(x)| \leq \Delta(q) \left[\omega_1\left(f, \frac{1}{n^\alpha}\right) + 2\|f\|_\infty T e^{-2\lambda n^{(1-\alpha)}} \right] =: \tau, \tag{9.51}$$

where $T := \max\left\{q, \frac{1}{q}\right\} e^{4\lambda}$

and

(ii)

$$\left\| H_n(f) - f \right\|_\infty \leq \tau. \tag{9.52}$$

We get that $\lim_{n\to\infty} H_n(f) = f$, *pointwise and uniformly.*

Next present

Theorem 9.12 (Anastassiou 2023a). *Let* $0 < \alpha, \beta < 1$, $q > 0$, $q \neq 1$, $f \in C^1([a,b], X)$, $x \in [a,b]$, $n \in \mathbb{N} : n^{1-\beta} > 2$. *Then*

(i)

$$\left| H_n(f,x) - f(x) \right| \leq \frac{\Delta(q)}{\Gamma(\alpha+1)} \left\{ \frac{\left(\omega_1\left(D_{x-}^\alpha f, \frac{1}{n^\beta}\right)_{[a,x]} + \omega_1\left(D_{*x}^\alpha f, \frac{1}{n^\beta}\right)_{[x,b]} \right)}{n^{\alpha\beta}} + Te^{-2\lambda n^{(1-\beta)}} \left(\left\| D_{x-}^\alpha f \right\|_{\infty,[a,x]} (x-a)^\alpha + \left\| D_{*x}^\alpha f \right\|_{\infty,[x,b]} (b-x)^\alpha \right) \right\} \tag{9.53}$$

and

(ii)

$$\left\| H_n f - f \right\|_\infty \leq \frac{\Delta(q)}{\Gamma(\alpha+1)} \left\{ \frac{\left(\sup_{x\in[a,b]} \omega_1\left(D_{x-}^\alpha f, \frac{1}{n^\beta}\right)_{[a,x]} + \sup_{x\in[a,b]} \omega_1\left(D_{*x}^\alpha f, \frac{1}{n^\beta}\right)_{[x,b]} \right)}{n^{\alpha\beta}} + (b-a)^\alpha Te^{-2\lambda n^{(1-\beta)}} \left(\sup_{x\in[a,b]} \left\| D_{x-}^\alpha f \right\|_{\infty,[a,x]} + \sup_{x\in[a,b]} \left\| D_{*x}^\alpha f \right\|_{\infty,[x,b]} \right) \right\}. \tag{9.54}$$

Here, $T := \max\left\{q, \frac{1}{q}\right\} e^{4\lambda}$.

9.3 Combine 2.1 and 2.2

Let $a, b \in \mathbb{R}, f \in C([a, b])$. Let also $q, \lambda > 0, A > 1, \gamma = 2\max\left\{q, \frac{1}{q}\right\}$ as in previous sections.

For the following theorems, we call

$$
\begin{aligned}
{}_1L_n(f, x) &:= L_n(f, x), x \in [a, b], \\
{}_2L_n(f, x) &:= H_n(f, x), x \in [a, b].
\end{aligned}
$$

Also we set

$$
\begin{aligned}
K_1(q) &= K(q), \\
K_2(q) &= \Delta(q).
\end{aligned}
$$

Furthermore we set

$$
\begin{aligned}
\hat{\beta}_{1,n}(\lambda, \alpha) &= A^{-\lambda\left(n^{1-\alpha}-2\right)}, \quad n \in \mathbb{N},\ 0 < \alpha < 1, \\
\hat{\beta}_{2,n}(\lambda, \alpha) &= e^{4\lambda - 2\lambda n^{1-\alpha}}, \quad n \in \mathbb{N},\ 0 < \alpha < 1.
\end{aligned}
$$

We present

Theorem 9.13. *Let $f \in C([a, b])$, $0 < \alpha < 1$, $n \in \mathbb{N} : n^{1-\alpha} > 2$, $x \in [a, b]$. Then for $i = 1, 2$*

(i)

$$
\left|{}_iL_n(f, x) - f(x)\right| \le K_i(q)\left[\omega_1\left(f, \frac{1}{n^\alpha}\right) + 2\|f\|_\infty \gamma\hat{\beta}_{i,n}(\lambda, \alpha)\right] =: \rho_i \tag{9.55}
$$

and

(ii)

$$
\left\|{}_iL_n(f) - f\right\|_\infty \le \rho_i. \tag{9.56}
$$

We get that $\lim_{n\to\infty} {}_iL_n(f) = f$, pointwise and uniformly.

Proof. From Theorems 9.5 and 9.11. □

Next we present

Theorem 9.14. *Let $0 < \alpha, \beta < 1$, $f \in C^1([a, b])$, $x \in [a, b]$, $n \in \mathbb{N} : n^{1-\beta} > 2$. Then for $i = 1, 2$*

(i)

$$\left|{}_iL_n(f,x)-f(x)\right|$$
$$\leq K_i(q)\frac{1}{\Gamma(\alpha+1)}\left\{\frac{\left(\omega_1\left(D_{x-}^{\alpha}f,\frac{1}{n^{\beta}}\right)_{[a,x]}+\omega_1\left(D_{*x}^{\alpha}f,\frac{1}{n^{\beta}}\right)_{[x,b]}\right)}{n^{\alpha\beta}}\right.$$
$$\left.+\gamma\hat{\beta}_{i,n}(\lambda,\beta)\left(\left\|D_{x-}^{\alpha}f\right\|_{\infty,[a,x]}(x-a)^{\alpha}+\left\|D_{*x}^{\alpha}f\right\|_{\infty,[x,b]}(b-x)^{\alpha}\right)\right\} \tag{9.57}$$

and

(ii)

$$\left\|{}_iL_nf-f\right\|_{\infty}\leq K_i(q)\frac{1}{\Gamma(\alpha+1)}\cdot$$
$$\left\{\frac{\left(\sup_{x\in[a,b]}\omega_1\left(D_{x-}^{\alpha}f,\frac{1}{n^{\beta}}\right)_{[a,x]}+\sup_{x\in[a,b]}\omega_1\left(D_{*x}^{\alpha}f,\frac{1}{n^{\beta}}\right)_{[x,b]}\right)}{n^{\alpha\beta}}\right.$$
$$\left.+(b-a)^{\alpha}\gamma\hat{\beta}_{i,n}(\lambda,\beta)\left(\sup_{x\in[a,b]}\left\|D_{x-}^{\alpha}f\right\|_{\infty,[a,x]}+\sup_{x\in[a,b]}\left\|D_{*x}^{\alpha}f\right\|_{\infty,[x,b]}\right)\right\}. \tag{9.58}$$

Proof. From Theorems 9.6 and 9.12. □

9.4 About Brownian Motion on Three-dimensional Sphere

(Kouloumpou and Papanicolaou 2013). The Brownian motion on S^n is a diffusion (Markov) process $W_t, t\geq 0$, on S^n whose transition density is a function $P(t,x,y)$ on $(0,\infty)\times S^n\times S^n$ satisfying

$$\frac{\partial P}{\partial t}=\frac{1}{2}\Delta_n P \tag{9.59}$$

and

$$P(t,x,y)\to\delta_x(y)\quad\text{as } t\to 0^+, \tag{9.60}$$

where Δ_n is the Laplace–Beltrami operator of S^n acting on the x-variables and $\delta_x(y)$ is the delta mass at x, i.e. $P(t,x,y)$ is the **heat kernel** of S^n. The heat kernel exists; it is unique, positive and smooth in (t,x,y).

Remark 9.3. The heat kernel $P(t,x,y)$ satisfies the following properties:

(1) Symmetry: $P(t,x,y) = P(t,y,x)$.
(2) The semigroup identity: For any $s \in (0,t)$,

$$P(t,x,y) = \int_{S^n} P(s,x,z)P(t-s,z,y)d\mu(z), \tag{9.61}$$

where $d\mu$ is the area measure element of S^n.
(3) For all $t > 0$ and $x \in S^n$,

$$\int_{S^n} P(t,x,y)d\mu(y) = 1. \tag{9.62}$$

(4) As $t \to \infty$, $P(t,x,y)$ approaches the uniform density on S^n, i.e.

$$\lim_{t\to\infty} P(t,x,y) = \frac{1}{A_n}, \tag{9.63}$$

where A_n is the area of the S^n with radius a. It is also well known that

$$A_n = \frac{2\pi^{\frac{n+1}{2}}a^n}{\left(\frac{n-1}{2}\right)!}, \quad \text{for } n \text{ odd},$$

$$A_n = \frac{2^n\left(\frac{n}{2}-1\right)!\pi^{\frac{n}{2}}a^n}{(n-1)!}, \quad \text{for } n \text{ even}. \tag{9.64}$$

Finally, the symmetry of S^n implies that $P(t,x,y)$ depends only on t and $d(x,y)$, the distance between x and y. Thus, in spherical coordinates, it depends on t and the angle φ between x and y. Hence,

$$P(t,x,y) = p(t,\varphi),$$

where $p(t,\varphi)$ satisfies

$$\frac{\partial p}{\partial t} = \frac{1}{2}\Delta_n p = \frac{1}{2a^2}\left[(n-1)\cot\varphi \cdot \frac{\partial p}{\partial \varphi} + \frac{\partial^2 p}{\partial \varphi^2}\right] \tag{9.65}$$

and

$$\lim_{t\to 0^+} aA_{n-1}p(t,\varphi)\cdot \sin^{n-1}\varphi = \delta(\varphi). \tag{9.66}$$

Here $\delta(\cdot)$ is the standard Dirac delta function on $\mathbb{R}$.

9.4.1 *Explicit form of the heat kernel of S^3*

Proposition 9.1. *Let W_t, $t \geq 0$, be the Brownian motion on a 3-dimensional sphere S^3 of radius a. The transition density function $p(t,\varphi)$ of W_t is given by*

$$p(t,\varphi) = \frac{\exp\left(\frac{t}{2a^2}\right)}{(2\pi t)^{3/2}\sin\varphi}\sum_{n\in\mathbb{Z}}(\varphi+2n\pi)\exp\left(-\frac{(\varphi+2n\pi)^2a^2}{2t}\right), \tag{9.67}$$

where $\mathbb{Z}$ is the set of all integers. Equivalently,

$$p(t,\varphi) = -\frac{i}{4\pi^2a^3\sin\varphi}\sum_{n\in\mathbb{Z}} n\exp\left(-\frac{t(n^2-1)}{2a^2}+i\varphi n\right) \tag{9.68}$$

and

$$p(t,\varphi) = \frac{1}{2\pi^2a^3\sin\varphi}\sum_{n\in\mathbb{N}} n\sin(n\varphi)\exp\left(-\frac{t(n^2-1)}{2a^2}\right). \tag{9.69}$$

Furthermore, $p(t,\varphi)$ is analytic about $\varphi = 0$ and $\varphi = \pi$. In fact,

$$p(t,0) = \lim_{\varphi\to 0+} p(t,\varphi) = \frac{1}{2\pi^2a^3}\sum_{n\in\mathbb{N}} n^2\exp\left(-\frac{t(n^2-1)}{2a^2}\right) \tag{9.70}$$

and

$$p(t,\pi) = \lim_{\varphi\to\pi-} p(t,\varphi) = \frac{1}{2\pi^2a^3}\sum_{n\in\mathbb{N}} n^2(-1)^n\exp\left(-\frac{t(n^2-1)}{2a^2}\right). \tag{9.71}$$

Theorem 9.15. *Consider function $g:\mathbb{R}\to\mathbb{R}$, which is bounded on $[0,\pi]$, i.e. there exists $M>0$ such that $|g(\phi)| \leq M$, for every $\phi\in[0,\pi]$, and Lebesgue measurable on $\mathbb{R}$. Let also $W(t,\phi)$ be the Brownian motion on S^3. Then the expectation*

$$E(|g(W)|)(t) = \int_0^\pi |g(\phi)|\, p(t,\phi)d\phi$$

is continuous in t, and

$$E(|g(W)|)(t) \leq \pi M p(t_o,\phi_0), \tag{9.72}$$

where

$$p(t_0,\phi_0) = \max_{(t,\phi)\in[t_1,t_2]\times[0,\pi]} p(t,\phi), \quad \text{with } 0<t_1<t_2.$$

Here $p(t, \phi)$ is the transition density function of the Brownian motion W_t, $t \geq 0$, on S^3 given by (9.67).

Proof. It is known that the transition density function of the Brownian motion $W_t, t \geq 0$, on $S^3, p(t, \phi)$ is continuous in $(t, \phi) \in [t_1, t_2] \times [0, \pi], t_1 > 0$.

By the extreme value theorem, there exists $(t_0, \phi_0) \in [t_1, t_2] \times [0, \pi]$ such that

$$p(t_0, \phi_0) = \max_{(t,\phi) \in [t_1, t_2] \times [0,\pi]} p(t, \phi).$$

So we have

$$0 \leq p(t, \phi) \leq p(t_0, \phi_0), \quad \text{for every } (t, \phi) \in [t_1, t_2] \times [0, \pi].$$

Let $N \in \mathbb{N}, t_N, t \in [t_1, t_2] : t_N \to t$, as $N \to \infty$.

Then, $p(t_N, \phi) \to p(t, \phi)$ for every $\phi \in [0, \pi]$.

The function $g : \mathbb{R} \to \mathbb{R}$ is bounded on $[0, \pi]$, i.e. there is a $M > 0$ such that $|g(\phi)| \leq M$, for every $\phi \in [0, \pi]$, and Lebesgue measurable on $\mathbb{R}$.

Furthermore, we have that

$$|g(\phi)| \, p(t_N, \phi) \to |g(\phi)| \, p(t, \phi), \quad \text{as } N \to \infty$$

and

$$|g(\phi)| \, p(t_N, \phi) \leq |g(\phi)| \, p(t_o, \phi_0), \quad \text{for all } \phi \in [0, \pi] \text{ and } N \in \mathbb{N}.$$

So, by dominated convergence theorem, we get that

$$E(|g(W)|)(t_N) \to E(|g(W)|)(t) \quad \text{as } N \to \infty.$$

Thus $E(|g(W)|)(t)$ is proved to be continuous in t.

Moreover,

$|g(\phi)| \, p(t, \phi) \leq M p(t_o, \phi_0)$ for all $t \in [t_1, t_2]$ and $\phi \in [0, \pi]$.

Thus,

$$E(|g(W)|)(t) = \int_0^\pi |g(\phi)| \, p(t, \phi) d\phi \leq \pi M p(t_o, \phi_0).$$

□

Proposition 9.2. *Consider function $g : \mathbb{R} \to \mathbb{R}$, which is bounded on $[0, \pi]$ and Lebesgue measurable on $\mathbb{R}$. Let also $W(t, \phi)$ be the Brownian motion on S^3. Then the expectation*

$$E(|g(W)|)(t) = \int_0^\pi |g(\phi)| \, p(t, \phi) d\phi$$

is differentiable in t*, and*

$$\frac{\partial E\left(|g(W)|\right)}{\partial t}=\int_0^{\pi}|g(\phi)|\,\frac{\partial\left(p(t,\phi)\right)}{\partial t}d\phi, \tag{9.73}$$

which is continuous in t.

Proof. As we said before, the transition density function of the Brownian motion $W_t, t\geq 0$ on $S^3, p(t,\phi)$ is continuous in $(t,\phi)\in[t_1,t_2]\times[0,\pi], t_1>0$.

We have

$$E\left(|g(W)|\right)=\int_0^{\pi}|g(\phi)|\,p(t,\phi)d\phi,\quad \text{for every } t\in[t_1,t_2].$$

We apply differentiation under the integral sign.

We note

$$|g(\phi)|\,\frac{\partial p(t,\phi)}{\partial t}\leq M\left\|\frac{\partial p(t,\phi)}{\partial t}\right\|_{\infty,[t_1,t_2]\times[0,\pi]}.$$

Therefore, there exists

$$\frac{\partial E\left(|g(W)|\right)}{\partial t}=\int_0^{\pi}|g(\phi)|\,\frac{\partial\left(p(t,\phi)\right)}{\partial t}d\phi,$$

which is continuous in t (same proof as in Theorem 9.15). ☐

9.5 Main Results

We start with

Proposition 9.3. *Let* $0<\alpha<1$*,* $n\in\mathbb{N}:n^{1-\alpha}>2$*,* $t\in[t_1,t_2]$*, where* $t_1>0$*. Then for* $i=1,2$

(i)

$$\left|{}_iL_n\left(E\left(|g(W)|\right)\right)(t)-E\left(|g(W)|\right)(t)\right|$$
$$\leq K_i(q)\left[\omega_1\left(E\left(|g(W)|\right),\frac{1}{n^{\alpha}}\right)+2\left\|E\left(|g(W)|\right)\right\|_{\infty}\gamma\hat{\beta}_{i,n}(\lambda,\alpha)\right]=:\rho_i \tag{9.74}$$

and

(ii)

$$\left\|{}_iL_n\left(E\left(|g(W)|\right)\right)-E\left(|g(W)|\right)\right\|_{\infty}\leq\rho_i. \tag{9.75}$$

We get that $\lim_{n\to\infty}{}_iL_n\left(E\left(|g(W)|\right)\right)=E\left(|g(W)|\right)$*, pointwise and uniformly.*

Proof. From Theorem 9.13. □

Next we present

Proposition 9.4. *Let* $0 < \alpha, \beta < 1$, $t \in [t_1, t_2]$, *where* $t_1 > 0$ $n \in \mathbb{N}$: $n^{1-\beta} > 2$. *Then for* $i = 1, 2$

(i)

$$\left|{}_iL_n\left(E\left(|g(W)|\right)\right)(t) - E\left(|g(W)|\right)(t)\right|$$
$$\leq K_i(q)\frac{1}{\Gamma(\alpha+1)}$$
$$\times\left\{\frac{\left(\omega_1\left(D_{t-}^{\alpha}E\left(|g(W)|\right),\frac{1}{n^{\beta}}\right)_{[t_1,t]} + \omega_1\left(D_{*t}^{\alpha}E\left(|g(W)|\right),\frac{1}{n^{\beta}}\right)_{[t,t_2]}\right)}{n^{\alpha\beta}}\right.$$
$$+\gamma\hat{\beta}_{i,n}(\lambda,\beta)\left(\left\|D_{t-}^{\alpha}E\left(|g(W)|\right)\right\|_{\infty,[t_1,t]}(t-t_1)^{\alpha}\right.$$
$$\left.\left.+\left\|D_{*t}^{\alpha}E\left(|g(W)|\right)\right\|_{\infty,[t,t_2]}(t_2-t)^{\alpha}\right)\right\} \tag{9.76}$$

and

(ii)

$$\left\|{}_iL_nE\left(|g(W)|\right) - E\left(|g(W)|\right)\right\|_{\infty} \leq K_i(q)\frac{1}{\Gamma(\alpha+1)}\cdot$$
$$\left\{\frac{\left(\sup_{t\in[t_1,t_2]}\omega_1\left(D_{t-}^{\alpha}E\left(|g(W)|\right),\frac{1}{n^{\beta}}\right)_{[t_1,t]} + \sup_{t\in[t_1,t_2]}\omega_1\left(D_{*t}^{\alpha}E\left(|g(W)|\right),\frac{1}{n^{\beta}}\right)_{[t,t_2]}\right)}{n^{\alpha\beta}}\right.$$
$$+\left(t_2-t_1\right)^{\alpha}\gamma\hat{\beta}_{i,n}(\lambda,\beta)$$
$$\left.\left(\sup_{t\in[t_1,t_2]}\left\|D_{t-}^{\alpha}E\left(|g(W)|\right)\right\|_{\infty,[t_1,t]} + \sup_{t\in[t_1,t_2]}\left\|D_{*t}^{\alpha}E\left(|g(W)|\right)\right\|_{\infty,[t,t_2]}\right)\right\}. \tag{9.77}$$

Proof. From Theorem 9.14. □

We continue with

Proposition 9.5. *Let* $0 < \alpha < 1$, $n \in \mathbb{N} : n^{1-\alpha} > 2$, $t \in [t_1, t_2]$, *where* $t_1 > 0$. *Then for* $i = 1, 2$

(i)

$$\left| {}_iL_n \left(\frac{\partial E(|g(W)|)}{\partial t} \right)(t) - \left(\frac{\partial E(|g(W)|)}{\partial t} \right)(t) \right|$$
$$\leq K_i(q) \left[\omega_1 \left(\frac{\partial E(|g(W)|)}{\partial t}, \frac{1}{n^\alpha} \right) + 2 \left\| \frac{\partial E(|g(W)|)}{\partial t} \right\|_\infty \gamma \hat{\beta}_{i,n}(\lambda, \alpha) \right] =: \rho_i \tag{9.78}$$

and

(ii)

$$\left\| {}_iL_n \left(\frac{\partial E(|g(W)|)}{\partial t} \right) - \frac{\partial E(|g(W)|)}{\partial t} \right\|_\infty \leq \rho_i. \tag{9.79}$$

We get that $\lim_{n \to \infty} {}_iL_n \left(\frac{\partial E(|g(W)|)}{\partial t} \right) = \frac{\partial E(|g(W)|)}{\partial t}$, *pointwise and uniformly.*

Proof. From Theorem 9.13. □

9.6 Applications

For a function $g : \mathbb{R} \to \mathbb{R}$, which is bounded on $[0, \pi]$ and Lebesgue measurable on $\mathbb{R}$ and $W(t, \phi)$ the Brownian motion on S^3. We use the following notations:

$$E(|g(W)|) := E(|g(W)|)^{(0)} \tag{9.80}$$

and

$$\frac{\partial E(|g(W)|)}{\partial t} := E(|g(W)|)^{(1)}. \tag{9.81}$$

We can apply our main results to the function $g(W) = W$. Consider the function $g : \mathbb{R} \to \mathbb{R}$, where $g(x) = x$ for every $x \in \mathbb{R}$. Let also $W(t, \phi)$ be the Brownian motion on S^3. Then the expectation

$$E(|W|)(t) = \int_0^\pi \phi p(t, \phi) d\phi$$

is continuous in t.

Moreover,

Corollary 9.1. *Let* $0 < \alpha < 1$, $n \in \mathbb{N} : n^{1-\alpha} > 2$, $t \in [t_1, t_2]$, *where* $t_1 > 0$. *Then for* $i = 1, 2$ *and* $j = 0, 1$

(i)

$$\left| {}_iL_n \left(E\left(|W|\right)^{(j)} \right)(t) - E\left(|W|\right)^{(j)}(t) \right|$$
$$\leq K_i(q) \left[\omega_1 \left(E\left(|W|\right)^{(j)}, \frac{1}{n^{\alpha}} \right) + 2 \left\| E\left(|W|\right)^{(j)} \right\|_{\infty} \gamma \hat{\beta}_{i,n}(\lambda, \alpha) \right] =: \rho_i \tag{9.82}$$

and

(ii)

$$\left\| {}_iL_n \left(E\left(|W|\right)^{(j)} \right) - E\left(|W|\right)^{(j)} \right\|_{\infty} \leq \rho_i \tag{9.83}$$

We get that $\lim_{n \to \infty} {}_iL_n \left(E\left(|W|\right)^{(j)} \right) = E\left(|W|\right)^{(j)}$, *pointwise and uniformly.*

Proof. From Propositions 9.3 and 9.5. □

Next we present

Corollary 9.2. *Let* $0 < \alpha, \beta < 1$, $t \in [t_1, t_2]$,*where* $t_1 > 0$ $n \in \mathbb{N}$: $n^{1-\beta} > 2$. *Then for* $i = 1, 2$

(i)

$$\left| {}_iL_n \left(E\left(|W|\right) \right)(t) - E\left(|W|\right)(t) \right|$$
$$\leq K_i(q) \frac{1}{\Gamma(\alpha+1)} \left\{ \frac{\left(\omega_1 \left(D_{t-}^{\alpha} E\left(|W|\right), \frac{1}{n^{\beta}} \right)_{[t_1, t]} + \omega_1 \left(D_{*t}^{\alpha} E\left(|W|\right), \frac{1}{n^{\beta}} \right)_{[t, t_2]} \right)}{n^{\alpha\beta}} \right.$$
$$+ \gamma \hat{\beta}_{i,n}(\lambda, \beta) \left(\left\| D_{t-}^{\alpha} E\left(|W|\right) \right\|_{\infty, [t_1, t]} (t - t_1)^{\alpha} \right.$$
$$\left. \left. + \left\| D_{*t}^{\alpha} E\left(|W|\right) \right\|_{\infty, [t, t_2]} (t_2 - t)^{\alpha} \right) \right\} \tag{9.84}$$

and

(ii)

$$\|{}_iL_nE(|W|) - E(|W|)\|_\infty \leq K_i(q)\frac{1}{\Gamma(\alpha+1)} \cdot$$

$$\left\{ \frac{\left(\sup_{t\in[t_1,t_2]} \omega_1\left(D_{t-}^{\alpha}E(|W|), \frac{1}{n^\beta}\right)_{[t_1,t]} + \sup_{t\in[t_1,t_2]} \omega_1\left(D_{*t}^{\alpha}E(|W|), \frac{1}{n^\beta}\right)_{[t,t_2]}\right)}{n^{\alpha\beta}} \right.$$

$$+ (t_2-t_1)^\alpha \gamma\hat{\beta}_{i,n}(\lambda,\beta)$$

$$\left. \left(\sup_{t\in[t_1,t_2]} \left\|D_{t-}^{\alpha}E(|W|)\right\|_{\infty,[t_1,t]} + \sup_{t\in[t_1,t_2]} \|D_{*t}^{\alpha}E(|W|)\|_{\infty,[t,t_2]}\right) \right\}. \tag{9.85}$$

Proof. From Proposition 9.4. □

For the following corollaries, we consider the function $g : \mathbb{R} \to \mathbb{R}$, where $g(x) = cosx$ for every $x \in \mathbb{R}$. Let also $W(t,\phi)$ be the Brownian motion on S^3. Then the expectation

$$E(|cosW|)(t) = \int_0^\pi |\cos\phi|\, p(t,\phi)d\phi$$

is continuous in t.
Moreover,

Corollary 9.3. *Let $0 < \alpha < 1$, $n \in \mathbb{N} : n^{1-\alpha} > 2$, $t \in [t_1, t_2]$, where $t_1 > 0$. Then for $i = 1, 2$ and $j = 0, 1$*

(i)

$$\left|{}_iL_n\left(E(|\cos W|)^{(j)}\right)(t) - E(|\cos W|)^{(j)}(t)\right|$$

$$\leq K_i(q)\left[\omega_1\left(E(|\cos W|)^{(j)}, \frac{1}{n^\alpha}\right) + 2\left\|E(|\cos W|)^{(j)}\right\|_\infty \gamma\hat{\beta}_{i,n}(\lambda,\alpha)\right] =: \rho_i \tag{9.86}$$

and

(ii)

$$\left\|{}_iL_n\left(E(|\cos W|)^{(j)}\right) - E(|\cos W|)^{(j)}\right\|_\infty \leq \rho_i. \tag{9.87}$$

We get that $\lim_{n\to\infty} {}_iL_n\left(E\left(\left|\cos W\right|\right)^{(j)}\right) = E\left(\left|\cos W\right|\right)^{(j)}$, *pointwise and uniformly.*

Proof. From Propositions 9.3 and 9.5. □

Next we present

Corollary 9.4. *Let* $0 < \alpha, \beta < 1$, $t \in [t_1, t_2]$, *where* $t_1 > 0$ $n \in \mathbb{N}$: $n^{1-\beta} > 2$. *Then for* $i = 1, 2$

(i)

$$\left|{}_iL_n\left(E\left(\left|\cos W\right|\right)\right)(t) - E\left(\left|\cos W\right|\right)(t)\right|$$
$$\leq K_i(q)\frac{1}{\Gamma(\alpha+1)}$$
$$\times\left\{\frac{\left(\omega_1\left(D_{t-}^{\alpha}E\left(\left|\cos W\right|\right), \frac{1}{n^{\beta}}\right)_{[t_1,t]} + \omega_1\left(D_{*t}^{\alpha}E\left(\left|\cos W\right|\right), \frac{1}{n^{\beta}}\right)_{[t,t_2]}\right)}{n^{\alpha\beta}}\right.$$
$$+\gamma\hat{\beta}_{i,n}(\lambda,\beta)\left(\left\|D_{t-}^{\alpha}E\left(\left|\cos W\right|\right)\right\|_{\infty,[t_1,t]}(t-t_1)^{\alpha}\right.$$
$$\left.\left.+\left\|D_{*t}^{\alpha}E\left(\left|\cos W\right|\right)\right\|_{\infty,[t,t_2]}(t_2-t)^{\alpha}\right)\right\} \tag{9.88}$$

and

(ii)

$$\left\|{}_iL_nE\left(\left|\cos W\right|\right) - E\left(\left|\cos W\right|\right)\right\|_{\infty} \leq K_i(q)\frac{1}{\Gamma(\alpha+1)}\cdot$$
$$\left\{\frac{\left(\sup_{t\in[t_1,t_2]}\omega_1\left(D_{t-}^{\alpha}E\left(\left|\cos W\right|\right), \frac{1}{n^{\beta}}\right)_{[t_1,t]} + \sup_{t\in[t_1,t_2]}\omega_1\left(D_{*t}^{\alpha}E\left(\left|\cos W\right|\right), \frac{1}{n^{\beta}}\right)_{[t,t_2]}\right)}{n^{\alpha\beta}}\right.$$
$$+(t_2-t_1)^{\alpha}\gamma\hat{\beta}_{i,n}(\lambda,\beta)$$
$$\left.\left(\sup_{t\in[t_1,t_2]}\left\|D_{t-}^{\alpha}E\left(\left|\cos W\right|\right)\right\|_{\infty,[t_1,t]} + \sup_{t\in[t_1,t_2]}\left\|D_{*t}^{\alpha}E\left(\left|\cos W\right|\right)\right\|_{\infty,[t,t_2]}\right)\right\}. \tag{9.89}$$

Proof. From Proposition 9.4. □

We can obtain similar results for the function $g : \mathbb{R} \to \mathbb{R}$, where $g(x) = \sin x$ for every $x \in \mathbb{R}$.

Let as consider now the function $g : \mathbb{R} \to \mathbb{R}$, where $g(x) = \tanh x$ for every $x \in \mathbb{R}$. Let also $W(t, \phi)$ be the Brownian motion on S^3. Then the expectation

$$E\left(|\tanh W|\right)(t) = \int_0^{\pi} |\tanh(\phi)|\, p(t, \phi) d\phi$$

is continuous in t.

Moreover,

Corollary 9.5. *Let $0 < \alpha < 1$, $n \in \mathbb{N} : n^{1-\alpha} > 2$, $t \in [t_1, t_2]$, where $t_1 > 0$. Then for $i = 1, 2$ and $j = 0, 1$*

(i)

$$\left|{}_iL_n\left(E\left(|\tanh W|\right)^{(j)}\right)(t) - E\left(|\tanh W|\right)^{(j)}(t)\right|$$
$$\leq K_i(q)\left[\omega_1\left(E\left(|\tanh W|\right)^{(j)}, \frac{1}{n^{\alpha}}\right) + 2\left\|E\left(|\tanh W|\right)^{(j)}\right\|_{\infty} \gamma\hat{\beta}_{i,n}(\lambda, \alpha)\right] =: \rho_i \tag{9.90}$$

and

(ii)

$$\left\|{}_iL_n\left(E\left(|\tanh W|\right)^{(j)}\right) - E\left(|\tanh W|\right)^{(j)}\right\|_{\infty} \leq \rho_i. \tag{9.91}$$

We get that $\lim_{n\to\infty} {}_iL_n\left(E\left(|\tanh W|\right)^{(j)}\right) = E\left(|\tanh W|\right)^{(j)}$, pointwise and uniformly.

Proof. From Propositions 9.3 and 9.5. □

Next we present

Corollary 9.6. *Let $0 < \alpha, \beta < 1$, $t \in [t_1, t_2]$, where $t_1 > 0$ $n \in \mathbb{N} : n^{1-\beta} > 2$. Then for $i = 1, 2$*

(i)

$$\left|{}_iL_n\left(E\left(|\tanh W|\right)\right)(t) - E\left(|\tanh W|\right)(t)\right| \leq K_i(q)\frac{1}{\Gamma(\alpha+1)}$$
$$\left\{\frac{\left(\omega_1\left(D_{t-}^{\alpha}E\left(|\tanh W|\right), \frac{1}{n^{\beta}}\right)_{[t_1,t]} + \omega_1\left(D_{*t}^{\alpha}E\left(|\tanh W|\right), \frac{1}{n^{\beta}}\right)_{[t,t_2]}\right)}{n^{\alpha\beta}}\right.$$

$$+\gamma\hat{\beta}_{i,n}(\lambda,\beta)$$

$$\left.\left(\left\|D_{t-}^{\alpha}E(|\tanh W|)\right\|_{\infty,[t_1,t]}(t-t_1)^{\alpha}+\left\|D_{*t}^{\alpha}E(|\tanh W|)\right\|_{\infty,[t,t_2]}(t_2-t)^{\alpha}\right)\right\} \tag{9.92}$$

and

(ii)

$$\left\|{}_iL_nE(|\tanh W|)-E(|\tanh W|)\right\|_{\infty}\leq K_i(q)\frac{1}{\Gamma(\alpha+1)}\cdot$$

$$\left\{\frac{\left(\sup_{t\in[t_1,t_2]}\omega_1\left(D_{t-}^{\alpha}E(|\tanh W|),\frac{1}{n^{\beta}}\right)_{[t_1,t]}+\sup_{t\in[t_1,t_2]}\omega_1\left(D_{*t}^{\alpha}E(|\tanh W|),\frac{1}{n^{\beta}}\right)_{[t,t_2]}\right)}{n^{\alpha\beta}}\right.$$

$$+(t_2-t_1)^{\alpha}\gamma\hat{\beta}_{i,n}(\lambda,\beta)$$

$$\left.\left(\sup_{t\in[t_1,t_2]}\left\|D_{t-}^{\alpha}E(|\tanh W|)\right\|_{\infty,[t_1,t]}+\sup_{t\in[t_1,t_2]}\left\|D_{*t}^{\alpha}E(|\tanh W|)\right\|_{\infty,[t,t_2]}\right)\right\}. \tag{9.93}$$

Proof. From Proposition 9.4. □

In the following, let as consider the function $g:\mathbb{R}\to\mathbb{R}$, where $g(x)=e^{-\ell x}, \ell>0$ for every $x\in\mathbb{R}$. Let also $W(t,\phi)$ be the Brownian motion on S^3. Then the expectation

$$E\left(e^{-\ell W}\right)(t)=\int_0^{\pi}e^{-\ell\phi}p(t,\phi)d\phi$$

is continuous in t.

Furthermore,

Corollary 9.7. *Let* $0<\alpha<1$, $n\in\mathbb{N}:n^{1-\alpha}>2$, $t\in[t_1,t_2]$, *where* $t_1>0$. *Then for* $i=1,2$ *and* $j=0,1$

(i)

$$\left|{}_iL_n\left(E\left(e^{-\ell W}\right)^{(j)}\right)(t)-E\left(e^{-\ell W}\right)^{(j)}(t)\right|$$

$$\leq K_i(q)\left[\omega_1\left(E\left(e^{-\ell W}\right)^{(j)},\frac{1}{n^{\alpha}}\right)+2\left\|E\left(e^{-\ell W}\right)^{(j)}\right\|_{\infty}\gamma\hat{\beta}_{i,n}(\lambda,\alpha)\right]=:\rho_i \tag{9.94}$$

and

(ii)

$$\left\|{}_iL_n\left(E\left(e^{-\ell W}\right)^{(j)}\right)-E\left(e^{-\ell W}\right)^{(j)}\right\|_\infty\le\rho_i. \tag{9.95}$$

We get that $\lim_{n\to\infty}\ {}_iL_n\left(E\left(e^{-\ell W}\right)^{(j)}\right)=E\left(e^{-\ell W}\right)^{(j)}$, *pointwise and uniformly.*

Proof. From Propositions 9.3 and 9.5. □

We continue with

Corollary 9.8. *Let* $0<\alpha,\beta<1$, $t\in[t_1,t_2]$,*where* $t_1>0$ $n\in\mathbb{N}:n^{1-\beta}>2$. *Then for* $i=1,2$

(i)

$$\begin{aligned}&\left|{}_iL_n\left(E\left(e^{-\ell W}\right)\right)(t)-E\left(e^{-\ell W}\right)(t)\right|\\&\le K_i(q)\frac{1}{\Gamma(\alpha+1)}\left\{\frac{\left(\omega_1\left(D_{t-}^{\alpha}E\left(e^{-\ell W}\right),\frac{1}{n^\beta}\right)_{[t_1,t]}+\omega_1\left(D_{*t}^{\alpha}E\left(e^{-\ell W}\right),\frac{1}{n^\beta}\right)_{[t,t_2]}\right)}{n^{\alpha\beta}}\right.\\&+\gamma\hat{\beta}_{i,n}(\lambda,\beta)\left(\left\|D_{t-}^{\alpha}E\left(e^{-\ell W}\right)\right\|_{\infty,[t_1,t]}(t-t_1)^\alpha\right.\\&\left.\left.+\left\|D_{*t}^{\alpha}E\left(e^{-\ell W}\right)\right\|_{\infty,[t,t_2]}(t_2-t)^\alpha\right)\right\}\end{aligned} \tag{9.96}$$

and

(ii)

$$\begin{aligned}&\left\|{}_iL_nE\left(e^{-\ell W}\right)-E\left(e^{-\ell W}\right)\right\|_\infty\le K_i(q)\frac{1}{\Gamma(\alpha+1)}\cdot\\&\left\{\frac{\left(\sup_{t\in[t_1,t_2]}\omega_1\left(D_{t-}^{\alpha}E\left(e^{-\ell W}\right),\frac{1}{n^\beta}\right)_{[t_1,t]}+\sup_{t\in[t_1,t_2]}\omega_1\left(D_{*t}^{\alpha}E\left(e^{-\ell W}\right),\frac{1}{n^\beta}\right)_{[t,t_2]}\right)}{n^{\alpha\beta}}\right.\\&+(t_2-t_1)^\alpha\gamma\hat{\beta}_{i,n}(\lambda,\beta)\\&\left.\left(\sup_{t\in[t_1,t_2]}\left\|D_{t-}^{\alpha}E\left(e^{-\ell W}\right)\right\|_{\infty,[t_1,t]}+\sup_{t\in[t_1,t_2]}\left\|D_{*t}^{\alpha}E\left(e^{-\ell W}\right)\right\|_{\infty,[t,t_2]}\right)\right\}.\end{aligned} \tag{9.97}$$

Proof. From Proposition 9.4. □

In the following, we consider the generalized logistic sigmoid function $g : \mathbb{R} \to \mathbb{R}$, where $g(x) = (1 + e^{-x})^{-\delta}$ for every $x \in \mathbb{R}$. Let also $W(t, \phi)$ be the Brownian motion on S^3. Then the expectation

$$E\left(\left(1+e^{-W}\right)^{-\delta}\right)(t) = \int_0^{\pi} \left(1+e^{-\phi}\right)^{-\delta} p(t,\phi)d\phi$$

is continuous in t.

Moreover,

Corollary 9.9. *Let $0 < \alpha < 1$, $n \in \mathbb{N} : n^{1-\alpha} > 2$, $t \in [t_1, t_2]$, where $t_1 > 0$. Then for $i = 1, 2$ and $j = 0, 1$*

(i)

$$\begin{aligned}
&\left| {}_iL_n\left(E\left(\left(1+e^{-W}\right)^{-\delta}\right)^{(j)}\right)(t) - E\left(\left(1+e^{-W}\right)^{-\delta}\right)^{(j)}(t)\right| \\
&\leq K_i(q)\left[\omega_1\left(E\left(\left(1+e^{-W}\right)^{-\delta}\right)^{(j)}, \frac{1}{n^{\alpha}}\right)\right. \\
&\left. +2\left\|E\left(\left(1+e^{-W}\right)^{-\delta}\right)^{(j)}\right\|_{\infty}\gamma\hat{\beta}_{i,n}(\lambda,\alpha)\right] =: \rho_i \qquad (9.98)
\end{aligned}$$

and

(ii)

$$\left\| {}_iL_n\left(E\left(\left(1+e^{-W}\right)^{-\delta}\right)^{(j)}\right) - E\left(\left(1+e^{-W}\right)^{-\delta}\right)^{(j)}\right\|_{\infty} \leq \rho_i \qquad (9.99)$$

We get that $\lim_{n\to\infty} {}_iL_n\left(E\left(\left(1+e^{-W}\right)^{-\delta}\right)^{(j)}\right) = E\left(\left(1+e^{-W}\right)^{-\delta}\right)^{(j)}$, *pointwise and uniformly.*

Proof. From Propositions 9.3 and 9.5. □

It follows

Corollary 9.10. *Let $0 < \alpha, \beta < 1$, $t \in [t_1, t_2]$, where $t_1 > 0$ $n \in \mathbb{N} : n^{1-\beta} > 2$. Then for $i = 1, 2$*

(i)

$$\left| {}_iL_n\left(E\left(\left(1+e^{-W}\right)^{-\delta}\right)\right)(t) - E\left(\left(1+e^{-W}\right)^{-\delta}\right)(t)\right| \leq K_i(q)\frac{1}{\Gamma(\alpha+1)}$$

$$\left\{\frac{\left(\omega_1\left(D_{t-}^{\alpha}E\left(\left(1+e^{-W}\right)^{-\delta}\right), \frac{1}{n^{\beta}}\right)_{[t_1,t]} + \omega_1\left(D_{*t}^{\alpha}E\left(\left(1+e^{-W}\right)^{-\delta}\right), \frac{1}{n^{\beta}}\right)_{[t,t_2]}\right)}{n^{\alpha\beta}}\right.$$

$$+\gamma\hat{\beta}_{i,n}(\lambda,\beta)$$

$$\left(\left\|D_{t-}^{\alpha}E\left(\left(1+e^{-W}\right)^{-\delta}\right)\right\|_{\infty,[t_1,t]}(t-t_1)^{\alpha}\right.$$

$$\left.\left.+\left\|D_{*t}^{\alpha}E\left(\left(1+e^{-W}\right)^{-\delta}\right)\right\|_{\infty,[t,t_2]}(t_2-t)^{\alpha}\right)\right\} \tag{9.100}$$

and

(ii)

$$\left\|{}_iL_nE\left(\left(1+e^{-W}\right)^{-\delta}\right)-E\left(\left(1+e^{-W}\right)^{-\delta}\right)\right\|_{\infty}\leq K_i(q)\frac{1}{\Gamma(\alpha+1)}\cdot$$

$$\left\{\frac{\left(\sup_{t\in[t_1,t_2]}\omega_1\left(D_{t-}^{\alpha}E\left(\left(1+e^{-W}\right)^{-\delta}\right),\frac{1}{n^{\beta}}\right)_{[t_1,t]}+\sup_{t\in[t_1,t_2]}\omega_1\left(D_{*t}^{\alpha}E\left(\left(1+e^{-W}\right)^{-\delta}\right),\frac{1}{n^{\beta}}\right)_{[t,t_2]}\right)}{n^{\alpha\beta}}\right.$$

$$+(t_2-t_1)^{\alpha}\gamma\hat{\beta}_{i,n}(\lambda,\beta)\left(\sup_{t\in[t_1,t_2]}\left\|D_{t-}^{\alpha}E\left(\left(1+e^{-W}\right)^{-\delta}\right)\right\|_{\infty,[t_1,t]}\right.$$

$$\left.\left.+\sup_{t\in[t_1,t_2]}\left\|D_{*t}^{\alpha}E\left(\left(1+e^{-W}\right)^{-\delta}\right)\right\|_{\infty,[t,t_2]}\right)\right\}. \tag{9.101}$$

Proof. From Proposition 9.4. □

When $\delta=1$, we have the usual logistic sigmoid function.

The Gompertz function $g:\mathbb{R}\to\mathbb{R}$, with $g(x)=e^{\mu e^{-x}},\mu<0$, is a sigmoid function which describes growth as being slowest at the start and end of a given time period. Let $W(t,\phi)$ be the Brownian motion on S^3. Then the expectation

$$E\left(e^{\mu e^{-W}}\right)(t)=\int_0^{\pi}e^{\mu e^{-\phi}}p(t,\phi)d\phi$$

is continuous in t.

It follows

Corollary 9.11. *Let* $0<\alpha<1$, $n\in\mathbb{N}:n^{1-\alpha}>2$, $t\in[t_1,t_2]$, *where* $t_1>0$. *Then for* $i=1,2$ *and* $j=0,1$

(i)

$$\left|{}_iL_n\left(E\left(e^{\mu e^{-W}}\right)^{(j)}\right)(t)-E\left(e^{\mu e^{-W}}\right)^{(j)}(t)\right|$$
$$\leq K_i(q)\left[\omega_1\left(E\left(e^{\mu e^{-W}}\right)^{(j)},\frac{1}{n^{\alpha}}\right)+2\left\|E\left(e^{\mu e^{-W}}\right)^{(j)}\right\|_{\infty}\gamma\hat{\beta}_{i,n}(\lambda,\alpha)\right]=:\rho_i \tag{9.102}$$

and

(ii)

$$\left\|{}_iL_n\left(E\left(e^{\mu e^{-W}}\right)^{(j)}\right)-E\left(e^{\mu e^{-W}}\right)^{(j)}\right\|_{\infty}\leq\rho_i \tag{9.103}$$

We get that $\lim_{n\to\infty}{}_iL_n\left(E\left(e^{\mu e^{-W}}\right)^{(j)}\right)=E\left(e^{\mu e^{-W}}\right)^{(j)}$*, pointwise and uniformly.*

Proof. From Propositions 9.3 and 9.5. □

We finish with

Corollary 9.12. *Let* $0<\alpha,\beta<1$, $t\in[t_1,t_2]$,*where* $t_1>0$ $n\in\mathbb{N}:n^{1-\beta}>2$. *Then for* $i=1,2$

(i)

$$\left|{}_iL_n\left(e^{\mu e^{-W}}\right)(t)-E\left(e^{\mu e^{-W}}\right)(t)\right|\leq K_i(q)\frac{1}{\Gamma(\alpha+1)}$$
$$\times\left\{\frac{\left(\omega_1\left(D_{t-}^{\alpha}E\left(e^{\mu e^{-W}}\right),\frac{1}{n^{\beta}}\right)_{[t_1,t]}+\omega_1\left(D_{*t}^{\alpha}E\left(e^{\mu e^{-W}}\right),\frac{1}{n^{\beta}}\right)_{[t,t_2]}\right)}{n^{\alpha\beta}}\right.$$
$$+\gamma\hat{\beta}_{i,n}(\lambda,\beta)$$
$$\left.\left(\left\|D_{t-}^{\alpha}E\left(e^{\mu e^{-W}}\right)\right\|_{\infty,[t_1,t]}(t-t_1)^{\alpha}+\left\|D_{*t}^{\alpha}E\left(e^{\mu e^{-W}}\right)\right\|_{\infty,[t,t_2]}(t_2-t)^{\alpha}\right)\right\} \tag{9.104}$$

and

(ii)

$$\left\| {}_iL_nE\left(e^{\mu e^{-W}}\right) - E\left(e^{\mu e^{-W}}\right)\right\|_\infty \leq K_i(q)\frac{1}{\Gamma(\alpha+1)}\cdot$$

$$\left\{\frac{\left(\sup_{t\in[t_1,t_2]}\omega_1\left(D_{t-}^\alpha E\left(e^{\mu e^{-W}}\right),\frac{1}{n^\beta}\right)_{[t_1,t]} + \sup_{t\in[t_1,t_2]}\omega_1\left(D_{*t}^\alpha E\left(e^{\mu e^{-W}}\right),\frac{1}{n^\beta}\right)_{[t,t_2]}\right)}{n^{\alpha\beta}}\right.$$

$$+ (t_2-t_1)^\alpha\gamma\hat{\beta}_{i,n}(\lambda,\beta)$$

$$\left.\left(\sup_{t\in[t_1,t_2]}\left\|D_{t-}^\alpha E\left(e^{\mu e^{-W}}\right)\right\|_{\infty,[t_1,t]} + \sup_{t\in[t_1,t_2]}\left\|D_{*t}^\alpha E\left(e^{\mu e^{-W}}\right)\right\|_{\infty,[t,t_2]}\right)\right\}. \tag{9.105}$$

Proof. From Proposition 9.4. □

Bibliography

G.A. Anastassiou, Rate of convergence of some neural network operators to the unit-univariate case, *Journal of Mathematical Analysis and Applications* 212 (1997), 237–262.

G.A. Anastassiou, *Quantitative Approximations*, Chapman and Hall / CRC Boca Raton, New York, 2001.

G.A. Anastassiou, On Right Fractional Calculus, *Chaos, Solitons and Fractals*, 42 (2009), 365–376.

G.A. Anastassiou, *Inteligent Systems: Approximation by Artificial Neural Networks*, Intelligent Systems Reference Library, Vol 19, Springer, Heidelberg, 2011a.

G.A. Anastassiou, Multivariate hyperbolic tangent neural network approximation, *Computer and Mathematics*, 61 (2011b), 809–821.

G.A. Anastassiou, Multivariate sigmoidal neural network approximation, *Neural Networks*, 24 (2011c), 378–386.

G.A. Anastassiou, Univariate hyperbolic tangent neural network approximation, *Mathematics and Computer Modeling*, 53 (2011d), 1111–1132.

G.A. Anastassiou, Fractional neural network approximation, *Computer and Mathematics with Applications*, 64 (2012a), 1655–1676.

G.A. Anastassiou, Univariate sigmoidal neural network approximation, *Journal of Computational Analysis and Applications*, 14(4), (2012b), 659–690.

G.A. Anastassiou, Inteligent Systems II: Complete Approximation by Neural Network Operators, Springer, Heidelberg, New York, 2016.

G.A. Anastassiou, A strong fractional calculus theory for Banach space valued functions, N*onlinear Functional Analysis and Applications (Korea)*, 22(3) (2017a), 495–524.

G.A. Anastassiou, Strong right fractional calculus for Banach space valued functions, *Revista Proyecciones*, 36(1) (2017b), 149–186.

G.A. Anastassiou, Vector fractional Korovkin type approximations, *Dynamic Systems and Applications*, 26 (2017c), 81–104.

G.A. Anastassiou, *Nonlinearity: Ordinary and Fractional Approximations by Sublinear and Max-Product Operators*, Springer, Heidelberg, New York, 2018a.

G.A. Anastassiou, *Intelligent Computations: Abstract Fractional Calculus, Inequalities, Approximations*, Springer, Heidelberg, New York, 2018b.

G.A. Anastasiou, *Abstract univariate neural network approximation using a q-deformed and λ-parametrized hyperbolic tangent activation function*, FRACT.FRACT./MDPI/ 2023a, 7, 208.

G.A. Anastassiou, *Banach Space Valued Neural Network*, Springer, Heidelberg, New York, 2023b.

G.A. Anastassiou, General sigmoid based Banach space valued neural network approximation, *Journal of Computational Analysis and Applications*, 31(4) (2023c), 520–534.

G.A. Anastassiou, *Parametrized, deformed and general neural networks*, accepted, Springer, Heidelberg, New York, 2023d.

G.A. Anastassiou, *q-Deformed and λ-parametrized hyperbolic tangent function based Banach space valued multivariate multi layer neural network approximations*, Annales Univ. Sci.Budapest, Sect. Comp., accepted, 2023e.

G.A. Anastassiou, *q-Deformed and λ-parametrized A-generalized logistic function based Banach space valued ordinary and fractional neural network approximations*, submitted 2023f.

G.A. Anastassiou, *q-Deformed and λ-parametrized A-generalized logistic function indused Banach space valued multivariate multi layer neural network approximations*, submitted 2023g.

G.A. Anastassiou, D. Kouloumpou, *Brownian motion approximation by parametrized and deformed neural networks*, RACSAM, 2024, 118:14.

Z. Chen and F. Cao, The approximation operators with sigmoidal functions, *Computers and Mathematics with Applications*, 58 (2009), 758–765.

K. Diethelm, The Analysis of Fractional Diferential Equations, *Lecture Notes in Mathematics* 2004, Springer-Verlag, Berlin, Heidelberg, 2010.

A.M.A El-sayed and M. Geber, On the finite Caputo and finite Riesz derivatives, *Electronic journal of Theoritical Physics*, 3(12) (2006), 81–95.

G.S. Frederico and D.F.M. Torres, Fractional optimal control in the sense of Caputo and the fractional Noether's theorem, *International Mathematical Forum*, 3(10) (2008), 479–493.

D. Kouloumpou and V.G. Papanicolaou, Certain Calculation Regarding the Brownian Motion on the Sphere, *Journal of Concrete and Applicable Mathematics,* 11(1) (2013).

S.G. Samko, A.A. Kilbas and O.I. Marichev, Fractional Integrals and Derivatives, Theory and Applications, (Gordon and Breach, Amsterdam, 1993)[English translation from the Russian, Integrals and Derivatives of Fractional Order and Some of Their Applications (Nauka i Tekhnika, Minsk,1987)].

Chapter 10

Approximation of Brownian Motion over Simple Graphs

The first author recently derived several approximation results by neural network operators, see his new monograph Anastassiou (2023d). There, the approximation methods were derived from the parametrized and deformed neural networks induced by the parametrized error and q-deformed and β-parametrized half-hyperbolic tangent activation functions. The results we apply here are univariate on a compact interval, regular and fractional. The outcome is the quantitative approximation of Brownian motion on simple graphs, in particular over a system S of semiaxes emanating from a common origin radially arranged and a particle moving randomly on S. We derive several Jackson-type inequalities estimating the degree of convergence of our neural network operators to a general expectation function of Brownian motion. We give a detailed list of approximation applications regarding the expectation of well-known functions of this Brownian motion. Smoothness of our functions is taken into account producing higher speeds of approximation. It follows Anastassiou and Kouloumpou (2023).

10.1 Introduction

The first author in Anastassiou (1997, 2001), see Chapters 2–5, was the first to establish neural network approximation to continuous functions with rates by very specifically defined neural network operators of Cardaliaguet–Euvrard and "Squashing" types, by employing the modulus of continuity of the engaged function or its high order derivative, and producing very tight Jackson-type inequalities. He treats there both the univariate and multivariate cases. The definitions of these operators' "bell-shaped" and "squashing" functions are assumed to be of compact support.

Also the first author inspired by Chen and Cao (2009) continued his studies on neural network approximation by introducing and using the proper quasi-interpolation operators of sigmoidal and hyperbolic tangent types which resulted into Anastassiou (2011a), Anastassiou (2011b), Anastassiou (2011c), Anastassiou (2011d) and Anastassiou (2012), by treating both the univariate and multivariate cases. He did also the corresponding fractional cases Anastassiou (2012), Anastassiou (2016) and Anastassiou (2018a).

In Anastassiou (2023a), Anastassiou (2023d), the first author continued similar studies for Banach space-valued functions for activation functions deriving from the parametrized error and q-deformed and β-parametrized half-hyperbolic tangent sigmoid functions. The authors based and inspired by Papanicolaou, Papageorgiou and Lepipas (2012) perform here neural network quantitative approximations to Brownian motion over a simple graph of a system of semiaxes.

They present a collection of Jackson-type inequalities estimating the error of approximation to a general expectation function of this Brownian motion and its derivative. They produce regular and fractional calculus results. They finish with a lot of important applications.

10.2 Basics

10.2.1 *About the parametrized (Gauss) error special activation function*

Here we follow Anastassiou (2023e).

We consider here the parametrized (Gauss) error special activation function

$$erf\lambda z = \frac{2}{\sqrt{\pi}} \int_0^{\lambda z} e^{-t^2} dt, \quad \lambda > 0,\ z \in \mathbb{R}, \tag{10.1}$$

which is a sigmoidal-type function and a strictly increasing function.

Of special interest in neural network theory is when $0 < \lambda < 1$, see Anastassiou (1997), Introduction.

It has the following basic properties:

$$\begin{gathered} erf\,(0) = 0, \quad erf\,(-\lambda x) = -erf\,(\lambda x)\,, \text{for every } 0 < \lambda < 1 \\ erf\,(+\infty) = 1, \quad erf\,(-\infty) = -1 \end{gathered} \tag{10.2}$$

and

$$(erf\,(\lambda x))' = \frac{2\lambda}{\sqrt{\pi}} e^{-(\lambda x)^2}, \quad x \in \mathbb{R}. \tag{10.3}$$

We consider the function

$$\chi(x) = \frac{1}{4}\left(erf\left(\lambda\left(x+1\right)\right) - erf\left(\lambda\left(x-1\right)\right)\right), \quad x \in \mathbb{R}, \tag{10.4}$$

and we note that

$$\chi(-x) = \chi(x). \tag{10.5}$$

Thus, χ is an even function.

Since $x+1 > x-1$, then $erf\left(\lambda\left(x+1\right)\right) > erf\left(\lambda\left(x-1\right)\right)$, and $\chi(x) > 0$, all $x \in \mathbb{R}$.

We see

$$\chi(0) = \frac{erf\lambda}{2}. \tag{10.6}$$

Let $x > 0$, we have that

$$\chi'(x) = \frac{\lambda}{2\sqrt{\pi}}\left(\frac{e^{\lambda^2(x-1)^2} - e^{\lambda^2(x+1)^2}}{e^{\lambda^2(x+1)^2}e^{\lambda^2(x-1)^2}}\right) < 0, \tag{10.7}$$

proving $\chi'(x) < 0$, for $x > 0$. That is, χ is strictly decreasing on $[0, \infty)$ and it is strictly increasing on $(-\infty, 0]$, and $\chi'(0) = 0$.

Clearly, the x-axis is the horizontal asymptote of χ.

Conclusion, χ is a bell symmetric function with maximum

$$\chi(0) = \frac{erf\lambda}{2}.$$

Theorem 10.1. *It holds*

$$\sum_{i=-\infty}^{\infty} \chi(x-i) = 1, \quad \forall\, x \in \mathbb{R}. \tag{10.8}$$

We have

Theorem 10.2. *We have that*

$$\int_{-\infty}^{\infty} \chi(x)\, dx = 1. \tag{10.9}$$

Hence, $\chi(x)$ is a density function on $\mathbb{R}$. We need

Theorem 10.3. *Let* $0 < \alpha < 1$, *and* $n \in \mathbb{N}$ *with* $n^{1-\alpha} > 2$, $\lambda > 0$. *It holds*

$$\sum_{\left\{\begin{array}{l} k=-\infty \\ : |nx-k| \geq n^{1-\alpha} \end{array}\right.}^{\infty} \chi(nx-k) < \left(1 - erf\left(\lambda\left(n^{1-\alpha}-2\right)\right)\right), \tag{10.10}$$

with

$$\lim_{n\to+\infty}\left(1 - erf\left(\lambda\left(n^{1-\alpha}-2\right)\right)\right) = 0.$$

Denote by $\lfloor \cdot \rfloor$ the integral part and by $\lceil \cdot \rceil$ the ceiling of a number. Furthermore we need

Theorem 10.4. *Let $x \in [a, b] \subset \mathbb{R}$, $\lambda > 0$ and $n \in \mathbb{N}$ so that $\lceil na \rceil \leq \lfloor nb \rfloor$. Then*

$$\frac{1}{\sum_{k=\lceil na \rceil}^{\lfloor nb \rfloor} \chi(nx - k)} < \frac{1}{\chi(1)} = \frac{4}{erf(2\lambda)}. \tag{10.11}$$

Remark 10.1. As in Anastassiou (2023b), we have that

$$\lim_{n \to \infty} \sum_{k=\lceil na \rceil}^{\lfloor nb \rfloor} \chi(nx - k) \neq 1. \tag{10.12}$$

Note 10.1. For large enough n, we always obtain $\lceil na \rceil \leq \lfloor nb \rfloor$. Also $a \leq \frac{k}{n} \leq b$, iff $\lceil na \rceil \leq k \leq \lfloor nb \rfloor$. As in Anastassiou (2023b), we obtain that

$$\sum_{k=\lceil na \rceil}^{\lfloor nb \rfloor} \chi(nx - k) \leq 1. \tag{10.13}$$

Definition 10.1. Let $f \in C([a, b])$ and $n \in \mathbb{N} : \lceil na \rceil \leq \lfloor nb \rfloor$. We introduce and define the X-valued linear neural network operators

$$A_n(f, x) := \frac{\sum_{k=\lceil na \rceil}^{\lfloor nb \rfloor} f\left(\frac{k}{n}\right) \chi(nx - k)}{\sum_{k=\lceil na \rceil}^{\lfloor nb \rfloor} \chi(nx - k)}, \quad x \in [a, b]. \tag{10.14}$$

Clearly here $A_n(f, x) \in C([a, b])$. We study here the pointwise and uniform convergence of $A_n(f, x)$ to $f(x)$ with rates.

For convenience also we call

$$A_n^*(f, x) := \sum_{k=\lceil na \rceil}^{\lfloor nb \rfloor} f\left(\frac{k}{n}\right) \chi(nx - k), \tag{10.15}$$

that is,

$$A_n(f, x) = \frac{A_n^*(f, x)}{\sum_{k=\lceil na \rceil}^{\lfloor nb \rfloor} \chi(nx - k)}. \tag{10.16}$$

So

$$A_n(f, x) - f(x) = \frac{A_n^*(f, x)}{\sum_{k=\lceil na \rceil}^{\lfloor nb \rfloor} \chi(nx - k)} - f(x)$$

$$= \frac{A_n^*(f, x) - f(x) \left(\sum_{k=\lceil na \rceil}^{\lfloor nb \rfloor} \chi(nx - k)\right)}{\sum_{k=\lceil na \rceil}^{\lfloor nb \rfloor} \chi(nx - k)}. \tag{10.17}$$

Consequently we derive

$$|A_n(f,x) - f(x)| \leq \frac{4}{erf(2\lambda)} \left| A_n^*(f,x) - f(x) \left(\sum_{k=\lceil na \rceil}^{\lfloor nb \rfloor} \chi(nx-k) \right) \right|. \tag{10.18}$$

That is,

$$|A_n(f,x) - f(x)| \leq \frac{4}{erf(2\lambda)} \left| \sum_{k=\lceil na \rceil}^{\lfloor nb \rfloor} \left(f\left(\frac{k}{n}\right) - f(x) \right) \chi(nx-k) \right|. \tag{10.19}$$

We estimate the right-hand side of (10.19).

For that we need, for $f \in C([a,b])$, the first modulus of continuity

$$\omega_1(f,\delta)_{[a,b]} := \omega_1(f,\delta) := \sup_{\substack{x,y \in [a,b] \\ |x-y| \leq \delta}} |f(x) - f(y)|, \quad \delta > 0. \tag{10.20}$$

The fact $f \in C([a,b])$ is equivalent to $\lim_{\delta \to 0} \omega_1(f,\delta) = 0$, see Anastassiou (2017c).

We present a series of real-valued neural network approximations to a function given with rates.

We first give

Theorem 10.5. *Let* $f \in C([a,b])$, $0 < \alpha < 1$, $n \in \mathbb{N} : n^{1-\alpha} > 2$, $\lambda > 0$, $x \in [a,b]$. *Then*

(i)

$$|A_n(f,x) - f(x)|$$
$$\leq \frac{4}{erf(2\lambda)} \left[\omega_1\left(f, \frac{1}{n^\alpha}\right) + 2\left(1 - erf\left(\lambda\left(n^{1-\alpha} - 2\right)\right)\right) \|f\|_\infty \right] =: \rho \tag{10.21}$$

and

(ii)

$$\|A_n(f) - f\|_\infty \leq \rho. \tag{10.22}$$

We note $\lim_{n \to \infty} A_n(f) = f$, *pointwise and uniformly.*

The speed of convergence is $\max\left\{\frac{1}{n^\alpha}, 2(1 - erf\left(\lambda\left(n^{1-\alpha} - 2\right)\right))\right\}$.

We need

Definition 10.2. (Anastassiou 2017a) Let $[a,b] \subset \mathbb{R}$, $\alpha > 0$; $m = \lceil \alpha \rceil \in \mathbb{N}$, ($\lceil \cdot \rceil$ is the ceiling of the number), $f : [a,b] \to \mathbb{R}$. We assume that $f^{(m)} \in L_1([a,b])$. We call the left Caputo fractional derivative of order α:

$$\left(D_{*a}^{\alpha} f\right)(x) := \frac{1}{\Gamma(m-\alpha)} \int_a^x (x-t)^{m-\alpha-1} f^{(m)}(t)\, dt, \quad \forall\, x \in [a,b]. \tag{10.23}$$

If $\alpha \in \mathbb{N}$, we set $D_{*a}^{\alpha} f := f^{(m)}$ the ordinary real-valued derivative (defined similar to numerical one, see Shilov 1996, p. 83) and also set $D_{*a}^{0} f := f$.

By Anastassiou (2017a), $\left(D_{*a}^{\alpha} f\right)(x)$ exists almost everywhere in $x \in [a,b]$ and $D_{*a}^{\alpha} f \in L_1([a,b])$.

If $\left\| f^{(m)} \right\|_{L_\infty([a,b])} < \infty$, then by Anastassiou (2017a), $D_{*a}^{\alpha} f \in C([a,b])$, hence $\left| D_{*a}^{\alpha} f \right| \in C([a,b])$.

We mention

Lemma 10.1 (Anastassiou 2017c). *Let $\alpha > 0$, $\alpha \notin \mathbb{N}$, $m = \lceil \alpha \rceil$, $f \in C^{m-1}([a,b])$ and $f^{(m)} \in L_\infty([a,b])$. Then $D_{*a}^{\alpha} f(a) = 0$.*

We mention

Definition 10.3 (Anastassiou 2017b). Let $[a,b] \subset \mathbb{R}$, $\alpha > 0$, $m := \lceil \alpha \rceil$. We assume that $f^{(m)} \in L_1([a,b])$, where $f : [a,b] \to \mathbb{R}$. We call the right Caputo fractional derivative of order α:

$$\left(D_{b-}^{\alpha} f\right)(x) := \frac{(-1)^m}{\Gamma(m-\alpha)} \int_x^b (z-x)^{m-\alpha-1} f^{(m)}(z)\, dz, \quad \forall\, x \in [a,b]. \tag{10.24}$$

We observe that $\left(D_{b-}^{m} f\right)(x) = (-1)^m f^{(m)}(x)$, for $m \in \mathbb{N}$, and $\left(D_{b-}^{0} f\right)(x) = f(x)$.

By Anastassiou (2017b), $\left(D_{b-}^{\alpha} f\right)(x)$ exists almost everywhere on $[a,b]$ and $\left(D_{b-}^{\alpha} f\right) \in L_1([a,b])$.

If $\left\| f^{(m)} \right\|_{L_\infty([a,b])} < \infty$, and $\alpha \notin \mathbb{N}$, by Anastassiou (2017b), $D_{b-}^{\alpha} f \in C([a,b])$, hence $\left| D_{b-}^{\alpha} f \right| \in C([a,b])$.

We need

Lemma 10.2 (Anastassiou 2017c). *Let $f \in C^{m-1}([a,b])$, $f^{(m)} \in L_\infty([a,b])$, $m = \lceil \alpha \rceil$, $\alpha > 0$, $\alpha \notin \mathbb{N}$. Then $D_{b-}^{\alpha} f(b) = 0$.*

We present the following real-valued fractional approximation result by $erf\lambda$ based neural networks.

Theorem 10.6. *Let* $0 < \alpha, \beta^* < 1$, $f \in C^1([a,b])$, $x \in [a,b]$, $n \in \mathbb{N}$: $n^{1-\beta^*} > 2$, $\lambda > 0$. *Then*

(i)

$$
\begin{aligned}
&|A_n(f,x) - f(x)| \\
&\quad \leq \frac{4}{erf(2\lambda)\Gamma(\alpha+1)} \left\{ \frac{\left(\omega_1\left(D_{x-}^{\alpha} f, \frac{1}{n^{\beta^*}}\right)_{[a,x]} + \omega_1\left(D_{*x}^{\alpha} f, \frac{1}{n^{\beta^*}}\right)_{[x,b]}\right)}{n^{\alpha\beta^*}} \right. \\
&\qquad + \left(1 - erf\left(\lambda\left(n^{1-\beta^*} - 2\right)\right)\right) \\
&\qquad \left. \times \left(\left\|D_{x-}^{\alpha} f\right\|_{\infty,[a,x]} (x-a)^{\alpha} + \left\|D_{*x}^{\alpha} f\right\|_{\infty,[x,b]} (b-x)^{\alpha}\right) \right\} \qquad (10.25)
\end{aligned}
$$

and
(ii)

$$
\begin{aligned}
&\|A_n f - f\|_{\infty} \leq \frac{4}{\Gamma(\alpha+1)\, erf(2\lambda)} \\
&\left\{ \frac{\left(\sup_{x\in[a,b]} \omega_1\left(D_{x-}^{\alpha} f, \frac{1}{n^{\beta^*}}\right)_{[a,x]} + \sup_{x\in[a,b]} \omega_1\left(D_{*x}^{\alpha} f, \frac{1}{n^{\beta^*}}\right)_{[x,b]}\right)}{n^{\alpha\beta^*}} \right. \\
&\quad + \left(1 - erf\left(\lambda\left(n^{1-\beta^*} - 2\right)\right)\right) \\
&\quad \left. \times (b-a)^{\alpha} \left(\sup_{x\in[a,b]} \left\|D_{x-}^{\alpha} f\right\|_{\infty,[a,x]} + \sup_{x\in[a,b]} \left\|D_{*x}^{\alpha} f\right\|_{\infty,[x,b]}\right) \right\}. \qquad (10.26)
\end{aligned}
$$

When $\alpha = \frac{1}{2}$, we derive

Corollary 10.1. *Let* $0 < \beta^* < 1$, $f \in C^1([a,b])$, $x \in [a,b]$, $n \in \mathbb{N}$: $n^{1-\beta^*} > 2$, $\lambda > 0$. *Then*

(i)

$$|A_n(f,x) - f(x)|$$

$$\leq \frac{8}{erf(2\lambda)\sqrt{\pi}} \left\{ \frac{\left(\omega_1\left(D_{x-}^{\frac{1}{2}}f, \frac{1}{n^{\beta^*}}\right)_{[a,x]} + \omega_1\left(D_{*x}^{\frac{1}{2}}f, \frac{1}{n^{\beta^*}}\right)_{[x,b]}\right)}{n^{\frac{\beta^*}{2}}} \right.$$

$$+ \left(1 - erf\left(\lambda\left(n^{1-\beta^*} - 2\right)\right)\right)$$

$$\left. \times \left(\left\|D_{x-}^{\frac{1}{2}}f\right\|_{\infty,[a,x]} \sqrt{(x-a)} + \left\|D_{*x}^{\frac{1}{2}}f\right\|_{\infty,[x,b]} \sqrt{(b-x)}\right) \right\} \quad (10.27)$$

and

(ii)

$$\|A_n f - f\|_\infty \leq \frac{8}{erf(2\lambda)\sqrt{\pi}}$$

$$\left\{ \frac{\left(\sup_{x\in[a,b]} \omega_1\left(D_{x-}^{\frac{1}{2}}f, \frac{1}{n^{\beta^*}}\right)_{[a,x]} + \sup_{x\in[a,b]} \omega_1\left(D_{*x}^{\frac{1}{2}}f, \frac{1}{n^{\beta^*}}\right)_{[x,b]}\right)}{n^{\frac{\beta}{2}}} \right.$$

$$+ \left(1 - erf\left(\lambda\left(n^{1-\beta^*} - 2\right)\right)\right)\sqrt{(b-a)}$$

$$\left. \times \left(\sup_{x\in[a,b]} \left\|D_{x-}^{\frac{1}{2}}f\right\|_{\infty,[a,x]} + \sup_{x\in[a,b]} \left\|D_{*x}^{\frac{1}{2}}f\right\|_{\infty,[x,b]}\right) \right\} < \infty. \quad (10.28)$$

10.2.2 *About q-deformed and β-parametrized half-hyperbolic tangent function φ_q*

All the following background comes from Anastassiou (2023f).

Here we describe the properties of the activation function

$$\varphi_q(t) := \frac{1 - qe^{-\beta t}}{1 + qe^{-\beta t}}, \quad \forall\, t \in \mathbb{R}, \quad (10.29)$$

where $q, \beta > 0$.

We have that

$$\varphi_q(0) = \frac{1-q}{1+q}$$

and

$$\varphi_q(-t) = -\varphi_{\frac{1}{q}}(t), \quad \forall\, t \in \mathbb{R}, \tag{10.30}$$

hence

$$\varphi'_{\frac{1}{q}}(t) = \varphi'_q(-t). \tag{10.31}$$

It is

$$\lim_{t\to+\infty} \varphi_q(t) = \varphi_q(+\infty) = 1 \tag{10.32}$$

and

$$\lim_{t\to-\infty} \varphi_q(t) = \varphi_q(-\infty) = -1. \tag{10.33}$$

Furthermore

$$\varphi'_q(t) = \frac{2\beta q e^{\beta t}}{\left(e^{\beta t}+q\right)^2} > 0, \quad \forall\, t \in \mathbb{R}, \tag{10.34}$$

therefore φ_q is strictly increasing. Moreover, in case of $t < \frac{\ln q}{\beta}$, we have that φ_q is strictly concave up, with $\varphi''_q\left(\frac{\ln q}{\beta}\right) = 0$.

And in case of $t > \frac{\ln q}{\beta}$, we have that φ_q is strictly concave down.

Clearly, φ_q is a shifted sigmoid function with $\varphi_q(0) = \frac{1-q}{1+q}$, and $\varphi_q(-x) = -\varphi_{q^{-1}}(x)$, $\forall\ x \in \mathbb{R}$ (a semi-odd function), see also Anastassiou 2023f.

We consider the function

$$\phi_q(x) := \frac{1}{4}\left(\varphi_q(x+1) - \varphi_q(x-1)\right) > 0, \tag{10.35}$$

$\forall x \in \mathbb{R}$; $\beta, q > 0$. Note that $\phi_q(\pm\infty) = 0$, so the x-axis is horizontal asymptote. We have that

$$\phi_q(-x) = \phi_{\frac{1}{q}}(x), \quad \forall\, x \in \mathbb{R}, \tag{10.36}$$

a deformed symmetry.

Next we have that φ_q is strictly increasing over $\left(-\infty, \frac{\ln q}{\beta} - 1\right)$ and strictly decreasing over $\left(\frac{\ln q}{\beta} + 1, +\infty\right)$.

Moreover, ϕ_q is concave down over $\left[\frac{\ln q}{\beta} - 1, \frac{\ln q}{\beta} + 1\right]$, and strictly concave down over $\left(\frac{\ln q}{\beta} - 1, \frac{\ln q}{\beta} + 1\right)$.

Consequently, ϕ_q has a bell-type shape over $\mathbb{R}$.

Of course, it holds $\phi_q''\left(\frac{\ln q}{\beta}\right) < 0$. Thus, at $x = \frac{\ln q}{\beta}$, we have the maximum value of ϕ_q which is

$$\phi_q\left(\frac{\ln q}{\beta}\right) = \frac{\left(1 - e^{-\beta}\right)}{2\left(1 + e^{-\beta}\right)} = \frac{\varphi_1(1)}{2}. \tag{10.37}$$

We mention

Theorem 10.7 (Anastassiou 2023c). *We have that*

$$\sum_{i=-\infty}^{\infty} \phi_q(x - i) = 1, \ \ \forall\, x \in \mathbb{R}, \forall\, q, \beta > 0. \tag{10.38}$$

It follows

Theorem 10.8. *(Anastassiou 2023c) It holds*

$$\int_{-\infty}^{\infty} \phi_q(x)\, dx = 1, \quad q, \beta > 0. \tag{10.39}$$

So ϕ_q is a density function on $\mathbb{R}$; $q, \beta > 0$.

We need the following result.

Theorem 10.9 (Anastassiou 2023c). *Let $0 < \alpha < 1$, and $n \in \mathbb{N}$ with $n^{1-\alpha} > 2$; $q, \beta > 0$. Then*

$$\sum_{\substack{k=-\infty \\ \left\{: |nx - k| \geq n^{1-\alpha}\right.}}^{\infty} \phi_q(nx - k) < 2 \max\left\{q, \frac{1}{q}\right\} e^{2\beta} e^{-\beta n^{(1-\alpha)}} = K e^{-\beta n^{(1-\alpha)}}, \tag{10.40}$$

where $K := 2\max\left\{q, \frac{1}{q}\right\} e^{2\beta}$.

Let $\lceil \cdot \rceil$ be the ceiling of the number and $\lfloor \cdot \rfloor$ be the integral part of the number.

We mention the following result.

Theorem 10.10 (Anastassiou 2023c). *Let $x \in [a, b] \subset \mathbb{R}$ and $n \in \mathbb{N}$ so that $\lceil na \rceil \leq \lfloor nb \rfloor$. For $q > 0$, we consider the number $\lambda_q > z_0 > 0$ with $\phi_q(z_0) = \phi_q(0)$ and $\beta, \lambda_q > 1$. Then*

$$\frac{1}{\sum_{k=\lceil na \rceil}^{\lfloor nb \rfloor} \phi_q(nx - k)} < \max\left\{\frac{1}{\phi_q(\lambda_q)}, \frac{1}{\phi_{\frac{1}{q}}\left(\lambda_{\frac{1}{q}}\right)}\right\} =: \theta(q). \tag{10.41}$$

We also mention

Remark 10.2 (Anastassiou 2023c). (i) We have that

$$\lim_{n \to +\infty} \sum_{k=\lceil na \rceil}^{\lfloor nb \rfloor} \phi_q (nx - k) \neq 1, \quad \text{for at least some } x \in [a, b], \tag{10.42}$$

where $\beta, q > 0$.

(ii) Let $[a, b] \subset \mathbb{R}$. For large n, we always have $\lceil na \rceil \leq \lfloor nb \rfloor$. Also $a \leq \frac{k}{n} \leq b$, iff $\lceil na \rceil \leq k \leq \lfloor nb \rfloor$. In general, it holds

$$\sum_{k=\lceil na \rceil}^{\lfloor nb \rfloor} \phi_q (nx - k) \leq 1. \tag{10.43}$$

We need

Definition 10.4. Let $f \in C([a, b])$ and $n \in \mathbb{N} : \lceil na \rceil \leq \lfloor nb \rfloor$. We introduce and define the real-valued linear neural network operators

$$H_n (f, x) := \frac{\sum_{k=\lceil na \rceil}^{\lfloor nb \rfloor} f\left(\frac{k}{n}\right) \Phi_q (nx - k)}{\sum_{k=\lceil na \rceil}^{\lfloor nb \rfloor} \Phi_q (nx - k)}, \quad x \in [a, b]; \, q, \beta > 0. \tag{10.44}$$

Clearly, $H_n (f) \in C([a, b])$.

We study here the pointwise and uniform convergence of $H_n (f, x)$ to $f(x)$ with rates.

For convenience, also we call

$$H_n^* (f, x) := \sum_{k=\lceil na \rceil}^{\lfloor nb \rfloor} f\left(\frac{k}{n}\right) \Phi_q (nx - k). \tag{10.45}$$

That is,

$$H_n (f, x) := \frac{H_n^* (f, x)}{\sum_{k=\lceil na \rceil}^{\lfloor nb \rfloor} \Phi_q (nx - k)}. \tag{10.46}$$

So

$$H_n(f,x) - f(x) = \frac{H_n^*(f,x)}{\sum_{k=\lceil na \rceil}^{\lfloor nb \rfloor} \Phi_q(nx-k)} - f(x)$$

$$= \frac{H_n^*(f,x) - f(x)\left(\sum_{k=\lceil na \rceil}^{\lfloor nb \rfloor} \Phi_q(nx-k)\right)}{\sum_{k=\lceil na \rceil}^{\lfloor nb \rfloor} \Phi_q(nx-k)}. \tag{10.47}$$

Consequently, we derive that

$$|H_n(f,x) - f(x)| \leq \theta(q) \left| H_n^*(f,x) - f(x)\left(\sum_{k=\lceil na \rceil}^{\lfloor nb \rfloor} \Phi_q(nx-k)\right)\right|$$

$$= \theta(q) \left| \sum_{k=\lceil na \rceil}^{\lfloor nb \rfloor} \left(f\left(\frac{k}{n}\right) - f(x)\right) \Phi_q(nx-k) \right|, \tag{10.48}$$

where $\theta(q)$ as in (10.41). We estimate the right-hand side of the last quantity.

We present a set of real-valued neural network approximations to a function given with rates.

Theorem 10.11. *Let $f \in C([a,b])$, $0 < \alpha < 1$, $n \in \mathbb{N} : n^{1-\alpha} > 2$, $q, \beta > 0$, $x \in [a,b]$. Then*

(i)

$$|H_n(f,x) - f(x)| \leq \theta(q) \left[\omega_1\left(f, \frac{1}{n^\alpha}\right) + 2\|f\|_\infty K e^{-\beta n^{(1-\alpha)}} \right] =: \tau, \tag{10.49}$$

*where K as in (**??**)*

and

(ii)

$$\|H_n(f) - f\|_\infty \leq \tau. \tag{10.50}$$

We get that $\lim_{n\to\infty} H_n(f) = f$, pointwise and uniformly.

Next we present

Theorem 10.12. *Let* $0 < \alpha, \beta^* < 1$, $q, \beta > 0$, $f \in C^1([a,b])$, $x \in [a,b]$, $n \in \mathbb{N} : n^{1-\beta^*} > 2$. *Then*

(i)

$$|H_n(f,x) - f(x)|$$
$$\leq \frac{\theta(q)}{\Gamma(\alpha+1)} \left\{ \frac{\left(\omega_1\left(D^{\alpha}_{x-}f, \frac{1}{n^{\beta^*}}\right)_{[a,x]} + \omega_1\left(D^{\alpha}_{*x}f, \frac{1}{n^{\beta^*}}\right)_{[x,b]}\right)}{n^{\alpha\beta^*}} \right.$$
$$\left. + Ke^{-\beta n^{(1-\beta^*)}} \left(\left\|D^{\alpha}_{x-}f\right\|_{\infty,[a,x]} (x-a)^{\alpha} + \left\|D^{\alpha}_{*x}f\right\|_{\infty,[x,b]} (b-x)^{\alpha}\right) \right\} \tag{10.51}$$

and
(ii)

$$\|H_n f - f\|_{\infty} \leq \frac{\theta(q)}{\Gamma(\alpha+1)}$$
$$\left\{ \frac{\left(\sup_{x\in[a,b]} \omega_1\left(D^{\alpha}_{x-}f, \frac{1}{n^{\beta^*}}\right)_{[a,x]} + \sup_{x\in[a,b]} \omega_1\left(D^{\alpha}_{*x}f, \frac{1}{n^{\beta^*}}\right)_{[x,b]}\right)}{n^{\alpha\beta^*}} \right.$$
$$\left. + (b-a)^{\alpha} Ke^{-\beta n^{(1-\beta^*)}} \left(\sup_{x\in[a,b]} \|D^{\alpha}_{x-}f\|_{\infty,[a,x]} + \sup_{x\in[a,b]} \|D^{\alpha}_{*x}f\|_{\infty,[x,b]}\right) \right\}. \tag{10.52}$$

When $\alpha = \frac{1}{2}$, we derive

Corollary 10.2. *Let* $0 < \beta^* < 1$, $q, \beta > 0$, $f \in C^1([a,b])$, $x \in [a,b]$, $n \in \mathbb{N} : n^{1-\beta^*} > 2$. *Then*

(i)

$$|H_n(f,x) - f(x)|$$
$$\leq \frac{2\theta(q)}{\sqrt{\pi}} \left\{ \frac{\left(\omega_1 \left(D^{\frac{1}{2}}_{x-} f, \frac{1}{n^{\beta^*}} \right)_{[a,x]} + \omega_1 \left(D^{\frac{1}{2}}_{*x} f, \frac{1}{n^{\beta^*}} \right)_{[x,b]} \right)}{n^{\frac{\beta^*}{2}}} \right.$$
$$\left. + K e^{-\beta n^{(1-\beta^*)}} \left(\left\| D^{\frac{1}{2}}_{x-} f \right\|_{\infty,[a,x]} \sqrt{(x-a)} + \left\| D^{\frac{1}{2}}_{*x} f \right\|_{\infty,[x,b]} \sqrt{(b-x)} \right) \right\} \tag{10.53}$$

and

(ii)

$$\|H_n f - f\|_\infty \leq \frac{2\theta(q)}{\sqrt{\pi}}$$
$$\left\{ \frac{\left(\sup_{x\in[a,b]} \omega_1 \left(D^{\frac{1}{2}}_{x-} f, \frac{1}{n^{\beta^*}} \right)_{[a,x]} + \sup_{x\in[a,b]} \omega_1 \left(D^{\frac{1}{2}}_{*x} f, \frac{1}{n^{\beta^*}} \right)_{[x,b]} \right)}{n^{\frac{\beta^*}{2}}} \right.$$
$$\left. + \sqrt{(b-a)} K e^{-\beta n^{(1-\beta^*)}} \left(\sup_{x\in[a,b]} \left\| D^{\frac{1}{2}}_{x-} f \right\|_{\infty,[a,x]} + \sup_{x\in[a,b]} \left\| D^{\frac{1}{2}}_{*x} f \right\|_{\infty,[x,b]} \right) \right\} < \infty. \tag{10.54}$$

10.3 Combine 2.1 and 2.2

Let $a, b \in \mathbb{R}$ with $a < b$, $f \in C([a,b])$. Let also $q, \lambda, \beta > 0, \gamma = \max\left\{q, \frac{1}{q}\right\}$.

For the following theorems, we call

$$_1L_n(f,x) := A_n(f,x), x \in [a,b]$$

$$_2L_n(f,x) := H_n(f,x), x \in [a,b].$$

Also we set

$$K_1 = K_1(\lambda) = \frac{4}{erf(2\lambda)},$$

$$K_2 = K_2(q) = \theta(q).$$

Furthermore we set

$$\hat{\beta}_{1,n} = \hat{\beta}_{1,n}(\lambda, \beta^*) = 2(1 - erf\left(\lambda\left(n^{1-\beta^*} - 2\right)\right)), n \in \mathbb{N}, \lambda > 0, 0 < \beta^* < 1,$$

$$\hat{\beta}_{2,n} = \hat{\beta}_{2,n}(q, \beta, \beta^*) = 2\gamma e^{2\beta - \beta n^{1-\beta^*}}, n \in \mathbb{N}, q, \beta > 0, 0 < \beta^* < 1.$$

We present

Theorem 10.13. *Let* $f \in C([a,b])$, $0 < \beta^* < 1$, $n \in \mathbb{N} : n^{1-\beta^*} > 2$, $q, \lambda, \beta > 0$, $x \in [a,b]$. *Then for* $i = 1, 2$

(i)

$$\left|{}_iL_n(f,x) - f(x)\right| \leq K_i \cdot \left[\omega_1\left(f, \frac{1}{n^{\beta^*}}\right) + \hat{\beta}_{i,n}\|f\|_\infty\right] =: \rho_i \tag{10.55}$$

and

(ii)

$$\left\|{}_iL_n(f) - f\right\|_\infty \leq \rho_i. \tag{10.56}$$

We get that $\lim_{n\to\infty} {}_iL_n(f) = f$, *pointwise and uniformly.*

Proof. From Theorems 10.5 and 10.11. □

Next we present

Theorem 10.14. *Let* $0 < \alpha, \beta^* < 1$, $q, \lambda, \beta > 0$, $f \in C^1([a,b])$, $x \in [a,b]$, $n \in \mathbb{N} : n^{1-\beta^*} > 2$. *Then for* $i = 1, 2$

(i)

$$\left|{}_iL_n(f,x) - f(x)\right|$$
$$\leq \frac{K_i}{\Gamma(\alpha+1)}\left\{\frac{\left(\omega_1\left(D_{x-}^\alpha f, \frac{1}{n^{\beta^*}}\right)_{[a,x]} + \omega_1\left(D_{*x}^\alpha f, \frac{1}{n^{\beta^*}}\right)_{[x,b]}\right)}{n^{\alpha\beta^*}}\right.$$
$$\left. + \frac{\hat{\beta}_{i,n}}{2}\left(\left\|D_{x-}^\alpha f\right\|_{\infty,[a,x]}(x-a)^\alpha + \left\|D_{*x}^\alpha f\right\|_{\infty,[x,b]}(b-x)^\alpha\right)\right\} \tag{10.57}$$

and
(ii)

$$\|{}_iL_n f - f\|_\infty \leq \frac{K_i}{\Gamma(\alpha+1)}$$

$$\left\{ \frac{\left(\sup_{x\in[a,b]} \omega_1 \left(D_{x-}^\alpha f, \frac{1}{n^{\beta^*}}\right)_{[a,x]} + \sup_{x\in[a,b]} \omega_1 \left(D_{*x}^\alpha f, \frac{1}{n^{\beta^*}}\right)_{[x,b]} \right)}{n^{\alpha\beta^*}} \right.$$

$$\left. + \frac{(b-a)^\alpha \hat{\beta}_{i,n}}{2} \left(\sup_{x\in[a,b]} \left\|D_{x-}^\alpha f\right\|_{\infty,[a,x]} + \sup_{x\in[a,b]} \left\|D_{*x}^\alpha f\right\|_{\infty,[x,b]} \right) \right\}. \tag{10.58}$$

Proof. From Theorems 10.6 and 10.12. □

When $\alpha = \frac{1}{2}$, we derive

Corollary 10.3. *Let* $0 < \beta^* < 1$, $q, \lambda, \beta > 0$, $f \in C^1([a,b])$, $x \in [a,b]$, $n \in \mathbb{N} : n^{1-\beta^*} > 2$. *Then for* $i = 1, 2$

(i)

$$|{}_iL_n(f,x) - f(x)|$$

$$\leq \frac{2K_i}{\sqrt{\pi}} \left\{ \frac{\left(\omega_1 \left(D_{x-}^{\frac{1}{2}} f, \frac{1}{n^{\beta^*}}\right)_{[a,x]} + \omega_1 \left(D_{*x}^{\frac{1}{2}} f, \frac{1}{n^{\beta^*}}\right)_{[x,b]} \right)}{n^{\frac{\beta^*}{2}}} \right.$$

$$\left. + \frac{\hat{\beta}_{i,n}}{2} \left(\left\|D_{x-}^{\frac{1}{2}} f\right\|_{\infty,[a,x]} \sqrt{(x-a)} + \left\|D_{*x}^{\frac{1}{2}} f\right\|_{\infty,[x,b]} \sqrt{(b-x)} \right) \right\} \tag{10.59}$$

and

(ii)

$$\|{}_iL_nf - f\|_\infty \leq \frac{2K_i}{\sqrt{\pi}}$$

$$\left\{ \frac{\left(\sup_{x\in[a,b]} \omega_1\left(D_{x-}^{\frac{1}{2}}f, \frac{1}{n^{\beta^*}}\right)_{[a,x]} + \sup_{x\in[a,b]} \omega_1\left(D_{*x}^{\frac{1}{2}}f, \frac{1}{n^{\beta^*}}\right)_{[x,b]} \right)}{n^{\frac{\beta^*}{2}}} \right.$$

$$\left. + \frac{\sqrt{(b-a)}\hat{\beta}_{i,n}}{2} \left(\sup_{x\in[a,b]} \left\| D_{x-}^{\frac{1}{2}}f \right\|_{\infty,[a,x]} + \sup_{x\in[a,b]} \left\| D_{*x}^{\frac{1}{2}}f \right\|_{\infty,[x,b]} \right) \right\} < \infty. \tag{10.60}$$

10.4 About Random Motion on Simple Graphs

Here we follow Papanicolaou, Papageorgiou and Lepipas (2012).

Suppose we have a system of S semiaxes with a common origin radially arranged and a particle moving randomly on S. Possible applications include spread of toxic particles in a system of channels or vessels or propagation of information in networks.

The mathematical model is the following: Let S be the set consisting of n semiaxes $S_1, \ldots, S_n, n \geq 2$, with a common origin 0 and X_t the Brownian motion process on S, namely the diffusion process on S whose infinitesimal generator L is

$$Lu = \frac{1}{2}u'', \tag{10.61}$$

where

$$u = (u_1, \ldots, u_n),$$

together with the continuity conditions (a total of $n-1$ equations),

$$u_1(0) = \cdots = u_n(0) \tag{10.62}$$

and the so-called "Kirchhoff condition"

$$u_1'(0) + \cdots + u_n'(0) = 0. \tag{10.63}$$

This is a Walsh's-type Brownian motion (see Barlow, Pitman and Yor 1989).

The process X_t does a standard Brownian motion on each of the semi-axes and, when it hits 0, it continues its motion on the jth semiaxis, $1 \leq j \leq n$, with probability $\frac{1}{n}$.

For each semiaxes $S_j, 1 \leq j \leq n$, it is convenient to use the coordinate $x_j, 0 \leq x_j \leq \infty$. Note that, if $u = (u_1, \ldots, u_n)$ is a function on S, then its jth component, u_j, is a function on S_j, thus $u_j = u_j(x_j)$.

The transition density of X_t is

$$p(t, x_k, y_j) = \frac{2}{n\sqrt{2\pi t}} e^{-\frac{(x_k+y_j)^2}{2t}}, \text{ if } k \neq j,$$

and

$$p(t, x_k, y_k) = \frac{1}{\sqrt{2\pi t}} \left(e^{-\frac{(x_k-y_k)^2}{2t}} - \frac{n-2}{n} e^{-\frac{(x_k+y_k)^2}{2t}} \right). \tag{10.64}$$

We need the following result.

Theorem 10.15. *Let $t \in [t_1, t_2]$, where $t_1, t_2 \in (0, \infty)$ with $t_1 < t_2$ are fixed. We consider the function $g : \mathbb{R} \to \mathbb{R}$, which is bounded on $(0, \infty)$, i.e. there exists $M > 0$ such that $|g(x)| \leq M$, for every $x \in (0, \infty)$, and Lebesgue measurable on $\mathbb{R}$. Let also X_t be the standard Brownian motion on each of the semiaxes $j = 1, \ldots, n$ as described above. Here x_k is fixed on S_k semiaxes, $k \in \{1, \ldots, n\}$. We consider the related expected value function*

$$r(t) := E_k(|g(X_t)|)$$

$$= \int_0^\infty |g(y_k)| \, p(t, x_k, y_k) dy_k + \sum_{j=1, j\neq k}^{n} \int_0^\infty |g(y_j)| \, p(t, x_k, y_j) dy_j, \ t \in [t_1, t_2].$$

The function $r(t)$ is continuous in t and differentiable.

Proof. First we observe that for $t \in [t_1, t_2]$ and $k, j \in \{1, \ldots, n\}$ with $k \neq j$

$$0 < p(t, x_k, y_j) < \frac{2}{\sqrt{2\pi t_1}}.$$

Also for $t \in [t_1, t_2]$, and $k \in \{1, \ldots, n\}$, it is

$$0 < p(t, x_k, y_k) < \frac{2}{\sqrt{2\pi t_1}}.$$

It is enough to prove that

$$I(t) := \int_0^\infty |g(y_k)| \, p(t, x_k, y_k) dy_k$$

is continuous in $t \in [t_1, t_2]$.

We have that

$$|g(y_k)| \leq M.$$

Thus,

$$|g(y_k)|\, p\,(t, x_k, y_k) \leq M \frac{2}{\sqrt{2\pi t_1}}.$$

Furthermore, as $0 < t_N \to t$, with $N \longrightarrow \infty$, we get

$$|g(y_k)|\, p\,(t_N, x_k, y_k) \longrightarrow |g(y_k)|\, p\,(t, x_k, y_k)\,, \text{ for every } y_k \geq 0.$$

By the dominating convergence theorem, $I\,(t_n) \longrightarrow I\,(t)$ and thus, $I\,(t)$ is continuous in t.

Consequently, the function

$$r\,(t) := E_k\,(|g(X_t)|)$$

is continuous in t. □

We also need the following theorem.

Theorem 10.16. *Let $t \in [t_1, t_2]\,,$ where $t_1, t_2 \in (0, \infty)$ with $t_1 < t_2$ are fixed. We consider function $g : \mathbb{R} \to \mathbb{R}$, which is bounded on $(0, \infty)$ and Lebesgue measurable on $\mathbb{R}$. Let also X_t be the standard Brownian motion on each of the semiaxes $j = 1, \ldots, n$ as described above. Here x_k is fixed on S_k semiaxes, $k \in \{1, \ldots, n\}$. Then the related expected value function*

$$r\,(t) := E_k\,(|g(X_t)|)$$

$$= \int_0^\infty |g(y_k)|\, p(t, x_k, y_k) dy_k + \sum_{j=1, j\neq k}^{n} \int_0^\infty |g(y_j)|\, p(t, x_k, y_j) dy_j,\ t \in [t_1, t_2]\,,$$

is differentiable in t, and

$$\frac{\partial r\,(t)}{\partial t} = \int_0^\infty |g(y_k)|\, \frac{\partial p(t, x_k, y_k)}{\partial t} dy_k$$

$$+ \sum_{j=1, j\neq k}^{n} \int_0^\infty |g(y_j)|\, \frac{\partial p(t, x_k, y_j)}{\partial t} dy_j,\ t \in [t_1, t_2]\,, \tag{10.65}$$

which is continuous in t.

Proof. First, we observe that for $t \in [t_1, t_2]$ and $k, j \in \{1, \ldots, n\}$ with $k \neq j$

$$\frac{\partial p\,(t, x_k, y_j)}{\partial t} = \frac{1}{nt\sqrt{2\pi t}} e^{-\frac{(x_k + y_j)^2}{2t}} \left(\frac{(x_k + y_j)^2}{t} - 1 \right).$$

Also for $t \in [t_1, t_2]$, and $k \in \{1, \ldots, n\}$, it is

$$\frac{\partial p(t, x_k, y_k)}{\partial t} = \frac{1}{2t\sqrt{2\pi t}}$$

$$\left[e^{-\frac{(x_k - y_k)^2}{2t}} \left(\frac{(x_k - y_k)^2}{t} - 1 \right) - \frac{(n-2)}{n} e^{-\frac{(x_k + y_k)^2}{2t}} \left(\frac{(x_k + y_k)^2}{t} - 1 \right) \right].$$

Furthermore, for $k \neq j$,

$$\left| \frac{\partial p(t, x_k, y_j)}{\partial t} \right| \leq \frac{1}{n t_1 \sqrt{2\pi t_1}} \left(\frac{(x_k + y_j)^2}{t_1} + 1 \right)$$

for every $y_j \in (0, \infty)$,
and

$$\left| \frac{\partial p(t, x_k, y_k)}{\partial t} \right| \leq \frac{1}{2t_1\sqrt{2\pi t_1}} \left[\left(\frac{(x_k - y_k)^2}{t_1} + 1 \right) + \frac{(n-2)}{n} \left(\frac{(x_k + y_k)^2}{t_1} + 1 \right) \right]$$

for every $y_k \in (0, \infty)$.

So $\frac{\partial p(t, x_k, y_j)}{\partial t}$ and $\frac{\partial p(t, x_k, y_k)}{\partial t}$ is bounded with respect to t. The bounds are integrable with respect to y_j and y_k, respectively.

We have

$$r(t) := E_k(|g(X_t)|) = \int_0^\infty |g(y_k)| \, p(t, x_k, y_k) dy_k$$
$$+ \sum_{j=1, j \neq k}^{n} \int_0^\infty |g(y_j)| \, p(t, x_k, y_j) dy_j, \; t \in [t_1, t_2].$$

We apply differentiation under the integral sign.

We note that

$$|g(y_k)| \, p(t, x_k, y_k) \leq M \frac{1}{2t_1\sqrt{2\pi t_1}} \left[\left(\frac{(x_k - y_k)^2}{t_1} + 1 \right) + \frac{(n-2)}{n} \left(\frac{(x_k + y_k)^2}{t_1} + 1 \right) \right]$$

and

$$|g(y_k)| \, p(t, x_k, y_j) \leq M \frac{1}{n t_1 \sqrt{2\pi t_1}} \left(\frac{(x_k + y_j)^2}{t_1} + 1 \right).$$

Therefore, there exists

$$\frac{\partial r(t)}{\partial t} = \int_0^\infty |g(y_k)| \frac{\partial p(t, x_k, y_k)}{\partial t} dy_k + \sum_{j=1, j \neq k}^{n} \int_0^\infty |g(y_j)| \frac{\partial p(t, x_k, y_j)}{\partial t} dy_j, \; t \in [t_1, t_2],$$

which is continuous in t (same proof as in Theorem 10.15). □

10.5 Main Results

We present the following general approximation results of Brownian Motion on simple graphs.

Theorem 10.17. *We consider function $g : \mathbb{R} \to \mathbb{R}$, which is bounded on $(0, \infty)$ and Lebesgue measurable on $\mathbb{R}$. Let also $r(t) := E_k[g(X_t)]$ the related expected value function.*

If $0 < \beta^* < 1$, $n \in \mathbb{N} : n^{1-\beta^*} > 2$, $q, \lambda, \beta > 0$, $t \in [t_1, t_2]$, where $t_1, t_2 \in (0, \infty)$ with $t_1 < t_2$, then for $i = 1, 2$

(i)

$$\left|{}_iL_n(r(t)) - r(t)\right| \leq K_i \cdot \left[\omega_1\left(r, \frac{1}{n^{\beta^*}}\right) + \hat{\beta}_{i,n}\|r\|_{\infty,[t_1,t_2]}\right] =: \rho_i \quad (10.66)$$

and

(ii)

$$\left\|{}_iL_n(r(t)) - r(t)\right\|_{\infty,[t_1,t_2]} \leq \rho_i. \quad (10.67)$$

We get that $\lim_{n\to\infty} {}_iL_n(r) = r$, pointwise and uniformly.

Proof. From Theorem 10.13. □

Next we present

Theorem 10.18. *We consider function $g : \mathbb{R} \to \mathbb{R}$, which is bounded on $(0, \infty)$ and Lebesgue measurable on $\mathbb{R}$. Let also $r(t) := E_k[g(X_t)]$ the related expected value function.*

If $0 < \alpha, \beta^* < 1$, $q, \lambda, \beta > 0$, $t \in [t_1, t_2]$, where $t_1, t_2 \in (0, \infty)$ with $t_1 < t_2$, and $n \in \mathbb{N} : n^{1-\beta^*} > 2$. Then for $i = 1, 2$

(i)

$$\begin{aligned} &\left|{}_iL_n(r(t)) - r(t)\right| \\ &\leq \frac{K_i}{\Gamma(\alpha+1)}\left\{\frac{\left(\omega_1\left(D_{t-}^{\alpha}r, \frac{1}{n^{\beta^*}}\right)_{[t_1,t]} + \omega_1\left(D_{*t}^{\alpha}r, \frac{1}{n^{\beta^*}}\right)_{[t,t_2]}\right)}{n^{\alpha\beta^*}}\right. \\ &\left.+ \frac{\hat{\beta}_{i,n}}{2}\left(\left\|D_{t-}^{\alpha}r\right\|_{\infty,[t_1,t]}(t-t_1)^{\alpha} + \left\|D_{*t}^{\alpha}r\right\|_{\infty,[t,t_2]}(t_2-t)^{\alpha}\right)\right\} \end{aligned} \quad (10.68)$$

and
(ii)

$$\left\|{}_iL_nr - r\right\|_{\infty,[t_1,t_2]} \leq \frac{K_i}{\Gamma(\alpha+1)}$$

$$\left\{\frac{\left(\sup_{t\in[t_1,t_2]}\omega_1\left(D_{t-}^{\alpha}r,\frac{1}{n^{\beta^*}}\right)_{[t_1,t]}+\sup_{t\in[t_1,t_2]}\omega_1\left(D_{*t}^{\alpha}r,\frac{1}{n^{\beta^*}}\right)_{[t,t_2]}\right)}{n^{\alpha\beta^*}}\right.$$

$$\left.+\frac{(t_2-t_1)^{\alpha}\hat{\beta}_{i,n}}{2}\left(\sup_{t\in[t_1,t_2]}\left\|D_{t-}^{\alpha}r\right\|_{\infty,[t_1,t]}+\sup_{t\in[t_1,t_2]}\left\|D_{*t}^{\alpha}r\right\|_{\infty,[t,t_2]}\right)\right\}. \tag{10.69}$$

Proof. From Theorem 10.14. □

When $\alpha = \frac{1}{2}$, we derive

Corollary 10.4. *We consider function $g : \mathbb{R} \to \mathbb{R}$, which is bounded on $(0,\infty)$ and Lebesgue measurable on $\mathbb{R}$. Let also $r(t) := E_k[g(X_t)]$ the related expected value function.*

If $0 < \beta^* < 1$, $q, \lambda, \beta > 0$, $t \in [t_1,t_2]$ and $n \in \mathbb{N} : n^{1-\beta^*} > 2$. Then for $i = 1,2$

(i)

$$\left|{}_iL_n(r(t)) - r(t)\right|$$

$$\leq \frac{2K_i}{\sqrt{\pi}}\left\{\frac{\left(\omega_1\left(D_{t-}^{\frac{1}{2}}r,\frac{1}{n^{\beta^*}}\right)_{[t_1,t]}+\omega_1\left(D_{*t}^{\frac{1}{2}}r,\frac{1}{n^{\beta^*}}\right)_{[t,t_2]}\right)}{n^{\frac{\beta^*}{2}}}\right.$$

$$\left.+\frac{\hat{\beta}_{i,n}}{2}\left(\left\|D_{t-}^{\frac{1}{2}}r\right\|_{\infty,[t_1,t]}\sqrt{(t-t_1)}+\left\|D_{*t}^{\frac{1}{2}}r\right\|_{\infty,[t,t_2]}\sqrt{(t_2-t)}\right)\right\} \tag{10.70}$$

and

(ii)

$$\|{}_iL_n r - r\|_{\infty,[t_1,t_2]} \leq \frac{2K_i}{\sqrt{\pi}}$$

$$\left\{ \frac{\left(\sup_{t\in[t_1,t_2]} \omega_1 \left(D_{t-}^{\frac{1}{2}} r, \frac{1}{n^{\beta^*}} \right)_{[t_1,t]} + \sup_{t\in[t_1,t_2]} \omega_1 \left(D_{*t}^{\frac{1}{2}} r, \frac{1}{n^{\beta^*}} \right)_{[t,t_2]} \right)}{n^{\frac{\beta^*}{2}}} \right.$$

$$\left. + \frac{\sqrt{(t_2-t_1)}\hat{\beta}_{i,n}}{2} \left(\sup_{t\in[t_1,t_2]} \left\| D_{t-}^{\frac{1}{2}} r \right\|_{\infty,[t_1,t]} + \sup_{t\in[t_1,t_2]} \left\| D_{*t}^{\frac{1}{2}} r \right\|_{\infty,[t,t_2]} \right) \right\} < \infty. \tag{10.71}$$

Proof. From Corollary 10.3. □

We continue with

Theorem 10.19. *We consider function $g : \mathbb{R} \to \mathbb{R}$, which is bounded on $(0,\infty)$ and Lebesgue measurable on $\mathbb{R}$. Let also $r(t) := E_k[g(X_t)]$ the related expected value function.*

If $0 < \beta^* < 1$, $n \in \mathbb{N} : n^{1-\beta^*} > 2$, $q, \lambda, \beta > 0$, $t \in [t_1, t_2]$, where $t_1, t_2 \in (0,\infty)$ with $t_1 < t_2$, then for $i = 1, 2$

(i)

$$\left| {}_iL_n \left(\frac{\partial r(t)}{\partial t} \right) - \frac{\partial r(t)}{\partial t} \right| \leq K_i \cdot \left[\omega_1 \left(\frac{\partial r}{\partial t}, \frac{1}{n^{\beta^*}} \right) + \hat{\beta}_{i,n} \left\| \frac{\partial r}{\partial t} \right\|_{\infty,[t_1,t_2]} \right] =: \rho_i \tag{10.72}$$

and

(ii)

$$\left\| {}_iL_n \left(\frac{\partial r}{\partial t} \right) - \frac{\partial r}{\partial t} \right\|_{\infty,[t_1,t_2]} \leq \rho_i. \tag{10.73}$$

We get that $\lim_{n\to\infty} {}_iL_n \left(\frac{\partial r}{\partial t} \right) = \frac{\partial r}{\partial t}$, pointwise and uniformly.

Proof. From Theorem 10.13. □

10.6 Applications

Let a function $g : \mathbb{R} \to \mathbb{R}$, which is bounded on $[t_1, t_2]$, where $t_1, t_2 > 0$ with $t_1 < t_2$ and Lebesgue measurable on $\mathbb{R}$. For the Brownian Motion on simple graphs X_t, we use the following notations:

$$r(t) := E_k(|g(X_t)|)$$
$$= \int_0^{\infty} |g(y_k)| \, p(t, x_k, y_k) dy_k + \sum_{j=1, j\neq k}^{n} \int_0^{\infty} |g(y_j)| \, p(t, x_k, y_j) dy_j := E_k(|g(X_t)|)^{(0)} \tag{10.74}$$

and

$$\frac{\partial r(t)}{\partial t} = \int_0^{\infty} |g(y_k)| \frac{\partial p(t, x_k, y_k)}{\partial t} dy_k + \sum_{j=1, j\neq k}^{n} \int_0^{\infty} |g(y_j)| \frac{\partial p(t, x_k, y_j)}{\partial t} dy_j := E_k(|g(X_t)|)^{(1)}. \tag{10.75}$$

We can apply our main results to the function $g(W) = W$. Consider the function $g : \mathbb{R} \to \mathbb{R}$, where $g(x) = x$ for every $x \in \mathbb{R}$. Let also $W = X_t$ be the Brownian motion on simple graphs. Then the expectation

$$E_k(|W|)(t) = \int_0^{\infty} |y_k| \, p(t, x_k, y_k) dy_k + \sum_{j=1, j\neq k}^{n} \int_0^{\infty} |y_j| \, p(t, x_k, y_j) dy_j$$

is continuous in t.

Moreover,

Corollary 10.5. *Let* $0 < \beta^* < 1$, $n \in \mathbb{N} : n^{1-\beta^*} > 2$, $q, \lambda, \beta > 0$, $t \in [t_1, t_2]$, *where* $t_1, t_2 \in (0, \infty)$ *with* $t_1 < t_2$, *then for* $i = 1, 2$ *and* $j = 0, 1$

(i)

$$\left| {}_iL_n\left(E_k(|W|)^{(j)}\right)(t) - E_k(|W|)^{(j)}(t)\right|$$
$$\leq K_i \cdot \left[\omega_1\left(E_k(|W|)^{(j)}, \frac{1}{n^{\beta^*}}\right) + \hat{\beta}_{i,n}\left\|E_k(|W|)^{(j)}\right\|_{\infty,[t_1,t_2]}\right] =: \rho_i \tag{10.76}$$

and

(ii)

$$\left\| {}_iL_n\left(E_k(|W|)^{(j)}\right)(t) - E_k(|W|)^{(j)}(t)\right\|_{\infty,[t_1,t_2]} \leq \rho_i. \tag{10.77}$$

We get that $\lim_{n\to\infty} {}_iL_n E_k(|W|)^{(j)}(t) = E_k(|W|)^{(j)}$, *pointwise and uniformly.*

Proof. From Theorems 10.17 and 10.19. □

Next we present

Corollary 10.6. *Let* $0 < \alpha, \beta^* < 1$, $q, \lambda, \beta > 0$, $t \in [t_1, t_2]$, *where* $t_1, t_2 \in (0, \infty)$ *with* $t_1 < t_2$, *and* $n \in \mathbb{N} : n^{1-\beta^*} > 2$. *Then for* $i = 1, 2$

(i)

$$\left|{}_iL_n\left(E_k\left(|W|\right)(t)\right) - E_k\left(|W|\right)(t)\right|$$

$$\leq \frac{K_i}{\Gamma(\alpha+1)} \left\{ \frac{\left(\omega_1\left(D_{t-}^{\alpha}E_k\left(|W|\right), \frac{1}{n^{\beta^*}}\right)_{[t_1,t]} + \omega_1\left(D_{*t}^{\alpha}E_k\left(|W|\right), \frac{1}{n^{\beta^*}}\right)_{[t,t_2]}\right)}{n^{\alpha\beta^*}} \right.$$

$$\left. + \frac{\hat{\beta}_{i,n}}{2}\left(\left\|D_{t-}^{\alpha}E_k\left(|W|\right)\right\|_{\infty,[t_1,t]}(t-t_1)^{\alpha} + \left\|D_{*t}^{\alpha}E_k\left(|W|\right)\right\|_{\infty,[t,t_2]}(t_2-t)^{\alpha}\right)\right\} \tag{10.78}$$

and
(ii)

$$\left\|{}_iL_nE_k\left(|W|\right) - E_k\left(|W|\right)\right\|_{\infty,[t_1,t_2]} \leq \frac{K_i}{\Gamma(\alpha+1)}$$

$$\left\{ \frac{\left(\sup_{t\in[t_1,t_2]}\omega_1\left(D_{t-}^{\alpha}E_k\left(|W|\right), \frac{1}{n^{\beta^*}}\right)_{[t_1,t]} + \sup_{t\in[t_1,t_2]}\omega_1\left(D_{*t}^{\alpha}E_k\left(|W|\right)\frac{1}{n^{\beta^*}}\right)_{[t,t_2]}\right)}{n^{\alpha\beta^*}} \right.$$

$$\left. + \frac{(t_2-t_1)^{\alpha}\hat{\beta}_{i,n}}{2}\left(\sup_{t\in[t_1,t_2]}\left\|D_{t-}^{\alpha}E_k\left(|W|\right)\right\|_{\infty,[t_1,t]} + \sup_{t\in[t_1,t_2]}\left\|D_{*t}^{\alpha}E_k\left(|W|\right)\right\|_{\infty,[t,t_2]}\right)\right\}. \tag{10.79}$$

Proof. From Theorem 10.18. □

When $\alpha = \frac{1}{2}$, we derive

Corollary 10.7. *Let* $0 < \beta^* < 1$, $q, \lambda, \beta > 0$, $t \in [t_1, t_2]$ *and* $n \in \mathbb{N}$: $n^{1-\beta^*} > 2$. *Then for* $i = 1, 2$

(i)

$$\left|{}_iL_n\left(E_k\left(|W|\right)(t)\right)-E_k\left(|W|\right)(t)\right|$$

$$\leq\frac{2K_i}{\sqrt{\pi}}\left\{\frac{\left(\omega_1\left(D_{t-}^{\frac{1}{2}}E_k\left(|W|\right),\frac{1}{n^{\beta^*}}\right)_{[t_1,t]}+\omega_1\left(D_{*t}^{\frac{1}{2}}E_k\left(|W|\right),\frac{1}{n^{\beta^*}}\right)_{[t,t_2]}\right)}{n^{\frac{\beta^*}{2}}}\right.$$

$$\left.+\frac{\hat{\beta}_{i,n}}{2}\left(\left\|D_{t-}^{\frac{1}{2}}E_k\left(|W|\right)\right\|_{\infty,[t_1,t]}\sqrt{(t-t_1)}+\left\|D_{*t}^{\frac{1}{2}}E_k\left(|W|\right)\right\|_{\infty,[t,t_2]}\sqrt{(t_2-t)}\right)\right\} \tag{10.80}$$

and
(ii)

$$\left\|{}_iL_nE_k\left(|W|\right)-E_k\left(|W|\right)\right\|_{\infty,[t_1,t_2]}\leq\frac{2K_i}{\sqrt{\pi}}$$

$$\left\{\frac{\left(\sup_{t\in[t_1,t_2]}\omega_1\left(D_{t-}^{\frac{1}{2}}E_k\left(|W|\right),\frac{1}{n^{\beta^*}}\right)_{[t_1,t]}+\sup_{t\in[t_1,t_2]}\omega_1\left(D_{*t}^{\frac{1}{2}}E_k\left(|W|\right),\frac{1}{n^{\beta^*}}\right)_{[t,t_2]}\right)}{n^{\frac{\beta^*}{2}}}\right.$$

$$+\frac{\sqrt{(t_2-t_1)}\hat{\beta}_{i,n}}{2}$$

$$\left.\left(\sup_{t\in[t_1,t_2]}\left\|D_{t-}^{\frac{1}{2}}E_k\left(|W|\right)\right\|_{\infty,[t_1,t]}+\sup_{t\in[t_1,t_2]}\left\|D_{*t}^{\frac{1}{2}}E_k\left(|W|\right)\right\|_{\infty,[t,t_2]}\right)\right\}<\infty. \tag{10.81}$$

Proof. From Corollary 10.4. □

For the next application, we consider the function $g:\mathbb{R}\to\mathbb{R}$, where $g(x)=\cos x$ for every $x\in\mathbb{R}$. Let also $W=X_t$ be the Brownian motion on simple graphs. Then the expectation

$$E_k\left(|\cos W|\right)(t)=\int_0^{\infty}\left|\cos\left(y_k\right)\right|p(t,x_k,y_k)dy_k+\sum_{j=1,j\neq k}^{n}\int_0^{\infty}\left|\cos\left(y_j\right)\right|p(t,x_k,y_j)dy_j$$

is continuous in t.

Moreover,

Corollary 10.8. *Let* $0 < \beta^* < 1$, $n \in \mathbb{N} : n^{1-\beta^*} > 2$, $q, \lambda, \beta > 0$, $t \in [t_1, t_2]$, *where* $t_1, t_2 \in (0, \infty)$ *with* $t_1 < t_2$, *then for* $i = 1, 2$ *and* $j = 0, 1$

(i)

$$\left| {}_iL_n\left(E_k\left(|\cos W|\right)^{(j)}\right)(t) - E_k\left(|\cos W|\right)^{(j)}(t)\right|$$
$$\leq K_i \cdot \left[\omega_1\left(E_k\left(|\cos W|\right)^{(j)}, \frac{1}{n^{\beta^*}}\right) + \hat{\beta}_{i,n}\left\|E_k\left(|\cos W|\right)^{(j)}\right\|_{\infty,[t_1,t_2]}\right] =: \rho_i \tag{10.82}$$

and

(ii)

$$\left\| {}_iL_n\left(E_k\left(|\cos W|\right)^{(j)}\right)(t) - E_k\left(|\cos W|\right)^{(j)}(t)\right\|_{\infty,[t_1,t_2]} \leq \rho_i. \tag{10.83}$$

We get that $\lim_{n\to\infty} {}_iL_nE_k\left(|\cos W|\right)^{(j)}(t) = E_k\left(|\cos W|\right)^{(j)}$, *pointwise and uniformly.*

Proof. From Theorems 10.17 and 10.19. ☐

Next we present

Corollary 10.9. *Let* $0 < \alpha, \beta^* < 1$, $q, \lambda, \beta > 0$, $t \in [t_1, t_2]$, *where* $t_1, t_2 \in (0, \infty)$ *with* $t_1 < t_2$, *and* $n \in \mathbb{N} : n^{1-\beta^*} > 2$. *Then for* $i = 1, 2$

(i)

$$\left| {}_iL_n\left(E_k\left(|\cos W|\right)(t)\right) - E_k\left(|\cos W|\right)(t)\right|$$
$$\leq \frac{K_i}{\Gamma(\alpha+1)}\left\{\frac{\left(\omega_1\left(D^\alpha_{t-}E_k\left(|\cos W|\right), \frac{1}{n^{\beta^*}}\right)_{[t_1,t]} + \omega_1\left(D^\alpha_{*t}E_k\left(|\cos W|\right), \frac{1}{n^{\beta^*}}\right)_{[t,t_2]}\right)}{n^{\alpha\beta^*}}\right.$$
$$\left. + \frac{\hat{\beta}_{i,n}}{2}\left(\left\|D^\alpha_{t-}E_k\left(|\cos W|\right)\right\|_{\infty,[t_1,t]}(t-t_1)^\alpha + \left\|D^\alpha_{*t}E_k\left(|\cos W|\right)\right\|_{\infty,[t,t_2]}(t_2-t)^\alpha\right)\right\} \tag{10.84}$$

and
(ii)

$$\left\|{}_iL_nE_k\left(|\cos W|\right)-E_k\left(|\cos W|\right)\right\|_{\infty,[t_1,t_2]}\leq\frac{K_i}{\Gamma\left(\alpha+1\right)}$$

$$\left\{\frac{\left(\sup_{t\in[t_1,t_2]}\omega_1\left(D_{t-}^{\alpha}E_k\left(|\cos W|\right),\frac{1}{n^{\beta^*}}\right)_{[t_1,t]}+\sup_{t\in[t_1,t_2]}\omega_1\left(D_{*t}^{\alpha}E_k\left(|\cos W|\right)\frac{1}{n^{\beta^*}}\right)_{[t,t_2]}\right)}{n^{\alpha\beta^*}}\right.$$

$$+\frac{\left(t_2-t_1\right)^{\alpha}\hat{\beta}_{i,n}}{2}$$

$$\left.\left(\sup_{t\in[t_1,t_2]}\left\|D_{t-}^{\alpha}E_k\left(|\cos W|\right)\right\|_{\infty,[t_1,t]}+\sup_{t\in[t_1,t_2]}\left\|D_{*t}^{\alpha}E_k\left(|\cos W|\right)\right\|_{\infty,[t,t_2]}\right)\right\}. \tag{10.85}$$

Proof. From Theorem 10.18. □

When $\alpha=\frac{1}{2}$, we derive

Corollary 10.10. *Let $0<\beta^*<1$, $q,\lambda,\beta>0$, $t\in[t_1,t_2]$ and $n\in\mathbb{N}$: $n^{1-\beta^*}>2$. Then for $i=1,2$*

(i)

$$\left|{}_iL_n\left(E_k\left(|\cos W|\right)(t)\right)-E_k\left(|\cos W|\right)(t)\right|$$

$$\leq\frac{2K_i}{\sqrt{\pi}}\left\{\frac{\left(\omega_1\left(D_{t-}^{\frac{1}{2}}E_k\left(|\cos W|\right),\frac{1}{n^{\beta^*}}\right)_{[t_1,t]}+\omega_1\left(D_{*t}^{\frac{1}{2}}E_k\left(|\cos W|\right),\frac{1}{n^{\beta^*}}\right)_{[t,t_2]}\right)}{n^{\frac{\beta^*}{2}}}\right.$$

$$\left.+\frac{\hat{\beta}_{i,n}}{2}\left(\left\|D_{t-}^{\frac{1}{2}}E_k\left(|\cos W|\right)\right\|_{\infty,[t_1,t]}\sqrt{(t-t_1)}+\left\|D_{*t}^{\frac{1}{2}}E_k\left(|\cos W|\right)\right\|_{\infty,[t,t_2]}\sqrt{(t_2-t)}\right)\right\} \tag{10.86}$$

and
(ii)

$$\left\|{}_iL_nE_k\left(|\cos W|\right)-E_k\left(|\cos W|\right)\right\|_{\infty,[t_1,t_2]}\leq\frac{2K_i}{\sqrt{\pi}}$$

$$\left\{\frac{\left(\sup_{t\in[t_1,t_2]}\omega_1\left(D_{t-}^{\frac{1}{2}}E_k\left(|\cos W|\right),\frac{1}{n^{\beta^*}}\right)_{[t_1,t]}+\sup_{t\in[t_1,t_2]}\omega_1\left(D_{*t}^{\frac{1}{2}}E_k\left(|\cos W|\right),\frac{1}{n^{\beta^*}}\right)_{[t,t_2]}\right)}{n^{\frac{\beta^*}{2}}}\right.$$

$$+\frac{\sqrt{(t_2-t_1)}\hat{\beta}_{i,n}}{2}$$

$$\left.\left(\sup_{t\in[t_1,t_2]}\left\|D_{t-}^{\frac{1}{2}}E_k\left(|\cos W|\right)\right\|_{\infty,[t_1,t]}+\sup_{t\in[t_1,t_2]}\left\|D_{*t}^{\frac{1}{2}}E_k\left(|\cos W|\right)\right\|_{\infty,[t,t_2]}\right)\right\}<\infty. \tag{10.87}$$

Proof. From Corollary 10.4. □

Let us consider now the function $g:\mathbb{R}\to\mathbb{R}$, where $g(x)=\tanh x$ for every $x\in\mathbb{R}$. Let also $W=X_t$ be the Brownian motion on simple graphs. Then the expectation

$$E_k\left(|\tanh W|\right)(t)$$

$$=\int_0^\infty|\tanh(y_k)|\,p(t,x_k,y_k)dy_k+\sum_{j=1,j\neq k}^{n}\int_0^\infty|\tanh(y_j)|\,p(t,x_k,y_j)dy_j$$

is continuous in t.

Moreover,

Corollary 10.11. *Let* $0<\beta^*<1$, $n\in\mathbb{N}:n^{1-\beta^*}>2$, $q,\lambda,\beta>0$, $t\in[t_1,t_2]$, *where* $t_1,t_2\in(0,\infty)$ *with* $t_1<t_2$, *then for* $i=1,2$ *and* $j=0,1$

(i)

$$\left|{}_iL_n\left(E_k\left(|\tanh W|\right)^{(j)}\right)(t)-E_k\left(|\tanh W|\right)^{(j)}(t)\right|$$

$$\leq K_i\cdot\left[\omega_1\left(E_k\left(|\tanh W|\right)^{(j)},\frac{1}{n^{\beta^*}}\right)+\hat{\beta}_{i,n}\left\|E_k\left(|\tanh W|\right)^{(j)}\right\|_{\infty,[t_1,t_2]}\right]=:\rho_i \tag{10.88}$$

and

(ii)

$$\left\|{}_iL_n\left(E_k\left(|\tanh W|\right)^{(j)}\right)(t)-E_k\left(|\tanh W|\right)^{(j)}(t)\right\|_{\infty,[t_1,t_2]}\le\rho_i. \quad (10.89)$$

We get that $\lim_{n\to\infty}{}_iL_nE_k\left(|\tanh W|\right)^{(j)}(t)=E_k\left(|\tanh W|\right)^{(j)}$, *pointwise and uniformly.*

Proof. From Theorems 10.17 and 10.19. □

Next we present

Corollary 10.12. *Let* $0<\alpha,\beta^*<1$, $q,\lambda,\beta>0$, $t\in[t_1,t_2]$, *where* $t_1,t_2\in(0,\infty)$ *with* $t_1<t_2$, *and* $n\in\mathbb{N}:n^{1-\beta^*}>2$. *Then for* $i=1,2$

(i)

$$\left|{}_iL_n\left(E_k\left(|\tanh W|\right)(t)\right)-E_k\left(|\tanh W|\right)(t)\right|$$

$$\le\frac{K_i}{\Gamma(\alpha+1)}\left\{\frac{\left(\omega_1\left(D_{t-}^{\alpha}E_k\left(|\tanh W|\right),\frac{1}{n^{\beta^*}}\right)_{[t_1,t]}+\omega_1\left(D_{*t}^{\alpha}E_k\left(|\tanh W|\right),\frac{1}{n^{\beta^*}}\right)_{[t,t_2]}\right)}{n^{\alpha\beta^*}}\right.$$

$$+\frac{\hat{\beta}_{i,n}}{2}$$

$$\left.\left(\left\|D_{t-}^{\alpha}E_k\left(|\tanh W|\right)\right\|_{\infty,[t_1,t]}(t-t_1)^{\alpha}+\left\|D_{*t}^{\alpha}E_k\left(|\tanh W|\right)\right\|_{\infty,[t,t_2]}(t_2-t)^{\alpha}\right)\right\} \quad (10.90)$$

and

(ii)

$$\left\|{}_iL_nE_k\left(|\tanh W|\right)-E_k\left(|\tanh W|\right)\right\|_{\infty,[t_1,t_2]}\le\frac{K_i}{\Gamma(\alpha+1)}$$

$$\left\{\frac{\left(\sup_{t\in[t_1,t_2]}\omega_1\left(D_{t-}^{\alpha}E_k\left(|\tanh W|\right),\frac{1}{n^{\beta^*}}\right)_{[t_1,t]}+\sup_{t\in[t_1,t_2]}\omega_1\left(D_{*t}^{\alpha}E_k\left(|\tanh W|\right)\frac{1}{n^{\beta^*}}\right)_{[t,t_2]}\right)}{n^{\alpha\beta^*}}\right.$$

$$+\frac{(t_2-t_1)^{\alpha}\hat{\beta}_{i,n}}{2}$$

$$\left.\left(\sup_{t\in[t_1,t_2]}\left\|D_{t-}^{\alpha}E_k\left(|\tanh W|\right)\right\|_{\infty,[t_1,t]}+\sup_{t\in[t_1,t_2]}\left\|D_{*t}^{\alpha}E_k\left(|\tanh W|\right)\right\|_{\infty,[t,t_2]}\right)\right\}. \quad (10.91)$$

Proof. From Theorem 10.18. □

When $\alpha = \frac{1}{2}$, we derive

Corollary 10.13. *Let* $0 < \beta^* < 1$, $q, \lambda, \beta > 0$, $t \in [t_1, t_2]$ *and* $n \in \mathbb{N} : n^{1-\beta^*} > 2$. *Then for* $i = 1, 2$

(i)

$$\left| {}_iL_n\left(E_k\left(|\tanh W|\right)(t)\right) - E_k\left(|\tanh W|\right)(t)\right| \leq \frac{2K_i}{\sqrt{\pi}}$$

$$\left\{ \frac{\left(\omega_1\left(D_{t-}^{\frac{1}{2}} E_k\left(|\tanh W|\right), \frac{1}{n^{\beta^*}}\right)_{[t_1,t]} + \omega_1\left(D_{*t}^{\frac{1}{2}} E_k\left(|\tanh W|\right), \frac{1}{n^{\beta^*}}\right)_{[t,t_2]}\right)}{n^{\frac{\beta^*}{2}}} \right.$$

$$+ \frac{\hat{\beta}_{i,n}}{2}$$

$$\left. \left(\left\| D_{t-}^{\frac{1}{2}} E_k\left(|\tanh W|\right)\right\|_{\infty,[t_1,t]} \sqrt{(t-t_1)} + \left\| D_{*t}^{\frac{1}{2}} E_k\left(|\tanh W|\right)\right\|_{\infty,[t,t_2]} \sqrt{(t_2-t)}\right) \right\} \tag{10.92}$$

and
(ii)

$$\left\| {}_iL_n E_k\left(|\tanh W|\right) - E_k\left(|\tanh W|\right)\right\|_{\infty,[t_1,t_2]} \leq \frac{2K_i}{\sqrt{\pi}}$$

$$\left\{ \frac{\left(\sup_{t\in[t_1,t_2]} \omega_1\left(D_{t-}^{\frac{1}{2}} E_k\left(|\tanh W|\right), \frac{1}{n^{\beta^*}}\right)_{[t_1,t]} + \sup_{t\in[t_1,t_2]} \omega_1\left(D_{*t}^{\frac{1}{2}} E_k\left(|\tanh W|\right), \frac{1}{n^{\beta^*}}\right)_{[t,t_2]}\right)}{n^{\frac{\beta^*}{2}}} \right.$$

$$+ \frac{\sqrt{(t_2-t_1)}\hat{\beta}_{i,n}}{2}$$

$$\left. \left(\sup_{t\in[t_1,t_2]} \left\| D_{t-}^{\frac{1}{2}} E_k\left(|\tanh W|\right)\right\|_{\infty,[t_1,t]} + \sup_{t\in[t_1,t_2]} \left\| D_{*t}^{\frac{1}{2}} E_k\left(|\tanh W|\right)\right\|_{\infty,[t,t_2]}\right) \right\} < \infty. \tag{10.93}$$

Proof. From Corollary 10.4. □

In the following, we consider the function $g:\mathbb{R}\to\mathbb{R}$, where $g(x)=e^{-\ell x},\ell>0$ for every $x\in\mathbb{R}$. Let also $W=X_t$ be the Brownian motion on simple graphs. Then the expectation

$$E_k\left(e^{-\ell W}\right)(t)=\int_0^{\infty}e^{-\ell y_k}p(t,x_k,y_k)dy_k+\sum_{j=1,j\neq k}^{n}\int_0^{\infty}e^{-\ell y_j}p(t,x_k,y_j)dy_j$$

is continuous in t.

Moreover,

Corollary 10.14. *Let* $0<\beta^*<1$, $n\in\mathbb{N}:n^{1-\beta^*}>2$, $q,\lambda,\beta>0$, $t\in[t_1,t_2]$, *where* $t_1,t_2\in(0,\infty)$ *with* $t_1<t_2$, *then for* $i=1,2$ *and* $j=0,1$

(i)

$$\left|{}_iL_n\left(E_k\left(e^{-\ell W}\right)^{(j)}\right)(t)-E_k\left(e^{-\ell W}\right)^{(j)}(t)\right|$$
$$\leq K_i\cdot\left[\omega_1\left(E_k\left(e^{-\ell W}\right)^{(j)},\frac{1}{n^{\beta^*}}\right)+\hat{\beta}_{i,n}\left\|E_k\left(e^{-\ell W}\right)^{(j)}\right\|_{\infty,[t_1,t_2]}\right]=:\rho_i \tag{10.94}$$

and

(ii)

$$\left\|{}_iL_n\left(E_k\left(e^{-\ell W}\right)^{(j)}\right)(t)-E_k\left(e^{-\ell W}\right)^{(j)}(t)\right\|_{\infty,[t_1,t_2]}\leq\rho_i. \tag{10.95}$$

We get that $\lim_{n\to\infty}{}_iL_nE_k\left(e^{-\ell W}\right)^{(j)}(t)=E_k\left(e^{-\ell W}\right)^{(j)}$, *pointwise and uniformly.*

Proof. From Theorems 10.17 and 10.19. □

Next we present

Corollary 10.15. *Let* $0<\alpha,\beta^*<1$, $q,\lambda,\beta>0$, $t\in[t_1,t_2]$, *where* $t_1,t_2\in(0,\infty)$ *with* $t_1<t_2$, *and* $n\in\mathbb{N}:n^{1-\beta^*}>2$. *Then for* $i=1,2$

(i)

$$\left|{}_iL_n\left(E_k\left(e^{-\ell W}\right)(t)\right)-E_k\left(e^{-\ell W}\right)(t)\right|$$
$$\leq\frac{K_i}{\Gamma(\alpha+1)}\left\{\frac{\left(\omega_1\left(D_{t-}^{\alpha}E_k\left(e^{-\ell W}\right),\frac{1}{n^{\beta^*}}\right)_{[t_1,t]}+\omega_1\left(D_{*t}^{\alpha}E_k\left(e^{-\ell W}\right),\frac{1}{n^{\beta^*}}\right)_{[t,t_2]}\right)}{n^{\alpha\beta^*}}\right.$$
$$\left.+\frac{\hat{\beta}_{i,n}}{2}\left(\left\|D_{t-}^{\alpha}E_k\left(e^{-\ell W}\right)\right\|_{\infty,[t_1,t]}(t-t_1)^{\alpha}+\left\|D_{*t}^{\alpha}E_k\left(e^{-\ell W}\right)\right\|_{\infty,[t,t_2]}(t_2-t)^{\alpha}\right)\right\} \tag{10.96}$$

and

(ii)

$$\left\| {}_iL_nE_k\left(e^{-\ell W}\right) - E_k\left(e^{-\ell W}\right)\right\|_{\infty,[t_1,t_2]} \leq \frac{K_i}{\Gamma(\alpha+1)}$$

$$\left\{\frac{\left(\sup_{t\in[t_1,t_2]}\omega_1\left(D^{\alpha}_{t-}E_k\left(e^{-\ell W}\right),\frac{1}{n^{\beta^*}}\right)_{[t_1,t]}+\sup_{t\in[t_1,t_2]}\omega_1\left(D^{\alpha}_{*t}E_k\left(e^{-\ell W}\right)\frac{1}{n^{\beta^*}}\right)_{[t,t_2]}\right)}{n^{\alpha\beta^*}}\right.$$

$$\left.+\frac{(t_2-t_1)^{\alpha}\hat{\beta}_{i,n}}{2}\left(\sup_{t\in[t_1,t_2]}\left\|D^{\alpha}_{t-}E_k\left(e^{-\ell W}\right)\right\|_{\infty,[t_1,t]}+\sup_{t\in[t_1,t_2]}\left\|D^{\alpha}_{*t}E_k\left(e^{-\ell W}\right)\right\|_{\infty,[t,t_2]}\right)\right\}. \tag{10.97}$$

Proof. From Theorem 10.18. □

When $\alpha = \frac{1}{2}$, we derive

Corollary 10.16. *Let* $0 < \beta^* < 1$, $q, \lambda, \beta > 0$, $t \in [t_1, t_2]$ *and* $n \in \mathbb{N}$: $n^{1-\beta^*} > 2$. *Then for* $i = 1, 2$

(i)

$$\left|{}_iL_n\left(E_k\left(e^{-\ell W}\right)(t)\right) - E_k\left(e^{-\ell W}\right)(t)\right|$$

$$\leq \frac{2K_i}{\sqrt{\pi}}\left\{\frac{\left(\omega_1\left(D^{\frac{1}{2}}_{t-}E_k\left(e^{-\ell W}\right),\frac{1}{n^{\beta^*}}\right)_{[t_1,t]}+\omega_1\left(D^{\frac{1}{2}}_{*t}E_k\left(e^{-\ell W}\right),\frac{1}{n^{\beta^*}}\right)_{[t,t_2]}\right)}{n^{\frac{\beta^*}{2}}}\right.$$

$$\left.+\frac{\hat{\beta}_{i,n}}{2}\left(\left\|D^{\frac{1}{2}}_{t-}E_k\left(e^{-\ell W}\right)\right\|_{\infty,[t_1,t]}\sqrt{(t-t_1)}+\left\|D^{\frac{1}{2}}_{*t}E_k\left(e^{-\ell W}\right)\right\|_{\infty,[t,t_2]}\sqrt{(t_2-t)}\right)\right\} \tag{10.98}$$

and
(ii)

$$\left\| {}_iL_nE_k\left(e^{-\ell W}\right) - E_k\left(e^{-\ell W}\right)\right\|_{\infty,[t_1,t_2]} \leq \frac{2K_i}{\sqrt{\pi}}$$

$$\left\{ \frac{\left(\sup_{t\in[t_1,t_2]} \omega_1\left(D_{t-}^{\frac{1}{2}}E_k\left(e^{-\ell W}\right), \frac{1}{n^{\beta^*}}\right)_{[t_1,t]} + \sup_{t\in[t_1,t_2]} \omega_1\left(D_{*t}^{\frac{1}{2}}E_k\left(e^{-\ell W}\right), \frac{1}{n^{\beta^*}}\right)_{[t,t_2]}\right)}{n^{\frac{\beta^*}{2}}} \right.$$

$$+ \frac{\sqrt{(t_2-t_1)}\hat{\beta}_{i,n}}{2}$$

$$\left.\left(\sup_{t\in[t_1,t_2]} \left\| D_{t-}^{\frac{1}{2}}E_k\left(e^{-\ell W}\right)\right\|_{\infty,[t_1,t]} + \sup_{t\in[t_1,t_2]} \left\| D_{*t}^{\frac{1}{2}}E_k\left(e^{-\ell W}\right)\right\|_{\infty,[t,t_2]}\right)\right\} < \infty. \tag{10.99}$$

Proof. From Corollary 10.4. □

Let the generalized logistic sigmoid function $g : \mathbb{R} \to \mathbb{R}$, where $g(x) = (1 + e^{-x})^{\delta}, \delta > 0$ for every $x \in \mathbb{R}$. Let also $W = X_t$ be the Brownian motion on simple graphs. Then the expectation

$$E_k\left(\left(1+e^{-W}\right)^{\delta}\right)(t)$$

$$= \int_0^{\infty} \left(1+e^{-y_k}\right)^{\delta} p(t,x_k,y_k)dy_k + \sum_{j=1,j\neq k}^{n} \int_0^{\infty} \left(1+e^{-y_j}\right)^{\delta} p(t,x_k,y_j)dy_j$$

is continuous in t.

Moreover,

Corollary 10.17. *Let* $0 < \beta^* < 1$, $n \in \mathbb{N} : n^{1-\beta^*} > 2$, $q, \lambda, \beta > 0$, $t \in [t_1, t_2]$, *where* $t_1, t_2 \in (0,\infty)$ *with* $t_1 < t_2$, *then for* $i = 1,2$ *and* $j = 0,1$

(i)

$$\left| {}_iL_n\left(E_k\left(\left(1+e^{-W}\right)^{\delta}\right)^{(j)}\right)(t) - E_k\left(\left(1+e^{-W}\right)^{\delta}\right)^{(j)}(t)\right|$$

$$\leq K_i \cdot \left[\omega_1\left(E_k\left(\left(1+e^{-W}\right)^{\delta}\right)^{(j)}, \frac{1}{n^{\beta^*}}\right) + \hat{\beta}_{i,n}\left\|E_k\left(\left(1+e^{-W}\right)^{\delta}\right)^{(j)}\right\|_{\infty,[t_1,t_2]}\right] =: \rho_i \tag{10.100}$$

and

(ii)

$$\left\| {}_iL_n\left(E_k\left(\left(1+e^{-W}\right)^{\delta}\right)^{(j)}\right)(t) - E_k\left(\left(1+e^{-W}\right)^{\delta}\right)^{(j)}(t)\right\|_{\infty,[t_1,t_2]} \leq \rho_i. \tag{10.101}$$

We get that $\lim_{n\to\infty} {}_iL_nE_k\left(\left(1+e^{-W}\right)^{\delta}\right)^{(j)}(t) = E_k\left(\left(1+e^{-W}\right)^{\delta}\right)^{(j)}$, *pointwise and uniformly.*

Proof. From Theorems 10.17 and 10.19. □

Next we present

Corollary 10.18. *Let* $0 < \alpha, \beta^* < 1$, $q, \lambda, \beta > 0$, $t \in [t_1, t_2]$, *where* $t_1, t_2 \in (0, \infty)$ *with* $t_1 < t_2$, *and* $n \in \mathbb{N} : n^{1-\beta^*} > 2$. *Then for* $i = 1, 2$

(i)

$$\left| {}_iL_n\left(E_k\left(\left(1+e^{-W}\right)^{\delta}\right)(t)\right) - E_k\left(\left(1+e^{-W}\right)^{\delta}\right)(t)\right| \leq \frac{K_i}{\Gamma(\alpha+1)}$$

$$\left\{ \frac{\left(\omega_1\left(D_{t-}^{\alpha}E_k\left(\left(1+e^{-W}\right)^{\delta}\right), \frac{1}{n^{\beta^*}}\right)_{[t_1,t]} + \omega_1\left(D_{*t}^{\alpha}E_k\left(\left(1+e^{-W}\right)^{\delta}\right), \frac{1}{n^{\beta^*}}\right)_{[t,t_2]}\right)}{n^{\alpha\beta^*}} \right.$$

$$+ \frac{\hat{\beta}_{i,n}}{2}$$

$$\left.\left(\left\|D_{t-}^{\alpha}E_k\left(\left(1+e^{-W}\right)^{\delta}\right)\right\|_{\infty,[t_1,t]}(t-t_1)^{\alpha} + \left\|D_{*t}^{\alpha}E_k\left(\left(1+e^{-W}\right)^{\delta}\right)\right\|_{\infty,[t,t_2]}(t_2-t)^{\alpha}\right)\right\} \tag{10.102}$$

and

(ii)

$$\left\| {}_iL_nE_k\left(\left(1+e^{-W}\right)^{\delta}\right) - E_k\left(\left(1+e^{-W}\right)^{\delta}\right)\right\|_{\infty,[t_1,t_2]} \leq \frac{K_i}{\Gamma(\alpha+1)}$$

$$\left\{ \frac{\left(\sup_{t\in[t_1,t_2]}\omega_1\left(D_{t-}^{\alpha}E_k\left(\left(1+e^{-W}\right)^{\delta}\right), \frac{1}{n^{\beta^*}}\right)_{[t_1,t]} + \sup_{t\in[t_1,t_2]}\omega_1\left(D_{*t}^{\alpha}E_k\left(\left(1+e^{-W}\right)^{\delta}\right)\frac{1}{n^{\beta^*}}\right)_{[t,t_2]}\right)}{n^{\alpha\beta^*}} \right.$$

$$+ \frac{(t_2-t_1)^{\alpha}\hat{\beta}_{i,n}}{2}$$

$$\left.\left(\sup_{t\in[t_1,t_2]}\left\|D_{t-}^{\alpha}E_k\left(\left(1+e^{-W}\right)^{\delta}\right)\right\|_{\infty,[t_1,t]} + \sup_{t\in[t_1,t_2]}\left\|D_{*t}^{\alpha}E_k\left(\left(1+e^{-W}\right)^{\delta}\right)\right\|_{\infty,[t,t_2]}\right)\right\}. \tag{10.103}$$

Proof. From Theorem 10.18. □

When $\alpha = \frac{1}{2}$, we derive

Corollary 10.19. *Let $0 < \beta^* < 1$, $q, \lambda, \beta > 0$, $t \in [t_1, t_2]$ and $n \in \mathbb{N}$: $n^{1-\beta^*} > 2$. Then for $i = 1, 2$*

(i)

$$\left|{}_iL_n\left(E_k\left(\left(1+e^{-W}\right)^{\delta}\right)(t)\right) - E_k\left(\left(1+e^{-W}\right)^{\delta}\right)(t)\right| \le \frac{2K_i}{\sqrt{\pi}}$$

$$\left\{\frac{\left(\omega_1\left(D_{t-}^{\frac{1}{2}}E_k\left(\left(1+e^{-W}\right)^{\delta}\right), \frac{1}{n^{\beta^*}}\right)_{[t_1,t]} + \omega_1\left(D_{*t}^{\frac{1}{2}}E_k\left(\left(1+e^{-W}\right)^{\delta}\right), \frac{1}{n^{\beta^*}}\right)_{[t,t_2]}\right)}{n^{\frac{\beta^*}{2}}}\right.$$

$$+ \frac{\hat{\beta}_{i,n}}{2}\left(\left\|D_{t-}^{\frac{1}{2}}E_k\left(\left(1+e^{-W}\right)^{\delta}\right)\right\|_{\infty,[t_1,t]}\sqrt{(t-t_1)}\right.$$

$$\left.\left. + \left\|D_{*t}^{\frac{1}{2}}E_k\left(\left(1+e^{-W}\right)^{\delta}\right)\right\|_{\infty,[t,t_2]}\sqrt{(t_2-t)}\right)\right\} \tag{10.104}$$

and

(ii)

$$\left\|{}_iL_nE_k\left(\left(1+e^{-W}\right)^{\delta}\right) - E_k\left(\left(1+e^{-W}\right)^{\delta}\right)\right\|_{\infty,[t_1,t_2]} \le \frac{2K_i}{\sqrt{\pi}}$$

$$\left\{\frac{\left(\sup_{t\in[t_1,t_2]}\omega_1\left(D_{t-}^{\frac{1}{2}}E_k\left(\left(1+e^{-W}\right)^{\delta}\right), \frac{1}{n^{\beta^*}}\right)_{[t_1,t]} + \sup_{t\in[t_1,t_2]}\omega_1\left(D_{*t}^{\frac{1}{2}}E_k\left(\left(1+e^{-W}\right)^{\delta}\right), \frac{1}{n^{\beta^*}}\right)_{[t,t_2]}\right)}{n^{\frac{\beta^*}{2}}}\right.$$

$$+ \frac{\sqrt{(t_2-t_1)}\hat{\beta}_{i,n}}{2}\left(\sup_{t\in[t_1,t_2]}\left\|D_{t-}^{\frac{1}{2}}E_k\left(\left(1+e^{-W}\right)^{\delta}\right)\right\|_{\infty,[t_1,t]}\right.$$

$$\left.\left. + \sup_{t\in[t_1,t_2]}\left\|D_{*t}^{\frac{1}{2}}E_k\left(\left(1+e^{-W}\right)^{\delta}\right)\right\|_{\infty,[t,t_2]}\right)\right\} < \infty. \tag{10.105}$$

Proof. From Corollary 10.4. □

When $\delta = 1$, we have the usual logistic sigmoid function.

For the last application, we consider the Gompertz function $g : \mathbb{R} \to \mathbb{R}$, where $g(x) = e^{\mu e^{-x}}$ $\mu < 0$ for every $x \in \mathbb{R}$. The Gompertz function is also a sigmoid function which describes growth as being slowest at the start and end of a given time period. Let also $W = X_t$ be the Brownian motion on simple graphs. Then the expectation

$$E_k\left(e^{\mu e^{-W}}\right)(t) = \int_0^\infty e^{\mu e^{-y_k}} p(t, x_k, y_k) dy_k + \sum_{j=1, j\neq k}^{n} \int_0^\infty e^{\mu e^{-y_j}} p(t, x_k, y_j) dy_j$$

is continuous in t.

Moreover,

Corollary 10.20. *Let* $0 < \beta^* < 1$, $n \in \mathbb{N} : n^{1-\beta^*} > 2$, $q, \lambda, \beta > 0$, $t \in [t_1, t_2]$, *where* $t_1, t_2 \in (0, \infty)$ *with* $t_1 < t_2$, *then for* $i = 1, 2$ *and* $j = 0, 1$

(i)

$$\left| {}_iL_n\left(E_k\left(e^{\mu e^{-W}}\right)^{(j)}\right)(t) - E_k\left(e^{\mu e^{-W}}\right)^{(j)}(t)\right|$$
$$\leq K_i \cdot \left[\omega_1\left(E_k\left(e^{\mu e^{-W}}\right)^{(j)}, \frac{1}{n^{\beta^*}}\right) + \hat{\beta}_{i,n}\left\|E_k\left(e^{\mu e^{-W}}\right)^{(j)}\right\|_{\infty,[t_1,t_2]}\right] =: \rho_i \tag{10.106}$$

and

(ii)

$$\left\| {}_iL_n\left(E_k\left(e^{\mu e^{-W}}\right)^{(j)}\right)(t) - E_k\left(e^{\mu e^{-W}}\right)^{(j)}(t)\right\|_{\infty,[t_1,t_2]} \leq \rho_i. \tag{10.107}$$

We get that $\lim_{n\to\infty} {}_iL_n E_k\left(e^{\mu e^{-W}}\right)^{(j)}(t) = E_k\left(e^{\mu e^{-W}}\right)^{(j)}$, *pointwise and uniformly.*

Proof. From Theorems 10.17 and 10.19. □

Next we present

Corollary 10.21. *Let* $0 < \alpha, \beta^* < 1$, $q, \lambda, \beta > 0$, $t \in [t_1, t_2]$, *where* $t_1, t_2 \in (0, \infty)$ *with* $t_1 < t_2$, *and* $n \in \mathbb{N} : n^{1-\beta^*} > 2$. *Then for* $i = 1, 2$

(i)

$$\left|{}_iL_n\left(E_k\left(e^{\mu e^{-W}}\right)(t)\right)-E_k\left(e^{\mu e^{-W}}\right)(t)\right|$$

$$\leq\frac{K_i}{\Gamma(\alpha+1)}\left\{\frac{\left(\omega_1\left(D_{t-}^{\alpha}E_k\left(e^{\mu e^{-W}}\right),\frac{1}{n^{\beta^*}}\right)_{[t_1,t]}+\omega_1\left(D_{*t}^{\alpha}E_k\left(e^{\mu e^{-W}}\right),\frac{1}{n^{\beta^*}}\right)_{[t,t_2]}\right)}{n^{\alpha\beta^*}}\right.$$

$$\left.+\frac{\hat{\beta}_{i,n}}{2}\left(\left\|D_{t-}^{\alpha}E_k\left(e^{\mu e^{-W}}\right)\right\|_{\infty,[t_1,t]}(t-t_1)^{\alpha}+\left\|D_{*t}^{\alpha}E_k\left(e^{\mu e^{-W}}\right)\right\|_{\infty,[t,t_2]}(t_2-t)^{\alpha}\right)\right\} \tag{10.108}$$

and
(ii)

$$\left\|{}_iL_nE_k\left(e^{\mu e^{-W}}\right)-E_k\left(e^{\mu e^{-W}}\right)\right\|_{\infty,[t_1,t_2]}\leq\frac{K_i}{\Gamma(\alpha+1)}$$

$$\left\{\frac{\left(\sup_{t\in[t_1,t_2]}\omega_1\left(D_{t-}^{\alpha}E_k\left(e^{\mu e^{-W}}\right),\frac{1}{n^{\beta^*}}\right)_{[t_1,t]}+\sup_{t\in[t_1,t_2]}\omega_1\left(D_{*t}^{\alpha}E_k\left(e^{\mu e^{-W}}\right)\frac{1}{n^{\beta^*}}\right)_{[t,t_2]}\right)}{n^{\alpha\beta^*}}\right.$$

$$+\frac{(t_2-t_1)^{\alpha}\hat{\beta}_{i,n}}{2}$$

$$\left.\left(\sup_{t\in[t_1,t_2]}\left\|D_{t-}^{\alpha}E_k\left(e^{\mu e^{-W}}\right)\right\|_{\infty,[t_1,t]}+\sup_{t\in[t_1,t_2]}\left\|D_{*t}^{\alpha}E_k\left(e^{\mu e^{-W}}\right)\right\|_{\infty,[t,t_2]}\right)\right\}. \tag{10.109}$$

Proof. From Theorem 10.18. ☐

When $\alpha=\frac{1}{2}$, we derive

Corollary 10.22. *Let* $0<\beta^*<1$, $q,\lambda,\beta>0$, $t\in[t_1,t_2]$ *and* $n\in\mathbb{N}$: $n^{1-\beta^*}>2$. *Then for* $i=1,2$

(i)

$$\left|{}_iL_n\left(E_k\left(e^{\mu e^{-W}}\right)(t)\right)-E_k\left(e^{\mu e^{-W}}\right)(t)\right|$$

$$\leq\frac{2K_i}{\sqrt{\pi}}\left\{\frac{\left(\omega_1\left(D_{t-}^{\frac{1}{2}}E_k\left(e^{\mu e^{-W}}\right),\frac{1}{n^{\beta^*}}\right)_{[t_1,t]}+\omega_1\left(D_{*t}^{\frac{1}{2}}E_k\left(e^{\mu e^{-W}}\right),\frac{1}{n^{\beta^*}}\right)_{[t,t_2]}\right)}{n^{\frac{\beta^*}{2}}}\right.$$

$$\left.+\frac{\hat{\beta}_{i,n}}{2}\left(\left\|D_{t-}^{\frac{1}{2}}E_k\left(e^{\mu e^{-W}}\right)\right\|_{\infty,[t_1,t]}\sqrt{(t-t_1)}+\left\|D_{*t}^{\frac{1}{2}}E_k\left(e^{\mu e^{-W}}\right)\right\|_{\infty,[t,t_2]}\sqrt{(t_2-t)}\right)\right\} \tag{10.110}$$

and

(ii)

$$\left\|{}_iL_nE_k\left(e^{\mu e^{-W}}\right)-E_k\left(e^{\mu e^{-W}}\right)\right\|_{\infty,[t_1,t_2]}\leq\frac{2K_i}{\sqrt{\pi}}$$

$$\left\{\frac{\left(\sup_{t\in[t_1,t_2]}\omega_1\left(D_{t-}^{\frac{1}{2}}E_k\left(e^{\mu e^{-W}}\right),\frac{1}{n^{\beta^*}}\right)_{[t_1,t]}+\sup_{t\in[t_1,t_2]}\omega_1\left(D_{*t}^{\frac{1}{2}}E_k\left(e^{\mu e^{-W}}\right),\frac{1}{n^{\beta^*}}\right)_{[t,t_2]}\right)}{n^{\frac{\beta^*}{2}}}\right.$$

$$+\frac{\sqrt{(t_2-t_1)}\hat{\beta}_{i,n}}{2}$$

$$\left.\left(\sup_{t\in[t_1,t_2]}\left\|D_{t-}^{\frac{1}{2}}E_k\left(e^{\mu e^{-W}}\right)\right\|_{\infty,[t_1,t]}+\sup_{t\in[t_1,t_2]}\left\|D_{*t}^{\frac{1}{2}}E_k\left(e^{\mu e^{-W}}\right)\right\|_{\infty,[t,t_2]}\right)\right\}<\infty. \tag{10.111}$$

Proof. From Corollary 10.4. □

Bibliography

G.A. Anastassiou, Rate of convergence of some neural network operators to the unit-univariate case, *Journal of Mathematical Analysis and Applications* 212 (1997), 237–262.

G.A. Anastassiou, *Quantitative Approximations*, Chapman and Hall/CRC Boca Raton, New York, 2001.

G.A. Anastassiou, On right fractional calculus, *Chaos, Solitons and Fractals*, 42 (2009), 365–376.

G.A. Anastassiou, *Inteligent Systems: Approximation by Artificial Neural Networks*, Intelligent Systems Reference Library, Vol 19, Springer, Heidelberg, 2011a.

G.A. Anastassiou, Multivariate hyperbolic tangent neural network approximation, *Computer and Mathematics*, 61 (2011b), 809–821.

G.A. Anastassiou, Multivariate sigmoidal neural network approximation, *Neural Networks*, 24 (2011c), 378–386.

G.A. Anastassiou, Univariate hyperbolic tangent neural network approximation, *Mathematics and Computer Modeling*, 53 (2011d), 1111–1132.

G.A. Anastassiou, Fractional neural network approximation, *Computer and Mathematics with Applications*, 64 (2012), 1655–1676.

G.A. Anastassiou, Univariate sigmoidal neural network approximation, *Journal of Computational Analysis and Applications*, 14(4) (2012), 659–690.

G.A. Anastassiou, *Inteligent Systems II: Complete Approximation by Neural Network Operators*, Springer, Heidelberg, New York, 2016.

G.A. Anastassiou, A strong Fractional Calculus Theory for Banach space valued functions, *Nonlinear Functional Analysis and Applications (Korea)*, 22(3) (2017a), 495–524.

G.A. Anastassiou, Strong Right Fractional Calculus for Banach space valued functions, *Revista Proyecciones*, 36(1) (2017b), 149–186.

G.A. Anastassiou, Vector fractional Korovkin type Approximations, *Dynamic Systems and Applications*, 26 (2017c), 81–104.

G.A. Anastassiou, *Nonlinearity: Ordinary and Fractional Approximations by Sublinear and Max-Product Operators*, Springer, Heidelberg, New York, 2018a.

G.A. Anastassiou, *Intelligent Computations: Abstract Fractional Calculus, Inequalities, Approximations*, Springer, Heidelberg, New York, 2018b.

G.A. Anastassiou, *Banach Space Valued Neural Network*, Springer, Heidelberg, New York, 2023a.

G.A. Anastassiou, General sigmoid based Banach space valued neural network approximation, *Journal of Computational Analysis and Applications*, 31(4) (2023b), 520–534.

G.A. Anastassiou, *q-Deformed and parametrized half hyperbolic tangent based Banach space valued multivariate multi layer neural network approximations*, Chapter 19, Parametrized, Deformed and General Neural Networks, accepted, Springer, Heidelberg New York 2023c.

G.A. Anastassiou, *Parametrized, deformed and general neural networks*, accepted, Springer, Heidelberg, New York, 2023d.

G.A. Anastassiou, *Parametrized error function based Banach space valued univariate neural network approximation*, Chapter 9, Parametrized, Deformed and General Neural Networks, accepted, Springer, Heidelberg New York 2023e.

G.A. Anastassiou, *q-Deformed and β-parametrized half hyperbolic tangent based Banach space valued ordinary and fractional neural network approximation*, Chapter 20, Parametrized, Deformed and General Neural Networks, accepted, Springer, Heidelberg New York 2023f.

G.A. Anastassiou and D. Kouloumpou, Approximation of Brownian motion on simple graphs, *Mathematics*, 11 (2023), 4329, https://doi.org/10.3390/math11204329.

Barlow MT, Pitman JW, Yor M(1989) On Walsh's Brownian motions. Sem. prob. XXIII. In: *Lecture notes in mathematics*, vol 1372. Springer, Berlin, pp. 275–293.

Z. Chen and F.Cao, The approximation operators with sigmoidal functions, *Computers and Mathematics with Applications*, 58 (2009), 758–765.

K. Diethelm, *The Analysis of Fractional Diferential Equations*, Lecture Notes in Mathematics 2004, Springer-Verlag, Berlin, Heidelberg, 2010.

A.M.A El-sayed and M. Geber, On the finite Caputo and finite Riesz derivatives, *Electronic journal of Theoritical Physics*, 3(12) (2006), 81–95.

G.S. Frederico and D.F.M. Torres, Fractional Optimal Control in the sense of Caputo and the fractional Noether's theorem, *International Mathematical Forum*, 3(10) (2008), 479–493.

V.G. Papanicolaou, E.G. Papageorgiou and D.C Lepipas, Random motion on simple graph, *Methodology and Computing in Applied Probability*, 14 (2012), 285–297

S.G. Samko, A.A. Kilbas and O.I. Marichev, Fractional Integrals and Derivatives, Theory and Applications, (Gordon and Breach, Amsterdam, 1993)[English translation from the Russian, Integrals and Derivatives of Fractional Order and Some of Their Applications (Nauka i Tekhnika, Minsk,1987)].

G.E. Shilov, *Elementary Functional Analysis*, Dover Publications, Inc., New York, 1996.

Chapter 11

Multivariate Fuzzy-Random with Perturbed Neural Network Approximation

In this chapter, we estimate the degree of approximation of multivariate pointwise and uniform convergences in the p-mean to the fuzzy-random unit operator of multivariate fuzzy-random quasi-interpolation perturbed activation functions-based neural network operators. These multivariate fuzzy-random operators arise in a natural way among multivariate fuzzy-random neural networks. The rates are given through multivariate probabilistic Jackson-type inequalities involving the multivariate fuzzy-random modulus of continuity of the engaged multivariate fuzzy-random function. It follows Anastassiou and Kouloumpou (2023e).

11.1 Fuzzy-Random Functions Background

See also Anastassiou (2016), Ch. 22, pp. 497-501.

We start with

Definition 11.1 (see Wu and Gong 2000b). Let $\mu : \mathbb{R} \to [0,1]$ with the following properties:

(i) It is normal, i.e., $\exists\, x_0 \in \mathbb{R} : \mu(x_0) = 1$.
(ii) $\mu(\lambda x + (1-\lambda) y) \geq \min\{\mu(x), \mu(y)\}$, $\forall\, x, y \in \mathbb{R}$, $\forall\, \lambda \in [0,1]$ (μ is called a convex fuzzy subset).
(iii) μ is upper semicontinuous on $\mathbb{R}$, i.e. $\forall\, x_0 \in \mathbb{R}$ and $\forall\, \varepsilon > 0$, $\exists$ neighborhood $V(x_0) : \mu(x) \leq \mu(x_0) + \varepsilon$, $\forall\, x \in V(x_0)$.
(iv) The set $\overline{\operatorname{supp}(\mu)}$ is compact in $\mathbb{R}$ (where $\operatorname{supp}(\mu) := \{x \in \mathbb{R}; \mu(x) > 0\}$).

We call μ a fuzzy real number. Denote the set of all μ with $\mathbb{R}_{\mathcal{F}}$.

For example, $\chi_{\{x_0\}} \in \mathbb{R}_{\mathcal{F}}$, for any $x_0 \in \mathbb{R}$, where $\chi_{\{x_0\}}$ is the characteristic function at x_0.

For $0 < r \leq 1$ and $\mu \in \mathbb{R}_{\mathcal{F}}$, define $[\mu]^r := \{x \in \mathbb{R} : \mu(x) \geq r\}$ and $[\mu]^0 := \overline{\{x \in \mathbb{R} : \mu(x) > 0\}}$.

Then it is well known that for each $r \in [0,1]$, $[\mu]^r$ is a closed and bounded interval of $\mathbb{R}$. For $u, v \in \mathbb{R}_{\mathcal{F}}$ and $\lambda \in \mathbb{R}$, we define uniquely the sum $u \oplus v$ and the product $\lambda \odot u$ by

$$[u \oplus v]^r = [u]^r + [v]^r, \quad [\lambda \odot u]^r = \lambda [u]^r, \quad \forall\, r \in [0,1],$$

where $[u]^r + [v]^r$ means the usual addition of two intervals (as subsets of $\mathbb{R}$) and $\lambda [u]^r$ means the usual product between a scalar and a subset of $\mathbb{R}$ (see, e.g. Wu and Gong (2000b)). Note $1 \odot u = u$ and it holds $u \oplus v = v \oplus u$, $\lambda \odot u = u \odot \lambda$. If $0 \leq r_1 \leq r_2 \leq 1$, then $[u]^{r_2} \subseteq [u]^{r_1}$. Actually, $[u]^r = \left[u_-^{(r)}, u_+^{(r)}\right]$, where $u_-^{(r)} < u_+^{(r)}$, $u_-^{(r)}, u_+^{(r)} \in \mathbb{R}$, $\forall\, r \in [0,1]$.

Define

$$D : \mathbb{R}_{\mathcal{F}} \times \mathbb{R}_{\mathcal{F}} \to \mathbb{R}_+ \cup \{0\}$$

by

$$D(u,v) := \sup_{r \in [0,1]} \max \left\{ \left| u_-^{(r)} - v_-^{(r)} \right|, \left| u_+^{(r)} - v_+^{(r)} \right| \right\},$$

where $[v]^r = \left[v_-^{(r)}, v_+^{(r)}\right]$; $u, v \in \mathbb{R}_{\mathcal{F}}$. We have that D is a metric on $\mathbb{R}_{\mathcal{F}}$. Then $(\mathbb{R}_{\mathcal{F}}, D)$ is a complete metric space, see Wu and Gong (2000b), with the properties

$$\begin{aligned} D(u \oplus w, v \oplus w) &= D(u,v), \quad \forall\, u, v, w \in \mathbb{R}_{\mathcal{F}}, \\ D(k \odot u, k \odot v) &= |k|\, D(u,v), \quad \forall\, u, v \in \mathbb{R}_{\mathcal{F}}, \forall\, k \in \mathbb{R}, \\ D(u \oplus v, w \oplus e) &\leq D(u,w) + D(v,e), \quad \forall\, u, v, w, e \in \mathbb{R}_{\mathcal{F}}. \end{aligned} \tag{11.1}$$

Let (M,d) metric space and $f, g : M \to \mathbb{R}_{\mathcal{F}}$ be fuzzy real number-valued functions. The distance between f, g is defined by

$$D^*(f,g) := \sup_{x \in M} D(f(x), g(x)).$$

On $\mathbb{R}_{\mathcal{F}}$, we define a partial order by "$\leq$": $u, v \in \mathbb{R}_{\mathcal{F}}$, $u \leq v$ iff $u_-^{(r)} \leq v_-^{(r)}$ and $u_+^{(r)} \leq v_+^{(r)}$, $\forall\, r \in [0,1]$.

$\sum^{*}$ denotes the fuzzy summation, $\widetilde{o} := \chi_{\{0\}} \in \mathbb{R}_{\mathcal{F}}$ the neutral element with respect to $\oplus$. For more, see also Wu and Gong (2001) and Wu and Ma (1991).

We need

Definition 11.2. (see also Gal (2000a), Definition 13.16, pp. 654) Let $(X, \mathcal{B}, P)$ be a probability space. A fuzzy-random variable is a $\mathcal{B}$-measurable mapping $g : X \to \mathbb{R}_{\mathcal{F}}$ (i.e. for any open set $U \subseteq \mathbb{R}_{\mathcal{F}}$, in the topology of $\mathbb{R}_{\mathcal{F}}$ generated by the metric D, we have

$$g^{-1}(U) = \{s \in X; g(s) \in U\} \in \mathcal{B}). \tag{11.2}$$

The set of all fuzzy-random variables is denoted by $\mathcal{L}_{\mathcal{F}}(X, \mathcal{B}, P)$. Let $g_n, g \in \mathcal{L}_{\mathcal{F}}(X, \mathcal{B}, P)$, $n \in \mathbb{N}$ and $0 < p < +\infty$. We say $g_n(s) \overset{\text{“}p\text{-mean”}}{\underset{n\to+\infty}{\to}} g(s)$ if

$$\lim_{n\to+\infty} \int_X D(g_n(s), g(s))^p \, P(ds) = 0. \tag{11.3}$$

Remark 11.1. (see Gal (2000a), p. 654) If $f, g \in \mathcal{L}_{\mathcal{F}}(X, \mathcal{B}, P)$, let us denote $F : X \to \mathbb{R}_+ \cup \{0\}$ by $F(s) = D(f(s), g(s))$, $s \in X$. Here, F is $\mathcal{B}$-measurable because $F = G \circ H$, where $G(u, v) = D(u, v)$ is continuous on $\mathbb{R}_{\mathcal{F}} \times \mathbb{R}_{\mathcal{F}}$, and $H : X \to \mathbb{R}_{\mathcal{F}} \times \mathbb{R}_{\mathcal{F}}$, $H(s) = (f(s), g(s))$, $s \in X$, is $\mathcal{B}$-measurable. This shows that the above convergence in q-mean makes sense.

Definition 11.3. (see Gal 2000a, Definition 13.17, p. 654) Let $(T, \mathcal{T})$ be a topological space. A mapping $f : T \to \mathcal{L}_{\mathcal{F}}(X, \mathcal{B}, P)$ will be called fuzzy-random function (or fuzzy-stochastic process) on T. We denote $f(t)(s) = f(t, s)$, $t \in T$, $s \in X$.

Remark 11.2. (see Gal 2000a, p. 655) Any usual fuzzy real function $f : T \to \mathbb{R}_{\mathcal{F}}$ can be identified with the degenerate fuzzy-random function $f(t, s) = f(t)$, $\forall\, t \in T$, $s \in X$.

Remark 11.3. (see Gal 2000a, p. 655) Fuzzy-random functions that coincide with probability one for each $t \in T$ will be consider equivalent.

Remark 11.4. (see Gal 2000a, p. 655) Let $f, g : T \to \mathcal{L}_{\mathcal{F}}(X, \mathcal{B}, P)$. Then $f \oplus g$ and $k \odot f$ are defined pointwise, i.e.

$$(f \oplus g)(t, s) = f(t, s) \oplus g(t, s),$$

$$(k \odot f)(t, s) = k \odot f(t, s), \quad t \in T, s \in X, \ k \in \mathbb{R}.$$

Definition 11.4 (see also Gal 2000a, Definition 13.18, pp. 655–656). For a fuzzy-random function $f : W \subseteq \mathbb{R}^N \to \mathcal{L}_{\mathcal{F}}(X, \mathcal{B}, P)$, $N \in \mathbb{N}$, we define the (first) fuzzy-random modulus of continuity

$$\Omega_1^{(\mathcal{F})}(f,\delta)_{L^p}$$

$$= \sup\left\{ \left(\int_X D^p\left(f(x,s), f(y,s)\right) P(ds) \right)^{\frac{1}{p}} : x, y \in W,\ \|x-y\|_\infty \le \delta \right\},$$

$0 < \delta$, $1 \le p < \infty$.

Definition 11.5 (Anastassiou 2014). Here $1 \le p < +\infty$. Let $f : W \subseteq \mathbb{R}^N \to \mathcal{L}_{\mathcal{F}}(X, \mathcal{B}, P)$, $N \in \mathbb{N}$, be a fuzzy random function. We call f a (p-mean) uniformly continuous fuzzy random function over W, iff $\forall\ \varepsilon > 0\ \exists\ \delta > 0$:whenever $\|x-y\|_\infty \le \delta$, $x, y \in W$, implies that

$$\int_X \left(D\left(f(x,s), f(y,s)\right)\right)^p P(ds) \le \varepsilon.$$

We denote it as $f \in C_{FR}^{U_p}(W)$.

Proposition 11.1 (Anastassiou 2014). *Let $f \in C_{FR}^{U_p}(W)$, where $W \subseteq \mathbb{R}^N$ is convex.*

Then $\Omega_1^{(\mathcal{F})}(f,\delta)_{L^p} < \infty$, any $\delta > 0$.

Proposition 11.2 (Anastassiou 2014). *Let $f, g : W \subseteq \mathbb{R}^N \to \mathcal{L}_{\mathcal{F}}(X, \mathcal{B}, P)$, $N \in \mathbb{N}$, be fuzzy random functions. It holds*

(i) *$\Omega_1^{(\mathcal{F})}(f,\delta)_{L^p}$ is nonnegative and nondecreasing in $\delta > 0$,*

(ii) *$\lim_{\delta \downarrow 0} \Omega_1^{(\mathcal{F})}(f,\delta)_{L^p} = \Omega_1^{(\mathcal{F})}(f,0)_{L^p} = 0$, iff $f \in C_{FR}^{U_p}(W)$.*

We need also

Proposition 11.3 (Anastassiou 2014). *Let f, g be fuzzy random variables from $\mathcal{S}$ into $\mathbb{R}_{\mathcal{F}}$. Then*

(i) *let $c \in \mathbb{R}$, then $c \odot f$ is a fuzzy random variable,*

(ii) *$f \oplus g$ is a fuzzy random variable.*

11.2 About Perturbed Neural Network Background

11.2.1 *About q-Deformed and λ-parametrized A-generalized logistic function-induced real space-valued multivariate multilayer neural network approximation*

Here we follow Anastassiou (2023b).

We consider the q-deformed and λ-parametrized function

$$\varphi_{q,\lambda}(x) = \frac{1}{1 + qA^{-\lambda x}}, \quad x \in \mathbb{R}, \quad \text{where } q, \lambda > 0, \ A > 1, \tag{11.4}$$

which is a sigmoid-type function and it is strictly increasing. This is an A-generalized logistic-type function. We easily observe that

$$\varphi_{q,\lambda}(+\infty) = 1, \quad \varphi_{q,\lambda}(-\infty) = 0. \tag{11.5}$$

Furthermore we have

$$\varphi_{q,\lambda}(x) = 1 - \varphi_{\frac{1}{q},\lambda}(-x) \tag{11.6}$$

and

$$\varphi_{q,\lambda}(0) = \frac{1}{1+q}. \tag{11.7}$$

Moreover, $\varphi''_{q,\lambda}(x) > 0$, for $x < \frac{\log_A q}{\lambda}$ and there $\varphi_{q,\lambda}$ is concave up. When $x > \frac{\log_A q}{\lambda}$, we have $\varphi''_{q,\lambda}(x) < 0$ and $\varphi_{q,\lambda}$ is concave down. Of course,

$$\varphi''_{q,\lambda}\left(\frac{\log_A q}{\lambda}\right) = 0.$$

So, $\varphi_{q,\lambda}$ is a sigmoid function, see Anastassiou (2023a).

We consider the activation function

$${}_1\mathcal{L}_q(x) := \frac{1}{2}\left(\varphi_{q,\lambda}(x+1) - \varphi_{q,\lambda}(x-1)\right), \quad x \in \mathbb{R}. \tag{11.8}$$

Then

$${}_1\mathcal{L}_q(-x) = {}_1\mathcal{L}_{\frac{1}{q},\lambda}(x), \quad \forall\, x \in \mathbb{R}. \tag{11.9}$$

We have that

$${}_1\mathcal{L}'_q(x) = \frac{1}{2}\left(\varphi'_{q,\lambda}(x+1) - \varphi'_{q,\lambda}(x-1)\right) < 0,$$

i.e. ${}_1\mathcal{L}_q$ is strictly decreasing over $\left(\frac{\log_A q}{\lambda}, +\infty\right)$. Furthermore, ${}_1\mathcal{L}_q$ is strictly concave down over $\left(\frac{\log_A q}{\lambda} - 1, \frac{\log_A q}{\lambda} + 1\right)$. Overall ${}_1\mathcal{L}_q$ is a bell-shaped function over $\mathbb{R}$. Of course, it holds ${}_1\mathcal{L}q, \lambda''\left(\frac{\log_A q}{\lambda}\right) < 0$. We have that the global maximum of ${}_1\mathcal{L}_q$ is

$$ {}_1\mathcal{L}_q\left(\frac{\log_A q}{\lambda}\right) = \frac{A^\lambda - 1}{2\left(A^\lambda + 1\right)}. \tag{11.10}$$

Finally, we have that

$$\lim_{x\to+\infty} {}_1\mathcal{L}_q\left(x\right) = \frac{1}{2}\left(\varphi_{q,\lambda}\left(+\infty\right) - \varphi_{q,\lambda}\left(+\infty\right)\right) = 0 \tag{11.11}$$

and

$$\lim_{x\to-\infty} {}_1\mathcal{L}_q\left(x\right) = \frac{1}{2}\left(\varphi_{q,\lambda}\left(-\infty\right) - \varphi_{q,\lambda}\left(-\infty\right)\right) = 0. \tag{11.12}$$

Consequently, the x-axis is the horizontal asymptote of ${}_1\mathcal{L}_q$. Of course, ${}_1\mathcal{L}_q\left(x\right) > 0$, $\forall\, x \in \mathbb{R}$.
We need

Theorem 11.1. *It holds*

$$\sum_{i=-\infty}^{\infty} {}_1\mathcal{L}_q\left(x - i\right) = 1, \quad \forall x \in \mathbb{R}, \quad \forall\, q, \lambda > 0, A > 1. \tag{11.13}$$

It follows

Theorem 11.2. *It holds*

$$\int_{-\infty}^{\infty} {}_1\mathcal{L}_q\left(x\right) dx = 1, \quad \lambda, q > 0, \ \ A > 1. \tag{11.14}$$

So ${}_1\mathcal{L}_q$ is a density function on $\mathbb{R}$; $\lambda, q > 0$, $A > 1$.
We need the following result.

Theorem 11.3. *Let $0 < \alpha < 1$, and $n \in \mathbb{N}$ with $n^{1-\alpha} > 2$. Then*

$$\sum_{\substack{k=-\infty \\ \left\{:\left|nx-k\right| \geq n^{1-\alpha}\right.}}^{\infty} {}_1\mathcal{L}_q\left(nx - k\right) < 2\max\left\{q, \frac{1}{q}\right\} \frac{1}{A^{\lambda\left(n^{1-\alpha}-2\right)}}$$

$$= \gamma A^{-\lambda\left(n^{1-\alpha}-2\right)} =: c_1(n, a), \tag{11.15}$$

where $q, \lambda > 0$, $A > 1$; $\gamma := 2\max\left\{q, \frac{1}{q}\right\}$.

Let $\lceil \cdot \rceil$ be the ceiling of the number and $\lfloor \cdot \rfloor$ be the integral part of the number.

Theorem 11.4. *Let $x \in [a,b] \subset \mathbb{R}$ and $n \in \mathbb{N}$ so that $\lceil na \rceil \leq \lfloor nb \rfloor$. For $q > 0$, $\lambda > 0$, $A > 1$, we consider the number $\lambda_q > z_0 > 0$ with ${}_1\mathcal{L}_q(z_0) = {}_1\mathcal{L}_q(0)$ and $\lambda_q > 1$. Then*

$$\frac{1}{\sum_{k=\lceil na \rceil}^{\lfloor nb \rfloor} {}_1\mathcal{L}_q(nx-k)} < max\left\{\frac{1}{{}_1\mathcal{L}_q(\lambda_q)}, \frac{1}{{}_1\mathcal{L}_{\frac{1}{q}}\left(\lambda_{\frac{1}{q}}\right)}\right\} =: \Psi_1(q). \tag{11.16}$$

We make

Remark 11.5. (i) We have that

$$\lim_{n\to+\infty} \sum_{k=\lceil na \rceil}^{\lfloor nb \rfloor} {}_1\mathcal{L}_q(nx-k) \neq 1, \quad \text{for at least some } x \in [a,b], \tag{11.17}$$

where $\lambda, q > 0$.

(ii) Let $[a,b] \subset \mathbb{R}$. For large n, we always have $\lceil na \rceil \leq \lfloor nb \rfloor$. Also $a \leq \frac{k}{n} \leq b$, iff $\lceil na \rceil \leq k \leq \lfloor nb \rfloor$. In general, it holds

$$\sum_{k=\lceil na \rceil}^{\lfloor nb \rfloor} {}_1\mathcal{L}_q(nx-k) \leq 1. \tag{11.18}$$

We introduce

$${}_1Z_q(x_1,\ldots,x_N) := {}_1Z_q(x) := \prod_{i=1}^{N} {}_1\mathcal{L}_q(x_i), \quad x = (x_1,\ldots,x_N) \in \mathbb{R}^N, \tag{11.19}$$

$\lambda, q > 0$, $A > 1$, $N \in \mathbb{N}$.

${}_1Z_q(x)$ it has the following properties:

(i) ${}_1Z_q(x) > 0, \ \forall\, x \in \mathbb{R}^N$,

(ii)

$$\sum_{k=-\infty}^{\infty} {}_1Z_q(x-k) := \sum_{k_1=-\infty}^{\infty}\sum_{k_2=-\infty}^{\infty}\cdots\sum_{k_N=-\infty}^{\infty} {}_1Z_q(x_1-k_1,\ldots,x_N-k_N) = 1, \tag{11.20}$$

where $k := (k_1,\ldots,k_n) \in \mathbb{Z}^N$, $\forall\, x \in \mathbb{R}^N$, hence

(iii)

$$\sum_{k=-\infty}^{\infty} {}_1Z_q(nx-k) = 1, \tag{11.21}$$

$\forall\, x \in \mathbb{R}^N$; $n \in \mathbb{N}$,

and

(iv)

$$\int_{\mathbb{R}^N} {}_1Z_q(x)\,dx = 1, \tag{11.22}$$

that is, ${}_1Z_q$ is a multivariate density function.

Here denote $\|x\|_\infty := \max\{|x_1|,\ldots,|x_N|\}$, $x \in \mathbb{R}^N$, also set $\infty := (\infty,\ldots,\infty)$, $-\infty := (-\infty,\ldots,-\infty)$ upon the multivariate context, and

$$\begin{aligned} \lceil na \rceil &:= (\lceil na_1 \rceil,\ldots,\lceil na_N \rceil), \\ \lfloor nb \rfloor &:= (\lfloor nb_1 \rfloor,\ldots,\lfloor nb_N \rfloor), \end{aligned} \tag{11.23}$$

where $a := (a_1,\ldots,a_N)$, $b := (b_1,\ldots,b_N)$.

We obviously see that for $0 < \beta^* < 1$ and $n \in \mathbb{N}$, a fixed $x \in \mathbb{R}^N$, we have that

$$\sum_{k=\lceil na \rceil}^{\lfloor nb \rfloor} {}_1Z_q(nx-k)$$

$$= \sum_{\begin{cases} k=\lceil na \rceil \\ \left\| \frac{k}{n}-x \right\|_\infty \le \frac{1}{n^{\beta^*}} \end{cases}}^{\lfloor nb \rfloor} {}_1Z_q(nx-k) + \sum_{\begin{cases} k=\lceil na \rceil \\ \left\| \frac{k}{n}-x \right\|_\infty > \frac{1}{n^{\beta^*}} \end{cases}}^{\lfloor nb \rfloor} {}_1Z_q(nx-k). \tag{11.24}$$

(v) We derive that

$$\sum_{\begin{cases} k=\lceil na \rceil \\ \left\| \frac{k}{n}-x \right\|_\infty > \frac{1}{n^{\beta^*}} \end{cases}}^{\lfloor nb \rfloor} {}_1Z_q(nx-k) < \gamma A^{-\lambda\left(n^{1-\beta^*}-2\right)} = c_1(n,\beta^*),\ \ 0 < \beta^* < 1, \tag{11.25}$$

with $n \in \mathbb{N} : n^{1-\beta^*} > 2$, $x \in \prod_{i=1}^N [a_i,b_i]$.

(vi) We get that

$$0 < \frac{1}{\sum_{k=\lceil na \rceil}^{\lfloor nb \rfloor} {}_1Z_q(nx-k)} < (\Psi_1(q))^N, \tag{11.26}$$

$\forall\, x \in \left(\prod_{i=1}^N [a_i, b_i]\right)$, $n \in \mathbb{N}$.

It is also clear that

(vii)

$$\sum_{\substack{k=-\infty \\ \left\{\left\| \frac{k}{n} - x \right\|_\infty > \frac{1}{n^{\beta^*}}\right.}}^{\infty} {}_1Z_q(nx-k) < \gamma A^{-\lambda\left(n^{1-\beta^*}-2\right)}, \tag{11.27}$$

$0 < \beta^* < 1$, $n \in \mathbb{N} : n^{1-\beta^*} > 2$, $x \in \mathbb{R}^N$.

Furthermore it holds

$$\lim_{n\to\infty} \sum_{k=\lceil na \rceil}^{\lfloor nb \rfloor} {}_1Z_q(nx-k) \neq 1, \tag{11.28}$$

for at least some $x \in \left(\prod_{i=1}^N [a_i, b_i]\right)$.

11.2.2 *About q-deformed and λ-parametrized hyperbolic tangent function $g_{q,\lambda}$*

Here we follow Anastassiou (2023c). Let us consider the function

$$g_{q,\lambda}(x) := \frac{e^{\lambda x} - qe^{-\lambda x}}{e^{\lambda x} + qe^{-\lambda x}}, \quad \lambda, q > 0, \;\; x \in \mathbb{R}. \tag{11.29}$$

We have that $g_{q,\lambda}$ is strictly increasing. We easily observe that

$$g_{q,\lambda}(+\infty) = 1 \quad \text{and} \quad g_{q,\lambda}(-\infty) = -1. \tag{11.30}$$

Furthermore,

$$g_{q,\lambda}(0) = \frac{1-q}{1+q} \tag{11.31}$$

and

$$g'_{\frac{1}{q},\lambda}(x) = g'_{q,\lambda}(-x). \tag{11.32}$$

Moreover, in case of $x < \frac{\ln q}{2\lambda}$, we have that $g_{q,\lambda}$ is strictly concave up, with $g''_{q,\lambda}\left(\frac{\ln q}{2\lambda}\right) = 0$.

And in case of $x > \frac{\ln q}{2\lambda}$, we have that $g_{q,\lambda}$ is strictly concave down.

Clearly, $g_{q,\lambda}$ is a shifted sigmoid function with $g_{q,\lambda}(0) = \frac{1-q}{1+q}$, and $g_{q,\lambda}(-x) = -g_{q^{-1},\lambda}(x)$ (a semi-odd function), see also [?].

By $1 > -1$, $x+1 > x-1$, we consider the activation function

$$ {}_{2}\mathcal{L}_{q}(x) := \frac{1}{4}\left(g_{q,\lambda}(x+1) - g_{q,\lambda}(x-1)\right) > 0, \tag{11.33} $$

$\forall x \in \mathbb{R}$; $q, \lambda > 0$. Note that ${}_{2}\mathcal{L}_{q}(\pm\infty) = 0$, so the x-axis is horizontal asymptote. We have that

$$ {}_{2}\mathcal{L}_{q}(-x) = {}_{2}\mathcal{L}_{\frac{1}{q}}(x), \quad \forall\, x \in \mathbb{R}; \quad q, \lambda > 0, \tag{11.34} $$

a deformed symmetry.

Next, we have that

$$ {}_{2}\mathcal{L}_{q}'(x) = \frac{1}{4}\left(g_{q,\lambda}'(x+1) - g_{q,\lambda}'(x-1)\right), \quad \forall\, x \in \mathbb{R}. \tag{11.35} $$

Moreover, ${}_{2}\mathcal{L}_{q}$ is strictly increasing over $\left(-\infty, \frac{\ln q}{2\lambda} - 1\right)$ and strictly decreasing over $\left(\frac{\ln q}{2\lambda} + 1, +\infty\right)$. Furthermore, ${}_{2}\mathcal{L}_{q}$ is concave down over $\left[\frac{\ln q}{2\lambda} - 1, \frac{\ln q}{2\lambda} + 1\right]$ and strictly concave down over $\left(\frac{\ln q}{2\lambda} - 1, \frac{\ln q}{2\lambda} + 1\right)$. Consequently, ${}_{2}\mathcal{L}_{q}$ has a bell-type shape over $\mathbb{R}$.

Of course, it holds ${}_{2}\mathcal{L}_{q}''\left(\frac{\ln q}{2\lambda}\right) < 0$. We also have that the maximum value of ${}_{2}\mathcal{L}_{q}$ is

$$ {}_{2}\mathcal{L}_{q}\left(\frac{\ln q}{2\lambda}\right) = \frac{\tanh(\lambda)}{2}, \quad \lambda > 0. \tag{11.36} $$

We give

Theorem 11.5. *We have that*

$$ \sum_{i=-\infty}^{\infty} {}_{2}\mathcal{L}_{q}(x-i) = 1, \quad \forall\, x \in \mathbb{R}, \quad \forall\, \lambda, q > 0. \tag{11.37} $$

We need

Theorem 11.6. *It holds*

$$ \int_{-\infty}^{\infty} {}_{2}\mathcal{L}_{q}(x)\, dx = 1, \quad \lambda, q > 0. \tag{11.38} $$

So ${}_{2}\mathcal{L}_{q}$ is a density function on $\mathbb{R}$; $\lambda, q > 0$.

We need the following result.

Theorem 11.7. *Let $0 < \alpha < 1$, and $n \in \mathbb{N}$ with $n^{1-\alpha} > 2$; $q, \lambda > 0$. Then*

$$\sum_{\substack{k=-\infty \\ \{:|nx-k| \geq n^{1-\alpha}}}^{\infty} {}_2\mathcal{L}_q (nx - k) < 2 \max \left\{ q, \frac{1}{q} \right\} e^{4\lambda} e^{-2\lambda n^{(1-\alpha)}}$$

$$= T e^{-2\lambda n^{(1-\alpha)}} =: c_2(n, a), \tag{11.39}$$

where $T := 2 \max \left\{ q, \frac{1}{q} \right\} e^{4\lambda}$.

Let $\lceil \cdot \rceil$ be the ceiling of the number and $\lfloor \cdot \rfloor$ be the integral part of the number.

Theorem 11.8. *Let $x \in [a, b] \subset \mathbb{R}$ and $n \in \mathbb{N}$ so that $\lceil na \rceil \leq \lfloor nb \rfloor$. For $q > 0$, $\lambda > 0$, we consider the number $\lambda_q > z_0 > 0$ with ${}_2\mathcal{L}_q (z_0) = {}_2\mathcal{L}_q (0)$ and $\lambda_q > 1$. Then*

$$\frac{1}{\sum_{k=\lceil na \rceil}^{\lfloor nb \rfloor} {}_2\mathcal{L}_q (nx - k)} < max \left\{ \frac{1}{{}_2\mathcal{L}_q (\lambda_q)}, \frac{1}{{}_2\mathcal{L}_{\frac{1}{q}} \left(\lambda_{\frac{1}{q}} \right)} \right\} =: \Psi_2 (q). \tag{11.40}$$

We make

Remark 11.6. (i) We have that

$$\lim_{n \to +\infty} \sum_{k=\lceil na \rceil}^{\lfloor nb \rfloor} {}_2\mathcal{L}_q (nx - k) \neq 1, \quad \text{for at least some } x \in [a, b], \tag{11.41}$$

where $\lambda, q > 0$.

(ii) Let $[a, b] \subset \mathbb{R}$. For large n we always have $\lceil na \rceil \leq \lfloor nb \rfloor$. Also $a \leq \frac{k}{n} \leq b$, iff $\lceil na \rceil \leq k \leq \lfloor nb \rfloor$. In general, it holds

$$\sum_{k=\lceil na \rceil}^{\lfloor nb \rfloor} {}_2\mathcal{L}_q (nx - k) \leq 1. \tag{11.42}$$

We introduce

$${}_2Z_q (x_1, \ldots, x_N) := {}_2Z_q (x) := \prod_{i=1}^{N} {}_2\mathcal{L}_q (x_i), \tag{11.43}$$

$x = (x_1, \ldots, x_N) \in \mathbb{R}^N$, $\lambda, q > 0$, $N \in \mathbb{N}$.

${}_2Z_q(x)$ it has the following properties:

(i) ${}_2Z_q(x) > 0, \ \forall \, x \in \mathbb{R}^N$,
(ii)

$$\sum_{k=-\infty}^{\infty} {}_2Z_q(x-k) := \sum_{k_1=-\infty}^{\infty} \sum_{k_2=-\infty}^{\infty} \cdots \sum_{k_N=-\infty}^{\infty} Z_q(x_1-k_1,\ldots,x_N-k_N) = 1, \tag{11.44}$$

where $k := (k_1,\ldots,k_n) \in \mathbb{Z}^N$, $\forall \, x \in \mathbb{R}^N$,
hence
(iii)

$$\sum_{k=-\infty}^{\infty} {}_2Z_q(nx-k) = 1, \tag{11.45}$$

$\forall \, x \in \mathbb{R}^N$; $n \in \mathbb{N}$,
and
(iv)

$$\int_{\mathbb{R}^N} {}_2Z_q(x)\, dx = 1, \tag{11.46}$$

that is, ${}_2Z_q$ is a multivariate density function.

We obviously see that

$$\sum_{k=\lceil na \rceil}^{\lfloor nb \rfloor} {}_2Z_q(nx-k) = \sum_{k=\lceil na \rceil}^{\lfloor nb \rfloor} \left(\prod_{i=1}^{N} {}_2\mathcal{L}_q(nx_i-k_i) \right)$$
$$= \sum_{k_1=\lceil na_1 \rceil}^{\lfloor nb_1 \rfloor} \cdots \sum_{k_N=\lceil na_N \rceil}^{\lfloor nb_N \rfloor} \left(\prod_{i=1}^{N} {}_2\mathcal{L}_q(nx_i-k_i) \right) = \prod_{i=1}^{N} \left(\sum_{k_i=\lceil na_i \rceil}^{\lfloor nb_i \rfloor} {}_2\mathcal{L}_q(nx_i-k_i) \right). \tag{11.47}$$

For $0 < \beta^* < 1$ and $n \in \mathbb{N}$, a fixed $x \in \mathbb{R}^N$, we have that

$$\sum_{k=\lceil na \rceil}^{\lfloor nb \rfloor} {}_2Z_q(nx-k)$$
$$= \sum_{\substack{k=\lceil na \rceil \\ \left\{ \left\| \frac{k}{n}-x \right\|_\infty \le \frac{1}{n^{\beta^*}} \right.}}^{\lfloor nb \rfloor} {}_2Z_q(nx-k) + \sum_{\substack{k=\lceil na \rceil \\ \left\{ \left\| \frac{k}{n}-x \right\|_\infty > \frac{1}{n^{\beta^*}} \right.}}^{\lfloor nb \rfloor} {}_2Z_q(nx-k). \tag{11.48}$$

In the last two sums, the counting is over disjoint vector sets of k's because the condition $\left\| \frac{k}{n} - x \right\|_{\infty} > \frac{1}{n^{\beta^*}}$ implies that there exists at least one $\left| \frac{k_r}{n} - x_r \right| > \frac{1}{n^{\beta^*}}$, where $r \in \{1, \ldots, N\}$.

(v) We also have that

$$\sum_{\substack{k=\lceil na \rceil \\ \left\{ \left\| \frac{k}{n} - x \right\|_{\infty} > \frac{1}{n^{\beta^*}} \right.}}^{\lfloor nb \rfloor} {}_2Z_q\left(nx - k\right) < Te^{-2\lambda n^{\left(1-\beta^*\right)}} = c_2(n, \beta^*), \quad 0 < \beta^* < 1, \tag{11.49}$$

with $n \in \mathbb{N} : n^{1-\beta^*} > 2$, $x \in \prod_{i=1}^{N} [a_i, b_i]$.

(vi) Moreover,

$$0 < \frac{1}{\sum_{k=\lceil na \rceil}^{\lfloor nb \rfloor} {}_2Z_q\left(nx - k\right)} < \left(\Psi_2\left(q\right)\right)^N, \tag{11.50}$$

$\forall\, x \in \left(\prod_{i=1}^{N} [a_i, b_i]\right)$, $n \in \mathbb{N}$.

It is also clear that

(vii)

$$\sum_{\substack{k=-\infty \\ \left\{ \left\| \frac{k}{n} - x \right\|_{\infty} > \frac{1}{n^{\beta^*}} \right.}}^{\infty} {}_2Z_q\left(nx - k\right) < Te^{-2\lambda n^{\left(1-\beta^*\right)}}, \tag{11.51}$$

$0 < \beta^* < 1$, $n \in \mathbb{N} : n^{1-\beta^*} > 2$, $x \in \mathbb{R}^N$.

Furthermore it holds

$$\lim_{n \to \infty} \sum_{k=\lceil na \rceil}^{\lfloor nb \rfloor} {}_2Z_q\left(nx - k\right) \neq 1, \tag{11.52}$$

for at least some $x \in \left(\prod_{i=1}^{N} [a_i, b_i]\right)$.

11.2.3 *About q-deformed and parametrized half-hyperbolic tangent function ϑ_q*

Here we follow Anastassiou (2023d). We introduce the function

$$\vartheta_q\left(t\right) := \frac{1 - qe^{-\beta t}}{1 + qe^{-\beta t}}, \quad \forall\, t \in \mathbb{R}, \tag{11.53}$$

where $q, \beta > 0$. ϑ_q is strictly increasing. We also observe that

$$\vartheta_q\left(-\infty\right) = -1 \quad \text{and} \quad \vartheta_q\left(+\infty\right) = 1. \tag{11.54}$$

Furthermore,

$$\vartheta_q(0) = \frac{1-q}{1+q}. \tag{11.55}$$

In case of $t < \frac{\ln q}{\beta}$, we have that ϑ_q is strictly concave up, with $\vartheta_q''\left(\frac{\ln q}{\beta}\right) = 0$.

And in case of $t > \frac{\ln q}{\beta}$, we have that ϑ_q is strictly concave down.

Clearly, ϑ_q is a shifted sigmoid function with $\vartheta_q(0) = \frac{1-q}{1+q}$, and $\vartheta_q(-x) = -\vartheta_{q^{-1}}(x)$, $\forall\, x \in \mathbb{R}$ (a semi-odd function), see also Anastassiou (2023a).

By $1 > -1$, $x+1 > x-1$, we consider the activation function

$${}_3\mathcal{L}_q(x) := \frac{1}{4}\left(\vartheta_q(x+1) - \vartheta_q(x-1)\right) > 0, \tag{11.56}$$

$\forall x \in \mathbb{R}$; $\beta, q > 0$. Note that $\mathcal{L}_q(\pm\infty) = 0$, so the x-axis is horizontal asymptote. Also it holds,

$${}_3\mathcal{L}_q(-x) = {}_3\mathcal{L}_{\frac{1}{q}}(x), \quad \forall\, x \in \mathbb{R}, \tag{11.57}$$

a deformed symmetry.

Next we have that

$${}_3\mathcal{L}_q'(x) = \frac{1}{4}\left(\vartheta_q'(x+1) - \vartheta_q'(x-1)\right), \quad \forall\, x \in \mathbb{R}. \tag{11.58}$$

Hence, ${}_3\mathcal{L}_q$ is strictly increasing over $\left(-\infty, \frac{\ln q}{\beta} - 1\right)$.

and strictly decreasing over $\left(\frac{\ln q}{\beta} + 1, +\infty\right)$.

Moreover, ${}_3\mathcal{L}_q$ is concave down over $\left[\frac{\ln q}{\beta} - 1, \frac{\ln q}{\beta} + 1\right]$, and strictly concave down over $\left(\frac{\ln q}{\beta} - 1, \frac{\ln q}{\beta} + 1\right)$.

Consequently, ${}_3\mathcal{L}_q$ has a bell-type shape over $\mathbb{R}$. Of course, it holds ${}_3\mathcal{L}_q''\left(\frac{\ln q}{\beta}\right) < 0$. The maximum value of ${}_3\mathcal{L}_q$ is

$${}_3\mathcal{L}_q\left(\frac{\ln q}{\beta}\right) = \frac{1-e^{-\beta}}{2\left(1+e^{-\beta}\right)}. \tag{11.59}$$

We give

Theorem 11.9. *We have that*

$$\sum_{i=-\infty}^{\infty} {}_3\mathcal{L}_q(x-i) = 1, \quad \forall\, x \in \mathbb{R}, \quad \forall\, q, \beta > 0. \tag{11.60}$$

It follows

Theorem 11.10. *It holds*

$$\int_{-\infty}^{\infty} {}_3\mathcal{L}_q(x)\,dx = 1, \quad q, \beta > 0. \tag{11.61}$$

So ${}_3\mathcal{L}_q$ is a density function on $\mathbb{R}$; $q, \beta > 0$.
We need the following result.

Theorem 11.11. *Let* $0 < \alpha < 1$, *and* $n \in \mathbb{N}$ *with* $n^{1-\alpha} > 2$; $q, \beta > 0$. *Then*

$$\sum_{\substack{k=-\infty \\ \left\{ :|nx-k| \geq n^{1-\alpha} \right.}}^{\infty} {}_3\mathcal{L}_q(nx-k) < 2\max\left\{q, \frac{1}{q}\right\} e^{2\beta} e^{-\beta n^{(1-\alpha)}}$$

$$= Ke^{-\beta n^{(1-\alpha)}} =: c_3(n, a), \tag{11.62}$$

where $K := 2\max\left\{q, \frac{1}{q}\right\} e^{2\beta}$.

We need

Theorem 11.12. *Let* $x \in [a, b] \subset \mathbb{R}$ *and* $n \in \mathbb{N}$ *so that* $\lceil na \rceil \leq \lfloor nb \rfloor$. *For* $q > 0$, *we consider the number* $\lambda_q > z_0 > 0$ *with* ${}_3\mathcal{L}_q(z_0) = {}_3\mathcal{L}\phi_q(0)$ *and* $\beta, \lambda_q > 1$. *Then*

$$\frac{1}{\sum_{k=\lceil na \rceil}^{\lfloor nb \rfloor} {}_3\mathcal{L}_q(nx-k)} < max\left\{\frac{1}{{}_3\mathcal{L}_q(\lambda_q)}, \frac{1}{{}_3\mathcal{L}_{\frac{1}{q}}\left(\lambda_{\frac{1}{q}}\right)}\right\} =: \Psi_3(q). \tag{11.63}$$

We make

Remark 11.7. (i) We have that

$$\lim_{n\to+\infty} \sum_{k=\lceil na \rceil}^{\lfloor nb \rfloor} {}_3\mathcal{L}_q(nx-k) \neq 1, \quad \text{for at least some } x \in [a, b], \tag{11.64}$$

where $\beta, q > 0$.

(ii) Let $[a, b] \subset \mathbb{R}$. For large n, we always have $\lceil na \rceil \leq \lfloor nb \rfloor$. Also $a \leq \frac{k}{n} \leq b$, iff $\lceil na \rceil \leq k \leq \lfloor nb \rfloor$. In general, it holds

$$\sum_{k=\lceil na \rceil}^{\lfloor nb \rfloor} {}_3\mathcal{L}_q(nx-k) \leq 1. \tag{11.65}$$

We introduce

$$
{}_3Z_q\left(x_1,\ldots,x_N\right) := {}_3Z_q\left(x\right) := \prod_{i=1}^{N} {}_3\mathcal{L}_q\left(x_i\right), \tag{11.66}
$$

$x = (x_1, \ldots, x_N) \in \mathbb{R}^N$, $\beta, q > 0$, $N \in \mathbb{N}$.

It has the following properties:

(i) ${}_3Z_q\left(x\right) > 0, \ \forall\, x \in \mathbb{R}^N$,

(ii)

$$
\sum_{k=-\infty}^{\infty} {}_3Z_q(x-k) := \sum_{k_1=-\infty}^{\infty} \sum_{k_2=-\infty}^{\infty} \cdots \sum_{k_N=-\infty}^{\infty} {}_3Z_q(x_1-k_1,\ldots,x_N-k_N) = 1, \tag{11.67}
$$

where $k := (k_1, \ldots, k_n) \in \mathbb{Z}^N$, $\forall\, x \in \mathbb{R}^N$,

hence

(iii)

$$
\sum_{k=-\infty}^{\infty} {}_3Z_q\left(nx-k\right) = 1, \tag{11.68}
$$

$\forall\, x \in \mathbb{R}^N$; $n \in \mathbb{N}$,

and

(iv)

$$
\int_{\mathbb{R}^N} {}_3Z_q\left(x\right) dx = 1, \tag{11.69}
$$

that is, ${}_3Z_q$ is a multivariate density function.

For $0 < \beta^* < 1$ and $n \in \mathbb{N}$, a fixed $x \in \mathbb{R}^N$, we have that

$$
\sum_{k=\lceil na \rceil}^{\lfloor nb \rfloor} {}_3Z_q\left(nx-k\right)
$$

$$
= \sum_{\substack{k=\lceil na \rceil \\ \left\{ \left\| \frac{k}{n}-x \right\|_\infty \le \frac{1}{n^{\beta^*}} \right.}}^{\lfloor nb \rfloor} {}_3Z_q\left(nx-k\right) + \sum_{\substack{k=\lceil na \rceil \\ \left\{ \left\| \frac{k}{n}-x \right\|_\infty > \frac{1}{n^{\beta^*}} \right.}}^{\lfloor nb \rfloor} {}_3Z_q\left(nx-k\right). \tag{11.70}
$$

In the last two sums, the counting is over disjoint vector sets of k's because the condition $\left\| \frac{k}{n} - x \right\|_\infty > \frac{1}{n^{\beta^*}}$ implies that there exists at least one $\left| \frac{k_r}{n} - x_r \right| > \frac{1}{n^{\beta^*}}$, where $r \in \{1, \ldots, N\}$.

(v) We also have that

$$\sum_{\substack{k=\lceil na\rceil \\ \left\{\left\|\frac{k}{n}-x\right\|_{\infty}>\frac{1}{n^{\beta^*}}\right.}}^{\lfloor nb\rfloor} {}_3Z_q\left(nx-k\right) < Ke^{-\beta n^{\left(1-\beta^*\right)}} = c_3(n,\beta^*), \quad 0<\beta^*<1, \tag{11.71}$$

with $n \in \mathbb{N} : n^{1-\beta^*} > 2$, $x \in \prod_{i=1}^{N}[a_i, b_i]$.

(vi) Moreover,

$$0 < \frac{1}{\sum_{k=\lceil na\rceil}^{\lfloor nb\rfloor} {}_3Z_q\left(nx-k\right)} < \left(\Psi_3\left(q\right)\right)^N, \tag{11.72}$$

$\forall\, x \in \left(\prod_{i=1}^{N}[a_i, b_i]\right)$, $n \in \mathbb{N}$.

It is also clear that

(vii)

$$\sum_{\substack{k=-\infty \\ \left\{\left\|\frac{k}{n}-x\right\|_{\infty}>\frac{1}{n^{\beta^*}}\right.}}^{\infty} {}_3Z_q\left(nx-k\right) < Ke^{-\beta n^{\left(1-\beta^*\right)}}, \tag{11.73}$$

$0<\beta^*<1$, $n \in \mathbb{N} : n^{1-\beta^*} > 2$, $x \in \mathbb{R}^N$. Furthermore it holds

$$\lim_{n\to\infty} \sum_{k=\lceil na\rceil}^{\lfloor nb\rfloor} {}_3Z_q\left(nx-k\right) \neq 1, \tag{11.74}$$

for at least some $x \in \left(\prod_{i=1}^{N}[a_i, b_i]\right)$. Let $f \in C\left(\prod_{i=1}^{N}[a_i, b_i]\right)$, and $n \in \mathbb{N}$ such that $\lceil na_i\rceil \leq \lfloor nb_i\rfloor$, $i = 1, \ldots, N$.

We define the multivariate-averaged positive linear quasi-interpolation neural network operators ($x := (x_1, \ldots, x_N) \in \left(\prod_{i=1}^{N}[a_i, b_i]\right)$); $j = 1, 2, 3$:

$${}_jF_n\left(f, x_1, \ldots, x_N\right) := {}_jF_n\left(f, x\right) := \frac{\sum_{k=\lceil na\rceil}^{\lfloor nb\rfloor} f\left(\frac{k}{n}\right) {}_jZ_q\left(nx-k\right)}{\sum_{k=\lceil na\rceil}^{\lfloor nb\rfloor} {}_jZ_q\left(nx-k\right)}$$

$$= \frac{\sum_{k_1=\lceil na_1\rceil}^{\lfloor nb_1\rfloor} \sum_{k_2=\lceil na_2\rceil}^{\lfloor nb_2\rfloor} \cdots \sum_{k_N=\lceil na_N\rceil}^{\lfloor nb_N\rfloor} f\left(\frac{k_1}{n}, \ldots, \frac{k_N}{n}\right)\left(\prod_{i=1}^{N} {}_j\mathcal{L}_q\left(nx_i-k_i\right)\right)}{\prod_{i=1}^{N}\left(\sum_{k_i=\lceil na_i\rceil}^{\lfloor nb_i\rfloor} {}_j\mathcal{L}_q\left(nx_i-k_i\right)\right)}. \tag{11.75}$$

For large enough $n \in \mathbb{N}$, we always obtain $\lceil na_i\rceil \leq \lfloor nb_i\rfloor$, $i = 1, \ldots, N$. Also $a_i \leq \frac{k_i}{n} \leq b_i$, iff $\lceil na_i\rceil \leq k_i \leq \lfloor nb_i\rfloor$, $i = 1, \ldots, N$.

For the next, we need, for $f \in C\left(\prod_{i=1}^{N}[a_i,b_i]\right)$, the first multivariate modulus of continuity

$$\omega_1(f,h) := \sup_{\substack{x,y\in\prod_{i=1}^{N}[a_i,b_i] \\ \|x-y\|_\infty \leq h}} |f(x)-f(y)|, \quad h>0. \tag{11.76}$$

It holds that

$$\lim_{h\to 0}\omega_1(f,h) = 0. \tag{11.77}$$

We mention

Theorem 11.13 (see Anastassiou 2023b, 2023c, 2023d). *Let* $f \in C\left(\prod_{i=1}^{N}[a_i,b_i]\right)$, $0<\beta^*<1$, $x \in \left(\prod_{i=1}^{N}[a_i,b_i]\right)$, $N,n\in\mathbb{N}$ *with* $n^{1-\beta^*}>2$; $j=1,2,3$. *Then*

(1)

$$|\,{}_jF_n(f,x) - f(x)| \leq (\Psi_j(q))^N\left[\omega_1\left(f,\frac{1}{n^\beta}\right) + 2c_j(n,\beta^*)\|f\|_\infty\right] =: \lambda_j \tag{11.78}$$

and

(2)

$$\|{}_jF_n(f) - f\|_\infty \leq \lambda_j. \tag{11.79}$$

We note that $\lim_{n\to\infty}{}_jF_n(f) = f$, *pointwise and uniformly.*

In this chapter, we extend Theorem 11.13 to the fuzzy-random level.

11.3 Main Result

11.3.1 *About p-mean approximation by fuzzy-random perturbed quasi-interpolation neural network operators*

All terms and assumptions here as in Sections 1, 2.

Let $f \in C_{\mathcal{FR}}^{U_p}\left(\prod_{i=1}^{N}[a_i,b_i]\right)$, $1\leq p<+\infty$, $n,N\in\mathbb{N}$, $0<\beta<1$, $\overrightarrow{x} \in \left(\prod_{i=1}^{N}[a_i,b_i]\right)$, $(X,\mathcal{B},P)$ probability space, $s\in X$; $j=1,2,3$.

We define the following multivariate fuzzy random perturbed quasi-interpolation linear neural network operators:

$$\left({}_jF_n^{\mathcal{FR}}(f)\right)(\overrightarrow{x},s) := \sum_{\overrightarrow{k}=\lceil na\rceil}^{\lfloor nb\rfloor *} f\left(\frac{\overrightarrow{k}}{n},s\right) \odot \frac{{}_jZ_q\left(n\overrightarrow{x}-\overrightarrow{k}\right)}{\sum_{\overrightarrow{k}=\lceil na\rceil}^{\lfloor nb\rfloor} {}_jZ_q\left(n\overrightarrow{x}-\overrightarrow{k}\right)} \tag{11.80}$$

(see also (11.75)).

We present

Theorem 11.14. *Let* $f \in C_{\mathcal{FR}}^{U_p}\left(\prod_{i=1}^N [a_i,b_i]\right)$, $0 < \beta^* < 1$, $\overrightarrow{x} \in \left(\prod_{i=1}^N [a_i,b_i]\right)$, $n, N \in \mathbb{N}$, *with* $n^{1-\beta^*} > 2$, $1 \le p < +\infty$. *Assume that* $\int_X (D^*(f(\cdot,s),\widetilde{o}))^p P(ds) < \infty$; $j = 1,2,3$. *Then*

(1)

$$\begin{aligned}\left(\int_X D^p\left(\left({}_jF_n^{\mathcal{FR}}(f)\right)(\overrightarrow{x},s), f(\overrightarrow{x},s)\right) P(ds)\right)^{\frac{1}{p}} &\le (\Psi_j(q))^N \left\{\Omega_1\left(f,\frac{1}{n^{\beta^*}}\right)_{L^p}\right. \\ &\quad \left. + 2c_j(n,\beta^*)\left(\int_X (D^*(f(\cdot,s),\widetilde{o}))^p P(ds)\right)^{\frac{1}{p}}\right\} =: \mu_j^{(\mathcal{FR})}, \end{aligned} \tag{11.81}$$

(2)

$$\left\| \left(\int_X D^p\left(\left({}_jFn^{\mathcal{FR}}(f)\right)(\overrightarrow{x},s), f(\overrightarrow{x},s)\right) P(ds)\right)^{\frac{1}{p}} \right\|_{\infty,\left(\prod_{i=1}^N [a_i,b_i]\right)} \le \mu_j^{(\mathcal{FR})}, \tag{11.82}$$

where $(\Psi_j(q))^N$ *as in (11.26), (11.50), (11.72) and* $c_j(n,\beta^*)$ *as in (11.25), (11.49), (**??**).*

Proof. We note that

$$\begin{aligned} D\left(f\left(\frac{\overrightarrow{k}}{n},s\right), f(\overrightarrow{x},s)\right) &\le D\left(f\left(\frac{\overrightarrow{k}}{n},s\right),\widetilde{o}\right) + D(f(\overrightarrow{x},s),\widetilde{o}) \\ &\le 2D^*(f(\cdot,s),\widetilde{o}). \end{aligned} \tag{11.83}$$

Hence,

$$D^p\left(f\left(\frac{\overrightarrow{k}}{n},s\right),f\left(\overrightarrow{x},s\right)\right)\leq 2^pD^{*p}\left(f\left(\cdot,s\right),\widetilde{o}\right) \tag{11.85}$$

and

$$\left(\int_X D^p\left(f\left(\frac{\overrightarrow{k}}{n},s\right),f\left(\overrightarrow{x},s\right)\right)P\left(ds\right)\right)^{\frac{1}{p}}\leq 2\left(\int_X\left(D^*\left(f\left(\cdot,s\right),\widetilde{o}\right)\right)^pP\left(ds\right)\right)^{\frac{1}{p}}. \tag{11.86}$$

We observe that

$$D\left(\left({}_jF_n^{\mathcal{FR}}\left(f\right)\right)\left(\overrightarrow{x},s\right),f\left(\overrightarrow{x},s\right)\right) \tag{11.87}$$

$$=D\left(\sum_{\overrightarrow{k}=\lceil na\rceil}^{\lfloor nb\rfloor *}f\left(\frac{\overrightarrow{k}}{n},s\right)\odot\frac{{}_jZ_q\left(nx-k\right)}{\sum_{\overrightarrow{k}=\lceil na\rceil}^{\lfloor nb\rfloor}{}_jZ_q\left(nx-k\right)},f\left(\overrightarrow{x},s\right)\odot 1\right)$$

$$=D\left(\sum_{\overrightarrow{k}=\lceil na\rceil}^{\lfloor nb\rfloor *}f\left(\frac{\overrightarrow{k}}{n},s\right)\odot\frac{{}_jZ_q\left(nx-k\right)}{\sum_{\overrightarrow{k}=\lceil na\rceil}^{\lfloor nb\rfloor}{}_jZ_q\left(nx-k\right)},f\left(\overrightarrow{x},s\right)\right.$$

$$\left.\odot\frac{\sum_{\overrightarrow{k}=\lceil na\rceil}^{\lfloor nb\rfloor}{}_jZ_q\left(nx-k\right)}{\sum_{\overrightarrow{k}=\lceil na\rceil}^{\lfloor nb\rfloor}{}_jZ_q\left(nx-k\right)}\right) \tag{11.88}$$

$$=D\left(\sum_{\overrightarrow{k}=\lceil na\rceil}^{\lfloor nb\rfloor *}f\left(\frac{\overrightarrow{k}}{n},s\right)\odot\frac{{}_jZ_q\left(nx-k\right)}{\sum_{\overrightarrow{k}=\lceil na\rceil}^{\lfloor nb\rfloor}{}_jZ_q\left(nx-k\right)},\right.$$

$$\left.\sum_{\overrightarrow{k}=\lceil na\rceil}^{\lfloor nb\rfloor *}f\left(\overrightarrow{x},s\right)\odot\frac{{}_jZ_q\left(nx-k\right)}{\sum_{\overrightarrow{k}=\lceil na\rceil}^{\lfloor nb\rfloor}{}_jZ_q\left(nx-k\right)}\right)$$

$$\leq\sum_{\overrightarrow{k}=\lceil na\rceil}^{\lfloor nb\rfloor}\left(\frac{{}_jZ_q\left(nx-k\right)}{\sum_{\overrightarrow{k}=\lceil na\rceil}^{\lfloor nb\rfloor}{}_jZ_q\left(nx-k\right)}\right)D\left(f\left(\frac{\overrightarrow{k}}{n},s\right),f\left(\overrightarrow{x},s\right)\right). \tag{11.89}$$

So

$$D\left(\left({}_jF_n^{\mathcal{FR}}(f)\right)(\overrightarrow{x},s), f(\overrightarrow{x},s)\right)$$

$$\leq \sum_{\overrightarrow{k}=\lceil na\rceil}^{\lfloor nb\rfloor} \left(\frac{{}_jZ_q(nx-k)}{\sum_{\overrightarrow{k}=\lceil na\rceil}^{\lfloor nb\rfloor} {}_jZ_q(nx-k)}\right) D\left(f\left(\frac{\overrightarrow{k}}{n},s\right), f(\overrightarrow{x},s)\right)$$

$$= \sum_{\substack{\overrightarrow{k}=\lceil na\rceil \\ \left\|\frac{\overrightarrow{k}}{n}-\overrightarrow{x}\right\|_\infty \leq \frac{1}{n^{\beta*}}}}^{\lfloor nb\rfloor} \left(\frac{{}_jZ_q(nx-k)}{\sum_{\overrightarrow{k}=\lceil na\rceil}^{\lfloor nb\rfloor} {}_jZ_q(nx-k)}\right) D\left(f\left(\frac{\overrightarrow{k}}{n},s\right), f(\overrightarrow{x},s)\right)$$

$$+ \sum_{\substack{\overrightarrow{k}=\lceil na\rceil \\ \left\|\frac{\overrightarrow{k}}{n}-\overrightarrow{x}\right\|_\infty > \frac{1}{n^{\beta*}}}}^{\lfloor nb\rfloor} \left(\frac{{}_jZ_q(nx-k)}{\sum_{\overrightarrow{k}=\lceil na\rceil}^{\lfloor nb\rfloor} {}_jZ_q(nx-k)}\right) D\left(f\left(\frac{\overrightarrow{k}}{n},s\right), f(\overrightarrow{x},s)\right). \tag{11.90}$$

Hence it holds

$$\left(\int_X D^p\left(\left({}_jF_n^{\mathcal{FR}}(f)\right)(\overrightarrow{x},s), f(\overrightarrow{x},s)\right) P(ds)\right)^{\frac{1}{p}} \tag{11.91}$$

$$\leq \sum_{\substack{\overrightarrow{k}=\lceil na\rceil \\ \left\|\frac{\overrightarrow{k}}{n}-\overrightarrow{x}\right\|_\infty \leq \frac{1}{n^{\beta*}}}}^{\lfloor nb\rfloor} \left(\frac{{}_jZ_q(nx-k)}{\sum_{\overrightarrow{k}=\lceil na\rceil}^{\lfloor nb\rfloor} {}_jZ_q(nx-k)}\right)$$

$$\times \left(\int_X D^p\left(f\left(\frac{\overrightarrow{k}}{n},s\right), f(\overrightarrow{x},s)\right) P(ds)\right)^{\frac{1}{p}}$$

$$+ \sum_{\substack{\overrightarrow{k}=\lceil na\rceil \\ \left\|\frac{\overrightarrow{k}}{n}-\overrightarrow{x}\right\|_\infty > \frac{1}{n^{\beta*}}}}^{\lfloor nb\rfloor} \left(\frac{{}_jZ_q(nx-k)}{\sum_{\overrightarrow{k}=\lceil na\rceil}^{\lfloor nb\rfloor} {}_jZ_q(nx-k)}\right)$$

$$\times \left(\int_X D^p\left(f\left(\frac{\overrightarrow{k}}{n},s\right), f(\overrightarrow{x},s)\right) P(ds)\right)^{\frac{1}{p}}$$

$$\leq \left(\frac{1}{\sum_{\overrightarrow{k}=\lceil na \rceil}^{\lfloor nb \rfloor} {}_jZ_q(nx-k)} \right) \cdot \left\{ \Omega_1^{(\mathcal{F})} \left(f, \frac{1}{n^{\beta^*}} \right)_{L^p} \right.$$

$$\left. + 2 \left(\int_X \left(D^* \left(f(\cdot, s), \widetilde{o} \right) \right)^p P(ds) \right)^{\frac{1}{p}} \left(\sum_{\substack{\overrightarrow{k}=\lceil na \rceil \\ \left\| \frac{\overrightarrow{k}}{n} - \overrightarrow{x} \right\|_\infty > \frac{1}{n^{\beta^*}}}}^{\lfloor nb \rfloor} {}_jZ_q(nx-k) \right) \right\} \tag{11.92}$$

(by (11.25), (11.26); (11.49), (11.50); (11.71), (11.72)

$$\leq (\Psi_j(q))^N \left\{ \Omega_1^{(\mathcal{F})} \left(f, \frac{1}{n^{\beta^*}} \right)_{L^p} + 2c_j(n, \beta^*) \left(\int_X \left(D^*(f(\cdot, s), \widetilde{o}) \right)^p P(ds) \right)^{\frac{1}{p}} \right\}. \tag{11.93}$$

We have proved claim. □

Conclusion 11.1. By Theorem 11.14, we obtain the pointwise and uniform convergences with rates in the p-mean and D-metric of the operator ${}_jF_n^{\mathcal{FR}}$ to the unit operator for $f \in C_{\mathcal{FR}}^{U_p}\left(\prod_{i=1}^N [a_i, b_i]\right)$, $j = 1, 2, 3$.

Bibliography

G.A. Anastassiou, Multivariate Fuzzy-Random Quasi-interpolation neural network approximation operators, *J. Fuzzy Mathematics*, 22(1), 2014, 167–184.

G.A. Anastassiou, *Intelligent Systems II: Complete Approximation by Neural Network Operators*, Springer, Heidelberg, New York, 2016.

G.A. Anastassiou, General sigmoid based Banach space valued neural network approximation, *Journal of Computational Analysis and Applications*, 31(4) (2023a), 520–534.

G.A. Anastassiou, *Parametrized Deformed and General Neural Networks*, Chapter 15, Springer, Heidelberg, New York, 2023b.

G.A. Anastassiou, *Parametrized Deformed and General Neural Networks*, Chapter 17, Springer, Heidelberg, New York, 2023c.

G.A. Anastassiou, *Parametrized Deformed and General Neural Networks*, Chapter 19, Springer, Heidelberg, New York, 2023d.

G.A. Anastassiou, D. Kouloumpou, *Multivariate fuzzy-random and perturbed neural network approximation*, Analele Universitatii Oradea, Fasc. Matematica, accepted, 2023e.

S. Gal, *Approximation Theory in Fuzzy Setting*, Chapter 13 in Handbook of Analytic-Computational Methods in Applied Mathematics, pp. 617–666, edited by G. Anastassiou, Chapman & Hall/CRC, 2000a, Boca Raton, New York.

C. Wu and Z. Gong, On Henstock integral of interval-valued functions and fuzzy valued functions, *Fuzzy Sets and Systems*, 115(3) (2000b), 377–391.

C. Wu and Z. Gong, On Henstock integral of fuzzy-number-valued functions (I), *Fuzzy Sets and Systems*, 120(3) (2001), 523–532.

C. Wu and M. Ma, On embedding problem of fuzzy number space: Part 1, *Fuzzy Sets and Systems*, 44 (1991), 33–38.

Chapter 12

Trigonometric and Hyperbolic Korovkin Properties

Based on trigonometric and hyperbolic Taylor-type formulae, we establish related Shisha–Mond form inequalities leading to interesting Korovkin theorems. We deal with the high order of approximation of positive linear operators to the unit operator. The results are quantitative via the modulus of continuity. We finish with applications to Bernstein operators. It follows Anastassiou (2023).

12.1 Introduction

In this chapter, mainly we are motivated by the following result.

Theorem 12.1 (Korovkin 1960, p. 14). *Let $[a, b]$ be a closed interval in $\mathbb{R}$ and $(L_n)_{n\in\mathbb{N}}$ be a sequence of positive linear operators mapping $C([a, b])$ into itself. Suppose that $(L_n f)$ converges uniformly to f for the three test functions $f = 1$, x, x^2. Then $(L_n f)$ converges uniformly to f on $[a, b]$ for all functions $f \in C([a, b])$.*

Let $f \in C([a, b])$ and $0 \leq h \leq b - a$. The first modulus of continuity of f at h is given by

$$\omega_1(f, h) = \sup_{\substack{x,y\in[a,b]\\ |x-y|\leq h}} |f(x) - f(y)|.$$

If $h > b - a$, then we define $\omega_1(f, h) = \omega_1(f, b - a)$.

Another motivation is the following.

Theorem 12.2 (Shisha and Mond 1968). *Let $[a,b] \subset \mathbb{R}$ a closed interval. Let $\{L_n\}_{n\in\mathbb{N}}$ be a sequence of positive linear operators acting on $C([a,b])$ into itself. For $n = 1,\ldots,$ suppose $L_n(1)$ is bounded. Let $f \in C([a,b])$. Then for $n = 1,2,\ldots,$ we have*

(i)

$$\|L_n f - f\|_\infty \leq \|f\|_\infty \|L_n 1 - 1\|_\infty + \|L_n 1 + 1\|_\infty \omega_1(f,\mu_n),$$

where

$$\mu_n = \left\|L_n\left((t-x)\right)^2(x)\right\|_\infty^{\frac{1}{2}}$$

and $\|\cdot\|_\infty$ stands for the sup-norm over $[a,b]$.

One can easily see, for $n = 1,2,\ldots$

(ii)

$$\mu_n^2 \leq \left\|L_n\left(t^2;x\right) - x^2\right\|_\infty + 2c\|L_n(t;x) - x\|_\infty + c^2\|L_n(1;x) - 1\|_\infty,$$

where $c = \max(|a|,|b|)$.

Thus, given the Korovkin assumptions (see Theorem 12.1) as $n \to \infty$, we get by (ii) that $\mu_n \to 0$ and by (i) that $\|L_n f - f\|_\infty \to 0$ for any $f \in C([a,b])$. That is, one derives the Korovkin conclusion in a quantitative way and with rates of convergence.

Here by the use of trigonometric and hyperbolic Taylor's formulae, we derive a higher order of quantitative approximation for sequences of positive linear operators converging to the unit operator. This is along with the related Shisha–Mond-type inequalities and corresponding Korovkin theorems.

12.2 Main Results

12.2.1 *Trigonometric approximation by positive linear operators*

We present the following quantitative results.

Theorem 12.3. *Let $f \in C^2([a,b])$, $x_0 \in [a,b]$, and a sequence of positive linear operators $(L_n)_{n\in\mathbb{N}}$ from $C([a,b])$ into itself, $r > 0$, and $D_3(x_0) := \left(L_n\left(|\cdot - x_0|^3\right)(x_0)\right)^{\frac{1}{3}}$.*

Then

(i)

$$|L_n(f)(x_0) - f(x_0)| \leq |f(x_0)| \, |L_n(1)(x_0) - 1|$$
$$+ |f'(x_0)| \, |L_n(\sin(\cdot - x_0))(x_0)| + 2|f''(x_0)| \left(L_n \left(\sin^2 \left(\frac{\cdot - x_0}{2} \right) \right) (x_0) \right)$$
$$+ \frac{\omega_1(f'' + f, rD_3(x_0))}{2} D_3^2(x_0) \left[(L_n(1)(x_0))^{\frac{1}{3}} + \frac{1}{3r} \right], \quad (12.1)$$

(ii)

$$\|L_n(f) - f\|_\infty \leq \|f\|_\infty \|L_n(1) - 1\|_\infty$$
$$+ \|f'\|_\infty \|L_n(\sin(\cdot - x_0))(x_0)\|_\infty + 2\|f''\|_\infty \left\| L_n \left(\sin^2 \left(\frac{\cdot - x_0}{2} \right) \right) (x_0) \right\|_\infty$$
$$+ \frac{\omega_1(f'' + f, r\|D_3\|_\infty)}{2} \|D_3\|_\infty^2 \left[\|L_n(1)\|_\infty^{\frac{1}{3}} + \frac{1}{3r} \right], \quad (12.2)$$

(iii) *if* $L_n(1) = 1$, *and* $f'(x_0) = f''(x_0) = 0$ *for some* $x_0 \in [a,b]$, *we obtain*

$$|L_n(f)(x_0) - f(x_0)|$$
$$\leq \frac{\omega_1(f'' + f, rD_3(x_0))}{2} D_3^2(x_0) \left[(L_n(1)(x_0))^{\frac{1}{3}} + \frac{1}{3r} \right], \quad (12.3)$$

and

(iv)

$$\Big| L_n(f)(x_0) - f(x_0) L_n(1)(x_0) - f'(x_0) L_n(\sin(\cdot - x_0))(x_0)$$
$$- 2f''(x_0) L_n \left(\sin^2 \left(\frac{\cdot - x_0}{2} \right) \right) (x_0) \Big|$$
$$\leq \frac{\omega_1(f'' + f, rD_3(x_0))}{2} D_3^2(x_0) \left[(L_n(1)(x_0))^{\frac{1}{3}} + \frac{1}{3r} \right]. \quad (12.4)$$

Proof. Here $f \in C^2([a,b])$, with $x, x_0 \in [a,b]$. By Anastassiou (2023), we have that

$$f(x) - f(x_0) = f'(x_0) \sin(x - x_0) + 2f''(x_0) \sin^2 \left(\frac{x - x_0}{2} \right)$$
$$+ \int_{x_0}^{x} [(f''(t) + f(t)) - (f''(x_0) + f(x_0))] \sin(x - t) \, dt. \quad (12.5)$$

Denote by

$$R(x, x_0) := \int_{x_0}^{x} [(f''(t) + f(t)) - (f''(x_0) + f(x_0))] \sin(x - t)\, dt. \tag{12.6}$$

Clearly, we have that

$$\begin{aligned} &|(f''(t) + f(t)) - (f''(x_0) + f(x_0))| \\ &\quad \leq \omega_1(f'' + f, h)\left(1 + \frac{|t - x_0|}{h}\right), \quad \text{all } t \in [a, b],\ h > 0. \end{aligned} \tag{12.7}$$

Let $x \geq x_0$, then

$$|R(x, x_0)| \leq \int_{x_0}^{x} |(f''(t) + f(t)) - (f''(x_0) + f(x_0))|\, |\sin(x - t)|\, dt \overset{(12.7)}{\leq} \tag{12.8}$$

(using $|\sin x| \leq |x|$, $\forall\ x \in \mathbb{R}$)

$$\begin{aligned} &\omega_1(f'' + f, h) \int_{x_0}^{x} \left(1 + \frac{(t - x_0)}{h}\right)(x - t)\, dt \\ &= \omega_1(f'' + f, h)\left[\int_{x_0}^{x} (x - t)\, dt + \frac{1}{h}\int_{x_0}^{x} (x - t)(t - x_0)\, dt\right] \\ &= \omega_1(f'' + f, h)\left[\frac{(x - x_0)^2}{2} + \frac{(x - x_0)^3}{6h}\right]. \end{aligned} \tag{12.9}$$

So, if $x \geq x_0$, we got that

$$|R(x, x_0)| \leq \frac{\omega_1(f'' + f, h)}{2}\left[(x - x_0)^2 + \frac{(x - x_0)^3}{3h}\right], \quad h > 0. \tag{12.10}$$

Let now $x < x_0$, then

$$\begin{aligned} |R(x, x_0)| &= \left|\int_{x}^{x_0} [(f''(t) + f(t)) - (f''(x_0) + f(x_0))] \sin(x - t)\, dt\right| \\ &\leq \int_{x}^{x_0} |(f''(t) + f(t)) - (f''(x_0) + f(x_0))|\, |\sin(x - t)|\, dt \qquad (12.11) \\ &\leq \omega_1(f'' + f, h) \int_{x}^{x_0} \left(1 + \frac{(x_0 - t)}{h}\right)(t - x)\, dt \end{aligned}$$

$$= \omega_1 (f'' + f, h) \left[\int_x^{x_0} (t - x)\, dt + \frac{1}{h} \int_x^{x_0} (x_0 - t)(t - x)\, dt \right]$$

$$= \omega_1 (f'' + f, h) \left[\frac{(x_0 - x)^2}{2} + \frac{(x_0 - x)^3}{6h} \right]. \tag{12.12}$$

So, if $x < x_0$, we got that

$$|R(x, x_0)| \le \frac{\omega_1 (f'' + f, h)}{2} \left[(x_0 - x)^2 + \frac{(x_0 - x)^3}{3h} \right]. \tag{12.13}$$

Thus, it holds

$$|R(x, x_0)| \le \frac{\omega_1 (f'' + f, h)}{2} \left[(x - x_0)^2 + \frac{|x - x_0|^3}{3h} \right], \quad h > 0, \tag{12.14}$$

$\forall\ x,\ x_0 \in [a, b]$.

Let a sequence of positive linear operators $(L_n)_{n \in \mathbb{N}}$ from $C([a, b])$ into itself.

We note that

$$|R(x, x_0)| \le |R(x, x_0)|, \quad \text{iff} \ - |R(x, x_0)| \le R(x, x_0) \le |R(x, x_0)|,$$

and

$$-L_n (|R(x, x_0)|) \le L_n (R(x, x_0)) \le L_n (|R(x, x_0)|).$$

That is,

$$|L_n (R(x, x_0))| \le L_n (|R(x, x_0)|). \tag{12.15}$$

By (12.5), we can write

$$f(\cdot) - f(x_0) = f'(x_0) \sin(\cdot - x_0) + 2f''(x_0) \sin^2 \left(\frac{\cdot - x_0}{2} \right) + R(\cdot, x_0). \tag{12.16}$$

Thus, it holds

$$L_n (f)(x_0) - f(x_0) L_n (1)(x_0)$$
$$= f'(x_0) L_n (\sin(\cdot - x_0))(x_0) + 2f''(x_0) L_n \left(\sin^2 \left(\frac{\cdot - x_0}{2} \right) \right)(x_0)$$
$$+ L_n (R(\cdot, x_0))(x_0), \tag{12.17}$$

$\forall\ x_0 \in [a, b]$.

Consequently, we have

$$\left|L_n(f)(x_0) - f(x_0) L_n(1)(x_0)\right|$$
$$\overset{(12.17)}{\leq} \left|f'(x_0)\right| \left|L_n(\sin(\cdot - x_0))(x_0)\right| + 2\left|f''(x_0)\right| \left(L_n\left(\sin^2\left(\frac{\cdot - x_0}{2}\right)\right)(x_0)\right)$$
$$+ L_n(|R(\cdot, x_0)|) \leq \tag{12.18}$$

(by (12.14))

$$\left|f'(x_0)\right| \left|L_n(\sin(\cdot - x_0))(x_0)\right| + 2\left|f''(x_0)\right| \left(L_n\left(\sin^2\left(\frac{\cdot - x_0}{2}\right)\right)(x_0)\right)$$
$$+ \frac{\omega_1(f'' + f, h)}{2}\left[L_n\left((\cdot - x_0)^2\right)(x_0) + \frac{L_n\left(|\cdot - x_0|^3\right)(x_0)}{3h}\right], \tag{12.19}$$

$\forall\, x_0 \in [a, b]$.

We observe that

$$\begin{aligned}
&\left|L_n(f)(x_0) - f(x_0)\right| \\
&\quad = \left|L_n(f)(x_0) - f(x_0) - f(x_0) L_n(1) x_0 + f(x_0) L_n(1)(x_0)\right| \\
&\quad = \left|(L_n(f)(x_0) - f(x_0) L_n(1)(x_0)) + f(x_0)(L_n(1)(x_0) - 1)\right| \\
&\quad \leq \left|L_n(f)(x_0) - f(x_0) L_n(1)(x_0)\right| + \left|f(x_0)\right| \left|L_n(1)(x_0) - 1\right|.
\end{aligned} \tag{12.20}$$

We have found that (by (12.19) and (12.20))

$$\left|L_n(f)(x_0) - f(x_0)\right| \leq \left|f(x_0)\right| \left|L_n(1)(x_0) - 1\right|$$
$$+ \left|f'(x_0)\right| \left|L_n(\sin(\cdot - x_0))(x_0)\right| + 2\left|f''(x_0)\right| \left(L_n\left(\sin^2\left(\frac{\cdot - x_0}{2}\right)\right)(x_0)\right)$$
$$+ \frac{\omega_1(f'' + f, h)}{2}\left[L_n\left((\cdot - x_0)^2\right)(x_0) + \frac{L_n\left(|\cdot - x_0|^3\right)(x_0)}{3h}\right] \tag{12.21}$$
$$\leq \left|f(x_0)\right| \left|L_n(1)(x_0) - 1\right| + \left|f'(x_0)\right| L_n(|\cdot - x_0|)(x_0)$$
$$+ \frac{\left|f''(x_0)\right|}{2} L_n\left((\cdot - x_0)^2\right)(x_0)$$
$$+ \frac{\omega_1(f'' + f, h)}{2}\left[L_n\left((\cdot - x_0)^2\right)(x_0) + \frac{L_n\left(|\cdot - x_0|^3\right)(x_0)}{3h}\right] =: (\psi), \tag{12.22}$$

$\forall\, x_0 \in [a, b]$, $h > 0$.

By Riesz representation theorem and Hölder's inequality, we get

$$L_n\left((\cdot - x_0)^2\right)(x_0) \leq \left(L_n\left(|\cdot - x_0|^3\right)(x_0)\right)^{\frac{2}{3}} (L_n(1)(x_0))^{\frac{1}{3}}, \tag{12.23}$$

$\forall\, x_0 \in [a, b]$.

Let $r > 0$, and take

$$h := r\left(L_n\left(|\cdot - x_0|^3\right)(x_0)\right)^{\frac{1}{3}} = rD_3(x_0), \tag{12.24}$$

where

$$D_3(x_0) := \left(L_n |\cdot - x_0|^3 (x_0)\right)^{\frac{1}{3}}. \tag{12.25}$$

That is,

$$\left(L_n(\cdot - x_0)^2\right)(x_0) \leq D_3^2(x_0)(L_n(1)(x_0))^{\frac{1}{3}}. \tag{12.26}$$

Momentarily, we assume that $D_3(x_0) > 0$. Therefore, it holds

$$\begin{aligned}
&\left[L_n\left((\cdot - x_0)^2\right)(x_0) + \frac{L_n\left(|\cdot - x_0|^3\right)(x_0)}{3h}\right] \\
&\leq \left[D_3^2(x_0)(L_n(1)(x_0))^{\frac{1}{3}} + \frac{D_3^3(x_0)}{3rD_3(x_0)}\right] \\
&= D_3^2(x_0)\left[(L_n(1)(x_0))^{\frac{1}{3}} + \frac{1}{3r}\right].
\end{aligned} \tag{12.27}$$

We continue as follows

$$\begin{aligned}
|L_n(f)(x_0) - f(x_0)| &\leq (\psi) \overset{(12.27)}{\leq} |f(x_0)|\,|L_n(1)(x_0) - 1| \\
&+ |f'(x_0)|\, L_n(|\cdot - x_0|)(x_0) + \frac{|f''(x_0)|}{2} L_n\left((\cdot - x_0)^2\right)(x_0) \\
&+ \frac{\omega_1(f'' + f, rD_3(x_0))}{2} D_3^2(x_0)\left[(L_n(1)(x_0))^{\frac{1}{3}} + \frac{1}{3r}\right].
\end{aligned} \tag{12.28}$$

Note also in general that

$$\begin{aligned}
L_n(|\cdot - x_0|)(x_0) &\leq \left(L_n\left((\cdot - x_0)^2\right)(x_0)\right)^{\frac{1}{2}} (L_n(1)(x_0))^{\frac{1}{2}} \\
&\leq D_3(x_0)(L_n(1)(x_0))^{\frac{2}{3}}.
\end{aligned} \tag{12.29}$$

In case of $D_3(x_0) = 0$, by (12.22), we have that

$$|L_n(f)(x_0) - f(x_0)| \leq |f(x_0)|\,|L_n(1)(x_0) - 1|. \tag{12.30}$$

By Riesz representation theorem, we have that

$$L_n(f)(x_0) = \int_{[a,b]} f(t)\, d\mu_{x_0}(t), \tag{12.31}$$

where μ_{x_0} is a positive finite Borel measure such that $\mu_{x_0}([a,b]) = M > 0$. Note that $M = L_n(1)(x_0)$.

So, here we have that

$$D_3^3(x_0) = L_n\left(|\cdot - x_0|^3\right)(x_0) = \int_{[a,b]} |t - x_0|^3\, d\mu_{x_0}(t) = 0, \tag{12.32}$$

which implies $t = x_0$, a.e, consequently we have that $\mu_{x_0} = \delta_{x_0} M$, where δ_{x_0} is the Dirac measure at x_0.

Thus,

$$L_n(f)(x_0) = f(x_0)\, M = f(x_0)\, L_n(1)(x_0). \tag{12.33}$$

Hence we derive

$$\begin{aligned} |L_n(f)(x_0) - f(x_0)| &= |f(x_0)\, L_n(1)(x_0) - f(x_0)| \\ &= |f(x_0)|\, |L_n(1)(x_0) - 1|. \end{aligned} \tag{12.34}$$

So (12.30) becomes equality. Thus, (12.28) is valid always.

By Anastassiou (2000, p. 388), $L_n\left(|\cdot - x_0|^3\right)(x_0)$ is continuous in $x_0 \in [a,b]$.

The theorem is proved. □

It follows a Korovkin-type theorem.

Theorem 12.4. *Here all as in Theorem 12.3. Assume that $L_n(1) \to 1$ and $L_n\left(|\cdot - x_0|^3\right)(x_0) \to 0$, both uniformly convergent, as $n \to \infty$. Then $L_n(f) \to f$, uniformly convergent, as $n \to \infty$, $\forall\, f \in C^2([a,b])$.*

Proof. By (12.2), using (12.23) and (12.29), and that every uniformly convergent sequence of bounded functions is uniformly bounded. Also that $\omega_1\left(f'' + f, r\, \|D_3\|_\infty\right) \to 0$, as $n \to \infty$. □

12.2.2 *Hyperbolic approximation by positive linear operators*

We give the following related results.

Theorem 12.5. *Let $f \in C^2([a,b])$, $x_0 \in [a,b]$, and a sequence of positive linear operators $(L_n)_{n\in\mathbb{N}}$ from $C([a,b])$ into itself, $r > 0$, and $D_3(x_0) := \left(L_n\left(|\cdot - x_0|^3\right)(x_0)\right)^{\frac{1}{3}}$. Then*

(i)

$$\begin{aligned}|L_n(f)(x_0) - f(x_0)| &\le |f(x_0)| \, |L_n(1)(x_0) - 1| \\ &+ |f'(x_0)| \, |L_n(\sinh(\cdot - x_0))(x_0)| + 2|f''(x_0)| \left(L_n\left(\sinh^2\left(\frac{\cdot - x_0}{2}\right)\right)(x_0)\right) \\ &+ \frac{\cosh(b-a)\,\omega_1(f'' - f, rD_3(x_0))}{2} D_3^2(x_0) \left[(L_n(1)(x_0))^{\frac{1}{3}} + \frac{1}{3r}\right], \end{aligned} \tag{12.35}$$

(ii)

$$\begin{aligned}\|L_n(f) - f\|_\infty &\le \|f\|_\infty \|L_n(1) - 1\|_\infty \\ &+ \|f'\|_\infty \|L_n(\sinh(\cdot - x_0))(x_0)\|_\infty + 2\|f''\|_\infty \left\|L_n\left(\sinh^2\left(\frac{\cdot - x_0}{2}\right)\right)(x_0)\right\|_\infty \\ &+ \frac{\cosh(b-a)\,\omega_1\left(f'' - f, r\|D_3\|_\infty\right)}{2} \|D_3\|_\infty^2 \left[\|L_n(1)\|_\infty^{\frac{1}{3}} + \frac{1}{3r}\right], \end{aligned} \tag{12.36}$$

(iii) *if* $L_n(1) = 1$, *and* $f'(x_0) = f''(x_0) = 0$ *for some* $x_0 \in [a,b]$, *we obtain*

$$\begin{aligned}&|L_n(f)(x_0) - f(x_0)| \\ &\le \frac{\cosh(b-a)\,\omega_1(f'' - f, rD_3(x_0))}{2} D_3^2(x_0) \left[(L_n(1)(x_0))^{\frac{1}{3}} + \frac{1}{3r}\right], \end{aligned} \tag{12.37}$$

and

(iv)

$$\begin{aligned}&\Big|L_n(f)(x_0) - f(x_0) L_n(1)(x_0) - f'(x_0) L_n(\sinh(\cdot - x_0))(x_0) \\ &\quad - 2f''(x_0) L_n\left(\sinh^2\left(\frac{\cdot - x_0}{2}\right)\right)(x_0)\Big| \\ &\le \frac{\cosh(b-a)\,\omega_1(f'' - f, rD_3(x_0))}{2} D_3^2(x_0) \left[(L_n(1)(x_0))^{\frac{1}{3}} + \frac{1}{3r}\right]. \end{aligned} \tag{12.38}$$

Proof. Here $f \in C^2([a,b])$, with $x, x_0 \in [a,b]$. By Anastassiou (2023), we have that

$$\begin{aligned}f(x) - f(x_0) &= f'(x_0)\sinh(x - x_0) + 2f''(x_0)\sinh^2\left(\frac{x - x_0}{2}\right) \\ &+ \int_{x_0}^{x} [(f''(t) - f(t)) - (f''(x_0) - f(x_0))] \sinh(x - t)\, dt. \end{aligned} \tag{12.39}$$

By the mean value theorem, we derive that

$$|\sinh x| \leq \cosh(b-a)\,|x|, \quad \forall\, x \in [-(b-a), b-a]. \tag{12.40}$$

Denote by

$$\overline{R}(x, x_0) := \int_{x_0}^{x} \left[(f''(t) - f(t)) - (f''(x_0) - f(x_0))\right] \sinh(x-t)\, dt. \tag{12.41}$$

Clearly, we have that

$$\begin{aligned} &|(f''(t) - f(t)) - (f''(x_0) - f(x_0))| \\ &\leq \omega_1(f'' - f, h)\left(1 + \frac{|t - x_0|}{h}\right), \quad \text{all } t \in [a,b],\ h > 0. \end{aligned} \tag{12.42}$$

Let $x \geq x_0$, then

$$\begin{aligned} \left|\overline{R}(x, x_0)\right| &\leq \int_{x_0}^{x} |(f''(t) - f(t)) - (f''(x_0) - f(x_0))|\, |\sinh(x-t)|\, dt \\ &\overset{\text{(by (12.41), (12.42))}}{\leq} \omega_1(f'' - f, h)\cosh(b-a)\int_{x_0}^{x}\left(1 + \frac{(t - x_0)}{h}\right)(x-t)\, dt \\ &= \cosh(b-a)\,\omega_1(f'' - f, h)\left[\int_{x_0}^{x}(x-t)\, dt + \frac{1}{h}\int_{x_0}^{x}(x-t)(t-x_0)\, dt\right] \\ &= \cosh(b-a)\,\omega_1(f'' - f, h)\left[\frac{(x-x_0)^2}{2} + \frac{(x-x_0)^3}{6h}\right]. \end{aligned} \tag{12.43}$$

So, if $x \geq x_0$, we got that

$$\left|\overline{R}(x, x_0)\right| \leq \frac{\cosh(b-a)\,\omega_1(f'' - f, h)}{2}\left[(x-x_0)^2 + \frac{(x-x_0)^3}{3h}\right], \quad h > 0. \tag{12.44}$$

In case of $x < x_0$, we find similarly that

$$\left|\overline{R}(x, x_0)\right| \leq \frac{\cosh(b-a)\,\omega_1(f'' - f, h)}{2}\left[(x_0-x)^2 + \frac{(x_0-x)^3}{3h}\right]. \tag{12.45}$$

Thus, it holds

$$\left|\overline{R}(x,x_0)\right| \leq \frac{\cosh(b-a)\,\omega_1(f''-f,h)}{2}\left[(x-x_0)^2+\frac{|x-x_0|^3}{3h}\right], \quad h>0, \tag{12.46}$$

$\forall\ x,\ x_0 \in [a,b]$.

Let a sequence of positive linear operators $(L_n)_{n\in\mathbb{N}}$ from $C([a,b])$ into itself.

Clearly we have

$$\left|L_n\left(\overline{R}(x,x_0)\right)\right| \leq L_n\left(\left|\overline{R}(x,x_0)\right|\right). \tag{12.47}$$

By (12.39), we can write

$$f(\cdot)-f(x_0)=f'(x_0)\sinh(\cdot-x_0)+2f''(x_0)\sinh^2\left(\frac{\cdot-x_0}{2}\right)+\overline{R}(\cdot,x_0). \tag{12.48}$$

Thus, it holds

$$\begin{aligned} &L_n(f)(x_0)-f(x_0)L_n(1)(x_0) \\ &\quad = f'(x_0)L_n(\sinh(\cdot-x_0))(x_0)+2f''(x_0)L_n\left(\sinh^2\left(\frac{\cdot-x_0}{2}\right)\right)(x_0) \\ &\qquad + L_n\left(\overline{R}(\cdot,x_0)\right)(x_0), \end{aligned} \tag{12.49}$$

$\forall\ x_0 \in [a,b]$.

Consequently, we have

$$\begin{aligned} &|L_n(f)(x_0)-f(x_0)L_n(1)(x_0)| \\ &\quad \overset{(12.49)}{\leq} |f'(x_0)|\,|L_n(\sinh(\cdot-x_0))(x_0)|+2|f''(x_0)|\left|L_n\left(\sinh^2\left(\frac{\cdot-x_0}{2}\right)\right)(x_0)\right| \\ &\qquad + L_n\left(\left|\overline{R}(\cdot,x_0)\right|\right) \end{aligned} \tag{12.50}$$

$$\begin{aligned} &\leq |f'(x_0)|\,|L_n(\sinh(\cdot-x_0))(x_0)|+2|f''(x_0)|\left|L_n\left(\sinh^2\left(\frac{\cdot-x_0}{2}\right)\right)(x_0)\right| \\ &\quad + \frac{\cosh(b-a)\,\omega_1(f''-f,h)}{2}\left[L_n\left((\cdot-x_0)^2\right)(x_0)+\frac{L_n\left(|\cdot-x_0|^3\right)(x_0)}{3h}\right], \end{aligned} \tag{12.51}$$

$\forall\ x_0 \in [a,b]$.

As in (12.20), we again have that

$$\begin{aligned} &|L_n(f)(x_0)-f(x_0)| \\ &\quad \leq |(L_n(f)(x_0)-f(x_0)L_n(1)(x_0))|+|f(x_0)|\,|L_n(1)(x_0)-1|. \end{aligned}$$

It turns out that (by (12.51) and (12.20))

$$\begin{aligned}
&\left|L_n(f)(x_0) - f(x_0)\right| \leq |f(x_0)|\,|L_n(1)(x_0) - 1| \\
&\quad + \left|f'(x_0)\right| |L_n(\sinh(\cdot - x_0))(x_0)| + 2\left|f''(x_0)\right| \left(L_n\left(\sinh^2\left(\frac{\cdot - x_0}{2}\right)\right)(x_0)\right) \\
&\quad + \frac{\cosh(b-a)\,\omega_1(f''-f,h)}{2}\left[L_n\left((\cdot - x_0)^2\right)(x_0) + \frac{L_n\left(|\cdot - x_0|^3\right)(x_0)}{3h}\right]
\end{aligned} \tag{12.52}$$

$$\begin{aligned}
&\leq |f(x_0)|\,|L_n(1)(x_0) - 1| + \cosh(b-a)\left|f'(x_0)\right| L_n(|\cdot - x_0|)(x_0) \\
&\quad + \frac{\cosh^2(b-a)\,|f''(x_0)|}{2} L_n\left((\cdot - x_0)^2\right)(x_0) \\
&\quad + \frac{\cosh(b-a)\,\omega_1(f''-f,h)}{2}\left[L_n\left((\cdot - x_0)^2\right)(x_0) + \frac{L_n\left(|\cdot - x_0|^3\right)(x_0)}{3h}\right] \\
&=: (\xi),
\end{aligned} \tag{12.53}$$

$\forall\, x_0 \in [a,b]$, $h > 0$.

As in (12.23) we have

$$L_n\left((\cdot - x_0)^2\right)(x_0) \leq \left(L_n(|\cdot - x_0|)^3(x_0)\right)^{\frac{2}{3}} \left(L_n(1)(x_0)\right)^{\frac{1}{3}}, \tag{12.54}$$

$\forall\, x_0 \in [a,b]$.

Let $r > 0$, and take

$$h := r\left(L_n\left(|\cdot - x_0|^3\right)(x_0)\right)^{\frac{1}{3}} = rD_3(x_0), \tag{12.55}$$

where

$$D_3(x_0) := \left(L_n\left(|\cdot - x_0|^3\right)(x_0)\right)^{\frac{1}{3}}. \tag{12.56}$$

As in (12.26) we have

$$\left(L_n(\cdot - x_0)^2\right)(x_0) \leq D_3^{\frac{2}{3}}(x_0)\left(L_n(1)(x_0)\right)^{\frac{1}{3}}. \tag{12.57}$$

At the moment, we assume that $D_3(x_0) > 0$. As in (12.27) we get

$$\begin{aligned}
&\left[L_n\left((\cdot - x_0)^2\right)(x_0) + \frac{L_n\left(|\cdot - x_0|^3\right)(x_0)}{3h}\right] \\
&\quad \leq D_3^{\frac{2}{3}}(x_0)\left[\left(L_n(1)(x_0)\right)^{\frac{1}{3}} + \frac{1}{3r}\right].
\end{aligned} \tag{12.58}$$

It follows that

$$|L_n(f)(x_0) - f(x_0)| \leq (\xi) \overset{(12.58)}{\leq} |f(x_0)| |L_n(1)(x_0) - 1|$$
$$+ \cosh(b-a) |f'(x_0)| L_n(|\cdot - x_0|)(x_0)$$
$$+ \frac{\cosh^2(b-a) |f''(x_0)|}{2} L_n\left((\cdot - x_0)^2\right)(x_0)$$
$$+ \frac{\cosh(b-a)\,\omega_1(f'' - f, rD_3(x_0))}{2} D_3^{\frac{2}{3}}(x_0) \left[(L_n(1)(x_0))^{\frac{1}{3}} + \frac{1}{3r}\right]. \quad (12.59)$$

As in (12.29) we get that

$$L_n(|\cdot - x_0|)(x_0) \leq D_3(x_0)(L_n(1)(x_0))^{\frac{2}{3}}. \quad (12.60)$$

In case of $D_3(x_0) = 0$, by (12.53), we have that

$$|L_n(f)(x_0) - f(x_0)| \leq |f(x_0)| |L_n(1)(x_0) - 1|. \quad (12.61)$$

The last is valid as equality as in the proof of Theorem 12.3.
The proof of this theorem now is completed. □

Next comes a Korovkin-type result.

Theorem 12.6. *All as in Theorem 12.5. Assume that $L_n(1) \to 1$ and $L_n\left(|\cdot - x_0|^3\right)(x_0) \to 0$, both uniformly convergent, as $n \to \infty$. Then $L_n(f) \to f$, uniformly convergent, as $n \to \infty$, $\forall\, f \in C^2([a,b])$.*

Proof. By (12.36), using (12.54) and (12.60), and that every uniformly convergent sequence of bounded functions is uniformly bounded. Also that $\omega_1(f'' - f, r\|D_3\|_\infty) \to 0$, as $n \to \infty$. □

12.3 Applications

Let $f \in C([0,1])$, $x \in [0,1]$, and the Bernstein polynomials

$$B_n(f)(x) = \sum_{k=0}^{n} f\left(\frac{k}{n}\right)\binom{n}{k} x^k (1-x)^{n-k}. \quad (12.62)$$

We have that $B_n(1) = 1$, and

$$B_n\left((\cdot - x)^2\right)(x) = \frac{x(1-x)}{n} \leq \frac{1}{4n}. \quad (12.63)$$

Clearly it holds

$$B_n\left(|\cdot - x|^3\right)(x) \leq B_n\left((\cdot - x)^2\right)(x) \leq \frac{1}{4n}, \tag{12.64}$$

$\forall\, n \in \mathbb{N}$.

We have that

$$D_3^*(x_0) := \left(B_n\left(|\cdot - x_0|^3\right)(x)\right)^{\frac{1}{3}} \leq \frac{1}{\sqrt[3]{4n}}, \tag{12.65}$$

$\forall\, x_0 \in [0,1]$, $n \in \mathbb{N}$.

By (12.29), we get that

$$B_n\left(|\cdot - x_0|\right)(x) \leq \frac{1}{\sqrt[3]{4n}}, \quad \forall\, x_0 \in [0,1],\ n \in \mathbb{N}. \tag{12.66}$$

Consequently, we obtain

$$\left\| B_n\left((\cdot - x_0)^2\right)(x_0)\right\|_\infty \leq \frac{1}{4n} \tag{12.67}$$

and

$$\|D_3^*\| \leq \frac{1}{\sqrt[3]{4n}}, \tag{12.68}$$

with

$$\|B_n\left(|\cdot - x_0|\right)(x_0)\|_\infty \leq \frac{1}{\sqrt[3]{4n}}, \quad \forall\, n \in \mathbb{N}. \tag{12.69}$$

By (12.28) and the above, we derive

Proposition 12.1. *Let $f \in C^2([0,1])$, $r > 0$. Then*

$$\|B_n(f) - f\|_\infty \leq \frac{\|f'\|_\infty}{\sqrt[3]{4n}} + \frac{\|f''\|_\infty}{8n} + \frac{\omega_1\left(f'' + f, \frac{r}{\sqrt[3]{4n}}\right)}{2(4n)^{\frac{2}{3}}}\left[1 + \frac{1}{3r}\right]. \tag{12.70}$$

By (12.59) and the above, we obtain

Proposition 12.2. *Let $f \in C^2([0,1])$, $r > 0$. Then*

$$\|B_n(f) - f\|_\infty$$
$$\leq \cosh 1\left[\frac{\|f'\|_\infty}{\sqrt[3]{4n}} + \frac{\cosh 1\,\|f''\|_\infty}{8n} + \frac{\omega_1\left(f'' - f, \frac{r}{\sqrt[3]{4n}}\right)}{2(4n)^{\frac{2}{3}}}\left[1 + \frac{1}{3r}\right]\right]. \tag{12.71}$$

From (12.70) and/or (12.71), we obtain that $B_n(f) \to f$, uniformly, as $n \to \infty$, $\forall\ f \in C^2([0,1])$.

We also get (use of (12.28)).

Corollary 12.1. *Let $f \in C^2([0,1])$, $r > 0$, with $f'(x_0) = f''(x_0) = 0$, for some $x_0 \in [0,1]$. Then*

$$|B_n(f)(x_0) - f(x_0)| \leq \frac{\omega_1\left(f'' + f, \frac{r}{\sqrt[3]{4n}}\right)}{2(4n)^{\frac{2}{3}}}\left(1 + \frac{1}{3r}\right). \tag{12.72}$$

We finally give (use of (12.59))

Corollary 12.2. *All as in Corollary 12.1. Then*

$$|B_n(f)(x_0) - f(x_0)| \leq \frac{\cosh 1\omega_1\left(f'' - f, \frac{r}{\sqrt[3]{4n}}\right)}{2(4n)^{\frac{2}{3}}}\left(1 + \frac{1}{3r}\right). \tag{12.73}$$

By (12.72), (12.73), we have $\lim_{n\to\infty} B_n(f)(x_0) = f(x_0)$, pointwise, at the high speed $\frac{1}{n}$.

Bibliography

G.A. Anastassiou, *Quantitative Approximations*, Chapman & Hall/CRC, Boca Raton, New York, 2000.

G.A. Anastassiou, *Opial and Ostrowski type inequalities based on trigonometric and hyperbolic type Taylor formulae*, submitted, 2023.

G.A. Anastassiou, *Trigonometric and hyperbolic Korovkin theory*, submitted, 2023.

P.P. Korovkin, *Linear Operators and Approximation Theory*, Hindustan Publ. Corp. Delhi, India, 1960.

O. Shisha and B. Mond, *The degree of convergence of sequences of linear positive operators, Proceedings of the National Academy of Sciences of the United States of America*, 60, (1968), 1196–1200.

Chapter 13

Integral Inequalities Using the New Conformable Derivatives

Here, we first describe the new conformable derivatives and integrals (Anderson and Ulness 2015). Then, we establish Opial, Ostrowski, Poincaré, Sobolev, Polya and Hilbert–Pachpatte-type integral inequalities. It follows Anastassiou (2023).

13.1 Background

All in this section comes from Anderson and Ulness (2015).

The motivation though comes from control theory Li, Ang and Chog (2006). A proportional derivative (PD) controller for controller output u at time t with two tuning parameters is given by

$$u(t) = k_p E(t) + k_d \frac{d}{dt} E(t), \tag{13.1}$$

where k_p is the proportional gain, k_d is the derivative gain and E is the error between the state variable and the process variable.

The above motivate the following:

Definition 13.1 (A class of New Conformable Derivatives) (Anderson and Ulness 2015). Let $\alpha \in [0,1]$, and let the functions $k_0, k_1 : [0,1] \times \mathbb{R} \to [0,\infty)$ be continuous such that

$$\begin{gathered} \lim_{\alpha \to 0+} k_1(\alpha,t) = 1, \quad \lim_{\alpha \to 0+} k_0(\alpha,t) = 0, \quad \forall\, t \in \mathbb{R}, \\ \lim_{\alpha \to 1-} k_1(\alpha,t) = 0, \quad \lim_{\alpha \to 1-} k_0(\alpha,t) = 1, \quad \forall\, t \in \mathbb{R}, \\ k_1(\alpha,t) \neq 0, \quad \alpha \in [0,1), \quad k_0(\alpha,t) \neq 0, \quad \alpha \in (0,1], \quad \forall\, t \in \mathbb{R}. \end{gathered} \tag{13.2}$$

Then, the following differentiable operator D^α, defined via

$$D^\alpha f(t) = k_1(\alpha,t) f(t) + k_0(\alpha,t) f'(t), \tag{13.3}$$

is the new conformable derivative given that $f'(t)$ exists for $t \in \mathbb{R}$.

Here, k_1 is a type of proportional gain k_p, k_0 is a type of derivative gain k_d, f is the error and $u = D^\alpha f$ is the controller output. For example, one may choose $k_1 := (1-\alpha)\,\omega^\alpha$ and $k_0 := \alpha\omega^{1-\alpha}$ for any $\omega \in (0,\infty)$, or $k_1 := (1-\alpha)\,|t|^\alpha$ and $k_0 := \alpha\,|t|^{1-\alpha}$ on $\mathbb{R} - \{0\}$, so that

$$D^\alpha f(t) = (1-\alpha)\,|t|^\alpha f(t) + \alpha\,|t|^{1-\alpha} f'(t). \tag{13.4}$$

Another class of conformable derivatives is

$$D^\alpha f(t) = \cos\left(\alpha\frac{\pi}{2}\right)|t|^\alpha f(t) + \sin\left(\alpha\frac{\pi}{2}\right)|t|^{1-\alpha} f'(t). \tag{13.5}$$

In general, we have that $D^\beta D^\alpha \neq D^\alpha D^\beta$ for $\alpha, \beta \in [0,1]$.

Definition 13.2 (Anderson and Ulness 2015). (Conformable Exponential Function). Let $\alpha \in (0,1]$, the points $s,t \in \mathbb{R}$ with $s \leq t$, and let the function $p : [s,t] \to \mathbb{R}$ be continuous. Let $k_0, k_1 : [0,1] \times \mathbb{R} \to [0,\infty)$ be continuous and satisfy (13.2) with $\frac{p}{k_0}$ and $\frac{k_1}{k_0}$ Riemann integrable on $[s,t]$. Then the exponential function with respect to D^α in (13.3) is defined to be

$$e_p(t,s) := e^{\int_s^t \frac{p(\tau)-k_1(\alpha,\tau)}{k_0(\alpha,\tau)}d\tau}, \quad e_0(t,s) := e^{-\int_s^t \frac{k_1(\alpha,\tau)}{k_0(\alpha,\tau)}d\tau}. \tag{13.6}$$

Using (13.3) and (13.6), we have the following basic results.

Lemma 13.1 (Anderson and Ulness 2015). *(Basic Derivatives). Let the conformable differential operator D^α be given as in (13.3), where $\alpha \in [0,1]$. Let the function $p : [s,t] \to \mathbb{R}$ be continuous. Let $k_0, k_1 : [0,1] \times \mathbb{R} \to [0,\infty)$ be continuous and satisfy (13.2), with $\frac{p}{k_0}$ and $\frac{k_1}{k_0}$ Riemann integrable on $[s,t]$. Assume the functions f and g are differentiable as needed. Then*

(i) $D^\alpha[af+bg] = aD^\alpha[f] + bD^\alpha[g]$ *for all* $a,b \in \mathbb{R}$,
(ii) $D^\alpha c = ck_1(\alpha,\cdot)$ *for all constants* $c \in \mathbb{R}$,
(iii) $D^\alpha[fg] = fD^\alpha[g] + gD^\alpha[f] - fgk_1(\alpha,\cdot)$,
(iv) $D^\alpha\left[\frac{f}{g}\right] = \frac{gD^\alpha[f]-fD^\alpha[g]}{g^2} + \frac{f}{g}k_1(\alpha,\cdot)$,
(v) *for* $\alpha \in (0,1]$ *and fixed* $s \in \mathbb{R}$, *the exponential function satisfies*

$$D_t^\alpha[e_p(t,s)] = p(t)\,e_p(t,s) \tag{13.7}$$

for $e_p(t,s)$ *given in (13.6),*
(vi) *for* $\alpha \in (0,1]$ *and for the exponential function* e_0 *given in (13.6), we have*

$$D^\alpha\left[\int_a^t \frac{f(s)\,e_0(t,s)}{k_0(\alpha,s)}ds\right] = f(t). \tag{13.8}$$

Definition 13.3 (Anderson and Ulness 2015). (Integrals). Let $\alpha \in (0,1]$ and $t_0 \in \mathbb{R}$. In the light of (13.6) and Lemmas 13.1 (v) and (vi), define the antiderivative via

$$\int D^{\alpha} f(t)\, d_{\alpha} t = f(t) + c e_0(t, t_0), \quad c \in \mathbb{R}.$$

Similarly, define the integral of f over $[a, b]$ as

$$\int_a^t f(s)\, e_0(t,s)\, d_{\alpha} s := \int_a^t \frac{f(s)\, e_0(t,s)}{k_0(t,s)} ds, \quad d_{\alpha} s := \frac{1}{k_0(\alpha, s)} ds; \tag{13.9}$$

recall that

$$e_0(t,s) := e^{-\int_s^t \frac{k_1(\alpha,\tau)}{k_0(\alpha,\tau)} d\tau} = e^{-\int_s^t k_1(\alpha,\tau) d_{\alpha}\tau}$$

from (13.6).

Lemma 13.2 (Anderson and Ulness 2015). *(Basic Integrals). Let the conformable differential operator D^{α} be given as in (13.3), and the integral be given as in (13.9) with $\alpha \in (0,1]$. Let the functions k_0, k_1 be continuous and satisfy (13.2), and let f and g be differentiable as needed. Then*

(i) *the derivative of the definite integral of f is given by*

$$D^{\alpha}\left[\int_a^t f(s)\, e_0(t,s)\, d_{\alpha} s\right] = f(t), \tag{13.10}$$

(ii) *the definite integral of the derivative of f is given by*

$$\int_a^t D^{\alpha}[f(s)]\, e_0(t,s)\, d_{\alpha} s = f(s)\, e_0(t,s)\, |_{s=a}^{t} := f(t) - f(a)\, e_0(t,a), \tag{13.11}$$

(iii) *an integration by parts formula is given by*

$$\int_a^b f(t)\, D^{\alpha}[g(t)]\, e_0(b,t)\, d_{\alpha} t = f(t)\, g(t)\, e_0(b,t)\, |_{t=a}^{b} - \int_a^b g(t)\left(D^{\alpha}[f(t)] - k_1(\alpha,t) f(t)\right) e_0(b,t)\, d_{\alpha} t, \tag{13.12}$$

(iv) *a version of the Leibniz rule for differentiation of an integral is given by*

$$D^{\alpha}\left[\int_a^t f(t,s)\, e_0(t,s)\, d_{\alpha} s\right] = \int_a^t \left(D_t^{\alpha}[f(t,s)] - k_1(\alpha,t) f(t,s)\right) e_0(t,s)\, d_{\alpha} s + f(t,t), \tag{13.13}$$

using (13.15), or, if e_0 *is absent,*

$$D^{\alpha}\left[\int_a^t f(t,s)\,d_{\alpha}s\right]=f(t,t)+\int_a^t D_t^{\alpha}\left[f(t,s)\right]d_{\alpha}s. \tag{13.14}$$

Definition 13.4 (Anderson and Ulness 2015). (Partial Conformable Derivatives). Let $\alpha\in[0,1]$, and let the functions $k_0,k_1:[0,1]\times\mathbb{R}\to[0,\infty)$ be continuous and satisfy (13.2). Given a function $f:\mathbb{R}^2\to\mathbb{R}$ such that $\frac{\partial}{\partial t}f(t,s)$ exists for each fixed $s\in\mathbb{R}$, define the partial differential operator D_t^{α} via

$$D_t^{\alpha}f(t,s)=k_1(\alpha,t)\,f(t,s)+k_0(\alpha,t)\,\frac{\partial}{\partial t}f(t,s). \tag{13.15}$$

Next comes the important Taylor's formula.

Theorem 13.1 (Anderson and Ulness 2015). *Let* $n\in\mathbb{N}$, $\alpha\in(0,1]$, *and suppose* f *is* n *times continuously differentiable on* $[t_0,\infty)$. *Let* $t,s\in[t_0,\infty)$, *and define the functions* h_k *by*

$$h_0(t,s)\equiv 1\quad\text{and}\quad h_{k+1}(t,s)=\int_s^t h_k(\tau,s)\,d_{\alpha}\tau\quad\text{for } k\in\mathbb{N}_0. \tag{13.16}$$

Then

$$f(t)=e_0(t,s)\sum_{k=0}^{n-1}(-1)^k h_k(s,t)\,(D^{\alpha})^k f(s)$$

$$+(-1)^{n-1}\int_s^t h_{n-1}(\tau,t)\,(D^{\alpha})^n f(\tau)\,e_0(t,\tau)\,d_{\alpha}\tau \tag{13.17}$$

for $t\in[t_0,\infty)$.

Example 13.1 (Anderson and Ulness 2015). For $\alpha\in(0,1]$, let $\omega_0,\omega_1\in(0,\infty)$, let k_1 satisfy (13.2), and take

$$k_0(\alpha,t)\equiv\alpha\omega_0^{1-\alpha}. \tag{13.18}$$

By (13.9),

$$d_{\alpha}\tau=\frac{1}{k_0(\alpha,\tau)}d\tau=\frac{1}{\alpha\omega_0^{1-\alpha}}d\tau.$$

Letting $h_0(t,s)\equiv 1$, we calculate h_1 via (13.16) to get

$$h_1(t,s)=\int_s^t h_0(\tau,s)\,d_{\alpha}\tau=\frac{1}{\alpha\omega_0^{1-\alpha}}\int_s^t 1d\tau=\frac{t-s}{\alpha\omega_0^{1-\alpha}};$$

additionally,

$$h_2(t,s) = \int_s^t h_1(\tau,s)\, d_\alpha \tau = \frac{1}{2!}\left(\frac{t-s}{\alpha\omega_0^{1-\alpha}}\right)^2.$$

In general, we have that

$$h_n(t,s) = \frac{1}{n!}\left(\frac{t-s}{\alpha\omega_0^{1-\alpha}}\right)^n. \tag{13.19}$$

Note that at $\alpha = 1$ we have

$$h_n(t,s) = \frac{1}{n!}(t-s)^n$$

as expected.

Example 13.2 (Anderson and Ulness 2015). For $\alpha \in (0,1]$, let $\omega_0, \omega_1 \in (0,\infty)$, let k_1 satisfy (13.2), and this time take

$$k_0(\alpha,t) = \alpha(\omega_0 t)^{1-\alpha}, \quad t \in [0,\infty). \tag{13.20}$$

By (13.9),

$$d_\alpha \tau = \frac{\tau^{\alpha-1}}{\alpha\omega_0^{1-\alpha}} d\tau.$$

Again starting with $h_0(t,s) \equiv 1$, we see that

$$h_1(t,s) = \int_s^t h_0(\tau,s)\, d_\alpha \tau = \frac{1}{\alpha\omega_0^{1-\alpha}} \int_s^t \tau^{\alpha-1} d\tau = \frac{t^\alpha - s^\alpha}{\alpha^2\omega_0^{1-\alpha}}$$

and

$$h_2(t,s) = \int_s^t h_1(\tau,s)\, d_\alpha \tau = \frac{1}{2!}\left(\frac{t^\alpha - s^\alpha}{\alpha^2\omega_0^{1-\alpha}}\right)^2.$$

Continuing, we find that

$$h_n(t,s) = \frac{1}{n!}\left(\frac{t^\alpha - s^\alpha}{\alpha^2\omega_0^{1-\alpha}}\right)^n, \tag{13.21}$$

which is just $\frac{1}{n!}(t-s)^n$ at $\alpha = 1$.

13.2 Main Results

Motivation comes from Anastassiou (2010)–Anastassiou (2016). We need

Remark 13.1. (to Lemma 13.2 (iii))

When $f = g$, we get

$$\int_a^b f(t) D^\alpha [f(t)] e_0(b,t) d_\alpha t = f^2(b) - f^2(a) e_0(b,a)$$
$$- \int_a^b f(t) D^\alpha [f(t)] e_0(b,t) d_\alpha t + \int_a^b k_1(\alpha,t) f^2(t) e_0(b,t) d_\alpha t. \tag{13.22}$$

Therefore, it holds

$$\int_a^b f(t) D^\alpha [f(t)] e_0(b,t) d_\alpha t = \left(\frac{f^2(b) - f^2(a) e_0(b,a)}{2}\right)$$
$$+ \frac{1}{2} \int_a^b f^2(t) k_1(\alpha,t) e_0(b,t) d_\alpha t. \tag{13.23}$$

We present the following Opial-type inequality:

Theorem 13.2. *All as in Theorem 13.1. Further assume that* $t \geq s$, $(D^\alpha)^k f(s) = 0$, *for* $k = 0, 1, \ldots, n-1$; $\alpha \in (0,1]$, $n \in \mathbb{N}$; *and* $p, q > 1 : \frac{1}{p} + \frac{1}{q} = 1$. *Then*

$$\int_s^t |f(w)| |(D^\alpha)^n f(w)| e_0(t,w) d_\alpha w$$
$$\leq \left[\int_s^t \left(\int_s^w |h_{n-1}(\tau,w)|^p e_0(w,\tau) d_\alpha \tau\right) e_0(t,w) d_\alpha w\right]^{\frac{1}{p}}$$
$$\left\{\frac{\left(\int_s^t |(D^\alpha)^n f(w)|^q e_0(t,w) d_\alpha w\right)^2}{2}\right.$$
$$\left. + \frac{1}{2}\left[\int_s^t \left(\int_s^w |(D^\alpha)^n f(\tau)|^q e_0(w,\tau) d_\alpha \tau\right)^2 k_1(\alpha,w) e_0(t,w) d_\alpha w\right]\right\}^{\frac{1}{q}}. \tag{13.24}$$

Proof. Since $(D^{\alpha})^k f(s) = 0$, for $k = 0, 1, \ldots, n-1$, $t \geq s$, we have

$$f(t) = (-1)^{n-1} \int_s^t h_{n-1}(\tau, t)(D^{\alpha})^n f(\tau) e_0(t, \tau) d_{\alpha}\tau \tag{13.25}$$

and

$$f(w) = (-1)^{n-1} \int_s^w h_{n-1}(\tau, w)(D^{\alpha})^n f(\tau) e_0(w, \tau) d_{\alpha}\tau, \tag{13.26}$$

for all $s \leq w \leq t$.

By Hölder's inequality, we obtain

$$\begin{aligned}
|f(w)| &\leq \int_s^w |h_{n-1}(\tau, w)| |(D^{\alpha})^n f(\tau)| e_0(w, \tau) d_{\alpha}\tau \\
&= \int_s^w |h_{n-1}(\tau, w)| (e_0(w, \tau))^{\frac{1}{p}} |(D^{\alpha})^n f(\tau)| (e_0(w, \tau))^{\frac{1}{q}} d_{\alpha}\tau \\
&\leq \left(\int_s^w |h_{n-1}(\tau, w)|^p (e_0(w, \tau)) d_{\alpha}\tau \right)^{\frac{1}{p}} \\
&\quad \times \left(\int_s^w |(D^{\alpha})^n f(\tau)|^q (e_0(w, \tau)) d_{\alpha}\tau \right)^{\frac{1}{q}}.
\end{aligned} \tag{13.27}$$

Call

$$z(w) := \int_s^w |(D^{\alpha})^n f(\tau)|^q (e_0(w, \tau)) d_{\alpha}\tau, \quad z(s) = 0, \tag{13.28}$$

$s \leq w \leq t$.

Then (by (13.10))

$$D^{\alpha} z(w) = |(D^{\alpha})^n f(w)|^q, \tag{13.29}$$

and

$$|(D^{\alpha})^n f(w)| = (D^{\alpha} z(w))^{\frac{1}{q}}, \quad \text{all } s \leq w \leq t. \tag{13.30}$$

Therefore, we have (all $s \leq w \leq t$)

$$\begin{aligned}
&|f(w)| |(D^{\alpha})^n f(w)| e_0(t, w) \\
&\leq \left[\left(\int_s^w |h_{n-1}(\tau, w)|^p e_0(w, \tau) d_{\alpha}\tau \right) e_0(t, w) \right]^{\frac{1}{p}} [z(w) D^{\alpha} z(w) e_0(t, w)]^{\frac{1}{q}}.
\end{aligned} \tag{13.31}$$

Next, we apply again Hölder's inequality and finally we use (13.23) to get that

$$\int_s^t |f(w)| |(D^\alpha)^n f(w)| e_0(t,w) \, d_\alpha w$$

$$\leq \int_s^t \left\{ \left[\left(\int_s^w |h_{n-1}(\tau,w)|^p e_0(w,\tau) \, d_\alpha \tau \right) e_0(t,w) \right]^{\frac{1}{p}} \left[z(w) D^\alpha z(w) e_0(t,w) \right]^{\frac{1}{q}} \right\} d_\alpha w$$

$$\leq \left(\int_s^t \left(\int_s^w |h_{n-1}(\tau,w)|^p e_0(w,\tau) \, d_\alpha \tau \right) e_0(t,w) \, d_\alpha w \right)^{\frac{1}{p}}$$

$$\left(\int_s^t z(w) D^\alpha z(w) e_0(t,w) \, d_\alpha w \right)^{\frac{1}{q}} =: (\xi). \tag{13.32}$$

By (13.23), we derive that

$$\int_s^t z(w) D^\alpha z(w) e_0(t,w) \, d_\alpha w$$

$$= \frac{z^2(t)}{2} + \frac{1}{2} \int_s^t z^2(w) k_1(\alpha,w) e_0(t,w) \, d_\alpha w$$

$$= \frac{\left(\int_s^t |(D^\alpha)^n f(w)|^q e_0(t,w) \, d_\alpha w \right)^2}{2}$$

$$+ \frac{1}{2} \left[\int_s^t \left(\int_s^w |(D^\alpha)^n f(\tau)|^q e_0(w,\tau) \, d_\alpha \tau \right)^2 k_1(\alpha,w) e_0(t,w) \, d_\alpha w \right]. \tag{13.33}$$

Consequently, we get that

$$(\xi) = \left[\int_s^t \left(\int_s^w |h_{n-1}(\tau,w)|^p e_0(w,\tau) \, d_\alpha \tau \right) e_0(t,w) \, d_\alpha w \right]^{\frac{1}{p}}$$

$$\left\{ \frac{\left(\int_s^t |(D^\alpha)^n f(w)|^q e_0(t,w) \, d_\alpha w \right)^2}{2} \right.$$

$$\left. + \frac{1}{2} \left[\int_s^t \left(\int_s^w |(D^\alpha)^n f(\tau)|^q e_0(w,\tau) \, d_\alpha \tau \right)^2 k_1(\alpha,w) e_0(t,w) \, d_\alpha w \right] \right\}^{\frac{1}{q}}. \tag{13.34}$$

The claim is proved. □

Corollary 13.1 (to Theorem 13.2, for $n = 1$). *Here f is continuously differentiable on $[t_0, \infty)$, $t, s \in [t_0, \infty)$, $t \geq s$, $f(s) = 0$, $\alpha \in (0, 1]$, and $p, q > 1 : \frac{1}{p} + \frac{1}{q} = 1$. Then*

$$\int_s^t |f(w)| |D^\alpha f(w)| e_0(t, w) d_\alpha w$$
$$\leq \left[\int_s^t \left(\int_s^w e_0(w, \tau) d_\alpha \tau\right) e_0(t, w) d_\alpha w\right]^{\frac{1}{p}}$$
$$\left\{\frac{\left(\int_s^t |D^\alpha f(w)|^q e_0(t, w) d_\alpha w\right)^2}{2}\right.$$
$$\left.+\frac{1}{2}\left[\int_s^t \left(\int_s^w |D^\alpha f(\tau)|^q e_0(w, \tau) d_\alpha \tau\right)^2 k_1(\alpha, w) e_0(t, w) d_\alpha w\right]\right\}^{\frac{1}{q}}. \tag{13.35}$$

Corollary 13.2 (to Theorem 13.2, for $n = 2$). *Let f be twice continuously differentiable on $[t_0, \infty)$, $t, s \in [t_0, \infty)$, $t \geq s$, $f(s) = D^\alpha f(s) = 0$, $\alpha \in (0, 1]$, and $p, q > 1 : \frac{1}{p} + \frac{1}{q} = 1$. Here it is $h_1(t, s) = \int_s^t \frac{d\tau}{k_0(\alpha, \tau)}$. Then*

$$\int_s^t |f(w)| \left|(D^\alpha)^2 f(w)\right| e_0(t, w) d_\alpha w$$
$$\leq \left[\int_s^t \left(\int_s^w |h_1(\tau, w)|^p e_0(w, \tau) d_\alpha \tau\right) e_0(t, w) d_\alpha w\right]^{\frac{1}{p}}$$
$$\left\{\frac{\left(\int_s^t \left|(D^\alpha)^2 f(w)\right|^q e_0(t, w) d_\alpha w\right)^2}{2}\right.$$
$$\left.+\frac{1}{2}\left[\int_s^t \left(\int_s^w \left|(D^\alpha)^2 f(\tau)\right|^q e_0(w, \tau) d_\alpha \tau\right)^2 k_1(\alpha, w) e_0(t, w) d_\alpha w\right]\right\}^{\frac{1}{q}}. \tag{13.36}$$

It follows an Ostrowski-like inequality.

Theorem 13.3. *Let $n \in \mathbb{N}$, and f is n times continuously differentiable over $[a.b] \subset \mathbb{R}$; h_k as in (13.16). For fixed $s \in [a, b]$, assume that*

$(D^{\alpha})^{k} f(s)=0,\ k=1,\ldots,n-1,\ \alpha\in(0,1]$, *and let* $p,q>1:\frac{1}{p}+\frac{1}{q}=1$. *Then*

$$\left|\int_a^b (f(t)-e_0(t,s)f(s))\,dt\right| \leq \int_a^b |f(t)-e_0(t,s)f(s)|\,dt$$

$$\leq \left[\|(D^{\alpha})^n f\|_{L_q([a,s],e_0(\alpha,\cdot)d_\alpha\tau)}\left(\int_a^s\left(\int_t^s (h_{n-1}(\tau,t))^p e_0(t,\tau)\,d_\alpha\tau\right)^{\frac{1}{p}} dt\right)\right.$$

$$\left.+\|(D^{\alpha})^n f\|_{L_q([s,b],d_\alpha\tau)}\left(\int_s^b\left(\int_s^t (|h_{n-1}(\tau,t)|)^p e_0(t,\tau)\,d_\alpha\tau\right)^{\frac{1}{p}} dt\right)\right]. \tag{13.37}$$

Proof. Here $(D^{\alpha})^{k} f(s)=0,\ k=1,\ldots,n-1;\ t,s\in[a,b]$, s is fixed.

We have by (13.17) that

$$f(t)-e_0(t,s)f(s)=(-1)^{n-1}\int_s^t h_{n-1}(\tau,t)(D^{\alpha})^n f(\tau)e_0(t,\tau)\,d_\alpha\tau.$$

Let $t\geq s$, then by Hölder's inequality, we get

$$|f(t)-e_0(t,s)f(s)|\leq\int_s^t |h_{n-1}(\tau,t)|\,|(D^{\alpha})^n f(\tau)|\,e_0(t,\tau)\,d_\alpha\tau$$

$$=\int_s^t\left(|h_{n-1}(\tau,t)|(e_0(t,\tau))^{\frac{1}{p}}\right)\left(|(D^{\alpha})^n f(\tau)|\,e_0(t,\tau)^{\frac{1}{q}}\right)d_\alpha\tau$$

$$\leq\left(\int_s^t |h_{n-1}(\tau,t)|^p e_0(t,\tau)\,d_\alpha\tau\right)^{\frac{1}{p}}\left(\int_s^t |(D^{\alpha})^n f(\tau)|^q e_0(t,\tau)\,d_\alpha\tau\right)^{\frac{1}{q}}$$

$$\leq\left(\int_s^t |h_{n-1}(\tau,t)|^p e_0(t,\tau)\,d_\alpha\tau\right)^{\frac{1}{p}}\left(\int_s^t |(D^{\alpha})^n f(\tau)|^q\,d_\alpha\tau\right)^{\frac{1}{q}}$$

$$=\|(D^{\alpha})^n f\|_{L_q([s,b],d_\alpha\tau)}\left(\int_s^t |h_{n-1}(\tau,t)|^p e_0(t,\tau)\,d_\alpha\tau\right)^{\frac{1}{p}}. \tag{13.38}$$

So when, $s\leq t\leq b$, we got that

$$|f(t)-e_0(t,s)f(s)|$$

$$\leq\|(D^{\alpha})^n f\|_{L_q([s,b],d_\alpha\tau)}\left(\int_s^t |h_{n-1}(\tau,t)|^p e_0(t,\tau)\,d_\alpha\tau\right)^{\frac{1}{p}}. \tag{13.39}$$

Let now $t \leq s$, then

$$- (f(t) - e_0(t,s) f(s)) = (-1)^{n-1} \int_t^s h_{n-1}(\tau,t) (D^{\alpha})^n f(\tau) e_0(t,\tau) d_{\alpha}\tau. \tag{13.40}$$

Hence, it holds (again by Hölder's inequality)

$$|f(t) - e_0(t,s) f(s)| \leq \int_t^s h_{n-1}(\tau,t) |(D^{\alpha})^n f(\tau)| e_0(t,\tau) d_{\alpha}\tau \tag{13.41}$$

$$= \int_t^s h_{n-1}(\tau,t) (e_0(t,\tau))^{\frac{1}{p}} |(D^{\alpha})^n f(\tau)| e_0(t,\tau)^{\frac{1}{q}} d_{\alpha}\tau$$

$$\leq \left(\int_t^s (h_{n-1}(\tau,t))^p e_0(t,\tau) d_{\alpha}\tau\right)^{\frac{1}{p}} \left(\int_t^s |(D^{\alpha})^n f(\tau)|^q e_0(t,\tau) d_{\alpha}\tau\right)^{\frac{1}{q}}$$

$$\leq \left(\int_t^s (h_{n-1}(\tau,t))^p e_0(t,\tau) d_{\alpha}\tau\right)^{\frac{1}{p}} \left(\int_a^s |(D^{\alpha})^n f(\tau)|^q e_0(\alpha,\tau) d_{\alpha}\tau\right)^{\frac{1}{q}}$$

$$= \left(\int_t^s (h_{n-1}(\tau,t))^p e_0(t,\tau) d_{\alpha}\tau\right)^{\frac{1}{p}} \|(D^{\alpha})^n f\|_{L_q([a,s],e_0(\alpha,\cdot)d_{\alpha}\tau)}. \tag{13.42}$$

Thus, when, $a \leq t \leq s$, we found that

$$|f(t) - e_0(t,s) f(s)| \leq \|(D^{\alpha})^n f\|_{L_q([a,s],e_0(\alpha,\cdot)d_{\alpha}\tau)} \left(\int_t^s (h_{n-1}(\tau,t))^p e_0(t,\tau) d_{\alpha}\tau\right)^{\frac{1}{p}}. \tag{13.43}$$

Consequently, it holds

$$\left|\int_a^b (f(t) - e_0(t,s) f(s)) dt\right| \leq \int_a^b |f(t) - e_0(t,s) f(s)| dt$$

$$= \int_a^s |f(t) - e_0(t,s) f(s)| dt + \int_s^b |f(t) - e_0(t,s) f(s)| dt$$

$$\leq \left[\left(\int_a^s \left(\int_t^s (h_{n-1}(\tau,t))^p e_0(t,\tau) d_{\alpha}\tau\right)^{\frac{1}{p}} dt\right) \|(D^{\alpha})^n f\|_{L_q([a,s],e_0(\alpha,\cdot)d_{\alpha}\tau)} + \left(\int_s^b \left(\int_s^t (|h_{n-1}(\tau,t)|)^p e_0(t,\tau) d_{\alpha}\tau\right)^{\frac{1}{p}} dt\right) \|(D^{\alpha})^n f\|_{L_q([s,b],d_{\alpha}\tau)}\right]. \tag{13.44}$$

The claim is proved. □

Next comes a Poincaré-type inequality.

Theorem 13.4. *Let all as in Theorem 13.3, including $f(s) = 0$, and $a \leq s \leq t \leq b$. Then*

$$\|f\|_{L_q([s,b])} \leq \left(\int_s^b \left(\int_s^t |h_{n-1}(\tau,t)|^p e_0(t,\tau) d_\alpha \tau \right)^{\frac{q}{p}} dt \right)^{\frac{1}{q}} \|(D^\alpha)^n f\|_{L_q([s,b],d_\alpha \tau)}. \tag{13.45}$$

Proof. Since $(D^\alpha)^k f(s) = 0$, for $k = 0, 1, \ldots, n-1$; $t \geq s$, we have

$$f(t) = (-1)^{n-1} \int_s^t h_{n-1}(\tau,t) (D^\alpha)^n f(\tau) e_0(t,\tau) d_\alpha \tau. \tag{13.46}$$

By Hölder's inequality, we obtain

$$\begin{aligned}
|f(t)| &\leq \int_s^t |h_{n-1}(\tau,t)| |(D^\alpha)^n f(\tau)| e_0(t,\tau) d_\alpha \tau \\
&= \int_s^t |h_{n-1}(\tau,t)| (e_0(t,\tau))^{\frac{1}{p}} |(D^\alpha)^n f(\tau)| e_0(t,\tau)^{\frac{1}{q}} d_\alpha \tau \\
&\leq \left(\int_s^t |h_{n-1}(\tau,t)|^p e_0(t,\tau) d_\alpha \tau \right)^{\frac{1}{p}} \left(\int_s^t |(D^\alpha)^n f(\tau)|^q e_0(t,\tau) d_\alpha \tau \right)^{\frac{1}{q}} \\
&\leq \left(\int_s^t |h_{n-1}(\tau,t)|^p e_0(t,\tau) d_\alpha \tau \right)^{\frac{1}{p}} \left(\int_s^b |(D^\alpha)^n f(\tau)|^q e_0(t,\tau) d_\alpha \tau \right)^{\frac{1}{q}},
\end{aligned} \tag{13.47}$$

$\forall\, t \in [s,b]$.

Hence it holds

$$\begin{aligned}
|f(t)|^q &\leq \left(\int_s^t |h_{n-1}(\tau,t)|^p e_0(t,\tau) d_\alpha \tau \right)^{\frac{q}{p}} \left(\int_s^b |(D^\alpha)^n f(\tau)|^q e_0(t,\tau) d_\alpha \tau \right) \\
&\leq \left(\int_s^t |h_{n-1}(\tau,t)|^p e_0(t,\tau) d_\alpha \tau \right)^{\frac{q}{p}} \left(\int_s^b |(D^\alpha)^n f(\tau)|^q d_\alpha \tau \right) \\
&= \left(\int_s^t |h_{n-1}(\tau,t)|^p e_0(t,\tau) d_\alpha \tau \right)^{\frac{q}{p}} \|(D^\alpha)^n f\|^q_{L_q([s,b],d_\alpha \tau)}.
\end{aligned} \tag{13.48}$$

Consequently, we get that

$$\int_s^b |f(t)|^q \, dt$$

$$\leq \left(\int_s^b \left(\int_s^t |h_{n-1}(\tau, t)|^p \, e_0(t, \tau) \, d_\alpha \tau \right)^{\frac{q}{p}} dt \right) \|(D^\alpha)^n f\|^q_{L_q([s,b], d_\alpha \tau)}. \tag{13.49}$$

Thus, we derive

$$\left(\int_s^b |f(t)|^q \, dt \right)^{\frac{1}{q}}$$

$$\leq \left(\int_s^b \left(\int_s^t |h_{n-1}(\tau, t)|^p \, e_0(t, \tau) \, d_\alpha \tau \right)^{\frac{q}{p}} dt \right)^{\frac{1}{q}} \|(D^\alpha)^n f\|_{L_q([s,b], d_\alpha \tau)}. \tag{13.50}$$

The claim is proved. □

It follows a Sobolev-type inequality.

Theorem 13.5. *All as in Theorem 13.4, $r \geq 1$. Then*

$$\|f\|_{L_r([s,b])}$$

$$\leq \left(\int_s^b \left(\int_s^t |h_{n-1}(\tau, t)|^p \, e_0(t, \tau) \, d_\alpha \tau \right)^{\frac{r}{p}} dt \right)^{\frac{1}{r}} \|(D^\alpha)^n f\|_{L_q([s,b], d_\alpha \tau)}. \tag{13.51}$$

Proof. As in (13.47) we obtain

$$|f(t)| \leq \left(\int_s^t |h_{n-1}(\tau, t)|^p \, e_0(t, \tau) \, d_\alpha \tau \right)^{\frac{1}{p}} \left(\int_s^b |(D^\alpha)^n f(\tau)|^q \, e_0(t, \tau) \, d_\alpha \tau \right)^{\frac{1}{q}} \tag{13.52}$$

$$\leq \left(\int_s^t |h_{n-1}(\tau, t)|^p \, e_0(t, \tau) \, d_\alpha \tau \right)^{\frac{1}{p}} \|(D^\alpha)^n f\|_{L_q([s,b], d_\alpha \tau)}, \tag{13.53}$$

$\forall \, t \in [s, b]$.

Hence, by $r \geq 1$, we obtain

$$|f(t)|^r \leq \left(\int_s^t |h_{n-1}(\tau, t)|^p \, e_0(t, \tau) \, d_\alpha \tau \right)^{\frac{r}{p}} \|(D^\alpha)^n f\|^r_{L_q([s,b], d_\alpha \tau)}, \tag{13.54}$$

$\forall \, t \in [s, b]$.

Consequently, it holds

$$\int_s^b |f(t)|^r dt$$
$$\leq \left(\int_s^b \left(\int_s^t |h_{n-1}(\tau,t)|^p e_0(t,\tau) d_\alpha \tau \right)^{\frac{r}{p}} dt \right) \|(D^\alpha)^n f\|^r_{L_q([s,b],d_\alpha \tau)}. \tag{13.55}$$

Finally, we get

$$\left(\int_s^b |f(t)|^r dt \right)^{\frac{1}{r}}$$
$$\leq \left(\int_s^b \left(\int_s^t |h_{n-1}(\tau,t)|^p e_0(t,\tau) d_\alpha \tau \right)^{\frac{r}{q}} dt \right)^{\frac{1}{r}} \|(D^\alpha)^n f\|_{L_q([s,b],d_\alpha \tau)}. \tag{13.56}$$

The claim is proved. □

We continue with Polya-type inequalities. From now on, $f \in C^n([a,b])$ *means in the sense of* D^α.

Corollary 13.3. *(to Theorem 13.3) Let* $n \in \mathbb{N}$, *and* $f \in C^n([a,b])$; h_k *as in (13.16). For fixed* $s \in [a,b]$, *assume that* $(D^\alpha)^k f(s) = 0$, $k = 0,1,\ldots,n-1$, $\alpha \in (0,1]$, *and let* $p,q > 1 : \frac{1}{p} + \frac{1}{q} = 1$. *Then*

$$\left| \int_a^b f(t) dt \right| \leq \int_a^b |f(t)| dt$$
$$\leq \left[\|(D^\alpha)^n f\|_{L_q([a,s],e_0(\alpha,\cdot)d_\alpha \tau)} \left(\int_a^s \left(\int_t^s (h_{n-1}(\tau,t))^p e_0(t,\tau) d_\alpha \tau \right)^{\frac{1}{p}} dt \right) \right.$$
$$\left. + \|(D^\alpha)^n f\|_{L_q([s,b],d_\alpha \tau)} \left(\int_s^b \left(\int_s^t (|h_{n-1}(\tau,t)|)^p e_0(t,\tau) d_\alpha \tau \right)^{\frac{1}{p}} dt \right) \right]. \tag{13.57}$$

Proof. By (13.37), just set $f(s) = 0$. □

We give the following result.

Theorem 13.6. *Let* $n \in \mathbb{N}$, $f \in C^n([a,b])$; h_k *as in (13.16). Assume that* $(D^\alpha)^k f(a) = (D^\alpha)^k f(b) = 0$, $k = 0,1,\ldots,n-1$, $\alpha \in (0,1]$, *and let*

$p, q > 1 : \frac{1}{p} + \frac{1}{q} = 1$. *Then*

$$\left| \int_a^b f(t)\, dt \right| \le \int_a^b |f(t)|\, dt$$

$$\le \|(D^\alpha)^n f\|_{L_q\left(\left[a, \frac{a+b}{2}\right], d_\alpha \tau\right)} \left(\int_a^{\frac{a+b}{2}} \left(\int_a^t (|h_{n-1}(\tau, t)|)^p e_0(t, \tau)\, d_\alpha \tau \right)^{\frac{1}{p}} dt \right)$$

$$+ \|(D^\alpha)^n f\|_{L_q\left(\left[\frac{a+b}{2}, b\right], e_0(\alpha, \cdot) d_\alpha \tau\right)} \left(\int_{\frac{a+b}{2}}^b \left(\int_t^b (h_{n-1}(\tau, t))^p e_0(t, \tau)\, d_\alpha \tau \right)^{\frac{1}{p}} dt \right).$$

Proof. Let $s = a$ and instead of b we choose $\frac{a+b}{2}$. Then, by (13.57), we derive

$$\left| \int_a^{\frac{a+b}{2}} f(t)\, dt \right| \le \int_a^{\frac{a+b}{2}} |f(t)|\, dt$$

$$\le \|(D^\alpha)^n f\|_{L_q\left(\left[a, \frac{a+b}{2}\right], d_\alpha \tau\right)} \left(\int_a^{\frac{a+b}{2}} \left(\int_a^t (|h_{n-1}(\tau, t)|)^p e_0(t, \tau)\, d_\alpha \tau \right)^{\frac{1}{p}} dt \right). \tag{13.59}$$

Next, assume that $s = b$ and instead of a we choose $\frac{a+b}{2}$.

Then, again by (13.57), we obtain

$$\left| \int_{\frac{a+b}{2}}^b f(t)\, dt \right| \le \int_{\frac{a+b}{2}}^b |f(t)|\, dt$$

$$\le \|(D^\alpha)^n f\|_{L_q\left(\left[\frac{a+b}{2}, b\right], e_0(\alpha, \cdot) d_\alpha \tau\right)} \left(\int_{\frac{a+b}{2}}^b \left(\int_t^b (h_{n-1}(\tau, t))^p e_0(t, \tau)\, d_\alpha \tau \right)^{\frac{1}{p}} dt \right). \tag{13.60}$$

Finally, we have that (by (13.59) and (13.60))

$$\left| \int_a^b f(t)\, dt \right| \le \int_a^b |f(t)|\, dt = \int_a^{\frac{a+b}{2}} |f(t)|\, dt + \int_{\frac{a+b}{2}}^b |f(t)|\, dt$$

$$\le \|(D^\alpha)^n f\|_{L_q\left(\left[a, \frac{a+b}{2}\right], d_\alpha \tau\right)} \left(\int_a^{\frac{a+b}{2}} \left(\int_a^t (|h_{n-1}(\tau, t)|)^p e_0(t, \tau)\, d_\alpha \tau \right)^{\frac{1}{p}} dt \right)$$

$$+ \|(D^\alpha)^n f\|_{L_q\left(\left[\frac{a+b}{2}, b\right], e_0(\alpha, \cdot) d_\alpha \tau\right)} \left(\int_{\frac{a+b}{2}}^b \left(\int_t^b (h_{n-1}(\tau, t))^p e_0(t, \tau) d_\alpha \tau \right)^{\frac{1}{p}} dt \right). \tag{13.61}$$

The claim is proved. ∎

We finish with a Hilbert–Pachpatte-type inequality.

Theorem 13.7. *Here* $j=1,2$. *Let* $n_j \in \mathbb{N}$, *and* $f_j \in C^{n_j}([a_j,b_j])$; h_{kj} *as in (13.16). For fixed* $s_j \in [a_j,b_j]$, *assume that* $(D^\alpha)^{k_j} f_j(s_j)=0$, $k_j = 0,1,\dots,n_j-1$, $\alpha \in (0,1]$, *and let* $p,q>1 : \frac{1}{p}+\frac{1}{q}=1$. *Then*

$$\int_{s_1}^{b_1}\int_{s_2}^{b_2} \frac{|f_1(t_1)|\,|f_2(t_2)|\,dt_1dt_2}{\left[\frac{\left(\int_{s_1}^{t_1}(h_{n_1-1}(\tau_1,t_1))^p e_0(t_1,\tau_1)d_\alpha\tau_1\right)}{p}+\frac{\left(\int_{s_2}^{t_2}(h_{n_2-1}(\tau_2,t_2))^q e_0(t_2,\tau_2)d_\alpha\tau_2\right)}{q}\right]}$$

$$\leq (b_1-s_1)(b_2-s_2)\left(\int_{s_1}^{b_1}|(D^\alpha)^{n_1}f_1(\tau_1)|^q\,d_\alpha\tau_1\right)^{\frac{1}{q}}\left(\int_{s_2}^{b_2}|(D^\alpha)^{n_2}f_2(\tau_2)|^p\,d_\alpha\tau_2\right)^{\frac{1}{p}}. \tag{13.62}$$

Proof. Here $j=1,2$. Since $(D^\alpha)^{k_j} f_j(s_j)=0$, $k_j=0,1,\dots,n_j-1$; $t_j \geq s_j$, we have

$$f_j(t_j) = (-1)^{n_j-1}\int_{s_j}^{t_j} h_{n_j-1}(\tau_j,t_j)\,(D^\alpha)^{n_j} f_j(\tau_j)\,e_0(t_j,\tau_j)\,d_\alpha\tau_j. \tag{13.63}$$

As in (13.27) we get

$$|f_1(t_1)| \leq \left(\int_{s_1}^{t_1}|h_{n_1-1}(\tau_1,t_1)|^p\,e_0(t_1,\tau_1)\,d_\alpha\tau_1\right)^{\frac{1}{p}}$$
$$\left(\int_{s_1}^{t_1}|(D^\alpha)^{n_1} f_1(\tau_1)|^q\,e_0(t_1,\tau_1)\,d_\alpha\tau_1\right)^{\frac{1}{q}} \tag{13.64}$$

and

$$|f_2(t_2)| \leq \left(\int_{s_2}^{t_2}|h_{n_2-1}(\tau_2,t_2)|^q\,e_0(t_2,\tau_2)\,d_\alpha\tau_2\right)^{\frac{1}{q}}$$
$$\left(\int_{s_2}^{t_2}|(D^\alpha)^{n_2} f_2(\tau_2)|^p\,e_0(t_2,\tau_2)\,d_\alpha\tau_2\right)^{\frac{1}{p}}. \tag{13.65}$$

Hence

$$|f_1(t_1)|\,|f_2(t_2)|$$
$$\leq \left(\int_{s_1}^{t_1}|h_{n_1-1}(\tau_1,t_1)|^p\,e_0(t_1,\tau_1)\,d_\alpha\tau_1\right)^{\frac{1}{p}}\left(\int_{s_2}^{t_2}|h_{n_2-1}(\tau_2,t_2)|^q\,e_0(t_2,\tau_2)\,d_\alpha\tau_2\right)^{\frac{1}{q}}$$
$$\left(\int_{s_1}^{t_1}|(D^\alpha)^{n_1}f_1(\tau_1)|^q\,e_0(t_1,\tau_1)\,d_\alpha\tau_1\right)^{\frac{1}{q}}\left(\int_{s_2}^{t_2}|(D^\alpha)^{n_2}f_2(\tau_2)|^p\,e_0(t_2,\tau_2)\,d_\alpha\tau_2\right)^{\frac{1}{p}} \tag{13.66}$$

(using Young's inequality for $a^*, b^* \geq 0$, $a^{*\frac{1}{p}} b^{*\frac{1}{q}} \leq \frac{a^*}{p} + \frac{b^*}{q}$)

$$\leq \left[\frac{\left(\int_{s_1}^{t_1} |h_{n_1-1}(\tau_1, t_1)|^p e_0(t_1, \tau_1)\, d_\alpha \tau_1\right)}{p} + \frac{\left(\int_{s_2}^{t_2} |h_{n_2-1}(\tau_2, t_2)|^q e_0(t_2, \tau_2)\, d_\alpha \tau_2\right)}{q}\right]$$

$$\left(\int_{s_1}^{t_1} |(D^\alpha)^{n_1} f_1(\tau_1)|^q e_0(t_1, \tau_1)\, d_\alpha \tau_1\right)^{\frac{1}{q}} \left(\int_{s_2}^{t_2} |(D^\alpha)^{n_2} f_2(\tau_2)|^p e_0(t_2, \tau_2)\, d_\alpha \tau_2\right)^{\frac{1}{p}}. \tag{13.67}$$

Thus, it holds

$$\int_{s_1}^{b_1}\int_{s_2}^{b_2} \frac{|f_1(t_1)|\,|f_2(t_2)|\, dt_1 dt_2}{\left[\frac{\left(\int_{s_1}^{t_1}\left(|h_{n_1-1}(\tau_1,t_1)|\right)^p e_0(t_1,\tau_1) d_\alpha \tau_1\right)}{p} + \frac{\left(\int_{s_2}^{t_2}\left(|h_{n_2-1}(\tau_2,t_2)|\right)^q e_0(t_2,\tau_2) d_\alpha \tau_2\right)}{q}\right]}$$

(denominator can be zero only when both $t_1 = s_1$ and $t_2 = s_2$)

$$\leq \left(\int_{s_1}^{b_1} \left(\int_{s_1}^{t_1} |(D^\alpha)^{n_1} f_1(\tau_1)|^q e_0(t_1, \tau_1)\, d_\alpha \tau_1\right)^{\frac{1}{q}} dt_1\right)$$

$$\times \left(\int_{s_2}^{b_2} \left(\int_{s_2}^{t_2} |(D^\alpha)^{n_2} f_2(\tau_2)|^p e_0(t_2, \tau_2)\, d_\alpha \tau_2\right)^{\frac{1}{p}} dt_2\right)$$

$$\leq \left(\int_{s_1}^{b_1} \left(\int_{s_1}^{t_1} |(D^\alpha)^{n_1} f_1(\tau_1)|^q d_\alpha \tau_1\right)^{\frac{1}{q}} dt_1\right)$$

$$\times \left(\int_{s_2}^{b_2} \left(\int_{s_2}^{t_2} |(D^\alpha)^{n_2} f_2(\tau_2)|^p d_\alpha \tau_2\right)^{\frac{1}{p}} dt_2\right)$$

$$\leq \left(\int_{s_1}^{b_1} \left(\int_{s_1}^{b_1} |(D^\alpha)^{n_1} f_1(\tau_1)|^q d_\alpha \tau_1\right)^{\frac{1}{q}} dt_1\right)$$

$$\times \left(\int_{s_2}^{b_2} \left(\int_{s_2}^{b_2} |(D^\alpha)^{n_2} f_2(\tau_2)|^p d_\alpha \tau_2\right)^{\frac{1}{p}} dt_2\right)$$

$$= (b_1 - s_1)(b_2 - s_2) \left(\int_{s_1}^{b_1} |(D^\alpha)^{n_1} f_1(\tau_1)|^q d_\alpha \tau_1\right)^{\frac{1}{q}}$$

$$\times \left(\int_{s_2}^{b_2} |(D^\alpha)^{n_2} f_2(\tau_2)|^p d_\alpha \tau_2\right)^{\frac{1}{p}}, \tag{13.68}$$

proving the claim. □

Bibliography

G.A. Anastassiou, *Probabilistic Inequalities*, World Scientific, New Jersey, Singapore, 2010.

G.A. Anastassiou, *Advanced Inequalities*, World Scientific, New Jersey, Singapore, 2011.

G.A. Anastassiou, *Intelligent Comparisons: Analytic Inequalities*, Springer, Heildelberg, New York, 2016.

G.A. Anastassiou, *Integral inequalities involving new conformable derivatives*, Progress in fractional differentiation and applications, accepted, 2023.

D.R. Anderson and D.J. Ulness, Newly defined conformable derivatives, *Advances in Dynamical Systems and Applications*, 10(2) (2015), 109–137.

Y. Li, K.H. Ang and G.C.Y. Chog, PID control systems analysis and design, *IEEE Control Syst. Mag.* 26(1) (2006), 32–41.

Chapter 14

About Proportional Fractional Calculus and Inequalities

Left and right Caputo fractional calculus is developed further with new interesting representation formulae. In that spirit, we introduce and develop in great generality the corresponding left and right proportional Caputo style fractional calculus. We establish the analogous proportional fractional representation formulae. Based on all the above, we derive left and right fractional integral inequalities of Opial, Poincaré, Sobolev and Hilbert–Pachpatte types. It follows Anastassiou (2023).

14.1 Background

14.1.1 *From left fractional calculus and more*

Let $\alpha \in (0,1)$ and $f \in C^1([a,b])$, $[a,b] \subset \mathbb{R}$.

The left Riemann–Liouville fractional integral of f is given by

$$(J_a^\alpha f)(x) := \frac{1}{\Gamma(\alpha)} \int_a^x (x-t)^{\alpha-1} f(t)\, dt, \tag{14.1}$$

for $a \le x \le b$, where Γ is the gamma function.

We set $J_a^0 := I$, the identity operator.

By Theorem 1.5, Anastassiou (2018), p. 3, $J_a^\alpha f \in C([a,b])$.

Here $\lceil \cdot \rceil$ denotes the ceiling of the number, i.e. $\lceil \alpha \rceil = 1$.

The left Caputo fractional derivative of f is given by

$$(D_{*a}^\alpha f)(x) := \frac{1}{\Gamma(1-\alpha)} \int_a^x (x-t)^{-\alpha} f'(t)\, dt, \tag{14.2}$$

$\forall\, x \in [a,b]$. Clearly, $D_{*a}^\alpha f \in C([a,b])$.

We set $D_{*a}^1 f := f'$ and $D_{*a}^0 f := f$.

By Anastassiou (2018, p. 11), we get that

$$\begin{aligned}\left(J_a^{\alpha} D_{*a}^{\alpha} f\right)(x) &= \left(J_a^1 f'\right)(x) \\ &= \int_a^x f'(t)\, dt = f(x) - f(a), \quad \forall\, x \in [a, b]. \qquad (14.3)\end{aligned}$$

When $f(a) = 0$, we obtain

$$\left(J_a^{\alpha} D_{*a}^{\alpha} f\right)(x) = f(x), \quad \forall\, x \in [a, b]. \qquad (14.4)$$

The left Riemann–Liouville fractional derivative of f is defined by

$$\begin{aligned}{}_L D_a^{\alpha} f(x) &= D J_a^{1-\alpha} f(x) \\ &= \frac{1}{\Gamma(1-\alpha)} \frac{d}{dx} \int_a^x (x-t)^{-\alpha} f(t)\, dt, \quad \forall\, x \in [a, b]. \qquad (14.5)\end{aligned}$$

By Lemma 3.5, Diethelm (2010), p. 53, we derive that

$${}_L D_a^{\alpha} f = D_{*a}^{\alpha} f,$$

iff

$$f(a) = 0.$$

And, by Theorem 3.7, Diethelm (2010), p. 53, we have that

$$D_{*a}^{\alpha} J_a^{\alpha} f = f, \qquad (14.6)$$

given f is continuous over $[a, b]$.

14.1.2 *From right fractional calculus and more*

Let $\alpha \in (0, 1)$ and $f \in C^1([a, b])$, $[a, b] \subset \mathbb{R}$.

The right Riemann–Liouville fractional integral of f is given by

$$I_{b-}^{\alpha} f(x) = \frac{1}{\Gamma(\alpha)} \int_x^b (J - x)^{\alpha - 1} f(J)\, dJ, \qquad (14.7)$$

$\forall\, x \in [a, b]$. We set $I_{b-}^0 := I$.

By Theorem 2.5, Anastassiou (2018, p. 37), $I_{b-}^{\alpha} f \in C([a, b])$.

The right Caputo fractional derivative of f of order α is given by

$$\begin{aligned}D_{b-}^{\alpha} f(x) &:= -I_{b-}^{1-\alpha} f'(x) \\ &= \frac{-1}{\Gamma(1-\alpha)} \int_x^b (J - x)^{-\alpha} f'(J)\, dJ, \qquad (14.8)\end{aligned}$$

$\forall\, x \in [a, b]$. Clearly $D_{b-}^{\alpha} f \in C([a, b])$.

The right Riemann–Liouville fractional derivative of f is defined by

$$_{R}D_{b-}^{\alpha}f(x)=\frac{-1}{\Gamma(1-\alpha)}\frac{d}{dx}\int_{x}^{b}(t-x)^{-\alpha}f(t)\,dt, \tag{14.9}$$

$\forall\, x\in[a,b]$.

We define $_{R}D_{b-}^{0}f:=I$.

That is,

$$_{R}D_{b-}^{\alpha}f=-DI_{b-}^{1-\alpha}f. \tag{14.10}$$

Given $f(b)=0$, by Theorem 23.9, Anastassiou (2011), p. 339, we obtain that

$$_{R}D_{b-}^{\alpha}f=\left(D_{b-}^{\alpha}f\right), \tag{14.11}$$

over $[a.b]$.

We have that

$$\begin{aligned}D_{b-}^{\alpha}I_{b-}^{\alpha}f&={}_{R}D_{b-}^{\alpha}I_{b-}^{\alpha}f\\&=(-1)\,DI_{b-}^{1-\alpha}I_{b-}^{\alpha}f=(-1)\,DI_{b-}^{1}f=(-1)^{2}f=f.\end{aligned}$$

So when $f(b)=0$, we have

$$D_{b-}^{\alpha}I_{b-}^{\alpha}f=f. \tag{14.12}$$

Also, by Anastassiou (2018, p. 43), we have

$$\begin{aligned}\left(I_{b-}^{\alpha}D_{b-}^{\alpha}f\right)(x)&=(-1)\left(I_{b-}^{1}f'\right)(x)\\&=-\int_{x}^{b}f'(J)\,dJ=\int_{b}^{x}f'(J)\,dJ=f(x)-f(b).\end{aligned} \tag{14.13}$$

Consequently, given $f(b)=0$, we obtain

$$I_{b-}^{\alpha}D_{b-}^{\alpha}f=f. \tag{14.14}$$

14.1.3 *Conclusion*

Let $\alpha\in(0,1)$, $f\in C^{1}([a,b])$.

(i) Given $f(a)=0$, we derive over $[a,b]$ that

$$J_{a}^{\alpha}D_{*a}^{\alpha}f=D_{*a}^{\alpha}J_{a}^{\alpha}f=f. \tag{14.15}$$

Hence, J_{a}^{α}, D_{*a}^{α} are functional inverses.

(ii) Given $f(b)=0$, we get over $[a,b]$ that

$$I_{b-}^{\alpha}D_{b-}^{\alpha}f=D_{b-}^{\alpha}I_{b-}^{\alpha}f=f. \tag{14.16}$$

Hence I_{b-}^{α}, D_{b-}^{α} are functional inverses.

So

$$\left(J_a^\alpha D_{*a}^\alpha\right)\left(J_a^\alpha D_{*a}^\alpha f\right) = \left(J_a^\alpha D_{*a}^\alpha\right)(f) = f, \tag{14.17}$$

and

$$J_a^\alpha \left(D_{*a}^\alpha \circ J_a^\alpha\right) D_{*a}^\alpha f = f, \tag{14.18}$$

and

$$J_a^\alpha \left(J_a^\alpha \circ D_{*a}^\alpha\right) D_{*a}^\alpha f = f. \tag{14.19}$$

That is,

$$\left(J_a^\alpha\right)^2 \left(D_{*a}^\alpha\right)^2 f = f. \tag{14.20}$$

Hence for any $N \in \mathbb{N}$, we obtain

$$\left(J_a^\alpha\right)^N \left(D_{*a}^\alpha\right)^N f = f, \tag{14.21}$$

given that $f(a) = 0$.

By similar reasoning, we derive that

$$\left(I_{b-}^\alpha\right)^N \left(D_{b-}^\alpha\right)^N f = f, \tag{14.22}$$

$\forall\ N \in \mathbb{N}$, given that $f(b) = 0$.

Consequently, if $\left(D_{*a}^\alpha\right)^{(i)} f \in C([a,b])$, $i = 1, \ldots, N$, we get that

$$J_a^\alpha \left(D_{*a}^\alpha\right)^N f = \left(D_{*a}^\alpha\right)^{N-1} f =: D_{*a}^{(N-1)\alpha} f, \tag{14.23}$$

given that $f(a) = 0$.

Furthermore, if $\left(D_{b-}^\alpha\right)^{(i)} f \in C([a,b])$, $i = 1, \ldots, N$, we obtain

$$I_{b-}^\alpha \left(D_{b-}^\alpha\right)^N f = \left(D_{b-}^\alpha\right)^{N-1} f =: D_{b-}^{(N-1)\alpha} f, \tag{14.24}$$

given that $f(b) = 0$.

So both cases are true for some $N \in \mathbb{N}$; $\alpha \in (0,1)$ and $f \in C^1([a,b])$.

14.1.4 *About new conformable calculus*

Here all comes from Anderson and Ulness (2015).

Definition 14.1 (A class of New Conformable Derivatives) (Anderson and Ulness 2015). Let $\alpha \in [0,1]$, and let the functions

$k_0, k_1 : [0,1] \times \mathbb{R} \to [0,\infty)$ be continuous such that

$$\lim_{\alpha \to 0+} k_1(\alpha,t) = 1, \quad \lim_{\alpha \to 0+} k_0(\alpha,t) = 0, \quad \forall\, t \in \mathbb{R},$$

$$\lim_{\alpha \to 1-} k_1(\alpha,t) = 0, \quad \lim_{\alpha \to 1-} k_0(\alpha,t) = 1, \quad \forall\, t \in \mathbb{R}, \tag{14.25}$$

$$k_1(\alpha,t) \neq 0,\ \alpha \in [0,1), \quad k_0(\alpha,t) \neq 0, \quad \alpha \in (0,1], \quad \forall\, t \in \mathbb{R}.$$

Then, the following differentiable operator D^α, defined via

$$D^\alpha f(t) = k_1(\alpha,t) f(t) + k_0(\alpha,t) f'(t), \tag{14.26}$$

is the new conformable derivative given that $f'(t)$ exists for $t \in \mathbb{R}$.

Definition 14.2 (Anderson and Ulness 2015). (Conformable Exponential Function). Let $\alpha \in (0,1]$, the points $s,t \in \mathbb{R}$ with $s \leq t$, and let the function $p : [s,t] \to \mathbb{R}$ be continuous. Let $k_0, k_1 : [0,1] \times \mathbb{R} \to [0,\infty)$ be continuous and satisfy (14.25) with $\frac{p}{k_0}$ and $\frac{k_1}{k_0}$ Riemann integrable on $[s,t]$. Then the exponential function with respect to D^α in (14.26) is defined to be

$$e_p(t,s) := e^{\int_s^t \frac{p(\tau)-k_1(\alpha,\tau)}{k_0(\alpha,\tau)} d\tau}, \quad e_0(t,s) := e^{-\int_s^t \frac{k_1(\alpha,\tau)}{k_0(\alpha,\tau)} d\tau}. \tag{14.27}$$

Using (14.26) and (14.27), we have the following basic results.

Lemma 14.1 (Anderson and Ulness 2015). *(Basic Derivatives). Let the conformable differential operator D^α be given as in (14.26), where $\alpha \in [0,1]$. Let the function $p : [s,t] \to \mathbb{R}$ be continuous. Let $k_0, k_1 : [0,1] \times \mathbb{R} \to [0,\infty)$ be continuous and satisfy (14.25), with $\frac{p}{k_0}$ and $\frac{k_1}{k_0}$ Riemann integrable on $[s,t]$. Assume the functions f and g are differentiable as needed. Then*

(i) $D^\alpha[af + bg] = aD^\alpha[f] + bD^\alpha[g]$ *for all* $a, b \in \mathbb{R}$,
(ii) $D^\alpha c = ck_1(\alpha,\cdot)$ *for all constants* $c \in \mathbb{R}$,
(iii) $D^\alpha[fg] = fD^\alpha[g] + gD^\alpha[f] - fgk_1(\alpha,\cdot)$,
(iv) $D^\alpha\left[\frac{f}{g}\right] = \frac{gD^\alpha[f] - fD^\alpha[g]}{g^2} + \frac{f}{g}k_1(\alpha,\cdot)$,
(v) *for* $\alpha \in (0,1]$ *and fixed* $s \in \mathbb{R}$, *the exponential function satisfies*

$$D_t^\alpha[e_p(t,s)] = p(t)\, e_p(t,s) \tag{14.28}$$

for $e_p(t,s)$ *given in (14.27),*
(vi) *for* $\alpha \in (0,1]$ *and for the exponential function* e_0 *given in (14.27), we have*

$$D^\alpha\left[\int_a^t \frac{f(s)\, e_0(t,s)}{k_0(\alpha,s)} ds\right] = f(t). \tag{14.29}$$

Definition 14.3 (Anderson and Ulness 2015). (Integrals). Let $\alpha \in (0,1]$ and $t_0 \in \mathbb{R}$. In the light of (14.27) and Lemmas 14.1 (v) and (vi), define the antiderivative via

$$\int D^{\alpha} f(t)\, d_{\alpha} t = f(t) + c e_0(t, t_0), \quad c \in \mathbb{R}.$$

Similarly, define the integral of f over $[a, b]$ as

$$\int_a^t f(s)\, e_0(t,s)\, d_{\alpha} s := \int_a^t \frac{f(s)\, e_0(t,s)}{k_0(t,s)} ds, \quad d_{\alpha} s := \frac{1}{k_0(\alpha, s)} ds; \tag{14.30}$$

recall that

$$e_0(t,s) := e^{-\int_s^t \frac{k_1(\alpha,\tau)}{k_0(\alpha,\tau)} d\tau} = e^{-\int_s^t k_1(\alpha,\tau) d_{\alpha}\tau}$$

from (14.27).

Lemma 14.2 (Anderson and Ulness 2015). *(Basic Integrals). Let the conformable differential operator D^{α} be given as in (14.26), and the integral be given as in (14.30) with $\alpha \in (0,1]$. Let the functions k_0, k_1 be continuous and satisfy (14.25), and let f and g be differentiable as needed. Then*

(i) *the derivative of the definite integral of f is given by*

$$D^{\alpha}\left[\int_a^t f(s)\, e_0(t,s)\, d_{\alpha} s\right] = f(t), \tag{14.31}$$

(ii) *the definite integral of the derivative of f is given by*

$$\int_a^t D^{\alpha}[f(s)]\, e_0(t,s)\, d_{\alpha} s = f(s)\, e_0(t,s)\,|_{s=a}^t := f(t) - f(a)\, e_0(t,a), \tag{14.32}$$

(iii) *an integration by parts formula is given by*

$$\int_a^b f(t)\, D^{\alpha}[g(t)]\, e_0(b,t)\, d_{\alpha} t = f(t)\, g(t)\, e_0(b,t)\,|_{t=a}^b - \int_{\alpha}^b g(t)\left(D^{\alpha}[f(t)] - k_1(\alpha,t) f(t)\right) e_0(b,t)\, d_{\alpha} t. \tag{14.33}$$

Remark 14.1. (to Lemma 14.2 (iii))

When $f = g$, we get

$$\int_a^b f(t)\, D^{\alpha}[f(t)]\, e_0(b,t)\, d_{\alpha} t = f^2(b) - f^2(a)\, e_0(b,a) - \int_a^b f(t)\, D^{\alpha}[f(t)]\, e_0(b,t)\, d_{\alpha} t + \int_a^b k_1(\alpha,t) f^2(t)\, e_0(b,t)\, d_{\alpha} t. \tag{14.34}$$

Therefore, it holds

$$\int_a^b f(t) D^\alpha [f(t)] e_0(b,t) d_\alpha t = \left(\frac{f^2(b) - f^2(a) e_0(b,a)}{2} \right) + \frac{1}{2} \int_a^b f^2(t) k_1(\alpha,t) e_0(b,t) d_\alpha t. \tag{14.35}$$

14.1.5 *Proportional left Caputo fractional calculus*

Proportional left Caputo fractional calculus was first introduced in Baleanu, Fernandez and Akgül (2020), $\alpha \in (0,1)$.

Here we go more generally.

We define for $f \in C^1([a,b])$ in the sense of D^α:

$$\left({}^P D_{*a}^\alpha f \right)(x) := \frac{1}{\Gamma(1-\alpha)} \int_a^x (x-t)^{-\alpha} D^\alpha f(t) dt, \tag{14.36}$$

$\forall\, x \in [a,b]$. Clearly, ${}^P D_{*a}^\alpha f \in C([a,b])$.

The function ${}^P D_{*a}^\alpha f$ is called the proportional left Caputo fractional derivative of f.

That is,

$$\left({}^P D_{*a}^\alpha f \right)(x) = J_a^{1-\alpha} (D^\alpha f)(x). \tag{14.37}$$

The left proportional integral is given by (14.30):

$$\left({}^P I_a f \right)(t) := \int_a^t f(s) e_0(t,s) d_\alpha s \tag{14.38}$$

and ${}^P I_a f \in C([a,b])$.

By (14.32), we get

$$\left({}^P I_a D^\alpha f \right)(t) = f(t) - f(a) e_0(t,a) \tag{14.39}$$

and ${}^P I_a D^\alpha f \in C([a,b])$.

We define the left Riemann–Liouville proportional integral as follows:

$$\left({}_L^P I_a^{RL} f \right)(t) := \int_a^t {}_L D_a^{1-\alpha} f(s) e_0(t,s) d_\alpha s. \tag{14.40}$$

Clearly, ${}_L^P I_a^{RL} f \in C([a,b])$ when $f(a) = 0$.

We have that

$$\left({}_L^P I_a^{RL}\left({}^P D_{*a}^{\alpha} f\right)\right)(t)=\int_a^t\left({}_L D_a^{1-\alpha}\left({}^P D_{*a}^{\alpha} f\right)\right)(s)\, e_0(t,s)\, d_\alpha s, \tag{14.41}$$

with $\left({}_L^P I_a^{RL}\left({}^P D_{*a}^{\alpha} f\right)\right) \in C([a,b])$.

We see that

$$\begin{aligned} {}^P D_{*a}^{\alpha}\, {}_L^P I_a^{RL} f(t) &= \left(J_a^{1-\alpha} \circ D^{\alpha}\right) \circ \left({}^P I_a \circ_L D_a^{1-\alpha}\right) f(t) \\ &\overset{(14.31)}{=} \left(J_a^{1-\alpha} \circ\, {}_L D_a^{1-\alpha}\right) f(t) \end{aligned} \tag{14.42}$$

(by [5], p. 39)

$$= f(t)-\frac{(t-a)^{-\alpha}}{\Gamma(\alpha)} \lim_{z \to a+} J_a^{\alpha} f(z)=f(t)-\frac{(t-a)^{-\alpha}}{\Gamma(\alpha)} \cdot 0=f(t), \quad t \neq a.$$

We have proved that

$${}^P D_{*a}^{\alpha}\, {}_L^P I_a^{RL} f=f, \quad t \neq a. \tag{14.43}$$

Next we study

$$\left({}_L^P I_a^{RL}\left({}^P D_{*a}^{\alpha} f\right)\right)(t)=\left({}^P I_a \circ\, {}_L D_a^{1-\alpha}\right) \circ\left(J_a^{1-\alpha} \circ D^{\alpha}\right) f(t) \tag{14.44}$$

(by Theorem 2.14, Diethelm (2010), p. 30)

$$=\left({}^P I_a \circ D^{\alpha}\right) f(t) \overset{(14.39)}{=} f(t)-f(a)\, e_0(t,a).$$

Assuming $f(a)=0$, we have

$$\left({}_L^P I_a^{RL}\, {}^P D_{*a}^{\alpha}\right) f=f. \tag{14.45}$$

Conclusion: When $f(a)=0$, we have

$$\left({}_L^P I_a^{RL}\, {}^P D_{*a}^{\alpha}\right) f=\left({}^P D_{*a}^{\alpha}\, {}_L^P I_a^{RL}\right) f=f, \quad t \neq a. \tag{14.46}$$

That is, ${}_L^P I_a^{RL}$, ${}^P D_{*a}^{\alpha}$ are functional inverses over (a,b].

Clearly, for any $N \in \mathbb{N}$, we get

$$\left({}_L^P I_a^{RL}\right)^N \left({}^P D_{*a}^{\alpha}\right)^N f=f, \quad t \neq a, \tag{14.47}$$

given that $f(a)=0$.

Thus, it holds

$${}_L^P I_a^{RL}\left({}^P D_{*a}^{\alpha}\right)^N f=\left({}^P D_{*a}^{\alpha}\right)^{N-1} f=:\left({}^P D_{*a}^{(N-1)\alpha}\right) f,\ t \neq a, \tag{14.48}$$

under the assumptions: $f(a) = 0$, and $\left({}^{P}D_{*a}^{\alpha}\right)^{(i)} f \in C([a,b])$, for $i = 1,\ldots,N$, where $f \in C^1([a,b])$, $\alpha \in (0,1)$.

14.1.6 *Proportional right Caputo fractional calculus*

We define for $f \in C^1([a,b])$ in the sense of D^α, $\alpha \in (0,1)$:

$$\left({}^{P}D_{b-}^{\alpha} f\right)(x) := \frac{-1}{\Gamma(1-\alpha)} \int_x^b (t-x)^{-\alpha} D^\alpha f(t)\, dt, \tag{14.49}$$

$\forall\, x \in [a,b]$. Clearly, ${}^{P}D_{b-}^{\alpha} f \in C([a,b])$.

That is,

$$\left({}^{P}D_{b-}^{\alpha} f\right)(x) = -I_{b-}^{1-\alpha}(D^\alpha f)(x). \tag{14.50}$$

We call ${}^{P}D_{b-}^{\alpha} f$ the proportional right Caputo fractional derivative of f.

The right proportional integral of f is given by

$$\left({}^{P}I_b f\right)(t) := \int_t^b f(s)\, e_0(b,s)\, d_\alpha s, \tag{14.51}$$

$\forall\, t \in [a,b]$, with ${}^{P}I_b f \in C([a,b])$.

We have that

$$\left({}^{P}I_b (D^\alpha f)\right)(t) = \int_t^b (D^\alpha f)(s)\, e_0(b,s)\, d_\alpha s \overset{(14.32)}{=} f(b) - f(t)\, e_0(b,t), \tag{14.52}$$

with $\left({}^{P}I_b (D^\alpha f)\right) \in C([a,b])$.

If $f(b) = 0$, we derive that

$$\left({}^{P}I_b \circ D^\alpha\right)(f(t)) = -f(t)\, e_0(b,t), \quad \forall t \in [a,b]. \tag{14.53}$$

Next we define

$$\left({}_{R}^{P}I_b^{RL} f\right)(t) := \int_t^b \left({}_{R}D_{b-}^{1-\alpha} f(s)\right) e_0(b,s)\, d_\alpha s, \tag{14.54}$$

$\forall\, t \in [a,b]$, where ${}_{R}D_{b-}^{1-\alpha} f$ the right Riemann–Liouville fractional derivative of f of order $1-\alpha$, $\alpha \in (0,1)$. When $f(b) = 0$, it is ${}_{R}^{P}I_b^{RL} f \in C([a,b])$.

Furthermore, we consider

$$\left({}_{R}^{P}I_b^{RL}\left({}^{P}D_{b-}^{\alpha} f\right)\right)(t) = \int_t^b \left({}_{R}D_{b-}^{1-\alpha}\left({}^{P}D_{b-}^{\alpha} f\right)\right)(s)\, e_0(b,s)\, d_\alpha s, \tag{14.55}$$

$\forall\, t \in [a,b]$, where $\left({}_{R}^{P}I_b^{RL}\left({}^{P}D_{b-}^{\alpha} f\right)\right) \in C([a,b])$.

Finally, we observe that

$$\left({}_{R}^{P}I_{b}^{RL}\circ^{P}D_{b-}^{\alpha}\right)f(t)\overset{\text{(by (14.50), (14.53))}}{=}-\left({}^{P}I_{b}\circ_{R}D_{b-}^{1-\alpha}\right)\circ\left(I_{b-}^{1-\alpha}\circ D^{\alpha}\right)f(t)$$
$$=-\left({}^{P}I_{b}\circ D^{\alpha}\right)f(t)=f(t)\,e_{0}(b,t),\qquad(14.56)$$

by assuming $f(b)=0$.

Above we used that (see (14.11) and (14.12))

$$D_{b-}^{1-\alpha}\circ I_{b-}^{1-\alpha}=I,\quad\text{the identity operator.}$$

Conclusion: When $f(b)=0$, we get that

$$\left({}_{R}^{P}I_{b}^{RL}\circ^{P}D_{b-}^{\alpha}\right)f(t)=f(t)\,e_{0}(b,t),\qquad(14.57)$$

$\forall\, t\in[a,b]$.

14.2 Main Results

We start with a collection of Opial-type inequalities.

Here $\alpha\in(0,1)$, $f\in C^{1}([a,b])$; $p,q>1:\frac{1}{p}+\frac{1}{q}=1$, $N\in\mathbb{N}$.

Theorem 14.1. *Assume that* $\alpha>\frac{1}{q}$; $f(a)=0$, $\left(D_{*a}^{i\alpha}f\right)\in C([a,b])$, $i=1,\ldots,N$. *Then*

$$\int_{a}^{x}\left|\left(D_{*a}^{(N-1)\alpha}f\right)(w)\right|\left|\left(D_{*a}^{N\alpha}f\right)(w)\right|dw$$
$$\leq\frac{(x-a)^{\alpha+\frac{1}{p}-\frac{1}{q}}}{2^{\frac{1}{q}}\Gamma(\alpha)\left[(p(\alpha-1)+1)(p(\alpha-1)+2)\right]^{\frac{1}{p}}}\left(\int_{a}^{x}\left|\left(D_{*a}^{N\alpha}f\right)(w)\right|^{q}dw\right)^{\frac{2}{q}},\qquad(14.58)$$

$\forall\, x\in[a,b]$.

Proof. By (14.23), we have

$$J_{a}^{\alpha}\left(D_{*a}^{N\alpha}\right)f(x)=D_{*a}^{(N-1)\alpha}f(x),\qquad(14.59)$$

given that $f(a)=0$, under the assumption $\left(D_{*a}^{i\alpha}f\right)\in C([a,b])$, $i=1,\ldots,N$.

That is,

$$\left(D_{*a}^{(N-1)\alpha}f\right)(w)=\frac{1}{\Gamma(\alpha)}\int_{a}^{w}(w-t)^{\alpha-1}\left(D_{*a}^{N\alpha}f\right)(t)\,dt,\qquad(14.60)$$

for all $a\leq w\leq x\leq b$.

Hence, by Hölder's inequality, we derive

$$\left|\left(D_{*a}^{(N-1)\alpha}f\right)(w)\right| \leq \frac{1}{\Gamma(\alpha)}\int_a^w (w-t)^{\alpha-1}\left|\left(D_{*a}^{N\alpha}f\right)(t)\right| dt$$

$$\leq \frac{1}{\Gamma(\alpha)}\left(\int_a^w (w-t)^{p(\alpha-1)}\,dt\right)^{\frac{1}{p}}\left(\int_a^w \left|\left(D_{*a}^{N\alpha}f\right)(t)\right|^q dt\right)^{\frac{1}{q}}$$

$$= \frac{1}{\Gamma(\alpha)}\left(\frac{(w-a)^{\frac{p(\alpha-1)+1}{p}}}{(p(\alpha-1)+1)^{\frac{1}{p}}}\right)\left(\int_a^w \left|\left(D_{*a}^{N\alpha}f\right)(t)\right|^q dt\right)^{\frac{1}{q}}. \quad (14.61)$$

Call

$$z(w) := \int_a^w \left|\left(D_{*a}^{N\alpha}f\right)(t)\right|^q dt, \quad z(a) = 0, \quad (14.62)$$

$a \leq w \leq x$.

Then

$$z'(w) := \left|\left(D_{*a}^{N\alpha}f\right)(w)\right|^q \quad (14.63)$$

and

$$\left|\left(D_{*a}^{N\alpha}f\right)(w)\right| = (z'(w))^{\frac{1}{q}}, \quad \text{all } a \leq w \leq x. \quad (14.64)$$

Therefore, we have (all $a \leq w \leq x$)

$$\left|\left(D_{*a}^{(N-1)\alpha}f\right)(w)\right|\left|\left(D_{*a}^{N\alpha}f\right)(w)\right| \leq \frac{1}{\Gamma(\alpha)}\left(\frac{(w-a)^{\frac{p(\alpha-1)+1}{p}}}{(p(\alpha-1)+1)^{\frac{1}{p}}}\right)(z(w)\,z'(w))^{\frac{1}{q}}. \quad (14.65)$$

Hence, it holds

$$\int_a^x \left|\left(D_{*a}^{(N-1)\alpha}f\right)(w)\right|\left|\left(D_{*a}^{N\alpha}f\right)(w)\right| dw$$

$$\leq \frac{1}{\Gamma(\alpha)(p(\alpha-1)+1)^{\frac{1}{p}}}\int_a^x (w-a)^{\frac{p(\alpha-1)+1}{p}}(z(w)\,z'(w))^{\frac{1}{q}}\,dw \leq$$

(by Hölder's inequality)

$$\frac{1}{\Gamma(\alpha)(p(\alpha-1)+1)^{\frac{1}{p}}}\left(\int_a^x (w-a)^{p(\alpha-1)+1}\,dw\right)^{\frac{1}{p}}\left(\int_a^x z(w)\,z'(w)\,dw\right)^{\frac{1}{q}}$$

$$= \frac{1}{\Gamma(\alpha)(p(\alpha-1)+1)^{\frac{1}{p}}}\frac{(x-a)^{\frac{p(\alpha-1)+2}{p}}}{(p(\alpha-1)+2)^{\frac{1}{p}}}\left(\int_a^x z(w)\,dz(w)\right)^{\frac{1}{q}}$$

$$= \frac{(x-a)^{\alpha+\frac{1}{p}-\frac{1}{q}}}{\Gamma(\alpha)\left[(p(\alpha-1)+1)(p(\alpha-1)+2)\right]^{\frac{1}{p}}} \frac{z^{\frac{2}{q}}(x)}{2^{\frac{1}{q}}}$$

$$= \frac{(x-a)^{\alpha+\frac{1}{p}-\frac{1}{q}}}{2^{\frac{1}{q}}\Gamma(\alpha)\left[(p(\alpha-1)+1)(p(\alpha-1)+2)\right]^{\frac{1}{p}}} \left(\int_a^x \left|\left(D_{*a}^{N\alpha} f\right)(t)\right|^q dt\right)^{\frac{2}{q}}. \tag{14.66}$$

The claim is proved. □

We continue with

Theorem 14.2. *Assume that* $\alpha > \frac{1}{q}$; $f(b) = 0$, $\left(D_{b-}^{i\alpha} f\right) \in C([a,b])$, $i = 1, \ldots, N$. *Then*

$$\int_x^b \left|\left(D_{b-}^{(N-1)\alpha} f\right)(w)\right| \left|\left(D_{b-}^{N\alpha} f\right)(w)\right| dw$$

$$\leq \frac{(b-x)^{\alpha+\frac{1}{p}-\frac{1}{q}}}{2^{\frac{1}{q}}\Gamma(\alpha)\left[(p(\alpha-1)+1)(p(\alpha-1)+2)\right]^{\frac{1}{p}}} \left(\int_x^b \left|\left(D_{b-}^{N\alpha} f\right)(w)\right|^q dw\right)^{\frac{2}{q}}, \tag{14.67}$$

$\forall\, x \in [a,b]$.

Proof. As similar to Theorem 14.1 is omitted. It is based on (14.24). □

Next comes a left proportional Caputo fractional Opial-type inequality.

Theorem 14.3. *Here* $f(a) = 0$ *and* ${}^P D_{*a}^{i\alpha} f \in C([a,b])$, *for* $i = 1, \ldots, N$. *Then, for every* $x \in [a,b]$, *we have*

$$\int_a^x \left|\left({}^P D_{*a}^{(N-1)\alpha} f\right)(w)\right| \left|{}_L D_a^{1-\alpha}\left({}^P D_{*a}^{N\alpha} f\right)(w)\right| e_0(x,w)\, d_\alpha w$$

$$\leq \int_a^x \left(\left(\int_a^w e_0(w,s)\, d_\alpha s\right) e_0(x,w)\, d_\alpha w\right)^{\frac{1}{p}}$$

$$\left\{\frac{\left(\int_a^x \left|{}_L D_a^{1-\alpha}\left({}^P D_{*a}^{N\alpha} f\right)(s)\right|^q e_0(x,s)\, d_\alpha s\right)^2}{2}\right.$$

$$\left. + \frac{1}{2}\left[\int_a^x \left(\int_a^w \left|{}_L D_a^{1-\alpha}\left({}^P D_{*a}^{N\alpha} f\right)(s)\right|^q e_0(w,s)\, d_\alpha s\right) k_1(\alpha,w)\, e_0(x,w)\, d_\alpha w\right]\right\}^{\frac{1}{q}}. \tag{14.68}$$

Proof. By (14.48), we have

$$ {}_{L}^{P}I_{a}^{RL}\left({}^{P}D_{*a}^{N\alpha}\right)f=\left({}^{P}D_{*a}^{(N-1)\alpha}\right)f,\quad on\ (a,b], \tag{14.69}$$

given that $f(a)=0$, under the assumption $\left(\left({}^{P}D_{*a}^{i\alpha}\right)f\right)\in C([a,b])$, $i=1,\ldots,N$.

That is,

$$\left({}^{P}D_{*a}^{(N-1)\alpha}f\right)(w)=\int_{a}^{w}\left({}_{L}D_{a}^{1-\alpha}\left({}^{P}D_{*a}^{N\alpha}f\right)\right)(s)\,e_{0}(w,s)\,d_{\alpha}s, \tag{14.70}$$

for all $a\lneqq w\leq x\leq b$.

Hence, by Hölder's inequality, we derive

$$\begin{aligned}\left|\left({}^{P}D_{*a}^{(N-1)\alpha}f\right)(w)\right| &\leq \int_{a}^{w}e_{0}(w,s)^{\frac{1}{p}}\left|{}_{L}D_{a}^{1-\alpha}\left({}^{P}D_{*a}^{N\alpha}f\right)(s)\right|e_{0}(w,s)^{\frac{1}{q}}\,d_{\alpha}s\\ &\leq\left(\int_{a}^{w}e_{0}(w,s)\,d_{\alpha}s\right)^{\frac{1}{p}}\left(\int_{a}^{w}\left|{}_{L}D_{a}^{1-\alpha}\left({}^{P}D_{*a}^{N\alpha}f\right)(s)\right|^{q}e_{0}(w,s)\,d_{\alpha}s\right)^{\frac{1}{q}}.\end{aligned} \tag{14.71}$$

Call

$$z(w):=\int_{a}^{w}\left|{}_{L}D_{a}^{1-\alpha}\left({}^{P}D_{*a}^{N\alpha}f\right)(s)\right|^{q}e_{0}(w,s)\,d_{\alpha}s,\quad z(a)=0, \tag{14.72}$$

$a\leq w\leq x$.

By (14.31), we have

$$D^{\alpha}z(w)=\left|{}_{L}D_{a}^{1-\alpha}\left({}^{P}D_{*a}^{N\alpha}f\right)(w)\right|^{q} \tag{14.73}$$

and

$$\left|{}_{L}D_{a}^{1-\alpha}\left({}^{P}D_{*a}^{N\alpha}f\right)(w)\right|=\left(D^{\alpha}z(w)\right)^{\frac{1}{q}}, \tag{14.74}$$

all $a\leq w\leq x$.

Therefore, we have (all $a\lneqq w\leq x$)

$$\begin{aligned}&\left|\left({}^{P}D_{*a}^{(N-1)\alpha}f\right)(w)\right|\left|{}_{L}D_{a}^{1-\alpha}\left({}^{P}D_{*a}^{N\alpha}f\right)(w)\right|e_{0}(x,w)\\ &\leq\left(\int_{a}^{w}e_{0}(w,s)\,d_{\alpha}s\right)^{\frac{1}{p}}\left(z(w)\,D^{\alpha}z(w)\right)^{\frac{1}{q}}e_{0}(x,w)\\ &=\left(\left(\int_{a}^{w}e_{0}(w,s)\,d_{\alpha}s\right)e_{0}(x,w)\right)^{\frac{1}{p}}\left(z(w)\,D^{\alpha}z(w)\,e_{0}(x,w)\right)^{\frac{1}{q}}.\end{aligned} \tag{14.75}$$

Next, we apply again Hölder's inequality and finally we use (14.35) to get that

$$\int_a^x \left|\left({}^P D_{*a}^{(N-1)\alpha} f\right)(w)\right| \left|{}_L D_a^{1-\alpha}\left({}^P D_{*a}^{N\alpha} f\right)(w)\right| e_0(x,w)\, d_\alpha w$$

$$\leq \int_a^x \left(\left(\int_a^w e_0(w,s)\, d_\alpha s\right) e_0(x,w)\right)^{\frac{1}{p}} \left(z(w)\, D^\alpha z(w)\, e_0(x,w)\right)^{\frac{1}{q}} d_\alpha w$$

$$\leq \left(\int_a^x \left(\left(\int_a^w e_0(w,s)\, d_\alpha s\right) e_0(x,w)\right) d_\alpha w\right)^{\frac{1}{p}}$$

$$\left(\int_a^x z(w)\, D^\alpha z(w)\, e_0(x,w)\, d_\alpha w\right)^{\frac{1}{q}} =: (\xi). \tag{14.76}$$

By (14.35), we derive that

$$\int_a^x z(w)\, D^\alpha z(w)\, e_0(x,w)\, d_\alpha w$$

$$= \frac{z^2(x)}{2} + \frac{1}{2}\int_a^x z^2(w)\, k_1(\alpha,w)\, e_0(x,w)\, d_\alpha w$$

$$= \frac{\left(\int_a^x \left|{}_L D_a^{1-\alpha}\left({}^P D_{*a}^{N\alpha} f\right)(s)\right|^q e_0(x,s)\, d_\alpha s\right)^2}{2}$$

$$+ \frac{1}{2}\left[\int_a^x \left(\int_a^w \left|{}_L D_a^{1-\alpha}\left({}^P D_{*a}^{N\alpha} f\right)(s)\right|^q e_0(w,s)\, d_\alpha s\right) k_1(\alpha,w)\, e_0(x,w)\, d_\alpha w\right]. \tag{14.77}$$

Consequently, we obtain that

$$(\xi) = \left(\int_a^x \left(\int_a^w e_0(w,s)\, d_\alpha s\right) e_0(x,w)\, d_\alpha w\right)^{\frac{1}{p}}$$

$$\left\{\frac{\left(\int_a^x \left|{}_L D_a^{1-\alpha}\left({}^P D_{*a}^{N\alpha} f\right)(s)\right|^q e_0(x,s)\, d_\alpha s\right)^2}{2}\right.$$

$$\left.+ \frac{1}{2}\left[\int_a^x \left(\int_a^w \left|{}_L D_a^{1-\alpha}\left({}^P D_{*a}^{N\alpha} f\right)(s)\right|^q e_0(w,s)\, d_\alpha s\right) k_1(\alpha,w)\, e_0(x,w)\, d_\alpha w\right]\right\}^{\frac{1}{q}}. \tag{14.78}$$

The claim is proved. □

We continue with a left Poincaré-type inequality.

Theorem 14.4. *Here* $\alpha \in (0,1)$, $f \in C^1([a,b])$; $p,q > 1 : \frac{1}{p} + \frac{1}{q} = 1$, $N \in \mathbb{N}$. *Assume that* $\alpha > \frac{1}{q}$; $f(a) = 0$, $\left(D_{*a}^{i\alpha} f\right) \in C([a,b])$, $i = 1, \ldots, N$.

Then

$$\left\| D_{*a}^{(N-1)\alpha} f \right\|_{L_q([a,b])} \leq \frac{(b-a)^{\alpha}}{\Gamma(\alpha)\,(p(\alpha-1)+1)^{\frac{1}{p}}\,(q\alpha)^{\frac{1}{q}}} \left\| D_{*a}^{N\alpha} f \right\|_{L_q([a,b])}. \tag{14.79}$$

Proof. As in (14.61) we have

$$\left| \left(D_{*a}^{(N-1)\alpha} f \right)(x) \right| \leq \frac{1}{\Gamma(\alpha)} \frac{(x-a)^{\frac{p(\alpha-1)+1}{p}}}{(p(\alpha-1)+1)^{\frac{1}{p}}} \left(\int_a^x \left| \left(D_{*a}^{N\alpha} f \right)(t) \right|^q dt \right)^{\frac{1}{q}}$$

$$\leq \frac{(x-a)^{\alpha-\frac{1}{q}}}{\Gamma(\alpha)\,(p(\alpha-1)+1)^{\frac{1}{p}}} \left\| D_{*a}^{N\alpha} f \right\|_{L_q([a,b])}, \tag{14.80}$$

$\forall\, x \in [a,b]$.

Hence it holds

$$\left| \left(D_{*a}^{(N-1)\alpha} f \right)(x) \right|^q \leq \frac{(x-a)^{q\alpha-1}}{\Gamma(\alpha)^q\,(p(\alpha-1)+1)^{\frac{q}{p}}} \left\| D_{*a}^{N\alpha} f \right\|_{L_q([a,b])}^q \tag{14.81}$$

and

$$\int_a^b \left| \left(D_{*a}^{(N-1)\alpha} f \right)(x) \right|^q dx \leq \frac{(b-a)^{q\alpha}}{\Gamma(\alpha)^q\,(p(\alpha-1)+1)^{\frac{q}{p}}\,q\alpha} \left\| D_{*a}^{N\alpha} f \right\|_{L_q([a,b])}^q. \tag{14.82}$$

The last results into

$$\left(\int_a^b \left| \left(D_{*a}^{(N-1)\alpha} f \right)(x) \right|^q dx \right)^{\frac{1}{q}}$$

$$\leq \frac{(b-a)^{\alpha}}{\Gamma(\alpha)\,(p(\alpha-1)+1)^{\frac{1}{p}}\,(q\alpha)^{\frac{1}{q}}} \left\| D_{*a}^{N\alpha} f \right\|_{L_q([a,b])}. \tag{14.83}$$

The claim is proved. □

Next comes a right Poincaré-type inequality.

Theorem 14.5. *Again* $\alpha \in (0,1)$, $f \in C^1([a,b])$; $p, q > 1 : \frac{1}{p} + \frac{1}{q} = 1$, $N \in \mathbb{N}$. *Assume that* $\alpha > \frac{1}{q}$; $f(b) = 0$, $\left(D_{b-}^{i\alpha} f \right) \in C([a,b])$, $i = 1, \ldots, N$. *Then*

$$\left\| D_{b-}^{(N-1)\alpha} f \right\|_{L_q([a,b])} \leq \frac{(b-a)^{\alpha}}{\Gamma(\alpha)\,(p(\alpha-1)+1)^{\frac{1}{p}}\,(q\alpha)^{\frac{1}{q}}} \left\| D_{b-}^{N\alpha} f \right\|_{L_q([a,b])}. \tag{14.84}$$

Proof. As similar to Theorem 14.4 is omitted. It is based on (14.24). □

Next comes a left Sobolev-type inequality.

Theorem 14.6. *Here all as in Theorem 14.4, $r > 0$. Then*

$$\left\| D_{*a}^{(N-1)\alpha} f \right\|_{L_r([a,b])} \leq \frac{(b-a)^{\alpha - \frac{1}{q} + \frac{1}{r}}}{\Gamma(\alpha)\left(p(\alpha-1)+1\right)^{\frac{1}{p}} \left(r\left(\alpha - \frac{1}{q}\right)+1\right)^{\frac{1}{r}}} \left\| D_{*a}^{N\alpha} f \right\|_{L_q([a,b])}. \quad (14.85)$$

Proof. From (14.80), we get ($r > 0$)

$$\left| \left(D_{*a}^{(N-1)\alpha} f \right)(x) \right|^r \leq \frac{(x-a)^{r\left(\alpha - \frac{1}{q}\right)}}{\Gamma(\alpha)^r \left(p(\alpha-1)+1\right)^{\frac{r}{p}}} \left\| D_{*a}^{N\alpha} f \right\|_{L_q([a,b])}^r, \quad (14.86)$$

$\forall\, x \in [a,b]$.

Hence it holds

$$\int_a^b \left| \left(D_{*a}^{(N-1)\alpha} f \right)(x) \right|^r dx \leq \frac{(b-a)^{r\left(\alpha - \frac{1}{q}\right)+1} \left\| D_{*a}^{N\alpha} f \right\|_{L_q([a,b])}^r}{\Gamma(\alpha)^r \left(p(\alpha-1)+1\right)^{\frac{r}{p}} \left(r\left(\alpha - \frac{1}{q}\right)+1\right)}. \quad (14.87)$$

That is,

$$\left(\int_a^b \left| \left(D_{*a}^{(N-1)\alpha} f \right)(x) \right|^r dx \right)^{\frac{1}{r}} \leq \frac{(b-a)^{\alpha - \frac{1}{q} + \frac{1}{r}}}{\Gamma(\alpha)\left(p(\alpha-1)+1\right)^{\frac{1}{p}} \left(r\left(\alpha - \frac{1}{q}\right)+1\right)^{\frac{1}{r}}} \left\| D_{*a}^{N\alpha} f \right\|_{L_q([a,b])}. \quad (14.88)$$

The claim is proved. □

It follows the right Sobolev inequality.

Theorem 14.7. *Here all as in Theorem 14.5, $r > 0$. Then*

$$\left\| D_{b-}^{(N-1)\alpha} f \right\|_{L_r([a,b])} \leq \frac{(b-a)^{\alpha - \frac{1}{q} + \frac{1}{r}}}{\Gamma(\alpha)\left(p(\alpha-1)+1\right)^{\frac{1}{p}} \left(r\left(\alpha - \frac{1}{q}\right)+1\right)^{\frac{1}{r}}} \left\| D_{b-}^{N\alpha} f \right\|_{L_q([a,b])}. \quad (14.89)$$

Proof. As similar to Theorem 14.6 is omitted. □

Next comes a left proportional Caputo fractional Poincaré-type inequality.

Theorem 14.8. *Here* $\alpha \in (0,1)$, $f \in C^1([a,b])$; $p,q > 1 : \frac{1}{p} + \frac{1}{q} = 1$, $N \in \mathbb{N}$. *We assume* $f(a) = 0$, ${}^P D_{*a}^{i\alpha} f \in C([a,b])$, $i = 1, \ldots, N$. *Then*

$$\left\| {}^P D_{*a}^{(N-1)\alpha} f \right\|_{L_q([a,b],d_\alpha x)} \leq \left(\int_a^b \left(\int_a^x e_0(x,s)\, d_\alpha s \right)^{\frac{q}{p}} d_\alpha x \right)^{\frac{1}{q}} \left\| {}_L D_a^{1-\alpha} \left({}^P D_{*a}^{N\alpha} f \right) \right\|_{L_q([a,b],d_\alpha x)}. \tag{14.90}$$

Proof. As in (14.71) we have

$$\begin{aligned} &\left| \left({}^P D_{*a}^{(N-1)\alpha} f \right)(x) \right| \\ &\leq \left(\int_a^x e_0(x,s)\, d_\alpha s \right)^{\frac{1}{p}} \left(\int_a^x \left| {}_L D_a^{1-\alpha} \left({}^P D_{*a}^{N\alpha} f \right)(s) \right|^q e_0(x,s)\, d_\alpha s \right)^{\frac{1}{q}} \\ &\leq \left(\int_a^x e_0(x,s)\, d_\alpha s \right)^{\frac{1}{p}} \left(\int_a^x \left| {}_L D_a^{1-\alpha} \left({}^P D_{*a}^{N\alpha} f \right)(s) \right|^q d_\alpha s \right)^{\frac{1}{q}} \\ &\leq \left(\int_a^x e_0(x,s)\, d_\alpha s \right)^{\frac{1}{p}} \left(\int_a^b \left| {}_L D_a^{1-\alpha} \left({}^P D_{*a}^{N\alpha} f \right)(s) \right|^q d_\alpha s \right)^{\frac{1}{q}} \\ &= \left(\int_a^x e_0(x,s)\, d_\alpha s \right)^{\frac{1}{p}} \left\| {}_L D_a^{1-\alpha} \left({}^P D_{*a}^{N\alpha} f \right) \right\|_{L_q([a,b],d_\alpha s)}. \end{aligned} \tag{14.91}$$

That is,

$$\left| \left({}^P D_{*a}^{(N-1)\alpha} f \right)(x) \right| \leq \left(\int_a^x e_0(x,s)\, d_\alpha s \right)^{\frac{1}{p}} \left\| {}_L D_a^{1-\alpha} \left({}^P D_{*a}^{N\alpha} f \right) \right\|_{L_q([a,b],d_\alpha s)}, \tag{14.92}$$

$\forall\, x \in (a,b]$.

Hence it holds

$$\left| \left({}^P D_{*a}^{(N-1)\alpha} f \right)(x) \right|^q \leq \left(\int_a^x e_0(x,s)\, d_\alpha s \right)^{\frac{q}{p}} \left\| {}_L D_a^{1-\alpha} \left({}^P D_{*a}^{N\alpha} f \right) \right\|_{L_q([a,b],d_\alpha s)}^q \tag{14.93}$$

and

$$\int_a^b \left| \left({}^P D_{*a}^{(N-1)\alpha} f \right)(x) \right|^q d_\alpha x$$

$$\leq \int_a^b \left(\int_a^x e_0(x,s)\, d_\alpha s \right)^{\frac{q}{p}} d_\alpha x \left\| {}_L D_a^{1-\alpha} \left({}^P D_{*a}^{N\alpha} f \right) \right\|_{L_q([a,b],d_\alpha s)}^q. \tag{14.94}$$

We derive

$$\left(\int_a^b \left| \left({}^P D_{*a}^{(N-1)\alpha} f \right)(x) \right|^q d_\alpha x \right)^{\frac{1}{q}}$$

$$\leq \left(\int_a^b \left(\int_a^x e_0(x,s)\, d_\alpha s \right)^{\frac{q}{p}} d_\alpha x \right)^{\frac{1}{q}} \left\| {}_L D_a^{1-\alpha} \left({}^P D_{*a}^{N\alpha} f \right) \right\|_{L_q([a,b],d_\alpha s)},$$

proving the claim. □

It follows a right proportional Caputo fractional Poincarê-type inequality.

Theorem 14.9. *Here* $\alpha \in (0,1)$, $f \in C^1([a,b])$; $p,q > 1 : \frac{1}{p} + \frac{1}{q} = 1$, $f(b) = 0$. *Then*

$$\left\| f e_0(b,\cdot) \right\|_{L_q([a,b],d_\alpha t)}$$

$$\leq \left(\int_a^b \left(\int_t^b e_0(b,s)\, d_\alpha s \right)^{\frac{q}{p}} d_\alpha x \right)^{\frac{1}{q}} \left\| {}_R D_{b-}^{1-\alpha} \left({}^P D_{b-}^{\alpha} f \right) \right\|_{L_q([a,b],e_0(b,\cdot)d_\alpha t)}. \tag{14.95}$$

Proof. By (14.57), when $f(b) = 0$, we have

$$f(t)\, e_0(b,t) = {}_R^P I_b^{RL} \left({}^P D_{b-}^{\alpha} (f(t)) \right)$$

$$= \int_t^b \left({}_R D_{b-}^{1-\alpha} \left({}^P D_{b-}^{\alpha} f \right) \right)(s)\, e_0(b,s)\, d_\alpha s, \tag{14.96}$$

$\forall\, t \in [a,b]$.

Hence by Hölder's inequality,

$$|f(t)| e_0(b,t) \leq \int_t^b e_0(b,s)^{\frac{1}{p}} \left| \left({}_R D_{b-}^{1-\alpha} \left({}^P D_{b-}^{\alpha} f \right) \right)(s) \right| e_0(b,s)^{\frac{1}{q}} d_\alpha s \tag{14.97}$$

$$\leq \left(\int_t^b e_0(b,s) d_\alpha s \right)^{\frac{1}{p}} \left(\int_t^b \left| \left({}_R D_{b-}^{1-\alpha} \left({}^P D_{b-}^{\alpha} f \right) \right)(s) \right|^q e_0(b,s) d_\alpha s \right)^{\frac{1}{q}}$$

$$\leq \left(\int_t^b e_0(b,s) d_\alpha s \right)^{\frac{1}{p}} \left\| {}_R D_{b-}^{1-\alpha} \left({}^P D_{b-}^{\alpha} f \right) \right\|_{L_q([a,b], e_0(b,\cdot) d_\alpha t)}, \tag{14.98}$$

$\forall\, t \in [a,b]$.

Hence

$$(|f(t)| e_0(b,t))^q \leq \left(\int_t^b e_0(b,s) d_\alpha s \right)^{\frac{q}{p}} \left\| {}_R D_{b-}^{1-\alpha} \left({}^P D_{b-}^{\alpha} f \right) \right\|_{L_q([a,b], e_0(b,\cdot) d_\alpha s)}^q \tag{14.99}$$

and

$$\int_a^b (|f(t)| e_0(b,t))^q d_\alpha t$$

$$\leq \left(\int_a^b \left(\int_t^b e_0(b,s) d_\alpha s \right)^{\frac{q}{p}} d_\alpha t \right) \left\| {}_R D_{b-}^{1-\alpha} \left({}^P D_{b-}^{\alpha} f \right) \right\|_{L_q([a,b], e_0(b,\cdot) d_\alpha t)}^q. \tag{14.100}$$

We obtain

$$\left(\int_a^b (|f(t)| e_0(b,t))^q d_\alpha t \right)^{\frac{1}{q}}$$

$$\leq \left(\int_a^b \left(\int_t^b e_0(b,s) d_\alpha s \right)^{\frac{q}{p}} d_\alpha t \right)^{\frac{1}{q}} \left\| {}_R D_{b-}^{1-\alpha} \left({}^P D_{b-}^{\alpha} f \right) \right\|_{L_q([a,b], e_0(b,\cdot) d_\alpha t)}. \tag{14.101}$$

The claim is proved. □

A left proportional fractional Sobolev-type inequality follows:

Theorem 14.10. *All as in Theorem 14.8, $r > 0$. Then*

$$\left\| {}^{P}D_{*a}^{(N-1)\alpha} f \right\|_{L_r([a,b])} \leq \left(\int_a^b \left(\int_a^x e_0(x,s)\, d_\alpha s \right)^{\frac{r}{p}} dx \right)^{\frac{1}{r}} \left\| {}_L D_a^{1-\alpha} \left({}^{P}D_{*a}^{N\alpha} f \right) \right\|_{L_q([a,b],d_\alpha s)}. \tag{14.102}$$

Proof. By (14.92), we get

$$\left| \left({}^{P}D_{*a}^{(N-1)\alpha} f \right)(x) \right|^r \leq \left(\int_a^x e_0(x,s)\, d_\alpha s \right)^{\frac{r}{p}} \left\| {}_L D_a^{1-\alpha} \left({}^{P}D_{*a}^{N\alpha} f \right) \right\|^r_{L_q([a,b],d_\alpha s)}, \tag{14.103}$$

$\forall\, x \in (a,b]$.

Hence it holds

$$\int_a^b \left| \left({}^{P}D_{*a}^{(N-1)\alpha} f \right)(x) \right|^r dx \leq \left(\int_a^b \left(\int_a^x e_0(x,s)\, d_\alpha s \right)^{\frac{r}{p}} dx \right) \left\| {}_L D_a^{1-\alpha} \left({}^{P}D_{*a}^{N\alpha} f \right) \right\|^r_{L_q([a,b],d_\alpha s)}. \tag{14.104}$$

That is,

$$\left(\int_a^b \left| \left({}^{P}D_{*a}^{(N-1)\alpha} f \right)(x) \right|^r dx \right)^{\frac{1}{r}} \leq \left(\int_a^b \left(\int_a^x e_0(x,s)\, d_\alpha s \right)^{\frac{r}{p}} dx \right)^{\frac{1}{r}} \left\| {}_L D_a^{1-\alpha} \left({}^{P}D_{*a}^{N\alpha} f \right) \right\|_{L_q([a,b],d_\alpha s)}. \tag{14.105}$$

The claim is established. □

A right proportional fractional Sobolev-type inequality follows:

Theorem 14.11. *All as in Theorem 14.9, $r > 0$. Then*

$$\left\| f e_0(b,\cdot) \right\|_{L_r([a,b])} \leq \left(\int_a^b \left(\int_t^b e_0(b,s)\, d_\alpha s \right)^{\frac{r}{p}} dt \right)^{\frac{1}{r}} \left\| {}_R D_{b-}^{1-\alpha} \left({}^{P}D_{b-}^{\alpha} f \right) \right\|_{L_q([a,b],e_0(b,\cdot)d_\alpha s)}. \tag{14.106}$$

Proof. From (14.98), we have

$$\left(|f(t)|\, e_0(b,t)\right)^r \le \left(\int_t^b e_0(b,s)\, d_\alpha s\right)^{\frac{r}{p}} \left\| {}_R D_{b-}^{1-\alpha}\left({}^P D_{b-}^{\alpha} f\right)\right\|_{L_q([a,b],e_0(b,\cdot)d_\alpha s)}^r, \tag{14.107}$$

$\forall\, t \in [a,b]$.

Hence it holds

$$\int_a^b \left(|f(t)|\, e_0(b,t)\right)^r dt \le \left(\int_a^b \left(\int_t^b e_0(b,s)\, d_\alpha s\right)^{\frac{r}{p}} dt\right) \left\| {}_R D_{b-}^{1-\alpha}\left({}^P D_{b-}^{\alpha} f\right)\right\|_{L_q([a,b],e_0(b,\cdot)d_\alpha s)}^r. \tag{14.108}$$

That is,

$$\left(\int_a^b \left(|f(t)|\, e_0(b,t)\right)^r dt\right)^{\frac{1}{r}} \le \left(\int_a^b \left(\int_t^b e_0(b,s)\, d_\alpha s\right)^{\frac{r}{p}} dt\right)^{\frac{1}{r}} \left\| {}_R D_{b-}^{1-\alpha}\left({}^P D_{b-}^{\alpha} f\right)\right\|_{L_q([a,b],e_0(b,\cdot)d_\alpha s)}. \tag{14.109}$$

□

Next we proceed with a left fractional Hilbert–Pachpatte-type inequality.

Theorem 14.12. *Here* $j = 1,2$; $\alpha_j \in (0,1)$, $f_j \in C^1([a_j,b_j])$; $p,q > 1$: $\frac{1}{p}+\frac{1}{q}=1$, $N_j \in \mathbb{N}$. *Assume* $\alpha_1 > \frac{1}{q}$, $\alpha_2 > \frac{1}{p}$; $f_j(a_j) = 0$, $\left(D_{*a_j}^{i_j\alpha_j} f_j\right) \in C([a_j,b_j])$, $i_j = 1,\ldots,N_j$. *Then*

$$\int_{a_1}^{b_1}\int_{a_2}^{b_2} \frac{\left|\left(D_{*a_1}^{(N_1-1)\alpha_1} f_1\right)(x_1)\right| \left|\left(D_{*a_2}^{(N_2-1)\alpha_2} f_2\right)(x_2)\right| dx_1 dx_2}{\left[\frac{(x_1-a_1)^{p(\alpha_1-1)+1}}{p(p(\alpha_1-1)+1)} + \frac{(x_2-a_2)^{q(x_2-1)+1}}{q(q(\alpha_2-1)+1)}\right]} \le \frac{(b_1-a_1)(b_2-a_2)}{\Gamma(\alpha_1)\Gamma(\alpha_2)} \left\|D_{*a_1}^{N_1\alpha_1} f_1\right\|_{L_q([a_1,b_1])} \left\|D_{*a_2}^{N_2\alpha_2} f_2\right\|_{L_p([a_2,b_2])}. \tag{14.110}$$

Proof. We have by (14.23) that $(j = 1, 2)$

$$\left(D_{*a_j}^{(N_j-1)\alpha_j} f_j\right)(x_j) = \frac{1}{\Gamma(\alpha_j)} \int_{a_j}^{x_j} (x_j - t_j)^{\alpha_j - 1} \left(D_{*a_j}^{N_j\alpha_j} f_j\right)(t_j)\, dt_j, \tag{14.111}$$

$\forall\ x_j \in [a_j, b_j]$.

Then

$$\left|\left(D_{*a_j}^{(N_j-1)\alpha_j} f_j\right)(x_j)\right| \leq \frac{1}{\Gamma(\alpha_j)} \int_{a_j}^{x_j} (x_j - t_j)^{\alpha_j - 1} \left|\left(D_{*a_j}^{N_j\alpha_j} f_j\right)(t_j)\right| dt_j, \tag{14.112}$$

$j = 1, 2$, $\forall\ x_j \in [a_j, b_j]$.

We get as before

$$\left|\left(D_{*a_1}^{(N_1-1)\alpha_1} f_1\right)(x_1)\right| \leq \frac{1}{\Gamma(\alpha_1)} \frac{(x_1 - a_1)^{\frac{p(\alpha_1-1)+1}{p}}}{(p(\alpha_1 - 1) + 1)^{\frac{1}{p}}} \left\|D_{*a_1}^{N_1\alpha_1} f_1\right\|_{L_q([a_1,b_1])} \tag{14.113}$$

and

$$\left|\left(D_{*a_2}^{(N_2-1)\alpha_2} f_2\right)(x_2)\right| \leq \frac{1}{\Gamma(\alpha_2)} \frac{(x_2 - a_2)^{\frac{q(\alpha_2-1)+1}{q}}}{(q(\alpha_2 - 1) + 1)^{\frac{1}{q}}} \left\|D_{*a_2}^{N_2\alpha_2} f_2\right\|_{L_p([a_2,b_2])}. \tag{14.114}$$

Hence we have

$$\begin{aligned}
&\left|\left(D_{*a_1}^{(N_1-1)\alpha_1} f_1\right)(x_1)\right| \left|\left(D_{*a_2}^{(N_2-1)\alpha_2} f_2\right)(x_2)\right| \\
&\leq \frac{(x_1 - a_1)^{\frac{p(\alpha_1-1)+1}{p}} (x_2 - a_2)^{\frac{q(\alpha_2-1)+1}{q}}}{\Gamma(\alpha_1)\Gamma(\alpha_2)(p(\alpha_1 - 1) + 1)^{\frac{1}{p}} (q(\alpha_2 - 1) + 1)^{\frac{1}{q}}} \\
&\quad \left\|D_{*a_1}^{N_1\alpha_1} f_1\right\|_{L_q([a_1,b_1])} \left\|D_{*a_2}^{N_2\alpha_2} f_2\right\|_{L_p([a_2,b_2])}
\end{aligned} \tag{14.115}$$

(using Young's inequality for $a^*, b^* \geq 0$, $a^{*\frac{1}{p}} b^{*\frac{1}{q}} \leq \frac{a^*}{p} + \frac{b^*}{q}$)

$$\begin{aligned}
&\leq \frac{1}{\Gamma(\alpha_1)\Gamma(\alpha_2)} \left(\frac{(x_1 - a_1)^{p(\alpha_1-1)+1}}{p(p(\alpha_1 - 1) + 1)} + \frac{(x_2 - a_2)^{q(\alpha_2-1)+1}}{q(q(\alpha_2 - 1) + 1)} \right) \\
&\quad \left\|D_{*a_1}^{N_1\alpha_1} f_1\right\|_{L_q([a_1,b_1])} \left\|D_{*a_2}^{N_2\alpha_2} f_2\right\|_{L_p([a_2,b_2])},
\end{aligned} \tag{14.116}$$

$\forall\ x_j \in [a_j, b_j]$; $j = 1, 2$.

So far we have

$$\frac{\left|\left(D_{*a_1}^{(N_1-1)\alpha_1} f_1\right)(x_1)\right| \left|\left(D_{*a_2}^{(N_2-1)\alpha_2} f_2\right)(x_2)\right|}{\left[\frac{(x_1-a_1)^{p(\alpha_1-1)+1}}{p(p(\alpha_1-1)+1)} + \frac{(x_2-a_2)^{q(\alpha_2-1)+1}}{q(q(\alpha_2-1)+1)}\right]}$$

$$\leq \frac{\left\|D_{*a_1}^{N_1\alpha_1} f_1\right\|_{L_q([a_1,b_1])} \left\|D_{*a_2}^{N_2\alpha_2} f_2\right\|_{L_p([a_2,b_2])}}{\Gamma(\alpha_1)\Gamma(\alpha_2)}, \tag{14.117}$$

$\forall\, x_j \in [a_j, b_j]$; $j = 1, 2$.

The denominator in (14.117) can be zero only when both $x_1 = a_1$ and $x_2 = a_2$. Therefore, we obtain (14.110), by integrating (14.117) over $[a_1, b_1] \times [a_2, b_2]$. □

The right fractional Hilbert–Pachpatte-type inequality follows:

Theorem 14.13. *Here* $j = 1, 2$; $\alpha_j \in (0,1)$, $f_j \in C^1([a_j, b_j])$; $p, q > 1$: $\frac{1}{p} + \frac{1}{q} = 1$, $N_j \in \mathbb{N}$. *Assume* $\alpha_1 > \frac{1}{q}$, $\alpha_2 > \frac{1}{p}$; $f_j(b_j) = 0$, $\left(D_{b_j-}^{i_j\alpha_j} f_j\right) \in C([a_j, b_j])$, $i_j = 1, \ldots, N_j$. *Then*

$$\int_{a_1}^{b_1}\int_{a_2}^{b_2} \frac{\left|\left(D_{b_1-}^{(N_1-1)\alpha_1} f_1\right)(x_1)\right| \left|\left(D_{b_2-}^{(N_2-1)\alpha_2} f_2\right)(x_2)\right| dx_1 dx_2}{\left[\frac{(b_1-x_1)^{p(\alpha_1-1)+1}}{p(p(\alpha_1-1)+1)} + \frac{(b_2-x_2)^{q(\alpha_2-1)+1}}{q(q(\alpha_2-1)+1)}\right]}$$

$$\leq \frac{(b_1-a_1)(b_2-a_2)}{\Gamma(\alpha_1)\Gamma(\alpha_2)} \left\|D_{b_1-}^{N_1\alpha_1} f_1\right\|_{L_q([a_1,b_1])} \left\|D_{b_2-}^{N_2\alpha_2} f_2\right\|_{L_p([a_2,b_2])}. \tag{14.118}$$

Proof. As similar to Theorem 14.12 is omitted. It is based on (14.24). □

Next we continue with a left proportional fractional Hilbert–Pachpatte inequality.

Theorem 14.14. *Here it is* $j = 1, 2$. *Let* $f_j \in C^1([a_j, b_j])$, $\alpha_j \in (0,1)$, $f_j(a_j) = 0$; *and* $\left({}^P D_{*a_j}^{i_j\alpha_j} f_j\right) \in C([a_j, b_j])$, *for* $i_j = 1, \ldots, N_j \in \mathbb{N}$. *Then*

$$\int_{a_1}^{b_1}\int_{a_2}^{b_2} \frac{\left|\left({}^P D_{*a_1}^{(N_1-1)\alpha_1} f_1\right)(x_1)\right| \left|\left({}^P D_{*a_2}^{(N_2-1)\alpha_2} f_2\right)(x_2)\right| dx_1 dx_2}{\left[\frac{\int_{a_1}^{x_1} e_0(x_1,s_1)d_{\alpha_1}s_1}{p} + \frac{\int_{a_2}^{x_2} e_0(x_2,s_2)d_{\alpha_2}s_2}{q}\right]}$$

$$\leq (b_1-a_1)(b_2-a_2) \left\|{}_L D_{a_1}^{1-\alpha_1}\left({}^P D_{*a_1}^{N_1\alpha_1} f_1\right)\right\|_{L_q\left([a_1,b_1],d_{\alpha_1}s_1\right)}$$

$$\left\|{}_L D_{a_2}^{1-\alpha_2}\left({}^P D_{*a_2}^{N_2\alpha_2} f_2\right)\right\|_{L_p\left([a_2,b_2],d_{\alpha_2}s_2\right)}. \tag{14.119}$$

Proof. Based on (14.48) ($j = 1, 2$) and as in (14.92), we have

$$\left|\left({}^P D_{*a_1}^{(N_1-1)\alpha_1} f_1\right)(x_1)\right|$$
$$\leq \left(\int_{a_1}^{x_1} e_0(x_1, s_1)\, d_{\alpha_1} s_1\right)^{\frac{1}{p}} \left\|{}_L D_{a_1}^{1-\alpha_1}\left({}^P D_{*a_1}^{N_1\alpha_1} f_1\right)\right\|_{L_q\left([a_1,b_1],d_{\alpha_1}s_1\right)}, \tag{14.120}$$

$\forall\, x_1 \in (a_1, b_1]$,

and

$$\left|\left({}^P D_{*a_2}^{(N_2-1)\alpha_2} f_2\right)(x_2)\right|$$
$$\leq \left(\int_{a_2}^{x_2} e_0(x_2, s_2)\, d_{\alpha_2} s_2\right)^{\frac{1}{q}} \left\|{}_L D_{a_2}^{1-\alpha_2}\left({}^P D_{*a_2}^{N_2\alpha_2} f_2\right)\right\|_{L_p\left([a_2,b_2],d_{\alpha_2}s_2\right)}, \tag{14.121}$$

$\forall\, x_2 \in (a_2, b_2]$.

Therefore, we have

$$\left|\left({}^P D_{*a_1}^{(N_1-1)\alpha_1} f_1\right)(x_1)\right| \left|\left({}^P D_{*a_2}^{(N_2-1)\alpha_2} f_2\right)(x_2)\right|$$
$$\leq \left[\left(\int_{a_1}^{x_1} e_0(x_1, s_1)\, d_{\alpha_1} s_1\right)^{\frac{1}{p}} \left(\int_{a_2}^{x_2} e_0(x_2, s_2)\, d_{\alpha_2} s_2\right)^{\frac{1}{q}}\right]$$
$$\times \left\|{}_L D_{a_1}^{1-\alpha_1}\left({}^P D_{*a_1}^{N_1\alpha_1} f_1\right)\right\|_{L_q\left([a_1,b_1],d_{\alpha_1}s_1\right)}$$
$$\times \left\|{}_L D_{a_2}^{1-\alpha_2}\left({}^P D_{*a_2}^{N_2\alpha_2} f_2\right)\right\|_{L_p\left([a_2,b_2],d_{\alpha_2}s_2\right)}$$
$$\leq \left[\frac{\int_{a_1}^{x_1} e_0(x_1, s_1)\, d_{\alpha_1} s_1}{p} + \frac{\int_{a_2}^{x_2} e_0(x_2, s_2)\, d_{\alpha_2} s_2}{q}\right]$$
$$\times \left\|{}_L D_{a_1}^{1-\alpha_1}\left({}^P D_{*a_1}^{N_1\alpha_1} f_1\right)\right\|_{L_q\left([a_1,b_1],d_{\alpha_1}s_1\right)}$$
$$\times \left\|{}_L D_{a_2}^{1-\alpha_2}\left({}^P D_{*a_2}^{N_2\alpha_2} f_2\right)\right\|_{L_p\left([a_2,b_2],d_{\alpha_2}s_2\right)}. \tag{14.122}$$

Consequently, it holds

$$\frac{\left|\left({}^{P}D_{*a_1}^{(N_1-1)\alpha_1} f_1\right)(x_1)\right| \left|\left({}^{P}D_{*a_2}^{(N_2-1)\alpha_2} f_2\right)(x_2)\right|}{\left[\frac{\int_{a_1}^{x_1} e_0(x_1,s_1) d_{\alpha_1} s_1}{p} + \frac{\int_{a_2}^{x_2} e_0(x_2,s_2) d_{\alpha_2} s_2}{q}\right]}$$
$$\leq \left\| {}_{L}D_{a_1}^{1-\alpha_1} \left({}^{P}D_{*a_1}^{N_1\alpha_1} f_1\right) \right\|_{L_q\left([a_1,b_1], d_{\alpha_1} s_1\right)}$$
$$\times \left\| {}_{L}D_{a_2}^{1-\alpha_2} \left({}^{P}D_{*a_2}^{N_2\alpha_2} f_2\right) \right\|_{L_p\left([a_2,b_2], d_{\alpha_2} s_2\right)}. \quad (14.123)$$

The denominator of (14.123) is zero only when both $a_1 = x_1$ and $a_2 = x_2$.

Inequality (14.119) comes after integrating (14.123) over $[a_1, b_1] \times [a_2, b_2]$. □

We finish with the right proportional fractional Hilbert–Pachpatte inequality.

Theorem 14.15. *Let $j = 1, 2$, and $\alpha_j \in (0, 1)$, $f_j \in C^1([a_j, b_j])$; $p, q > 1$: $\frac{1}{p} + \frac{1}{q} = 1$, $f_j(b_j) = 0$. Then*

$$\int_{a_1}^{b_1} \int_{a_2}^{b_2} \frac{|f_1(t_1)| |f_2(t_2)| e_0(b_1, t_1) e_0(b_2, t_2) dt_1 dt_2}{\left[\frac{\int_{t_1}^{b_1} e_0(b_1,s_1) d_{\alpha_1} s_1}{p} + \frac{\int_{t_2}^{b_2} e_0(b_2,s_2) d_{\alpha_2} s_2}{q}\right]}$$
$$\leq (b_1 - a_1)(b_2 - a_2) \left\| {}_{R}D_{b_1-}^{1-\alpha_1} \left({}^{P}D_{b_1-}^{\alpha_1} f_1\right) \right\|_{L_q\left([a_1,b_1], e_0(b_1,\cdot) d_{\alpha_1} s_1\right)}$$
$$\left\| {}_{R}D_{b_2-}^{1-\alpha_2} \left({}^{P}D_{b_2-}^{\alpha_2} f_2\right) \right\|_{L_q\left([a_2,b_2], e_0(b_2,\cdot) d_{\alpha_2} s_2\right)}. \quad (14.124)$$

Proof. Similar to (14.98), based on (14.57), we derive

$$|f_1(t_1)| e_0(b_1, t_1)$$
$$\leq \left(\int_{t_1}^{b_1} e_0(b_1, s_1) d_{\alpha_1} s_1\right)^{\frac{1}{p}} \left\| {}_{R}D_{b_1-}^{1-\alpha_1} \left({}^{P}D_{b_1-}^{\alpha_1} f_1\right) \right\|_{L_q\left([a_1,b_1], e_0(b_1,\cdot) d_{\alpha_1} s_1\right)} \quad (14.125)$$

and

$$|f_2(t_2)| e_0(b_2, t_2)$$
$$\leq \left(\int_{t_2}^{b_2} e_0(b_2, s_2) d_{\alpha_2} s_2\right)^{\frac{1}{q}} \left\| {}_{R}D_{b_2-}^{1-\alpha_2} \left({}^{P}D_{b_2-}^{\alpha_2} f_2\right) \right\|_{L_q\left([a_2,b_2], e_0(b_2,\cdot) d_{\alpha_2} s_2\right)}. \quad (14.126)$$

Hence it holds

$$\left|f_1(t_1)\right|\left|f_2(t_2)\right| e_0(b_1,t_1)\, e_0(b_2,t_2)$$
$$\leq \left(\int_{t_1}^{b_1} e_0(b_1,s_1)\, d_{\alpha_1} s_1\right)^{\frac{1}{p}} \left(\int_{t_2}^{b_2} e_0(b_2,s_2)\, d_{\alpha_2} s_2\right)^{\frac{1}{q}}$$
$$\left\| {}_R D_{b_1-}^{1-\alpha_1} \left({}^P D_{b_1-}^{\alpha_1} f_1\right)\right\|_{L_q\left([a_1,b_1], e_0(b_1,\cdot) d_{\alpha_1} s_1\right)}$$
$$\left\| {}_R D_{b_2-}^{1-\alpha_2} \left({}^P D_{b_2-}^{\alpha_2} f_2\right)\right\|_{L_q\left([a_2,b_2], e_0(b_2,\cdot) d_{\alpha_2} s_2\right)} \tag{14.127}$$
$$\leq \left[\frac{\int_{t_1}^{b_1} e_0(b_1,s_1)\, d_{\alpha_1} s_1}{p} + \frac{\int_{t_2}^{b_2} e_0(b_2,s_2)\, d_{\alpha_2} s_2}{q}\right]$$
$$\left\| {}_R D_{b_1-}^{1-\alpha_1} \left({}^P D_{b_1-}^{\alpha_1} f_1\right)\right\|_{L_q\left([a_1,b_1], e_0(b_1,\cdot) d_{\alpha_1} s_1\right)}$$
$$\left\| {}_R D_{b_2-}^{1-\alpha_2} \left({}^P D_{b_2-}^{\alpha_2} f_2\right)\right\|_{L_q\left([a_2,b_2], e_0(b_2,\cdot) d_{\alpha_2} s_2\right)}. \tag{14.128}$$

Consequently, we get

$$\frac{\left|f_1(t_1)\right|\left|f_2(t_2)\right| e_0(b_1,t_1)\, e_0(b_2,t_2)}{\left[\frac{\int_{t_1}^{b_1} e_0(b_1,s_1) d_{\alpha_1} s_1}{p} + \frac{\int_{t_2}^{b_2} e_0(b_2,s_2) d_{\alpha_2} s_2}{q}\right]}$$
$$\leq \left\| {}_R D_{b_1-}^{1-\alpha_1} \left({}^P D_{b_1-}^{\alpha_1} f_1\right)\right\|_{L_q\left([a_1,b_1], e_0(b_1,\cdot) d_{\alpha_1} s_1\right)}$$
$$\left\| {}_R D_{b_2-}^{1-\alpha_2} \left({}^P D_{b_2-}^{\alpha_2} f_2\right)\right\|_{L_q\left([a_2,b_2], e_0(b_2,\cdot) d_{\alpha_2} s_2\right)}. \tag{14.129}$$

The denominator in (14.129) is zero only when both $t_1 = b_1$ and $t_2 = b_2$.

Inequality (14.124) is implied by integration of (14.129) over $[a_1,b_1] \times [a_2,b_2]$. □

Bibliography

G.A. Anastassiou, *Intelligent Mathematics: Computational Analysis*, Springer, Heidelberg, New York, 2011.

G.A. Anastassiou, *Intelligent Computations: Abstract Fractional Calculus, Inequalities, Approximations*, Springer, Heildelberg, New York, 2018.

G.A. Anastassiou, *Towards proportional fractional calculus and inequalities*, Progress in fractional differentiation and applications, accepted, 2023.

D.R. Anderson and D.J. Ulness, Newly defined conformable derivatives, *Advances in Dynamical Systems and Applications*, 10(2) (2015), 109–137.

D. Baleanu, A. Fernandez and A. Akgül, On a fractional operator combining proportional and classical differintegrals, *Mathematics* 8 (2020), 360; doi: 10.3390/math8030360.

K. Diethelm, *The Analysis of Fractional Differential Equations*, Springer, New York, Heidelberg, 2010.

Chapter 15

New Radial Ostrowski Inequalities on the Ball

Here we present general multivariate radial mixed Ostrowski type inequalities over balls. The proofs are derived by implementation of some essential estimates out of some new trigonometric and hyperbolic Taylor's formulae (Anastassiou 2023a) and reducing the multivariate problem to a univariate one via general polar coordinates. It follows Anastassiou (2023b).

15.1 Introduction

We are motivated by the following:

Ostrowski (1938) proved the following famous inequality.

Theorem 15.1. *Let $f : [a,b] \to \mathbb{R}$ be continuous on $[a,b]$ and differentiable on (a,b) whose derivative $f' : (a,b) \to \mathbb{R}$ is bounded on (a,b), i.e., $\|f'\|_\infty = \sup_{t\in(a,b)} |f'(t)| < +\infty$. Then*

$$\left| \frac{1}{b-a} \int_a^b f(t)\, dt - f(x) \right| \leq \left[\frac{1}{4} + \frac{\left(x - \frac{a+b}{2}\right)^2}{(b-a)^2} \right] (b-a) \|f'\|_\infty, \qquad (15.1)$$

for any $x \in [a,b]$. The constant $\frac{1}{4}$ is the best.

Ostrowski-type inequalities have great applications to numerical analysis and probability and their literature is enormous.

Here $K = \mathbb{R}$ or $\mathbb{C}$.

Recently the author proved the following:

Theorem 15.2 (Anastassiou 2023a). *Let $f \in C_K^3([c,d])$, $a \in [c,d]$, such that $f'(a) = f''(a) = 0$. Then*

(i)

$$\left|\frac{1}{d-c}\int_{c}^{d} f(x)\,dx - f(a)\right| \leq \|f''' + f'\|_{\infty} \frac{\left[(d-a)^{3} + (a-c)^{3}\right]}{6(d-c)}, \tag{15.2}$$

(ii) *when* $f'\left(\frac{c+d}{2}\right) = f''\left(\frac{c+d}{2}\right) = 0$, *and* $a = \frac{c+d}{2}$, *we get*

$$\left|\frac{1}{d-c}\int_{c}^{d} f(x)\,dx - f\left(\frac{c+d}{2}\right)\right| \leq \|f' + f'''\|_{\infty} \frac{(d-c)^{2}}{24}. \tag{15.3}$$

We are also motivated by author's monograph, see Chapters 5 and 6.

This work is based on author's recent article Anastassiou (2023a), where we developed some new trigonometric and hyperbolic-type Taylor's formulae.

We prove here a collection of multivariate Ostrowski-type inequalities related to radial functions over a ball in $\mathbb{R}^N$, with respect to all norms $\|\cdot\|_p$, $1 \leq p \leq \infty$, and we give also their generalizations.

15.2 Main Results

We make

Remark 15.1. We define the ball

$$B(0,R) := \left\{x \in \mathbb{R}^N : |x| < R\right\} \subseteq \mathbb{R}^N, \quad N \geq 2, \ R > 0,$$

and the sphere

$$S^{N-1} := \left\{x \in \mathbb{R}^N : |x| = 1\right\},$$

where $|\cdot|$ is the Euclidean norm.

Let $d\omega$ be the element of surface measure on $\mathbb{S}^{N-1}$ and let

$$\omega_N = \int_{S^{N-1}} d\omega = \frac{2\pi^{\frac{N}{2}}}{\Gamma\left(\frac{N}{2}\right)}. \tag{15.4}$$

For $x \in \mathbb{R}^N - \{0\}$, we can write uniquely $x = r\omega$, where $r = |x| > 0$ and $\omega = \frac{x}{r} \in S^{N-1}$, $|\omega| = 1$.

Note that

$$\int_{B(0,R)} dy = \frac{\omega_N R^N}{N} \tag{15.5}$$

is the Lebesgue measure of the ball.

Following Rudin (1970, exercise 6, pp. 149–150) and Stroock (1999, Theorem 5.2.2, pp. 87–88), we can write for $F : \overline{B(0,R)} \to \mathbb{R}$ a Lebesgue integrable function that

$$\int_{B(0,R)} F(x)\,dx = \int_{S^{N-1}} \left(\int_0^R F(r\omega)\, r^{N-1} dr \right) d\omega; \tag{15.6}$$

a formula to be used next.

We present the following multivariate radial Ostrowski-type inequality.

Theorem 15.3. *Let the function $f : \overline{B(0,R)} \to \mathbb{R}$ be radial, that is, there exists a function g such that $f(x) = g(r)$, where $r = |x|$, $r \in [0,R]$, $\forall$ $x \in \overline{B(0,R)}$. We further assume that $g \in C^3([0,R])$, and $g^{(k)}(r_0) = 0$, $k = 1,2$, where $r_0 \in [0,R]$ is fixed. Then ($\forall$ $\omega \in S^{N-1}$)*

$$\left| f(r_0\omega) - \frac{\int_{B(0,R)} f(y)\,dy}{Vol(B(0,R))} \right| = \left| g(r_0) - \frac{N}{R^N} \int_0^R g(s)\, s^{N-1} ds \right|$$

$$\le \frac{N!}{R^N} \left[\|g''' + g'\|_{\infty,[0,r_0]} \frac{r_0^{3+N}}{(3+N)!} \right.$$

$$\left. + \|g''' + g'\|_{\infty,[r_0,R]} \sum_{k=0}^{N-1} \frac{(-1)^{N+k-1}}{k!\,(N-k+3)!} R^k (R-r_0)^{N-k+3} \right]. \tag{15.7}$$

Proof. As in Anastassiou (2023a), we get that

$$|g(r) - g(r_0)| \le \|g''' + g'\|_{\infty,[r_0,R]} \frac{(r-r_0)^3}{3!}, \tag{15.8}$$

$\forall$ $r \in [r_0, R]$,

and

$$|g(r) - g(r_0)| \le \|g''' + g'\|_{\infty,[0,r_0]} \frac{(r_0-r)^3}{3!}, \tag{15.9}$$

$\forall$ $r \in [0, r_0]$.

Next we observe

$$\left| f(r_0\omega) - \frac{\int_{B(0,R)} f(y)\,dy}{Vol(B(0,R))} \right|$$

$$= \left| g(r_0) - \frac{\int_{S^{N-1}} \left(\int_0^R g(s)\, s^{N-1} ds \right) d\omega}{\int_{S^{N-1}} \left(\int_0^R s^{N-1} ds \right) d\omega} \right|$$

$$= \left| g(r_0) - \frac{N}{R^N} \int_0^R g(s)\, s^{N-1} ds \right| = \frac{N}{R^N} \left| \int_0^R s^{N-1} \left(g(r_0) - g(s) \right) ds \right|$$

$$\leq \frac{N}{R^N} \int_0^R s^{N-1} \left| g(r_0) - g(s) \right| ds \tag{15.10}$$

$$= \frac{N}{R^N} \left[\int_0^{r_0} s^{N-1} \left| g(r_0) - g(s) \right| ds + \int_{r_0}^R s^{N-1} \left| g(r_0) - g(s) \right| ds \right]$$

$$\leq \frac{N}{6R^N} \left[\left\| g''' + g' \right\|_{\infty,[0,r_0]} \int_0^{r_0} s^{N-1} (r_0 - s)^3\, ds \right.$$

$$\left. + \left\| g''' + g' \right\|_{\infty,[r_0,R]} \int_{r_0}^R s^{N-1} (s - r_0)^3\, ds \right]$$

$$= \frac{N}{6R^N} \left[\left\| g''' + g' \right\|_{\infty,[0,r_0]} \int_0^{r_0} (r_0 - s)^{4-1} (s - 0)^{N-1}\, ds \right. \tag{15.11}$$

$$\left. + \left\| g''' + g' \right\|_{\infty,[r_0,R]} (-1)^{N-1} \int_{r_0}^R \left((R - s) - R \right)^{N-1} (s - r_0)^3\, ds \right]$$

$$= \frac{N}{6R^N} \left[\left\| g''' + g' \right\|_{\infty,[0,r_0]} \frac{\Gamma(4)\,\Gamma(N)}{\Gamma(4+N)} r_0^{3+N} \right.$$

$$+ \left\| g''' + g' \right\|_{\infty,[r_0,R]} (-1)^{N-1}$$

$$\left. \times \sum_{k=0}^{N-1} (-1)^k R^k \binom{N-1}{k} \int_{r_0}^R (R - s)^{N-k-1} (s - r_0)^{4-1}\, ds \right]$$

$$= \frac{N}{6R^N} \left[\left\| g''' + g' \right\|_{\infty,[0,r_0]} \frac{3!\,(N-1)!}{(3+N)!} r_0^{3+N} \right. \tag{15.12}$$

$$+ \|g''' + g'\|_{\infty,[r_0,R]} (-1)^{N-1}$$

$$\left. \times \sum_{k=0}^{N-1} (-1)^k R^k \binom{N-1}{k} \frac{(N-k-1)!3!}{(N-k+3)!} (R-r_0)^{N-k+3} \right]$$

$$= \frac{N}{R^N} \left[\|g''' + g'\|_{\infty,[0,r_0]} \frac{(N-1)!}{(3+N)!} r_0^{3+N} + \|g''' + g'\|_{\infty,[r_0,R]} \right.$$

$$\left. \times \sum_{k=0}^{N-1} (-1)^{k+N-1} R^k \binom{N-1}{k} \frac{(N-k-1)!}{(N-k+3)!} (R-r_0)^{N-k+3} \right]$$

$$= \frac{N!}{R^N} \left[\|g''' + g'\|_{\infty,[0,r_0]} \frac{r_0^{3+N}}{(3+N)!} + \|g''' + g'\|_{\infty,[r_0,R]} \right.$$

$$\left. \times \sum_{k=0}^{N-1} (-1)^{k+N-1} \frac{R^k}{k!(N-k-1)!} \frac{(N-k-1)!}{(N-k+3)!} (R-r_0)^{N-k+3} \right]$$

$$= \frac{N!}{R^N} \left[\|g''' + g'\|_{\infty,[0,r_0]} \frac{r_0^{3+N}}{(3+N)!} \right.$$

$$\left. + \|g''' + g'\|_{\infty,[r_0,R]} \sum_{k=0}^{N-1} \frac{(-1)^{N+k-1}}{k!(N-k+3)!} R^k (R-r_0)^{N-k+3} \right]. \quad (15.13)$$

□

We continue with a more general multivariate radial Ostrowski inequality.

Theorem 15.4. *Let the function $f : \overline{B(0,R)} \to \mathbb{R}$ be radial, that is, there exists a function g such that $f(x) = g(r)$, where $r = |x|$, $r \in [0,R]$, $\forall$ $x \in \overline{B(0,R)}$. We further assume that $g \in C^5([0,R])$, and $g^{(k)}(r_0) = 0$, $k = 1,2,3,4$, where $r_0 \in [0,R]$ is fixed. Here $\alpha, \beta \in \mathbb{R} : \alpha\beta(\alpha^2 - \beta^2) \neq 0$. Then ($\forall$ $\omega \in S^{N-1}$)*

$$\left| f(r_0\omega) - \frac{\int_{B(0,R)} f(y)\,dy}{Vol(B(0,R))} \right| = \left| g(r_0) - \frac{N}{R^N} \int_0^R g(s) s^{N-1} ds \right| \quad (15.14)$$

$$\leq \frac{2N!}{R^N |\beta^2 - \alpha^2|} \left[\left\| g^{(5)} + (\alpha^2 + \beta^2) g^{(3)} + \alpha^2\beta^2 g' \right\|_{\infty,[0,r_0]} \frac{r_0^{3+N}}{(3+N)!} \right.$$

$$\left. \left\| g^{(5)} + (\alpha^2 + \beta^2) g^{(3)} + \alpha^2\beta^2 g' \right\|_{\infty,[r_0,R]} \sum_{k=0}^{N-1} \frac{(-1)^{N+k-1}}{k!(N-k+3)!} R^k (R-r_0)^{N-k+3} \right]. \quad (15.15)$$

Proof. As in Anastassiou (2023a), we get that

$$\begin{aligned}
&|g(r)-g(r_0)| \\
&\leq \frac{1}{3\left|\beta^2-\alpha^2\right|}\left\|g^{(5)}+\left(\alpha^2+\beta^2\right)g^{(3)}+\alpha^2\beta^2 g'\right\|_{\infty,[r_0,R]}(r-r_0)^3 \\
&=: A(r-r_0)^3, \quad \forall\, r\in[r_0,R],
\end{aligned} \tag{15.16}$$

and

$$\begin{aligned}
&|g(r)-g(r_0)| \\
&\leq \frac{1}{3\left|\beta^2-\alpha^2\right|}\left\|g^{(5)}+\left(\alpha^2+\beta^2\right)g^{(3)}+\alpha^2\beta^2 g'\right\|_{\infty,[0,r_0]}(r_0-r)^3 \\
&=: B(r_0-r)^3, \quad \forall\, r\in[0,r_0].
\end{aligned} \tag{15.17}$$

Next we observe

$$\left|f(r_0\omega)-\frac{\int_{B(0,R)}f(y)\,dy}{Vol(B(0,R))}\right|\leq$$

(as in the proof of Theorem 15.1)

$$\frac{N}{R^N}\left[\int_0^{r_0}s^{N-1}|g(r_0)-g(s)|\,ds+\int_{r_0}^{R}s^{N-1}|g(r_0)-g(s)|\,ds\right]$$

$$\leq\frac{N}{R^N}\left[B\int_0^{r_0}s^{N-1}(r_0-s)^3\,ds+A\int_{r_0}^{R}s^{N-1}(s-r_0)^3\,ds\right] = \tag{15.18}$$

(as earlier)

$$\begin{aligned}
&\frac{N}{R^N}\Bigg[B\frac{3!\,(N-1)!}{(3+N)!}r_0^{3+N} \\
&\quad +A(-1)^{N-1}\sum_{k=0}^{N-1}(-1)^k R^k\binom{N-1}{k}\frac{(N-k-1)!3!}{(N-k+3)!}(R-r_0)^{N-k+3}\Bigg] \\
&=\frac{6N!}{R^N}\left[B\frac{r_0^{3+N}}{(3+N)!}+A\sum_{k=0}^{N-1}\frac{(-1)^{N+k-1}}{k!\,(N-k+3)!}R^k(R-r_0)^{N-k+3}\right].
\end{aligned}$$

□

It follows an L_1 Ostrowski inequality.

Theorem 15.5. *All as in Theorem 15.3, except now $g \in C^2([0,R])$. Then $(\forall\ \omega \in S^{N-1})$*

$$\left| f(r_0\omega) - \frac{\int_{B(0,R)} f(y)\,dy}{Vol(B(0,R))} \right| = \left| g(r_0) - \frac{N}{R^N}\int_0^R g(s)\,s^{N-1}ds \right|$$

$$\leq \frac{1}{R^N}\left[\left\| g'' + g - g(r_0) \right\|_{L_1([0,r_0])} \frac{r_0^{N+1}}{(N+1)} \right.$$

$$\left. + \left\| g'' + g - g(r_0) \right\|_{L_1([r_0,R])} \left[\left(\frac{N}{N+1}\right)\left(R^{N+1} - r_0^{N+1}\right) - r_0\left(R^N - r_0^N\right)\right]\right]. \tag{15.19}$$

Proof. As in Anastassiou (2023a), we get that

$$|g(r) - g(r_0)| \leq \|g'' + g - g(r_0)\|_{L_1([0,r_0])}(r_0 - r), \tag{15.20}$$

$\forall\ r \in [0, r_0]$,

and

$$|g(r) - g(r_0)| \leq \|g'' + g - g(r_0)\|_{L_1([r_0,R])}(r - r_0), \tag{15.21}$$

$\forall\ r \in [r_0, R]$.

Next we observe

$$\left| f(r_0\omega) - \frac{\int_{B(0,R)} f(y)\,dy}{Vol(B(0,R))} \right| = \left| g(r_0) - \frac{N}{R^N}\int_0^R g(s)\,s^{N-1}ds \right|$$

$$\leq \frac{N}{R^N}\left[\int_0^{r_0} s^{N-1}|g(r_0) - g(s)|\,ds + \int_{r_0}^R s^{N-1}|g(r_0) - g(s)|\,ds \right]$$

$$\leq \frac{N}{R^N}\left[\|g'' + g - g(r_0)\|_{L_1([0,r_0])} \int_0^{r_0} s^{N-1}(r_0 - s)^{2-1}\,ds \right. \tag{15.22}$$

$$\left. + \|g'' + g - g(r_0)\|_{L_1([r_0,R])} \int_{r_0}^R s^{N-1}(s - r_0)\,ds \right]$$

$$= \frac{N}{R^N}\left[\|g'' + g - g(r_0)\|_{L_1([0,r_0])} \frac{r_0^{N+1}}{N(N+1)} \right.$$

$$\left. + \|g'' + g - g(r_0)\|_{L_1([r_0,R])} \left[\left(\frac{R^{N+1} - r_0^{N+1}}{N+1}\right) - r_0\left(\frac{R^N - r_0^N}{N}\right)\right]\right]$$

$$= \frac{1}{R^N} \left[\|g'' + g - g(r_0)\|_{L_1([0,r_0])} \frac{r_0^{N+1}}{(N+1)} \right.$$

$$\left. + \|g'' + g - g(r_0)\|_{L_1([r_0,R])} \left[\left(\frac{N}{N+1} \right) \left(R^{N+1} - r_0^{N+1} \right) - r_0 \left(R^N - r_0^N \right) \right] \right]. \tag{15.23}$$

□

Next comes an Ostrowski multivariate radial inequality for $\|\cdot\|_p$, $p > 1$.

Theorem 15.6. *Let $p, q > 1 : \frac{1}{p} + \frac{1}{q} = 1$. Let all as in Theorem 15.5. Then ($\forall\, \omega \in S^{N-1}$)*

$$\left| f(r_0\omega) - \frac{\int_{B(0,R)} f(y)\, dy}{Vol(B(0,R))} \right| = \left| g(r_0) - \frac{N}{R^N} \int_0^R g(s)\, s^{N-1} ds \right|$$

$$\leq \frac{N!\Gamma\left(2 + \frac{1}{q}\right)}{R^N (q+1)^{\frac{1}{q}}} \left[\|g'' + g - g(r_0)\|_{L_p([0,r_0])} \frac{r_0^{N+1+\frac{1}{q}}}{\Gamma\left(N + 2 + \frac{1}{q}\right)} \right.$$

$$\left. + \|g'' + g - g(r_0)\|_{L_p([r_0,R])} \sum_{k=0}^{N-1} \frac{(-1)^{N+k-1} R^k (R - r_0)^{N-k+1+\frac{1}{q}}}{k!\Gamma\left(N - k + 2 + \frac{1}{q}\right)} \right]. \tag{15.24}$$

Proof. As in Anastassiou (2023a), we get that

$$|g(r) - g(r_0)| \leq \|g'' + g - g(r_0)\|_{L_p([r_0,R])} \frac{(r - r_0)^{\frac{q+1}{1}}}{(q+1)^{\frac{1}{q}}}, \tag{15.25}$$

$\forall\, r \in [r_0, R]$,

and

$$|g(r) - g(r_0)| \leq \|g'' + g - g(r_0)\|_{L_p([0,r_0])} \frac{(r_0 - r)^{\frac{q+1}{q}}}{(q+1)^{\frac{1}{q}}}, \tag{15.26}$$

$\forall\, r \in [0, r_0]$.

Next we observe

$$\left| f(r_0\omega) - \frac{\int_{B(0,R)} f(y)\, dy}{Vol(B(0,R))} \right| =$$

(as earlier)

$$\left|g(r_0) - \frac{N}{R^N}\int_0^R g(s)\, s^{N-1} ds\right|$$

$$\le \frac{N}{R^N}\left[\int_0^{r_0} s^{N-1}\left|g(r_0) - g(s)\right| ds + \int_{r_0}^R s^{N-1}\left|g(r_0) - g(s)\right| ds\right]$$

$$\le \frac{N}{R^N (q+1)^{\frac{1}{q}}}\left[\left\|g'' + g - g(r_0)\right\|_{L_p([0,r_0])}\int_0^{r_0} s^{N-1}(r_0 - s)^{\left(2+\frac{1}{q}\right)-1} ds\right. \tag{15.27}$$

$$\left. + \left\|g'' + g - g(r_0)\right\|_{L_p([r_0,R])}\int_{r_0}^R s^{N-1}(s - r_0)^{1+\frac{1}{q}} ds\right]$$

$$= \frac{N}{R^N (q+1)^{\frac{1}{q}}}\left[\left\|g'' + g - g(r_0)\right\|_{L_p([0,r_0])}\frac{(N-1)!\Gamma\left(2+\frac{1}{q}\right)}{\Gamma\left(N+2+\frac{1}{q}\right)} r_0^{N+1+\frac{1}{q}}\right.$$

$$+ \left\|g'' + g - g(r_0)\right\|_{L_p([r_0,R])}(-1)^{N-1}$$

$$\left.\times \int_{r_0}^R ((R-s) - R)^{N-1}(s - r_0)^{1+\frac{1}{q}} ds\right] \tag{15.28}$$

$$= \frac{N}{R^N (q+1)^{\frac{1}{q}}}\left[\left\|g'' + g - g(r_0)\right\|_{L_p([0,r_0])}\frac{(N-1)!\Gamma\left(2+\frac{1}{q}\right)}{\Gamma\left(N+2+\frac{1}{q}\right)} r_0^{N+1+\frac{1}{q}}\right.$$

$$+ \left\|g'' + g - g(r_0)\right\|_{L_p([r_0,R])}\sum_{k=0}^{N-1}\frac{(N-1)!}{k!(N-k-1)!}(-1)^{N+k-1} R^k$$

$$\left.\times \frac{(N-k-1)!\Gamma\left(2+\frac{1}{q}\right)}{\Gamma\left(N-k+2+\frac{1}{q}\right)}(R - r_0)^{N-k+1+\frac{1}{q}}\right]$$

$$= \frac{N!\Gamma\left(2+\frac{1}{q}\right)}{R^N (q+1)^{\frac{1}{q}}}\left[\left\|g'' + g - g(r_0)\right\|_{L_p([0,r_0])}\frac{r_0^{N+1+\frac{1}{q}}}{\Gamma\left(N+2+\frac{1}{q}\right)}\right.$$

$$\left. + \left\|g'' + g - g(r_0)\right\|_{L_p([r_0,R])}\sum_{k=0}^{N-1}(-1)^{N+k-1} R^k \frac{(R - r_0)^{N-k+1+\frac{1}{q}}}{k!\Gamma\left(N-k+2+\frac{1}{q}\right)}\right]. \tag{15.29}$$

□

It follows a general L_1 estimate.

Theorem 15.7. *Let the function $f : \overline{B(0,R)} \to \mathbb{R}$ be radial, that is, there exists a function g such that $f(x) = g(r)$, where $r = |x|$, $r \in [0,R]$, $\forall\, x \in \overline{B(0,R)}$. We further assume that $g \in C^4([0,R])$, and $g^{(k)}(r_0) = 0$, $k = 1,2,3,4$, where $r_0 \in [0,R]$ is fixed. Let also $\alpha, \beta \in \mathbb{R} : \alpha\beta(\alpha^2 - \beta^2) \neq 0$. Then ($\forall\, \omega \in S^{N-1}$)*

$$\left| f(r_0\omega) - \frac{\int_{B(0,R)} f(y)\, dy}{Vol(B(0,R))} \right| = \left| g(r_0) - \frac{N}{R^N} \int_0^R g(s)\, s^{N-1} ds \right|$$

$$\leq \frac{2}{|\beta^2 - \alpha^2| R^N} \left[\left\| g'''' + (\alpha^2 + \beta^2) g'' + \alpha^2\beta^2 g - \alpha^2\beta^2 g(r_0) \right\|_{L_1([0,r_0])} \frac{r_0^{N+1}}{(N+1)} \right.$$

$$+ \left\| g'''' + (\alpha^2 + \beta^2) g'' + \alpha^2\beta^2 g - \alpha^2\beta^2 g(r_0) \right\|_{L_1([r_0,R])}$$

$$\left. \left[\left(\frac{N}{N+1} \right) \left(R^{N+1} - r_0^{N+1} \right) - r_0 \left(R^N - r_0^N \right) \right] \right]. \tag{15.30}$$

Proof. As in Anastassiou (2023a), we get that

$$|g(r) - g(r_0)|$$

$$\leq \frac{2}{|\beta^2 - \alpha^2|} \left\| g'''' + (\alpha^2 + \beta^2) g'' + \alpha^2\beta^2 g - \alpha^2\beta^2 g(r_0) \right\|_{L_1([0,r_0])} (r_0 - r), \tag{15.31}$$

$\forall\, r \in [0, r_0]$,

and

$$|g(r) - g(r_0)|$$

$$\leq \frac{2}{|\beta^2 - \alpha^2|} \left\| g'''' + (\alpha^2 + \beta^2) g'' + \alpha^2\beta^2 g - \alpha^2\beta^2 g(r_0) \right\|_{L_1([r_0,R])} (r - r_0), \tag{15.32}$$

$\forall\, r \in [r_0, R]$.

The rest of the proof goes as in Theorem 15.5. ☐

Next comes a general L_p, $p > 1$, estimate.

Theorem 15.8. *All as in Theorem 15.7 and let* $p, q > 1 : \frac{1}{p} + \frac{1}{q} = 1$. *Then* $(\forall\, \omega \in S^{N-1})$

$$\left| f(r_0\omega) - \frac{\int_{B(0,R)} f(y)\,dy}{Vol\,(B(0,R))} \right| = \left| g(r_0) - \frac{N}{R^N} \int_0^R g(s)\, s^{N-1} ds \right|$$

$$\leq \frac{2N!\Gamma\left(2+\frac{1}{q}\right)}{\left|\beta^2-\alpha^2\right| R^N (q+1)^{\frac{1}{q}}}$$

$$\left[\left\| g'''' + \left(\alpha^2+\beta^2\right) g'' + \alpha^2\beta^2 g - \alpha^2\beta^2 g(r_0) \right\|_{L_p([0,r_0])} \frac{r_0^{N+1+\frac{1}{q}}}{\Gamma\left(N+2+\frac{1}{q}\right)} \right.$$

$$+ \left\| g'''' + \left(\alpha^2+\beta^2\right) g'' + \alpha^2\beta^2 g - \alpha^2\beta^2 g(r_0) \right\|_{L_p([r_0,R])}$$

$$\left. \sum_{k=0}^{N-1} \frac{(-1)^{N+k-1} R^k (R-r_0)^{N-k+1+\frac{1}{q}}}{k!\Gamma\left(N-k+2+\frac{1}{q}\right)} \right]. \tag{15.33}$$

Proof. As in Anastassiou (2023a), we have that

$$|g(r) - g(r_0)| \leq \frac{2}{\left|\beta^2-\alpha^2\right|}$$

$$\left\| g'''' + \left(\alpha^2+\beta^2\right) g'' + \alpha^2\beta^2 g - \alpha^2\beta^2 g(r_0) \right\|_{L_p([0,r_0])} \frac{(r_0-r)^{\frac{q+1}{q}}}{(q+1)^{\frac{1}{q}}}, \tag{15.34}$$

$\forall\, r \in [0, r_0]$,

and

$$|g(r) - g(r_0)| \leq \frac{2}{\left|\beta^2-\alpha^2\right|}$$

$$\left\| g'''' + \left(\alpha^2+\beta^2\right) g'' + \alpha^2\beta^2 g - \alpha^2\beta^2 g(r_0) \right\|_{L_p([r_0,R])} \frac{(r-r_0)^{\frac{q+1}{q}}}{(q+1)^{\frac{1}{q}}}, \tag{15.35}$$

$\forall\, r \in [r_0, R]$.

The rest of the proof goes as in Theorem 15.6. □

Bibliography

G.A. Anastassiou, *Advances on Fractional Inequalities*, Springer, Heidelberg, New York, 2011.

G.A. Anastassiou, Opial and Ostrowski type inequalities based on trigonometric and hyperbolic Taylor formulae, *Malaya Journal of Matematik*, 11(S) (2023a), 1–26.

G.A. Anastassiou, Updated radial Ostrowski inequalities over a ball, *Romai Journal*, 19(1) (2023b), 25–34.

W. Rudin, *Real and Complex Analysis*, International student edition, McGraw Hill, London, New York, 1970.

D. Stroock, *A Concise Introduction in the Theory of Integration*, Third Edition, Birkhaüser, Boston, Basel, Berlin, 1999.

A. Ostrowski, Über die Absolutabweichung einer differentiebaren Funktion von ihrem Integralmittelwert, *Commentarii Mathematici Helvetici*, 10, (1938), 226–227.

Chapter 16

New Ostrowski Inequalities on a Spherical Shell

Here we present general multivariate mixed Ostrowski-type inequalities over spherical shells and balls. We cover the radial and not necessarily radial cases. The proofs are derived by the use of some estimates coming out of some new trigonometric and hyperbolic Taylor's formulae (Anastassiou 2023a) and reducing the multivariate problem to a univariate one via general polar coordinates. It follows Anastassiou (2023c).

16.1 Introduction

We are motivated by the following:

Ostrowski (1938) proved the following famous inequality.

Theorem 16.1. *Let $f : [a, b] \to \mathbb{R}$ be continuous on $[a, b]$ and differentiable on (a, b) whose derivative $f' : (a, b) \to \mathbb{R}$ is bounded on (a, b), i.e., $\|f'\|_\infty = \sup_{t \in (a,b)} |f'(t)| < +\infty$. Then*

$$\left| \frac{1}{b-a} \int_a^b f(t)\, dt - f(x) \right| \leq \left[\frac{1}{4} + \frac{\left(x - \frac{a+b}{2}\right)^2}{(b-a)^2} \right] (b-a) \|f'\|_\infty, \quad (16.1)$$

for any $x \in [a, b]$. The constant $\frac{1}{4}$ is the best.

Ostrowski-type inequalities have great applications to numerical analysis and probability and their literature is enormous.

Here $K = \mathbb{R}$ or $\mathbb{C}$.

Recently the author proved the following:

Theorem 16.2. *(Anastassiou 2023a) Let $f \in C_K^3([c, d])$, $a \in [c, d]$, such that $f'(a) = f''(a) = 0$. Then*

(i)

$$\left|\frac{1}{d-c}\int_c^d f(x)\,dx - f(a)\right| \leq \|f''' + f'\|_\infty \frac{\left[(d-a)^3 + (a-c)^3\right]}{6(d-c)}, \quad (16.2)$$

(ii) *when* $f'\left(\frac{c+d}{2}\right) = f''\left(\frac{c+d}{2}\right) = 0$, *and* $a = \frac{c+d}{2}$, *we get*

$$\left|\frac{1}{d-c}\int_c^d f(x)\,dx - f\left(\frac{c+d}{2}\right)\right| \leq \|f' + f'''\|_\infty \frac{(d-c)^2}{24}. \quad (16.3)$$

We are also motivated by author's monograph, see Chapter 6.

This work is based on author's recent article Anastassiou (2023a), where we developed some new trigonometric and hyperbolic-type Taylor's formulae.

We prove here a collection of multivariate Ostrowski-type inequalities related to functions over a spherical shell in $\mathbb{R}^N$, with respect to all norms $\|\cdot\|_p$, $1 \leq p \leq \infty$, and we give also their generalizations.

16.2 Main Results

We need

Remark 16.1. Let the spherical shell

$$A := B(0, R_2) - \overline{B(0, R_1)},$$

$0 < R_1 < R_2$, $A \subseteq \mathbb{R}^N$, $N \geq 2$, $x \in \overline{A}$. Consider that $f : \overline{A} \to \mathbb{R}$ is radial, that is, there exists g such that $f(x) = g(r)$, $r = |x|$, $r \in [R_1, R_2]$, $\forall\, x \in \overline{A}$, $|\cdot|$ the Euclidean norm. Here x can be written uniquely as $x = r\omega$, where $r = |x| > 0$ and $\omega = \frac{x}{r} \in S^{N-1}$, $|\omega| = 1$, see (Rudin 1970, pp. 149–150 and Stroock 1999, p. 421).

Furthermore, for $F : \overline{A} \to \mathbb{R}$ a Lebesgue integrable function we have that

$$\int_A F(x)\,dx = \int_{S^{N-1}} \left(\int_{R_1}^{R_2} F(r\omega)\, r^{N-1} dr\right) d\omega, \quad (16.4)$$

where $S^{N-1} := \left\{x \in \mathbb{R}^N : |x| = 1\right\}$.

Let $d\omega$ be the element of surface measure on $\mathbb{S}^{N-1}$ with surface area

$$\omega_N = \int_{S^{N-1}} d\omega = \frac{2\pi^{\frac{N}{2}}}{\Gamma\left(\frac{N}{2}\right)}. \quad (16.5)$$

Here it is volume of A,

$$vol\,(A) = \frac{\omega_N \left(R_2^N - R_1^N\right)}{N}. \tag{16.6}$$

We present the following multivariate radial Ostrowski-type inequality on the spherical shell.

Theorem 16.3. *Let the function $f : \overline{A} \to \mathbb{R}$ be radial, that is, there exists g such that $f(x) = g(r)$, $r = |x|$, $\forall\, x \in \overline{A}$; $\omega \in S^{N-1}$. Assume that $g \in C^3([R_1, R_2])$, and $g^{(k)}(r_0) = 0$, $k = 1, 2$, where $r_0 \in [R_1, R_2]$ is fixed. Then ($\forall\, \omega \in S^{N-1}$)*

$$\left| f(r_0\omega) - \frac{\int_A f(y)\,dy}{Vol\,(A)} \right| = \left| g(r_0) - \left(\frac{N}{R_2^N - R_1^N}\right) \int_{R_1}^{R_2} g(s)\, s^{N-1} ds \right|$$
$$\leq \left(\frac{N!}{R_2^N - R_1^N}\right) \left\{ \|g''' + g'\|_{\infty,[R_1,r_0]} \left(\sum_{k=0}^{N-1} \frac{R_1^k (r_0 - R_1)^{3+N-k}}{k!\,(N-k+3)!} \right) \right.$$
$$\left. + \|g''' + g'\|_{\infty,[r_0,R_2]} \left(\sum_{k=0}^{N-1} \frac{(-1)^{N+k-1}}{k!\,(N-k+3)!} R_2^k (R_2 - r_0)^{3+N-k} \right) \right\}. \tag{16.7}$$

Proof. As in Anastassiou (2023a), we get that

$$|g(r) - g(r_0)| \leq \|g''' + g'\|_{\infty,[r_0,R_2]} \frac{(r - r_0)^3}{3!}, \tag{16.8}$$

$\forall\, r \in [r_0, R_2]$,

and

$$|g(r) - g(r_0)| \leq \|g''' + g'\|_{\infty,[R_1,r_0]} \frac{(r_0 - r)^3}{3!}, \tag{16.9}$$

$\forall\, r \in [R_1, r_0]$.

Next we observe that

$$\left| f(r_0\omega) - \frac{\int_A f(y)\,dy}{Vol\,(A)} \right|$$
$$= \left| g(r_0) - \left(\frac{N}{R_2^N - R_1^N}\right) \int_{R_1}^{R_2} g(s)\, s^{N-1} ds \right|$$
$$= \left(\frac{N}{R_2^N - R_1^N}\right) \left| \int_{R_1}^{R_2} (g(r_0) - g(s))\, s^{N-1} ds \right| \tag{16.10}$$

$$\leq \left(\frac{N}{R_2^N - R_1^N}\right) \int_{R_1}^{R_2} |g(r_0) - g(s)| \, s^{N-1} ds$$
$$= \left(\frac{N}{R_2^N - R_1^N}\right) \left\{ \int_{R_1}^{r_0} |g(r_0) - g(s)| \, s^{N-1} ds \right.$$
$$\left. + \int_{r_0}^{R_2} |g(r_0) - g(s)| \, s^{N-1} ds \right\}$$
$$\leq \left(\frac{N}{6\left(R_2^N - R_1^N\right)}\right) \left\{ \|g''' + g'\|_{\infty,[R_1,r_0]} \int_{R_1}^{r_0} (r_0 - s)^3 \, s^{N-1} ds \right.$$
$$\left. + \|g''' + g'\|_{\infty,[r_0,R_2]} \int_{r_0}^{R_2} (s - r_0)^3 \, s^{N-1} ds \right\} =: (*). \tag{16.11}$$

Here we calculate

$$I_1 := \int_{R_1}^{r_0} (r_0 - s)^3 \, s^{N-1} ds$$
$$= \int_{R_1}^{r_0} (r_0 - s)^3 \left((s - R_1) + R_1\right)^{N-1} ds$$
$$= \sum_{k=0}^{N-1} \binom{N-1}{k} R_1^k \int_{R_1}^{r_0} (r_0 - s)^{4-1} (s - R_1)^{N-k-1} ds$$
$$= \sum_{k=0}^{N-1} \frac{(N-1)!}{k!\,(N-k-1)!} R_1^k \frac{3!\,(N-k-1)!}{(N-k+3)!} (r_0 - R_1)^{3+N-k}$$
$$= \sum_{k=0}^{N-1} \frac{(N-1)!3!}{k!\,(N-k+3)!} R_1^k (r_0 - R_1)^{3+N-k}. \tag{16.12}$$

That is,

$$I_1 = 3!\,(N-1)! \sum_{k=0}^{N-1} \frac{R_1^k (r_0 - R_1)^{3+N-k}}{k!\,(N-k+3)!}. \tag{16.13}$$

Also

$$I_2 := \int_{r_0}^{R_2} (s - r_0)^3 \, s^{N-1} ds$$
$$= (-1)^{N-1} \int_{r_0}^{R_2} \left((R_2 - s) - R_2\right)^{N-1} (s - r_0)^3 \, ds$$

$$= (-1)^{N-1} \sum_{k=0}^{N-1} \frac{(N-1)!}{k!\,(N-k-1)!} (-1)^k R_2^k \int_{r_0}^{R_2} (R_2 - s)^{(N-k)-1} (s - r_0)^{4-1} ds$$

$$= \sum_{k=0}^{N-1} \frac{(N-1)!}{k!\,(N-k-1)!} (-1)^{N+k-1} R_2^k \frac{(N-k-1)!3!}{(N-k+3)!} (R_2 - r_0)^{3+N-k}. \tag{16.14}$$

That is,

$$I_2 = 3!\,(N-1)! \sum_{k=0}^{N-1} \frac{(-1)^{N+k-1}}{k!\,(N-k+3)!} R_2^k (R_2 - r_0)^{3+N-k}. \tag{16.15}$$

Consequently, we obtain

$$(*) = \left(\frac{N!}{R_2^N - R_1^N}\right) \left\{ \|g''' + g'\|_{\infty,[R_1,r_0]} \sum_{k=0}^{N-1} \frac{R_1^k (r_0 - R_1)^{3+N-k}}{k!\,(N-k+3)!} \right.$$
$$\left. + \|g''' + g'\|_{\infty,[r_0,R_2]} \sum_{k=0}^{N-1} \frac{(-1)^{N+k-1}}{k!\,(N-k+3)!} R_2^k (R_2 - r_0)^{3+N-k} \right\}. \tag{16.16}$$

The proof is completed. □

It follows an L_1 Ostrowski inequality.

Theorem 16.4. *All as in Theorem 16.3, except now $g \in C^2([R_1, R_2])$. Then ($\forall\, \omega \in S^{N-1}$)*

$$\left| f(r_0\omega) - \frac{\int_A f(y)\,dy}{Vol(A)} \right| = \left| g(r_0) - \left(\frac{N}{R_2^N - R_1^N}\right) \int_{R_1}^{R_2} g(s)\, s^{N-1} ds \right|$$
$$\leq \left(\frac{N!}{R_2^N - R_1^N}\right) \left\{ \|g'' + g - g(r_0)\|_{L_1([R_1,r_0])} \left(\sum_{k=0}^{N-1} \frac{R_1^k (r_0 - R_1)^{N-k+1}}{k!\,(N-k+1)!} \right) \right.$$
$$\left. + \|g'' + g - g(r_0)\|_{L_1([r_0,R_2])} \left(\sum_{k=0}^{N-1} \frac{(-1)^{N+k-1}}{k!\,(N-k+1)!} R_2^k (R_2 - r_0)^{N-k+1} \right) \right\}. \tag{16.17}$$

Proof. As in Anastassiou (2023a), we get that

$$|g(r) - g(r_0)| \leq \|g'' + g - g(r_0)\|_{L_1([r_0,R_2])} (r - r_0), \tag{16.18}$$

$\forall\, r \in [r_0, R_2]$,

and

$$|g(r) - g(r_0)| \leq \|g'' + g - g(r_0)\|_{L_1([R_1,r_0])} (r_0 - r), \tag{16.19}$$

$\forall\, r \in [R_1, r_0]$.

Next we observe

$$\left| f(r_0\omega) - \frac{\int_A f(y)\,dy}{Vol(A)} \right| = \left| g(r_0) - \left(\frac{N}{R_2^N - R_1^N}\right) \int_{R_1}^{R_2} g(s)\, s^{N-1} ds \right|$$

$\leq$ (as earlier)

$$\leq \left(\frac{N}{R_2^N - R_1^N}\right) \left\{ \int_{R_1}^{r_0} |g(r_0) - g(s)|\, s^{N-1} ds + \int_{r_0}^{R_2} |g(r_0) - g(s)|\, s^{N-1} ds \right\}$$

$$\leq \left(\frac{N}{R_2^N - R_1^N}\right) \left\{ \left\| g'' + g - g(r_0) \right\|_{L_1([R_1, r_0])} \int_{R_1}^{r_0} (r_0 - s)\, s^{N-1} ds \right.$$

$$\left. + \left\| g'' + g - g(r_0) \right\|_{L_1([r_0, R_2])} \int_{r_0}^{R_2} (s - r_0)\, s^{N-1} ds \right\}. \tag{16.20}$$

Here we calculate

$$I_1^* := \int_{R_1}^{r_0} (r_0 - s)\, s^{N-1} ds$$

$$= \int_{R_1}^{r_0} (r_0 - s) \left((s - R_1) + R_1\right)^{N-1} ds$$

$$= \sum_{k=0}^{N-1} \binom{N-1}{k} R_1^k \int_{R_1}^{r_0} (r_0 - s)^{2-1} (s - R_1)^{N-k-1} ds$$

$$= \sum_{k=0}^{N-1} \frac{(N-1)!}{k!\,(N-k-1)!} R_1^k \frac{(N-k-1)!}{(N-k+1)!} (r_0 - R_1)^{N-k+1}$$

$$= \sum_{k=0}^{N-1} \frac{(N-1)!}{k!\,(N-k+1)!} R_1^k (r_0 - R_1)^{N-k+1}. \tag{16.21}$$

That is,

$$I_1^* = (N-1)! \sum_{k=0}^{N-1} \frac{R_1^k (r_0 - R_1)^{N-k+1}}{k!\,(N-k+1)!}. \tag{16.22}$$

Also

$$I_2^* := \int_{r_0}^{R_2} (s - r_0)\, s^{N-1} ds$$

$$= (-1)^{N-1} \int_{r_0}^{R_2} \left((R_2 - s) - R_2\right)^{N-1} (s - r_0)^{2-1} ds$$

$$= (-1)^{N-1} \sum_{k=0}^{N-1} \frac{(N-1)!}{k!\,(N-k-1)!} (-1)^k R_2^k \int_{r_0}^{R_2} (R_2 - s)^{(N-k)-1} (s - r_0)^{2-1} ds$$

$$= \sum_{k=0}^{N-1} \frac{(N-1)!}{k!\,(N-k-1)!} (-1)^{N+k-1} R_2^k \frac{(N-k-1)!}{(N-k+1)!} (R_2 - r_0)^{N-k+1}$$

$$= \sum_{k=0}^{N-1} \frac{(-1)^{N+k-1}(N-1)!}{k!\,(N-k+1)!} R_2^k (R_2 - r_0)^{N-k+1}. \tag{16.23}$$

That is,

$$I_2^* = (N-1)! \sum_{k=0}^{N-1} \frac{(-1)^{N+k-1}}{k!\,(N-k+1)!} R_2^k (R_2 - r_0)^{N-k+1}. \tag{16.24}$$

The theorem is proved. □

Next comes an Ostrowski multivariate radial inequality for $\|\cdot\|_p$, $p > 1$.

Theorem 16.5. *Let $p, q > 1 : \frac{1}{p} + \frac{1}{q} = 1$. The rest as in Theorem 16.4. Then ($\forall\, \omega \in S^{N-1}$)*

$$\left| f(r_0\omega) - \frac{\int_A f(y)\,dy}{Vol(A)} \right| = \left| g(r_0) - \left(\frac{N}{R_2^N - R_1^N} \right) \int_{R_1}^{R_2} g(s)\, s^{N-1} ds \right|$$

$$\leq \frac{N!\,\Gamma\left(2+\frac{1}{q}\right)}{\left(R_2^N - R_1^N\right)(q+1)^{\frac{1}{q}}}$$

$$\left\{ \|g'' + g - g(r_0)\|_{L_p([R_1, r_0])} \left(\sum_{k=0}^{N-1} \frac{1}{k!\,\Gamma\left(2+\frac{1}{q}+N-k\right)} \right.\right.$$

$$\left. \times R_1^k (r_0 - R_1)^{N-k+1+\frac{1}{q}} \right) + \|g'' + g - g(r_0)\|_{L_p([r_0, R_2])}$$

$$\left.\times \left(\sum_{k=0}^{N-1} \frac{(-1)^{N+k-1}}{k!\,\Gamma\left(N-k+2+\frac{1}{q}\right)} R_2^k (R_2 - r_0)^{N-k+1+\frac{1}{q}} \right) \right\}. \tag{16.25}$$

Proof. As in Anastassiou (2023a), we get that

$$|g(r) - g(r_0)| \leq \|g'' + g - g(r_0)\|_{L_p([r_0, R_2])} \frac{(r - r_0)^{\frac{q+1}{q}}}{(q+1)^{\frac{1}{q}}}, \tag{16.26}$$

$\forall\, r \in [r_0, R_2]$,

and

$$|g(r) - g(r_0)| \leq \|g'' + g - g(r_0)\|_{L_p([R_1, r_0])} \frac{(r_0 - r)^{\frac{q+1}{q}}}{(q+1)^{\frac{1}{q}}}, \tag{16.27}$$

$\forall\, r \in [R_1, r_0]$.

Next we observe

$$\left| f(r_0\omega) - \frac{\int_A f(y)\, dy}{Vol(A)} \right|$$

$$= \left| g(r_0) - \left(\frac{N}{R_2^N - R_1^N} \right) \int_{R_1}^{R_2} g(s)\, s^{N-1} ds \right|$$

$\leq$ (as earlier)

$$\leq \left(\frac{N}{R_2^N - R_1^N} \right) \left\{ \int_{R_1}^{r_0} |g(r_0) - g(s)|\, s^{N-1} ds + \int_{r_0}^{R_2} |g(r_0) - g(s)|\, s^{N-1} ds \right\}$$

$$\leq \left(\frac{N}{R_2^N - R_1^N} \right) \frac{1}{(q+1)^{\frac{1}{q}}} \left\{ \left\| g'' + g - g(r_0) \right\|_{L_p([R_1, r_0])} \int_{R_1}^{r_0} (r_0 - s)^{\frac{q+1}{q}} s^{N-1} ds \right.$$

$$\left. + \left\| g'' + g - g(r_0) \right\|_{L_p([r_0, R_2])} \int_{r_0}^{R_2} (s - r_0)^{\frac{q+1}{q}} s^{N-1} ds \right\}. \tag{16.28}$$

Here we calculate

$$J_1 := \int_{R_1}^{r_0} (r_0 - s)^{\frac{q+1}{q}} s^{N-1} ds$$

$$= \int_{R_1}^{r_0} (r_0 - s)^{\frac{q+1}{q}} ((s - R_1) + R_1)^{N-1} ds$$

$$= \sum_{k=0}^{N-1} \binom{N-1}{k} R_1^k \int_{R_1}^{r_0} (r_0 - s)^{\left(2+\frac{1}{q}\right)-1} (s - R_1)^{N-k-1} ds$$

$$= \sum_{k=0}^{N-1} \binom{N-1}{k} R_1^k \frac{\Gamma\left(2 + \frac{1}{q}\right) (N-k-1)!}{\Gamma\left(2 + \frac{1}{q} + N - k\right)} (r_0 - R_1)^{1+\frac{1}{q}+N-k}$$

$$= \sum_{k=0}^{N-1} \frac{(N-1)!}{k!} \frac{\Gamma\left(2 + \frac{1}{q}\right)}{\Gamma\left(2 + \frac{1}{q} + N - k\right)} R_1^k (r_0 - R_1)^{1+\frac{1}{q}+N-k}. \tag{16.29}$$

We have found that

$$J_1 = (N-1)! \sum_{k=0}^{N-1} \frac{\Gamma\left(2 + \frac{1}{q}\right)}{k!\Gamma\left(2 + \frac{1}{q} + N - k\right)} R_1^k (r_0 - R_1)^{1+\frac{1}{q}+N-k}. \tag{16.30}$$

Also we have

$$\begin{aligned}
J_2 &:= \int_{r_0}^{R_2} (s-r_0)^{\left(1+\frac{1}{q}\right)} s^{N-1} ds \\
&= (-1)^{N-1} \int_{r_0}^{R_2} ((R_2-s)-R_2)^{N-1} (s-r_0)^{\left(1+\frac{1}{q}\right)} ds \\
&= (-1)^{N-1} \sum_{k=0}^{N-1} \frac{(N-1)!}{k!\,(N-k-1)!} (-1)^k R_2^k \\
&\quad \times \int_{r_0}^{R_2} (R_2-s)^{(N-k)-1} (s-r_0)^{\left(2+\frac{1}{q}\right)-1} ds \\
&= \sum_{k=0}^{N-1} \frac{(N-1)!}{k!\,(N-k-1)!} (-1)^{k+N-1} R_2^k \frac{(N-k-1)!\Gamma\left(2+\frac{1}{q}\right)}{\Gamma\left(N-k+2+\frac{1}{q}\right)} \\
&\quad \times (R_2-r_0)^{N-k+1+\frac{1}{q}} \\
&= (N-1)! \sum_{k=0}^{N-1} \frac{(-1)^{k+N-1}\Gamma\left(2+\frac{1}{q}\right)}{k!\Gamma\left(N-k+2+\frac{1}{q}\right)} R_2^k (R_2-r_0)^{N-k+1+\frac{1}{q}}.
\end{aligned} \tag{16.31}$$

That is,

$$J_2 = (N-1)! \sum_{k=0}^{N-1} \frac{(-1)^{k+N-1}\Gamma\left(2+\frac{1}{q}\right)}{k!\Gamma\left(N-k+2+\frac{1}{q}\right)} R_2^k (R_2-r_0)^{N-k+1+\frac{1}{q}}. \tag{16.32}$$

The proof of the theorem is finished. □

We make

Remark 16.2. We treat here the general, not necessarily radial case of f. We apply Theorem 16.3 to $f(r\omega)$, ω fixed, $r \in [R_1, R_2]$, under the following assumptions:

$f(\cdot\omega) \in C^3([R_1, R_2])$, $\forall\ \omega \in S^{N-1}$, where $f : \overline{A} \to \mathbb{R}$ is Lebesgue integrable; $\frac{\partial^k f}{\partial r^k}$, $k = 1, 2$, vanish on $\partial B(0, r_0)$, r_0 is fixed in $[R_1, R_2]$; $\left(\frac{\partial^3 f}{\partial r^3} + \frac{\partial f}{\partial r}\right) \in B\left(\overline{A_1}\right)$ (bounded functions), where $A_1 := B(0, R_2) - \overline{B(0, r_0)}$; and $\left(\frac{\partial^3 f}{\partial r^3} + \frac{\partial f}{\partial r}\right) \in B\left(\overline{A_2}\right)$, where $A_2 := B(0, r_0) - \overline{B(0, R_1)}$.

Then $(\forall\ \omega \in S^{N-1})$

$$\left| f(r_0\omega) - \left(\frac{N}{R_2^N - R_1^N}\right) \int_{R_1}^{R_2} f(s\omega)\, s^{N-1} ds \right|$$

$$\leq \left(\frac{N!}{R_2^N - R_1^N}\right) \left\{ \left\| \frac{\partial^3 f}{\partial r^3} + \frac{\partial f}{\partial r} \right\|_{\infty, \overline{A_2}} \left(\sum_{k=0}^{N-1} \frac{R_1^k (r_0 - R_1)^{N-k+3}}{k!\,(N-k+3)!} \right) \right.$$

$$\left. + \left\| \frac{\partial^3 f}{\partial r^3} + \frac{\partial f}{\partial r} \right\|_{\infty, \overline{A_1}} \left(\sum_{k=0}^{N-1} \frac{(-1)^{N+k-1}}{k!\,(N-k+3)!} R_2^k (R_2 - r_0)^{N-k+3} \right) \right\} =: \lambda_1. \tag{16.33}$$

Therefore,

$$\left| \frac{\int_{S^{N-1}} f(r_0\omega)\, d\omega}{\omega_N} - \frac{N}{(R_2^N - R_1^N)\,\omega_N} \int_{S^{N-1}} \left(\int_{R_1}^{R_2} f(s\omega)\, s^{N-1} ds \right) d\omega \right| \leq \lambda_1. \tag{16.34}$$

That is,

$$\left| \frac{\Gamma\left(\frac{N}{2}\right)}{2\pi^{\frac{N}{2}}} \int_{S^{N-1}} f(r_0\omega)\, d\omega - \frac{\int_A f(x)\, dx}{Vol(A)} \right| \leq \lambda_1. \tag{16.35}$$

Therefore, it holds for $x \in \overline{A}$ that

$$\left| f(x) - \frac{\int_A f(x)\, dx}{Vol(A)} \right| \leq \left| f(x) - \frac{\Gamma\left(\frac{N}{2}\right)}{2\pi^{\frac{N}{2}}} \int_{S^{N-1}} f(r_0\omega)\, d\omega \right| + \lambda_1. \tag{16.36}$$

We have proved the following:

Theorem 16.6. *Let $f : \overline{A} \to \mathbb{R}$ be Lebesgue integrable with $f(\cdot\omega) \in C^3([R_1, R_2])$, $\forall\ \omega \in S^{N-1}$; $\frac{\partial^k f}{\partial r^k}$, $k = 1, 2$, vanish on $\partial B(0, r_0)$, r_0 is fixed in $[R_1, R_2]$; and $\left(\frac{\partial^3 f}{\partial r^3} + \frac{\partial f}{\partial r}\right) \in B\left(\overline{A_1}\right)$, $\left(\frac{\partial^3 f}{\partial r^3} + \frac{\partial f}{\partial r}\right) \in B\left(\overline{A_2}\right)$. Then for $x \in \overline{A}$ we have*

$$\left| f(x) - \frac{\int_A f(x)\, dx}{Vol(A)} \right| \leq \left| f(x) - \frac{\Gamma\left(\frac{N}{2}\right)}{2\pi^{\frac{N}{2}}} \int_{S^{N-1}} f(r_0\omega)\, d\omega \right|$$

$$+ \left(\frac{N!}{R_2^N - R_1^N}\right) \left\{ \left\| \frac{\partial^3 f}{\partial r^3} + \frac{\partial f}{\partial r} \right\|_{\infty, \overline{A_2}} \left(\sum_{k=0}^{N-1} \frac{R_1^k (r_0 - R_1)^{N-k+3}}{k!\,(N-k+3)!} \right) \right.$$

$$\left. + \left\| \frac{\partial^3 f}{\partial r^3} + \frac{\partial f}{\partial r} \right\|_{\infty, \overline{A_1}} \left(\sum_{k=0}^{N-1} \frac{(-1)^{N+k-1}}{k!\,(N-k+3)!} R_2^k (R_2 - r_0)^{N-k+3} \right) \right\}. \tag{16.37}$$

We need the following multivariate radial Ostrowski-type inequality.

Theorem 16.7 (Anastassiou 2023b). *Let the function* $f : \overline{B(0,R)} \to \mathbb{R}$ *be radial, that is, there exists a function* g *such that* $f(x) = g(r)$, *where* $r = |x|$, $r \in [0,R]$, $\forall\, x \in \overline{B(0,R)}$. *We further assume that* $g \in C^3([0,R])$, *and* $g^{(k)}(r_0) = 0$, $k = 1, 2$, *where* $r_0 \in [0,R]$ *is fixed. Then* ($\forall\, \omega \in S^{N-1}$)

$$\left| f(r_0\omega) - \frac{\int_{B(0,R)} f(y)\,dy}{Vol(B(0,R))} \right| = \left| g(r_0) - \frac{N}{R^N}\int_0^R g(s)\,s^{N-1}ds \right|$$
$$\leq \frac{N!}{R^N}\left[\|g''' + g'\|_{\infty,[0,r_0]} \frac{r_0^{3+N}}{(3+N)!} \right.$$
$$\left. + \|g''' + g'\|_{\infty,[r_0,R]} \sum_{k=0}^{N-1} \frac{(-1)^{N+k-1}}{k!\,(N-k+3)!} R^k (R-r_0)^{N-k+3} \right]. \tag{16.38}$$

We make

Remark 16.3. Let $f : \overline{B(0,R)} \to \mathbb{R}$ be a Lebesgue integrable function, that is, not necessarily a radial function. We apply Theorem 16.7 to $f(r\omega)$, where ω is fixed, $\forall\, r \in [0,R]$, under the following assumptions:

$f(\cdot\omega) \in C^3([0,R])$, $\forall\, \omega \in S^{N-1}$. Furthermore, $\frac{\partial^k f(r_0\omega)}{\partial r^k} = 0$, $k = 1, 2$, $\forall\, \omega \in S^{N-1}$, where $r_0 \in [0,R]$ is fixed. Finally, we assume that

$$\left\| \left(\frac{\partial^3 f(t\omega)}{\partial r^3} + \frac{\partial f(t\omega)}{\partial r} \right) \right\|_{\infty,(t\in[0,r_0])}, \left\| \left(\frac{\partial^3 f(t\omega)}{\partial r^3} + \frac{\partial f(t\omega)}{\partial r} \right) \right\|_{\infty,(t\in[r_0,R])} \leq K, \tag{16.39}$$

$\forall\, \omega \in S^{N-1}$, where $K > 0$.

By (16.38), we obtain ($\forall\, \omega \in S^{N-1}$)

$$\left| f(r_0\omega) - \frac{N}{R^N}\int_0^R f(s\omega)\,s^{N-1}ds \right|$$
$$\leq \frac{KN!}{R^N}\left[\frac{r_0^{3+N}}{(3+N)!} + \sum_{k=0}^{N-1} \frac{(-1)^{N+k-1}}{k!\,(N-k+3)!} R^k (R-r_0)^{N-k+3} \right] =: \lambda_2. \tag{16.40}$$

Therefore,

$$\left| \frac{\int_{S^{N-1}} f(r_0\omega)\,d\omega}{\omega_N} - \frac{N}{R^N\omega_N}\int_{S^{N-1}} \left(\int_0^R f(s\omega)\,s^{N-1}ds \right) d\omega \right| \leq \lambda_2. \tag{16.41}$$

That is,

$$\left| \frac{\Gamma\left(\frac{N}{2}\right)}{2\pi^{\frac{N}{2}}} \int_{S^{N-1}} f\left(r_0\omega\right) d\omega - \frac{\int_{B(0,R)} f\left(x\right) dx}{Vol\left(B\left(0,R\right)\right)} \right| \leq \lambda_2. \tag{16.42}$$

Therefore, it holds for $x \in \overline{B\left(0,R\right)}$ that

$$\left| f\left(x\right) - \frac{\int_{B(0,R)} f\left(x\right) dx}{Vol\left(B\left(0,R\right)\right)} \right| \leq \left| f\left(x\right) - \frac{\Gamma\left(\frac{N}{2}\right)}{2\pi^{\frac{N}{2}}} \int_{S^{N-1}} f\left(r_0\omega\right) d\omega \right| + \lambda_2. \tag{16.43}$$

We have proved on the ball the following:

Theorem 16.8. *Let $f : \overline{B\left(0,R\right)} \to \mathbb{R}$ be a Lebesgue integrable function, that is, not necessarily a radial function. Here $f\left(\cdot\omega\right) \in C^3\left(\left[0,R\right]\right)$, $R > 0$, $\forall\, \omega \in S^{N-1}$. Furthermore, $\frac{\partial^k f(r_0\omega)}{\partial r^k} = 0$, $k = 1,2$, $\forall\, \omega \in S^{N-1}$, where $r_0 \in \left[0,R\right]$ is fixed. Finally, we assume that*

$$\left\| \left(\frac{\partial^3 f\left(t\omega\right)}{\partial r^3} + \frac{\partial f\left(t\omega\right)}{\partial r} \right) \right\|_{\infty,(t\in[0,r_0])}, \left\| \left(\frac{\partial^3 f\left(t\omega\right)}{\partial r^3} + \frac{\partial f\left(t\omega\right)}{\partial r} \right) \right\|_{\infty,(t\in[r_0,R])} \leq K, \tag{16.44}$$

$\forall\, \omega \in S^{N-1}$, where $K > 0$.

Then, for $x \in \overline{B\left(0,R\right)}$ we have

$$\left| f\left(x\right) - \frac{\int_{B(0,R)} f\left(x\right) dx}{Vol\left(B\left(0,R\right)\right)} \right| \leq \left| f\left(x\right) - \frac{\Gamma\left(\frac{N}{2}\right)}{2\pi^{\frac{N}{2}}} \int_{S^{N-1}} f\left(r_0\omega\right) d\omega \right|$$
$$+ \frac{KN!}{R^N} \left[\frac{r_0^{3+N}}{(3+N)!} + \left(\sum_{k=0}^{N-1} \frac{\left(-1\right)^{N+k-1}}{k!\left(N-k+3\right)!} R^k \left(R-r_0\right)^{N-k+3} \right) \right]. \tag{16.45}$$

Next we generalize Theorem 16.3.

Theorem 16.9. *Let the function $f : \overline{A} \to \mathbb{R}$ be radial, that is, there exists g such that $f\left(x\right) = g\left(r\right)$, $r = \left|x\right|$, $\forall\, x \in \overline{A}$; $\omega \in S^{N-1}$. Assume that $g \in C^5\left(\left[R_1,R_2\right]\right)$, and $g^{(k)}\left(r_0\right) = 0$, $k = 1,2,3,4$, where $r_0 \in \left[R_1,R_2\right]$*

is fixed. Here $\alpha, \beta \in \mathbb{R} : \alpha\beta\left(\alpha^2 - \beta^2\right) \neq 0$. *Then* $(\forall\ \omega \in S^{N-1})$

$$\left|f(r_0\omega) - \frac{\int_A f(y)\,dy}{Vol(A)}\right| = \left|g(r_0) - \left(\frac{N}{R_2^N - R_1^N}\right)\int_{R_1}^{R_2} g(s)\,s^{N-1}ds\right|$$
$$\leq \left(\frac{N!}{R_2^N - R_1^N}\right)\frac{2}{|\beta^2 - \alpha^2|}$$
$$\left\{\left\|g^{(5)} + \left(\alpha^2 + \beta^2\right)g^{(3)} + \alpha^2\beta^2 g'\right\|_{\infty,[R_1,r_0]}\left(\sum_{k=0}^{N-1}\frac{R_1^k\left(r_0 - R_1\right)^{N-k+3}}{k!\,(N-k+3)!}\right)\right.$$
$$+\left\|g^{(5)} + \left(\alpha^2 + \beta^2\right)g^{(3)} + \alpha^2\beta^2 g'\right\|_{\infty,[r_0,R_2]}$$
$$\left.\times\left(\sum_{k=0}^{N-1}\frac{(-1)^{N+k-1}R_2^k\left(R_2 - r_0\right)^{N-k+3}}{k!\,(N-k+3)!}\right)\right\}. \tag{16.46}$$

Proof. As in Anastassiou (2023a), we get that

$$|g(r) - g(r_0)| \leq \frac{1}{3\,|\beta^2 - \alpha^2|}\left\|g^{(5)} + \left(\alpha^2 + \beta^2\right)g^{(3)} + \alpha^2\beta^2 g'\right\|_{\infty,[r_0,R_2]}(r - r_0)^3$$
$$=: A^*\,(r - r_0)^3, \quad \forall\ r \in [r_0, R_2], \tag{16.47}$$

and

$$|g(r) - g(r_0)| \leq \frac{1}{3\,|\beta^2 - \alpha^2|}\left\|g^{(5)} + \left(\alpha^2 + \beta^2\right)g^{(3)} + \alpha^2\beta^2 g'\right\|_{\infty,[R_1,r_0]}(r_0 - r)^3$$
$$=: B\,(r_0 - r)^3, \quad \forall\ r \in [R_1, r_0]. \tag{16.48}$$

We observe that

$$\left|f(r_0\omega) - \frac{\int_A f(y)\,dy}{Vol(A)}\right| \leq \text{(as earlier)}$$
$$\leq \left(\frac{N}{R_2^N - R_1^N}\right)\left\{\int_{R_1}^{r_0}|g(r_0) - g(s)|\,s^{N-1}ds + \int_{r_0}^{R_2}|g(r_0) - g(s)|\,s^{N-1}ds\right\} \tag{16.49}$$
$$\leq \left(\frac{N}{R_2^N - R_1^N}\right)\left\{B\int_{R_1}^{r_0}(r_0 - s)^3\,s^{N-1}ds + A^*\int_{r_0}^{R_2}(s - r_0)^3\,s^{N-1}ds\right\}$$
$$\overset{\text{(by (16.13), (16.15))}}{=} \left(\frac{N}{R_2^N - R_1^N}\right)\left\{B3!\,(N-1)!\left(\sum_{k=0}^{N-1}\frac{R_1^k\left(r_0 - R_1\right)^{3+N-k}}{k!\,(N-k+3)!}\right)\right.$$
$$\left.+\,A^*3!\,(N-1)!\left(\sum_{k=0}^{N-1}\frac{(-1)^{N+k-1}}{k!\,(N-k+3)!}R_2^k\left(R_2 - r_0\right)^{3+N-k}\right)\right\}$$

$$= \left(\frac{N!3!}{R_2^N - R_1^N}\right) \left\{ B \left(\sum_{k=0}^{N-1} \frac{R_1^k (r_0 - R_1)^{N-k+3}}{k!(N-k+3)!} \right) \right.$$

$$\left. + A^* \left(\sum_{k=0}^{N-1} \frac{(-1)^{N+k-1}}{k!(N-k+3)!} R_2^k (R_2 - r_0)^{N-k+3} \right) \right\}. \tag{16.50}$$

The theorem is proved. □

A generalized radial L_1 result follows.

Theorem 16.10. *All as in Theorem 16.9, except now $f \in C^4([R_1, R_2])$. Then ($\forall\, \omega \in S^{N-1}$)*

$$\left| f(r_0\omega) - \frac{\int_A f(y)\,dy}{Vol(A)} \right| = \left| g(r_0) - \left(\frac{N}{R_2^N - R_1^N}\right) \int_{R_1}^{R_2} g(s)\, s^{N-1} ds \right|$$

$$\leq \left(\frac{N!}{R_2^N - R_1^N}\right) \frac{2}{|\beta^2 - \alpha^2|}$$

$$\left\{ \left\| g'''' + (\alpha^2 + \beta^2) g'' + \alpha^2\beta^2 g - \alpha^2\beta^2 g(r_0) \right\|_{L_1([R_1, r_0])} \right.$$

$$\left(\sum_{k=0}^{N-1} \frac{R_1^k (r_0 - R_1)^{N-k+1}}{k!(N-k+1)!} \right)$$

$$+ \left\| g'''' + (\alpha^2 + \beta^2) g'' + \alpha^2\beta^2 g - \alpha^2\beta^2 g(r_0) \right\|_{L_1([r_0, R_2])}$$

$$\left. \left(\sum_{k=0}^{N-1} \frac{(-1)^{N+k-1} R_2^k (R_2 - r_0)^{N-k+1}}{k!(N-k+1)!} \right) \right\}. \tag{16.51}$$

Proof. As in Anastassiou (2023a), we get that

$$|g(r) - g(r_0)|$$

$$\leq \frac{2}{|\beta^2 - \alpha^2|} \left\| g'''' + (\alpha^2 + \beta^2) g'' + \alpha^2\beta^2 g - \alpha^2\beta^2 g(r_0) \right\|_{L_1([R_1, r_0])} (r_0 - r), \tag{16.52}$$

$\forall\, r \in [R_1, r_0]$,

and

$$|g(r) - g(r_0)|$$

$$\leq \frac{2}{|\beta^2 - \alpha^2|} \left\| g'''' + (\alpha^2 + \beta^2) g'' + \alpha^2\beta^2 g - \alpha^2\beta^2 g(r_0) \right\|_{L_1([r_0, R_2])} (r - r_0), \tag{16.53}$$

$\forall\, r \in [r_0, R_2]$.

The rest of the proof goes as in Theorem 16.4. □

It follows a generalized Ostrowski multivariate radial inequality for $\|\cdot\|_p$, $p > 1$.

Theorem 16.11. *All as in Theorem 16.10, plus let* $p, q > 1 : \frac{1}{p} + \frac{1}{q} = 1$. *Then* $(\forall\ \omega \in S^{N-1})$

$$\left| f(r_0\omega) - \frac{\int_A f(y)\,dy}{Vol(A)} \right| = \left| g(r_0) - \left(\frac{N}{R_2^N - R_1^N} \right) \int_{R_1}^{R_2} g(s)\, s^{N-1} ds \right|$$

$$\leq \left(\frac{N!}{R_2^N - R_1^N} \right) \frac{\Gamma\left(2 + \frac{1}{q}\right)}{(q+1)^{\frac{1}{q}}}$$

$$\left\{ \left\| g'''' + \left(\alpha^2 + \beta^2\right) g'' + \alpha^2\beta^2 g - \alpha^2\beta^2 g(r_0) \right\|_{L_p([R_1, r_0])} \right.$$

$$\left(\sum_{k=0}^{N-1} \frac{R_1^k (r_0 - R_1)^{N-k+1+\frac{1}{q}}}{k!\Gamma\left(2 + \frac{1}{q} + N - k\right)} \right)$$

$$+ \left\| g'''' + \left(\alpha^2 + \beta^2\right) g'' + \alpha^2\beta^2 g - \alpha^2\beta^2 g(r_0) \right\|_{L_p([r_0, R_2])}$$

$$\left. \left(\sum_{k=0}^{N-1} \frac{(-1)^{N+k-1} R_2^k (R_2 - r_0)^{N-k+1+\frac{1}{q}}}{k!\Gamma\left(N - k + 2 + \frac{1}{q}\right)} \right) \right\}. \tag{16.54}$$

Proof. As in Anastassiou (2023a), we have that

$$|g(r) - g(r_0)| \leq \frac{2}{|\beta^2 - \alpha^2| (q+1)^{\frac{1}{q}}}$$

$$\left\| g'''' + \left(\alpha^2 + \beta^2\right) g'' + \alpha^2\beta^2 g - \alpha^2\beta^2 g(r_0) \right\|_{L_p([R_1, r_0])} (r_0 - r)^{\frac{q+1}{q}}, \tag{16.55}$$

$\forall\ r \in [R_1, r_0]$,

and

$$|g(r) - g(r_0)| \leq \frac{2}{|\beta^2 - \alpha^2| (q+1)^{\frac{1}{q}}}$$

$$\left\| g'''' + \left(\alpha^2 + \beta^2\right) g'' + \alpha^2\beta^2 g - \alpha^2\beta^2 g(r_0) \right\|_{L_p([r_0, R_2])} (r - r_0)^{\frac{q+1}{q}}, \tag{16.56}$$

$\forall\ r \in [r_0, R_2]$.

The rest of the proof goes as in Theorem 16.5. □

It follows a non-radial generalization of Theorem 16.6.

Theorem 16.12. *Let $f : \overline{A} \to \mathbb{R}$ be Lebesgue measurable with $f(\cdot\omega) \in C^5([R_1, R_2])$, $\forall\ \omega \in S^{N-1}$; $\frac{\partial^k f}{\partial r^k}$, $k = 1,2,3,4$, vanish on $\partial B(0, r_0)$, r_0 is fixed in $[R_1, R_2]$; and $\left(\frac{\partial^5 f}{\partial r^5} + (\alpha^2+\beta^2)\frac{\partial^3 f}{\partial r^3} + \alpha^2\beta^2\frac{\partial f}{\partial r}\right) \in B(\overline{A_1})$, $\left(\frac{\partial^5 f}{\partial r^5} + (\alpha^2+\beta^2)\frac{\partial^3 f}{\partial r^3} + \alpha^2\beta^2\frac{\partial f}{\partial r}\right) \in B(\overline{A_2})$. Here $\alpha, \beta \in \mathbb{R}$: $\alpha\beta(\alpha^2-\beta^2) \neq 0$. Then for $x \in \overline{A}$ we have*

$$\left|f(x) - \frac{\int_A f(x)\,dx}{Vol(A)}\right| \leq \left|f(x) - \frac{\Gamma\left(\frac{N}{2}\right)}{2\pi^{\frac{N}{2}}}\int_{S^{N-1}} f(r_0\omega)\,d\omega\right|$$

$$+\left(\frac{N!}{R_2^N - R_1^N}\right)\frac{2}{|\beta^2-\alpha^2|}$$

$$\left\{\left\|\frac{\partial^5 f}{\partial r^5} + (\alpha^2+\beta^2)\frac{\partial^3 f}{\partial r^3} + \alpha^2\beta^2\frac{\partial f}{\partial r}\right\|_{\infty,\overline{A_2}}\left(\sum_{k=0}^{N-1}\frac{R_1^k(r_0-R_1)^{N-k+3}}{k!(N-k+3)!}\right)\right.$$

$$+\left\|\frac{\partial^5 f}{\partial r^5} + (\alpha^2+\beta^2)\frac{\partial^3 f}{\partial r^3} + \alpha^2\beta^2\frac{\partial f}{\partial r}\right\|_{\infty,\overline{A_1}}$$

$$\left.\times\left(\sum_{k=0}^{N-1}\frac{(-1)^{N+k-1}R_2^k(R_2-r_0)^{N-k+3}}{k!(N-k+3)!}\right)\right\}. \tag{16.57}$$

Proof. As very similar to Theorem 16.6 is omitted. Now it is based on Theorem 16.9. ☐

We need the following multivariate radial Ostrowski-type general inequality.

Theorem 16.13. *(Anastassiou 2023b) Let the function $f : \overline{B(0,R)} \to \mathbb{R}$ be radial, that is there exists a function g such that $f(x) = g(r)$, where $r = |x|$, $r \in [0, R]$, $\forall\ x \in \overline{B(0,R)}$. We further assume that $g \in C^5([0,R])$, and $g^{(k)}(r_0) = 0$, $k = 1,2,3,4$, where $r_0 \in [0,R]$ is fixed. Here $\alpha, \beta \in \mathbb{R}$: $\alpha\beta(\alpha^2-\beta^2) \neq 0$. Then ($\forall\ \omega \in S^{N-1}$)*

$$\left|f(r_0\omega) - \frac{\int_{B(0,R)} f(y)\,dy}{Vol(B(0,R))}\right| = \left|g(r_0) - \frac{N}{R^N}\int_0^R g(s)\,s^{N-1}ds\right|$$

$$\leq \frac{2N!}{R^N|\beta^2-\alpha^2|}\left\{\left\|g^{(5)} + (\alpha^2+\beta^2)g^{(3)} + \alpha^2\beta^2 g'\right\|_{\infty,[0,r_0]}\left(\frac{r_0^{3+N}}{(3+N)!}\right)\right.$$

$$\left\| g^{(5)} + \left(\alpha^2+\beta^2\right) g^{(3)} + \alpha^2\beta^2 g' \right\|_{\infty,[r_0,R]}$$

$$\times \left(\sum_{k=0}^{N-1} \frac{(-1)^{N+k-1}}{k!\,(N-k+3)!} R^k \left(R-r_0\right)^{N-k+3} \right) \Bigg\}. \tag{16.58}$$

We finish with a non-radial necessarily Ostrowski-type general inequality on the ball.

Theorem 16.14. *Let $f : \overline{B(0,R)} \to \mathbb{R}$ be a Lebesgue integrable function, that is, not necessarily a radial function. Here $f(\cdot\omega) \in C^5([0,R])$, $R > 0$, $\forall\ \omega \in S^{N-1}$. Furthermore $\frac{\partial^k f(r_0\omega)}{\partial r^k} = 0$, $k = 1,2,3,4$, $\forall\ \omega \in S^{N-1}$, where $r_0 \in [0,R]$ is fixed. Assume that $\left\| \left(\frac{\partial^5 f(t\omega)}{\partial r^5} + \left(\alpha^2+\beta^2\right) \frac{\partial^3 f(t\omega)}{\partial r^3} + \alpha^2\beta^2 \frac{\partial f(t\omega)}{\partial r} \right) \right\|_{\infty,(t\in[0,r_0])}$, $\left\| \left(\frac{\partial^5 f(t\omega)}{\partial r^5} + \left(\alpha^2+\beta^2\right) \frac{\partial^3 f(t\omega)}{\partial r^3} + \alpha^2\beta^2 \frac{\partial f(t\omega)}{\partial r} \right) \right\|_{\infty,(t\in[r_0,R])} \leq K^*, \forall\ \omega \in S^{N-1}$, where $K^* > 0$. Here $\alpha, \beta \in \mathbb{R} : \alpha\beta\left(\alpha^2-\beta^2\right) \neq 0$.*

Then, for $x \in \overline{B(0,R)}$ we have

$$\left| f(x) - \frac{\int_{B(0,R)} f(x)\,dx}{Vol\,(B(0,R))} \right| \leq \left| f(x) - \frac{\Gamma\left(\frac{N}{2}\right)}{2\pi^{\frac{N}{2}}} \int_{S^{N-1}} f(r_0\omega)\,d\omega \right|$$

$$+ \frac{2K^* N!}{R^N \left|\beta^2-\alpha^2\right|} \left\{ \left(\frac{r_0^{3+N}}{(3+N)!} \right) + \left(\sum_{k=0}^{N-1} \frac{(-1)^{N+k-1}}{k!\,(N-k+3)!} R^k \left(R-r_0\right)^{N-k+3} \right) \right\}. \tag{16.59}$$

Proof. Very similar to Theorem 16.8, now it is based on Theorem 16.13. □

Bibliography

G.A. Anastassiou, *Advances on Fractional Inequalities*, Springer, Heidelberg, New York, 2011.

G.A. Anastassiou, *Opial and Ostrowski type inequalities based on trigonometric and hyperbolic Taylor formulae*, submitted, 2023a.

G.A. Anastassiou, *Updated radial Ostrowski inequalities over a ball*, submitted, 2023b.

G.A. Anastassiou, Updated Ostrowski inequalities over a spherical shell, *Bulletin of the Transilvania University of Brasov, Series III: Mathematics and Computer Science*, 3(65)(2) (2023c), 31–52.

W. Rudin, *Real and Complex Analysis*, International student edition, McGraw Hill, London, New York, 1970.

D. Stroock, *A Concise Introduction in the Theory of Integration*, Third Edition, Birkhaüser, Boston, Basel, Berlin, 1999.

A. Ostrowski, Über die Absolutabweichung einer differentiebaren Funktion von ihrem Integralmittelwert, *Comment. Math. Helv.* 10, (1938), 226–227.

Chapter 17

Trigonometric and Hyperbolic Polya Inequalities

Here based on trigonometric and hyperbolic-type Taylor formulae, we derive Polya-type inequalities in a number of cases. It follows Anastassiou (2024).

17.1 Main Results

We present a collection of Polya-type inequalities.

Theorem 17.1. *Let $f \in C^2([a,b], K)$, where $K = \mathbb{R}$ or $\mathbb{C}$, such that $f^{(k)}(a) = f^{(k)}(b) = 0$, $k = 0, 1$, and $p, q > 1 : \frac{1}{p} + \frac{1}{q} = 1$. We set*

$$M_1 := \max\left\{ \|f'' + f\|_{\infty,\left[a,\frac{a+b}{2}\right]}, \|f'' + f\|_{\infty,\left[\frac{a+b}{2},b\right]} \right\}, \tag{17.1}$$

$$M_2 := \max\left\{ \|f'' + f\|_{L_1\left(\left[a,\frac{a+b}{2}\right]\right)}, \|f'' + f\|_{L_1\left(\left[\frac{a+b}{2},b\right]\right)} \right\}, \tag{17.2}$$

and

$$M_3 := \max\left\{ \|f'' + f\|_{L_q\left(\left[a,\frac{a+b}{2}\right]\right)}, \|f'' + f\|_{L_q\left(\left[\frac{a+b}{2},b\right]\right)} \right\}. \tag{17.3}$$

Then

$$\left| \int_a^b f(x)\,dx \right| \leq \int_a^b |f(x)|\,dx \leq \min \left\{ \begin{array}{c} \frac{(b-a)^3}{8} M_1, \\ \\ \frac{(b-a)^2}{4} M_2, \\ \\ \frac{(b-a)^{2+\frac{1}{p}}}{2^{1+\frac{1}{p}}(p+1)^{\frac{1}{p}}\left(2+\frac{1}{p}\right)} M_3 \end{array} \right\}. \tag{17.4}$$

Proof. Here $f \in C^2([a,b], K)$, such that $f^{(k)}(a) = f^{(k)}(b) = 0$, $k = 0, 1$. By Corollary 3.4 of Ali and Páles (2022), we have

$$f(x) = \int_a^x (f''(t) + f(t)) \sin(x - t)\,dt \tag{17.5}$$

and

$$f(x) = \int_x^b (f''(t) + f(t)) \sin(x - t) \, dt. \tag{17.6}$$

By using $|\sin x| \leq |x|$, $\forall \, x \in \mathbb{R}$, we obtain

$$\begin{aligned} |f(x)| &\leq \int_a^x |f''(t) + f(t)| \, |\sin(x - t)| \, dt \\ &\leq \int_a^x |f''(t) + f(t)| \, (x - t) \, dt \leq \left(\int_a^x (x - t) \, dt \right) \|f'' + f\|_{\infty, \left[a, \frac{a+b}{2}\right]} \\ &= \frac{(x - a)^2}{2} \|f'' + f\|_{\infty, \left[a, \frac{a+b}{2}\right]}, \quad \forall \, x \in \left[a, \frac{a+b}{2}\right]. \end{aligned} \tag{17.7}$$

Also it holds

$$\begin{aligned} |f(x)| &\leq (x - a) \left(\int_a^x |f''(t) + f(t)| \, dt \right) \\ &\leq (x - a) \|f'' + f\|_{L_1\left(\left[a, \frac{a+b}{2}\right]\right)}, \quad \forall \, x \in \left[a, \frac{a+b}{2}\right]. \end{aligned} \tag{17.8}$$

Furthermore, by Hölder's inequality, we have $(p, q > 1 : \frac{1}{p} + \frac{1}{q} = 1)$

$$\begin{aligned} |f(x)| &\leq \left(\int_a^x |f''(t) + f(t)|^q \, dt \right)^{\frac{1}{q}} \left(\int_a^x (x - t)^p \, dt \right)^{\frac{1}{p}} \\ &\leq \|f'' + f\|_{L_q\left(\left[a, \frac{a+b}{2}\right]\right)} \frac{(x - a)^{\frac{p+1}{p}}}{(p + 1)^{\frac{1}{p}}}, \quad \forall \, x \in \left[a, \frac{a+b}{2}\right]. \end{aligned} \tag{17.9}$$

We have found that

$$|f(x)| \leq \left\{ \begin{array}{l} \frac{(x-a)^2}{2} \|f'' + f\|_{\infty, \left[a, \frac{a+b}{2}\right]}, \\ (x - a) \|f'' + f\|_{L_1\left(\left[a, \frac{a+b}{2}\right]\right)}, \\ \frac{(x-a)^{1+\frac{1}{p}}}{(p+1)^{\frac{1}{p}}} \|f'' + f\|_{L_q\left(\left[a, \frac{a+b}{2}\right]\right)}, \quad p, q > 1 : \frac{1}{p} + \frac{1}{q} = 1 \end{array} \right\}, \tag{17.10}$$

$\forall \, x \in \left[a, \frac{a+b}{2}\right]$.

Similarly acting, we get that

$$\begin{aligned} |f(x)| &= \left| \int_x^b (f''(t) + f(t)) \sin(x-t)\, dt \right| \\ &\leq \int_x^b |f''(t) + f(t)| \, |\sin(t-x)| \, dt \\ &\leq \int_x^b |f''(t) + f(t)| (t-x)\, dt, \end{aligned} \tag{17.11}$$

and

$$|f(x)| \leq \left\{ \begin{array}{l} \frac{(b-x)^2}{2} \|f'' + f\|_{\infty, \left[\frac{a+b}{2}, b\right]}, \\ (b-x) \|f'' + f\|_{L_1\left(\left[\frac{a+b}{2}, b\right]\right)}, \\ \frac{(b-x)^{1+\frac{1}{p}}}{(p+1)^{\frac{1}{p}}} \|f'' + f\|_{L_q\left(\left[\frac{a+b}{2}, b\right]\right)}, \quad p, q > 1 : \frac{1}{p} + \frac{1}{q} = 1 \end{array} \right\}, \tag{17.12}$$

$\forall\, x \in \left[\frac{a+b}{2}, b\right]$.

Consequently, we obtain

$$\int_a^{\frac{a+b}{2}} |f(x)|\, dx \leq \left\{ \begin{array}{l} \frac{(b-a)^3}{16} \|f'' + f\|_{\infty, \left[a, \frac{a+b}{2}\right]}, \\ \frac{(b-a)^2}{8} \|f'' + f\|_{L_1\left(\left[a, \frac{a+b}{2}\right]\right)}, \\ \frac{(b-a)^{2+\frac{1}{p}}}{2^{2+\frac{1}{p}} (p+1)^{\frac{1}{p}} \left(2+\frac{1}{p}\right)} \|f'' + f\|_{L_q\left(\left[a, \frac{a+b}{2}\right]\right)}, \\ \qquad p, q > 1 : \frac{1}{p} + \frac{1}{q} = 1. \end{array} \right\} \tag{17.13}$$

Similarly, we derive that

$$\int_{\frac{a+b}{2}}^b |f(x)|\, dx \leq \left\{ \begin{array}{l} \frac{(b-a)^3}{16} \|f'' + f\|_{\infty, \left[\frac{a+b}{2}, b\right]}, \\ \frac{(b-a)^2}{8} \|f'' + f\|_{L_1\left(\left[\frac{a+b}{2}, b\right]\right)}, \\ \frac{(b-a)^{2+\frac{1}{p}}}{2^{2+\frac{1}{p}} (p+1)^{\frac{1}{p}} \left(2+\frac{1}{p}\right)} \|f'' + f\|_{L_q\left(\left[\frac{a+b}{2}, b\right]\right)}, \quad p, q > 1 : \frac{1}{p} + \frac{1}{q} = 1. \end{array} \right\} \tag{17.14}$$

We have that

$$\int_a^b |f(x)|\,dx = \int_a^{\frac{a+b}{2}} |f(x)|\,dx + \int_{\frac{a+b}{2}}^b |f(x)|\,dx$$

$$\leq \begin{cases} \frac{(b-a)^3}{16}\left[\|f''+f\|_{\infty,\left[a,\frac{a+b}{2}\right]} + \|f''+f\|_{\infty,\left[\frac{a+b}{2},b\right]}\right], \\ \frac{(b-a)^2}{8}\left[\|f''+f\|_{L_1\left(\left[a,\frac{a+b}{2}\right]\right)} + \|f''+f\|_{L_1\left(\left[\frac{a+b}{2},b\right]\right)}\right], \\ \frac{(b-a)^{2+\frac{1}{p}}}{2^{2+\frac{1}{p}}(p+1)^{\frac{1}{p}}\left(2+\frac{1}{p}\right)}\left[\|f''+f\|_{L_q\left(\left[a,\frac{a+b}{2}\right]\right)}, + \|f''+f\|_{L_q\left(\left[\frac{a+b}{2},b\right]\right)}\right]. \end{cases} \tag{17.15}$$

$$\leq \left\{\begin{array}{l} \frac{(b-a)^3}{8}M_1, \\ \frac{(b-a)^2}{4}M_2, \\ \frac{(b-a)^{2+\frac{1}{p}}}{2^{1+\frac{1}{p}}(p+1)^{\frac{1}{p}}\left(2+\frac{1}{p}\right)}M_3 \end{array}\right\}. \tag{17.16}$$

The claim is proved. □

We continue with

Theorem 17.2. *All as in Theorem 17.1. Denote*

$$M_1^* := \max\left\{\|f''-f\|_{\infty,\left[a,\frac{a+b}{2}\right]}, \|f''-f\|_{\infty,\left[\frac{a+b}{2},b\right]}\right\}, \tag{17.17}$$

$$M_2^* := \max\left\{\|f''-f\|_{L_1\left(\left[a,\frac{a+b}{2}\right]\right)}, \|f''-f\|_{L_1\left(\left[\frac{a+b}{2},b\right]\right)}\right\}, \tag{17.18}$$

and

$$M_3^* := \max\left\{\|f''-f\|_{L_q\left(\left[a,\frac{a+b}{2}\right]\right)}, \|f''-f\|_{L_q\left(\left[\frac{a+b}{2},b\right]\right)}\right\}. \tag{17.19}$$

Then

$$\left|\int_a^b f(x)\,dx\right| \leq \int_a^b |f(x)|\,dx \leq \min\cosh(b-a)\times\left\{\begin{array}{l} \frac{(b-a)^3}{8}M_1^*, \\ \frac{(b-a)^2}{4}M_2^*, \\ \frac{(b-a)^{2+\frac{1}{p}}}{2^{1+\frac{1}{p}}(p+1)^{\frac{1}{p}}\left(2+\frac{1}{p}\right)}M_3^* \end{array}\right\}. \tag{17.20}$$

Proof. As similar to Theorem 17.1 it is omitted. It is based on Corollary 3.5 of Ali and Páles (2022). Also we use that $|\sinh x| \leq \cosh(b-a)|x|$, $\forall$ $x \in [-(b-a), b-a]$, by the mean value theorem. $\square$

It follows

Theorem 17.3. *Let $f \in C^4([a,b], K)$, where $K = \mathbb{R}$ or $\mathbb{C}$, such that $f^{(k)}(a) = f^{(k)}(b) = 0$, $k = 0,1,2,3$, and $p, q > 1 : \frac{1}{p} + \frac{1}{q} = 1$. We set*

$$A_1 := \max\left\{ \left\| f^{(4)} - f \right\|_{\infty,\left[a,\frac{a+b}{2}\right]}, \left\| f^{(4)} - f \right\|_{\infty,\left[\frac{a+b}{2},b\right]} \right\}, \tag{17.21}$$

$$A_2 := \max\left\{ \left\| f^{(4)} - f \right\|_{L_1\left(\left[a,\frac{a+b}{2}\right]\right)}, \left\| f^{(4)} - f \right\|_{L_1\left(\left[\frac{a+b}{2},b\right]\right)} \right\}, \tag{17.22}$$

and

$$A_3 := \max\left\{ \left\| f^{(4)} - f \right\|_{L_q\left(\left[a,\frac{a+b}{2}\right]\right)}, \left\| f^{(4)} - f \right\|_{L_q\left(\left[\frac{a+b}{2},b\right]\right)} \right\}. \tag{17.23}$$

Then

$$\left| \int_a^b f(x)\, dx \right| \leq \int_a^b |f(x)|\, dx$$

$$\leq \min \frac{(\cosh(b-a)+1)}{4} \times \left\{ \begin{array}{l} \frac{(b-a)^3}{4} A_1, \\ \frac{(b-a)^2}{2} A_2, \\ \frac{(b-a)^{2+\frac{1}{p}}}{2^{\frac{1}{p}}(p+1)^{\frac{1}{p}}\left(2+\frac{1}{p}\right)} A_3 \end{array} \right\}. \tag{17.24}$$

Proof. As similar to Theorem 17.1 it is omitted. It is based on Corollary 3.6 of Ali and Páles (2022). $\square$

We continue with

Theorem 17.4. *Let $f \in C^4([a,b], K)$, where $K = \mathbb{R}$ or $\mathbb{C}$, such that $f^{(k)}(a) = f^{(k)}(b) = 0$, $k = 0,1,2,3$, and $p, q > 1 : \frac{1}{p} + \frac{1}{q} = 1$. Let also $\alpha, \beta \in \mathbb{R}$ with $\alpha\beta\left(\alpha^2 - \beta^2\right) \neq 0$. We set*

$$B_1 := \max\left\{ \left\| f^{(4)} + \left(\alpha^2 + \beta^2\right) f'' + \alpha^2\beta^2 f \right\|_{\infty,\left[a,\frac{a+b}{2}\right]}, \right.$$

$$\left. \left\| f^{(4)} + \left(\alpha^2 + \beta^2\right) f'' + \alpha^2\beta^2 f \right\|_{\infty,\left[\frac{a+b}{2},b\right]} \right\}, \tag{17.25}$$

$$B_2 := \max\left\{\left\|f^{(4)} + \left(\alpha^2+\beta^2\right)f'' + \alpha^2\beta^2 f\right\|_{L_1\left(\left[a,\frac{a+b}{2}\right]\right)},\right.$$
$$\left.\left\|f^{(4)} + \left(\alpha^2+\beta^2\right)f'' + \alpha^2\beta^2 f\right\|_{L_1\left(\left[\frac{a+b}{2},b\right]\right)}\right\}, \tag{17.26}$$

and

$$B_3 := \max\left\{\left\|f^{(4)} + \left(\alpha^2+\beta^2\right)f'' + \alpha^2\beta^2 f\right\|_{L_q\left(\left[a,\frac{a+b}{2}\right]\right)},\right.$$
$$\left.\left\|f^{(4)} + \left(\alpha^2+\beta^2\right)f'' + \alpha^2\beta^2 f\right\|_{L_q\left(\left[\frac{a+b}{2},b\right]\right)}\right\}. \tag{17.27}$$

Then

$$\left|\int_a^b f(x)\,dx\right| \le \int_a^b |f(x)|\,dx$$
$$\le \min \frac{1}{|\beta^2-\alpha^2|} \times \begin{cases} \frac{(b-a)^3}{4} B_1, \\ \frac{(b-a)^2}{2} B_2, \\ \frac{(b-a)^{2+\frac{1}{p}}}{2^{\frac{1}{p}}(p+1)^{\frac{1}{p}}\left(2+\frac{1}{p}\right)} B_3 \end{cases}. \tag{17.28}$$

Proof. As similar to Theorem 17.1 it is omitted. It is based on Corollary 3.7 of Ali and Páles (2022). □

We finish with

Theorem 17.5. *All as in Theorem 17.4, $|\alpha|, |\beta| < 1$. However here, instead of B_1, B_2, B_3, we set*

$$D_1 := \max\left\{\left\|f^{(4)} - \left(\alpha^2+\beta^2\right)f'' + \alpha^2\beta^2 f\right\|_{\infty,\left[a,\frac{a+b}{2}\right]},\right.$$
$$\left.\left\|f^{(4)} - \left(\alpha^2+\beta^2\right)f'' + \alpha^2\beta^2 f\right\|_{\infty,\left[\frac{a+b}{2},b\right]}\right\}, \tag{17.29}$$

$$D_2 := \max\left\{\left\|f^{(4)} - \left(\alpha^2+\beta^2\right)f'' + \alpha^2\beta^2 f\right\|_{L_1\left(\left[a,\frac{a+b}{2}\right]\right)},\right.$$
$$\left.\left\|f^{(4)} - \left(\alpha^2+\beta^2\right)f'' + \alpha^2\beta^2 f\right\|_{L_1\left(\left[\frac{a+b}{2},b\right]\right)}\right\}, \tag{17.30}$$

and

$$D_3 := \max\left\{\left\|f^{(4)} - \left(\alpha^2+\beta^2\right)f'' + \alpha^2\beta^2 f\right\|_{L_q\left(\left[a,\frac{a+b}{2}\right]\right)},\right.$$
$$\left.\left\|f^{(4)} - \left(\alpha^2+\beta^2\right)f'' + \alpha^2\beta^2 f\right\|_{L_q\left(\left[\frac{a+b}{2},b\right]\right)}\right\}. \tag{17.31}$$

Then

$$\left|\int_a^b f(x)\,dx\right| \leq \int_a^b |f(x)|\,dx$$
$$\leq \min\frac{\cosh(b-a)}{|\beta^2-\alpha^2|} \times \begin{cases} \frac{(b-a)^3}{4}D_1, \\ \frac{(b-a)^2}{2}D_2, \\ \frac{(b-a)^{2+\frac{1}{p}}}{2^{\frac{1}{p}}(p+1)^{\frac{1}{p}}\left(2+\frac{1}{p}\right)}D_3 \end{cases}. \tag{17.32}$$

Proof. As similar to Theorem 17.1 it is omitted. It is based on Corollary 3.9 of Ali and Páles (2022). □

Bibliography

Ali Hasan Ali and Zsolt Páles, Taylor-type expansions in terms of exponential polynomials, *Mathematical Inequalities & Applications*, 25(4) (2022), 1123–1141.

G.A. Anastassiou, Trigonometric and hyperbolic Polya type inequalities, *Journal of Computational Analysis and Applications*, 32(1) (2024), 186–191.

Chapter 18

New Opial and Polya Inequalities Over a Spherical Shell

Here we present general multivariate Opial and Polya-type inequalities over spherical shells. The proofs are derived by the use of some estimates coming out of some new trigonometric and hyperbolic Taylor's formulae (Ali and Pales 2022, Anastassiou 2023) and reducing the multivariate problem to a univariate one via general polar coordinates. It follows Anastassiou (2024).

18.1 Background

We need

Remark 18.1. Let the spherical shell

$$A := B(0, R_2) - \overline{B(0, R_1)},$$

$0 < R_1 < R_2$, $A \subseteq \mathbb{R}^N$, $N \geq 2$, $x \in \overline{A}$; $r = |x|$, $r \in [R_1, R_2]$, $\forall\, x \in \overline{A}$, $|\cdot|$ the Euclidean norm. Here x can be written uniquely as $x = r\omega$, where $r = |x| > 0$ and $\omega = \frac{x}{r} \in S^{N-1}$, $|\omega| = 1$, see (Rudin 1970, pp. 149–150 and Stroock 1999, p. 421).

Furthermore, for $F : \overline{A} \to \mathbb{R}$ a Lebesgue integrable function we have that

$$\int_A F(x)\,dx = \int_{S^{N-1}} \left(\int_{R_1}^{R_2} F(r\omega)\, r^{N-1} dr \right) d\omega, \tag{18.1}$$

where $S^{N-1} := \left\{ x \in \mathbb{R}^N : |x| = 1 \right\}$.

Let $d\omega$ be the element of surface measure on $\mathbb{S}^{N-1}$ with surface area

$$\omega_N = \int_{S^{N-1}} d\omega = \frac{2\pi^{\frac{N}{2}}}{\Gamma\left(\frac{N}{2}\right)}. \tag{18.2}$$

Here it is volume of A,

$$vol\,(A) = \frac{\omega_N \left(R_2^N - R_1^N\right)}{N}. \tag{18.3}$$

Above it is $B(0,r) := \left\{x \in \mathbb{R}^N : |x| < r\right\}$, $r > 0$.

Here K is either $\mathbb{R}$ or $\mathbb{C}$, and $C_K^n(I)$ denotes functions n-times continuously differentiable on an interval $I \subset \mathbb{R}$ with values in K.

From Anastassiou (2023), we need to mention the following Opial-type inequalities.

Theorem 18.1. *Let $f \in C_K^2(I)$, with interval $I \subset \mathbb{R}$, $a, x \in I$, $a < x$, and $f(a) = f'(a) = 0$, with $p, q > 1 : \frac{1}{p} + \frac{1}{q} = 1$. Then*

$$\int_a^x |f(w)|\,|f''(w) + f(w)|\,dw$$
$$\leq 2^{-\frac{1}{q}} \left(\int_a^x \left(\int_a^w |\sin(w-t)|^p\,dt\right) dw\right)^{\frac{1}{p}} \left(\int_a^x |f''(w) + f(w)|^q\,dw\right)^{\frac{2}{q}}. \tag{18.4}$$

Theorem 18.2. *Let $f \in C_K^2(I)$, $a, x \in I$, $a < x$, and $f(a) = f'(a) = 0$, with $p, q > 1 : \frac{1}{p} + \frac{1}{q} = 1$. Then*

$$\int_a^x |f(w)|\,|f''(w) - f(w)|\,dw$$
$$\leq 2^{-\frac{1}{q}} \left(\int_a^x \left(\int_a^w |\sinh(w-t)|^p\,dt\right) dw\right)^{\frac{1}{p}}$$
$$\times \left(\int_a^x |f''(w) - f(w)|^q\,dw\right)^{\frac{2}{q}}. \tag{18.5}$$

Theorem 18.3. *Let $f \in C_K^4(I)$, interval $I \subset \mathbb{R}$, let $a, x \in I$, $a < x$, $f(a) = f'(a) = f''(a) = f'''(a) = 0$, with $p, q > 1 : \frac{1}{p} + \frac{1}{q} = 1$. Then*

$$\int_a^x |f(w)| \left|f^{(iv)}(w) - f(w)\right| dw$$
$$\leq 2^{-\left(1+\frac{1}{q}\right)} \left(\int_a^x \left(\int_a^w |\sinh(w-t) - \sin(w-t)|^p\,dt\right) dw\right)^{\frac{1}{p}}$$
$$\left(\int_a^x \left|f^{(iv)}(w) - f(w)\right|^q dw\right)^{\frac{2}{q}}. \tag{18.6}$$

Theorem 18.4. *All as in Theorem 18.3. Let* $\alpha, \beta \in \mathbb{R} : \alpha\beta\left(\alpha^2 - \beta^2\right) \neq 0$. *Then*

$$\int_a^x |f(w)| \left| f^{(4)}(w) + \left(\alpha^2 + \beta^2\right) f''(w) + \alpha^2\beta^2 f(w) \right| dw$$
$$\leq \frac{1}{2^{\frac{1}{q}} |\alpha\beta| \left|\beta^2 - \alpha^2\right|} \left(\int_a^x \left(\int_a^w |\beta \sin(\alpha(w-t)) - \alpha \sin(\beta(w-t))|^p \, dt \right) dw \right)^{\frac{1}{p}}$$
$$\left(\int_a^x \left| f^{(4)}(w) + \left(\alpha^2 + \beta^2\right) f''(w) + \alpha^2\beta^2 f(w) \right|^q dw \right)^{\frac{2}{q}}. \quad (18.7)$$

Theorem 18.5. *All as in Theorem 18.3. Let* $\alpha \in \mathbb{R}$, $\alpha \neq 0$. *Then*

$$\int_a^x |f(w)| \left| f^{(4)}(w) + 2\alpha^2 f''(w) + \alpha^4 f(w) \right| dw$$
$$\leq \frac{1}{2^{\frac{1}{q}+1} |\alpha|^3} \left(\int_a^x \left(\int_a^w |\sin(\alpha(w-t)) - \alpha(w-t)\cos(\alpha(w-t))|^p \, dt \right) dw \right)^{\frac{1}{p}}$$
$$\left(\int_a^x \left| f^{(4)}(w) + 2\alpha^2 f''(w) + \alpha^4 f(w) \right|^q dw \right)^{\frac{2}{q}}. \quad (18.8)$$

Theorem 18.6. *All as in Theorem 18.4. Then*

$$\int_a^x |f(w)| \left| f^{(4)}(w) - \left(\alpha^2 + \beta^2\right) f''(w) + \alpha^2\beta^2 f(w) \right| dw$$
$$\leq \frac{1}{2^{\frac{1}{q}} |\alpha\beta| \left|\beta^2 - \alpha^2\right|}$$
$$\times \left(\int_a^x \left(\int_a^w |\alpha \sinh(\beta(w-t)) - \beta \sinh(\alpha(w-t))|^p \, dt \right) dw \right)^{\frac{1}{p}}$$
$$\left(\int_a^x \left| f^{(4)}(w) - \left(\alpha^2 + \beta^2\right) f''(w) + \alpha^2\beta^2 f(w) \right|^q dw \right)^{\frac{2}{q}}. \quad (18.9)$$

Theorem 18.7. *All as in Theorem 18.5. Then*

$$\int_a^x |f(w)| \left| f^{(4)}(w) - 2\alpha^2 f''(w) + \alpha^4 f(w) \right| dw$$
$$\leq \frac{1}{2^{\frac{1}{q}+1} |\alpha|^3}$$

$$\times\left(\int_a^x\left(\int_a^w|\alpha(w-t)\cosh(\alpha(w-t))-\sinh(\alpha(w-t))|^p\,dt\right)dw\right)^{\frac{1}{p}}$$

$$\left(\int_a^x\left|f^{(4)}(w)-2\alpha^2 f''(w)+\alpha^4 f(w)\right|^q dw\right)^{\frac{2}{q}}. \tag{18.10}$$

We use the above Opial-type inequalities in the case of $p=q=2$.

The motivation came from the following famous Opial's inequality.

Theorem 18.8 (Opial 1960). *Let $c>0$ and $y(x)$ be real, continuously differentiable on $[0,c]$, with $y(0)=y(c)=0$. Then*

$$\int_0^c|y(x)\,y'(x)|\,dx\le\frac{c}{4}\int_0^c\left(y'(x)\right)^2dx. \tag{18.11}$$

Equality holds for the function $y(x)=x$ on $\left[0,\frac{c}{2}\right]$ and $y(x)=c-x$ on $\left[\frac{c}{2},c\right]$.

18.2 Results

First, we present a collection of Opial-type inequalities on the spherical shell A.

Theorem 18.9. *Let $F:\overline{A}\to\mathbb{R}$ Lebesgue integrable function with $F(\cdot\omega)\in C^2([R_1,R_2])$, with $F(R_1\omega)=\frac{\partial F}{\partial r}(R_1\omega)=0$, $\forall\,\omega\in S^{N-1}$. Then*

$$\int_A|F(x)|\left|F(x)+\frac{\partial^2F(x)}{\partial r^2}\right|dx$$

$$\le 2^{-\frac{1}{2}}\left(\frac{R_2}{R_1}\right)^{N-1}\left(\int_{R_1}^{R_2}\left(\int_{R_1}^{r}(\sin(r-t))^2\,dt\right)dr\right)^{\frac{1}{2}}$$

$$\times\int_A\left(F(x)+\frac{\partial^2F(x)}{\partial r^2}\right)^2dx. \tag{18.12}$$

Proof. Here we apply Theorem 18.1 to $F(\cdot\omega)$ for $p=q=2$. So, for every $\omega\in S^{N-1}$, we have that

$$\int_{R_1}^{R_2}|F(r\omega)|\left|\frac{\partial^2F(r\omega)}{\partial r^2}+F(r\omega)\right|dr$$

$$\leq 2^{-\frac{1}{2}} \left(\int_{R_1}^{R_2} \left(\int_{R_1}^{r} (\sin (r-t))^2 \, dt \right) dr \right)^{\frac{1}{2}}$$
$$\times \left(\int_{R_1}^{R_2} \left(\frac{\partial^2 F(r\omega)}{\partial r^2} + F(r\omega) \right)^2 dr \right). \quad (18.13)$$

We have $R_1 \leq r \leq R_2$ and $R_1^{N-1} \leq r^{N-1} \leq R_2^{N-1}$, and $R_2^{1-N} \leq r^{1-N} \leq R_1^{1-N}$.

We observe the following:

$$R_2^{1-N} \int_{R_1}^{R_2} |F(r\omega)| \left| \frac{\partial^2 F(r\omega)}{\partial r^2} + F(r\omega) \right| r^{N-1} dr$$
$$\leq \int_{R_1}^{R_2} |F(r\omega)| \left| \frac{\partial^2 F(r\omega)}{\partial r^2} + F(r\omega) \right| r^{N-1} r^{1-N} dr$$
$$= \int_{R_1}^{R_2} |F(r\omega)| \left| \frac{\partial^2 F(r\omega)}{\partial r^2} + F(r\omega) \right| dr$$
$$\overset{(18.13)}{\leq} 2^{-\frac{1}{2}} \left(\int_{R_1}^{R_2} \left(\int_{R_1}^{r} (\sin (r-t))^2 \, dt \right) dr \right)^{\frac{1}{2}} \quad (18.14)$$
$$\times \left(\int_{R_1}^{R_2} \left(\frac{\partial^2 F(r\omega)}{\partial r^2} + F(r\omega) \right)^2 r^{N-1} r^{1-N} dr \right)$$
$$\leq R_1^{1-N} 2^{-\frac{1}{2}} \left(\int_{R_1}^{R_2} \left(\int_{R_1}^{r} (\sin (r-t))^2 \, dt \right) dr \right)^{\frac{1}{2}}$$
$$\times \left(\int_{R_1}^{R_2} \left(\frac{\partial^2 F(r\omega)}{\partial r^2} + F(r\omega) \right)^2 r^{N-1} dr \right). \quad (18.15)$$

Therefore, it holds

$$\int_{R_1}^{R_2} |F(r\omega)| \left| \frac{\partial^2 F(r\omega)}{\partial r^2} + F(r\omega) \right| r^{N-1} dr$$
$$\leq \left(\frac{R_1}{R_2} \right)^{1-N} 2^{-\frac{1}{2}} \left(\int_{R_1}^{R_2} \left(\int_{R_1}^{r} (\sin (r-t))^2 \, dt \right) dr \right)^{\frac{1}{2}}$$
$$\times \left(\int_{R_1}^{R_2} \left(\frac{\partial^2 F(r\omega)}{\partial r^2} + F(r\omega) \right)^2 r^{N-1} dr \right). \quad (18.16)$$

Consequently, we obtain

$$\int_{S^{N-1}} \left(\int_{R_1}^{R_2} |F(r\omega)| \left| \frac{\partial^2 F(r\omega)}{\partial r^2} + F(r\omega) \right| r^{N-1} dr \right) d\omega$$

$$\leq 2^{-\frac{1}{2}} \left(\frac{R_2}{R_1} \right)^{N-1} \left(\int_{R_1}^{R_2} \left(\int_{R_1}^{r} (\sin(r-t))^2 \, dt \right) dr \right)^{\frac{1}{2}}$$

$$\times \int_{S^{N-1}} \left(\int_{R_1}^{R_2} \left(\frac{\partial^2 F(r\omega)}{\partial r^2} + F(r\omega) \right)^2 r^{N-1} dr \right) d\omega. \tag{18.17}$$

Applying (18.1), we obtain (18.12). □

Next, we present more Opial-type inequalities on spherical shell. Their proofs are similar to the proof of Theorem 18.9 and are based on Theorems 18.2–18.7. Use also of (18.1).

Theorem 18.10. *Same assumptions as in Theorem 18.9. Then*

$$\int_A |F(x)| \left| F(x) - \frac{\partial^2 F(x)}{\partial r^2} \right| dx$$

$$\leq 2^{-\frac{1}{2}} \left(\frac{R_2}{R_1} \right)^{N-1} \left(\int_{R_1}^{R_2} \left(\int_{R_1}^{r} (\sinh(r-t))^2 \, dt \right) dr \right)^{\frac{1}{2}}$$

$$\times \int_A \left(F(x) - \frac{\partial^2 F(x)}{\partial r^2} \right)^2 dx. \tag{18.18}$$

Proof. Based on (18.5). □

Theorem 18.11. *Let $F : \overline{A} \to \mathbb{R}$ Lebesgue integrable function with $F(\cdot\omega) \in C^4([R_1, R_2])$, with $F(R_1\omega) = \frac{\partial^{(i)} F}{\partial r^{(i)}}(R_1\omega) = 0$, $i = 1, 2, 3$; $\forall$ $\omega \in S^{N-1}$. Then*

$$\int_A |F(x)| \left| \frac{\partial^4 F(x)}{\partial r^4} - F(x) \right| dx$$

$$\leq 2^{-\frac{3}{2}} \left(\frac{R_2}{R_1} \right)^{N-1} \left(\int_{R_1}^{R_2} \left(\int_{R_1}^{r} (\sinh(r-t) - \sin(r-t))^2 \, dt \right) dr \right)^{\frac{1}{2}}$$

$$\times \int_A \left(\frac{\partial^4 F(x)}{\partial r^4} - F(x) \right)^2 dx. \tag{18.19}$$

Proof. Based on (18.6). □

Theorem 18.12. *All as in Theorem 18.11. Let* $\alpha, \beta \in \mathbb{R} : \alpha\beta\left(\alpha^2 - \beta^2\right) \neq 0$. *Then*

$$\int_A |F(x)| \left| \frac{\partial^4 F(x)}{\partial r^4} + \left(\alpha^2 + \beta^2\right) \frac{\partial^2 F(x)}{\partial r^2} + \alpha^2 \beta^2 F(x) \right| dx$$

$$\leq \frac{1}{\sqrt{2}\left|\alpha\beta\left(\beta^2 - \alpha^2\right)\right|} \left(\frac{R_2}{R_1}\right)^{N-1}$$

$$\times \left(\int_{R_1}^{R_2} \left(\int_{R_1}^{r} \left(\beta \sin\left(\alpha\left(r-t\right)\right) - \alpha \sin\left(\beta\left(r-t\right)\right)\right)^2 dt \right) dr \right)^{\frac{1}{2}}$$

$$\times \int_A \left(\frac{\partial^4 F(x)}{\partial r^4} + \left(\alpha^2 + \beta^2\right) \frac{\partial^2 F(x)}{\partial r^2} + \alpha^2 \beta^2 F(x) \right)^2 dx. \quad (18.20)$$

Proof. Based on (18.7). □

Theorem 18.13. *All as in Theorem 18.11. Let* $\alpha \in \mathbb{R}$, $\alpha \neq 0$. *Then*

$$\int_A |F(x)| \left| \frac{\partial^4 F(x)}{\partial r^4} + 2\alpha^2 \frac{\partial^2 F(x)}{\partial r^2} + \alpha^4 F(x) \right| dx$$

$$\leq \frac{1}{2^{\frac{3}{2}}\left|\alpha^3\right|} \left(\frac{R_2}{R_1}\right)^{N-1}$$

$$\times \left(\int_{R_1}^{R_2} \left(\int_{R_1}^{r} \left(\sin\left(\alpha\left(r-t\right)\right) - \alpha\left(r-t\right)\cos\left(\alpha\left(r-t\right)\right)\right)^2 dt \right) dr \right)^{\frac{1}{2}}$$

$$\times \int_A \left(\frac{\partial^4 F(x)}{\partial r^4} + 2\alpha^2 \frac{\partial^2 F(x)}{\partial r^2} + \alpha^4 F(x) \right)^2 dx. \quad (18.21)$$

Proof. Based on (18.8). □

Theorem 18.14. *All as in Theorem 18.12. Then*

$$\int_A |F(x)| \left| \frac{\partial^4 F(x)}{\partial r^4} - \left(\alpha^2 + \beta^2\right) \frac{\partial^2 F(x)}{\partial r^2} + \alpha^2 \beta^2 F(x) \right| dx$$

$$\leq \frac{1}{\sqrt{2}\left|\alpha\beta\left(\beta^2 - \alpha^2\right)\right|} \left(\frac{R_2}{R_1}\right)^{N-1}$$

$$\times\left(\int_{R_1}^{R_2}\left(\int_{R_1}^{r}(\alpha\sinh(\beta(r-t))-\beta\sinh(\alpha(r-t)))^2\,dt\right)dr\right)^{\frac{1}{2}}$$

$$\times\int_A\left(\frac{\partial^4 F(x)}{\partial r^4}-\left(\alpha^2+\beta^2\right)\frac{\partial^2 F(x)}{\partial r^2}+\alpha^2\beta^2 F(x)\right)^2 dx. \quad (18.22)$$

Proof. By (18.9). □

Finally, we give the following Opial-type inequality.

Theorem 18.15. *All as in Theorem 18.13. Then*

$$\int_A |F(x)|\left|\frac{\partial^4 F(x)}{\partial r^4}-2\alpha^2\frac{\partial^2 F(x)}{\partial r^2}+\alpha^4 F(x)\right| dx$$

$$\leq \frac{1}{2^{\frac{3}{2}}|\alpha|^3}\left(\frac{R_2}{R_1}\right)^{N-1}$$

$$\times\left(\int_{R_1}^{R_2}\left(\int_{R_1}^{r}(\alpha(r-t)\cosh(\alpha(r-t))-\sinh(\alpha(r-t)))^2\,dt\right)dr\right)^{\frac{1}{2}}$$

$$\times\int_A\left(\frac{\partial^4 F(x)}{\partial r^4}-2\alpha^2\frac{\partial^2 F(x)}{\partial r^2}+\alpha^4 F(x)\right)^2 dx. \quad (18.23)$$

Proof. Based on (18.10). □

We need the following results.

Theorem 18.16. *(Ali and Pales 2022) For $f\in C_K^2([a,b])$ and $x\in[a,b]$: $f(a)=f'(a)=0$, we have that*

$$f(x)=\int_a^x\left(f''(t)+f(t)\right)\sin(x-t)\,dt \quad (18.24)$$

and

$$f(x)=\int_a^x\left(f''(t)-f(t)\right)\sinh(x-t)\,dt. \quad (18.25)$$

Theorem 18.17. *(Ali and Pales 2022) For $f\in C_K^4([a,b])$ and $x\in[a,b]$: $f(a)=f'(a)=f''(a)=f'''(a)=0$, we have that*

$$f(x)=\int_a^x\left(f''''(t)-f(t)\right)\left(\frac{\sinh(x-t)-\sin(x-t)}{2}\right)dt. \quad (18.26)$$

Theorem 18.18. *(Ali and Pales 2022) Let $\alpha, \beta \in \mathbb{R}$ with $\alpha\beta\left(\beta^2 - \alpha^2\right) \neq 0$, and $f \in C_K^4([a,b])$, $x \in [a,b] : f(a) = f'(a) = f''(a) = f'''(a) = 0$. Then*

$$f(x) = \frac{1}{\alpha\beta\left(\beta^2 - \alpha^2\right)} \int_a^x \left(f''''(t) + \left(\alpha^2 + \beta^2\right) f''(t) + \alpha^2\beta^2 f(t)\right)$$

$$\left(\beta \sin\left(\alpha\left(x - t\right)\right) - \alpha \sin\left(\beta\left(x - t\right)\right)\right) dt. \tag{18.27}$$

Theorem 18.19. *(Ali and Pales 2022) Let $\alpha, \beta \in \mathbb{R}$ with $\alpha\beta\left(\alpha^2 - \beta^2\right) \neq 0$, and $f \in C_K^4([a,b])$, $x \in [a,b] : f(a) = f'(a) = f''(a) = f'''(a) = 0$. Then*

$$f(x) = \frac{1}{\alpha\beta\left(\beta^2 - \alpha^2\right)} \int_a^x \left(f''''(t) - \left(\alpha^2 + \beta^2\right) f''(t) + \alpha^2\beta^2 f(t)\right)$$

$$\left(\alpha \sinh\left(\beta\left(x - t\right)\right) - \beta \sinh\left(\alpha\left(x - t\right)\right) - \right) dt. \tag{18.28}$$

We use

$$|\sin x| \leq |x|, \quad \forall\, x \in \mathbb{R}, \tag{18.29}$$

$$|\sinh x| \leq \cosh(b - a)\,|x|, \quad \forall\, x \in [-(b-a), b - a]. \tag{18.30}$$

Both of the above come by applications of mean value theorem.

We give the following Polya-type univariate inequalities.

Theorem 18.20. *For $f \in C_K^2([a,b])$ and $x \in [a,b] : f(a) = f'(a) = 0$, it holds*

$$\int_a^b |f(x)|\, dx \leq \frac{(b-a)^2}{2} \int_a^b |f''(t) + f(t)|\, dt \tag{18.31}$$

and

$$\int_a^b |f(x)|\, dx \leq \cosh(b-a) \frac{(b-a)^2}{2} \int_a^b |f''(t) - f(t)|\, dt. \tag{18.32}$$

Proof. (i) By (18.24), we have that

$$\begin{aligned} |f(x)| &\leq \int_a^x |f''(t) + f(t)|\, |\sin(x - t)|\, dt \\ &\leq \int_a^x |f''(t) + f(t)|\, (x - t)\, dt \\ &\leq (x - a) \int_a^x |f''(t) + f(t)|\, dt \leq (x - a) \int_a^b |f''(t) + f(t)|\, dt. \end{aligned} \tag{18.33}$$

Therefore, it holds

$$\int_a^b |f(x)|\,dx \leq \left(\int_a^b (x-a)\,dx\right)\int_a^b |f''(t)+f(t)|\,dt$$
$$= \frac{(b-a)^2}{2}\int_a^b |f''(t)+f(t)|\,dt. \tag{18.34}$$

(ii) By (18.25), we have that

$$|f(x)| \leq \int_a^x |f''(t)-f(t)|\,|\sinh(x-t)|\,dt$$
$$\leq \cosh(b-a)\int_a^x |f''(t)-f(t)|\,(x-t)\,dt$$
$$\leq \cosh(b-a)\,(x-a)\int_a^x |f''(t)-f(t)|\,dt$$
$$\leq \cosh(b-a)\,(x-a)\int_a^b |f''(t)-f(t)|\,dt. \tag{18.35}$$

Therefore, we get

$$\int_a^b |f(x)|\,dx \leq \cosh(b-a)\left(\int_a^b (x-a)\,dx\right)\int_a^b |f''(t)-f(t)|\,dt$$
$$= \cosh(b-a)\,\frac{(b-a)^2}{2}\int_a^b |f''(t)-f(t)|\,dt. \tag{18.36}$$

□

Theorem 18.21. *All as in Theorem 18.17. Then*

$$\int_a^b |f(x)|\,dx \leq (\cosh(b-a)+1)\,\frac{(b-a)^2}{4}\int_a^b |f''''(t)-f(t)|\,dt. \tag{18.37}$$

Theorem 18.22. *All as in Theorem 18.18. Then*

$$\int_a^b |f(x)|\,dx \leq \frac{(b-a)^2}{|\beta^2-\alpha^2|}\int_a^b \left|f''''(t)+\left(\alpha^2+\beta^2\right)f''(t)+\alpha^2\beta^2 f(t)\right|\,dt. \tag{18.38}$$

Theorem 18.23. *All as in Theorem 18.19, plus* $|\alpha|, |\beta| < 1$. *Then*

$$\int_a^b |f(x)|\,dx \leq \frac{\cosh(b-a)\,(b-a)^2}{|\beta^2-\alpha^2|}\int_a^b \left|f''''(t)\right.$$
$$\left.-\left(\alpha^2+\beta^2\right)f''(t)+\alpha^2\beta^2 f(t)\right|\,dt. \tag{18.39}$$

Next comes a collection of Polya-type inequalities on the spherical shell. Their proofs are based on Theorems 18.20–18.23, (18.1) and they are similar to the proof of Theorem 18.9, and as such details are omitted.

Theorem 18.24. *Same assumptions as in Theorem 18.9. Then*

$$\int_A |F(x)| \leq \left(\frac{R_2}{R_1}\right)^{N-1} \frac{(R_2 - R_1)^2}{2} \int_A \left| \frac{\partial^2 F(x)}{\partial r^2} + F(x) \right| dx \tag{18.40}$$

and

$$\int_A |F(x)| \leq \left(\frac{R_2}{R_1}\right)^{N-1} \cosh(R_2 - R_1) \frac{(R_2 - R_1)^2}{2} \int_A \left| \frac{\partial^2 F(x)}{\partial r^2} - F(x) \right| dx. \tag{18.41}$$

Proof. Based on Theorem 18.20. □

Theorem 18.25. *Same assumptions as in Theorem 18.11. Then*

$$\int_A |F(x)| \leq \left(\frac{R_2}{R_1}\right)^{N-1} (\cosh(R_2 - R_1) + 1) \frac{(R_2 - R_1)^2}{4} \int_A \left| \frac{\partial^4 F(x)}{\partial r^4} - F(x) \right| dx. \tag{18.42}$$

Proof. Based on Theorem 18.21. □

Theorem 18.26. *All as in Theorem 18.12. Then*

$$\int_A |F(x)| \leq \left(\frac{R_2}{R_1}\right)^{N-1} \frac{(R_2 - R_1)^2}{|\beta^2 - \alpha^2|} \int_A \left| \frac{\partial^4 F(x)}{\partial r^4} + (\alpha^2 + \beta^2) \frac{\partial^2 F(x)}{\partial r^2} + \alpha^2 \beta^2 F(x) \right| dx. \tag{18.43}$$

Proof. Based on Theorem 18.22. □

We finish with

Theorem 18.27. *All as in Theorem 18.12, plus $|\alpha|, |\beta| < 1$. Then*

$$\int_A |F(x)| \leq \left(\frac{R_2}{R_1}\right)^{N-1} \frac{\cosh(R_2 - R_1)(R_2 - R_1)^2}{|\beta^2 - \alpha^2|} \int_A \left| \frac{\partial^4 F(x)}{\partial r^4} - (\alpha^2 + \beta^2) \frac{\partial^2 F(x)}{\partial r^2} + \alpha^2 \beta^2 F(x) \right| dx. \tag{18.44}$$

Proof. Based on Theorem 18.23. □

Bibliography

Ali Hasan Ali, Zsolt Pales, Taylor-type expansions in terms of exponential polynomials, *Mathematical Inequalities and Applications*, 25(4) (2022), 1123–1141.

G.A. Anastassiou, Opial and Ostrowski type inequalities based on trigonometric and hyperbolic Taylor formulae, *Malaya Journal of Matematik*, 11(S) 2023, 1–26.

G.A. Anastassiou, New Opial and Polya type inequalities over a spherical shell, *Journal of Computational Analysis and Applications*, 32(1) (2024), 211–221.

W. Rudin, *Real and Complex Analysis*, International student edition, McGraw Hill, London, New York, 1970.

Z. Opial, Sur une inegalité, *Annales Polonici Mathematici*, 8 (1960), 29–32.

D. Stroock, *A Concise Introduction in the Theory of Integration*, Third Edition, Birkhaüser, Boston, Basel, Berlin, 1999.

Chapter 19

Trigonometric and Hyperbolic Poincaré, Sobolev and Hilbert–Pachpatte Inequalities

In this chapter, based on trigonometric and hyperbolic-type Taylor formulae, we establish Poincaré, Sobolev and Hilbert–Pachpatte-type inequalities of different kinds, specific and general. It follows Anastassiou (2023).

19.1 Main Results

We start with a collection of Poincaré-type inequalities.

Theorem 19.1. *Let $f \in C^2([a,b], K)$, where $K = \mathbb{R}$ or $\mathbb{C}$, such that $f(a) = f'(a) = 0$, and $p, q > 1 : \frac{1}{p} + \frac{1}{q} = 1$. Then*

$$\|f\|_{L_q([a,b])} \leq \left(\int_a^b \left(\int_a^x |\sin(x-t)|^p \, dt \right)^{\frac{q}{p}} dx \right)^{\frac{1}{q}} \|f + f''\|_{L_q([a,b])} . \tag{19.1}$$

Proof. Since $f(a) = f'(a) = 0$, by Corollary 3.4 of Ali and Páles (2022), we have

$$f(x) = \int_a^x (f''(t) + f(t)) \sin(x-t) \, dt, \tag{19.2}$$

$\forall \, x \in [a,b]$.

It follows by Hölder's inequality that

$$\begin{aligned} |f(x)| &\leq \int_a^x |f''(t) + f(t)| \, |\sin(x-t)| \, dt \\ &\leq \left(\int_a^x |\sin(x-t)|^p \, dt \right)^{\frac{1}{p}} \left(\int_a^x |(f'' + f)(t)|^q \, dt \right)^{\frac{1}{q}} \\ &\leq \left(\int_a^x |\sin(x-t)|^p \, dt \right)^{\frac{1}{p}} \|f'' + f\|_{L_q([a,b])}. \end{aligned} \tag{19.3}$$

Hence,

$$\left|f(x)\right|^q \leq \left(\int_a^x \left|\sin(x-t)\right|^p dt\right)^{\frac{q}{p}} \left\|f''+f\right\|_{L_q([a,b])}^q \tag{19.4}$$

and

$$\int_a^b \left|f(x)\right|^q dx \leq \left(\int_a^b \left(\int_a^x \left|\sin(x-t)\right|^p dt\right)^{\frac{q}{p}} dx\right) \left\|f''+f\right\|_{L_q([a,b])}^q. \tag{19.5}$$

We have proved that

$$\left(\int_a^b \left|f(x)\right|^q dx\right)^{\frac{1}{q}} \leq \left(\int_a^b \left(\int_a^x \left|\sin(x-t)\right|^p dt\right)^{\frac{q}{p}} dx\right)^{\frac{1}{q}} \left\|f''+f\right\|_{L_q([a,b])}. \tag{19.6}$$

□

It follows

Theorem 19.2. *All as in Theorem 19.1. Then*

$$\left\|f\right\|_{L_q([a,b])} \leq \left(\int_a^b \left(\int_a^x \left|\sinh(x-t)\right|^p dt\right)^{\frac{q}{p}} dx\right)^{\frac{1}{q}} \left\|f-f''\right\|_{L_q([a,b])}. \tag{19.7}$$

Proof. As similar to Theorem 19.1 is omitted. It is based on Corollary 3.5 of Ali and Páles (2022). □

We continue with

Theorem 19.3. *Let $f \in C^4([a,b], K)$, where $K = \mathbb{R}$ or $\mathbb{C}$, such that $f(a) = f'(a) = f''(a) = f'''(a) = 0$, and $p, q > 1 : \frac{1}{p} + \frac{1}{q} = 1$. Then*

$$\left\|f\right\|_{L_q([a,b])}$$
$$\leq \frac{1}{2}\left(\int_a^b \left(\int_a^x \left|\sinh(x-t) - \sin(x-t)\right|^p dt\right)^{\frac{q}{p}} dx\right)^{\frac{1}{q}} \left\|f-f''''\right\|_{L_q([a,b])}. \tag{19.8}$$

Proof. As similar to Theorem 19.1 is omitted. It is based on Corollary 3.6 of Ali and Páles (2022). □

It follows

Theorem 19.4. *Let $f \in C^4([a,b],K)$, where $K = \mathbb{R}$ or $\mathbb{C}$, such that $f(a) = f'(a) = f''(a) = f'''(a) = 0$, and $p,q > 1 : \frac{1}{p} + \frac{1}{q} = 1$. Also let $\alpha, \beta \in \mathbb{R}$ with $\alpha\beta(\alpha^2 - \beta^2) \neq 0$. Then*

$$\|f\|_{L_q([a,b])} \leq \frac{1}{|\alpha||\beta||\beta^2 - \alpha^2|} \left(\int_a^b \left(\int_a^x |\beta \sin(\alpha(x-t)) - \alpha \sin(\beta(x-t))|^p dt \right)^{\frac{q}{p}} dx \right)^{\frac{1}{q}} \left\| f'''' + (\alpha^2 + \beta^2) f'' + \alpha^2 \beta^2 f \right\|_{L_q([a,b])}. \tag{19.9}$$

Proof. As similar to Theorem 19.1 is omitted. It is based on Corollary 3.7 of Ali and Páles (2022). □

Next comes

Theorem 19.5. *Let $f \in C^4([a,b],K)$, where $K = \mathbb{R}$ or $\mathbb{C}$, such that $f^{(i)}(a) = 0$, $i = 0,1,2,3$, and $p,q > 1 : \frac{1}{p} + \frac{1}{q} = 1$; $\alpha \in \mathbb{R}$, $\alpha \neq 0$. Then*

$$\|f\|_{L_q([a,b])} \leq \frac{1}{2|\alpha|^3} \left(\int_a^b \left(\int_a^x |\sin(\alpha(x-t)) - \alpha(x-t)\cos(\alpha(x-t))|^p dt \right)^{\frac{q}{p}} dx \right)^{\frac{1}{q}} \left\| f^{(4)} + 2\alpha^2 f'' + \alpha^4 f \right\|_{L_q([a,b])}. \tag{19.10}$$

Proof. As similar to Theorem 19.1 is omitted. It is based on Corollary 3.8 of Ali and Páles (2022). □

We continue with

Theorem 19.6. *All as in Theorem 19.4. Then*

$$\|f\|_{L_q([a,b])} \leq \frac{1}{|\alpha||\beta||\beta^2 - \alpha^2|} \times \left(\int_a^b \left(\int_a^x |\alpha \sinh(\beta(x-t)) - \beta \sinh(\alpha(x-t))|^p dt \right)^{\frac{q}{p}} dx \right)^{\frac{1}{q}} \left\| f^{(4)} - (\alpha^2 + \beta^2) f'' + \alpha^2 \beta^2 f \right\|_{L_q([a,b])}. \tag{19.11}$$

Proof. As similar to Theorem 19.1 is omitted. It is based on Corollary 3.9 of Ali and Páles (2022). □

We also give

Theorem 19.7. *All as in Theorem 19.5. Then*

$$\|f\|_{L_q([a,b])}$$
$$\leq \frac{1}{2|\alpha|^3}\left(\int_a^b\left(\int_a^x |\alpha(x-t)\cosh(\alpha(x-t))-\sinh(\alpha(x-t))|^p\,dt\right)^{\frac{q}{p}}dx\right)^{\frac{1}{q}}$$
$$\left\|f^{(4)}-2\alpha^2 f''+\alpha^4 f\right\|_{L_q([a,b])}. \tag{19.12}$$

Proof. As similar to Theorem 19.1 is omitted. It is based on Corollary 3.10 of Ali and Páles (2022). □

We make

Remark 19.1. The following comes from Ali and Páles (2022). Let K denote $\mathbb{R}$ or $\mathbb{C}$, $[a,b]\subset\mathbb{R}$. For $c=(c_0,\ldots,c_n)\in K^{n+1}$ with $c_n=1$, let the nth-order linear differential operator

$$D_c : C^n([a,b],K)\to C([a,b],K)$$

be defined by the formula

$$D_c(f) := c_n f^{(n)}+\cdots+c_1 f'+c_0 f \quad (f\in C^n([a,b],K)).$$

Let $\omega_c\in C^n(\mathbb{R},\mathbb{C})$ denote the unique solution of the initial value problem

$$D_c(\omega_c)=0,\quad \omega_c^{(l)}(0)=\delta_{l,n-1}\quad (l\in\{0,\ldots,n-1\}). \tag{19.13}$$

The function ω_c is be called the characteristic solution of $D_c(\omega)=0$.

Define

$$(T_{a,c}f)(x) := \sum_{j=0}^{n-1}\left(f^{(j)}(a)\sum_{i=0}^{n-1-j} c_{i+j+1}\omega_c^{(i)}(x-a)\right), \tag{19.14}$$

if $f^{(j)}(a)=0$, $j=0,1,\ldots,n-1$, then

$$(T_{a,c}f)(x)=0,\ \forall\, x\in[a,b].$$

Theorem 19.8. *(Ali and Páles 2022) Let $n\in\mathbb{N}$, $c=(c_0,\ldots,c_n)\in K^{n+1}$ with $c_n=1$. Then for all $f\in C^n([a,b],K)$, $x\in[a,b]$, we have*

$$f(x)=(T_{a,c}f)(x)+\int_a^x D_c(f)(t)\,\omega_c(x-t)\,dt. \tag{19.15}$$

If $f^{(j)}(a) = 0,\ j = 0, 1, \ldots, n-1,$ *then*

$$f(x) = \int_a^x D_c(f)(t)\,\omega_c(x-t)\,dt, \tag{19.16}$$

$\forall\ x \in [a, b]$.

We give the following general Poincaré-type inequality.

Theorem 19.9. *All as in Remark 19.1 and* $f^{(j)}(a) = 0,\ j = 0, 1, \ldots, n-1,$ $p, q > 1 : \frac{1}{p} + \frac{1}{q} = 1$. *Then*

$$\|f\|_{L_q([a,b])} \leq \left(\int_a^b \left(\int_a^x |\omega_c(x-t)|^p\,dt\right)^{\frac{q}{p}} dx\right)^{\frac{1}{q}} \|D_c(f)\|_{L_q([a,b])}. \tag{19.17}$$

Proof. As similar to Theorem 19.1 is omitted. It is based on Remark 19.1 and Theorem 19.8. □

Next follow Sobolev-type inequalities.

Theorem 19.10. *All as in Theorem 19.1,* $r > 0$. *Then*

$$\|f\|_{L_r([a,b])} \leq \left(\int_a^b \left(\int_a^x |\sin(x-t)|^p\,dt\right)^{\frac{r}{p}} dx\right)^{\frac{1}{r}} \|f'' + f\|_{L_q([a,b])}. \tag{19.18}$$

Proof. As in (19.3) we have

$$|f(x)| \leq \left(\int_a^x |\sin(x-t)|^p\,dt\right)^{\frac{1}{p}} \|f'' + f\|_{L_q([a,b])}, \tag{19.19}$$

and (by $r > 0$)

$$|f(x)|^r \leq \left(\int_a^x |\sin(x-t)|^p\,dt\right)^{\frac{r}{p}} \|f'' + f\|_{L_q([a,b])}^r, \tag{19.20}$$

$\forall\ x \in [a, b]$.

Thus,

$$\int_a^b |f(x)|^r\,dx \leq \left(\int_a^b \left(\int_a^x |\sin(x-t)|^p\,dt\right)^{\frac{r}{p}} dx\right) \|f'' + f\|_{L_q([a,b])}^r \tag{19.21}$$

and

$$\left(\int_a^b |f(x)|^r dx\right)^{\frac{1}{r}} \leq \left(\int_a^b \left(\int_a^x |\sin(x-t)|^p dt\right)^{\frac{r}{p}} dx\right)^{\frac{1}{r}} \|f''+f\|_{L_q([a,b])}, \tag{19.22}$$

proving the claim. □

Theorem 19.11. *All as in Theorem 19.2, $r>0$. Then*

$$\|f\|_{L_r([a,b])} \leq \left(\int_a^b \left(\int_a^x |\sinh(x-t)|^p dt\right)^{\frac{r}{p}} dx\right)^{\frac{1}{r}} \|f-f''\|_{L_q([a,b])}. \tag{19.23}$$

Proof. As similar to Theorem 19.10 is omitted. □

Theorem 19.12. *All as in Theorem 19.3, $r>0$. Then*

$$\|f\|_{L_r([a,b])} \leq \frac{1}{2}\left(\int_a^b \left(\int_a^x |\sinh(x-t)-\sin(x-t)|^p dt\right)^{\frac{r}{p}} dx\right)^{\frac{1}{r}} \|f-f''''\|_{L_q([a,b])}. \tag{19.24}$$

Proof. As similar to Theorem 19.10 is omitted. □

Theorem 19.13. *All as in Theorem 19.4, $r>0$. Then*

$$\|f\|_{L_r([a,b])} \leq \frac{1}{|\alpha||\beta||\beta^2-\alpha^2|} \times \left(\int_a^b \left(\int_a^x |\beta\sin(\alpha(x-t))-\alpha\sin(\beta(x-t))|^p dt\right)^{\frac{r}{p}} dx\right)^{\frac{1}{r}} \left\|f''''+\left(\alpha^2+\beta^2\right)f''+\alpha^2\beta^2 f\right\|_{L_q([a,b])}. \tag{19.25}$$

Proof. As similar to Theorem 19.10 is omitted. □

Theorem 19.14. *All as in Theorem 19.5, $r>0$. Then*

$$\|f\|_{L_r([a,b])} \leq \frac{1}{2|\alpha|^3}\left(\int_a^b \left(\int_a^x |\sin(\alpha(x-t))-\alpha(x-t)\cos(\alpha(x-t))|^p dt\right)^{\frac{r}{p}} dx\right)^{\frac{1}{r}} \left\|f^{(4)}+2\alpha^2 f''+\alpha^4 f\right\|_{L_q([a,b])}. \tag{19.26}$$

Proof. As similar to Theorem 19.10 is omitted. □

Theorem 19.15. *All as in Theorem 19.6, $r > 0$. Then*

$$\|f\|_{L_r([a,b])} \leq \frac{1}{|\alpha| |\beta| |\beta^2 - \alpha^2|} \left(\int_a^b \left(\int_a^x |\alpha \sinh(\beta(x-t)) - \beta \sinh(\alpha(x-t))|^p dt \right)^{\frac{r}{p}} dx \right)^{\frac{1}{r}} \left\| f^{(4)} - (\alpha^2 + \beta^2) f'' + \alpha^2 \beta^2 f \right\|_{L_q([a,b])}. \tag{19.27}$$

Proof. As similar to Theorem 19.10 is omitted. □

Theorem 19.16. *All as in Theorem 19.7, $r > 0$. Then*

$$\|f\|_{L_r([a,b])} \leq \frac{1}{2|\alpha|^3} \left(\int_a^b \left(\int_a^x |\alpha(x-t)\cosh(\alpha(x-t)) - \sinh(\alpha(x-t))|^p dt \right)^{\frac{r}{p}} dx \right)^{\frac{1}{r}} \left\| f^{(4)} - 2\alpha^2 f'' + \alpha^4 f \right\|_{L_q([a,b])}. \tag{19.28}$$

Proof. It is omitted. □

Theorem 19.17. *All as in Theorem 19.9, $r > 0$. Then*

$$\|f\|_{L_r([a,b])} \leq \left(\int_a^b \left(\int_a^x |\omega_c(x-t)|^p dt \right)^{\frac{r}{p}} dx \right)^{\frac{1}{r}} \|D_c(f)\|_{L_q([a,b])}. \tag{19.29}$$

Proof. It is omitted. □

We continue with a collection of Hilbert–Pachpatte inequalities.

Theorem 19.18. *Here $j = 1, 2$. Let $f_j \in C^2([a_j, b_j], K)$, where $K = \mathbb{R}$ or $\mathbb{C}$, such that $f_j(a_j) = f_j'(a_j) = 0$, and $p, q > 1 : \frac{1}{p} + \frac{1}{q} = 1$. Then*

$$\int_{a_1}^{b_1} \int_{a_2}^{b_2} \frac{|f_1(x_1)| |f_2(x_2)| dx_1 dx_2}{\left[\frac{\int_{a_1}^{x_1} |\sin(x_1 - t_1)|^p dt_1}{p} + \frac{\int_{a_2}^{x_2} |\sin(x_2 - t_2)|^q dt_2}{q} \right]} \leq (b_1 - a_1)(b_2 - a_2) \|f_1'' + f_1\|_{L_q([a_1,b_1])} \|f_2'' + f_2\|_{L_p([a_2,b_2])}. \tag{19.30}$$

Proof. As in (19.3) we have

$$|f_1(x_1)| \leq \left(\int_{a_1}^{x_1} |\sin(x_1 - t_1)|^p \, dt_1 \right)^{\frac{1}{p}} \|f_1'' + f_1\|_{L_q([a_1,b_1])}, \tag{19.31}$$

$\forall \, x_1 \in [a_1, b_1]$,

and

$$|f_2(x_2)| \leq \left(\int_{a_2}^{x_2} |\sin(x_2 - t_2)|^q \, dt_2 \right)^{\frac{1}{q}} \|f_2'' + f_2\|_{L_p([a_2,b_2])}. \tag{19.32}$$

$\forall \, x_2 \in [a_2, b_2]$.

Hence we have

$$|f_1(x_1)| \, |f_2(x_2)| \leq \left(\int_{a_1}^{x_1} |\sin(x_1 - t_1)|^p \, dt_1 \right)^{\frac{1}{p}} \left(\int_{a_2}^{x_2} |\sin(x_2 - t_2)|^q \, dt_2 \right)^{\frac{1}{q}}$$

$$\|f_1'' + f_1\|_{L_q([a_1,b_1])} \, \|f_2'' + f_2\|_{L_p([a_2,b_2])} \leq \tag{19.33}$$

(using Young's inequality for $a, b \geq 0$, $a^{\frac{1}{p}} b^{\frac{1}{q}} \leq \frac{a}{p} + \frac{b}{q}$)

$$\left[\frac{\int_{a_1}^{x_1} |\sin(x_1 - t_1)|^p \, dt_1}{p} + \frac{\int_{a_2}^{x_2} |\sin(x_2 - t_2)|^q \, dt_2}{q} \right]$$

$$\|f_1'' + f_1\|_{L_q([a_1,b_1])} \, \|f_2'' + f_2\|_{L_p([a_2,b_2])}. \tag{19.34}$$

So far we have

$$\frac{|f_1(x_1)| \, |f_2(x_2)|}{\left[\frac{\int_{a_1}^{x_1} |\sin(x_1 - t_1)|^p dt_1}{p} + \frac{\int_{a_2}^{x_2} |\sin(x_2 - t_2)|^q dt_2}{q} \right]}$$

$$\leq \|f_1'' + f_1\|_{L_q([a_1,b_1])} \, \|f_2'' + f_2\|_{L_p([a_2,b_2])}, \tag{19.35}$$

$\forall \, (x_1, x_2) \in [a_1, b_1] \times [a_2, b_2]$.

The denominator in (19.35) can be zero only when both $x_1 = a_1$ and $x_2 = a_2$.

Therefore, we obtain (19.30), by integrating (19.35) over $[a_1, b_1] \times [a_2, b_2]$. □

Theorem 19.19. *All as in Theorem 19.18. Then*

$$\int_{a_1}^{b_1} \int_{a_2}^{b_2} \frac{|f_1(x_1)| \, |f_2(x_2)| \, dx_1 dx_2}{\left[\frac{\int_{a_1}^{x_1} |\sinh(x_1 - t_1)|^p dt_1}{p} + \frac{\int_{a_2}^{x_2} |\sinh(x_2 - t_2)|^q dt_2}{q} \right]}$$

$$\leq (b_1 - a_1)(b_2 - a_2) \|f_1 - f_1''\|_{L_q([a_1,b_1])} \, \|f_2 - f_2''\|_{L_p([a_2,b_2])}. \tag{19.36}$$

Proof. As similar to Theorem 19.18 is omitted. □

Theorem 19.20. *Here $j = 1, 2$. Let $f_j \in C^4([a_j, b_j], K)$, where $K = \mathbb{R}$ or $\mathbb{C}$, such that $f_j^{(i)}(a_j) = 0$, $i = 0, 1, 2, 3$, and $p, q > 1 : \frac{1}{p} + \frac{1}{q} = 1$. Then*

$$\int_{a_1}^{b_1}\int_{a_2}^{b_2} \frac{|f_1(x_1)|\,|f_2(x_2)|\,dx_1 dx_2}{\left[\frac{\int_{a_1}^{x_1}|\sinh(x_1-t_1)-\sin(x_1-t_1)|^p dt_1}{p} + \frac{\int_{a_2}^{x_2}|\sinh(x_2-t_2)-\sin(x_2-t_2)|^q dt_2}{q}\right]}$$

$$\leq \frac{(b_1-a_1)(b_2-a_2)}{4}\left\|f_1 - f_1^{(4)}\right\|_{L_q([a_1,b_1])}\left\|f_2 - f_2^{(4)}\right\|_{L_p([a_2,b_2])}. \tag{19.37}$$

Proof. As similar to Theorem 19.18 is omitted. □

Theorem 19.21. *Here $j = 1, 2$. Let $f_j \in C^4([a_j, b_j], K)$, where $K = \mathbb{R}$ or $\mathbb{C}$, such that $f_j^{(i)}(a_j) = 0$, $i = 0, 1, 2, 3$, and $p, q > 1 : \frac{1}{p} + \frac{1}{q} = 1$. Also let $\alpha, \beta \in \mathbb{R}$ with $\alpha\beta(\alpha^2 - \beta^2) \neq 0$. Then*

$$\int_{a_1}^{b_1}\int_{a_2}^{b_2} \frac{|f_1(x_1)|\,|f_2(x_2)|\,dx_1 dx_2}{\left[\begin{array}{l}\frac{\int_{a_1}^{x_1}|\beta\sin(\alpha(x_1-t_1))-\alpha\sin(\beta(x_1-t_1))|^p dt_1}{p}\\ \quad + \frac{\int_{a_2}^{x_2}|\beta\sin(\alpha(x_2-t_2))-\alpha\sin(\beta(x_2-t_2))|^q dt_2}{q}\end{array}\right]}$$

$$\leq \frac{(b_1-a_1)(b_2-a_2)}{\alpha^2\beta^2(\beta^2-\alpha^2)^2}\left\|f_1^{(4)} + (\alpha^2+\beta^2)f_1'' + \alpha^2\beta^2 f_1\right\|_{L_q([a_1,b_1])}$$

$$\left\|f_2^{(4)} + (\alpha^2+\beta^2)f_2'' + \alpha^2\beta^2 f_2\right\|_{L_p([a_2,b_2])}. \tag{19.38}$$

Proof. It is omitted. □

Theorem 19.22. *Here $j = 1, 2$. Let $f_j \in C^4([a_j, b_j], K)$, where $K = \mathbb{R}$ or $\mathbb{C}$, such that $f_j^{(i)}(a_j) = 0$, $i = 0, 1, 2, 3$, and $p, q > 1 : \frac{1}{p} + \frac{1}{q} = 1$; $\alpha \in \mathbb{R}$, $\alpha \neq 0$. Then*

$$\int_{a_1}^{b_1}\int_{a_2}^{b_2} \frac{|f_1(x_1)|\,|f_2(x_2)|\,dx_1 dx_2}{\left[\begin{array}{l}\frac{\int_{a_1}^{x_1}|\sin(\alpha(x_1-t_1))-\alpha(x_1-t_1)\cos(\alpha(x_1-t_1))|^p dt_1}{p}\\ \quad + \frac{\int_{a_2}^{x_2}|\sin(\alpha(x_2-t_2))-\alpha(x_2-t_2)\cos(\alpha(x_2-t_2))|^q dt_2}{q}\end{array}\right]}$$

$$\leq \frac{(b_1-a_1)(b_2-a_2)}{4\alpha^6}\left\|f_1^{(4)} + 2\alpha^2 f_1'' + \alpha^4 f_1\right\|_{L_q([a_1,b_1])}$$

$$\left\|f_2^{(4)} + 2\alpha^2 f_2'' + \alpha^4 f_2\right\|_{L_p([a_2,b_2])}. \tag{19.39}$$

Proof. It is omitted. □

Theorem 19.23. *All as in Theorem 19.21. Then*

$$\int_{a_1}^{b_1}\int_{a_2}^{b_2}\frac{|f_1(x_1)||f_2(x_2)|\,dx_1dx_2}{\left[\frac{\int_{a_1}^{x_1}|\alpha\sinh(\beta(x_1-t_1))-\beta\sinh(\alpha(x_1-t_1))|^p dt_1}{p}+\frac{\int_{a_2}^{x_2}|\alpha\sinh(\beta(x_2-t_2))-\beta\sinh(\alpha(x_2-t_2))|^q dt_2}{q}\right]}$$

$$\leq\frac{(b_1-a_1)(b_2-a_2)}{\alpha^2\beta^2(\beta^2-\alpha^2)^2}\left\|f_1^{(4)}-(\alpha^2+\beta^2)f_1''+\alpha^2\beta^2f_1\right\|_{L_q([a_1,b_1])}$$

$$\left\|f_2^{(4)}-(\alpha^2+\beta^2)f_2''+\alpha^2\beta^2f_2\right\|_{L_p([a_2,b_2])}. \tag{19.40}$$

Proof. It is omitted. □

Theorem 19.24. *All as in Theorem 19.22. Then*

$$\int_{a_1}^{b_1}\int_{a_2}^{b_2}\frac{|f_1(x_1)||f_2(x_2)|\,dx_1dx_2}{\left[\frac{\int_{a_1}^{x_1}|\alpha(x_1-t_1)\cosh(\alpha(x_1-t_1))-\sinh(\alpha(x_1-t_1))|^p dt_1}{p}+\frac{\int_{a_2}^{x_2}|\alpha(x_2-t_2)\cosh(\alpha(x_2-t_2))-\sinh(\alpha(x_2-t_2))|^q dt_2}{q}\right]}$$

$$\leq\frac{(b_1-a_1)(b_2-a_2)}{4\alpha^6}\left\|f_1^{(4)}-2\alpha^2f_1''+\alpha^4f_1\right\|_{L_q([a_1,b_1])}$$

$$\left\|f_2^{(4)}-2\alpha^2f_2''+\alpha^4f_2\right\|_{L_p([a_2,b_2])}. \tag{19.41}$$

Proof. It is omitted. □

We finish with

Theorem 19.25. *Let $j=1,2$. Here $f_j\in C^4([a_j,b_j],K)$, where $K=\mathbb{R}$ or $\mathbb{C}$, and $f_j^{(i)}(a_j)=0$, $i=0,1,\ldots,n-1$. All the rest are as in Remark 19.1 and $p,q>1:\frac{1}{p}+\frac{1}{q}=1$. Then*

$$\int_{a_1}^{b_1}\int_{a_2}^{b_2}\frac{|f_1(x_1)||f_2(x_2)|\,dx_1dx_2}{\left[\frac{\int_{a_1}^{x_1}|\omega_c(x_1-t_1)|^p dt_1}{p}+\frac{\int_{a_2}^{x_2}|\omega_c(x_2-t_2)|^q dt_2}{q}\right]}$$

$$\leq(b_1-a_1)(b_2-a_2)\|D_c(f_1)\|_{L_q([a_1,b_1])}\|D_c(f_2)\|_{L_p([a_2,b_2])}. \tag{19.42}$$

Proof. It is omitted. □

Bibliography

A.H. Ali and Z. Páles, Taylor-type expansions in terms of exponential polynomials, *Mathematical Inequalities & Applications*, 25(4) (2022), 1123–1141.

G.A. Anastassiou, Trigonometric and hyperbolic Poincaré, Sobolev and Hilbert-Pachpatte type inequalities, *Annals of Communications in Mathematics*, 6(3) (2023), 191–198.

Chapter 20

Trigonometric-Induced Degree of Approximation by Smooth Picard Singular Integral Operators

In this chapter, we continue the study of smooth Picard singular integral operators that started in Anastassiou (2011), see there Chapters 10–14. This time, the foundation of our research is a trigonometric Taylor's formula. We establish the convergence of our operators to the unit operator with rates via Jackson-type inequalities engaging the first modulus of continuity. Of interest, here is a residual appearing term. Note that our operators are not positive. Our results are pointwise and uniform. It follows Anastassiou (2023b).

20.1 Introduction

We are motivated by Anastassiou (2006, 2011) Chapters 10–14, Anastassiou (2023a), and Ali and Pales (2022). We use a trigonometric new Taylor formula from 20.3, see also 20.4. Here we consider some very general operators, the smooth Picard singular integral operators over the real line and we study further their convergence properties quantitatively. We establish related inequalities involving the first modulus of continuity with respect to uniform norm and the estimates are pointwise and uniform. We provide a detailed proof.

20.2 Results

By Anastassiou (2023a) and Ali and Pales (2022), for $f \in C^2(\mathbb{R})$ and $a, x \in \mathbb{R}$, we have by trigonometric Taylor formula

$$f(x) - f(a) = f'(a)\sin(x-a) + 2f''(a)\sin^2\left(\frac{x-a}{2}\right) + \int_a^x [(f''(t) + f(t)) - (f''(a) + f(a))]\sin(x-t)\,dt. \tag{20.1}$$

For $r \in \mathbb{N}$ and $n \in \mathbb{Z}^+$, we set

$$\alpha_j := \begin{cases} (-1)^{r-j}\binom{r}{j} j^{-n}, & j = 1, \ldots, r. \\ 1 - \sum_{j=1}^{r}(-1)^{r-j}\binom{r}{j} j^{-n}, & j = 0, \end{cases} \tag{20.2}$$

that is,

$$\sum_{j=0}^{r} \alpha_j = 1. \tag{20.3}$$

$C_U(\mathbb{R})$ denotes the space of uniformly continuous functions on $\mathbb{R}$, and $C_B(\mathbb{R})$ denotes the space of bounded continuous functions on $\mathbb{R}$.

Here we consider both $f, f'' \in C_U(\mathbb{R}) \cup C_B(\mathbb{R})$.

For $x \in \mathbb{R}$, $\xi > 0$, we consider the Lebesgue integrals, so-called smooth Picard operators

$$P_{r,\xi}(f,x) = \frac{1}{2\xi}\int_{-\infty}^{\infty}\left(\sum_{j=0}^{r}\alpha_j f(x+jt)\right)e^{-\frac{|t|}{\xi}}\,dt, \tag{20.4}$$

see Anastassiou (2006); $P_{r,\xi}$ are not in general positive operators, see Anastassiou (2011).

We note by

$$\frac{1}{2\xi}\int_{-\infty}^{\infty} e^{-\frac{|t|}{\xi}}\,dt = 1 \tag{20.5}$$

that

$$P_{r,\xi}(c,x) = c, \quad \text{where } c \text{ is a constant}, \tag{20.6}$$

and

$$P_{r,\xi}(f,x)-f(x)=\frac{1}{2\xi}\left(\sum_{j=0}^{r}\alpha_j\int_{-\infty}^{\infty}(f(x+jt)-f(x))\,e^{-\frac{|t|}{\xi}}dt\right). \tag{20.7}$$

Denote by

$$\omega_1(f,\delta):=\sup_{\substack{x,y\in\mathbb{R}\\|x-y|\le\delta}}|f(x)-f(y)|,\quad \delta>0, \tag{20.8}$$

the first modulus of continuity of f.

We set

$$\Delta(x):=P_{r,\xi}(f,x)-f(x)-f''(x)\left(\sum_{j=0}^{r}\frac{\alpha_j j^2}{j^2\xi^2+1}\right)\xi^2;\quad \xi>0,\ x\in\mathbb{R}. \tag{20.9}$$

We present our uniform approximation result.

Theorem 20.1. *It holds*

$$|\Delta(x)|\le\|\Delta(x)\|_\infty\le\xi^2\omega_1(f''+f,\xi)\left(\sum_{j=0}^{r}|\alpha_j|\,j^2(j+1)\right)=:A;$$

$$\xi>0,\ x\in\mathbb{R}. \tag{20.10}$$

And $\|\Delta(x)\|_\infty\to0,$ *as* $\xi\to0$. *If* $f''(x)=0,$ *then* $|P_{r,\xi}(f,x)-f(x)|\le A.$

Proof. By (20.1), we get that

$$f(x+jt)-f(x)=f'(x)\sin(jt)+2f''(x)\sin^2\left(\frac{jt}{2}\right)$$
$$+\int_x^{x+jt}[(f''(s)+f(s))-(f''(x)+f(x))]\sin(x+jt-s)\,ds, \tag{20.11}$$

or better

$$f(x+jt)-f(x)=f'(x)\sin(jt)+2f''(x)\sin^2\left(\frac{jt}{2}\right)$$
$$+\int_0^{jt}[(f''(x+z)+f(x+z))-(f''(x)+f(x))]\sin(jt-z)\,dz. \tag{20.12}$$

Furthermore, it holds

$$\sum_{j=0}^{r} \alpha_j \left[f(x+jt) - f(x)\right]$$
$$= f'(x) \sum_{j=0}^{r} \alpha_j \sin(jt) + 2f''(x) \sum_{j=0}^{r} \alpha_j \sin^2\left(\frac{jt}{2}\right)$$
$$+ \sum_{j=0}^{r} \alpha_j \int_0^{jt} \left[(f''(x+z) + f(x+z)) - (f''(x) + f(x))\right] \sin(jt - z)\, dz, \tag{20.13}$$

or better

$$\sum_{j=0}^{r} \alpha_j \left[f(x+jt) - f(x)\right]$$
$$= f'(x) \sum_{j=0}^{r} \alpha_j \sin(jt) + 2f''(x) \sum_{j=0}^{r} \alpha_j \sin^2\left(\frac{jt}{2}\right)$$
$$+ \sum_{j=0}^{r} \alpha_j j \int_0^{t} \left[(f''(x+jw) + f(x+jw)) - (f''(x) + f(x))\right]$$
$$\times \sin j(t-w)\, dw. \tag{20.14}$$

Call

$$R := R(t)$$
$$:= \sum_{j=0}^{r} \alpha_j j \int_0^{t} \left[(f''(x+jw) + f(x+jw)) - (f''(x) + f(x))\right]$$
$$\times \sin j(t-w)\, dw, \tag{20.15}$$

$\forall\, t \in \mathbb{R}$.

Then, for $t \geq 0$,

$$|R| \leq \sum_{j=0}^{r} |\alpha_j|\, j \int_0^{t} \left|(f''(x+jw) + f(x+jw)) - (f''(x) + f(x))\right|$$
$$\times \left|\sin j(t-w)\right| dw$$
$$\leq \sum_{j=0}^{r} |\alpha_j|\, j \int_0^{t} \omega_1(f'' + f, jw)\, j(t-w)\, dw =$$

$(\xi > 0)$

$$\sum_{j=0}^{r} |\alpha_j| j^2 \int_0^t \omega_1 \left(f'' + f, \frac{\xi j w}{\xi}\right)(t-w)\,dw$$

$$\leq \sum_{j=0}^{r} |\alpha_j| j^2 \omega_1 (f'' + f, \xi) \int_0^t \left(1 + \frac{jw}{\xi}\right)(t-w)\,dw$$

$$= \omega_1 (f'' + f, \xi) \sum_{j=0}^{r} |\alpha_j| j^2 \left[\int_0^t (t-w)\,dw + \frac{j}{\xi}\int_0^t w^{2-1}(t-w)^{2-1}\,dw\right]$$

$$= \omega_1 (f'' + f, \xi) \sum_{j=0}^{r} |\alpha_j| j^2 \left[\left.\frac{(t-w)^2}{2}\right|_t^0 + \frac{j}{\xi}\frac{1}{6}t^3\right]$$

$$= \omega_1 (f'' + f, \xi) \sum_{j=0}^{r} |\alpha_j| j^2 \left[\frac{t^2}{2} + j\frac{t^3}{6\xi}\right] 1. \qquad (20.16)$$

Hence $(t \geq 0)$

$$|R| \leq \omega_1 (f'' + f, \xi) \sum_{j=0}^{r} |\alpha_j| j^2 \left[\frac{t^2}{2} + j\frac{t^3}{6\xi}\right]. \qquad (20.17)$$

Let now $t < 0$, then

$$|R| \leq \sum_{j=0}^{r} |\alpha_j| j \left| \int_0^t \left[(f''(x+jw) + f(x+jw)) - (f''(x) + f(x))\right] \times \sin j(t-w)\,dw \right|$$

$$\leq \sum_{j=0}^{r} |\alpha_j| j \int_t^0 \left|(f''(x+jw) + f(x+jw)) - (f''(x) + f(x))\right| \times |\sin j(t-w)|\,dw$$

$$\leq \sum_{j=0}^{r} |\alpha_j| j \int_t^0 \omega_1 (f'' + f, -jw)\, j(w-t)\,dw$$

$$= \sum_{j=0}^{r} |\alpha_j| j \int_t^0 \omega_1 \left(f'' + f, -jw\frac{\xi}{\xi}\right) j(w-t)\,dw$$

$$= \omega_1 (f'' + f, \xi) \sum_{j=0}^{r} |\alpha_j| j^2 \int_t^0 \left(1 - \frac{j}{\xi} w\right)(w - t)\, dw$$

$$= \omega_1 (f'' + f, \xi) \sum_{j=0}^{r} |\alpha_j| j^2$$

$$\times \left[\int_t^0 (w - t)\, dw + \frac{j}{\xi} \int_t^0 (0 - w)^{2-1} (w - t)^{2-1} dw\right]$$

$$= \omega_1 (f'' + f, \xi) \sum_{j=0}^{r} |\alpha_j| j^2 \left[\left.\frac{(w - t)^2}{2}\right|_t^0 + \frac{j}{\xi} \frac{1}{6} (-t)^3\right]$$

$$= \omega_1 (f'' + f, \xi) \sum_{j=0}^{r} |\alpha_j| j^2 \left[\frac{t^2}{2} + j \frac{(-t)^3}{6\xi}\right]. \tag{20.18}$$

We found that $(t < 0)$

$$|R| \leq \omega_1 (f'' + f, \xi) \sum_{j=0}^{r} |\alpha_j| j^2 \left[\frac{t^2}{2} + j \frac{(-t)^3}{6\xi}\right]. \tag{20.19}$$

Consequently, for $t \in \mathbb{R}$, we obtain

$$|R(t)| \leq \omega_1 (f'' + f, \xi) \sum_{j=0}^{r} |\alpha_j| j^2 \left[\frac{t^2}{2} + \frac{j}{6\xi} |t|^3\right], \quad \xi > 0. \tag{20.20}$$

So, we have

$$\sum_{j=0}^{r} \alpha_j [f(x + jt) - f(x)]$$

$$- f'(x) \sum_{j=0}^{r} \alpha_j \sin(jt) - 2f''(x) \sum_{j=0}^{r} \alpha_j \sin^2\left(\frac{jt}{2}\right) = R(t). \tag{20.21}$$

Therefore, it holds

$$\Delta_1 (x) := P_{r,\xi} (f, x) - f(x) - f'(x) \sum_{j=0}^{r} \alpha_j \frac{1}{2\xi} \left(\int_{-\infty}^{\infty} \sin(jt)\, e^{-\frac{|t|}{\xi}} dt\right)$$

$$- 2f''(x) \sum_{j=0}^{r} \alpha_j \frac{1}{2\xi} \left(\int_{-\infty}^{\infty} \sin^2\left(\frac{jt}{2}\right) e^{-\frac{|t|}{\xi}} dt\right)$$

$$= \frac{1}{2\xi} \int_{-\infty}^{\infty} R(t)\, e^{-\frac{|t|}{\xi}} dt. \tag{20.22}$$

Hence we get

$$|\Delta_1(x)| \le \frac{1}{2\xi}\int_{-\infty}^{\infty}|R(t)|\,e^{-\frac{|t|}{\xi}}dt$$

$$\le \frac{1}{2\xi}\int_{-\infty}^{\infty}\left[\omega_1(f''+f,\xi)\sum_{j=0}^{r}|\alpha_j|\,j^2\left[\frac{t^2}{2}+\frac{j}{6\xi}|t|^3\right]\right]e^{-\frac{|t|}{\xi}}dt \quad (20.23)$$

(we use

$$\int_{-\infty}^{\infty}t^k e^{-\frac{|t|}{\xi}}dt = \begin{cases}0, & k \text{ odd},\\ 2k!\xi^{k+1}, & k \text{ even}\end{cases}\Bigg) \quad (20.24)$$

$$= \omega_1(f''+f,\xi)\sum_{j=0}^{r}|\alpha_j|\,j^2\frac{1}{2\xi}\int_{-\infty}^{\infty}\left[\frac{t^2}{2}+\frac{j}{6\xi}|t|^3\right]e^{-\frac{|t|}{\xi}}dt$$

$$= \omega_1(f''+f,\xi)\sum_{j=0}^{r}|\alpha_j|\,j^2\left[\frac{1}{4\xi}\int_{-\infty}^{\infty}t^2e^{-\frac{|t|}{\xi}}dt+\frac{j}{12\xi^2}\int_{-\infty}^{\infty}|t|^3e^{-\frac{|t|}{\xi}}dt\right]$$

$$= \omega_1(f''+f,\xi)\sum_{j=0}^{r}|\alpha_j|\,j^2\left[\frac{1}{4\xi}4\xi^3+\frac{j\xi^4}{6\xi^2}\int_0^{\infty}\left(\frac{t}{\xi}\right)^3e^{-\frac{|t|}{\xi}}d\frac{t}{\xi}\right]$$

$$= \omega_1(f''+f,\xi)\sum_{j=0}^{r}|\alpha_j|\,j^2\left[\xi^2+\xi^2\frac{j}{6}\int_0^{\infty}z^3e^{-z}dz\right]$$

$$= \omega_1(f''+f,\xi)\sum_{j=0}^{r}|\alpha_j|\,j^2\left[\xi^2+\xi^2\frac{j}{6}6\right]$$

$$= \omega_1(f''+f,\xi)\sum_{j=0}^{r}|\alpha_j|\,j^2\left[\xi^2+\xi^2 j\right]$$

$$= \omega_1(f''+f,\xi)\,\xi^2\left(\sum_{j=0}^{r}|\alpha_j|\,j^2(1+j)\right). \quad (20.25)$$

We have proved that

$$|\Delta_1(x)| \le \xi^2\omega_1(f''+f,\xi)\left(\sum_{j=0}^{r}|\alpha_j|\,j^2(j+1)\right);\quad \xi>0,\ x\in\mathbb{R}. \quad (20.26)$$

Note that $\Delta_1(x)\to 0$, as $\xi\to 0$.

Next we simplify the left-hand side (20.22).

We observe that

$$\int_{-\infty}^{\infty} \sin(jt)\, e^{-\frac{|t|}{\xi}} dt = \int_{-\infty}^{0} \sin(jt)\, e^{-\frac{|t|}{\xi}} dt + \int_{0}^{\infty} \sin(jt)\, e^{-\frac{|t|}{\xi}} dt. \quad (20.27)$$

Note $-\infty \leq t \leq 0 \Rightarrow \infty \geq -t \geq 0$. So

$$-\int_{-\infty}^{0} \sin(j(-(-t)))\, e^{-\frac{|t|}{\xi}} d(-t) = -\int_{-\infty}^{0} (-\sin(j(-t)))\, e^{-\frac{|t|}{\xi}} d(-t)$$

$$= \int_{-\infty}^{0} \sin j(-t)\, e^{-\frac{|t|}{\xi}} d(-t) = \int_{\infty}^{0} \sin j(t)\, e^{-\frac{|t|}{\xi}} dt$$

$$= -\int_{0}^{\infty} \sin(jt)\, e^{-\frac{|t|}{\xi}} dt. \quad (20.28)$$

Therefore, it is

$$\int_{-\infty}^{\infty} \sin(jt)\, e^{-\frac{|t|}{\xi}} dt = 0, \quad j = 0, 1, \ldots, r. \quad (20.29)$$

Furthermore, we have that

$$\int_{-\infty}^{\infty} \sin^2\left(\frac{jt}{2}\right) e^{-\frac{|t|}{\xi}} dt = 2\int_{0}^{\infty} \sin^2\left(\frac{jt}{2}\right) e^{-\frac{t}{\xi}} dt, \quad j = 0, 1, \ldots, r. \quad (20.30)$$

The last follows by

$$\int_{-\infty}^{0} \sin^2\left(\frac{jt}{2}\right) e^{-\frac{|t|}{\xi}} dt = -\int_{-\infty}^{0} \left(-\sin\left(\frac{j(-t)}{2}\right)\right)^2 e^{-\frac{|-t|}{\xi}} d(-t)$$

$$\overset{(z=-t)}{=} -\int_{\infty}^{0} \sin^2\left(\frac{jz}{2}\right) e^{-\frac{|z|}{\xi}} dz = \int_{0}^{\infty} \sin^2\left(\frac{jz}{2}\right) e^{-\frac{z}{\xi}} dz. \quad (20.31)$$

Next, we calculate

$$\int_{0}^{\infty} \sin^2\left(\frac{jt}{2}\right) e^{-\frac{|t|}{\xi}} dt = \xi \int_{0}^{\infty} \sin^2\left(\left(\frac{j\xi}{2}\right)\frac{t}{\xi}\right) e^{-\frac{t}{\xi}} d\frac{t}{\xi} =$$

(call $\frac{t}{\xi} =: x$ and $\frac{j\xi}{2} =: a_1$)

$$\xi \int_{0}^{\infty} \sin^2(a_1 x)\, e^{-x} dx = \quad (20.32)$$

(by Wolfram Alpha Computational Intelligence)

$$\xi\left(\frac{2a_1^2}{4a_1^2+1}\right)=\frac{j^2\xi^3}{2\left(j^2\xi^2+1\right)}.$$

Thus,

$$\int_{-\infty}^{\infty}\sin^2\left(\frac{jt}{2}\right)e^{-\frac{|t|}{\xi}}dt=\frac{j^2\xi^3}{j^2\xi^2+1},\quad j=0,1,\ldots,r. \tag{20.33}$$

Consequently, it holds

$$\frac{1}{2\xi}\int_{-\infty}^{\infty}\sin^2\left(\frac{jt}{2}\right)e^{-\frac{|t|}{\xi}}dt=\frac{j^2\xi^2}{2\left(j^2\xi^2+1\right)}\to 0,\text{ as }\xi\to 0,\quad j=0,1,\ldots,r. \tag{20.34}$$

Finally, we obtain

$$-2f''(x)\sum_{j=0}^{r}\alpha_j\frac{1}{2\xi}\left(\int_{-\infty}^{\infty}\sin^2\left(\frac{jt}{2}\right)e^{-\frac{|t|}{\xi}}dt\right)$$

$$=-f''(x)\left(\sum_{j=0}^{r}\alpha_j\frac{j^2}{\left(j^2\xi^2+1\right)}\right)\xi^2. \tag{20.35}$$

Clearly now it is $\Delta_1(x)=\Delta(x)$.
The proof of the theorem is finally completed. □

We finish with

Corollary 20.1. *Additionally assume $f''\in C_B(\mathbb{R})$. It follows ($\xi>0$, $x\in\mathbb{R}$)*

$$\left\|P_{r,\xi}(f,x)-f(x)\right\|_\infty\leq\omega_1\left(f''+f.\xi\right)\left(\sum_{j=0}^{r}|\alpha_j|\,j^2\,(j+1)\right)\xi^2$$

$$+\left\|f''(x)\right\|_\infty\left(\sum_{j=0}^{r}|\alpha_j|\frac{j^2}{1+j^2\xi^2}\right)\xi^2\to 0,\quad as\ \xi\to 0. \tag{20.36}$$

Proof. Easy, by (20.9) and (20.10). □

Bibliography

A.H. Ali, Z. Pales, Taylor-type expansions in terms of exponential polynomials, *Mathematical Inequalities and Applications*, 25(4) (2022), 1123–1141.

G.A. Anastassiou, Basic convergence with rates of smooth Picard singular integral operators, *Journal of Computational Analysis and Applications*, 8(4) (2006), 313–334.

G.A. Anastassiou, *Intelligent Mathematics: Computational Analysis*, Springer, Heidelberg, New York, Chapter 10, 2011.

G.A. Anastassiou, Opial and Ostrowski type inequalities based on trigonometric and hyperbolic type Taylor formulae, *Malaya Journal Mathematik*, 11(S) 2023a, 1–26.

G.A. Anastassiou, *Trigonometric generated rate of convergence of smooth Picard singular integral operators*, *Journal of Applied and Pure Mathematics*, accepted, 2023b.

Chapter 21

Trigonometric-Derived L_p Degree of Approximation by Smooth Picard Singular Integral Operators

In this chapter, we continue the study of smooth Picard singular integral operators that started in Anastassiou (2011), see there Chapters 10–14. This time, the foundation of our research is a trigonometric Taylor's formula. We establish the L_p convergence of our operators to the unit operator with rates via Jackson-type inequalities engaging the first L_p modulus of continuity. Of interest, here is a residual appearing term. Note that our operators are not positive. It follows Anastassiou (2023).

21.1 Introduction

We are motivated by Anastassiou (2006, 2011) Chapters 10–14, Anastassiou (2023), and Ali and Pales (2022). We use a trigonometric new Taylor formula from Anastassiou (2023), see also Ali and Pales (2022). Here we consider some very general operators, the smooth Picard singular integral operators over the real line, and we study further their L_p, $p \geq 1$, convergence properties quantitatively. We establish related inequalities involving the first L_p, $p \geq 1$, modulus of continuity with respect to L_p, $p \geq 1$, norm. We provide detailed proofs.

21.2 Results

By Anastassiou (2023) and Ali and Pales (2022), for $f \in C^2(\mathbb{R})$ and $a, x \in \mathbb{R}$, we have by the trigonometric Taylor formula

$$f(x) - f(a) = f'(a)\sin(x-a) + 2f''(a)\sin^2\left(\frac{x-a}{2}\right) + \int_a^x \left[\left(f''(t) + f(t)\right) - \left(f''(a) + f(a)\right)\right]\sin(x-t)\,dt. \tag{21.1}$$

For $r \in \mathbb{N}$ and $n \in \mathbb{Z}^+$, we set

$$\alpha_j := \begin{cases} (-1)^{r-j} \binom{r}{j} j^{-n}, & j = 1, \ldots, r, \\ 1 - \sum_{j=1}^{r} (-1)^{r-j} \binom{r}{j} j^{-n}, & j = 0, \end{cases} \tag{21.2}$$

that is,

$$\sum_{j=0}^{r} \alpha_j = 1. \tag{21.3}$$

Here we consider both $f, f'' \in L_p(\mathbb{R}) \cap C(\mathbb{R})$, $1 \leq p < \infty$.

For $x \in \mathbb{R}$, $\xi > 0$, we consider the Lebesgue integrals, so-called smooth Picard operators,

$$P_{r,\xi}(f, x) = \frac{1}{2\xi} \int_{-\infty}^{\infty} \left(\sum_{j=0}^{r} \alpha_j f(x + jt) \right) e^{-\frac{|t|}{\xi}} dt, \tag{21.4}$$

see Anastassiou (2006); $P_{r,\xi}$ are not in general positive operators, see Anastassiou (2011).

We note by

$$\frac{1}{2\xi} \int_{-\infty}^{\infty} e^{-\frac{|t|}{\xi}} dt = 1 \tag{21.5}$$

that

$$P_{r,\xi}(c, x) = c, \quad \text{where } c \text{ is a constant}, \tag{21.6}$$

and

$$P_{r,\xi}(f, x) - f(x) = \frac{1}{2\xi} \left(\sum_{j=0}^{r} \alpha_j \int_{-\infty}^{\infty} (f(x + jt) - f(x)) e^{-\frac{|t|}{\xi}} dt \right). \tag{21.7}$$

Denote by

$$\omega_1(f, h)_p := \sup_{\substack{t \in \mathbb{R} \\ |t| \leq h}} \|f(x + t) - f(x)\|_{p,x} \tag{21.8}$$

the first L_p modulus of smoothness of f, $1 \leq p < \infty$.

By (21.1), we get that

$$f(x + jt) - f(x) = f'(x) \sin(jt) + 2f''(x) \sin^2\left(\frac{jt}{2}\right) + \int_x^{x+jt} [(f''(s) + f(s)) - (f''(x) + f(x))] \sin(x + jt - s) \, ds, \tag{21.9}$$

or better,

$$f(x+jt)-f(x)=f'(x)\sin(jt)+2f''(x)\sin^2\left(\frac{jt}{2}\right)$$
$$+\int_0^{jt}\left[\left(f''(x+z)+f(x+z)\right)-\left(f''(x)+f(x)\right)\right]\sin(jt-z)\,dz. \tag{21.10}$$

Furthermore, it holds

$$\sum_{j=0}^{r}\alpha_j\left[f(x+jt)-f(x)\right]$$
$$=f'(x)\sum_{j=0}^{r}\alpha_j\sin(jt)+2f''(x)\sum_{j=0}^{r}\alpha_j\sin^2\left(\frac{jt}{2}\right)$$
$$+\sum_{j=0}^{r}\alpha_j\int_0^{jt}\left[\left(f''(x+z)+f(x+z)\right)-\left(f''(x)+f(x)\right)\right]\sin(jt-z)\,dz, \tag{21.11}$$

or better

$$\sum_{j=0}^{r}\alpha_j\left[f(x+jt)-f(x)\right]$$
$$=f'(x)\sum_{j=0}^{r}\alpha_j\sin(jt)+2f''(x)\sum_{j=0}^{r}\alpha_j\sin^2\left(\frac{jt}{2}\right)$$
$$+\sum_{j=0}^{r}\alpha_j j\int_0^{t}\left[\left(f''(x+jw)+f(x+jw)\right)-\left(f''(x)+f(x)\right)\right]$$
$$\times\sin j(t-w)\,dw. \tag{21.12}$$

Call

$$R:=R(t)$$
$$:=\sum_{j=0}^{r}\alpha_j j\int_0^{t}\left[\left(f''(x+jw)+f(x+jw)\right)-\left(f''(x)+f(x)\right)\right]$$
$$\times\sin j(t-w)\,dw, \tag{21.13}$$

$\forall\, t\in\mathbb{R}$.

We note that

$$\Delta(x) := P_{r,\xi}(f,x) - f(x) - f'(x)\sum_{j=0}^{r}\alpha_j \frac{1}{2\xi}\left(\int_{-\infty}^{\infty}\sin(jt)\,e^{-\frac{|t|}{\xi}}dt\right)$$

$$- 2f''(x)\sum_{j=0}^{r}\alpha_j \frac{1}{2\xi}\left(\int_{-\infty}^{\infty}\sin^2\left(\frac{jt}{2}\right)e^{-\frac{|t|}{\xi}}dt\right)$$

$$= \frac{1}{2\xi}\int_{-\infty}^{\infty} R(t)\,e^{-\frac{|t|}{\xi}}dt. \tag{21.14}$$

Next we simplify (21.14).

We observe that

$$\int_{-\infty}^{\infty}\sin(jt)\,e^{-\frac{|t|}{\xi}}dt = \int_{-\infty}^{0}\sin(jt)\,e^{-\frac{|t|}{\xi}}dt + \int_{0}^{\infty}\sin(jt)\,e^{-\frac{|t|}{\xi}}dt. \tag{21.15}$$

Note $-\infty \le t \le 0 \Rightarrow \infty \ge -t \ge 0$. So

$$-\int_{-\infty}^{0}\sin(j(-(-t)))\,e^{-\frac{|t|}{\xi}}d(-t) = -\int_{-\infty}^{0}(-\sin(j(-t)))\,e^{-\frac{|t|}{\xi}}d(-t)$$

$$= \int_{-\infty}^{0}(\sin j(-t))\,e^{-\frac{|t|}{\xi}}d(-t) = \int_{\infty}^{0}\sin j(t)\,e^{-\frac{|t|}{\xi}}dt$$

$$= -\int_{0}^{\infty}\sin(jt)\,e^{-\frac{|t|}{\xi}}dt. \tag{21.16}$$

Therefore, it is

$$\int_{-\infty}^{\infty}\sin(jt)\,e^{-\frac{|t|}{\xi}}dt = 0, \quad j = 0, 1, \ldots, r. \tag{21.17}$$

Furthermore, we have that

$$\int_{-\infty}^{\infty}\sin^2\left(\frac{jt}{2}\right)e^{-\frac{|t|}{\xi}}dt = 2\int_{0}^{\infty}\sin^2\left(\frac{jt}{2}\right)e^{-\frac{t}{\xi}}dt, \quad j = 0, 1, \ldots, r. \tag{21.18}$$

The last follows by

$$\int_{-\infty}^{0}\sin^2\left(\frac{jt}{2}\right)e^{-\frac{|t|}{\xi}}dt = -\int_{-\infty}^{0}\left(-\sin\left(\frac{j(-t)}{2}\right)\right)^2 e^{-\frac{|-t|}{\xi}}d(-t)$$

$$\overset{(z=-t)}{=} -\int_{\infty}^{0}\sin^2\left(\frac{jz}{2}\right)e^{-\frac{|z|}{\xi}}dz = \int_{0}^{\infty}\sin^2\left(\frac{jz}{2}\right)e^{-\frac{z}{\xi}}dz. \tag{21.19}$$

Next, we calculate

$$\int_0^{\infty} \sin^2\left(\frac{jt}{2}\right) e^{-\frac{|t|}{\xi}} dt = \xi \int_0^{\infty} \sin^2\left(\left(\frac{j\xi}{2}\right)\frac{t}{\xi}\right) e^{-\frac{t}{\xi}} d\frac{t}{\xi} =$$

(call $\frac{t}{\xi} =: x$ and $\frac{j\xi}{2} =: a$)

$$\xi \int_0^{\infty} \sin^2(ax) e^{-x} dx = \tag{21.20}$$

(by Wolfram Alpha Computational Intelligence)

$$\xi\left(\frac{2a^2}{4a^2+1}\right) = \frac{j^2\xi^3}{2(j^2\xi^2+1)}.$$

Thus, we find that

$$\int_{-\infty}^{\infty} \sin^2\left(\frac{jt}{2}\right) e^{-\frac{|t|}{\xi}} dt = \frac{j^2\xi^3}{j^2\xi^2+1}, \quad j = 0, 1, \ldots, r. \tag{21.21}$$

Consequently, it holds ($\xi > 0$)

$$\frac{1}{2\xi}\int_{-\infty}^{\infty} \sin^2\left(\frac{jt}{2}\right) e^{-\frac{|t|}{\xi}} dt = \frac{j^2\xi^2}{2(j^2\xi^2+1)} \to 0, \quad \text{as } \xi \to 0, \quad j = 0, 1, \ldots, r. \tag{21.22}$$

Eventually, we obtain

$$\begin{aligned} & -2f''(x)\sum_{j=0}^{r} \alpha_j \frac{1}{2\xi}\left(\int_{-\infty}^{\infty} \sin^2\left(\frac{jt}{2}\right) e^{-\frac{|t|}{\xi}} dt\right) \\ & = -f''(x)\left(\sum_{j=0}^{r} \alpha_j \frac{j^2}{(j^2\xi^2+1)}\right)\xi^2. \end{aligned} \tag{21.23}$$

Consequently, based on the above, it is

$$\Delta(x) = P_{r,\xi}(f,x) - f(x) - f''(x)\left(\sum_{j=0}^{r} \alpha_j \frac{j^2}{(j^2\xi^2+1)}\right)\xi^2, \quad \xi > 0. \tag{21.24}$$

We present our first result, L_p approximation, $p > 1$.

Theorem 21.1. *Let $p, q > 1 : \frac{1}{p} + \frac{1}{q} = 1$, $\xi > 0$, and both $f, f'' \in L_p(\mathbb{R}) \cap C(\mathbb{R})$, $x \in \mathbb{R}$. Then*

$$\|\Delta(x)\|_p = \left\| P_{r,\xi}(f) - f - f'' \left(\sum_{j=0}^{r} \alpha_j \frac{j^2}{(j^2\xi^2 + 1)} \right) \xi^2 \right\|_p$$

$$\leq \left(\frac{4(r+1)}{q(q+1)} \right)^{\frac{1}{q}} \left(\frac{2}{p} \right)^{2+\frac{1}{p}} \omega_1 (f'' + f, \xi)_p \, \xi^2$$

$$\left\{ \sum_{j=0}^{r} |\alpha_j|^p j^{2p} \left[\Gamma(qp - q + p + 1) + \frac{(2j)^p}{p^p (p+1)} \Gamma(qp - q + 2p + 1) \right] \right\}^{\frac{1}{p}}$$

$$\to 0, \tag{21.25}$$

as $\xi \to 0$.

Above Γ stands for the gamma function.

Proof. Let

$$I := \int_0^t [(f''(x + jw) + f(x + jw)) - (f''(x) + f(x))] \sin j(t - w) \, dw,$$

$$\forall \, t \in \mathbb{R}. \tag{21.26}$$

For $t < 0$, we have that

$$|I| = \left| \int_t^0 [(f''(x + jw) + f(x + jw)) - (f''(x) + f(x))] \sin j(t - w) \, dw \right|$$

$$\leq \int_t^0 |(f''(x + jw) + f(x + jw)) - (f''(x) + f(x))| \, |\sin j(t - w)| \, dw$$

$$\leq j \int_t^0 |(f''(x + jw) + f(x + jw)) - (f''(x) + f(x))| \, (w - t) \, dw$$

$$= -j \int_t^0 |(f''(x - j(-w)) + f(x - j(-w))) - (f''(x) + f(x))|$$

$$\times (-t - (-w)) \, d(-w) \tag{21.27}$$

$(t \leq w \leq 0 \Rightarrow -t \geq -w =: \theta \geq 0)$

$$= -j \int_{-t}^{0} \left| \left(f''(x - j\theta) + f(x - j\theta) \right) - \left(f''(x) + f(x) \right) \right| (-t - \theta)\, d\theta$$

$$= j \int_{0}^{-t} \left| \left(f''(x - j\theta) + f(x - j\theta) \right) - \left(f''(x) + f(x) \right) \right| (-t - \theta)\, d\theta$$

$$= j \int_{0}^{|t|} \left| \left(f''(x + sign(t)\, j\theta) + f(x + sign(t)\, j\theta) \right) - \left(f''(x) + f(x) \right) \right| (|t| - \theta)\, d\theta. \quad (21.28)$$

So, we have proved that

$$|I| \leq j \int_{0}^{|t|} \left| \left(f''(x + sign(t)\, j\theta) + f(x + sign(t)\, j\theta) \right) - \left(f''(x) + f(x) \right) \right| \times (|t| - \theta)\, d\theta, \quad (21.29)$$

$\forall\, t \in \mathbb{R}$,

and by (21.13),

$$|R(t)| \leq \sum_{j=0}^{r} |\alpha_j|\, j^2 \int_{0}^{|t|} \left| \left(f''(x + j sign(t)\, \theta) + f(x + j sign(t)\, \theta) \right) - \left(f''(x) + f(x) \right) \right| \times (|t| - \theta)\, d\theta, \quad (21.30)$$

$\forall\, t \in \mathbb{R}$.

By (21.14), we have

$$\Delta(x) = \frac{1}{2\xi} \int_{-\infty}^{\infty} R(t)\, e^{-\frac{|t|}{\xi}}\, dt. \quad (21.31)$$

Hence, it holds $(p, q > 1 : \frac{1}{p} + \frac{1}{q} = 1)$,

$$\int_{-\infty}^{\infty} |\Delta(x)|^p\, dx = \frac{1}{(2\xi)^p} \int_{-\infty}^{\infty} \left| \int_{-\infty}^{\infty} R(t)\, e^{-\frac{|t|}{\xi}}\, dt \right|^p dx$$

$$\leq \frac{1}{(2\xi)^p} \int_{-\infty}^{\infty} \left(\int_{-\infty}^{\infty} |R(t)|\, e^{-\frac{|t|}{\xi}}\, dt \right)^p dx$$

$$= \frac{1}{(2\xi)^p} \int_{-\infty}^{\infty} \left(\int_{-\infty}^{\infty} |R(t)| \, e^{-\frac{|t|}{2\xi}} e^{-\frac{|t|}{2\xi}} dt \right)^p dx$$

$$\leq \frac{1}{(2\xi)^p} \int_{-\infty}^{\infty} \left(\int_{-\infty}^{\infty} |R(t)|^p \, e^{-\frac{|pt|}{2\xi}} dt \right) \left(\int_{-\infty}^{\infty} e^{-\frac{|qt|}{2\xi}} dt \right)^{\frac{p}{q}} dx$$

$$= \frac{1}{(2\xi)^p} \left(\frac{4\xi}{q} \right)^{\frac{p}{q}} \int_{-\infty}^{\infty} \left(\int_{-\infty}^{\infty} |R(t)|^p \, e^{-\frac{|pt|}{2\xi}} dt \right) dx$$

$$= \frac{2^{p-2}\xi^{-1}}{q^{p-1}} \left(\int_{-\infty}^{\infty} \left(\int_{-\infty}^{\infty} |R(t)|^p \, e^{-\frac{|pt|}{2\xi}} dt \right) dx \right) =: (*). \tag{21.32}$$

But, we need to treat

$$|R(t)| \leq \sum_{j=0}^{r} |\alpha_j| \, j^2 \tag{21.33}$$

$$\times \left(\int_0^{|t|} \left| (f''(x + jsign(t)\theta) + f(x + jsign(t)\theta)) - (f''(x) + f(x)) \right|^p d\theta \right)^{\frac{1}{p}}$$

$$\times \left(\int_0^{|t|} (|t| - \theta)^q \, d\theta \right)^{\frac{1}{q}}$$

$$= \sum_{j=0}^{r} |\alpha_j| \, j^2 \left(\int_0^{|t|} \left| (f''(x + jsign(t)\theta) + f(x + jsign(t)\theta)) \right. \right.$$

$$\left. \left. - (f''(x) + f(x)) \right|^p d\theta \right)^{\frac{1}{p}}$$

$$\frac{|t|^{q+\frac{1}{q}}}{(q+1)^{\frac{1}{q}}}$$

$$= \left\{ \sum_{j=0}^{r} \left((|\alpha_j| \, j^2)^p \right)^{\frac{1}{p}} \right.$$

$$\left. \left(\int_0^{|t|} \left| (f''(x + jsign(t)\theta) + f(x + jsign(t)\theta)) - (f''(x) + f(x)) \right|^p d\theta \right)^{\frac{1}{p}} \right\}$$

$$\frac{|t|^{q+\frac{1}{q}}}{(q+1)^{\frac{1}{q}}} \leq \tag{21.34}$$

$(0 < \frac{1}{p} < 1)$

$$\left\{ (r+1)^{\frac{1}{q}} \left(\sum_{j=0}^{r} |\alpha_j|^p j^{2p} \right. \right.$$

$$\left. \left. \int_0^{|t|} \left| \left(f''(x + jsign(t)\theta) + f(x + jsign(t)\theta)\right) - \left(f''(x) + f(x)\right) \right|^p d\theta \right)^{\frac{1}{p}} \right\}$$

$$\frac{|t|^{q+\frac{1}{q}}}{(q+1)^{\frac{1}{q}}}.$$

Hence, we find that

$$|R(t)|^p \leq \left(\frac{r+1}{q+1}\right)^{\frac{p}{q}} \left[\sum_{j=0}^{r} |\alpha_j|^p j^{2p} \right.$$

$$\left. \left(\int_0^{|t|} \left| \left(f''(x + jsign(t)\theta) + f(x + jsign(t)\theta)\right) - \left(f''(x) + f(x)\right) \right|^p d\theta \right) \right]$$

$$\times |t|^{\frac{(q+1)p}{q}}. \tag{21.35}$$

Therefore, we get

$$(*) \leq \left(\frac{2^{p-2}\xi^{-1}}{q^{p-1}} \left(\frac{r+1}{q+1}\right)^{\frac{p}{q}} \right) \left\{ \int_{-\infty}^{\infty} \left(\int_{-\infty}^{\infty} \left[\sum_{j=0}^{r} |\alpha_j|^p j^{2p} \right. \right. \right.$$

$$\left. \int_0^{|t|} \left| \left(f''(x + jsign(t)\theta) + f(x + jsign(t)\theta)\right) - \left(f''(x) + f(x)\right) \right|^p d\theta \right]$$

$$\left. \left. |t|^{\frac{(q+1)p}{q}} e^{-\frac{|pt|}{2\xi}} dt \right) dx \right\} \tag{21.36}$$

(set $c_1 = \frac{2^{p-2}\xi^{-1}}{q^{p-1}} \left(\frac{r+1}{q+1}\right)^{\frac{p}{q}}$)

$$= c_1 \left\{ \int_{-\infty}^{\infty} \left(\int_{-\infty}^{\infty} \left[\sum_{j=0}^{r} |\alpha_j|^p j^{2p} \right. \right. \right.$$

$$\left. \left. \int_0^{|t|} \left| \left(f''(x + jsign(t)\theta) + f(x + jsign(t)\theta)\right) - \left(f''(x) + f(x)\right) \right|^p d\theta \right] dx \right)$$

$$\left. |t|^{\frac{(q+1)p}{q}} e^{-\frac{|pt|}{2\xi}} dt \right\}$$

$$= c_1 \left\{ \int_{-\infty}^{\infty} \left[\sum_{j=0}^{r} |\alpha_j|^p \, j^{2p} \right. \right. \tag{21.37}$$

$$\int_0^{|t|} \left(\int_{-\infty}^{\infty} |(f''(x + jsign(t)\theta) + f(x + jsign(t)\theta))$$

$$\left. - (f''(x) + f(x))|^p \, dx \right) d\theta \Bigg]$$

$$\left. |t|^{\frac{(q+1)p}{q}} e^{-\frac{|pt|}{2\xi}} dt \right\} \leq$$

$(\xi > 0)$

$$c_1 \left\{ \int_{-\infty}^{\infty} \left[\sum_{j=0}^{r} |\alpha_j|^p \, j^{2p} \left(\int_0^{|t|} \omega_1 \left(f'' + f, \xi \frac{j\theta}{\xi} \right)_p^p d\theta \right) \right] \right.$$

$$\left. |t|^{\frac{(q+1)p}{q}} e^{-\frac{|pt|}{2\xi}} dt \right\}$$

$$\leq \omega_1 \left(f'' + f, \xi \right)_p^p c_1 \left\{ \int_{-\infty}^{\infty} \left[\sum_{j=0}^{r} |\alpha_j|^p \, j^{2p} \left(\int_0^{|t|} \left(1 + \frac{j}{\xi} \theta \right)^p d\theta \right) \right] \right.$$

$$\left. |t|^{(q+1)(p-1)} e^{-\frac{|pt|}{2\xi}} dt \right\} \tag{21.38}$$

$$\leq c_1 \omega_1 \left(f'' + f, \xi \right)_p^p \left\{ \int_{-\infty}^{\infty} \left[\sum_{j=0}^{r} |\alpha_j|^p \, j^{2p} 2^{p-1} \left(\int_0^{|t|} \left(1 + \frac{j^p}{\xi^p} \theta^p \right) d\theta \right) \right] \right.$$

$$\left. |t|^{(q+1)(p-1)} e^{-\frac{|pt|}{2\xi}} dt \right\}$$

$$= 2^{p-1} c_1 \omega_1 \left(f'' + f, \xi \right)_p^p$$

$$\left\{ \int_{-\infty}^{\infty} \left[\sum_{j=0}^{r} |\alpha_j|^p \, j^{2p} \left(|t| + \frac{j^p}{\xi^p} \frac{|t|^{p+1}}{p+1} \right) \right] |t|^{(q+1)(p-1)} e^{-\frac{|pt|}{2\xi}} dt \right\}$$

$$= 2^{p-1} c_1 \omega_1 \left(f'' + f, \xi \right)_p^p$$

$$\left\{ \sum_{j=0}^{r} |\alpha_j|^p \, j^{2p} \int_{-\infty}^{\infty} \left(|t| + \frac{j^p}{\xi^p} \frac{|t|^{p+1}}{p+1} \right) |t|^{(q+1)(p-1)} e^{-\frac{|pt|}{2\xi}} dt \right\}$$

$$= 2^p c_1 \omega_1 \left(f'' + f, \xi\right)_p^p \tag{21.39}$$

$$\left\{ \sum_{j=0}^{r} |\alpha_j|^p j^{2p} \int_0^\infty \left(t + \frac{j^p}{\xi^p (p+1)} t^{p+1}\right) t^{(q+1)(p-1)} e^{-\frac{pt}{2\xi}} dt \right\}$$

$$= 2^p c_1 \omega_1 \left(f'' + f, \xi\right)_p^p$$

$$\left\{ \sum_{j=0}^{r} |\alpha_j|^p j^{2p} \left[\int_0^\infty t^{(q+1)(p-1)+1} e^{-\frac{pt}{2\xi}} dt \right.\right.$$

$$\left.\left. + \frac{j^p}{\xi^p (p+1)} \int_0^\infty t^{(q+1)(p-1)+p+1} e^{-\frac{pt}{2\xi}} dt \right] \right\} \tag{21.40}$$

$$= 2^p c_1 \omega_1 \left(f'' + f, \xi\right)_p^p \left\{ \sum_{j=0}^{r} |\alpha_j|^p j^{2p} \right.$$

$$\times \left[\xi^{qp-q+p+1} \int_0^\infty \left(\frac{t}{\xi}\right)^{(q+1)(p-1)+1} e^{-\left(\frac{p}{2}\right)\frac{t}{\xi}} d\frac{t}{\xi} \right.$$

$$\left.\left. + \frac{j^p}{\xi^p (p+1)} \xi^{qp-q+2p+1} \int_0^\infty \left(\frac{t}{\xi}\right)^{qp-q+2p} e^{-\left(\frac{p}{2}\right)\frac{t}{\xi}} d\frac{t}{\xi} \right] \right\} =$$

(above it is $(q+1)(p-1)+1 = qp - q + p > 0$, $(q+1)(p-1)+p+1 = qp - q + 2p > 0$)

$$2^p c_1 \omega_1 \left(f'' + f, \xi\right)_p^p \xi^{qp-q+p+1} \left[\sum_{j=0}^{r} |\alpha_j|^p j^{2p} \left\{ \int_0^\infty x^{qp-q+p} e^{-\left(\frac{p}{2}\right)x} dx \right.\right.$$

$$\left.\left. + \frac{j^p}{(p+1)} \int_0^\infty x^{qp-q+2p} e^{-\left(\frac{p}{2}\right)x} dx \right] \right\} = \tag{21.41}$$

(next we use, for $a, b > 0$, that it holds $\int_0^\infty x^a e^{-bx} dx = b^{-a-1}\Gamma(a+1)$, by Wolfram Alpha Computational Intelligence)

$$2^p c_1 \omega_1 \left(f'' + f, \xi\right)_p^p \xi^{qp-q+p+1} \left[\sum_{j=0}^{r} |\alpha_j|^p j^{2p} \left\{ \left(\frac{p}{2}\right)^{-qp+q-p-1} \Gamma(qp - q + p + 1) \right.\right.$$

$$\left.\left. + \frac{j^p}{(p+1)} \left(\frac{p}{2}\right)^{-qp+q-2p-1} \Gamma(qp - q + 2p + 1) \right] \right\}$$

$$= \frac{2^{2(p-1)}}{q^{p-1}} \left(\frac{r+1}{q+1}\right)^{\frac{p}{q}} \omega_1 \left(f'' + f, \xi\right)_p^p \xi^{qp-q+p} \left[\sum_{j=0}^{r} |\alpha_j|^p j^{2p}\right.$$

$$\left\{\left(\frac{p}{2}\right)^{-qp+q-p-1} \left(\Gamma\left(qp-q+p+1\right) + \frac{j^p}{(p+1)} \left(\frac{p}{2}\right)^{-p}\right.\right.$$

$$\left.\left.\left.\times \Gamma\left(qp-q+2p+1\right)\right)\right\}\right]. \tag{21.42}$$

So for $p, q > 1 : \frac{1}{p} + \frac{1}{q} = 1$, $\xi > 0$, we have proved the following:

$$\int_{-\infty}^{\infty} |\Delta(x)|^p \, dx \leq \frac{2^{2(p-1)}}{q^{p-1}} \left(\frac{r+1}{q+1}\right)^{\frac{p}{q}} \omega_1 \left(f'' + f, \xi\right)_p^p \xi^{qp-q+p} \left(\frac{p}{2}\right)^{-qp+q-p-1}$$

$$\left\{\sum_{j=0}^{r} |\alpha_j|^p j^{2p} \left[\Gamma\left(qp-q+p+1\right) + \frac{(2j)^p}{p^p (p+1)} \Gamma\left(qp-q+2p+1\right)\right]\right\}. \tag{21.43}$$

The proof of the theorem is now completed. □

Next comes the L_1-related approximation.

Theorem 21.2. *Let $\xi > 0$, both $f, f'' \in L_1(\mathbb{R}) \cap C(\mathbb{R})$, $x \in \mathbb{R}$. Then*

$$\|\Delta(x)\|_1 \leq \left[\sum_{j=0}^{r} |\alpha_j| j^2 (2+3j)\right] \omega_1 \left(f'' + f, \xi\right)_1 \xi^2 \to 0, \quad as\ \xi \to 0. \tag{21.44}$$

Proof. We have that

$$|\Delta(x)| = \frac{1}{2\xi} \left| \int_{-\infty}^{\infty} R(t) e^{-\frac{|t|}{\xi}} dt \right| \leq \frac{1}{2\xi} \int_{-\infty}^{\infty} |R(t)| e^{-\frac{|t|}{\xi}} dt$$

$$\leq \frac{1}{2\xi} \sum_{j=0}^{r} |\alpha_j| j^2 \int_{-\infty}^{\infty} \left(\int_0^{|t|} |(f''(x + j sign(t)\theta) + f(x + j sign(t)\theta))\right.$$

$$\left. - (f''(x) + f(x))| (|t| - \theta) \, d\theta \right) e^{-\frac{|t|}{\xi}} dt. \tag{21.45}$$

Thus, we get

$$\|\Delta(x)\|_1 = \int_{-\infty}^{\infty} |\Delta(x)|\, dx$$

$$\leq \frac{1}{2\xi} \sum_{j=0}^{r} |\alpha_j| j^2 \left(\int_{-\infty}^{\infty} \left(\int_{-\infty}^{\infty} \left(\int_0^{|t|} \left| (f''(x + j sign(t)\theta) + f(x + j sign(t)\theta)) \right.\right.\right.\right.$$

$$\left.\left.\left.\left. - (f''(x) + f(x)) \right| (|t| - \theta)\, d\theta \right) e^{-\frac{|t|}{\xi}} dt \right) dx \right) \tag{21.46}$$

$$\leq \frac{1}{2\xi} \sum_{j=0}^{r} |\alpha_j| j^2 \left(\int_{-\infty}^{\infty} \left(\int_{-\infty}^{\infty} \left(|t| \int_0^{|t|} \left| (f''(x + j sign(t)\theta) + f(x + j sign(t)\theta)) \right.\right.\right.\right.$$

$$\left.\left.\left.\left. - (f''(x) + f(x)) \right| d\theta \right) e^{-\frac{|t|}{\xi}} dt \right) dx \right)$$

$$= \frac{1}{2\xi} \sum_{j=0}^{r} |\alpha_j| j^2 \left(\int_{-\infty}^{\infty} \left(\int_{-\infty}^{\infty} \left(|t| \int_0^{|t|} \left| (f''(x + j sign(t)\theta) + f(x + j sign(t)\theta)) \right.\right.\right.\right.$$

$$\left.\left.\left.\left. - (f''(x) + f(x)) \right| d\theta \right) e^{-\frac{|t|}{\xi}} dx \right) dt \right) \tag{21.47}$$

$$= \frac{1}{2\xi} \sum_{j=0}^{r} |\alpha_j| j^2 \left(\int_{-\infty}^{\infty} \left(\int_{-\infty}^{\infty} \left(\int_0^{|t|} \left| (f''(x + j sign(t)\theta) + f(x + j sign(t)\theta)) \right.\right.\right.\right.$$

$$\left.\left.\left.\left. - (f''(x) + f(x)) \right| d\theta \right) dx \right) |t|\, e^{-\frac{|t|}{\xi}} dt \right)$$

$$= \frac{1}{2\xi} \sum_{j=0}^{r} |\alpha_j| j^2 \left(\int_{-\infty}^{\infty} \left(\int_0^{|t|} \left(\int_{-\infty}^{\infty} \left| (f''(x + j sign(t)\theta) + f(x + j sign(t)\theta)) \right.\right.\right.\right.$$

$$\left.\left.\left.\left. - (f''(x) + f(x)) \right| dx \right) d\theta \right) |t|\, e^{-\frac{|t|}{\xi}} dt \right)$$

$$\leq \frac{1}{2\xi} \sum_{j=0}^{r} |\alpha_j| j^2 \left(\int_{-\infty}^{\infty} \left(\int_0^{|t|} \omega_1 (f'' + f, j\theta)_1\, d\theta \right) |t|\, e^{-\frac{|t|}{\xi}} dt \right)$$

$$\leq \frac{\omega_1 (f'' + f, \xi)_1}{2\xi} \sum_{j=0}^{r} |\alpha_j| j^2 \left(\int_{-\infty}^{\infty} \left(\int_0^{|t|} \left(1 + \frac{j}{\xi}\theta \right) d\theta \right) |t|\, e^{-\frac{|t|}{\xi}} dt \right)$$

$$= \frac{\omega_1 (f'' + f, \xi)_1}{2\xi} \sum_{j=0}^{r} |\alpha_j| j^2 \left(\int_{-\infty}^{\infty} \left[|t| + \frac{j}{\xi} \frac{|t|^2}{2} \right] |t|\, e^{-\frac{|t|}{\xi}} dt \right) \tag{21.48}$$

$$= \xi^2 \omega_1 \left(f'' + f, \xi\right)_1 \sum_{j=0}^{r} |\alpha_j| \, j^2 \left(\int_0^{\infty} \left[\frac{t}{\xi} + \frac{j}{2} \left(\frac{t}{\xi} \right)^2 \right] \frac{t}{\xi} e^{-\frac{t}{\xi}} d\frac{t}{\xi} \right)$$

$$= \xi^2 \omega_1 \left(f'' + f, \xi\right)_1 \sum_{j=0}^{r} |\alpha_j| \, j^2 \left[\int_0^{\infty} \left(t + \frac{j}{2} t^2 \right) t e^{-t} dt \right]$$

$$= \xi^2 \omega_1 \left(f'' + f, \xi\right)_1 \sum_{j=0}^{r} |\alpha_j| \, j^2 \left[\int_0^{\infty} t^2 e^{-t} dt + \frac{j}{2} \int_0^{\infty} t^3 e^{-t} dt \right]$$

$$= \xi^2 \omega_1 \left(f'' + f, \xi\right)_1 \sum_{j=0}^{r} |\alpha_j| \, j^2 \left[2 + \frac{j}{2} \cdot 6 \right]$$

$$= \xi^2 \omega_1 \left(f'' + f, \xi\right)_1 \sum_{j=0}^{r} |\alpha_j| \, j^2 \left[2 + 3j \right]. \tag{21.49}$$

Above we used that

$$\int_0^{\infty} t^2 e^{-t} dt = 2 \quad \text{and} \quad \int_0^{\infty} t^3 e^{-t} dt = 6.$$

The theorem is proved. ☐

We finish with the following.

Corollary 21.1. *(to Theorem 21.1) It holds*

$$\left\| P_{r,\xi}(f) - f \right\|_p \le \left\| f'' \right\|_p \left(\sum_{j=0}^{r} |\alpha_j| \frac{j^2}{(j^2 \xi^2 + 1)} \right) \xi^2$$

$$+ \left(\frac{4(r+1)}{q(q+1)} \right)^{\frac{1}{q}} \left(\frac{2}{p} \right)^{2+\frac{1}{p}} \omega_1 \left(f'' + f, \xi\right)_p \xi^2$$

$$\times \left\{ \sum_{j=0}^{r} |\alpha_j|^p \, j^{2p} \left[\Gamma(qp - q + p + 1) + \frac{(2j)^p}{p^p (p+1)} \Gamma(qp - q + 2p + 1) \right] \right\}^{\frac{1}{p}}$$

$$\to 0, \tag{21.50}$$

as $\xi \to 0$.

Corollary 21.2 (to Theorem 21.2). *It holds*

$$\left\| P_{r,\xi}(f) - f \right\|_1 \leq \left\| f'' \right\|_1 \left(\sum_{j=0}^{r} |\alpha_j| \frac{j^2}{(j^2\xi^2+1)} \right) \xi^2$$

$$+ \left[\sum_{j=0}^{r} |\alpha_j| \, j^2 (2+3j) \right] \omega_1 (f'' + f, \xi)_1 \, \xi^2 \to 0, \quad as\ \xi \to 0. \qquad (21.51)$$

Bibliography

A.H. Ali, Z. Pales, Taylor-type expansions in terms of exponential polynomials, *Mathematical Inequalities and Applications*, 25(4) (2022), 1123–1141.

G.A. Anastassiou, Basic convergence with rates of smooth Picard singular integral operators, *Journal of Computational Analysis and Applications*, 8(4) (2006), 313–334.

G.A. Anastassiou, *Intelligent Mathematics: Computational Analysis*, Springer, Heidelberg, New York, Chapter 12, 2011.

G.A. Anastassiou, Opial and Ostrowski type inequalities based on trigonometric and hyperbolic type Taylor formulae, *Malaya Journal Mathematik*, 11(S) 2023, 1–26.

G.A. Anastassiou, Trigonometric generated L_p degree of approximation by smooth Picard singular integral operators, *Annals of Communications in Mathematics*, 6(4) (2023), 209–219.

Chapter 22

Parametrized and Trigonometric-Induced Quantitative Convergence of Smooth Picard Singular Integral Operators

In this chapter, we continue the research on the smooth Picard singular integral operators that started in Anastassiou (2011), see there Chapters 10–14. This time, the foundation of our research is a parametrized trigonometric Taylor's formula. We establish the convergence of our operators to the unit operator with rates via Jackson-type inequalities engaging the first modulus of continuity. Of interest, here is a parametrized residual appearing term. Note that our operators are not positive. Our results are pointwise and uniform. It follows Anastassiou (2023b).

22.1 Introduction

We are motivated by Anastassiou (2006, 2011) Chapters 10–14, Anastassiou (2023a), and Ali and Pales (2022). We use a parametrized trigonometric new Taylor formula from Anastassiou (2023a), see also Ali and Pales (2022). Here we consider some very general operators, the smooth Picard singular integral operators over the real line, and we study further their parametrized convergence properties quantitatively. We establish related inequalities involving the first modulus of continuity with respect to uniform norm and the estimates are pointwise and uniform. We provide a detailed proof.

22.2 Results

By Anastassiou (2023a) and Ali and Pales (2022), for $\alpha, \beta \in \mathbb{R}$ with $\alpha\beta\left(\alpha^2 - \beta^2\right) \neq 0$, and $f \in C^4(\mathbb{R})$, $a, x \in \mathbb{R}$, we have the following general trigonometric Taylor formula:

$$\begin{aligned}
f(x) - f(a) = {} & f'(a)\left(\frac{\beta^3 \sin(\alpha(x-a)) - \alpha^3 \sin(\beta(x-a))}{\alpha\beta(\beta^2-\alpha^2)}\right) \\
& + f''(a)\left(\frac{\cos(\alpha(x-a)) - \cos(\beta(x-a))}{\beta^2-\alpha^2}\right) \\
& + f'''(a)\left(\frac{\beta \sin(\alpha(x-a)) - \alpha \sin(\beta(x-a))}{\alpha\beta(\beta^2-\alpha^2)}\right) \\
& + \frac{2\left(f''''(a) + (\alpha^2+\beta^2) f''(a)\right)}{(\alpha\beta)^2(\beta^2-\alpha^2)} \\
& \times \left(\beta^2 \sin^2\left(\frac{\alpha(x-a)}{2}\right) - \alpha^2 \sin^2\left(\frac{\beta(x-a)}{2}\right)\right) \\
& + \frac{1}{\alpha\beta(\beta^2-\alpha^2)} \int_a^x \left[\left(f''''(t) + (\alpha^2+\beta^2) f''(t) + \alpha^2\beta^2 f(t)\right)\right. \\
& - \left.\left(f''''(a) + (\alpha^2+\beta^2) f''(a) + \alpha^2\beta^2 f(a)\right)\right] \\
& \times \left[\beta \sin(\alpha(x-t)) - \alpha \sin(\beta(x-t))\right] dt. \qquad (22.1)
\end{aligned}$$

For $r \in \mathbb{N}$ and $n \in \mathbb{Z}^+$, we set

$$\alpha_j := \begin{cases} (-1)^{r-j} \binom{r}{j} j^{-n}, & j = 1, \ldots, r, \\ 1 - \sum_{j=1}^{r} (-1)^{r-j} \binom{r}{j} j^{-n}, & n = 0, \end{cases} \qquad (22.2)$$

that is,

$$\sum_{j=0}^{r} \alpha_j = 1. \qquad (22.3)$$

$C_U(\mathbb{R})$ denotes the space of uniformly continuous functions on $\mathbb{R}$, and $C_B(\mathbb{R})$ denotes the space of bounded continuous functions on $\mathbb{R}$.

Here we consider all $f, f', f'', f''', f^{(4)} \in C_U(\mathbb{R}) \cup C_B(\mathbb{R})$.

For $x \in \mathbb{R}$, $\xi > 0$, we consider the Lebesgue integrals, so-called smooth Picard operators,

$$P_{r,\xi}(f,x) = \frac{1}{2\xi} \int_{-\infty}^{\infty} \left(\sum_{j=0}^{r} \alpha_j f(x+jt)\right) e^{-\frac{|t|}{\xi}} dt, \qquad (22.4)$$

see Anastassiou (2006); $P_{r,\xi}$ are not in general positive operators, see Anastassiou (2011).

We note by

$$\frac{1}{2\xi}\int_{-\infty}^{\infty}e^{-\frac{|t|}{\xi}}dt=1 \tag{22.5}$$

that

$$P_{r,\xi}\left(c,x\right)=c, \text{ where } c \text{ is a constant,}$$

and

$$P_{r,\xi}\left(f,x\right)-f\left(x\right)=\frac{1}{2\xi}\left(\sum_{j=0}^{r}\alpha_{j}\int_{-\infty}^{\infty}\left(f\left(x+jt\right)-f\left(x\right)\right)e^{-\frac{|t|}{\xi}}dt\right). \tag{22.6}$$

Denote by

$$\omega_{1}\left(f,\delta\right):=\sup_{\substack{x,y\in\mathbb{R}\\|x-y|\leq\delta}}\left|f\left(x\right)-f\left(y\right)\right|,\ \ \delta>0,$$

the first modulus of continuity of f.

We set

$$E\left(x\right):=P_{r,\xi}\left(f,x\right)-f\left(x\right)-f''\left(x\right)\left\{\sum_{j=0}^{r}\alpha_{j}j^{2}\left(\frac{1}{\left(\alpha^{2}j^{2}\xi^{2}+1\right)\left(\beta^{2}j^{2}\xi^{2}+1\right)}\right)\right\}\xi^{2}$$
$$-\left(f^{(4)}\left(x\right)+\left(\alpha^{2}+\beta^{2}\right)f''\left(x\right)\right)\left[\sum_{j=0}^{r}\alpha_{j}j^{4}\left(\frac{1}{\left(\alpha^{2}j^{2}\xi^{2}+1\right)\left(\beta^{2}j^{2}\xi^{2}+1\right)}\right)\right]\xi^{4}, \tag{22.7}$$

$\xi>0$, $x\in\mathbb{R}$.

We present our main uniform approximation result.

Theorem 22.1. *It holds*

$$\left|E\left(x\right)\right|\leq\left\|E\left(x\right)\right\|_{\infty}\leq\frac{2\xi^{2}\omega_{1}\left(\left(f^{(4)}+\left(\alpha^{2}+\beta^{2}\right)f''+\alpha^{2}\beta^{2}f\right),\xi\right)}{\left|\beta^{2}-\alpha^{2}\right|}$$
$$\left(\sum_{j=0}^{r}\left|\alpha_{j}\right|j^{2}\left(j+1\right)\right)=:B;\quad \xi>0,\ \ x\in\mathbb{R}, \tag{22.8}$$

and $\left\|E\left(x\right)\right\|_{\infty}\rightarrow0,$ *as* $\xi\rightarrow0.$

If $f''\left(x\right)=f^{(4)}\left(x\right)=0,$ *then*

$$\left|P_{r,\xi}\left(f,x\right)-f\left(x\right)\right|\leq B.$$

Proof. By (22.1), we get that

$$f(x+jt)-f(x)=f'(x)\left(\frac{\beta^3\sin(\alpha jt)-\alpha^3\sin(\beta jt)}{\alpha\beta(\beta^2-\alpha^2)}\right)$$

$$+f''(x)\left(\frac{\cos(\alpha jt)-\cos(\beta jt)}{\beta^2-\alpha^2}\right)$$

$$+f'''(x)\left(\frac{\beta\sin(\alpha jt)-\alpha\sin(\beta jt)}{\alpha\beta(\beta^2-\alpha^2)}\right)$$

$$+\frac{2\left(f^{(4)}(x)+\left(\alpha^2+\beta^2\right)f''(x)\right)}{(\alpha\beta)^2(\beta^2-\alpha^2)}\left(\beta^2\sin^2\left(\frac{\alpha jt}{2}\right)-\alpha^2\sin^2\left(\frac{\beta jt}{2}\right)\right)$$

$$+\frac{1}{\alpha\beta(\beta^2-\alpha^2)}\int_x^{x+jt}\left[\left(f^{(4)}(s)+\left(\alpha^2+\beta^2\right)f''(s)+\alpha^2\beta^2f(s)\right)\right.$$

$$\left.-\left(f^{(4)}(x)+\left(\alpha^2+\beta^2\right)f''(x)+\alpha^2\beta^2f(x)\right)\right]$$

$$[\beta\sin(\alpha(x+jt-s))-\alpha\sin(\beta(x+jt-s))]\,ds, \tag{22.9}$$

or better,

$$f(x+jt)-f(x)=f'(x)\left(\frac{\beta^3\sin(\alpha jt)-\alpha^3\sin(\beta jt)}{\alpha\beta(\beta^2-\alpha^2)}\right)$$

$$+f''(x)\left(\frac{\cos(\alpha jt)-\cos(\beta jt)}{\beta^2-\alpha^2}\right)$$

$$+f'''(x)\left(\frac{\beta\sin(\alpha jt)-\alpha\sin(\beta jt)}{\alpha\beta(\beta^2-\alpha^2)}\right)$$

$$+\frac{2\left(f^{(4)}(x)+\left(\alpha^2+\beta^2\right)f''(x)\right)}{(\alpha\beta)^2(\beta^2-\alpha^2)}\left(\beta^2\sin^2\left(\frac{\alpha jt}{2}\right)-\alpha^2\sin^2\left(\frac{\beta jt}{2}\right)\right)$$

$$+\frac{1}{\alpha\beta(\beta^2-\alpha^2)}\int_0^{jt}\left[\left(f^{(4)}(x+z)+\left(\alpha^2+\beta^2\right)f''(x+z)+\alpha^2\beta^2f(x+z)\right)\right.$$

$$\left.-\left(f^{(4)}(x)+\left(\alpha^2+\beta^2\right)f''(x)+\alpha^2\beta^2f(x)\right)\right]$$

$$[\beta\sin(\alpha(jt-z))-\alpha\sin(\beta(jt-z))]\,dz. \tag{22.10}$$

Furthermore, it holds

$$\sum_{j=0}^{r} \alpha_j \left[f(x+jt) - f(x)\right]$$

$$= \left(\frac{f'(x)}{\alpha\beta(\beta^2-\alpha^2)}\right)\left[\beta^3 \sum_{j=0}^{r} \alpha_j \sin(\alpha jt) - \alpha^3 \sum_{j=0}^{r} \alpha_j \sin(\beta jt)\right]$$

$$+ \left(\frac{f''(x)}{\beta^2-\alpha^2}\right)\left[\sum_{j=0}^{r} \alpha_j \cos(\alpha jt) - \sum_{j=0}^{r} \alpha_j \cos(\beta jt)\right]$$

$$+ \left(\frac{f^{(3)}(x)}{\alpha\beta(\beta^2-\alpha^2)}\right)\left[\beta \sum_{j=0}^{r} \alpha_j \sin(\alpha jt) - \alpha \sum_{j=0}^{r} \alpha_j \sin(\beta jt)\right]$$

$$+ \left(\frac{2\left(f^{(4)}(x) + \left(\alpha^2+\beta^2\right) f''(x)\right)}{(\alpha\beta)^2(\beta^2-\alpha^2)}\right)$$

$$\times \left(\beta^2 \sum_{j=0}^{r} \alpha_j \sin^2\left(\frac{\alpha jt}{2}\right) - \alpha^2 \sum_{j=0}^{r} \alpha_j \sin^2\left(\frac{\beta jt}{2}\right)\right) + \frac{1}{\alpha\beta(\beta^2-\alpha^2)}$$

$$\times \sum_{j=0}^{r} \alpha_j \int_0^{jt} \left[\left(f^{(4)}(x+z) + \left(\alpha^2+\beta^2\right) f''(x+z) + \alpha^2\beta^2 f(x+z)\right)\right.$$

$$\left. - \left(f^{(4)}(x) + \left(\alpha^2+\beta^2\right) f''(x) + \alpha^2\beta^2 f(x)\right)\right]$$

$$\times \left[\beta \sin(\alpha(jt-z)) - \alpha \sin(\beta(jt-z))\right] dz, \qquad (22.11)$$

or better

$$\sum_{j=0}^{r} \alpha_j \left[f(x+jt) - f(x)\right]$$

$$= \left(\frac{f'(x)}{\alpha\beta(\beta^2-\alpha^2)}\right)\left[\beta^3 \sum_{j=0}^{r} \alpha_j \sin(\alpha jt) - \alpha^3 \sum_{j=0}^{r} \alpha_j \sin(\beta jt)\right]$$

$$+ \left(\frac{f''(x)}{\beta^2-\alpha^2}\right)\left[\sum_{j=0}^{r} \alpha_j \cos(\alpha jt) - \sum_{j=0}^{r} \alpha_j \cos(\beta jt)\right]$$

$$+ \left(\frac{f^{(3)}(x)}{\alpha\beta(\beta^2-\alpha^2)}\right)\left[\beta \sum_{j=0}^{r} \alpha_j \sin(\alpha jt) - \alpha \sum_{j=0}^{r} \alpha_j \sin(\beta jt)\right]$$

$$+\left(\frac{2\left(f^{(4)}(x)+\left(\alpha^{2}+\beta^{2}\right)f''(x)\right)}{(\alpha\beta)^{2}\left(\beta^{2}-\alpha^{2}\right)}\right)$$

$$\times\left(\beta^{2}\sum_{j=0}^{r}\alpha_{j}\sin^{2}\left(\frac{\alpha jt}{2}\right)-\alpha^{2}\sum_{j=0}^{r}\alpha_{j}\sin^{2}\left(\frac{\beta jt}{2}\right)\right)+\frac{1}{\alpha\beta\left(\beta^{2}-\alpha^{2}\right)}$$

$$\times\sum_{j=0}^{r}\alpha_{j}j\int_{0}^{t}\left[\left(f^{(4)}(x+jw)+\left(\alpha^{2}+\beta^{2}\right)f''(x+jw)+\alpha^{2}\beta^{2}f(x+jw)\right)\right.$$

$$\left.-\left(f^{(4)}(x)+\left(\alpha^{2}+\beta^{2}\right)f''(x)+\alpha^{2}\beta^{2}f(x)\right)\right]$$

$$\times\left[\beta\sin\left(\alpha j(t-w)\right)-\alpha\sin\left(\beta j(t-w)\right)\right]dw. \tag{22.12}$$

We call

$$R:=R(t):=\frac{1}{\alpha\beta\left(\beta^{2}-\alpha^{2}\right)}\sum_{j=0}^{r}\alpha_{j}j$$

$$\times\int_{0}^{t}\left[\left(f^{(4)}(x+jw)+\left(\alpha^{2}+\beta^{2}\right)f''(x+jw)+\alpha^{2}\beta^{2}f(x+jw)\right)\right.$$

$$\left.-\left(f^{(4)}(x)+\left(\alpha^{2}+\beta^{2}\right)f''(x)+\alpha^{2}\beta^{2}f(x)\right)\right]$$

$$\left[\beta\sin\left(\alpha j(t-w)\right)-\alpha\sin\left(\beta j(t-w)\right)\right]dw,\quad\forall\, t\in\mathbb{R}. \tag{22.13}$$

Then, for $t\geq 0$,

$$|R|\leq\frac{1}{\left|\alpha\beta\left(\beta^{2}-\alpha^{2}\right)\right|}\sum_{j=0}^{r}\left|\alpha_{j}\right|j$$

$$\times\int_{0}^{t}\left|\left(f^{(4)}(x+jw)+\left(\alpha^{2}+\beta^{2}\right)f''(x+jw)+\alpha^{2}\beta^{2}f(x+jw)\right)\right.$$

$$\left.-\left(f^{(4)}(x)+\left(\alpha^{2}+\beta^{2}\right)f''(x)+\alpha^{2}\beta^{2}f(x)\right)\right|$$

$$\left|\beta\sin\left(\alpha j(t-w)\right)-\alpha\sin\left(\beta j(t-w)\right)\right|dw\leq$$

(by $|\sin x|\leq|x|$, $\forall\, x\in\mathbb{R}$)

$$\frac{2|\alpha\beta|}{\left|\alpha\beta\left(\beta^{2}-\alpha^{2}\right)\right|}\sum_{j=0}^{r}\left|\alpha_{j}\right|j^{2}$$

$$\times\int_{0}^{t}\omega_{1}\left(\left(f^{(4)}+\left(\alpha^{2}+\beta^{2}\right)f''+\alpha^{2}\beta^{2}f\right),jw\right)(t-w)\,dw$$

$(\xi > 0)$

$$= \frac{2}{|\beta^2 - \alpha^2|} \sum_{j=0}^{r} |\alpha_j| j^2$$

$$\times \int_0^t \omega_1 \left(\left(f^{(4)} + \left(\alpha^2 + \beta^2\right) f'' + \alpha^2 \beta^2 f \right), \frac{\xi j w}{\xi} \right) (t - w)\, dw \quad (22.14)$$

$$\leq \frac{2}{|\beta^2 - \alpha^2|} \sum_{j=0}^{r} |\alpha_j| j^2 \omega_1 \left(\left(f^{(4)} + \left(\alpha^2 + \beta^2\right) f'' + \alpha^2 \beta^2 f \right), \xi \right)$$

$$\times \int_0^t \left(1 + \frac{jw}{\xi} \right) (t - w)\, dw$$

$$= \frac{2\omega_1 \left(\left(f^{(4)} + \left(\alpha^2 + \beta^2\right) f'' + \alpha^2 \beta^2 f \right), \xi \right)}{|\beta^2 - \alpha^2|}$$

$$\times \sum_{j=0}^{r} |\alpha_j| j^2 \left[\int_0^t (t - w)\, dw + \frac{j}{\xi} \int_0^t w^{2-1} (t - w)^{2-1}\, dw \right]$$

$$= \frac{2\omega_1 \left(\left(f^{(4)} + \left(\alpha^2 + \beta^2\right) f'' + \alpha^2 \beta^2 f \right), \xi \right)}{|\beta^2 - \alpha^2|} \sum_{j=0}^{r} |\alpha_j| j^2 \left[\frac{t^2}{2} + j \frac{t^3}{6\xi} \right]$$

$$= \frac{\omega_1 \left(\left(f^{(4)} + \left(\alpha^2 + \beta^2\right) f'' + \alpha^2 \beta^2 f \right), \xi \right)}{|\beta^2 - \alpha^2|} \sum_{j=0}^{r} |\alpha_j| j^2 \left[t^2 + j \frac{t^3}{3\xi} \right]. \quad (22.15)$$

Hence it holds $(t \geq 0)$

$$|R| \leq \frac{\omega_1 \left(\left(f^{(4)} + \left(\alpha^2 + \beta^2\right) f'' + \alpha^2 \beta^2 f \right), \xi \right)}{|\beta^2 - \alpha^2|} \sum_{j=0}^{r} |\alpha_j| j^2 \left[t^2 + \frac{j}{3\xi} t^3 \right], \quad (22.16)$$

$\xi > 0$.

Let now $t < 0$, then

$$|R| \leq \frac{1}{|\alpha\beta\left(\beta^2 - \alpha^2\right)|} \sum_{j=0}^{r} |\alpha_j| j$$

$$\times \int_t^0 \left| \left(f^{(4)}(x + jw) + \left(\alpha^2 + \beta^2\right) f''(x + jw) + \alpha^2 \beta^2 f(x + jw) \right) \right.$$

$$\left. - \left(f^{(4)}(x) + \left(\alpha^2 + \beta^2\right) f''(x) + \alpha^2 \beta^2 f(x) \right) \right|$$

$$\times |\beta \sin(\alpha j (t - w)) - \alpha \sin(\beta j (t - w))|\, dw$$

$$\leq \frac{2}{|\beta^2-\alpha^2|}\sum_{j=0}^{r}|\alpha_j|\,j^2$$

$$\times \int_t^0 \omega_1\left(\left(f^{(4)}+\left(\alpha^2+\beta^2\right)f''+\alpha^2\beta^2 f\right),-j\frac{w\xi}{\xi}\right)(w-t)\,dw$$

$$\leq \frac{2\omega_1\left(\left(f^{(4)}+\left(\alpha^2+\beta^2\right)f''+\alpha^2\beta^2 f\right),\xi\right)}{|\beta^2-\alpha^2|}$$

$$\times \sum_{j=0}^{r}|\alpha_j|\,j^2\left(\int_t^0\left(1-\frac{j}{\xi}w\right)(w-t)\,dw\right)$$

$$= \frac{2\omega_1\left(\left(f^{(4)}+\left(\alpha^2+\beta^2\right)f''+\alpha^2\beta^2 f\right),\xi\right)}{|\beta^2-\alpha^2|}$$

$$\times \sum_{j=0}^{r}|\alpha_j|\,j^2\left[\int_t^0 (w-t)\,dw+\frac{j}{\xi}\int_t^0 (0-w)^{2-1}(w-t)^{2-1}\,dw\right]$$

$$= \frac{\omega_1\left(\left(f^{(4)}+\left(\alpha^2+\beta^2\right)f''+\alpha^2\beta^2 f\right),\xi\right)}{|\beta^2-\alpha^2|}\sum_{j=0}^{r}|\alpha_j|\,j^2\left[t^2+\frac{j}{3\xi}(-t)^3\right]. \tag{22.17}$$

We found that (for $t<0$)

$$|R|\leq \frac{\omega_1\left(\left(f^{(4)}+\left(\alpha^2+\beta^2\right)f''+\alpha^2\beta^2 f\right),\xi\right)}{|\beta^2-\alpha^2|}\sum_{j=0}^{r}|\alpha_j|\,j^2\left[t^2+\frac{j}{3\xi}(-t)^3\right]. \tag{22.18}$$

Consequently, for $t\in\mathbb{R}$, we obtain

$$|R(t)|\leq \frac{\omega_1\left(\left(f^{(4)}+\left(\alpha^2+\beta^2\right)f''+\alpha^2\beta^2 f\right),\xi\right)}{|\beta^2-\alpha^2|}$$
$$\times \sum_{j=0}^{r}|\alpha_j|\,j^2\left[t^2+\frac{j}{3\xi}|t|^3\right],\quad \xi>0. \tag{22.19}$$

So, we have (by (22.12))

$$\sum_{j=0}^{r}\alpha_j\left[f(x+jt)-f(x)\right]$$

$$-\left(\frac{f'(x)}{\alpha\beta\left(\beta^2-\alpha^2\right)}\right)\left[\beta^3\sum_{j=0}^{r}\alpha_j\sin(\alpha jt)-\alpha^3\sum_{j=0}^{r}\alpha_j\sin(\beta jt)\right]$$

$$-\left(\frac{f''(x)}{\beta^2-\alpha^2}\right)\left[\sum_{j=0}^{r}\alpha_j\cos(\alpha jt)-\sum_{j=0}^{r}\alpha_j\cos(\beta jt)\right]$$

$$-\left(\frac{f^{(3)}(x)}{\alpha\beta(\beta^2-\alpha^2)}\right)\left[\beta\sum_{j=0}^{r}\alpha_j\sin(\alpha jt)-\alpha\sum_{j=0}^{r}\alpha_j\sin(\beta jt)\right]$$

$$-\left(\frac{2\left(f^{(4)}(x)+\left(\alpha^2+\beta^2\right)f''(x)\right)}{(\alpha\beta)^2(\beta^2-\alpha^2)}\right)$$

$$\times\left(\beta^2\sum_{j=0}^{r}\alpha_j\sin^2\left(\frac{\alpha jt}{2}\right)-\alpha^2\sum_{j=0}^{r}\alpha_j\sin^2\left(\frac{\beta jt}{2}\right)\right)$$

$$=R(t). \tag{22.20}$$

Therefore, it holds

$$E_1(x):=P_{r,\xi}(f,x)-f(x)-\left(\frac{f'(x)}{\alpha\beta(\beta^2-\alpha^2)}\right)$$

$$\left[\beta^3\sum_{j=0}^{r}\alpha_j\frac{1}{2\xi}\left(\int_{-\infty}^{\infty}\sin(\alpha jt)\,e^{-\frac{|t|}{\xi}}dt\right)-\alpha^3\sum_{j=0}^{r}\alpha_j\frac{1}{2\xi}\left(\int_{-\infty}^{\infty}\sin(\beta jt)\,e^{-\frac{|t|}{\xi}}dt\right)\right]$$

$$-\left(\frac{f''(x)}{\beta^2-\alpha^2}\right)$$

$$\left[\sum_{j=0}^{r}\alpha_j\frac{1}{2\xi}\left(\int_{-\infty}^{\infty}\cos(\alpha jt)\,e^{-\frac{|t|}{\xi}}dt\right)-\sum_{j=0}^{r}\alpha_j\frac{1}{2\xi}\left(\int_{-\infty}^{\infty}\cos(\beta jt)\,e^{-\frac{|t|}{\xi}}dt\right)\right]$$

$$-\left(\frac{f^{(3)}(x)}{\alpha\beta(\beta^2-\alpha^2)}\right)$$

$$\left[\beta\sum_{j=0}^{r}\alpha_j\frac{1}{2\xi}\left(\int_{-\infty}^{\infty}\sin(\alpha jt)\,e^{-\frac{|t|}{\xi}}dt\right)-\alpha\sum_{j=0}^{r}\alpha_j\frac{1}{2\xi}\left(\int_{-\infty}^{\infty}\sin(\beta jt)\,e^{-\frac{|t|}{\xi}}dt\right)\right]$$

$$-\left(\frac{2\left(f^{(4)}(x)+\left(\alpha^2+\beta^2\right)f''(x)\right)}{(\alpha\beta)^2(\beta^2-\alpha^2)}\right)$$

$$\left[\beta^2\sum_{j=0}^{r}\alpha_j\frac{1}{2\xi}\left(\int_{-\infty}^{\infty}\sin^2\left(\frac{\alpha jt}{2}\right)e^{-\frac{|t|}{\xi}}dt\right)\right.$$

$$\left.-\alpha^2\sum_{j=0}^{r}\alpha_j\frac{1}{2\xi}\left(\int_{-\infty}^{\infty}\sin^2\left(\frac{\beta jt}{2}\right)e^{-\frac{|t|}{\xi}}dt\right)\right]$$

$$=\frac{1}{2\xi}\int_{-\infty}^{\infty}R(t)\,e^{-\frac{|t|}{\xi}}dt. \tag{22.21}$$

Hence we get

$$\begin{aligned}|E_1(x)| &\leq \frac{1}{2\xi}\int_{-\infty}^{\infty}|R(t)|\,e^{-\frac{|t|}{\xi}}dt\\ &\overset{(22.19)}{\leq} \frac{\omega_1\left(\left(f^{(4)}+\left(\alpha^2+\beta^2\right)f''+\alpha^2\beta^2 f\right),\xi\right)}{|\beta^2-\alpha^2|}\\ &\quad \sum_{j=0}^{r}|\alpha_j|\,j^2\left[\frac{1}{2\xi}\int_{-\infty}^{\infty}t^2e^{-\frac{|t|}{\xi}}dt+\frac{j}{3\xi}\frac{1}{2\xi}\int_{-\infty}^{\infty}|t|^3\,e^{-\frac{|t|}{\xi}}dt\right]=\end{aligned} \tag{22.22}$$

$$\left(\text{we use } \int_{-\infty}^{\infty}t^ke^{-\frac{|t|}{\xi}}dt=\begin{Bmatrix}0, & k \text{ odd}\\ 2k!\xi^{k+1}, & k \text{ even}\end{Bmatrix}\right)$$

$$\frac{\omega_1\left(\left(f^{(4)}+\left(\alpha^2+\beta^2\right)f''+\alpha^2\beta^2 f\right),\xi\right)}{|\beta^2-\alpha^2|}$$

$$\sum_{j=0}^{r}|\alpha_j|\,j^2\left[\frac{1}{2\xi}4\xi^3+\frac{j\xi^4}{3\xi^2}\int_0^{\infty}\left(\frac{t}{\xi}\right)^3e^{-\frac{t}{\xi}}d\frac{t}{\xi}\right] \tag{22.23}$$

$$\begin{aligned}&=\frac{\omega_1\left(\left(f^{(4)}+\left(\alpha^2+\beta^2\right)f''+\alpha^2\beta^2 f\right),\xi\right)}{|\beta^2-\alpha^2|}\\ &\quad\sum_{j=0}^{r}|\alpha_j|\,j^2\left[2\xi^2+\xi^2\frac{j}{3}\int_0^{\infty}z^3e^{-z}dz\right]\\ &=\frac{\omega_1\left(\left(f^{(4)}+\left(\alpha^2+\beta^2\right)f''+\alpha^2\beta^2 f\right),\xi\right)}{|\beta^2-\alpha^2|}\sum_{j=0}^{r}|\alpha_j|\,j^2\left[2\xi^2+\xi^2\frac{j}{3}6\right]\\ &=\frac{2\omega_1\left(\left(f^{(4)}+\left(\alpha^2+\beta^2\right)f''+\alpha^2\beta^2 f\right),\xi\right)\xi^2}{|\beta^2-\alpha^2|}\left(\sum_{j=0}^{r}|\alpha_j|\,j^2(j+1)\right).\end{aligned} \tag{22.24}$$

We have proved that

$$|E_1(x)|\leq\frac{2\xi^2\omega_1\left(\left(f^{(4)}+\left(\alpha^2+\beta^2\right)f''+\alpha^2\beta^2 f\right),\xi\right)}{|\beta^2-\alpha^2|}\left(\sum_{j=0}^{r}|\alpha_j|\,j^2(j+1)\right), \tag{22.25}$$

$\xi>0$, $\forall\, x\in\mathbb{R}$.

Note that

$$E_1(x) \to 0, \quad \text{as } \xi \to 0.$$

Next, we simplify $E_1(x)$:
We observe that

$$\int_{-\infty}^{\infty} \sin(\alpha j t)\, e^{-\frac{|t|}{\xi}} dt = \int_{-\infty}^{0} \sin(\alpha j t)\, e^{-\frac{|t|}{\xi}} dt + \int_{0}^{\infty} \sin(\alpha j t)\, e^{-\frac{|t|}{\xi}} dt. \tag{22.26}$$

Note $-\infty \leq t \leq 0 \Rightarrow \infty \geq -t \geq 0$. So

$$\int_{-\infty}^{0} \sin(\alpha j t)\, e^{-\frac{|t|}{\xi}} dt = -\int_{-\infty}^{0} \sin(\alpha j(-(-t)))\, e^{-\frac{|t|}{\xi}} d(-t)$$

$$= -\int_{-\infty}^{0} (-\sin(\alpha j(-t)))\, e^{-\frac{|t|}{\xi}} d(-t) = \int_{-\infty}^{0} \sin(\alpha j(-t))\, e^{-\frac{|t|}{\xi}} d(-t)$$

$$= \int_{\infty}^{0} \sin(\alpha j t)\, e^{-\frac{|t|}{\xi}} dt = -\int_{0}^{\infty} \sin(\alpha j t)\, e^{-\frac{|t|}{\xi}} dt. \tag{22.27}$$

So

$$\int_{-\infty}^{\infty} \sin(\alpha j t)\, e^{-\frac{|t|}{\xi}} dt = 0,$$

and all sine integrals in (22.21) are zeros.

Furthermore, we have that

$$\int_{-\infty}^{\infty} \sin^2\left(\frac{\alpha j t}{2}\right) e^{-\frac{|t|}{\xi}} dt = 2\int_{0}^{\infty} \sin^2\left(\frac{\alpha j t}{2}\right) e^{-\frac{|t|}{\xi}} dt, \quad j = 0, 1, \ldots, r. \tag{22.28}$$

The last follows by

$$\int_{-\infty}^{0} \sin^2\left(\frac{\alpha j t}{2}\right) e^{-\frac{|t|}{\xi}} dt = -\int_{-\infty}^{0} \left(-\sin\left(\frac{-\alpha j t}{2}\right)\right)^2 e^{-\frac{|-t|}{\xi}} d(-t)$$

$$\overset{(z=-t)}{=} -\int_{\infty}^{0} \sin^2\left(\frac{\alpha j z}{2}\right) e^{-\frac{|z|}{\xi}} dz = \int_{0}^{\infty} \sin^2\left(\frac{\alpha j z}{2}\right) e^{-\frac{z}{\xi}} dz. \tag{22.29}$$

Next, we calculate

$$\int_{0}^{\infty} \sin^2\left(\frac{\alpha j}{2} z\right) e^{-\frac{|z|}{\xi}} dz = \xi \int_{0}^{\infty} \sin^2\left(\left(\frac{\alpha j \xi}{2}\right)\frac{z}{\xi}\right) e^{-\frac{z}{\xi}} d\frac{z}{\xi} = \tag{22.30}$$

(call $\frac{t}{\xi} =: x$ and $\frac{\alpha j \xi}{2} =: a_1$)

$$\xi \int_0^{\infty} \sin^2 (a_1 x) e^{-x} dx =$$

(by Wolfram Alpha Computational Intelligence)

$$\xi \left(\frac{2a_1^2}{4a_1^2 + 1} \right) = \frac{\alpha^2 j^2 \xi^3}{2 (\alpha^2 j^2 \xi^2 + 1)}.$$

Thus, it is

$$\int_{-\infty}^{\infty} \sin^2 \left(\frac{\alpha j t}{2} \right) e^{-\frac{|t|}{\xi}} dt = \frac{\alpha^2 j^2 \xi^3}{\alpha^2 j^2 \xi^2 + 1}, \quad j = 0, 1, \ldots, r. \tag{22.31}$$

Consequently, it holds

$$\frac{1}{2\xi} \int_{-\infty}^{\infty} \sin^2 \left(\frac{\alpha j t}{2} \right) e^{-\frac{|t|}{\xi}} dt = \frac{\alpha^2 j^2 \xi^2}{2 (\alpha^2 j^2 \xi^2 + 1)} \to 0, \text{ as } \xi \to 0, \ j = 0, 1, \ldots, r. \tag{22.32}$$

We note that

$$\int_{-\infty}^{\infty} \cos (\alpha j t) e^{-\frac{|t|}{\xi}} dt = 2 \int_0^{\infty} \cos (\alpha j t) e^{-\frac{t}{\xi}} dt$$

$$= 2\xi \int_0^{\infty} \cos \left((\alpha j \xi) \frac{t}{\xi} \right) e^{-\frac{t}{\xi}} d\frac{t}{\xi} = \tag{22.33}$$

(call $\alpha j \xi =: \alpha_2$, $\frac{t}{\xi} =: x$)

$$2\xi \int_0^{\infty} \cos (\alpha_2 x) e^{-x} dx = 2\xi \left(\frac{1}{\alpha_2^2 + 1} \right) = 2\xi \left(\frac{1}{\alpha^2 j^2 \xi^2 + 1} \right),$$

and

$$\frac{1}{2\xi} \int_{-\infty}^{\infty} \cos (\alpha j t) e^{-\frac{|t|}{\xi}} dt = \frac{1}{\alpha^2 j^2 \xi^2 + 1}, \tag{22.34}$$

$j = 0, 1, \ldots, r$.

So, based on the above we write down the simplified version of $E_1(x)$ as follows:

$$E_1(x) = P_{r,\xi}(f,x) - f(x) - \left(\frac{f''(x)}{\beta^2-\alpha^2}\right)\left\{\sum_{j=0}^{r}\alpha_j\left[\left(\frac{1}{\alpha^2 j^2\xi^2+1}\right)-\left(\frac{1}{\beta^2 j^2\xi^2+1}\right)\right]\right\} - \left(\frac{2\left(f^{(4)}(x)+\left(\alpha^2+\beta^2\right)f''(x)\right)}{(\alpha\beta)^2\left(\beta^2-\alpha^2\right)}\right) \times\left[\beta^2\sum_{j=0}^{r}\alpha_j\left(\frac{\alpha^2 j^2\xi^2}{2\left(\alpha^2 j^2\xi^2+1\right)}\right)-\alpha^2\sum_{j=0}^{r}\alpha_j\left(\frac{\beta^2 j^2\xi^2}{2\left(\beta^2 j^2\xi^2+1\right)}\right)\right] \tag{22.35}$$

$$= P_{r,\xi}(f,x) - f(x) - f''(x)\left\{\sum_{j=0}^{r}\alpha_j\left[\frac{j^2\xi^2}{\left(\alpha^2 j^2\xi^2+1\right)\left(\beta^2 j^2\xi^2+1\right)}\right]\right\} \tag{22.36}$$

$$- \left(\frac{2\left(f^{(4)}(x)+\left(\alpha^2+\beta^2\right)f''(x)\right)}{(\alpha\beta)^2\left(\beta^2-\alpha^2\right)}\right) \times\left[\beta^2\sum_{j=0}^{r}\alpha_j\left(\frac{\alpha^2 j^2\xi^2}{2\left(\alpha^2 j^2\xi^2+1\right)}\right)-\alpha^2\sum_{j=0}^{r}\alpha_j\left(\frac{\beta^2 j^2\xi^2}{2\left(\beta^2 j^2\xi^2+1\right)}\right)\right]$$

$$= P_{r,\xi}(f,x) - f(x) - f''(x)\left\{\sum_{j=0}^{r}\alpha_j j^2\left[\frac{1}{\left(\alpha^2 j^2\xi^2+1\right)\left(\beta^2 j^2\xi^2+1\right)}\right]\right\}\xi^2 - \left(f^{(4)}(x)+\left(\alpha^2+\beta^2\right)f''(x)\right) \times\left[\sum_{j=0}^{r}\alpha_j j^4\left(\frac{1}{\left(\alpha^2 j^2\xi^2+1\right)\left(\beta^2 j^2\xi^2+1\right)}\right)\right]\xi^4 = E(x). \tag{22.37}$$

The proof of the theorem is now complete. □

In the special case of $\alpha = 2$, $\beta = 1$, we set

$$\overline{E(x)} := P_{r,\xi}(f,x) - f(x) - f''(x)\left\{\sum_{j=0}^{r} \alpha_j \left[\frac{j^2}{(4j^2\xi^2+1)(j^2\xi^2+1)}\right]\right\}\xi^2 - \left(f^{(4)}(x) + 5f''(x)\right)\left[\sum_{j=0}^{r} \alpha_j \frac{j^4}{(4j^2\xi^2+1)(j^2\xi^2+1)}\right]\xi^4, \tag{22.38}$$

$\xi > 0$, $x \in \mathbb{R}$.

That is $\overline{E(x)}$ is $E(x)$ for $\alpha = 2$, $\beta = 1$.

We mention

Corollary 22.1. *It holds*

$$\left\|\overline{E(x)}\right\|_{\infty} \leq \frac{2\xi^2\omega_1\left(\left(f^{(4)} + 5f'' + 4f\right), \xi\right)}{3}\left(\sum_{j=0}^{r} |\alpha_j| j^2 (j+1)\right), \tag{22.39}$$

$\xi > 0$, $x \in \mathbb{R}$.

And $\left\|\overline{E(x)}\right\| \to 0$, *as* $\xi \to 0$.

We finish with

Corollary 22.2. *(to Theorem 22.1) Additionally let* $f'', f^{(4)} \in C_B(\mathbb{R})$. *It holds*

$$\begin{aligned}
|P_{r,\xi}(f,x) - f(x)| &\leq \|P_{r,\xi}(f,x) - f(x)\|_{\infty} \\
&\leq \|f''\|_{\infty}\left\{\sum_{j=0}^{r} |\alpha_j| j^2 \left[\frac{1}{(\alpha^2 j^2\xi^2+1)(\beta^2 j^2\xi^2+1)}\right]\right\}\xi^2 \\
&\quad + \left(\left\|f^{(4)}\right\|_{\infty} + (\alpha^2+\beta^2)\|f''\|_{\infty}\right) \\
&\quad \times \left[\sum_{j=0}^{r} |\alpha_j| j^4 \left(\frac{1}{(\alpha^2 j^2\xi^2+1)(\beta^2 j^2\xi^2+1)}\right)\right]\xi^4 \\
&\quad + \frac{2\xi^2\omega_1\left(\left(f^{(4)} + (\alpha^2+\beta^2)f'' + \alpha^2\beta^2 f\right), \xi\right)}{|\beta^2-\alpha^2|} \\
&\quad \times \left(\sum_{j=0}^{r} |\alpha_j| j^2 (j+1)\right) \to 0, \quad \text{as } \xi \to 0.
\end{aligned} \tag{22.40}$$

Bibliography

A.H. Ali, Z. Pales, Taylor-type expansions in terms of exponential polynomials, *Mathematical Inequalities and Applications*, 25(4) (2022), 1123–1141.

G.A. Anastassiou, Basic convergence with rates of smooth Picard singular integral operators, *Journal Computational Analysis and Application*, 8(4) (2006), 313–334.

G.A. Anastassiou, *Intelligent Mathematics: Computational Analysis*, Springer, Heidelberg, New York, Chapter 10, 2011.

G.A. Anastassiou, Opial and Ostrowski type inequalities based on trigonometric and hyperbolic type Taylor formulae, *Malaya Journal Mathematik*, 11(S) (2023a), 1–26.

G.A. Anastassiou, *Parametrized and Trigonometric generated quantitative convergence of smooth Picard singular integral operators*, submitted, 2023b.

Chapter 23

Parametrized and Trigonometric L_p Approximation with Rates of Smooth Picard Singular Integral Operators

In this chapter, we continue the research on the smooth Picard singular integral operators that started in Anastassiou (2011), see there Chapters 10–14. This time, the foundation of our research is a parametrized trigonometric Taylor's formula. We establish the L_p, $p \geq 1$, convergence of our operators to the unit operator with rates via Jackson-type inequalities engaging the first L_p modulus of smoothness. Of interest, here is a parametrized residual appearing term. Note that our operators are not positive. It follows Anastassiou (2023b).

23.1 Introduction

We are motivated by Anastassiou (2006, 2011) Chapters 10–14, Anastassiou (2023a), and Ali and Pales (2022). We use a parametrized trigonometric new Taylor formula from Anastassiou (2023a), see also Ali and Pales (2022). Here we consider some very general operators, the smooth Picard singular integral operators over the real line, and we study further their parametrized L_p convergence properties quantitatively. We establish related L_p, $p \geq 1$, inequalities involving the first modulus of smoothness with respect to the L_p, $p \geq 1$, norm. We provide detailed proofs.

23.2 Results

By Anastassiou (2023a) and Ali and Pales (2022), for $\alpha, \beta \in \mathbb{R}$ with $\alpha\beta\left(\alpha^2 - \beta^2\right) \neq 0$, and $f \in C^4(\mathbb{R})$, $a, x \in \mathbb{R}$, we have the following general trigonometric Taylor formula:

$$\begin{aligned}
f(x) - f(a) &= f'(a)\left(\frac{\beta^3 \sin(\alpha(x-a)) - \alpha^3 \sin(\beta(x-a))}{\alpha\beta(\beta^2-\alpha^2)}\right) \\
&\quad + f''(a)\left(\frac{\cos(\alpha(x-a)) - \cos(\beta(x-a))}{\beta^2-\alpha^2}\right) \\
&\quad + f'''(a)\left(\frac{\beta \sin(\alpha(x-a)) - \alpha \sin(\beta(x-a))}{\alpha\beta(\beta^2-\alpha^2)}\right) \\
&\quad + \frac{2\left(f''''(a) + (\alpha^2+\beta^2) f''(a)\right)}{(\alpha\beta)^2(\beta^2-\alpha^2)} \\
&\quad \times \left(\beta^2 \sin^2\left(\frac{\alpha(x-a)}{2}\right) - \alpha^2 \sin^2\left(\frac{\beta(x-a)}{2}\right)\right) \\
&\quad + \frac{1}{\alpha\beta(\beta^2-\alpha^2)}\int_a^x \left[\left(f''''(t) + (\alpha^2+\beta^2) f''(t) + \alpha^2\beta^2 f(t)\right)\right. \\
&\quad \left. - \left(f''''(a) + (\alpha^2+\beta^2) f''(a) + \alpha^2\beta^2 f(a)\right)\right] \\
&\quad \left[\beta \sin(\alpha(x-t)) - \alpha \sin(\beta(x-t))\right] dt. \qquad (23.1)
\end{aligned}$$

For $r \in \mathbb{N}$ and $n \in \mathbb{Z}^+$, we set

$$\alpha_j := \begin{cases} (-1)^{r-j}\binom{r}{j} j^{-n}, & j = 1, \ldots, r. \\ 1 - \sum_{j=1}^{r} (-1)^{r-j}\binom{r}{j} j^{-n}, & n = 0, \end{cases} \qquad (23.2)$$

that is,

$$\sum_{j=0}^{r} \alpha_j = 1. \qquad (23.3)$$

Here we consider all $f, f', f'', f''', f^{(4)} \in L_p(\mathbb{R}) \cap C(\mathbb{R})$, $1 \leq p < \infty$.

For $x \in \mathbb{R}$, $\xi > 0$, we consider the Lebesgue integrals, so-called smooth Picard operators,

$$P_{r,\xi}(f,x) = \frac{1}{2\xi}\int_{-\infty}^{\infty}\left(\sum_{j=0}^{r} \alpha_j f(x+jt)\right) e^{-\frac{|t|}{\xi}} dt, \qquad (23.4)$$

see Anastassiou (2006); $P_{r,\xi}$ are not in general positive operators, see Anastassiou (2011).

We note by

$$\frac{1}{2\xi}\int_{-\infty}^{\infty} e^{-\frac{|t|}{\xi}}dt = 1 \tag{23.5}$$

that

$$P_{r,\xi}(c,x) = c, \text{ where } c \text{ is a constant,} \tag{23.6}$$

and

$$P_{r,\xi}(f,x) - f(x) = \frac{1}{2\xi}\left(\sum_{j=0}^{r}\alpha_j \int_{-\infty}^{\infty}(f(x+jt) - f(x))\, e^{-\frac{|t|}{\xi}}dt\right). \tag{23.7}$$

Denote by

$$\omega_1(f,h)_p := \sup_{\substack{t\in\mathbb{R} \\ |t|\le h}} \|f(x+t) - f(x)\|_{p,x} \tag{23.8}$$

the first L_p modulus of smoothness of f, $1 \le p < \infty$.

By (23.1), we get that

$$\begin{aligned} f(x+jt) - f(x) &= f'(x)\left(\frac{\beta^3\sin(\alpha jt) - \alpha^3\sin(\beta jt)}{\alpha\beta(\beta^2-\alpha^2)}\right) \\ &+ f''(x)\left(\frac{\cos(\alpha jt) - \cos(\beta jt)}{\beta^2-\alpha^2}\right) \\ &+ f'''(x)\left(\frac{\beta\sin(\alpha jt) - \alpha\sin(\beta jt)}{\alpha\beta(\beta^2-\alpha^2)}\right) \\ &+ \frac{2\left(f^{(4)}(x) + \left(\alpha^2+\beta^2\right)f''(x)\right)}{(\alpha\beta)^2(\beta^2-\alpha^2)}\left(\beta^2\sin^2\left(\frac{\alpha jt}{2}\right) - \alpha^2\sin^2\left(\frac{\beta jt}{2}\right)\right) \\ &+ \frac{1}{\alpha\beta(\beta^2-\alpha^2)}\int_x^{x+jt}\Big[\left(f^{(4)}(s) + \left(\alpha^2+\beta^2\right)f''(s) + \alpha^2\beta^2 f(s)\right) \\ &- \left(f^{(4)}(x) + \left(\alpha^2+\beta^2\right)f''(x) + \alpha^2\beta^2 f(x)\right)\Big] \\ &[\beta\sin(\alpha(x+jt-s)) - \alpha\sin(\beta(x+jt-s))]\, ds, \end{aligned} \tag{23.9}$$

or better,

$$f(x+jt)-f(x)=f'(x)\left(\frac{\beta^3\sin(\alpha jt)-\alpha^3\sin(\beta jt)}{\alpha\beta(\beta^2-\alpha^2)}\right)$$

$$+f''(x)\left(\frac{\cos(\alpha jt)-\cos(\beta jt)}{\beta^2-\alpha^2}\right)$$

$$+f'''(x)\left(\frac{\beta\sin(\alpha jt)-\alpha\sin(\beta jt)}{\alpha\beta(\beta^2-\alpha^2)}\right)$$

$$+\frac{2\left(f^{(4)}(x)+\left(\alpha^2+\beta^2\right)f''(x)\right)}{(\alpha\beta)^2\left(\beta^2-\alpha^2\right)}\left(\beta^2\sin^2\left(\frac{\alpha jt}{2}\right)-\alpha^2\sin^2\left(\frac{\beta jt}{2}\right)\right)$$

$$+\frac{1}{\alpha\beta(\beta^2-\alpha^2)}\int_0^{jt}\left[\left(f^{(4)}(x+z)+\left(\alpha^2+\beta^2\right)f''(x+z)+\alpha^2\beta^2f(x+z)\right)\right.$$

$$\left.-\left(f^{(4)}(x)+\left(\alpha^2+\beta^2\right)f''(x)+\alpha^2\beta^2f(x)\right)\right]$$

$$\times\left[\beta\sin(\alpha(jt-z))-\alpha\sin(\beta(jt-z))\right]dz. \tag{23.10}$$

Furthermore, it holds

$$\sum_{j=0}^{r}\alpha_j\left[f(x+jt)-f(x)\right]$$

$$=\left(\frac{f'(x)}{\alpha\beta(\beta^2-\alpha^2)}\right)\left[\beta^3\sum_{j=0}^{r}\alpha_j\sin(\alpha jt)-\alpha^3\sum_{j=0}^{r}\alpha_j\sin(\beta jt)\right]$$

$$+\left(\frac{f''(x)}{\beta^2-\alpha^2}\right)\left[\sum_{j=0}^{r}\alpha_j\cos(\alpha jt)-\sum_{j=0}^{r}\alpha_j\cos(\beta jt)\right]$$

$$+\left(\frac{f^{(3)}(x)}{\alpha\beta(\beta^2-\alpha^2)}\right)\left[\beta\sum_{j=0}^{r}\alpha_j\sin(\alpha jt)-\alpha\sum_{j=0}^{r}\alpha_j\sin(\beta jt)\right]$$

$$+\left(\frac{2\left(f^{(4)}(x)+\left(\alpha^2+\beta^2\right)f''(x)\right)}{(\alpha\beta)^2\left(\beta^2-\alpha^2\right)}\right)$$

$$\times \left(\beta^2 \sum_{j=0}^{r} \alpha_j \sin^2\left(\frac{\alpha jt}{2}\right) - \alpha^2 \sum_{j=0}^{r} \alpha_j \sin^2\left(\frac{\beta jt}{2}\right)\right) + \frac{1}{\alpha\beta\left(\beta^2-\alpha^2\right)} \sum_{j=0}^{r} \alpha_j$$

$$\times \int_0^{jt} \left[\left(f^{(4)}(x+z) + \left(\alpha^2+\beta^2\right) f''(x+z) + \alpha^2\beta^2 f(x+z)\right)\right.$$

$$\left. - \left(f^{(4)}(x) + \left(\alpha^2+\beta^2\right) f''(x) + \alpha^2\beta^2 f(x)\right)\right]$$

$$\times \left[\beta \sin\left(\alpha\left(jt-z\right)\right) - \alpha \sin\left(\beta\left(jt-z\right)\right)\right] dz, \tag{23.11}$$

or better

$$\sum_{j=0}^{r} \alpha_j \left[f(x+jt) - f(x)\right]$$

$$= \left(\frac{f'(x)}{\alpha\beta\left(\beta^2-\alpha^2\right)}\right) \left[\beta^3 \sum_{j=0}^{r} \alpha_j \sin(\alpha jt) - \alpha^3 \sum_{j=0}^{r} \alpha_j \sin(\beta jt)\right]$$

$$+ \left(\frac{f''(x)}{\beta^2-\alpha^2}\right) \left[\sum_{j=0}^{r} \alpha_j \cos(\alpha jt) - \sum_{j=0}^{r} \alpha_j \cos(\beta jt)\right]$$

$$+ \left(\frac{f^{(3)}(x)}{\alpha\beta\left(\beta^2-\alpha^2\right)}\right) \left[\beta \sum_{j=0}^{r} \alpha_j \sin(\alpha jt) - \alpha \sum_{j=0}^{r} \alpha_j \sin(\beta jt)\right]$$

$$+ \left(\frac{2\left(f^{(4)}(x) + \left(\alpha^2+\beta^2\right) f''(x)\right)}{(\alpha\beta)^2\left(\beta^2-\alpha^2\right)}\right)$$

$$\times \left(\beta^2 \sum_{j=0}^{r} \alpha_j \sin^2\left(\frac{\alpha jt}{2}\right) - \alpha^2 \sum_{j=0}^{r} \alpha_j \sin^2\left(\frac{\beta jt}{2}\right)\right) + \frac{1}{\alpha\beta\left(\beta^2-\alpha^2\right)} \sum_{j=0}^{r} \alpha_j j$$

$$\times \int_0^{t} \left[\left(f^{(4)}(x+jw) + \left(\alpha^2+\beta^2\right) f''(x+jw) + \alpha^2\beta^2 f(x+jw)\right)\right.$$

$$\left. - \left(f^{(4)}(x) + \left(\alpha^2+\beta^2\right) f''(x) + \alpha^2\beta^2 f(x)\right)\right]$$

$$\times \left[\beta \sin\left(\alpha j\left(t-w\right)\right) - \alpha \sin\left(\beta j\left(t-w\right)\right)\right] dw. \tag{23.12}$$

We call

$$R := R(t) := \frac{1}{\alpha\beta(\beta^2-\alpha^2)} \sum_{j=0}^{r} \alpha_j j$$

$$\int_0^t \Big[\Big(f^{(4)}(x+jw) + (\alpha^2+\beta^2) f''(x+jw) + \alpha^2\beta^2 f(x+jw)\Big)$$

$$- \Big(f^{(4)}(x) + (\alpha^2+\beta^2) f''(x) + \alpha^2\beta^2 f(x)\Big)\Big]$$

$$\times [\beta \sin(\alpha j(t-w)) - \alpha \sin(\beta j(t-w))]\, dw, \quad \forall\, t \in \mathbb{R}. \tag{23.13}$$

We set

$$E(x) := P_{r,\xi}(f,x) - f(x) - \left(\frac{f'(x)}{\alpha\beta(\beta^2-\alpha^2)}\right)$$

$$\left[\beta^3 \sum_{j=0}^{r} \alpha_j \frac{1}{2\xi}\left(\int_{-\infty}^{\infty} \sin(\alpha j t)\, e^{-\frac{|t|}{\xi}} dt\right) - \alpha^3 \sum_{j=0}^{r} \alpha_j \frac{1}{2\xi}\left(\int_{-\infty}^{\infty} \sin(\beta j t)\, e^{-\frac{|t|}{\xi}} dt\right)\right]$$

$$-\left(\frac{f''(x)}{\beta^2-\alpha^2}\right)$$

$$\times \left[\sum_{j=0}^{r} \alpha_j \frac{1}{2\xi}\left(\int_{-\infty}^{\infty} \cos(\alpha j t)\, e^{-\frac{|t|}{\xi}} dt\right) - \sum_{j=0}^{r} \alpha_j \frac{1}{2\xi}\left(\int_{-\infty}^{\infty} \cos(\beta j t)\, e^{-\frac{|t|}{\xi}} dt\right)\right]$$

$$-\left(\frac{f^{(3)}(x)}{\alpha\beta(\beta^2-\alpha^2)}\right)$$

$$\times \left[\beta \sum_{j=0}^{r} \alpha_j \frac{1}{2\xi}\left(\int_{-\infty}^{\infty} \sin(\alpha j t)\, e^{-\frac{|t|}{\xi}} dt\right) - \alpha \sum_{j=0}^{r} \alpha_j \frac{1}{2\xi}\left(\int_{-\infty}^{\infty} \sin(\beta j t)\, e^{-\frac{|t|}{\xi}} dt\right)\right]$$

$$-\left(\frac{2\left(f^{(4)}(x) + (\alpha^2+\beta^2) f''(x)\right)}{(\alpha\beta)^2(\beta^2-\alpha^2)}\right)$$

$$\times \left[\beta^2 \sum_{j=0}^{r} \alpha_j \frac{1}{2\xi}\left(\int_{-\infty}^{\infty} \sin^2\left(\frac{\alpha j t}{2}\right) e^{-\frac{|t|}{\xi}} dt\right)\right.$$

$$\left. -\alpha^2 \sum_{j=0}^{r} \alpha_j \frac{1}{2\xi}\left(\int_{-\infty}^{\infty} \sin^2\left(\frac{\beta j t}{2}\right) e^{-\frac{|t|}{\xi}} dt\right)\right]$$

$$= \frac{1}{2\xi}\int_{-\infty}^{\infty} R(t)\, e^{-\frac{|t|}{\xi}} dt. \tag{23.14}$$

Next we simplify $E(x)$:

We observe that

$$\int_{-\infty}^{\infty} \sin(\alpha jt) e^{-\frac{|t|}{\xi}} dt = \int_{-\infty}^{0} \sin(\alpha jt) e^{-\frac{|t|}{\xi}} dt + \int_{0}^{\infty} \sin(\alpha jt) e^{-\frac{|t|}{\xi}} dt. \tag{23.15}$$

Note $-\infty \leq t \leq 0 \Rightarrow \infty \geq -t \geq 0$. So

$$\int_{-\infty}^{0} \sin(\alpha jt) e^{-\frac{|t|}{\xi}} dt = -\int_{-\infty}^{0} \sin(\alpha j(-(-t))) e^{-\frac{|t|}{\xi}} d(-t)$$

$$= -\int_{-\infty}^{0} (-\sin(\alpha j(-t))) e^{-\frac{|t|}{\xi}} d(-t) = \int_{-\infty}^{0} \sin(\alpha j(-t)) e^{-\frac{|t|}{\xi}} d(-t)$$

$$= \int_{\infty}^{0} \sin(\alpha jt) e^{-\frac{|t|}{\xi}} dt = -\int_{0}^{\infty} \sin(\alpha jt) e^{-\frac{|t|}{\xi}} dt. \tag{23.16}$$

So

$$\int_{-\infty}^{\infty} \sin(\alpha jt) e^{-\frac{|t|}{\xi}} dt = 0,$$

and all sine integrals in (23.14) are zeros.

Furthermore, we have that

$$\int_{-\infty}^{\infty} \sin^2\left(\frac{\alpha jt}{2}\right) e^{-\frac{|t|}{\xi}} dt = 2\int_{0}^{\infty} \sin^2\left(\frac{\alpha jt}{2}\right) e^{-\frac{|t|}{\xi}} dt, \quad j = 0, 1, \ldots, r. \tag{23.17}$$

The last follows by

$$\int_{-\infty}^{0} \sin^2\left(\frac{\alpha jt}{2}\right) e^{-\frac{|t|}{\xi}} dt = -\int_{-\infty}^{0} \left(-\sin\left(\frac{-\alpha jt}{2}\right)\right)^2 e^{-\frac{|-t|}{\xi}} d(-t)$$

$$\overset{(z=-t)}{=} -\int_{\infty}^{0} \sin^2\left(\frac{\alpha jz}{2}\right) e^{-\frac{|z|}{\xi}} dz = \int_{0}^{\infty} \sin^2\left(\frac{\alpha jz}{2}\right) e^{-\frac{z}{\xi}} dz. \tag{23.18}$$

Next, we calculate

$$\int_{0}^{\infty} \sin^2\left(\frac{\alpha j}{2} z\right) e^{-\frac{|z|}{\xi}} dz = \xi \int_{0}^{\infty} \sin^2\left(\left(\frac{\alpha j\xi}{2}\right) \frac{z}{\xi}\right) e^{-\frac{z}{\xi}} d\frac{z}{\xi} = \tag{23.19}$$

(call $\frac{t}{\xi} =: x$ and $\frac{\alpha j\xi}{2} =: a_1$)

$$\xi \int_{0}^{\infty} \sin^2(a_1 x) e^{-x} dx =$$

(by Wolfram Alpha Computational Intelligence)

$$\xi\left(\frac{2a_1^2}{4a_1^2+1}\right)=\frac{\alpha^2 j^2\xi^3}{2\left(\alpha^2 j^2\xi^2+1\right)}.$$

Thus, it is

$$\int_{-\infty}^{\infty}\sin^2\left(\frac{\alpha jt}{2}\right)e^{-\frac{|t|}{\xi}}dt=\frac{\alpha^2 j^2\xi^3}{\alpha^2 j^2\xi^2+1},\quad j=0,1,\ldots,r, \tag{23.20}$$

and

$$\frac{1}{2\xi}\int_{-\infty}^{\infty}\cos\left(\alpha jt\right)e^{-\frac{|t|}{\xi}}dt=\frac{1}{\alpha^2 j^2\xi^2+1}, \tag{23.21}$$

$j=0,1,\ldots,r.$

So, based on the above we write down the simplified version of $E\left(x\right)$ as follows:

$$\begin{aligned}E\left(x\right)&=P_{r,\xi}\left(f,x\right)-f\left(x\right)\\&\quad-\left(\frac{f''\left(x\right)}{\beta^2-\alpha^2}\right)\left\{\sum_{j=0}^{r}\alpha_j\left[\left(\frac{1}{\alpha^2 j^2\xi^2+1}\right)-\left(\frac{1}{\beta^2 j^2\xi^2+1}\right)\right]\right\}\\&\quad-\left(\frac{2\left(f^{(4)}\left(x\right)+\left(\alpha^2+\beta^2\right)f''\left(x\right)\right)}{\left(\alpha\beta\right)^2\left(\beta^2-\alpha^2\right)}\right)\\&\quad\times\left[\beta^2\sum_{j=0}^{r}\alpha_j\left(\frac{\alpha^2 j^2\xi^2}{2\left(\alpha^2 j^2\xi^2+1\right)}\right)-\alpha^2\sum_{j=0}^{r}\alpha_j\left(\frac{\beta^2 j^2\xi^2}{2\left(\beta^2 j^2\xi^2+1\right)}\right)\right]\end{aligned} \tag{23.22}$$

$$\begin{aligned}&=P_{r,\xi}\left(f,x\right)-f\left(x\right)-f''\left(x\right)\left\{\sum_{j=0}^{r}\alpha_j\left[\frac{j^2\xi^2}{\left(\alpha^2 j^2\xi^2+1\right)\left(\beta^2 j^2\xi^2+1\right)}\right]\right\}\\&\quad-\left(\frac{2\left(f^{(4)}\left(x\right)+\left(\alpha^2+\beta^2\right)f''\left(x\right)\right)}{\left(\alpha\beta\right)^2\left(\beta^2-\alpha^2\right)}\right)\\&\quad\times\left[\beta^2\sum_{j=0}^{r}\alpha_j\left(\frac{\alpha^2 j^2\xi^2}{2\left(\alpha^2 j^2\xi^2+1\right)}\right)-\alpha^2\sum_{j=0}^{r}\alpha_j\left(\frac{\beta^2 j^2\xi^2}{2\left(\beta^2 j^2\xi^2+1\right)}\right)\right].\end{aligned} \tag{23.23}$$

Some more simplifications lead to

$$E(x) = P_{r,\xi}(f,x) - f(x) - f''(x)\left\{\sum_{j=0}^{r} \alpha_j j^2 \left[\frac{1}{(\alpha^2 j^2 \xi^2 + 1)(\beta^2 j^2 \xi^2 + 1)}\right]\right\}\xi^2$$
$$- \left(f^{(4)}(x) + (\alpha^2 + \beta^2) f''(x)\right)\left[\sum_{j=0}^{r} \alpha_j j^4 \left(\frac{1}{(\alpha^2 j^2 \xi^2 + 1)(\beta^2 j^2 \xi^2 + 1)}\right)\right]\xi^4, \tag{23.24}$$

$\xi > 0$, $x \in \mathbb{R}$.

We present our first result, about L_p approximation, $p > 1$.

Theorem 23.1. *Let $\alpha, \beta \in \mathbb{R}$ with $\alpha\beta(\alpha^2 - \beta^2) \neq 0$, and let all $f, f', f'', f^{(3)}, f^{(4)} \in L_p(\mathbb{R}) \cap C(\mathbb{R})$, $1 < p < \infty$. Here $q > 1 : \frac{1}{p} + \frac{1}{q} = 1$; $\xi > 0$, $x \in \mathbb{R}$. Then*

$$\|E(x)\|_p = \Bigg\| P_{r,\xi}(f,x) - f(x) - f''(x)\left\{\sum_{j=0}^{r} \alpha_j j^2 \left(\frac{1}{(\alpha^2 j^2 \xi^2 + 1)(\beta^2 j^2 \xi^2 + 1)}\right)\right\}\xi^2$$
$$- \left(f^{(4)}(x) + (\alpha^2 + \beta^2) f''(x)\right) \times \left(\sum_{j=0}^{r} \alpha_j j^4 \left(\frac{1}{(\alpha^2 j^2 \xi^2 + 1)(\beta^2 j^2 \xi^2 + 1)}\right)\right)\xi^4 \Bigg\|_p$$
$$\leq \frac{2^{1+\frac{2}{q}}}{q^{\frac{1}{q}}|\beta^2 - \alpha^2|}\left(\frac{r+1}{q+1}\right)^{\frac{1}{q}} \omega_1\left(f^{(4)} + (\alpha^2 + \beta^2) f'' + \alpha^2\beta^2 f, \xi\right)_p \xi^2$$
$$\left[\sum_{j=0}^{r} |\alpha_j|^p j^{2p} \left\{\left(\frac{p}{2}\right)^{-qp+q-p-1} \left(\Gamma(qp - q + p + 1) + \frac{j^p}{(p+1)}\left(\frac{p}{2}\right)^{-p} \Gamma(qp - q + 2p + 1)\right)\right\}\right]^{\frac{1}{p}} \to 0, \tag{23.25}$$

as $\xi \to 0$.

Above Γ is the gamma function.

Proof. Call the function

$$F := f^{(4)} + (\alpha^2 + \beta^2) f'' + \alpha^2\beta^2 f. \tag{23.26}$$

Then, we get

$$R = R(t) = \frac{1}{\alpha\beta\left(\beta^2 - \alpha^2\right)} \sum_{j=0}^{r} \alpha_j j \int_0^t \left[F(x + jw) - F(x)\right]$$
$$\left[\beta \sin(\alpha j (t - w)) - \alpha \sin(\beta j (t - w))\right] dw, \quad \forall\, t \in \mathbb{R}. \tag{23.27}$$

We isolate and study

$$I := \int_0^t \left[F(x + jw) - F(x)\right]$$
$$\left[\beta \sin(\alpha j (t - w)) - \alpha \sin(\beta j (t - w))\right] dw, \quad \forall\, t \in \mathbb{R}. \tag{23.28}$$

For $t < 0$, we have that

$$|I| = \left| \int_t^0 \left[F(x + jw) - F(x)\right] \left[\beta \sin(\alpha j (t - w)) - \alpha \sin(\beta j (t - w))\right] dw \right|$$
$$\leq \int_t^0 |F(x + jw) - F(x)| \left|\beta \sin(\alpha j (t - w)) - \alpha \sin(\beta j (t - w))\right| dw$$
$$\leq 2|\alpha||\beta| j \int_t^0 |F(x + jw) - F(x)| (w - t)\, dw$$
$$= -2|\alpha||\beta| j \int_t^0 |F(x - j(-w)) - F(x)| (-t - (-w))\, d(-w) = \tag{23.29}$$

$(t \leq w \leq 0 \Rightarrow -t \geq -w =: \theta \geq 0)$

$$-2|\alpha||\beta| j \int_{-t}^0 |F(x - j\theta) - F(x)| (-t - \theta)\, d\theta$$
$$= 2|\alpha||\beta| j \int_0^{-t} |F(x - j\theta) - F(x)| (-t - \theta)\, d\theta$$
$$= 2|\alpha\beta| j \int_0^{|t|} |F(x + sign(t)\, j\theta) - F(x)| (|t| - \theta)\, d\theta.$$

So, we have proved that

$$|I| \leq 2|\alpha\beta| j \int_0^{|t|} |F(x + sign(t)\, j\theta) - F(x)| (|t| - \theta)\, d\theta, \quad \forall\, t \in \mathbb{R}, \tag{23.30}$$

and, by (23.27),

$$|R(t)| \leq \frac{2}{|\beta^2 - \alpha^2|} \sum_{j=0}^{r} |\alpha_j| j^2 \int_0^{|t|} |F(x + j sign(t)\, \theta) - F(x)| (|t| - \theta)\, d\theta, \tag{23.31}$$

$\forall\, t \in \mathbb{R}$.

By (23.14), we have

$$E(x) = \frac{1}{2\xi} \int_{-\infty}^{\infty} R(t) e^{-\frac{|t|}{\xi}} dt. \tag{23.32}$$

Hence it holds ($p, q > 1 : \frac{1}{p} + \frac{1}{q} = 1$),

$$\begin{aligned}
\int_{-\infty}^{\infty} |E(x)|^p dx &= \frac{1}{(2\xi)^p} \int_{-\infty}^{\infty} \left| \int_{-\infty}^{\infty} R(t) e^{-\frac{|t|}{\xi}} dt \right|^p dx \\
&\leq \frac{1}{(2\xi)^p} \int_{-\infty}^{\infty} \left(\int_{-\infty}^{\infty} |R(t)| e^{-\frac{|t|}{\xi}} dt \right)^p dx \qquad (23.33) \\
&= \frac{1}{(2\xi)^p} \int_{-\infty}^{\infty} \left(\int_{-\infty}^{\infty} |R(t)| e^{-\frac{|t|}{2\xi}} e^{-\frac{|t|}{2\xi}} dt \right)^p dx \\
&\leq \frac{1}{(2\xi)^p} \int_{-\infty}^{\infty} \left(\int_{-\infty}^{\infty} |R(t)|^p e^{-\frac{|pt|}{2\xi}} dt \right) \left(\int_{-\infty}^{\infty} e^{-\frac{|qt|}{2\xi}} dt \right)^{\frac{p}{q}} dx \\
&= \frac{1}{(2\xi)^p} \left(\frac{4\xi}{q} \right)^{\frac{p}{q}} \int_{-\infty}^{\infty} \left(\int_{-\infty}^{\infty} |R(t)|^p e^{-\frac{|pt|}{2\xi}} dt \right) dx \\
&= \frac{2^{p-2}\xi^{-1}}{q^{p-1}} \left(\int_{-\infty}^{\infty} \left(\int_{-\infty}^{\infty} |R(t)|^p e^{-\frac{|pt|}{2\xi}} dt \right) dx \right) =: (*).
\end{aligned} \tag{23.34}$$

But, we need to treat

$$\begin{aligned}
|R(t)| &\leq \frac{2}{|\beta^2 - \alpha^2|} \sum_{j=0}^{r} |\alpha_j| j^2 \\
&\quad \times \left(\int_0^{|t|} |F(x + j\, sign(t)\, \theta) - F(x)|^p d\theta \right)^{\frac{1}{p}} \left(\int_0^{|t|} (|t| - \theta)^q d\theta \right)^{\frac{1}{q}} \\
&= \frac{2}{|\beta^2 - \alpha^2|} \sum_{j=0}^{r} |\alpha_j| j^2 \left(\int_0^{|t|} |F(x + j\, sign(t)\, \theta) - F(x)|^p d\theta \right)^{\frac{1}{p}} \frac{|t|^{q+\frac{1}{q}}}{(q+1)^{\frac{1}{q}}} \\
&= \frac{2}{|\beta^2 - \alpha^2|} \left\{ \sum_{j=0}^{r} \left(\left(|\alpha_j| j^2 \right)^p \right)^{\frac{1}{p}} \left(\int_0^{|t|} |F(x + j\, sign(t)\, \theta) - F(x)|^p d\theta \right)^{\frac{1}{p}} \right\} \\
&\quad \frac{|t|^{q+\frac{1}{q}}}{(q+1)^{\frac{1}{q}}} \leq
\end{aligned} \tag{23.35}$$

$(0 < \frac{1}{p} < 1)$

$$\frac{2(r+1)^{\frac{1}{q}}}{|\beta^2 - \alpha^2|} \left(\sum_{j=0}^{r} |\alpha_j|^p j^{2p} \int_0^{|t|} |F(x + jsign(t)\theta) - F(x)|^p d\theta \right)^{\frac{1}{p}} \frac{|t|^{q+\frac{1}{q}}}{(q+1)^{\frac{1}{q}}}.$$

Hence, we find that

$$|R(t)|^p \leq \frac{2^p}{|\beta^2 - \alpha^2|^p} \left(\frac{r+1}{q+1} \right)^{\frac{p}{q}}$$
$$\times \left[\sum_{j=0}^{r} |\alpha_j|^p j^{2p} \int_0^{|t|} |F(x + jsign(t)\theta) - F(x)|^p d\theta \right] |t|^{\frac{(q+1)p}{q}}. \tag{23.36}$$

Therefore, we get

$$(*) \leq \left(\frac{2^{p-2}\xi^{-1}}{q^{p-1}} \frac{2^p}{|\beta^2 - \alpha^2|^p} \left(\frac{r+1}{q+1} \right)^{\frac{p}{q}} \right) \left\{ \int_{-\infty}^{\infty} \left(\int_{-\infty}^{\infty} \left[\sum_{j=0}^{r} |\alpha_j|^p j^{2p} \right.\right.\right.$$
$$\left.\left.\left. \int_0^{|t|} |F(x + jsign(t)\theta) - F(x)|^p d\theta \right] |t|^{\frac{(q+1)p}{q}} e^{-\frac{|pt|}{2\xi}} dt \right) dx \right\}$$

$$= \left(\text{set } c_1 = \frac{2^{2(p-1)}\xi^{-1}}{q^{p-1}|\beta^2 - \alpha^2|^p} \left(\frac{r+1}{q+1} \right)^{\frac{p}{q}} \right)$$

$$c_1 \left\{ \int_{-\infty}^{\infty} \left(\int_{-\infty}^{\infty} \left[\sum_{j=0}^{r} |\alpha_j|^p j^{2p} \right.\right.\right.$$
$$\left.\left.\left. \int_0^{|t|} |F(x + jsign(t)\theta) - F(x)|^p d\theta \right] dx \right) |t|^{\frac{(q+1)p}{q}} e^{-\frac{|pt|}{2\xi}} dt \right\}$$

$$= c_1 \left\{ \int_{-\infty}^{\infty} \left[\sum_{j=0}^{r} |\alpha_j|^p j^{2p} \int_0^{|t|} \left(\int_{-\infty}^{\infty} |F(x + jsign(t)\theta) - F(x)|^p dx \right) d\theta \right] \right.$$
$$\left. |t|^{\frac{(q+1)p}{q}} e^{-\frac{|pt|}{2\xi}} dt \right\} \leq \tag{23.37}$$

$(\xi > 0)$

$$c_1 \left\{ \int_{-\infty}^{\infty} \left[\sum_{j=0}^{r} |\alpha_j|^p j^{2p} \left(\int_0^{|t|} \omega_1 \left(F, \xi \frac{j\theta}{\xi} \right)_p^p d\theta \right) \right] |t|^{\frac{(q+1)p}{q}} e^{-\frac{|pt|}{2\xi}} dt \right\} \tag{23.38}$$

$$\leq c_1 \omega_1 (F, \xi)_p^p \left\{ \int_{-\infty}^{\infty} \left[\sum_{j=0}^{r} |\alpha_j|^p j^{2p} \left(\int_0^{|t|} \left(1 + \frac{j}{\xi}\theta \right)^p d\theta \right) \right] \right.$$

$$\left. |t|^{(q+1)(p-1)} e^{-\frac{|pt|}{2\xi}} dt \right\}$$

$$\leq c_1 \omega_1 (F, \xi)_p^p \left\{ \int_{-\infty}^{\infty} \left[\sum_{j=0}^{r} |\alpha_j|^p j^{2p} 2^{p-1} \left(\int_0^{|t|} \left(1 + \frac{j^p}{\xi^p}\theta^p \right) d\theta \right) \right] \right.$$

$$\left. |t|^{(q+1)(p-1)} e^{-\frac{|pt|}{2\xi}} dt \right\}$$

$$= 2^{p-1} c_1 \omega_1 (F, \xi)_p^p$$

$$\left\{ \int_{-\infty}^{\infty} \left[\sum_{j=0}^{r} |\alpha_j|^p j^{2p} \left(|t| + \frac{j^p}{\xi^p} \frac{|t|^{p+1}}{p+1} \right) \right] |t|^{(q+1)(p-1)} e^{-\frac{|pt|}{2\xi}} dt \right\}$$

$$= 2^{p-1} c_1 \omega_1 (F, \xi)_p^p$$

$$\left\{ \sum_{j=0}^{r} |\alpha_j|^p j^{2p} \int_{-\infty}^{\infty} \left(|t| + \frac{j^p}{\xi^p} \frac{|t|^{p+1}}{p+1} \right) |t|^{(q+1)(p-1)} e^{-\frac{|pt|}{2\xi}} dt \right\} \tag{23.39}$$

$$= 2^p c_1 \omega_1 (F, \xi)_p^p$$

$$\left\{ \sum_{j=0}^{r} |\alpha_j|^p j^{2p} \int_0^{\infty} \left(t + \frac{j^p}{\xi^p (p+1)} t^{p+1} \right) t^{(q+1)(p-1)} e^{-\frac{pt}{2\xi}} dt \right\}$$

$$= 2^p c_1 \omega_1 (F, \xi)_p^p$$

$$\left\{ \sum_{j=0}^{r} |\alpha_j|^p j^{2p} \left[\int_0^{\infty} t^{(q+1)(p-1)+1} e^{-\frac{pt}{2\xi}} dt \right. \right.$$

$$+ \frac{j^p}{\xi^p (p+1)} \int_0^\infty t^{(q+1)(p-1)+p+1} e^{-\frac{pt}{2\xi}} dt \Bigg] \Bigg\}$$

$$= 2^p c_1 \omega_1 (F, \xi)_p^p \Bigg\{ \sum_{j=0}^{r} |\alpha_j|^p j^{2p} \Bigg[\xi^{qp-q+p+1} \int_0^\infty \left(\frac{t}{\xi}\right)^{(q+1)(p-1)+1} e^{-\left(\frac{p}{2}\right)\frac{t}{\xi}} d\frac{t}{\xi}$$

$$+ \frac{j^p}{\xi^p (p+1)} \xi^{qp-q+2p+1} \int_0^\infty \left(\frac{t}{\xi}\right)^{qp-q+2p} e^{-\left(\frac{p}{2}\right)\frac{t}{\xi}} d\frac{t}{\xi} \Bigg] \Bigg\} = \qquad (23.40)$$

(above it is $(q+1)(p-1)+1 = qp-q+p > 0$, $(q+1)(p-1)+p+1 = qp-q+2p > 0$)

$$2^p c_1 \omega_1 (F, \xi)_p^p \xi^{qp-q+p+1} \Bigg[\sum_{j=0}^{r} |\alpha_j|^p j^{2p} \Bigg\{ \int_0^\infty x^{qp-q+p} e^{-\left(\frac{p}{2}\right)x} dx$$

$$+ \frac{j^p}{(p+1)} \int_0^\infty x^{qp-q+2p} e^{-\left(\frac{p}{2}\right)x} dx \Bigg] \Bigg\} = \qquad (23.41)$$

(next we use, for $a, b > 0$ that it holds $\int_0^\infty x^a e^{-bx} dx = b^{-a-1}\Gamma(a+1)$, by Wolfram Alpha Computational Intelligence)

$$2^p c_1 \omega_1 (F, \xi)_p^p \xi^{qp-q+p+1} \Bigg[\sum_{j=0}^{r} |\alpha_j|^p j^{2p} \Bigg\{ \left(\frac{p}{2}\right)^{-qp+q-p-1} \Gamma(qp-q+p+1)$$

$$+ \frac{j^p}{(p+1)} \left(\frac{p}{2}\right)^{-qp+q-2p-1} \Gamma(qp-q+2p+1) \Bigg\} \Bigg]$$

$$= \frac{2^{3p-2}}{q^{p-1} |\beta^2 - \alpha^2|^p} \left(\frac{r+1}{q+1}\right)^{\frac{p}{q}} \omega_1 (F, \xi)_p^p \xi^{qp-q+p}$$

$$\times \Bigg[\sum_{j=0}^{r} |\alpha_j|^p j^{2p} \Bigg\{ \left(\frac{p}{2}\right)^{-qp+q-p-1}$$

$$\left(\Gamma(qp-q+p+1) + \frac{j^p}{(p+1)} \left(\frac{p}{2}\right)^{-p} \Gamma(qp-q+2p+1) \right) \Bigg\} \Bigg]. \qquad (23.42)$$

So for $p, q > 1 : \frac{1}{p} + \frac{1}{q} = 1$, $\xi > 0$, we have proved the following:

$$\int_{-\infty}^{\infty} |E(x)|^p \, dx \le \frac{2^{3p-2}}{q^{p-1} |\beta^2 - \alpha^2|^p} \left(\frac{r+1}{q+1} \right)^{\frac{p}{q}}$$
$$\omega_1 \left(f^{(4)} + (\alpha^2 + \beta^2) f'' + \alpha^2 \beta^2 f, \xi \right)_p^p \xi^{qp-q+p}$$
$$\times \left[\sum_{j=0}^{r} |\alpha_j|^p j^{2p} \left\{ \left(\frac{p}{2} \right)^{-qp+q-p-1} \right. \right.$$
$$\left. \left. \left(\Gamma(qp - q + p + 1) + \frac{j^p}{(p+1)} \left(\frac{p}{2} \right)^{-p} \Gamma(qp - q + 2p + 1) \right) \right\} \right]. \tag{23.43}$$

The proof of the theorem is now completed. □

Next comes the L_1-related approximation.

Theorem 23.2. *Let $\alpha, \beta \in \mathbb{R}$ with $\alpha\beta(\alpha^2 - \beta^2) \neq 0$, and let all $f, f', f'', f^{(3)}, f^{(4)} \in L_1(\mathbb{R}) \cap C(\mathbb{R})$, $\xi > 0$, $x \in \mathbb{R}$. Then*

$$\|E(x)\|_1 = \| P_{r,\xi}(f, x) - f(x)$$
$$- f''(x) \left\{ \sum_{j=0}^{r} \alpha_j j^2 \left(\frac{1}{(\alpha^2 j^2 \xi^2 + 1)(\beta^2 j^2 \xi^2 + 1)} \right) \right\} \xi^2$$
$$- \left(f^{(4)}(x) + (\alpha^2 + \beta^2) f''(x) \right)$$
$$\times \left(\sum_{j=0}^{r} \alpha_j j^4 \left(\frac{1}{(\alpha^2 j^2 \xi^2 + 1)(\beta^2 j^2 \xi^2 + 1)} \right) \right) \xi^4 \Bigg\|_1$$
$$\le \frac{2\xi^2 \omega_1 \left(f^{(4)} + (\alpha^2 + \beta^2) f'' + \alpha^2 \beta^2 f, \xi \right)_1}{|\beta^2 - \alpha^2|} \left(\sum_{j=0}^{r} |\alpha_j| j^2 (2 + 3j) \right) \to 0, \tag{23.44}$$

as $\xi \to 0$.

Proof. We have that (by (23.14))

$$\begin{aligned}
|E(x)| &= \frac{1}{2\xi}\left|\int_{-\infty}^{\infty} R(t)\, e^{-\frac{|t|}{\xi}} dt\right| \leq \frac{1}{2\xi}\int_{-\infty}^{\infty} |R(t)|\, e^{-\frac{|t|}{\xi}} dt \\
&\overset{(23.31)}{\leq} \frac{1}{|\beta^2-\alpha^2|\,\xi}\sum_{j=0}^{r}|\alpha_j|\, j^2 \\
&\quad \int_{-\infty}^{\infty}\left(\int_0^{|t|}|F(x+j\,sign(t)\,\theta)-F(x)|\,(|t|-\theta)\, d\theta\right) e^{-\frac{|t|}{\xi}} dt.
\end{aligned} \tag{23.45}$$

Thus, we get

$$\begin{aligned}
\|E(x)\|_1 &= \int_{-\infty}^{\infty}|E(x)|\, dx \\
&\leq \frac{1}{|\beta^2-\alpha^2|\,\xi}\sum_{j=0}^{r}|\alpha_j|\, j^2 \\
&\quad \times\left(\int_{-\infty}^{\infty}\left(\int_{-\infty}^{\infty}\left(\int_0^{|t|}|F(x+j\,sign(t)\,\theta)-F(x)|\,(|t|-\theta)\, d\theta\right) e^{-\frac{|t|}{\xi}} dt\right) dx\right)
\end{aligned} \tag{23.46}$$

$$\begin{aligned}
&\leq \frac{1}{|\beta^2-\alpha^2|\,\xi}\sum_{j=0}^{r}|\alpha_j|\, j^2 \\
&\quad \times\left(\int_{-\infty}^{\infty}\left(\int_{-\infty}^{\infty}\left(|t|\int_0^{|t|}|F(x+j\,sign(t)\,\theta)-F(x)|\, d\theta\right) e^{-\frac{|t|}{\xi}} dt\right) dx\right) \\
&= \frac{1}{|\beta^2-\alpha^2|\,\xi}\sum_{j=0}^{r}|\alpha_j|\, j^2 \\
&\quad \times\left(\int_{-\infty}^{\infty}\left(\int_{-\infty}^{\infty}\left(|t|\int_0^{|t|}|F(x+j\,sign(t)\,\theta)-F(x)|\, d\theta\right) e^{-\frac{|t|}{\xi}} dx\right) dt\right) \\
&= \frac{1}{|\beta^2-\alpha^2|\,\xi}\sum_{j=0}^{r}|\alpha_j|\, j^2 \\
&\quad \times\left(\int_{-\infty}^{\infty}\left(\int_{-\infty}^{\infty}\left(\int_0^{|t|}|F(x+j\,sign(t)\,\theta)-F(x)|\, d\theta\right) dx\right)|t|\, e^{-\frac{|t|}{\xi}} dt\right)
\end{aligned}$$

$$= \frac{1}{|\beta^2 - \alpha^2| \xi} \sum_{j=0}^{r} |\alpha_j| j^2$$

$$\times \left(\int_{-\infty}^{\infty} \left(\int_{0}^{|t|} \left(\int_{-\infty}^{\infty} |F(x + j sign(t) \theta) - F(x)| dx \right) d\theta \right) |t| e^{-\frac{|t|}{\xi}} dt \right) \tag{23.47}$$

$$\leq \frac{1}{|\beta^2 - \alpha^2| \xi} \sum_{j=0}^{r} |\alpha_j| j^2 \left(\int_{-\infty}^{\infty} \left(\int_{0}^{|t|} \omega_1 (F, j\theta)_1 d\theta \right) |t| e^{-\frac{|t|}{\xi}} dt \right)$$

$$\leq \frac{\omega_1 (F, \xi)_1}{|\beta^2 - \alpha^2| \xi} \sum_{j=0}^{r} |\alpha_j| j^2 \left(\int_{-\infty}^{\infty} \left(\int_{0}^{|t|} \left(1 + \frac{j}{\xi} \theta \right) d\theta \right) |t| e^{-\frac{|t|}{\xi}} dt \right)$$

$$= \frac{\omega_1 (F, \xi)_1}{|\beta^2 - \alpha^2| \xi} \sum_{j=0}^{r} |\alpha_j| j^2 \left(\int_{-\infty}^{\infty} \left[|t| + \frac{j}{\xi} \frac{|t|^2}{2} \right] |t| e^{-\frac{|t|}{\xi}} dt \right)$$

$$= \frac{2\xi^2 \omega_1 (F, \xi)_1}{|\beta^2 - \alpha^2|} \sum_{j=0}^{r} |\alpha_j| j^2 \left(\int_{0}^{\infty} \left[\frac{t}{\xi} + \frac{j}{2} \left(\frac{t}{\xi} \right)^2 \right] \frac{t}{\xi} e^{-\frac{t}{\xi}} d\frac{t}{\xi} \right) \tag{23.48}$$

$$= \frac{2\xi^2 \omega_1 (F, \xi)_1}{|\beta^2 - \alpha^2|} \sum_{j=0}^{r} |\alpha_j| j^2 \left(\int_{0}^{\infty} \left(t + \frac{j}{2} t^2 \right) t e^{-t} dt \right)$$

$$= \frac{2\xi^2 \omega_1 (F, \xi)_1}{|\beta^2 - \alpha^2|} \sum_{j=0}^{r} |\alpha_j| j^2 \left[\int_{0}^{\infty} t^2 e^{-t} dt + \frac{j}{2} \int_{0}^{\infty} t^3 e^{-t} dt \right]$$

$$= \frac{2\xi^2 \omega_1 (F, \xi)_1}{|\beta^2 - \alpha^2|} \sum_{j=0}^{r} |\alpha_j| j^2 \left[2 + \frac{j}{2} \cdot 6 \right]$$

$$= \frac{2\xi^2 \omega_1 (F, \xi)_1}{|\beta^2 - \alpha^2|} \sum_{j=0}^{r} |\alpha_j| j^2 (2 + 3j). \tag{23.49}$$

Above we used that

$$\int_{0}^{\infty} t^2 e^{-t} dt = 2 \quad \text{and} \quad \int_{0}^{\infty} t^3 e^{-t} dt = 6.$$

The theorem is proved. □

We finish with the following:

Corollary 23.1. *(to Theorem 23.1, $p>1$) It holds*

$$\left\|P_{r,\xi}(f)-f\right\|_{p}$$

$$\leq \left\|f''\right\|_{p}\left\{\sum_{j=0}^{r}\left|\alpha_{j}\right|j^{2}\left(\frac{1}{\left(\alpha^{2}j^{2}\xi^{2}+1\right)\left(\beta^{2}j^{2}\xi^{2}+1\right)}\right)\right\}\xi^{2}$$

$$+\left(\left\|f^{(4)}\right\|_{p}+\left(\alpha^{2}+\beta^{2}\right)\left\|f''\right\|_{p}\right)$$

$$\times\left(\sum_{j=0}^{r}\left|\alpha_{j}\right|j^{4}\left(\frac{1}{\left(\alpha^{2}j^{2}\xi^{2}+1\right)\left(\beta^{2}j^{2}\xi^{2}+1\right)}\right)\right)\xi^{4}$$

$$+\frac{2^{1+\frac{2}{q}}}{q^{\frac{1}{q}}\left|\beta^{2}-\alpha^{2}\right|}\left(\frac{r+1}{q+1}\right)^{\frac{1}{q}}\omega_{1}\left(f^{(4)}+\left(\alpha^{2}+\beta^{2}\right)f''+\alpha^{2}\beta^{2}f,\xi\right)_{p}\xi^{2}$$

$$\times\left[\sum_{j=0}^{r}\left|\alpha_{j}\right|^{p}j^{2p}\left\{\left(\frac{p}{2}\right)^{-qp+q-p-1}\right.\right.$$

$$\left.\left.\left(\Gamma\left(qp-q+p+1\right)+\frac{j^{p}}{(p+1)}\left(\frac{p}{2}\right)^{-p}\Gamma\left(qp-q+2p+1\right)\right)\right\}\right]^{\frac{1}{p}}\to 0, \tag{23.50}$$

as $\xi\to 0$.

Corollary 23.2. *(to Theorem 23.2, $p=1$) It holds*

$$\left\|P_{r,\xi}(f)-f\right\|_{1}$$

$$\leq \left\|f''\right\|_{1}\left\{\sum_{j=0}^{r}\left|\alpha_{j}\right|j^{2}\left(\frac{1}{\left(\alpha^{2}j^{2}\xi^{2}+1\right)\left(\beta^{2}j^{2}\xi^{2}+1\right)}\right)\right\}\xi^{2}$$

$$+\left(\left\|f^{(4)}\right\|_{1}+\left(\alpha^{2}+\beta^{2}\right)\left\|f''\right\|_{1}\right)$$

$$\times\left(\sum_{j=0}^{r}|\alpha_j|\,j^4\left(\frac{1}{(\alpha^2j^2\xi^2+1)(\beta^2j^2\xi^2+1)}\right)\right)\xi^4$$

$$+\frac{2\xi^2\omega_1\left(f^{(4)}+\left(\alpha^2+\beta^2\right)f''+\alpha^2\beta^2f,\xi\right)_1}{|\beta^2-\alpha^2|}\left(\sum_{j=0}^{r}|\alpha_j|\,j^2\,(2+3j)\right)\to 0, \tag{23.51}$$

as $\xi\to 0$.

Bibliography

A.H. Ali, Z. Pales, Taylor-type expansions in terms of exponential polynomials, *Mathematical Inequalities and Applications*, 25(4) (2022), 1123–1141.

G.A. Anastassiou, Basic convergence with rates of smooth Picard singular integral operators, *Journal of Computational Analysis and Applications*, 8(4) (2006), 313–334.

G.A. Anastassiou, *Intelligent Mathematics: Computational Analysis*, Springer, Heidelberg, New York, Chapter 12, 2011.

G.A. Anastassiou, Opial and Ostrowski type inequalities based on trigonometric and hyperbolic type Taylor formulae, *Malaya Journal Mathematik*, 11(S) (2023a), 1–26.

G.A. Anastassiou, Parametrized and Trigonometric L_p quantitative convergence of smooth Picard singular integral operators, *Indian Journal of Mathematics*, accepted, 2023b.

Chapter 24

Update on Uniform Approximation by Smooth Picard Multivariate Singular Integral Operators Revisited

In this chapter, we reexamine the uniform approximation properties of smooth Picard multivariate singular integral operators over $\mathbb{R}^N$, $N \geq 1$. We establish their convergence to the unit operator with rates. The estimates are pointwise and uniform. The established inequalities involve the multivariate first modulus of continuity. Our approach is based on a new multivariate trigonometric Taylor formula. At first, we present in detail the general theory of uniform approximation by general smooth multivariate singular integral operators, which then is applied to the Picard operator case. It follows Anastassiou (2023b).

24.1 Introduction

The rate of convergence of univariate and multivariate singular integral operators has been studied extensively in Anastassiou (2011a)–Anastassiou (2021) and Anastassiou and Gal (2000), Anastassiou and Mezei (2012) and Mohapatra and Rodriguez (1990). All these motivate our current work. In particular, we studied the smooth singular integral operators in Anastassiou (2011a)–Anastassiou (2021) and Anastassiou and Mezei (2012), which are not in general positive ones.

Here we continue the study of the last ones at the multivariate level, at first in general, and then apply our theory to the smooth Picard ones. The main tool here, we are based on, is a new trigonometric multivariate

Taylor formula from Anastassiou (2023a). Our quantitative estimates are pointwise and uniform, using the multivariate first modulus of continuity.

24.2 Results

Here $r \in \mathbb{N}$, $m \in \mathbb{Z}_+$, and define

$$\alpha_j := \alpha_{j,r}^{[m]} := \begin{cases} (-1)^{r-j} \binom{r}{j} j^{-m}, & \text{if } j = 1, 2, \ldots, r, \\ 1 - \sum_{j=1}^{r} (-1)^{r-j} \binom{r}{j} j^{-m}, & \text{if } j = 0. \end{cases} \tag{24.1}$$

See that

$$\sum_{j=0}^{r} \alpha_{j,r}^{[m]} = 1 \tag{24.2}$$

and

$$-\sum_{j=1}^{r} (-1)^{r-j} \binom{r}{j} = (-1)^r \binom{r}{0}. \tag{24.3}$$

Let μ_{ξ_n} be a probability Borel measure on $\mathbb{R}^N$, $N \geq 1$, $\xi_n > 0$, $n \in \mathbb{N}$. We now define the multiple smooth singular integral operators

$$\theta_n (f; x_1, \ldots, x_N) := \theta_{r,n}^{[m]} (f; x_1, \ldots, x_N)$$

$$:= \sum_{j=0}^{r} \alpha_{j,r}^{[m]} \int_{\mathbb{R}^N} f (x_1 + s_1 j, x_2 + s_2 j, \ldots, x_N + s_N j) \, d\mu_{\xi_n} (s), \tag{24.4}$$

where $s := (s_1, \ldots, s_N)$, $x := (x_1, \ldots, x_N) \in \mathbb{R}^N$; $n, r \in \mathbb{Z}$, $m \in \mathbb{Z}_+$, $f : \mathbb{R}^N \to \mathbb{R}$ is a Borel measurable function, and also $(\xi_n)_{n \in \mathbb{N}}$ is a bounded sequence of positive real numbers, we take $0 < \xi_n \leq 1$.

Remark 24.1. The operators $\theta_{r,n}^{[m]}$ are not in general positive, see Anastassiou (2011b), p. 2.

We observe that

Lemma 24.1. *It holds*

$$\theta_{r,n}^{[m]} (c; x_1, \ldots, x_n) = c,$$

where c is a constant.

We need

Definition 24.1. Let $f \in C\left(\mathbb{R}^N\right)$, $N \geq 1$. We define the first modulus of continuity of f as

$$\omega_1(f,\delta) := \sup_{\substack{x,y \in \mathbb{R}^N: \\ \|x-y\|_\infty \leq \delta}} |f(x) - f(y)|, \quad \delta > 0, \tag{24.5}$$

where $\|\cdot\|_\infty$ is the max norm in $\mathbb{R}^N$. The functional $\omega_1(f,\delta)$ is bounded for f being bounded or uniformly continuous, and $\omega_1(f,\delta) \to 0$ as $\delta \to 0$, in the case of f being uniformly continuous.

We present the main general approximation result regarding the operator θ_n.

Theorem 24.1. *Here $f \in C^2\left(\mathbb{R}^N\right)$ and let all $\alpha_i \in \mathbb{Z}^+$, $i = 1,\ldots,N$, $N \geq 1$, $|\alpha| := \sum_{i=1}^N \alpha_i = 2$; $x \in \mathbb{R}^N$, and all the partials f_α of order 2, along with $f \in C_B\left(\mathbb{R}^N\right)$ (continuous and bounded functions); or all f_α of order 2, $f \in C_U\left(\mathbb{R}^N\right)$ (uniformly continuous functions). Let μ_{ξ_n} be a Borel probability measure on $\mathbb{R}^N$, for $0 < \xi_n \leq 1$, $n \in \mathbb{N}$.*

Suppose that for all $\alpha := (\alpha_1,\ldots,\alpha_N)$, $\alpha_i \in \mathbb{Z}^+$, $i = 1,\ldots,N$, $|\alpha| = \sum_{i=1}^N \alpha_i = 2$, $j = 0,1,\ldots,r$, we have that both

$$I_{1j}(\alpha) := \int_{\mathbb{R}^N} \left(1 + \frac{j\|s\|_1}{3\xi_n}\right)\left(\prod_{i=1}^N |s_i|^{\alpha_i}\right) d\mu_{\xi_n}(s), \tag{24.6}$$

$$I_{2j}(\alpha) := \int_{\mathbb{R}^N} \left(1 + \frac{j\|s\|_1}{3\xi_n}\right) d\mu_{\xi_n}(s) \tag{24.7}$$

are uniformly bounded in $\xi_n \in (0,1]$.

Denote ($n \in \mathbb{N}$)

$$\begin{aligned}
\Delta_n(x) &:= \theta_n(f,x) - f(x) \\
&- \left(\sum_{j=0}^r \alpha_j j\right) \sin(1)\left[\sum_{i=1}^N \frac{\partial f(x)}{\partial x_i}\left(\int_{\mathbb{R}^N} s_i d\mu_{\xi_n}(s)\right)\right] \\
&- 2\left(\sum_{j=0}^r \alpha_j j^2\right)\sin^2\left(\frac{1}{2}\right)\left\{\sum_{i=1}^N \left(\int_{\mathbb{R}^N} s_i^2 d\mu_{\xi_n}(s)\right)\frac{\partial^2 f(x)}{\partial x_i^2}\right. \\
&\left. + \sum_{\substack{i \neq j^*, \\ i,j^* \in \{1,\ldots,N\}}} \left(\int_{\mathbb{R}^N} s_i s_{j^*} d\mu_{\xi_n}(s)\right)\frac{\partial^2 f(x)}{\partial x_i \partial x_{j^*}}\right\}.
\end{aligned} \tag{24.8}$$

Then

(i)

$$|\Delta_n(x)| \leq \|\Delta_n(x)\|_\infty \leq \sum_{j=0}^{r} |\alpha_j|$$

$$\left[\left[j^2 \sum_{\substack{\alpha_i \in \mathbb{Z}^+, \\ \alpha:|\alpha|=2}} \left(\frac{1}{\prod_{i=1}^{N} \alpha_i!}\right) \omega_1(f_\alpha, \xi_n) \int_{\mathbb{R}^N} \left(1 + \frac{j\|s\|_\infty}{3\xi_n}\right) \left(\prod_{i=1}^{N} |s_i|^{\alpha_i}\right) d\mu_{\xi_n}(s)\right]\right.$$

$$\left. + \frac{1}{2}\omega_1(f, \xi_n) \int_{\mathbb{R}^N} \left(1 + \frac{j\|s\|_\infty}{3\xi_n}\right) d\mu_{\xi_n}(s)\right] =: \varphi_{\xi_n}. \tag{24.9}$$

In case of all f_α of order 2 and $f \in C_U(\mathbb{R}^N)$ and $\xi_n \to 0$, as $n \to \infty$, then $\Delta_n(x)$, $\|\Delta_n(x)\|_\infty \to 0$ with rates.

(ii) *If $\frac{\partial f(x)}{\partial x_i} = 0$, $i = 1, \ldots, N$, and $f_\alpha(x) = 0$, $\alpha_i \in \mathbb{Z}^+$, $i = 1, \ldots, N$, with $|\alpha| = 2$, then*

$$|\theta_n(f, x) - f(x)| \leq \varphi_{\xi_n}. \tag{24.10}$$

And $\theta_n(f, x) \to f(x)$ in the uniformly continuous case.

(iii) *Additionally assume all partials of order ≤ 2 are bounded. Hence*

$$\|\theta_n(f) - f\|_\infty \leq \left(\sum_{j=0}^{r} |\alpha_j| j\right) (0.8414) \left[\sum_{i=1}^{N} \left\|\frac{\partial f}{\partial x_i}\right\|_\infty \left(\int_{\mathbb{R}^N} |s_i| d\mu_{\xi_n}(s)\right)\right]$$

$$+ \left(\sum_{j=0}^{r} |\alpha_j| j^2\right) (0.4596) \left\{\sum_{i=1}^{N} \left(\int_{\mathbb{R}^N} s_i^2 d\mu_{\xi_n}(s)\right) \left\|\frac{\partial^2 f}{\partial x_i^2}\right\|_\infty\right.$$

$$\left. + \sum_{\substack{i \neq j^*, \\ i, j^* \in \{1, \ldots, N\}}} \left(\int_{\mathbb{R}^N} |s_i| |s_{j^*}| d\mu_{\xi_n}(s)\right) \left\|\frac{\partial^2 f}{\partial x_i \partial x_{j^*}}\right\|_\infty\right\} + \varphi_{\xi_n}. \tag{24.11}$$

If all $\int_{\mathbb{R}^N} s_i^2 d\mu_{\xi_n}(s)$ and $\int_{\mathbb{R}^N} |s_i| |s_{j^*}| d\mu_{\xi_n}(s)$ converge to zero, as $n \to \infty$, with $\xi_n \to 0$, and all f_α of order 2, $f \in C_U(\mathbb{R}^N)$, then

$$\|\theta_n(f) - f\|_\infty \to 0 \quad \text{with rates, as } \xi_n \to 0,\ n \to +\infty.$$

Proof. Let $s := (s_1, \ldots, s_N)$, $x := (x_1, \ldots, x_N)$, $z := (z_1, \ldots, z_N) := (x_1 + s_1 j,\ x_2 + s_2 j, \ldots x_N + s_N j) = x + sj$; $j = 0, 1, \ldots, r$, and $x := x_0 = (x_{01}, \ldots, x_{0N}) = (x_1, \ldots, x_N)$, all in $\mathbb{R}^N$.

Here $f \in C^2(\mathbb{R}^N)$, $N \in \mathbb{N}$, and clearly all the mixed partials commute.

Consider

$$g_{x+sj}(t) := f(x + t(sj)), \quad 0 \le t \le 1. \tag{24.12}$$

Note that $g_{x+sj}(0) = f(x)$, $g_{x+sj}(1) = f(x + sj)$.

We have (by Anastassiou (2023a))

$$\begin{aligned}
f(x+sj) - f(x) &= g_{x+sj}(1) - g_{x+sj}(0) \\
&= g'_{x+sj}(0)\sin(1) + 2g''_{x+sj}(0)\sin^2\left(\frac{1}{2}\right) \\
&\quad + \int_0^1 \left[\left(g''_{x+sj}(t) + g_{x+sj}(t)\right) - \left(g''_{x+sj}(0) + g_{x+sj}(0)\right)\right]\sin(1-t)\,dt \\
&= \left(\sum_{i=1}^N (s_i j)\frac{\partial f}{\partial x_i}(x)\right)\sin(1) \\
&\quad + 2\left\{\left[\left(\sum_{i=1}^N (s_i j)\frac{\partial}{\partial x_i}\right)^2 f\right](x)\right\}\sin^2\left(\frac{1}{2}\right) \\
&\quad + \int_0^1 \left\{\left\{\left[\left(\sum_{i=1}^N (s_i j)\frac{\partial}{\partial x_i}\right)^2 f\right](x + t(sj)) + f(x + t(sj))\right\}\right. \\
&\quad \left. - \left\{\left[\left(\sum_{i=1}^N (s_i j)\frac{\partial}{\partial x_i}\right)^2 f\right](x) + f(x)\right\}\right\}\sin(1-t)\,dt.
\end{aligned} \tag{24.13}$$

Denote the remainder $(j = 0, 1, \ldots, r)$

$$\begin{aligned}
R_j := \int_0^1 &\left\{\left\{\left[\left(\sum_{i=1}^N (s_i j)\frac{\partial}{\partial x_i}\right)^2 f\right](x + tsj) + f(x + tsj)\right\}\right. \\
&\left. - \left\{\left[\left(\sum_{i=1}^N (s_i j)\frac{\partial}{\partial x_i}\right)^2 f\right](x) + f(x)\right\}\right\}\sin(1-t)\,dt
\end{aligned}$$

$$= \int_0^1 \left\{ \sum_{\substack{\alpha:=(\alpha_1,\ldots,\alpha_N),\alpha_i\in\mathbb{Z}_+ \\ i=1,\ldots,N,|\alpha|:=\sum_{i=1}^N \alpha_i=2}} \left(\frac{2}{\prod_{i=1}^N \alpha_i!} \right) \left(\prod_{i=1}^N (js_i)^{\alpha_i} \right) [f_\alpha (x+t(js)) \right.$$

$$\left. -f_\alpha (x)] + (f(x+t(js)) - f(x)) \right\} \sin(1-t)\, dt. \tag{24.14}$$

Therefore, it holds

$$|R_j| \leq \int_0^1 \left\{ \sum_{\substack{\alpha:=(\alpha_1,\ldots,\alpha_N),\alpha_i\in\mathbb{Z}_+ \\ i=1,\ldots,N,|\alpha|:=\sum_{i=1}^N \alpha_i=2}} \left(\frac{2}{\prod_{i=1}^N \alpha_i!} \right) \right.$$

$$\left. \left(\prod_{i=1}^N (j|s_i|)^{\alpha_i} \right) |f_\alpha (x+tsj) - f_\alpha (x)| + |f(x+tsj) - f(x)| \right\} |\sin(1-t)|\, dt \tag{24.15}$$

$$\leq \int_0^1 \left\{ \sum_{\substack{\alpha:=(\alpha_1,\ldots,\alpha_N),\alpha_i\in\mathbb{Z}_+ \\ i=1,\ldots,N,|\alpha|:=\sum_{i=1}^N \alpha_i=2}} \left(\frac{2}{\prod_{i=1}^N \alpha_i!} \right) \left(\prod_{i=1}^N (j|s_i|)^{\alpha_i} \right) \omega_1 \left(f_\alpha, tj\|s\|_\infty\right) \right.$$

$$\left. +\omega_1 \left(f, tj\|s\|_\infty\right) \right\} |\sin(1-t)|\, dt \leq \tag{24.16}$$

$(0 < \xi_n \leq 1)$

$$\int_0^1 \left\{ \sum_{\substack{\alpha:=(\alpha_1,\ldots,\alpha_N),\alpha_i\in\mathbb{Z}_+ \\ i=1,\ldots,N,|\alpha|:=\sum_{i=1}^N \alpha_i=2}} \left(\frac{2}{\prod_{i=1}^N \alpha_i!} \right) \left(\prod_{i=1}^N (j|s_i|)^{\alpha_i} \right) \omega_1 (f_\alpha, \xi_n) \left(1 + \frac{tj\|s\|_\infty}{\xi_n}\right) \right.$$

$$\left. +\omega_1 (f, \xi_n) \left(1 + \frac{tj\|s\|_\infty}{\xi_n}\right) \right\} |\sin(1-t)|\, dt$$

$$= \left\{ \left[2 \sum_{\substack{\alpha:=(\alpha_1,\ldots,\alpha_N),\alpha_i\in\mathbb{Z}_+ \\ i=1,\ldots,N,|\alpha|:=\sum_{i=1}^{N}\alpha_i=2}} \left(\frac{1}{\prod_{i=1}^{N}\alpha_i!} \right) \left(\prod_{i=1}^{N} (j\,|s_i|)^{\alpha_i} \right) \omega_1(f_\alpha,\xi_n) \right] \right.$$

$$\left. + \omega_1(f,\xi_n) \right\} \left[\int_0^1 \left(1 + \frac{tj\,\|s\|_\infty}{\xi_n} \right) |\sin(1-t)|\, dt \right]$$

$$\leq \left\{ \left[2 \sum_{\substack{\alpha:=(\alpha_1,\ldots,\alpha_N),\alpha_i\in\mathbb{Z}_+ \\ i=1,\ldots,N,|\alpha|:=\sum_{i=1}^{N}\alpha_i=2}} \left(\frac{1}{\prod_{i=1}^{N}\alpha_i!} \right) \left(\prod_{i=1}^{N} (j\,|s_i|)^{\alpha_i} \right) \omega_1(f_\alpha,\xi_n) \right] \right.$$

$$\left. + \omega_1(f,\xi_n) \right\} \left[\int_0^1 \left(1 + \frac{tj\,\|s\|_\infty}{\xi_n} \right) (1-t)\, dt \right]. \tag{24.17}$$

So far we have proved that

$$|R_j| \leq \left\{ \left[2 \sum_{\substack{\alpha_i\in\mathbb{Z}_+, \\ |\alpha|:=\sum_{i=1}^{N}\alpha_i=2}} \left(\frac{1}{\prod_{i=1}^{N}\alpha_i!} \right) \left(\prod_{i=1}^{N} (j\,|s_i|)^{\alpha_i} \right) \omega_1(f_\alpha,\xi_n) \right] \right.$$

$$\left. + \omega_1(f,\xi_n) \right\} \left[\int_0^1 \left(1 + \frac{tj\,\|s\|_\infty}{\xi_n} \right) (1-t)\, dt \right], \tag{24.18}$$

$j = 0, 1, \ldots, r;\ 0 < \xi_n \leq 1$.

So, we have found that

$$|R_j| \leq \left[\left[j^2 \sum_{\substack{\alpha_i\in\mathbb{Z}_+, \\ |\alpha|=2}} \left(\frac{1}{\prod_{i=1}^{N}\alpha_i!} \right) \left(\prod_{i=1}^{N} |s_i|^{\alpha_i} \right) \omega_1(f_\alpha,\xi_n) \right] + \frac{1}{2}\omega_1(f,\xi_n) \right]$$

$$\left[1 + \frac{j\,\|s\|_\infty}{3\xi_n} \right], \quad j = 0, 1, \ldots, r;\ 0 < \xi_n \leq 1. \tag{24.19}$$

Next we can write

$$\sum_{j=0}^{r}\alpha_j\left[f\left(x+sj\right)-f\left(x\right)\right]-\left(\sum_{j=0}^{r}\alpha_j j\right)\left(\sum_{i=1}^{N}s_i\frac{\partial f}{\partial x_i}\left(x\right)\right)\sin\left(1\right)$$
$$-2\left(\sum_{j=0}^{r}\alpha_j j^2\right)\left\{\left[\left(\sum_{i=1}^{N}s_i\frac{\partial}{\partial x_i}\right)^2 f\right]\left(x\right)\right\}\sin^2\left(\frac{1}{2}\right)=\sum_{j=0}^{r}\alpha_j R_j. \tag{24.20}$$

Call

$$R:=\sum_{j=0}^{r}\alpha_j R_j. \tag{24.21}$$

Hence it holds

$$|R|\leq\sum_{j=0}^{r}|\alpha_j|\,|R_j|\leq\sum_{j=0}^{r}|\alpha_j|$$
$$\left[\left[j^2\sum_{\left\{\substack{\alpha:=(\alpha_1,\ldots,\alpha_N),\alpha_i\in\mathbb{Z}_+\\ i=1,\ldots,N,|\alpha|:=\sum_{i=1}^{N}\alpha_i=2}\right\}}\left(\frac{1}{\prod_{i=1}^{N}\alpha_i!}\right)\left(\prod_{i=1}^{N}|s_i|^{\alpha_i}\right)\omega_1\left(f_\alpha,\xi_n\right)\right]\right.$$
$$\left.+\frac{1}{2}\omega_1\left(f,\xi_n\right)\right] \tag{24.22}$$
$$\left[1+\frac{j\left\|s\right\|_\infty}{3\xi_n}\right]$$
$$=\sum_{j=0}^{r}|\alpha_j|\left[\left[j^2\sum_{\alpha:|\alpha|=2}\left(\frac{1}{\prod_{i=1}^{N}\alpha_i!}\right)\omega_1\left(f_\alpha,\xi_n\right)\left[1+\frac{j\left\|s\right\|_\infty}{3\xi_n}\right]\left(\prod_{i=1}^{N}|s_i|^{\alpha_i}\right)\right]\right.$$
$$\left.+\frac{1}{2}\omega_1\left(f,\xi_n\right)\left[1+\frac{j\left\|s\right\|_\infty}{3\xi_n}\right]\right], \tag{24.23}$$

$0<\xi_n\leq 1$.

See that

$$\sin 1 \cong 0.8414$$
$$(\sin 0.5)^2 \cong (0.4794)^2 \cong 0.2298.$$

Clearly, it holds

$$\theta_n(f,x) - f(x) = \sum_{j=0}^{r} \alpha_j \int_{\mathbb{R}^N} (f(x+sj) - f(x))\, d\mu_{\xi_n}(s). \tag{24.24}$$

We observe that

$$\Delta_n(x) = \sum_{j=0}^{r} \alpha_j \int_{\mathbb{R}^N} R_j d\mu_{\xi_n}(s). \tag{24.25}$$

We have that

$$\begin{aligned}
&\sum_{j=0}^{r} |\alpha_j| \int_{\mathbb{R}^N} |R_j|\, d\mu_{\xi_n}(s) \\
&\quad \leq \sum_{j=0}^{r} |\alpha_j| \left[\left[j^2 \sum_{\alpha:|\alpha|=2} \left(\frac{1}{\prod_{i=1}^{N} \alpha_i!} \right) \omega_1(f_\alpha, \xi_n) \int_{\mathbb{R}^N} \right.\right. \\
&\quad \left. \times \left(1 + \frac{j\|s\|_\infty}{3\xi_n}\right) \left(\prod_{i=1}^{N} |s_i|^{\alpha_i} \right) d\mu_{\xi_n}(s) \right] \\
&\quad \left. + \frac{1}{2}\omega_1(f,\xi_n) \int_{\mathbb{R}^N} \left(1 + \frac{j\|s\|_\infty}{3\xi_n}\right) d\mu_{\xi_n}(s) \right] = \varphi_{\xi_n}.
\end{aligned} \tag{24.26}$$

To remind (see also (24.6) and (24.7))

$$\|s\|_\infty \leq \sum_{i=1}^{N} |s_i| =: \|s\|_1, \tag{24.27}$$

hence the integrals in (24.9) and (24.26) are uniformly bounded in $\xi_n \in (0,1]$.

Note also that $(j = 0, 1, \ldots, r)$

$$\begin{aligned}
&\int_{\mathbb{R}^N} |s_i|\, |s_{j^*}|\, d\mu_{\xi_n}(s) \leq I_{1j}(\alpha) < \infty, \\
&\int_{\mathbb{R}^N} s_i^2 d\mu_{\xi_n}(s) \leq I_{1j}(\alpha) < \infty,
\end{aligned} \tag{24.28}$$

and

$$\int_{\mathbb{R}^N} |s_i| \, d\mu_{\xi_n}(s) \le \left(\int_{\mathbb{R}^N} s_i^2 d\mu_{\xi_n}(s) \right)^{\frac{1}{2}} < \infty, \tag{24.29}$$

by Hölder's inequality, and all of them are uniformly bounded in $\xi_n \in (0, 1]$.

Thus, in the uniformly continuous case of f_α, $|\alpha| = 2$, and f we get $\varphi_{\xi_n} \to 0$, as $\xi_n \to 0$.

That is, $\|\Delta_n(x)\|_\infty \to 0$, as $\xi_n \to 0$.

The proof of the theorem is now completed. □

We make

Remark 24.2. Next we apply Theorem 24.1 to specific multivariate smooth Picard singular integral operators

$$P_n(f; x_1, \ldots, x_N) := P_{r,n}^{[m]}(f; x_1, \ldots, x_N)$$

$$:= \frac{1}{(2\xi_n)^N} \sum_{j=0}^{r} \alpha_{j,r}^{[m]} \int_{\mathbb{R}^N} f(x_1 + s_1 j, x_2 + s_2 j, \ldots, x_N + s_N j) \, e^{-\frac{\sum_{i=1}^N |s_i|}{\xi_n}}$$

$$\times ds_1 \ldots ds_N, \tag{24.30}$$

$r, n \in \mathbb{N}$, $m \in \mathbb{Z}_+$, $0 < \xi_n \le 1$.

Clearly here it is

$$d\mu_{\xi_n}(s) = \frac{1}{(2\xi_n)^N} e^{-\frac{\sum_{i=1}^N |s_i|}{\xi_n}} ds_1 \cdots ds_N, \quad s \in \mathbb{R}^N. \tag{24.31}$$

We observe that

$$\frac{1}{(2\xi_n)^N} \int_{\mathbb{R}^N} e^{-\frac{\sum_{i=1}^N |s_i|}{\xi_n}} ds_1 \cdots ds_N = 1, \tag{24.32}$$

see Anastassiou (2021), Chapter 22, p. 356.

Here $\alpha_i \in \mathbb{Z}^+$, $i = 1, \ldots, N : |\alpha| = \sum_{i=1}^N \alpha_i = 2$. We note that

$$\int_{\mathbb{R}^N} \left(\prod_{i=1}^N |s_i|^{\alpha_i} \right) e^{-\frac{\sum_{i=1}^N |s_i|}{\xi_n}} ds_1 \cdots ds_N \le 4^N \xi_n^{N+2} \le 4^N, \tag{24.33}$$

by Anastassiou (2021), p. 364.

So (24.6) and (24.7) are confirmed for $j = 0$ when $d\mu_{\xi_n}$ is as in (24.31).

We need

Theorem 24.2. *Let $N \in \mathbb{N}$, $\alpha_i \in \mathbb{Z}^+$, $i = 1, \ldots, N : |\alpha| = \sum_{i=1}^N \alpha_i = 2$, $\xi_n \in (0,1]$, $n \in \mathbb{N}$; $j = 1, 2, \ldots, r$. Then*

$$I_{1j}^*(\alpha) := \frac{1}{(2\xi_n)^N} \int_{\mathbb{R}^N} \left(1 + \frac{j\,\|s\|_1}{3\xi_n}\right) \left(\prod_{i=1}^N |s_i|^{\alpha_i}\right) e^{-\frac{\sum_{i=1}^N |s_i|}{\xi_n}} ds_1 \cdots ds_N$$

$$\leq \xi_n^2 \left[\left(1 + \frac{j}{3}N\right) + \left(1 + \frac{j}{3}\right) \prod_{i=1}^N \left(\frac{\lfloor e(\alpha_i + 1)! \rfloor}{e}\right)\right]$$

$$\leq \left[\left(1 + \frac{j}{3}N\right) + \left(1 + \frac{j}{3}\right)\left(\frac{16}{e}\right)^N\right] < +\infty, \tag{24.34}$$

are uniformly bounded in $\xi_n \in (0,1]$, fulfilling (24.6). Above $\lfloor \cdot \rfloor$ is the integral part of the number symbol.

Proof. Let $j = 1, \ldots, r$, then

$$I_{1j}^*(\alpha) = \frac{1}{(2\xi_n)^N} \int_{\mathbb{R}^N} \left(1 + \frac{j\,\|s\|_1}{3\xi_n}\right) \left(\prod_{i=1}^N |s_i|^{\alpha_i}\right) e^{-\frac{\sum_{i=1}^N |s_i|}{\xi_n}} ds_1 \cdots ds_N$$

$$= \frac{1}{\xi_n^N} \int_{\mathbb{R}_+^N} \left(1 + \frac{j\left(\sum_{i=1}^N s_i\right)}{3\xi_n}\right) \left(\prod_{i=1}^N s_i^{\alpha_i}\right) e^{-\frac{\sum_{i=1}^N s_i}{\xi_n}} ds_1 \cdots ds_N$$

$$= \xi_n^2 \int_{\mathbb{R}_+^N} \left(1 + \frac{j}{3}\left(\sum_{i=1}^N \frac{s_i}{\xi_n}\right)\right) \left(\prod_{i=1}^N \left(\frac{s_i}{\xi_n}\right)^{\alpha_i}\right)$$

$$\times e^{-\sum_{i=1}^N \frac{s_i}{\xi_n}} d\left(\frac{s_1}{\xi_n}\right) \cdots d\left(\frac{s_N}{\xi_n}\right) \tag{24.35}$$

$$= \xi_n^2 \int_{\mathbb{R}_+^N} \left(1 + \frac{j}{3}\left(\sum_{i=1}^N z_i\right)\right) \left(\prod_{i=1}^N z_i^{\alpha_i}\right) e^{-\sum_{i=1}^N z_i} dz_1 \cdots dz_N$$

$$= \xi_n^2 \left[\int_{[0,1]^N} \left(1 + \frac{j}{3}\left(\sum_{i=1}^N z_i\right)\right) \left(\prod_{i=1}^N z_i^{\alpha_i}\right) e^{-\sum_{i=1}^N z_i} dz_1 \cdots dz_N\right.$$

$$\left. + \int_{(\mathbb{R}_+ - [0,1])^N} \left(1 + \frac{j}{3}\left(\sum_{i=1}^N z_i\right)\right) \left(\prod_{i=1}^N z_i^{\alpha_i}\right) e^{-\sum_{i=1}^N z_i} dz_1 \cdots dz_N\right]$$

$$\leq \xi_n^2 \left[\left(1 + \frac{j}{3}N\right) + \int_{(\mathbb{R}_+ - [0,1])^N} \left(1 + \frac{j}{3}\left(\sum_{i=1}^N z_i\right)\right)\right.$$

$$\times \left(\prod_{i=1}^{N} z_i^{\alpha_i} \right) e^{-\sum_{i=1}^{N} z_i} dz_1 \cdots dz_N \Bigg]$$

$$\leq \xi_n^2 \left[\left(1 + \frac{j}{3} N \right) + \left(1 + \frac{j}{3} \right) \int_{(\mathbb{R}_+ - [0,1])^N} \left(\sum_{i=1}^{N} z_i \right) \right.$$

$$\left. \times \left(\prod_{i=1}^{N} z_i^{\alpha_i} \right) e^{-\sum_{i=1}^{N} z_i} dz_1 \cdots dz_N \right]$$

$$\leq \xi_n^2 \left[\left(1 + \frac{j}{3} N \right) + \left(1 + \frac{j}{3} \right) \int_{(\mathbb{R}_+ - [0,1])^N} \left(\prod_{i=1}^{N} z_i \right) \left(\prod_{i=1}^{N} z_i^{\alpha_i} \right) \right.$$

$$\left. \times \left(\prod_{i=1}^{N} e^{-z_i} \right) \left(\prod_{i=1}^{N} dz_i \right) \right]$$

$$= \xi_n^2 \left[\left(1 + \frac{j}{3} N \right) + \left(1 + \frac{j}{3} \right) \prod_{i=1}^{N} \int_1^{\infty} z_i^{\alpha_i + 1} e^{-z_i} dz_i \right]$$

$$= \xi_n^2 \left[\left(1 + \frac{j}{3} N \right) + \left(1 + \frac{j}{3} \right) \prod_{i=1}^{N} \int_1^{\infty} z_i^{(\alpha_i + 2) - 1} e^{-z_i} dz_i \right] \tag{24.36}$$

(by Gradshteyn and Ryzhik (2015), p. 348)

$$\xi_n^2 \left[\left(1 + \frac{j}{3} N \right) + \left(1 + \frac{j}{3} \right) \prod_{i=1}^{N} \Gamma\left((\alpha_i + 2), 1 \right) \right], \tag{24.37}$$

where $\Gamma(\cdot, \cdot)$ is the upper incomplete gamma function.

We have proved that, $j = 1, \ldots, r$, that

$$I_{1j}^*(\alpha) \leq \xi_n^2 \left[\left(1 + \frac{j}{3} N \right) + \left(1 + \frac{j}{3} \right) \prod_{i=1}^{N} \Gamma\left((\alpha_i + 2), 1 \right) \right]$$

$$\leq \xi_n^2 \left[\left(1 + \frac{j}{3} N \right) + \left(1 + \frac{j}{3} \right) \prod_{i=1}^{N} \left(\frac{\lfloor e (\alpha_i + 1)! \rfloor}{e} \right) \right]$$

$$\leq \left[\left(1 + \frac{j}{3} N \right) + \left(1 + \frac{j}{3} \right) \left(\frac{16}{e} \right)^N \right] < +\infty, \tag{24.38}$$

therefore $I_{1j}^*(\alpha)$ are uniformly bounded.

Above we used the formula

$$\Gamma(s + 1, 1) = \frac{\lfloor es! \rfloor}{e}, \quad s \in \mathbb{N}. \tag{24.39}$$

Here $\alpha_i + 2 \in \mathbb{N}$, hence

$$\Gamma\left((\alpha_i+2),1\right) = \frac{\lfloor e\,(\alpha_i+1)!\rfloor}{e} \leq \frac{\lfloor e3!\rfloor}{e} = \frac{\lfloor 6e\rfloor}{e} = \frac{\lfloor 16.30968\rfloor}{e} = \frac{16}{e}. \tag{24.40}$$

The claim is proved. □

It follows

Theorem 24.3. *Let $N \in \mathbb{N}$, $\xi_n \in (0,1]$, $n \in \mathbb{N}$; $j = 1,2,\ldots,r$. Then*

$$\begin{aligned} I^*_{2j}(\alpha) &:= \frac{1}{(2\xi_n)^N}\int_{\mathbb{R}^N}\left(1+\frac{j\left(\sum_{i=1}^N |s_i|\right)}{3\xi_n}\right)e^{-\frac{\sum_{i=1}^N |s_i|}{\xi_n}}ds_1\cdots ds_N \\ &\leq \left[\left(1+\frac{j}{3}N\right)+\left(1+\frac{j}{3}\right)\left(\frac{2}{e}\right)^N\right] < +\infty \end{aligned} \tag{24.41}$$

are uniformly bounded in $\xi_n \in (0,1]$, fulfilling (24.7).

Proof. We have

$$\begin{aligned} I^*_{2j}(\alpha) &= \frac{1}{(2\xi_n)^N}\int_{\mathbb{R}^N}\left(1+\frac{j\left(\sum_{i=1}^N |s_i|\right)}{3\xi_n}\right)e^{-\frac{\sum_{i=1}^N |s_i|}{\xi_n}}ds_1\cdots ds_N \\ &= \frac{1}{\xi_n^N}\int_{\mathbb{R}^N_+}\left(1+\frac{j\left(\sum_{i=1}^N s_i\right)}{3\xi_n}\right)e^{-\frac{\sum_{i=1}^N s_i}{\xi_n}}ds_1\cdots ds_N \qquad (24.42) \\ &= \int_{\mathbb{R}^N_+}\left(1+\frac{j}{3}\left(\sum_{i=1}^N z_i\right)\right)e^{-\sum_{i=1}^N z_i}dz_1\cdots dz_N \\ &= \int_{[0,1]^N}\left(1+\frac{j}{3}\left(\sum_{i=1}^N z_i\right)\right)e^{-\sum_{i=1}^N z_i}dz_1\cdots dz_N \\ &\quad + \int_{(\mathbb{R}_+-[0,1])^N}\left(1+\frac{j}{3}\left(\sum_{i=1}^N z_i\right)\right)e^{-\sum_{i=1}^N z_i}dz_1\cdots dz_N \\ &\leq \left(1+\frac{j}{3}N\right)+\int_{(\mathbb{R}_+-[0,1])^N}\left(1+\frac{j}{3}\left(\sum_{i=1}^N z_i\right)\right)e^{-\sum_{i=1}^N z_i}dz_1\cdots dz_N \end{aligned}$$

$$\leq \left(1+\frac{j}{3}N\right)+\left(1+\frac{j}{3}\right)\int_{(\mathbb{R}_+-[0,1])^N}\left(\sum_{i=1}^N z_i\right)e^{-\sum_{i=1}^N z_i}dz_1\cdots dz_N$$

$$\leq \left(1+\frac{j}{3}N\right)+\left(1+\frac{j}{3}\right)\int_{(\mathbb{R}_+-[0,1])^N}\left(\prod_{i=1}^N z_i\right)\left(\prod_{i=1}^N e^{-z_i}\right)\left(\prod_{i=1}^N dz_i\right)$$

$$=\left(1+\frac{j}{3}N\right)+\left(1+\frac{j}{3}\right)\prod_{i=1}^N\int_1^\infty z_i e^{-z_i}dz_i$$

$$=\left(1+\frac{j}{3}N\right)+\left(1+\frac{j}{3}\right)\prod_{i=1}^N\int_1^\infty z_i^{2-1}e^{-z_i}dz_i$$

$$=\left(1+\frac{j}{3}N\right)+\left(1+\frac{j}{3}\right)\prod_{i=1}^N\Gamma(2,1)$$

$$=\left(1+\frac{j}{3}N\right)+\left(1+\frac{j}{3}\right)\left(\frac{\lfloor e\rfloor}{e}\right)^N$$

$$=\left(1+\frac{j}{3}N\right)+\left(1+\frac{j}{3}\right)\left(\frac{2}{e}\right)^N<+\infty. \tag{24.43}$$

□

We make

Remark 24.3. By (24.28), (24.29), Remark 24.2 and Theorem 24.2, we observe that $(j=0,1,\ldots,r)$

$$\frac{1}{(2\xi_n)^N}\int_{\mathbb{R}^N}|s_i|\,|s_{j^*}|\,e^{-\frac{\sum_{i=1}^N|s_i|}{\xi_n}}ds_1\cdots ds_N,$$

$$\frac{1}{(2\xi_n)^N}\int_{\mathbb{R}^N}s_i^2 e^{-\frac{\sum_{i=1}^N|s_i|}{\xi_n}}ds_1\cdots ds_N,$$

$$\frac{1}{(2\xi_n)^N}\int_{\mathbb{R}^N}|s_i|\,e^{-\frac{\sum_{i=1}^N|s_i|}{\xi_n}}ds_1\cdots ds_N \tag{24.44}$$

are uniformly bounded in $\xi_n\in(0,1]$ and they converge to zero as $\xi_n\to 0$.

Based on Theorem 24.1, Remark 24.2, Theorem 24.2, Theorem 24.7 and Remark 24.3, we present our major result about approximation properties of smooth Picard singular integral operators P_n, see (24.30).

Theorem 24.4. *Here* $f\in C^2(\mathbb{R}^N)$ *and let* $\alpha_i\in\mathbb{Z}^+$, $i=1,\ldots,N$, $N\geq 1$, $|\alpha|:=\sum_{i=1}^N\alpha_i=2$; $x\in\mathbb{R}^N$, *and* f_α *of order* 2, $f\in C_B(\mathbb{R}^N)\cup C_U(\mathbb{R}^N)$, *and* $0<\xi_n\leq 1$, $n\in\mathbb{N}$.

Denote $(n \in \mathbb{N})$

$$\overline{\Delta}_n(x) := P_n(f,x) - f(x)$$

$$- \left(\sum_{j=0}^{r} \alpha_j j\right) \sin(1) \left[\sum_{i=1}^{N} \frac{\partial f(x)}{\partial x_i} \left(\frac{1}{(2\xi_n)^N} \int_{\mathbb{R}^N} s_i e^{-\frac{\sum_{i=1}^{N}|s_i|}{\xi_n}} ds_1 \cdots ds_N\right)\right]$$

$$- 2\left(\sum_{j=0}^{r} \alpha_j j^2\right) \sin^2\left(\frac{1}{2}\right) \left\{\sum_{i=1}^{N} \left(\frac{1}{(2\xi_n)^N} \int_{\mathbb{R}^N} s_i^2 e^{-\frac{\sum_{i=1}^{N}|s_i|}{\xi_n}} ds_1 \cdots ds_N\right) \frac{\partial^2 f(x)}{\partial x_i^2}\right.$$

$$\left. + \sum_{\substack{i \neq j^*, \\ i,j^* \in \{1,\ldots,N\}}} \left(\frac{1}{(2\xi_n)^N} \int_{\mathbb{R}^N} s_i s_{j^*} e^{-\frac{\sum_{i=1}^{N}|s_i|}{\xi_n}} ds_1 \cdots ds_N\right) \frac{\partial^2 f(x)}{\partial x_i \partial x_{j^*}}\right\}. \tag{24.45}$$

Then

(i)

$$\left|\overline{\Delta}_n(x)\right| \leq \left\|\overline{\Delta}_n(x)\right\|_\infty$$

$$\leq \sum_{j=0}^{r} |\alpha_j| \left[\left[j^2 \sum_{\substack{\alpha_i \in \mathbb{Z}^+, \\ \alpha : |\alpha| = 2}} \left(\frac{1}{\prod_{i=1}^{N} \alpha_i!}\right) \omega_1(f_\alpha, \xi_n) \frac{1}{(2\xi_n)^N}\right.\right.$$

$$\left.\int_{\mathbb{R}^N} \left(1 + \frac{j\|s\|_\infty}{3\xi_n}\right) \left(\prod_{i=1}^{N} |s_i|^{\alpha_i}\right) e^{-\frac{\sum_{i=1}^{N}|s_i|}{\xi_n}} ds_1 \cdots ds_N\right]$$

$$\left. + \frac{1}{2}\omega_1(f,\xi_n) \frac{1}{(2\xi_n)^N} \int_{\mathbb{R}^N} \left(1 + \frac{j\|s\|_\infty}{3\xi_n}\right) e^{-\frac{\sum_{i=1}^{N}|s_i|}{\xi_n}} ds_1 \cdots ds_N\right] =: \overline{\varphi}_{\xi_n}. \tag{24.46}$$

In case of all f_α *of order* 2 *and* $f \in C_U(\mathbb{R}^N)$ *and* $\xi_n \to 0$, *as* $n \to \infty$, *then* $\overline{\Delta}_n(x)$, $\left\|\overline{\Delta}_n(x)\right\|_\infty \to 0$ *with rates.*

(ii) *If* $\frac{\partial f(x)}{\partial x_i} = 0$, $i = 1,\ldots,N$, *and* $f_\alpha(x) = 0$, $\alpha_i \in \mathbb{Z}^+$, $i = 1,\ldots,N$, *with* $|\alpha| = 2$, *then*

$$|P_n(f,x) - f(x)| \leq \overline{\varphi}_{\xi_n}. \tag{24.47}$$

And $P_n(f,x) \to f(x)$ *in the uniformly continuous case.*

(iii) *Additionally assume that all partials of order* ≤ 2 *are bounded. Hence*

$$\|P_n(f) - f\|_\infty \leq \left(\sum_{j=0}^{r} |\alpha_j| j\right)(0.8414)$$

$$\left[\sum_{i=1}^{N} \left\|\frac{\partial f}{\partial x_i}\right\|_\infty \frac{1}{(2\xi_n)^N} \left(\int_{\mathbb{R}^N} |s_i| e^{-\frac{\sum_{i=1}^{N}|s_i|}{\xi_n}} ds_1 \cdots ds_N\right)\right]$$

$$+ \left(\sum_{j=0}^{r} |\alpha_j| j^2\right)(0.4596)$$

$$\left\{\sum_{i=1}^{N} \frac{1}{(2\xi_n)^N} \left(\int_{\mathbb{R}^N} s_i^2 e^{-\frac{\sum_{i=1}^{N}|s_i|}{\xi_n}} ds_1 \cdots ds_N\right) \left\|\frac{\partial^2 f}{\partial x_i^2}\right\|_\infty\right.$$

$$\left. + \sum_{\substack{i \neq j^*, \\ i,j^* \in \{1,\ldots,N\}}} \frac{1}{(2\xi_n)^N} \left(\int_{\mathbb{R}^N} |s_i| |s_{j^*}| e^{-\frac{\sum_{i=1}^{N}|s_i|}{\xi_n}} ds_1 \cdots ds_N\right) \left\|\frac{\partial^2 f}{\partial x_i \partial x_{j^*}}\right\|_\infty\right\}$$

$$+ \overline{\varphi}_{\xi_n}. \tag{24.48}$$

If all f_α *of order* 2, $f \in C_U(\mathbb{R}^N)$, *then*

$$\|P_n(f) - f\|_\infty \to 0 \text{ with rates, as } \xi_n \to 0,\ n \to +\infty.$$

Bibliography

G.A. Anastassiou, *Intelligent Mathematics: Computational Analysis*, Springer, Heidelberg, New York, 2011a.

G.A. Anastassiou, *Approximation by Multivariate Singular Integrals*, Springer, New York, 2011b.

G.A. Anastassiou, *Constructive Fractional Analysis with Applications*, Springer, Heidelberg, New York, 2021.

G.A. Anastassiou, *General multiple sigmoid functions relied complex valued multivariate trigonometric and hyperbolic neural network approximations*, submitted, 2023a.

G.A. Anastassiou and S. Gal, *Approximation Theory*, Birkhaüser, Boston, 2000.

G.A. Anastassiou and R. Mezei, *Approximation by Singular Integrals*, Cambridge Scientific Publishers, Cambridge, UK, 2012.

G.A. Anastassiou, *Uniform Approximation by smooth Picard Multivariate Singular Integral Operators*, submitted, 2023b.

I.S. Gradshteyn and I.M. Ryzhik, *Table of Integrals, Series and Products*, 8th edn. Elsevier, Amsterdam, 2015.

R.N. Mohapatra and R.S. Rodriguez, On the rate of convergence of singular integrals for Hölder continuous functions, *Mathematische Nachrichten*, 149 (1990), 117–124.

Chapter 25

Trigonometric-Derived Multivariate Smooth Picard Singular Integrals L_p Quantitative Approximation

In this chapter, we reexamine the L_p, $1 \leq p < \infty$, approximation properties of general smooth multivariate singular integral operators over $\mathbb{R}^N$, $N \geq 1$. It is a trigonometric-based approach with detailed applications to the corresponding smooth multivariate Picard singular integral operators. The results are quantitative via Jackson-type inequalities involving their first L_p modulus of continuity. It follows Anastassiou (2023a).

25.1 Introduction

The rate of convergence of univariate and multivariate singular integral operators has been studied extensively in Anastassiou (2011a)–Anastassiou (2021) and Anastassiou and Gal (2000), Anastassiou and Mezei (2012) and Mohapatra and Rodriguez (1990). All these motivate our current work. In particular, we studied the smooth singular integral operators in Anastassiou (2011a)–Anastassiou (2021) and Anastassiou and Mezei (2012), which are not in general positive ones, their L_p approximation properties.

Here we continue the L_p study of the last ones at the multivariate level, at first in general, and then apply our theory to the smooth Picard ones. The main tool, we are based on, is a new trigonometric multivariate Taylor formula from Anastassiou (2023b). Our quantitative estimates are related to L_p approximations, using the multivariate first L_p modulus of continuity.

25.2 Results

Here $r \in \mathbb{N}$, $m \in \mathbb{Z}_+$, we define

$$\alpha_j := \alpha_{j,r}^{[m]} := \begin{cases} (-1)^{r-j} \binom{r}{j} j^{-m}, & \text{if } j = 1, 2, \ldots, r, \\ 1 - \sum_{j=1}^{r} (-1)^{r-j} \binom{r}{j} j^{-m}, & \text{if } j = 0, \end{cases} \tag{25.1}$$

and

$$\delta_k := \delta_{k,r}^{[m]} := \sum_{j=0}^{r} \alpha_{j,r}^{[m]} j^k, \quad k = 1, 2, \ldots, m \in \mathbb{N}. \tag{25.2}$$

See that

$$\sum_{j=0}^{r} \alpha_{j,r}^{[m]} = 1 \tag{25.3}$$

and

$$-\sum_{j=1}^{r} (-1)^{r-j} \binom{r}{j} = (-1)^r \binom{r}{0}. \tag{25.4}$$

Let μ_{ξ_n} be a probability Borel measure on $\mathbb{R}^N$, $N \geq 1$, $\xi_n > 0$, $n \in \mathbb{N}$. We now define the multiple smooth singular integral operators

$$\theta_n (f; x_1, \ldots, x_N) := \theta_{r,n}^{[m]} (f; x_1, \ldots, x_N)$$
$$:= \sum_{j=0}^{r} \alpha_{j,r}^{[m]} \int_{\mathbb{R}^N} f (x_1 + s_1 j, x_2 + s_2 j, \ldots, x_N + s_N j) \, d\mu_{\xi_n} (s), \tag{25.5}$$

where $s := (s_1, \ldots, s_N)$, $x := (x_1, \ldots, x_N) \in \mathbb{R}^N$; $n, r \in \mathbb{Z}$, $m \in \mathbb{Z}_+$, $f : \mathbb{R}^N \to \mathbb{R}$ is a Borel measurable function, and also $(\xi_n)_{n \in \mathbb{N}}$ is a bounded sequence of positive real numbers.

The operators $\theta_{r,n}^{[m]}$ preserve constants and are not in general positive operators, see Anastassiou (2011b), pp. 2–3.

Here we deal with $f \in C^m (\mathbb{R}^N)$, $m \in \mathbb{Z}^+$, with $f_\alpha \in L_p (\mathbb{R}^N)$, $|\alpha| = m \in \mathbb{Z}^+$, $p \geq 1$, where f_α denotes the mixed partial $\frac{\partial^{\widetilde{j}} f(\cdot,\ldots,\cdot)}{\partial x_1^{\alpha_1} \cdots \partial x_N^{\alpha_N}}$, $\alpha_j \in \mathbb{Z}^+$, $j = 1, \ldots, N : |\alpha| := \sum_{j=1}^{N} \alpha_j = \widetilde{j}$, $\widetilde{j} = 1, \ldots, m$.

We need

Definition 25.1. (see also Anastassiou (2011b), p. 20) We call

$$\Delta_u^r f(x) := \Delta_{u_1,u_2,\ldots,u_N}^r f(x_1,\ldots,x_N)$$
$$:= \sum_{j=0}^{r} (-1)^{r-j} \binom{r}{j} f(x_1+ju_1, x_2+ju_2,\ldots,x_N+ju_N). \tag{25.6}$$

Let $p \geq 1$, the modulus of smoothness of order r is given by

$$\omega_r(f;h)_p := \sup_{\|u\|_2 \leq h} \|\Delta_u^r(f)\|_p, \tag{25.7}$$

$h > 0$.

(I) We consider the general case of $m \in \mathbb{N}$, $p > 1$.
We make

Remark 25.1. For $\widetilde{j} = 1,\ldots,m$, and $\alpha := (\alpha_1,\ldots,\alpha_N)$, $\alpha_i \in \mathbb{Z}^+$, $i = 1,\ldots,N$, $|\alpha| := \sum_{i=1}^N \alpha_i = \widetilde{j}$, under the assumption of Theorem 25.1,

$$c_{\alpha,n,\widetilde{j}} := c_{\alpha,n} := \int_{\mathbb{R}^N} \prod_{i=1}^N s_i^{\alpha_i} d\mu_{\xi_n}(s_1,\ldots,s_N) \in \mathbb{R}, \tag{25.8}$$

see also Anastassiou (2011b).
From Anastassiou (2011b), we get

$$E_{r,n}^{[m]}(x) := \theta_{r,n}^{[m]}(f;x) - f(x) - \sum_{\widetilde{j}=1}^{m} \delta_{\widetilde{j},r}^{[m]} \left(\sum_{|\alpha|=\widetilde{j}} \frac{c_{\alpha,n} f_\alpha(x)}{\prod_{i=1}^N \alpha_i!} \right) \tag{25.9}$$

$$= m \sum_{|\alpha|=\widetilde{j}} \left(\frac{1}{\prod_{i=1}^N \alpha_i!} \right) \left(\int_{\mathbb{R}^N} \left(\prod_{i=1}^N s_i^{\alpha_i} \int_0^1 (1-\theta)^{m-1} \left(\Delta_{\theta s}^r f_\alpha(x)\right) d\theta \right) d\mu_{\xi_n}(s) \right)$$

$$=: R_{r,n}^{[m]}(x), \quad \forall\, x \in \mathbb{R}^N. \tag{25.10}$$

Let $p,q > 1 : \frac{1}{p} + \frac{1}{q} = 1$. Then

$$\left| E_{r,n}^{[m]}(x) \right|^p = \left| R_{r,n}^{[m]}(x) \right|^p$$

$$\left(\text{set } c_1 := m \sum_{|\alpha|=\widetilde{j}} \left(\frac{1}{\prod_{i=1}^N \alpha_i!}\right)\right)$$

$$= \left| c_1 \int_{\mathbb{R}^N} \left(\prod_{i=1}^N s_i^{\alpha_i} \int_0^1 (1-\theta)^{m-1} \left(\Delta_{\theta s}^r f_\alpha(x)\right) d\theta \right) d\mu_{\xi_n}(s) \right|^p$$

$$\leq \left(c_1 \int_{\mathbb{R}^N} \left(\prod_{i=1}^N |s_i|^{\alpha_i} \int_0^1 (1-\theta)^{m-1} \left|\Delta_{\theta s}^r f_\alpha(x)\right| d\theta \right) d\mu_{\xi_n}(s) \right)^p . \tag{25.11}$$

Hence we have

$$I_1 := \int_{\mathbb{R}^N} \left| E_{r,n}^{[m]}(x) \right|^p dx$$

$$\leq \int_{\mathbb{R}^N} \left(c_1 \int_{\mathbb{R}^N} \left(\prod_{i=1}^N |s_i|^{\alpha_i} \int_0^1 (1-\theta)^{m-1} \left|\Delta_{\theta s}^r f_\alpha(x)\right| d\theta \right) d\mu_{\xi_n}(s) \right)^p dx = \tag{25.12}$$

(Call $0 \leq \gamma_\alpha(s,x) := \prod_{i=1}^N |s_i|^{\alpha_i} \int_0^1 (1-\theta)^{m-1} \left|\Delta_{\theta s}^r f_\alpha(x)\right| d\theta$)

$$\int_{\mathbb{R}^N} \left(c_1 \int_{\mathbb{R}^N} \gamma_\alpha(s,x) \, d\mu_{\xi_n}(s) \right)^p dx =: I_2. \tag{25.13}$$

Therefore, it holds

$$I_2 \leq \int_{\mathbb{R}^N} \left(\left(N^m/(m-1)!\right)^{\frac{p}{q}} c_1 \int_{\mathbb{R}^N} \gamma_\alpha^p(s,x) \, d\mu_{\xi_n}(s) \right) dx =: I_3. \tag{25.14}$$

But we have

$$\gamma_\alpha(s,x) \leq \prod_{i=1}^N |s_i|^{\alpha_i} \left(\int_0^1 \left((1-\theta)^{m-1} \right)^q d\theta \right)^{\frac{1}{q}} \left(\int_0^1 \left|\Delta_{\theta s}^r f_\alpha(x)\right|^p d\theta \right)^{\frac{1}{p}}$$

$$= \prod_{i=1}^N |s_i|^{\alpha_i} \frac{1}{(q(m-1)+1)^{\frac{1}{q}}} \left(\int_0^1 \left|\Delta_{\theta s}^r f_\alpha(x)\right|^p d\theta \right)^{\frac{1}{p}} . \tag{25.15}$$

Hence, we get

$$\gamma_\alpha^p(s,x) \leq \prod_{i=1}^N |s_i|^{\alpha_i p} \frac{1}{(q(m-1)+1)^{\frac{p}{q}}} \left(\int_0^1 \left|\Delta_{\theta s}^r f_\alpha(x)\right|^p d\theta \right). \tag{25.16}$$

Thus, we obtain

$$I_3 \leq \int_{\mathbb{R}^N} \left(c_1^* \int_{\mathbb{R}^N} \prod_{i=1}^N |s_i|^{\alpha_i p} \left(\int_0^1 \left|\Delta_{\theta s}^r f_\alpha(x)\right|^p d\theta \right) d\mu_{\xi_n}(s) \right) dx \tag{25.17}$$

(where $c_1^* := (N^m/(m-1)!)^{\frac{p}{q}} \frac{1}{(q(m-1)+1)^{\frac{p}{q}}} c_1$)

$$= c_1^* \int_{\mathbb{R}^N} \left(\int_0^1 \left(\int_{\mathbb{R}^N} \prod_{i=1}^N |s_i|^{\alpha_i p} \left| \Delta_{\theta s}^r f_\alpha(x) \right|^p dx \right) d\theta \right) d\mu_{\xi_n}(s)$$

$$= c_1^* \int_{\mathbb{R}^N} \prod_{i=1}^N |s_i|^{\alpha_i p} \left(\int_0^1 \left(\int_{\mathbb{R}^N} \left| \Delta_{\theta s}^r f_\alpha(x) \right|^p dx \right) d\theta \right) d\mu_{\xi_n}(s) =: I_4. \tag{25.18}$$

Consequently, we derive

$$I_4 \leq c_1^* \int_{\mathbb{R}^N} \prod_{i=1}^N |s_i|^{\alpha_i p} \left(\int_0^1 \left(\omega_r \left(f_\alpha; \theta \|s\|_2 \right)_p \right)^p d\theta \right) d\mu_{\xi_n}(s)$$

$$\leq c_1^* \int_{\mathbb{R}^N} \left(\prod_{i=1}^N |s_i|^{\alpha_i p} \right) \omega_r \left(f_\alpha; \|s\|_2 \right)_p^p d\mu_{\xi_n}(s)$$

$$= c_1^* \int_{\mathbb{R}^N} \left(\prod_{i=1}^N |s_i|^{\alpha_i p} \right) \omega_r \left(f_\alpha; \xi_n \frac{\|s\|_2}{\xi_n} \right)_p^p d\mu_{\xi_n}(s) \tag{25.19}$$

(by $\omega_r(f, \lambda h)_p \leq (1+\lambda)^r \omega_r(f, h)_p$, for any $h, \lambda > 0$, $p \geq 1$)

$$\leq c_1^* \omega_r \left(f_\alpha; \xi_n \right)_p^p \left(\int_{\mathbb{R}^N} \left(\prod_{i=1}^N |s_i|^{\alpha_i p} \right) \left(1 + \frac{\|s\|_2}{\xi_n} \right)^{rp} d\mu_{\xi_n}(s) \right). \tag{25.20}$$

We have proved that

$$\int_{\mathbb{R}^N} \left| E_{r,n}^{[m]}(x) \right|^p dx$$

$$\leq c_1^* \omega_r \left(f_\alpha; \xi_n \right)_p^p \left(\int_{\mathbb{R}^N} \left[\left(\prod_{i=1}^N |s_i|^{\alpha_i} \right) \left(1 + \frac{\|s\|_2}{\xi_n} \right)^r \right]^p d\mu_{\xi_n}(s) \right). \tag{25.21}$$

Thus, we get ($p > 1$)

$$\left\| E_{r,n}^{[m]}(x) \right\|_p$$

$$\leq \left[c_1^* \left[\int_{\mathbb{R}^N} \left[\left(\prod_{i=1}^N |s_i|^{\alpha_i} \right) \left(1 + \frac{\|s\|_2}{\xi_n} \right)^r \right]^p d\mu_{\xi_n}(s) \right] \omega_r \left(f_\alpha; \xi_n \right)_p^p \right]^{\frac{1}{p}}. \tag{25.22}$$

We make

Remark 25.2. Note that $(p > 1)$

$$\int_{\mathbb{R}^N}\left(\prod_{i=1}^{N}|s_i|^{\alpha_i}\right)\left(1+\frac{\|s\|_2}{\xi_n}\right)^r d\mu_{\xi_n}(s)$$

$$\leq\left[\int_{\mathbb{R}^N}\left(\left(\prod_{i=1}^{N}|s_i|^{\alpha_i}\right)\left(1+\frac{\|s\|_2}{\xi_n}\right)^r\right)^p d\mu_{\xi_n}(s)\right]^{\frac{1}{p}}<\infty, \quad (25.23)$$

by assumption of Theorem 25.1.

As in Anastassiou (2011b) then we get that

$$\int_{\mathbb{R}^N}\prod_{i=1}^{N}|s_i|^{\alpha_i}\, d\mu_{\xi_n}(s)<\infty. \quad (25.24)$$

Hence $c_{\alpha,n,\widetilde{j}}\in\mathbb{R}$.

From the above, we have proved

Theorem 25.1. *Let $f\in C^m\left(\mathbb{R}^N\right)$, $m\in\mathbb{N}$, $N\geq 1$, with $f_\alpha\in L_p\left(\mathbb{R}^N\right)$, $|\alpha|=m$, $x\in\mathbb{R}^N$. Let $p,q>1:\frac{1}{p}+\frac{1}{q}=1$. Here μ_{ξ_n} is a Borel probability measure on $\mathbb{R}^N$ for $\xi_n>0$, $(\xi_n)_{n\in\mathbb{N}}$ bounded sequence. Assume for all $\alpha:=(\alpha_1,\ldots,\alpha_N)$, $\alpha_i\in\mathbb{Z}^+$, $i=1,\ldots,N$, $|\alpha|:=\sum_{i=1}^{N}\alpha_i=m$ that we have*

$$\int_{\mathbb{R}^N}\left(\left(\prod_{i=1}^{N}|s_i|^{\alpha_i}\right)\left(1+\frac{\|s\|_2}{\xi_n}\right)^r\right)^p d\mu_{\xi_n}(s)<\infty. \quad (25.25)$$

For $\widetilde{j}=1,\ldots,m$, and $\alpha:=(\alpha_1,\ldots,\alpha_N)$, $\alpha_i\in\mathbb{Z}^+$, $i=1,\ldots,N$, $|\alpha|:=\sum_{i=1}^{N}\alpha_i=\widetilde{j}$, call

$$c_{\alpha,n,\widetilde{j}}:=\int_{\mathbb{R}^N}\prod_{i=1}^{N}s_i^{\alpha_i}\, d\mu_{\xi_n}(s). \quad (25.26)$$

Then

$$\left\|E_{r,n}^{[m]}\right\|_p=\left\|\theta_{r,n}^{[m]}(f;x)-f(x)-\sum_{\widetilde{j}=1}^{m}\delta_{\widetilde{j},r}^{[m]}\left(\sum_{|\alpha|=\widetilde{j}}\frac{c_{\alpha,n,\widetilde{j}}f_\alpha(x)}{\left(\prod_{i=1}^{N}\alpha_i!\right)}\right)\right\|_{p,x}$$

$$\leq\left[\left(N^m/(m-1)!\right)^{\frac{p}{q}}\left(\frac{m}{(q(m-1)+1)^{\frac{p}{q}}}\right)\left(\sum_{|\alpha|=m}\frac{1}{\prod_{i=1}^{N}\alpha_i!}\right.\right.$$

$$\left.\left.\cdot\left[\int_{\mathbb{R}^N}\left[\left(\prod_{i=1}^{N}|s_i|^{\alpha_i}\right)\left(1+\frac{\|s\|_2}{\xi_n}\right)^r\right]^p d\mu_{\xi_n}(s)\right]\omega_r\left(f_\alpha,\xi_n\right)_p^p\right)\right]^{\frac{1}{p}}. \quad (25.27)$$

As $n \to \infty$ *and* $\xi_n \to 0$, *by (25.27), we obtain that* $\left\| E_{r,n}^{[m]} \right\|_p \to 0$ *with rates.*

One also gets by (25.27) that

$$\left\| \theta_{r,n}^{[m]} (f; x) - f(x) \right\|_{p,x}$$

$$\leq \sum_{\widetilde{j}=1}^{m} \left| \delta_{\widetilde{j},r}^{[m]} \right| \left(\sum_{|\alpha|=\widetilde{j}} \frac{\left| c_{\alpha,n,\widetilde{j}} \right|}{\prod_{i=1}^{N} \alpha_i!} \| f_\alpha \|_p \right) + R.H.S.\ (25.27), \tag{25.28}$$

given that $\| f_\alpha \|_p < \infty$, $|\alpha| = \widetilde{j}$, $\widetilde{j} = 1, \ldots, m$.

Assuming that $c_{\alpha,n,\widetilde{j}} \to 0$, $\xi_n \to 0$, *as* $n \to \infty$, *we get* $\left\| \theta_{r,n}^{[m]} (f) - f \right\|_p \to 0$, *that is,* $\theta_{r,n}^{[m]} \to I$ *the unit operator, in* L_p *norm, with rates.*

Inequality (25.27) provides a correction in the constants of the inequality in Theorem 4 in Anastassiou (2011b), p. 25.

(II) Here we treat the case of $f \in C^2(\mathbb{R}^N)$ with $f_\alpha \in L_p(\mathbb{R}^N)$, $|\alpha| = 2$, $p \geq 1$.

We make

Remark 25.3. The following are based on a new multivariate trigonometric Taylor formula (Anastassiou 2023b).

Let $s := (s_1, \ldots, s_N)$, $x := (x_1, \ldots, x_N)$, $z := (z_1, \ldots, z_N) := (x_1 + s_1 j,\ x_2 + s_2 j, \ldots x_N + s_N j) = x + sj$; $j = 0, 1, \ldots, r$, and $x := x_0 = (x_{01}, \ldots, x_{0N}) = (x_1, \ldots, x_N)$, all in $\mathbb{R}^N$; $p > 1$, $0 < \xi_n \leq 1$, $n \in \mathbb{N}$.

Here $f \in C^2(\mathbb{R}^N)$, $N \in \mathbb{N}$, and clearly all the mixed partials commute.

Consider

$$g_{x+sj}(t) := f(x + t(sj)), \quad 0 \leq t \leq 1. \tag{25.29}$$

Note that $g_{x+sj}(0) = f(x)$, $g_{x+sj}(1) = f(x + sj)$.

We have (by Anastassiou (2023b))

$$f(x+sj) - f(x) = g_{x+sj}(1) - g_{x+sj}(0)$$

$$= g'_{x+sj}(0) \sin(1) + 2 g''_{x+sj}(0) \sin^2\left(\frac{1}{2}\right)$$

$$+ \int_0^1 \left[\left(g''_{x+sj}(t) + g_{x+sj}(t) \right) - \left(g''_{x+sj}(0) + g_{x+sj}(0) \right) \right] \sin(1-t)\, dt$$

$$= \left(\sum_{i=1}^{N} (s_i j) \frac{\partial f}{\partial x_i}(x) \right) \sin(1)$$

$$+2\left\{\left[\left(\sum_{i=1}^{N}(s_i j)\frac{\partial}{\partial x_i}\right)^2 f\right](x)\right\}\sin^2\left(\frac{1}{2}\right)$$

$$+\int_0^1\left\{\left\{\left[\left(\sum_{i=1}^{N}(s_i j)\frac{\partial}{\partial x_i}\right)^2 f\right](x+t(sj))+f(x+t(sj))\right\}\right.$$

$$\left.-\left\{\left[\left(\sum_{i=1}^{N}(s_i j)\frac{\partial}{\partial x_i}\right)^2 f\right](x)+f(x)\right\}\right\}\sin(1-t)\,dt. \tag{25.30}$$

Denote the remainder $(j=0,1,\ldots,r)$

$$R_j:=\int_0^1\left\{\left\{\left[\left(\sum_{i=1}^{N}(s_i j)\frac{\partial}{\partial x_i}\right)^2 f\right](x+tsj)+f(x+tsj)\right\}\right.$$

$$\left.-\left\{\left[\left(\sum_{i=1}^{N}(s_i j)\frac{\partial}{\partial x_i}\right)^2 f\right](x)+f(x)\right\}\right\}\sin(1-t)\,dt$$

$$=\int_0^1\left\{\sum_{\substack{\alpha:=(\alpha_1,\ldots,\alpha_N),\alpha_i\in\mathbb{Z}_+\\ i=1,\ldots,N,|\alpha|:=\sum_{i=1}^N\alpha_i=2}}\left(\frac{2}{\prod_{i=1}^N\alpha_i!}\right)\left(\prod_{i=1}^N(js_i)^{\alpha_i}\right)\right.$$

$$\left.\times\left[f_\alpha(x+t(js))-f_\alpha(x)\right]+\left(f(x+t(js))-f(x)\right)\right\}\sin(1-t)\,dt. \tag{25.31}$$

Therefore, it holds

$$|R_j|\le\int_0^1\left\{\sum_{\substack{\alpha:=(\alpha_1,\ldots,\alpha_N),\alpha_i\in\mathbb{Z}_+\\ i=1,\ldots,N,|\alpha|:=\sum_{i=1}^N\alpha_i=2}}\left(\frac{2}{\prod_{i=1}^N\alpha_i!}\right)\left(\prod_{i=1}^N(j|s_i|)^{\alpha_i}\right)\right.$$

$$\left.|f_\alpha(x+tsj)-f_\alpha(x)|+|f(x+tsj)-f(x)|\right\}|\sin(1-t)|\,dt. \tag{25.32}$$

By $|\sin x| \leq |x|$, $x \in \mathbb{R}$, we get

$$|R_j| \leq \int_0^1 (1-t) \left\{ \left(\sum_{\substack{\alpha:=(\alpha_1,\ldots,\alpha_N),\alpha_i\in\mathbb{Z}_+ \\ i=1,\ldots,N,|\alpha|:=\sum_{i=1}^N \alpha_i=2}} \left(\frac{2}{\prod_{i=1}^N \alpha_i!} \right) \left(\prod_{i=1}^N (j\,|s_i|)^{\alpha_i} \right) |f_\alpha(x+tsj) - f_\alpha(x)| \right) + |f(x+tsj) - f(x)| \right\} dt. \tag{25.33}$$

Next we can write

$$\sum_{j=0}^r \alpha_j [f(x+sj) - f(x)] - \left(\sum_{j=0}^r \alpha_j j \right) \left(\sum_{i=1}^N s_i \frac{\partial f}{\partial x_i}(x) \right) \sin(1) - 2 \left(\sum_{j=0}^r \alpha_j j^2 \right) \left\{ \left[\left(\sum_{i=1}^N s_i \frac{\partial}{\partial x_i} \right)^2 f \right](x) \right\} \sin^2\left(\frac{1}{2}\right) = \sum_{j=0}^r \alpha_j R_j. \tag{25.34}$$

Call

$$R := \sum_{j=0}^r \alpha_j R_j. \tag{25.35}$$

Clearly, it holds

$$\theta_n(f,x) - f(x) = \sum_{j=0}^r \alpha_j \int_{\mathbb{R}^N} (f(x+sj) - f(x))\, d\mu_{\xi_n}(s). \tag{25.36}$$

Denote ($n \in \mathbb{N}$)

$$\Delta_n(x) := \theta_n(f,x) - f(x) - \left(\sum_{j=0}^r \alpha_j j \right) \sin(1) \left[\sum_{i=1}^N \frac{\partial f(x)}{\partial x_i} \left(\int_{\mathbb{R}^N} s_i\, d\mu_\xi(s) \right) \right] - 2 \left(\sum_{j=0}^r \alpha_j j^2 \right) \sin^2\left(\frac{1}{2}\right) \left\{ \sum_{i=1}^N \left(\int_{\mathbb{R}^N} s_i^2\, d\mu_\xi(s) \right) \frac{\partial^2 f(x)}{\partial x_i^2} + \sum_{\substack{i\neq j^*, \\ i,j^*\in\{1,\ldots,N\}}} \left(\int_{\mathbb{R}^N} s_i s_{j^*}\, d\mu_\xi(s) \right) \frac{\partial^2 f(x)}{\partial x_i \partial x_{j^*}} \right\}. \tag{25.37}$$

We observe that

$$\Delta_n(x) = \sum_{j=0}^{r} \alpha_j \int_{\mathbb{R}^N} R_j d\mu_{\xi_n}(s) = \int_{\mathbb{R}^N} R d\mu_{\xi_n}(s). \tag{25.38}$$

Let $p, q > 1 : \frac{1}{p} + \frac{1}{q} = 1$. Then

$$\begin{aligned} |\Delta_n(x)|^p &= \left| \int_{\mathbb{R}^N} R d\mu_{\xi_n}(s) \right|^p \leq \left(\int_{\mathbb{R}^N} |R| \, d\mu_{\xi_n}(s) \right)^p \\ &\leq \left(\int_{\mathbb{R}^N} \left(\sum_{j=0}^{r} |\alpha_j| \, |R_j| \right) d\mu_{\xi_n}(s) \right)^p \\ &= \left(\sum_{j=0}^{r} |\alpha_j| \left(\int_{\mathbb{R}^N} |R_j| \, d\mu_{\xi_n}(s) \right) \right)^p \leq \end{aligned} \tag{25.39}$$

(by Hölder's inequality)

$$\left(\sum_{j=0}^{r} |\alpha_j|^q \right)^{\frac{p}{q}} \left(\sum_{j=0}^{r} \left(\int_{\mathbb{R}^N} |R_j| \, d\mu_{\xi_n}(s) \right)^p \right) \leq$$

(again by Hölder's inequality)

$$\left(\sum_{j=0}^{r} |\alpha_j|^q \right)^{\frac{p}{q}} \left(\sum_{j=0}^{r} \left(\int_{\mathbb{R}^N} |R_j|^p \, d\mu_{\xi_n}(s) \right) \right).$$

Thus, we have

$$|\Delta_n(x)|^p \leq \left(\sum_{j=0}^{r} |\alpha_j|^q \right)^{\frac{p}{q}} \left(\sum_{j=0}^{r} \left(\int_{\mathbb{R}^N} |R_j|^p \, d\mu_{\xi_n}(s) \right) \right). \tag{25.40}$$

By (25.33), we get that

$$|R_j| \leq 2 \left(\int_0^1 (1-t)^q \, dt \right)^{\frac{1}{q}} \left\{ \int_0^1 \left[\left(\sum_{|\alpha|=2} \left(\frac{1}{\prod_{i=1}^{N} \alpha_i!} \right) \left(\prod_{i=1}^{N} (j |s_i|)^{\alpha_i} \right) \right. \right. \right.$$

$$\left. |f_\alpha(x+tsj) - f_\alpha(x)| \right) + |f(x+tsj) - f(x)| \Big]^p \, dt \Big\}^{\frac{1}{p}}$$

$$\leq \frac{2}{(q+1)^{\frac{1}{q}}} \left\{ \left\{ \sum_{|\alpha|=2} \left(\left(\left(\frac{1}{\prod_{i=1}^{N} \alpha_i!} \right) \left(\prod_{i=1}^{N} (j|s_i|)^{\alpha_i} \right) \right) \right)^p \right)^{\frac{1}{p}} \right.$$

$$\left. \left(\int_0^1 |f_\alpha(x+tsj) - f_\alpha(x)|^p \, dt \right)^{\frac{1}{p}} \right\} + \left(\int_0^1 |f(x+tsj) - f(x)|^p \, dt \right)^{\frac{1}{p}} \right\}$$

$$\leq \frac{2}{(q+1)^{\frac{1}{q}}} \left(\frac{N(N+1)+2}{2} \right)^{\frac{1}{q}} \left\{ \sum_{|\alpha|=2} \left(\left(\frac{1}{\prod_{i=1}^{N} \alpha_i!} \right) \left(\prod_{i=1}^{N} (j|s_i|)^{\alpha_i} \right) \right)^p \right.$$

$$\left. \left(\int_0^1 |f_\alpha(x+tsj) - f_\alpha(x)|^p \, dt \right) + \left(\int_0^1 |f(x+tsj) - f(x)|^p \, dt \right) \right\}^{\frac{1}{p}}. \tag{25.41}$$

Hence

$$|R_j|^p \leq \frac{2^p}{(q+1)^{\frac{p}{q}}} \left(\frac{N(N+1)+2}{2} \right)^{\frac{p}{q}}$$

$$\left\{ \sum_{|\alpha|=2} \left(\left(\frac{1}{\prod_{i=1}^{N} \alpha_i!} \right) \left(\prod_{i=1}^{N} (j|s_i|)^{\alpha_i} \right) \right)^p \right.$$

$$\left. \left(\int_0^1 |f_\alpha(x+tsj) - f_\alpha(x)|^p \, dt \right) + \left(\int_0^1 |f(x+tsj) - f(x)|^p \, dt \right) \right\}. \tag{25.42}$$

By (25.40) and (25.42), we continue as follows:

$$|\Delta_n(x)|^p \leq \left(\sum_{j=0}^{r} |\alpha_j|^q \right)^{\frac{p}{q}} \left\{ \frac{2^p}{(q+1)^{\frac{p}{q}}} \left(\frac{N(N+1)+2}{2} \right)^{\frac{p}{q}} \right\}$$

$$\left\{ \sum_{j=0}^{r} \left[\sum_{|\alpha|=2} \left(\frac{1}{\prod_{i=1}^{N} \alpha_i!} \right)^p \right. \right.$$

$$\left(\int_{\mathbb{R}^N} \left(\prod_{i=1}^{N} (j|s_i|)^{\alpha_i} \right)^p \left(\int_0^1 |f_\alpha(x+tsj) - f_\alpha(x)|^p \, dt \right) d\mu_{\xi_n}(s) \right)$$

$$\left. \left. + \left(\int_{\mathbb{R}^N} \left(\int_0^1 |f(x+tsj) - f(x)|^p \, dt \right) d\mu_{\xi_n}(s) \right) \right] \right\}. \tag{25.43}$$

Furthermore, it holds

$$\int_{\mathbb{R}^N} |\Delta_n(x)|^p dx \leq \left(\sum_{j=0}^{r} |\alpha_j|^q\right)^{\frac{p}{q}} \left\{\frac{2^p}{(q+1)^{\frac{p}{q}}} \left(\frac{N(N+1)+2}{2}\right)^{\frac{p}{q}}\right\}$$

$$\left\{\sum_{j=0}^{r} \left[\sum_{|\alpha|=2} \left(\frac{1}{\prod_{i=1}^{N} \alpha_i!}\right)^p\right.\right.$$

$$\left(\int_{\mathbb{R}^N} \left(\prod_{i=1}^{N} (j|s_i|)^{\alpha_i}\right)^p \left(\int_0^1 \left(\int_{\mathbb{R}^N} |f_\alpha(x+tsj) - f_\alpha(x)|^p dx\right) dt\right) d\mu_{\xi_n}(s)\right)$$

$$\left.\left. + \left(\int_{\mathbb{R}^N} \left(\int_0^1 \left(\int_{\mathbb{R}^N} |f(x+tsj) - f(x)|^p dx\right) dt\right) d\mu_{\xi_n}(s)\right)\right]\right\} \tag{25.44}$$

$$\leq \left(\sum_{j=0}^{r} |\alpha_j|^q\right)^{\frac{p}{q}} \left\{\frac{2^p}{(q+1)^{\frac{p}{q}}} \left(\frac{N(N+1)+2}{2}\right)^{\frac{p}{q}}\right\}$$

$$\left\{\sum_{j=0}^{r} \left[\sum_{|\alpha|=2} \left(\frac{1}{\prod_{i=1}^{N} \alpha_i!}\right)^p\right.\right.$$

$$\left(\int_{\mathbb{R}^N} \left(\prod_{i=1}^{N} (j|s_i|)^{\alpha_i}\right)^p \left(\int_0^1 \omega_1(f_\alpha, jt\|s\|_2)_p^p dt\right) d\mu_{\xi_n}(s)\right)$$

$$\left.\left. + \left(\int_{\mathbb{R}^N} \left(\int_0^1 \omega_1(f_\alpha, jt\|s\|_2)_p^p dt\right) d\mu_{\xi_n}(s)\right)\right]\right\}$$

$$\leq \left(\sum_{j=0}^{r} |\alpha_j|^q\right)^{\frac{p}{q}} \left\{\frac{2^p}{(q+1)^{\frac{p}{q}}} \left(\frac{N(N+1)+2}{2}\right)^{\frac{p}{q}}\right\}$$

$$\left\{\sum_{j=0}^{r} \left[\sum_{|\alpha|=2} \left(\frac{1}{\prod_{i=1}^{N} \alpha_i!}\right)^p \left(\int_{\mathbb{R}^N} \left(\prod_{i=1}^{N} (j|s_i|)^{\alpha_i}\right)^p\right.\right.\right.$$

$$\times \omega_1\left(f_\alpha, \frac{\xi_n j\|s\|_2}{\xi_n}\right)_p^p d\mu_{\xi_n}(s)\right) \tag{25.45}$$

$$+\left(\int_{\mathbb{R}^N}\omega_1\left(f,\frac{\xi_n j\|s\|_2}{\xi_n}{}_2\right)_p^p d\mu_{\xi_n}(s)\right)\Bigg]\Bigg\}$$

$$\leq\left(\sum_{j=0}^r|\alpha_j|^q\right)^{\frac{p}{q}}\left\{\frac{2^p}{(q+1)^{\frac{p}{q}}}\left(\frac{N(N+1)+2}{2}\right)^{\frac{p}{q}}\right\}$$

$$\left\{\sum_{j=0}^r\left[\sum_{|\alpha|=2}\left(\frac{1}{\prod_{i=1}^N\alpha_i!}\right)^p\omega_1(f_\alpha,\xi_n)_p^p\right.\right.$$

$$\times\left(\int_{\mathbb{R}^N}\left(\prod_{i=1}^N(j|s_i|)^{\alpha_i}\right)^p\left(1+\frac{j\|s\|_2}{\xi_n}\right)^p d\mu_{\xi_n}(s)\right) \tag{25.46}$$

$$\left.\left.+\omega_1(f,\xi_n)_p^p\left(\int_{\mathbb{R}^N}\left(1+\frac{j\|s\|_2}{\xi_n}\right)^p d\mu_{\xi_n}(s)\right)\right]\right\}$$

$$\leq\left(\sum_{j=0}^r|\alpha_j|^q\right)^{\frac{p}{q}}\left\{\frac{2^p}{(q+1)^{\frac{p}{q}}}\left(\frac{N(N+1)+2}{2}\right)^{\frac{p}{q}}\right\}$$

$$\left\{\sum_{j=0}^r\left[\sum_{|\alpha|=2}\left(\frac{1}{\prod_{i=1}^N\alpha_i!}\right)^p\omega_1(f_\alpha,\xi_n)_p^p\right.\right.$$

$$\times\left(\int_{\mathbb{R}^N}\left(\prod_{i=1}^N(j|s_i|)^{\alpha_i}\right)^p\left(1+\frac{j^p\|s\|_2^p}{\xi_n^p}\right)d\mu_{\xi_n}(s)\right)$$

$$\left.\left.+\omega_1(f,\xi_n)_p^p\left(\int_{\mathbb{R}^N}\left(1+\frac{j^p\|s\|_2^p}{\xi_n^p}\right)d\mu_{\xi_n}(s)\right)\right]\right\}. \tag{25.47}$$

Then it holds

$$\|\Delta_n(x)\|_p\leq\left(\sum_{j=0}^r|\alpha_j|^q\right)^{\frac{1}{q}}\left\{\frac{2}{(q+1)^{\frac{1}{q}}}\left(\frac{N(N+1)+2}{2}\right)^{\frac{1}{q}}\right\}$$

$$\left\{\sum_{j=0}^r\left[j^{2p}\sum_{|\alpha|=2}\left(\frac{1}{\prod_{i=1}^N\alpha_i!}\right)^p\omega_1(f_\alpha,\xi_n)_p^p\right.\right.$$

$$\times\left(\int_{\mathbb{R}^N}\left(\prod_{i=1}^{N}|s_i|^{p\alpha_i}\right)\left(1+\frac{j^p\|s\|_2^p}{\xi_n^p}\right)d\mu_{\xi_n}(s)\right)$$

$$+\omega_1(f,\xi_n)_p^p\left(\int_{\mathbb{R}^N}\left(1+\frac{j^p\|s\|_2^p}{\xi_n^p}\right)d\mu_{\xi_n}(s)\right)\Bigg]\Bigg\}^{\frac{1}{p}}. \quad (25.48)$$

We have proved the following trigonometric-induced alternative L_p approximation result for θ_n operators.

Theorem 25.2. *Let $p,q>1:\frac{1}{p}+\frac{1}{q}=1$, $0<\xi_n\le 1$, $n\in\mathbb{N}$. Here we deal with $f\in C^2(\mathbb{R}^N)$, $N\ge 1$, with f, $f_\alpha\in L_p(\mathbb{R}^N)$, $|\alpha|=2$, where $\alpha_i\in\mathbb{Z}^+$, $i=1,\ldots,N$, and $|\alpha|=\sum_{i=1}^{N}\alpha_i$; $x\in\mathbb{R}^N$. Let μ_{ξ_n} be a Borel probability measure on $\mathbb{R}^N$. Suppose that for all $\alpha:|\alpha|=2$, $j=0,1,\ldots,r$, we have that both*

$$I_{1j}(\alpha):=\int_{\mathbb{R}^N}\left(\prod_{i=1}^{N}|s_i|^{p\alpha_i}\right)\left(1+\frac{j^p\|s\|_2^p}{\xi_n^p}\right)d\mu_{\xi_n}(s), \quad (25.49)$$

$$I_2:=\int_{\mathbb{R}^N}\left(\frac{\|s\|_2}{\xi_n}\right)^p d\mu_{\xi_n}(s) \quad (25.50)$$

are uniformly bounded in $\xi_n\in(0,1]$.

Denote ($n\in\mathbb{N}$)

$$\Delta_n(x):=\theta_n(f,x)-f(x)-\left(\sum_{j=0}^{r}\alpha_j j\right)\sin(1)\left[\sum_{i=1}^{N}\frac{\partial f(x)}{\partial x_i}\left(\int_{\mathbb{R}^N}s_i d\mu_{\xi_n}(s)\right)\right]$$

$$-2\left(\sum_{j=0}^{r}\alpha_j j^2\right)\sin^2\left(\frac{1}{2}\right)\left\{\sum_{i=1}^{N}\left(\int_{\mathbb{R}^N}s_i^2 d\mu_{\xi_n}(s)\right)\frac{\partial^2 f(x)}{\partial x_i^2}\right.$$

$$\left.+\sum_{\substack{i\neq j^*,\\ i,j^*\in\{1,\ldots,N\}}}\left(\int_{\mathbb{R}^N}s_i s_{j^*}d\mu_{\xi_n}(s)\right)\frac{\partial^2 f(x)}{\partial x_i\partial x_{j^*}}\right\}. \quad (25.51)$$

Then, it holds

$$\|\Delta_n(x)\|_p \leq \left(\sum_{j=0}^{r} |\alpha_j|^q\right)^{\frac{1}{q}} \left\{\frac{2}{(q+1)^{\frac{1}{q}}} \left(\frac{N(N+1)+2}{2}\right)^{\frac{1}{q}}\right\}$$

$$\left\{\sum_{j=0}^{r} \left[j^{2p} \sum_{|\alpha|=2} \left(\frac{1}{\prod_{i=1}^{N} \alpha_i!}\right)^p \omega_1(f_\alpha, \xi_n)_p^p \right.\right.$$

$$\times \left(\int_{\mathbb{R}^N} \left(\prod_{i=1}^{N} |s_i|^{p\alpha_i}\right) \left(1 + \frac{j^p \|s\|_2^p}{\xi_n^p}\right) d\mu_{\xi_n}(s)\right)$$

$$\left.\left. + \omega_1(f, \xi_n)_p^p \left(\int_{\mathbb{R}^N} \left(1 + \frac{j^p \|s\|_2^p}{\xi_n^p}\right) d\mu_{\xi_n}(s)\right)\right]\right\}^{\frac{1}{p}}. \quad (25.52)$$

As $n \to \infty$ and $\xi_n \to 0$, by (25.52), we obtain that $\|\Delta_n\|_p \to 0$ with rates. One also gets by (25.52) that

$$\|\theta_n(f, x) - f(x)\|_{p,x}$$

$$\leq \left(\sum_{j=0}^{r} |\alpha_j| j\right) \sin(1) \left[\sum_{i=1}^{N} \left\|\frac{\partial f}{\partial x_i}\right\|_p \left(\int_{\mathbb{R}^N} |s_i| d\mu_{\xi_n}(s)\right)\right]$$

$$+ 2 \left(\sum_{j=0}^{r} |\alpha_j| j^2\right) \sin^2\left(\frac{1}{2}\right) \left\{\sum_{i=1}^{N} \left(\int_{\mathbb{R}^N} s_i^2 d\mu_{\xi_n}(s)\right) \left\|\frac{\partial^2 f}{\partial x_i^2}\right\|_p\right.$$

$$\left. + \sum_{\substack{i \neq j^*, \\ i, j^* \in \{1, \ldots, N\}}} \left(\int_{\mathbb{R}^N} |s_i s_{j^*}| d\mu_{\xi_n}(s)\right) \left\|\frac{\partial^2 f}{\partial x_i \partial x_{j^*}}\right\|_p\right\} + R.H.S.\ (25.52), \quad (25.53)$$

given that $\|f_\alpha\|_p < \infty$, $|\alpha| = \widetilde{j}$, $\widetilde{j} = 1, 2$. Assuming that $\int_{\mathbb{R}^N} s_i^2 d\mu_{\xi_n}(s)$, $\int_{\mathbb{R}^N} |s_i s_{j^*}| d\mu_{\xi_n}(s)$, $i, j^* \in \{1, \ldots, N\}$, $i \neq j^*$, converge to zero as $\xi_n \to 0$, we get $\|\theta_n(f, x) - f(x)\|_p \to 0$, that is $\theta_n \to I$ the unit operator, in L_p norm, with rates.

We make

Remark 25.4. It follows the L_1 approximation by θ_n operators, i.e. $p = 1$ case. They are all the same, as in Remark 25.3, till including (25.38):

$$\Delta_n(x) = \sum_{j=0}^{r} \alpha_j \int_{\mathbb{R}^N} R_j d\mu_{\xi_n}(s) = \int_{\mathbb{R}^N} R d\mu_{\xi_n}(s).$$

Hence it holds

$$|\Delta_n(x)| \le \left(\sum_{j=0}^{r} |\alpha_j| \left(\int_{\mathbb{R}^N} |R_j| \, d\mu_{\xi_n}(s) \right) \right)$$

$$\overset{\text{(by (25.33))}}{\le} \left\{ \sum_{j=0}^{r} |\alpha_j| \int_{\mathbb{R}^N} \left(\int_0^1 \left\{ \sum_{|\alpha|=2} \left(\frac{2}{\prod_{i=1}^{N} \alpha_i!} \right) \left(\prod_{i=1}^{N} (j|s_i|)^{\alpha_i} \right) \right. \right. \right.$$

$$\left. \left. \left. |f_\alpha(x+tsj) - f_\alpha(x)| \right) + |f(x+tsj) - f(x)| \right\} dt \right) d\mu_{\xi_n}(s) \Big\}. \tag{25.54}$$

Therefore, we have

$$\int_{\mathbb{R}^N} |\Delta_n(x)| \, dx$$

$$\le \left\{ \sum_{j=0}^{r} |\alpha_j| \left[\left(\int_{\mathbb{R}^N} \left(\int_0^1 \left\{ \sum_{|\alpha|=2} \left(\frac{2}{\prod_{i=1}^{N} \alpha_i!} \right) \left(\prod_{i=1}^{N} (j|s_i|)^{\alpha_i} \right) \right. \right. \right. \right. \right.$$

$$\left. \left. \left. \left(\int_{\mathbb{R}^N} |f_\alpha(x+tsj) - f_\alpha(x)| \, dx \right) dt \right\} d\mu_{\xi_n}(s) \right)$$

$$\left. \left. + \left(\int_{\mathbb{R}^N} \left(\int_0^1 \left(\int_{\mathbb{R}^N} |f(x+tsj) - f(x)| \, dx \right) dt \right) d\mu_{\xi_n}(s) \right) \right] \right\}$$

$$\le \left\{ \sum_{j=0}^{r} |\alpha_j| \left[\left(\int_{\mathbb{R}^N} \left(\int_0^1 \left\{ \sum_{|\alpha|=2} \left(\frac{2}{\prod_{i=1}^{N} \alpha_i!} \right) \left(\prod_{i=1}^{N} (j|s_i|)^{\alpha_i} \right) \right. \right. \right. \right. \right.$$

$$\omega_1\left(f_\alpha, jt\|s\|_2\right)_1 dt \Big\} d\mu_{\xi_n}(s) \Big) \tag{25.55}$$

$$\left. \left. + \left(\int_{\mathbb{R}^N} \left(\int_0^1 \omega_1\left(f, jt\|s\|_2\right)_1 dt \right) d\mu_{\xi_n}(s) \right) \right] \right\}$$

$$\leq \left\{ \sum_{j=0}^{r} |\alpha_j| \left[\sum_{|\alpha|=2} \left(\frac{2}{\prod_{i=1}^{N} \alpha_i!} \right) \left(\int_{\mathbb{R}^N} \left(\prod_{i=1}^{N} (j |s_i|)^{\alpha_i} \right) \right. \right. \right.$$

$$\left. \left. \left. \omega_1 \left(f_\alpha, \frac{\xi_n j \|s\|_2}{\xi_n} \right)_1 d\mu_{\xi_n}(s) \right) + \left(\int_{\mathbb{R}^N} \omega_1 \left(f, \frac{\xi_n j \|s\|_2}{\xi_n} \right)_1 d\mu_{\xi_n}(s) \right) \right] \right\}$$

$$\leq \left\{ \sum_{j=0}^{r} |\alpha_j| \left[\sum_{|\alpha|=2} \left(\frac{2}{\prod_{i=1}^{N} \alpha_i!} \right) \omega_1 (f_\alpha, \xi_n)_1 \left(\int_{\mathbb{R}^N} \left(\prod_{i=1}^{N} (j |s_i|)^{\alpha_i} \right) \right. \right. \right.$$

$$\left. \left. \left. \left(1 + \frac{j \|s\|_2}{\xi_n} \right) d\mu_{\xi_n}(s) \right) + \omega_1 (f, \xi_n)_1 \left(\int_{\mathbb{R}^N} \left(1 + \frac{j \|s\|_2}{\xi_n} \right) d\mu_{\xi_n}(s) \right) \right] \right\}. \tag{25.56}$$

Then, it holds

$$\|\Delta_n(x)\|_1 \leq \left\{ \sum_{j=0}^{r} |\alpha_j| \left[j^2 \sum_{|\alpha|=2} \left(\frac{2}{\prod_{i=1}^{N} \alpha_i!} \right) \omega_1(f_\alpha, \xi_n)_1 \left(\int_{\mathbb{R}^N} \left(\prod_{i=1}^{N} (|s_i|)^{\alpha_i} \right) \right. \right. \right.$$

$$\left. \left. \left. \left(1 + \frac{j \|s\|_2}{\xi_n} \right) d\mu_{\xi_n}(s) \right) + \omega_1 (f, \xi_n)_1 \left(\int_{\mathbb{R}^N} \left(1 + \frac{j \|s\|_2}{\xi_n} \right) d\mu_{\xi_n}(s) \right) \right] \right\}. \tag{25.57}$$

We have proved the following trigonometric-based alternative L_1 approximation result for θ_n operators.

Theorem 25.3. *Let $0 < \xi_n \leq 1$, $n \in \mathbb{N}$, $x \in \mathbb{R}^N$. Here we deal with $f \in C^2(\mathbb{R}^N)$, $N \geq 1$, with $f, f_\alpha \in L_1(\mathbb{R}^N)$, $|\alpha| = 2$, where $\alpha_i \in \mathbb{Z}^+$, $i = 1, \ldots, N$, and $|\alpha| = \sum_{i=1}^{N} \alpha_i$. Let $\mu_{\xi n}$ be a Borel probability measure on $\mathbb{R}^N$. Suppose that for all $\alpha : |\alpha| = 2$, $j = 0, 1, \ldots, r$, we have that both*

$$I_{1j}^*(\alpha) := \int_{\mathbb{R}^N} \left(\prod_{i=1}^{N} |s_i|^{\alpha_i} \right) \left(1 + \frac{j \|s\|_2}{\xi_n} \right) d\mu_{\xi_n}(s) \tag{25.58}$$

and

$$I_2^* := \int_{\mathbb{R}^N} \frac{\|s\|_2}{\xi_n} d\mu_{\xi_n}(s) \tag{25.59}$$

are uniformly bounded in $\xi_n \in (0, 1]$.

Here $\Delta_n(x)$ is as in (25.51).

Then, it holds

$$\|\Delta_n(x)\|_1 \leq \left\{\sum_{j=0}^{r}|\alpha_j|\left[j^2\sum_{|\alpha|=2}\left(\frac{2}{\prod_{i=1}^{N}\alpha_i!}\right)\omega_1(f_\alpha,\xi_n)_1\left(\int_{\mathbb{R}^N}\left(\prod_{i=1}^{N}(|s_i|)^{\alpha_i}\right)\right.\right.\right.$$

$$\left.\left.\left.\left(1+\frac{j\|s\|_2}{\xi_n}\right)d\mu_{\xi_n}(s)\right)+\omega_1(f,\xi_n)_1\left(\int_{\mathbb{R}^N}\left(1+\frac{j\|s\|_2}{\xi_n}\right)d\mu_{\xi_n}(s)\right)\right]\right\}. \tag{25.60}$$

As $n\to\infty$ and $\xi_n\to 0$, by (25.60), we obtain that $\|\Delta_n\|_1\to 0$ with rates. One also obtains by (25.60) that

$$\|\theta_n(f)-f\|_1$$

$$\leq\left(\sum_{j=0}^{r}|\alpha_j|\,j\right)\sin(1)\left[\sum_{i=1}^{N}\left\|\frac{\partial f}{\partial x_i}\right\|_1\left(\int_{\mathbb{R}^N}|s_i|\,d\mu_{\xi_n}(s)\right)\right]$$

$$+2\left(\sum_{j=0}^{r}|\alpha_j|\,j^2\right)\sin^2\left(\frac{1}{2}\right)\left\{\sum_{i=1}^{N}\left(\int_{\mathbb{R}^N}s_i^2d\mu_{\xi_n}(s)\right)\left\|\frac{\partial^2 f}{\partial x_i^2}\right\|_1\right.$$

$$\left.+\sum_{\substack{i\neq j^*,\\ i,j^*\in\{1,\ldots,N\}}}\left(\int_{\mathbb{R}^N}|s_is_{j^*}|\,d\mu_{\xi_n}(s)\right)\left\|\frac{\partial^2 f}{\partial x_i\partial x_{j^*}}\right\|_1\right\}+R.H.S.\ (25.60), \tag{25.61}$$

given that $\|f_\alpha\|_1<\infty$, $|\alpha|=\widetilde{j}$, $\widetilde{j}=1,2$. Assuming that $\int_{\mathbb{R}^N}s_i^2d\mu_{\xi_n}(s)$, $\int_{\mathbb{R}^N}|s_is_{j^*}|\,d\mu_{\xi_n}(s)$, $i,j^*\in\{1,\ldots,N\}$, $i\neq j^*$, converge to zero as $\xi_n\to 0$, we derive $\|\theta_n(f)-f\|_1\to 0$, that is, $\theta_n\to I$ in L_1 norm, with rates.

We make

Remark 25.5. Next we apply Theorems 4, 25.2 and 25.3 to the specific multivariate smooth Picard singular integral operators

$$P_n(f;x_1,\ldots,x_N):=P_{r,n}^{[m]}(f;x_1,\ldots,x_N):=\frac{1}{(2\xi_n)^N}\sum_{j=0}^{r}\alpha_{j,r}^{[m]}$$

$$\times\int_{\mathbb{R}^N}f(x_1+s_1j,x_2+s_2j,\ldots,x_N+s_Nj)\,e^{-\frac{\sum_{i=1}^{N}|s_i|}{\xi_n}}ds_1\cdots ds_N, \tag{25.62}$$

$r,n\in\mathbb{N}$, $m\in\mathbb{Z}_+$, $0<\xi_n\leq 1$.

Clearly here it is

$$d\mu_{\xi_n}(s) = \frac{1}{(2\xi_n)^N} e^{-\frac{\sum_{i=1}^N |s_i|}{\xi_n}} ds_1 \cdots ds_N, \quad s \in \mathbb{R}^N. \tag{25.63}$$

We observe that

$$\frac{1}{(2\xi_n)^N} \int_{\mathbb{R}^N} e^{-\frac{\sum_{i=1}^N |s_i|}{\xi_n}} ds_1 \cdots ds_N = 1, \tag{25.64}$$

see Anastassiou (2021), Chap. 22, p. 356.

We need

Theorem 25.4. *(Anastassiou 2021, p. 361) Let $r, N \in \mathbb{N}$, $p > 1$, $\alpha_i \in \mathbb{Z}^+$, $i = 1, \ldots, N : |\alpha| := \sum_{i=1}^N \alpha_i = m \in \mathbb{N}$, $\xi_n \in (0, 1]$, $n \in \mathbb{N}$. Then*

$$\begin{aligned} A_{\xi_n}(\alpha) &:= \frac{1}{(2\xi_n)^N} \int_{\mathbb{R}^N} \left(\left(\prod_{i=1}^N |s_i|^{\alpha_i} \right) \left(1 + \frac{\|s\|_2}{\xi_n} \right)^r \right)^p e^{-\frac{\sum_{i=1}^N |s_i|}{\xi_n}} ds_1 \cdots ds_N \\ &\leq \xi^{mp} \left[(1+N)^{rp} + 2^{rp} \Gamma^N ((m+r)p+1, 1) \right] \\ &\leq (1+N)^{rp} + 2^{rp} \Gamma^N ((m+r)p+1, 1) < +\infty \end{aligned} \tag{25.65}$$

are uniformly bounded.

Above $\Gamma(\cdot, \cdot)$ is the upper incomplete gamma function.

Clearly, it holds

$$B_{\xi_n}(\alpha) := \frac{1}{(2\xi_n)^N} \int_{\mathbb{R}^N} \left(\prod_{i=1}^N |s_i|^{\alpha_i} \right) e^{-\frac{\sum_{i=1}^N |s_i|}{\xi_n}} ds_1 \cdots ds_N < \infty, \tag{25.66}$$

uniformly bounded in $\xi_n \in (0, 1]$. And clearly $B_{\xi_n}(\alpha) \to 0$, as $\xi_n \to 0$, $n \to \infty$, by (25.65). It follows an application of Theorem 25.1.

Theorem 25.5. *Let $f \in C^m(\mathbb{R}^N)$, $m \in \mathbb{N}$, $N \geq 1$, with $f_\alpha \in L_p(\mathbb{R}^N)$, $|\alpha| = m$, $x \in \mathbb{R}^N$. Let $p, q > 1 : \frac{1}{p} + \frac{1}{q} = 1$; $\xi_n \in (0, 1]$, $n \in \mathbb{N}$. For $\widetilde{j} = 1, \ldots, m$, and $\alpha := (\alpha_1, \ldots, \alpha_N)$, $\alpha_i \in \mathbb{Z}^+$, $i = 1, \ldots, N$, $|\alpha| := \sum_{i=1}^N \alpha_i = \widetilde{j}$, call*

$$\overline{c}_{\alpha,n,\widetilde{j}} := \frac{1}{(2\xi_n)^N} \int_{\mathbb{R}^N} \left(\prod_{i=1}^N s_i^{\alpha_i} \right) e^{-\frac{\sum_{i=1}^N |s_i|}{\xi_n}} ds_1 \cdots ds_N. \tag{25.67}$$

Then

$$\left\|\overline{E}_{r,n}^{[m]}\right\|_p := \left\| P_{r,n}^{[m]}(f;x) - f(x) - \sum_{\widetilde{j}=1}^{m} \delta_{\widetilde{j},r}^{[m]} \left(\sum_{|\alpha|=\widetilde{j}} \frac{\overline{c}_{\alpha,n,\widetilde{j}} f_\alpha(x)}{\left(\prod_{i=1}^N \alpha_i!\right)} \right) \right\|_{p,x}$$

$$\leq \left[\left(N^m/(m-1)!\right)^{\frac{p}{q}} \left(\frac{m}{\left(q(m-1)+1\right)^{\frac{p}{q}}} \right) \left(\sum_{|\alpha|=m} \frac{1}{\prod_{i=1}^N \alpha_i!} \right.\right.$$

$$\left.\left. \cdot A_{\xi_n}(\alpha)\, \omega_r(f_\alpha, \xi_n)_p^p \right) \right]^{\frac{1}{p}}, \tag{25.68}$$

where $A_{\xi_n}(\alpha)$ as in (25.65).

As $n \to \infty$ and $\xi_n \to 0$, by (25.68), we obtain that $\left\|\overline{E}_{r,n}^{[m]}\right\|_p \to 0$ with rates.

One also gets by (25.68) that

$$\left\| P_{r,n}^{[m]}(f;x) - f(x) \right\|_{p,x}$$

$$\leq \sum_{\widetilde{j}=1}^{m} \left|\delta_{\widetilde{j},r}^{[m]}\right| \left(\sum_{|\alpha|=\widetilde{j}} \frac{\left|\overline{c}_{\alpha,n,\widetilde{j}}\right|}{\prod_{i=1}^N \alpha_i!} \|f_\alpha\|_p \right) + R.H.S.\ (25.68), \tag{25.69}$$

given that $\|f_\alpha\|_p < \infty$, $|\alpha| = \widetilde{j}$, $\widetilde{j} = 1, \ldots, m$.

Furthermore, it holds $\left\| P_{r,n}^{[m]}(f) - f \right\|_p \to 0$, that is, $P_{r,n}^{[m]} \to I$, in L_p norm, with rates.

We need the following auxiliary results.

Theorem 25.6. *Let $N \in \mathbb{N}$, $\alpha_i \in \mathbb{Z}^+$, $i = 1, \ldots, N : |\alpha| = \sum_{i=1}^N \alpha_i = 2$, $\xi_n \in (0,1]$, $n \in \mathbb{N}$, $p \geq 1$, $j = 0, 1, \ldots, r$. Then*

$$I_{1jn}^*(\alpha) := \frac{1}{(2\xi_n)^N} \int_{\mathbb{R}^N} \left(\prod_{i=1}^N |s_i|^{p\alpha_i} \right) \left(1 + \frac{j^p \|s\|_2^p}{\xi_n^p} \right) e^{-\frac{\sum_{i=1}^N |s_i|}{\xi_n}} ds_1 \cdots ds_N$$

$$\leq \xi_n^{2p} \left[(1 + N j^p) + (1 + j^p) \Gamma^N(3p+1, 1) \right]$$

$$\leq \left[(1 + N j^p) + (1 + j^p) \Gamma^N(3p+1, 1) \right] < +\infty, \tag{25.70}$$

uniformly bounded in $\xi_n \in (0,1]$.

Also $I_{1jn}^*(\alpha) \to 0$, as $\xi_n \to 0$, $n \to \infty$.

Above $\Gamma(\cdot,\cdot)$ denotes the upper incomplete gamma function. Denote $I_{1jn}^{**}(\alpha)$ the $I_{1jn}^*(\alpha)$ when $p = 1$.

Proof. We have that

$$I_{1jn}^{*}(\alpha)=\frac{1}{(2\xi_n)^N}\int_{\mathbb{R}^N}\left(\prod_{i=1}^N |s_i|^{p\alpha_i}\right)\left(1+\frac{j^p\|s\|_2^p}{\xi_n^p}\right)e^{-\frac{\sum_{i=1}^N|s_i|}{\xi_n}}ds_1\cdots ds_N$$

$$=\frac{1}{\xi_n^N}\int_{\mathbb{R}_+^N}\left(\prod_{i=1}^N s_i^{p\alpha_i}\right)\left(1+\frac{j^p\|s\|_2^p}{\xi_n^p}\right)e^{-\frac{\sum_{i=1}^N s_i}{\xi_n}}ds_1\cdots ds_N \tag{25.71}$$

$$\leq\frac{1}{\xi_n^N}\int_{\mathbb{R}_+^N}\left(\prod_{i=1}^N s_i^{p\alpha_i}\right)\left(1+j^p\frac{\left(\sum_{i=1}^N s_i\right)^p}{\xi_n^p}\right)e^{-\frac{\sum_{i=1}^N s_i}{\xi_n}}ds_1\cdots ds_N$$

$$=\xi_n^{2p}\int_{\mathbb{R}_+^N}\left(\prod_{i=1}^N\left(\frac{s_i}{\xi_n}\right)^{p\alpha_i}\right)\left(1+j^p\sum_{i=1}^N\left(\frac{s_i}{\xi_n}\right)^p\right)e^{-\sum_{i=1}^N\frac{s_i}{\xi_n}}d\frac{s_1}{\xi_n}\cdots d\frac{s_N}{\xi_n}$$

$$=\xi_n^{2p}\int_{\mathbb{R}_+^N}\left(\prod_{i=1}^N z_i^{p\alpha_i}\right)\left(1+j^p\sum_{i=1}^N z_i^p\right)e^{-\sum_{i=1}^N z_i}dz_1\cdots dz_N$$

$$=\xi_n^{2p}\left[\int_{[0,1]^N}\left(\prod_{i=1}^N z_i^{p\alpha_i}\right)\left(1+j^p\sum_{i=1}^N z_i^p\right)e^{-\sum_{i=1}^N z_i}dz_1\cdots dz_N\right. \tag{25.72}$$

$$\left.+\int_{(\mathbb{R}_+-[0,1])^N}\left(\prod_{i=1}^N z_i^{p\alpha_i}\right)\left(1+j^p\sum_{i=1}^N z_i^p\right)e^{-\sum_{i=1}^N z_i}dz_1\cdots dz_N\right]$$

$$\leq\xi_n^{2p}\left[(1+j^pN)+\int_{(\mathbb{R}_+-[0,1])^N}\left(\prod_{i=1}^N z_i^{p\alpha_i}\right)\left(1+j^p\sum_{i=1}^N z_i^p\right)e^{-\sum_{i=1}^N z_i}dz_1\cdots dz_N\right]$$

$$\leq\xi_n^{2p}\left[(1+j^pN)+(1+j^p)\int_{(\mathbb{R}_+-[0,1])^N}\left(\sum_{i=1}^N z_i^p\right)\left(\prod_{i=1}^N z_i^{p\alpha_i}\right)e^{-\sum_{i=1}^N z_i}dz\cdots dz_N\right]$$

$$\leq\xi_n^{2p}\left[(1+j^pN)+(1+j^p)\int_{(\mathbb{R}_+-[0,1])^N}\left(\prod_{i=1}^N z_i^p\right)\left(\prod_{i=1}^N z_i^{p\alpha_i}\right)e^{-\sum_{i=1}^N z_i}dz\cdots dz_N\right]$$

$$=\xi_n^{2p}\left[(1+j^pN)+(1+j^p)\prod_{i=1}^N\int_1^\infty z_i^{p(\alpha_i+1)}e^{-z_i}dz_i\right]$$

$$=\xi_n^{2p}\left[(1+j^pN)+(1+j^p)\prod_{i=1}^N\int_1^\infty z_i^{(p(\alpha_i+1)+1)-1}e^{-z_i}dz_i\right]= \tag{25.73}$$

(by Gradshteyn and Ryzhik (2015), p. 348)

$$\xi_n^{2p}\left[(1+j^pN)+(1+j^p)\prod_{i=1}^N\Gamma\left((p(\alpha_i+1)+1),1\right)\right]\leq$$

(where $\Gamma(\cdot,\cdot)$ is the upper incomplete gamma function) (by Gradshteyn and Ryzhik 2015, p. 909)

$$\xi_n^{2p}\left[(1+j^pN)+(1+j^p)\Gamma^N(3p+1,1)\right]. \tag{25.74}$$

We have proved that

$$\begin{aligned} I_{1jn}^*(\alpha) &\leq \xi_n^{2p}\left[(1+Nj^p)+(1+j^p)\Gamma^N(3p+1,1)\right] \\ &\leq \left[(1+Nj^p)+(1+j^p)\Gamma^N(3p+1,1)\right] < +\infty, \end{aligned} \tag{25.75}$$

uniformly bounded in $\xi_n \in (0,1]$.

Above we used (Gradshteyn and Ryzhik 2015, p. 909) the formula

$$\Gamma(\alpha, xy) = y^\alpha e^{-xy}\int_0^\infty e^{-ty}(t+x)^{\alpha-1}\,dt, \tag{25.76}$$

where $\mathrm{Re}y > 0$, $x > 0$, $\mathrm{Re}\alpha > 1$.

We note that ($x = y = 1$, $\alpha_i \leq 2$)

$$\begin{aligned} \Gamma((\alpha_i+1)p+1,1) &= e^{-1}\int_0^\infty e^{-t}(t+1)^{(\alpha_i+1)p}\,dt \\ &\leq e^{-1}\int_0^\infty e^{-t}(t+1)^{(2+1)p}\,dt = \Gamma(3p+1,1). \end{aligned} \tag{25.77}$$

That is,

$$\Gamma((\alpha_i+1)p+1,1) \leq \Gamma(3p+1,1), \tag{25.78}$$

for all $i = 1,\ldots,N$.

The theorem is proved. □

Corollary 25.1. *(to Theorem 25.6) It holds*

$$K_{1n}(\alpha) := \frac{1}{(2\xi_n)^N}\int_{\mathbb{R}^N}\left(\prod_{i=1}^N |s_i|^{\alpha_i}\right)e^{-\frac{\sum_{i=1}^N |s_i|}{\xi_n}}\,ds_1\cdots ds_N < +\infty, \tag{25.79}$$

uniformly bounded in $\xi_n \in (0,1]$. *Also* $K_{1n}(\alpha) \to 0$, *as* $\xi_n \to 0$, $n \to \infty$.

Proof. By (25.70). □

Proposition 25.1. *Let* $N \in \mathbb{N}$, $\alpha_i \in \mathbb{Z}^+$, $i = 1, \ldots, N : |\alpha| = \sum_{i=1}^{N} \alpha_i = 2$, $\xi_n \in (0,1]$, $n \in \mathbb{N}$, $p \geq 1$. *Then*

$$M_n := \frac{1}{(2\xi_n)^N} \int_{\mathbb{R}^N} \left(\frac{\|s\|_2}{\xi_n} \right)^p e^{-\frac{\sum_{i=1}^{N} |s_i|}{\xi_n}} ds_1 \cdots ds_N$$
$$\leq N^p + \Gamma^N (p+1, 1) < +\infty, \tag{25.80}$$

uniformly bounded in ξ_n. *Denote* M_n^* *the* M_n *when* $p = 1$.

Proof. We have

$$M_n = \frac{1}{(2\xi_n)^N} \int_{\mathbb{R}^N} \left(\frac{\|s\|_2}{\xi_n} \right)^p e^{-\frac{\sum_{i=1}^{N} |s_i|}{\xi_n}} ds_1 \cdots ds_N$$
$$= \frac{1}{\xi_n^N} \int_{\mathbb{R}_+^N} \left(\frac{\|s\|_2}{\xi_n} \right)^p e^{-\frac{\sum_{i=1}^{N} s_i}{\xi_n}} ds_1 \cdots ds_N$$
$$\leq \frac{1}{\xi_n^N} \int_{\mathbb{R}_+^N} \left(\frac{\sum_{i=1}^{N} s_i}{\xi_n} \right)^p e^{-\frac{\sum_{i=1}^{N} s_i}{\xi_n}} ds_1 \cdots ds_N \tag{25.81}$$
$$= \int_{\mathbb{R}_+^N} \left(\sum_{i=1}^{N} \left(\frac{s_i}{\xi_n} \right)^p \right) e^{-\sum_{i=1}^{N} \frac{s_i}{\xi_n}} d\frac{s_1}{\xi_n} \cdots d\frac{s_N}{\xi_n}$$
$$= \int_{\mathbb{R}_+^N} \left(\sum_{i=1}^{N} z_i \right)^p e^{-\sum_{i=1}^{N} z_i} dz_1 \cdots dz_N$$
$$= \int_{[0,1]^N} \left(\sum_{i=1}^{N} z_i \right)^p e^{-\sum_{i=1}^{N} z_i} dz_1 \cdots dz_N$$
$$+ \int_{(\mathbb{R}_+ - [0,1])^N} \left(\sum_{i=1}^{N} z_i \right)^p e^{-\sum_{i=1}^{N} z_i} dz_1 \cdots dz_N \Bigg]$$
$$\leq N^p + \int_{(\mathbb{R}_+ - [0,1])^N} \prod_{i=1}^{N} z_i^p \prod_{i=1}^{N} e^{-z_i} \prod_{i=1}^{N} dz_i$$
$$= N^p + \prod_{i=1}^{N} \int_1^{\infty} z_i^p e^{-z_i} dz_i = N^p + \prod_{i=1}^{N} \int_1^{\infty} z_i^{(p+1)-1} e^{-z_i} dz_i$$
$$= N^p + \Gamma^N (p+1, 1) < +\infty, \tag{25.82}$$

uniformly bounded in $\xi_n \in (0,1]$. ◻

Next, we apply Theorem 25.2 to multivariate Picard operators P_n, L_p approximation $p > 1$.

Theorem 25.7. *Let $p, q > 1 : \frac{1}{p} + \frac{1}{q} = 1$, $0 < \xi_n \leq 1$, $n \in \mathbb{N}$. We consider $f \in C^2\left(\mathbb{R}^N\right)$, $N \geq 1$, with f, $f_\alpha \in L_p\left(\mathbb{R}^N\right)$, $|\alpha| = 2$, $\alpha_i \in \mathbb{Z}^+$, $i = 1, \ldots, N$, and $|\alpha| = \sum_{i=1}^N \alpha_i$; $x \in \mathbb{R}^N$; $j = 0, 1, \ldots, r$. Denote by*

$$\psi_{in}^{(1)} := \frac{1}{(2\xi_n)^N} \int_{\mathbb{R}^N} |s_i| \, e^{-\frac{\sum_{i=1}^N |s_i|}{\xi_n}} ds_1 \cdots ds_N, \tag{25.83}$$

$$\psi_{in}^{(2)} := \frac{1}{(2\xi_n)^N} \int_{\mathbb{R}^N} s_i^2 e^{-\frac{\sum_{i=1}^N |s_i|}{\xi_n}} ds_1 \cdots ds_N, \tag{25.84}$$

and

$$\psi_{ij^*n} := \frac{1}{(2\xi_n)^N} \int_{\mathbb{R}^N} |s_i s_{j^*}| \, e^{-\frac{\sum_{i=1}^N |s_i|}{\xi_n}} ds_1 \cdots ds_N, \tag{25.85}$$

where $i, j^* \in \{1, \ldots, N\}$, $i \neq j^*$; $n \in \mathbb{N}$.

Also denote ($n \in \mathbb{N}$)

$$\begin{aligned}
&\overline{\Delta}_n(x) := P_n(f, x) - f(x) \\
&- \left(\sum_{j=0}^r \alpha_j j\right) \sin(1) \left[\sum_{i=1}^N \frac{\partial f(x)}{\partial x_i} \left(\frac{1}{(2\xi_n)^N} \int_{\mathbb{R}^N} s_i e^{-\frac{\sum_{i=1}^N |s_i|}{\xi_n}} ds_1 \cdots ds_N\right)\right] \\
&- 2\left(\sum_{j=0}^r \alpha_j j^2\right) \sin^2\left(\frac{1}{2}\right) \left\{\sum_{i=1}^N \left(\frac{1}{(2\xi_n)^N} \int_{\mathbb{R}^N} s_i^2 e^{-\frac{\sum_{i=1}^N |s_i|}{\xi_n}} ds_1 \cdots ds_N\right) \frac{\partial^2 f(x)}{\partial x_i^2}\right. \\
&\left.+ \sum_{\substack{i \neq j^*, \\ i, j^* \in \{1, \ldots, N\}}} \left(\frac{1}{(2\xi_n)^N} \int_{\mathbb{R}^N} s_i s_{j^*} e^{-\frac{\sum_{i=1}^N |s_i|}{\xi_n}} ds_1 \cdots ds_N\right) \frac{\partial^2 f(x)}{\partial x_i \partial x_{j^*}}\right\}.
\end{aligned} \tag{25.86}$$

Then, it holds

$$\begin{aligned}
&\left\|\overline{\Delta}_n(x)\right\|_p \leq \left(\sum_{j=0}^r |\alpha_j|^q\right)^{\frac{1}{q}} \left\{\frac{2}{(q+1)^{\frac{1}{q}}} \left(\frac{N(N+1)+2}{2}\right)^{\frac{1}{q}}\right\} \\
&\left\{\sum_{j=0}^r \left[j^{2p} \sum_{|\alpha|=2} \left(\frac{1}{\prod_{i=1}^N \alpha_i!}\right)^p \omega_1(f_\alpha, \xi_n)_p^p I_{1jn}^*(\alpha) + \omega_1(f, \xi_n)_p^p \left(1 + j^p M_n\right)\right]\right\}^{\frac{1}{p}},
\end{aligned} \tag{25.87}$$

where $I_{1jn}^*(\alpha)$ as in (25.70) and M_n as in (25.80).

As $n \to \infty$ and $\xi_n \to 0$, by (25.87), we obtain that $\left\|\overline{\Delta}_n\right\|_p \to 0$ with rates.

One also gets by (25.87) that

$$\|P_n(f) - f\|_p \leq \left(\sum_{j=0}^{r} |\alpha_j| \, j\right) \sin(1) \left[\sum_{i=1}^{N} \left\|\frac{\partial f}{\partial x_i}\right\|_p \psi_{in}^{(1)}\right]$$
$$+ 2\left(\sum_{j=0}^{r} |\alpha_j| \, j^2\right) \sin^2\left(\frac{1}{2}\right) \left\{\sum_{i=1}^{N} \psi_{in}^{(2)} \left\|\frac{\partial^2 f}{\partial x_i^2}\right\|_p\right.$$
$$\left. + \sum_{\substack{i \neq j^*, \\ i, j^* \in \{1, \ldots, N\}}} \psi_{ij^*n} \left\|\frac{\partial^2 f}{\partial x_i \partial x_{j^*}}\right\|_p \right\} + R.H.S. \ (25.87), \tag{25.88}$$

given that $\|f_\alpha\|_p < \infty$, $|\alpha| = \widetilde{j}$, $\widetilde{j} = 1, 2$. Also we obtain $\|P_n(f) - f\|_p \to 0$, i.e. $P_n \to I$ the unit operator, in L_p norm, with rates.

Proof. By Theorems 25.2 and 25.6, Corollary 25.1 and Proposition 25.1. □

Finally, we apply Theorem 25.3 to an alternative L_1 approximation by the P_n multivariate operators, $p = 1$ case.

Theorem 25.8. *Let $0 < \xi_n \leq 1$, $n \in \mathbb{N}$, $x \in \mathbb{R}^N$. Here we consider $f \in C^2(\mathbb{R}^N)$, $N \geq 1$, with f, $f_\alpha \in L_1(\mathbb{R}^N)$, $|\alpha| = 2$, where $\alpha_i \in \mathbb{Z}^+$, $i = 1, \ldots, N$, and $|\alpha| = \sum_{i=1}^{N} \alpha_i$; $j = 0, 1, \ldots, r$. Here $\overline{\Delta}_n(x)$ is as in (25.86), $\psi_{in}^{(1)}$ as in (25.83), $\psi_{in}^{(2)}$ as in (25.84), ψ_{ij^*n} as in (25.85).*

Then

$$\left\|\overline{\Delta}_n(x)\right\|_1$$
$$\leq \left\{\sum_{j=0}^{r} |\alpha_j| \left[j^2 \sum_{|\alpha|=2} \left(\frac{2}{\prod_{i=1}^{N} \alpha_i!}\right) \omega_1(f_\alpha, \xi_n)_1 \, I_{1jn}^{**}(\alpha) + \omega_1(f, \xi_n)_1 \left(1 + j M_n^*\right)\right]\right\}, \tag{25.89}$$

where $I_{1jn}^{**}(\alpha)$ as in Theorem 25.6, and M_n^* as in Proposition 25.1.

As $n \to \infty$ and $\xi_n \to 0$, by (25.89), we obtain that $\left\|\overline{\Delta}_n\right\|_1 \to 0$ with rates.

One also obtains by (25.89) that

$$\|P_n(f)-f\|_1 \le \left(\sum_{j=0}^{r}|\alpha_j|\, j\right)\sin(1)\left[\sum_{i=1}^{N}\left\|\frac{\partial f}{\partial x_i}\right\|_1 \psi_{in}^{(1)}\right]$$
$$+2\left(\sum_{j=0}^{r}|\alpha_j|\, j^2\right)\sin^2\left(\frac{1}{2}\right)\left\{\sum_{i=1}^{N}\psi_{in}^{(2)}\left\|\frac{\partial^2 f}{\partial x_i^2}\right\|_1\right.$$
$$\left.+\sum_{\substack{i\neq j^*,\\ i,j^*\in\{1,\ldots,N\}}}\psi_{ij^*n}\left\|\frac{\partial^2 f}{\partial x_i \partial x_{j^*}}\right\|_1\right\}+R.H.S.\ (25.89), \tag{25.90}$$

given that $\|f_\alpha\|_1<\infty$, $|\alpha|=\widetilde{j}$, $\widetilde{j}=1,2$. *Also we obtain* $\|P_n(f)-f\|_1\to 0$, *i.e.* $P_n\to I$ *in* L_1 *norm with rates.*

Proof. By Theorems 25.3 and 25.6, Corollary 25.1 and Proposition ??. □

Bibliography

G.A. Anastassiou, *Intelligent Mathematics: Computational Analysis*, Springer, Heidelberg, New York, 2011a.

G.A. Anastassiou, *Approximation by Multivariate Singular Integrals*, Springer, New York, 2011b.

G.A. Anastassiou, *Constructive Fractional Analysis with Applications*, Springer, Heidelberg, New York, 2021.

G.A. Anastassiou, Trigonometric based multivariate smooth Picard singular integrals L_p approximation, *Bulletin of The Allahabad Mathematical Society*, accepted, 2023a.

G.A. Anastassiou, *General multiple sigmoid functions relied complex valued multivariate trigonometric and hyperbolic neural network approximations*, submitted, 2023b.

G.A. Anastassiou, and S. Gal, *Approximation Theory*, Birkhaüser, Boston, 2000.

G.A. Anastassiou, and R. Mezei, *Approximation by Singular Integrals*, Cambridge Scientific Publishers, Cambridge, UK, 2012.

I.S. Gradshteyn, and I.M. Ryzhik, *Table of Integrals, Series and Product*, 8th edn. Elsevier, Amsterdam, 2015.

R.N. Mohapatra and R.S. Rodriguez, On the rate of convergence of singular integrals for Hölder continuous functions, *Mathematische Nachrichten*, 149 (1990), 117–124.

Chapter 26

Trigonometric Produced Rate of Approximation of Various Smooth Singular Integral Operators

In this chapter, we continue the study of approximation of various smooth singular integral operators. This time, the foundation of our research is a trigonometric Taylor's formula. We establish the convergence of our operators to the unit operator with rates via Jackson-type inequalities engaging the first modulus of continuity. Of interest, here is a residual appearing term. Note that our operators are not positive. Our results are pointwise and uniform. The studied operators here are of the following types: Gauss–Weierstrass, Poisson–Cauchy and trigonometric. It follows Anastassiou (2024).

26.1 Introduction

We are motivated by Anastassiou (2006, 2011) Chapters 10–14, Ali and Pales (2022), and Anastassiou (2023). We use a trigonometric new Taylor formula from Ali and Pales (2022), see also Anastassiou (2023). Here we consider some very general operators, the smooth Gauss–Weierstrass, Poisson–Cauchy and trigonometric singular integral operators over the real line and we study further their convergence properties quantitatively. We establish related inequalities involving the first modulus of continuity with respect to uniform norm and the estimates are pointwise and uniform. We demonstrate detailed proofs.

26.2 Results

By Ali and Pales (2022) and Anastassiou (2023), for $f \in C^2(\mathbb{R})$ and $a, x \in \mathbb{R}$, we have by trigonometric Taylor formula

$$f(x) - f(a) = f'(a)\sin(x-a) + 2f''(a)\sin^2\left(\frac{x-a}{2}\right) + \int_a^x [(f''(t) + f(t)) - (f''(a) + f(a))]\sin(x-t)\,dt. \tag{26.1}$$

For $r \in \mathbb{N}$ and $n \in \mathbb{Z}^+$, we set

$$\alpha_j := \begin{cases} (-1)^{r-j}\binom{r}{j} j^{-n}, & j = 1, \ldots, r, \\ 1 - \sum_{j=1}^{r} (-1)^{r-j}\binom{r}{j} j^{-n}, & j = 0, \end{cases} \tag{26.2}$$

that is,

$$\sum_{j=0}^{r} \alpha_j = 1. \tag{26.3}$$

$C_U(\mathbb{R})$ denotes the space of uniformly continuous functions on $\mathbb{R}$, and $C_B(\mathbb{R})$ denotes the space of bounded continuous functions on $\mathbb{R}$.

Here we consider both $f, f'' \in C_U(\mathbb{R}) \cup C_B(\mathbb{R})$.

Denote by

$$\omega_1(f, \delta) := \sup_{\substack{x,y \in \mathbb{R} \\ |x-y| \leq \delta}} |f(x) - f(y)|, \quad \delta > 0, \tag{26.4}$$

the first modulus of continuity of f.

(I) We define the smooth Gauss–Weierstrass singular integral operators (Anastassiou and Mezei 2012).

Let $f \in C^2(\mathbb{R})$, we define for $x \in \mathbb{R}$, $\xi > 0$ the Lebesgue integral

$$W_{r,\xi}(f; x) = \frac{1}{\sqrt{\pi\xi}} \int_{-\infty}^{\infty} \left(\sum_{j=0}^{r} \alpha_j f(x + jt)\right) e^{-\frac{t^2}{\xi}}\,dt. \tag{26.5}$$

We assume that $W_{r,\xi}(f;x) \in \mathbb{R}$ for all $x \in \mathbb{R}$. We use also that

$$W_{r,\xi}(f;x) = \frac{1}{\sqrt{\pi\xi}} \sum_{j=0}^{r} \alpha_j \left(\int_{-\infty}^{\infty} f(x+jt) e^{-\frac{t^2}{\xi}} dt \right), \tag{26.6}$$

note by $\frac{1}{\sqrt{\pi\xi}} \int_{-\infty}^{\infty} e^{-\frac{t^2}{\xi}} dt = 1$ that $W_{r,\xi}(c;x) = c$, c constant and

$$W_{r,\xi}(f;x) - f(x) = \frac{1}{\sqrt{\pi\xi}} \left(\sum_{j=0}^{r} \alpha_j \int_{-\infty}^{\infty} (f(x+jt) - f(x)) e^{-\frac{t^2}{\xi}} dt \right). \tag{26.7}$$

We set

$$\Delta_2(x) := W_{r,\xi}(f;x) - f(x) - f''(x) \sum_{j=0}^{r} \alpha_j \left(1 - e^{-\frac{j^2}{4}\xi}\right), \quad x \in \mathbb{R}, \tag{26.8}$$

$j = 0, 1, \ldots, r \in \mathbb{N}$.

(II) We define the smooth Poisson–Cauchy singular integral operators (Anastassiou and Mezei 2012).

Let $\alpha \in \mathbb{N}$, $\beta > \frac{1}{2\alpha}$ and $f \in C^2(\mathbb{R})$. We define for $x \in \mathbb{R}$, $\xi > 0$ the Lebesgue integral

$$M_{r,\xi}(f;x) = W \int_{-\infty}^{\infty} \frac{\sum_{j=0}^{r} \alpha_j f(x+jt)}{(t^{2\alpha} + \xi^{2\alpha})^{\beta}} dt, \tag{26.9}$$

where the constant is defined as

$$W = \frac{\Gamma(\beta)\, \alpha \beta^{2\alpha\beta-1}}{\Gamma\left(\frac{1}{2\alpha}\right) \Gamma\left(\beta - \frac{1}{2\alpha}\right)}. \tag{26.10}$$

We assume that $M_{r,\xi}(f;x) \in \mathbb{R}$ for all $x \in \mathbb{R}$. We use also that

$$M_{r,\xi}(f;x) = W \sum_{j=0}^{r} \alpha_j \left(\int_{-\infty}^{\infty} f(x+jt) \frac{1}{(t^{2\alpha} + \xi^{2\alpha})^{\beta}} dt \right). \tag{26.11}$$

We note by $W \int_{-\infty}^{\infty} \frac{1}{(t^{2\alpha}+\xi^{2\alpha})^{\beta}} dt = 1$ that $M_{r,\xi}(c;x) = c$, c constant and

$$M_{r,\xi}(f;x) - f(x) = W \left(\sum_{j=0}^{r} \alpha_j \int_{-\infty}^{\infty} [f(x+jt) - f(x)] \frac{1}{(t^{2\alpha} + \xi^{2\alpha})^{\beta}} dt \right). \tag{26.12}$$

We set

$$\Delta_3(x) := M_{r,\xi}(f;x) - f(x) - 4f''(x)\frac{\alpha\Gamma(\beta)}{\Gamma\left(\frac{1}{2\alpha}\right)\Gamma\left(\beta-\frac{1}{2\alpha}\right)}$$

$$\times\sum_{j=0}^{r}\alpha_j\int_0^{\infty}\frac{\sin^2\left(\frac{j\xi}{2}t\right)}{(1+t^{2\alpha})^{\beta}}dt, \tag{26.13}$$

$\xi > 0$, $x \in \mathbb{R}$; $\beta > \frac{1}{2\alpha}$, $\alpha \in \mathbb{N}$; $j = 0, 1, \ldots, r \in \mathbb{N}$.

(III) We define the smooth trigonometric singular integral operators (Anastassiou and Mezei 2012) as follows:

Let $\xi > 0$, $f \in C^2(\mathbb{R})$, $x \in \mathbb{R}$, $\beta \in \mathbb{N}$; we set

$$T_{r,\xi}(f;x) := \frac{1}{\lambda}\int_{-\infty}^{\infty}\left(\sum_{j=0}^{r}\alpha_j f(x+jt)\right)\left(\frac{\sin\left(\frac{t}{\xi}\right)}{t}\right)^{2\beta}dt, \tag{26.14}$$

where

$$\lambda := \int_{-\infty}^{\infty}\left(\frac{\sin\left(\frac{t}{\xi}\right)}{t}\right)^{2\beta}dt = 2\xi^{1-2\beta}\int_0^{\infty}\left(\frac{\sin t}{t}\right)^{2\beta}dt \tag{26.15}$$

(by [Edwards (1954)], p. 210, item 1033)

$$= 2\xi^{1-2\beta}\pi(-1)^{\beta}\beta\sum_{k=1}^{\beta}(-1)^k\frac{k^{2\beta-1}}{(\beta-k)!\,(\beta+k)!}.$$

Denote

$$\lambda_1 := 2\pi(-1)^{\beta}\beta\sum_{k=1}^{\beta}(-1)^k\frac{k^{2\beta-1}}{(\beta-k)!\,(\beta+k)!}, \tag{26.16}$$

that is,

$$\lambda = \lambda_1\xi^{1-2\beta}. \tag{26.17}$$

We suppose that $T_{r,\xi}(f;x) \in \mathbb{R}$ for all $x \in \mathbb{R}$. Clearly, again it is

$$\frac{1}{\lambda}\int_{-\infty}^{\infty}\left(\frac{\sin\left(\frac{t}{\xi}\right)}{t}\right)^{2\beta}dt = 1, \tag{26.18}$$

and $T_{r,\xi}(c;x) = c$, c constant.

We set

$$\Delta_4(x) := T_{r,\xi}(f;x) - f(x) - 4f''(x)\lambda_1^{-1}$$

$$\times \sum_{j=0}^{r} \alpha_j \int_0^\infty \sin^2\left(\frac{j\xi}{2}t\right)\left(\frac{\sin t}{t}\right)^{2\beta} dt, \qquad (26.19)$$

where $\xi > 0$, $x \in \mathbb{R}$; $\beta \in \mathbb{N} - \{1,2\}$; $j = 0,1,\ldots,r \in \mathbb{N}$.

Next we present uniform approximation results regarding the smooth singular operators $W_{r,\xi}$, $M_{r,\xi}$ and $T_{r,\xi}$.

Theorem 26.1. *It holds*

$$|\Delta_2(x)| \le \|\Delta_2\|_\infty \le \omega_1(f''+f,\xi)\sum_{j=0}^{r}|\alpha_j| j^2\left[\frac{\xi}{4} + \frac{j\sqrt{\xi}}{6\sqrt{\pi}}\right] =: A_2;$$

$$\xi > 0,\ x \in \mathbb{R}. \qquad (26.20)$$

And $\|\Delta_2\|_\infty \to 0$, *as* $\xi \to 0$. *If* $f''(x) = 0$, *then* $|W_{r,\xi}(f,x) - f(x)| \le A_2$ *and* $W_{r,\xi}(f,x) \to f(x)$, *as* $\xi \to 0$.

Proof. By (26.1), we get that

$$f(x+jt) - f(x) = f'(x)\sin(jt) + 2f''(x)\sin^2\left(\frac{jt}{2}\right)$$

$$+ \int_x^{x+jt} \left[(f''(s)+f(s)) - (f''(x)+f(x))\right]\sin(x+jt-s)\,ds, \qquad (26.21)$$

or better

$$f(x+jt) - f(x) = f'(x)\sin(jt) + 2f''(x)\sin^2\left(\frac{jt}{2}\right)$$

$$+ \int_0^{jt} \left[(f''(x+z)+f(x+z)) - (f''(x)+f(x))\right]\sin(jt-z)\,dz. \qquad (26.22)$$

Furthermore, it holds

$$\begin{aligned}
&\sum_{j=0}^{r} \alpha_j \left[f(x+jt) - f(x) \right] \\
&= f'(x) \sum_{j=0}^{r} \alpha_j \sin(jt) + 2f''(x) \sum_{j=0}^{r} \alpha_j \sin^2\left(\frac{jt}{2}\right) \\
&\quad + \sum_{j=0}^{r} \alpha_j \int_0^{jt} \left[\left(f''(x+z) + f(x+z)\right) - \left(f''(x) + f(x)\right) \right] \sin(jt - z)\, dz,
\end{aligned} \tag{26.23}$$

or better

$$\begin{aligned}
&\sum_{j=0}^{r} \alpha_j \left[f(x+jt) - f(x) \right] \\
&= f'(x) \sum_{j=0}^{r} \alpha_j \sin(jt) + 2f''(x) \sum_{j=0}^{r} \alpha_j \sin^2\left(\frac{jt}{2}\right) \\
&\quad + \sum_{j=0}^{r} \alpha_j j \int_0^{t} \left[\left(f''(x+jw) + f(x+jw)\right) - \left(f''(x) + f(x)\right) \right] \sin j(t - w)\, dw.
\end{aligned} \tag{26.24}$$

Call

$$\begin{aligned}
R &:= R(t) \\
&:= \sum_{j=0}^{r} \alpha_j j \int_0^{t} \left[\left(f''(x+jw) + f(x+jw)\right) - \left(f''(x) + f(x)\right) \right] \sin j(t - w)\, dw,
\end{aligned} \tag{26.25}$$

$\forall\, t \in \mathbb{R}$.

Then, for $t \geq 0$,

$$\begin{aligned}
|R| \leq &\sum_{j=0}^{r} |\alpha_j|\, j \int_0^{t} \left| \left(f''(x+jw) + f(x+jw)\right) - \left(f''(x) + f(x)\right) \right| \\
&\times \left| \sin j(t - w) \right| dw \\
&\leq \sum_{j=0}^{r} |\alpha_j|\, j \int_0^{t} \omega_1\left(f'' + f, jw\right) j (t - w)\, dw =
\end{aligned}$$

$(\xi > 0)$

$$\sum_{j=0}^{r} |\alpha_j| j^2 \int_0^t \omega_1 \left(f'' + f, \frac{\xi j w}{\xi} \right) (t-w)\, dw$$

$$\leq \sum_{j=0}^{r} |\alpha_j| j^2 \omega_1 (f'' + f, \xi) \int_0^t \left(1 + \frac{jw}{\xi} \right) (t-w)\, dw$$

$$= \omega_1 (f'' + f, \xi) \sum_{j=0}^{r} |\alpha_j| j^2 \left[\int_0^t (t-w)\, dw + \frac{j}{\xi} \int_0^t w^{2-1} (t-w)^{2-1}\, dw \right]$$

$$= \omega_1 (f'' + f, \xi) \sum_{j=0}^{r} |\alpha_j| j^2 \left[\frac{(t-w)^2}{2} \Big|_t^0 + \frac{j}{\xi} \frac{1}{6} t^3 \right]$$

$$= \omega_1 (f'' + f, \xi) \sum_{j=0}^{r} |\alpha_j| j^2 \left[\frac{t^2}{2} + j \frac{t^3}{6\xi} \right]. \tag{26.26}$$

Hence $(t \geq 0)$

$$|R| \leq \omega_1 (f'' + f, \xi) \sum_{j=0}^{r} |\alpha_j| j^2 \left[\frac{t^2}{2} + j \frac{t^3}{6\xi} \right]. \tag{26.27}$$

Let now $t < 0$, then

$$|R| \leq \sum_{j=0}^{r} |\alpha_j| j \left| \int_0^t \left[\left(f''(x+jw) + f(x+jw) \right) - \left(f''(x) + f(x) \right) \right] \sin j (t-w) dw \right|$$

$$\leq \sum_{j=0}^{r} |\alpha_j| j \int_t^0 \left| \left(f''(x+jw) + f(x+jw) \right) - \left(f''(x) + f(x) \right) \right| |\sin j (t-w)|\, dw$$

$$\leq \sum_{j=0}^{r} |\alpha_j| j \int_t^0 \omega_1 (f'' + f, -jw)\, j (w-t)\, dw$$

$$= \sum_{j=0}^{r} |\alpha_j| j \int_t^0 \omega_1 \left(f'' + f, -jw \frac{\xi}{\xi} \right) j (w-t)\, dw$$

$$= \omega_1 (f'' + f, \xi) \sum_{j=0}^{r} |\alpha_j| j^2 \int_t^0 \left(1 - \frac{j}{\xi} w \right) (w-t)\, dw$$

$$= \omega_1 (f'' + f, \xi) \sum_{j=0}^{r} |\alpha_j| j^2 \left[\int_t^0 (w-t)\, dw + \frac{j}{\xi} \int_t^0 (0-w)^{2-1} (w-t)^{2-1}\, dw \right]$$

$$= \omega_1 \left(f'' + f, \xi\right) \sum_{j=0}^{r} \left|\alpha_j\right| j^2 \left[\left.\frac{(w-t)^2}{2}\right|_t^0 + \frac{j}{\xi}\frac{1}{6}(-t)^3 \right]$$

$$= \omega_1 \left(f'' + f, \xi\right) \sum_{j=0}^{r} \left|\alpha_j\right| j^2 \left[\frac{t^2}{2} + j\frac{(-t)^3}{6\xi} \right]. \tag{26.28}$$

We found that $(t < 0)$

$$|R| \leq \omega_1 \left(f'' + f, \xi\right) \sum_{j=0}^{r} \left|\alpha_j\right| j^2 \left[\frac{t^2}{2} + j\frac{(-t)^3}{6\xi} \right]. \tag{26.29}$$

Consequently, for $t \in \mathbb{R}$, we obtain

$$|R(t)| \leq \omega_1 \left(f'' + f, \xi\right) \sum_{j=0}^{r} \left|\alpha_j\right| j^2 \left[\frac{t^2}{2} + \frac{j}{6\xi}|t|^3 \right], \quad \xi > 0. \tag{26.30}$$

So, we have

$$\sum_{j=0}^{r} \alpha_j \left[f(x+jt) - f(x)\right] - f'(x) \sum_{j=0}^{r} \alpha_j \sin(jt)$$

$$- 2f''(x) \sum_{j=0}^{r} \alpha_j \sin^2\left(\frac{jt}{2}\right) = R(t). \tag{26.31}$$

We call

$$\overline{\Delta}_2(x) := W_{r,\xi}(f,x) - f(x) - f'(x) \sum_{j=0}^{r} \alpha_j \frac{1}{\sqrt{\pi\xi}} \left(\int_{-\infty}^{\infty} \sin(jt)\, e^{-\frac{t^2}{\xi}} dt \right)$$

$$- 2f''(x) \sum_{j=0}^{r} \alpha_j \frac{1}{\sqrt{\pi\xi}} \left(\int_{-\infty}^{\infty} \sin^2\left(\frac{jt}{2}\right) e^{-\frac{t^2}{\xi}} dt \right)$$

$$= \frac{1}{\sqrt{\pi\xi}} \int_{-\infty}^{\infty} R(t)\, e^{-\frac{t^2}{\xi}} dt. \tag{26.32}$$

Hence we get that

$$\left|\overline{\Delta}_2(x)\right| \leq \frac{1}{\sqrt{\pi\xi}} \int_{-\infty}^{\infty} R(t)\, e^{-\frac{t^2}{\xi}} dt$$

$$\leq \frac{1}{\sqrt{\pi\xi}} \int_{-\infty}^{\infty} \left[\omega_1 \left(f'' + f, \xi\right) \sum_{j=0}^{r} \left|\alpha_j\right| j^2 \left[\frac{t^2}{2} + \frac{j}{6\xi}|t|^3 \right] \right] e^{-\frac{t^2}{\xi}} dt \tag{26.33}$$

$$= \omega_1 \left(f'' + f, \xi\right) \sum_{j=0}^{r} \left|\alpha_j\right| j^2 \frac{1}{\sqrt{\pi\xi}} \int_{-\infty}^{\infty} \left[\frac{t^2}{2} + \frac{j}{6\xi}|t|^3 \right] e^{-\frac{t^2}{\xi}} dt$$

$$= \omega_1 \left(f'' + f, \xi\right) \sum_{j=0}^{r} |\alpha_j| j^2 \left[\frac{1}{2\sqrt{\pi\xi}} \int_{-\infty}^{\infty} t^2 e^{-\frac{t^2}{\xi}} dt + \frac{j}{6\xi\sqrt{\pi\xi}} \int_{-\infty}^{\infty} |t|^3 e^{-\frac{t^2}{\xi}} dt\right] \tag{26.34}$$

$$= \omega_1 \left(f'' + f, \xi\right) \sum_{j=0}^{r} |\alpha_j| j^2 \left[\frac{1}{\sqrt{\pi\xi}} \int_{0}^{\infty} t^2 e^{-\frac{t^2}{\xi}} dt + \frac{j}{3\xi\sqrt{\pi\xi}} \int_{0}^{\infty} t^3 e^{-\frac{t^2}{\xi}} dt\right]$$

$$= \omega_1 \left(f'' + f, \xi\right) \sum_{j=0}^{r} |\alpha_j| j^2 \left[\frac{\left(\sqrt{\xi}\right)^3}{\sqrt{\pi\xi}} \int_{0}^{\infty} \left(\frac{t}{\sqrt{\xi}}\right)^2 e^{-\left(\frac{t}{\sqrt{\xi}}\right)^2} d\left(\frac{t}{\sqrt{\xi}}\right)\right.$$

$$\left. + \frac{j\left(\sqrt{\xi}\right)^4}{3\xi\sqrt{\pi\xi}} \int_{0}^{\infty} \left(\frac{t}{\sqrt{\xi}}\right)^3 e^{-\left(\frac{t}{\sqrt{\xi}}\right)^2} d\left(\frac{t}{\sqrt{\xi}}\right)\right] \tag{26.35}$$

$$= \omega_1 \left(f'' + f, \xi\right) \sum_{j=0}^{r} |\alpha_j| j^2 \left[\frac{\xi}{\sqrt{\pi}} \int_{0}^{\infty} z^2 e^{-z^2} dz + \frac{j\left(\sqrt{\xi}\right)^3 \xi^{-1}}{3\sqrt{\pi}} \int_{0}^{\infty} z^3 e^{-z^2} dz\right]$$

$$= \omega_1 \left(f'' + f, \xi\right) \sum_{j=0}^{r} |\alpha_j| j^2 \left[\frac{\xi}{\sqrt{\pi}} \frac{\sqrt{\pi}}{4} + \frac{j\sqrt{\xi}}{3\sqrt{\pi}} \frac{1}{2}\right]$$

$$= \omega_1 \left(f'' + f, \xi\right) \sum_{j=0}^{r} |\alpha_j| j^2 \left[\frac{\xi}{4} + \frac{j\sqrt{\xi}}{6\sqrt{\pi}}\right]. \tag{26.36}$$

We have proved that

$$\left|\overline{\Delta}_2(x)\right| \leq \omega_1 \left(f'' + f, \xi\right) \sum_{j=0}^{r} |\alpha_j| j^2 \left[\frac{\xi}{4} + \frac{j\sqrt{\xi}}{6\sqrt{\pi}}\right]; \quad \xi > 0, \ x \in \mathbb{R}. \tag{26.37}$$

Note that $\overline{\Delta}_2(x) \to 0$, as $\xi \to 0$.

Next we simplify the left-hand side of (26.32).

We observe the following:

Clearly, it is

$$\int_{-\infty}^{\infty} \sin(jt)\, e^{-\frac{t^2}{\xi}} dt = 0, \quad j = 0, 1, \ldots, r. \tag{26.38}$$

Furthermore, we have that

$$\int_{-\infty}^{\infty} \sin^2\left(\frac{jt}{2}\right) e^{-\frac{t^2}{\xi}} dt = 2\int_{0}^{\infty} \sin^2\left(\frac{jt}{2}\right) e^{-\frac{t^2}{\xi}} dt$$

$$= 2\sqrt{\xi} \int_{0}^{\infty} \sin^2\left(\left(\frac{j\sqrt{\xi}}{2}\right) \frac{t}{\sqrt{\xi}}\right) e^{-\left(\frac{t}{\sqrt{\xi}}\right)^2} d\frac{t}{\sqrt{\xi}}$$

($\frac{t}{\sqrt{\xi}} =: x$ and $\frac{j\sqrt{\xi}}{2} =: \beta_1$)

$$= 2\sqrt{\xi}\int_0^{\infty} \sin^2(\beta_1 x)\, e^{-x^2} dx$$

$$= \sqrt{\xi}\frac{1}{2}\sqrt{\pi}e^{-\beta_1^2}\left(e^{\beta_1^2} - 1\right)$$

$$= \frac{\sqrt{\xi}\sqrt{\pi}}{2}e^{-\frac{j^2}{4}\xi}\left(e^{\frac{j^2}{4}\xi} - 1\right). \tag{26.39}$$

Consequently, we derive

$$\overline{\Delta}_2(x) \overset{(26.32)}{=} W_{r,\xi}(f,x) - f(x) - f''(x)\sum_{j=0}^{r}\alpha_j\left(1 - e^{-\frac{j^2}{4}\xi}\right) = \Delta_2(x). \tag{26.40}$$

The theorem is proved. □

We derive the following.

Corollary 26.1. *(to Theorem 26.1) Additionally assume $f'' \in C_B(\mathbb{R})$. It follows ($\xi > 0$, $x \in \mathbb{R}$)*

$$\|W_{r,\xi}(f) - f\|_{\infty} \leq \omega_1(f'' + f, \xi)\sum_{j=0}^{r}|\alpha_j| j^2\left[\frac{\xi}{4} + \frac{j\sqrt{\xi}}{6\sqrt{\pi}}\right]$$
$$+ \|f''\|_{\infty}\left(\sum_{j=0}^{r}|\alpha_j|\left(1 - e^{-\frac{j^2}{4}\xi}\right)\right) \to 0, \quad \text{as } \xi \to 0. \tag{26.41}$$

Proof. By (26.20) and (26.8). □

We continue with the following result.

Theorem 26.2. *It holds*

$$|\Delta_3(x)| \leq \|\Delta_3\|_{\infty} \leq \frac{\omega_1(f'' + f, \xi)\,\xi^2}{2\Gamma\left(\frac{1}{2\alpha}\right)\Gamma\left(\beta - \frac{1}{2\alpha}\right)}$$
$$\sum_{j=0}^{r}|\alpha_j| j^2\left[\Gamma\left(\frac{3}{2\alpha}\right)\Gamma\left(\beta - \frac{3}{2\alpha}\right) + \frac{j}{3}\Gamma\left(\frac{2}{\alpha}\right)\Gamma\left(\beta - \frac{2}{\alpha}\right)\right] =: A_3; \tag{26.42}$$

$\xi > 0$, $x \in \mathbb{R}$; and $\beta > \frac{2}{\alpha}$, $\alpha \in \mathbb{N}$.

Note that $\|\Delta_3\|_{\infty} \to 0$, as $\xi \to 0$. If $f''(x) = 0$, then $|M_{r,\xi}(f,x) - f(x)| \leq A_3$ and $M_{r,\xi}(f,x) \to f(x)$, as $\xi \to 0$.

Proof. The proof is the same as the proof of Theorem 26.1 from start till equation (26.31).

We call

$$\overline{\Delta}_3(x) := M_{r,\xi}(f,x) - f(x) - f'(x) \sum_{j=0}^{r} \alpha_j W \left(\int_{-\infty}^{\infty} \sin(jt) \frac{1}{(t^{2\alpha} + \xi^{2\alpha})^{\beta}} dt \right)$$
$$- 2f''(x) \sum_{j=0}^{r} \alpha_j W \left(\int_{-\infty}^{\infty} \sin^2\left(\frac{jt}{2}\right) \frac{1}{(t^{2\alpha} + \xi^{2\alpha})^{\beta}} dt \right)$$
$$= W \int_{-\infty}^{\infty} R(t) \frac{1}{(t^{2\alpha} + \xi^{2\alpha})^{\beta}} dt. \tag{26.43}$$

Hence we get

$$|\overline{\Delta}_3(x)| \le W \int_{-\infty}^{\infty} |R(t)| \frac{1}{(t^{2\alpha} + \xi^{2\alpha})^{\beta}} dt$$
$$\le W \int_{-\infty}^{\infty} \left[\omega_1(f'' + f, \xi) \sum_{j=0}^{r} |\alpha_j| j^2 \left[\frac{t^2}{2} + \frac{j}{6\xi} |t|^3 \right] \right] \frac{1}{(t^{2\alpha} + \xi^{2\alpha})^{\beta}} dt \tag{26.44}$$
$$= \omega_1(f'' + f, \xi) \sum_{j=0}^{r} |\alpha_j| j^2 W \int_{-\infty}^{\infty} \left[\frac{t^2}{2} + \frac{j}{6\xi} |t|^3 \right] \frac{1}{(t^{2\alpha} + \xi^{2\alpha})^{\beta}} dt$$
$$= \omega_1(f'' + f, \xi) \sum_{j=0}^{r} |\alpha_j| j^2 \left[\frac{W}{2} \int_{-\infty}^{\infty} t^2 \frac{1}{(t^{2\alpha} + \xi^{2\alpha})^{\beta}} dt + \frac{jW}{6\xi} \int_{-\infty}^{\infty} |t|^3 \frac{1}{(t^{2\alpha} + \xi^{2\alpha})^{\beta}} dt \right]$$
$$= \omega_1(f'' + f, \xi) \sum_{j=0}^{r} |\alpha_j| j^2 \left[W \int_0^{\infty} t^2 \frac{1}{(t^{2\alpha} + \xi^{2\alpha})^{\beta}} dt + \frac{jW}{3\xi} \int_0^{\infty} t^3 \frac{1}{(t^{2\alpha} + \xi^{2\alpha})^{\beta}} dt \right]$$
$$= \omega_1(f'' + f, \xi) \sum_{j=0}^{r} |\alpha_j| j^2 \left[W \xi^{-2\alpha\beta} \xi^2 \xi \int_0^{\infty} \left(\frac{t}{\xi}\right)^2 \frac{1}{\left(\left(\frac{t}{\xi}\right)^{2\alpha} + 1\right)^{\beta}} d\left(\frac{t}{\xi}\right) \right.$$
$$\left. + \frac{jW}{3\xi} \xi^{-2\alpha\beta} \xi^4 \int_0^{\infty} \left(\frac{t}{\xi}\right)^3 \frac{1}{\left(\left(\frac{t}{\xi}\right)^{2\alpha} + 1\right)^{\beta}} d\left(\frac{t}{\xi}\right) \right]$$
$$= \omega_1(f'' + f, \xi) \sum_{j=0}^{r} |\alpha_j| j^2$$
$$\left[W \xi^{3-2\alpha\beta} \int_0^{\infty} \frac{z^2}{(z^{2\alpha} + 1)^{\beta}} dz + \frac{jW}{3} \xi^{3-2\alpha\beta} \int_0^{\infty} \frac{z^3}{(z^{2\alpha} + 1)^{\beta}} dz \right] = \tag{26.45}$$

(by Zwillinger (1995), p. 397, formula 595)

$$\omega_1(f''+f,\xi)\sum_{j=0}^{r}|\alpha_j|\,j^2 W\xi^{3-2\alpha\beta}\left[\frac{\Gamma(\frac{3}{2\alpha})\,\Gamma(\beta-(\frac{3}{2\alpha}))}{2\alpha\Gamma(\beta)}+\frac{j}{3}\frac{\Gamma(\frac{2}{\alpha})\,\Gamma(\beta-\frac{2}{\alpha})}{2\alpha\Gamma(\beta)}\right] \tag{26.46}$$

$$=\omega_1(f''+f,\xi)\sum_{j=0}^{r}|\alpha_j|\,j^2\frac{\Gamma(\beta)\,\alpha\xi^{2\alpha\beta-1}\xi^{3-2\alpha\beta}}{\Gamma\left(\frac{1}{2\alpha}\right)\Gamma\left(\beta-\frac{1}{2\alpha}\right)}$$

$$\left[\frac{\Gamma\left(\frac{3}{2\alpha}\right)\Gamma\left(\beta-\left(\frac{3}{2\alpha}\right)\right)}{2\alpha\Gamma(\beta)}+\frac{j}{3}\frac{\Gamma\left(\frac{2}{\alpha}\right)\Gamma\left(\beta-\frac{2}{\alpha}\right)}{2\alpha\Gamma(\beta)}\right]$$

$$=\frac{\omega_1(f''+f,\xi)}{2}\sum_{j=0}^{r}|\alpha_j|\,j^2\frac{\xi^2}{\Gamma\left(\frac{1}{2\alpha}\right)\Gamma\left(\beta-\frac{1}{2\alpha}\right)}$$

$$\left[\Gamma\left(\frac{3}{2\alpha}\right)\Gamma\left(\beta-\frac{3}{2\alpha}\right)+\frac{j}{3}\Gamma\left(\frac{2}{\alpha}\right)\Gamma\left(\beta-\frac{2}{\alpha}\right)\right]. \tag{26.47}$$

We have proved that

$$\left|\overline{\Delta}_3(x)\right|\leq\frac{\omega_1(f''+f,\xi)\,\xi^2}{2\Gamma\left(\frac{1}{2\alpha}\right)\Gamma\left(\beta-\frac{1}{2\alpha}\right)}$$

$$\sum_{j=0}^{r}|\alpha_j|\,j^2\left[\Gamma\left(\frac{3}{2\alpha}\right)\Gamma\left(\beta-\frac{3}{2\alpha}\right)+\frac{j}{3}\Gamma\left(\frac{2}{\alpha}\right)\Gamma\left(\beta-\frac{2}{\alpha}\right)\right]; \tag{26.48}$$

$\xi>0$, $x\in\mathbb{R}$; and $\beta>\frac{2}{\alpha}$, $\alpha\in\mathbb{N}$.

Note that $\overline{\Delta}_3(x)\to 0$, as $\xi\to 0$.

Next we simplify the left-hand side of (26.43).

We observe the following:

Clearly, it is

$$\int_{-\infty}^{\infty}\sin(jt)\frac{1}{\left(t^{2\alpha}+\xi^{2\alpha}\right)^{\beta}}dt=0,\quad j=0,1,\ldots,r. \tag{26.49}$$

Furthermore, we have that

$$W\int_{-\infty}^{\infty}\sin^2\left(\frac{jt}{2}\right)\frac{1}{\left(t^{2\alpha}+\xi^{2\alpha}\right)^{\beta}}dt=W2\int_{0}^{\infty}\sin^2\left(\frac{jt}{2}\right)\frac{1}{\left(t^{2\alpha}+\xi^{2\alpha}\right)^{\beta}}dt$$

$$=W2\xi^{-2\alpha\beta}\xi\int_{0}^{\infty}\sin^2\left(\left(\frac{j\xi}{2}\right)\frac{t}{\xi}\right)\frac{1}{\left(\left(\frac{t}{\xi}\right)^{2\alpha}+1\right)^{\beta}}d\left(\frac{t}{\xi}\right)= \tag{26.50}$$

($\frac{t}{\xi} =: x$ and $\frac{j\xi}{2} =: \gamma$)

$$= 2W\xi^{1-2\alpha\beta} \int_0^\infty \sin^2(\gamma x) \frac{1}{(x^{2\alpha}+1)^\beta} dx$$

$$= 2\frac{\Gamma(\beta)\,\alpha\xi^{2\alpha\beta-1}}{\Gamma\left(\frac{1}{2\alpha}\right)\Gamma\left(\beta-\frac{1}{2\alpha}\right)}\xi^{1-2\alpha\beta} \int_0^\infty \frac{\sin^2(\gamma x)}{(x^{2\alpha}+1)^\beta} dx$$

$$= \frac{2\alpha\Gamma(\beta)}{\Gamma\left(\frac{1}{2\alpha}\right)\Gamma\left(\beta-\frac{1}{2\alpha}\right)} \int_0^\infty \frac{\sin^2(\gamma x)}{(x^{2\alpha}+1)^\beta} dx$$

$$\leq \frac{2\alpha\Gamma(\beta)}{\Gamma\left(\frac{1}{2\alpha}\right)\Gamma\left(\beta-\frac{1}{2\alpha}\right)} \int_0^\infty \frac{dx}{(1+x^{2\alpha})^\beta} = \tag{26.51}$$

(by Zwillinger (1995), p. 397, formula 595)

$$\frac{2\alpha\Gamma(\beta)}{\Gamma\left(\frac{1}{2\alpha}\right)\Gamma\left(\beta-\frac{1}{2\alpha}\right)} \frac{\Gamma\left(\frac{1}{2\alpha}\right)\Gamma\left(\beta-\frac{1}{2\alpha}\right)}{2\alpha\Gamma(\beta)} = 1, \tag{26.52}$$

for $\beta > \frac{1}{2\alpha}$, $\alpha \in \mathbb{N}$.

Therefore, it holds ($j = 0, 1, \ldots, r$)

$$W \int_{-\infty}^\infty \sin^2\left(\frac{jt}{2}\right) \frac{1}{(t^{2\alpha}+\xi^{2\alpha})^\beta} dt \leq 1, \tag{26.53}$$

where $\beta > \frac{1}{2\alpha}$, $\alpha \in \mathbb{N}$.

Consequently, we get that

$$\overline{\Delta}_3(x) \overset{(26.43)}{=} M_{r,\xi}(f,x) - f(x)$$

$$- 4f''(x) \frac{\alpha\Gamma(\beta)}{\Gamma\left(\frac{1}{2\alpha}\right)\Gamma\left(\beta-\frac{1}{2\alpha}\right)} \sum_{j=0}^r \alpha_j \int_0^\infty \frac{\sin^2\left(\frac{j\xi}{2}t\right)}{(t^{2\alpha}+1)^\beta} dt = \Delta_3(x), \tag{26.54}$$

$\xi > 0$, $x \in \mathbb{R}$; $\beta > \frac{1}{2\alpha}$, $\alpha \in \mathbb{N}$.

The theorem is proved. □

We give the following.

Corollary 26.2. *(to Theorem 26.2) Additionally assume $f'' \in C_B(\mathbb{R})$. It follows ($\xi > 0$, $x \in \mathbb{R}$)*

$$\|M_{r,\xi}(f) - f\|_\infty \leq \frac{\omega_1(f'' + f, \xi)\,\xi^2}{2\Gamma\left(\frac{1}{2\alpha}\right)\Gamma\left(\beta - \frac{1}{2\alpha}\right)}$$

$$\sum_{j=0}^{r} |\alpha_j|\, j^2 \left[\Gamma\left(\frac{3}{2\alpha}\right)\Gamma\left(\beta - \frac{3}{2\alpha}\right) + \frac{j}{3}\Gamma\left(\frac{2}{\alpha}\right)\Gamma\left(\beta - \frac{2}{\alpha}\right)\right]$$

$$+ 4\|f''\|_\infty \frac{\alpha\Gamma(\beta)}{\Gamma\left(\frac{1}{2\alpha}\right)\Gamma\left(\beta - \frac{1}{2\alpha}\right)} \sum_{j=0}^{r} |\alpha_j| \int_0^\infty \frac{\sin^2\left(\frac{j\xi}{2}t\right)}{(1+t^{2\alpha})^\beta} dt \to 0, \quad \text{as } \xi \to 0, \tag{26.55}$$

$\xi > 0$, $\beta > \frac{1}{2\alpha}$, $\alpha \in \mathbb{N}$.

Proof. By (26.42) and (26.13). And we note that

$$\left|\sin\left(\frac{j\xi}{2}t\right)\right| \leq \frac{j\xi}{2}t, \quad \forall\, t \in \mathbb{R}_+.$$

Hence we get

$$\sin^2\left(\frac{j\xi}{2}t\right) \leq \frac{j^2\xi^2}{4}t^2, \quad \forall\, t \in \mathbb{R}_+.$$

Consequently, it holds

$$\int_0^\infty \frac{\sin^2\left(\frac{j\xi}{2}t\right)}{(1+t^{2\alpha})^\beta} dt \leq \frac{j^2\xi^2}{4} \int_0^\infty \frac{t^2}{(1+t^{2\alpha})^\beta} dt$$

$$= \frac{j^2\xi^2}{8} \frac{\Gamma\left(\frac{3}{2\alpha}\right)\Gamma\left(\beta - \left(\frac{3}{2\alpha}\right)\right)}{\alpha\Gamma(\beta)} \to 0, \quad \text{as } \xi \to 0.$$

□

The last main result comes next.

Theorem 26.3. *It holds*

$$|\Delta_4(x)| \le \|\Delta_4\|_\infty \le \frac{\omega_1(f''+f,\xi)}{\lambda_1}\xi^2$$

$$\times \left\{ \sum_{j=0}^{r} |\alpha_j| j^2 \left[\frac{\pi(-1)^{\beta-1}(2\beta)!}{8(2\beta-3)!} \left(\sum_{k=1}^{\beta} (-1)^k \frac{k^{2\beta-3}}{(\beta-k)!(\beta+k)!} \right) \right.\right.$$

$$\left.\left. + \frac{j}{3} \left(\frac{-(2\beta)!}{8(2\beta-4)!} \left(\sum_{k=1}^{\beta} \frac{(-1)^{\beta-k} k^{2\beta-k} \ln(2k)}{(\beta-k)!(\beta+k)!} \right) \right) \right] \right\} =: A_4, \tag{26.56}$$

$\xi > 0$, $x \in \mathbb{R}$ *and* $\beta \in \mathbb{N} - \{1, 2\}$.

Note that $\|\Delta_4\|_\infty \to 0$, *as* $\xi \to 0$. *If* $f''(x) = 0$, *then* $|T_{r,\xi}(f,x) - f(x)| \le A_4$ *and* $T_{r,\xi}(f,x) \to f(x)$, *as* $\xi \to 0$.

Proof. The proof is the same as the proof of Theorem 26.1 from start till equation (26.31).

We call

$$\overline{\Delta}_4(x) := T_{r,\xi}(f,x) - f(x) - f'(x)\sum_{j=0}^{r} \alpha_j \lambda^{-1} \left(\int_{-\infty}^{\infty} \sin(jt) \left(\frac{\sin\left(\frac{t}{\xi}\right)}{t} \right)^{2\beta} dt \right)$$

$$- 2f''(x) \sum_{j=0}^{r} \alpha_j \lambda^{-1} \left(\int_{-\infty}^{\infty} \sin^2\left(\frac{jt}{2}\right) \left(\frac{\sin\left(\frac{t}{\xi}\right)}{t} \right)^{2\beta} dt \right)$$

$$= \lambda^{-1} \int_{-\infty}^{\infty} R(t) \left(\frac{\sin\left(\frac{t}{\xi}\right)}{t} \right)^{2\beta} dt. \tag{26.57}$$

Hence we get

$$\left|\overline{\Delta}_4(x)\right| \le \lambda^{-1} \int_{-\infty}^{\infty} |R(t)| \left(\frac{\sin\left(\frac{t}{\xi}\right)}{t} \right)^{2\beta} dt$$

$$\le \lambda^{-1} \int_{-\infty}^{\infty} \left[\omega_1(f''+f,\xi) \sum_{j=0}^{r} |\alpha_j| j^2 \left[\frac{t^2}{2} + \frac{j}{6\xi}|t|^3 \right] \right] \left(\frac{\sin\left(\frac{t}{\xi}\right)}{t} \right)^{2\beta} dt \tag{26.58}$$

$$= \omega_1 (f'' + f, \xi) \sum_{j=0}^{r} |\alpha_j| j^2 \lambda^{-1} \int_{-\infty}^{\infty} \left[\frac{t^2}{2} + \frac{j}{6\xi} |t|^3 \right] \left(\frac{\sin\left(\frac{t}{\xi}\right)}{t} \right)^{2\beta} dt$$

$$= \omega_1 (f'' + f, \xi) \sum_{j=0}^{r} |\alpha_j| j^2$$

$$\left[\frac{\lambda^{-1}}{2} \int_{-\infty}^{\infty} t^2 \left(\frac{\sin\left(\frac{t}{\xi}\right)}{t} \right)^{2\beta} dt + \frac{j\lambda^{-1}}{6\xi} \int_{-\infty}^{\infty} |t|^3 \left(\frac{\sin\left(\frac{t}{\xi}\right)}{t} \right)^{2\beta} dt \right] \tag{26.59}$$

$$= \omega_1 (f'' + f, \xi) \sum_{j=0}^{r} |\alpha_j| j^2$$

$$\times \left[\lambda^{-1} \int_{0}^{\infty} t^2 \left(\frac{\sin\left(\frac{t}{\xi}\right)}{t} \right)^{2\beta} dt + \frac{j\lambda^{-1}}{3\xi} \int_{0}^{\infty} t^3 \left(\frac{\sin\left(\frac{t}{\xi}\right)}{t} \right)^{2\beta} dt \right]$$

$$= \omega_1 (f'' + f, \xi) \sum_{j=0}^{r} |\alpha_j| j^2 \left[\lambda^{-1} \xi^{3-2\beta} \int_{0}^{\infty} \left(\frac{t}{\xi} \right)^2 \left(\frac{\sin\left(\frac{t}{\xi}\right)}{\left(\frac{t}{\xi}\right)} \right)^{2\beta} d\left(\frac{t}{\xi} \right) \right.$$

$$\left. + \frac{j\lambda^{-1}}{3\xi} \xi^{4-2\beta} \int_{0}^{\infty} \left(\frac{t}{\xi} \right)^3 \left(\frac{\sin\left(\frac{t}{\xi}\right)}{\left(\frac{t}{\xi}\right)} \right)^{2\beta} d\left(\frac{t}{\xi} \right) \right]$$

$$= \omega_1 (f'' + f, \xi) \sum_{j=0}^{r} |\alpha_j| j^2$$

$$\left[\lambda^{-1} \xi^{3-2\beta} \int_{0}^{\infty} x^2 \left(\frac{\sin x}{x} \right)^{2\beta} dx + \frac{j\lambda^{-1}}{3} \xi^{3-2\beta} \int_{0}^{\infty} x^3 \left(\frac{\sin x}{x} \right)^{2\beta} dx \right] \tag{26.60}$$

(by Edwards (1954), p. 210, item 1033)

$$= \omega_1 (f'' + f, \xi) \sum_{j=0}^{r} |\alpha_j| j^2$$

$$\times \left[\lambda_1^{-1} \xi^2 \left(\frac{\pi (-1)^{\beta-1} (2\beta)!}{8 (2\beta - 3)!} \left(\sum_{k=1}^{\beta} (-1)^k \frac{k^{2\beta-3}}{(\beta - k)! (\beta + k)!} \right) \right) \right.$$

$$+\frac{j\lambda_1^{-1}}{3}\xi^2\left(\frac{-(2\beta)!}{8(2\beta-4)!}\left(\sum_{k=1}^{\beta}\frac{(-1)^{\beta-k}k^{2\beta-k}\ln(2k)}{(\beta-k)!(\beta+k)!}\right)\right)\Bigg]$$

$$=\frac{\omega_1(f''+f,\xi)}{\lambda_1}\xi^2\left\{\sum_{j=0}^{r}|\alpha_j|j^2\left[\frac{\pi(-1)^{\beta-1}(2\beta)!}{8(2\beta-3)!}\left(\sum_{k=1}^{\beta}(-1)^k\frac{k^{2\beta-3}}{(\beta-k)!(\beta+k)!}\right)\right.\right.$$

$$\left.\left.+\frac{j}{3}\left(\frac{-(2\beta)!}{8(2\beta-4)!}\left(\sum_{k=1}^{\beta}\frac{(-1)^{\beta-k}k^{2\beta-4}\ln(2k)}{(\beta-k)!(\beta+k)!}\right)\right)\right]\right\}. \tag{26.61}$$

We have proved that

$$\left|\overline{\Delta}_4(x)\right|\le\frac{\omega_1(f''+f,\xi)}{\lambda_1}\xi^2\left\{\sum_{j=0}^{r}|\alpha_j|j^2\right.$$

$$\left[\frac{\pi(-1)^{\beta-1}(2\beta)!}{8(2\beta-3)!}\left(\sum_{k=1}^{\beta}(-1)^k\frac{k^{2\beta-3}}{(\beta-k)!(\beta+k)!}\right)\right.$$

$$\left.\left.+\frac{j}{3}\left(\frac{-(2\beta)!}{8(2\beta-4)!}\left(\sum_{k=1}^{\beta}\frac{(-1)^{\beta-k}k^{2\beta-4}\ln(2k)}{(\beta-k)!(\beta+k)!}\right)\right)\right]\right\}, \tag{26.62}$$

$\xi>0$, $x\in\mathbb{R}$; and $\beta\in\mathbb{N}-\{1,2\}$.

Note that $\overline{\Delta}_4(x)\to 0$, as $\xi\to 0$.

Next we simplify the left-hand side of (26.57).

We observe the following:

Clearly, it is

$$\int_{-\infty}^{\infty}\sin(jt)\left(\frac{\sin\left(\frac{t}{\xi}\right)}{t}\right)^{2\beta}dt=0,\quad j=0,1,\ldots,r. \tag{26.63}$$

Furthermore, we have that

$$\lambda^{-1}\int_{-\infty}^{\infty}\sin^2\left(\frac{jt}{2}\right)\left(\frac{\sin\left(\frac{t}{\xi}\right)}{t}\right)^{2\beta}dt$$

$$=2\lambda_1^{-1}\xi^{2\beta-1}\xi^{1-2\beta}\int_0^{\infty}\sin^2\left(\left(\frac{j\xi}{2}\right)\frac{t}{\xi}\right)\left(\frac{\sin\left(\frac{t}{\xi}\right)}{\frac{t}{\xi}}\right)^{2\beta}d\left(\frac{t}{\xi}\right)=$$

($\frac{t}{\xi} =: x$ and $\frac{j\xi}{2} =: \gamma$)

$$= 2\lambda_1^{-1} \int_0^{\infty} \sin^2(\gamma x) \left(\frac{\sin x}{x}\right)^{2\beta} dx$$

$$\leq 2\lambda_1^{-1} \int_0^{\infty} \left(\frac{\sin x}{x}\right)^{2\beta} dx \tag{26.64}$$

(by Edwards (1954), p. 210, item 1033)

$$= 2\lambda_1^{-1}\pi(-1)^{\beta}\beta\left(\sum_{k=1}^{\beta}(-1)^k \frac{k^{2\beta-1}}{(\beta-k)!(\beta+k)!}\right) = 1, \quad \text{for } \beta \in \mathbb{N}. \tag{26.65}$$

Therefore, it holds ($j = 0, 1, \ldots, r$)

$$\lambda^{-1} \int_{-\infty}^{\infty} \sin^2\left(\frac{jt}{2}\right) \left(\frac{\sin\left(\frac{t}{\xi}\right)}{t}\right)^{2\beta} dt \leq 1, \tag{26.66}$$

where $\beta \in \mathbb{N}$.

Consequently, we get that

$$\overline{\Delta}_4(x) \overset{(26.57)}{=} T_{r,\xi}(f,x) - f(x)$$
$$- 4f''(x)\lambda_1^{-1}\sum_{j=0}^{r}\alpha_j \int_0^{\infty} \sin^2\left(\frac{j\xi}{2}t\right)\left(\frac{\sin t}{t}\right)^{2\beta} dt = \Delta_4(x), \tag{26.67}$$

where $\xi > 0$, $x \in \mathbb{R}$; $\beta \in \mathbb{N} - \{1, 2\}$.

The theorem is proved. □

We finish with the following.

Corollary 26.3. *(to Theorem 26.3) Additionally assume $f'' \in C_B(\mathbb{R})$. It follows ($\xi > 0$, $x \in \mathbb{R}$)*

$$\|T_{r,\xi}(f) - f\|_{\infty} \leq \frac{\omega_1(f'' + f, \xi)}{\lambda_1}\xi^2$$
$$\times \left\{\sum_{j=0}^{r}|\alpha_j| j^2 \left[\frac{\pi(-1)^{\beta-1}(2\beta)!}{8(2\beta-3)!}\left(\sum_{k=1}^{\beta}(-1)^k \frac{k^{2\beta-3}}{(\beta-k)!(\beta+k)!}\right)\right.\right.$$

$$+\frac{j}{3}\left(\frac{-(2\beta)!}{8(2\beta-4)!}\left(\sum_{k=1}^{\beta}\frac{(-1)^{\beta-k}k^{2\beta-4}\ln(2k)}{(\beta-k)!(\beta+k)!}\right)\right)\Bigg]\Bigg\}$$

$$+4\|f''\|_{\infty}\lambda_1^{-1}\sum_{j=0}^{r}|\alpha_j|\int_0^{\infty}\sin^2\left(\frac{j\xi}{2}t\right)\left(\frac{\sin t}{t}\right)^{2\beta}dt\to 0,\quad as\ \xi\to 0, \tag{26.68}$$

where $\xi>0$, $\beta\in\mathbb{N}-\{1,2\}$.

Proof. By (26.56) and (26.19). And from

$$\sin^2\left(\frac{j\xi}{2}t\right)\le\frac{j^2\xi^2}{4}t^2,\ \ \forall\, t\in\mathbb{R}_+,$$

we obtain that

$$\int_0^{\infty}\sin^2\left(\frac{j\xi}{2}t\right)\left(\frac{\sin t}{t}\right)^{2\beta}dt\le\frac{j^2\xi^2}{4}\int_0^{\infty}t^2\left(\frac{\sin t}{t}\right)^{2\beta}dt$$

$$=\frac{j^2\xi^2}{4}\left(\frac{\pi(-1)^{\beta-1}(2\beta)!}{8(2\beta-3)!}\left(\sum_{k=1}^{\beta}(-1)^k\frac{k^{2\beta-3}}{(\beta-k)!(\beta+k)!}\right)\right)\to 0,\ \text{as}\ \xi\to 0.$$

□

Bibliography

A.H. Ali, Z. Pales, Taylor-type expansions in terms of exponential polynomials, *Mathematical Inequalities and Applications*, 25(4) (2022), 1123–1141.

G.A. Anastassiou, Basic convergence with rates of smooth Picard singular integral operators, *Journal of Computational Analysis and Applications*, 8(4) (2006), 313-334.

G.A. Anastassiou, *Intelligent Mathematics: Computational Analysis*, Springer, Heidelberg, New York, Chapter 10, 2011.

G.A. Anastassiou, Opial and Ostrowski type inequalities based on trigonometric and hyperbolic type Taylor formulae, *Mathematische Nachrichten*, 11(S) (2023), 1–26.

G.A. Anastassiou, Trigonometric derived rate of convergence of various smooth singular integral operators, *Modern Mathematical Methods*, 2 (2024) (1), 27–40.

G.A. Anastassiou and R. Mezei, *Approximation by Singular Integrals*, Cambridge Scientific Publishers, Cambridge, UK, 2012.

J. Edwards, *A Treatise on the Integral Calculus, Vol II*, Chelsea, New York, 1954.

D. Zwillinger, *CRC Standard Mathematical Tables and Formulae*, 30th edn. Chapman & Hall/CRC, Boca Raton (1995).

Chapter 27

Trigonometric Derived L_p Quantitative Approximation by Various Smooth Singular Integral Operators

In this chapter, we continue with the study of smooth Gauss–Weierstrass, Poisson–Cauchy and trigonometric singular integral operators that started in Anastassiou (2011), see there Chapters 10–14. This time, the foundation of our research is a trigonometric Taylor's formula. We establish the univariate L_p convergence of our operators to the unit operator with rates via Jackson-type inequalities involving the first L_p modulus of continuity. Of interest, here is a residual appearing term. Note that our operators are not positive. It follows Anastassiou (2024).

27.1 Introduction

We are motivated by Anastassiou (2006, 2011) Chapters 10–14, Anastassiou (2023), and Ali and Pales (2022). We use a trigonometric new Taylor formula from Anastassiou (2023), see also Ali and Pales (2022). Here we consider some very general operators, the smooth Gauss–Weierstrass, Poisson–Cauchy and trigonometric singular integral operators over the real line and we study further their L_p, $p \geq 1$, convergence properties quantitatively. We establish related inequalities involving the first L_p, $p \geq 1$, modulus of continuity with respect to L_p, $p \geq 1$, norm. We provide detailed proofs.

For recent related works, see Bardaro *et al.* (2021), Tapiawala *et al.* (2019)–Uysal and Dutta (2020). Other important articles on the topic are Aral (2005)–Aral and Gal (2008) and Mohapatra and Rodriguez (1990).

For the history of the topic, we mention about the monograph Anastassiou and Mezei (2012) of 2012, which was the first complete source to deal exclusively with the classic theory of the approximation of singular integrals to the identity-unit operator. The authors there studied quantitatively the basic approximation properties of the general Picard, Gauss–Weierstrass and Poisson–Cauchy singular integral operators over the real line, which are not positive linear operators. In particular, they studied the rate of convergence of these operators to the unit operator, as well as the related simultaneous approximation. This is given via inequalities and with the use of higher-order modulus of smoothness of the high order derivative of the involved function. Some of these inequalities are proven to be sharp. Also, they studied the global smoothness preservation property of these operators. Furthermore, they gave asymptotic expansions of Voronovskaya type for the error of approximation. They continued with the study of related properties of the general fractional Gauss–Weierstrass and Poisson–Cauchy singular integral operators. These properties were studied with respect to L_p norm, $1 \leq p \leq \infty$. The case of Lipschitz-type function approximation was studied separately and in detail. Furthermore, they presented the corresponding general approximation theory of general singular integral operators with lots of applications to, the under focused till then, trigonometric singular integral.

27.2 Basics

By Ali and Pales (2022) and Anastassiou (2023), for $f \in C^2(\mathbb{R})$ and $a, x \in \mathbb{R}$, we have by trigonometric Taylor formula

$$f(x) - f(a) = f'(a)\sin(x-a) + 2f''(a)\sin^2\left(\frac{x-a}{2}\right)$$

$$+ \int_a^x [(f''(t) + f(t)) - (f''(a) + f(a))]\sin(x-t)\,dt. \tag{27.1}$$

For $r \in \mathbb{N}$ and $n \in \mathbb{Z}^+$, we set

$$\alpha_j := \begin{cases} (-1)^{r-j}\binom{r}{j} j^{-n}, & j = 1, \ldots, r, \\ 1 - \sum\limits_{j=1}^{r} (-1)^{r-j}\binom{r}{j} j^{-n}, & j = 0, \end{cases} \tag{27.2}$$

that is,

$$\sum_{j=0}^{r} \alpha_j = 1. \tag{27.3}$$

Here we consider both $f, f'' \in L_p(\mathbb{R}) \cap C(\mathbb{R})$, $1 \leq p < \infty$.

Denote by

$$\omega_1(f, h)_p := \sup_{\substack{t \in \mathbb{R} \\ |t| \leq h}} \|f(x+t) - f(x)\|_{p,x}, \quad h > 0, \tag{27.4}$$

the first L_p modulus of continuity of f, $1 \leq p < \infty$.

(I) We define the smooth Gauss–Weierstrass singular integral operators (Anastassiou and Mezei 2012).

Let $f \in C^2(\mathbb{R})$, we define for $x \in \mathbb{R}$, $\xi > 0$ the Lebesgue integral

$$W_{r,\xi}(f; x) = \frac{1}{\sqrt{\pi\xi}} \int_{-\infty}^{\infty} \left(\sum_{j=0}^{r} \alpha_j f(x + jt) \right) e^{-\frac{t^2}{\xi}} dt. \tag{27.5}$$

We assume that $W_{r,\xi}(f; x) \in \mathbb{R}$ for all $x \in \mathbb{R}$. We use also that

$$W_{r,\xi}(f; x) = \frac{1}{\sqrt{\pi\xi}} \sum_{j=0}^{r} \alpha_j \left(\int_{-\infty}^{\infty} f(x + jt) e^{-\frac{t^2}{\xi}} dt \right), \tag{27.6}$$

note by $\frac{1}{\sqrt{\pi\xi}} \int_{-\infty}^{\infty} e^{-\frac{t^2}{\xi}} dt = 1$ that $W_{r,\xi}(c; x) = c$, c constant and

$$W_{r,\xi}(f; x) - f(x) = \frac{1}{\sqrt{\pi\xi}} \left(\sum_{j=0}^{r} \alpha_j \int_{-\infty}^{\infty} (f(x + jt) - f(x)) e^{-\frac{t^2}{\xi}} dt \right). \tag{27.7}$$

We set

$$\Delta_2(x) := W_{r,\xi}(f; x) - f(x) - f''(x) \sum_{j=0}^{r} \alpha_j \left(1 - e^{-\frac{j^2}{4}\xi} \right), \quad x \in \mathbb{R}, \tag{27.8}$$

$j = 0, 1, \ldots, r \in \mathbb{N}$.

(II) We define the smooth Poisson-Cauchy singular integral operators (Anastassiou and Mezei 2012).

Let $\alpha \in \mathbb{N}$, $\beta > \frac{1}{2\alpha}$ and $f \in C^2(\mathbb{R})$. We define for $x \in \mathbb{R}$, $\xi > 0$ the Lebesgue integral

$$M_{r,\xi}(f; x) = W \int_{-\infty}^{\infty} \frac{\sum_{j=0}^{r} \alpha_j f(x + jt)}{(t^{2\alpha} + \xi^{2\alpha})^{\beta}} dt, \tag{27.9}$$

where the constant is defined as

$$W=\frac{\Gamma(\beta)\,\alpha\xi^{2\alpha\beta-1}}{\Gamma\left(\frac{1}{2\alpha}\right)\Gamma\left(\beta-\frac{1}{2\alpha}\right)}. \tag{27.10}$$

We assume that $M_{r,\xi}(f;x)\in\mathbb{R}$ for all $x\in\mathbb{R}$. We use also that

$$M_{r,\xi}(f;x)=W\sum_{j=0}^{r}\alpha_j\left(\int_{-\infty}^{\infty}f(x+jt)\frac{1}{\left(t^{2\alpha}+\xi^{2\alpha}\right)^{\beta}}dt\right). \tag{27.11}$$

We note by $W\int_{-\infty}^{\infty}\frac{1}{(t^{2\alpha}+\xi^{2\alpha})^{\beta}}dt=1$ that $M_{r,\xi}(c;x)=c$, c constant and

$$M_{r,\xi}(f;x)-f(x)=W\left(\sum_{j=0}^{r}\alpha_j\int_{-\infty}^{\infty}[f(x+jt)-f(x)]\frac{1}{\left(t^{2\alpha}+\xi^{2\alpha}\right)^{\beta}}dt\right). \tag{27.12}$$

We set

$$\Delta_3(x):=M_{r,\xi}(f;x)-f(x)$$
$$-4f''(x)\frac{\alpha\Gamma(\beta)}{\Gamma\left(\frac{1}{2\alpha}\right)\Gamma\left(\beta-\frac{1}{2\alpha}\right)}\sum_{j=0}^{r}\alpha_j\int_0^{\infty}\frac{\sin^2\left(\frac{j\xi}{2}t\right)}{\left(1+t^{2\alpha}\right)^{\beta}}dt, \tag{27.13}$$

$\xi>0$, $x\in\mathbb{R}$; $\beta>\frac{1}{2\alpha}$, $\alpha\in\mathbb{N}$; $j=0,1,\ldots,r\in\mathbb{N}$.

(III) We define the smooth trigonometric singular integral operators (Anastassiou and Mezei 2012) as follows:

Let $\xi>0$, $f\in C^2(\mathbb{R})$, $x\in\mathbb{R}$, $\beta\in\mathbb{N}$; we set

$$T_{r,\xi}(f;x):=\frac{1}{\lambda}\int_{-\infty}^{\infty}\left(\sum_{j=0}^{r}\alpha_j f(x+jt)\right)\left(\frac{\sin\left(\frac{t}{\xi}\right)}{t}\right)^{2\beta}dt, \tag{27.14}$$

where

$$\lambda:=\int_{-\infty}^{\infty}\left(\frac{\sin\left(\frac{t}{\xi}\right)}{t}\right)^{2\beta}dt=2\xi^{1-2\beta}\int_0^{\infty}\left(\frac{\sin t}{t}\right)^{2\beta}dt \tag{27.15}$$

(by Edwards (1954), p. 210, item 1033)

$$=2\xi^{1-2\beta}\pi(-1)^{\beta}\beta\sum_{k=1}^{\beta}(-1)^k\frac{k^{2\beta-1}}{(\beta-k)!\,(\beta+k)!}.$$

Denote

$$\lambda_1 := 2\pi (-1)^{\beta} \beta \sum_{k=1}^{\beta} (-1)^k \frac{k^{2\beta-1}}{(\beta-k)!\,(\beta+k)!}, \tag{27.16}$$

that is,

$$\lambda = \lambda_1 \xi^{1-2\beta}. \tag{27.17}$$

We suppose that $T_{r,\xi}(f;x) \in \mathbb{R}$ for all $x \in \mathbb{R}$. Clearly, again it is

$$\frac{1}{\lambda} \int_{-\infty}^{\infty} \left(\frac{\sin\left(\frac{t}{\xi}\right)}{t} \right)^{2\beta} dt = 1, \tag{27.18}$$

and $T_{r,\xi}(c;x) = c$, c constant.

We set

$$\Delta_4(x) := T_{r,\xi}(f;x) - f(x) - 4f''(x)\lambda_1^{-1} \sum_{j=0}^{r} \alpha_j \int_0^{\infty} \sin^2\left(\frac{j\xi}{2}t\right) \left(\frac{\sin t}{t}\right)^{2\beta} dt, \tag{27.19}$$

where $\xi > 0$, $x \in \mathbb{R}$; $\beta \in \mathbb{N} - \{1,2\}$; $j = 0,1,\ldots,r \in \mathbb{N}$.

We need the following set of calculations.

Remark 27.1. By (27.1) we get that

$$f(x+jt) - f(x) = f'(x)\sin(jt) + 2f''(x)\sin^2\left(\frac{jt}{2}\right) + \int_x^{x+jt} \left[(f''(s)+f(s)) - (f''(x)+f(x))\right] \sin(x+jt-s)\,ds, \tag{27.20}$$

or better

$$f(x+jt) - f(x) = f'(x)\sin(jt) + 2f''(x)\sin^2\left(\frac{jt}{2}\right) + \int_0^{jt} \left[(f''(x+z)+f(x+z)) - (f''(x)+f(x))\right] \sin(jt-z)\,dz. \tag{27.21}$$

Furthermore, it holds

$$\sum_{j=0}^{r}\alpha_j\left[f\left(x+jt\right)-f\left(x\right)\right]$$
$$=f'\left(x\right)\sum_{j=0}^{r}\alpha_j\sin\left(jt\right)+2f''\left(x\right)\sum_{j=0}^{r}\alpha_j\sin^2\left(\frac{jt}{2}\right)$$
$$+\sum_{j=0}^{r}\alpha_j\int_0^{jt}\left[\left(f''\left(x+z\right)+f\left(x+z\right)\right)-\left(f''\left(x\right)+f\left(x\right)\right)\right]\sin\left(jt-z\right)dz, \tag{27.22}$$

or better

$$\sum_{j=0}^{r}\alpha_j\left[f\left(x+jt\right)-f\left(x\right)\right]$$
$$=f'\left(x\right)\sum_{j=0}^{r}\alpha_j\sin\left(jt\right)+2f''\left(x\right)\sum_{j=0}^{r}\alpha_j\sin^2\left(\frac{jt}{2}\right)$$
$$+\sum_{j=0}^{r}\alpha_j j\int_0^{t}\left[\left(f''\left(x+jw\right)+f\left(x+jw\right)\right)-\left(f''\left(x\right)+f\left(x\right)\right)\right]\sin j\left(t-w\right)dw. \tag{27.23}$$

Call

$$R:=R\left(t\right)$$
$$:=\sum_{j=0}^{r}\alpha_j j\int_0^{t}\left[\left(f''\left(x+jw\right)+f\left(x+jw\right)\right)-\left(f''\left(x\right)+f\left(x\right)\right)\right]\sin j\left(t-w\right)dw, \tag{27.24}$$

$\forall\, t\in\mathbb{R}$.

We note that

$$\Delta_2^*\left(x\right):=W_{r,\xi}\left(f;x\right)-f\left(x\right)-f'\left(x\right)\sum_{j=0}^{r}\alpha_j\frac{1}{\sqrt{\pi\xi}}\left(\int_{-\infty}^{\infty}\sin\left(jt\right)e^{-\frac{t^2}{\xi}}dt\right)$$
$$-2f''\left(x\right)\sum_{j=0}^{r}\alpha_j\frac{1}{\sqrt{\pi\xi}}\left(\int_{-\infty}^{\infty}\sin^2\left(\frac{jt}{2}\right)e^{-\frac{t^2}{\xi}}dt\right)$$
$$=\frac{1}{\sqrt{\pi\xi}}\int_{-\infty}^{\infty}R\left(t\right)e^{-\frac{t^2}{\xi}}dt, \tag{27.25}$$

and

$$\begin{aligned}\Delta_3^*(x) &:= M_{r,\xi}(f;x) - f(x)\\ &\quad - f'(x)\sum_{j=0}^{r}\alpha_j W\left(\int_{-\infty}^{\infty}\sin(jt)\frac{1}{\left(t^{2\alpha}+\xi^{2\alpha}\right)^{\beta}}dt\right)\\ &\quad - 2f''(x)\sum_{j=0}^{r}\alpha_j W\left(\int_{-\infty}^{\infty}\sin^2\left(\frac{jt}{2}\right)\frac{1}{\left(t^{2\alpha}+\xi^{2\alpha}\right)^{\beta}}dt\right)\\ &= W\int_{-\infty}^{\infty}R(t)\frac{1}{\left(t^{2\alpha}+\xi^{2\alpha}\right)^{\beta}}dt, \end{aligned} \tag{27.26}$$

and

$$\begin{aligned}\Delta_4^*(x) &:= T_{r,\xi}(f;x) - f(x) - f'(x)\sum_{j=0}^{r}\alpha_j\frac{1}{\lambda}\left(\int_{-\infty}^{\infty}\sin(jt)\left(\frac{\sin\left(\frac{t}{\xi}\right)}{t}\right)^{2\beta}dt\right)\\ &\quad - 2f''(x)\sum_{j=0}^{r}\alpha_j\frac{1}{\lambda}\left(\int_{-\infty}^{\infty}\sin^2\left(\frac{jt}{2}\right)\left(\frac{\sin\left(\frac{t}{\xi}\right)}{t}\right)^{2\beta}dt\right)\\ &= \frac{1}{\lambda}\int_{-\infty}^{\infty}R(t)\left(\frac{\sin\left(\frac{t}{\xi}\right)}{t}\right)^{2\beta}dt. \end{aligned} \tag{27.27}$$

Clearly, it holds

$$\int_{-\infty}^{\infty}\sin(jt)\,e^{-\frac{t^2}{\xi}}dt = 0, \tag{27.28}$$

$$\int_{-\infty}^{\infty}\sin(jt)\frac{1}{\left(t^{2\alpha}+\xi^{2\alpha}\right)^{\beta}}dt = 0 \tag{27.29}$$

and

$$\int_{-\infty}^{\infty}\sin(jt)\left(\frac{\sin\left(\frac{t}{\xi}\right)}{t}\right)^{2\beta}dt = 0, \tag{27.30}$$

for all $j = 0, 1, \ldots, r$.

Furthermore, we have that

$$\int_{-\infty}^{\infty} \sin^2\left(\frac{jt}{2}\right) e^{-\frac{t^2}{\xi}} dt = 2\int_0^{\infty} \sin^2\left(\frac{jt}{2}\right) e^{-\frac{t^2}{\xi}} dt, \tag{27.31}$$

$$\int_{-\infty}^{\infty} \sin^2\left(\frac{jt}{2}\right) \frac{1}{\left(t^{2\alpha}+\xi^{2\alpha}\right)^{\beta}} dt = 2\int_0^{\infty} \sin^2\left(\frac{jt}{2}\right) \frac{1}{\left(t^{2\alpha}+\xi^{2\alpha}\right)^{\beta}} dt, \tag{27.32}$$

$$\int_{-\infty}^{\infty} \sin^2\left(\frac{jt}{2}\right) \left(\frac{\sin\left(\frac{t}{\xi}\right)}{t}\right)^{2\beta} dt = 2\int_0^{\infty} \sin^2\left(\frac{jt}{2}\right) \left(\frac{\sin\left(\frac{t}{\xi}\right)}{t}\right)^{2\beta} dt, \tag{27.33}$$

$j = 0, 1, \ldots, r$.

More specifically we have that

$$\int_{-\infty}^{\infty} \sin^2\left(\frac{jt}{2}\right) e^{-\frac{t^2}{\xi}} dt = 2\int_0^{\infty} \sin^2\left(\frac{jt}{2}\right) e^{-\frac{t^2}{\xi}} dt$$

$$= 2\sqrt{\xi}\int_0^{\infty} \sin^2\left(\left(\frac{j\sqrt{\xi}}{2}\right)\frac{t}{\sqrt{\xi}}\right) e^{-\left(\frac{t}{\sqrt{\xi}}\right)^2} d\frac{t}{\sqrt{\xi}}$$

($\frac{t}{\sqrt{\xi}} =: x$ and $\frac{j\sqrt{\xi}}{2} =: \beta_1$)

$$= 2\sqrt{\xi}\int_0^{\infty} \sin^2\left(\beta_1 x\right) e^{-x^2} dx$$

$$= \sqrt{\xi}\frac{1}{2}\sqrt{\pi} e^{-\beta_1^2}\left(e^{\beta_1^2} - 1\right)$$

$$= \frac{\sqrt{\xi}\sqrt{\pi}}{2} e^{-\frac{j^2}{4}\xi}\left(e^{\frac{j^2}{4}\xi} - 1\right). \tag{27.34}$$

Consequently, we derive

$$\Delta_2^*(x) \overset{\text{(by (27.8))}}{=} W_{r,\xi}(f,x) - f(x) - f''(x)\sum_{j=0}^{r}\alpha_j\left(1 - e^{-\frac{j^2}{4}\xi}\right) = \Delta_2(x). \tag{27.35}$$

Furthermore, we obtain

$$W\int_{-\infty}^{\infty} \sin^2\left(\frac{jt}{2}\right) \frac{1}{\left(t^{2\alpha}+\xi^{2\alpha}\right)^{\beta}} dt \le W\int_{-\infty}^{\infty} \frac{1}{\left(t^{2\alpha}+\xi^{2\alpha}\right)^{\beta}} dt = 1, \tag{27.36}$$

$\beta > \frac{1}{2\alpha}$, $\alpha \in \mathbb{N}$; $j = 0, 1, \ldots, r$.

Hence we get that

$$\Delta_3^*(x) \overset{(27.13)}{=} M_{r,\xi}(f;x) - f(x)$$

$$- 4f''(x) \frac{\alpha\Gamma(\beta)}{\Gamma\left(\frac{1}{2\alpha}\right)\Gamma\left(\beta - \frac{1}{2\alpha}\right)} \sum_{j=0}^{r} \alpha_j \int_0^\infty \frac{\sin^2\left(\frac{j\xi t}{2}\right)}{(1+t^{2\alpha})^\beta} dt = \Delta_3(x), \quad (27.37)$$

$\xi > 0$, $x \in \mathbb{R}$; $\beta > \frac{1}{2\alpha}$, $\alpha \in \mathbb{N}$; $j = 0, 1, \ldots, r \in \mathbb{N}$.

Similarly, we observe that

$$\lambda^{-1} \int_{-\infty}^{\infty} \sin^2\left(\frac{jt}{2}\right) \left(\frac{\sin\left(\frac{t}{\xi}\right)}{t}\right)^{2\beta} dt \leq \lambda^{-1} \int_{-\infty}^{\infty} \left(\frac{\sin\left(\frac{t}{\xi}\right)}{t}\right)^{2\beta} dt = 1. \quad (27.38)$$

Consequently, it holds

$$\Delta_4^*(x) = T_{r,\xi}(f;x) - f(x)$$

$$- 4f''(x)\lambda_1^{-1} \sum_{j=0}^{r} \alpha_j \int_0^\infty \sin^2\left(\frac{j\xi}{2}t\right)\left(\frac{\sin t}{t}\right)^{2\beta} dt = \Delta_4(x), \quad (27.39)$$

where $\xi > 0$, $x \in \mathbb{R}$; $\beta \in \mathbb{N} - \{1, 2\}$; $j = 0, 1, \ldots, r \in \mathbb{N}$.

Above it is $\lambda_1 \overset{(27.17)}{=} \lambda\xi^{2\beta-1}$.

27.3 Main Results

We present our first main result, about Gauss–Weierstrass operators, L_p approximation, $p > 1$.

Theorem 27.1. *Let $p, q > 1 : \frac{1}{p} + \frac{1}{q} = 1$, $\xi > 0$, and both $f, f'' \in L_p(\mathbb{R}) \cap C(\mathbb{R})$, $x \in \mathbb{R}$. Then*

$$\|\Delta_2(x)\|_p = \left\| W_{r,\xi}(f) - f - f''\left(\sum_{j=0}^{r} \alpha_j \left(1 - e^{-\frac{j^2}{4}\xi}\right)\right)\right\|_p$$

$$\leq \frac{2}{(\sqrt{\pi})^{\frac{1}{p}}} \left(\frac{2}{q}\right)^{\frac{1}{2q}} \left(\frac{r+1}{q+1}\right)^{\frac{1}{q}} \omega_1\left(f'' + f, \sqrt{\xi}\right)_p \xi$$

$$\left\{\sum_{j=0}^{r} |\alpha_j|^p j^{2p} \left\{\left[\frac{1}{(2p+1)} + \left(\frac{p}{2}\right)^{-(2p+1)} \Gamma(2p+1)\right]\right.\right.$$

$$+\frac{j^{p}}{(p+1)}\left[\frac{1}{(3p+1)}+\left(\frac{p}{2}\right)^{-(3p+1)}\Gamma(3p+1)\right]\Bigg\}\Bigg\}^{\frac{1}{p}}$$

$$:=\theta_{1}(\xi)\to 0,\ \ as\ \xi\to 0. \tag{27.40}$$

Above Γ *stands for the gamma function.*

Proof. From (27.24), let

$$I:=\int_{0}^{t}\left[\left(f''(x+jw)+f(x+jw)\right)-\left(f''(x)+f(x)\right)\right]\sin j(t-w)\,dw,$$

$$\forall\, t\in\mathbb{R}. \tag{27.41}$$

For $t<0$, we have that

$$|I|=\left|\int_{0}^{t}\left[\left(f''(x+jw)+f(x+jw)\right)-\left(f''(x)+f(x)\right)\right]\sin j(t-w)\,dw\right|$$

$$\leq\int_{0}^{t}\left|\left(f''(x+jw)+f(x+jw)\right)-\left(f''(x)+f(x)\right)\right|\left|\sin j(t-w)\right|dw$$

$$\leq j\int_{t}^{0}\left|\left(f''(x+jw)+f(x+jw)\right)-\left(f''(x)+f(x)\right)\right|(w-t)\,dw$$

$$=-j\int_{t}^{0}\left|\left(f''(x-j(-w))+f(x-j(-w))\right)-\left(f''(x)+f(x)\right)\right|$$

$$\times\left(-t-(-w)\right)d(-w) \tag{27.42}$$

$(t\leq w\leq 0\Rightarrow -t\geq -w=:\theta\geq 0)$

$$=-j\int_{-t}^{0}\left|\left(f''(x-j\theta)+f(x-j\theta)\right)-\left(f''(x)+f(x)\right)\right|(-t-\theta)\,d\theta$$

$$=j\int_{0}^{-t}\left|\left(f''(x-j\theta)+f(x-j\theta)\right)-\left(f''(x)+f(x)\right)\right|(-t-\theta)\,d\theta$$

$$=j\int_{0}^{|t|}\left|\left(f''(x+sign(t)\,j\theta)+f(x+sign(t)\,j\theta)\right)\right.$$

$$\left.-\left(f''(x)+f(x)\right)\right|(|t|-\theta)\,d\theta. \tag{27.43}$$

So, we have proved that

$$|I|\leq j\int_{0}^{|t|}\left|\left(f''(x+sign(t)\,j\theta)+f(x+sign(t)\,j\theta)\right)\right.$$

$$\left.-\left(f''(x)+f(x)\right)\right|(|t|-\theta)\,d\theta, \tag{27.44}$$

$\forall\, t\in\mathbb{R}$,

and by (27.24),

$$|R(t)| \le \sum_{j=0}^{r} |\alpha_j| j^2$$
$$\int_0^{|t|} |(f''(x + jsign(t)\theta) + f(x + jsign(t)\theta))$$
$$-(f''(x) + f(x))| (|t| - \theta) d\theta, \tag{27.45}$$

$\forall\, t \in \mathbb{R}$.

Let now $p, q > 1 : \frac{1}{p} + \frac{1}{q} = 1$.

We observe that

$$|R(t)| \le \sum_{j=0}^{r} |\alpha_j| j^2 \tag{27.46}$$
$$\left(\int_0^{|t|} |(f''(x + jsign(t)\theta) + f(x + jsign(t)\theta)) - (f''(x) + f(x))|^p d\theta\right)^{\frac{1}{p}}$$
$$\left(\int_0^{|t|} (|t| - \theta)^q d\theta\right)^{\frac{1}{q}} = \sum_{j=0}^{r} |\alpha_j| j^2$$
$$\left(\int_0^{|t|} |(f''(x + jsign(t)\theta) + f(x + jsign(t)\theta)) - (f''(x) + f(x))|^p d\theta\right)^{\frac{1}{p}}$$
$$\frac{|t|^{q+\frac{1}{q}}}{(q+1)^{\frac{1}{q}}}$$
$$= \left(\int_0^{|t|} |(f''(x + jsign(t)\theta) + f(x + jsign(t)\theta)) - (f''(x) + f(x))|^p d\theta\right)^{\frac{1}{p}}\Bigg\}$$
$$\frac{|t|^{q+\frac{1}{q}}}{(q+1)^{\frac{1}{q}}} \le \Bigg\{ (r+1)^{\frac{1}{q}} \left(\sum_{j=0}^{r} |\alpha_j|^p j^{2p}\right.$$

$(0 < \frac{1}{p} < 1)$

$$\int_0^{|t|} |(f''(x + jsign(t)\theta) + f(x + jsign(t)\theta)) - (f''(x) + f(x))|^p d\theta\Bigg)^{\frac{1}{p}}\Bigg\}$$
$$\frac{|t|^{q+\frac{1}{q}}}{(q+1)^{\frac{1}{q}}}. \tag{27.47}$$

Hence, we find that

$$|R(t)|^{p} \leq \left(\frac{r+1}{q+1}\right)^{\frac{p}{q}} \left[\sum_{j=0}^{r} |\alpha_j|^{p} j^{2p} \left(\int_0^{|t|} \left|(f''(x+jsign(t)\theta)+f(x+jsign(t)\theta)) - (f''(x)+f(x))\right|^{p} d\theta\right)\right] \times |t|^{\frac{(q+1)p}{q}}. \quad (27.48)$$

By (27.25), we have

$$\Delta_2^{*}(x) = \frac{1}{\sqrt{\pi\xi}} \int_{-\infty}^{\infty} R(t)\, e^{-\frac{t^2}{\xi}} dt. \quad (27.49)$$

We observe that

$$\begin{aligned}
\int_{-\infty}^{\infty} |\Delta_2^{*}(x)|^{p} dx &= \frac{1}{(\sqrt{\pi\xi})^{p}} \int_{-\infty}^{\infty} \left|\int_{-\infty}^{\infty} R(t)\, e^{-\frac{t^2}{\xi}} dt\right|^{p} dx \\
&\leq \frac{1}{(\sqrt{\pi\xi})^{p}} \int_{-\infty}^{\infty} \left(\int_{-\infty}^{\infty} |R(t)|\, e^{-\frac{t^2}{\xi}} dt\right)^{p} dx \\
&= \frac{1}{(\sqrt{\pi\xi})^{p}} \int_{-\infty}^{\infty} \left(\int_{-\infty}^{\infty} |R(t)|\, e^{-\frac{t^2}{2\xi}} e^{-\frac{t^2}{2\xi}} dt\right)^{p} dx \\
&\leq \frac{1}{(\sqrt{\pi\xi})^{p}} \int_{-\infty}^{\infty} \left(\int_{-\infty}^{\infty} |R(t)|^{p} e^{-\frac{pt^2}{2\xi}} dt\right) \left(\int_{-\infty}^{\infty} e^{-\frac{qt^2}{2\xi}} dt\right)^{\frac{p}{q}} dx \\
&= \left(\frac{2\pi\xi}{q}\right)^{\frac{p}{2q}} \frac{1}{(\pi\xi)^{\frac{p}{2}}} \int_{-\infty}^{\infty} \left(\int_{-\infty}^{\infty} |R(t)|^{p} e^{-\frac{pt^2}{2\xi}} dt\right) dx \\
&= \left(\frac{2}{q}\right)^{\frac{p}{2q}} \frac{1}{\sqrt{\pi\xi}} \int_{-\infty}^{\infty} \left(\int_{-\infty}^{\infty} |R(t)|^{p} e^{-\frac{pt^2}{2\xi}} dt\right) dx \\
&\overset{(27.48)}{\leq} \left(\frac{2}{q}\right)^{\frac{p}{2q}} \frac{1}{\sqrt{\pi\xi}} \left(\frac{r+1}{q+1}\right)^{\frac{p}{q}} \int_{-\infty}^{\infty} \left[\sum_{j=0}^{r} |\alpha_j|^{p} j^{2p} \left(\int_{-\infty}^{\infty} \left(\int_0^{|t|} \left|(f''(x+jsign(t)\theta)+f(x+jsign(t)\theta)) \right.\right.\right.\right. \\
&\left.\left.\left.\left. - (f''(x)+f(x))\right|^{p} d\theta\right)\right) |t|^{\frac{(q+1)p}{q}} e^{-\frac{pt^2}{2\xi}}\right] dx \quad (27.50)
\end{aligned}$$

(call

$$c_1 := \left(\frac{2}{q}\right)^{\frac{p}{2q}} \frac{1}{\sqrt{\pi\xi}} \left(\frac{r+1}{q+1}\right)^{\frac{p}{q}}) \tag{27.51}$$

$$= c_1 \left\{ \sum_{j=0}^{r} |\alpha_j|^p j^{2p} \int_{-\infty}^{\infty} \left[\int_0^{|t|} \right.\right.$$

$$\left(\int_{-\infty}^{\infty} \left| \left(f''(x + jsign(t)\theta) + f(x + jsign(t)\theta)\right) - \left(f''(x) + f(x)\right) \right|^p dx \right)$$

$$\left.\left. d\theta \right] |t|^{\frac{(q+1)p}{q}} e^{-\frac{pt^2}{2\xi}} \right\} dt \leq$$

$(\xi > 0)$

$$c_1 \left\{ \sum_{j=0}^{r} |\alpha_j|^p j^{2p} \int_{-\infty}^{\infty} \left(\int_0^{|t|} \omega_1 \left(f'' + f, \sqrt{\xi} \frac{j\theta}{\sqrt{\xi}}\right)_p^p d\theta \right) |t|^{\frac{(q+1)p}{q}} e^{-\frac{pt^2}{2\xi}} dt \right\}$$

$$\leq \omega_1 \left(f'' + f, \sqrt{\xi}\right)_p^p c_1$$

$$\left\{ \sum_{j=0}^{r} |\alpha_j|^p j^{2p} \int_{-\infty}^{\infty} \left(\int_0^{|t|} \left(1 + \frac{j}{\sqrt{\xi}}\theta\right)^p d\theta \right) |t|^{\frac{(q+1)p}{q}} e^{-\frac{pt^2}{2\xi}} dt \right\}$$

$$\leq c_1 \omega_1 \left(f'' + f, \sqrt{\xi}\right)_p^p \tag{27.52}$$

$$\left\{ \sum_{j=0}^{r} |\alpha_j|^p j^{2p} 2^{p-1} \int_{-\infty}^{\infty} \left(\int_0^{|t|} \left(1 + \frac{j^p}{\sqrt{\xi}^p}\theta^p\right) d\theta \right) |t|^{\frac{(q+1)p}{q}} e^{-\frac{pt^2}{2\xi}} dt \right\}$$

$$= c_1 \omega_1 \left(f'' + f, \sqrt{\xi}\right)_p^p$$

$$\left\{ \sum_{j=0}^{r} |\alpha_j|^p j^{2p} 2^{p-1} \int_{-\infty}^{\infty} \left(|t| + \frac{j^p}{\sqrt{\xi}^p} \frac{|t|^{p+1}}{(p+1)} \right) |t|^{\frac{(q+1)p}{q}} e^{-\frac{pt^2}{2\xi}} dt \right\}$$

$$= 2^p c_1 \omega_1 \left(f'' + f, \sqrt{\xi}\right)_p^p$$

$$\left\{ \sum_{j=0}^{r} |\alpha_j|^p j^{2p} \int_0^{\infty} \left(t + \frac{j^p}{\sqrt{\xi}^p} \frac{t^{p+1}}{(p+1)} \right) t^{(q+1)(p-1)} e^{-\frac{pt^2}{2\xi}} dt \right\}$$

$$= 2^p c_1 \omega_1 \left(f'' + f, \sqrt{\xi}\right)_p^p \left\{ \sum_{j=0}^{r} |\alpha_j|^p j^{2p} \right.$$

$$\left. \left[\int_0^{\infty} t^{(q+1)(p-1)+1} e^{-\frac{pt^2}{2\xi}} dt + \frac{j^p}{\sqrt{\xi}^p (p+1)} \int_0^{\infty} t^{(q+1)(p-1)+(p+1)} e^{-\frac{pt^2}{2\xi}} dt \right] \right\}$$

$$= 2^p c_1 \omega_1 \left(f'' + f, \sqrt{\xi}\right)_p^p \tag{27.53}$$

$$\left\{\sum_{j=0}^{r}|\alpha_j|^p j^{2p}\left[\left(\sqrt{\xi}\right)^{(q+1)(p-1)+2}\int_0^\infty\left(\frac{t}{\sqrt{\xi}}\right)^{(q+1)(p-1)+1}e^{-\frac{p}{2}\left(\frac{t}{\sqrt{\xi}}\right)^2}d\left(\frac{t}{\sqrt{\xi}}\right)\right.\right.$$

$$+\frac{j^p}{\sqrt{\xi}^p\,(p+1)}\left(\sqrt{\xi}\right)^{(q+1)(p-1)+(p+2)}$$

$$\left.\left.\times\int_0^\infty\left(\frac{t}{\sqrt{\xi}}\right)^{(q+1)(p-1)+(p+1)}e^{-\frac{p}{2}\left(\frac{t}{\sqrt{\xi}}\right)^2}d\left(\frac{t}{\sqrt{\xi}}\right)\right]\right\} \tag{27.54}$$

$$=2^p c_1\omega_1\left(f''+f,\sqrt{\xi}\right)_p^p$$

$$\left\{\sum_{j=0}^{r}|\alpha_j|^p j^{2p}\left[\left(\sqrt{\xi}\right)^{(q+1)(p-1)+2}\int_0^\infty x^{(q+1)(p-1)+1}e^{-\frac{p}{2}x^2}dx\right.\right.$$

$$\left.\left.+\frac{j^p}{(p+1)}\left(\sqrt{\xi}\right)^{(q+1)(p-1)+2}\int_0^\infty x^{(q+1)(p-1)+(p+1)}e^{-\frac{p}{2}x^2}dx\right]\right\} \tag{27.55}$$

(above it is $(q+1)(p-1)+1=qp-q+p>0$,

$$(q+1)(p-1)+(p+1)=qp-q+2p>0)$$

$$=2^p c_1\omega_1\left(f''+f,\sqrt{\xi}\right)_p^p$$

$$\left\{\sum_{j=0}^{r}|\alpha_j|^p j^{2p}\left[\xi^{\left(\frac{qp-q+p+1}{2}\right)}\int_0^\infty x^{(qp-q+p)}e^{-\frac{p}{2}x^2}dx\right.\right.$$

$$\left.\left.+\frac{j^p}{(p+1)}\xi^{\frac{(qp-q+p+1)}{2}}\int_0^\infty x^{(qp-q+2p)}e^{-\frac{p}{2}x^2}dx\right]\right\}=:(*). \tag{27.56}$$

We estimate the following integrals:

$$\int_0^\infty x^{(qp-q+p)}e^{-\frac{p}{2}x^2}dx=\int_0^1 x^{(qp-q+p)}e^{-\frac{p}{2}x^2}dx+\int_1^\infty x^{(qp-q+p)}e^{-\frac{p}{2}x^2}dx$$

$$\leq\int_0^1 x^{(qp-q+p)}dx+\int_1^\infty x^{(qp-q+p)}e^{-\frac{p}{2}x}dx$$

$$\leq\frac{1}{(qp-q+p+1)}+\int_0^\infty x^{(qp-q+p)}e^{-\frac{p}{2}x}dx$$

$$=\frac{1}{(qp-q+p+1)}+\left(\frac{p}{2}\right)^{-qp+q-p-1}\Gamma(qp-q+p+1), \tag{27.57}$$

and similarly, it holds

$$\int_0^{\infty} x^{(qp-q+2p)} e^{-\frac{p}{2}x^2} dx \leq \frac{1}{(qp-q+2p+1)} + \int_0^{\infty} x^{(qp-q+2p)} e^{-\frac{p}{2}x} dx$$

$$= \frac{1}{(qp-q+2p+1)} + \left(\frac{p}{2}\right)^{-qp+q-2p-1} \Gamma(qp-q+2p+1), \qquad (27.58)$$

where Γ is the gamma function.

We found that

$$\int_{-\infty}^{\infty} |\Delta_2^*(x)|^p dx \leq 2^p c_1 \omega_1 \left(f'' + f, \sqrt{\xi}\right)_p^p \xi^{\left(\frac{qp-q+p+1}{2}\right)}$$

$$\left\{\sum_{j=0}^{r} |\alpha_j|^p j^{2p} \left\{\left[\frac{1}{(qp-q+p+1)} + \left(\frac{p}{2}\right)^{-qp+q-p-1} \Gamma(qp-q+p+1)\right]\right.\right.$$

$$\left.\left.+ \frac{j^p}{(p+1)} \left[\frac{1}{(qp-q+2p+1)} + \left(\frac{p}{2}\right)^{-qp+q-2p-1} \Gamma(qp-q+2p+1)\right]\right]\right\}$$

$$= 2^p \left(\frac{2}{q}\right)^{\frac{p}{2q}} \frac{1}{\sqrt{\pi}} \left(\frac{r+1}{q+1}\right)^{\frac{p}{q}} \omega_1 \left(f'' + f, \sqrt{\xi}\right)_p^p \xi^{\frac{(qp-q+p)}{2}} \qquad (27.59)$$

$$\left\{\sum_{j=0}^{r} |\alpha_j|^p j^{2p} \left\{\left[\frac{1}{(qp-q+p+1)} + \left(\frac{p}{2}\right)^{-qp+q-p-1} \Gamma(qp-q+p+1)\right]\right.\right.$$

$$\left.\left.+ \frac{j^p}{(p+1)} \left[\frac{1}{(qp-q+2p+1)} + \left(\frac{p}{2}\right)^{-qp+q-2p-1} \Gamma(qp-q+2p+1)\right]\right]\right\}$$

$$= \frac{2^p}{\sqrt{\pi}} \left(\frac{2}{q}\right)^{\frac{p}{2q}} \left(\frac{r+1}{q+1}\right)^{\frac{p}{q}} \omega_1 \left(f'' + f, \sqrt{\xi}\right)_p^p \xi^p \qquad (27.60)$$

$$\left\{\sum_{j=0}^{r} |\alpha_j|^p j^{2p} \left\{\left[\frac{1}{(2p+1)} + \left(\frac{p}{2}\right)^{-(2p+1)} \Gamma(2p+1)\right]\right.\right.$$

$$\left.\left.+ \frac{j^p}{(p+1)} \left[\frac{1}{(3p+1)} + \left(\frac{p}{2}\right)^{-(3p+1)} \Gamma(3p+1)\right]\right]\right\}. \qquad (27.61)$$

The proof of the theorem is now completed. □

We give the following.

Corollary 27.1. *(to Theorem 27.1) It holds $(p>1)$*

$$\left\|W_{r,\xi}(f)-f\right\|_{p}\leq\theta_{1}(\xi)$$
$$+\left\|f''\right\|_{p}\left(\sum_{j=0}^{r}\left|\alpha_{j}\right|\left(1-e^{-\frac{j^{2}}{4}\xi}\right)\right)\rightarrow 0,\ \ as\ \xi\rightarrow 0. \tag{27.62}$$

Above $\theta_1(\xi)$ is as in (27.40).

It follows L_p, $p>1$, approximation by Poisson–Cauchy operators.

Theorem 27.2. *Let $\alpha\in\mathbb{N}$, $\beta>\frac{4}{\alpha}$, $\xi>0$, $p,q>1:\frac{1}{p}+\frac{1}{q}=1$, and $2\left(1+\frac{1}{\alpha\beta}\right)\geq q>\frac{3p+1}{\alpha\beta}$. Call*

$$\delta_{2}:=\frac{\Gamma(\beta)\,\Gamma\left(\frac{q\beta}{2}-\frac{1}{2\alpha}\right)^{\frac{1}{q}}(r+1)^{\frac{1}{q}}}{\Gamma^{\frac{1}{p}}\left(\frac{1}{2\alpha}\right)\Gamma\left(\beta-\frac{1}{2\alpha}\right)\Gamma\left(\frac{q\beta}{2}\right)(q+1)^{\frac{1}{q}}}.$$

Here $f,f''\in L_p(\mathbb{R})\cap C(\mathbb{R})$. Then

$$\left\|\Delta_{3}\right\|_{p}=\left\|M_{r,\xi}(f)-f-4f''\frac{\alpha\Gamma(\beta)}{\Gamma\left(\frac{1}{2\alpha}\right)\Gamma\left(\beta-\frac{1}{2\alpha}\right)}\sum_{j=0}^{r}\alpha_{j}\left(\int_{0}^{\infty}\frac{\sin^{2}\left(\frac{j\xi t}{2}\right)}{\left(1+t^{2\alpha}\right)^{\beta}}dt\right)\right\|_{p}$$
$$\leq 2^{\frac{1}{q}}\delta_{2}\omega_{1}\left(f''+f,\xi\right)_{p}\xi^{2+\alpha\beta(2-q)}$$
$$\left\{\sum_{j=0}^{r}\left|\alpha_{j}\right|^{p}j^{2p}\left[\Gamma\left(\frac{2p+1}{2\alpha}\right)\Gamma\left(\frac{q\beta}{2}-\left(\frac{2p+1}{2\alpha}\right)\right)\right.\right.$$
$$\left.\left.+\frac{j^{p}}{(p+1)}\Gamma\left(\frac{3p+1}{2\alpha}\right)\Gamma\left(\frac{q\beta}{2}-\left(\frac{3p+1}{2\alpha}\right)\right)\right]\right\}^{\frac{1}{p}}$$
$$:=\theta_{2}(\xi)\rightarrow 0,\quad as\ \xi\rightarrow 0. \tag{27.63}$$

Proof. Equations/inequalities (27.41)–(27.48) are all used here.

By (27.26), we have that

$$\Delta_{3}^{*}(x)=W\int_{-\infty}^{\infty}R(t)\,\frac{1}{\left(t^{2\alpha}+\xi^{2\alpha}\right)^{\beta}}dt, \tag{27.64}$$

where W as in (27.10); $\alpha\in\mathbb{N}$, $\beta>\frac{4}{\alpha}$; $\xi>0$.

We observe that $(p,q>1:\frac{1}{p}+\frac{1}{q}=1)$

$$\int_{-\infty}^{\infty}|\Delta_3^*(x)|^p\,dx = W^p\int_{-\infty}^{\infty}\left|\int_{-\infty}^{\infty}R(t)\frac{1}{(t^{2\alpha}+\xi^{2\alpha})^{\beta}}dt\right|^p dx$$

$$\leq W^p\int_{-\infty}^{\infty}\left(\int_{-\infty}^{\infty}|R(t)|\frac{1}{(t^{2\alpha}+\xi^{2\alpha})^{\beta}}dt\right)^p dx$$

$$= W^p\int_{-\infty}^{\infty}\left(\int_{-\infty}^{\infty}|R(t)|\frac{1}{(t^{2\alpha}+\xi^{2\alpha})^{\frac{\beta}{2}}}\frac{1}{(t^{2\alpha}+\xi^{2\alpha})^{\frac{\beta}{2}}}dt\right)^p dx$$

$$\leq W^p\int_{-\infty}^{\infty}\left(\int_{-\infty}^{\infty}|R(t)|^p\frac{1}{(t^{2\alpha}+\xi^{2\alpha})^{\frac{p\beta}{2}}}dt\right)\left(\int_{-\infty}^{\infty}\frac{1}{(t^{2\alpha}+\xi^{2\alpha})^{\frac{q\beta}{2}}}dt\right)^{\frac{p}{q}} dx$$

$$= \left[\frac{\alpha\Gamma(\beta)^p\,\Gamma\left(\frac{q\beta}{2}-\frac{1}{2\alpha}\right)^{p-1}\xi^{\alpha\beta p-1}}{\Gamma\left(\frac{1}{2\alpha}\right)\Gamma^p\left(\beta-\frac{1}{2\alpha}\right)\Gamma\left(\frac{q\beta}{2}\right)^{p-1}}\right]$$

$$\times\int_{-\infty}^{\infty}\left(\int_{-\infty}^{\infty}|R(t)|^p\frac{1}{(t^{2\alpha}+\xi^{2\alpha})^{\frac{p\beta}{2}}}dt\right)dx = \tag{27.65}$$

(call

$$\rho := \frac{\alpha\Gamma(\beta)^p\,\Gamma\left(\frac{q\beta}{2}-\frac{1}{2\alpha}\right)^{p-1}\xi^{\alpha\beta p-1}}{\Gamma\left(\frac{1}{2\alpha}\right)\Gamma^p\left(\beta-\frac{1}{2\alpha}\right)\Gamma\left(\frac{q\beta}{2}\right)^{p-1}})$$

$$\rho\int_{-\infty}^{\infty}\left(\int_{-\infty}^{\infty}|R(t)|^p\frac{1}{(t^{2\alpha}+\xi^{2\alpha})^{\frac{p\beta}{2}}}dt\right)dx$$

$$\leq \rho\left(\frac{r+1}{q+1}\right)^{(p-1)}\left[\sum_{j=0}^{r}|\alpha_j|^p j^{2p}\left(\int_{-\infty}^{\infty}\left(\int_{-\infty}^{\infty}\right.\right.\right.$$

$$\left(\int_0^{|t|}|(f''(x+j\,sign(t)\,\theta)+f(x+j\,sign(t)\,\theta))\right.$$

$$\left.\left.\left.\left.-(f''(x)+f(x))|^p\,d\theta\right)|t|^{\frac{(q+1)p}{q}}\frac{dt}{(t^{2\alpha}+\xi^{2\alpha})^{\frac{p\beta}{2}}}\right)dx\right)\right] \tag{27.66}$$

(call

$$c_2 := \rho \left(\frac{r+1}{q+1}\right)^{(p-1)})$$

$$= c_2 \left[\sum_{j=0}^{r} |\alpha_j|^p j^{2p} \left(\int_{-\infty}^{\infty} \left[\int_0^{|t|} \left(\int_{-\infty}^{\infty} |(f''(x + jsign(t)\theta) + f(x + jsign(t)\theta)) - (f''(x) + f(x))|^p dx\right) d\theta\right] |t|^{(q+1)(p-1)} \frac{dt}{(t^{2\alpha} + \xi^{2\alpha})^{\frac{p\beta}{2}}}\right)\right] \tag{27.67}$$

$$\leq c_2 \left[\sum_{j=0}^{r} |\alpha_j|^p j^{2p} \int_{-\infty}^{\infty} \left(\int_0^{|t|} \omega_1 \left(f'' + f, \xi\frac{j\theta}{\xi}\right)_p^p d\theta\right) \times |t|^{(q+1)(p-1)} \frac{dt}{(t^{2\alpha} + \xi^{2\alpha})^{\frac{p\beta}{2}}}\right] \tag{27.68}$$

$$\leq c_2 \omega_1 (f'' + f, \xi)_p^p \left[\sum_{j=0}^{r} |\alpha_j|^p j^{2p} \int_{-\infty}^{\infty} \left(\int_0^{|t|} \left(1 + \frac{j}{\xi}\theta\right)^p d\theta\right) |t|^{(q+1)(p-1)} \frac{dt}{(t^{2\alpha} + \xi^{2\alpha})^{\frac{p\beta}{2}}}\right]$$

$$\leq 2^{p-1} c_2 \omega_1 (f'' + f, \xi)_p^p \left[\sum_{j=0}^{r} |\alpha_j|^p j^{2p} \int_{-\infty}^{\infty} \left(\int_0^{|t|} \left(1 + \frac{j^p}{\xi^p}\theta^p\right) d\theta\right) |t|^{(q+1)(p-1)} \frac{dt}{(t^{2\alpha} + \xi^{2\alpha})^{\frac{p\beta}{2}}}\right] \tag{27.69}$$

$$= 2^{p-1} c_2 \omega_1 (f'' + f, \xi)_p^p \left[\sum_{j=0}^{r} |\alpha_j|^p j^{2p} \int_{-\infty}^{\infty} \left(|t| + \frac{j^p}{\xi^p} \frac{|t|^{p+1}}{(p+1)}\right) |t|^{(q+1)(p-1)} \frac{dt}{(t^{2\alpha} + \xi^{2\alpha})^{\frac{p\beta}{2}}}\right]$$

$$= 2^{p} c_2 \omega_1 (f'' + f, \xi)_p^p \left[\sum_{j=0}^{r} |\alpha_j|^p j^{2p} \int_0^{\infty} \left(t + \frac{j^p}{\xi^p} \frac{t^{p+1}}{(p+1)}\right) \frac{t^{(q+1)(p-1)}}{(t^{2\alpha} + \xi^{2\alpha})^{\frac{q\beta}{2}}} dt\right]$$

$$= 2^{p} c_2 \omega_1 (f'' + f, \xi)_p^p \left\{\sum_{j=0}^{r} |\alpha_j|^p j^{2p} \left[\int_0^{\infty} \frac{t^{(q+1)(p-1)+1}}{(t^{2\alpha} + \xi^{2\alpha})^{\frac{q\beta}{2}}} dt + \frac{j^p}{\xi^p (p+1)} \int_0^{\infty} \frac{t^{(q+1)(p-1)+(p+1)}}{(t^{2\alpha} + \xi^{2\alpha})^{\frac{q\beta}{2}}} dt\right]\right\} \tag{27.70}$$

$$= 2^p c_2 \omega_1 \left(f'' + f, \xi\right)_p^p$$
$$\times \left\{ \sum_{j=0}^{r} |\alpha_j|^p j^{2p} \left[\xi^{(q+1)(p-1)+2-\alpha\beta q} \int_0^\infty \frac{\left(\frac{t}{\xi}\right)^{(q+1)(p-1)+1}}{\left(\left(\frac{t}{\xi}\right)^{2\alpha}+1\right)^{\frac{q\beta}{2}}} d\left(\frac{t}{\xi}\right) \right.\right.$$
$$\left.\left. + \frac{j^p}{\xi^p (p+1)} \xi^{(q+1)(p-1)+(p+2)-\alpha\beta q} \int_0^\infty \frac{\left(\frac{t}{\xi}\right)^{(q+1)(p-1)+(p+1)}}{\left(\left(\frac{t}{\xi}\right)^{2\alpha}+1\right)^{\frac{q\beta}{2}}} d\left(\frac{t}{\xi}\right) \right]\right\}$$
$$= 2^p c_2 \omega_1 \left(f'' + f, \xi\right)_p^p \xi^{(q+1)(p-1)+2-\alpha\beta q}$$
$$\times \left\{ \sum_{j=0}^{r} |\alpha_j|^p j^{2p} \left[\int_0^\infty \frac{x^{(q+1)(p-1)+1}}{\left(1+x^{2\alpha}\right)^{\frac{q\beta}{2}}} dx \right.\right.$$
$$\left.\left. + \frac{j^p}{(p+1)} \int_0^\infty \frac{x^{(q+1)(p-1)+(p+1)}}{\left(1+x^{2\alpha}\right)^{\frac{q\beta}{2}}} dx \right]\right\} \tag{27.71}$$
$$= 2^p c_2 \omega_1 \left(f'' + f, \xi\right)_p^p \xi^{qp-q+p+1-\alpha\beta q}$$
$$\times \left\{ \sum_{j=0}^{r} |\alpha_j|^p j^{2p} \left[\int_0^\infty \frac{x^{qp-q+p}}{\left(1+x^{2\alpha}\right)^{\frac{q\beta}{2}}} dx + \frac{j^p}{(p+1)} \int_0^\infty \frac{x^{qp-q+2p}}{\left(1+x^{2\alpha}\right)^{\frac{q\beta}{2}}} dx \right]\right\}$$
$$= 2^p c_2 \omega_1 \left(f'' + f, \xi\right)_p^p \xi^{2p+1-\alpha\beta q}$$
$$\times \left\{ \sum_{j=0}^{r} |\alpha_j|^p j^{2p} \left[\int_0^\infty \frac{x^{2p}}{\left(1+x^{2\alpha}\right)^{\frac{q\beta}{2}}} dx + \frac{j^p}{(p+1)} \int_0^\infty \frac{x^{3p}}{\left(1+x^{2\alpha}\right)^{\frac{q\beta}{2}}} dx \right]\right\} = \tag{27.72}$$

(by Zwillinger (1995), p. 397, formula 595)

$$2^p c_2 \omega_1 \left(f'' + f, \xi\right)_p^p \xi^{2p+1-\alpha\beta q}$$
$$\times \left\{ \sum_{j=0}^{r} |\alpha_j|^p j^{2p} \left[\frac{\Gamma\left(\frac{2p+1}{2\alpha}\right)\Gamma\left(\frac{q\beta}{2} - \left(\frac{2p+1}{2\alpha}\right)\right)}{2\alpha\Gamma\left(\frac{q\beta}{2}\right)} \right.\right.$$
$$\left.\left. + \frac{j^p}{(p+1)} \frac{\Gamma\left(\frac{3p+1}{2\alpha}\right)\Gamma\left(\frac{q\beta}{2} - \left(\frac{3p+1}{2\alpha}\right)\right)}{2\alpha\Gamma\left(\frac{q\beta}{2}\right)} \right]\right\}$$
$$= 2^p c_2 \omega_1 \left(f'' + f, \xi\right)_p^p \xi^{2p+1-\alpha\beta q}$$
$$\times \left\{ \sum_{j=0}^{r} |\alpha_j|^p j^{2p} \left[\frac{\Gamma\left(\frac{2p+1}{2\alpha}\right)\Gamma\left(\frac{q\beta}{2} - \left(\frac{2p+1}{2\alpha}\right)\right) + \frac{j^p}{(p+1)}\Gamma\left(\frac{3p+1}{2\alpha}\right)\Gamma\left(\frac{q\beta}{2} - \left(\frac{3p+1}{2\alpha}\right)\right)}{2\alpha\Gamma\left(\frac{q\beta}{2}\right)} \right]\right\} \tag{27.73}$$

(under the assumption $q > \frac{3p+1}{\alpha\beta}$)
(it is

$$c_2 = \rho\left(\frac{r+1}{q+1}\right)^{p-1} = \frac{\alpha\Gamma(\beta)^p\,\Gamma\left(\frac{q\beta}{2}-\frac{1}{2\alpha}\right)^{(p-1)}(r+1)^{(p-1)}\,\xi^{\alpha\beta p-1}}{\Gamma\left(\frac{1}{2\alpha}\right)\Gamma^p\left(\beta-\frac{1}{2\alpha}\right)\Gamma\left(\frac{q\beta}{2}\right)^{(p-1)}(q+1)^{(p-1)}}$$

and call

$$\psi_2 := \frac{\alpha\Gamma(\beta)^p\,\Gamma\left(\frac{q\beta}{2}-\frac{1}{2\alpha}\right)^{(p-1)}(r+1)^{(p-1)}}{\Gamma\left(\frac{1}{2\alpha}\right)\Gamma^p\left(\beta-\frac{1}{2\alpha}\right)\Gamma\left(\frac{q\beta}{2}\right)^{(p-1)}(q+1)^{(p-1)}})$$

$$= 2^p\psi_2\omega_1\left(f''+f,\xi\right)_p^p\,\xi^{2p+\alpha\beta(p-q)}$$

$$\left\{\sum_{j=0}^{r}|\alpha_j|^p\,j^{2p}\left[\frac{\Gamma\left(\frac{2p+1}{2\alpha}\right)\Gamma\left(\frac{q\beta}{2}-\left(\frac{2p+1}{2\alpha}\right)\right)+\frac{j^p}{(p+1)}\Gamma\left(\frac{3p+1}{2\alpha}\right)\Gamma\left(\frac{q\beta}{2}-\left(\frac{3p+1}{2\alpha}\right)\right)}{2\alpha\Gamma\left(\frac{q\beta}{2}\right)}\right]\right\}. \tag{27.74}$$

The claim is proved. □

We give the following.

Corollary 27.2. *(to Theorem 27.2) Assume that $\beta > \frac{3}{2\alpha}$. It holds $(p>1)$*

$$\|M_{r,\xi}(f)-f\|_p \le \theta_2(\xi)$$

$$+4\|f''\|_p\frac{\alpha\Gamma(\beta)}{\Gamma\left(\frac{1}{2\alpha}\right)\Gamma\left(\beta-\frac{1}{2\alpha}\right)}\sum_{j=0}^{r}|\alpha_j|\left(\int_0^{\infty}\frac{\sin^2\left(\frac{j\xi t}{2}\right)}{\left(1+t^{2\alpha}\right)^{\beta}}dt\right)\to 0,$$

$$as\ \xi\to 0. \tag{27.75}$$

Above $\theta_2(\xi)$ is as in (27.63).

Proof. By Theorem 27.2, and by $|\sin x|\le|x|$, $x\in\mathbb{R}$:

$$\int_0^{\infty}\frac{\sin^2\left(\frac{j\xi t}{2}\right)}{\left(1+t^{2\alpha}\right)^{\beta}}dt \le \int_0^{\infty}\frac{\frac{j^2\xi^2t^2}{4}}{\left(1+t^{2\alpha}\right)^{\beta}}dt = \left(\frac{j^2}{4}\right)\xi^2\int_0^{\infty}\frac{t^2}{\left(1+t^{2\alpha}\right)^{\beta}}dt \tag{27.76}$$

(by Zwillinger (1995), p. 397, formula 595)

$$= \xi^2\left(\frac{j^2}{4}\right)\frac{\Gamma\left(\frac{3}{2\alpha}\right)\Gamma\left(\beta-\frac{3}{2\alpha}\right)}{2\alpha\Gamma(\beta)}\to 0,\quad \text{as } \xi\to 0.$$

□

We continue with L_p $(p > 1)$ approximation by trigonometric singular operators.

Theorem 27.3. *Let $p, q > 1 : \frac{1}{p} + \frac{1}{q} = 1$, $\beta \in \mathbb{N}$ and $\beta > \frac{\lceil 3p \rceil + 1}{2}$ ($\lceil \cdot \rceil$ the ceiling of the number), $\xi > 0$, λ_1 is as in (27.16).*

When $\overline{\lambda} \in \mathbb{N}$ is even we define

$$\psi_{1\overline{\lambda}} := \frac{\pi (-1)^{\frac{2\beta - \overline{\lambda}}{2}} (2\beta)!}{2^{\overline{\lambda}+1} (2\beta - \overline{\lambda} - 1)!} \left(\sum_{k=1}^{\beta} (-1)^k \frac{k^{2\beta - \overline{\lambda} - 1}}{(\beta - k)! (\beta + k)!} \right), \tag{27.77}$$

and when $\overline{\lambda}$ is odd we define

$$\psi_{2\overline{\lambda}} := \frac{(-1)^{\frac{\overline{\lambda}-1}{2}} (2\beta)!}{2^{\overline{\lambda}} (2\beta - \overline{\lambda} - 1)!} \left(\sum_{k=1}^{\beta} (-1)^{\beta - k} \frac{k^{2\beta - \overline{\lambda} - 1} \ln (2k)}{(\beta - k)! (\beta + k)!} \right), \tag{27.78}$$

and we set

$$\psi_{\overline{\lambda}} := \begin{Bmatrix} \psi_{1\overline{\lambda}}, & \text{if } \overline{\lambda} \text{ is even,} \\ \psi_{2\overline{\lambda}}, & \text{if } \overline{\lambda} \text{ is odd} \end{Bmatrix}. \tag{27.79}$$

Then

$$\|\Delta_4\|_p = \left\| T_{r,\xi}(f) - f - 4f'' \left(\lambda_1^{-1} \sum_{j=0}^{r} \alpha_j \int_0^{\infty} \sin^2 \left(\frac{j\xi t}{2} \right) \left(\frac{\sin t}{t} \right)^{2\beta} dt \right) \right\|_p$$

$$\leq \frac{2^{1+\frac{1}{p}}}{\lambda_1^{\frac{1}{p}} (p+1)^{\frac{1}{p}}} \left(\frac{r+1}{q+1} \right)^{\frac{1}{q}} \omega_1 (f'' + f, \xi)_p \, \xi^2 \left[\sum_{j=1}^{r} |\alpha_j|^p j^{3p} \right]^{\frac{1}{p}}$$

$$\left\{ \left(\frac{-(2\beta)!}{8(2\beta - 4)!} \sum_{k=1}^{\beta} (-1)^{\beta - k} \frac{k^{2\beta - 4} \ln (2k)}{(\beta - k)! (\beta + k)!} \right) + \psi_{\lceil 3p \rceil} \right\}^{\frac{1}{p}}$$

$$=: \theta_3 (\xi) \to 0, \quad \text{as } \xi \to 0. \tag{27.80}$$

Proof. Equations/inequalities (27.41)–(27.48) are all used here.

By (27.27), we have that

$$\Delta_4^* (x) = \int_{-\infty}^{\infty} R(t) \frac{1}{\lambda} \left(\frac{\sin \left(\frac{t}{\xi} \right)}{t} \right)^{2\beta} dt, \tag{27.81}$$

where $\lambda > 0$ is as in (27.15); in fact $\lambda = \lambda_1 \xi^{1-2\beta}$ by (27.17), where λ_1 is as in (27.16); $\xi > 0$.

By (27.18), we have

$$\int_{-\infty}^{\infty}\left(\frac{1}{\lambda}\left(\frac{\sin\left(\frac{t}{\xi}\right)}{t}\right)^{2\beta}\right)dt=1. \tag{27.82}$$

We observe that $(p,q>1:\frac{1}{p}+\frac{1}{q}=1)$

$$\int_{-\infty}^{\infty}|\Delta_4^*(x)|^p\,dx=\int_{-\infty}^{\infty}\left|\int_{-\infty}^{\infty}R(t)\,\frac{1}{\lambda}\left(\frac{\sin\left(\frac{t}{\xi}\right)}{t}\right)^{2\beta}dt\right|^p dx$$

$$\leq\int_{-\infty}^{\infty}\left(\int_{-\infty}^{\infty}|R(t)|\,\frac{1}{\lambda}\left(\frac{\sin\left(\frac{t}{\xi}\right)}{t}\right)^{2\beta}dt\right)^p dx$$

$$\leq\int_{-\infty}^{\infty}\left(\int_{-\infty}^{\infty}|R(t)|^p\,\frac{1}{\lambda}\left(\frac{\sin\left(\frac{t}{\xi}\right)}{t}\right)^{2\beta}dt\right)dx$$

$$=\lambda^{-1}\int_{-\infty}^{\infty}\left(\int_{-\infty}^{\infty}|R(t)|^p\left(\frac{\sin\left(\frac{t}{\xi}\right)}{t}\right)^{2\beta}dt\right)dx$$

$$\overset{(27.48)}{\leq}\lambda^{-1}\left(\frac{r+1}{q+1}\right)^{\frac{p}{q}}\left[\sum_{j=0}^{r}|\alpha_j|^p\,j^{2p}\left(\int_{-\infty}^{\infty}\left(\int_{-\infty}^{\infty}\right.\right.\right.$$

$$\left(\int_0^{|t|}\left|\left(f''(x+jsign(t)\,\theta)+f(x+jsign(t)\,\theta)\right)-\left(f''(x)+f(x)\right)\right|^p d\theta\right)$$

$$\left.\left.\left.|t|^{\frac{(q+1)p}{q}}\left(\frac{\sin\left(\frac{t}{\xi}\right)}{t}\right)^{2\beta}dt\right)dx\right)\right] \tag{27.83}$$

$$=\lambda^{-1}\left(\frac{r+1}{q+1}\right)^{\frac{p}{q}}\left[\sum_{j=0}^{r}|\alpha_j|^p\,j^{2p}\left(\int_{-\infty}^{\infty}\left(\int_0^{|t|}\right.\right.\right.$$

$$\left(\int_{-\infty}^{\infty}\left|\left(f''(x+jsign(t)\,\theta)+f(x+jsign(t)\,\theta)\right)-\left(f''(x)+f(x)\right)\right|^p dx\right)$$

$$\left.\left.\left.d\theta\right)|t|^{\frac{(q+1)q}{p}}\left(\frac{\sin\left(\frac{t}{\xi}\right)}{t}\right)^{2\beta}dt\right)\right]\leq\lambda^{-1}\left(\frac{r+1}{q+1}\right)^{\frac{p}{q}}$$

$$\left[\sum_{j=0}^{r}|\alpha_j|^p j^{2p}\int_{-\infty}^{\infty}\left(\int_0^{|t|}\omega_1\left(f''+f,\frac{\xi j\theta}{\xi}\right)_p^p d\theta\right)|t|^{\frac{(q+1)q}{p}}\left(\frac{\sin\left(\frac{t}{\xi}\right)}{t}\right)^{2\beta}dt\right]$$

$$\le \lambda^{-1}\left(\frac{r+1}{q+1}\right)^{\frac{p}{q}}\omega_1\left(f''+f,\xi\right)_p^p$$

$$\left[\sum_{j=0}^{r}|\alpha_j|^p j^{2p}\int_{-\infty}^{\infty}\left(\int_0^{|t|}\left(1+\frac{j}{\xi}\theta\right)^p d\theta\right)|t|^{\frac{(q+1)p}{q}}\left(\frac{\sin\left(\frac{t}{\xi}\right)}{t}\right)^{2\beta}dt\right] \tag{27.84}$$

$$=\frac{\lambda^{-1}\xi}{(p+1)}\left(\frac{r+1}{q+1}\right)^{\frac{p}{q}}\omega_1\left(f''+f,\xi\right)_p^p$$

$$\left[\sum_{j=1}^{r}|\alpha_j|^p j^{2p-1}\int_{-\infty}^{\infty}\left(\left(1+\frac{j}{\xi}|t|\right)^{p+1}-1\right)|t|^{\frac{(q+1)p}{q}}\left(\frac{\sin\left(\frac{t}{\xi}\right)}{t}\right)^{2\beta}dt\right]$$

$$\le \frac{2^p\lambda^{-1}\xi}{(p+1)}\left(\frac{r+1}{q+1}\right)^{\frac{p}{q}}\omega_1\left(f''+f,\xi\right)_p^p$$

$$\left[\sum_{j=1}^{r}|\alpha_j|^p j^{2p-1}\left(\int_{-\infty}^{\infty}\left(\frac{j^{(p+1)}}{\xi^{(p+1)}}|t|^{p+1}\right)|t|^{(q+1)(p-1)}\left(\frac{\sin\left(\frac{t}{\xi}\right)}{t}\right)^{2\beta}dt\right)\right]$$

$$=\frac{2^p\lambda^{-1}\xi^{-p}}{(p+1)}\left(\frac{r+1}{q+1}\right)^{(p-1)}\omega_1\left(f''+f,\xi\right)_p^p\left[\sum_{j=1}^{r}|\alpha_j|^p j^{3p}\right]$$

$$\left(\int_{-\infty}^{\infty}|t|^{(q+1)(p-1)+(p+1)}\left(\frac{\sin\left(\frac{t}{\xi}\right)}{t}\right)^{2\beta}dt\right) \tag{27.85}$$

$$=\frac{2^p}{\lambda\xi^p(p+1)}\left(\frac{r+1}{q+1}\right)^{(p-1)}\omega_1\left(f''+f,\xi\right)_p^p\left[\sum_{j=1}^{r}|\alpha_j|^p j^{3p}\right]$$

$$\left(\int_{-\infty}^{\infty}|t|^{3p}\left(\frac{\sin\left(\frac{t}{\xi}\right)}{t}\right)^{2\beta}dt\right)$$

$$=\frac{2^{p+1}}{\lambda\xi^p(p+1)}\left(\frac{r+1}{q+1}\right)^{(p-1)}\omega_1\left(f''+f,\xi\right)_p^p\left[\sum_{j=1}^{r}|\alpha_j|^p j^{3p}\right]$$

$$\left(\int_0^\infty t^{3p}\left(\frac{\sin\left(\frac{t}{\xi}\right)}{t}\right)^{2\beta}dt\right)$$

$$=\frac{2^{p+1}\xi^{2p+1-2\beta}}{\lambda\,(p+1)}\left(\frac{r+1}{q+1}\right)^{(p-1)}\omega_1\left(f''+f,\xi\right)_p^p\left[\sum_{j=1}^r|\alpha_j|^p\,j^{3p}\right]$$

$$\left(\int_0^\infty x^{3p}\left(\frac{\sin x}{x}\right)^{2\beta}dx\right) \tag{27.86}$$

$$=\frac{2^{p+1}\xi^{2p}}{\lambda_1\,(p+1)}\left(\frac{r+1}{q+1}\right)^{(p-1)}\omega_1\left(f''+f,\xi\right)_p^p\left[\sum_{j=1}^r|\alpha_j|^p\,j^{3p}\right]$$

$$\left(\int_0^1 x^{3p}\left(\frac{\sin x}{x}\right)^{2\beta}dx+\int_1^\infty x^{3p}\left(\frac{\sin x}{x}\right)^{2\beta}dx\right)$$

$$\leq\frac{2^{p+1}\xi^{2p}}{(p+1)\,\lambda_1}\left(\frac{r+1}{q+1}\right)^{(p-1)}\omega_1\left(f''+f,\xi\right)_p^p\left[\sum_{j=1}^r|\alpha_j|^p\,j^{3p}\right]$$

$$\left(\int_0^1 x^{3}\left(\frac{\sin x}{x}\right)^{2\beta}dx+\int_1^\infty x^{\lceil 3p\rceil}\left(\frac{\sin x}{x}\right)^{2\beta}dx\right)$$

$$\leq\frac{2^{p+1}\xi^{2p}}{(p+1)\,\lambda_1}\left(\frac{r+1}{q+1}\right)^{(p-1)}\omega_1\left(f''+f,\xi\right)_p^p\left[\sum_{j=1}^r|\alpha_j|^p\,j^{3p}\right] \tag{27.87}$$

$$\left(\int_0^\infty x^{3}\left(\frac{\sin x}{x}\right)^{2\beta}dx+\int_0^\infty x^{\lceil 3p\rceil}\left(\frac{\sin x}{x}\right)^{2\beta}dx\right)=$$

(we use [Edwards (1954)], p. 210, item 1033)

$$\frac{2^{p+1}}{(p+1)}\frac{\xi^{2p}}{\lambda_1}\left(\frac{r+1}{q+1}\right)^{(p-1)}\omega_1\left(f''+f,\xi\right)_p^p\left[\sum_{j=1}^r|\alpha_j|^p\,j^{3p}\right]$$

$$\left(\left(\frac{(-1)\,(2\beta)!}{8\,(2\beta-4)!}\sum_{k=1}^{\beta}(-1)^{\beta-k}\,\frac{k^{2\beta-4}\ln(2k)}{(\beta-k)!\,(\beta+k)!}\right)+\psi_{\lceil 3p\rceil}\right)<\infty.$$

The theorem is proved. □

We give the following.

Corollary 27.3. *(to Theorem 27.3) It holds $(p > 1)$*

$$\|T_{r,\xi}(f) - f\|_p \leq \theta_3(\xi) + 4\|f''\|_p \lambda_1^{-1} \left(\sum_{j=0}^{r} |\alpha_j| \left(\int_0^\infty \sin^2\left(\frac{j\xi t}{2}\right) \left(\frac{\sin t}{t}\right)^{2\beta} dt \right) \right) \to 0, \quad \text{as } \xi \to 0. \tag{27.88}$$

Above $\theta_3(\xi)$ is as in (27.80).

Proof. By Theorem 27.3, and by $|\sin x| \leq |x|$, $x \in \mathbb{R}$:

$$\int_0^\infty \sin^2\left(\frac{j\xi t}{2}\right) \left(\frac{\sin t}{t}\right)^{2\beta} dt \leq \left(\frac{j^2}{4}\right) \xi^2 \left(\int_0^\infty t^2 \left(\frac{\sin t}{t}\right)^{2\beta} dt \right)$$

(by Edwards (1954), p. 210, item 1033)

$$= \left(\frac{j^2}{4}\right) \left\{ \frac{\pi (-1)^{\beta-1} (2\beta)!}{8(2\beta-3)!} \left(\sum_{k=1}^{\beta} (-1)^k \frac{k^{2\beta-3}}{(\beta-k)!(\beta+k)!} \right) \right\} \xi^2 \to 0, \quad \text{as } \xi \to 0. \tag{27.89}$$

□

L_1 approximation by Gauss–Weierstrass operators follows.

Theorem 27.4. *Let $\xi > 0$, both $f, f'' \in L_1(\mathbb{R}) \cap C(\mathbb{R})$. Then*

$$\|\Delta_2\|_1 = \left\| W_{r,\xi}(f) - f - f'' \left(\sum_{j=0}^{r} \alpha_j \left(1 - e^{-\frac{j^2}{4}\xi}\right) \right) \right\|_1 \leq \frac{2}{\sqrt{\pi}} \omega_1\left(f'' + f, \sqrt{\xi}\right)_1 \xi \left\{ \sum_{j=1}^{r} |\alpha_j| j^2 \left[\frac{7}{3} + j\frac{25}{8}\right] \right\} := E_1(\xi) \to 0, \quad \text{as } \xi \to 0. \tag{27.90}$$

Proof. We have that

$$\Delta_2^*(x) = \frac{1}{\sqrt{\pi\xi}} \int_{-\infty}^{\infty} R(t) e^{-\frac{t^2}{\xi}} dt. \tag{27.91}$$

We observe that

$$\|\Delta_2^*\|_1 = \int_{-\infty}^{\infty} |\Delta_2^*(x)| dx = \frac{1}{\sqrt{\pi\xi}} \int_{-\infty}^{\infty} \left| \int_{-\infty}^{\infty} R(t) e^{-\frac{t^2}{\xi}} dt \right| dx$$

$$\leq \frac{1}{\sqrt{\pi\xi}} \int_{-\infty}^{\infty} \left(\int_{-\infty}^{\infty} |R(t)| e^{-\frac{t^2}{\xi}} dt \right) dx$$

$$\overset{(27.45)}{\leq} \frac{1}{\sqrt{\pi\xi}} \left[\sum_{j=0}^{r} |\alpha_j| j^2 \left(\int_{-\infty}^{\infty} \left(\int_{-\infty}^{\infty} \right.\right.\right.$$

$$\left(\int_0^{|t|} \left| (f''(x + jsign(t)\theta) + f(x + jsign(t)\theta)) \right.\right.$$

$$\left.\left.\left.\left. - (f''(x) + f(x)) \right| (|t| - \theta) d\theta \right) e^{-\frac{t^2}{\xi}} dt \right) dx \right] \tag{27.92}$$

$$\leq \frac{1}{\sqrt{\pi\xi}} \left[\sum_{j=0}^{r} |\alpha_j| j^2 \left(\int_{-\infty}^{\infty} \left(\int_{-\infty}^{\infty} \left(|t| \int_0^{|t|} \right.\right.\right.\right.$$

$$\left| (f''(x + jsign(t)\theta) + f(x + jsign(t)\theta)) \right.$$

$$\left.\left.\left.\left. - (f''(x) + f(x)) \right| d\theta \right) e^{-\frac{t^2}{\xi}} dt \right) dx \right) \right]$$

$$= \frac{1}{\sqrt{\pi\xi}} \left[\sum_{j=0}^{r} |\alpha_j| j^2 \left(\int_{-\infty}^{\infty} \left(\int_{-\infty}^{\infty} \left(\int_0^{|t|} \right.\right.\right.\right.$$

$$\left| (f''(x + jsign(t)\theta) + f(x + jsign(t)\theta)) \right.$$

$$\left.\left.\left.\left. - (f''(x) + f(x)) \right| d\theta \right) dx \right) |t| e^{-\frac{t^2}{\xi}} dt \right) \right]$$

$$= \frac{1}{\sqrt{\pi\xi}} \left[\sum_{j=0}^{r} |\alpha_j| j^2 \left(\int_{-\infty}^{\infty} \left(\int_0^{|t|} \left(\int_{-\infty}^{\infty} \right.\right.\right.\right.$$

$$\left| (f''(x + jsign(t)\theta) + f(x + jsign(t)\theta)) \right.$$

$$\left.\left.\left.\left. - (f''(x) + f(x)) \right| dx \right) d\theta \right) |t| e^{-\frac{t^2}{\xi}} dt \right) \right]$$

$$\leq \frac{1}{\sqrt{\pi\xi}}\left[\sum_{j=0}^{r}|\alpha_j|\,j^2\left(\int_{-\infty}^{\infty}\left(\int_0^{|t|}\omega_1\left(f''+f,\frac{\sqrt{\xi}}{\sqrt{\xi}}j\theta\right)_1 d\theta\right)|t|\,e^{-\frac{t^2}{\xi}}dt\right)\right]$$

$$\leq \frac{1}{\sqrt{\pi\xi}}\omega_1\left(f''+f,\sqrt{\xi}\right)_1$$

$$\left[\sum_{j=0}^{r}|\alpha_j|\,j^2\left(\int_{-\infty}^{\infty}\left(\int_0^{|t|}\left(1+\frac{j}{\sqrt{\xi}}\theta\right)d\theta\right)|t|\,e^{-\frac{t^2}{\xi}}dt\right)\right] \tag{27.93}$$

$$=\frac{1}{\sqrt{\pi\xi}}\omega_1\left(f''+f,\sqrt{\xi}\right)_1\left[\sum_{j=0}^{r}|\alpha_j|\,j^2\left(\int_{-\infty}^{\infty}\left(|t|+\frac{j}{\sqrt{\xi}}\frac{t^2}{2}\right)|t|\,e^{-\frac{t^2}{\xi}}dt\right)\right]$$

$$=\frac{2}{\sqrt{\pi\xi}}\omega_1\left(f''+f,\sqrt{\xi}\right)_1\left[\sum_{j=0}^{r}|\alpha_j|\,j^2\left(\int_0^{\infty}\left(t+\frac{j}{\sqrt{\xi}}\frac{t^2}{2}\right)te^{-\frac{t^2}{\xi}}dt\right)\right]$$

$$=\frac{2}{\sqrt{\pi}\sqrt{\xi}}\omega_1\left(f''+f,\sqrt{\xi}\right)_1$$

$$\left[\sum_{j=0}^{r}|\alpha_j|\,j^2\left[\int_0^{\infty}t^2e^{-\frac{t^2}{\xi}}dt+\frac{j}{2\sqrt{\xi}}\int_0^{\infty}t^3e^{-\frac{t^2}{\xi}}dt\right]\right]$$

$$=\frac{2}{\sqrt{\pi}\sqrt{\xi}}\omega_1\left(f''+f,\sqrt{\xi}\right)_1 \tag{27.94}$$

$$\left[\sum_{j=0}^{r}|\alpha_j|\,j^2\left[\left(\sqrt{\xi}\right)^3\int_0^{\infty}\left(\frac{t}{\sqrt{\xi}}\right)^2e^{-\left(\frac{t}{\sqrt{\xi}}\right)^2}d\frac{t}{\sqrt{\xi}}\right.\right.$$

$$\left.\left.+\frac{j}{2\sqrt{\xi}}\left(\sqrt{\xi}\right)^4\int_0^{\infty}\left(\frac{t}{\sqrt{\xi}}\right)^3e^{-\left(\frac{t}{\sqrt{\xi}}\right)^2}d\frac{t}{\sqrt{\xi}}\right]\right]$$

$$=\frac{2\xi}{\sqrt{\pi}}\omega_1\left(f''+f,\sqrt{\xi}\right)_1$$

$$\left\{\sum_{j=0}^{r}|\alpha_j|\,j^2\left[\int_0^{\infty}x^2e^{-x^2}dx+\frac{j}{2}\int_0^{\infty}x^3e^{-x^2}dx\right]\right\}$$

$$=\frac{2\xi}{\sqrt{\pi}}\omega_1\left(f''+f,\sqrt{\xi}\right)_1\left\{\sum_{j=0}^{r}|\alpha_j|\,j^2\right.$$

$$\left.\left[\left[\int_0^1x^2e^{-x^2}dx+\int_1^{\infty}x^2e^{-x^2}dx\right]+\frac{j}{2}\left[\int_0^1x^3e^{-x^2}dx+\int_1^{\infty}x^3e^{-x^2}dx\right]\right]\right\}$$

$$\leq\frac{2\xi}{\sqrt{\pi}}\omega_1\left(f''+f,\sqrt{\xi}\right)_1$$

$$\left\{\sum_{j=0}^{r}|\alpha_j|\,j^2\left[\left[\frac{1}{3}+\int_1^{\infty}x^2e^{-x}dx\right]+\frac{j}{2}\left[\frac{1}{4}+\int_1^{\infty}x^3e^{-x}dx\right]\right]\right\}$$

$$\leq \frac{2\xi}{\sqrt{\pi}}\omega_1\left(f''+f,\sqrt{\xi}\right)_1$$

$$\left\{\sum_{j=0}^{r}|\alpha_j|\,j^2\left[\left[\frac{1}{3}+\int_0^{\infty}x^2e^{-x}dx\right]+\frac{j}{2}\left[\frac{1}{4}+\int_0^{\infty}x^3e^{-x}dx\right]\right]\right\} \tag{27.95}$$

$$= \frac{2\xi}{\sqrt{\pi}}\omega_1\left(f''+f,\sqrt{\xi}\right)_1\left\{\sum_{j=0}^{r}|\alpha_j|\,j^2\left[\left[\frac{1}{3}+2\right]+\frac{j}{2}\left[\frac{1}{4}+6\right]\right]\right\}$$

$$= \frac{2\xi}{\sqrt{\pi}}\omega_1\left(f''+f,\sqrt{\xi}\right)_1\left\{\sum_{j=0}^{r}|\alpha_j|\,j^2\left[\frac{7}{3}+j\frac{25}{8}\right]\right\}.$$

The theorem is proved. □

We give

Corollary 27.4. *(to Theorem 27.4) It holds*

$$\|W_{r,\xi}(f)-f\|_1 \leq E_1(\xi)$$

$$+\|f''\|_1\left(\sum_{j=0}^{r}|\alpha_j|\left(1-e^{-\frac{j^2}{4}\xi}\right)\right)\to 0, \quad as\ \xi\to 0. \tag{27.96}$$

Above $E_1(\xi)$ *is as in (27.90).*

It follows the L_1 approximation by Poisson–Cauchy operators.

Theorem 27.5. *Let* $\xi>0$, $f,f''\in L_1(\mathbb{R})\cap C(\mathbb{R})$ *and* $\beta>\frac{2}{\alpha}$, $\alpha\in\mathbb{N}$. *Then*

$$\|\Delta_3\|_1$$

$$=\left\|M_{r,\xi}(f)-f-4f''\left[\left(\frac{\alpha\Gamma(\beta)}{\Gamma\left(\frac{1}{2\alpha}\right)\Gamma\left(\beta-\frac{1}{2\alpha}\right)}\right)\left(\sum_{j=0}^{r}\alpha_j\int_0^{\infty}\frac{\sin^2\left(\frac{j\xi t}{2}\right)}{(1+t^{2\alpha})^{\beta}}dt\right)\right]\right\|_1$$

$$\leq \frac{\omega_1(f''+f,\xi)_1\,\xi^2}{\Gamma\left(\frac{1}{2\alpha}\right)\Gamma\left(\beta-\frac{1}{2\alpha}\right)}$$

$$\left[\sum_{j=0}^{r}|\alpha_j|\,j^2\left(\Gamma\left(\frac{3}{2\alpha}\right)\Gamma\left(\beta-\frac{3}{2\alpha}\right)+\frac{j}{2}\Gamma\left(\frac{2}{\alpha}\right)\Gamma\left(\beta-\frac{2}{\alpha}\right)\right)\right]$$

$$:= E_2(\xi)\to 0, \tag{27.97}$$

as $\xi\to 0$.

Proof. We have that

$$\Delta_3^*(x) = W \int_{-\infty}^{\infty} R(t) \frac{1}{(t^{2\alpha} + \xi^{2\alpha})^{\beta}} dt. \tag{27.98}$$

We observe that

$$\|\Delta_3^*\|_1 = \int_{-\infty}^{\infty} |\Delta_3^*(x)| \, dx = W \int_{-\infty}^{\infty} \left(\left| \int_{-\infty}^{\infty} R(t) \frac{1}{(t^{2\alpha} + \xi^{2\alpha})^{\beta}} dt \right| \right) dx$$

$$\leq W \int_{-\infty}^{\infty} \left(\int_{-\infty}^{\infty} |R(t)| \frac{1}{(t^{2\alpha} + \xi^{2\alpha})^{\beta}} dt \right) dx$$

$$\overset{(27.45)}{\leq} W \left[\sum_{j=0}^{r} |\alpha_j| \, j^2 \left(\int_{-\infty}^{\infty} \left(\int_{-\infty}^{\infty} \right.\right.\right.$$

$$\left(\int_0^{|t|} \left| \left(f''(x + jsign(t)\theta) + f(x + jsign(t)\theta) \right) \right.\right.$$

$$\left.\left.\left.\left. - \left(f''(x) + f(x) \right) \right| (|t| - \theta) \, d\theta \right) \frac{1}{(t^{2\alpha} + \xi^{2\alpha})^{\beta}} dt \right) dx \right) \right] \tag{27.99}$$

$$\leq W \left[\sum_{j=0}^{r} |\alpha_j| \, j^2 \left(\int_{-\infty}^{\infty} \left(\int_{-\infty}^{\infty} \left(|t| \int_0^{|t|} \right.\right.\right.\right.$$

$$\left| \left(f''(x + jsign(t)\theta) + f(x + jsign(t)\theta) \right) - \left(f''(x) + f(x) \right) \right| d\theta \Big)$$

$$\left.\left.\left. \frac{1}{(t^{2\alpha} + \xi^{2\alpha})^{\beta}} dt \right) dx \right) \right]$$

$$= W \left[\sum_{j=0}^{r} |\alpha_j| \, j^2 \left(\int_{-\infty}^{\infty} \left(\int_{-\infty}^{\infty} \left(|t| \int_0^{|t|} \right.\right.\right.\right.$$

$$\left| \left(f''(x + jsign(t)\theta) + f(x + jsign(t)\theta) \right) - \left(f''(x) + f(x) \right) \right| d\theta \Big) \, dx \Big)$$

$$\left.\left. \frac{1}{(t^{2\alpha} + \xi^{2\alpha})^{\beta}} dt \right) \right]$$

$$= W \left[\sum_{j=0}^{r} |\alpha_j| \, j^2 \left(\int_{-\infty}^{\infty} \left(\int_0^{|t|} \left(\int_{-\infty}^{\infty} \right.\right.\right.\right.$$

$$\left| \left(f''(x + jsign(t)\theta) + f(x + jsign(t)\theta) \right) - \left(f''(x) + f(x) \right) \right| dx \Big) \, d\theta \Big)$$

$$\left.\left. \frac{|t|}{(t^{2\alpha} + \xi^{2\alpha})^{\beta}} dt \right) \right]$$

$$\leq W\left[\sum_{j=0}^{r}|\alpha_j|j^2\left(\int_{-\infty}^{\infty}\left(\int_0^{|t|}\omega_1\left(f''+f,\frac{\xi}{\xi}j\theta\right)_1 d\theta\right)\frac{|t|}{(t^{2\alpha}+\xi^{2\alpha})^{\beta}}dt\right)\right]$$

$$\leq W\omega_1\left(f''+f,\xi\right)_1 \tag{27.100}$$

$$\left[\sum_{j=0}^{r}|\alpha_j|j^2\left(\int_{-\infty}^{\infty}\left(\int_0^{|t|}\left(1+\frac{j}{\xi}\theta\right)d\theta\right)\frac{|t|}{(t^{2\alpha}+\xi^{2\alpha})^{\beta}}dt\right)\right]$$

$$=W\omega_1\left(f''+f,\xi\right)_1\left[\sum_{j=0}^{r}|\alpha_j|j^2\left(\int_{-\infty}^{\infty}\left(|t|+\frac{j}{\xi}\frac{t^2}{2}\right)\frac{|t|}{(t^{2\alpha}+\xi^{2\alpha})^{\beta}}dt\right)\right]$$

$$=2W\omega_1\left(f''+f,\xi\right)_1\left[\sum_{j=0}^{r}|\alpha_j|j^2\left(\int_0^{\infty}\left(t+\frac{j}{\xi}\frac{t^2}{2}\right)\frac{t}{(t^{2\alpha}+\xi^{2\alpha})^{\beta}}dt\right)\right]$$

$$=2W\omega_1\left(f''+f,\xi\right)_1$$

$$\left[\sum_{j=0}^{r}|\alpha_j|j^2\left(\int_0^{\infty}\frac{t^2}{(t^{2\alpha}+\xi^{2\alpha})^{\beta}}dt+\frac{j}{2\xi}\int_0^{\infty}\frac{t^3}{(t^{2\alpha}+\xi^{2\alpha})^{\beta}}dt\right)\right]$$

$$=2W\omega_1\left(f''+f,\xi\right)_1 \tag{27.101}$$

$$\left[\sum_{j=0}^{r}|\alpha_j|j^2\left(\xi^{3-2\alpha\beta}\int_0^{\infty}\frac{\left(\frac{t}{\xi}\right)^2}{\left(\left(\frac{t}{\xi}\right)^{2\alpha}+1\right)^{\beta}}d\frac{t}{\xi}+\right.\right.$$

$$\left.\left.\frac{j}{2\xi}\xi^{4-2\alpha\beta}\int_0^{\infty}\frac{\left(\frac{t}{\xi}\right)^3}{\left(\left(\frac{t}{\xi}\right)^{2\alpha}+1\right)^{\beta}}d\frac{t}{\xi}\right)\right]$$

$$=2W\xi^{3-2\alpha\beta}\omega_1\left(f''+f,\xi\right)_1$$

$$\left[\sum_{j=0}^{r}|\alpha_j|j^2\left[\int_0^{\infty}\frac{x^2}{(1+x^{2\alpha})^{2\beta}}dx+\frac{j}{2}\int_0^{\infty}\frac{x^3}{(1+x^{2\alpha})^{2\beta}}dx\right]\right]\overset{(\text{by }(27.10))}{=}$$

(and by Zwillinger (1995), p. 397, formula 595)

$$\frac{2\alpha\Gamma(\beta)\,\xi^2}{\Gamma\left(\frac{1}{2\alpha}\right)\Gamma\left(\beta-\frac{1}{2\alpha}\right)}\omega_1\left(f''+f,\xi\right)_1$$

$$\left[\sum_{j=0}^{r}|\alpha_j|j^2\left[\frac{\Gamma\left(\frac{3}{2\alpha}\right)\Gamma\left(\beta-\frac{3}{2\alpha}\right)}{2\alpha\Gamma(\beta)}+\frac{j}{2}\frac{\Gamma\left(\frac{2}{\alpha}\right)\Gamma\left(\beta-\frac{2}{\alpha}\right)}{2\alpha\Gamma(\beta)}\right]\right]$$

$$= \frac{\xi^2}{\Gamma\left(\frac{1}{2\alpha}\right)\Gamma\left(\beta-\frac{1}{2\alpha}\right)}\omega_1\left(f''+f,\xi\right)_1$$
$$\left[\sum_{j=0}^{r}|\alpha_j|\,j^2\left(\Gamma\left(\frac{3}{2\alpha}\right)\Gamma\left(\beta-\frac{3}{2\alpha}\right)+\frac{j}{2}\Gamma\left(\frac{2}{\alpha}\right)\Gamma\left(\beta-\frac{2}{\alpha}\right)\right)\right], \tag{27.102}$$

under the assumption $\beta > \frac{2}{\alpha}$.

The proof is completed. □

We also give

Corollary 27.5. *(to Theorem 27.5) It holds ($\beta > \frac{2}{\alpha}$, $\alpha \in \mathbb{N}$)*

$$\left\|M_{r,\xi}(f)-f\right\|_1 \leq E_2(\xi)+4\left\|f''\right\|_1$$
$$\left[\left(\frac{\alpha\Gamma(\beta)}{\Gamma\left(\frac{1}{2\alpha}\right)\Gamma\left(\beta-\frac{1}{2\alpha}\right)}\right)\left(\sum_{j=0}^{r}|\alpha_j|\left(\int_0^\infty\frac{\sin^2\left(\frac{j\xi t}{2}\right)}{\left(1+t^{2\alpha}\right)^\beta}dt\right)\right)\right]$$
$$\to 0, \quad \text{as } \xi \to 0. \tag{27.103}$$

Above $E_2(\xi)$ is as in (27.97).

We continue with L_1 approximation by trigonometric singular operators.

Theorem 27.6. *Let $\xi > 0$, $f, f'' \in L_1(\mathbb{R}) \cap C(\mathbb{R})$, and $\beta \in \mathbb{N} - \{1,2\}$. Then*

$$\|\Delta_4\|_1 = \left\|T_{r,\xi}(f)-f-4f''\left(\lambda_1^{-1}\sum_{j=0}^{r}\alpha_j\int_0^\infty\sin^2\left(\frac{j\xi t}{2}\right)\left(\frac{\sin t}{t}\right)^{2\beta}dt\right)\right\|_1$$
$$\leq \lambda_1^{-1}\omega_1\left(f''+f,\xi\right)_1\xi^2\left\{\sum_{j=1}^{r}|\alpha_j|\,j^2\right.$$
$$\left[\left(\frac{\pi(-1)^{(\beta-1)}(2\beta)!}{4(2\beta-3)!}\sum_{k=1}^{\beta}(-1)^k\frac{k^{2\beta-3}}{(\beta-k)!(\beta+k)!}\right)\right.$$
$$\left.+\frac{j}{2}\left(\frac{-(2\beta)!}{4(2\beta-4)!}\sum_{k=1}^{\beta}(-1)^{\beta-k}\frac{k^{2\beta-4}\ln(2k)}{(\beta-k)!(\beta+k)!}\right)\right]$$
$$=: E_3(\xi) \to 0, \quad \text{as } \xi \to 0. \tag{27.104}$$

Proof. We have that

$$\Delta_4^*(x) = \lambda^{-1} \int_{-\infty}^{\infty} R(t) \left(\frac{\sin\left(\frac{t}{\xi}\right)}{t} \right)^{2\beta} dt. \tag{27.105}$$

We observe that

$$\|\Delta_4^*\| = \int_{-\infty}^{\infty} |\Delta_4^*(x)|\, dx = \lambda^{-1} \int_{-\infty}^{\infty} \left| \int_{-\infty}^{\infty} R(t) \left(\frac{\sin\left(\frac{t}{\xi}\right)}{t} \right)^{2\beta} dt \right| dx$$

$$\leq \lambda^{-1} \int_{-\infty}^{\infty} \left(\int_{-\infty}^{\infty} |R(t)| \left(\frac{\sin\left(\frac{t}{\xi}\right)}{t} \right)^{2\beta} dt \right) dx$$

$$\overset{(27.45)}{\leq} \lambda^{-1} \left[\sum_{j=0}^{r} |\alpha_j|\, j^2 \left(\int_{-\infty}^{\infty} \left(\int_{-\infty}^{\infty} \right.\right.\right.$$

$$\left(\int_0^{|t|} \left| (f''(x + jsign(t)\theta) + f(x + jsign(t)\theta)) \right.\right.$$

$$\left.\left.\left.\left.\left. - (f''(x) + f(x)) \right| (|t| - \theta)\, d\theta \right) \left(\frac{\sin\left(\frac{t}{\xi}\right)}{t} \right)^{2\beta} dt \right) dx \right) \right]$$

$$\leq \lambda^{-1} \left[\sum_{j=0}^{r} |\alpha_j|\, j^2 \left(\int_{-\infty}^{\infty} \left(\int_{-\infty}^{\infty} \left(|t| \int_0^{|t|} \right.\right.\right.\right.$$

$$\left. \left| (f''(x + jsign(t)\theta) + f(x + jsign(t)\theta)) - (f''(x) + f(x)) \right| d\theta \right)$$

$$\left.\left.\left. \left(\frac{\sin\left(\frac{t}{\xi}\right)}{t} \right)^{2\beta} dt \right) dx \right) \right]$$

$$= \lambda^{-1} \left[\sum_{j=0}^{r} |\alpha_j|\, j^2 \left(\int_{-\infty}^{\infty} \left(\int_{-\infty}^{\infty} \left(|t| \int_0^{|t|} \right.\right.\right.\right.$$

$$\left|\left(f''\left(x+jsign\left(t\right)\theta\right)+f\left(x+jsign\left(t\right)\theta\right)\right)-\left(f''\left(x\right)+f\left(x\right)\right)\right|d\theta\Bigg)dx\Bigg)$$

$$\left(\frac{\sin\left(\frac{t}{\xi}\right)}{t}\right)^{2\beta}dt\Bigg)\Bigg]$$

$$=\lambda^{-1}\left[\sum_{j=0}^{r}\left|\alpha_j\right|j^2\left(\int_{-\infty}^{\infty}\left(\int_0^{|t|}\left(\int_{-\infty}^{\infty}\right.\right.\right.\right.$$

$$\left|\left(f''\left(x+jsign\left(t\right)\theta\right)+f\left(x+jsign\left(t\right)\theta\right)\right)-\left(f''\left(x\right)+f\left(x\right)\right)\right|dx\Bigg)d\theta\Bigg)$$

$$|t|\left(\frac{\sin\left(\frac{t}{\xi}\right)}{t}\right)^{2\beta}dt\Bigg)\Bigg]$$

$$\leq\lambda^{-1}\left[\sum_{j=0}^{r}\left|\alpha_j\right|j^2\left(\int_{-\infty}^{\infty}\left(\int_0^{|t|}\omega_1\left(f''+f,\frac{\xi}{\xi}j\theta\right)_1d\theta\right)\right.\right.$$

$$|t|\left(\frac{\sin\left(\frac{t}{\xi}\right)}{t}\right)^{2\beta}dt\Bigg)\Bigg] \tag{27.107}$$

$$\leq\lambda^{-1}\omega_1\left(f''+f,\xi\right)_1$$

$$\left[\sum_{j=0}^{r}\left|\alpha_j\right|j^2\left(\int_{-\infty}^{\infty}\left(\int_0^{|t|}\left(1+\frac{j}{\xi}\theta\right)d\theta\right)|t|\left(\frac{\sin\left(\frac{t}{\xi}\right)}{t}\right)^{2\beta}dt\right)\right]$$

$$=\lambda^{-1}\omega_1\left(f''+f,\xi\right)_1\left[\sum_{j=0}^{r}\left|\alpha_j\right|j^2\left(\int_{-\infty}^{\infty}\left(|t|+\frac{j}{\xi}\frac{t^2}{2}\right)|t|\left(\frac{\sin\left(\frac{t}{\xi}\right)}{t}\right)^{2\beta}dt\right)\right]$$

$$=2\lambda^{-1}\omega_1\left(f''+f,\xi\right)_1\left[\sum_{j=0}^{r}\left|\alpha_j\right|j^2\left(\int_0^{\infty}\left(t+\frac{j}{\xi}\frac{t^2}{2}\right)t\left(\frac{\sin\left(\frac{t}{\xi}\right)}{t}\right)^{2\beta}dt\right)\right]$$

$$=2\lambda^{-1}\omega_1\left(f''+f,\xi\right)_1$$

$$\left[\sum_{j=0}^{r}|\alpha_j|\,j^2\left(\int_0^\infty t^2\left(\frac{\sin\left(\frac{t}{\xi}\right)}{t}\right)^{2\beta}dt+\frac{j}{2\xi}\int_0^\infty t^3\left(\frac{\sin\left(\frac{t}{\xi}\right)}{t}\right)^{2\beta}dt\right)\right]$$

$$=2\lambda^{-1}\omega_1\left(f''+f,\xi\right)_1$$

$$\left[\sum_{j=0}^{r}|\alpha_j|\,j^2\left(\xi^{3-2\beta}\int_0^\infty\left(\frac{t}{\xi}\right)^2\left(\frac{\sin\left(\frac{t}{\xi}\right)}{\frac{t}{\xi}}\right)^{2\beta}d\frac{t}{\xi}\right.\right.$$

$$\left.\left.+\frac{j}{2\xi}\xi^{4-2\beta}\int_0^\infty\left(\frac{t}{\xi}\right)^3\left(\frac{\sin\left(\frac{t}{\xi}\right)}{\frac{t}{\xi}}\right)^{2\beta}d\frac{t}{\xi}\right)\right]$$

$$=2\lambda^{-1}\omega_1\left(f''+f,\xi\right)_1\xi^{3-2\beta}$$

$$\left[\sum_{j=0}^{r}|\alpha_j|\,j^2\left(\int_0^\infty x^2\left(\frac{\sin x}{x}\right)^{2\beta}dx+\frac{j}{2}\int_0^\infty x^3\left(\frac{\sin x}{x}\right)^{2\beta}dx\right)\right]$$

$$=2\lambda_1^{-1}\omega_1\left(f''+f,\xi\right)_1\xi^2$$

$$\left[\sum_{j=0}^{r}|\alpha_j|\,j^2\left(\int_0^\infty x^2\left(\frac{\sin x}{x}\right)^{2\beta}dx+\frac{j}{2}\int_0^\infty x^3\left(\frac{\sin x}{x}\right)^{2\beta}dx\right)\right] \tag{27.108}$$

(we use Edwards (1954), p. 210, item 1033)

$$=2\lambda_1^{-1}\omega_1\left(f''+f,\xi\right)_1\xi^2\left\{\sum_{j=1}^{r}|\alpha_j|\,j^2\right.$$

$$\left[\left(\frac{\pi\,(-1)^{(\beta-1)}\,(2\beta)!}{8\,(2\beta-3)!}\sum_{k=1}^{\beta}(-1)^k\,\frac{k^{2\beta-3}}{(\beta-k)!\,(\beta+k)!}\right)\right.$$

$$\left.+\frac{j}{2}\left(\frac{(-1)\,(2\beta)!}{8\,(2\beta-4)!}\sum_{k=1}^{\beta}(-1)^{\beta-k}\,\frac{k^{2\beta-4}\,(\ln(2k))}{(\beta-k)!\,(\beta+k)!}\right)\right], \tag{27.109}$$

under $\beta>2$, $\beta\in\mathbb{N}$.

The theorem is proved. □

We finish with the following.

Corollary 27.6 (to Theorem 27.6). *It holds* ($\beta \in \mathbb{N} - \{1, 2\}$),

$$\|T_{r,\xi}(f) - f\|_1 \leq E_3(\xi)$$
$$+ 4\|f''\|_1 \lambda_1^{-1} \left(\sum_{j=0}^{r} |\alpha_j| \left(\int_0^\infty \sin^2\left(\frac{j\xi t}{2}\right) \left(\frac{\sin t}{t}\right)^{2\beta} dt \right) \right)$$
$$\to 0, \quad as\ \xi \to 0. \tag{27.110}$$

Above $E_3(\xi)$ *is as in (27.104).*

Bibliography

A.H. Ali and Z. Pales, Taylor-type expansions in terms of exponential polynomials, *Mathematical Inequalities and Applications*, 25(4) (2022), 1123–1141.

G.A. Anastassiou, Basic convergence with rates of smooth Picard singular integral operators, *Journal of Computational Analysis and Applications*, 8(4) (2006), 313–334.

G.A. Anastassiou, *Intelligent Mathematics: Computational Analysis*, Springer, Heidelberg, New York, Chapter 12, 2011.

G.A. Anastassiou, Opial and Ostrowski type inequalities based on trigonometric and hyperbolic type Taylor formulae, *Malaya Journal Mathematik*, 11(S) (2023), 1–26.

G.A. Anastassiou, *Trigonometric deduced L_p degree of approximation by various smooth singular integral operators*, submitted, 2024.

G.A. Anastassiou, R. Mezei, *Approximation by Singular Integrals*, Cambridge Scientific Publishers, Cambridge, UK, 2012.

A. Aral, On a generalized Gauss Weierstrass singular integral, *Fasciculi Mathematici*. No. 35 (2005), 23–33.

A. Aral, Pointwise approximation by the generalization of Picard and Gauss-Weierstrass singular integrals, *Journal of Concrete and Applicable Mathematics*, 6(4) (2008), 327–339.

A. Aral, On generalized Picard integral operators, *Advances in Summability and Approximation Theory*, 157–168, Springer, Singapore, 2018.

A. Aral, E. Deniz and H. Erbay, The Picard and Gauss-Weiertrass singular integrals in (p, q)-calculus, *Bulletin of the Malaysian Mathematical Sciences Society*, 43(2) (2020), 1569–1583.

A. Aral and S.G. Gal, q-generalizations of the Picard and Gauss-Weierstrass singular integrals, *Taiwanese Journal of Mathematics*, 12(9) (2008), 2501–2515.

Carlo Bardaro, Ilaria Mantellini, Gumrah Uysal, Basar Yilmaz, A class of integral operators that fix exponential functions, *Mediterranean Journal of Mathematics*, 18(5) (2021), Paper No. 179, 21.

J. Edwards, *A Treatise on the Integral Calculus, Vol II*, Chelsea, New York, 1954.

R.N. Mohapatra and R.S. Rodriguez, On the rate of convergence of singular integrals for Hölder continuous functions, *Mathematische Nachrichten*, 149 (1990), 117–124.

D. Tapiawala, G. Uysal and V.N. Mishra, Recent observations on nonlinear two-parameter singular integral operators, *Journal of Inequalities and Special Functions*, 10(2) (2019), 1–9.

G.O. Ozalp and G. Uysal, On certain multidimensional nonlinear integrals, *Communications Faculty of Sciences University of Ankara Series A1 Mathematics and Statistics*, 69(2) (2020), 1356–1367.

G. Uysal, On nonlinear bivariate $[m_1, m_2]$-singular integral operators, *Mathematical Methods in the Applied Sciences*, 42(16) (2019), 5455–5467.

G. Uysal and H. Dutta, On weighted convergence of double singular integral operators involving summation, *Applied Mathematical Analysis: Theory, Methods, and Applications*, 579-603, Stud. Syst. Decis. Control, 177, Springer, Cham, 2020.

D. Zwillinger, *CRC Standard Mathematical Tables and Formulae*, 30th edn. Chapman & Hall/CRC, Boca Raton (1995).

Chapter 28

Parametrized and Trigonometric-Produced Uniform Quantitative Approximation by Various Smooth Singular Integral Operators

In this chapter, we continue with the study of smooth Gauss–Weierstrass, Poisson–Cauchy and trigonometric singular integral operators that started in Anastassiou (2011), see there Chapters 10–14. This time, the foundation of our research is a trigonometric Taylor's formula. We prove the parametrized univariate uniform convergence of our operators to the unit operator with rates via Jackson-type parametrized inequalities involving the first modulus of continuity. Of interest, here is a residual appearing term. Note that our operators are not in general positive. It follows Anastassiou (2024).

28.1 Introduction

We are motivated by Anastassiou (2006, 2011) Chapters 10–14, Anastassiou (2023) and Ali and Pales (2022). We use a trigonometric new Taylor formula from Anastassiou (2023), see also Ali and Pales (2022). Here we consider some very general operators, the smooth Gauss–Weierstrass, Poisson–Cauchy and trigonometric singular integral operators over the real line, and we study further their parametrized and uniform convergence properties quantitatively. We establish related parametrized inequalities involving the first modulus of continuity. We provide detailed proofs.

Other important motivating articles on the topic are Aral (2005)–Ali and Gal (2008).

For the history of the topic, we mention about the monograph Anastassiou and Mezei (2012), which was the first complete source to deal exclusively with the classic theory of the approximation of singular integrals to the identity-unit operator. The authors there studied quantitatively the basic approximation properties of the general Picard, Gauss–Weierstrass and Poisson–Cauchy singular integral operators over the real line, which are not positive linear operators. In particular, they studied the rate of convergence of these operators to the unit operator, as well as the related simultaneous approximation. This is given via inequalities and with the use of higher-order modulus of smoothness of the high order derivative of the involved function. Some of these inequalities are proven to be sharp. Also, they studied the global smoothness preservation property of these operators. Furthermore, they gave asymptotic expansions of Voronovskaya type for the error of approximation. They continued with the study of related properties of the general fractional Gauss–Weierstrass and Poisson–Cauchy singular integral operators. These properties were studied with respect to L_p norm, $1 \leq p \leq \infty$. The case of Lipschitz-type function approximation was studied separately and in detail. Furthermore, they presented the corresponding general approximation theory of general singular integral operators with lots of applications to, the under focused till then, trigonometric singular integral.

28.2 About Part I: On Gauss–Weierstrass Smooth Singular Integrals

By Ali and Pales (2022) and Anastassiou (2023), for $\alpha, \beta \in \mathbb{R}$ with $\alpha\beta\left(\alpha^2 - \beta^2\right) \neq 0$, and $f \in C^4(\mathbb{R})$, $a, x \in \mathbb{R}$, we have the following general trigonometric Taylor formula:

$$
\begin{aligned}
f(x) - f(a) = {} & f'(a)\left(\frac{\beta^3 \sin(\alpha(x-a)) - \alpha^3 \sin(\beta(x-a))}{\alpha\beta\left(\beta^2-\alpha^2\right)}\right) \\
& + f''(a)\left(\frac{\cos(\alpha(x-a)) - \cos(\beta(x-a))}{\beta^2-\alpha^2}\right) \\
& + f'''(a)\left(\frac{\beta \sin(\alpha(x-a)) - \alpha \sin(\beta(x-a))}{\alpha\beta\left(\beta^2-\alpha^2\right)}\right) \\
& + \frac{2\left(f''''(a) + \left(\alpha^2+\beta^2\right) f''(a)\right)}{(\alpha\beta)^2\left(\beta^2-\alpha^2\right)}
\end{aligned}
$$

$$\times\left(\beta^2\sin^2\left(\frac{\alpha(x-a)}{2}\right)-\alpha^2\sin^2\left(\frac{\beta(x-a)}{2}\right)\right)$$
$$+\frac{1}{\alpha\beta(\beta^2-\alpha^2)}\int_a^x\left[\left(f''''(t)+\left(\alpha^2+\beta^2\right)f''(t)+\alpha^2\beta^2f(t)\right)\right.$$
$$\left.-\left(f''''(a)+\left(\alpha^2+\beta^2\right)f''(a)+\alpha^2\beta^2f(a)\right)\right]$$
$$\times\left[\beta\sin(\alpha(x-t))-\alpha\sin(\beta(x-t))\right]dt. \tag{28.1}$$

For $r\in\mathbb{N}$ and $n\in\mathbb{Z}^+$, we set

$$\alpha_j:=\begin{cases}(-1)^{r-j}\binom{r}{j}j^{-n}, & j=1,\ldots,r.\\ 1-\sum_{j=1}^{r}(-1)^{r-j}\binom{r}{j}j^{-n}, & n=0,\end{cases} \tag{28.2}$$

that is,

$$\sum_{j=0}^{r}\alpha_j=1. \tag{28.3}$$

Here we consider all $f,f',f'',f''',f^{(4)}\in C_U(\mathbb{R})\cap C_B(\mathbb{R})$.

For $x\in\mathbb{R}$, $\xi>0$ we consider the Lebesgue integrals, so-called smooth Gauss–Weierstrass operators

$$W_{r,\xi}(f,x)=\frac{1}{\sqrt{\pi\xi}}\int_{-\infty}^{\infty}\left(\sum_{j=0}^{r}\alpha_jf(x+jt)\right)e^{-\frac{t^2}{\xi}}dt, \tag{28.4}$$

see Anastassiou and Mezei (2012), $W_{r,\xi}$ are not in general positive operators, see Anastassiou and Mezei (2012).

We note by

$$\frac{1}{\sqrt{\pi\xi}}\int_{-\infty}^{\infty}e^{-\frac{t^2}{\xi}}dt=1 \tag{28.5}$$

that

$$W_{r,\xi}(c,x)=c,\text{ where } c \text{ is a constant}, \tag{28.6}$$

and

$$W_{r,\xi}(f,x)-f(x)=\frac{1}{\sqrt{\pi\xi}}\left(\sum_{j=0}^{r}\alpha_j\int_{-\infty}^{\infty}(f(x+jt)-f(x))e^{-\frac{t^2}{\xi}}dt\right). \tag{28.7}$$

Denote by

$$\omega_1(f,h) := \sup_{\substack{x,y\in\mathbb{R}\\ |x-y|\le\delta}} |f(x)-f(y)|, \tag{28.8}$$

the first modulus of continuity of f.

By (28.1), we get that

$$\begin{aligned}
f(x+jt)-f(x) &= f'(x)\left(\frac{\beta^3\sin(\alpha jt)-\alpha^3\sin(\beta jt)}{\alpha\beta(\beta^2-\alpha^2)}\right)\\
&+ f''(x)\left(\frac{\cos(\alpha jt)-\cos(\beta jt)}{\beta^2-\alpha^2}\right)\\
&+ f'''(x)\left(\frac{\beta\sin(\alpha jt)-\alpha\sin(\beta jt)}{\alpha\beta(\beta^2-\alpha^2)}\right)\\
&+ \frac{2\left(f^{(4)}(x)+(\alpha^2+\beta^2)f''(x)\right)}{(\alpha\beta)^2(\beta^2-\alpha^2)}\left(\beta^2\sin^2\left(\frac{\alpha jt}{2}\right)-\alpha^2\sin^2\left(\frac{\beta jt}{2}\right)\right)\\
&+ \frac{1}{\alpha\beta(\beta^2-\alpha^2)}\int_x^{x+jt}\Big[\left(f^{(4)}(s)+(\alpha^2+\beta^2)f''(s)+\alpha^2\beta^2 f(s)\right)\\
&- \left(f^{(4)}(x)+(\alpha^2+\beta^2)f''(x)+\alpha^2\beta^2 f(x)\right)\Big]\\
&[\beta\sin(\alpha(x+jt-s))-\alpha\sin(\beta(x+jt-s))]\,ds,
\end{aligned} \tag{28.9}$$

or better,

$$\begin{aligned}
f(x+jt)-f(x) &= f'(x)\left(\frac{\beta^3\sin(\alpha jt)-\alpha^3\sin(\beta jt)}{\alpha\beta(\beta^2-\alpha^2)}\right)\\
&+ f''(x)\left(\frac{\cos(\alpha jt)-\cos(\beta jt)}{\beta^2-\alpha^2}\right)\\
&+ f'''(x)\left(\frac{\beta\sin(\alpha jt)-\alpha\sin(\beta jt)}{\alpha\beta(\beta^2-\alpha^2)}\right)\\
&+ \frac{2\left(f^{(4)}(x)+(\alpha^2+\beta^2)f''(x)\right)}{(\alpha\beta)^2(\beta^2-\alpha^2)}\left(\beta^2\sin^2\left(\frac{\alpha jt}{2}\right)-\alpha^2\sin^2\left(\frac{\beta jt}{2}\right)\right)\\
&+ \frac{1}{\alpha\beta(\beta^2-\alpha^2)}\int_0^{jt}\Big[\left(f^{(4)}(x+z)+(\alpha^2+\beta^2)f''(x+z)+\alpha^2\beta^2 f(x+z)\right)\\
&- \left(f^{(4)}(x)+(\alpha^2+\beta^2)f''(x)+\alpha^2\beta^2 f(x)\right)\Big]\\
&[\beta\sin(\alpha(jt-z))-\alpha\sin(\beta(jt-z))]\,dz.
\end{aligned} \tag{28.10}$$

Furthermore, it holds

$$\sum_{j=0}^{r}\alpha_j\left[f\left(x+jt\right)-f\left(x\right)\right]$$

$$=\left(\frac{f'\left(x\right)}{\alpha\beta\left(\beta^2-\alpha^2\right)}\right)\left[\beta^3\sum_{j=0}^{r}\alpha_j\sin\left(\alpha jt\right)-\alpha^3\sum_{j=0}^{r}\alpha_j\sin\left(\beta jt\right)\right]$$

$$+\left(\frac{f''\left(x\right)}{\beta^2-\alpha^2}\right)\left[\sum_{j=0}^{r}\alpha_j\cos\left(\alpha jt\right)-\sum_{j=0}^{r}\alpha_j\cos\left(\beta jt\right)\right]$$

$$+\left(\frac{f^{(3)}\left(x\right)}{\alpha\beta\left(\beta^2-\alpha^2\right)}\right)\left[\beta\sum_{j=0}^{r}\alpha_j\sin\left(\alpha jt\right)-\alpha\sum_{j=0}^{r}\alpha_j\sin\left(\beta jt\right)\right]$$

$$+\left(\frac{2\left(f^{(4)}\left(x\right)+\left(\alpha^2+\beta^2\right)f''\left(x\right)\right)}{\left(\alpha\beta\right)^2\left(\beta^2-\alpha^2\right)}\right)$$

$$\times\left(\beta^2\sum_{j=0}^{r}\alpha_j\sin^2\left(\frac{\alpha jt}{2}\right)-\alpha^2\sum_{j=0}^{r}\alpha_j\sin^2\left(\frac{\beta jt}{2}\right)\right)+\frac{1}{\alpha\beta\left(\beta^2-\alpha^2\right)}$$

$$\times\sum_{j=0}^{r}\alpha_j\int_0^{jt}\left[\left(f^{(4)}\left(x+z\right)+\left(\alpha^2+\beta^2\right)f''\left(x+z\right)+\alpha^2\beta^2f\left(x+z\right)\right)\right.$$

$$\left.-\left(f^{(4)}\left(x\right)+\left(\alpha^2+\beta^2\right)f''\left(x\right)+\alpha^2\beta^2f\left(x\right)\right)\right]$$

$$\times\left[\beta\sin\left(\alpha\left(jt-z\right)\right)-\alpha\sin\left(\beta\left(jt-z\right)\right)\right]dz, \tag{28.11}$$

or better

$$\sum_{j=0}^{r}\alpha_j\left[f\left(x+jt\right)-f\left(x\right)\right]$$

$$=\left(\frac{f'\left(x\right)}{\alpha\beta\left(\beta^2-\alpha^2\right)}\right)\left[\beta^3\sum_{j=0}^{r}\alpha_j\sin\left(\alpha jt\right)-\alpha^3\sum_{j=0}^{r}\alpha_j\sin\left(\beta jt\right)\right]$$

$$+\left(\frac{f''\left(x\right)}{\beta^2-\alpha^2}\right)\left[\sum_{j=0}^{r}\alpha_j\cos\left(\alpha jt\right)-\sum_{j=0}^{r}\alpha_j\cos\left(\beta jt\right)\right]$$

$$
+\left(\frac{f^{(3)}(x)}{\alpha\beta\left(\beta^2-\alpha^2\right)}\right)\left[\beta\sum_{j=0}^{r}\alpha_j\sin(\alpha jt)-\alpha\sum_{j=0}^{r}\alpha_j\sin(\beta jt)\right]
$$

$$
+\left(\frac{2\left(f^{(4)}(x)+\left(\alpha^2+\beta^2\right)f''(x)\right)}{(\alpha\beta)^2\left(\beta^2-\alpha^2\right)}\right)
$$

$$
\times\left(\beta^2\sum_{j=0}^{r}\alpha_j\sin^2\left(\frac{\alpha jt}{2}\right)-\alpha^2\sum_{j=0}^{r}\alpha_j\sin^2\left(\frac{\beta jt}{2}\right)\right)+\frac{1}{\alpha\beta\left(\beta^2-\alpha^2\right)}
$$

$$
\times\sum_{j=0}^{r}\alpha_j j\int_0^t\left[\left(f^{(4)}(x+jw)+\left(\alpha^2+\beta^2\right)f''(x+jw)+\alpha^2\beta^2f(x+jw)\right)\right.
$$

$$
\left.-\left(f^{(4)}(x)+\left(\alpha^2+\beta^2\right)f''(x)+\alpha^2\beta^2f(x)\right)\right]
$$

$$
\times\left[\beta\sin(\alpha j(t-w))-\alpha\sin(\beta j(t-w))\right]dw. \tag{28.12}
$$

We call

$$
R:=R(t):=\frac{1}{\alpha\beta\left(\beta^2-\alpha^2\right)}\sum_{j=0}^{r}\alpha_j j
$$

$$
\int_0^t\left[\left(f^{(4)}(x+jw)+\left(\alpha^2+\beta^2\right)f''(x+jw)+\alpha^2\beta^2f(x+jw)\right)\right.
$$

$$
\left.-\left(f^{(4)}(x)+\left(\alpha^2+\beta^2\right)f''(x)+\alpha^2\beta^2f(x)\right)\right]
$$

$$
\left[\beta\sin(\alpha j(t-w))-\alpha\sin(\beta j(t-w))\right]dw,\quad\forall\, t\in\mathbb{R}. \tag{28.13}
$$

We set

$$
E_1(x):=W_{r,\xi}(f,x)-f(x)-\left(\frac{f'(x)}{\alpha\beta\left(\beta^2-\alpha^2\right)}\right)
$$

$$
\left[\beta^3\sum_{j=0}^{r}\alpha_j\frac{1}{\sqrt{\pi\xi}}\left(\int_{-\infty}^{\infty}\sin(\alpha jt)\,e^{-\frac{t^2}{\xi}}dt\right)-\alpha^3\sum_{j=0}^{r}\alpha_j\frac{1}{\sqrt{\pi\xi}}\left(\int_{-\infty}^{\infty}\sin(\beta jt)\,e^{-\frac{t^2}{\xi}}dt\right)\right]
$$

$$
-\left(\frac{f''(x)}{\beta^2-\alpha^2}\right)
$$

$$
\left[\sum_{j=0}^{r}\alpha_j\frac{1}{\sqrt{\pi\xi}}\left(\int_{-\infty}^{\infty}\cos(\alpha jt)\,e^{-\frac{t^2}{\xi}}dt\right)-\sum_{j=0}^{r}\alpha_j\frac{1}{\sqrt{\pi\xi}}\left(\int_{-\infty}^{\infty}\cos(\beta jt)\,e^{-\frac{t^2}{\xi}}dt\right)\right]
$$

$$
-\left(\frac{f^{(3)}(x)}{\alpha\beta\left(\beta^2-\alpha^2\right)}\right)
$$

$$\left[\beta\sum_{j=0}^{r}\alpha_j\frac{1}{\sqrt{\pi\xi}}\left(\int_{-\infty}^{\infty}\sin(\alpha jt)\,e^{-\frac{t^2}{\xi}}dt\right)-\alpha\sum_{j=0}^{r}\alpha_j\frac{1}{\sqrt{\pi\xi}}\left(\int_{-\infty}^{\infty}\sin(\beta jt)\,e^{-\frac{t^2}{\xi}}dt\right)\right]$$

$$-\left(\frac{2\left(f^{(4)}(x)+\left(\alpha^2+\beta^2\right)f''(x)\right)}{(\alpha\beta)^2\left(\beta^2-\alpha^2\right)}\right)$$

$$\left[\beta^2\sum_{j=0}^{r}\alpha_j\frac{1}{\sqrt{\pi\xi}}\left(\int_{-\infty}^{\infty}\sin^2\left(\frac{\alpha jt}{2}\right)e^{-\frac{t^2}{\xi}}dt\right)\right.$$

$$\left.-\alpha^2\sum_{j=0}^{r}\alpha_j\frac{1}{\sqrt{\pi\xi}}\left(\int_{-\infty}^{\infty}\sin^2\left(\frac{\beta jt}{2}\right)e^{-\frac{t^2}{\xi}}dt\right)\right]$$

$$=\frac{1}{\sqrt{\pi\xi}}\int_{-\infty}^{\infty}R(t)\,e^{-\frac{t^2}{\xi}}dt. \tag{28.14}$$

Next we simplify $E_1(x)$:

We observe that

$$\int_{-\infty}^{\infty}\sin(\alpha jt)\,e^{-\frac{t^2}{\xi}}dt=\int_{-\infty}^{0}\sin(\alpha jt)\,e^{-\frac{t^2}{\xi}}dt+\int_{0}^{\infty}\sin(\alpha jt)\,e^{-\frac{t^2}{\xi}}dt. \tag{28.15}$$

Note $-\infty\leq t\leq 0\Rightarrow\infty\geq -t\geq 0$. So

$$\int_{-\infty}^{0}\sin(\alpha jt)\,e^{-\frac{\left|t^2\right|}{\xi}}dt=-\int_{-\infty}^{0}\sin(\alpha j(-(-t)))\,e^{-\frac{t^2}{\xi}}d(-t)$$

$$=-\int_{-\infty}^{0}(-\sin(\alpha j(-t)))\,e^{-\frac{t^2}{\xi}}d(-t)=\int_{-\infty}^{0}\sin(\alpha j(-t))\,e^{-\frac{t^2}{\xi}}d(-t)$$

$$=\int_{\infty}^{0}\sin(\alpha jt)\,e^{-\frac{t^2}{\xi}}dt=-\int_{0}^{\infty}\sin(\alpha jt)\,e^{-\frac{t^2}{\xi}}dt. \tag{28.16}$$

So

$$\int_{-\infty}^{\infty}\sin(\alpha jt)\,e^{-\frac{t^2}{\xi}}dt=0,$$

and all sine integrals in (28.14) are zeros.

Furthermore, we have that

$$\int_{-\infty}^{\infty}\sin^2\left(\frac{\alpha jt}{2}\right)e^{-\frac{t^2}{\xi}}dt=2\int_{0}^{\infty}\sin^2\left(\frac{\alpha jt}{2}\right)e^{-\frac{t^2}{\xi}}dt$$

$$=2\sqrt{\xi}\int_{0}^{\infty}\sin^2\left(\left(\frac{\alpha j\sqrt{\xi}}{2}\right)\frac{t}{\sqrt{\xi}}\right)e^{-\left(\frac{t}{\sqrt{\xi}}\right)^2}d\frac{t}{\sqrt{\xi}}$$

($\frac{t}{\sqrt{\xi}} =: x$, and $\frac{\alpha j \sqrt{\xi}}{2} = \beta_1$)

$$= 2\sqrt{\xi} \int_0^{\infty} \sin^2(\beta_1 x) e^{-x^2} dx = 2\sqrt{\xi}\frac{1}{4}\sqrt{\pi} e^{-\beta_1^2}\left(e^{\beta_1^2} - 1\right)$$

$$= \frac{\sqrt{\pi\xi}}{2}\left(1 - e^{-\beta_1^2}\right) = \frac{\sqrt{\pi\xi}}{2}\left(1 - e^{-\frac{\alpha^2 j^2 \xi}{4}}\right). \tag{28.17}$$

That is,

$$\int_{-\infty}^{\infty} \sin^2\left(\frac{\alpha j t}{2}\right) e^{-\frac{t^2}{\xi}} dt = \frac{\sqrt{\pi\xi}}{2}\left(1 - e^{-\frac{\alpha^2 j^2 \xi}{4}}\right), \tag{28.18}$$

and similarly,

$$\int_{-\infty}^{\infty} \sin^2\left(\frac{\beta j t}{2}\right) e^{-\frac{t^2}{\xi}} dt = \frac{\sqrt{\pi\xi}}{2}\left(1 - e^{-\frac{\beta^2 j^2 \xi}{4}}\right). \tag{28.19}$$

Next, we treat

$$\int_{-\infty}^{\infty} \cos(\alpha j t) e^{-\frac{t^2}{\xi}} dt = 2\int_0^{\infty} \cos(\alpha j t) e^{-\frac{t^2}{\xi}} dt$$

$$= 2\sqrt{\xi}\int_0^{\infty} \cos\left(\left(\frac{\alpha j \sqrt{\xi}}{2}\right)\frac{t}{\sqrt{\xi}}\right) e^{-\left(\frac{t}{\sqrt{\xi}}\right)^2} d\frac{t}{\sqrt{\xi}} \tag{28.20}$$

($\frac{t}{\sqrt{\xi}} =: x$, and $\frac{\alpha j \sqrt{\xi}}{2} =: \beta_1$)

$$= 2\sqrt{\xi}\int_0^{\infty} \cos(\beta_1 x) e^{-x^2} dx = 2\sqrt{\xi}\frac{1}{2}\sqrt{\pi} e^{-\frac{\beta_1^2}{4}} = \sqrt{\pi\xi} e^{-\frac{\alpha^2 j^2 \xi}{16}}.$$

That is,

$$\int_{-\infty}^{\infty} \cos(\alpha j t) e^{-\frac{t^2}{\xi}} dt = \sqrt{\pi\xi} e^{-\frac{\alpha^2 j^2 \xi}{16}}, \tag{28.21}$$

and similarly,

$$\int_{-\infty}^{\infty} \cos(\beta j t) e^{-\frac{t^2}{\xi}} dt = \sqrt{\pi\xi} e^{-\frac{\beta^2 j^2 \xi}{16}}. \tag{28.22}$$

Hence, we have the simplified expression,

$$E_1(x) = W_{r,\xi}(f,x) - f(x)$$

$$- \left(\frac{f''(x)}{\beta^2 - \alpha^2}\right) \left[\sum_{j=0}^{r} \alpha_j \left(e^{-\frac{\alpha^2 j^2 \xi}{16}} - e^{-\frac{\beta^2 j^2 \xi}{16}}\right)\right]$$

$$- \left(\frac{f^{(4)}(x) + (\alpha^2 + \beta^2) f''(x)}{(\alpha\beta)^2 (\beta^2 - \alpha^2)}\right)$$

$$\left[\beta^2 \sum_{j=0}^{r} \alpha_j \left(1 - e^{-\frac{\alpha^2 j^2 \xi}{4}}\right) - \alpha^2 \sum_{j=0}^{r} \alpha_j \left(1 - e^{-\frac{\beta^2 j^2 \xi}{4}}\right)\right]$$

$$= \frac{1}{\sqrt{\pi\xi}} \int_{-\infty}^{\infty} R(t) e^{-\frac{t^2}{\xi}} dt. \tag{28.23}$$

Remark 28.1. Call the function

$$F := f^{(4)} + (\alpha^2 + \beta^2) f'' + \alpha^2 \beta^2 f. \tag{28.24}$$

Then, we get

$$R = R(t) = \frac{1}{\alpha\beta(\beta^2 - \alpha^2)} \sum_{j=0}^{r} \alpha_j j \int_0^t [F(x + jw) - F(x)]$$

$$[\beta \sin(\alpha j (t - w)) - \alpha \sin(\beta j (t - w))] dw, \quad \forall\, t \in \mathbb{R}. \tag{28.25}$$

We isolate and study

$$I := \int_0^t [F(x + jw) - F(x)]$$

$$[\beta \sin(\alpha j (t - w)) - \alpha \sin(\beta j (t - w))] dw, \quad \forall\, t \in \mathbb{R}. \tag{28.26}$$

For $t < 0$, we have that

$$|I| = \left| \int_t^0 [F(x + jw) - F(x)] [\beta \sin(\alpha j (t - w)) - \alpha \sin(\beta j (t - w))] dw \right|$$

$$\leq \int_t^0 |F(x + jw) - F(x)| |\beta \sin(\alpha j (t - w)) - \alpha \sin(\beta j (t - w))| dw \tag{28.27}$$

(by $|\sin x| \leq |x|$, $\forall\, x \in \mathbb{R}$)

$$\leq 2\,|\alpha|\,|\beta|\,j \int_t^0 |F(x+jw) - F(x)|\,(w-t)\,dw$$

$$= -2\,|\alpha|\,|\beta|\,j \int_t^0 |F(x - j(-w)) - F(x)|\,(-t-(-w))\,d(-w) =$$

$(t \leq w \leq 0 \Rightarrow -t \geq -w =: \theta \geq 0)$

$$-\,2\,|\alpha|\,|\beta|\,j \int_{-t}^0 |F(x-j\theta) - F(x)|\,(-t-\theta)\,d\theta$$

$$= 2\,|\alpha|\,|\beta|\,j \int_0^{-t} |F(x-j\theta) - F(x)|\,(-t-\theta)\,d\theta$$

$$= 2\,|\alpha\beta|\,j \int_0^{|t|} |F(x + sign(t)\,j\theta) - F(x)|\,(|t|-\theta)\,d\theta.$$

So, we have proved that

$$|I| \leq 2\,|\alpha\beta|\,j \int_0^{|t|} |F(x + sign(t)\,j\theta) - F(x)|\,(|t|-\theta)\,d\theta, \quad \forall\, t \in \mathbb{R}, \tag{28.28}$$

and, by (28.25),

$$|R(t)| \leq \frac{2}{|\beta^2-\alpha^2|} \sum_{j=0}^{r} |\alpha_j|\,j^2 \int_0^{|t|} |F(x + jsign(t)\,\theta) - F(x)|\,(|t|-\theta)\,d\theta, \tag{28.29}$$

$\forall\, t \in \mathbb{R}$.

By (28.14), we have

$$E_1(x) = \frac{1}{\sqrt{\pi\xi}} \int_{-\infty}^{\infty} R(t)\,e^{-\frac{t^2}{\xi}}\,dt. \tag{28.30}$$

About Part II: On the smooth Poisson–Cauchy singular integral operators (Anastassiou and Mezei 2012)

Let $\overline{\alpha} \in \mathbb{N}$, $\overline{\beta} > \frac{1}{2\overline{\alpha}}$ and $f \in C^2(\mathbb{R})$. We define for $x \in \mathbb{R}$, $\xi > 0$ the Lebesgue integral

$$M_{r,\xi}(f;x) = W \int_{-\infty}^{\infty} \frac{\sum\limits_{j=0}^{r} \alpha_j f(x+jt)}{\left(t^{2\overline{\alpha}} + \xi^{2\overline{\alpha}}\right)^{\overline{\beta}}}\,dt, \tag{28.31}$$

where the constant is defined as

$$W = \frac{\Gamma\left(\overline{\beta}\right)\overline{\alpha}\xi^{2\overline{\alpha}\overline{\beta}-1}}{\Gamma\left(\frac{1}{2\overline{\alpha}}\right)\Gamma\left(\overline{\beta}-\frac{1}{2\overline{\alpha}}\right)}. \tag{28.32}$$

We assume that $M_{r,\xi}(f;x) \in \mathbb{R}$ for all $x \in \mathbb{R}$. We use also that

$$M_{r,\xi}(f;x) = W\sum_{j=0}^{r}\alpha_j\left(\int_{-\infty}^{\infty} f(x+jt)\frac{1}{\left(t^{2\overline{\alpha}}+\xi^{2\overline{\alpha}}\right)^{\overline{\beta}}}dt\right). \tag{28.33}$$

We note by $W\int_{-\infty}^{\infty}\frac{1}{\left(t^{2\overline{\alpha}}+\xi^{2\overline{\alpha}}\right)^{\overline{\beta}}}dt = 1$ that $M_{r,\xi}(c;x) = c$, c constant, and

$$M_{r,\xi}(f;x) - f(x) = W\left(\sum_{j=0}^{r}\alpha_j\int_{-\infty}^{\infty}\left[f(x+jt)-f(x)\right]\frac{1}{\left(t^{2\overline{\alpha}}+\xi^{2\overline{\alpha}}\right)^{\overline{\beta}}}dt\right). \tag{28.34}$$

We set

$$E_2(x) := M_{r,\xi}(f,x) - f(x) - \left(\frac{f'(x)}{\alpha\beta\left(\beta^2-\alpha^2\right)}\right)$$
$$\left[\beta^3\sum_{j=0}^{r}\alpha_j W\left(\int_{-\infty}^{\infty}\sin(\alpha jt)\frac{dt}{\left(t^{2\overline{\alpha}}+\xi^{2\overline{\alpha}}\right)^{\overline{\beta}}}\right)\right.$$
$$\left.-\alpha^3\sum_{j=0}^{r}\alpha_j W\left(\int_{-\infty}^{\infty}\sin(\beta jt)\frac{dt}{\left(t^{2\overline{\alpha}}+\xi^{2\overline{\alpha}}\right)^{\overline{\beta}}}\right)\right] - \left(\frac{f''(x)}{\beta^2-\alpha^2}\right) \tag{28.35}$$
$$\left[\sum_{j=0}^{r}\alpha_j W\left(\int_{-\infty}^{\infty}\cos(\alpha jt)\frac{dt}{\left(t^{2\overline{\alpha}}+\xi^{2\overline{\alpha}}\right)^{\overline{\beta}}}\right)\right.$$
$$\left.-\sum_{j=0}^{r}\alpha_j W\left(\int_{-\infty}^{\infty}\cos(\beta jt)\frac{dt}{\left(t^{2\overline{\alpha}}+\xi^{2\overline{\alpha}}\right)^{\overline{\beta}}}\right)\right] - \left(\frac{f^{(3)}(x)}{\alpha\beta\left(\beta^2-\alpha^2\right)}\right)$$
$$\left[\beta\sum_{j=0}^{r}\alpha_j W\left(\int_{-\infty}^{\infty}\sin(\alpha jt)\frac{dt}{\left(t^{2\overline{\alpha}}+\xi^{2\overline{\alpha}}\right)^{\overline{\beta}}}\right)\right.$$
$$\left.-\alpha\sum_{j=0}^{r}\alpha_j W\left(\int_{-\infty}^{\infty}\sin(\beta jt)\frac{dt}{\left(t^{2\overline{\alpha}}+\xi^{2\overline{\alpha}}\right)^{\overline{\beta}}}\right)\right]$$

$$-\left(\frac{2\left(f^{(4)}(x)+\left(\alpha^2+\beta^2\right)f''(x)\right)}{(\alpha\beta)^2\left(\beta^2-\alpha^2\right)}\right)$$

$$\left[\beta^2\sum_{j=0}^{r}\alpha_j W\left(\int_{-\infty}^{\infty}\sin^2\left(\frac{\alpha jt}{2}\right)\frac{dt}{\left(t^{2\overline{\alpha}}+\xi^{2\overline{\alpha}}\right)^{\overline{\beta}}}\right)\right.$$

$$\left.-\alpha^2\sum_{j=0}^{r}\alpha_j W\left(\int_{-\infty}^{\infty}\sin^2\left(\frac{\beta jt}{2}\right)\frac{dt}{\left(t^{2\overline{\alpha}}+\xi^{2\overline{\alpha}}\right)^{\overline{\beta}}}\right)\right]$$

$$=M_{r,\xi}(f,x)-f(x)-\left(\frac{2f''(x)}{\beta^2-\alpha^2}\right)W \tag{28.36}$$

$$\left[\sum_{j=0}^{r}\alpha_j\left(\int_0^{\infty}\left(\cos(\alpha jt)-\cos(\beta jt)\right)\frac{dt}{\left(t^{2\overline{\alpha}}+\xi^{2\overline{\alpha}}\right)^{\overline{\beta}}}\right)\right]$$

$$-\left(\frac{4\left(f^{(4)}(x)+\left(\alpha^2+\beta^2\right)f''(x)\right)}{(\alpha\beta)^2\left(\beta^2-\alpha^2\right)}\right)W$$

$$\left[\beta^2\sum_{j=0}^{r}\alpha_j\left(\int_0^{\infty}\sin^2\left(\frac{\alpha jt}{2}\right)\frac{dt}{\left(t^{2\overline{\alpha}}+\xi^{2\overline{\alpha}}\right)^{\overline{\beta}}}\right)\right.$$

$$\left.-\alpha^2\sum_{j=0}^{r}\alpha_j\left(\int_0^{\infty}\sin^2\left(\frac{\beta jt}{2}\right)\frac{dt}{\left(t^{2\overline{\alpha}}+\xi^{2\overline{\alpha}}\right)^{\overline{\beta}}}\right)\right]$$

$$=W\int_{-\infty}^{\infty}R(t)\frac{dt}{\left(t^{2\overline{\alpha}}+\xi^{2\overline{\alpha}}\right)^{\overline{\beta}}}. \tag{28.37}$$

That is,

$$E_2(x)=W\int_{-\infty}^{\infty}R(t)\frac{dt}{\left(t^{2\overline{\alpha}}+\xi^{2\overline{\alpha}}\right)^{\overline{\beta}}}. \tag{28.38}$$

About Part III: On the smooth trigonometric singular integral operators (Anastassiou and Mezei 2012)

Let $\xi>0$, $f\in C^2(\mathbb{R})$, $x\in\mathbb{R}$, $\overline{\beta}\in\mathbb{N}$; we set

$$T_{r,\xi}(f;x):=\frac{1}{\lambda}\int_{-\infty}^{\infty}\left(\sum_{j=0}^{r}\alpha_j f(x+jt)\right)\left(\frac{\sin\left(\frac{t}{\xi}\right)}{t}\right)^{2\overline{\beta}}dt, \tag{28.39}$$

where

$$\lambda := \int_{-\infty}^{\infty} \left(\frac{\sin\left(\frac{t}{\xi}\right)}{t} \right)^{2\overline{\beta}} dt = 2\xi^{1-2\overline{\beta}} \int_{0}^{\infty} \left(\frac{\sin t}{t} \right)^{2\overline{\beta}} dt \tag{28.40}$$

(by Edwards (1954), p. 210, item 1033)

$$= 2\xi^{1-2\overline{\beta}} \pi (-1)^{\overline{\beta}} \overline{\beta} \sum_{k=1}^{\overline{\beta}} (-1)^k \frac{k^{2\overline{\beta}-1}}{(\overline{\beta}-k)!\,(\overline{\beta}+k)!}.$$

Denote

$$\lambda_1 := 2\pi (-1)^{\overline{\beta}} \overline{\beta} \sum_{k=1}^{\overline{\beta}} (-1)^k \frac{k^{2\overline{\beta}-1}}{(\overline{\beta}-k)!\,(\overline{\beta}+k)!}, \tag{28.41}$$

that is,

$$\lambda = \lambda_1 \xi^{1-2\overline{\beta}}. \tag{28.42}$$

We suppose that $T_{r,\xi}(f;x) \in \mathbb{R}$ for all $x \in \mathbb{R}$. Clearly, again it is

$$\frac{1}{\lambda} \int_{-\infty}^{\infty} \left(\frac{\sin\left(\frac{t}{\xi}\right)}{t} \right)^{2\overline{\beta}} dt = 1, \tag{28.43}$$

and $T_{r,\xi}(c;x) = c$, c constant, and

$$T_{r,\xi}(f;x) - f(x) = \frac{1}{\lambda} \left(\sum_{j=0}^{r} \alpha_j \int_{-\infty}^{\infty} [f(x+jt) - f(x)] \left(\frac{\sin\left(\frac{t}{\xi}\right)}{t} \right)^{2\overline{\beta}} dt \right). \tag{28.44}$$

We set

$$E_3(x) := T_{r,\xi}(f,x) - f(x) - \left(\frac{f'(x)}{\alpha\beta(\beta^2-\alpha^2)} \right)$$

$$\left[\beta^3 \sum_{j=0}^{r} \alpha_j \frac{1}{\lambda} \left(\int_{-\infty}^{\infty} \sin(\alpha j t) \left(\frac{\sin\left(\frac{t}{\xi}\right)}{t} \right)^{2\overline{\beta}} dt \right) \right.$$

$$\left. - \alpha^3 \sum_{j=0}^{r} \alpha_j \frac{1}{\lambda} \left(\int_{-\infty}^{\infty} \sin(\beta j t) \left(\frac{\sin\left(\frac{t}{\xi}\right)}{t} \right)^{2\overline{\beta}} dt \right) \right]$$

$$-\left(\frac{f''(x)}{\beta^2-\alpha^2}\right)\left[\sum_{j=0}^{r}\alpha_j\frac{1}{\lambda}\left(\int_{-\infty}^{\infty}\cos(\alpha jt)\left(\frac{\sin\left(\frac{t}{\xi}\right)}{t}\right)^{2\overline{\beta}}dt\right)\right.$$

$$\left.-\sum_{j=0}^{r}\alpha_j\frac{1}{\lambda}\left(\int_{-\infty}^{\infty}\cos(\beta jt)\left(\frac{\sin\left(\frac{t}{\xi}\right)}{t}\right)^{2\overline{\beta}}dt\right)\right]$$

$$-\left(\frac{f^{(3)}(x)}{\alpha\beta(\beta^2-\alpha^2)}\right)\left[\beta\sum_{j=0}^{r}\alpha_j\frac{1}{\lambda}\left(\int_{-\infty}^{\infty}\sin(\alpha jt)\left(\frac{\sin\left(\frac{t}{\xi}\right)}{t}\right)^{2\overline{\beta}}dt\right)\right. \tag{28.45}$$

$$\left.-\alpha\sum_{j=0}^{r}\alpha_j\frac{1}{\lambda}\left(\int_{-\infty}^{\infty}\sin(\beta jt)\left(\frac{\sin\left(\frac{t}{\xi}\right)}{t}\right)^{2\overline{\beta}}dt\right)\right]$$

$$-\left(\frac{2\left(f^{(4)}(x)+\left(\alpha^2+\beta^2\right)f''(x)\right)}{(\alpha\beta)^2\left(\beta^2-\alpha^2\right)}\right)$$

$$\left[\beta^2\sum_{j=0}^{r}\alpha_j\frac{1}{\lambda}\left(\int_{-\infty}^{\infty}\sin^2\left(\frac{\alpha jt}{2}\right)\left(\frac{\sin\left(\frac{t}{\xi}\right)}{t}\right)^{2\overline{\beta}}dt\right)\right.$$

$$\left.-\alpha^2\sum_{j=0}^{r}\alpha_j\frac{1}{\lambda}\left(\int_{-\infty}^{\infty}\sin^2\left(\frac{\beta jt}{2}\right)\left(\frac{\sin\left(\frac{t}{\xi}\right)}{t}\right)^{2\overline{\beta}}dt\right)\right]$$

$$=T_{r,\xi}(f,x)-f(x)-\left(\frac{2f''(x)}{\beta^2-\alpha^2}\right)\frac{1}{\lambda}$$

$$\left[\sum_{j=0}^{r}\alpha_j\left(\int_{-\infty}^{\infty}\left(\cos(\alpha jt)-\cos(\beta jt)\right)\left(\frac{\sin\left(\frac{t}{\xi}\right)}{t}\right)^{2\overline{\beta}}dt\right)\right]$$

$$-\left(\frac{4\left(f^{(4)}(x)+\left(\alpha^2+\beta^2\right)f''(x)\right)}{(\alpha\beta)^2\left(\beta^2-\alpha^2\right)}\right)\frac{1}{\lambda} \tag{28.46}$$

$$\left[\beta^2 \sum_{j=0}^{r} \alpha_j \left(\int_{-\infty}^{\infty} \sin^2\left(\frac{\alpha j t}{2}\right) \left(\frac{\sin\left(\frac{t}{\xi}\right)}{t}\right)^{2\overline{\beta}} dt \right) \right.$$

$$\left. - \alpha^2 \sum_{j=0}^{r} \alpha_j \left(\int_{-\infty}^{\infty} \sin^2\left(\frac{\beta j t}{2}\right) \left(\frac{\sin\left(\frac{t}{\xi}\right)}{t}\right)^{2\overline{\beta}} dt \right) \right]$$

$$= \frac{1}{\lambda} \int_{-\infty}^{\infty} R(t) \left(\frac{\sin\left(\frac{t}{\xi}\right)}{t}\right)^{2\overline{\beta}} dt. \tag{28.47}$$

That is,

$$E_3(x) = \frac{1}{\lambda} \int_{-\infty}^{\infty} R(t) \left(\frac{\sin\left(\frac{t}{\xi}\right)}{t}\right)^{2\overline{\beta}} dt. \tag{28.48}$$

We need

Remark 28.2. By (28.29), we get $(t \in \mathbb{R})$

$$|R(t)| \leq \frac{2}{|\beta^2 - \alpha^2|} \sum_{j=0}^{r} |\alpha_j| j^2 \int_0^{|t|} |F(x + j sign(t)\theta) - F(x)| (|t| - \theta) d\theta \leq$$

$(\xi > 0)$

$$\frac{2}{|\beta^2 - \alpha^2|} \sum_{j=0}^{r} |\alpha_j| j^2 \int_0^{|t|} \omega_1\left(F, \xi \frac{j\theta}{\xi}\right) (|t| - \theta) d\theta$$

$$\leq \frac{2\omega_1(F,\xi)}{|\beta^2 - \alpha^2|} \left[\sum_{j=0}^{r} |\alpha_j| j^2 \int_0^{|t|} \left(1 + \frac{j\theta}{\xi}\right) (|t| - \theta) d\theta \right]$$

$$= \frac{2\omega_1(F,\xi)}{|\beta^2 - \alpha^2|} \left[\sum_{j=0}^{r} |\alpha_j| j^2 \left[\int_0^{|t|} (|t| - \theta) d\theta + \frac{j}{\xi} \int_0^{|t|} (|t| - \theta)^{2-1} (\theta - 0)^{2-1} d\theta \right] \right]$$

$$= \frac{2\omega_1(F,\xi)}{|\beta^2 - \alpha^2|} \left[\sum_{j=0}^{r} |\alpha_j| j^2 \left[\frac{t^2}{2} + \frac{j}{\xi} \frac{|t|^3}{6} \right] \right]$$

$$= \frac{\omega_1(F,\xi)}{|\beta^2 - \alpha^2|} \left[\sum_{j=0}^{r} |\alpha_j| j^2 \left[t^2 + j \frac{|t|^3}{3\xi} \right] \right]. \tag{28.49}$$

Consequently, for $t \in \mathbb{R}$, we obtain

$$|R(t)| \leq \frac{\omega_1(F,\xi)}{|\beta^2-\alpha^2|}\left[\sum_{j=0}^{r}|\alpha_j|\,j^2\left[t^2+\frac{j}{3\xi}|t|^3\right]\right]. \tag{28.50}$$

28.3 Main Results

Next we present uniform approximation by the $W_{r,\xi}$ operators.

Theorem 28.1. *It holds ($\xi > 0$, $x \in \mathbb{R}$)*

$$\begin{aligned}
|E_1(x)| &\leq \|E_1\|_\infty = \Bigg\| W_{r,\xi}(f,x) - f(x) \\
&\quad - \frac{f''(x)}{(\beta^2-\alpha^2)}\left[\sum_{j=0}^{r}\alpha_j\left(e^{-\frac{\alpha^2 j^2\xi}{16}} - e^{-\frac{\beta^2 j^2\xi}{16}}\right)\right] \\
&\quad - \left(\frac{f^{(4)}(x) + (\alpha^2+\beta^2)f''(x)}{(\alpha\beta)^2(\beta^2-\alpha^2)}\right) \\
&\quad \left[\beta^2\sum_{j=0}^{r}\alpha_j\left(1-e^{-\frac{\alpha^2 j^2\xi}{4}}\right) - \alpha^2\sum_{j=0}^{r}\alpha_j\left(1-e^{-\frac{\beta^2 j^2\xi}{4}}\right)\right]\Bigg\|_{\infty,x} \\
&\leq \frac{\sum_{j=0}^{r}|\alpha_j|\,j^2\left(\frac{\sqrt{\pi}}{2}+\frac{j}{3}\right)}{|\beta^2-\alpha^2|\sqrt{\pi}}\omega_1\left(f^{(4)} + (\alpha^2+\beta^2)f'' + \alpha^2\beta^2 f, \sqrt{\xi}\right)\cdot\xi \\
&=: B_1 \to 0,
\end{aligned} \tag{28.51}$$

as $\xi \to 0$.

If $f''(x) = f^{(4)}(x) = 0$, then

$$|W_{r,\xi}(f,x) - f(x)| \leq B_1, \tag{28.52}$$

and $W_{r,\xi}(f,x) \to f(x)$, as $\xi \to 0$.

Proof. By (28.30), we get

$$|E_1(x)| \leq \frac{1}{\sqrt{\pi\xi}}\int_{-\infty}^{\infty}|R(t)|\,e^{-\frac{t^2}{\xi}}\,dt$$

(apply (28.50) with ξ to be $\sqrt{\xi}$)

$$\leq \frac{\omega_1\left(F,\sqrt{\xi}\right)}{\left|\beta^2-\alpha^2\right|\sqrt{\pi\xi}}\left[\sum_{j=0}^{r}\left|\alpha_j\right| j^2\left[\int_{-\infty}^{\infty} t^2 e^{-\frac{t^2}{\xi}}dt+\frac{j}{3\sqrt{\xi}}\int_{-\infty}^{\infty}|t|^3 e^{-\frac{t^2}{\xi}}dt\right]\right]$$

$$=\frac{2\omega_1\left(F,\sqrt{\xi}\right)}{\left|\beta^2-\alpha^2\right|\sqrt{\pi\xi}}\left[\sum_{j=0}^{r}\left|\alpha_j\right| j^2\left[\int_{0}^{\infty} t^2 e^{-\frac{t^2}{\xi}}dt+\frac{j}{3\sqrt{\xi}}\int_{0}^{\infty} t^3 e^{-\frac{t^2}{\xi}}dt\right]\right]$$

$$=\frac{2\omega_1\left(F,\sqrt{\xi}\right)}{\left|\beta^2-\alpha^2\right|\sqrt{\pi\xi}}\left[\sum_{j=0}^{r}\left|\alpha_j\right| j^2\left[\left(\sqrt{\xi}\right)^3\int_{0}^{\infty}\left(\frac{t}{\sqrt{\xi}}\right)^2 e^{-\left(\frac{t}{\sqrt{\xi}}\right)^2}d\left(\frac{t}{\sqrt{\xi}}\right)\right.\right.$$

$$\left.\left.+\frac{j}{3\sqrt{\xi}}\left(\sqrt{\xi}\right)^4\int_{0}^{\infty}\left(\frac{t}{\sqrt{\xi}}\right)^3 e^{-\left(\frac{t}{\sqrt{\xi}}\right)^2}d\left(\frac{t}{\sqrt{\xi}}\right)\right]\right]$$

$$=\frac{2\omega_1\left(F,\sqrt{\xi}\right)\xi}{\left|\beta^2-\alpha^2\right|\sqrt{\pi}}\left[\sum_{j=0}^{r}\left|\alpha_j\right| j^2\left[\int_{0}^{\infty} x^2 e^{-x^2}dx+\frac{j}{3}\int_{0}^{\infty} x^3 e^{-x^2}dx\right]\right]$$

$$=\frac{2\omega_1\left(F,\sqrt{\xi}\right)\xi}{\left|\beta^2-\alpha^2\right|\sqrt{\pi}}\left[\sum_{j=0}^{r}\left|\alpha_j\right| j^2\left[\frac{\sqrt{\pi}}{4}+\frac{j}{6}\right]\right]. \tag{28.53}$$

Thus, we obtain

$$\left|E_1(x)\right|\leq\frac{\omega_1\left(F,\sqrt{\xi}\right)\xi}{\left|\beta^2-\alpha^2\right|\sqrt{\pi}}\left[\sum_{j=0}^{r}\left|\alpha_j\right| j^2\left[\frac{\sqrt{\pi}}{2}+\frac{j}{3}\right]\right]. \tag{28.54}$$

The claim is proved. □

Next we present uniform approximation by the $M_{r,\xi}$ operators.

Theorem 28.2. *It holds* ($\xi>0$, $x\in\mathbb{R}$, $\overline{\beta}>\frac{2}{\overline{\alpha}}$, $\overline{\alpha}\in\mathbb{N}$)

$$\left|E_2(x)\right|\leq\left\|E_2\right\|_\infty=\left\| M_{r,\xi}(f,x)-f(x)\right.$$

$$-\left(\frac{2f''(x)}{\beta^2-\alpha^2}\right)W\left[\sum_{j=0}^{r}\alpha_j\left(\int_0^\infty\left(\cos(\alpha jt)-\cos(\beta jt)\right)\frac{dt}{\left(t^{2\overline{\alpha}}+\xi^{2\overline{\alpha}}\right)^{\overline{\beta}}}\right)\right]$$

$$-\left(\frac{4\left(f^{(4)}(x)+\left(\alpha^2+\beta^2\right)f''(x)\right)}{(\alpha\beta)^2\left(\beta^2-\alpha^2\right)}\right)W$$

$$\left[\beta^2 \sum_{j=0}^{r} \alpha_j \left(\int_0^\infty \sin^2\left(\frac{\alpha j t}{2}\right) \frac{dt}{\left(t^{2\overline{\alpha}} + \xi^{2\overline{\alpha}}\right)^{\overline{\beta}}}\right)\right.$$

$$\left.\left. - \alpha^2 \sum_{j=0}^{r} \alpha_j \left(\int_0^\infty \sin^2\left(\frac{\beta j t}{2}\right) \frac{dt}{\left(t^{2\overline{\alpha}} + \xi^{2\overline{\alpha}}\right)^{\overline{\beta}}}\right)\right]\right\|_{\infty,x}$$

$$\leq \left\{ \frac{\left[\sum_{j=1}^{r} |\alpha_j| j^2 \left(\Gamma\left(\frac{3}{2\overline{\alpha}}\right)\Gamma\left(\overline{\beta} - \frac{3}{2\overline{\alpha}}\right) + \frac{j}{3}\Gamma\left(\frac{2}{\overline{\alpha}}\right)\Gamma\left(\overline{\beta} - \frac{2}{\overline{\alpha}}\right)\right)\right]}{|\beta^2 - \alpha^2| \Gamma\left(\frac{1}{2\overline{\alpha}}\right)\Gamma\left(\overline{\beta} - \frac{1}{2\overline{\alpha}}\right)} \right\}$$

$$\omega_1\left(f^{(4)} + \left(\alpha^2 + \beta^2\right) f'' + \alpha^2\beta^2 f, \xi\right) \cdot \xi^2 =: B_2 \to 0, \tag{28.55}$$

as $\xi \to 0$.

If $f''(x) = f^{(4)}(x) = 0$, *then*

$$\left|M_{r,\xi}(f,x) - f(x)\right| \leq B_2, \tag{28.56}$$

and $M_{r,\xi}(f,x) \to f(x)$, *as* $\xi \to 0$.

Proof. By (28.38), we get

$$|E_2(x)| \leq W \int_{-\infty}^{\infty} |R(t)| \frac{dt}{\left(t^{2\overline{\alpha}} + \xi^{2\overline{\alpha}}\right)^{\overline{\beta}}}$$

(by (28.50))

$$\leq \frac{W\omega_1(F,\xi)}{|\beta^2 - \alpha^2|} \left[\sum_{j=0}^{r} |\alpha_j| j^2 \left[\int_{-\infty}^{\infty} t^2 \frac{dt}{\left(t^{2\overline{\alpha}} + \xi^{2\overline{\alpha}}\right)^{\overline{\beta}}} + \frac{j}{3\xi} \int_{-\infty}^{\infty} |t|^3 \frac{dt}{\left(t^{2\overline{\alpha}} + \xi^{2\overline{\alpha}}\right)^{\overline{\beta}}}\right]\right] \tag{28.57}$$

$$= \frac{2W\omega_1(F,\xi)}{|\beta^2 - \alpha^2|} \left[\sum_{j=0}^{r} |\alpha_j| j^2 \left[\int_0^\infty t^2 \frac{dt}{\left(t^{2\overline{\alpha}} + \xi^{2\overline{\alpha}}\right)^{\overline{\beta}}} + \frac{j}{3\xi} \int_0^\infty t^3 \frac{dt}{\left(t^{2\overline{\alpha}} + \xi^{2\overline{\alpha}}\right)^{\overline{\beta}}}\right]\right]$$

$$= \frac{2W\omega_1(F,\xi)}{|\beta^2-\alpha^2|}\left[\sum_{j=0}^{r}|\alpha_j|\,j^2\xi^{3-2\overline{\alpha}\overline{\beta}}\left[\int_0^\infty\left(\frac{t}{\xi}\right)^2\frac{d\left(\frac{t}{\xi}\right)}{\left(\left(\frac{t}{\xi}\right)^{2\overline{\alpha}}+1\right)^{\overline{\beta}}}\right.\right.$$

$$\left.\left.+\frac{j}{3}\int_0^\infty\left(\frac{t}{\xi}\right)^3\frac{d\left(\frac{t}{\xi}\right)}{\left(\left(\frac{t}{\xi}\right)^{2\overline{\alpha}}+1\right)^{\overline{\beta}}}\right]\right]$$

$$= \left(\frac{2\Gamma\left(\overline{\beta}\right)\overline{\alpha}\xi^2}{\Gamma\left(\frac{1}{2\overline{\alpha}}\right)\Gamma\left(\overline{\beta}-\frac{1}{2\overline{\alpha}}\right)}\right)\frac{\omega_1(F,\xi)}{|\beta^2-\alpha^2|}$$

$$\left[\sum_{j=0}^{r}|\alpha_j|\,j^2\left[\int_0^\infty x^2\frac{dx}{\left(1+x^{2\overline{\alpha}}\right)^{\overline{\beta}}}+\frac{j}{3}\int_0^\infty x^3\frac{dx}{\left(1+x^{2\overline{\alpha}}\right)^{\overline{\beta}}}\right]\right] = \tag{28.58}$$

(by Zwillinger (1995), p. 397, formula 595)

$$\left(\frac{2\overline{\alpha}\Gamma\left(\overline{\beta}\right)}{\Gamma\left(\frac{1}{2\overline{\alpha}}\right)\Gamma\left(\overline{\beta}-\frac{1}{2\overline{\alpha}}\right)|\beta^2-\alpha^2|}\right)\xi^2\omega_1(F,\xi)$$

$$\left[\sum_{j=0}^{r}|\alpha_j|\,j^2\left[\frac{\Gamma\left(\frac{3}{2\overline{\alpha}}\right)\Gamma\left(\overline{\beta}-\frac{3}{2\overline{\alpha}}\right)}{2\overline{\alpha}\Gamma\left(\overline{\beta}\right)}+\frac{j}{3}\left(\frac{\Gamma\left(\frac{2}{\overline{\alpha}}\right)\Gamma\left(\overline{\beta}-\frac{2}{\overline{\alpha}}\right)}{2\overline{\alpha}\Gamma\left(\overline{\beta}\right)}\right)\right]\right]$$

$$= \left(\frac{\xi^2\omega_1(F,\xi)}{\Gamma\left(\frac{1}{2\overline{\alpha}}\right)\Gamma\left(\overline{\beta}-\frac{1}{2\overline{\alpha}}\right)|\beta^2-\alpha^2|}\right)$$

$$\left[\sum_{j=0}^{r}|\alpha_j|\,j^2\left[\Gamma\left(\frac{3}{2\overline{\alpha}}\right)\Gamma\left(\overline{\beta}-\frac{3}{2\overline{\alpha}}\right)+\frac{j}{3}\Gamma\left(\frac{2}{\overline{\alpha}}\right)\Gamma\left(\overline{\beta}-\frac{2}{\overline{\alpha}}\right)\right]\right].$$

Thus, we derive

$$|E_2(x)| \leq \frac{\omega_1(F,\xi)\,\xi^2}{|\beta^2-\alpha^2|\,\Gamma\left(\frac{1}{2\overline{\alpha}}\right)\Gamma\left(\overline{\beta}-\frac{1}{2\overline{\alpha}}\right)}$$

$$\left[\sum_{j=0}^{r}|\alpha_j|\,j^2\left[\Gamma\left(\frac{3}{2\overline{\alpha}}\right)\Gamma\left(\overline{\beta}-\frac{3}{2\overline{\alpha}}\right)+\frac{j}{3}\Gamma\left(\frac{2}{\overline{\alpha}}\right)\Gamma\left(\overline{\beta}-\frac{2}{\overline{\alpha}}\right)\right]\right]. \tag{28.60}$$

The claim is proved. □

It follows the uniform approximation by the $T_{r,\xi}$ operators.

Theorem 28.3. *It holds* ($\xi > 0$, $x \in \mathbb{R}$, $\overline{\beta} \in \mathbb{N} - \{1, 2\}$)

$$\left|E_3(x)\right| \leq \left\|E_3\right\|_\infty = \Bigg\| T_{r,\xi}(f,x) - f(x) - \left(\frac{2f''(x)}{\beta^2 - \alpha^2}\right)\frac{1}{\lambda}$$

$$\times \left[\sum_{j=0}^{r} \alpha_j \left(\int_0^\infty (\cos(\alpha j t) - \cos(\beta j t)) \left(\frac{\sin\left(\frac{t}{\xi}\right)}{t}\right)^{2\overline{\beta}} dt\right)\right]$$

$$- \left(\frac{4\left(f^{(4)}(x) + \left(\alpha^2 + \beta^2\right) f''(x)\right)}{(\alpha\beta)^2\left(\beta^2 - \alpha^2\right)}\right)\frac{1}{\lambda}$$

$$\left[\beta^2 \sum_{j=0}^{r} \alpha_j \left(\int_0^\infty \sin^2\left(\frac{\alpha j t}{2}\right) \left(\frac{\sin\left(\frac{t}{\xi}\right)}{t}\right)^{2\overline{\beta}} dt\right)\right.$$

$$\left.- \alpha^2 \sum_{j=0}^{r} \alpha_j \left(\int_0^\infty \sin^2\left(\frac{\beta j t}{2}\right) \left(\frac{\sin\left(\frac{t}{\xi}\right)}{t}\right)^{2\overline{\beta}} dt\right)\right]\Bigg\|_{\infty,x}$$

$$\leq \frac{2\omega_1\left(f^{(4)} + \left(\alpha^2 + \beta^2\right) f'' + \alpha^2\beta^2 f, \xi\right)\xi^2}{\lambda_1\left|\beta^2 - \alpha^2\right|}$$

$$\left[\sum_{j=1}^{r} \left|\alpha_j\right| j^2 \left[\left(\frac{\pi(-1)^{\overline{\beta}-1}(2\overline{\beta})!}{8(2\overline{\beta} - 3)!}\right)\left(\sum_{k=1}^{\overline{\beta}} (-1)^k \frac{k^{2\overline{\beta}-3}}{(\overline{\beta} - k)!(\overline{\beta} + k)!}\right)\right)\right.$$

$$\left.\left. + \frac{j}{3}\left(\frac{-(2\overline{\beta})!}{8(2\overline{\beta} - 4)!}\left(\sum_{k=1}^{\overline{\beta}} \frac{(-1)^{\overline{\beta}-k} k^{2\overline{\beta}-k} \ln(2k)}{(\overline{\beta} - k)!(\overline{\beta} + k)!}\right)\right)\right]\right] =: B_3 \to 0, \tag{28.61}$$

as $\xi \to 0$.

If $f''(x) = f^{(4)}(x) = 0$, *then*

$$\left|T_{r,\xi}(f,x) - f(x)\right| \leq B_3, \tag{28.62}$$

and $T_{r,\xi}(f,x) \to f(x)$, *as* $\xi \to 0$.

Proof. By (28.48), we get

$$|E_3(x)| \leq \frac{1}{\lambda}\int_{-\infty}^{\infty}|R(t)|\left(\frac{\sin\left(\frac{t}{\xi}\right)}{t}\right)^{2\overline{\beta}}dt$$

(by (28.50))

$$\leq \frac{\omega_1(F,\xi)}{\lambda|\beta^2-\alpha^2|}\left[\sum_{j=0}^{r}|\alpha_j|\,j^2\left[\int_{-\infty}^{\infty}t^2\left(\frac{\sin\left(\frac{t}{\xi}\right)}{t}\right)^{2\overline{\beta}}dt+\frac{j}{3\xi}\int_{-\infty}^{\infty}|t|^3\left(\frac{\sin\left(\frac{t}{\xi}\right)}{t}\right)^{2\overline{\beta}}dt\right]\right]$$

$$=\frac{2\omega_1(F,\xi)}{\lambda|\beta^2-\alpha^2|}\left[\sum_{j=0}^{r}|\alpha_j|\,j^2\left[\int_{0}^{\infty}t^2\left(\frac{\sin\left(\frac{t}{\xi}\right)}{t}\right)^{2\overline{\beta}}dt+\frac{j}{3\xi}\int_{0}^{\infty}t^3\left(\frac{\sin\left(\frac{t}{\xi}\right)}{t}\right)^{2\overline{\beta}}dt\right]\right]$$

$$=\frac{2\omega_1(F,\xi)}{\lambda|\beta^2-\alpha^2|}\left[\sum_{j=0}^{r}|\alpha_j|\,j^2\xi^{3-2\overline{\beta}}\left[\int_{0}^{\infty}\left(\frac{t}{\xi}\right)^2\left(\frac{\sin\left(\frac{t}{\xi}\right)}{\frac{t}{\xi}}\right)^{2\overline{\beta}}d\left(\frac{t}{\xi}\right)\right.\right.$$

$$\left.\left.+\frac{j}{3}\int_{0}^{\infty}\left(\frac{t}{\xi}\right)^3\left(\frac{\sin\left(\frac{t}{\xi}\right)}{\frac{t}{\xi}}\right)^{2\overline{\beta}}d\left(\frac{t}{\xi}\right)\right]\right]$$

$$=\frac{2\omega_1(F,\xi)\,\xi^2}{\lambda_1|\beta^2-\alpha^2|}\left[\sum_{j=0}^{r}|\alpha_j|\,j^2\left[\int_{0}^{\infty}x^2\left(\frac{\sin x}{x}\right)^{2\overline{\beta}}dx+\frac{j}{3}\int_{0}^{\infty}x^3\left(\frac{\sin x}{x}\right)^{2\overline{\beta}}dx\right]\right]= \tag{28.63}$$

(by Edwards (1954), p. 210, formula 1033)

$$\frac{2\omega_1(F,\xi)\,\xi^2}{\lambda_1|\beta^2-\alpha^2|}$$

$$\left[\sum_{j=1}^{r}|\alpha_j|\,j^2\left[\left(\frac{\pi(-1)^{\overline{\beta}-1}(2\overline{\beta})!}{8(2\overline{\beta}-3)!}\left(\sum_{k=1}^{\overline{\beta}}(-1)^k\frac{k^{2\overline{\beta}-3}}{(\overline{\beta}-k)!(\overline{\beta}+k)!}\right)\right)\right.\right.$$

$$\left.\left.+\frac{j}{3}\left(\frac{-(2\overline{\beta})!}{8(2\overline{\beta}-4)!}\left(\sum_{k=1}^{\overline{\beta}}\frac{(-1)^{\overline{\beta}-k}k^{2\overline{\beta}-k}\ln(2k)}{(\overline{\beta}-k)!(\overline{\beta}+k)!}\right)\right)\right]\right]. \tag{28.64}$$

The claim is proved. □

We make

Remark 28.3. Let $\overline{E_i(x)}$ be $E_i(x)$, $i = 1, 2, 3$, when $\alpha = 2$ and $\beta = 1$.

We have

$$\overline{E_1(x)} = W_{r,\xi}(f, x) - f(x) + \frac{f''(x)}{3}\left[\sum_{j=0}^{r} \alpha_j \left(e^{-\frac{j^2\xi}{4}} - e^{-\frac{j^2\xi}{16}}\right)\right]$$

$$+ \left(\frac{f^{(4)}(x) + 5f''(x)}{12}\right)\left[\sum_{j=0}^{r} \alpha_j \left(1 - e^{-j^2\xi}\right) - 4\sum_{j=0}^{r} \alpha_j \left(1 - e^{-\frac{j^2\xi}{4}}\right)\right]. \tag{28.65}$$

We also have

$$\overline{E_2(x)} := M_{r,\xi}(f, x) - f(x) + \frac{2f''(x)}{3} W$$

$$\left[\sum_{j=0}^{r} \alpha_j \left(\int_0^{\infty} (\cos(2jt) - \cos(jt)) \frac{dt}{\left(t^{2\overline{\alpha}} + \xi^{2\overline{\alpha}}\right)^{\overline{\beta}}}\right)\right]$$

$$+ \left(\frac{4\left(f^{(4)}(x) + 5f''(x)\right)}{12}\right) W$$

$$\left[\sum_{j=0}^{r} \alpha_j \left(\int_0^{\infty} \sin^2(jt) \frac{dt}{\left(t^{2\overline{\alpha}} + \xi^{2\overline{\alpha}}\right)^{\overline{\beta}}}\right)\right.$$

$$\left. - 4\sum_{j=0}^{r} \alpha_j \left(\int_0^{\infty} \sin^2\left(\frac{jt}{2}\right) \frac{dt}{\left(t^{2\overline{\alpha}} + \xi^{2\overline{\alpha}}\right)^{\overline{\beta}}}\right)\right]. \tag{28.66}$$

Furthermore, it is

$$\overline{E_3(x)} = T_{r,\xi}(f, x) - f(x)$$

$$+ \frac{2f''(x)}{3\lambda}\left[\sum_{j=0}^{r} \alpha_j \left(\int_{-\infty}^{\infty} (\cos(2jt) - \cos(jt)) \left(\frac{\sin\left(\frac{t}{\xi}\right)}{t}\right)^{2\overline{\beta}} dt\right)\right]$$

$$+ \left(\frac{4\left(f^{(4)}(x) + 5f''(x)\right)}{12\lambda}\right)$$

$$\left[\sum_{j=0}^{r}\alpha_j\left(\int_{-\infty}^{\infty}\sin^2(jt)\left(\frac{\sin\left(\frac{t}{\xi}\right)}{t}\right)^{2\overline{\beta}}dt\right)\right.$$
$$\left.-4\sum_{j=0}^{r}\alpha_j\left(\int_{-\infty}^{\infty}\sin^2\left(\frac{jt}{2}\right)\left(\frac{\sin\left(\frac{t}{\xi}\right)}{t}\right)^{2\overline{\beta}}dt\right)\right]. \tag{28.67}$$

We finish with the following special results.

Corollary 28.1. *(to Theorem 28.1) It holds*

$$\left|\overline{E_1(x)}\right| \leq \left\|\overline{E_1}\right\|_\infty$$
$$\leq \frac{\sum_{j=0}^{r}|\alpha_j|\, j^2\left(\frac{\sqrt{\pi}}{2}+\frac{j}{3}\right)}{3\sqrt{\pi}}\omega_1\left(f^{(4)}+5f''+4f,\sqrt{\xi}\right)\cdot\xi \to 0, \tag{28.68}$$

as $\xi \to 0$.

Corollary 28.2. *(to Theorem 28.2) It holds*

$$\left|\overline{E_2(x)}\right| \leq \left\|\overline{E_2}\right\|_\infty$$
$$\leq \left\{\frac{\left[\sum_{j=1}^{r}|\alpha_j|\, j^2\left(\Gamma\left(\frac{3}{2\overline{\alpha}}\right)\Gamma\left(\overline{\beta}-\frac{3}{2\overline{\alpha}}\right)+\frac{j}{3}\Gamma\left(\frac{2}{\overline{\alpha}}\right)\Gamma\left(\overline{\beta}-\frac{2}{\overline{\alpha}}\right)\right)\right]}{3\Gamma\left(\frac{1}{2\overline{\alpha}}\right)\Gamma\left(\overline{\beta}-\frac{1}{2\overline{\alpha}}\right)}\right\}$$
$$\omega_1\left(f^{(4)}+5f''+4f,\xi\right)\xi^2 \to 0, \tag{28.69}$$

as $\xi \to 0$.

Corollary 28.3. *(to Theorem 28.3) It holds*

$$\left|\overline{E_3(x)}\right| \leq \left\|\overline{E_3}\right\|_\infty$$
$$\leq \frac{2\omega_1\left(f^{(4)}+5f''+4f,\xi\right)\xi^2}{3\lambda_1}$$
$$\left[\sum_{j=1}^{r}|\alpha_j|\, j^2\left[\left(\frac{\pi(-1)^{\overline{\beta}-1}(2\overline{\beta})!}{8(2\overline{\beta}-3)!}\left(\sum_{k=1}^{\overline{\beta}}(-1)^k\frac{k^{2\overline{\beta}-3}}{(\overline{\beta}-k)!(\overline{\beta}+k)!}\right)\right)\right.\right.$$
$$\left.\left.+\frac{j}{3}\left(\frac{-(2\overline{\beta})!}{8(2\overline{\beta}-4)!}\left(\sum_{k=1}^{\overline{\beta}}\frac{(-1)^{\overline{\beta}-k}k^{2\overline{\beta}-k}\ln(2k)}{(\overline{\beta}-k)!(\overline{\beta}+k)!}\right)\right)\right]\right] \to 0, \tag{28.70}$$

as $\xi \to 0$.

Bibliography

A.H. Ali, Z. Pales, Taylor-type expansions in terms of exponential polynomials, *Mathematical Inequalities and Applications*, 25(4) (2022), 1123–1141.

G.A. Anastassiou, Basic convergence with rates of smooth Picard singular integral operators, *Journal of Computational Analysis and Applications*, 8(4) (2006), 313–334.

G.A. Anastassiou, *Intelligent Mathematics: Computational Analysis*, Springer, Heidelberg, New York, Chapter 12, 2011.

G.A. Anastassiou, Opial and Ostrowski type inequalities based on trigonometric and hyperbolic type Taylor formulae, *Malaya Journal of Matematik*, 11(S) (2023), 1–26.

G.A. Anastassiou, *Parametrized and Trigonometric derived uniform approximation by various smooth singular integral operators*, submitted, 2024.

G.A. Anastassiou and R. Mezei, *Approximation by Singular Integrals*, Cambridge Scientific Publishers, Cambridge, UK, 2012.

A. Aral, On a generalized Gauss Weierstrass singular integral, *Fasciculi Mathematici*, (35) (2005), 23–33.

A. Aral, Pointwise approximation by the generalization of Picard and Gauss-Weierstrass singular integrals, *Journal of Concrete Applicable Mathematics*, 6(4) (2008), 327–339.

A. Aral, On generalized Picard integral operators, *Advances in Summability and Approximation Theory*, 157–168, Springer, Singapore, 2018.

A. Aral, E. Deniz and H. Erbay, The Picard and Gauss-Weiertrass singular integrals in (p, q)-calculus, *Bulletin of the Malaysian Mathematical Sciences Society*, 43(2) (2020), 1569–1583.

A. Aral and S.G. Gal, q-generalizations of the Picard and Gauss-Weierstrass singular integrals, *Taiwanese Journal of Mathematics*, 12(9) (2008), 2501–2515.

J. Edwards, *A Treatise on the Integral Calculus, Vol II*, Chelsea, New York, 1954.

R.N. Mohapatra and R.S. Rodriguez, On the rate of convergence of singular integrals for Hölder continuous functions, *Mathematische Nachricnten*, 149 (1990), 117–124.

D. Zwillinger, *CRC Standard Mathematical Tables and Formulae*, 30th ed. Chapman & Hall/CRC, Boca Raton (1995).

Chapter 29

Parametrized Trigonometric-Implied L_p Degree of Quantitative Approximation by Various Smooth Singular Integral Operators

In this chapter, we continue with the study of smooth Gauss–Weierstrass, Poisson–Cauchy and trigonometric singular integral operators that started in Anastassiou (2011), see there Chapters 10–14. This time, the foundation of our research is a trigonometric Taylor's formula. We prove the parametrized univariate L_p convergence of our operators to the unit operator with rates via Jackson-type parametrized inequalities involving the first L_p modulus of continuity. Of interest, here is a residual appearing term. Note that our operators are not in general positive. It follows Anastassiou (2024).

29.1 Introduction

We are motivated by Anastassiou (2006, 2011) Chapters 10–14, Anastassiou (2023)and Ali and Pales (2022). We use a trigonometric new Taylor formula from Anastassiou (2023), see also Ali and Pales (2022). Here we consider some very general operators, the smooth Gauss–Weierstrass, Poisson–Cauchy and trigonometric singular integral operators over the real line, and we study further their $L_p, p \geq 1$, parametrized convergence properties quantitatively. We establish related parametrized inequalities involving the first L_p, $p \geq 1$, modulus of continuity with respect to L_p, $p \geq 1$, norm. We provide detailed proofs.

Other important motivating articles on the topic are Aral (2005)–Aral and Gal (2008).

For the history of the topic, we mention about the monograph Anastassiou and Mezei (2012), which was the first complete source to deal exclusively with the classic theory of the approximation of singular integrals to the identity-unit operator. The authors there studied quantitatively the basic approximation properties of the general Picard, Gauss–Weierstrass and Poisson–Cauchy singular integral operators over the real line, which are not positive linear operators. In particular, they studied the rate of convergence of these operators to the unit operator, as well as the related simultaneous approximation. This is given via inequalities and with the use of higher-order modulus of smoothness of the high order derivative of the involved function. Some of these inequalities are proven to be sharp. Also, they studied the global smoothness preservation property of these operators. Furthermore, they gave asymptotic expansions of Voronovskaya type for the error of approximation. They continued with the study of related properties of the general fractional Gauss–Weierstrass and Poisson–Cauchy singular integral operators. These properties were studied with respect to L_p norm, $1 \leq p \leq \infty$. The case of Lipschitz-type function approximation was studied separately and in detail. Furthermore, they presented the corresponding general approximation theory of general singular integral operators with lots of applications to, the under focused till then, trigonometric singular integral.

29.2 Part I: On Gauss–Weierstrass Smooth Singular Integrals

By Ali and Pales (2022) and Anastassiou (2023), for $\alpha, \beta \in \mathbb{R}$ with $\alpha\beta\left(\alpha^2 - \beta^2\right) \neq 0$, and $f \in C^4(\mathbb{R})$, $a, x \in \mathbb{R}$, we have the following general trigonometric Taylor formula:

$$f(x) - f(a) = f'(a)\left(\frac{\beta^3 \sin(\alpha(x-a)) - \alpha^3 \sin(\beta(x-a))}{\alpha\beta\left(\beta^2-\alpha^2\right)}\right)$$

$$+ f''(a)\left(\frac{\cos(\alpha(x-a)) - \cos(\beta(x-a))}{\beta^2-\alpha^2}\right)$$

$$+ f'''(a)\left(\frac{\beta \sin(\alpha(x-a)) - \alpha \sin(\beta(x-a))}{\alpha\beta\left(\beta^2-\alpha^2\right)}\right)$$

$$+ \frac{2\left(f''''(a) + \left(\alpha^2+\beta^2\right) f''(a)\right)}{(\alpha\beta)^2\left(\beta^2-\alpha^2\right)}\left(\beta^2 \sin^2\left(\frac{\alpha(x-a)}{2}\right)\right.$$

$$\left. - \alpha^2 \sin^2\left(\frac{\beta(x-a)}{2}\right)\right)$$

$$+\frac{1}{\alpha\beta\left(\beta^2-\alpha^2\right)}\int_a^x\left[\left(f''''(t)+\left(\alpha^2+\beta^2\right)f''(t)+\alpha^2\beta^2 f(t)\right)\right.$$
$$\left.-\left(f''''(a)+\left(\alpha^2+\beta^2\right)f''(a)+\alpha^2\beta^2 f(a)\right)\right]$$
$$\left[\beta\sin\left(\alpha\left(x-t\right)\right)-\alpha\sin\left(\beta\left(x-t\right)\right)\right]dt. \tag{29.1}$$

For $r\in\mathbb{N}$ and $n\in\mathbb{Z}^+$, we set

$$\alpha_j := \begin{cases} (-1)^{r-j}\binom{r}{j} j^{-n}, & j=1,\ldots,r, \\ 1-\sum\limits_{j=1}^{r}(-1)^{r-j}\binom{r}{j} j^{-n}, & n=0, \end{cases} \tag{29.2}$$

that is,

$$\sum_{j=0}^{r}\alpha_j = 1. \tag{29.3}$$

Here we consider all $f, f', f'', f''', f^{(4)} \in L_p(\mathbb{R})\cap C(\mathbb{R})$, $1\le p<\infty$.

For $x\in\mathbb{R}$, $\xi>0$, we consider the Lebesgue integrals, so-called smooth Gauss–Weierstrass operators,

$$W_{r,\xi}(f,x) = \frac{1}{\sqrt{\pi\xi}}\int_{-\infty}^{\infty}\left(\sum_{j=0}^{r}\alpha_j f(x+jt)\right)e^{-\frac{t^2}{\xi}}dt, \tag{29.4}$$

see Anastassiou and Mezei 2012; $W_{r,\xi}$ are not in general positive operators, see Anastassiou and Mezei 2012.

We note by

$$\frac{1}{\sqrt{\pi\xi}}\int_{-\infty}^{\infty}e^{-\frac{t^2}{\xi}}dt = 1 \tag{29.5}$$

that

$$W_{r,\xi}(c,x) = c, \quad \text{where } c \text{ is a constant}, \tag{29.6}$$

and

$$W_{r,\xi}(f,x)-f(x) = \frac{1}{\sqrt{\pi\xi}}\left(\sum_{j=0}^{r}\alpha_j\int_{-\infty}^{\infty}\left(f(x+jt)-f(x)\right)e^{-\frac{t^2}{\xi}}dt\right). \tag{29.7}$$

Denote by

$$\omega_1(f,h)_p := \sup_{\substack{t\in\mathbb{R}\\ |t|\le h}} \|f(x+t)-f(x)\|_{p,x}, \tag{29.8}$$

the first L_p modulus of smoothness of f, $1 \le p < \infty$.

By (29.1), we get that

$$\begin{aligned}
f(x+jt)-f(x) &= f'(x)\left(\frac{\beta^3\sin(\alpha jt)-\alpha^3\sin(\beta jt)}{\alpha\beta(\beta^2-\alpha^2)}\right)\\
&+ f''(x)\left(\frac{\cos(\alpha jt)-\cos(\beta jt)}{\beta^2-\alpha^2}\right)\\
&+ f'''(x)\left(\frac{\beta\sin(\alpha jt)-\alpha\sin(\beta jt)}{\alpha\beta(\beta^2-\alpha^2)}\right)\\
&+ \frac{2\left(f^{(4)}(x)+(\alpha^2+\beta^2)f''(x)\right)}{(\alpha\beta)^2(\beta^2-\alpha^2)}\left(\beta^2\sin^2\left(\frac{\alpha jt}{2}\right)-\alpha^2\sin^2\left(\frac{\beta jt}{2}\right)\right)\\
&+ \frac{1}{\alpha\beta(\beta^2-\alpha^2)}\int_x^{x+jt}\left[\left(f^{(4)}(s)+(\alpha^2+\beta^2)f''(s)+\alpha^2\beta^2 f(s)\right)\right.\\
&- \left.\left(f^{(4)}(x)+(\alpha^2+\beta^2)f''(x)+\alpha^2\beta^2 f(x)\right)\right]\\
&[\beta\sin(\alpha(x+jt-s))-\alpha\sin(\beta(x+jt-s))]\,ds,
\end{aligned} \tag{29.9}$$

or better,

$$\begin{aligned}
f(x+jt)-f(x) &= f'(x)\left(\frac{\beta^3\sin(\alpha jt)-\alpha^3\sin(\beta jt)}{\alpha\beta(\beta^2-\alpha^2)}\right)\\
&+ f''(x)\left(\frac{\cos(\alpha jt)-\cos(\beta jt)}{\beta^2-\alpha^2}\right)\\
&+ f'''(x)\left(\frac{\beta\sin(\alpha jt)-\alpha\sin(\beta jt)}{\alpha\beta(\beta^2-\alpha^2)}\right)\\
&+ \frac{2\left(f^{(4)}(x)+(\alpha^2+\beta^2)f''(x)\right)}{(\alpha\beta)^2(\beta^2-\alpha^2)}\left(\beta^2\sin^2\left(\frac{\alpha jt}{2}\right)-\alpha^2\sin^2\left(\frac{\beta jt}{2}\right)\right)\\
&+ \frac{1}{\alpha\beta(\beta^2-\alpha^2)}\int_0^{jt}\left[\left(f^{(4)}(x+z)+(\alpha^2+\beta^2)f''(x+z)\right.\right.\\
&+ \left.\alpha^2\beta^2 f(x+z)\right)-\left.\left(f^{(4)}(x)+(\alpha^2+\beta^2)f''(x)+\alpha^2\beta^2 f(x)\right)\right]\\
&[\beta\sin(\alpha(jt-z))-\alpha\sin(\beta(jt-z))]\,dz.
\end{aligned} \tag{29.10}$$

Furthermore, it holds

$$\sum_{j=0}^{r} \alpha_j \left[f(x+jt) - f(x) \right]$$

$$= \left(\frac{f'(x)}{\alpha\beta(\beta^2-\alpha^2)} \right) \left[\beta^3 \sum_{j=0}^{r} \alpha_j \sin(\alpha jt) - \alpha^3 \sum_{j=0}^{r} \alpha_j \sin(\beta jt) \right]$$

$$+ \left(\frac{f''(x)}{\beta^2-\alpha^2} \right) \left[\sum_{j=0}^{r} \alpha_j \cos(\alpha jt) - \sum_{j=0}^{r} \alpha_j \cos(\beta jt) \right]$$

$$+ \left(\frac{f^{(3)}(x)}{\alpha\beta(\beta^2-\alpha^2)} \right) \left[\beta \sum_{j=0}^{r} \alpha_j \sin(\alpha jt) - \alpha \sum_{j=0}^{r} \alpha_j \sin(\beta jt) \right]$$

$$+ \left(\frac{2\left(f^{(4)}(x) + (\alpha^2+\beta^2) f''(x)\right)}{(\alpha\beta)^2(\beta^2-\alpha^2)} \right) \left(\beta^2 \sum_{j=0}^{r} \alpha_j \sin^2\left(\frac{\alpha jt}{2}\right) \right.$$

$$\left. - \alpha^2 \sum_{j=0}^{r} \alpha_j \sin^2\left(\frac{\beta jt}{2}\right) \right) + \frac{1}{\alpha\beta(\beta^2-\alpha^2)} \sum_{j=0}^{r} \alpha_j \int_0^{jt}$$

$$\times \left[\left(f^{(4)}(x+z) + (\alpha^2+\beta^2) f''(x+z) + \alpha^2\beta^2 f(x+z) \right) \right.$$

$$\left. - \left(f^{(4)}(x) + (\alpha^2+\beta^2) f''(x) + \alpha^2\beta^2 f(x) \right) \right]$$

$$\left[\beta \sin(\alpha(jt-z)) - \alpha \sin(\beta(jt-z)) \right] dz, \tag{29.11}$$

or better

$$\sum_{j=0}^{r} \alpha_j \left[f(x+jt) - f(x) \right]$$

$$= \left(\frac{f'(x)}{\alpha\beta(\beta^2-\alpha^2)} \right) \left[\beta^3 \sum_{j=0}^{r} \alpha_j \sin(\alpha jt) - \alpha^3 \sum_{j=0}^{r} \alpha_j \sin(\beta jt) \right]$$

$$+ \left(\frac{f''(x)}{\beta^2-\alpha^2} \right) \left[\sum_{j=0}^{r} \alpha_j \cos(\alpha jt) - \sum_{j=0}^{r} \alpha_j \cos(\beta jt) \right]$$

$$+ \left(\frac{f^{(3)}(x)}{\alpha\beta(\beta^2-\alpha^2)} \right) \left[\beta \sum_{j=0}^{r} \alpha_j \sin(\alpha jt) - \alpha \sum_{j=0}^{r} \alpha_j \sin(\beta jt) \right]$$

$$+\left(\frac{2\left(f^{(4)}(x)+\left(\alpha^2+\beta^2\right)f''(x)\right)}{(\alpha\beta)^2\left(\beta^2-\alpha^2\right)}\right)\left(\beta^2\sum_{j=0}^{r}\alpha_j\sin^2\left(\frac{\alpha jt}{2}\right)\right.$$

$$\left.-\alpha^2\sum_{j=0}^{r}\alpha_j\sin^2\left(\frac{\beta jt}{2}\right)\right)+\frac{1}{\alpha\beta\left(\beta^2-\alpha^2\right)}\sum_{j=0}^{r}\alpha_j j\int_0^t$$

$$\times\left[\left(f^{(4)}(x+jw)+\left(\alpha^2+\beta^2\right)f''(x+jw)+\alpha^2\beta^2 f(x+jw)\right)\right.$$

$$\left.-\left(f^{(4)}(x)+\left(\alpha^2+\beta^2\right)f''(x)+\alpha^2\beta^2 f(x)\right)\right]$$

$$[\beta\sin(\alpha j(t-w))-\alpha\sin(\beta j(t-w))]\,dw. \tag{29.12}$$

We call

$$R:=R(t):=\frac{1}{\alpha\beta\left(\beta^2-\alpha^2\right)}\sum_{j=0}^{r}\alpha_j j$$

$$\int_0^t\left[\left(f^{(4)}(x+jw)+\left(\alpha^2+\beta^2\right)f''(x+jw)+\alpha^2\beta^2 f(x+jw)\right)\right.$$

$$\left.-\left(f^{(4)}(x)+\left(\alpha^2+\beta^2\right)f''(x)+\alpha^2\beta^2 f(x)\right)\right]$$

$$[\beta\sin(\alpha j(t-w))-\alpha\sin(\beta j(t-w))]\,dw,\quad\forall\, t\in\mathbb{R}. \tag{29.13}$$

We set

$$E_1(x):=W_{r,\xi}(f,x)-f(x)-\left(\frac{f'(x)}{\alpha\beta\left(\beta^2-\alpha^2\right)}\right)$$

$$\left[\beta^3\sum_{j=0}^{r}\alpha_j\frac{1}{\sqrt{\pi\xi}}\left(\int_{-\infty}^{\infty}\sin(\alpha jt)\,e^{-\frac{t^2}{\xi}}dt\right)\right.$$

$$\left.-\alpha^3\sum_{j=0}^{r}\alpha_j\frac{1}{\sqrt{\pi\xi}}\left(\int_{-\infty}^{\infty}\sin(\beta jt)\,e^{-\frac{t^2}{\xi}}dt\right)\right]-\left(\frac{f''(x)}{\beta^2-\alpha^2}\right)$$

$$\left[\sum_{j=0}^{r}\alpha_j\frac{1}{\sqrt{\pi\xi}}\left(\int_{-\infty}^{\infty}\cos(\alpha jt)\,e^{-\frac{t^2}{\xi}}dt\right)-\sum_{j=0}^{r}\alpha_j\frac{1}{\sqrt{\pi\xi}}\left(\int_{-\infty}^{\infty}\cos(\beta jt)\,e^{-\frac{t^2}{\xi}}dt\right)\right]$$

$$-\left(\frac{f^{(3)}(x)}{\alpha\beta\left(\beta^2-\alpha^2\right)}\right)$$

$$\left[\beta\sum_{j=0}^{r}\alpha_j\frac{1}{\sqrt{\pi\xi}}\left(\int_{-\infty}^{\infty}\sin(\alpha jt)\,e^{-\frac{t^2}{\xi}}dt\right)\right.$$

$$\left.-\alpha\sum_{j=0}^{r}\alpha_j\frac{1}{\sqrt{\pi\xi}}\left(\int_{-\infty}^{\infty}\sin(\beta jt)\,e^{-\frac{t^2}{\xi}}dt\right)\right]$$

$$-\left(\frac{2\left(f^{(4)}(x)+\left(\alpha^2+\beta^2\right)f''(x)\right)}{(\alpha\beta)^2\left(\beta^2-\alpha^2\right)}\right)$$

$$\left[\beta^2\sum_{j=0}^{r}\alpha_j\frac{1}{\sqrt{\pi\xi}}\left(\int_{-\infty}^{\infty}\sin^2\left(\frac{\alpha jt}{2}\right)e^{-\frac{t^2}{\xi}}dt\right)\right.$$

$$\left.-\alpha^2\sum_{j=0}^{r}\alpha_j\frac{1}{\sqrt{\pi\xi}}\left(\int_{-\infty}^{\infty}\sin^2\left(\frac{\beta jt}{2}\right)e^{-\frac{t^2}{\xi}}dt\right)\right]$$

$$=\frac{1}{\sqrt{\pi\xi}}\int_{-\infty}^{\infty}R(t)\,e^{-\frac{t^2}{\xi}}dt. \tag{29.14}$$

Next we simplify $E_1(x)$:
We observe that

$$\int_{-\infty}^{\infty}\sin(\alpha jt)\,e^{-\frac{t^2}{\xi}}dt=\int_{-\infty}^{0}\sin(\alpha jt)\,e^{-\frac{t^2}{\xi}}dt+\int_{0}^{\infty}\sin(\alpha jt)\,e^{-\frac{t^2}{\xi}}dt. \tag{29.15}$$

Note $-\infty\le t\le 0\Rightarrow\infty\ge -t\ge 0$. So

$$\int_{-\infty}^{0}\sin(\alpha jt)\,e^{-\frac{\left|t^2\right|}{\xi}}dt=-\int_{-\infty}^{0}\sin(\alpha j(-(-t)))\,e^{-\frac{t^2}{\xi}}d(-t)$$

$$=-\int_{-\infty}^{0}(-\sin(\alpha j(-t)))\,e^{-\frac{t^2}{\xi}}d(-t)=\int_{-\infty}^{0}\sin(\alpha j(-t))\,e^{-\frac{t^2}{\xi}}d(-t)$$

$$=\int_{\infty}^{0}\sin(\alpha jt)\,e^{-\frac{t^2}{\xi}}dt=-\int_{0}^{\infty}\sin(\alpha jt)\,e^{-\frac{t^2}{\xi}}dt. \tag{29.16}$$

So

$$\int_{-\infty}^{\infty}\sin(\alpha jt)\,e^{-\frac{t^2}{\xi}}dt=0,$$

and all sine integrals in (29.14) are zeros.

Furthermore, we have that

$$\int_{-\infty}^{\infty} \sin^2\left(\frac{\alpha jt}{2}\right) e^{-\frac{t^2}{\xi}} dt = 2\int_{0}^{\infty} \sin^2\left(\frac{\alpha jt}{2}\right) e^{-\frac{t^2}{\xi}} dt$$

$$= 2\sqrt{\xi}\int_{0}^{\infty} \sin^2\left(\left(\frac{\alpha j\sqrt{\xi}}{2}\right)\frac{t}{\sqrt{\xi}}\right) e^{-\left(\frac{t}{\sqrt{\xi}}\right)^2} d\frac{t}{\sqrt{\xi}}$$

($\frac{t}{\sqrt{\xi}} =: x$, and $\frac{\alpha j\sqrt{\xi}}{2} = \beta_1$)

$$= 2\sqrt{\xi}\int_{0}^{\infty} \sin^2(\beta_1 x) e^{-x^2} dx = 2\sqrt{\xi}\frac{1}{4}\sqrt{\pi}e^{-\beta_1^2}\left(e^{\beta_1^2} - 1\right)$$

$$= \frac{\sqrt{\pi\xi}}{2}\left(1 - e^{-\beta_1^2}\right) = \frac{\sqrt{\pi\xi}}{2}\left(1 - e^{-\frac{\alpha^2 j^2 \xi}{4}}\right). \tag{29.17}$$

That is,

$$\int_{-\infty}^{\infty} \sin^2\left(\frac{\alpha jt}{2}\right) e^{-\frac{t^2}{\xi}} dt = \frac{\sqrt{\pi\xi}}{2}\left(1 - e^{-\frac{\alpha^2 j^2 \xi}{4}}\right), \tag{29.18}$$

and similarly,

$$\int_{-\infty}^{\infty} \sin^2\left(\frac{\beta jt}{2}\right) e^{-\frac{t^2}{\xi}} dt = \frac{\sqrt{\pi\xi}}{2}\left(1 - e^{-\frac{\beta^2 j^2 \xi}{4}}\right). \tag{29.19}$$

Next, we treat

$$\int_{-\infty}^{\infty} \cos(\alpha jt) e^{-\frac{t^2}{\xi}} dt = 2\int_{0}^{\infty} \cos(\alpha jt) e^{-\frac{t^2}{\xi}} dt =$$

$$= 2\sqrt{\xi}\int_{0}^{\infty} \cos\left(\left(\frac{\alpha j\sqrt{\xi}}{2}\right)\frac{t}{\sqrt{\xi}}\right) e^{-\left(\frac{t}{\sqrt{\xi}}\right)^2} d\frac{t}{\sqrt{\xi}} \tag{29.20}$$

($\frac{t}{\sqrt{\xi}} =: x$, and $\frac{\alpha j\sqrt{\xi}}{2} =: \beta_1$)

$$= 2\sqrt{\xi}\int_{0}^{\infty} \cos(\beta_1 x) e^{-x^2} dx = 2\sqrt{\xi}\frac{1}{2}\sqrt{\pi}e^{-\frac{\beta_1^2}{4}} = \sqrt{\pi\xi}e^{-\frac{\alpha^2 j^2 \xi}{16}}.$$

That is,

$$\int_{-\infty}^{\infty} \cos(\alpha jt) e^{-\frac{t^2}{\xi}} dt = \sqrt{\pi\xi}e^{-\frac{\alpha^2 j^2 \xi}{16}}, \tag{29.21}$$

and similarly,

$$\int_{-\infty}^{\infty} \cos\left(\beta j t\right) e^{-\frac{t^2}{\xi}} dt = \sqrt{\pi\xi} e^{-\frac{\beta^2 j^2 \xi}{16}}. \tag{29.22}$$

Hence, we have the simplified expression,

$$\begin{aligned} E_1(x) &= W_{r,\xi}(f,x) - f(x) \\ &\quad - \left(\frac{f''(x)}{\beta^2-\alpha^2}\right)\left[\sum_{j=0}^{r} \alpha_j \left(e^{-\frac{\alpha^2 j^2 \xi}{16}} - e^{-\frac{\beta^2 j^2 \xi}{16}}\right)\right] \\ &\quad - \left(\frac{f^{(4)}(x) + \left(\alpha^2+\beta^2\right) f''(x)}{(\alpha\beta)^2\left(\beta^2-\alpha^2\right)}\right) \\ &\quad \left[\beta^2 \sum_{j=0}^{r} \alpha_j \left(1 - e^{-\frac{\alpha^2 j^2 \xi}{4}}\right) - \alpha^2 \sum_{j=0}^{r} \alpha_j \left(1 - e^{-\frac{\beta^2 j^2 \xi}{4}}\right)\right] \\ &= \frac{1}{\sqrt{\pi\xi}} \int_{-\infty}^{\infty} R(t) e^{-\frac{t^2}{\xi}} dt. \end{aligned} \tag{29.23}$$

It follows a parametrized L_p $(p > 1)$ approximation result for $W_{r,\xi}$, $\xi > 0$.

Theorem 29.1. *Let $\alpha, \beta \in \mathbb{R}$ with $\alpha\beta\left(\alpha^2-\beta^2\right) \neq 0$, and let all $f, f', f'', f^{(3)}, f^{(4)} \in L_p(\mathbb{R}) \cap C(\mathbb{R})$, $1 < p < \infty$. Here $q > 1 : \frac{1}{p} + \frac{1}{q} = 1$; $\xi > 0$, $x \in \mathbb{R}$. Then*

$$\begin{aligned} \left\|E_1(x)\right\|_p &= \Bigg\| W_{r,\xi}(f,x) - f(x) \\ &\quad - \frac{f''(x)}{\left(\beta^2-\alpha^2\right)}\left[\sum_{j=0}^{r} \alpha_j \left(e^{-\frac{\alpha^2 j^2 \xi}{16}} - e^{-\frac{\beta^2 j^2 \xi}{16}}\right)\right] \\ &\quad - \left(\frac{f^{(4)}(x) + \left(\alpha^2+\beta^2\right) f''(x)}{(\alpha\beta)^2\left(\beta^2-\alpha^2\right)}\right) \\ &\quad \left[\beta^2 \sum_{j=0}^{r} \alpha_j \left(1 - e^{-\frac{\alpha^2 j^2 \xi}{4}}\right) - \alpha^2 \sum_{j=0}^{r} \alpha_j \left(1 - e^{-\frac{\beta^2 j^2 \xi}{4}}\right)\right]\Bigg\|_p \\ &\leq \frac{\left(\sum_{j=1}^{r} |\alpha_j| j^2\right)^{\frac{1}{q}} 4 \cdot 2^{\frac{1}{\beta}}}{\left(\sqrt{\pi}\right)^{\frac{1}{p}} \left|\beta^2-\alpha^2\right| (q+1)^{\frac{1}{q}} (p+1)^{\frac{1}{p}}} \end{aligned}$$

$$\left[\sum_{j=1}^{r}|\alpha_j|\,j^{p+2}\left(\frac{1}{(3p+1)}+\Gamma(3p+1)\right)\right]^{\frac{1}{p}}$$

$$\omega_1\left(f^{(4)}+\left(\alpha^2+\beta^2\right)f''+\alpha^2\beta^2 f,\sqrt{\xi}\right)_p\cdot\xi=:\Phi_1(\xi)\to 0, \quad (29.24)$$

as $\xi\to 0$.

Above Γ *stands for the gamma function.*

Proof. Call the function

$$F:=f^{(4)}+\left(\alpha^2+\beta^2\right)f''+\alpha^2\beta^2 f. \quad (29.25)$$

Then, we get

$$R=R(t)=\frac{1}{\alpha\beta\left(\beta^2-\alpha^2\right)}\sum_{j=0}^{r}\alpha_j j\int_0^t\left[F(x+jw)-F(x)\right]$$

$$\left[\beta\sin(\alpha j(t-w))-\alpha\sin(\beta j(t-w))\right]dw, \quad \forall\, t\in\mathbb{R}. \quad (29.26)$$

We isolate and study

$$I:=\int_0^t\left[F(x+jw)-F(x)\right]$$

$$\left[\beta\sin(\alpha j(t-w))-\alpha\sin(\beta j(t-w))\right]dw, \quad \forall\, t\in\mathbb{R}. \quad (29.27)$$

For $t<0$, we have that

$$|I|=\left|\int_t^0\left[F(x+jw)-F(x)\right]\left[\beta\sin(\alpha j(t-w))-\alpha\sin(\beta j(t-w))\right]dw\right|$$

$$\leq\int_t^0\left|F(x+jw)-F(x)\right|\left|\beta\sin(\alpha j(t-w))-\alpha\sin(\beta j(t-w))\right|dw$$

$$\leq 2|\alpha||\beta|\,j\int_t^0\left|F(x+jw)-F(x)\right|(w-t)\,dw$$

$$=-2|\alpha||\beta|\,j\int_t^0\left|F(x-j(-w))-F(x)\right|(-t-(-w))\,d(-w)= \quad (29.28)$$

$(t \leq w \leq 0 \Rightarrow -t \geq -w =: \theta \geq 0)$

$$- 2\left|\alpha\right|\left|\beta\right| j \int_{-t}^{0} \left|F\left(x - j\theta\right) - F\left(x\right)\right| \left(-t - \theta\right) d\theta$$

$$= 2\left|\alpha\right|\left|\beta\right| j \int_{0}^{-t} \left|F\left(x - j\theta\right) - F\left(x\right)\right| \left(-t - \theta\right) d\theta$$

$$= 2\left|\alpha\beta\right| j \int_{0}^{|t|} \left|F\left(x + sign\left(t\right) j\theta\right) - F\left(x\right)\right| \left(\left|t\right| - \theta\right) d\theta.$$

So, we have proved that

$$\left|I\right| \leq 2\left|\alpha\beta\right| j \int_{0}^{|t|} \left|F\left(x + sign\left(t\right) j\theta\right) - F\left(x\right)\right| \left(\left|t\right| - \theta\right) d\theta, \quad \forall\, t \in \mathbb{R}, \tag{29.29}$$

and, by (29.26),

$$\left|R\left(t\right)\right| \leq \frac{2}{\left|\beta^2 - \alpha^2\right|} \sum_{j=0}^{r} \left|\alpha_j\right| j^2 \int_{0}^{|t|} \left|F\left(x + jsign\left(t\right)\theta\right) - F\left(x\right)\right| \left(\left|t\right| - \theta\right) d\theta, \tag{29.30}$$

$\forall\, t \in \mathbb{R}$.

By (29.14), we have

$$E_1\left(x\right) = \frac{1}{\sqrt{\pi\xi}} \int_{-\infty}^{\infty} R\left(t\right) e^{-\frac{t^2}{\xi}} dt. \tag{29.31}$$

Hence it holds $(p, q > 1 : \frac{1}{p} + \frac{1}{q} = 1)$,

$$\int_{-\infty}^{\infty} \left|E_1\left(x\right)\right|^p dx = \int_{-\infty}^{\infty} \left(\frac{1}{\sqrt{\pi\xi}} \left|\int_{-\infty}^{\infty} R\left(t\right) e^{-\frac{t^2}{\xi}} dt\right|\right)^p dx$$

$$\leq \int_{-\infty}^{\infty} \left(\int_{-\infty}^{\infty} \left|R\left(t\right)\right| \frac{e^{-\frac{t^2}{\xi}}}{\sqrt{\pi\xi}} dt\right)^p dx$$

$$\leq \frac{1}{\sqrt{\pi\xi}} \int_{-\infty}^{\infty} \left(\int_{-\infty}^{\infty} \left|R\left(t\right)\right|^p e^{-\frac{t^2}{\xi}} dt\right) dx \overset{(29.30)}{\leq} \frac{1}{\sqrt{\pi\xi}} \frac{2^p}{\left|\beta^2 - \alpha^2\right|^p} \tag{29.32}$$

$$\int_{-\infty}^{\infty} \left(\int_{-\infty}^{\infty} \left[\sum_{j=1}^{r} \left|\alpha_j\right| j^2 \right.\right.$$

$$\left.\left.\times \left(\int_{0}^{|t|} \left|F\left(x + jsign\left(t\right)\theta\right) - F\left(x\right)\right| \left(\left|t\right| - \theta\right) d\theta\right)\right]^p e^{-\frac{t^2}{\xi}} dt\right) dx$$

$$\overset{\text{(by Jensen inequality)}}{\leq} \frac{1}{\sqrt{\pi\xi}} \frac{2^p}{|\beta^2-\alpha^2|^p} \int_{-\infty}^{\infty} \left(\int_{-\infty}^{\infty} \left(\sum_{j=1}^{r} |\alpha_j| j^2 \right)^{p-1} \right.$$

$$\left. \left. \left[\sum_{j=1}^{r} |\alpha_j| j^2 \left(\int_0^{|t|} |F(x + jsign(t)\theta) - F(x)| (|t| - \theta) d\theta \right)^p \right] e^{-\frac{t^2}{\xi}} dt \right) dx$$

$$\leq \frac{2^p \left(\sum_{j=1}^{r} |\alpha_j| j^2 \right)^{p-1}}{\sqrt{\pi\xi} |\beta^2-\alpha^2|^p} \left[\sum_{j=1}^{r} |\alpha_j| j^2 \int_{-\infty}^{\infty} \left(\int_{-\infty}^{\infty} \right. \right.$$

$$\left. \left. \left(\int_0^{|t|} |F(x + jsign(t)\theta) - F(x)|^p d\theta \right) \left(\int_0^{|t|} (|t| - \theta)^q d\theta \right)^{\frac{p}{q}} e^{-\frac{t^2}{\xi}} dt \right) dx \right]$$

$$= \frac{2^p \left(\sum_{j=1}^{r} |\alpha_j| j^2 \right)^{p-1}}{\sqrt{\pi\xi} |\beta^2-\alpha^2|^p (q+1)^{p-1}} \left[\sum_{j=1}^{r} |\alpha_j| j^2 \int_{-\infty}^{\infty} \left(\int_{-\infty}^{\infty} \right. \right.$$

$$\left. \left. \left(\int_0^{|t|} |F(x + jsign(t)\theta) - F(x)|^p d\theta \right) |t|^{(q+1)(p-1)} e^{-\frac{t^2}{\xi}} dt \right) dx \right] = \tag{29.33}$$

(call $c := \frac{2^p \left(\sum_{j=1}^{r} |\alpha_j| j^2 \right)^{p-1}}{\sqrt{\pi\xi} |\beta^2-\alpha^2|^p (q+1)^{p-1}}$)

$$c \left[\sum_{j=1}^{r} |\alpha_j| j^2 \int_{-\infty}^{\infty} \left(\int_0^{|t|} \left(\int_{-\infty}^{\infty} \right. \right. \right.$$

$$\left. \left. \left. |F(x + jsign(t)\theta) - F(x)|^p dx \right) d\theta \right) |t|^{(q+1)(p-1)} e^{-\frac{t^2}{\xi}} dt \right]$$

$$\leq c \left[\sum_{j=1}^{r} |\alpha_j| j^2 \int_{-\infty}^{\infty} \left(\int_0^{|t|} \omega_1 \left(F, \sqrt{\xi} \frac{j\theta}{\sqrt{\xi}} \right)_p^p d\theta \right) |t|^{(q+1)(p-1)} e^{-\frac{t^2}{\xi}} dt \right] \tag{29.34}$$

$$\leq c\omega_1 \left(F, \sqrt{\xi} \right)_p^p \left[\sum_{j=1}^{r} |\alpha_j| j^2 \int_{-\infty}^{\infty} \left(\int_0^{|t|} \left(1 + \frac{j\theta}{\sqrt{\xi}} \right)^p d\theta \right) |t|^{(q+1)(p-1)} e^{-\frac{t^2}{\xi}} dt \right]$$

$$= c\omega_1 \left(F, \sqrt{\xi} \right)_p^p \frac{\sqrt{\xi}}{(p+1)} \left[\sum_{j=1}^{r} |\alpha_j| j \int_{-\infty}^{\infty} \left[\left(1 + \frac{j}{\sqrt{\xi}} |t| \right)^{p+1} - 1 \right] \right.$$

$$\left. \times |t|^{(q+1)(p-1)} e^{-\frac{t^2}{\xi}} dt \right]$$

$$\leq \frac{2^p c\sqrt{\xi}}{(p+1)}\omega_1\left(F,\sqrt{\xi}\right)_p^p\left[\sum_{j=1}^r|\alpha_j|\,j\frac{j^{p+1}}{\left(\sqrt{\xi}\right)^{p+1}}\int_{-\infty}^{\infty}|t|^{p+1}\,|t|^{(q+1)(p-1)}\,e^{-\frac{t^2}{\xi}}dt\right]$$

$$= \frac{2^{p+1} c\left(\sqrt{\xi}\right)^{-p}}{(p+1)}\omega_1\left(F,\sqrt{\xi}\right)_p^p\left[\sum_{j=1}^r|\alpha_j|\,j^{p+2}\int_0^{\infty}t^{(q+1)(p-1)+(p+1)}e^{-\frac{t^2}{\xi}}dt\right] \tag{29.35}$$

$$= \frac{2^{p+1} c\left(\sqrt{\xi}\right)^{-p}}{(p+1)}\omega_1\left(F,\sqrt{\xi}\right)_p^p\left(\sqrt{\xi}\right)^{3p+1}$$

$$\left[\sum_{j=1}^r|\alpha_j|\,j^{p+2}\int_0^{\infty}\left(\frac{t}{\sqrt{\xi}}\right)^{3p}e^{-\left(\frac{t}{\sqrt{\xi}}\right)^2}d\frac{t}{\sqrt{\xi}}\right]$$

$$= \frac{2^{p+1} c\left(\sqrt{\xi}\right)^{2p+1}}{(p+1)}\omega_1\left(F,\sqrt{\xi}\right)_p^p\left[\sum_{j=1}^r|\alpha_j|\,j^{p+2}\int_0^{\infty}x^{3p}e^{-x^2}dx\right]$$

$$= \frac{2^{p+1} c\left(\sqrt{\xi}\right)^{2p+1}}{(p+1)}\omega_1\left(F,\sqrt{\xi}\right)_p^p$$

$$\times\left[\sum_{j=1}^r|\alpha_j|\,j^{p+2}\left(\int_0^1 x^{3p}e^{-x^2}dx+\int_1^{\infty}x^{3p}e^{-x^2}dx\right)\right] \tag{29.36}$$

$$\leq \frac{2^{p+1} c\left(\sqrt{\xi}\right)^{2p+1}}{(p+1)}\omega_1\left(F,\sqrt{\xi}\right)_p^p\left[\sum_{j=1}^r|\alpha_j|\,j^{p+2}\left(\int_0^1 x^{3p}dx+\int_1^{\infty}x^{3p}e^{-x}dx\right)\right]$$

$$\leq \frac{2^{p+1} c\left(\sqrt{\xi}\right)^{2p+1}}{(p+1)}\omega_1\left(F,\sqrt{\xi}\right)_p^p\left[\sum_{j=1}^r|\alpha_j|\,j^{p+2}\left(\frac{1}{(3p+1)}+\int_0^{\infty}x^{3p}e^{-x}dx\right)\right]$$

$$= \frac{2^{p+1} c\left(\sqrt{\xi}\right)^{2p+1}}{(p+1)}\omega_1\left(F,\sqrt{\xi}\right)_p^p\left[\sum_{j=1}^r|\alpha_j|\,j^{p+2}\left(\frac{1}{(3p+1)}+\Gamma(3p+1)\right)\right] \tag{29.37}$$

$$= \frac{\left(\sum_{j=1}^r|\alpha_j|\,j^2\right)^{p-1}2^{2p+1}}{\sqrt{\pi}\,|\beta^2-\alpha^2|^p\,(q+1)^{p-1}\,(p+1)}\xi^p\omega_1\left(F,\sqrt{\xi}\right)_p^p$$

$$\left[\sum_{j=1}^r|\alpha_j|\,j^{p+2}\left(\frac{1}{(3p+1)}+\Gamma(3p+1)\right)\right]$$

$$= \frac{\left(\sum_{j=1}^r|\alpha_j|\,j^2\right)^{p-1}2^{2p+1}}{\sqrt{\pi}\,|\beta^2-\alpha^2|^p\,(q+1)^{p-1}\,(p+1)}\left[\sum_{j=1}^r|\alpha_j|\,j^{p+2}\left(\frac{1}{(3p+1)}+\Gamma(3p+1)\right)\right]$$

$$\omega_1\left(f^{(4)}+\left(\alpha^2+\beta^2\right)f''+\alpha^2\beta^2 f,\sqrt{\xi}\right)_p^p\xi^p, \tag{29.38}$$

proving the claim. ☐

It follows a consequence.

Corollary 29.1. *(to Theorem 29.1) It holds*

$$\left\| W_{r,\xi}(f) - f \right\|_p \leq \Phi_1(\xi)$$

$$+ \frac{\|f''\|_p}{|\beta^2 - \alpha^2|} \left[\sum_{j=0}^{r} |\alpha_j| \left| e^{-\frac{\alpha^2 j^2 \xi}{16}} - e^{-\frac{\beta^2 j^2 \xi}{16}} \right| \right]$$

$$+ \left(\frac{\left\|f^{(4)}\right\|_p + (\alpha^2 + \beta^2) \|f''\|_p}{(\alpha\beta)^2 |\beta^2 - \alpha^2|} \right)$$

$$\left| \beta^2 \sum_{j=0}^{r} \alpha_j \left(1 - e^{-\frac{\alpha^2 j^2 \xi}{4}} \right) - \alpha^2 \sum_{j=0}^{r} \alpha_j \left(1 - e^{-\frac{\beta^2 j^2 \xi}{4}} \right) \right| \to 0, \quad (29.39)$$

as $\xi \to 0$.

Above $\Phi_1(\xi)$ *is as in (29.24).*

Proof. Directly from (29.24). □

Next comes the L_1 approximation by $W_{r,\xi}$.

Theorem 29.2. *Let* $\alpha, \beta \in \mathbb{R}$ *with* $\alpha\beta(\alpha^2 - \beta^2) \neq 0$, *and let all* $f, f', f'', f^{(3)}, f^{(4)} \in L_1(\mathbb{R}) \cap C(\mathbb{R})$; $\xi > 0$, $x \in \mathbb{R}$. *Then*

$$\|E_1(x)\|_1 = \| W_{r,\xi}(f, x) - f(x)$$

$$- \frac{f''(x)}{(\beta^2 - \alpha^2)} \left[\sum_{j=0}^{r} \alpha_j \left(e^{-\frac{\alpha^2 j^2 \xi}{16}} - e^{-\frac{\beta^2 j^2 \xi}{16}} \right) \right]$$

$$- \left(\frac{f^{(4)}(x) + (\alpha^2 + \beta^2) f''(x)}{(\alpha\beta)^2 (\beta^2 - \alpha^2)} \right)$$

$$\left[\beta^2 \sum_{j=0}^{r} \alpha_j \left(1 - e^{-\frac{\alpha^2 j^2 \xi}{4}} \right) - \alpha^2 \sum_{j=0}^{r} \alpha_j \left(1 - e^{-\frac{\beta^2 j^2 \xi}{4}} \right) \right] \Bigg\|_1$$

$$\leq \left(\frac{4}{\sqrt{\pi} |\beta^2 - \alpha^2|} \right) \left[\sum_{j=1}^{r} |\alpha_j| j^2 \left(\frac{7}{3} + j \frac{25}{8} \right) \right]$$

$$\omega_1 \left(f^{(4)} + (\alpha^2 + \beta^2) f'' + \alpha^2 \beta^2 f, \sqrt{\xi} \right)_1 \cdot \xi =: \Psi_1(\xi) \to 0, \quad (29.40)$$

as $\xi \to 0$.

Proof. We have (F is as in (29.25))

$$\int_{-\infty}^{\infty}|E_1(x)|\,dx \overset{(29.31)}{=} \int_{-\infty}^{\infty}\left(\frac{1}{\sqrt{\pi\xi}}\left|\int_{-\infty}^{\infty}R(t)\,e^{-\frac{t^2}{\xi}}dt\right|\right)dx$$

$$\leq \int_{-\infty}^{\infty}\left(\int_{-\infty}^{\infty}|R(t)|\,\frac{e^{-\frac{t^2}{\xi}}}{\sqrt{\pi\xi}}dt\right)dx \overset{(29.30)}{\leq} \frac{1}{\sqrt{\pi\xi}}\frac{2}{|\beta^2-\alpha^2|}$$

$$\int_{-\infty}^{\infty}\left(\int_{-\infty}^{\infty}\left[\sum_{j=1}^{r}|\alpha_j|\,j^2\left(\int_0^{|t|}|F(x+jsign(t)\,\theta)-F(x)|\,(|t|-\theta)\,d\theta\right)\right]\right.$$

$$\left.\times e^{-\frac{t^2}{\xi}}dt\right)dx$$

$$\leq \frac{2}{\sqrt{\pi\xi}\,|\beta^2-\alpha^2|}\left[\sum_{j=1}^{r}|\alpha_j|\,j^2\int_{-\infty}^{\infty}\left(\int_{-\infty}^{\infty}\right.\right.$$

$$\left.\left.\left(\int_0^{|t|}|F(x+jsign(t)\,\theta)-F(x)|\,d\theta\right)|t|\,e^{-\frac{t^2}{\xi}}dt\right)dx\right] \tag{29.41}$$

$$=\frac{2}{\sqrt{\pi\xi}\,|\beta^2-\alpha^2|}\left[\sum_{j=1}^{r}|\alpha_j|\,j^2\int_{-\infty}^{\infty}\left(\int_0^{|t|}\left(\int_{-\infty}^{\infty}\right.\right.\right.$$

$$\left.\left.\left.|F(x+jsign(t)\,\theta)-F(x)|\,dx\right)d\theta\right)|t|\,e^{-\frac{t^2}{\xi}}dt\right]$$

$$\leq \frac{2}{\sqrt{\pi\xi}\,|\beta^2-\alpha^2|}\left[\sum_{j=1}^{r}|\alpha_j|\,j^2\int_{-\infty}^{\infty}\left(\int_0^{|t|}\omega_1\left(F,\sqrt{\xi}\frac{j\theta}{\sqrt{\xi}}\right)_1d\theta\right)|t|\,e^{-\frac{t^2}{\xi}}dt\right]$$

$$\leq \frac{2}{\sqrt{\pi\xi}\,|\beta^2-\alpha^2|}\omega_1\left(F,\sqrt{\xi}\right)_1\left[\sum_{j=1}^{r}|\alpha_j|\,j^2\int_{-\infty}^{\infty}\left(\int_0^{|t|}\left(1+\frac{j\theta}{\sqrt{\xi}}\right)d\theta\right)|t|\,e^{-\frac{t^2}{\xi}}dt\right]$$

$$=\frac{2}{\sqrt{\pi\xi}\,|\beta^2-\alpha^2|}\omega_1\left(F,\sqrt{\xi}\right)_1\left[\sum_{j=1}^{r}|\alpha_j|\,j^2\int_{-\infty}^{\infty}\left(|t|+\frac{j}{2\sqrt{\xi}}t^2\right)|t|\,e^{-\frac{t^2}{\xi}}dt\right]$$

$$=\frac{4}{\sqrt{\pi\xi}\,|\beta^2-\alpha^2|}\omega_1\left(F,\sqrt{\xi}\right)_1\left[\sum_{j=1}^{r}|\alpha_j|\,j^2\int_0^{\infty}\left(t+\frac{j}{2\sqrt{\xi}}t^2\right)te^{-\frac{t^2}{\xi}}dt\right] \tag{29.42}$$

$$=\frac{4}{\sqrt{\pi\xi}\,|\beta^2-\alpha^2|}\omega_1\left(F,\sqrt{\xi}\right)_1\left[\sum_{j=1}^{r}|\alpha_j|\,j^2\left(\int_0^{\infty}t^2e^{-\frac{t^2}{\xi}}dt+\frac{j}{2\sqrt{\xi}}\int_0^{\infty}t^3e^{-\frac{t^2}{\xi}}dt\right)\right]$$

$$= \frac{4}{\sqrt{\pi\xi}\,|\beta^2-\alpha^2|}\omega_1\left(F,\sqrt{\xi}\right)_1\left[\sum_{j=1}^{r}|\alpha_j|\,j^2\left(\left(\sqrt{\xi}\right)^3\int_0^\infty\left(\frac{t}{\sqrt{\xi}}\right)^2 e^{-\left(\frac{t}{\sqrt{\xi}}\right)^2}d\frac{t}{\sqrt{\xi}}\right.\right.$$

$$\left.\left.+\frac{j}{2\sqrt{\xi}}\left(\sqrt{\xi}\right)^4\int_0^\infty\left(\frac{t}{\sqrt{\xi}}\right)^3 e^{-\left(\frac{t}{\sqrt{\xi}}\right)^2}d\frac{t}{\sqrt{\xi}}\right)\right]$$

$$= \frac{4\xi}{\sqrt{\pi}\,|\beta^2-\alpha^2|}\omega_1\left(F,\sqrt{\xi}\right)_1\left[\sum_{j=1}^{r}|\alpha_j|\,j^2\left(\int_0^\infty x^2e^{-x^2}dx\right.\right. \tag{29.43}$$

$$\left.\left.+\frac{j}{2}\int_0^\infty x^3e^{-x^2}dx\right)\right]$$

$$\leq \frac{4\xi}{\sqrt{\pi}\,|\beta^2-\alpha^2|}\omega_1\left(F,\sqrt{\xi}\right)_1\left[\sum_{j=1}^{r}|\alpha_j|\,j^2\left[\left(\int_0^1 x^2dx+\int_0^\infty x^2e^{-x}dx\right)\right.\right.$$

$$\left.\left.+\frac{j}{2}\left(\int_0^1 x^3dx+\int_0^\infty x^3e^{-x}dx\right)\right]\right]$$

$$= \frac{4\xi}{\sqrt{\pi}\,|\beta^2-\alpha^2|}\omega_1\left(F,\sqrt{\xi}\right)_1\left[\sum_{j=1}^{r}|\alpha_j|\,j^2\left[\left(\frac{1}{3}+\Gamma(3)\right)+\frac{j}{2}\left(\frac{1}{4}+\Gamma(4)\right)\right]\right]$$

$$= \frac{4\xi}{\sqrt{\pi}\,|\beta^2-\alpha^2|}\omega_1\left(F,\sqrt{\xi}\right)_1\left[\sum_{j=1}^{r}|\alpha_j|\,j^2\left(\frac{7}{3}+j\frac{25}{8}\right)\right]$$

$$= \frac{4\xi}{\sqrt{\pi}\,|\beta^2-\alpha^2|}\omega_1\left(f^{(4)}+\left(\alpha^2+\beta^2\right)f''+\alpha^2\beta^2 f,\sqrt{\xi}\right)_1$$

$$\times\left[\sum_{j=1}^{r}|\alpha_j|\,j^2\left(\frac{7}{3}+j\frac{25}{8}\right)\right]. \tag{29.44}$$

The claim is proved. ☐

We give

Corollary 29.2. *(to Theorem 29.2) It holds*

$$\|W_{r,\xi}(f)-f\|_1 \leq \Psi_1(\xi)$$

$$+\frac{\|f''\|_1}{|\beta^2-\alpha^2|}\left[\sum_{j=0}^{r}|\alpha_j|\left(e^{-\frac{\alpha^2j^2\xi}{16}}-e^{-\frac{\beta^2j^2\xi}{16}}\right)\right]$$

$$+\left(\frac{\left\|f^{(4)}\right\|_1+\left(\alpha^2+\beta^2\right)\left\|f''\right\|_1}{(\alpha\beta)^2\left|\beta^2-\alpha^2\right|}\right)$$

$$\left|\beta^2\sum_{j=0}^{r}\alpha_j\left(1-e^{-\frac{\alpha^2 j^2\xi}{4}}\right)-\alpha^2\sum_{j=0}^{r}\alpha_j\left(1-e^{-\frac{\beta^2 j^2\xi}{4}}\right)\right|\to 0, \qquad (29.45)$$

as $\xi\to 0$.

Above $\Psi_1(\xi)$ *is as in (29.40).*

Proof. Directly from (29.40). □

29.3 Part II: On the Smooth Poisson–Cauchy Singular Integral Operators (Anastassiou and Mezei (2012))

Let $\overline{\alpha}\in\mathbb{N}$, $\overline{\beta}>\frac{1}{2\overline{\alpha}}$ and $f\in C^2(\mathbb{R})$. We define for $x\in\mathbb{R}$, $\xi>0$ the Lebesgue integral

$$M_{r,\xi}(f;x)=W\int_{-\infty}^{\infty}\frac{\sum_{j=0}^{r}\alpha_j f(x+jt)}{\left(t^{2\overline{\alpha}}+\xi^{2\overline{\alpha}}\right)^{\overline{\beta}}}dt, \qquad (29.46)$$

where the constant is defined as

$$W=\frac{\Gamma\left(\overline{\beta}\right)\overline{\alpha}\xi^{2\overline{\alpha}\overline{\beta}-1}}{\Gamma\left(\frac{1}{2\overline{\alpha}}\right)\Gamma\left(\overline{\beta}-\frac{1}{2\overline{\alpha}}\right)}. \qquad (29.47)$$

We assume that $M_{r,\xi}(f;x)\in\mathbb{R}$ for all $x\in\mathbb{R}$. We will use also that

$$M_{r,\xi}(f;x)=W\sum_{j=0}^{r}\alpha_j\left(\int_{-\infty}^{\infty}f(x+jt)\frac{1}{\left(t^{2\overline{\alpha}}+\xi^{2\overline{\alpha}}\right)^{\overline{\beta}}}dt\right). \qquad (29.48)$$

We note by $W\int_{-\infty}^{\infty}\frac{1}{\left(t^{2\overline{\alpha}}+\xi^{2\overline{\alpha}}\right)^{\overline{\beta}}}dt=1$ that $M_{r,\xi}(c;x)=c$, c constant, and

$$M_{r,\xi}(f;x)-f(x)=W\left(\sum_{j=0}^{r}\alpha_j\int_{-\infty}^{\infty}\left[f(x+jt)-f(x)\right]\frac{1}{\left(t^{2\overline{\alpha}}+\xi^{2\overline{\alpha}}\right)^{\overline{\beta}}}dt\right). \qquad (29.49)$$

We set

$$E_2(x) := M_{r,\xi}(f,x) - f(x) - \left(\frac{f'(x)}{\alpha\beta(\beta^2-\alpha^2)}\right)$$

$$\times \left[\beta^3 \sum_{j=0}^{r} \alpha_j W \left(\int_{-\infty}^{\infty} \sin(\alpha j t) \frac{dt}{\left(t^{2\overline{\alpha}} + \xi^{2\overline{\alpha}}\right)^{\overline{\beta}}}\right)\right.$$

$$\left. - \alpha^3 \sum_{j=0}^{r} \alpha_j W \left(\int_{-\infty}^{\infty} \sin(\beta j t) \frac{dt}{\left(t^{2\overline{\alpha}} + \xi^{2\overline{\alpha}}\right)^{\overline{\beta}}}\right)\right] - \left(\frac{f''(x)}{\beta^2-\alpha^2}\right) \tag{29.50}$$

$$\times \left[\sum_{j=0}^{r} \alpha_j W \left(\int_{-\infty}^{\infty} \cos(\alpha j t) \frac{dt}{\left(t^{2\overline{\alpha}} + \xi^{2\overline{\alpha}}\right)^{\overline{\beta}}}\right)\right.$$

$$\left. - \sum_{j=0}^{r} \alpha_j W \left(\int_{-\infty}^{\infty} \cos(\beta j t) \frac{dt}{\left(t^{2\overline{\alpha}} + \xi^{2\overline{\alpha}}\right)^{\overline{\beta}}}\right)\right] - \left(\frac{f^{(3)}(x)}{\alpha\beta(\beta^2-\alpha^2)}\right)$$

$$\times \left[\beta \sum_{j=0}^{r} \alpha_j W \left(\int_{-\infty}^{\infty} \sin(\alpha j t) \frac{dt}{\left(t^{2\overline{\alpha}} + \xi^{2\overline{\alpha}}\right)^{\overline{\beta}}}\right)\right.$$

$$\left. - \alpha \sum_{j=0}^{r} \alpha_j W \left(\int_{-\infty}^{\infty} \sin(\beta j t) \frac{dt}{\left(t^{2\overline{\alpha}} + \xi^{2\overline{\alpha}}\right)^{\overline{\beta}}}\right)\right]$$

$$- \left(\frac{2\left(f^{(4)}(x) + \left(\alpha^2+\beta^2\right) f''(x)\right)}{(\alpha\beta)^2 (\beta^2-\alpha^2)}\right)$$

$$\times \left[\beta^2 \sum_{j=0}^{r} \alpha_j W \left(\int_{-\infty}^{\infty} \sin^2\left(\frac{\alpha j t}{2}\right) \frac{dt}{\left(t^{2\overline{\alpha}} + \xi^{2\overline{\alpha}}\right)^{\overline{\beta}}}\right)\right.$$

$$\left. - \alpha^2 \sum_{j=0}^{r} \alpha_j W \left(\int_{-\infty}^{\infty} \sin^2\left(\frac{\beta j t}{2}\right) \frac{dt}{\left(t^{2\overline{\alpha}} + \xi^{2\overline{\alpha}}\right)^{\overline{\beta}}}\right)\right]$$

$$= M_{r,\xi}(f,x) - f(x) - \left(\frac{2f''(x)}{\beta^2-\alpha^2}\right) W$$

$$\times\left[\sum_{j=0}^{r}\alpha_j\left(\int_0^{\infty}\left(\cos\left(\alpha jt\right)-\cos\left(\beta jt\right)\right)\frac{dt}{\left(t^{2\overline{\alpha}}+\xi^{2\overline{\alpha}}\right)^{\overline{\beta}}}\right)\right]$$

$$-\left(\frac{4\left(f^{(4)}\left(x\right)+\left(\alpha^2+\beta^2\right)f''\left(x\right)\right)}{\left(\alpha\beta\right)^2\left(\beta^2-\alpha^2\right)}\right)W \tag{29.51}$$

$$\times\left[\beta^2\sum_{j=0}^{r}\alpha_j\left(\int_0^{\infty}\sin^2\left(\frac{\alpha jt}{2}\right)\frac{dt}{\left(t^{2\overline{\alpha}}+\xi^{2\overline{\alpha}}\right)^{\overline{\beta}}}\right)\right.$$

$$\left.-\alpha^2\sum_{j=0}^{r}\alpha_j\left(\int_0^{\infty}\sin^2\left(\frac{\beta jt}{2}\right)\frac{dt}{\left(t^{2\overline{\alpha}}+\xi^{2\overline{\alpha}}\right)^{\overline{\beta}}}\right)\right]$$

$$=W\int_{-\infty}^{\infty}R\left(t\right)\frac{dt}{\left(t^{2\overline{\alpha}}+\xi^{2\overline{\alpha}}\right)^{\overline{\beta}}}. \tag{29.52}$$

It follows a parametrized L_p ($p>1$) approximation result for $M_{r,\xi}$, $\xi>0$.

Theorem 29.3. *Let $\alpha,\beta\in\mathbb{R}$ with $\alpha\beta\left(\alpha^2-\beta^2\right)\neq0$, and let all $f,f',f'',f^{(3)},f^{(4)}\in L_p\left(\mathbb{R}\right)\cap C\left(\mathbb{R}\right)$, $1<p<\infty$. Here $q>1:\frac{1}{p}+\frac{1}{q}=1$; $\xi>0$, $x\in\mathbb{R}$. Assume $\overline{\alpha}\in\mathbb{N}$, $\overline{\beta}>\frac{3p+1}{2\overline{\alpha}}$. Then*

$$\left\|E_2\left(x\right)\right\|_p=\left\|M_{r,\xi}\left(f,x\right)-f\left(x\right)\right.$$

$$-\left(\frac{2f''\left(x\right)}{\beta^2-\alpha^2}\right)W\left[\sum_{j=0}^{r}\alpha_j\left(\int_0^{\infty}\left(\cos\left(\alpha jt\right)-\cos\left(\beta jt\right)\right)\frac{dt}{\left(t^{2\overline{\alpha}}+\xi^{2\overline{\alpha}}\right)^{\overline{\beta}}}\right)\right]$$

$$-\left(\frac{4\left(f^{(4)}\left(x\right)+\left(\alpha^2+\beta^2\right)f''\left(x\right)\right)}{\left(\alpha\beta\right)^2\left(\beta^2-\alpha^2\right)}\right)W$$

$$\times\left[\beta^2\sum_{j=0}^{r}\alpha_j\left(\int_0^{\infty}\sin^2\left(\frac{\alpha jt}{2}\right)\frac{dt}{\left(t^{2\overline{\alpha}}+\xi^{2\overline{\alpha}}\right)^{\overline{\beta}}}\right)\right.$$

$$\left.\left.-\alpha^2\sum_{j=0}^{r}\alpha_j\left(\int_0^{\infty}\sin^2\left(\frac{\beta jt}{2}\right)\frac{dt}{\left(t^{2\overline{\alpha}}+\xi^{2\overline{\alpha}}\right)^{\overline{\beta}}}\right)\right]\right\|_p$$

$$\leq\left[\frac{4\Gamma^{\frac{1}{p}}\left(\frac{3p+1}{2\overline{\alpha}}\right)\Gamma^{\frac{1}{p}}\left(\overline{\beta}-\left(\frac{3p+1}{2\overline{\alpha}}\right)\right)}{\left(p+1\right)^{\frac{1}{p}}\left(q+1\right)^{\frac{1}{q}}\left|\beta^2-\alpha^2\right|\Gamma^{\frac{1}{p}}\left(\frac{1}{2\overline{\alpha}}\right)\Gamma^{\frac{1}{p}}\left(\overline{\beta}-\frac{1}{2\overline{\alpha}}\right)}\right]$$

$$\left(\sum_{j=1}^{r}|\alpha_j|\,j^2\right)^{\frac{1}{q}}\left(\sum_{j=1}^{r}|\alpha_j|\,j^{p+2}\right)^{\frac{1}{p}}$$

$$\omega_1\left(f^{(4)}+\left(\alpha^2+\beta^2\right)f''+\alpha^2\beta^2 f,\xi\right)_p\cdot\xi^2=:\Phi_2(\xi)\to 0, \tag{29.53}$$

as $\xi\to 0$.

Proof. By (29.52), we have ($\overline{\alpha}\in\mathbb{N}$, $\overline{\beta}>\frac{3p+1}{2\overline{\alpha}}$)

$$E_2(x)=W\int_{-\infty}^{\infty}R(t)\frac{1}{\left(t^{2\overline{\alpha}}+\xi^{2\overline{\alpha}}\right)^{\overline{\beta}}}dt.$$

Hence it holds ($p,q>1:\frac{1}{p}+\frac{1}{q}=1$),

$$\int_{-\infty}^{\infty}|E_2(x)|^p\,dx=\int_{-\infty}^{\infty}\left(W\left|\int_{-\infty}^{\infty}R(t)\frac{1}{\left(t^{2\overline{\alpha}}+\xi^{2\overline{\alpha}}\right)^{\overline{\beta}}}dt\right|\right)^p dx$$

$$\leq\int_{-\infty}^{\infty}\left(W\int_{-\infty}^{\infty}|R(t)|\frac{1}{\left(t^{2\overline{\alpha}}+\xi^{2\overline{\alpha}}\right)^{\overline{\beta}}}dt\right)^p dx$$

$$\leq W\int_{-\infty}^{\infty}\left(\int_{-\infty}^{\infty}|R(t)|^p\frac{1}{\left(t^{2\overline{\alpha}}+\xi^{2\overline{\alpha}}\right)^{\overline{\beta}}}dt\right)dx\overset{(29.30)}{\leq}\frac{2^pW}{|\beta^2-\alpha^2|^p}$$

$$\int_{-\infty}^{\infty}\left(\int_{-\infty}^{\infty}\left[\sum_{j=1}^{r}|\alpha_j|\,j^2\left(\int_0^{|t|}|F(x+j\,sign(t)\,\theta)-F(x)|\,(|t|-\theta)\,d\theta\right)\right]^p\right.$$

$$\left.\frac{1}{\left(t^{2\overline{\alpha}}+\xi^{2\overline{\alpha}}\right)^{\overline{\beta}}}dt\right)dx \tag{29.54}$$

(by Jensen inequality)

$$\leq\frac{2^pW}{|\beta^2-\alpha^2|^p}\int_{-\infty}^{\infty}\left(\int_{-\infty}^{\infty}\left(\sum_{j=1}^{r}|\alpha_j|\,j^2\right)^{p-1}\right.$$

$$\left[\sum_{j=1}^{r}|\alpha_j|\,j^2\left(\int_0^{|t|}|F(x+j\,sign(t)\,\theta)-F(x)|\,(|t|-\theta)\,d\theta\right)^p\right]$$

$$\left.\frac{1}{\left(t^{2\overline{\alpha}}+\xi^{2\overline{\alpha}}\right)^{\overline{\beta}}}dt\right)dx$$

$$\leq W\left(\sum_{j=1}^{r}|\alpha_j|\,j^2\right)^{p-1}\frac{2^p}{|\beta^2-\alpha^2|^p}\left[\sum_{j=1}^{r}|\alpha_j|\,j^2\int_{-\infty}^{\infty}\left(\int_{-\infty}^{\infty}\right.\right.$$

$$\left(\int_0^{|t|} |F(x + jsign(t)\theta) - F(x)|^p d\theta\right) \left(\int_0^{|t|} (|t| - \theta)^q d\theta\right)^{\frac{p}{q}}$$

$$\left.\left.\frac{1}{(t^{2\overline{\alpha}} + \xi^{2\overline{\alpha}})^{\overline{\beta}}} dt\right) dx\right] \tag{29.55}$$

$$= W\left(\sum_{j=1}^{r} |\alpha_j| j^2\right)^{p-1} \frac{2^p}{|\beta^2 - \alpha^2|^p (q+1)^{p-1}} \left[\sum_{j=1}^{r} |\alpha_j| j^2 \int_{-\infty}^{\infty} \left(\int_{-\infty}^{\infty}\right.\right.$$

$$\left.\left.\left(\int_0^{|t|} |F(x + jsign(t)\theta) - F(x)|^p d\theta\right) |t|^{(q+1)(p-1)} \frac{1}{(t^{2\overline{\alpha}} + \xi^{2\overline{\alpha}})^{\overline{\beta}}} dt\right) dx\right] =$$

(call $\rho := W\left(\sum_{j=1}^{r} |\alpha_j| j^2\right)^{p-1} \frac{2^p}{|\beta^2 - \alpha^2|^p (q+1)^{p-1}}$)

$$\rho\left[\sum_{j=1}^{r} |\alpha_j| j^2 \int_{-\infty}^{\infty} \left(\int_0^{|t|} \left(\int_{-\infty}^{\infty}\right.\right.\right.$$

$$\left.\left.\left.|F(x + jsign(t)\theta) - F(x)|^p dx\right) d\theta\right) |t|^{(q+1)(p-1)} \frac{1}{(t^{2\overline{\alpha}} + \xi^{2\overline{\alpha}})^{\overline{\beta}}} dt\right]$$

$$\leq \rho\left[\sum_{j=1}^{r} |\alpha_j| j^2 \int_{-\infty}^{\infty} \left(\int_0^{|t|} \omega_1\left(F, \xi\frac{j\theta}{\xi}\right)_p^p d\theta\right) |t|^{(q+1)(p-1)} \frac{1}{(t^{2\overline{\alpha}} + \xi^{2\overline{\alpha}})^{\overline{\beta}}} dt\right] \tag{29.56}$$

$$\leq \rho\omega_1(F, \xi)_p^p \left[\sum_{j=1}^{r} |\alpha_j| j^2 \int_{-\infty}^{\infty} \left(\int_0^{|t|} \left(1 + \frac{j\theta}{\xi}\right)^p d\theta\right)\right.$$

$$\left.\times |t|^{(q+1)(p-1)} \frac{1}{(t^{2\overline{\alpha}} + \xi^{2\overline{\alpha}})^{\overline{\beta}}} dt\right]$$

$$= \rho\omega_1(F, \xi)_p^p \frac{\xi}{(p+1)}$$

$$\left[\sum_{j=1}^{r} |\alpha_j| j \int_{-\infty}^{\infty} \left[\left(1 + \frac{j}{\xi}|t|\right)^{p+1} - 1\right] |t|^{(q+1)(p-1)} \frac{1}{(t^{2\overline{\alpha}} + \xi^{2\overline{\alpha}})^{\overline{\beta}}} dt\right]$$

$$\leq \frac{2^p \rho\xi\omega_1(F, \xi)_p^p}{(p+1)} \left[\sum_{j=1}^{r} |\alpha_j| j \frac{j^{p+1}}{\xi^{p+1}} \int_{-\infty}^{\infty} |t|^{p+1} |t|^{(q+1)(p-1)} \frac{1}{(t^{2\overline{\alpha}} + \xi^{2\overline{\alpha}})^{\overline{\beta}}} dt\right] \tag{29.57}$$

$$= \frac{2^{p+1} \rho\xi^{-p}\omega_1(F, \xi)_p^p}{(p+1)} \left[\sum_{j=1}^{r} |\alpha_j| j^{p+2} \int_0^{\infty} t^{(q+1)(p-1)+(p+1)} \frac{1}{(t^{2\overline{\alpha}} + \xi^{2\overline{\alpha}})^{\overline{\beta}}} dt\right]$$

$$= \frac{2^{p+1}\rho\xi^{-p}\omega_1(F,\xi)_p^p}{(p+1)}$$

$$\times\left[\sum_{j=1}^{r}|\alpha_j|\,j^{p+2}\xi^{3p-2\overline{\alpha}\overline{\beta}+1}\int_0^\infty\left(\frac{t}{\xi}\right)^{3p}\frac{1}{\left(\left(\frac{t}{\xi}\right)^{2\overline{\alpha}}+1\right)^{\overline{\beta}}}d\left(\frac{t}{\xi}\right)\right]$$

$$= \frac{2^{p+1}\rho\xi^{2p-2\overline{\alpha}\overline{\beta}+1}\omega_1(F,\xi)_p^p}{(p+1)}\left[\sum_{j=1}^{r}|\alpha_j|\,j^{p+2}\right]\left[\int_0^\infty x^{3p}\frac{1}{(x^{2\overline{\alpha}}+1)^{\overline{\beta}}}dx\right]= \tag{29.58}$$

(by Zwillinger (1995), p. 397 formula 595)

$$\frac{2^{p+1}\rho\xi^{2p-2\overline{\alpha}\overline{\beta}+1}\omega_1(F,\xi)_p^p}{(p+1)}\left[\sum_{j=1}^{r}|\alpha_j|\,j^{p+2}\right]\left[\frac{\Gamma\left(\frac{3p+2}{2\overline{\alpha}}\right)\Gamma\left(\overline{\beta}-\left(\frac{3p+1}{2\overline{\alpha}}\right)\right)}{2\overline{\alpha}\Gamma\left(\overline{\beta}\right)}\right]$$

$$= \frac{2^{p+1}\xi^{2p-2\overline{\alpha}\overline{\beta}+1}}{(p+1)}\left(W\left(\sum_{j=1}^{r}|\alpha_j|\,j^2\right)^{p-1}\frac{2^p}{|\beta^2-\alpha^2|^p(q+1)^{p-1}}\right)\omega_1(F,\xi)_p^p$$

$$\left[\sum_{j=1}^{r}|\alpha_j|\,j^{p+2}\right]\left[\frac{\Gamma\left(\frac{3p+2}{2\overline{\alpha}}\right)\Gamma\left(\overline{\beta}-\left(\frac{3p+1}{2\overline{\alpha}}\right)\right)}{2\overline{\alpha}\Gamma\left(\overline{\beta}\right)}\right]$$

$$\overset{(29.47)}{=}\frac{2^{p+1}\xi^{2p-2\overline{\alpha}\overline{\beta}+1}}{(q+1)^{p-1}(p+1)|\beta^2-\alpha^2|^p}\left(\frac{\Gamma\left(\overline{\beta}\right)\overline{\alpha}\xi^{2\overline{\alpha}\overline{\beta}-1}}{\Gamma\left(\frac{1}{2\overline{\alpha}}\right)\Gamma\left(\overline{\beta}-\frac{1}{2\overline{\alpha}}\right)}\right)\left(\sum_{j=1}^{r}|\alpha_j|\,j^2\right)^{p-1} \tag{29.59}$$

$$\left[\sum_{j=1}^{r}|\alpha_j|\,j^{p+2}\right]\left(\frac{\Gamma\left(\frac{3p+2}{2\overline{\alpha}}\right)\Gamma\left(\overline{\beta}-\left(\frac{3p+1}{2\overline{\alpha}}\right)\right)}{2\overline{\alpha}\Gamma\left(\overline{\beta}\right)}\right)\omega_1(F,\xi)_p^p$$

$$= \frac{4^p\xi^{2p}}{(q+1)^{p-1}(p+1)|\beta^2-\alpha^2|^p\,\Gamma\left(\frac{1}{2\overline{\alpha}}\right)\Gamma\left(\overline{\beta}-\frac{1}{2\overline{\alpha}}\right)}\left(\sum_{j=1}^{r}|\alpha_j|\,j^2\right)^{p-1}$$

$$\left(\sum_{j=1}^{r}|\alpha_j|\,j^{p+2}\right)\Gamma\left(\frac{3p+1}{2\overline{\alpha}}\right)\Gamma\left(\overline{\beta}-\left(\frac{3p+1}{2\overline{\alpha}}\right)\right)\omega_1(F,\xi)_p^p$$

$$= \left[\frac{4^p\Gamma\left(\frac{3p+1}{2\overline{\alpha}}\right)\Gamma\left(\overline{\beta}-\left(\frac{3p+1}{2\overline{\alpha}}\right)\right)}{(p+1)(q+1)^{(p-1)}|\beta^2-\alpha^2|^p\,\Gamma\left(\frac{1}{2\overline{\alpha}}\right)\Gamma\left(\overline{\beta}-\frac{1}{2\overline{\alpha}}\right)}\right]\left(\sum_{j=1}^{r}|\alpha_j|\,j^2\right)^{p-1}$$

$$\left(\sum_{j=1}^{r}|\alpha_j|\,j^{p+2}\right)\omega_1\left(f^{(4)}+\left(\alpha^2+\beta^2\right)f''+\alpha^2\beta^2 f,\xi\right)_p^p\cdot\xi^{2p}, \tag{29.60}$$

the claim is proved. □

Next we give a consequence of Theorem 29.3.

Corollary 29.3. *(to Theorem 29.3) It holds*

$$\|M_{r,\xi}(f) - f\|_p \leq \Phi_2(\xi)$$

$$+ \|f''\|_p \left(\frac{2W}{|\beta^2 - \alpha^2|}\right) \left[\sum_{j=0}^{r} |\alpha_j| \int_0^\infty |\cos(\alpha j t) - \cos(\beta j t)| \frac{dt}{\left(t^{2\overline{\alpha}} + \xi^{2\overline{\alpha}}\right)^{\overline{\beta}}}\right]$$

$$+ \left(\frac{4\left\|f^{(4)}\right\|_p + \left(\alpha^2 + \beta^2\right)\|f''\|_p}{(\alpha\beta)^2 |\beta^2 - \alpha^2|}\right) W$$

$$\left[\beta^2 \sum_{j=0}^{r} |\alpha_j| \left(\int_0^\infty \sin^2\left(\frac{\alpha j t}{2}\right) \frac{dt}{\left(t^{2\overline{\alpha}} + \xi^{2\overline{\alpha}}\right)^{\overline{\beta}}}\right)\right.$$

$$\left. + \alpha^2 \sum_{j=0}^{r} |\alpha_j| \left(\int_0^\infty \sin^2\left(\frac{\beta j t}{2}\right) \frac{dt}{\left(t^{2\overline{\alpha}} + \xi^{2\overline{\alpha}}\right)^{\overline{\beta}}}\right)\right] \to 0, \tag{29.61}$$

as $\xi \to 0$.

Above $\Phi_2(\xi)$ *is as in (29.53).*

Proof. We have

$$W \int_0^\infty |\cos(\alpha j t) - \cos(\beta j t)| \frac{dt}{\left(t^{2\overline{\alpha}} + \xi^{2\overline{\alpha}}\right)^{\overline{\beta}}}$$

$$= 2W \int_0^\infty \left|\sin\left(\frac{(\alpha+\beta) j t}{2}\right)\right| \left|\sin\left(\frac{(\alpha-\beta) j t}{2}\right)\right| \frac{dt}{\left(t^{2\overline{\alpha}} + \xi^{2\overline{\alpha}}\right)^{\overline{\beta}}}$$

$$\leq 2W \int_0^\infty \frac{|\alpha+\beta| j t}{2} \frac{|\alpha-\beta| j t}{2} \frac{dt}{\left(t^{2\overline{\alpha}} + \xi^{2\overline{\alpha}}\right)^{\overline{\beta}}}$$

$$= \frac{W |\alpha^2 - \beta^2| j^2}{2} \int_0^\infty \frac{t^2 dt}{\left(t^{2\overline{\alpha}} + \xi^{2\overline{\alpha}}\right)^{\overline{\beta}}}$$

$$= \frac{W |\alpha^2 - \beta^2| j^2}{2} \xi^{3-2\overline{\alpha}\overline{\beta}} \int_0^\infty \frac{x^2 dx}{\left(1 + x^{2\overline{\alpha}}\right)^{\overline{\beta}}} = \tag{29.62}$$

(by Zwillinger (1995), p. 397, formula 595)

$$\frac{\left|\alpha^2-\beta^2\right| j^2}{2}\left(\frac{\Gamma\left(\overline{\beta}\right)\overline{\alpha}\xi^{2\overline{\alpha}\overline{\beta}-1}}{\Gamma\left(\frac{1}{2\overline{\alpha}}\right)\Gamma\left(\overline{\beta}-\frac{1}{2\overline{\alpha}}\right)}\right)\xi^{3-2\overline{\alpha}\overline{\beta}}\frac{\Gamma\left(\frac{3}{2\overline{\alpha}}\right)\Gamma\left(\overline{\beta}-\frac{3}{2\overline{\alpha}}\right)}{2\overline{\alpha}\Gamma\left(\overline{\beta}\right)}$$

$$=\frac{\left|\beta^2-\alpha^2\right| j^2}{4}\frac{\Gamma\left(\frac{3}{2\overline{\alpha}}\right)\Gamma\left(\overline{\beta}-\frac{3}{2\overline{\alpha}}\right)}{\Gamma\left(\frac{1}{2\overline{\alpha}}\right)\Gamma\left(\overline{\beta}-\frac{1}{2\overline{\alpha}}\right)}\xi^2\to 0,\quad \text{as } \xi\to 0. \tag{29.63}$$

Furthermore, we have that

$$W\int_0^{\infty}\sin^2\left(\frac{\alpha jt}{2}\right)\frac{dt}{\left(t^{2\overline{\alpha}}+\xi^{2\overline{\alpha}}\right)^{\overline{\beta}}}\le W\frac{\alpha^2 j^2}{4}\int_0^{\infty}t^2\frac{dt}{\left(t^{2\overline{\alpha}}+\xi^{2\overline{\alpha}}\right)^{\overline{\beta}}} \tag{29.64}$$

$$=\left(\frac{\Gamma\left(\overline{\beta}\right)\overline{\alpha}\xi^{2\overline{\alpha}\overline{\beta}-1}}{\Gamma\left(\frac{1}{2\overline{\alpha}}\right)\Gamma\left(\overline{\beta}-\frac{1}{2\overline{\alpha}}\right)}\right)\frac{\alpha^2 j^2}{4}\xi^{3-2\overline{\alpha}\overline{\beta}}\left(\int_0^{\infty}\frac{x^2dx}{\left(1+x^{2\overline{\alpha}}\right)^{\overline{\beta}}}\right)$$

$$=\left(\frac{\Gamma\left(\overline{\beta}\right)\overline{\alpha}\xi^{2}}{\Gamma\left(\frac{1}{2\overline{\alpha}}\right)\Gamma\left(\overline{\beta}-\frac{1}{2\overline{\alpha}}\right)}\right)\frac{\alpha^2 j^2}{4}\frac{\Gamma\left(\frac{3}{2\overline{\alpha}}\right)\Gamma\left(\overline{\beta}-\frac{3}{2\overline{\alpha}}\right)}{2\overline{\alpha}\Gamma\left(\overline{\beta}\right)}$$

$$=\frac{\alpha^2 j^2}{8}\left(\frac{\Gamma\left(\frac{3}{2\overline{\alpha}}\right)\Gamma\left(\overline{\beta}-\frac{3}{2\overline{\alpha}}\right)}{\Gamma\left(\frac{1}{2\overline{\alpha}}\right)\Gamma\left(\overline{\beta}-\frac{1}{2\overline{\alpha}}\right)}\right)\xi^2\to 0,\quad \text{as } \xi\to 0. \tag{29.65}$$

The proof now is clear. □

Next comes the L_1 approximation by $M_{r,\xi}$.

Theorem 29.4. *Let $\alpha,\beta\in\mathbb{R}$ with $\alpha\beta\left(\alpha^2-\beta^2\right)\neq 0$, and let all $f,f',f'',f^{(3)},f^{(4)}\in L_1\left(\mathbb{R}\right)\cap C\left(\mathbb{R}\right)$; $\xi>0$, $x\in\mathbb{R}$. Here $\overline{\alpha}\in\mathbb{N}$, $\overline{\beta}>\frac{2}{\overline{\alpha}}$. Then*

$$\left\|E_2\left(x\right)\right\|_1=\| M_{r,\xi}\left(f,x\right)-f\left(x\right)$$

$$-\left(\frac{2f''\left(x\right)}{\beta^2-\alpha^2}\right)W\left[\sum_{j=0}^{r}\alpha_j\left(\int_0^{\infty}\left(\cos\left(\alpha jt\right)-\cos\left(\beta jt\right)\right)\frac{dt}{\left(t^{2\overline{\alpha}}+\xi^{2\overline{\alpha}}\right)^{\overline{\beta}}}\right)\right]$$

$$-\left(\frac{4\left(f^{(4)}\left(x\right)+\left(\alpha^2+\beta^2\right)f''\left(x\right)\right)}{\left(\alpha\beta\right)^2\left(\beta^2-\alpha^2\right)}\right)W$$

$$\left[\beta^2\sum_{j=0}^{r}\alpha_j\left(\int_0^{\infty}\sin^2\left(\frac{\alpha jt}{2}\right)\frac{dt}{\left(t^{2\overline{\alpha}}+\xi^{2\overline{\alpha}}\right)^{\overline{\beta}}}\right)\right.$$

$$-\alpha^2 \sum_{j=0}^{r} \alpha_j \left(\int_0^{\infty} \sin^2\left(\frac{\beta j t}{2}\right) \frac{dt}{\left(t^{2\overline{\alpha}} + \xi^{2\overline{\alpha}}\right)^{\overline{\beta}}} \right) \Bigg] \Bigg\|_1$$

$$\leq \left(\frac{2}{\left|\beta^2 - \alpha^2\right| \Gamma\left(\frac{1}{2\overline{\alpha}}\right) \Gamma\left(\overline{\beta} - \frac{1}{2\overline{\alpha}}\right)} \right)$$

$$\left[\sum_{j=1}^{r} |\alpha_j| j^2 \left[\Gamma\left(\frac{3}{2\overline{\alpha}}\right) \Gamma\left(\overline{\beta} - \frac{3}{2\overline{\alpha}}\right) + \frac{j}{2} \Gamma\left(\frac{2}{\overline{\alpha}}\right) \Gamma\left(\overline{\beta} - \frac{2}{\overline{\alpha}}\right) \right] \right]$$

$$\omega_1 \left(f^{(4)} + \left(\alpha^2 + \beta^2\right) f'' + \alpha^2 \beta^2 f, \xi \right)_1 \cdot \xi^2 =: \Psi_2(\xi) \to 0, \tag{29.66}$$

as $\xi \to 0$.

Proof. We have that (F is as in (29.25))

$$\int_{-\infty}^{\infty} |E_2(x)| \, dx \overset{(29.52)}{=} W \int_{-\infty}^{\infty} \left| \int_{-\infty}^{\infty} R(t) \frac{dt}{\left(t^{2\overline{\alpha}} + \xi^{2\overline{\alpha}}\right)^{\overline{\beta}}} \right| dx$$

$$\leq W \int_{-\infty}^{\infty} \left(\int_{-\infty}^{\infty} |R(t)| \frac{dt}{\left(t^{2\overline{\alpha}} + \xi^{2\overline{\alpha}}\right)^{\overline{\beta}}} \right) dx \overset{(29.30)}{\leq} \frac{2W}{\left|\beta^2 - \alpha^2\right|}$$

$$\int_{-\infty}^{\infty} \left(\int_{-\infty}^{\infty} \left[\sum_{j=1}^{r} |\alpha_j| j^2 \left(\int_0^{|t|} |F(x + j \, sign(t) \theta) - F(x)| \left(|t| - \theta\right) d\theta \right) \right] \right.$$

$$\left. \frac{dt}{\left(t^{2\overline{\alpha}} + \xi^{2\overline{\alpha}}\right)^{\overline{\beta}}} \right) dx \tag{29.67}$$

$$\leq \frac{2W}{\left|\beta^2 - \alpha^2\right|} \int_{-\infty}^{\infty} \left(\int_{-\infty}^{\infty} \left[\sum_{j=1}^{r} |\alpha_j| j^2 \right.\right.$$

$$\left.\left. \left(\int_0^{|t|} |F(x + j \, sign(t) \theta) - F(x)| \, d\theta \right) |t| \frac{dt}{\left(t^{2\overline{\alpha}} + \xi^{2\overline{\alpha}}\right)^{\overline{\beta}}} \right) dx \right]$$

$$= \frac{2W}{\left|\beta^2 - \alpha^2\right|} \left[\sum_{j=1}^{r} |\alpha_j| j^2 \int_{-\infty}^{\infty} \left(\int_0^{|t|} \left(\int_{-\infty}^{\infty} \right.\right.\right.$$

$$\left.\left.\left. |F(x + j \, sign(t) \theta) - F(x)| \, dx \right) d\theta \right) \frac{|t| \, dt}{\left(t^{2\overline{\alpha}} + \xi^{2\overline{\alpha}}\right)^{\overline{\beta}}} \right]$$

$$\leq \frac{2W}{|\beta^2-\alpha^2|}\left[\sum_{j=1}^{r}|\alpha_j|\,j^2\int_{-\infty}^{\infty}\left(\int_0^{|t|}\omega_1\left(F,\xi\frac{j\theta}{\xi}\right)_1 d\theta\right)\frac{|t|\,dt}{\left(t^{2\overline{\alpha}}+\xi^{2\overline{\alpha}}\right)^{\overline{\beta}}}\right]$$

$$\leq \frac{2W}{|\beta^2-\alpha^2|}\omega_1(F,\xi)_1\left[\sum_{j=1}^{r}|\alpha_j|\,j^2\int_{-\infty}^{\infty}\left(\int_0^{|t|}\left(1+\frac{j\theta}{\xi}\right)d\theta\right)\frac{|t|\,dt}{\left(t^{2\overline{\alpha}}+\xi^{2\overline{\alpha}}\right)^{\overline{\beta}}}\right]$$

$$= \frac{2W}{|\beta^2-\alpha^2|}\omega_1(F,\xi)_1\left[\sum_{j=1}^{r}|\alpha_j|\,j^2\int_{-\infty}^{\infty}\left(|t|+\frac{j}{2\xi}t^2\right)\frac{|t|\,dt}{\left(t^{2\overline{\alpha}}+\xi^{2\overline{\alpha}}\right)^{\overline{\beta}}}\right] \quad (29.68)$$

$$= \frac{4W}{|\beta^2-\alpha^2|}\omega_1(F,\xi)_1\left[\sum_{j=1}^{r}|\alpha_j|\,j^2\int_0^{\infty}\left(t+\frac{j}{2\xi}t^2\right)\frac{tdt}{\left(t^{2\overline{\alpha}}+\xi^{2\overline{\alpha}}\right)^{\overline{\beta}}}\right]$$

$$= \frac{4W}{|\beta^2-\alpha^2|}\omega_1(F,\xi)_1\left[\sum_{j=1}^{r}|\alpha_j|\,j^2\left[\int_0^{\infty}\frac{t^2dt}{\left(t^{2\overline{\alpha}}+\xi^{2\overline{\alpha}}\right)^{\overline{\beta}}}+\frac{j}{2\xi}\int_0^{\infty}\frac{t^3dt}{\left(t^{2\overline{\alpha}}+\xi^{2\overline{\alpha}}\right)^{\overline{\beta}}}\right]\right]$$

$$= \frac{4W}{|\beta^2-\alpha^2|}\omega_1(F,\xi)_1\left[\sum_{j=1}^{r}|\alpha_j|\,j^2\left[\xi^{3-2\overline{\alpha}\overline{\beta}}\int_0^{\infty}\frac{x^2dx}{\left(1+x^{2\overline{\alpha}}\right)^{\overline{\beta}}}\right.\right.$$

$$\left.\left.+\frac{j}{2\xi}\xi^{4-2\overline{\alpha}\overline{\beta}}\int_0^{\infty}\frac{x^3dx}{\left(1+x^{2\overline{\alpha}}\right)^{\overline{\beta}}}\right]\right] \quad (29.69)$$

$$= \frac{4W\xi^{3-2\overline{\alpha}\overline{\beta}}}{|\beta^2-\alpha^2|}\omega_1(F,\xi)_1\left[\sum_{j=1}^{r}|\alpha_j|\,j^2\left[\int_0^{\infty}\frac{x^2dx}{\left(1+x^{2\overline{\alpha}}\right)^{\overline{\beta}}}\right.\right.$$

$$\left.\left.+\frac{j}{2}\int_0^{\infty}\frac{x^3dx}{\left(1+x^{2\overline{\alpha}}\right)^{\overline{\beta}}}\right]\right] =$$

(by Zwillinger (1995), p. 397, formula 595)

$$\frac{4}{|\beta^2-\alpha^2|}\left(\frac{\Gamma\left(\overline{\beta}\right)\overline{\alpha}\xi^2}{\Gamma\left(\frac{1}{2\overline{\alpha}}\right)\Gamma\left(\overline{\beta}-\frac{1}{2\overline{\alpha}}\right)}\right)\omega_1(F,\xi)_1\left[\sum_{j=1}^{r}|\alpha_j|\,j^2\left[\frac{\Gamma\left(\frac{3}{2\overline{\alpha}}\right)\Gamma\left(\overline{\beta}-\frac{3}{2\overline{\alpha}}\right)}{2\overline{\alpha}\Gamma\left(\overline{\beta}\right)}\right.\right.$$

$$\left.\left.+\frac{j}{2}\frac{\Gamma\left(\frac{2}{\overline{\alpha}}\right)\Gamma\left(\overline{\beta}-\frac{2}{\overline{\alpha}}\right)}{2\overline{\alpha}\Gamma\left(\overline{\beta}\right)}\right]\right]$$

$$= \frac{2}{|\beta^2-\alpha^2|\,\Gamma\left(\frac{1}{2\overline{\alpha}}\right)\Gamma\left(\overline{\beta}-\frac{1}{2\overline{\alpha}}\right)}\omega_1(F,\xi)_1\,\xi^2$$

$$\left[\sum_{j=1}^{r}|\alpha_j|\,j^2\left[\Gamma\left(\frac{3}{2\overline{\alpha}}\right)\Gamma\left(\overline{\beta}-\frac{3}{2\overline{\alpha}}\right)+\frac{j}{2}\Gamma\left(\frac{2}{\overline{\alpha}}\right)\Gamma\left(\overline{\beta}-\frac{2}{\overline{\alpha}}\right)\right]\right]. \quad (29.70)$$

The claim is proved. □

It follows the related result.

Corollary 29.4. *(to Theorem 29.4) It holds*

$$\left\|M_{r,\xi}(f)-f\right\|_1 \leq \Psi_2(\xi)+\left\|f''\right\|_1\left(\frac{2W}{\left|\beta^2-\alpha^2\right|}\right)$$

$$\left[\sum_{j=0}^{r}\left|\alpha_j\right|\int_0^{\infty}\left|\cos(\alpha jt)-\cos(\beta jt)\right|\frac{dt}{\left(t^{2\overline{\alpha}}+\xi^{2\overline{\alpha}}\right)^{\overline{\beta}}}\right]$$

$$+\left(\frac{4\left\|f^{(4)}\right\|_1+\left(\alpha^2+\beta^2\right)\left\|f''\right\|_1}{(\alpha\beta)^2\left|\beta^2-\alpha^2\right|}\right)W$$

$$\left[\beta^2\sum_{j=0}^{r}\left|\alpha_j\right|\left(\int_0^{\infty}\sin^2\left(\frac{\alpha jt}{2}\right)\frac{dt}{\left(t^{2\overline{\alpha}}+\xi^{2\overline{\alpha}}\right)^{\overline{\beta}}}\right)\right.$$

$$\left.+\alpha^2\sum_{j=0}^{r}\left|\alpha_j\right|\left(\int_0^{\infty}\sin^2\left(\frac{\beta jt}{2}\right)\frac{dt}{\left(t^{2\overline{\alpha}}+\xi^{2\overline{\alpha}}\right)^{\overline{\beta}}}\right)\right]\to 0 \tag{29.71}$$

as $\xi\to 0$. Above $\Psi_2(\xi)$ is as in (29.66).

Proof. Directly from (29.66), and see also the proof of Corollary 29.3. □

29.4 Part III: On the Smooth Trigonometric Singular Integral Operators (Anastassiou and Mezei (2012))

Let $\xi>0$, $f\in C^2(\mathbb{R})$, $x\in\mathbb{R}$, $\overline{\beta}\in\mathbb{N}$; we set

$$T_{r,\xi}(f;x):=\frac{1}{\lambda}\int_{-\infty}^{\infty}\left(\sum_{j=0}^{r}\alpha_j f(x+jt)\right)\left(\frac{\sin\left(\frac{t}{\xi}\right)}{t}\right)^{2\overline{\beta}}dt, \tag{29.72}$$

where

$$\lambda:=\int_{-\infty}^{\infty}\left(\frac{\sin\left(\frac{t}{\xi}\right)}{t}\right)^{2\overline{\beta}}dt=2\xi^{1-2\overline{\beta}}\int_0^{\infty}\left(\frac{\sin t}{t}\right)^{2\overline{\beta}}dt \tag{29.73}$$

(by Edwards (1954), p. 210, item 1033)

$$=2\xi^{1-2\overline{\beta}}\pi(-1)^{\overline{\beta}}\,\overline{\beta}\sum_{k=1}^{\overline{\beta}}(-1)^k\frac{k^{2\overline{\beta}-1}}{(\overline{\beta}-k)!\,(\overline{\beta}+k)!}.$$

Denote

$$\lambda_1 := 2\pi \left(-1\right)^{\overline{\beta}} \overline{\beta} \sum_{k=1}^{\overline{\beta}} \left(-1\right)^k \frac{k^{2\overline{\beta}-1}}{\left(\overline{\beta}-k\right)! \left(\overline{\beta}+k\right)!}, \tag{29.74}$$

that is,

$$\lambda = \lambda_1 \xi^{1-2\overline{\beta}}. \tag{29.75}$$

We suppose that $T_{r,\xi}\left(f;x\right) \in \mathbb{R}$ for all $x \in \mathbb{R}$. Clearly, again it is

$$\frac{1}{\lambda} \int_{-\infty}^{\infty} \left(\frac{\sin\left(\frac{t}{\xi}\right)}{t} \right)^{2\overline{\beta}} dt = 1, \tag{29.76}$$

and $T_{r,\xi}\left(c;x\right) = c$, c constant, and

$$T_{r,\xi}\left(f;x\right) - f\left(x\right) = \frac{1}{\lambda} \left(\sum_{j=0}^{r} \alpha_j \int_{-\infty}^{\infty} \left[f\left(x+jt\right) - f\left(x\right)\right] \left(\frac{\sin\left(\frac{t}{\xi}\right)}{t} \right)^{2\overline{\beta}} dt \right). \tag{29.77}$$

We set

$$\begin{aligned} E_3\left(x\right) :=\; & T_{r,\xi}\left(f,x\right) - f\left(x\right) - \left(\frac{f'\left(x\right)}{\alpha\beta\left(\beta^2-\alpha^2\right)} \right) \\ & \left[\beta^3 \sum_{j=0}^{r} \alpha_j \frac{1}{\lambda} \left(\int_{-\infty}^{\infty} \sin\left(\alpha j t\right) \left(\frac{\sin\left(\frac{t}{\xi}\right)}{t} \right)^{2\overline{\beta}} dt \right) \right. \\ & \left. - \alpha^3 \sum_{j=0}^{r} \alpha_j \frac{1}{\lambda} \left(\int_{-\infty}^{\infty} \sin\left(\beta j t\right) \left(\frac{\sin\left(\frac{t}{\xi}\right)}{t} \right)^{2\overline{\beta}} dt \right) \right] \\ & - \left(\frac{f''\left(x\right)}{\beta^2-\alpha^2} \right) \left[\sum_{j=0}^{r} \alpha_j \frac{1}{\lambda} \left(\int_{-\infty}^{\infty} \cos\left(\alpha j t\right) \left(\frac{\sin\left(\frac{t}{\xi}\right)}{t} \right)^{2\overline{\beta}} dt \right) \right. \\ & \left. - \sum_{j=0}^{r} \alpha_j \frac{1}{\lambda} \left(\int_{-\infty}^{\infty} \cos\left(\beta j t\right) \left(\frac{\sin\left(\frac{t}{\xi}\right)}{t} \right)^{2\overline{\beta}} dt \right) \right] \end{aligned}$$

$$-\left(\frac{f^{(3)}(x)}{\alpha\beta(\beta^2-\alpha^2)}\right)\left[\beta\sum_{j=0}^{r}\alpha_j\frac{1}{\lambda}\left(\int_{-\infty}^{\infty}\sin(\alpha jt)\left(\frac{\sin\left(\frac{t}{\xi}\right)}{t}\right)^{2\overline{\beta}}dt\right)\right. \tag{29.78}$$

$$\left.-\alpha\sum_{j=0}^{r}\alpha_j\frac{1}{\lambda}\left(\int_{-\infty}^{\infty}\sin(\beta jt)\left(\frac{\sin\left(\frac{t}{\xi}\right)}{t}\right)^{2\overline{\beta}}dt\right)\right]$$

$$-\left(\frac{2\left(f^{(4)}(x)+\left(\alpha^2+\beta^2\right)f''(x)\right)}{(\alpha\beta)^2(\beta^2-\alpha^2)}\right)$$

$$\left[\beta^2\sum_{j=0}^{r}\alpha_j\frac{1}{\lambda}\left(\int_{-\infty}^{\infty}\sin^2\left(\frac{\alpha jt}{2}\right)\left(\frac{\sin\left(\frac{t}{\xi}\right)}{t}\right)^{2\overline{\beta}}dt\right)\right.$$

$$\left.-\alpha^2\sum_{j=0}^{r}\alpha_j\frac{1}{\lambda}\left(\int_{-\infty}^{\infty}\sin^2\left(\frac{\beta jt}{2}\right)\left(\frac{\sin\left(\frac{t}{\xi}\right)}{t}\right)^{2\overline{\beta}}dt\right)\right]$$

$$=T_{r,\xi}(f,x)-f(x)-\left(\frac{2f''(x)}{\beta^2-\alpha^2}\right)\frac{1}{\lambda}$$

$$\left[\sum_{j=0}^{r}\alpha_j\left(\int_{-\infty}^{\infty}(\cos(\alpha jt)-\cos(\beta jt))\left(\frac{\sin\left(\frac{t}{\xi}\right)}{t}\right)^{2\overline{\beta}}dt\right)\right]$$

$$-\left(\frac{4\left(f^{(4)}(x)+\left(\alpha^2+\beta^2\right)f''(x)\right)}{(\alpha\beta)^2(\beta^2-\alpha^2)}\right)\frac{1}{\lambda} \tag{29.79}$$

$$\left[\beta^2\sum_{j=0}^{r}\alpha_j\left(\int_{-\infty}^{\infty}\sin^2\left(\frac{\alpha jt}{2}\right)\left(\frac{\sin\left(\frac{t}{\xi}\right)}{t}\right)^{2\overline{\beta}}dt\right)\right.$$

$$\left.-\alpha^2\sum_{j=0}^{r}\alpha_j\left(\int_{-\infty}^{\infty}\sin^2\left(\frac{\beta jt}{2}\right)\left(\frac{\sin\left(\frac{t}{\xi}\right)}{t}\right)^{2\overline{\beta}}dt\right)\right]$$

$$=\frac{1}{\lambda}\int_{-\infty}^{\infty}R(t)\left(\frac{\sin\left(\frac{t}{\xi}\right)}{t}\right)^{2\overline{\beta}}dt. \tag{29.80}$$

It follows a parametrized L_p $(p>1)$ approximation result for $T_{r,\xi}$, $\xi>0$.

Theorem 29.5. *Let $p,q>1: \frac{1}{p}+\frac{1}{q}=1$, $\overline{\beta}\in\mathbb{N}$ and $\overline{\beta}>\frac{\lceil 3p\rceil+1}{2}$ ($\lceil\cdot\rceil$ the ceiling of the number), $\xi>0$, λ_1 is as in (29.74).*

When $\overline{\lambda}\in\mathbb{N}$ is even we define

$$\psi_{1\overline{\lambda}}:=\frac{\pi(-1)^{\frac{2\overline{\beta}-\overline{\lambda}}{2}}(2\overline{\beta})!}{2^{\overline{\lambda}+1}(2\overline{\beta}-\overline{\lambda}-1)!}\left(\sum_{k=1}^{\overline{\beta}}(-1)^k\frac{k^{2\overline{\beta}-\overline{\lambda}-1}}{(\overline{\beta}-k)!(\overline{\beta}+k)!}\right),$$

and when $\overline{\lambda}$ is odd we define

$$\psi_{2\overline{\lambda}}:=\frac{(-1)^{\frac{\overline{\lambda}-1}{2}}(2\overline{\beta})!}{2^{\overline{\lambda}}(2\overline{\beta}-\overline{\lambda}-1)!}\left(\sum_{k=1}^{\overline{\beta}}(-1)^{\overline{\beta}-k}\frac{k^{2\overline{\beta}-\overline{\lambda}-1}\ln(2k)}{(\overline{\beta}-k)!(\overline{\beta}+k)!}\right),$$

and we set

$$\psi_{\overline{\lambda}}:=\left\{\begin{matrix}\psi_{1\overline{\lambda}}, & \text{if } \overline{\lambda} \text{ is even,}\\ \psi_{2\overline{\lambda}}, & \text{if } \overline{\lambda} \text{ is odd}\end{matrix}\right\}.$$

Let $\alpha,\beta\in\mathbb{R}:\alpha\beta(\alpha^2-\beta^2)\neq0$; $f,f',f'',f^{(3)},f^{(4)}\in L_p(\mathbb{R})\cap C(\mathbb{R})$; $\xi>0$, $x\in\mathbb{R}$.

Then

$$\begin{aligned}\|E_3\|_p=\Bigg\|&T_{r,\xi}(f,x)-f(x)-\left(\frac{2f''(x)}{\beta^2-\alpha^2}\right)\frac{1}{\lambda}\\&\left[\sum_{j=0}^{r}\alpha_j\left(\int_0^\infty(\cos(\alpha jt)-\cos(\beta jt))\left(\frac{\sin\left(\frac{t}{\xi}\right)}{t}\right)^{2\overline{\beta}}dt\right)\right]\\&-\left(\frac{4\left(f^{(4)}(x)+(\alpha^2+\beta^2)f''(x)\right)}{(\alpha\beta)^2(\beta^2-\alpha^2)}\right)\frac{1}{\lambda}\\&\left[\beta^2\sum_{j=0}^{r}\alpha_j\left(\int_0^\infty\sin^2\left(\frac{\alpha jt}{2}\right)\left(\frac{\sin\left(\frac{t}{\xi}\right)}{t}\right)^{2\overline{\beta}}dt\right)\right.\\&\left.-\alpha^2\sum_{j=0}^{r}\alpha_j\left(\int_0^\infty\sin^2\left(\frac{\beta jt}{2}\right)\left(\frac{\sin\left(\frac{t}{\xi}\right)}{t}\right)^{2\overline{\beta}}dt\right)\right]\Bigg\|_p\leq\end{aligned}\qquad(29.81)$$

$$\leq \frac{4 \cdot 2^{\frac{1}{p}}}{\lambda_1^{\frac{1}{p}} \left|\beta^2 - \alpha^2\right| (q+1)^{\frac{1}{q}} (p+1)^{\frac{1}{p}}} \left(\sum_{j=1}^{r} |\alpha_j| j^2 \right)^{\frac{1}{q}} \left(\sum_{j=1}^{r} |\alpha_j| j^{p+2} \right)^{\frac{1}{p}}$$

$$\left[\left(\frac{-\left(2\overline{\beta}\right)!}{8\left(2\overline{\beta}-4\right)!} \sum_{k=1}^{\overline{\beta}} (-1)^{\overline{\beta}-k} \frac{k^{2\overline{\beta}-4} \ln(2k)}{\left(\overline{\beta}-k\right)! \left(\overline{\beta}+k\right)!} \right) + \psi_{\lceil 3p \rceil} \right]^{\frac{1}{p}}$$

$$\omega_1 \left(f^{(4)} + \left(\alpha^2 + \beta^2\right) f'' + \alpha^2 \beta^2 f, \xi \right)_p \cdot \xi^2 =: \Phi_3(\xi) \to 0, \quad as\ \xi \to 0.$$

Proof. By (29.80), we have $\left(\overline{\beta} \in \mathbb{N} : \overline{\beta} > \frac{\lceil 3p \rceil + 1}{2}\right)$

$$E_3(x) = \frac{1}{\lambda} \int_{-\infty}^{\infty} R(t) \left(\frac{\sin\left(\frac{t}{\xi}\right)}{t} \right)^{2\overline{\beta}} dt. \tag{29.82}$$

Hence it holds $\left(p, q > 1 : \frac{1}{p} + \frac{1}{q} = 1\right)$

$$\int_{-\infty}^{\infty} |E_3(x)|^p \, dx = \int_{-\infty}^{\infty} \left(\frac{1}{\lambda} \left| \int_{-\infty}^{\infty} R(t) \left(\frac{\sin\left(\frac{t}{\xi}\right)}{t} \right)^{2\overline{\beta}} dt \right| \right)^p dx$$

$$\leq \int_{-\infty}^{\infty} \left(\frac{1}{\lambda} \int_{-\infty}^{\infty} |R(t)| \left(\frac{\sin\left(\frac{t}{\xi}\right)}{t} \right)^{2\overline{\beta}} dt \right)^p dx \tag{29.83}$$

$$\leq \frac{1}{\lambda} \int_{-\infty}^{\infty} \left(\int_{-\infty}^{\infty} |R(t)|^p \left(\frac{\sin\left(\frac{t}{\xi}\right)}{t} \right)^{2\overline{\beta}} dt \right) dx$$

$$\overset{(29.30)}{\leq} \frac{2^p}{\left|\beta^2 - \alpha^2\right|^p \lambda} \left(\int_{-\infty}^{\infty} \left(\int_{-\infty}^{\infty} \left[\sum_{j=0}^{r} |\alpha_j| j^2 \right. \right. \right.$$

$$\left. \left. \left. \left(\int_0^{|t|} |F(x + j\,sign(t)\,\theta) - F(x)| \left(|t| - \theta\right) d\theta \right) \right]^p \left(\frac{\sin\left(\frac{t}{\xi}\right)}{t} \right)^{2\overline{\beta}} dt \right) dx \right.$$

$$\overset{\text{(by Jensen's inequality)}}{\leq} \frac{2^p}{\left|\beta^2 - \alpha^2\right|^p \lambda} \int_{-\infty}^{\infty} \left(\int_{-\infty}^{\infty} \left(\sum_{j=0}^{r} |\alpha_j| j^2 \right)^{p-1} \right.$$

$$\left[\sum_{j=0}^{r}|\alpha_j|\,j^2\left(\int_0^{|t|}|F(x+jsign(t)\theta)-F(x)|\,(|t|-\theta)\,d\theta\right)^p\right]$$

$$\times\left(\frac{\sin\left(\frac{t}{\xi}\right)}{t}\right)^{2\overline{\beta}}dt\Bigg)\,dx$$

$$\leq\frac{1}{\lambda}\left(\sum_{j=0}^{r}|\alpha_j|\,j^2\right)^{p-1}\frac{2^p}{|\beta^2-\alpha^2|^p}$$

$$\left[\sum_{j=0}^{r}|\alpha_j|\,j^2\int_{-\infty}^{\infty}\left(\int_{-\infty}^{\infty}\left(\int_0^{|t|}|F(x+jsign(t)\theta)-F(x)|^p\,d\theta\right)\right.\right.$$

$$\left.\left.\left(\int_0^{|t|}(|t|-\theta)^q\,d\theta\right)^{\frac{p}{q}}\left(\frac{\sin\left(\frac{t}{\xi}\right)}{t}\right)^{2\overline{\beta}}dt\right)dx\right]$$

$$=\frac{1}{\lambda}\left(\sum_{j=0}^{r}|\alpha_j|\,j^2\right)^{p-1}\frac{2^p}{|\beta^2-\alpha^2|^p\,(q+1)^{p-1}}$$

$$\left[\sum_{j=0}^{r}|\alpha_j|\,j^2\int_{-\infty}^{\infty}\left(\int_{-\infty}^{\infty}\left(\int_0^{|t|}|F(x+jsign(t)\theta)-F(x)|^p\,d\theta\right)\right.\right.$$

$$\left.\left.|t|^{(q+1)(p-1)}\left(\frac{\sin\left(\frac{t}{\xi}\right)}{t}\right)^{2\overline{\beta}}dt\right)dx\right] \tag{29.84}$$

(call $\mu:=\frac{1}{\lambda}\left(\sum_{j=0}^{r}|\alpha_j|\,j^2\right)^{p-1}\frac{2^p}{|\beta^2-\alpha^2|^p(q+1)^{p-1}}$)

$$=\mu\left[\sum_{j=0}^{r}|\alpha_j|\,j^2\int_{-\infty}^{\infty}\left(\int_0^{|t|}\left(\int_{-\infty}^{\infty}|F(x+jsign(t)\theta)-F(x)|^p\,dx\right)d\theta\right)\right.$$

$$\left.|t|^{(q+1)(p-1)}\left(\frac{\sin\left(\frac{t}{\xi}\right)}{t}\right)^{2\overline{\beta}}dt\right]$$

$$\leq\mu\left[\sum_{j=0}^{r}|\alpha_j|\,j^2\int_{-\infty}^{\infty}\left(\int_0^{|t|}\omega_1\left(F,\xi\frac{j\theta}{\xi}\right)_p^p d\theta\right)|t|^{(q+1)(p-1)}\left(\frac{\sin\left(\frac{t}{\xi}\right)}{t}\right)^{2\overline{\beta}}dt\right] \tag{29.85}$$

$$\leq \mu\omega_1(F,\xi)_p^p$$

$$\left[\sum_{j=0}^{r}|\alpha_j|\,j^2\int_{-\infty}^{\infty}\left(\int_0^{|t|}\left(1+\frac{j\theta}{\xi}\right)^p d\theta\right)|t|^{(q+1)(p-1)}\left(\frac{\sin\left(\frac{t}{\xi}\right)}{t}\right)^{2\overline{\beta}}dt\right]$$

$$=\mu\omega_1(F,\xi)_p^p\,\frac{\xi}{(p+1)}$$

$$\left[\sum_{j=1}^{r}|\alpha_j|\,j\int_{-\infty}^{\infty}\left(\left(1+\frac{j}{\xi}|t|\right)^{p+1}-1\right)|t|^{(q+1)(p-1)}\left(\frac{\sin\left(\frac{t}{\xi}\right)}{t}\right)^{2\overline{\beta}}dt\right]$$

$$\leq\frac{2^p\mu\xi\omega_1(F,\xi)_p^p}{(p+1)}$$

$$\left[\sum_{j=1}^{r}|\alpha_j|\,j\frac{j^{p+1}}{\xi^{p+1}}\int_{-\infty}^{\infty}|t|^{p+1}|t|^{(q+1)(p-1)}\left(\frac{\sin\left(\frac{t}{\xi}\right)}{t}\right)^{2\overline{\beta}}dt\right] \tag{29.86}$$

$$=\frac{2^{p+1}\mu\xi^{-p}\omega_1(F,\xi)_p^p}{(p+1)}\left[\sum_{j=1}^{r}|\alpha_j|\,j^{p+2}\right.$$

$$\left.\int_0^{\infty}t^{(q+1)(p-1)+(p+1)}\left(\frac{\sin\left(\frac{t}{\xi}\right)}{t}\right)^{2\overline{\beta}}dt\right]$$

$$=\frac{2^{p+1}\mu\xi^{-p}\omega_1(F,\xi)_p^p}{(p+1)}\left[\sum_{j=1}^{r}|\alpha_j|\,j^{p+2}\xi^{3p-2\overline{\beta}+1}\right.$$

$$\left.\int_0^{\infty}\left(\frac{t}{\xi}\right)^{3p}\left(\frac{\sin\left(\frac{t}{\xi}\right)}{\left(\frac{t}{\xi}\right)}\right)^{2\overline{\beta}}d\frac{t}{\xi}\right]$$

$$=\frac{2^{p+1}\mu\xi^{2p-2\overline{\beta}+1}\omega_1(F,\xi)_p^p}{(p+1)}\left[\sum_{j=1}^{r}|\alpha_j|\,j^{p+2}\right]$$

$$\left[\int_0^{\infty}x^{3p}\left(\frac{\sin x}{x}\right)^{2\overline{\beta}}dx\right] \tag{29.87}$$

$$= \frac{2^{p+1}\mu\xi^{2p-2\overline{\beta}+1}\omega_1(F,\xi)_p^p}{(p+1)}\left[\sum_{j=1}^{r}|\alpha_j|\,j^{p+2}\right]$$

$$\left[\int_0^1 x^{3p}\left(\frac{\sin x}{x}\right)^{2\overline{\beta}}dx+\int_1^{\infty}x^{3p}\left(\frac{\sin x}{x}\right)^{2\overline{\beta}}dx\right]$$

$$\leq \frac{2^{p+1}\mu\xi^{2p-2\overline{\beta}+1}\omega_1(F,\xi)_p^p}{(p+1)}\left(\sum_{j=1}^{r}|\alpha_j|\,j^{p+2}\right)$$

$$\left[\int_0^1 x^{3}\left(\frac{\sin x}{x}\right)^{2\overline{\beta}}dx+\int_1^{\infty}x^{\lceil 3p\rceil}\left(\frac{\sin x}{x}\right)^{2\overline{\beta}}dx\right]$$

$$\leq \frac{2^{p+1}\mu\xi^{2p-2\overline{\beta}+1}\omega_1(F,\xi)_p^p}{(p+1)}\left(\sum_{j=1}^{r}|\alpha_j|\,j^{p+2}\right)$$

$$\left[\int_0^{\infty} x^{3}\left(\frac{\sin x}{x}\right)^{2\overline{\beta}}dx+\int_0^{\infty}x^{\lceil 3p\rceil}\left(\frac{\sin x}{x}\right)^{2\overline{\beta}}dx\right] \tag{29.88}$$

(we use Edwards (1954), p. 210, item 1033)

$$= \frac{2^{p+1}\mu\xi^{2p-2\overline{\beta}+1}\omega_1(F,\xi)_p^p}{(p+1)}\left(\sum_{j=1}^{r}|\alpha_j|\,j^{p+2}\right)$$

$$\left[\left(\frac{-(2\overline{\beta})!}{8(2\overline{\beta}-4)!}\sum_{k=1}^{\overline{\beta}}(-1)^{\overline{\beta}-k}\frac{k^{2\overline{\beta}-4}\ln(2k)}{(\overline{\beta}-k)!(\overline{\beta}+k)!}\right)+\psi_{\lceil 3p\rceil}\right]$$

$$= \frac{2^{p+1}\xi^{2p}\omega_1(F,\xi)_p^p}{(p+1)}\left(\sum_{j=1}^{r}|\alpha_j|\,j^{p+2}\right)$$

$$\left[\left(\frac{-(2\overline{\beta})!}{8(2\overline{\beta}-4)!}\sum_{k=1}^{\overline{\beta}}(-1)^{\overline{\beta}-k}\frac{k^{2\overline{\beta}-4}\ln(2k)}{(\overline{\beta}-k)!(\overline{\beta}+k)!}\right)+\psi_{\lceil 3p\rceil}\right]$$

$$\frac{1}{\lambda_1}\left(\sum_{j=0}^{r}|\alpha_j|\,j^2\right)^{p-1}\frac{2^p}{|\beta^2-\alpha^2|^p(q+1)^{p-1}}$$

$$= \frac{2^{2p+1}}{\lambda_1|\beta^2-\alpha^2|^p(q+1)^{p-1}(p+1)}\left(\sum_{j=0}^{r}|\alpha_j|\,j^2\right)^{p-1}\left(\sum_{j=1}^{r}|\alpha_j|\,j^{p+2}\right)$$

$$\left[\left(\frac{-(2\overline{\beta})!}{8(2\overline{\beta}-4)!}\sum_{k=1}^{\overline{\beta}}(-1)^{\overline{\beta}-k}\frac{k^{2\overline{\beta}-4}\ln(2k)}{(\overline{\beta}-k)!(\overline{\beta}+k)!}\right)+\psi_{\lceil 3p\rceil}\right]\omega_1(F,\xi)_p^p\,\xi^{2p}<\infty. \tag{29.89}$$

The claim is proved. □

Next we give a consequence of Theorem 29.5.

Corollary 29.5. *(to Theorem 29.5) It holds*

$$\left\|T_{r,\xi}(f)-f\right\|_p \leq \Phi_3(\xi)+\left\|f''\right\|_p\left(\frac{2\xi^{2\overline{\beta}-1}}{\lambda_1\left|\beta^2-\alpha^2\right|}\right)$$

$$\left[\sum_{j=0}^{r}\left|\alpha_j\right|\int_0^{\infty}\left|\cos(\alpha jt)-\cos(\beta jt)\right|\left(\frac{\sin\left(\frac{t}{\xi}\right)}{t}\right)^{2\overline{\beta}}dt\right]$$

$$+\left(\frac{4\left\|f^{(4)}\right\|_p+\left(\alpha^2+\beta^2\right)\left\|f''\right\|_p}{(\alpha\beta)^2\left|\beta^2-\alpha^2\right|}\right)\frac{\xi^{2\overline{\beta}-1}}{\lambda_1}$$

$$\left[\beta^2\sum_{j=0}^{r}\left|\alpha_j\right|\left(\int_0^{\infty}\sin^2\left(\frac{\alpha jt}{2}\right)\left(\frac{\sin\left(\frac{t}{\xi}\right)}{t}\right)^{2\overline{\beta}}dt\right)\right.$$

$$\left.+\alpha^2\sum_{j=0}^{r}\left|\alpha_j\right|\left(\int_0^{\infty}\sin^2\left(\frac{\beta jt}{2}\right)\left(\frac{\sin\left(\frac{t}{\xi}\right)}{t}\right)^{2\overline{\beta}}dt\right)\right]\to 0, \quad (29.90)$$

as $\xi\to 0$. Above $\Phi_3(\xi)$ is as in (29.81).

Proof. We have that

$$\frac{\xi^{2\overline{\beta}-1}}{\lambda_1}\int_0^{\infty}\left|\cos(\alpha jt)-\cos(\beta jt)\right|\left(\frac{\sin\left(\frac{t}{\xi}\right)}{t}\right)^{2\overline{\beta}}dt$$

$$=\frac{2\xi^{2\overline{\beta}-1}}{\lambda_1}\int_0^{\infty}\left|\sin\left(\frac{(\alpha+\beta)jt}{2}\right)\right|\left|\sin\left(\frac{(\alpha-\beta)jt}{2}\right)\right|\left(\frac{\sin\left(\frac{t}{\xi}\right)}{t}\right)^{2\overline{\beta}}dt$$

$$\leq\frac{2\xi^{2\overline{\beta}-1}}{\lambda_1}\int_0^{\infty}\frac{|\alpha+\beta|\,jt}{2}\frac{|\alpha-\beta|\,jt}{2}\left(\frac{\sin\left(\frac{t}{\xi}\right)}{t}\right)^{2\overline{\beta}}dt$$

$$=\frac{\xi^{2\overline{\beta}-1}}{\lambda_1}\frac{\left|\alpha^2-\beta^2\right|}{2}j^2\int_0^{\infty}t^2\left(\frac{\sin\left(\frac{t}{\xi}\right)}{t}\right)^{2\overline{\beta}}dt$$

$$= \frac{\xi^{2\overline{\beta}-1}}{\lambda_1} \xi^{3-2\overline{\beta}} \frac{\left|\alpha^2 - \beta^2\right|}{2} j^2 \int_0^\infty \left(\frac{t}{\xi}\right)^2 \left(\frac{\sin\left(\frac{t}{\xi}\right)}{\left(\frac{t}{\xi}\right)}\right)^{2\overline{\beta}} d\frac{t}{\xi}$$

$$= \frac{\xi^2}{\lambda_1} \frac{\left|\alpha^2 - \beta^2\right|}{2} j^2 \int_0^\infty x^2 \left(\frac{\sin x}{x}\right)^{2\overline{\beta}} dx$$

$$= \frac{\xi^2}{\lambda_1} \frac{\left|\alpha^2 - \beta^2\right|}{2} j^2 \psi_{12} \to 0, \quad \text{as } \xi \to 0. \tag{29.91}$$

Furthermore, we have that

$$\frac{\xi^{2\overline{\beta}-1}}{\lambda_1} \int_0^\infty \sin^2\left(\frac{\alpha j t}{2}\right) \left(\frac{\sin\left(\frac{t}{\xi}\right)}{t}\right)^{2\overline{\beta}} dt \leq \frac{\xi^{2\overline{\beta}-1}}{\lambda_1} \frac{\alpha^2 j^2}{4} \int_0^\infty t^2 \left(\frac{\sin\left(\frac{t}{\xi}\right)}{t}\right)^{2\overline{\beta}} dt$$

$$= \frac{\xi^2}{\lambda_1} \frac{\alpha^2 j^2}{4} \psi_{12} \to 0, \quad \text{as } \xi \to 0. \tag{29.92}$$

The proof now is clear. □

Next comes the corresponding L_1 approximation result.

Theorem 29.6. *Let $\overline{\beta} \in \mathbb{N} : \overline{\beta} > 2$, and $\alpha, \beta \in \mathbb{R} : \alpha\beta\left(\alpha^2 - \beta^2\right) \neq 0$; $f, f', f'', f^{(3)}, f^{(4)} \in L_1(\mathbb{R}) \cap C(\mathbb{R})$; $\xi > 0$, $x \in \mathbb{R}$. Then*

$$\|E_3\|_1 = \Bigg\| T_{r,\xi}(f, x) - f(x)$$

$$- \left(\frac{2f''(x)}{\beta^2 - \alpha^2}\right) \frac{1}{\lambda} \left[\sum_{j=0}^r \alpha_j \left(\int_0^\infty \left(\cos(\alpha j t) - \cos(\beta j t)\right) \left(\frac{\sin\left(\frac{t}{\xi}\right)}{t}\right)^{2\overline{\beta}} dt\right)\right]$$

$$- \left(\frac{4\left(f^{(4)}(x) + \left(\alpha^2 + \beta^2\right) f''(x)\right)}{(\alpha\beta)^2 \left(\beta^2 - \alpha^2\right)}\right) \frac{1}{\lambda}$$

$$\left[\beta^2 \sum_{j=0}^r \alpha_j \left(\int_0^\infty \sin^2\left(\frac{\alpha j t}{2}\right) \left(\frac{\sin\left(\frac{t}{\xi}\right)}{t}\right)^{2\overline{\beta}} dt\right)\right.$$

$$\left.- \alpha^2 \sum_{j=0}^r \alpha_j \left(\int_0^\infty \sin^2\left(\frac{\beta j t}{2}\right) \left(\frac{\sin\left(\frac{t}{\xi}\right)}{t}\right)^{2\overline{\beta}} dt\right)\right] \Bigg\|_1$$

$$\leq \left[\frac{4}{|\beta^2-\alpha^2|\,\lambda_1}\right]$$

$$\left[\sum_{j=1}^{r}|\alpha_j|\,j^2\left[\psi_{12}+\frac{j}{2}\psi_{23}\right]\right]$$

$$\omega_1\left(f^{(4)}+\left(\alpha^2+\beta^2\right)f''+\alpha^2\beta^2 f,\xi\right)_1\cdot\xi^2=:\Psi_3(\xi)\to 0, \tag{29.93}$$

as $\xi\to 0$.

Proof. We have that (F is as in (29.25))

$$\int_{-\infty}^{\infty}|E_3(x)|\,dx \overset{(29.82)}{=} \int_{-\infty}^{\infty}\left(\frac{1}{\lambda}\left|\int_{-\infty}^{\infty}R(t)\left(\frac{\sin\left(\frac{t}{\xi}\right)}{t}\right)^{2\overline{\beta}}dt\right|\right)dx$$

$$\leq \int_{-\infty}^{\infty}\left(\frac{1}{\lambda}\int_{-\infty}^{\infty}|R(t)|\left(\frac{\sin\left(\frac{t}{\xi}\right)}{t}\right)^{2\overline{\beta}}dt\right)dx \overset{(29.30)}{\leq} \frac{2}{|\beta^2-\alpha^2|\,\lambda}$$

$$\int_{-\infty}^{\infty}\left(\int_{-\infty}^{\infty}\left[\sum_{j=1}^{r}|\alpha_j|\,j^2\left(\int_0^{|t|}|F(x+j\,sign(t)\,\theta)-F(x)|\,(|t|-\theta)\,d\theta\right)\right]\right.$$

$$\left.\left(\frac{\sin\left(\frac{t}{\xi}\right)}{t}\right)^{2\overline{\beta}}dt\right)dx \tag{29.94}$$

$$\leq \frac{2}{|\beta^2-\alpha^2|\,\lambda}\int_{-\infty}^{\infty}\left(\int_{-\infty}^{\infty}\left[\sum_{j=1}^{r}|\alpha_j|\,j^2\right.\right.$$

$$\left.\left.\left(\int_0^{|t|}|F(x+j\,sign(t)\,\theta)-F(x)|\,d\theta\right)|t|\left(\frac{\sin\left(\frac{t}{\xi}\right)}{t}\right)^{2\overline{\beta}}dt\right)dx\right]$$

$$=\frac{2}{|\beta^2-\alpha^2|\,\lambda}\left[\sum_{j=1}^{r}|\alpha_j|\,j^2\int_{-\infty}^{\infty}\left(\int_0^{|t|}\left(\int_{-\infty}^{\infty}\right.\right.\right.$$

$$\left.\left.\left.|F(x+j\,sign(t)\,\theta)-F(x)|\,dx\right)d\theta\right)|t|\left(\frac{\sin\left(\frac{t}{\xi}\right)}{t}\right)^{2\overline{\beta}}dt\right]$$

$$\leq \frac{2}{|\beta^2-\alpha^2|\lambda}\left[\sum_{j=1}^{r}|\alpha_j|j^2\int_{-\infty}^{\infty}\left(\int_0^{|t|}\omega_1\left(F,\xi\frac{j\theta}{\xi}\right)_1 d\theta\right)|t|\left(\frac{\sin\left(\frac{t}{\xi}\right)}{t}\right)^{2\overline{\beta}}dt\right]$$

$$\leq \frac{2}{|\beta^2-\alpha^2|\lambda}\omega_1(F,\xi)_1\left[\sum_{j=1}^{r}|\alpha_j|j^2\int_{-\infty}^{\infty}\left(\int_0^{|t|}\left(1+\frac{j\theta}{\xi}\right)d\theta\right)\right.$$

$$\left.\times|t|\left(\frac{\sin\left(\frac{t}{\xi}\right)}{t}\right)^{2\overline{\beta}}dt\right]$$

$$=\frac{2}{|\beta^2-\alpha^2|\lambda}\omega_1(F,\xi)_1\left[\sum_{j=1}^{r}|\alpha_j|j^2\int_{-\infty}^{\infty}\left(|t|+\frac{j}{2\xi}t^2\right)|t|\left(\frac{\sin\left(\frac{t}{\xi}\right)}{t}\right)^{2\overline{\beta}}dt\right]$$

$$=\frac{4}{|\beta^2-\alpha^2|\lambda}\omega_1(F,\xi)_1\left[\sum_{j=1}^{r}|\alpha_j|j^2\int_{0}^{\infty}\left(t+\frac{j}{2\xi}t^2\right)t\left(\frac{\sin\left(\frac{t}{\xi}\right)}{t}\right)^{2\overline{\beta}}dt\right]$$

$$=\frac{4}{|\beta^2-\alpha^2|\lambda}\omega_1(F,\xi)_1\sum_{j=1}^{r}|\alpha_j|j^2$$

$$\left[\int_0^{\infty}t^2\left(\frac{\sin\left(\frac{t}{\xi}\right)}{t}\right)^{2\overline{\beta}}dt+\frac{j}{2\xi}\int_0^{\infty}t^3\left(\frac{\sin\left(\frac{t}{\xi}\right)}{t}\right)^{2\overline{\beta}}dt\right]$$

$$=\frac{4\xi^{3-2\overline{\beta}}}{|\beta^2-\alpha^2|\lambda}\omega_1(F,\xi)_1\left[\sum_{j=1}^{r}|\alpha_j|j^2\left[\int_0^{\infty}x^2\left(\frac{\sin x}{x}\right)^{2\overline{\beta}}dx\right.\right.$$

$$\left.\left.+\frac{j}{2}\int_0^{\infty}x^3\left(\frac{\sin x}{x}\right)^{2\overline{\beta}}dx\right]\right]= \tag{29.95}$$

(we use Edwards (1954), p. 210, formula 1033)

$$\frac{4\xi^2}{|\beta^2-\alpha^2|\lambda_1}\omega_1(F,\xi)_1\left[\sum_{j=1}^{r}|\alpha_j|j^2\left[\psi_{12}+\frac{j}{2}\psi_{23}\right]\right]<\infty. \tag{29.96}$$

The claim is proved. □

We finish this work with the following result.

Corollary 29.6 (to Theorem 29.6). *It holds*

$$\|T_{r,\xi}(f)-f\|_1 \leq \Psi_3(\xi)+\|f''\|_1\left(\frac{2\xi^{2\overline{\beta}-1}}{\lambda_1\left|\beta^2-\alpha^2\right|}\right)$$
$$\times\left[\sum_{j=0}^{r}|\alpha_j|\int_0^{\infty}|\cos(\alpha jt)-\cos(\beta jt)|\left(\frac{\sin\left(\frac{t}{\xi}\right)}{t}\right)^{2\overline{\beta}}dt\right]$$
$$+\left(\frac{4\left\|f^{(4)}\right\|_1+\left(\alpha^2+\beta^2\right)\|f''\|_1}{(\alpha\beta)^2\left|\beta^2-\alpha^2\right|}\right)\frac{\xi^{2\overline{\beta}-1}}{\lambda_1}$$
$$\times\left[\beta^2\sum_{j=0}^{r}|\alpha_j|\left(\int_0^{\infty}\sin^2\left(\frac{\alpha jt}{2}\right)\left(\frac{\sin\left(\frac{t}{\xi}\right)}{t}\right)^{2\overline{\beta}}dt\right)\right.$$
$$\left.+\alpha^2\sum_{j=0}^{r}|\alpha_j|\left(\int_0^{\infty}\sin^2\left(\frac{\beta jt}{2}\right)\left(\frac{\sin\left(\frac{t}{\xi}\right)}{t}\right)^{2\overline{\beta}}dt\right)\right]\to 0, \quad (29.97)$$

as $\xi \to 0$*. Above* $\Psi_3(\xi)$ *is as in (29.93).*

Proof. As similar to Corollary 29.5 is omitted. □

Bibliography

A.H. Ali and Z. Pales, Taylor-type expansions in terms of exponential polynomials, *Mathematical Inequalities and Applications*, 25(4) (2022), 1123–1141.

G.A. Anastassiou, Basic convergence with rates of smooth Picard singular integral operators, *Journal of Computational Analysis and Applications*, 8(4) (2006), 313–334.

G.A. Anastassiou, *Intelligent Mathematics: Computational Analysis*, Springer, Heidelberg, New York, Chapter 12, 2011.

G.A. Anastassiou, Opial and Ostrowski type inequalities based on trigonometric and hyperbolic type Taylor formulae, *Malaya Journal of Matematik*, 11(S) (2023), 1–26.

G.A. Anastassiou, *Parametrized Trigonometric Derived L_p Degree of Approximation by Various Smooth Singular Integral Operators*, submitted, 2024.

G.A. Anastassiou and R. Mezei, *Approximation by Singular Integrals*, Cambridge Scientific Publishers, Cambridge, UK, 2012.

A. Aral, On a generalized Gauss Weierstrass singular integral, *Fasciculi Mathematici*, (35) (2005), 23–33.

A. Aral, *Pointwise approximation by the generalization of Picard and Gauss-Weierstrass singular integrals*, *Journal of Concrete & Applicable Mathematics*, 6(4) (2008), 327–339.

A. Aral, On generalized Picard integral operators, *Advances in Summability and Approximation Theory*, 157–168, Springer, Singapore, 2018.

A. Aral, E. Deniz and H. Erbay, The Picard and Gauss-Weiertrass singular integrals in (p, q)-calculus, *Bulletin of the Malaysian Mathematical Sciences Society*, 43(2) (2020), 1569–1583.

A. Aral and S.G. Gal, q-generalizations of the Picard and Gauss-Weierstrass singular integrals, *Taiwanese Journal of Mathematics*, 12(9) (2008), 2501–2515.

J. Edwards, *A Treatise on the Integral Calculus, Vol II*, Chelsea, New York, 1954.

R.N. Mohapatra and R.S. Rodriguez,On the rate of convergence of singular integrals for Hölder continuous functions, *Mathematische Nachrichten*, 149 (1990), 117–124.

D. Zwillinger, *CRC standard Mathematical Tables and Formulae*, 30th ed. Chapman & Hall/CRC, Boca Raton (1995).

Chapter 30

Trigonometric-Implied Multivariate Smooth Gauss–Weierstrass Singular Integrals Quantitative Approximation

In this chapter, we employ the uniform and L_p, $1 \leq p < \infty$, approximation properties of general smooth multivariate singular integral operators over $\mathbb{R}^N$, $N \geq 1$. It is a trigonometric-relied approach with detailed applications to the corresponding smooth multivariate Gauss–Weierstrass singular integral operators. The results are quantitative via Jackson-type inequalities involving the first uniform and L_p moduli of continuity. It follows Anastassiou (2024).

30.1 Introduction

The degree of approximation by univariate and multivariate singular integral operators has been researched extensively in Anastassiou (2011)–Anastassiou (2021) and Anastassiou and Gal (2000), Anastassiou and Mezei (2012) and Mohapatra and Rodriguez (1990). All these sources motivate our current work. In particular, we studied the approximation properties of the smooth singular integral operators in Anastassiou (2011)–Anastassiou (2021) and Anastassiou and Mezei (2012). These are not in general positive operators. Here we use the uniform and L_p, $p \geq 1$, results of our multivariate general theory Anastassiou (2023) and Anastassiou (2023) to establish approximation properties of the smooth Gauss–Weierstrass singular integral operators. The degrees of approximation are given quantitatively by employing the uniform and L_p first moduli of continuity. The fundamental tool here comes from Anastassiou (2023), where a multivariate trigonometric Taylor formula is presented.

30.2 Background on General Theory

Here $r \in \mathbb{N}$, $m \in \mathbb{Z}_+$, we define

$$\alpha_j := \alpha_{j,r}^{[m]} := \begin{cases} (-1)^{r-j} \binom{r}{j} j^{-m}, & \text{if } j = 1, 2, \ldots, r, \\ 1 - \sum_{j=1}^{r} (-1)^{r-j} \binom{r}{j} j^{-m}, & \text{if } j = 0, \end{cases} \tag{30.1}$$

and

$$\delta_k := \delta_{k,r}^{[m]} := \sum_{j=0}^{r} \alpha_{j,r}^{[m]} j^k, \quad k = 1, 2, \ldots, m \in \mathbb{N}. \tag{30.2}$$

See that

$$\sum_{j=0}^{r} \alpha_{j,r}^{[m]} = 1 \tag{30.3}$$

and

$$-\sum_{j=1}^{r} (-1)^{r-j} \binom{r}{j} = (-1)^r \binom{r}{0}.$$

Let μ_{ξ_n} be a probability Borel measure on $\mathbb{R}^N$, $N \geq 1$, $\xi_n > 0$, $n \in \mathbb{N}$. We now define the multiple smooth singular integral operators

$$\theta_n (f; x_1, \ldots, x_N) := \theta_{r,n}^{[m]} (f; x_1, \ldots, x_N)$$
$$:= \sum_{j=0}^{r} \alpha_{j,r}^{[m]} \int_{\mathbb{R}^N} f (x_1 + s_1 j, x_2 + s_2 j, \ldots, x_N + s_N j) \, d\mu_{\xi_n} (s), \tag{30.4}$$

where $s := (s_1, \ldots, s_N)$, $x := (x_1, \ldots, x_N) \in \mathbb{R}^N$; $n, r \in \mathbb{Z}$, $m \in \mathbb{Z}_+$, $f : \mathbb{R}^N \to \mathbb{R}$ is a Borel measurable function, and also $(\xi_n)_{n \in \mathbb{N}}$ is a bounded sequence of positive real numbers, we take $0 < \xi_n \leq 1$.

Remark 30.1. The operators $\theta_{r,n}^{[m]}$ are not in general positive, see Anastassiou (2011, p. 2).

We observe that

Lemma 30.1. *It holds*

$$\theta_{r,n}^{[m]} (c; x_1, \ldots, x_n) = c,$$

where c is a constant.

We need

Definition 30.1. Let $f \in C\left(\mathbb{R}^N\right)$, $N \geq 1$. We define the first uniform modulus of continuity of f as

$$\omega_1(f, \delta) := \sup_{\substack{x, y \in \mathbb{R}^N: \\ \|x-y\|_\infty \leq \delta}} |f(x) - f(y)|, \quad \delta > 0, \tag{30.5}$$

where $\|\cdot\|_\infty$ is the max norm in $\mathbb{R}^N$. The functional $\omega_1(f, \delta)$ is bounded for f being bounded or uniformly continuous, and $\omega_1(f, \delta) \to 0$ as $\delta \to 0$, in the case of f being uniformly continuous.

We mention the main uniform general approximation result regarding the operator θ_n.

Theorem 30.1. *(Anastassiou 2023) Here $f \in C^2\left(\mathbb{R}^N\right)$ and let all $\alpha_i \in \mathbb{Z}^+$, $i = 1, \ldots, N$, $N \geq 1$, $|\alpha| := \sum_{i=1}^N \alpha_i = 2$; $x \in \mathbb{R}^N$, and all the partials f_α of order 2, along with $f \in C_B\left(\mathbb{R}^N\right)$ (continuous and bounded functions); or all f_α of order 2, $f \in C_U\left(\mathbb{R}^N\right)$ (uniformly continuous functions). Let μ_{ξ_n} be a Borel probability measure on $\mathbb{R}^N$, for $0 < \xi_n \leq 1$, $n \in \mathbb{N}$.*

Suppose that for all $\alpha := (\alpha_1, \ldots, \alpha_N)$, $\alpha_i \in \mathbb{Z}^+$, $i = 1, \ldots, N$, $|\alpha| = \sum_{i=1}^N \alpha_i = 2$, $j = 0, 1, \ldots, r$, we have that both

$$I_{1j}(\alpha) := \int_{\mathbb{R}^N} \left(1 + \frac{j\|s\|_1}{3\xi_n}\right) \left(\prod_{i=1}^N |s_i|^{\alpha_i}\right) d\mu_{\xi_n}(s), \tag{30.6}$$

$$I_{2j}(\alpha) := \int_{\mathbb{R}^N} \left(1 + \frac{j\|s\|_1}{3\xi_n}\right) d\mu_{\xi_n}(s) \tag{30.7}$$

are uniformly bounded in $\xi_n \in (0, 1]$.

Denote $(n \in \mathbb{N})$

$$\Delta_n(x) := \theta_n(f, x) - f(x) - \left(\sum_{j=0}^r \alpha_j j\right) \sin(1) \times \left[\sum_{i=1}^N \frac{\partial f(x)}{\partial x_i} \left(\int_{\mathbb{R}^N} s_i d\mu_{\xi_n}(s)\right)\right] - 2\left(\sum_{j=0}^r \alpha_j j^2\right) \sin^2\left(\frac{1}{2}\right) \left\{\sum_{i=1}^N \left(\int_{\mathbb{R}^N} s_i^2 d\mu_{\xi_n}(s)\right) \frac{\partial^2 f(x)}{\partial x_i^2} + \sum_{\substack{i \neq j^*, \\ i, j^* \in \{1, \ldots, N\}}} \left(\int_{\mathbb{R}^N} s_i s_{j^*} d\mu_{\xi_n}(s)\right) \frac{\partial^2 f(x)}{\partial x_i \partial x_{j^*}}\right\}. \tag{30.8}$$

Then

(i)

$$|\Delta_n(x)| \leq \|\Delta_n(x)\|_\infty \leq \sum_{j=0}^{r} |\alpha_j|$$

$$\times \left[\left[j^2 \sum_{\substack{\alpha_i \in \mathbb{Z}^+, \\ \alpha:|\alpha|=2}} \left(\frac{1}{\prod_{i=1}^{N} \alpha_i!} \right) \omega_1(f_\alpha, \xi_n) \right.\right.$$

$$\left. \times \int_{\mathbb{R}^N} \left(1 + \frac{j\|s\|_\infty}{3\xi_n}\right) \left(\prod_{i=1}^{N} |s_i|^{\alpha_i}\right) d\mu_{\xi_n}(s) \right]$$

$$\left. + \frac{1}{2}\omega_1(f, \xi_n) \int_{\mathbb{R}^N} \left(1 + \frac{j\|s\|_\infty}{3\xi_n}\right) d\mu_{\xi_n}(s) \right] =: \varphi_{\xi_n}. \tag{30.9}$$

In case of all f_α of order 2 and $f \in C_U(\mathbb{R}^N)$ and $\xi_n \to 0$, as $n \to \infty$, then $\Delta_n(x)$, $\|\Delta_n(x)\|_\infty \to 0$ with rates.

(ii) *If $\frac{\partial f(x)}{\partial x_i} = 0$, $i = 1, \ldots, N$, and $f_\alpha(x) = 0$, $\alpha_i \in \mathbb{Z}^+$, $i = 1, \ldots, N$, with $|\alpha| = 2$, then*

$$|\theta_n(f, x) - f(x)| \leq \varphi_{\xi_n}. \tag{30.10}$$

And $\theta_n(f, x) \to f(x)$ in the uniformly continuous case.

(iii) *Additionally, assume all partials of order ≤ 2 are bounded. Hence*

$$\|\theta_n(f) - f\|_\infty \leq \left(\sum_{j=0}^{r} |\alpha_j| j\right) (0.8414) \left[\sum_{i=1}^{N} \left\|\frac{\partial f}{\partial x_i}\right\|_\infty \left(\int_{\mathbb{R}^N} |s_i| d\mu_{\xi_n}(s)\right)\right]$$

$$+ \left(\sum_{j=0}^{r} |\alpha_j| j^2\right) (0.4596) \left\{ \sum_{i=1}^{N} \left(\int_{\mathbb{R}^N} s_i^2 d\mu_{\xi_n}(s)\right) \left\|\frac{\partial^2 f}{\partial x_i^2}\right\|_\infty \right.$$

$$\left. + \sum_{\substack{i \neq j^*, \\ i, j^* \in \{1, \ldots, N\}}} \left(\int_{\mathbb{R}^N} |s_i| |s_{j^*}| d\mu_{\xi_n}(s)\right) \left\|\frac{\partial^2 f}{\partial x_i \partial x_{j^*}}\right\|_\infty \right\} + \varphi_{\xi_n}. \tag{30.11}$$

If all $\int_{\mathbb{R}^N} s_i^2 d\mu_{\xi_n}(s)$ *and* $\int_{\mathbb{R}^N} |s_i| |s_{j^*}| d\mu_{\xi_n}(s)$ *converge to zero, as* $n \to \infty$, *with* $\xi_n \to 0$, *and all* f_α *of order* 2, $f \in C_U(\mathbb{R}^N)$, *then*

$$\|\theta_n(f) - f\|_\infty \to 0 \text{ with rates, as } \xi_n \to 0,\ n \to +\infty.$$

Next we deal with $f \in C^m(\mathbb{R}^N)$, $m \in \mathbb{Z}^+$, with $f_\alpha \in L_p(\mathbb{R}^N)$, $|\alpha| = m \in \mathbb{Z}^+$, $p \geq 1$, where f_α denotes the mixed partial $\frac{\partial^{\widetilde{j}} f(\cdot,\ldots,\cdot)}{\partial x_1^{\alpha_1} \ldots \partial x_N^{\alpha_N}}$, $\alpha_j \in \mathbb{Z}^+$, $j = 1, \ldots, N : |\alpha| := \sum_{j=1}^N \alpha_j = \widetilde{j}$, $\widetilde{j} = 1, \ldots, m$.

We need

Definition 30.2. (see also Anastassiou (2011), p. 20) We call

$$\Delta_u^r f(x) := \Delta_{u_1,u_2,\ldots,u_N}^r f(x_1, \ldots, x_N) := \sum_{j=0}^r (-1)^{r-j} \binom{r}{j} f(x_1 + ju_1, x_2 + ju_2, \ldots, x_N + ju_N). \tag{30.12}$$

Let $p \geq 1$, the L_p modulus of smoothness of order r is given by

$$\omega_r(f; h)_p := \sup_{\|u\|_2 \leq h} \|\Delta_u^r(f)\|_p, \tag{30.13}$$

$h > 0$.

We mention

Theorem 30.2. *(Anastassiou 2023) Let* $f \in C^m(\mathbb{R}^N)$, $m \in \mathbb{N}$, $N \geq 1$, *with* $f_\alpha \in L_p(\mathbb{R}^N)$, $|\alpha| = m$, $x \in \mathbb{R}^N$. *Let* $p, q > 1 : \frac{1}{p} + \frac{1}{q} = 1$. *Here* μ_{ξ_n} *is a Borel probability measure on* $\mathbb{R}^N$ *for* $\xi_n > 0$, $(\xi_n)_{n \in \mathbb{N}}$ *bounded sequence. Assume for all* $\alpha := (\alpha_1, \ldots, \alpha_N)$, $\alpha_i \in \mathbb{Z}^+$, $i = 1, \ldots, N$, $|\alpha| := \sum_{i=1}^N \alpha_i = m$ *that we have*

$$\int_{\mathbb{R}^N} \left(\left(\prod_{i=1}^N |s_i|^{\alpha_i} \right) \left(1 + \frac{\|s\|_2}{\xi_n} \right)^r \right)^p d\mu_{\xi_n}(s) < \infty. \tag{30.14}$$

For $\widetilde{j} = 1, \ldots, m$, *and* $\alpha := (\alpha_1, \ldots, \alpha_N)$, $\alpha_i \in \mathbb{Z}^+$, $i = 1, \ldots, N$, $|\alpha| := \sum_{i=1}^N \alpha_i = \widetilde{j}$, *call*

$$c_{\alpha,n,\widetilde{j}} := \int_{\mathbb{R}^N} \prod_{i=1}^N s_i^{\alpha_i} d\mu_{\xi_n}(s). \tag{30.15}$$

Then

$$\left\| E_{r,n}^{[m]} \right\|_p := \left\| \theta_{r,n}^{[m]} (f;x) - f(x) - \sum_{\widetilde{j}=1}^{m} \delta_{\widetilde{j},r}^{[m]} \left(\sum_{|\alpha|=\widetilde{j}} \frac{c_{\alpha,n,\widetilde{j}} f_\alpha (x)}{\left(\prod_{i=1}^{N} \alpha_i!\right)} \right) \right\|_{p,x}$$

$$\leq \left[\left(N^m/(m-1)!\right)^{\frac{p}{q}} \left(\frac{m}{(q(m-1)+1)^{\frac{p}{q}}} \right) \left(\sum_{|\alpha|=m} \frac{1}{\prod_{i=1}^{N} \alpha_i!} \cdot \right.\right.$$

$$\left.\left. \left[\int_{\mathbb{R}^N} \left[\left(\prod_{i=1}^{N} |s_i|^{\alpha_i} \right) \left(1 + \frac{\|s\|_2}{\xi_n} \right)^r \right]^p d\mu_{\xi_n}(s) \right] \omega_r (f_\alpha, \xi_n)_p^p \right) \right]^{\frac{1}{p}}. \tag{30.16}$$

As $n \to \infty$ and $\xi_n \to 0$, by (30.16), we obtain that $\left\| E_{r,n}^{[m]} \right\|_p \to 0$ with rates.

One also gets by (30.16) that

$$\left\| \theta_{r,n}^{[m]} (f;x) - f(x) \right\|_{p,x}$$

$$\leq \sum_{\widetilde{j}=1}^{m} \left| \delta_{\widetilde{j},r}^{[m]} \right| \left(\sum_{|\alpha|=\widetilde{j}} \frac{\left| c_{\alpha,n,\widetilde{j}} \right|}{\prod_{i=1}^{N} \alpha_i!} \|f_\alpha\|_p \right) + R.H.S.(30.16), \tag{30.17}$$

given that $\|f_\alpha\|_p < \infty$, $|\alpha| = \widetilde{j}$, $\widetilde{j} = 1, \ldots, m$.

Assuming that $c_{\alpha,n,\widetilde{j}} \to 0$, $\xi_n \to 0$, as $n \to \infty$, we get $\left\| \theta_{r,n}^{[m]} (f) - f \right\|_p \to 0$, that is, $\theta_{r,n}^{[m]} \to I$ the unit operator, in L_p norm, with rates.

We make

Remark 30.2. Note that $(p > 1)$

$$\int_{\mathbb{R}^N} \left(\prod_{i=1}^{N} |s_i|^{\alpha_i} \right) \left(1 + \frac{\|s\|_2}{\xi_n} \right)^r d\mu_{\xi_n}(s)$$

$$\leq \left[\int_{\mathbb{R}^N} \left(\left(\prod_{i=1}^{N} |s_i|^{\alpha_i} \right) \left(1 + \frac{\|s\|_2}{\xi_n} \right)^r \right)^p d\mu_{\xi_n}(s) \right]^{\frac{1}{p}} < \infty, \tag{30.18}$$

by assumption of Theorem 30.2.

By (30.18), we get that

$$\int_{\mathbb{R}^N} \prod_{i=1}^{N} |s_i|^{\alpha_i} d\mu_{\xi_n}(s) < \infty. \tag{30.19}$$

Hence $c_{\alpha,n,\widetilde{j}} \in \mathbb{R}$.

We mention also the following trigonometric induced alternative L_p approximation result for θ_n operators.

Theorem 30.3. *(Anastassiou 2023) Let $p, q > 1 : \frac{1}{p} + \frac{1}{q} = 1$, $0 < \xi_n \leq 1$, $n \in \mathbb{N}$. Here we deal with $f \in C^2(\mathbb{R}^N)$, $N \geq 1$, with f, $f_\alpha \in L_p(\mathbb{R}^N)$, $|\alpha| = 2$, where $\alpha_i \in \mathbb{Z}^+$, $i = 1, \ldots, N$, and $|\alpha| = \sum_{i=1}^N \alpha_i$; $x \in \mathbb{R}^N$. Let μ_{ξ_n} be a Borel probability measure on $\mathbb{R}^N$. Suppose that for all $\alpha : |\alpha| = 2$, $j = 0, 1, \ldots, r$, we have that both*

$$I_{1j}(\alpha) := \int_{\mathbb{R}^N} \left(\prod_{i=1}^N |s_i|^{p\alpha_i} \right) \left(1 + \frac{j^p \|s\|_2^p}{\xi_n^p} \right) d\mu_{\xi_n}(s), \tag{30.20}$$

$$I_2 := \int_{\mathbb{R}^N} \left(\frac{\|s\|_2}{\xi_n} \right)^p d\mu_{\xi_n}(s) \tag{30.21}$$

are uniformly bounded in $\xi_n \in (0, 1]$.

Denote ($n \in \mathbb{N}$)

$$\begin{aligned} \Delta_n(x) &:= \theta_n(f, x) - f(x) - \left(\sum_{j=0}^r \alpha_j j \right) \sin(1) \\ &\times \left[\sum_{i=1}^N \frac{\partial f(x)}{\partial x_i} \left(\int_{\mathbb{R}^N} s_i d\mu_{\xi_n}(s) \right) \right] \\ &- 2 \left(\sum_{j=0}^r \alpha_j j^2 \right) \sin^2\left(\frac{1}{2}\right) \left\{ \sum_{i=1}^N \left(\int_{\mathbb{R}^N} s_i^2 d\mu_{\xi_n}(s) \right) \frac{\partial^2 f(x)}{\partial x_i^2} \right. \\ &+ \left. \sum_{\substack{i \neq j^*, \\ i, j^* \in \{1, \ldots, N\}}} \left(\int_{\mathbb{R}^N} s_i s_{j^*} d\mu_{\xi_n}(s) \right) \frac{\partial^2 f(x)}{\partial x_i \partial x_{j^*}} \right\}. \end{aligned} \tag{30.22}$$

Then, it holds

$$\|\Delta_n(x)\|_p \leq \left(\sum_{j=0}^r |\alpha_j|^q \right)^{\frac{1}{q}} \left\{ \frac{2}{(q+1)^{\frac{1}{q}}} \left(\frac{N(N+1)+2}{2} \right)^{\frac{1}{q}} \right\}$$

$$\left\{ \sum_{j=0}^r \left[j^{2p} \sum_{|\alpha|=2} \left(\frac{1}{\prod_{i=1}^N \alpha_i!} \right)^p \omega_1(f_\alpha, \xi_n)_p^p \left(\int_{\mathbb{R}^N} \left(\prod_{i=1}^N |s_i|^{p\alpha_i} \right) \right. \right. \right.$$

$$\times \left(1 + \frac{j^p \|s\|_2^p}{\xi_n^p}\right) d\mu_{\xi_n}(s)\Bigg)$$

$$+ \omega_1(f, \xi_n)_p^p \left(\int_{\mathbb{R}^N} \left(1 + \frac{j^p \|s\|_2^p}{\xi_n^p}\right) d\mu_{\xi_n}(s)\right)\Bigg]\Bigg\}^{\frac{1}{p}}. \tag{30.23}$$

As $n \to \infty$ and $\xi_n \to 0$, by (30.23), we obtain that $\|\Delta_n\|_p \to 0$ with rates. One also gets by (30.23) that

$$\|\theta_n(f,x) - f(x)\|_{p,x}$$

$$\leq \left(\sum_{j=0}^{r} |\alpha_j| j\right) \sin(1) \left[\sum_{i=1}^{N} \left\|\frac{\partial f}{\partial x_i}\right\|_p \left(\int_{\mathbb{R}^N} |s_i| \, d\mu_{\xi_n}(s)\right)\right]$$

$$+ 2\left(\sum_{j=0}^{r} |\alpha_j| j^2\right) \sin^2\left(\frac{1}{2}\right) \left\{\sum_{i=1}^{N} \left(\int_{\mathbb{R}^N} s_i^2 d\mu_{\xi_n}(s)\right) \left\|\frac{\partial^2 f}{\partial x_i^2}\right\|_p\right.$$

$$\left. + \sum_{\substack{i \neq j^*, \\ i,j^* \in \{1,\dots,N\}}} \left(\int_{\mathbb{R}^N} |s_i s_{j^*}| \, d\mu_{\xi_n}(s)\right) \left\|\frac{\partial^2 f}{\partial x_i \partial x_{j^*}}\right\|_p\right\} + R.H.S.\,(30.23), \tag{30.24}$$

given that $\|f_\alpha\|_p < \infty$, $|\alpha| = \widetilde{j}$, $\widetilde{j} = 1,2$. Assuming that $\int_{\mathbb{R}^N} s_i^2 d\mu_{\xi_m}(s)$, $\int_{\mathbb{R}^N} |s_i s_{j^*}| \, d\mu_{\xi_n}(s)$, $i, j^* \in \{1,\dots,N\}$, $i \neq j^*$, converge to zero as $\xi_n \to 0$, we get $\|\theta_n(f,x) - f(x)\|_p \to 0$, that is, $\theta_n \to I$ the unit operator, in L_p norm, with rates.

Furthermore, we mention the following trigonometric-based alternative L_1 approximation result for θ_n operators.

Theorem 30.4. *(Anastassiou 2023) Let $0 < \xi_n \leq 1$, $n \in \mathbb{N}$, $x \in \mathbb{R}^N$. Here we deal with $f \in C^2(\mathbb{R}^N)$, $N \geq 1$, with $f, f_\alpha \in L_1(\mathbb{R}^N)$, $|\alpha| = 2$, where $\alpha_i \in \mathbb{Z}^+$, $i = 1,\dots,N$, and $|\alpha| = \sum_{i=1}^N \alpha_i$. Let μ_{ξ_n} be a Borel probability measure on $\mathbb{R}^N$. Suppose that for all $\alpha : |\alpha| = 2$, $j = 0,1,\dots,r$, we have that both*

$$I_{1j}^*(\alpha) := \int_{\mathbb{R}^N} \left(\prod_{i=1}^{N} |s_i|^{\alpha_i}\right) \left(1 + \frac{j\|s\|_2}{\xi_n}\right) d\mu_{\xi_n}(s) \tag{30.25}$$

and

$$I_2^* := \int_{\mathbb{R}^N} \frac{\|s\|_2}{\xi_n} d\mu_{\xi_n}(s) \tag{30.26}$$

are uniformly bounded in $\xi_n \in (0,1]$.

Here $\Delta_n(x)$ *is as in (30.22).*

Then, it holds

$$\|\Delta_n(x)\|_1 \leq \left\{ \sum_{j=0}^{r} |\alpha_j| \left[j^2 \sum_{|\alpha|=2} \left(\frac{2}{\prod_{i=1}^{N} \alpha_i!} \right) \omega_1(f_\alpha, \xi_n)_1 \right.\right.$$
$$\times \left(\int_{\mathbb{R}^N} \left(\prod_{i=1}^{N} (|s_i|)^{\alpha_i} \right) \left(1 + \frac{j\|s\|_2}{\xi_n} \right) d\mu_{\xi_n}(s) \right)$$
$$\left.\left. + \omega_1(f, \xi_n)_1 \left(\int_{\mathbb{R}^N} \left(1 + \frac{j\|s\|_2}{\xi_n} \right) d\mu_{\xi_n}(s) \right) \right] \right\}. \tag{30.27}$$

As $n \to \infty$ *and* $\xi_n \to 0$, *by (30.27), we obtain that* $\|\Delta_n\|_1 \to 0$ *with rates. One also obtains by (30.27) that*

$$\|\theta_n(f) - f\|_1 \leq \left(\sum_{j=0}^{r} |\alpha_j| j \right) \sin(1) \left[\sum_{i=1}^{N} \left\| \frac{\partial f}{\partial x_i} \right\|_1 \left(\int_{\mathbb{R}^N} |s_i| d\mu_{\xi_n}(s) \right) \right]$$
$$+ 2 \left(\sum_{j=0}^{r} |\alpha_j| j^2 \right) \sin^2\left(\frac{1}{2}\right) \left\{ \sum_{i=1}^{N} \left(\int_{\mathbb{R}^N} s_i^2 d\mu_{\xi_n}(s) \right) \left\| \frac{\partial^2 f}{\partial x_i^2} \right\|_1 \right.$$
$$\left. + \sum_{\substack{i \neq j^*, \\ i,j^* \in \{1,\ldots,N\}}} \left(\int_{\mathbb{R}^N} |s_i s_{j^*}| d\mu_{\xi_n}(s) \right) \left\| \frac{\partial^2 f}{\partial x_i \partial x_{j^*}} \right\|_1 \right\} + R.H.S.\,(30.27), \tag{30.28}$$

given that $\|f_\alpha\|_1 < \infty$, $|\alpha| = \widetilde{j}$, $\widetilde{j} = 1, 2$. *Assuming that* $\int_{\mathbb{R}^N} s_i^2 d\mu_{\xi_n}(s)$, $\int_{\mathbb{R}^N} |s_i s_{j^*}| d\mu_{\xi_n}(s)$, $i, j^* \in \{1, \ldots, N\}$, $i \neq j^*$, *converge to zero as* $\xi_n \to 0$, *we derive* $\|\theta_n(f) - f\|_1 \to 0$, *that is,* $\theta_n \to I$ *in* L_1 *norm, with rates.*

30.3 Auxiliary Essential Results

We need

Theorem 30.5. *Let $N \in \mathbb{N}$; $\alpha_i \in \mathbb{Z}^+$, $i = 1, \ldots, N : |\alpha| := \sum_{i=1}^{N} \alpha_i = 2$, $\xi_n \in (0,1]$, $n \in \mathbb{N}$; $j = 0, 1, \ldots, r \in \mathbb{N}$. Then*

$$I_{1j}^{*}(\alpha) := \frac{1}{\left(\sqrt{\pi \xi_n}\right)^N} \int_{\mathbb{R}^N} \left(1 + \frac{j \|s\|_1}{3\xi_n}\right) \left(\prod_{i=1}^{N} |s_i|^{\alpha_i}\right) e^{-\frac{\sum_{i=1}^{N} s_i^2}{\xi_n}} ds_1 \ldots ds_N$$

$$\leq \sqrt{\xi_n} \left(\frac{2}{\sqrt{\pi}}\right)^N \left[(1 + jN) + (1 + j)\left(\frac{\lfloor 6e \rfloor}{e}\right)^N\right]$$

$$\leq \left(\frac{2}{\sqrt{\pi}}\right)^N \left[(1 + jN) + (1 + j)\left(\frac{\lfloor 6e \rfloor}{e}\right)^N\right] < +\infty \tag{30.29}$$

are uniformly bounded, and $I_{1j}^{*}(\alpha) \to 0$, as $\xi_n \to 0$.

Above $\lfloor \cdot \rfloor$ denotes the integral part.

Proof. We have that

$$I_{1j}^{*}(\alpha) = \frac{1}{\left(\sqrt{\pi \xi_n}\right)^N} \int_{\mathbb{R}^N} \left(1 + \frac{j \|s\|_1}{3\xi_n}\right) \left(\prod_{i=1}^{N} |s_i|^{\alpha_i}\right) e^{-\frac{\sum_{i=1}^{N} s_i^2}{\xi_n}} ds_1 \ldots ds_N$$

$$= \frac{2^N}{\left(\sqrt{\pi}\right)^N \left(\sqrt{\xi_n}\right)^N} \int_{\mathbb{R}_+^N} \left(1 + \frac{j \|s\|_1}{3\xi_n}\right) \left(\prod_{i=1}^{N} s_i^{\alpha_i}\right) e^{-\frac{\sum_{i=1}^{N} s_i^2}{\xi_n}} ds_1 \ldots ds_N$$

$$= \frac{2^N}{\left(\sqrt{\pi}\right)^N \left(\sqrt{\xi_n}\right)^N} \int_{\mathbb{R}_+^N} \left(1 + \frac{j \left(\sum_{i=1}^{N} s_i\right)}{3\xi_n}\right) \left(\prod_{i=1}^{N} s_i^{\alpha_i}\right) e^{-\frac{\sum_{i=1}^{N} s_i^2}{\xi_n}} ds_1 \ldots ds_N$$

$$\leq \frac{2^N \left(\sqrt{\xi_n}\right)^2}{\left(\sqrt{\pi}\right)^N} \int_{\mathbb{R}_+^N} \left(\frac{1}{\sqrt{\xi_n}} + j \frac{1}{\sqrt{\xi_n}} \left(\sum_{i=1}^{N} \frac{s_i}{\sqrt{\xi_n}}\right)\right) \tag{30.30}$$

$$\left(\prod_{i=1}^{N} \left(\frac{s_i}{\sqrt{\xi_n}}\right)^{\alpha_i}\right) e^{-\sum_{i=1}^{N} \left(\frac{s_i}{\sqrt{\xi_n}}\right)^2} \frac{ds_1}{\sqrt{\xi_n}} \cdots \frac{ds_N}{\sqrt{\xi_n}}$$

$$= \frac{2^N \sqrt{\xi_n}}{\left(\sqrt{\pi}\right)^N} \int_{\mathbb{R}_+^N} \left(1 + j \left(\sum_{i=1}^{N} z_i\right)\right) \left(\prod_{i=1}^{N} z_i^{\alpha_i}\right) e^{-\left(\sum_{i=1}^{N} z_i^2\right)} dz_1 \ldots dz_N \tag{30.31}$$

$$= \frac{2^N \sqrt{\xi_n}}{\left(\sqrt{\pi}\right)^N} \left[\int_{[0,1]^N} \left(1 + j \left(\sum_{i=1}^{N} z_i\right)\right) \left(\prod_{i=1}^{N} z_i^{\alpha_i}\right) e^{-\left(\sum_{i=1}^{N} z_i^2\right)} dz_1 \ldots dz_N\right.$$

$$+ \int_{(\mathbb{R}_+-[0,1])^N} \left(1 + j\left(\sum_{i=1}^N z_i\right)\right)\left(\prod_{i=1}^N z_i^{\alpha_i}\right) e^{-\left(\sum_{i=1}^N z_i^2\right)} dz_1 \ldots dz_N\Bigg]$$

$$\leq \frac{2^N\sqrt{\xi_n}}{\left(\sqrt{\pi}\right)^N}\Bigg[(1+jN) + (1+j)$$

$$\times \int_{(\mathbb{R}_+-[0,1])^N} \left(\sum_{i=1}^N z_i\right)\left(\prod_{i=1}^N z_i^{\alpha_i}\right) e^{-\sum_{i=1}^N z_i} dz_1 \ldots dz_N\Bigg]$$

$$\leq \frac{2^N\sqrt{\xi_n}}{\left(\sqrt{\pi}\right)^N}\Bigg[(1+jN) + (1+j)$$

$$\times \int_{(\mathbb{R}_+-[0,1])^N} \left(\prod_{i=1}^N z_i\right)\left(\prod_{i=1}^N z_i^{\alpha_i}\right) \prod_{i=1}^N e^{-z_i} \prod_{i=1}^N dz_i\Bigg]$$

$$= \frac{2^N\sqrt{\xi_n}}{\left(\sqrt{\pi}\right)^N}\left[(1+jN) + (1+j)\prod_{i=1}^N \int_1^\infty z_i^{(\alpha_i+2)-1} e^{-z_i} dz_i\right] = \tag{30.32}$$

(by Gradshteyn and Ryzhik (2015), p. 348)

$$\frac{2^N\sqrt{\xi_n}}{\left(\sqrt{\pi}\right)^N}\left[(1+jN) + (1+j)\prod_{i=1}^N \Gamma\left((\alpha_i+2), 1\right)\right] =$$

(where $\Gamma(\cdot,\cdot)$ is the upper incomplete gamma function)

$$\frac{2^N\sqrt{\xi_n}}{\left(\sqrt{\pi}\right)^N}\left[(1+jN) + (1+j)\prod_{i=1}^N \frac{\lfloor e(\alpha_i+1)!\rfloor}{e}\right]$$

$$\leq \frac{2^N\sqrt{\xi_n}}{\left(\sqrt{\pi}\right)^N}\left[(1+jN) + (1+j)\left(\frac{\lfloor 6e\rfloor}{e}\right)^N\right] < +\infty.$$

That is,

$$I_{1j}^*(\alpha) \leq \sqrt{\xi_n}\left(\frac{2}{\sqrt{\pi}}\right)^N\left[(1+jN) + (1+j)\left(\frac{\lfloor 6e\rfloor}{e}\right)^N\right]$$

$$\leq \left(\frac{2}{\sqrt{\pi}}\right)^N\left[(1+jN) + (1+j)\left(\frac{\lfloor 6e\rfloor}{e}\right)^N\right] < +\infty \tag{30.33}$$

are uniformly bounded, furthermore $I_{1j}^*(\alpha) \to 0$, as $\xi_n \to 0$. □

We make

Remark 30.3. By Theorem 30.5, $j = 0, 1, \ldots, r \in \mathbb{N}$; $i, j^* \in \{1, \ldots, N\}$, $i \neq j^*$, we have that

$$\frac{1}{\left(\sqrt{\pi\xi_n}\right)^N} \int_{\mathbb{R}^N} |s_i| \, |s_{j^*}| \, e^{-\frac{\sum_{i=1}^N s_i^2}{\xi_n}} ds_1 \cdots ds_N,$$

$$\frac{1}{\left(\sqrt{\pi\xi_n}\right)^N} \int_{\mathbb{R}^N} s_i^2 e^{-\frac{\sum_{i=1}^N s_i^2}{\xi_n}} ds_1 \cdots ds_N \leq I_{1j}^* < \infty, \tag{30.34}$$

and

$$\frac{1}{\left(\sqrt{\pi\xi_n}\right)^N} \int_{\mathbb{R}^N} |s_i| \, e^{-\frac{\sum_{i=1}^N s_i^2}{\xi_n}} ds_1 \cdots ds_N \leq \tag{30.35}$$

$$\frac{1}{\left(\sqrt{\pi\xi_n}\right)^N} \int_{\mathbb{R}^N} s_i^2 e^{-\frac{\sum_{i=1}^N s_i^2}{\xi_n}} ds_1 \cdots ds_N < \infty,$$

and all these integrals are uniformly bounded in $\xi_n \in (0, 1]$. And all integrals converge to zero, as $\xi_n \to 0$.

We continue with

Theorem 30.6. *Let $N \geq 1$, $\xi_n \in (0, 1]$, $n \in \mathbb{N}$. Then*

$$I_2^* := \frac{1}{\left(\sqrt{\pi\xi_n}\right)^N} \int_{\mathbb{R}^N} \frac{\|s\|_1}{\xi_n} e^{-\frac{\sum_{i=1}^N s_i^2}{\xi_n}} ds_1 \cdots ds_N$$

$$\leq \frac{1}{\sqrt{\xi_n}} \left(\frac{2}{\sqrt{\pi}}\right)^N \left[N + \left(\frac{\lfloor e \rfloor}{e}\right)^N\right]. \tag{30.36}$$

Proof. We observe that

$$I_2^* = \frac{1}{\left(\sqrt{\pi\xi_n}\right)^N} \int_{\mathbb{R}^N} \frac{\|s\|_1}{\xi_n} e^{-\frac{\sum_{i=1}^N s_i^2}{\xi_n}} ds_1 \cdots ds_N$$

$$= \frac{2^N}{\left(\sqrt{\pi}\right)^N \left(\sqrt{\xi_n}\right)^N} \int_{\mathbb{R}_+^N} \frac{\sum_{i=1}^N s_i}{\xi_n} e^{-\frac{\sum_{i=1}^N s_i^2}{\xi_n}} ds_1 \cdots ds_N$$

$$= \frac{2^N}{\left(\sqrt{\pi}\right)^N} \frac{1}{\sqrt{\xi_n}} \int_{\mathbb{R}_+^N} \left(\sum_{i=1}^N \left(\frac{s_i}{\sqrt{\xi_n}}\right)\right) e^{-\sum_{i=1}^N \left(\frac{s_i}{\sqrt{\xi_n}}\right)^2} d\frac{s_1}{\sqrt{\xi_n}} \cdots d\frac{s_N}{\sqrt{\xi_n}} = \tag{30.37}$$

$$\frac{2^N}{\left(\sqrt{\pi}\right)^N} \frac{1}{\sqrt{\xi_n}} \int_{\mathbb{R}_+^N} \left(\sum_{i=1}^N z_i\right) e^{-\sum_{i=1}^N z_i^2} dz_1 \cdots dz_N$$

$$= \left(\frac{2}{\sqrt{\pi}}\right)^N \frac{1}{\sqrt{\xi_n}} \left[\int_{[0,1]^N} \left(\sum_{i=1}^N z_i \right) e^{-\sum_{i=1}^N z_i^2} dz_1 \cdots dz_N \right.$$

$$\left. + \int_{(\mathbb{R}_+ - [0,1])^N} \left(\sum_{i=1}^N z_i \right) e^{-\sum_{i=1}^N z_i^2} dz_1 \cdots dz_N \right]$$

$$\leq \left(\frac{2}{\sqrt{\pi}}\right)^N \frac{1}{\sqrt{\xi_n}} \left[N + \int_{(\mathbb{R}_+ - [0,1])^N} \left(\prod_{i=1}^N z_i \right) e^{-\sum_{i=1}^N z_i} dz_1 \cdots dz_N \right]$$

$$= \left(\frac{2}{\sqrt{\pi}}\right)^N \frac{1}{\sqrt{\xi_n}} \left[N + \prod_{i=1}^N \int_1^\infty z_i e^{-z_i} dz_i \right]$$

$$= \left(\frac{2}{\sqrt{\pi}}\right)^N \frac{1}{\sqrt{\xi_n}} \left[N + \left(\int_1^\infty z^{2-1} e^{-z} dz \right)^N \right] = \tag{30.38}$$

(by Gradshteyn and Ryzhik (2015), p. 348)

$$\left(\frac{2}{\sqrt{\pi}}\right)^N \frac{1}{\sqrt{\xi_n}} \left[N + \Gamma^N (2,1) \right]$$

$$= \left(\frac{2}{\sqrt{\pi}}\right)^N \frac{1}{\sqrt{\xi_n}} \left[N + \left(\frac{\lfloor e \rfloor}{e} \right)^N \right],$$

proving the claim. □

We need the following.

Theorem 30.7. *(Anastassiou (2021), p. 403) Let $r, N, m \in \mathbb{N}$, with $m > r$; $\alpha_i \in \mathbb{Z}^+$, $i = 1, \ldots, N : |\alpha| := \sum_{i=1}^N \alpha_i = m$, $\xi_n \in (0,1]$, $n \in \mathbb{N}$; $p > 1$. Then*

$$\widetilde{A}_{\xi_n}(\alpha) := \frac{1}{\left(\sqrt{\pi \xi_n}\right)^N} \int_{\mathbb{R}^N} \left(\left(\prod_{i=1}^N |s_i|^{\alpha_i} \right) \left(1 + \frac{\|s\|_2}{\xi_n} \right)^r \right)^p e^{-\frac{\sum_{i=1}^N s_i^2}{\xi_n}} ds_1 \cdots ds_N$$

$$\leq \sqrt{\xi_n}^{p(m-r)} \left(\frac{2}{\sqrt{\pi}}\right)^N \left[(1+N)^{rp} + 2^{rp} \Gamma^N \left((m+r)p + 1, 1 \right) \right]$$

$$\leq \left(\frac{2}{\sqrt{\pi}}\right)^N \left[(1+N)^{rp} + 2^{rp} \Gamma^N \left((m+r)p + 1, 1 \right) \right] < +\infty \tag{30.39}$$

are uniformly bounded, where $m > r$.

Also $\widetilde{A}_{\xi_n}(\alpha) \to 0$, as $\xi_n \to 0$, $n \to +\infty$.

Above $\Gamma(\cdot,\cdot)$ is the upper incomplete gamma function.

Clearly, it holds $(m > r)$

$$B_{\xi_n}(\alpha) := \frac{1}{\left(\sqrt{\pi\xi_n}\right)^N} \int_{\mathbb{R}^N} \left(\prod_{i=1}^N |s_i|^{\alpha_i}\right) e^{-\frac{\sum_{i=1}^N s_i^2}{\xi_n}} ds_1 \cdots ds_N < \infty, \quad (30.40)$$

uniformly bounded in $\xi_n \in (0,1]$. *And, clearly* $B_{\xi_n}(\alpha) \to 0$, *as* $\xi_n \to 0$, $n \to \infty$, *by (30.39).*

We need the following.

Theorem 30.8. *Let* $\alpha_i \in \mathbb{Z}^+$, $i = 1, \ldots, N \in \mathbb{N} : |\alpha| := \sum_{i=1}^N \alpha_i = 2$, $\xi_n \in (0,1]$, $n \in \mathbb{N}$, $p \geq 1$, *and* $j = 0, 1, \ldots, r \in \mathbb{N}$. *Then*

$$\widetilde{A}_{j\xi_n}(\alpha) := \frac{1}{\left(\sqrt{\pi\xi_n}\right)^N} \int_{\mathbb{R}^N} \left(\prod_{i=1}^N |s_i|^{p\alpha_i}\right) \left(1 + \frac{j^p \|s\|_2^p}{\xi_n^p}\right) e^{-\frac{\sum_{i=1}^N s_i^2}{\xi_n}} ds_1 \cdots ds_N$$

$$\leq \left(\sqrt{\xi_n}\right)^p \left(\frac{2}{\sqrt{\pi}}\right)^N \left[(1 + j^p N^p) + (1 + j^p)\Gamma^N(3p+1, 1)\right]$$

$$\leq \left(\frac{2}{\sqrt{\pi}}\right)^N \left[(1 + j^p N^p) + (1 + j^p)\Gamma^N(3p+1, 1)\right] < +\infty \quad (30.41)$$

are uniformly bounded, furthermore $\widetilde{A}_{j\xi_n}(\alpha) \to 0$, *as* $\xi_n \to 0$, $n \to \infty$.

Call $\widetilde{A}_{j\xi_n}(\alpha)$ *as* $\widetilde{\widetilde{A}}_{j\xi_n}(\alpha)$, *when* $p = 1$.

Proof. We estimate $(p \geq 1)$

$$\widetilde{A}_{j\xi_n}(\alpha) = \frac{1}{\left(\sqrt{\pi\xi_n}\right)^N} \int_{\mathbb{R}^N} \left(\prod_{i=1}^N |s_i|^{p\alpha_i}\right) \left(1 + \frac{j^p \|s\|_2^p}{\xi_n^p}\right) e^{-\frac{\sum_{i=1}^N s_i^2}{\xi_n}} ds_1 \cdots ds_N$$

$$= \frac{2^N}{\left(\sqrt{\pi}\right)^N \left(\sqrt{\xi_n}\right)^N} \int_{\mathbb{R}_+^N} \left(\prod_{i=1}^N s_i^{p\alpha_i}\right) \left(1 + \frac{j^p \|s\|_2^p}{\xi_n^p}\right) e^{-\frac{\sum_{i=1}^N s_i^2}{\xi_n}} ds_1 \cdots ds_N$$

$$\leq \frac{2^N \left(\sqrt{\xi_n}\right)^{2p}}{\left(\sqrt{\pi}\right)^N} \int_{\mathbb{R}_+^N} \left(\prod_{i=1}^N \left(\frac{s_i}{\sqrt{\xi_n}}\right)^{p\alpha_i}\right) \quad (30.42)$$

$$\left(\frac{1}{\left(\sqrt{\xi_n}\right)^p} + \left(\frac{j}{\sqrt{\xi_n}}\right)^p \left(\sum_{i=1}^N \frac{s_i}{\sqrt{\xi_n}}\right)^p\right) e^{-\sum_{i=1}^N \left(\frac{s_i}{\sqrt{\xi_n}}\right)^2} d\frac{s_1}{\sqrt{\xi_n}} \cdots d\frac{s_N}{\sqrt{\xi_n}}$$

$$= \left(\frac{2}{\sqrt{\pi}}\right)^N \left(\sqrt{\xi_n}\right)^p \int_{\mathbb{R}_+^N} \left(\prod_{i=1}^N z_i^{p\alpha_i}\right) \left(1 + j^p \left(\sum_{i=1}^N z_i\right)^p\right)$$

$$\times\ e^{-\sum_{i=1}^N z_i^2} dz_1 \cdots dz_N$$

$$= \left(\sqrt{\xi_n}\right)^p \left(\frac{2}{\sqrt{\pi}}\right)^N \left[\int_{[0,1]^N} \left(\prod_{i=1}^N z_i^{p\alpha_i}\right)\left(1 + j^p \left(\sum_{i=1}^N z_i\right)^p\right)\right.$$
$$\times e^{-\sum_{i=1}^N z_i^2} dz_1 \cdots dz_N$$
$$\left. + \int_{(\mathbb{R}_+ - [0,1])^N} \left(\prod_{i=1}^N z_i^{p\alpha_i}\right)\left(1 + j^p \left(\sum_{i=1}^N z_i\right)^p\right) e^{-\sum_{i=1}^N z_i^2} dz_1 \cdots dz_N\right] \tag{30.43}$$

$$\leq \left(\sqrt{\xi_n}\right)^p \left(\frac{2}{\sqrt{\pi}}\right)^N \left[(1 + j^p N^p) + (1 + j^p) \int_{(\mathbb{R}_+ - [0,1])^N} \left(\prod_{i=1}^N z_i^{p\alpha_i}\right)\right.$$
$$\left.\times \left(\sum_{i=1}^N z_i\right)^p e^{-\sum_{i=1}^N z_i} dz_1 \cdots dz_N\right] \leq \left(\sqrt{\xi_n}\right)^p \left(\frac{2}{\sqrt{\pi}}\right)^N$$
$$\times \left[(1 + j^p N^p) + (1 + j^p) \int_{(\mathbb{R}_+ - [0,1])^N} \left(\prod_{i=1}^N z_i^{p\alpha_i}\right)\left(\prod_{i=1}^N z_i^p\right)\right.$$
$$\left.\times \left(\prod_{i=1}^N e^{-z_i}\right) \prod_{i=1}^N dz_i\right] = \left(\sqrt{\xi_n}\right)^p \left(\frac{2}{\sqrt{\pi}}\right)^N \left[(1 + j^p N^p) + (1 + j^p)\right.$$
$$\left.\times \int_{(\mathbb{R}_+ - [0,1])^N} \left(\prod_{i=1}^N z_i^{p(\alpha_i+1)}\right)\left(\prod_{i=1}^N e^{-z_i}\right) \prod_{i=1}^N dz_i\right] = \left(\sqrt{\xi_n}\right)^p \left(\frac{2}{\sqrt{\pi}}\right)^N$$
$$\times \left[(1 + j^p N^p) + (1 + j^p) \prod_{i=1}^N \left(\int_1^\infty z_i^{p(\alpha_i+1)} e^{-z_i} dz_i\right)\right] = \tag{30.44}$$

(by Gradshteyn and Ryzhik (2015), p. 348)

$$\left(\sqrt{\xi_n}\right)^p \left(\frac{2}{\sqrt{\pi}}\right)^N \left[(1 + j^p N^p) + (1 + j^p) \prod_{i=1}^N \Gamma\left(\left(p\left(\alpha_i + 1\right) + 1\right), 1\right)\right] \leq$$

(by Gradshteyn and Ryzhik (2015), p. 909)

$$\left(\sqrt{\xi_n}\right)^p \left(\frac{2}{\sqrt{\pi}}\right)^N \left[(1 + j^p N^p) + (1 + j^p) \Gamma^N (3p + 1, 1)\right] \leq \tag{30.45}$$

$$\left(\frac{2}{\sqrt{\pi}}\right)^N \left[(1 + j^p N^p) + (1 + j^p) \Gamma^N (3p + 1, 1)\right] < +\infty$$

are uniformly bounded. □

We need the following.

Theorem 30.9. *Let $p \geq 1$, $\xi_n \in (0,1]$, $n \in \mathbb{N}$, $N \geq 1$. Then*

$$K_{p\xi_n} := \frac{1}{\left(\sqrt{\pi \xi_n}\right)^N} \int_{\mathbb{R}^N} \left(\frac{\|s\|_2}{\xi_n}\right)^p e^{-\frac{\sum_{i=1}^N s_i^2}{\xi_n}} ds_1 \dots ds_N$$

$$\leq \frac{1}{\left(\sqrt{\xi_n}\right)^p} \left(\frac{2}{\sqrt{\pi}}\right)^N \left[N^p + \Gamma^N (p+1,1)\right]. \tag{30.46}$$

We set $K_{p\xi_n}$ as $K^*_{\xi_n}$, when $p = 1$.

Proof. We have $(p \geq 1)$

$$K_{p\xi_n} = \frac{1}{\left(\sqrt{\pi \xi_n}\right)^N} \int_{\mathbb{R}^N} \left(\frac{\|s\|_2}{\xi_n}\right)^p e^{-\frac{\sum_{i=1}^N s_i^2}{\xi_n}} ds_1 \cdots ds_N$$

$$= \frac{2^N}{\left(\sqrt{\pi}\right)^N \left(\sqrt{\xi_n}\right)^N} \int_{\mathbb{R}_+^N} \left(\frac{\|s\|_2}{\xi_n}\right)^p e^{-\frac{\sum_{i=1}^N s_i^2}{\xi_n}} ds_1 \cdots ds_N$$

$$\leq \frac{2^N}{\left(\sqrt{\pi}\right)^N \left(\sqrt{\xi_n}\right)^N} \int_{\mathbb{R}_+^N} \left(\frac{\sum_{i=1}^N s_i}{\xi_n}\right)^p e^{-\frac{\sum_{i=1}^N s_i^2}{\xi_n}} ds_1 \cdots ds_N = \tag{30.47}$$

$$\frac{2^N}{\left(\sqrt{\pi}\right)^N} \frac{1}{\left(\sqrt{\xi_n}\right)^p} \int_{\mathbb{R}_+^N} \left(\sum_{i=1}^N \left(\frac{s_i}{\sqrt{\xi_n}}\right)\right)^p e^{-\sum_{i=1}^N \left(\frac{s_i}{\sqrt{\xi_n}}\right)^2} d\frac{s_1}{\sqrt{\xi_n}} \cdots d\frac{s_N}{\sqrt{\xi_n}}$$

$$= \frac{2^N}{\left(\sqrt{\pi}\right)^N} \left(\sqrt{\xi_n}\right)^{-p} \int_{\mathbb{R}_+^N} \left(\sum_{i=1}^N z_i\right)^p e^{-\sum_{i=1}^N z_i^2} dz_1 \cdots dz_N$$

$$= \frac{2^N}{\left(\sqrt{\pi}\right)^N} \left(\sqrt{\xi_n}\right)^{-p} \left[\int_{[0,1]^N} \left(\sum_{i=1}^N z_i\right)^p e^{-\sum_{i=1}^N z_i^2} dz_1 \cdots dz_N\right.$$

$$\left. + \int_{(\mathbb{R}_+ - [0,1])^N} \left(\sum_{i=1}^N z_i\right)^p e^{-\sum_{i=1}^N z_i^2} dz_1 \cdots dz_N\right]$$

$$\leq \frac{2^N}{\left(\sqrt{\pi}\right)^N} \left(\sqrt{\xi_n}\right)^{-p} \left[N^p + \int_{(\mathbb{R}_+ - [0,1])^N} \left(\prod_{i=1}^N z_i^p\right) e^{-\sum_{i=1}^N z_i} dz_1 \cdots dz_N\right]$$

$$= \frac{2^N}{\left(\sqrt{\pi}\right)^N} \left(\sqrt{\xi_n}\right)^{-p} \left[N^p + \prod_{i=1}^N \int_1^\infty z_i^p e^{-z_i} dz_i\right] \tag{30.48}$$

$$= \frac{2^N}{\left(\sqrt{\pi}\right)^N} \left(\sqrt{\xi_n}\right)^{-p} \left[N^p + \left(\int_1^\infty z^p e^{-z} dz\right)^N\right] =$$

(by Gradshteyn and Ryzhik (2015), p. 348)

$$\frac{2^N}{\left(\sqrt{\pi}\right)^N}\left(\sqrt{\xi_n}\right)^{-p}\left[N^p+\Gamma^N(p+1,1)\right].$$

□

30.4 Main Results

The general smooth multivariate Gauss–Weierstrass singular integral operators are defined as follows:

$$W_n(f;x_1,\ldots,x_N):=W_{r,n}^{[m]}(f;x_1,\ldots,x_N)$$
$$:=\frac{1}{\left(\sqrt{\pi\xi_n}\right)^N}\sum_{j=0}^{r}\alpha_{j,r}^{[m]}\int_{\mathbb{R}^N}f(x_1+s_1j,x_2+s_2j,\ldots,x_N+s_Nj)$$
$$\times e^{-\frac{\sum_{i=1}^N s_i^2}{\xi_n}}ds_1\cdots ds_N. \tag{30.49}$$

Observe that

$$\frac{1}{\left(\sqrt{\pi\xi_n}\right)^N}\int_{\mathbb{R}^N}e^{-\frac{\sum_{i=1}^N s_i^2}{\xi_n}}ds_1\ldots ds_N=1, \tag{30.50}$$

see Anastassiou (2011), p. 15.

That is, $W_{r,n}^{[m]}$ are the $\theta_{r,n}^{[m]}$ operators applied for the Borel probability measures on $\mathbb{R}^N$, $N\geq 1$,

$$d\mu_{\xi_n}(s)=\frac{1}{\left(\sqrt{\pi\xi_n}\right)^N}e^{-\frac{\sum_{i=1}^N s_i^2}{\xi_n}}ds_1\ldots ds_N,\quad s\in\mathbb{R}^N, \tag{30.51}$$

where $0<\xi_n\leq 1$, $n\in\mathbb{N}$.

We apply to $W_{r,n}^{[m]}$ the Theorems 30.1, 30.2, 30.3 and 30.4, with the help of all Section 3. That is a presentation of approximation properties by W_n.

Here apply first Theorem 30.1 to W_n operators.

Theorem 30.10. *We consider $f\in C^2(\mathbb{R}^N)$ and let all $\alpha_i\in\mathbb{Z}^+$, $i=1,\ldots,N$, $N\geq 1$, $|\alpha|:=\sum_{i=1}^N\alpha_i=2$; $x\in\mathbb{R}^N$, and all the partials f_α of order 2, along with $f\in C_B(\mathbb{R}^N)$; or all f_α of order 2, $f\in C_U(\mathbb{R}^N)$; $0<\xi_n\leq 1$, $n\in\mathbb{N}$.*

Denote ($n \in \mathbb{N}$)

$$\Delta_n^*(x) := W_n(f,x) - f(x) - \left(\sum_{j=0}^{r} \alpha_j j\right) \sin(1)$$

$$\times \left[\sum_{i=1}^{N} \frac{\partial f(x)}{\partial x_i} \frac{1}{\left(\sqrt{\pi \xi_n}\right)^N} \left(\int_{\mathbb{R}^N} s_i e^{-\frac{\sum_{i=1}^N s_i^2}{\xi_n}} ds_1 \ldots ds_N\right)\right]$$

$$- 2\left(\sum_{j=0}^{r} \alpha_j j^2\right) \sin^2\left(\frac{1}{2}\right) \left\{\sum_{i=1}^{N} \frac{1}{\left(\sqrt{\pi \xi_n}\right)^N}\right.$$

$$\times \left(\int_{\mathbb{R}^N} s_i^2 e^{-\frac{\sum_{i=1}^N s_i^2}{\xi_n}} ds_1 \cdots ds_N\right) \frac{\partial^2 f(x)}{\partial x_i^2}$$

$$\left. + \sum_{\substack{i \neq j^*, \\ i,j^* \in \{1,\ldots,N\}}} \frac{1}{\left(\sqrt{\pi \xi_n}\right)^N} \left(\int_{\mathbb{R}^N} s_i s_{j^*} e^{-\frac{\sum_{i=1}^N s_i^2}{\xi_n}} ds_1 \cdots ds_N\right) \frac{\partial^2 f(x)}{\partial x_i \partial x_{j^*}}\right\}. \tag{30.52}$$

Then

(i)

$$|\Delta_n^*(x)| \leq \|\Delta_n^*\|_\infty \leq \sum_{j=0}^{r} |\alpha_j|$$

$$\times \left[\left[j^2 \sum_{\substack{\alpha_i \in \mathbb{Z}^+, \\ \alpha : |\alpha| = 2}} \left(\frac{1}{\prod_{i=1}^N \alpha_i!}\right) \omega_1(f_\alpha, \xi_n) I_{1j}^*(\alpha)\right] + \frac{1}{2}\omega_1(f, \xi_n)\left[1 + \frac{j}{3} I_2^*\right]\right]$$

$$=: \varphi_{\xi_n}^*, \tag{30.53}$$

where $I_{1j}^*(\alpha)$ *is as in (30.29), and* I_2^* *is as in (30.36).*

In case of all f_α *of order* 2 *and* $f \in C_U\left(\mathbb{R}^N\right)$ *and* $\xi_n \to 0$, *as* $n \to \infty$, *then* $\Delta_n^*(x)$, $\|\Delta_n^*\|_\infty \to 0$ *with rates.*

(ii) *If* $\frac{\partial f(x)}{\partial x_i} = 0$, $i = 1, \ldots, N$, *and* $f_\alpha(x) = 0$, $\alpha_i \in \mathbb{Z}^+$, $i = 1, \ldots, N$, *with* $|\alpha| = 2$, *then*

$$|W_n(f,x) - f(x)| \leq \varphi_{\xi_n}^*. \tag{30.54}$$

And $W_n(f,x) \to f(x)$ *in the uniformly continuous case.*

(iii) *Additionally, assume all partials of order* ≤ 2 *are bounded. Hence*

$$\|W_n(f) - f\|_\infty \leq \left(\sum_{j=0}^{r} |\alpha_j|\, j\right)(0.8414)$$

$$\times \left[\sum_{i=1}^{N} \left\|\frac{\partial f}{\partial x_i}\right\|_\infty \frac{1}{\left(\sqrt{\pi\xi_n}\right)^N} \left(\int_{\mathbb{R}^N} |s_i|\, e^{-\frac{\sum_{i=1}^N s_i^2}{\xi_n}} ds_1 \cdots ds_N\right)\right]$$

$$+ \left(\sum_{j=0}^{r} |\alpha_j|\, j^2\right)(0.4596) \left\{\sum_{i=1}^{N} \frac{1}{\left(\sqrt{\pi\xi_n}\right)^N}\right.$$

$$\times \left(\int_{\mathbb{R}^N} s_i^2 e^{-\frac{\sum_{i=1}^N s_i^2}{\xi_n}} ds_1 \cdots ds_N\right) \left\|\frac{\partial^2 f}{\partial x_i^2}\right\|_\infty + \sum_{\substack{i \neq j^*,\\ i, j^* \in \{1,\ldots,N\}}} \frac{1}{\left(\sqrt{\pi\xi_n}\right)^N}$$

$$\times \left.\left(\int_{\mathbb{R}^N} |s_i|\, |s_{j^*}|\, e^{-\frac{\sum_{i=1}^N s_i^2}{\xi_n}} ds_1 \cdots ds_N\right) \left\|\frac{\partial^2 f}{\partial x_i \partial x_{j^*}}\right\|_\infty \right\} + \varphi_{\xi_n}^*. \quad (30.55)$$

If all f_α *of order* 2, $f \in C_U\left(\mathbb{R}^N\right)$, *then*

$$\|W_n(f) - f\|_\infty \to 0 \text{ with rates, as } \xi_n \to 0,\ n \to +\infty.$$

Proof. By Theorems 30.1, 30.5, 30.6 and Remark 30.3. □

Next we apply Theorem 30.2 to W_n operators.

Theorem 30.11. *Let* $f \in C^m\left(\mathbb{R}^N\right)$, $m \in \mathbb{N}$, $m > r$, $N \geq 1$, *with* $f_\alpha \in L_p\left(\mathbb{R}^N\right)$, $|\alpha| = m$, $x \in \mathbb{R}^N$. *Let* $p, q > 1 : \frac{1}{p} + \frac{1}{q} = 1$; $0 < \xi_n \leq 1$, $n \in \mathbb{N}$. *For* $\widetilde{j} = 1, \ldots, m$, *and* $\alpha := (\alpha_1, \ldots, \alpha_N)$, $\alpha_i \in \mathbb{Z}^+$, $i = 1, \ldots, N$, $|\alpha| := \sum_{i=1}^N \alpha_i = \widetilde{j}$, *call*

$$c_{\alpha,n,\widetilde{j}}^* := \frac{1}{\left(\sqrt{\pi\xi_n}\right)^N} \int_{\mathbb{R}^N} \left(\prod_{i=1}^N s_i^{\alpha_i}\right) e^{-\frac{\sum_{i=1}^N s_i^2}{\xi_n}} ds_1 \cdots ds_N. \quad (30.56)$$

Then

$$\left\|E_{r,n}^{*[m]}\right\|_p := \left\|W_{r,n}(f;x) - f(x) - \sum_{\widetilde{j}=1}^{m} \delta_{\widetilde{j},r}^{[m]} \left(\sum_{|\alpha|=\widetilde{j}} \frac{c_{\alpha,n,\widetilde{j}} f_\alpha(x)}{\left(\prod_{i=1}^N \alpha_i!\right)}\right)\right\|_{p,x}$$

$$\leq \left[\left(N^m/(m-1)!\right)^{\frac{p}{q}} \left(\frac{m}{(q(m-1)+1)^{\frac{p}{q}}} \right) \right.$$

$$\left. \times \left(\sum_{|\alpha|=m} \frac{1}{\prod_{i=1}^{N} \alpha_i!} \widetilde{A}_{\xi_n}(\alpha)\, \omega_r(f_\alpha, \xi_n)_p^p \right) \right]^{\frac{1}{p}}, \tag{30.57}$$

where $\widetilde{A}_{\xi_n}(\alpha)$ *as in (30.39).*

As $n \to \infty$ *and* $\xi_n \to 0$, *by (30.57), we obtain that* $\left\| E_{r,n}^{*[m]} \right\|_p \to 0$ *with rates.*

One also obtains by (30.57) that

$$\|W_n(f;x) - f(x)\|_{p,x}$$

$$\leq \sum_{\widetilde{j}=1}^{m} \left| \delta_{\widetilde{j},r}^{[m]} \right| \left(\sum_{|\alpha|=\widetilde{j}} \frac{\left| c_{\alpha,n,\widetilde{j}}^{*} \right|}{\prod_{i=1}^{N} \alpha_i!} \|f_\alpha\|_p \right) + R.H.S.(30.57), \tag{30.58}$$

given that $\|f_\alpha\|_p < \infty$, $|\alpha| = \widetilde{j}$, $\widetilde{j} = 1, \ldots, m$.

Finally, we get $\|W_n(f) - f\|_p \to 0$ *as* $\xi_n \to 0$, *and* $n \to \infty$. *That is,* $W_n \to I$ *the unit operator, in* L_p *norm, with rates.*

Proof. By Theorems 30.2 and 30.7 and (30.40). ☐

Next we apply Theorem 30.3 to W_n operators.

Theorem 30.12. *Let* $p, q > 1 : \frac{1}{p} + \frac{1}{q} = 1$, $0 < \xi_n \leq 1$, $n \in \mathbb{N}$. *Here we deal with* $f \in C^2(\mathbb{R}^N)$, $N \geq 1$, *with* f, $f_\alpha \in L_p(\mathbb{R}^N)$, $|\alpha| = 2$, *where* $\alpha_i \in \mathbb{Z}^+$, $i = 1, \ldots, N$, *and* $|\alpha| = \sum_{i=1}^{N} \alpha_i$; $x \in \mathbb{R}^N$; $j = 0, 1, \ldots, r$.

Denote $(n \in \mathbb{N})$

$$\overline{\Delta}_n(x) := W_n(f,x) - f(x) - \left(\sum_{j=0}^{r} \alpha_j j \right) \sin(1)$$

$$\times \left[\sum_{i=1}^{N} \frac{\partial f(x)}{\partial x_i} \frac{1}{\left(\sqrt{\pi \xi_n}\right)^N} \left(\int_{\mathbb{R}^N} s_i e^{-\frac{\sum_{i=1}^{N} s_i^2}{\xi_n}} ds_1 \cdots ds_N \right) \right]$$

$$-2 \left(\sum_{j=0}^{r} \alpha_j j^2 \right) \sin^2\left(\frac{1}{2}\right) \left\{ \sum_{i=1}^{N} \frac{1}{\left(\sqrt{\pi \xi_n}\right)^N} \right.$$

$$\times \left(\int_{\mathbb{R}^N} s_i^2 e^{-\frac{\sum_{i=1}^N s_i^2}{\xi_n}} ds_1 \cdots ds_N \right) \frac{\partial^2 f(x)}{\partial x_i^2} + \sum_{\substack{i \neq j^*, \\ i,j^* \in \{1,\ldots,N\}}} \frac{1}{\left(\sqrt{\pi \xi_n}\right)^N}$$

$$\left. \times \left(\int_{\mathbb{R}^N} s_i s_{j^*} e^{-\frac{\sum_{i=1}^N s_i^2}{\xi_n}} ds_1 \cdots ds_N \right) \frac{\partial^2 f(x)}{\partial x_i \partial x_{j^*}} \right\}. \tag{30.59}$$

Then

$$\left\|\overline{\Delta}_n\right\|_p \leq \left(\sum_{j=0}^r |\alpha_j|^q \right)^{\frac{1}{q}} \left\{ \frac{2}{(q+1)^{\frac{1}{q}}} \left(\frac{N(N+1)+2}{2} \right)^{\frac{1}{q}} \right\}$$

$$\left\{ \sum_{j=0}^r \left[j^{2p} \sum_{|\alpha|=2} \left(\frac{1}{\prod_{i=1}^N \alpha_i!} \right)^p \omega_1 (f_\alpha, \xi_n)_p^p \widetilde{A}_{j\xi_n}(\alpha) \right.\right.$$

$$\left.\left. + \ \omega_1 (f, \xi_n)_p^p \left[1 + j^p K_{p\xi_n}\right] \right] \right\}^{\frac{1}{p}}, \tag{30.60}$$

where $\widetilde{A}_{j\xi_n}(\alpha)$ *as in (30.41), and* $K_{p\xi_n}$ *as in (30.46).*

One also obtains from (30.60) that

$$\|W_n(f) - f\|_p \leq \left(\sum_{j=0}^r |\alpha_j| j \right) \sin(1) \left[\sum_{i=1}^N \left\| \frac{\partial f}{\partial x_i} \right\|_p \frac{1}{\left(\sqrt{\pi \xi_n}\right)^N} \right.$$

$$\left. \times \left(\int_{\mathbb{R}^N} |s_i| e^{-\frac{\sum_{i=1}^N s_i^2}{\xi_n}} ds_1 \cdots ds_N \right) \right] + 2 \left(\sum_{j=0}^r |\alpha_j| j^2 \right) \sin^2\left(\frac{1}{2}\right)$$

$$\times \left\{ \sum_{i=1}^N \frac{1}{\left(\sqrt{\pi \xi_n}\right)^N} \left(\int_{\mathbb{R}^N} s_i^2 e^{-\frac{\sum_{i=1}^N s_i^2}{\xi_n}} ds_1 \cdots ds_N \right) \left\| \frac{\partial^2 f}{\partial x_i^2} \right\|_p + \sum_{\substack{i \neq j^*, \\ i,j^* \in \{1,\ldots,N\}}} \right.$$

$$\left. \frac{1}{\left(\sqrt{\pi \xi_n}\right)^N} \left(\int_{\mathbb{R}^N} |s_i s_{j^*}| e^{-\frac{\sum_{i=1}^N s_i^2}{\xi_n}} ds_1 \cdots ds_N \right) \left\| \frac{\partial^2 f}{\partial x_i \partial x_{j^*}} \right\|_p \right\}$$

$$+ \text{R.H.S.}\,(30.60), \tag{30.61}$$

given that $\|f_\alpha\|_p < \infty$, $|\alpha| = \widetilde{j}$, $\widetilde{j} = 1, 2$.

Additionally assume that $\omega_1(f,\xi_n)_p \leq \lambda\xi_n$, $\lambda > 0$, *then as* $n \to \infty$ *and* $\xi_n \to 0$, *by (30.60), we obtain that* $\left\|\overline{\Delta}_n\right\|_p \to 0$ *with rates, and furthermore by (30.61), we derive that* $\|W_n(f) - f\|_p \to 0$, *that is,* $W_n \to I$ *the unit operator, in* L_p *norm, with rates.*

Proof. By Theorems 30.3, 30.8 and 30.9 and Remark 30.3. □

We finish with an application of Theorem 30.4 to W_n operators.

Theorem 30.13. *Let* $0 < \xi_n \leq 1$, $n \in \mathbb{N}$, $x \in \mathbb{R}^N$. *Here we deal with* $f \in C^2(\mathbb{R}^N)$, $N \geq 1$, *with* $f, f_\alpha \in L_1(\mathbb{R}^N)$, $|\alpha| = 2$, *where* $\alpha_i \in \mathbb{Z}^+$, $i = 1,\ldots,N$, *and* $|\alpha| = \sum_{i=1}^N \alpha_i$; $j = 0,1,\ldots,r$. *Here* $\overline{\Delta}_n$ *as in (30.59). Then*

$$\left\|\overline{\Delta}_n\right\|_1 \leq \left\{\sum_{j=0}^r |\alpha_j| \left[j^2 \sum_{|\alpha|=2} \left(\frac{2}{\prod_{i=1}^N \alpha_i!}\right) \omega_1(f_\alpha,\xi_n)_1 \widetilde{\widetilde{A}}_{j\xi_n}(\alpha) + \omega_1(f,\xi_n)_1 \left[1 + jK^*_{\xi_n}\right]\right]\right\}, \tag{30.62}$$

where $\widetilde{\widetilde{A}}_{j\xi_n}(\alpha)$ *as in Theorem 30.8, and* $K^*_{\xi_n}$ *as in Theorem 30.9.*

We derive also that

$$\|W_n(f) - f\|_1 \leq \left(\sum_{j=0}^r |\alpha_j| j\right) \sin(1) \left[\sum_{i=1}^N \left\|\frac{\partial f}{\partial x_i}\right\|_1 \frac{1}{\left(\sqrt{\pi\xi_n}\right)^N} \times \left(\int_{\mathbb{R}^N} |s_i| e^{-\frac{\sum_{i=1}^N s_i^2}{\xi_n}} ds_1 \cdots ds_N\right)\right]$$
$$+ 2\left(\sum_{j=0}^r |\alpha_j| j^2\right) \sin^2\left(\frac{1}{2}\right) \left\{\sum_{i=1}^N \frac{1}{\left(\sqrt{\pi\xi_n}\right)^N} \times \left(\int_{\mathbb{R}^N} s_i^2 e^{-\frac{\sum_{i=1}^N s_i^2}{\xi_n}} ds_1 \cdots ds_N\right) \left\|\frac{\partial^2 f}{\partial x_i^2}\right\|_1 + \sum_{\substack{i \neq j^*,\\ i,j^* \in \{1,\ldots,N\}}} \frac{1}{\left(\sqrt{\pi\xi_n}\right)^N} \times \left(\int_{\mathbb{R}^N} |s_i s_{j^*}| e^{-\frac{\sum_{i=1}^N s_i^2}{\xi_n}} ds_1 \cdots ds_N\right) \left\|\frac{\partial^2 f}{\partial x_i \partial x_{j^*}}\right\|_1\right\}$$
$$+ R.H.S.\,(30.62), \tag{30.63}$$

given that $\|f_\alpha\|_1 < \infty$, $|\alpha| = \widetilde{j}$, $\widetilde{j} = 1, 2$. *Additionally assume that* $\omega_1(f, \xi_n)_1 \leq \lambda^* \xi_n$, $\lambda^* > 0$, *then as* $n \to \infty$ *and* $\xi_n \to 0$, *by (30.62), we obtain that* $\|\overline{\Delta}_n\|_1 \to 0$ *with rates, and furthermore by (30.63), we derive that* $\|W_n(f) - f\|_1 \to 0$, *that is,* $W_n \to I$ *the unit operator, in* L_1 *norm, with rates.*

Proof. By Theorems 30.4, 30.8 and 30.9 and Remark 30.3. □

Bibliography

G.A. Anastassiou, *Intelligent Mathematics: Computational Analysis*, Springer, Heidelberg, New York, 2011a.

G.A. Anastassiou, *Approximation by Multivariate Singular Integrals*, Springer, New York, 2011b.

G.A. Anastassiou, *Constructive Fractional Analysis with Applications*, Springer, Heidelberg, New York, 2021.

G.A. Anastassiou, Trigonometric induced multivariate smooth Gauss-Weierstrass singular integrals approximation, *Mathematics*, 12 (2024), 115, https//doi.org/10.3390/math12010115.

G.A. Anastassiou, *Trigonometric based multivariate smooth Picard singular integrals L_p approximation*, Bulletin of the Allahabad Mathematical Society, accepted, 2023a.

G.A. Anastassiou, General multiple sigmoid functions relied complex valued multivariate trigonometric and hyperbolic neural network approximations, submitted, 2023b.

G.A. Anastassiou, *Uniform Approximation by smooth Picard Multivariate Singular Integral Operators revisited*, 2023c.

G.A. Anastassiou and S. Gal, *Approximation Theory*, Birkhaüser, Boston, 2000.

G.A. Anastassiou and R. Mezei, *Approximation by Singular Integrals*, Cambridge Scientific Publishers, Cambridge, UK, 2012.

I.S. Gradshteyn and I.M. Ryzhik, *Table of Integrals, Series and Products*, 8th edn. Elsevier, Amsterdam, 2015.

R.N. Mohapatra and R.S. Rodriguez, On the rate of convergence of singular integrals for Hölder continuous functions, *Mathematische Nachrichten*, 149 (1990), 117–124.

Chapter 31

Trigonometric-Based Multivariate Smooth Poisson–Cauchy Singular Integrals Quantitative Approximation

In this chapter, we apply the uniform and L_p, $1 \leq p < \infty$, approximation properties of general smooth multivariate singular integral operators over $\mathbb{R}^N$, $N \geq 1$. It is a trigonometric-based approach with detailed applications to the corresponding smooth multivariate Poisson–Cauchy singular integral operators. The results are quantitative via Jackson-type inequalities involving the first uniform and L_p moduli of continuity. It follows Anastassiou (2023).

31.1 Introduction

The degree of approximation by univariate and multivariate singular integral operators has been studied extensively in Anastassiou (2011)–Anastassiou (2021) and [Anastassiou and Gal (2000)], Anastassiou and Mezei (2012) and Mohapatra and Rodriguez (1990). All these sources motivate our current work. In particular, we studied the approximation properties of the smooth singular integral operators in Anastassiou (2011)–Anastassiou (2021) and Anastassiou and Mezei (2012). These are not in general positive operators. Here we use the uniform and L_p, $p \geq 1$, results of our multivariate general theory Anastassiou (2023) and Anastassiou (2023) to establish approximation properties of the smooth Poisson–Cauchy singular integral operators. The degrees of approximation are given quantitatively by using the uniform and L_p first moduli of continuity. The essential tool here comes from Anastassiou (2023), where a multivariate trigonometric Taylor formula is proved.

31.2 Background of General Theory

Here $r \in \mathbb{N}$, $m \in \mathbb{Z}_+$, we define

$$\alpha_j := \alpha_{j,r}^{[m]} := \begin{cases} (-1)^{r-j} \binom{r}{j} j^{-m}, & \text{if } j = 1, 2, \ldots, r, \\ 1 - \sum_{j=1}^{r} (-1)^{r-j} \binom{r}{j} j^{-m}, & \text{if } j = 0, \end{cases} \tag{31.1}$$

and

$$\delta_k := \delta_{k,r}^{[m]} := \sum_{j=0}^{r} \alpha_{j,r}^{[m]} j^k, \quad k = 1, 2, \ldots, m \in \mathbb{N}. \tag{31.2}$$

See that

$$\sum_{j=0}^{r} \alpha_{j,r}^{[m]} = 1 \tag{31.3}$$

and

$$-\sum_{j=1}^{r} (-1)^{r-j} \binom{r}{j} = (-1)^r \binom{r}{0}.$$

Let μ_{ξ_n} be a probability Borel measure on $\mathbb{R}^N$, $N \geq 1$, $\xi_n > 0$, $n \in \mathbb{N}$. We now define the multiple smooth singular integral operators

$$\theta_n (f; x_1, \ldots, x_N) := \theta_{r,n}^{[m]} (f; x_1, \ldots, x_N)$$

$$:= \sum_{j=0}^{r} \alpha_{j,r}^{[m]} \int_{\mathbb{R}^N} f (x_1 + s_1 j, x_2 + s_2 j, \ldots, x_N + s_N j) \, d\mu_{\xi_n} (s), \tag{31.4}$$

where $s := (s_1, \ldots, s_N)$, $x := (x_1, \ldots, x_N) \in \mathbb{R}^N$; $n, r \in \mathbb{Z}$, $m \in \mathbb{Z}_+$, $f : \mathbb{R}^N \to \mathbb{R}$ is a Borel measurable function, and also $(\xi_n)_{n \in \mathbb{N}}$ is a bounded sequence of positive real numbers, we take $0 < \xi_n \leq 1$.

Remark 31.1. The operators $\theta_{r,n}^{[m]}$ are not in general positive, see Anastassiou (2011), p. 2.

We observe that

Lemma 31.1. *It holds*

$$\theta_{r,n}^{[m]} (c; x_1, \ldots, x_n) = c,$$

where c is a constant.

We need

Definition 31.1. Let $f \in C\left(\mathbb{R}^N\right)$, $N \geq 1$. We define the first uniform modulus of continuity of f as

$$\omega_1(f, \delta) := \sup_{\substack{x, y \in \mathbb{R}^N: \\ \|x-y\|_\infty \leq \delta}} |f(x) - f(y)|, \quad \delta > 0, \tag{31.5}$$

where $\|\cdot\|_\infty$ is the max norm in $\mathbb{R}^N$. The functional $\omega_1(f, \delta)$ is bounded for f being bounded or uniformly continuous, and $\omega_1(f, \delta) \to 0$ as $\delta \to 0$, in the case of f being uniformly continuous.

We mention the main uniform general approximation result regarding the operator θ_n.

Theorem 31.1. *(Anastassiou 2023) Here $f \in C^2\left(\mathbb{R}^N\right)$ and let all $\alpha_i \in \mathbb{Z}^+$, $i = 1, \ldots, N$, $N \geq 1$, $|\alpha| := \sum_{i=1}^N \alpha_i = 2$; $x \in \mathbb{R}^N$, and all the partials f_α of order 2, along with $f \in C_B\left(\mathbb{R}^N\right)$ (continuous and bounded functions); or all f_α of order 2, $f \in C_U\left(\mathbb{R}^N\right)$ (uniformly continuous functions). Let μ_{ξ_n} be a Borel probability measure on $\mathbb{R}^N$, for $0 < \xi_n \leq 1$, $n \in \mathbb{N}$.*

Suppose that for all $\alpha := (\alpha_1, \ldots, \alpha_N)$, $\alpha_i \in \mathbb{Z}^+$, $i = 1, \ldots, N$, $|\alpha| = \sum_{i=1}^N \alpha_i = 2$, $j = 0, 1, \ldots, r$, we have that both

$$I_{1j}(\alpha) := \int_{\mathbb{R}^N} \left(1 + \frac{j\|s\|_1}{3\xi_n}\right) \left(\prod_{i=1}^N |s_i|^{\alpha_i}\right) d\mu_{\xi_n}(s), \tag{31.6}$$

$$I_{2j}(\alpha) := \int_{\mathbb{R}^N} \left(1 + \frac{j\|s\|_1}{3\xi_n}\right) d\mu_{\xi_n}(s) \tag{31.7}$$

are uniformly bounded in $\xi_n \in (0, 1]$.

Denote ($n \in \mathbb{N}$)

$$\Delta_n(x) := \theta_n(f, x) - f(x) - \left(\sum_{j=0}^r \alpha_j j\right) \sin(1)$$
$$\times \left[\sum_{i=1}^N \frac{\partial f(x)}{\partial x_i} \left(\int_{\mathbb{R}^N} s_i d\mu_{\xi_n}(s)\right)\right]$$
$$-2\left(\sum_{j=0}^r \alpha_j j^2\right) \sin^2\left(\frac{1}{2}\right) \left\{\sum_{i=1}^N \left(\int_{\mathbb{R}^N} s_i^2 d\mu_{\xi_n}(s)\right) \frac{\partial^2 f(x)}{\partial x_i^2}\right.$$
$$\left. + \sum_{\substack{i \neq j^*, \\ i, j^* \in \{1, \ldots, N\}}} \left(\int_{\mathbb{R}^N} s_i s_{j^*} d\mu_{\xi_n}(s)\right) \frac{\partial^2 f(x)}{\partial x_i \partial x_{j^*}}\right\}. \tag{31.8}$$

Then

(i)

$$|\Delta_n(x)| \leq \|\Delta_n(x)\|_\infty \leq \sum_{j=0}^{r} |\alpha_j|$$

$$\times \left[\left[j^2 \sum_{\substack{\alpha_i \in \mathbb{Z}^+, \\ \alpha : |\alpha| = 2}} \left(\frac{1}{\prod_{i=1}^{N} \alpha_i!} \right) \omega_1(f_\alpha, \xi_n) \right.\right.$$

$$\left. \int_{\mathbb{R}^N} \left(1 + \frac{j\|s\|_\infty}{3\xi_n}\right) \left(\prod_{i=1}^{N} |s_i|^{\alpha_i} \right) d\mu_{\xi_n}(s) \right]$$

$$\left. + \frac{1}{2}\omega_1(f, \xi_n) \int_{\mathbb{R}^N} \left(1 + \frac{j\|s\|_\infty}{3\xi_n}\right) d\mu_{\xi_n}(s) \right] =: \varphi_{\xi_n}. \tag{31.9}$$

In case of all f_α of order 2 and $f \in C_U(\mathbb{R}^N)$ and $\xi_n \to 0$, as $n \to \infty$, then $\Delta_n(x)$, $\|\Delta_n(x)\|_\infty \to 0$ with rates.

(ii) *If $\frac{\partial f(x)}{\partial x_i} = 0$, $i = 1, \ldots, N$, and $f_\alpha(x) = 0$, $\alpha_i \in \mathbb{Z}^+$, $i = 1, \ldots, N$, with $|\alpha| = 2$, then*

$$|\theta_n(f, x) - f(x)| \leq \varphi_{\xi_n}. \tag{31.10}$$

And $\theta_n(f, x) \to f(x)$ in the uniformly continuous case.

(iii) *Additionally assume, all partials of order ≤ 2 are bounded. Hence*

$$\|\theta_n(f) - f\|_\infty \leq \left(\sum_{j=0}^{r} |\alpha_j| j \right) (0.8414) \left[\sum_{i=1}^{N} \left\| \frac{\partial f}{\partial x_i} \right\|_\infty \left(\int_{\mathbb{R}^N} |s_i| d\mu_{\xi_n}(s) \right) \right]$$

$$+ \left(\sum_{j=0}^{r} |\alpha_j| j^2 \right) (0.4596) \left\{ \sum_{i=1}^{N} \left(\int_{\mathbb{R}^N} s_i^2 d\mu_{\xi_n}(s) \right) \left\| \frac{\partial^2 f}{\partial x_i^2} \right\|_\infty \right.$$

$$\left. + \sum_{\substack{i \neq j^*, \\ i, j^* \in \{1, \ldots, N\}}} \left(\int_{\mathbb{R}^N} |s_i| |s_{j^*}| d\mu_{\xi_n}(s) \right) \left\| \frac{\partial^2 f}{\partial x_i \partial x_{j^*}} \right\|_\infty \right\} + \varphi_{\xi_n}. \tag{31.11}$$

If all $\int_{\mathbb{R}^N} s_i^2 d\mu_{\xi_n}(s)$ *and* $\int_{\mathbb{R}^N} |s_i| |s_{j^*}| d\mu_{\xi_n}(s)$ *converge to zero, as* $n \to \infty$*, with* $\xi_n \to 0$*, and all* f_α *of order* 2*,* $f \in C_U(\mathbb{R}^N)$*, then*

$$\|\theta_n(f) - f\|_\infty \to 0 \text{ with rates, as } \xi_n \to 0, n \to +\infty.$$

Next we deal with $f \in C^m(\mathbb{R}^N)$, $m \in \mathbb{Z}^+$, with $f_\alpha \in L_p(\mathbb{R}^N)$, $|\alpha| = m \in \mathbb{Z}^+$, $p \geq 1$, where f_α denotes the mixed partial $\frac{\partial^{\widetilde{j}} f(\cdot,\ldots,\cdot)}{\partial x_1^{\alpha_1} \ldots \partial x_N^{\alpha_N}}$, $\alpha_j \in \mathbb{Z}^+$, $j = 1, \ldots, N : |\alpha| := \sum_{j=1}^N \alpha_j = \widetilde{j}$, $\widetilde{j} = 1, \ldots, m$.

We need

Definition 31.2. (see also Anastassiou (2011), p. 20) We call

$$\Delta_u^r f(x) := \Delta_{u_1,u_2,\ldots,u_N}^r f(x_1, \ldots, x_N)$$
$$:= \sum_{j=0}^r (-1)^{r-j} \binom{r}{j} f(x_1 + ju_1, x_2 + ju_2, \ldots, x_N + ju_N). \tag{31.12}$$

Let $p \geq 1$, the L_p modulus of smoothness of order r is given by

$$\omega_r(f; h)_p := \sup_{\|u\|_2 \leq h} \|\Delta_u^r(f)\|_p, \tag{31.13}$$

$h > 0$.

We mention

Theorem 31.2. *(Anastassiou 2023) Let* $f \in C^m(\mathbb{R}^N)$*,* $m \in \mathbb{N}$*,* $N \geq 1$*, with* $f_\alpha \in L_p(\mathbb{R}^N)$*,* $|\alpha| = m$*,* $x \in \mathbb{R}^N$*. Let* $p, q > 1 : \frac{1}{p} + \frac{1}{q} = 1$*. Here* μ_{ξ_n} *is a Borel probability measure on* $\mathbb{R}^N$ *for* $\xi_n > 0$*,* $(\xi_n)_{n\in\mathbb{N}}$ *bounded sequence. Assume for all* $\alpha := (\alpha_1, \ldots, \alpha_N)$*,* $\alpha_i \in \mathbb{Z}^+$*,* $i = 1, \ldots, N$*,* $|\alpha| := \sum_{i=1}^N \alpha_i = m$ *that we have*

$$\int_{\mathbb{R}^N} \left(\left(\prod_{i=1}^N |s_i|^{\alpha_i} \right) \left(1 + \frac{\|s\|_2}{\xi_n} \right)^r \right)^p d\mu_{\xi_n}(s) < \infty. \tag{31.14}$$

For $\widetilde{j} = 1, \ldots, m$*, and* $\alpha := (\alpha_1, \ldots, \alpha_N)$*,* $\alpha_i \in \mathbb{Z}^+$*,* $i = 1, \ldots, N$*,* $|\alpha| := \sum_{i=1}^N \alpha_i = \widetilde{j}$*, call*

$$c_{\alpha,n,\widetilde{j}} := \int_{\mathbb{R}^N} \prod_{i=1}^N s_i^{\alpha_i} d\mu_{\xi_n}(s). \tag{31.15}$$

Then

$$\left\|E_{r,n}^{[m]}\right\|_p := \left\|\theta_{r,n}^{[m]}(f;x) - f(x) - \sum_{\widetilde{j}=1}^{m} \delta_{\widetilde{j},r}^{[m]} \left(\sum_{|\alpha|=\widetilde{j}} \frac{c_{\alpha,n,\widetilde{j}} f_\alpha(x)}{\left(\prod_{i=1}^N \alpha_i!\right)}\right)\right\|_{p,x}$$

$$\leq \left[(N^m/(m-1)!)^{\frac{p}{q}} \left(\frac{m}{(q(m-1)+1)^{\frac{p}{q}}}\right) \left(\sum_{|\alpha|=m} \frac{1}{\prod_{i=1}^N \alpha_i!}\right.\right.$$

$$\left.\left. \cdot \left[\int_{\mathbb{R}^N} \left[\left(\prod_{i=1}^N |s_i|^{\alpha_i}\right)\left(1+\frac{\|s\|_2}{\xi_n}\right)^r\right]^p d\mu_{\xi_n}(s)\right] \omega_r(f_\alpha,\xi_n)_p^p\right)\right]^{\frac{1}{p}}. \tag{31.16}$$

As $n \to \infty$ and $\xi_n \to 0$, by (31.16), we obtain that $\left\|E_{r,n}^{[m]}\right\|_p \to 0$ with rates.

One also gets by (31.16) that

$$\left\|\theta_{r,n}^{[m]}(f;x) - f(x)\right\|_{p,x}$$

$$\leq \sum_{\widetilde{j}=1}^{m} \left|\delta_{\widetilde{j},r}^{[m]}\right| \left(\sum_{|\alpha|=\widetilde{j}} \frac{\left|c_{\alpha,n,\widetilde{j}}\right|}{\prod_{i=1}^N \alpha_i!} \|f_\alpha\|_p\right) + R.H.S.(31.16), \tag{31.17}$$

given that $\|f_\alpha\|_p < \infty$, $|\alpha| = \widetilde{j}$, $\widetilde{j} = 1, \ldots, m$.

Assuming that $c_{\alpha,n,\widetilde{j}} \to 0$, $\xi_n \to 0$, as $n \to \infty$, we get $\left\|\theta_{r,n}^{[m]}(f) - f\right\|_p \to 0$, that is, $\theta_{r,n}^{[m]} \to I$ the unit operator, in L_p norm, with rates.

We make

Remark 31.2. Note that $(p > 1)$

$$\int_{\mathbb{R}^N} \left(\prod_{i=1}^N |s_i|^{\alpha_i}\right)\left(1+\frac{\|s\|_2}{\xi_n}\right)^r d\mu_{\xi_n}(s)$$

$$\leq \left[\int_{\mathbb{R}^N} \left(\left(\prod_{i=1}^N |s_i|^{\alpha_i}\right)\left(1+\frac{\|s\|_2}{\xi_n}\right)^r\right)^p d\mu_{\xi_n}(s)\right]^{\frac{1}{p}} < \infty, \tag{31.18}$$

by assumption of Theorem 31.2.

By (31.18), we get that

$$\int_{\mathbb{R}^N} \prod_{i=1}^N |s_i|^{\alpha_i} d\mu_{\xi_n}(s) < \infty. \tag{31.19}$$

Hence $c_{\alpha,n,\widetilde{j}} \in \mathbb{R}$.

We mention also the following trigonometric-induced alternative L_p approximation result for θ_n operators.

Theorem 31.3. *(Anastassiou 2023) Let $p, q > 1 : \frac{1}{p} + \frac{1}{q} = 1$, $0 < \xi_n \leq 1$, $n \in \mathbb{N}$. Here we deal with $f \in C^2(\mathbb{R}^N)$, $N \geq 1$, with f, $f_\alpha \in L_p(\mathbb{R}^N)$, $|\alpha| = 2$, where $\alpha_i \in \mathbb{Z}^+$, $i = 1, \ldots, N$, and $|\alpha| = \sum_{i=1}^N \alpha_i$; $x \in \mathbb{R}^N$. Let μ_{ξ_n} be a Borel probability measure on $\mathbb{R}^N$. Suppose that for all $\alpha : |\alpha| = 2$, $j = 0, 1, \ldots, r$, we have that both*

$$I_{1j}(\alpha) := \int_{\mathbb{R}^N} \left(\prod_{i=1}^N |s_i|^{p\alpha_i} \right) \left(1 + \frac{j^p \|s\|_2^p}{\xi_n^p} \right) d\mu_{\xi_n}(s), \tag{31.20}$$

$$I_2 := \int_{\mathbb{R}^N} \left(\frac{\|s\|_2}{\xi_n} \right)^p d\mu_{\xi_n}(s) \tag{31.21}$$

are uniformly bounded in $\xi_n \in (0, 1]$.

Denote ($n \in \mathbb{N}$)

$$\Delta_n(x) := \theta_n(f, x) - f(x) - \left(\sum_{j=0}^r \alpha_j j \right) \sin(1)$$
$$\times \left[\sum_{i=1}^N \frac{\partial f(x)}{\partial x_i} \left(\int_{\mathbb{R}^N} s_i d\mu_{\xi_n}(s) \right) \right]$$
$$- 2 \left(\sum_{j=0}^r \alpha_j j^2 \right) \sin^2\left(\frac{1}{2}\right) \left\{ \sum_{i=1}^N \left(\int_{\mathbb{R}^N} s_i^2 d\mu_{\xi_n}(s) \right) \frac{\partial^2 f(x)}{\partial x_i^2} \right.$$
$$\left. + \sum_{\substack{i \neq j^*, \\ i, j^* \in \{1, \ldots, N\}}} \left(\int_{\mathbb{R}^N} s_i s_{j^*} d\mu_{\xi_n}(s) \right) \frac{\partial^2 f(x)}{\partial x_i \partial x_{j^*}} \right\}. \tag{31.22}$$

Then, it holds

$$\|\Delta_n(x)\|_p \leq \left(\sum_{j=0}^r |\alpha_j|^q \right)^{\frac{1}{q}} \left\{ \frac{2}{(q+1)^{\frac{1}{q}}} \left(\frac{N(N+1)+2}{2} \right)^{\frac{1}{q}} \right\}$$
$$\times \left\{ \sum_{j=0}^r \left[j^{2p} \sum_{|\alpha|=2} \left(\frac{1}{\prod_{i=1}^N \alpha_i!} \right)^p \right. \right.$$

$$\times\, \omega_1 \left(f_\alpha, \xi_n\right)_p^p \left(\int_{\mathbb{R}^N} \left(\prod_{i=1}^N |s_i|^{p\alpha_i}\right) \left(1 + \frac{j^p \|s\|_2^p}{\xi_n^p}\right) d\mu_{\xi_n}(s)\right)$$

$$\left. \left. + \,\omega_1 \left(f, \xi_n\right)_p^p \left(\int_{\mathbb{R}^N} \left(1 + \frac{j^p \|s\|_2^p}{\xi_n^p}\right) d\mu_{\xi_n}(s)\right)\right]\right\}^{\frac{1}{p}}. \quad (31.23)$$

As $n \to \infty$ and $\xi_n \to 0$, by (31.23), we obtain that $\|\Delta_n\|_p \to 0$ with rates. One also gets by (31.23) that

$$\|\theta_n(f,x) - f(x)\|_{p,x}$$

$$\leq \left(\sum_{j=0}^r |\alpha_j|\, j\right) \sin(1) \left[\sum_{i=1}^N \left\|\frac{\partial f}{\partial x_i}\right\|_p \left(\int_{\mathbb{R}^N} |s_i|\, d\mu_{\xi_n}(s)\right)\right]$$

$$+ 2 \left(\sum_{j=0}^r |\alpha_j|\, j^2\right) \sin^2\left(\frac{1}{2}\right) \left\{\sum_{i=1}^N \left(\int_{\mathbb{R}^N} s_i^2 d\mu_{\xi_n}(s)\right) \left\|\frac{\partial^2 f}{\partial x_i^2}\right\|_p\right.$$

$$\left. + \sum_{\substack{i \neq j^*, \\ i, j^* \in \{1,\ldots,N\}}} \left(\int_{\mathbb{R}^N} |s_i s_{j^*}|\, d\mu_{\xi_n}(s)\right) \left\|\frac{\partial^2 f}{\partial x_i \partial x_{j^*}}\right\|_p \right\} + R.H.S.\,(31.23), \quad (31.24)$$

given that $\|f_\alpha\|_p < \infty$, $|\alpha| = \widetilde{j}$, $\widetilde{j} = 1, 2$. Assuming that $\int_{\mathbb{R}^N} s_i^2 d\mu_{\xi_m}(s)$, $\int_{\mathbb{R}^N} |s_i s_{j^*}|\, d\mu_{\xi_n}(s)$, $i, j^* \in \{1, \ldots, N\}$, $i \neq j^*$, converge to zero as $\xi_n \to 0$, we get $\|\theta_n(f,x) - f(x)\|_p \to 0$, that is, $\theta_n \to I$ the unit operator, in L_p norm, with rates.

Furthermore, we mention the following trigonometric-based alternative L_1 approximation result for θ_n operators.

Theorem 31.4. *([Anastassiou (2023)]) Let $0 < \xi_n \leq 1$, $n \in \mathbb{N}$, $x \in \mathbb{R}^N$. Here we deal with $f \in C^2(\mathbb{R}^N)$, $N \geq 1$, with $f, f_\alpha \in L_1(\mathbb{R}^N)$, $|\alpha| = 2$, where $\alpha_i \in \mathbb{Z}^+$, $i = 1, \ldots, N$, and $|\alpha| = \sum_{i=1}^N \alpha_i$. Let μ_{ξ_n} be a Borel probability measure on $\mathbb{R}^N$. Suppose that for all $\alpha : |\alpha| = 2$, $j = 0, 1, \ldots, r$, we have that both*

$$I_{1j}^*(\alpha) := \int_{\mathbb{R}^N} \left(\prod_{i=1}^N |s_i|^{\alpha_i}\right) \left(1 + \frac{j\|s\|_2}{\xi_n}\right) d\mu_{\xi_n}(s) \quad (31.25)$$

and

$$I_2^* := \int_{\mathbb{R}^N} \frac{\|s\|_2}{\xi_n} d\mu_{\xi_n}(s) \tag{31.26}$$

are uniformly bounded in $\xi_n \in (0,1]$.

Here $\Delta_n(x)$ *is as in (31.22).*

Then, it holds

$$\|\Delta_n(x)\|_1 \leq \left\{ \sum_{j=0}^{r} |\alpha_j| \left[j^2 \sum_{|\alpha|=2} \left(\frac{2}{\prod_{i=1}^N \alpha_i!} \right) \right. \right.$$
$$\times\, \omega_1(f_\alpha, \xi_n)_1 \left(\int_{\mathbb{R}^N} \left(\prod_{i=1}^N (|s_i|)^{\alpha_i} \right) \left(1 + \frac{j\|s\|_2}{\xi_n} \right) d\mu_{\xi_n}(s) \right)$$
$$\left. \left. +\, \omega_1(f, \xi_n)_1 \left(\int_{\mathbb{R}^N} \left(1 + \frac{j\|s\|_2}{\xi_n} \right) d\mu_{\xi_n}(s) \right) \right] \right\}. \tag{31.27}$$

As $n \to \infty$ *and* $\xi_n \to 0$, *by (31.27), we obtain that* $\|\Delta_n\|_1 \to 0$ *with rates. One also obtains by (31.27) that*

$$\|\theta_n(f) - f\|_1 \leq \left(\sum_{j=0}^{r} |\alpha_j| j \right) \sin(1) \left[\sum_{i=1}^{N} \left\| \frac{\partial f}{\partial x_i} \right\|_1 \left(\int_{\mathbb{R}^N} |s_i| \, d\mu_{\xi_n}(s) \right) \right]$$
$$+ 2 \left(\sum_{j=0}^{r} |\alpha_j| j^2 \right) \sin^2\left(\frac{1}{2}\right) \left\{ \sum_{i=1}^{N} \left(\int_{\mathbb{R}^N} s_i^2 d\mu_{\xi_n}(s) \right) \left\| \frac{\partial^2 f}{\partial x_i^2} \right\|_1 \right.$$
$$\left. + \sum_{\substack{i \neq j^*, \\ i, j^* \in \{1, \ldots, N\}}} \left(\int_{\mathbb{R}^N} |s_i s_{j^*}| \, d\mu_{\xi_n}(s) \right) \left\| \frac{\partial^2 f}{\partial x_i \partial x_{j^*}} \right\|_1 \right\} + R.H.S.\,(31.27), \tag{31.28}$$

given that $\|f_\alpha\|_1 < \infty$, $|\alpha| = \widetilde{j}$, $\widetilde{j} = 1, 2$. *Assuming that* $\int_{\mathbb{R}^N} s_i^2 d\mu_{\xi_n}(s)$, $\int_{\mathbb{R}^N} |s_i s_{j^*}| \, d\mu_{\xi_n}(s)$, $i, j^* \in \{1, \ldots, N\}$, $i \neq j^*$, *converge to zero as* $\xi_n \to 0$, *we derive* $\|\theta_n(f) - f\|_1 \to 0$, *that is,* $\theta_n \to I$ *in* L_1 *norm, with rates.*

We need

Definition 31.3. The general multivariate Poisson–Cauchy singular integral operators are defined as follows:

$$U_n := U_{r,n}^{[m]}(f; x_1, \ldots, x_N) := W_n^N \sum_{j=0}^{r} \alpha_{j,r}^{[m]} \cdot \tag{31.29}$$

$$\int_{\mathbb{R}^N} f(x_1 + s_1 j, \ldots, x_N + s_N j) \prod_{i=1}^{N} \frac{1}{\left(s_i^{2\alpha} + \xi_n^{2\alpha}\right)^{\beta}} ds_1 \ldots ds_N,$$

with $\alpha \in \mathbb{N}$, $\beta > \frac{1}{2\alpha}$, and

$$W_n := \frac{\Gamma(\beta)\, \alpha \xi_n^{2\alpha\beta - 1}}{\Gamma\left(\frac{1}{2\alpha}\right) \Gamma\left(\beta - \frac{1}{2\alpha}\right)}, \tag{31.30}$$

see Anastassiou (2011), p. 15.

Note that

$$W_n^N \int_{\mathbb{R}^N} \prod_{i=1}^{N} \frac{1}{\left(s_i^{2\alpha} + \xi_n^{2\alpha}\right)^{\beta}} ds_1 \ldots ds_N = 1, \tag{31.31}$$

see Anastassiou and Mezei (2009), Zwillinger (1995), p. 397, formula 595.

That is, U_n are the θ_n operators applied for the Borel probability measures on $\mathbb{R}^N$, $N \geq 1$,

$$d\mu_{\xi_n}(s) = W_n^N \left(\prod_{i=1}^{N} \frac{1}{\left(s_i^{2\alpha} + \xi_n^{2\alpha}\right)^{\beta}} \right) ds_1 \ldots ds_N, \quad s \in \mathbb{R}^N, \tag{31.32}$$

where $0 < \xi_n \leq 1$, $n \in \mathbb{N}$.

We apply to U_n Theorems 31.1, 31.2, 31.3 and 31.4, with the help of all of Section 3. That is a presentation of approximation properties by U_n.

31.3 Auxiliary Results

We need

Theorem 31.5. *Let* $N \in \mathbb{N}$; $\alpha_i \in \mathbb{Z}^+$, $i = 1, \ldots, N : |\alpha| := \sum_{i=1}^{N} \alpha_i = 2$, $\xi_n \in (0,1]$, $n \in \mathbb{N}$; $j = 0, 1, \ldots, r \in \mathbb{N}$; $\alpha \in \mathbb{N}$, $\beta > \frac{1}{2\alpha}$, W_n *as in (31.30). Then*

$$\overline{I}_{1j}(\alpha) := W_n^N \int_{\mathbb{R}^N} \left(1 + \frac{j\,\|s\|_1}{3\xi_n}\right) \left(\prod_{i=1}^{N} |s_i|^{\alpha_i}\right) \prod_{i=1}^{N} \frac{1}{\left(s_i^{2\alpha} + \xi_n^{2\alpha}\right)^{\beta}} ds_1 \ldots ds_N \tag{31.33}$$

$$\leq \frac{\xi_n^{2\alpha\beta(N-1)+2}}{\Gamma^N\left(\frac{1}{2\alpha}\right) \Gamma^N\left(\beta - \frac{1}{2\alpha}\right)}$$

$$\left[(2\alpha\Gamma(\beta))^N\left(1+\frac{j}{3}N\right)+\left(1+\frac{j}{3}\right)\Gamma^N\left(\frac{2}{\alpha}\right)\Gamma^N\left(\beta-\frac{2}{\alpha}\right)\right]$$

$$\leq \frac{1}{\Gamma^N\left(\frac{1}{2\alpha}\right)\Gamma^N\left(\beta-\frac{1}{2\alpha}\right)}$$

$$\left[(2\alpha\Gamma(\beta))^N\left(1+\frac{j}{3}N\right)+\left(1+\frac{j}{3}\right)\Gamma^N\left(\frac{2}{\alpha}\right)\Gamma^N\left(\beta-\frac{2}{\alpha}\right)\right]<+\infty \tag{31.34}$$

are uniformly bounded in ξ_n, *and* $\overline{I}_{1j}(\alpha)\to 0$, *as* $\xi_n\to 0$.

Proof. We have that

$$\overline{I}_{1j}(\alpha)=W_n^N\int_{\mathbb{R}^N}\left(1+\frac{j\|s\|_1}{3\xi_n}\right)\left(\prod_{i=1}^N|s_i|^{\alpha_i}\right)\prod_{i=1}^N\frac{1}{\left(s_i^{2\alpha}+\xi_n^{2\alpha}\right)^\beta}ds_1\ldots ds_N$$

$$=2^NW_n^N\int_{\mathbb{R}_+^N}\left(1+\frac{j\left(\sum_{i=1}^N s_i\right)}{3\xi_n}\right)\left(\prod_{i=1}^N s_i^{\alpha_i}\right)\prod_{i=1}^N\frac{1}{\left(s_i^{2\alpha}+\xi_n^{2\alpha}\right)^\beta}ds_1\ldots ds_N \tag{31.35}$$

$$\left(c:=\frac{\Gamma(\beta)\,\alpha}{\Gamma\left(\frac{1}{2\alpha}\right)\Gamma\left(\beta-\frac{1}{2\alpha}\right)},\text{ i.e. } W_n=c\xi_n^{2\alpha\beta-1}\right)$$

$$=(2c)^N\xi_n^{2\alpha\beta N-N}\int_{\mathbb{R}_+^N}\left(1+j\frac{\sum_{i=1}^N s_i}{3\xi_n}\right)\left(\prod_{i=1}^N s_i^{\alpha_i}\right)\prod_{i=1}^N\frac{1}{\left(s_i^{2\alpha}+\xi_n^{2\alpha}\right)^\beta}\prod_{i=1}^N ds_i$$

$$=(2c)^N\xi_n^{2\alpha\beta N-N}\xi_n^2\xi_n^{-2\alpha\beta}\xi_n^N\int_{\mathbb{R}_+^N}\left(1+\frac{j}{3}\sum_{i=1}^N\left(\frac{s_i}{\xi_n}\right)\right)$$

$$\left(\prod_{i=1}^N\left(\frac{s_i}{\xi_n}\right)^{\alpha_i}\right)\prod_{i=1}^N\frac{1}{\left(\left(\frac{s_i}{\xi_n}\right)^{2\alpha}+1\right)^\beta}\prod_{i=1}^N d\left(\frac{s_i}{\xi_n}\right)$$

$$=(2c)^N\xi_n^{2\alpha\beta(N-1)+2}\int_{\mathbb{R}_+^N}\left(1+\frac{j}{3}\sum_{i=1}^N z_i\right)\left(\prod_{i=1}^N z_i^{\alpha_i}\right)\prod_{i=1}^N\frac{1}{\left(z_i^{2\alpha}+1\right)^\beta}\prod_{i=1}^N dz_i$$

$$=(2c)^N\xi_n^{2\alpha\beta(N-1)+2}\left[\int_{[0,1]^N}\left(1+\frac{j}{3}\sum_{i=1}^N z_i\right)\right.$$

$$\times\left(\prod_{i=1}^N z_i^{\alpha_i}\right)\prod_{i=1}^N\frac{1}{\left(z_i^{2\alpha}+1\right)^\beta}\prod_{i=1}^N dz_i$$

$$+\int_{(\mathbb{R}_+-[0,1])^N}\left(1+\frac{j}{3}\left(\sum_{i=1}^N z_i\right)\right)\left(\prod_{i=1}^N z_i^{\alpha_i}\right)\prod_{i=1}^N\frac{1}{\left(z_i^{2\alpha}+1\right)^{\beta}}\prod_{i=1}^N dz_i\Bigg] \tag{31.36}$$

$$\leq (2c)^N\,\xi_n^{2\alpha\beta(N-1)+2}\left[\left(1+\frac{j}{3}N\right)\right.$$

$$\left.+\left(1+\frac{j}{3}\right)\int_{(\mathbb{R}_+-[0,1])^N}\left(\sum_{i=1}^N z_i\right)\left(\prod_{i=1}^N z_i^{\alpha_i}\right)\prod_{i=1}^N\frac{1}{\left(z_i^{2\alpha}+1\right)^{\beta}}\prod_{i=1}^N dz_i\right]$$

$$\leq (2c)^N\,\xi_n^{2\alpha\beta(N-1)+2}\left[\left(1+\frac{j}{3}N\right)\right.$$

$$\left.+\left(1+\frac{j}{3}\right)\int_{(\mathbb{R}_+-[0,1])^N}\left(\prod_{i=1}^N z_i\right)\left(\prod_{i=1}^N z_i^{\alpha_i}\right)\prod_{i=1}^N\frac{1}{\left(z_i^{2\alpha}+1\right)^{\beta}}\prod_{i=1}^N dz_i\right]$$

$$= (2c)^N\,\xi_n^{2\alpha\beta(N-1)+2}$$

$$\left[\left(1+\frac{j}{3}N\right)+\left(1+\frac{j}{3}\right)\int_{(\mathbb{R}_+-[0,1])^N}\left(\prod_{i=1}^N z_i^{\alpha_i+1}\right)\prod_{i=1}^N\frac{1}{\left(z_i^{2\alpha}+1\right)^{\beta}}\prod_{i=1}^N dz_i\right] \tag{31.37}$$

$$= (2c)^N\,\xi_n^{2\alpha\beta(N-1)+2}\left[\left(1+\frac{j}{3}N\right)+\left(1+\frac{j}{3}\right)\prod_{i=1}^N\right.$$

$$\left.\times\int_1^{\infty} z_i^{\alpha_i+1}\frac{1}{\left(z_i^{2\alpha}+1\right)^{\beta}}dz_i\right]$$

$$\leq (2c)^N\,\xi_n^{2\alpha\beta(N-1)+2}\left[\left(1+\frac{j}{3}N\right)+\left(1+\frac{j}{3}\right)\left(\int_1^{\infty}\frac{z^{2+1}}{\left(z^{2\alpha}+1\right)^{\beta}}dz\right)^N\right]$$

$$\leq (2c)^N\,\xi_n^{2\alpha\beta(N-1)+2}\left[\left(1+\frac{j}{3}N\right)+\left(1+\frac{j}{3}\right)\left(\int_0^{\infty}\frac{z^{3}}{\left(z^{2\alpha}+1\right)^{\beta}}dz\right)^N\right] \tag{31.38}$$

(by Zwillinger (1995), p. 397, formula 595)

$$= (2c)^N\,\xi_n^{2\alpha\beta(N-1)+2}\left[\left(1+\frac{j}{3}N\right)+\left(1+\frac{j}{3}\right)\left(\frac{\Gamma\left(\frac{2}{\alpha}\right)\Gamma\left(\beta-\frac{2}{\alpha}\right)}{2\alpha\Gamma(\beta)}\right)^N\right]$$

$$= \left(\frac{2\Gamma(\beta)\alpha}{\Gamma\left(\frac{1}{2\alpha}\right)\Gamma\left(\beta - \frac{1}{2\alpha}\right)} \right)^N \xi_n^{2\alpha\beta(N-1)+2}$$

$$\times \left[\left(1 + \frac{j}{3}N\right) + \left(1 + \frac{j}{3}\right) \left(\frac{\Gamma\left(\frac{2}{\alpha}\right)\Gamma\left(\beta - \frac{2}{\alpha}\right)}{2\alpha\Gamma(\beta)} \right)^N \right],$$

where $N \in \mathbb{N}$ and $\beta > \frac{1}{2\alpha}$. The claim is proved. □

Remark 31.3. By Theorem 31.5, with $i, j^* \in \{1, \ldots, N\}$, $i \neq j^*$, we have that

$$W_n^N \int_{\mathbb{R}^N} |s_i| \, |s_{j^*}| \left(\prod_{i=1}^N \frac{1}{\left(s_i^{2\alpha} + \xi_n^{2\alpha}\right)^\beta} \right) ds_1 \ldots ds_N,$$

$$W_n^N \int_{\mathbb{R}^N} s_i^2 \left(\prod_{i=1}^N \frac{1}{\left(s_i^{2\alpha} + \xi_n^{2\alpha}\right)^\beta} \right) ds_1 \ldots ds_N \leq \overline{I}_{1j}(\alpha) < \infty, \quad (31.39)$$

$(j = 0, 1, \ldots, r \in \mathbb{N})$,

and

$$W_n^N \int_{\mathbb{R}^N} |s_i| \left(\prod_{i=1}^N \frac{1}{\left(s_i^{2\alpha} + \xi_n^{2\alpha}\right)^\beta} \right) ds_1 \ldots ds_N$$

$$\leq \left(W_n^N \int_{\mathbb{R}^N} s_i^2 \left(\prod_{i=1}^N \frac{1}{\left(s_i^{2\alpha} + \xi_n^{2\alpha}\right)^\beta} \right) ds_1 \ldots ds_N \right)^{1/2} < \infty, \quad (31.40)$$

and all these integrals are uniformly bounded in $\xi_n \in (0, 1]$. And all integrals converge to zero, as $\xi_n \to 0$.

We continue.

Theorem 31.6. *Let $N \geq 1$, $\xi_n \in (0, 1]$, $n \in \mathbb{N}$, $\beta > \frac{1}{\alpha}$. Then*

$$\overline{I}_2 := W_n^N \int_{\mathbb{R}^N} \frac{\|s\|_1}{\xi_n} \prod_{i=1}^N \frac{1}{\left(s_i^{2\alpha} + \xi_n^{2\alpha}\right)^\beta} ds_1 \ldots ds_N$$

$$\leq \frac{\xi^{2\alpha\beta(N-1)}}{\Gamma^N\left(\frac{1}{2\alpha}\right)\Gamma^N\left(\beta - \frac{1}{2\alpha}\right)} \left[(2\alpha\Gamma(\beta))^N N + \Gamma^N\left(\frac{1}{\alpha}\right)\Gamma^N\left(\beta - \frac{1}{\alpha}\right) \right]$$

$$\leq \frac{1}{\Gamma^N\left(\frac{1}{2\alpha}\right)\Gamma^N\left(\beta - \frac{1}{2\alpha}\right)} \left[(2\alpha\Gamma(\beta))^N N + \Gamma^N\left(\frac{1}{\alpha}\right)\Gamma^N\left(\beta - \frac{1}{\alpha}\right) \right]$$

$$< +\infty, \quad (31.41)$$

uniformly bounded in ξ_n, and as $\xi_n \to 0$, $\overline{I}_2 \to 0$, when $N > 1$.

Proof. We observe that

$$\overline{I}_2 = W_n^N \int_{\mathbb{R}^N} \frac{\|s\|_1}{\xi_n} \prod_{i=1}^{N} \frac{1}{\left(s_i^{2\alpha} + \xi_n^{2\alpha}\right)^{\beta}} ds_1 \dots ds_N$$

$$= 2^N W_n^N \int_{\mathbb{R}_+^N} \left(\frac{\sum_{i=1}^N s_i}{\xi_n}\right) \prod_{i=1}^{N} \frac{1}{\left(s_i^{2\alpha} + \xi_n^{2\alpha}\right)^{\beta}} ds_1 \dots ds_N$$

$$= \left(c := \frac{\Gamma(\beta)\,\alpha}{\Gamma\left(\frac{1}{2\alpha}\right)\Gamma\left(\beta - \frac{1}{2\alpha}\right)}, \text{ i.e. } W_n = c\xi_n^{2\alpha\beta-1}\right)$$

$$= (2c)^N \xi_n^{2\alpha\beta N - N} \xi_n^{-2\alpha\beta} \xi_n^N \int_{\mathbb{R}_+^N} \left(\sum_{i=1}^N \left(\frac{s_i}{\xi_n}\right)\right)$$

$$\times \prod_{i=1}^{N} \frac{1}{\left(\left(\frac{s_i}{\xi_n}\right)^{2\alpha} + 1\right)^{\beta}} \prod_{i=1}^{N} d\left(\frac{s_i}{\xi_n}\right) \tag{31.42}$$

$$(2c)^N \xi_n^{2\alpha\beta(N-1)} \int_{\mathbb{R}_+^N} \left(\sum_{i=1}^N z_i\right) \left(\prod_{i=1}^{N} \frac{1}{\left(z_i^{2\alpha}+1\right)^{\beta}}\right) \prod_{i=1}^{N} dz_i$$

$$= (2c)^N \xi_n^{2\alpha\beta(N-1)} \left[\int_{[0,1]^N} \left(\sum_{i=1}^N z_i\right) \left(\prod_{i=1}^{N} \frac{1}{\left(z_i^{2\alpha}+1\right)^{\beta}}\right) \prod_{i=1}^{N} dz_i\right.$$

$$\left. + \int_{(\mathbb{R}_+ - [0,1])^N} \left(\sum_{i=1}^N z_i\right) \left(\prod_{i=1}^{N} \frac{1}{\left(z_i^{2\alpha}+1\right)^{\beta}}\right) \prod_{i=1}^{N} dz_i\right]$$

$$\leq (2c)^N \xi_n^{2\alpha\beta(N-1)} \left[N + \int_{(\mathbb{R}_+ - [0,1])^N} \left(\prod_{i=1}^N z_i\right) \left(\prod_{i=1}^{N} \frac{1}{\left(z_i^{2\alpha}+1\right)^{\beta}}\right) \prod_{i=1}^{N} dz_i\right] \tag{31.43}$$

$$(2c)^N \xi_n^{2\alpha\beta(N-1)} \left[N + \prod_{i=1}^{N} \int_1^{\infty} \frac{z_i}{\left(z_i^{2\alpha}+1\right)^{\beta}} dz_i\right]$$

$$= (2c)^N \xi_n^{2\alpha\beta(N-1)} \left[N + \left(\int_1^{\infty} \frac{z}{\left(z^{2\alpha}+1\right)^{\beta}} dz\right)^N\right]$$

$$\leq (2c)^N \xi_n^{2\alpha\beta(N-1)} \left[N + \left(\int_0^{\infty} \frac{z}{\left(z^{2\alpha}+1\right)^{\beta}} dz\right)^N\right]$$

(by Zwillinger (1995), p. 397, formula 595)

$$
\begin{aligned}
&= (2c)^N \xi_n^{2\alpha\beta(N-1)} \left[N + \left(\frac{\Gamma\left(\frac{1}{\alpha}\right)\Gamma\left(\beta - \frac{1}{\alpha}\right)}{2\alpha\Gamma(\beta)} \right)^N \right] \\
&= \left(\frac{2\alpha\Gamma(\beta)}{\Gamma\left(\frac{1}{2\alpha}\right)\Gamma\left(\beta - \frac{1}{2\alpha}\right)} \right)^N \xi_n^{2\alpha\beta(N-1)} \left[N + \left(\frac{\Gamma\left(\frac{1}{\alpha}\right)\Gamma\left(\beta - \frac{1}{\alpha}\right)}{2\alpha\Gamma(\beta)} \right)^N \right]
\end{aligned} \tag{31.44}
$$

$$
\begin{aligned}
&= \frac{\xi_n^{2\alpha\beta(N-1)}}{\Gamma^N\left(\frac{1}{2\alpha}\right)\Gamma^N\left(\beta - \frac{1}{2\alpha}\right)} \left[(2\alpha\Gamma(\beta))^N N + \Gamma^N\left(\frac{1}{\alpha}\right)\Gamma^N\left(\beta - \frac{1}{\alpha}\right) \right] \\
&\leq \frac{1}{\Gamma^N\left(\frac{1}{2\alpha}\right)\Gamma^N\left(\beta - \frac{1}{2\alpha}\right)} \left[(2\alpha\Gamma(\beta))^N N + \Gamma^N\left(\frac{1}{\alpha}\right)\Gamma^N\left(\beta - \frac{1}{\alpha}\right) \right] < +\infty,
\end{aligned}
$$

uniformly bounded in ξ_n, and as $\xi_n \to 0$, we get $\overline{I}_2 \to 0$, for $N > 1$, $\beta > \frac{1}{\alpha}$. ☐

We apply

Theorem 31.7. *(Anastassiou 2021, p. 447) Let $r, m \in \mathbb{N}$, $N \in \mathbb{N}$-$\{1\}$, $\alpha \in \mathbb{N}$, $\beta > \frac{(m+r)p+1}{2\alpha}$, $\alpha_j \in \mathbb{Z}^+$, $p > 1$, $j = 1, \ldots, N : |\overline{\alpha}| := \sum_{j=1}^N \alpha_j = m$, $\xi_n \in (0,1]$, $n \in \mathbb{N}$. Denote $W_n := \frac{\alpha\Gamma(\beta)\xi_n^{2\alpha\beta-1}}{\Gamma\left(\frac{1}{2\alpha}\right)\Gamma\left(\beta-\frac{1}{2\alpha}\right)}$. Then*

$$
\begin{aligned}
\overline{\theta}_{\xi_n}(\overline{\alpha}) &:= W_n^N \int_{\mathbb{R}^N} \left(\left(\prod_{i=1}^N |s_i|^{\alpha_i} \right) \left(1 + \frac{\|s\|_2}{\xi_n} \right)^r \right)^p \prod_{i=1}^N \frac{1}{\left(s_i^{2\alpha} + \xi_n^{2\alpha}\right)^\beta} \prod_{i=1}^N ds_i \\
&\leq \xi_n^{2\alpha\beta(N-1)+mp} \left(\frac{2\alpha\Gamma(\beta)}{\Gamma\left(\frac{1}{2\alpha}\right)\Gamma\left(\beta - \frac{1}{2\alpha}\right)} \right)^N \\
&\quad \times \left[(1+N)^{rp} + 2^{rp} \left(\frac{\Gamma\left(\frac{(m+r)p+1}{2\alpha}\right)\Gamma\left(\beta - \left(\frac{(m+r)p+1}{2\alpha}\right)\right)}{2\alpha\Gamma(\beta)} \right)^N \right] \\
&\leq \left(\frac{2\alpha\Gamma(\beta)}{\Gamma\left(\frac{1}{2\alpha}\right)\Gamma\left(\beta - \frac{1}{2\alpha}\right)} \right)^N
\end{aligned}
$$

$$\times \left[(1+N)^{rp} + 2^{rp} \left(\frac{\Gamma\left(\frac{(m+r)p+1}{2\alpha}\right)\Gamma\left(\beta - \left(\frac{(m+r)p+1}{2\alpha}\right)\right)}{2\alpha\Gamma(\beta)} \right)^N \right] < +\infty, \tag{31.45}$$

uniformly bounded in ξ_n.

Also $\overline{\theta}_{\xi_n}(\overline{\alpha}) \to 0$, *as* $\xi_n \to 0$, $n \to +\infty$.

Remark 31.4. (on Theorem 31.7) Clearly, it holds

$$\overline{B}_{\xi_n}(\overline{\alpha}) := W_n^N \int_{\mathbb{R}^N} \left(\prod_{i=1}^N |s_i|^{\alpha_i} \right) \prod_{i=1}^N \frac{1}{\left(s_i^{2\alpha} + \xi_n^{2\alpha}\right)^\beta} \prod_{i=1}^N ds_i < \infty, \tag{31.46}$$

uniformly bounded in $\xi_n \in (0,1]$. And, clearly $\overline{B}_{\xi_n}(\overline{\alpha}) \to 0$, as $\xi_n \to 0$, $n \to \infty$, by (31.45).

We need the following.

Theorem 31.8. *Let* $\alpha_i \in \mathbb{Z}^+$, $i = 1, \ldots, N \in \mathbb{N} : |\alpha| := \sum_{i=1}^N \alpha_i = 2$, $\xi_n \in (0,1]$, $n \in \mathbb{N}$, $p \geq 1$, *and* $j = 0, 1, \ldots, r \in \mathbb{N}$. *We assume that* $\beta > \frac{3p+1}{2\alpha}$. *Then*

$$\overline{A}_{j\xi_n}(\alpha) := W_n^N \int_{\mathbb{R}^N} \left(\prod_{i=1}^N |s_i|^{p\alpha_i} \right) \left(1 + \frac{j^p \|s\|_2^p}{\xi_n^p} \right) \prod_{i=1}^N \frac{1}{\left(s_i^{2\alpha} + \xi_n^{2\alpha}\right)^\beta} \prod_{i=1}^N ds_i \tag{31.47}$$

$$\leq \frac{\xi_n^{2\alpha\beta(N-1)+2p}}{\Gamma^N\left(\frac{1}{2\alpha}\right)\Gamma^N\left(\beta - \frac{1}{2\alpha}\right)}$$

$$\times \left[(2\alpha\Gamma(\beta))^N (1 + j^p N^p) + (1+j^p)\Gamma^N\left(\frac{3p+1}{2\alpha}\right)\Gamma^N\left(\beta - \left(\frac{3p+1}{2\alpha}\right)\right) \right]$$

$$\leq \frac{1}{\Gamma^N\left(\frac{1}{2\alpha}\right)\Gamma^N\left(\beta - \frac{1}{2\alpha}\right)} \tag{31.48}$$

$$\times \left[(2\alpha\Gamma(\beta))^N (1 + j^p N^p) + (1+j^p)\Gamma^N\left(\frac{3p+1}{2\alpha}\right)\Gamma^N\left(\beta - \left(\frac{3p+1}{2\alpha}\right)\right) \right]$$

$$< +\infty,$$

uniformly bounded in $\xi_n \in (0,1]$, *and* $\overline{A}_{j\xi_n}(\alpha) \to 0$, *as* $\xi_n \to 0$, $n \to \infty$. *When* $p = 1$, *we denote* $\overline{A}_{j\xi_n}(\alpha)$ *as* $\overline{\overline{A}}_{j\xi_n}(\alpha)$.

Proof. We estimate ($p \geq 1$)

$$\overline{A}_{j\xi_n}(\alpha) = W_n^N \int_{\mathbb{R}^N} \left(\prod_{i=1}^N |s_i|^{p\alpha_i} \right) \left(1 + \frac{j^p \|s\|_2^p}{\xi_n^p} \right) \prod_{i=1}^N \frac{1}{\left(s_i^{2\alpha} + \xi_n^{2\alpha}\right)^\beta} \prod_{i=1}^N ds_i$$

$$= 2^N W_n^N \int_{\mathbb{R}_+^N} \left(\prod_{i=1}^N s_i^{p\alpha_i} \right) \left(1 + \frac{j^p \|s\|_2^p}{\xi_n^p} \right) \prod_{i=1}^N \frac{1}{\left(s_i^{2\alpha} + \xi_n^{2\alpha}\right)^\beta} \prod_{i=1}^N ds_i$$

$$\leq 2^N W_n^N \int_{\mathbb{R}_+^N} \left(\prod_{i=1}^N s_i^{p\alpha_i} \right) \left(1 + j^p \left(\sum_{i=1}^N \frac{s_i}{\xi_n} \right)^p \right) \prod_{i=1}^N \frac{1}{\left(s_i^{2\alpha} + \xi_n^{2\alpha}\right)^\beta} \prod_{i=1}^N ds_i \tag{31.49}$$

$$\left(c := \frac{\Gamma(\beta)\,\alpha}{\Gamma\left(\frac{1}{2\alpha}\right) \Gamma\left(\beta - \frac{1}{2\alpha}\right)}, \text{ i.e. } W_n = c\xi_n^{2\alpha\beta-1} \right)$$

$$= (2c)^N \xi_n^{2\alpha\beta N - N} \xi_n^{2p} \xi_n^{-2\alpha\beta} \xi_n^N \int_{\mathbb{R}_+^N} \left(\prod_{i=1}^N \left(\frac{s_i}{\xi_n} \right)^{p\alpha_i} \right)$$

$$\times \left(1 + j^p \left(\sum_{i=1}^N \frac{s_i}{\xi_n} \right)^p \right) \prod_{i=1}^N \frac{1}{\left(\left(\frac{s_i}{\xi_n} \right)^{2\alpha} + 1 \right)^\beta} \prod_{i=1}^N d\left(\frac{s_i}{\xi_n} \right)$$

$$= (2c)^N \xi_n^{2\alpha\beta(N-1)+2p} \int_{\mathbb{R}_+^N} \left(\prod_{i=1}^N z_i^{p\alpha_i} \right) \left(1 + j^p \left(\sum_{i=1}^N z_i \right)^p \right)$$

$$\times \prod_{i=1}^N \frac{1}{\left(z_i^{2\alpha} + 1\right)^\beta} \prod_{i=1}^N dz_i$$

$$= (2c)^N \xi_n^{2\alpha\beta(N-1)+2p} \left[\int_{[0,1]^N} \left(\prod_{i=1}^N z_i^{p\alpha_i} \right) \left(1 + j^p \left(\sum_{i=1}^N z_i \right)^p \right) \right.$$

$$\times \prod_{i=1}^N \frac{1}{\left(z_i^{2\alpha} + 1\right)^\beta} \prod_{i=1}^N dz_i \tag{31.50}$$

$$\left. + \int_{(\mathbb{R}_+ - [0,1])^N} \left(\prod_{i=1}^N z_i^{p\alpha_i} \right) \left(1 + j^p \left(\sum_{i=1}^N z_i \right)^p \right) \prod_{i=1}^N \frac{1}{\left(z_i^{2\alpha} + 1\right)^\beta} \prod_{i=1}^N dz_i \right]$$

$$\leq (2c)^N \xi_n^{2\alpha\beta(N-1)+2p} \left[(1 + j^p N^p) + \int_{(\mathbb{R}_+ - [0,1])^N} \left(\prod_{i=1}^N z_i^{p\alpha_i} \right) \right.$$

$$\times \left(\left(\sum_{i=1}^{N} z_i \right)^p + j^p \left(\sum_{i=1}^{N} z_i \right)^p \right) \prod_{i=1}^{N} \frac{1}{\left(z_i^{2\alpha}+1\right)^{\beta}} \prod_{i=1}^{N} dz_i \Bigg]$$

$$= (2c)^N \xi_n^{2\alpha\beta(N-1)+2p} \left[(1+j^p N^p) + (1+j^p) \int_{(\mathbb{R}_+-[0,1])^N} \left(\prod_{i=1}^{N} z_i^{p\alpha_i} \right) \right.$$

$$\left. \times \left(\sum_{i=1}^{N} z_i \right)^p \prod_{i=1}^{N} \frac{1}{\left(z_i^{2\alpha}+1\right)^{\beta}} \prod_{i=1}^{N} dz_i \right]$$

$$\leq (2c)^N \xi_n^{2\alpha\beta(N-1)+2p} \left[(1+j^p N^p) + (1+j^p) \int_{(\mathbb{R}_+-[0,1])^N} \left(\prod_{i=1}^{N} z_i^{p\alpha_i} \right) \right.$$

$$\left. \times \left(\prod_{i=1}^{N} z_i^{p} \right) \prod_{i=1}^{N} \frac{1}{\left(z_i^{2\alpha}+1\right)^{\beta}} \prod_{i=1}^{N} dz_i \right]$$

$$= (2c)^N \xi_n^{2\alpha\beta(N-1)+2p} [(1+j^p N^p) + (1+j^p)$$

$$\times \int_{(\mathbb{R}_+-[0,1])^N} \prod_{i=1}^{N} z_i^{p(\alpha_i+1)} \prod_{i=1}^{N} \frac{1}{\left(z_i^{2\alpha}+1\right)^{\beta}} \prod_{i=1}^{N} dz_i \Bigg] \tag{31.51}$$

$$= (2c)^N \xi_n^{2\alpha\beta(N-1)+2p} \left[(1+j^p N^p) + (1+j^p) \prod_{i=1}^{N} \right.$$

$$\left. \times \int_{1}^{\infty} z_i^{p(\alpha_i+1)} \frac{1}{\left(z_i^{2\alpha}+1\right)^{\beta}} dz_i \right]$$

$$\leq (2c)^N \xi_n^{2\alpha\beta(N-1)+2p} \left[(1+j^p N^p) + (1+j^p) \left(\int_{1}^{\infty} \frac{z^{3p}}{\left(z^{2\alpha}+1\right)^{\beta}} dz \right)^N \right]$$

$$\leq (2c)^N \xi_n^{2\alpha\beta(N-1)+2p} \left[(1+j^p N^p) + (1+j^p) \left(\int_{0}^{\infty} \frac{z^{3p}}{\left(z^{2\alpha}+1\right)^{\beta}} dz \right)^N \right] \tag{31.52}$$

(by Zwillinger (1995), p. 397, formula 595, by assuming $\beta > \frac{3p+1}{2\alpha}$)

$$(2c)^N \xi_n^{2\alpha\beta(N-1)+2p}$$

$$\times \left[(1+j^p N^p) + (1+j^p) \left(\frac{\Gamma\left(\frac{3p+1}{2\alpha}\right) \Gamma\left(\beta - \left(\frac{3p+1}{2\alpha}\right)\right)}{2\alpha \Gamma(\beta)} \right)^N \right]$$

$$= \left(\frac{2\Gamma(\beta)\alpha}{\Gamma\left(\frac{1}{2\alpha}\right)\Gamma\left(\beta - \frac{1}{2\alpha}\right)} \right)^N \xi_n^{2\alpha\beta(N-1)+2p}$$

$$\times \left[(1 + j^p N^p) + (1 + j^p) \left(\frac{\Gamma\left(\frac{3p+1}{2\alpha}\right)\Gamma\left(\beta - \left(\frac{3p+1}{2\alpha}\right)\right)}{2\alpha\Gamma(\beta)} \right)^N \right]$$

$$= \frac{\xi_n^{2\alpha\beta(N-1)+2p}}{\Gamma^N\left(\frac{1}{2\alpha}\right)\Gamma^N\left(\beta - \frac{1}{2\alpha}\right)} \left[(2\alpha\Gamma(\beta))^N (1 + j^p N^p) \right.$$

$$\left. + (1 + j^p)\Gamma^N\left(\frac{3p+1}{2\alpha}\right)\Gamma^N\left(\beta - \left(\frac{3p+1}{2\alpha}\right)\right) \right]. \tag{31.53}$$

The claim is proved. ☐

We need the following.

Theorem 31.9. *Let $p \geq 1$, $\xi_n \in (0,1]$, $n \in \mathbb{N}$, $N \geq 1$; and $\beta > \frac{p+1}{2\alpha}$. Then*

$$\overline{K}_{p\xi_n} := W_n^N \int_{\mathbb{R}^N} \left(\frac{\|s\|_2}{\xi_n} \right)^p \prod_{i=1}^N \frac{1}{\left(s_i^{2\alpha} + \xi_n^{2\alpha}\right)^\beta} \prod_{i=1}^N ds_i \tag{31.54}$$

$$\leq \frac{\xi^{2\alpha\beta(N-1)}}{\Gamma^N\left(\frac{1}{2\alpha}\right)\Gamma^N\left(\beta - \frac{1}{2\alpha}\right)}$$

$$\times \left[(2\alpha\Gamma(\beta))^N N^p + \Gamma^N\left(\frac{p+1}{2\alpha}\right)\Gamma^N\left(\beta - \left(\frac{p+1}{2\alpha}\right)\right) \right]$$

$$\leq \frac{1}{\Gamma^N\left(\frac{1}{2\alpha}\right)\Gamma^N\left(\beta - \frac{1}{2\alpha}\right)}$$

$$\times \left[(2\alpha\Gamma(\beta))^N N^p + \Gamma^N\left(\frac{p+1}{2\alpha}\right)\Gamma^N\left(\beta - \left(\frac{p+1}{2\alpha}\right)\right) \right] < +\infty, \tag{31.55}$$

uniformly bounded in $\xi_n \in (0,1]$, and $\overline{K}_{p\xi_n} \to 0$, as $\xi_n \to 0$, $n \to \infty$, for $N > 1$. When $p = 1$, we denote $\overline{K}_{p\xi_n}$ as $\overline{\overline{K}}_{1\xi_n}$.

Proof. We estimate ($p \geq 1$)

$$\overline{K}_{p\xi_n} = W_n^N \int_{\mathbb{R}^N} \left(\frac{\|s\|_2}{\xi_n} \right)^p \prod_{i=1}^N \frac{1}{\left(s_i^{2\alpha} + \xi_n^{2\alpha}\right)^\beta} \prod_{i=1}^N ds_i$$

$$= 2^N W_n^N \int_{\mathbb{R}_+^N} \left(\frac{\|s\|_2}{\xi_n} \right)^p \prod_{i=1}^N \frac{1}{\left(s_i^{2\alpha} + \xi_n^{2\alpha}\right)^\beta} \prod_{i=1}^N ds_i$$

$$\leq 2^N W_n^N \int_{\mathbb{R}_+^N} \left(\frac{\sum_{i=1}^N s_i}{\xi_n} \right)^p \prod_{i=1}^N \frac{1}{\left(s_i^{2\alpha} + \xi_n^{2\alpha}\right)^\beta} \prod_{i=1}^N ds_i = \tag{31.56}$$

$$\left(c := \frac{\Gamma(\beta)\,\alpha}{\Gamma\left(\frac{1}{2\alpha}\right)\Gamma\left(\beta - \frac{1}{2\alpha}\right)}, \text{ i.e. } W_n = c\xi_n^{2\alpha\beta-1} \right)$$

$$(2c)^N \xi_n^{2\alpha\beta N - N} \xi_n^{-2\alpha\beta} \xi_n^N \int_{\mathbb{R}_+^N} \left(\sum_{i=1}^N \left(\frac{s_i}{\xi_n} \right) \right)^p \prod_{i=1}^N \frac{1}{\left(\left(\frac{s_i}{\xi_n} \right)^{2\alpha} + 1 \right)^\beta} \prod_{i=1}^N d\left(\frac{s_i}{\xi_n} \right)$$

$$= (2c)^N \xi_n^{2\alpha\beta(N-1)} \int_{\mathbb{R}_+^N} \left(\sum_{i=1}^N z_i \right)^p \prod_{i=1}^N \frac{1}{\left(z_i^{2\alpha}+1\right)^\beta} \prod_{i=1}^N dz_i$$

$$= (2c)^N \xi_n^{2\alpha\beta(N-1)} \left[\int_{[0,1]^N} \left(\sum_{i=1}^N z_i \right)^p \prod_{i=1}^N \frac{1}{\left(z_i^{2\alpha}+1\right)^\beta} \prod_{i=1}^N dz_i \right.$$

$$\left. + \int_{(\mathbb{R}_+ - [0,1])^N} \left(\sum_{i=1}^N z_i \right)^p \prod_{i=1}^N \frac{1}{\left(z_i^{2\alpha}+1\right)^\beta} \prod_{i=1}^N dz_i \right]$$

$$\leq (2c)^N \xi_n^{2\alpha\beta(N-1)} \left[N^p + \int_{(\mathbb{R}_+ - [0,1])^N} \left(\prod_{i=1}^N z_i^p \right) \prod_{i=1}^N \frac{1}{\left(z_i^{2\alpha}+1\right)^\beta} \prod_{i=1}^N dz_i \right] \tag{31.57}$$

$$= (2c)^N \xi_n^{2\alpha\beta(N-1)} \left[N^p + \left(\int_1^\infty z^p \frac{1}{\left(z^{2\alpha}+1\right)^\beta} dz \right)^N \right]$$

$$\leq (2c)^N \xi_n^{2\alpha\beta(N-1)} \left[N^p + \left(\int_0^\infty \frac{z^p}{\left(z^{2\alpha}+1\right)^\beta} dz \right)^N \right]$$

(by Zwillinger (1995), p. 397, formula 595, by $\beta > \frac{p+1}{2\alpha}$)

$$(2c)^N \xi_n^{2\alpha\beta(N-1)} \left[N^p + \left(\frac{\Gamma\left(\frac{p+1}{2\alpha}\right)\Gamma\left(\beta - \left(\frac{p+1}{2\alpha}\right)\right)}{2\alpha\Gamma(\beta)} \right)^N \right]$$

$$= \left(\frac{2\alpha\Gamma(\beta)}{\Gamma\left(\frac{1}{2\alpha}\right)\Gamma\left(\beta - \frac{1}{2\alpha}\right)} \right)^N \xi_n^{2\alpha\beta(N-1)}$$

$$\times \left[N^p + \left(\frac{\Gamma\left(\frac{p+1}{2\alpha}\right)\Gamma\left(\beta - \left(\frac{p+1}{2\alpha}\right)\right)}{2\alpha\Gamma(\beta)} \right)^N \right]$$

$$= \frac{\xi_n^{2\alpha\beta(N-1)}}{\Gamma^N\left(\frac{1}{2\alpha}\right)\Gamma^N\left(\beta-\frac{1}{2\alpha}\right)}$$
$$\times\left[\left(2\alpha\Gamma\left(\beta\right)\right)^N N^p+\Gamma^N\left(\frac{p+1}{2\alpha}\right)\Gamma^N\left(\beta-\left(\frac{p+1}{2\alpha}\right)\right)\right], \quad (31.58)$$

proving the claim. □

31.4 Main Results

Here we apply Theorem 31.1 to U_n operators, regarding uniform convergence.

Theorem 31.10. *We consider $f \in C^2\left(\mathbb{R}^N\right)$ and let all $\alpha_i \in \mathbb{Z}^+$, $i = 1,\ldots,N$, $N \geq 1$, $|\alpha| := \sum_{i=1}^N \alpha_i = 2$; $x \in \mathbb{R}^N$, and all the partials f_α of order 2, along with $f \in C_B\left(\mathbb{R}^N\right)$, or all f_α of order 2, $f \in C_U\left(\mathbb{R}^N\right)$; $0 < \xi_n \leq 1$, $n \in \mathbb{N}$. Assume that $\beta > \frac{1}{\alpha}$.*

Denote ($n \in \mathbb{N}$)

$$\overline{\Delta}_n(x) := U_n(f,x) - f(x) - \left(\sum_{j=0}^r \alpha_j j\right)\sin(1)$$
$$\times\left[\sum_{i=1}^N \frac{\partial f(x)}{\partial x_i} W_n^N\left(\int_{\mathbb{R}^N} s_i \prod_{i=1}^N \frac{1}{\left(s_i^{2\alpha}+\xi_n^{2\alpha}\right)^\beta} ds_1\ldots ds_N\right)\right]$$
$$-2\left(\sum_{j=0}^r \alpha_j j^2\right)\sin^2\left(\frac{1}{2}\right)$$
$$\times\left\{\sum_{i=1}^N W_n^N\left(\int_{\mathbb{R}^N} s_i^2 \prod_{i=1}^N \frac{1}{\left(s_i^{2\alpha}+\xi_n^{2\alpha}\right)^\beta} ds_1\ldots ds_N\right)\frac{\partial^2 f(x)}{\partial x_i^2}\right.$$
$$\left.+\sum_{\substack{i\neq j^*,\\ i,j^*\in\{1,\ldots,N\}}} W_n^N\left(\int_{\mathbb{R}^N} s_i s_{j^*} \prod_{i=1}^N \frac{1}{\left(s_i^{2\alpha}+\xi_n^{2\alpha}\right)^\beta} ds_1\ldots ds_N\right)\frac{\partial^2 f(x)}{\partial x_i \partial x_{j^*}}\right\}. \quad (31.59)$$

Then

(i)

$$\left|\overline{\Delta}_n(x)\right| \leq \left\|\overline{\Delta}_n\right\|_\infty \leq \sum_{j=0}^{r} |\alpha_j|$$

$$\times \left[\left[j^2 \sum_{\substack{\alpha_i \in \mathbb{Z}^+, \\ \alpha:|\alpha|=2}} \left(\frac{1}{\prod_{i=1}^{N} \alpha_i!}\right) \omega_1(f_\alpha, \xi_n) \overline{I}_{1j}(\alpha)\right]\right.$$

$$\left. + \frac{1}{2}\omega_1(f, \xi_n)\left[1 + \frac{j}{3}\overline{I}_2\right]\right] =: \overline{\varphi}_{\xi_n}, \tag{31.60}$$

where $\overline{I}_{1j}(\alpha)$ is as in (31.33) and $\overline{I}_2$ is as in (31.41).

In case of all f_α of order 2 and $f \in C_U(\mathbb{R}^N)$ and $\xi_n \to 0$, as $n \to \infty$, then $\overline{\Delta}_n(x)$, $\left\|\overline{\Delta}_n\right\| \to 0$ with rates.

(ii) *If $\frac{\partial f}{\partial x_i} = 0$, $i = 1, \ldots, N$, and $f_\alpha(x) = 0$, $\alpha_i \in \mathbb{Z}^+$, $i = 1, \ldots, N$, with $|\alpha| = 2$, then*

$$|U_n(f, x) - f(x)| \leq \overline{\varphi}_{\xi_n}. \tag{31.61}$$

And $U_n(f, x) \to f(x)$ in the uniformly continuous case.

(iii) *Additionally, assume all partials of order ≤ 2 are bounded. Hence*

$$\|U_n(f) - f\|_\infty \leq \left(\sum_{j=0}^{r} |\alpha_j| j\right)(0.8414)$$

$$\times \left[\sum_{i=1}^{N} \left\|\frac{\partial f}{\partial x_i}\right\|_\infty W_n^N \left(\int_{\mathbb{R}^N} |s_i| \prod_{i=1}^{N} \frac{1}{\left(s_i^{2\alpha} + \xi_n^{2\alpha}\right)^\beta} ds_1 \ldots ds_N\right)\right]$$

$$+ \left(\sum_{j=0}^{r} |\alpha_j| j^2\right)(0.4596)\left\{\sum_{i=1}^{N} W_n^N\right.$$

$$\times \left(\int_{\mathbb{R}^N} s_i^2 \prod_{i=1}^{N} \frac{1}{\left(s_i^{2\alpha} + \xi_n^{2\alpha}\right)^\beta} ds_1 \ldots ds_N\right) \left\|\frac{\partial^2 f}{\partial x_i^2}\right\|_\infty + \sum_{\substack{i \neq j^*, \\ i,j^* \in \{1,\ldots,N\}}} W_n^N$$

$$\times \left(\int_{\mathbb{R}^N} |s_i| \, |s_{j^*}| \prod_{i=1}^{N} \frac{1}{\left(s_i^{2\alpha} + \xi_n^{2\alpha}\right)^{\beta}} ds_1 \ldots ds_N \right) \left\| \frac{\partial^2 f}{\partial x_i \partial x_{j^*}} \right\|_{\infty} \Bigg\} + \overline{\varphi}_{\xi_n}. \tag{31.62}$$

If all f_α of order 2, $f \in C_U\left(\mathbb{R}^N\right)$

$$\|U_n(f) - f\|_{\infty} \to 0 \textit{ with rates, as } \xi_n \to 0, n \to +\infty.$$

Proof. By Theorems 31.1, 31.5 and 31.6 and Remark 31.3. □

Next we apply Theorem 31.2 to U_n operators.

Theorem 31.11. *Let $f \in C^m\left(\mathbb{R}^N\right)$, $m \in \mathbb{N}$, $N > 1$, with $f_{\overline{\alpha}} \in L_p\left(\mathbb{R}^N\right)$, $|\overline{\alpha}| = m$, $x \in \mathbb{R}^N$. Let $p, q > 1 : \frac{1}{p} + \frac{1}{q} = 1$; $0 < \xi_n \leq 1$, $n \in \mathbb{N}$. For $\overline{j} = 1, \ldots, m$, and $\overline{\alpha} := (\alpha_1, \ldots, \alpha_N)$, $\alpha_i \in \mathbb{Z}^+$, $i = 1, \ldots, N$, $|\overline{\alpha}| := \sum_{i=1}^{N} \alpha_i = \overline{j}$, call*

$$\overline{c}_{\overline{\alpha},n,\overline{j}} := W_n^N \int_{\mathbb{R}^N} \left(\prod_{i=1}^{N} s_i^{\alpha_i} \right) \prod_{i=1}^{N} \frac{1}{\left(s_i^{2\alpha} + \xi_n^{2\alpha}\right)^{\beta}} ds_1 \ldots ds_N. \tag{31.63}$$

Here we assume $\beta > \frac{(m+r)p+1}{2\alpha}$.

Then

$$\left\| \overline{E}_{r,n}^{[m]} \right\|_p := \left\| U_{r,n}(f; x) - f(x) - \sum_{\overline{j}=1}^{m} \delta_{\overline{j},r}^{[m]} \left(\sum_{|\overline{\alpha}|=\overline{j}} \frac{\overline{c}_{\overline{\alpha},n,\overline{j}} f_{\overline{\alpha}}(x)}{\left(\prod_{i=1}^{N} \alpha_i!\right)} \right) \right\|_{p,x}$$

$$\leq \left[\left(N^m/(m-1)!\right)^{\frac{p}{q}} \left(\frac{m}{\left(q(m-1)+1\right)^{\frac{p}{q}}} \right) \right.$$

$$\left. \times \left(\sum_{|\overline{\alpha}|=m} \frac{1}{\prod_{i=1}^{N} \alpha_i!} \overline{\theta}_{\xi_n}(\overline{\alpha}) \, \omega_r\left(f_{\overline{\alpha}}, \xi_n\right)_p^p \right) \right]^{\frac{1}{p}}, \tag{31.64}$$

where $\overline{\theta}_{\xi_n}(\overline{\alpha})$ as in (31.45).

As $n \to \infty$ and $\xi_n \to 0$, by (31.64), we obtain that $\left\| \overline{E}_{r,n}^{[m]} \right\|_p \to 0$ with rates.

One also obtains by (31.64) that

$$\|U_n(f;x) - f(x)\|_{p,x}$$
$$\leq \sum_{\overline{j}=1}^{m} \left|\delta_{\overline{j},r}^{[m]}\right| \left(\sum_{|\overline{\alpha}|=\overline{j}} \frac{\left|\overline{c}_{\overline{\alpha},n,\overline{j}}\right|}{\prod_{i=1}^{N} \alpha_i!} \|f_{\overline{\alpha}}\|_p \right) + R.H.S.(31.64), \qquad (31.65)$$

given that $\|f_{\overline{\alpha}}\|_p < \infty$, $|\overline{\alpha}| = \overline{j}$, $\overline{j} = 1, \ldots, m$.

Finally we get $\|U_n(f) - f\|_p \to 0$ *as* $\xi_n \to 0$, *and* $n \to \infty$. *That is,* $U_n \to I$ *the unit operator, in* L_p *norm, with rates.*

Proof. By Theorems 31.2 and 31.7 and (31.46). □

Next we apply Theorem 31.3 to U_n operators.

Theorem 31.12. *Let* $p, q > 1 : \frac{1}{p} + \frac{1}{q} = 1$, $0 < \xi_n \leq 1$, $n \in \mathbb{N}$. *Here we deal with* $f \in C^2(\mathbb{R}^N)$, $N \geq 1$, *with* f, $f_\alpha \in L_p(\mathbb{R}^N)$, $|\alpha| = 2$, *where* $\alpha_i \in \mathbb{Z}^+$, $i = 1, \ldots, N$, *and* $|\alpha| = \sum_{i=1}^{N} \alpha_i$; $x \in \mathbb{R}^N$; $j = 0, 1, \ldots, r$. *Assume* $\beta > \frac{3p+1}{2\alpha}$.

Denote $(n \in \mathbb{N})$

$$\overline{\Delta}_n(x) := U_n(f,x) - f(x) - \left(\sum_{j=0}^{r} \alpha_j j\right) \sin(1)$$
$$\times \left[\sum_{i=1}^{N} \frac{\partial f(x)}{\partial x_i} W_n^N \left(\int_{\mathbb{R}^N} s_i \prod_{i=1}^{N} \frac{1}{\left(s_i^{2\alpha} + \xi_n^{2\alpha}\right)^\beta} ds_1 \ldots ds_N \right) \right]$$
$$- 2 \left(\sum_{j=0}^{r} \alpha_j j^2\right) \sin^2\left(\frac{1}{2}\right) \left\{ \sum_{i=1}^{N} W_n^N \right.$$
$$\times \left(\int_{\mathbb{R}^N} s_i^2 \prod_{i=1}^{N} \frac{1}{\left(s_i^{2\alpha} + \xi_n^{2\alpha}\right)^\beta} ds_1 \ldots ds_N \right) \frac{\partial^2 f(x)}{\partial x_i^2}$$
$$\left. + \sum_{\substack{i \neq j^*, \\ i,j^* \in \{1,\ldots,N\}}} W_n^N \left(\int_{\mathbb{R}^N} s_i s_{j^*} \prod_{i=1}^{N} \frac{1}{\left(s_i^{2\alpha} + \xi_n^{2\alpha}\right)^\beta} ds_1 \ldots ds_N \right) \frac{\partial^2 f(x)}{\partial x_i \partial x_{j^*}} \right\}. \qquad (31.66)$$

Then

$$\left\|\overline{\Delta}_n\right\|_p \leq \left(\sum_{j=0}^{r} |\alpha_j|^q\right)^{\frac{1}{q}} \left\{\frac{2}{(q+1)^{\frac{1}{q}}} \left(\frac{N(N+1)+2}{2}\right)^{\frac{1}{q}}\right\}$$

$$\times \left\{\sum_{j=0}^{r} \left[j^{2p} \sum_{|\alpha|=2} \left(\frac{1}{\prod_{i=1}^{N} \alpha_i!}\right)^p \omega_1 (f_\alpha, \xi_n)_p^p \overline{A}_{j\xi_n}(\alpha)\right.\right.$$

$$\left.\left. + \,\omega_1 (f, \xi_n)_p^p \left[1 + j^p \overline{K}_{p\xi_n}\right]\right]\right\}^{\frac{1}{p}}, \tag{31.67}$$

where $\overline{A}_{j\xi_n}(\alpha)$ *as in (31.47) and* $\overline{K}_{p\xi_n}$ *as in (31.55). Note that* $\left\|\overline{\Delta}_n\right\|_p \to 0$ *with rates as* $\xi_n \to 0$, $n \to \infty$, *by (31.67).*

One also obtains from (31.67) that

$$\|U_n(f) - f\|_p$$

$$\leq \left(\sum_{j=0}^{r} |\alpha_j| \, j\right) \sin(1) \left[\sum_{i=1}^{N} \left\|\frac{\partial f}{\partial x_i}\right\|_p W_n^N\right.$$

$$\left.\times \left(\int_{\mathbb{R}^N} |s_i| \prod_{i=1}^{N} \frac{1}{\left(s_i^{2\alpha} + \xi_n^{2\alpha}\right)^\beta} ds_1 \ldots ds_N\right)\right]$$

$$+ 2 \left(\sum_{j=0}^{r} |\alpha_j| \, j^2\right) \sin^2\left(\frac{1}{2}\right) \left\{\sum_{i=1}^{N} W_n^N\right.$$

$$\times \left(\int_{\mathbb{R}^N} s_i^2 \prod_{i=1}^{N} \frac{1}{\left(s_i^{2\alpha} + \xi_n^{2\alpha}\right)^\beta} ds_1 \ldots ds_N\right) \left\|\frac{\partial^2 f}{\partial x_i^2}\right\|_p + \sum_{\substack{i \neq j^*, \\ i, j^* \in \{1, \ldots, N\}}} W_n^N$$

$$\left.\times \left(\int_{\mathbb{R}^N} |s_i s_{j^*}| \prod_{i=1}^{N} \frac{1}{\left(s_i^{2\alpha} + \xi_n^{2\alpha}\right)^\beta} ds_1 \ldots ds_N\right) \left\|\frac{\partial^2 f}{\partial x_i \partial x_{j^*}}\right\|_p\right\}$$

$$+ \,R.H.S.\,(31.67), \tag{31.68}$$

given that $\|f_\alpha\|_p < \infty$, $|\alpha| = \widetilde{j}$, $\widetilde{j} = 1, 2$.

Clearly $\|U_n(f) - f\|_p \to 0$, *as* $n \to \infty$ *and* $\xi_n \to 0$. *That is,* $U_n \to I$, *the unit operator, in* L_p *norm with rates.*

Proof. By Theorems 31.3, 31.8 and 31.9 and Remark 31.3. □

We finish with an application of Theorem 31.4 to U_n operators.

Theorem 31.13. *Let $0 < \xi_n \leq 1$, $n \in \mathbb{N}$, $x \in \mathbb{R}^N$. Here we deal with $f \in C^2(\mathbb{R}^N)$, $N \geq 1$, with f, $f_\alpha \in L_1(\mathbb{R}^N)$, $|\alpha| = 2$, where $\alpha_i \in \mathbb{Z}^+$, $i = 1, \ldots, N$, and $|\alpha| = \sum_{i=1}^N \alpha_i$; $j = 0, 1, \ldots, r$. Assume that $\beta > \frac{2}{\alpha}$. Here $\overline{\Delta}_n$ as in (31.66). Then*

$$\left\|\overline{\Delta}_n\right\|_1 \leq \left\{\sum_{j=0}^r |\alpha_j| \left[j^2 \sum_{|\alpha|=2} \left(\frac{2}{\prod_{i=1}^N \alpha_i!}\right) \omega_1(f_\alpha, \xi_n)_1 \overline{\overline{A}}_{j\xi_n}(\alpha) + \omega_1(f, \xi_n)_1 \left[1 + j\overline{\overline{K}}_{\xi_n}\right]\right]\right\}, \tag{31.69}$$

where $\overline{\overline{A}}_{j\xi_n}(\alpha)$ as in Theorem 31.8 and $\overline{\overline{K}}_{\xi_n}$ as in Theorem 31.9.

Note that $\left\|\overline{\Delta}_n\right\|_1 \to 0$ with rates as $\xi_n \to 0$, $n \to \infty$, by (31.69).

We derive also that

$$\begin{aligned}
&\|U_n(f) - f\|_1 \\
&\leq \left(\sum_{j=0}^r |\alpha_j| j\right) \sin(1) \left[\sum_{i=1}^N \left\|\frac{\partial f}{\partial x_i}\right\|_1 W_n^N \right.\\
&\times \left.\left(\int_{\mathbb{R}^N} |s_i| \prod_{i=1}^N \frac{1}{(s_i^{2\alpha} + \xi_n^{2\alpha})^\beta} ds_1 \ldots ds_N\right)\right] \\
&+ 2\left(\sum_{j=0}^r |\alpha_j| j^2\right) \sin^2\left(\frac{1}{2}\right) \left\{\sum_{i=1}^N W_n^N \right.\\
&\times \left(\int_{\mathbb{R}^N} s_i^2 \prod_{i=1}^N \frac{1}{(s_i^{2\alpha} + \xi_n^{2\alpha})^\beta} ds_1 \ldots ds_N\right) \left\|\frac{\partial^2 f}{\partial x_i^2}\right\|_1 + \sum_{\substack{i \neq j^*, \\ i, j^* \in \{1, \ldots, N\}}} \\
&\times \left. W_n^N \left(\int_{\mathbb{R}^N} |s_i s_{j^*}| \prod_{i=1}^N \frac{1}{(s_i^{2\alpha} + \xi_n^{2\alpha})^\beta} ds_1 \ldots ds_N\right) \left\|\frac{\partial^2 f}{\partial x_i \partial x_{j^*}}\right\|_1 \right\} \\
&+ R.H.S.\,(31.69),
\end{aligned} \tag{31.70}$$

given that $\|f_\alpha\|_1 < \infty$, $|\alpha| = \widetilde{j}$, $\widetilde{j} = 1, 2$. *As* $n \to \infty$ *and* $\xi_n \to 0$, *by (31.70), we derive that* $\|U_n(f) - f\|_1 \to 0$, *that is,* $U_n \to I$ *the unit operator, in* L_1 *norm, with rates.*

Proof. By Theorems 31.4, 31.8 and 31.9 and Remark 31.3. □

Bibliography

G.A. Anastassiou, *Intelligent Mathematics: Computational Analysis*, Springer, Heidelberg, New York, 2011a.

G.A. Anastassiou, *Approximation by Multivariate Singular Integrals*, Springer, New York, 2011b.

G.A. Anastassiou, *Constructive Fractional Analysis with Applications*, Springer, Heidelberg, New York, 2021.

G.A. Anastassiou, Trigonometric based multivariate smooth Picard singular integrals L_p approximation, *Bulletin of the Allahabad Mathematical Society*, accepted, 2023a.

G.A. Anastassiou, *General multiple sigmoid functions relied complex valued multivariate trigonometric and hyperbolic neural network approximations*, submitted, 2023b.

G.A. Anastassiou, *Trigonometric background multivariate smooth Poisson-Cauchy singular integrals approximation*, submitted, 2023c.

G.A. Anastassiou, *Uniform Approximation by Smooth Picard Multivariate Singular Integral Operators revisited*, submitted, 2023d.

G.A. Anastassiou and S. Gal, *Approximation Theory*, Birkhaüser, Boston, 2000.

G.A. Anastassiou and R. Mezei, Uniform convergence with rates of smooth Poisson-Cauchy type singular integral operators, *Mathematical and Computer Modelling*, 50 (2009), 1553–1570.

G.A. Anastassiou, R. Mezei, *Approximation by Singular Integrals*, Cambridge Scientific Publishers, Cambridge, UK, 2012.

R.N. Mohapatra and R.S. Rodriguez, On the rate of convergence of singular integrals for Hölder continuous functions, *Mathematische Nachrichten*, 149 (1990), 117–124.

D. Zwillinger, *CRC Standard Mathematical Tables and Formulae*, 30th ed. Chapman & Hall/CRC, Boca Raton (1995).

Chapter 32

Trigonometric-Implied Multivariate Smooth Trigonometric Singular Integrals Quantitative Approximation

In this chapter, we apply the uniform and L_p, $1 \leq p < \infty$ approximation properties of general smooth multivariate singular integral operators over $\mathbb{R}^N$, $N \geq 1$. It is a trigonometric-based approach with detailed applications to the corresponding smooth multivariate trigonometric singular integral operators. The results are quantitative via Jackson-type inequalities involving the first uniform and L_p moduli of continuity. It follows Anastassiou (2024).

32.1 Introduction

The degree of approximation by univariate and multivariate singular integral operators has been studied extensively in Anastassiou (2011a)–Anastassiou (2021) and Anastassiou and Gal (2000), Anastassiou and Mezei (2012) and Mohapatra and Rodriguez (1990). All these sources motivate our current work. In particular, we studied the approximation properties of the smooth singular integral operators in Anastassiou (2011a)–Anastassiou (2021) and Anastassiou and Mezei (2012). These are not in general positive operators. Here we use the uniform and L_p, $p \geq 1$, results of our multivariate general theory Anastassiou (2023c) and Anastassiou (2023a)to establish approximation properties of the smooth trigonometric singular integral operators. The degrees of approximation are given quantitatively by using the uniform and L_p first moduli of continuity. The essential tool here comes from Anastassiou (2023b), where a multivariate trigonometric Taylor formula is proved.

32.2 Background of General Theory

Here $r \in \mathbb{N}$, $m \in \mathbb{Z}_+$, we define

$$\alpha_j := \alpha_{j,r}^{[m]} := \begin{cases} (-1)^{r-j} \binom{r}{j} j^{-m}, & \text{if } j = 1, 2, \ldots, r, \\ 1 - \sum_{j=1}^{r} (-1)^{r-j} \binom{r}{j} j^{-m}, & \text{if } j = 0, \end{cases} \tag{32.1}$$

and

$$\delta_k := \delta_{k,r}^{[m]} := \sum_{j=0}^{r} \alpha_{j,r}^{[m]} j^k, \quad k = 1, 2, \ldots, m \in \mathbb{N}. \tag{32.2}$$

See that

$$\sum_{j=0}^{r} \alpha_{j,r}^{[m]} = 1 \tag{32.3}$$

and

$$-\sum_{j=1}^{r} (-1)^{r-j} \binom{r}{j} = (-1)^r \binom{r}{0}.$$

Let μ_{ξ_n} be a probability Borel measure on $\mathbb{R}^N$, $N \geq 1$, $\xi_n > 0$, $n \in \mathbb{N}$. We now define the multiple smooth singular integral operators

$$\begin{aligned} &\theta_n(f; x_1, \ldots, x_N) := \theta_{r,n}^{[m]}(f; x_1, \ldots, x_N) \\ &:= \sum_{j=0}^{r} \alpha_{j,r}^{[m]} \int_{\mathbb{R}^N} f(x_1 + s_1 j, x_2 + s_2 j, \ldots, x_N + s_N j)\, d\mu_{\xi_n}(s), \end{aligned} \tag{32.4}$$

where $s := (s_1, \ldots, s_N)$, $x := (x_1, \ldots, x_N) \in \mathbb{R}^N$; $n, r \in \mathbb{Z}$, $m \in \mathbb{Z}_+$, $f : \mathbb{R}^N \to \mathbb{R}$ is a Borel measurable function, and also $(\xi_n)_{n\in\mathbb{N}}$ is a bounded sequence of positive real numbers, we take $0 < \xi_n \leq 1$.

Remark 32.1. The operators $\theta_{r,n}^{[m]}$ are not in general positive, see Anastassiou (2011b), p. 2.

We observe that

Lemma 32.1. *It holds*

$$\theta_{r,n}^{[m]}(c; x_1, \ldots, x_n) = c,$$

where c is a constant.

We need

Definition 32.1. Let $f \in C(\mathbb{R}^N)$, $N \geq 1$. We define the first uniform modulus of continuity of f as

$$\omega_1(f,\delta) := \sup_{\substack{x,y\in\mathbb{R}^N: \\ \|x-y\|_\infty \leq \delta}} |f(x)-f(y)|, \ \delta > 0, \tag{32.5}$$

where $\|\cdot\|_\infty$ is the max norm in $\mathbb{R}^N$. The functional $\omega_1(f,\delta)$ is bounded for f being bounded or uniformly continuous, and $\omega_1(f,\delta) \to 0$ as $\delta \to 0$, in the case of f being uniformly continuous.

We mention the main uniform general approximation result regarding the operator θ_n.

Theorem 32.1. *(Anastassiou 2023c) Here $f \in C^2(\mathbb{R}^N)$ and let all $\alpha_i \in \mathbb{Z}^+$, $i = 1,\ldots,N$, $N \geq 1$, $|\alpha| := \sum_{i=1}^N \alpha_i = 2$; $x \in \mathbb{R}^N$, and all the partials f_α of order 2, along with $f \in C_B(\mathbb{R}^N)$ (continuous and bounded functions); or all f_α of order 2, $f \in C_U(\mathbb{R}^N)$ (uniformly continuous functions). Let μ_{ξ_n} be a Borel probability measure on $\mathbb{R}^N$, for $0 < \xi_n \leq 1$, $n \in \mathbb{N}$.*

Suppose that for all $\alpha := (\alpha_1,\ldots,\alpha_N)$, $\alpha_i \in \mathbb{Z}^+$, $i = 1,\ldots,N$, $|\alpha| = \sum_{i=1}^N \alpha_i = 2$, $j = 0,1,\ldots,r$, we have that both

$$I_{1j}(\alpha) := \int_{\mathbb{R}^N} \left(1 + \frac{j\,\|s\|_1}{3\xi_n}\right)\left(\prod_{i=1}^N |s_i|^{\alpha_i}\right) d\mu_{\xi_n}(s), \tag{32.6}$$

$$I_{2j}(\alpha) := \int_{\mathbb{R}^N} \left(1 + \frac{j\,\|s\|_1}{3\xi_n}\right) d\mu_{\xi_n}(s) \tag{32.7}$$

are uniformly bounded in $\xi_n \in (0,1]$.

Denote ($n \in \mathbb{N}$)

$$\begin{aligned}\Delta_n(x) := {} & \theta_n(f,x) - f(x) - \left(\sum_{j=0}^r \alpha_j j\right)\sin(1)\left[\sum_{i=1}^N \frac{\partial f(x)}{\partial x_i}\left(\int_{\mathbb{R}^N} s_i d\mu_{\xi_n}(s)\right)\right] \\ & - 2\left(\sum_{j=0}^r \alpha_j j^2\right)\sin^2\left(\frac{1}{2}\right)\left\{\sum_{i=1}^N \left(\int_{\mathbb{R}^N} s_i^2 d\mu_{\xi_n}(s)\right)\frac{\partial^2 f(x)}{\partial x_i^2}\right. \\ & \left. + \sum_{\substack{i\neq j^*, \\ i,j^*\in\{1,\ldots,N\}}} \left(\int_{\mathbb{R}^N} s_i s_{j^*} d\mu_{\xi_n}(s)\right)\frac{\partial^2 f(x)}{\partial x_i \partial x_{j^*}}\right\}.\end{aligned} \tag{32.8}$$

Then

(i)

$$|\Delta_n(x)| \leq \|\Delta_n(x)\|_\infty \leq \sum_{j=0}^{r} |\alpha_j|$$

$$\left[\left[j^2 \sum_{\substack{\alpha_i \in \mathbb{Z}^+, \\ \alpha:|\alpha|=2}} \left(\frac{1}{\prod_{i=1}^N \alpha_i!}\right) \omega_1(f_\alpha, \xi_n) \int_{\mathbb{R}^N} \left(1 + \frac{j\|s\|_\infty}{3\xi_n}\right) \left(\prod_{i=1}^N |s_i|^{\alpha_i}\right) d\mu_{\xi_n}(s) \right]\right.$$

$$\left. + \frac{1}{2}\omega_1(f, \xi_n) \int_{\mathbb{R}^N} \left(1 + \frac{j\|s\|_\infty}{3\xi_n}\right) d\mu_{\xi_n}(s) \right] =: \varphi_{\xi_n}. \tag{32.9}$$

In case of all f_α of order 2 and $f \in C_U(\mathbb{R}^N)$ and $\xi_n \to 0$, as $n \to \infty$, then $\Delta_n(x)$, $\|\Delta_n(x)\|_\infty \to 0$ with rates.

(ii) *If $\frac{\partial f(x)}{\partial x_i} = 0$, $i = 1, \ldots, N$, and $f_\alpha(x) = 0$, $\alpha_i \in \mathbb{Z}^+$, $i = 1, \ldots, N$, with $|\alpha| = 2$, then*

$$|\theta_n(f, x) - f(x)| \leq \varphi_{\xi_n}. \tag{32.10}$$

And $\theta_n(f, x) \to f(x)$ in the uniformly continuous case.

(iii) *Additionally, assume all partials of order ≤ 2 are bounded. Hence*

$$\|\theta_n(f) - f\|_\infty \leq \left(\sum_{j=0}^{r} |\alpha_j| j\right) (0.8414) \left[\sum_{i=1}^{N} \left\|\frac{\partial f}{\partial x_i}\right\|_\infty \left(\int_{\mathbb{R}^N} |s_i| d\mu_{\xi_n}(s)\right)\right]$$

$$+ \left(\sum_{j=0}^{r} |\alpha_j| j^2\right) (0.4596) \left\{\sum_{i=1}^{N} \left(\int_{\mathbb{R}^N} s_i^2 d\mu_{\xi_n}(s)\right) \left\|\frac{\partial^2 f}{\partial x_i^2}\right\|_\infty\right.$$

$$\left. + \sum_{\substack{i \neq j^*, \\ i,j^* \in \{1,\ldots,N\}}} \left(\int_{\mathbb{R}^N} |s_i| |s_{j^*}| d\mu_{\xi_n}(s)\right) \left\|\frac{\partial^2 f}{\partial x_i \partial x_{j^*}}\right\|_\infty \right\} + \varphi_{\xi_n}. \tag{32.11}$$

If all $\int_{\mathbb{R}^N} s_i^2 d\mu_{\xi_n}(s)$ and $\int_{\mathbb{R}^N} |s_i| |s_{j^*}| d\mu_{\xi_n}(s)$ converge to zero, as $n \to \infty$, with $\xi_n \to 0$, and all f_α of order 2, $f \in C_U(\mathbb{R}^N)$, then

$$\|\theta_n(f) - f\|_\infty \to 0 \text{ with rates, as } \xi_n \to 0, n \to +\infty.$$

Next we deal with $f \in C^m(\mathbb{R}^N)$, $m \in \mathbb{Z}^+$, with $f_\alpha \in L_p(\mathbb{R}^N)$, $|\alpha| = m \in \mathbb{Z}^+$, $p \geq 1$; where f_α denotes the mixed partial $\frac{\partial^{\widetilde{j}} f(\cdot,\ldots,\cdot)}{\partial x_1^{\alpha_1} \ldots \partial x_N^{\alpha_N}}$, $\alpha_j \in \mathbb{Z}^+$, $j = 1, \ldots, N : |\alpha| := \sum_{j=1}^{N} \alpha_j = \widetilde{j}$, $\widetilde{j} = 1, \ldots, m$.

We need

Definition 32.2. (see also Anastassiou (2011b), p. 20) We call

$$\Delta_u^r f(x) := \Delta_{u_1,u_2,\ldots,u_N}^{\ r} f(x_1,\ldots,x_N)$$
$$:= \sum_{j=0}^{r}(-1)^{r-j}\binom{r}{j} f(x_1+ju_1, x_2+ju_2,\ldots,x_N+ju_N). \tag{32.12}$$

Let $p \geq 1$, the L_p modulus of smoothness of order r is given by

$$\omega_r(f;h)_p := \sup_{\|u\|_2 \leq h} \|\Delta_u^r(f)\|_p, \tag{32.13}$$

$h > 0$.

We mention

Theorem 32.2. *(Anastassiou 2023a) Let $f \in C^m(\mathbb{R}^N)$, $m \in \mathbb{N}$, $N \geq 1$, with $f_\alpha \in L_p(\mathbb{R}^N)$, $|\alpha| = m$, $x \in \mathbb{R}^N$. Let $p, q > 1 : \frac{1}{p} + \frac{1}{q} = 1$. Here μ_{ξ_n} is a Borel probability measure on $\mathbb{R}^N$ for $\xi_n > 0$, $(\xi_n)_{n\in\mathbb{N}}$ bounded sequence. Assume for all $\alpha := (\alpha_1,\ldots,\alpha_N)$, $\alpha_i \in \mathbb{Z}^+$, $i = 1,\ldots,N$, $|\alpha| := \sum_{i=1}^N \alpha_i = m$ that we have*

$$\int_{\mathbb{R}^N}\left(\left(\prod_{i=1}^N |s_i|^{\alpha_i}\right)\left(1+\frac{\|s\|_2}{\xi_n}\right)^r\right)^p d\mu_{\xi_n}(s) < \infty. \tag{32.14}$$

For $\widetilde{j} = 1,\ldots,m$, and $\alpha := (\alpha_1,\ldots,\alpha_N)$, $\alpha_i \in \mathbb{Z}^+$, $i = 1,\ldots,N$, $|\alpha| := \sum_{i=1}^N \alpha_i = \widetilde{j}$, call

$$c_{\alpha,n,\widetilde{j}} := \int_{\mathbb{R}^N} \prod_{i=1}^N s_i^{\alpha_i} d\mu_{\xi_n}(s). \tag{32.15}$$

Then

$$\left\|E_{r,n}^{[m]}\right\|_p := \left\|\theta_{r,n}^{[m]}(f;x) - f(x) - \sum_{\widetilde{j}=1}^m \delta_{\widetilde{j},r}^{[m]}\left(\sum_{|\alpha|=\widetilde{j}} \frac{c_{\alpha,n,\widetilde{j}} f_\alpha(x)}{\left(\prod_{i=1}^N \alpha_i!\right)}\right)\right\|_{p,x}$$
$$\leq \left[(N^m/(m-1)!)^{\frac{p}{q}}\left(\frac{m}{(q(m-1)+1)^{\frac{p}{q}}}\right)\left(\sum_{|\alpha|=m}\frac{1}{\prod_{i=1}^N \alpha_i!}\cdot\right.\right.$$
$$\left.\left.\left[\int_{\mathbb{R}^N}\left[\left(\prod_{i=1}^N |s_i|^{\alpha_i}\right)\left(1+\frac{\|s\|_2}{\xi_n}\right)^r\right]^p d\mu_{\xi_n}(s)\right]\omega_r(f_\alpha,\xi_n)_p^p\right)\right]^{\frac{1}{p}}. \tag{32.16}$$

As $n \to \infty$ and $\xi_n \to 0$, by (32.16), we obtain that $\left\| E_{r,n}^{[m]} \right\|_p \to 0$ with rates.

One also gets by (32.16) that

$$\left\| \theta_{r,n}^{[m]}(f;x) - f(x) \right\|_{p,x}$$

$$\leq \sum_{\widetilde{j}=1}^{m} \left| \delta_{\widetilde{j},r}^{[m]} \right| \left(\sum_{|\alpha|=\widetilde{j}} \frac{\left| c_{\alpha,n,\widetilde{j}} \right|}{\prod_{i=1}^{N} \alpha_i!} \left\| f_\alpha \right\|_p \right) + R.H.S.\ (32.16), \quad (32.17)$$

given that $\|f_\alpha\|_p < \infty$, $|\alpha| = \widetilde{j}$, $\widetilde{j} = 1, \ldots, m$.

Assuming that $c_{\alpha,n,\widetilde{j}} \to 0$, $\xi_n \to 0$, as $n \to \infty$, we get $\|\theta_{r,n}^{[m]}(f) - f\|_p \to 0$, that is, $\theta_{r,n}^{[m]} \to I$ the unit operator, in L_p norm, with rates.

We make

Remark 32.2. Note that $(p > 1)$

$$\int_{\mathbb{R}^N} \left(\prod_{i=1}^{N} |s_i|^{\alpha_i} \right) \left(1 + \frac{\|s\|_2}{\xi_n} \right)^r d\mu_{\xi_n}(s)$$

$$\leq \left[\int_{\mathbb{R}^N} \left(\left(\prod_{i=1}^{N} |s_i|^{\alpha_i} \right) \left(1 + \frac{\|s\|_2}{\xi_n} \right)^r \right)^p d\mu_{\xi_n}(s) \right]^{\frac{1}{p}} < \infty, \quad (32.18)$$

by assumption of Theorem 32.2.

By (32.18), we get that

$$\int_{\mathbb{R}^N} \prod_{i=1}^{N} |s_i|^{\alpha_i} \, d\mu_{\xi_n}(s) < \infty. \quad (32.19)$$

Hence $c_{\alpha,n,\widetilde{j}} \in \mathbb{R}$.

We mention also the following trigonometric-induced alternative L_p approximation result for θ_n operators.

Theorem 32.3. *(Anastassiou 2023a) Let $p, q > 1 : \frac{1}{p} + \frac{1}{q} = 1$, $0 < \xi_n \leq 1$, $n \in \mathbb{N}$. Here we deal with $f \in C^2(\mathbb{R}^N)$, $N \geq 1$, with f, $f_\alpha \in L_p(\mathbb{R}^N)$, $|\alpha| = 2$, where $\alpha_i \in \mathbb{Z}^+$, $i = 1, \ldots, N$, and $|\alpha| = \sum_{i=1}^{N} \alpha_i$; $x \in \mathbb{R}^N$. Let μ_{ξ_n} be a Borel probability measure on $\mathbb{R}^N$. Suppose that for all $\alpha : |\alpha| = 2$, $j = 0, 1, \ldots, r$, we have that both*

$$I_{1j}(\alpha) := \int_{\mathbb{R}^N} \left(\prod_{i=1}^{N} |s_i|^{p\alpha_i} \right) \left(1 + \frac{j^p \|s\|_2^p}{\xi_n^p} \right) d\mu_{\xi_n}(s), \quad (32.20)$$

$$I_2 := \int_{\mathbb{R}^N} \left(\frac{\|s\|_2}{\xi_n} \right)^p d\mu_{\xi_n}(s) \quad (32.21)$$

are uniformly bounded in $\xi_n \in (0,1]$.

Denote $(n \in \mathbb{N})$

$$\Delta_n(x) := \theta_n(f,x) - f(x) - \left(\sum_{j=0}^{r} \alpha_j j\right) \sin(1) \left[\sum_{i=1}^{N} \frac{\partial f(x)}{\partial x_i} \left(\int_{\mathbb{R}^N} s_i d\mu_{\xi_n}(s)\right)\right]$$

$$- 2\left(\sum_{j=0}^{r} \alpha_j j^2\right) \sin^2\left(\frac{1}{2}\right) \left\{\sum_{i=1}^{N} \left(\int_{\mathbb{R}^N} s_i^2 d\mu_{\xi_n}(s)\right) \frac{\partial^2 f(x)}{\partial x_i^2}\right.$$

$$\left. + \sum_{\substack{i \neq j^*, \\ i,j^* \in \{1,\dots,N\}}} \left(\int_{\mathbb{R}^N} s_i s_{j^*} d\mu_{\xi_n}(s)\right) \frac{\partial^2 f(x)}{\partial x_i \partial x_{j^*}}\right\}. \tag{32.22}$$

Then, it holds

$$\|\Delta_n(x)\|_p \leq \left(\sum_{j=0}^{r} |\alpha_j|^q\right)^{\frac{1}{q}} \left\{\frac{2}{(q+1)^{\frac{1}{q}}} \left(\frac{N(N+1)+2}{2}\right)^{\frac{1}{q}}\right\}$$

$$\left\{\sum_{j=0}^{r} \left[j^{2p} \sum_{|\alpha|=2} \left(\frac{1}{\prod_{i=1}^{N} \alpha_i!}\right)^p \omega_1(f_\alpha, \xi_n)_p^p \left(\int_{\mathbb{R}^N} \left(\prod_{i=1}^{N} |s_i|^{p\alpha_i}\right)\right.\right.\right.$$

$$\left. \times \left(1 + \frac{j^p \|s\|_2^p}{\xi_n^p}\right) d\mu_{\xi_n}(s)\right)$$

$$\left.\left. + \omega_1(f,\xi_n)_p^p \left(\int_{\mathbb{R}^N} \left(1 + \frac{j^p \|s\|_2^p}{\xi_n^p}\right) d\mu_{\xi_n}(s)\right)\right]\right\}^{\frac{1}{p}}. \tag{32.23}$$

As $n \to \infty$ *and* $\xi_n \to 0$, *by (32.23), we obtain that* $\|\Delta_n\|_p \to 0$ *with rates. One also gets by (32.23) that*

$$\|\theta_n(f,x) - f(x)\|_{p,x}$$

$$\leq \left(\sum_{j=0}^{r} |\alpha_j| j\right) \sin(1) \left[\sum_{i=1}^{N} \left\|\frac{\partial f}{\partial x_i}\right\|_p \left(\int_{\mathbb{R}^N} |s_i| \, d\mu_{\xi_n}(s)\right)\right]$$

$$+2\left(\sum_{j=0}^{r}|\alpha_j|\,j^2\right)\sin^2\left(\frac{1}{2}\right)\left\{\sum_{i=1}^{N}\left(\int_{\mathbb{R}^N}s_i^2 d\mu_{\xi_n}(s)\right)\left\|\frac{\partial^2 f}{\partial x_i^2}\right\|_p\right.$$

$$\left.+\sum_{\substack{i\neq j^*,\\ i,j^*\in\{1,\ldots,N\}}}\left(\int_{\mathbb{R}^N}|s_i s_{j^*}|\,d\mu_{\xi_n}(s)\right)\left\|\frac{\partial^2 f}{\partial x_i\partial x_{j^*}}\right\|_p\right\}+R.H.S.\ (32.23), \tag{32.24}$$

given that $\|f_\alpha\|_p<\infty$, $|\alpha|=\widetilde{j}$, $\widetilde{j}=1,2$. *Assuming that* $\int_{\mathbb{R}^N}s_i^2 d\mu_{\xi_m}(s)$, $\int_{\mathbb{R}^N}|s_i s_{j^*}|\,d\mu_{\xi_n}(s)$, $i,j^*\in\{1,\ldots,N\}$, $i\neq j^*$, *converge to zero as* $\xi_n\to 0$, *we get* $\|\theta_n(f,x)-f(x)\|_p\to 0$, *that is,* $\theta_n\to I$ *the unit operator, in* L_p *norm, with rates.*

Furthermore, we mention the following trigonometric-based alternative L_1 approximation result for θ_n operators.

Theorem 32.4. *(Anastassiou 2023a) Let* $0<\xi_n\leq 1$, $n\in\mathbb{N}$, $x\in\mathbb{R}^N$. *Here we deal with* $f\in C^2(\mathbb{R}^N)$, $N\geq 1$, *with* f, $f_\alpha\in L_1(\mathbb{R}^N)$, $|\alpha|=2$, *where* $\alpha_i\in\mathbb{Z}^+$, $i=1,\ldots,N$, *and* $|\alpha|=\sum_{i=1}^N\alpha_i$. *Let* μ_{ξ_n} *be a Borel probability measure on* $\mathbb{R}^N$. *Suppose that for all* $\alpha:|\alpha|=2$, $j=0,1,\ldots,r$, *we have that both*

$$I_{1j}^*(\alpha):=\int_{\mathbb{R}^N}\left(\prod_{i=1}^N|s_i|^{\alpha_i}\right)\left(1+\frac{j\,\|s\|_2}{\xi_n}\right)d\mu_{\xi_n}(s) \tag{32.25}$$

and

$$I_2^*:=\int_{\mathbb{R}^N}\frac{\|s\|_2}{\xi_n}d\mu_{\xi_n}(s) \tag{32.26}$$

are uniformly bounded in $\xi_n\in(0,1]$.

Here $\Delta_n(x)$ *is as in (32.22).*

Then, it holds

$$\|\Delta_n(x)\|_1\leq\left\{\sum_{j=0}^{r}|\alpha_j|\left[j^2\sum_{|\alpha|=2}\left(\frac{2}{\prod_{i=1}^N\alpha_i!}\right)\omega_1(f_\alpha,\xi_n)_1\left(\int_{\mathbb{R}^N}\left(\prod_{i=1}^N(|s_i|)^{\alpha_i}\right)\right.\right.\right.$$

$$\left.\left.\left.\left(1+\frac{j\,\|s\|_2}{\xi_n}\right)d\mu_{\xi_n}(s)\right)+\omega_1(f,\xi_n)_1\left(\int_{\mathbb{R}^N}\left(1+\frac{j\,\|s\|_2}{\xi_n}\right)d\mu_{\xi_n}(s)\right)\right]\right\}. \tag{32.27}$$

As $n \to \infty$ and $\xi_n \to 0$, by (32.27), we obtain that $\|\Delta_n\|_1 \to 0$ with rates. One also obtains by (32.27) that

$$\|\theta_n(f) - f\|_1$$
$$\leq \left(\sum_{j=0}^{r} |\alpha_j| j\right) \sin(1) \left[\sum_{i=1}^{N} \left\|\frac{\partial f}{\partial x_i}\right\|_1 \left(\int_{\mathbb{R}^N} |s_i| \, d\mu_{\xi_n}(s)\right)\right]$$
$$+ 2\left(\sum_{j=0}^{r} |\alpha_j| j^2\right) \sin^2\left(\frac{1}{2}\right) \left\{\sum_{i=1}^{N} \left(\int_{\mathbb{R}^N} s_i^2 d\mu_{\xi_n}(s)\right) \left\|\frac{\partial^2 f}{\partial x_i^2}\right\|_1\right.$$
$$\left. + \sum_{\substack{i \neq j^*, \\ i, j^* \in \{1, \dots, N\}}} \left(\int_{\mathbb{R}^N} |s_i s_{j^*}| \, d\mu_{\xi_n}(s)\right) \left\|\frac{\partial^2 f}{\partial x_i \partial x_{j^*}}\right\|_1 \right\} + R.H.S.\ (32.27), \tag{32.28}$$

given that $\|f_\alpha\|_1 < \infty$, $|\alpha| = \widetilde{j}$, $\widetilde{j} = 1, 2$. Assuming that $\int_{\mathbb{R}^N} s_i^2 d\mu_{\xi_n}(s)$, $\int_{\mathbb{R}^N} |s_i s_{j^*}| \, d\mu_{\xi_n}(s)$, $i, j^* \in \{1, \dots, N\}$, $i \neq j^*$, converge to zero as $\xi_n \to 0$, we derive $\|\theta_n(f) - f\|_1 \to 0$, that is, $\theta_n \to I$ in L_1 norm, with rates.

We need

Definition 32.3. The general multivariate trigonometric singular integral operators are

$$T_n(f; x_1, \dots, x_N) := T_{r,n}^{[m]}(f; x_1, \dots, x_N) := \lambda_n^{-N} \sum_{j=0}^{r} \alpha_{j,r}^{[m]} \cdot$$
$$\int_{\mathbb{R}^N} f(x_1 + s_1 j, \dots, x_N + s_N j) \prod_{i=1}^{N} \left(\frac{\sin\left(\frac{s_i}{\xi_n}\right)}{s_i}\right)^{2\beta} ds_1 \dots ds_N, \tag{32.29}$$

with $\beta \in \mathbb{N}$, and

$$\lambda_n := 2\xi_n^{1-2\beta} \pi (-1)^\beta \beta \sum_{k=1}^{\beta} (-1)^k \frac{k^{2\beta-1}}{(\beta - k)!(\beta + k)!}, \tag{32.30}$$

see [Anastassiou (2011b); ?], p. 210, item 1033.

Note that

$$\lambda_n^{-N} \int_{\mathbb{R}^N} \prod_{i=1}^{N} \left(\frac{\sin\left(\frac{s_i}{\xi_n}\right)}{s_i} \right)^{2\beta} ds_1, \ldots, ds_N = 1, \tag{32.31}$$

see also [Anastassiou (2011b); ?, p. 210, item 1033].

So here it is

$$d\mu_{\xi_n}(s) = \lambda_n^{-N} \prod_{i=1}^{N} \left(\frac{\sin\left(\frac{s_i}{\xi_n}\right)}{s_i} \right)^{2\beta} ds_1, \ldots, ds_N, \tag{32.32}$$

$s \in \mathbb{R}^N$, $0 < \xi_n \leq 1$, $n \in \mathbb{N}$.

We apply to T_n Theorems 32.1, 32.2, 32.3 and 32.4 with the help of section 3. That is a study of approximation by T_n.

32.3 Auxiliary Results

We mention

Lemma 32.2. *(Anastassiou 2021, p. 486) Let $N \in \mathbb{N}$, $r > 0$, $z_i \in \mathbb{R}_+$, $i = 1, \ldots, N$. Then*

$$\left(1 + \sum_{i=1}^{N} z_i\right)^r \leq \prod_{i=1}^{N} (1 + z_i)^r. \tag{32.33}$$

In Definition 32.3, we call

$$\gamma := 2\pi(-1)^\beta \beta \sum_{k=1}^{\beta} (-1)^k \frac{k^{2\beta-1}}{(\beta-k)!\,(\beta+k)!}, \tag{32.34}$$

that is,

$$\lambda_n = \gamma \xi_n^{1-2\beta}. \tag{32.35}$$

We need

Theorem 32.5. *Let $N \in \mathbb{N}$; $\alpha_i \in \mathbb{Z}^+$, $i = 1, \ldots, N : |\alpha| := \sum_{i=1}^{N} \alpha_i = 2$, $\xi_n \in (0, 1]$, $n \in \mathbb{N}$; $j = 0, 1, \ldots, r \in \mathbb{N}$; $\beta \in \mathbb{N} - \{1, 2\}$. We define*

$$\psi_{10} := \pi (-1)^\beta \beta \sum_{k=1}^{\beta} (-1)^k \frac{k^{2\beta-1}}{(\beta-k)!\,(\beta+k)!}, \tag{32.36}$$

$$\psi_{12} := \frac{\pi (-1)^{\beta-1} (2\beta)!}{8\,(2\beta-3)!} \sum_{k=1}^{\beta} (-1)^k \frac{k^{2\beta-3}}{(\beta-k)!\,(\beta+k)!}, \tag{32.37}$$

$$\psi_{21} := \frac{(2\beta)!}{2\,(2\beta-2)!} \sum_{k=1}^{\beta} (-1)^{\beta-k} \frac{k^{2(\beta-1)} \ln 2k}{(\beta-k)!\,(\beta+k)!} \tag{32.38}$$

and

$$\psi_{23} := \frac{-(2\beta)!}{8(2\beta-4)!} \sum_{k=1}^{\beta} (-1)^{\beta-k} \frac{k^{2(\beta-2)} \ln 2k}{(\beta-k)!\,(\beta+k)!}. \tag{32.39}$$

Then

$$J_{1j}(\alpha) := \lambda_n^{-N} \int_{\mathbb{R}^N} \left(1 + \frac{j\,\|s\|_1}{3\xi_n}\right) \left(\prod_{i=1}^{N} |s_i|^{\alpha_i}\right) \prod_{i=1}^{N} \left(\frac{\sin\left(\frac{s_i}{\xi_n}\right)}{s_i}\right)^{2\beta} ds_1 \ldots ds_N \tag{32.40}$$

$$\leq \xi_n^2 2^N \gamma^{-N} \left[\psi_{10} + \psi_{12} + \frac{j}{3}(\psi_{21} + \psi_{23})\right]^N$$

$$\leq 2^N \gamma^{-N} \left[\psi_{10} + \psi_{12} + \frac{j}{3}(\psi_{21} + \psi_{23})\right]^N < +\infty \tag{32.41}$$

is uniformly bounded with respect to ξ_n*, and* $J_{1j}(\alpha) \to 0$*, as* $\xi_n \to 0$*,* $n \to \infty$.

Proof. We have that

$$J_{1j}(\alpha) = \lambda_n^{-N} \int_{\mathbb{R}^N} \left(1 + \frac{j\,\|s\|_1}{3\xi_n}\right) \left(\prod_{i=1}^{N} |s_i|^{\alpha_i}\right) \prod_{i=1}^{N} \left(\frac{\sin\left(\frac{s_i}{\xi_n}\right)}{s_i}\right)^{2\beta} ds_1 \ldots ds_N$$

$$= 2^N \lambda_n^{-N} \int_{\mathbb{R}_+^N} \left(1 + \frac{j\,\|s\|_1}{3\xi_n}\right) \left(\prod_{i=1}^{N} s_i^{\alpha_i}\right) \prod_{i=1}^{N} \left(\frac{\sin\left(\frac{s_i}{\xi_n}\right)}{s_i}\right)^{2\beta} \prod_{i=1}^{N} ds_i$$

$$= 2^N \lambda_n^{-N} \int_{\mathbb{R}_+^N} \left(1 + \frac{j\sum_{i=1}^{N} s_i}{3\xi_n}\right) \left(\prod_{i=1}^{N} s_i^{\alpha_i}\right) \prod_{i=1}^{N} \left(\frac{\sin\left(\frac{s_i}{\xi_n}\right)}{s_i}\right)^{2\beta} \prod_{i=1}^{N} ds_i \tag{32.42}$$

$$= 2^N \lambda_n^{-N} \xi_n^2 \xi_n^{-2\beta N} \xi_n^N \int_{\mathbb{R}_+^N} \left(1 + \frac{j}{3} \sum_{i=1}^{N} \left(\frac{s_i}{\xi_n}\right)\right)$$

$$\left(\prod_{i=1}^{N} \left(\frac{s_i}{\xi_n}\right)^{\alpha_i}\right) \prod_{i=1}^{N} \left(\frac{\sin\left(\frac{s_i}{\xi_n}\right)}{\frac{s_i}{\xi_n}}\right)^{2\beta} \prod_{i=1}^{N} d\left(\frac{s_i}{\xi_n}\right)$$

$$= 2^N \lambda_n^{-N} \xi_n^{(1-2\beta)N+2} \int_{\mathbb{R}_+^N} \left(1 + \frac{j}{3} \sum_{i=1}^{N} z_i\right) \left(\prod_{i=1}^{N} z_i^{\alpha_i}\right) \prod_{i=1}^{N} \left(\frac{\sin z_i}{z_i}\right)^{2\beta} \prod_{i=1}^{N} dz_i =$$

$$
\begin{aligned}
&\left(\text{by } \lambda_n = \gamma \xi_n^{1-2\beta}\right) \\
&= 2^N \gamma^{-N} \xi_n^2 \int_{\mathbb{R}_+^N} \left(1 + \frac{j}{3} \sum_{i=1}^{N} z_i\right) \left(\prod_{i=1}^{N} z_i^{\alpha_i}\right) \prod_{i=1}^{N} \left(\frac{\sin z_i}{z_i}\right)^{2\beta} \prod_{i=1}^{N} dz_i \\
&\leq 2^N \gamma^{-N} \xi_n^2 \int_{\mathbb{R}_+^N} \prod_{i=1}^{N} \left(1 + \frac{j}{3} z_i\right) \left(\prod_{i=1}^{N} z_i^{\alpha_i}\right) \prod_{i=1}^{N} \left(\frac{\sin z_i}{z_i}\right)^{2\beta} \prod_{i=1}^{N} dz_i \\
&= 2^N \gamma^{-N} \xi_n^2 \prod_{i=1}^{N} \left(\int_0^{\infty} \left(1 + \frac{j}{3} z_i\right) z_i^{\alpha_i} \left(\frac{\sin z_i}{z_i}\right)^{2\beta} dz_i\right) =
\end{aligned} \tag{32.43}
$$

$$
\begin{aligned}
&= 2^N \gamma^{-N} \xi_n^2 \prod_{i=1}^{N} \left[\int_0^{\infty} z_i^{\alpha_i} \left(\frac{\sin z_i}{z_i}\right)^{2\beta} dz_i + \frac{j}{3} \int_0^{\infty} z_i^{1+\alpha_i} \left(\frac{\sin z_i}{z_i}\right)^{2\beta} dz_i\right] \\
&= 2^N \gamma^{-N} \xi_n^2 \prod_{i=1}^{N} \left[\int_0^{1} z_i^{\alpha_i} \left(\frac{\sin z_i}{z_i}\right)^{2\beta} dz_i + \int_1^{\infty} z_i^{\alpha_i} \left(\frac{\sin z_i}{z_i}\right)^{2\beta} dz_i \right. \\
&\quad \left. + \frac{j}{3} \int_0^{1} z_i^{1+\alpha_i} \left(\frac{\sin z_i}{z_i}\right)^{2\beta} dz_i + \frac{j}{3} \int_1^{\infty} z_i^{1+\alpha_i} \left(\frac{\sin z_i}{z_i}\right)^{2\beta} dz_i\right] \\
&\leq 2^N \gamma^{-N} \xi_n^2 \left[\int_0^{1} \left(\frac{\sin z}{z}\right)^{2\beta} dz + \frac{j}{3} \int_0^{1} z \left(\frac{\sin z}{z}\right)^{2\beta} dz \right. \\
&\quad \left. + \int_1^{\infty} z^2 \left(\frac{\sin z}{z}\right)^{2\beta} dz + \frac{j}{3} \int_1^{\infty} z^3 \left(\frac{\sin z}{z}\right)^{2\beta} dz\right]^N \\
&\leq 2^N \gamma^{-N} \xi_n^2 \left[\int_0^{\infty} \left(\frac{\sin z}{z}\right)^{2\beta} dz + \frac{j}{3} \int_0^{\infty} z \left(\frac{\sin z}{z}\right)^{2\beta} dz \right. \\
&\quad \left. + \int_0^{\infty} z^2 \left(\frac{\sin z}{z}\right)^{2\beta} dz + \frac{j}{3} \int_0^{\infty} z^3 \left(\frac{\sin z}{z}\right)^{2\beta} dz\right]^N =: I.
\end{aligned} \tag{32.44}
$$

Based on Edwards (1954), p. 210, item 1033 and Anastassiou and Mezei (2012), and by Anastassiou (2021), p. 489, we have that

$$
\int_0^{\infty} \left(\frac{\sin z}{z}\right)^{2\beta} dz = \pi (-1)^{\beta} \beta \sum_{k=1}^{\beta} (-1)^k \frac{k^{2\beta-1}}{(\beta-k)!\,(\beta+k)!} = \psi_{10}, \tag{32.45}
$$

$$\int_0^\infty z^2 \left(\frac{\sin z}{z}\right)^{2\beta} dz = \frac{\pi (-1)^{\beta-1} (2\beta)!}{8(2\beta-3)!} \sum_{k=1}^{\beta} (-1)^k \frac{k^{2\beta-3}}{(\beta-k)!(\beta+k)!} = \psi_{12}, \tag{32.46}$$

$$\int_0^\infty z \left(\frac{\sin z}{z}\right)^{2\beta} dz = \frac{(2\beta)!}{2(2\beta-2)!} \sum_{k=1}^{\beta} (-1)^{\beta-k} \frac{k^{2\beta-2}\ln(2k)}{(\beta-k)!(\beta+k)!} = \psi_{21} \tag{32.47}$$

and

$$\int_0^\infty z^3 \left(\frac{\sin z}{z}\right)^{2\beta} dz = \frac{-(2\beta)!}{8(2\beta-4)!} \sum_{k=1}^{\beta} \frac{(-1)^{\beta-k} k^{2\beta-4}\ln(2k)}{(\beta-k)!(\beta+k)!} = \psi_{23}. \tag{32.48}$$

Hence, it holds

$$\begin{aligned} I &= 2^N \gamma^{-N} \xi_n^2 \left[\psi_{10} + \frac{j}{3}\psi_{21} + \psi_{12} + \frac{j}{3}\psi_{23}\right]^N \\ &\le 2^N \gamma^{-N} \left[\psi_{10} + \psi_{12} + \frac{j}{3}(\psi_{21} + \psi_{23})\right]^N < +\infty. \end{aligned} \tag{32.49}$$

That is, $J_{1j}(\alpha)$ is uniformly bounded with respect to ξ_n, and $J_{1j}(\alpha) \to 0$, as $\xi_n \to 0$, $n \to \infty$. The claim is proved. ☐

Remark 32.3. (to Theorem 32.5) If $i, j^* \in \{1, \ldots, N\}$, $i \neq j^*$, we have that

$$\lambda_n^{-N} \int_{\mathbb{R}^N} |s_i| \, |s_{j^*}| \prod_{i=1}^N \left(\frac{\sin\left(\frac{s_i}{\xi_n}\right)}{s_i}\right)^{2\beta} \prod_{i=1}^N ds_i,$$

$$\lambda_n^{-N} \int_{\mathbb{R}^N} s_i^2 \prod_{i=1}^N \left(\frac{\sin\left(\frac{s_i}{\xi_n}\right)}{s_i}\right)^{2\beta} \prod_{i=1}^N ds_i \le J_{1j}(\alpha) < \infty, \tag{32.50}$$

$(j = 0, 1, \ldots, r \in \mathbb{N})$,
and

$$\lambda_n^{-N} \int_{\mathbb{R}^N} |s_i| \prod_{i=1}^N \left(\frac{\sin\left(\frac{s_i}{\xi_n}\right)}{s_i}\right)^{2\beta} \prod_{i=1}^N ds_i \le$$

$$\times \left(\lambda_n^{-N} \int_{\mathbb{R}^N} s_i^2 \prod_{i=1}^N \left(\frac{\sin\left(\frac{s_i}{\xi_n}\right)}{s_i}\right)^{2\beta} \prod_{i=1}^N ds_i\right)^{1/2} < \infty, \tag{32.51}$$

and all these integrals are uniformly bounded in $\xi_n \in (0,1]$. And all integrals converge to zero, as $\xi_n \to 0$, $n \to \infty$.

Remark 32.4. (to Theorem 32.5) By (32.50) and (32.51), we obtain $(j = 0,1,\ldots,r \in \mathbb{N})$

$$\begin{aligned} J_2 &:= \lambda_n^{-N} \int_{\mathbb{R}^N} \frac{\|s\|_1}{\xi_n} \prod_{i=1}^{N} \left(\frac{\sin\left(\frac{s_i}{\xi_n}\right)}{s_i} \right)^{2\beta} \prod_{i=1}^{N} ds_i \\ &= \frac{1}{\xi_n} \sum_{i=1}^{N} \left(\lambda_n^{-N} \int_{\mathbb{R}^N} |s_i| \prod_{i=1}^{N} \left(\frac{\sin\left(\frac{s_i}{\xi_n}\right)}{s_i} \right)^{2\beta} \prod_{i=1}^{N} ds_i \right) \\ &\leq \frac{1}{\xi_n} \sum_{i=1}^{N} \left(J_{1j}(\alpha)\right)^{\frac{1}{2}} = \frac{N}{\xi_n} \left(J_{1j}(\alpha)\right)^{\frac{1}{2}} \\ &\overset{(32.40)}{\leq} \frac{N}{\xi_n} \xi_n 2^{\frac{N}{2}} \gamma^{-\frac{N}{2}} \left[\psi_{10} + \psi_{12} + \frac{j}{3}\left(\psi_{21} + \psi_{23}\right) \right]^{\frac{N}{2}} \\ &= N 2^{\frac{N}{2}} \gamma^{-\frac{N}{2}} \left[\psi_{10} + \psi_{12} + \frac{j}{3}\left(\psi_{21} + \psi_{23}\right) \right]^{\frac{N}{2}} < \infty, \end{aligned} \tag{32.52}$$

uniformly bounded with respect to ξ_n.

We have proved the following result.

Theorem 32.6. *All as in Theorem 32.5. Then* $(j = 0,1,\ldots,r \in \mathbb{N})$

$$\begin{aligned} J_2 &:= \lambda_n^{-N} \int_{\mathbb{R}^N} \frac{\|s\|_1}{\xi_n} \prod_{i=1}^{N} \left(\frac{\sin\left(\frac{s_i}{\xi_n}\right)}{s_i} \right)^{2\beta} \prod_{i=1}^{N} ds_i \\ &\leq N 2^{\frac{N}{2}} \gamma^{-\frac{N}{2}} \left[\psi_{10} + \psi_{12} + \frac{j}{3}\left(\psi_{21} + \psi_{23}\right) \right]^{\frac{N}{2}} < \infty, \end{aligned} \tag{32.53}$$

uniformly bounded with respect to $\xi_n \in (0,1]$.

We need the following result.

Theorem 32.7. *(Anastassiou 2021, p. 491) Let* $p > 1$; $r, \beta, N \in \mathbb{N}$, $\alpha_j \in \mathbb{Z}^+$, $j = 1,\ldots,N$: $|\overline{\alpha}| := \sum_{j=1}^{N} \alpha_j = m$, $\xi_n \in (0,1]$, $n \in \mathbb{N}$. *Here we take* $\beta > \frac{\lceil rp \rceil + \lceil mp \rceil + 1}{2}$, *where* $\lceil \cdot \rceil$ *is the ceiling of the number and* γ, λ_n *are as in*

(32.34) and (32.35), respectively, and λ *runs as* $\lambda = 0, 1, \ldots, \lceil rp \rceil$. *When* λ *is even, we define*

$$\psi_{1\lambda} := \frac{\pi (-1)^{\frac{2\beta-\lambda}{2}} (2\beta)!}{2^{\lambda+1} (2\beta - \lambda - 1)!} \left(\sum_{k=1}^{\beta} (-1)^k \frac{k^{2\beta-\lambda-1}}{(\beta - k)! (\beta + k)!} \right), \tag{32.54}$$

and when λ *is odd, we define*

$$\psi_{2\lambda} := \frac{(-1)^{\frac{\lambda-1}{2}} (2\beta)!}{2^{\lambda} (2\beta - \lambda - 1)!} \left(\sum_{k=1}^{\beta} (-1)^{\beta-k} \frac{k^{2\beta-\lambda-1} \ln (2k)}{(\beta - k)! (\beta + k)!} \right), \tag{32.55}$$

and we set

$$\psi_\lambda := \begin{cases} \psi_{1\lambda}, & \text{if } \lambda \text{ is even}, \\ \psi_{2\lambda}, & \text{if } \lambda \text{ is odd}. \end{cases} \tag{32.56}$$

Similarly, it is defined $\psi_{\lambda+\lceil mp \rceil}$ *with* $\lambda + \lceil mp \rceil$ *in the place of* λ.

Then

$$C_{\xi_n} (\overline{\alpha}) := \lambda_n^{-N} \int_{\mathbb{R}^N} \left(\left(\prod_{i=1}^{N} |s_i|^{\alpha_i} \right) \left(1 + \frac{\|s\|_2}{\xi_n} \right)^r \right)^p \prod_{i=1}^{N} \left(\frac{\sin \left(\frac{s_i}{\xi_n} \right)}{s_i} \right)^{2\beta} \prod_{i=1}^{N} ds_i \tag{32.57}$$

$$\leq \xi_n^{2\beta(N-1)+mp} 2^N \gamma^{-N} \left\{ \sum_{\lambda=0}^{\lceil rp \rceil} \binom{\lceil rp \rceil}{\lambda} \left(\psi_\lambda + \psi_{\lambda+\lceil mp \rceil} \right) \right\}^N$$

$$\leq 2^N \gamma^{-N} \left\{ \sum_{\lambda=0}^{\lceil rp \rceil} \binom{\lceil rp \rceil}{\lambda} \left(\psi_\lambda + \psi_{\lambda+\lceil mp \rceil} \right) \right\}^N < \infty, \tag{32.58}$$

uniformly bounded with respect to $\xi_n \in (0, 1]$, *and convergent to zero as* $\xi_n \to 0$, *when* $n \to \infty$.

Remark 32.5. (to Theorem 32.7) Clearly, it holds

$$\widetilde{B}_{\xi_n} (\overline{\alpha}) := \lambda_n^{-N} \int_{\mathbb{R}^N} \left(\prod_{i=1}^{N} |s_i|^{\alpha_i} \right) \prod_{i=1}^{N} \left(\frac{\sin \left(\frac{s_i}{\xi_n} \right)}{s_i} \right)^{2\beta} \prod_{i=1}^{N} ds_i < \infty, \tag{32.59}$$

uniformly bounded in $\xi_n \in (0, 1]$. And, clearly $\widetilde{B}_{\xi_n} (\overline{\alpha}) \to 0$, as $\xi_n \to 0$, $n \to \infty$, by (32.57).

We need the following result.

Theorem 32.8. *Let* $N \in \mathbb{N}$, $\alpha_i \in \mathbb{Z}^+$, $i = 1, \ldots, N : |\alpha| := \sum_{i=1}^{N} \alpha_i = 2$, $\xi_n \in (0, 1]$, $n \in \mathbb{N}$*;* $j = 0, 1, \ldots, r \in \mathbb{N}$*;* $p \geq 1$*;* $\beta > \frac{3\lceil p \rceil + 1}{2}$ *and* $\beta \in \mathbb{N}$*;* $\lambda = 0, 1, \ldots, \lceil p \rceil$.

When λ *is even, define*

$$\psi_{1\lambda} := \frac{\pi (-1)^{\frac{2\beta-\lambda}{2}} (2\beta)!}{2^{\lambda+1} (2\beta - \lambda - 1)!} \sum_{k=1}^{\beta} (-1)^k \frac{k^{2\beta-\lambda-1}}{(\beta - k)! (\beta + k)!}, \tag{32.60}$$

for λ *odd, define*

$$\psi_{2\lambda} := \frac{(-1)^{\frac{\lambda-1}{2}} (2\beta)!}{2^{\lambda} (2\beta - \lambda - 1)!} \sum_{k=1}^{\beta} (-1)^{\beta-k} \frac{k^{2\beta-\lambda-1} \ln (2k)}{(\beta - k)! (\beta + k)!}, \tag{32.61}$$

and

$$\psi_{\lambda} := \begin{cases} \psi_{1\lambda}, & \text{when } \lambda \text{ is even,} \\ \psi_{2\lambda}, & \text{when } \lambda \text{ is odd.} \end{cases} \tag{32.62}$$

When $\lambda + 2\lceil p \rceil$ *is even, define*

$$\psi_{1,\lambda+2\lceil p \rceil} := \frac{\pi (-1)^{\frac{2\beta-\lambda-2\lceil p \rceil}{2}} (2\beta)!}{2^{\lambda+2\lceil p \rceil+1} (2\beta - \lambda - 2\lceil p \rceil - 1)!} \sum_{k=1}^{\beta} (-1)^k \frac{k^{2\beta-\lambda-2\lceil p \rceil-1}}{(\beta - k)! (\beta + k)!}, \tag{32.63}$$

for $\lambda + 2\lceil p \rceil$ *odd, define*

$$\psi_{2,\lambda+2\lceil p \rceil} := \frac{(-1)^{\frac{\lambda+2\lceil p \rceil-1}{2}} (2\beta)!}{2^{\lambda+2\lceil p \rceil} (2\beta - \lambda - 2\lceil p \rceil - 1)!} \sum_{k=1}^{\beta} (-1)^{\beta-k} \frac{k^{2\beta-\lambda-2\lceil p \rceil-1} \ln (2k)}{(\beta - k)! (\beta + k)!}, \tag{32.64}$$

and

$$\psi_{\lambda+2\lceil p \rceil} := \begin{cases} \psi_{1,\lambda+2\lceil p \rceil}, & \text{when } \lambda + 2\lceil p \rceil \text{ is even,} \\ \psi_{2,\lambda+2\lceil p \rceil}, & \text{when } \lambda + 2\lceil p \rceil \text{ is odd.} \end{cases} \tag{32.65}$$

Then

$$\widetilde{A}_{j\xi_n}(\alpha) := \lambda_n^{-N} \int_{\mathbb{R}^N} \left(\prod_{i=1}^{N} |s_i|^{p\alpha_i} \right) \left(1 + \frac{j^p \|s\|_2^p}{\xi_n^p} \right) \prod_{i=1}^{N} \left(\frac{\sin\left(\frac{s_i}{\xi_n}\right)}{s_i} \right)^{2\beta} \prod_{i=1}^{N} ds_i \tag{32.66}$$

$$\leq \xi_n^{2p} 2^N \gamma^{-N} \left\{ \sum_{\lambda=0}^{\lceil p \rceil} \binom{\lceil p \rceil}{\lambda} j^{\lambda} \left(\psi_{\lambda} + \psi_{\lambda+2\lceil p \rceil} \right) \right\}^N$$

$$\leq 2^N \gamma^{-N} \left\{ \sum_{\lambda=0}^{\lceil p \rceil} \binom{\lceil p \rceil}{\lambda} j^{\lambda} \left(\psi_{\lambda} + \psi_{\lambda+2\lceil p \rceil} \right) \right\}^N < \infty. \tag{32.67}$$

That is, $\widetilde{A}_{j\xi_n}(\alpha)$ is uniformly bounded with respect to $\xi_n \in (0,1]$, and $\widetilde{A}_{j\xi_n}(\alpha) \to 0$, as $\xi_n \to 0$, $n \to \infty$.

When $p = 1$, we denote $\widetilde{A}_{j\xi_n}(\alpha)$ as $\widetilde{\widetilde{A}}_{j\xi_n}(\alpha)$.

Proof. We have that $(p \geq 1)$

$$\widetilde{A}_{j\xi_n}(\alpha) = \lambda_n^{-N} \int_{\mathbb{R}^N} \left(\prod_{i=1}^N |s_i|^{p\alpha_i}\right)\left(1 + \frac{j^p \|s\|_2^p}{\xi_n^p}\right)\prod_{i=1}^N \left(\frac{\sin\left(\frac{s_i}{\xi_n}\right)}{s_i}\right)^{2\beta} \prod_{i=1}^N ds_i$$

$$= 2^N \lambda_n^{-N} \int_{\mathbb{R}_+^N} \left(\prod_{i=1}^N s_i^{p\alpha_i}\right)\left(1 + \frac{j^p \|s\|_2^p}{\xi_n^p}\right)\prod_{i=1}^N \left(\frac{\sin\left(\frac{s_i}{\xi_n}\right)}{s_i}\right)^{2\beta} \prod_{i=1}^N ds_i$$

$$\leq 2^N \lambda_n^{-N} \int_{\mathbb{R}_+^N} \left(\prod_{i=1}^N s_i^{p\alpha_i}\right)\left(1 + j^p \left(\sum_{i=1}^N \frac{s_i}{\xi_n}\right)^p\right)\prod_{i=1}^N \left(\frac{\sin\left(\frac{s_i}{\xi_n}\right)}{s_i}\right)^{2\beta} \prod_{i=1}^N ds_i$$

$$= 2^N \lambda_n^{-N} \xi_n^{2p} \xi_n^{-2\beta N} \xi_n^N \tag{32.68}$$

$$\int_{\mathbb{R}_+^N} \left(\prod_{i=1}^N \left(\frac{s_i}{\xi_n}\right)^{p\alpha_i}\right)\left(1 + j^p \left(\sum_{i=1}^N \frac{s_i}{\xi_n}\right)^p\right)\prod_{i=1}^N \left(\frac{\sin\left(\frac{s_i}{\xi_n}\right)}{\frac{s_i}{\xi_n}}\right)^{2\beta} \prod_{i=1}^N d\left(\frac{s_i}{\xi_n}\right)$$

$$= 2^N \lambda_n^{-N} \xi_n^{(1-2\beta)N+2p} \int_{\mathbb{R}_+^N} \left(\prod_{i=1}^N z_i^{p\alpha_i}\right)\left(1 + j^p \left(\sum_{i=1}^N z_i\right)^p\right)\prod_{i=1}^N \left(\frac{\sin z_i}{z_i}\right)^{2\beta} \prod_{i=1}^N dz_i$$

$= (\text{by } \lambda_n = \gamma \xi_n^{1-2\beta})$

$$2^N \gamma^{-N} \xi_n^{2p} \int_{\mathbb{R}_+^N} \left(\prod_{i=1}^N z_i^{p\alpha_i}\right)\left(1 + j^p \left(\sum_{i=1}^N z_i\right)^p\right)\prod_{i=1}^N \left(\frac{\sin z_i}{z_i}\right)^{2\beta} \prod_{i=1}^N dz_i \tag{32.69}$$

$$= 2^N \gamma^{-N} \xi_n^{2p} \int_{\mathbb{R}_+^N} \left(\prod_{i=1}^N z_i^{p\alpha_i}\right)\left(1 + \left(\sum_{i=1}^N j z_i\right)^p\right)\prod_{i=1}^N \left(\frac{\sin z_i}{z_i}\right)^{2\beta} \prod_{i=1}^N dz_i$$

$$\leq 2^N \gamma^{-N} \xi_n^{2p} \int_{\mathbb{R}_+^N} \left(\prod_{i=1}^N z_i^{p\alpha_i}\right)\left(1 + \sum_{i=1}^N j z_i\right)^p \prod_{i=1}^N \left(\frac{\sin z_i}{z_i}\right)^{2\beta} \prod_{i=1}^N dz_i$$

$$\overset{(\text{by } (32.33))}{\leq} 2^N \gamma^{-N} \xi_n^{2p} \int_{\mathbb{R}_+^N} \left(\prod_{i=1}^N z_i^{p\alpha_i}\right)\prod_{i=1}^N (1 + j z_i)^p \prod_{i=1}^N \left(\frac{\sin z_i}{z_i}\right)^{2\beta} \prod_{i=1}^N dz_i$$

$$= 2^N \gamma^{-N} \xi_n^{2p} \prod_{i=1}^N \left(\int_0^\infty z_i^{p\alpha_i} (1 + j z_i)^p \left(\frac{\sin z_i}{z_i}\right)^{2\beta} dz_i\right) \tag{32.70}$$

$$\leq 2^N \gamma^{-N} \xi_n^{2p} \prod_{i=1}^{N} \left(\int_0^{\infty} z_i^{p\alpha_i} (1 + jz_i)^{\lceil p \rceil} \left(\frac{\sin z_i}{z_i} \right)^{2\beta} dz_i \right)$$

$$= 2^N \gamma^{-N} \xi_n^{2p} \prod_{i=1}^{N} \left[\sum_{\lambda=0}^{\lceil p \rceil} \binom{\lceil p \rceil}{\lambda} j^{\lambda} \left(\int_0^{\infty} z_i^{\lambda + p\alpha_i} \left(\frac{\sin z_i}{z_i} \right)^{2\beta} dz_i \right) \right]$$

$$= 2^N \gamma^{-N} \xi_n^{2p} \prod_{i=1}^{N} \left\{ \sum_{\lambda=0}^{\lceil p \rceil} \binom{\lceil p \rceil}{\lambda} j^{\lambda} \right.$$

$$\left. \left[\int_0^{1} z_i^{\lambda + p\alpha_i} \left(\frac{\sin z_i}{z_i} \right)^{2\beta} dz_i + \int_1^{\infty} z_i^{\lambda + p\alpha_i} \left(\frac{\sin z_i}{z_i} \right)^{2\beta} dz_i \right] \right\}$$

$$\leq 2^N \gamma^{-N} \xi_n^{2p} \left\{ \sum_{\lambda=0}^{\lceil p \rceil} \binom{\lceil p \rceil}{\lambda} j^{\lambda} \right.$$

$$\left. \left[\int_0^{1} z^{\lambda} \left(\frac{\sin z}{z} \right)^{2\beta} dz + \int_1^{\infty} z_i^{\lambda + 2\lceil p \rceil} \left(\frac{\sin z}{z} \right)^{2\beta} dz \right] \right\}^N \tag{32.71}$$

$$\leq 2^N \gamma^{-N} \xi_n^{2p} \left\{ \sum_{\lambda=0}^{\lceil p \rceil} \binom{\lceil p \rceil}{\lambda} j^{\lambda} \right.$$

$$\left. \left[\int_0^{\infty} z^{\lambda} \left(\frac{\sin z}{z} \right)^{2\beta} dz + \int_0^{\infty} z_i^{\lambda + 2\lceil p \rceil} \left(\frac{\sin z}{z} \right)^{2\beta} dz \right] \right\}^N =: I.$$

Based on Edwards (1954), p. 210, item 1033 and Anastassiou and Mezei (2012), we have the following calculations for $\lambda = 0, 1, \ldots, \lceil p \rceil$:

Let λ be even, then

$$\int_0^{\infty} z^{\lambda} \left(\frac{\sin z}{z} \right)^{2\beta} dz = \frac{\pi (-1)^{\frac{2\beta - \lambda}{2}} (2\beta)!}{2^{\lambda+1} (2\beta - \lambda - 1)!} \sum_{k=1}^{\beta} (-1)^k \frac{k^{2\beta - \lambda - 1}}{(\beta - k)! (\beta + k)!} = \psi_{1\lambda}. \tag{32.72}$$

Let λ be odd, then

$$\int_0^{\infty} z^{\lambda} \left(\frac{\sin z}{z} \right)^{2\beta} dz = \frac{(-1)^{\frac{\lambda - 1}{2}} (2\beta)!}{2^{\lambda} (2\beta - \lambda - 1)!} \sum_{k=1}^{\beta} (-1)^{\beta - k} \frac{k^{2\beta - \lambda - 1} \ln (2k)}{(\beta - k)! (\beta + k)!} = \psi_{2\lambda}. \tag{32.73}$$

Therefore, it holds

$$\int_0^{\infty} z^{\lambda} \left(\frac{\sin z}{z} \right)^{2\beta} dz = \psi_{\lambda} = \begin{cases} \psi_{1\lambda}, \text{ when } \lambda \text{ is even}, \\ \psi_{2\lambda}, \text{ when } \lambda \text{ is odd}. \end{cases} \tag{32.74}$$

Similarly, for $\lambda + 2\lceil p \rceil$ being even, we get

$$\int_0^\infty z^{\lambda+2\lceil p\rceil}\left(\frac{\sin z}{z}\right)^{2\beta} dz = \frac{\pi(-1)^{\frac{2\beta-\lambda-2\lceil p\rceil}{2}}(2\beta)!}{2^{\lambda+2\lceil p\rceil+1}(2\beta-\lambda-2\lceil p\rceil-1)!} \times \sum_{k=1}^{\beta}(-1)^k \frac{k^{2\beta-\lambda-2\lceil p\rceil-1}}{(\beta-k)!(\beta+k)!} = \psi_{1,\lambda+2\lceil p\rceil}. \tag{32.75}$$

When $\lambda + 2\lceil p \rceil$ is odd, we obtain

$$\int_0^\infty z^{\lambda+2\lceil p\rceil}\left(\frac{\sin z}{z}\right)^{2\beta} dz = \frac{(-1)^{\frac{\lambda+2\lceil p\rceil-1}{2}}(2\beta)!}{2^{\lambda+2\lceil p\rceil}(2\beta-\lambda-2\lceil p\rceil-1)!} \times \sum_{k=1}^{\beta}(-1)^{\beta-k} \frac{k^{2\beta-\lambda-2\lceil p\rceil-1}\ln(2k)}{(\beta-k)!(\beta+k)!} = \psi_{2,\lambda+2\lceil p\rceil}. \tag{32.76}$$

Therefore, it holds

$$\int_0^\infty z^{\lambda+2\lceil p\rceil}\left(\frac{\sin z}{z}\right)^{2\beta} dz = \psi_{\lambda+2\lceil p\rceil} = \begin{cases} \psi_{1,\lambda+2\lceil p\rceil}, & \text{when } \lambda+2\lceil p\rceil \text{ is even}, \\ \psi_{2,\lambda+2\lceil p\rceil}, & \text{when } \lambda+2\lceil p\rceil \text{ is odd}. \end{cases} \tag{32.77}$$

That is,

$$I = 2^N \gamma^{-N} \xi_n^{2p} \left\{ \sum_{\lambda=0}^{\lceil p\rceil} \binom{\lceil p\rceil}{\lambda} j^\lambda \left[\psi_\lambda + \psi_{\lambda+2\lceil p\rceil}\right] \right\}^N \tag{32.78}$$

$$\leq 2^N \gamma^{-N} \left\{ \sum_{\lambda=0}^{\lceil p\rceil} \binom{\lceil p\rceil}{\lambda} j^\lambda \left[\psi_\lambda + \psi_{\lambda+2\lceil p\rceil}\right] \right\}^N < \infty. \tag{32.79}$$

That is, $\widetilde{A}_{j\xi_n}(\alpha)$ is uniformly bounded with respect to $\xi_n \in (0,1]$, and $\widetilde{A}_{j\xi_n}(\alpha) \to 0$, as $\xi_n \to 0$, $n \to \infty$.

The claim is proved. $\square$

The last auxiliary result follows.

Theorem 32.9. *Let $N \in \mathbb{N}$; $\xi_n \in (0,1]$, $n \in \mathbb{N}$; $p \geq 1$; $\lambda = 0, 1, \ldots, \lceil p \rceil$; $\beta > \frac{\lceil p\rceil+1}{2}$ and $\beta \in \mathbb{N}$. Here $\psi_{1\lambda}$ is as in (32.60), $\psi_{2\lambda}$ as in (32.61) and ψ_λ*

as in (32.62). Then

$$\widetilde{K}_{p\xi_n} := \lambda_n^{-N} \int_{\mathbb{R}^N} \left(\frac{\|s\|_2}{\xi_n}\right)^p \prod_{i=1}^N \left(\frac{\sin\left(\frac{s_i}{\xi_n}\right)}{s_i}\right)^{2\beta} \prod_{i=1}^N ds_i$$

$$\leq 2^N \gamma^{-N} \left\{\sum_{\lambda=0}^{\lceil p\rceil} \binom{\lceil p\rceil}{\lambda} \psi_\lambda\right\}^N < \infty \tag{32.80}$$

are uniformly bounded with respect to $\xi_n \in (0,1]$. *When* $p = 1$, *we denote* $\widetilde{K}_{p\xi_n}$ *as* $\widetilde{\widetilde{K}}_{1\xi_n}$.

Proof. We estimate ($p \geq 1$)

$$\widetilde{K}_{p\xi_n} = \lambda_n^{-N} \int_{\mathbb{R}^N} \left(\frac{\|s\|_2}{\xi_n}\right)^p \prod_{i=1}^N \left(\frac{\sin\left(\frac{s_i}{\xi_n}\right)}{s_i}\right)^{2\beta} \prod_{i=1}^N ds_i$$

$$= 2^N \lambda_n^{-N} \int_{\mathbb{R}_+^N} \left(\frac{\|s\|_2}{\xi_n}\right)^p \prod_{i=1}^N \left(\frac{\sin\left(\frac{s_i}{\xi_n}\right)}{s_i}\right)^{2\beta} \prod_{i=1}^N ds_i$$

$$\leq 2^N \lambda_n^{-N} \int_{\mathbb{R}_+^N} \left(\frac{\sum_{i=1}^N s_i}{\xi_n}\right)^p \prod_{i=1}^N \left(\frac{\sin\left(\frac{s_i}{\xi_n}\right)}{s_i}\right)^{2\beta} \prod_{i=1}^N ds_i \tag{32.81}$$

$$= 2^N \lambda_n^{-N} \xi_n^{-2\beta N} \xi_n^N \int_{\mathbb{R}_+^N} \left(\sum_{i=1}^N \frac{s_i}{\xi_n}\right)^p \prod_{i=1}^N \left(\frac{\sin\left(\frac{s_i}{\xi_n}\right)}{\frac{s_i}{\xi_n}}\right)^{2\beta} \prod_{i=1}^N d\frac{s_i}{\xi_n}$$

$$= 2^N \lambda_n^{-N} \xi_n^{N(1-2\beta)} \int_{\mathbb{R}_+^N} \left(\sum_{i=1}^N z_i\right)^p \prod_{i=1}^N \left(\frac{\sin z_i}{z_i}\right)^{2\beta} \prod_{i=1}^N dz_i$$

$$= (\text{by } \lambda_n = \gamma \xi_n^{1-2\beta})$$

$$= 2^N \gamma^{-N} \int_{\mathbb{R}_+^N} \left(\sum_{i=1}^N z_i\right)^p \prod_{i=1}^N \left(\frac{\sin z_i}{z_i}\right)^{2\beta} \prod_{i=1}^N dz_i$$

$$\leq 2^N \gamma^{-N} \int_{\mathbb{R}_+^N} \left(1 + \sum_{i=1}^N z_i\right)^p \prod_{i=1}^N \left(\frac{\sin z_i}{z_i}\right)^{2\beta} \prod_{i=1}^N dz_i$$

$$\overset{(\text{by } (32.33))}{\leq} 2^N \gamma^{-N} \int_{\mathbb{R}_+^N} \prod_{i=1}^N (1+z_i)^p \prod_{i=1}^N \left(\frac{\sin z_i}{z_i}\right)^{2\beta} \prod_{i=1}^N dz_i$$

$$= 2^N \gamma^{-N} \left(\int_0^{\infty} (1+z)^p \left(\frac{\sin z}{z} \right)^{2\beta} dz \right)^N$$

$$\leq 2^N \gamma^{-N} \left(\int_0^{\infty} (1+z)^{\lceil p \rceil} \left(\frac{\sin z}{z} \right)^{2\beta} dz \right)^N$$

$$= 2^N \gamma^{-N} \left\{ \sum_{\lambda=0}^{\lceil p \rceil} \binom{\lceil p \rceil}{\lambda} \int_0^{\infty} z^{\lambda} \left(\frac{\sin z}{z} \right)^{2\beta} dz \right\}^N$$

$$= 2^N \gamma^{-N} \left\{ \sum_{\lambda=0}^{\lceil p \rceil} \binom{\lceil p \rceil}{\lambda} \psi_{\lambda} \right\}^N < \infty, \tag{32.82}$$

ψ_{λ} is as in (32.74) and (32.62).

Thus, $\widetilde{K}_{p\xi_n}$ are uniformly bounded with respect to $\xi_n \in (0,1]$. □

32.4 Main Results

Here we apply Theorem 32.1 to T_n operators, regarding uniform approximation.

Theorem 32.10. *We consider $f \in C^2(\mathbb{R}^N)$, $N \in \mathbb{N}$, and let all $\alpha_i \in \mathbb{Z}^+$, $i = 1, \ldots, N$, $|\alpha| := \sum_{i=1}^N \alpha_i = 2$; $x \in \mathbb{R}^N$, and all the partials f_α of order 2, along with $f \in C_B(\mathbb{R}^N)$; or all f_α of order 2, $f \in C_U(\mathbb{R}^N)$; $0 < \xi_n \leq 1$, $n \in \mathbb{N}$; $\beta \in \mathbb{N} - \{1,2\}$; $j = 0, 1, \ldots, r \in \mathbb{N}$.*

Denote ($n \in \mathbb{N}$)

$$\widetilde{\Delta}_n(x) := T_n(f,x) - f(x) - \left(\sum_{j=0}^{r} \alpha_j j \right) \sin(1)$$

$$\left[\sum_{i=1}^{N} \frac{\partial f(x)}{\partial x_i} \lambda_n^{-N} \left(\int_{\mathbb{R}^N} s_i \prod_{i=1}^{N} \left(\frac{\sin\left(\frac{s_i}{\xi_n}\right)}{s_i} \right)^{2\beta} ds_1, \ldots, ds_N \right) \right]$$

$$- \left(\sum_{j=0}^{r} \alpha_j j^2 \right) \sin^2\left(\frac{1}{2}\right)$$

$$2\left\{\sum_{i=1}^{N}\lambda_n^{-N}\left(\int_{\mathbb{R}^N} s_i^2 \prod_{i=1}^{N}\left(\frac{\sin\left(\frac{s_i}{\xi_n}\right)}{s_i}\right)^{2\beta} ds_1,\ldots,ds_N\right)\frac{\partial^2 f(x)}{\partial x_i^2}\right.$$

$$\left.+\sum_{\substack{i\neq j^*,\\ i,j^*\in\{1,\ldots,N\}}}\lambda_n^{-N}\left(\int_{\mathbb{R}^N} s_i s_{j^*}\prod_{i=1}^{N}\left(\frac{\sin\left(\frac{s_i}{\xi_n}\right)}{s_i}\right)^{2\beta} ds_1\cdots ds_N\right)\frac{\partial^2 f(x)}{\partial x_i \partial x_{j^*}}\right\}. \tag{32.83}$$

Then

(i)

$$\left|\widetilde{\Delta}_n(x)\right| \leq \left\|\widetilde{\Delta}_n\right\|_\infty \leq \sum_{j=0}^{r}|\alpha_j|$$

$$\left[\left[j^2\sum_{\substack{\alpha_i\in\mathbb{Z}^+,\\ \alpha:|\alpha|=2}}\left(\frac{1}{\prod_{i=1}^{N}\alpha_i!}\right)\omega_1\left(f_\alpha,\xi_n\right)J_{1j}\left(\alpha\right)\right]\right.$$

$$\left.+\frac{1}{2}\omega_1\left(f,\xi_n\right)\left[1+\frac{j}{3}J_2\right]\right] =: \widetilde{\varphi}_{\xi_n}, \tag{32.84}$$

where $J_{1j}(\alpha)$ is as in (32.40) and J_2 is as in (32.53).

In case of all f_α of order 2 and $f \in C_U\left(\mathbb{R}^N\right)$ and $\xi_n \to 0$, as $n \to \infty$, then $\widetilde{\Delta}_n(x)$, $\left\|\widetilde{\Delta}_n\right\|_\infty \to 0$ with rates.

(ii) *If $\frac{\partial f(x)}{\partial x_i} = 0$, $i = 1, \ldots, N$, and $f_\alpha(x) = 0$, $\alpha_i \in \mathbb{Z}^+$, $i = 1, \ldots, N$, with $|\alpha| = 2$, then*

$$\left|T_n\left(f,x\right)-f(x)\right| \leq \widetilde{\varphi}_{\xi_n}. \tag{32.85}$$

And $T_n(f,x) \to f(x)$ in the uniformly continuous case, as $\xi_n \to 0$, $n \to \infty$.

(iii) *Additionally, assume all partials of order ≤ 2 are bounded. Hence*

$$\left\|T_n\left(f\right)-f\right\|_\infty \leq \left(\sum_{j=0}^{r}|\alpha_j|\,j\right)(0.8414)$$

$$\left[\sum_{i=1}^{N}\left\|\frac{\partial f}{\partial x_i}\right\|_\infty \lambda_n^{-N}\left(\int_{\mathbb{R}^N}|s_i|\prod_{i=1}^{N}\left(\frac{\sin\left(\frac{s_i}{\xi_n}\right)}{s_i}\right)^{2\beta}\prod_{i=1}^{N}ds_i\right)\right]$$

$$+\left(\sum_{j=0}^{r}|\alpha_j|\,j^2\right)(0.4596)\left\{\sum_{i=1}^{N}\lambda_n^{-N}\left(\int_{\mathbb{R}^N}s_i^2\prod_{i=1}^{N}\left(\frac{\sin\left(\frac{s_i}{\xi_n}\right)}{s_i}\right)^{2\beta}\prod_{i=1}^{N}ds_i\right)\left\|\frac{\partial^2 f}{\partial x_i^2}\right\|_\infty\right.$$

$$\left.+\sum_{\substack{i\neq j^*,\\ i,j^*\in\{1,\ldots,N\}}}\lambda_n^{-N}\left(\int_{\mathbb{R}^N}|s_i||s_{j^*}|\prod_{i=1}^{N}\left(\frac{\sin\left(\frac{s_i}{\xi_n}\right)}{s_i}\right)^{2\beta}\prod_{i=1}^{N}ds_i\right)\left\|\frac{\partial^2 f}{\partial x_i\partial x_{j^*}}\right\|_\infty\right\}+\widetilde{\varphi}_{\xi_n}. \tag{32.86}$$

If all f_α of order 2, $f\in C_U\left(\mathbb{R}^N\right)$, then

$$\|T_n(f)-f\|_\infty\to 0 \text{ with rates, as } \xi_n\to 0,\ n\to+\infty.$$

Proof. By Theorems 32.1, 32.5 and 32.6 and Remark 32.3. □

Next we apply Theorem 32.2 to T_n operators.

Theorem 32.11. *Let $f\in C^m\left(\mathbb{R}^N\right)$, $m\in\mathbb{N}$, $N>1$, with $f_{\overline{\alpha}}\in L_p\left(\mathbb{R}^N\right)$, $|\overline{\alpha}|=m$, $x\in\mathbb{R}^N$. Let $p,q>1:\frac{1}{p}+\frac{1}{q}=1$; $0<\xi_n\leq 1$, $n\in\mathbb{N}$. For $\overline{j}=1,\ldots,m$, and $\overline{\alpha}:=(\alpha_1,\ldots,\alpha_N)$, $\alpha_i\in\mathbb{Z}^+$, $i=1,\ldots,N$, $|\overline{\alpha}|:=\sum_{i=1}^{N}\alpha_i=\overline{j}$, call*

$$\widetilde{c}_{\overline{\alpha},n,\overline{j}}:=\lambda_n^{-N}\int_{\mathbb{R}^N}\left(\prod_{i=1}^{N}s_i^{\alpha_i}\right)\prod_{i=1}^{N}\left(\frac{\sin\left(\frac{s_i}{\xi_n}\right)}{s_i}\right)^{2\beta}ds_1\cdots ds_N. \tag{32.87}$$

Here we assume $\beta>\frac{\lceil rp\rceil+\lceil mp\rceil+1}{2}$, $\beta\in\mathbb{N}$.

Then

$$\left\|\widetilde{E}_{r,n}^{[m]}\right\|_p:=\left\|T_{r,n}(f;x)-f(x)-\sum_{\overline{j}=1}^{m}\delta_{\overline{j},r}^{[m]}\left(\sum_{|\overline{\alpha}|=\overline{j}}\frac{\widetilde{c}_{\overline{\alpha},n,\overline{j}}f_{\overline{\alpha}}(x)}{\left(\prod_{i=1}^{N}\alpha_i!\right)}\right)\right\|_{p,x}$$

$$\leq\left[(N^m/(m-1)!)^{\frac{p}{q}}\left(\frac{m}{(q(m-1)+1)^{\frac{p}{q}}}\right)\right.$$

$$\left.\times\left(\sum_{|\overline{\alpha}|=m}\frac{1}{\prod_{i=1}^{N}\alpha_i!}C_{\xi_n}(\overline{\alpha})\,\omega_r(f_{\overline{\alpha}},\xi_n)_p^p\right)\right]^{\frac{1}{p}}, \tag{32.88}$$

where $C_{\xi_n}(\overline{\alpha})$ as in (32.57).

As $n \to \infty$ and $\xi_n \to 0$, by (32.88), we obtain that $\left\|\widetilde{E}_{r,n}^{[m]}\right\|_p \to 0$ with rates.

One also obtains by (32.88) that

$$\|T_n(f;x) - f(x)\|_{p,x}$$
$$\leq \sum_{\overline{j}=1}^{m} \left|\delta_{\overline{j},r}^{[m]}\right| \left(\sum_{|\overline{\alpha}|=\overline{j}} \frac{\left|\widetilde{c}_{\overline{\alpha},n,\overline{j}}\right|}{\prod_{i=1}^{N} \alpha_i!} \|f_{\overline{\alpha}}\|_p \right) + R.H.S.(32.88), \quad (32.89)$$

given that $\|f_{\overline{\alpha}}\|_p < \infty$, $|\overline{\alpha}| = \overline{j}$, $\overline{j} = 1, \ldots, m$.

Finally, we get $\|T_n(f) - f\|_p \to 0$ as $\xi_n \to 0$, and $n \to \infty$. That is, $T_n \to I$ the unit operator, in L_p norm, with rates.

Proof. By Theorems 32.2 and 32.7 and (32.59). □

Next we apply Theorem 32.3 to T_n operators.

Theorem 32.12. *Let $p, q > 1 : \frac{1}{p} + \frac{1}{q} = 1$, $0 < \xi_n \leq 1$, $n \in \mathbb{N}$. Here we deal with $f \in C^2(\mathbb{R}^N)$, $N \geq 1$, with f, $f_\alpha \in L_p(\mathbb{R}^N)$, $|\alpha| = 2$, where $\alpha_i \in \mathbb{Z}^+$, $i = 1, \ldots, N$, and $|\alpha| = \sum_{i=1}^{N} \alpha_i$; $x \in \mathbb{R}^N$; $j = 0, 1, \ldots, r$. Assume $\beta > \frac{3[p]+1}{2}$, $\beta \in \mathbb{N}$.*

Denote ($n \in \mathbb{N}$)

$$\widetilde{\Delta}_n(x) := T_n(f,x) - f(x) - \left(\sum_{j=0}^{r} \alpha_j j\right) \sin(1)$$
$$\left[\sum_{i=1}^{N} \frac{\partial f(x)}{\partial x_i} \lambda_n^{-N} \left(\int_{\mathbb{R}^N} s_i \prod_{i=1}^{N} \left(\frac{\sin\left(\frac{s_i}{\xi_n}\right)}{s_i}\right)^{2\beta} ds_1, \ldots, ds_N\right)\right]$$
$$-2\left(\sum_{j=0}^{r} \alpha_j j^2\right) \sin^2\left(\frac{1}{2}\right)$$
$$\left\{\sum_{i=1}^{N} \lambda_n^{-N} \left(\int_{\mathbb{R}^N} s_i^2 \prod_{i=1}^{N} \left(\frac{\sin\left(\frac{s_i}{\xi_n}\right)}{s_i}\right)^{2\beta} ds_1, \ldots, ds_N\right) \frac{\partial^2 f(x)}{\partial x_i^2}\right.$$

$$+\sum_{\substack{i\neq j^*,\\ i,j^*\in\{1,\ldots,N\}}}\lambda_n^{-N}\left(\int_{\mathbb{R}^N}s_is_{j^*}\prod_{i=1}^{N}\left(\frac{\sin\left(\frac{s_i}{\xi_n}\right)}{s_i}\right)^{2\beta}ds_1\cdots ds_N\right)\frac{\partial^2 f(x)}{\partial x_i\partial x_{j^*}}\Bigg\}. \tag{32.90}$$

Then

$$\left\|\widetilde{\Delta}_n\right\|_p\leq\left(\sum_{j=0}^{r}|\alpha_j|^q\right)^{\frac{1}{q}}\left\{\frac{2}{(q+1)^{\frac{1}{q}}}\left(\frac{N(N+1)+2}{2}\right)^{\frac{1}{q}}\right\}$$

$$\times\left\{\sum_{j=0}^{r}\left[j^{2p}\sum_{|\alpha|=2}\left(\frac{1}{\prod_{i=1}^{N}\alpha_i!}\right)^p\omega_1(f_\alpha,\xi_n)_p^p\widetilde{A}_{j\xi_n}(\alpha)+\omega_1(f,\xi_n)_p^p\left[1+j^p\widetilde{K}_{p\xi_n}\right]\right]\right\}^{\frac{1}{p}}, \tag{32.91}$$

where $\widetilde{A}_{j\xi_n}(\alpha)$ *as in (32.66) and* $\widetilde{K}_{p\xi_n}$ *as in (32.80). Note that* $\left\|\widetilde{\Delta}_n\right\|_p\to 0$ *with rates as* $\xi_n\to 0$, $n\to\infty$, *by (32.91).*

One also obtains from (32.91) that

$$\|T_n(f)-f\|_p$$

$$\leq\left(\sum_{j=0}^{r}|\alpha_j|\,j\right)\sin(1)\left[\sum_{i=1}^{N}\left\|\frac{\partial f}{\partial x_i}\right\|_p\lambda_n^{-N}\left(\int_{\mathbb{R}^N}|s_i|\prod_{i=1}^{N}\left(\frac{\sin\left(\frac{s_i}{\xi_n}\right)}{s_i}\right)^{2\beta}ds_1,\ldots,ds_N\right)\right]$$

$$+2\left(\sum_{j=0}^{r}|\alpha_j|\,j^2\right)\sin^2\left(\frac{1}{2}\right)$$

$$\left\{\sum_{i=1}^{N}\lambda_n^{-N}\left(\int_{\mathbb{R}^N}s_i^2\prod_{i=1}^{N}\left(\frac{\sin\left(\frac{s_i}{\xi_n}\right)}{s_i}\right)^{2\beta}ds_1\cdots ds_N\right)\left\|\frac{\partial^2 f}{\partial x_i^2}\right\|_p\right.$$

$$\left.+\sum_{\substack{i\neq j^*,\\ i,j^*\in\{1,\ldots,N\}}}\lambda_n^{-N}\left(\int_{\mathbb{R}^N}|s_is_{j^*}|\prod_{i=1}^{N}\left(\frac{\sin\left(\frac{s_i}{\xi_n}\right)}{s_i}\right)^{2\beta}ds_1\cdots ds_N\right)\left\|\frac{\partial^2 f}{\partial x_i\partial x_{j^*}}\right\|_p\right\}$$

$$+R.H.S.(32.91), \tag{32.92}$$

given that $\|f_\alpha\|_p<\infty$, $|\alpha|=\widetilde{j}$, $\widetilde{j}=1,2$.

Clearly, $\|T_n(f)-f\|_p\to 0$, *as* $n\to\infty$ *and* $\xi_n\to 0$. *That is,* $T_n\to I$, *the unit operator, in* L_p *norm with rates.*

Proof. By Theorems 32.3, 32.8 and 32.9 and Remark 32.3. □

We finish with an application of Theorem 32.4 to T_n operators.

Theorem 32.13. *Let* $0 < \xi_n \leq 1$, $n \in \mathbb{N}$, $x \in \mathbb{R}^N$. *Here we deal with* $f \in C^2(\mathbb{R}^N)$, $N \geq 1$, *with* $f, f_\alpha \in L_1(\mathbb{R}^N)$, $|\alpha| = 2$, *where* $\alpha_i \in \mathbb{Z}^+$, $i = 1, \ldots, N$, *and* $|\alpha| = \sum_{i=1}^N \alpha_i$; $j = 0, 1, \ldots, r$. *Here* $\widetilde{\Delta}_n$ *is as in (32.90),* $\beta \in \mathbb{N} - \{1, 2\}$. *Then*

$$\left\|\widetilde{\Delta}_n\right\|_1 \leq \left\{\sum_{j=0}^{r} |\alpha_j| \left[j^2 \sum_{|\alpha|=2} \left(\frac{2}{\prod_{i=1}^{N} \alpha_i!}\right) \omega_1(f_\alpha, \xi_n)_1 \widetilde{\widetilde{A}}_{j\xi_n}(\alpha)\right.\right.$$

$$\left.\left. + \omega_1(f, \xi_n)_1 \left[1 + j\widetilde{\widetilde{K}}_{1\xi_n}\right]\right]\right\}, \tag{32.93}$$

where $\widetilde{\widetilde{A}}_{j\xi_n}(\alpha)$ *as in Theorem 32.8 and* $\widetilde{\widetilde{K}}_{\xi_n}$ *as in Theorem 32.9.*

Note that $\left\|\widetilde{\Delta}_n\right\|_1 \to 0$ *with rates as* $\xi_n \to 0$, $n \to \infty$, *by (32.93).*

We derive also that

$$\|T_n(f) - f\|_1$$

$$\leq \left(\sum_{j=0}^{r} |\alpha_j| j\right) \sin(1) \left[\sum_{i=1}^{N} \left\|\frac{\partial f}{\partial x_i}\right\|_1 \lambda_n^{-N} \left(\int_{\mathbb{R}^N} |s_i| \prod_{i=1}^{N} \left(\frac{\sin\left(\frac{s_i}{\xi_n}\right)}{s_i}\right)^{2\beta} ds_1, \ldots, ds_N\right)\right]$$

$$+ 2\left(\sum_{j=0}^{r} |\alpha_j| j^2\right) \sin^2\left(\frac{1}{2}\right)$$

$$\left\{\sum_{i=1}^{N} \lambda_n^{-N} \left(\int_{\mathbb{R}^N} s_i^2 \prod_{i=1}^{N} \left(\frac{\sin\left(\frac{s_i}{\xi_n}\right)}{s_i}\right)^{2\beta} ds_1, \ldots, ds_N\right) \left\|\frac{\partial^2 f}{\partial x_i^2}\right\|_1\right.$$

$$\left. + \sum_{\substack{i \neq j^*, \\ i, j^* \in \{1, \ldots, N\}}} \lambda_n^{-N} \left(\int_{\mathbb{R}^N} |s_i s_{j^*}| \prod_{i=1}^{N} \left(\frac{\sin\left(\frac{s_i}{\xi_n}\right)}{s_i}\right)^{2\beta} ds_1, \ldots, ds_N\right) \left\|\frac{\partial^2 f}{\partial x_i \partial x_{j^*}}\right\|_1\right\}$$

$$+ R.H.S.(32.93), \tag{32.94}$$

given that $\|f_\alpha\|_1 < \infty$, $|\alpha| = \widetilde{j}$, $\widetilde{j} = 1, 2$. *As* $n \to \infty$ *and* $\xi_n \to 0$, *by (32.94), we derive that* $\|T_n(f) - f\|_1 \to 0$, *that is,* $T_n \to I$ *the unit operator, in* L_1 *norm, with rates.*

Proof. By Theorems 32.4, 32.8 and 32.9 and Remark 32.3. □

Bibliography

G.A. Anastassiou, *Intelligent Mathematics: Computational Analysis*, Springer, Heidelberg, New York, 2011a.

G.A. Anastassiou, *Approximation by Multivariate Singular Integrals*, Springer, New York, 2011b.

G.A. Anastassiou, *Constructive Fractional Analysis with Applications*, Springer, Heidelberg, New York, 2021.

G.A. Anastassiou, *Trigonometric based multivariate smooth Picard singular integrals L_p approximation*, Bulletin Allah. Math. Soc., accepted, 2023a.

G.A. Anastassiou, *Trigonometric background multivariate smooth Trigonometric singular integrals approximations*, RACSAM, accepted, 2024.

G.A. Anastassiou, *General multiple sigmoid functions relied complex valued multivariate trigonometric and hyperbolic neural network approximations*, 2023b.

G.A. Anastassiou, *Uniform Approximation by Smooth Picard Multivariate Singular Integral Operators revisited*, 2023c.

G.A. Anastassiou, S. Gal, *Approximation Theory*, Birkhaüser, Boston, 2000.

G.A. Anastassiou, R. Mezei, *Convergence of complex general singular integral operators, Journal of Concrete and Applicable Mathematics*, 10 (3–4) (2012), 259–283.

G.A. Anastassiou, R. Mezei, *Approximation by Singular Integrals*, Cambridge Scientific Publishers, Cambridge, UK, 2012.

J. Edwards, *A Treatise on the Integral Calculus, Vol II*, Chelsea, New York, 1954.

R.N. Mohapatra and R.S. Rodriguez, On the rate of convergence of singular integrals for Hölder continuous functions, *Mathematische Nachrichten*, 149 (1990), 117–124.

Conclusion

This monograph is based on new types of Taylor's formulae, of the trigonometric and hyperbolic forms. We covered perturbed neural network approximations by themselves and with their connections to Brownian motion and stochastic processes, univariate and multivariate analytical inequalities ordinary and fractional, about Korovkin theory, and approximations by singular integrals, univariate and multivariate.

Differentiability of functions is a strong aspect of our work.

The above material is contained in 32 chapters, and at the end of each chapter, a detailed list of references is added. All exposed here are under the umbrella of Computational Analysis and Applications. The presented approaches are original, and chapters are self-contained and can be read independently.

Index

A

activation function, 38, 184, 286
Ali, A.H., 1–2, 633, 669
Anastassiou, G.A., 633
asymptotic expansion, 249

B

ball, 446
bell symmetric function, 185, 319
bell-shaped, 77
bell-shaped function, 81
Borel measurable function, 588, 786
Borel probability measure, 759
bounded continuous functions, 534
brain asymmetry, 112
Brownian motion, 285, 298, 300–301, 337, 813

C

Caputo left fractional derivative, 290, 417
Caputo right fractional derivative, 291
compact subset, 226
companion operator, 158
complete metric space, 360
complex Banach space, 158
concave down, 325
conformable calculus, 420
conformable derivatives, 399
conformable differential operator, 400
conformable exponential function, 400
continuous and bounded functions, 787
contraction property, 109
critical number, 116

D

density function, 38, 44, 83, 293
differential operator, 1
diffusion process, 333
Dirac delta function, 299

E

expectation, 311
expected value function, 334
exponential function, 400

F

feed-forward, 78
fractional approximation, 323
fractional Hilbert–Pachpatte, 437
fuzzy real number, 359
fuzzy summation, 361
fuzzy-random function, 362
fuzzy-random variable, 361
fuzzy-stochastic process, 361

G

gamma function, 522
Gauss–Weierstrass singular integral operators, 614
Gauss–Weierstrass smooth singular integrals, 670
Gaussian distributions, 278

generalized logistic-type, 78
global maximum, 116
Gompertz function, 353

H

half-hyperbolic, 324
Haykin, S., 39
heat kernel, 299
Hilbert–Pachpatte-type inequality, 414
horizontal asymptote, 185, 219
hyperbolic, 56
hyperbolic approximation, 241
hyperbolic Polya inequalities, 475
hyperbolic tangent function, 111
hyperbolic Taylor's formula, 1, 57, 198
hyperbolic-trigonometric, 67, 139

I

inner product, 113
integral part, 44, 252
iterated multilayer, 149

J

Jackson-type inequalities, 38
Jensen inequality, 704

K

Korovkin properties, 383
Korovkin theory, 813

L

L_1 approximation, 602
L_1 approximation by Poisson–Cauchy operators, 660
L_1-related approximation, 563
L_p approximation, 521, 557
L_p convergence, 549
L_p first moduli of continuity, 757
L_p modulus of continuity, 587, 785
Lebesgue integrable function, 458
Lebesgue integrals, 508
Lebesgue measurable, 301, 337
Lebesgue measure, 447
linear differential operators, 1
logistic function, 38
logistic sigmoid function, 78, 350

M

maximum value, 117
McCulloch, W., 39
mean value theorem, 50
Mitchell, T.M., 39
modulus of continuity, 46, 119, 509, 735
multilayer approximations, 150
multivariate approximation, 249
multivariate density function, 156, 253, 366
multivariate fuzzy-random, 359
multivariate moduli of continuity, 77, 218
multivariate Picard singular integral operators, 587
multivariate positive linear, 255
multivariate radial, 445
multivariate smooth Picard singular integral operators, 604
multivariate Taylor formulae, 149
multivariate-averaged positive, 375

N

network approximation, 56
neural network operators, 37, 78
neutral element, 361
normalized neural network, 158

O

open convex subset, 227
Opial-type, 484
Opial-type inequalities, 15
Ostrowski-type inequality, 16, 445, 447

P

Pales, Z., 1–2, 633, 669
λ-parametrized, 39
parametrized, 533

parametrized (Gauss) error, 318
partial conformable derivatives, 402
partial derivative, 250
partial order, 360
perturbed neural network, 363
perturbed quasi-interpolation, 376
Picard multivariate singular integral operators, 569
Picard singular integral, 507
Pitts, W., 39
Poincaré-type inequalities, 495
pointwise, 45
pointwise and uniformly, 47, 331
Poisson distributions, 277
Poisson–Cauchy singular integral operators, 615
Polya inequalities, 483
positive linear operators mapping, 383
probability Borel measure, 786
probability space, 279
proportional derivative, 399
proportional fractional calculus, 417

Q

q-deformed, 39
quantitative approximation, 217
quasi-interpolation, 78
quasi-interpolation operators, 183

R

radial, 447
random motion, 333
rate of convergence, 120
Riemann–Liouville fractional integral, 417
Riesz representation, 389
right Riemann–Liouville fractional, 418

S

scaled function, 219
shifted sigmoid function, 293
Shisha–Mond-type inequalities, 384
sigmoid function, 41, 311
simple graphs, 333
smooth Gauss–Weierstrass singular integral operators, 635
Sobolev-type inequality, 432, 499
speed of convergence, 66
spherical shell, 457
squashing functions, 37
standard Brownian motion, 335
stationary Gaussian, 250
stationary process, 250
stochastic approximation, 271
stochastic process, 278, 813
strictly convex, 184
strictly decreasing, 41
strictly increasing, 41, 325
supremum norm, 226
surface measure, 446

T

Taylor's formulae, 2, 112
time separating, 270
time separating process, 270
transition density, 334
transition density function, 300
trigonometric, 4
trigonometric approximation, 94, 120, 187
trigonometric singular integral operators, 616
trigonometric Taylor's formula, 517, 614, 669, 694
trigonometric-induced, 790

U

uniform convergence, 45
uniformly bounded, 577
uniformly continuous, 572
uniformly convergent, 390
unit operator, 95
upper incomplete gamma function, 605, 608

www.ingramcontent.com/pod-product-compliance
Lightning Source LLC
LaVergne TN
LVHW010337220926
848484LV00004BB/917
9789811293856